General, Organic, & Biological
CHEMISTRY

THE ELEMENTS

Element	Symbol	Atomic Number	Relative Atomic Mass*	Element	Symbol	Atomic Number	Relative Atomic Mass*
Actinium	Ac	89	(227)	Mercury	Hg	80	200.59
Aluminum	Al	13	26.9815	Molybdenum	Mo	42	95.94
Americium	Am	95	(243)	Neodymium	Nd	60	144.24
Antimony	Sb	51	121.760	Neon	Ne	10	20.1797
Argon	Ar	18	39.948	Neptunium	Np	93	(237)
Arsenic	As	33	74.9216	Nickel	Ni	28	58.693
Astatine	At	85	(210)	Niobium	Nb	41	92.9064
Barium	Ba	56	137.327	Nitrogen	N	7	14.0067
Berkelium	Bk	97	(247)	Nobelium	No	102	(259)
Beryllium	Be	4	9.0122	Osmium	Os	76	190.2
Bismuth	Bi	83	208.9804	Oxygen	O	8	15.9994
Bohrium	Bh	107	(272)	Palladium	Pd	46	106.42
Boron	B	5	10.811	Phosphorus	P	15	30.9738
Bromine	Br	35	79.904	Platinum	Pt	78	195.08
Cadmium	Cd	48	112.411	Plutonium	Pu	94	(244)
Calcium	Ca	20	40.078	Polonium	Po	84	(209)
Californium	Cf	98	(251)	Potassium	K	19	39.0983
Carbon	C	6	12.011	Praseodymium	Pr	59	140.9076
Cerium	Ce	58	140.115	Promethium	Pm	61	(145)
Cesium	Cs	55	132.9054	Protactinium	Pa	91	231.03588
Chlorine	Cl	17	35.453	Radium	Ra	88	(226)
Chromium	Cr	24	51.9961	Radon	Rn	86	(222)
Cobalt	Co	27	58.9332	Rhenium	Re	75	186.207
Copper	Cu	29	63.546	Rhodium	Rh	45	102.9055
Curium	Cm	96	(247)	Roentgenium	Rg	111	(280)
Darmstadtium	Ds	110	(281)	Rubidium	Rb	37	85.4678
Dubnium	Db	105	(268)	Ruthenium	Ru	44	101.07
Dysprosium	Dy	66	162.50	Rutherfordium	Rf	104	(267)
Einsteinium	Es	99	(252)	Samarium	Sm	62	150.36
Erbium	Er	68	167.26	Scandium	Sc	21	44.9559
Europium	Eu	63	151.964	Seaborgium	Sg	106	(271)
Fermium	Fm	100	(257)	Selenium	Se	34	78.96
Fluorine	F	9	18.9984	Silicon	Si	14	28.0855
Francium	Fr	87	(223)	Silver	Ag	47	107.8682
Gadolinium	Gd	64	157.25	Sodium	Na	11	22.9898
Gallium	Ga	31	69.723	Strontium	Sr	38	87.62
Germanium	Ge	32	72.64	Sulfur	S	16	32.066
Gold	Au	79	196.9665	Tantalum	Ta	73	180.9479
Hafnium	Hf	72	178.49	Technetium	Tc	43	(98)
Hassium	Hs	108	(270)	Tellurium	Te	52	127.60
Helium	He	2	4.0026	Terbium	Tb	65	158.9253
Holmium	Ho	67	164.9303	Thallium	Tl	81	204.3833
Hydrogen	H	1	1.0079	Thorium	Th	90	232.0381
Indium	In	49	114.82	Thulium	Tm	69	168.9342
Iodine	I	53	126.9045	Tin	Sn	50	118.710
Iridium	Ir	77	192.22	Titanium	Ti	22	47.88
Iron	Fe	26	55.845	Tungsten	W	74	183.84
Krypton	Kr	36	83.80	Uranium	U	92	238.0289
Lanthanum	La	57	138.9055	Vanadium	V	23	50.9415
Lawrencium	Lr	103	(262)	Xenon	Xe	54	131.29
Lead	Pb	82	207.2	Ytterbium	Yb	70	173.04
Lithium	Li	3	6.941	Yttrium	Y	39	88.9059
Lutetium	Lu	71	174.967	Zinc	Zn	30	65.41
Magnesium	Mg	12	24.3050	Zirconium	Zr	40	91.224
Manganese	Mn	25	54.9380			112**	(285)
Meitnerium	Mt	109	(276)			114	(289)
Mendelevium	Md	101	(258)			116	(293)

*Values in parentheses represent the mass number of the most stable isotope.

**The names and symbols for elements 112, 114, and 116 have not been chosen.

General, Organic, & Biological
CHEMISTRY

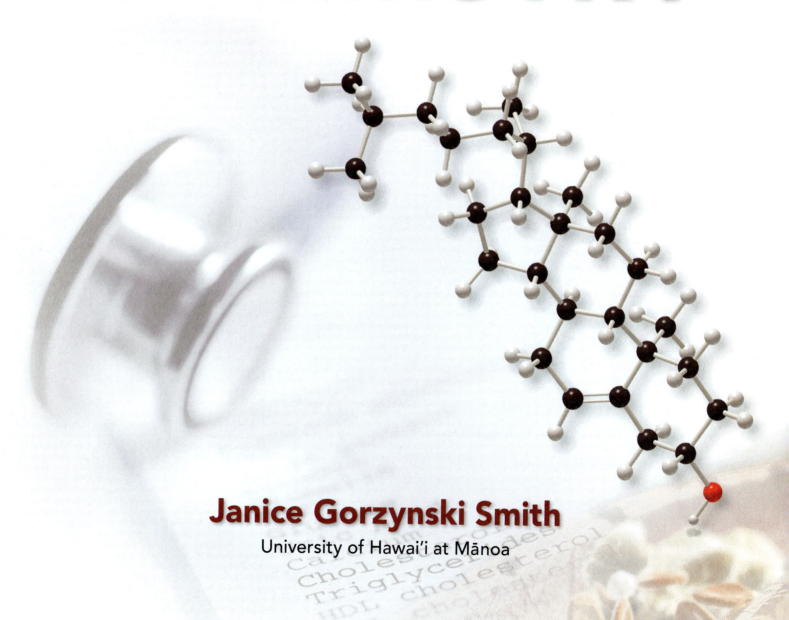

Janice Gorzynski Smith

University of Hawai'i at Mānoa

Higher Education

Boston   Burr Ridge, IL   Dubuque, IA   New York   San Francisco   St. Louis
Bangkok   Bogotá   Caracas   Kuala Lumpur   Lisbon   London   Madrid   Mexico City
Milan   Montreal   New Delhi   Santiago   Seoul   Singapore   Sydney   Taipei   Toronto

GENERAL, ORGANIC, AND BIOLOGICAL CHEMISTRY

1 2 3 4 5 6 7 8 9 0 DOW/DOW 0 9

ISBN 978–0–07–302657–2
MHID 0–07–302657–3

Publisher: *Thomas D. Timp*
Senior Sponsoring Editor: *Tamara L. Hodge*
Vice-President New Product Launches: *Michael Lange*
Senior Developmental Editor: *Donna Nemmers*
Marketing Manager: *Todd L. Turner*
Senior Project Manager: *Jayne L. Klein*
Senior Production Supervisor: *Laura Fuller*
Senior Media Project Manager: *Tammy Juran*
Designer: *Laurie B. Janssen*
(USE) Cover Image: *Lower right: ©3D4Medical.com/gettyimages, background: ©Andrew Brookes/Corbis*
Lead Photo Research Coordinator: *Carrie K. Burger*
Photo Research: *Mary Reeg*
Supplement Producer: *Mary Jane Lampe*
Compositor: *Precision Graphics*
Typeface: *10/12.5 Times LT Std*
Printer: *R. R. Donnelley Willard, OH*

The credits section for this book begins on page C-1 and is considered an extension of the copyright page.

Library of Congress Cataloging-in-Publication Data

Smith, Janice G.
    General, organic, and biological chemistry / Janice Gorzynski Smith. — 1st ed.
        p. cm.
    Includes index.
    ISBN 978–0–07–302657–2 — ISBN 0–07–302657–3 (hard copy : alk. paper) 1. Chemistry—Textbooks.
I. Title.

    QD31.3.S63 2010
    540—dc22

                                                                        2008044484

www.mhhe.com

Dedication

To my husband Dan, children Erin, Jenna, Matthew, and Zachary, and father Stanley, and in memory of my mother Dorothea and daughter Megan.

About the Author

Janice Gorzynski Smith was born in Schenectady, New York, and grew up following the Yankees, listening to the Beatles, and water skiing on Sacandaga Reservoir. She became interested in chemistry in high school, and went on to major in chemistry at Cornell University where she received an A.B. degree *summa cum laude*. Jan earned a Ph.D. in Organic Chemistry from Harvard University under the direction of Nobel Laureate E.J. Corey, and she also spent a year as a National Science Foundation National Needs Postdoctoral Fellow at Harvard. During her tenure with the Corey group, she completed the total synthesis of the plant growth hormone gibberellic acid.

Following her postdoctoral work, Jan joined the faculty of Mount Holyoke College where she was employed for 21 years. During this time she was active in teaching chemistry lecture and lab courses, conducting a research program in organic synthesis, and serving as department chair. Her organic chemistry class was named one of Mount Holyoke's "Don't-miss courses" in a survey by *Boston* magazine. After spending two sabbaticals amidst the natural beauty and diversity in Hawai'i in the 1990s, Jan and her family moved there permanently in 2000. She is currently a faculty member at the University of Hawai'i at Mānoa, where she teaches a one-semester organic and biological chemistry course for nursing students, as well as the two-semester organic chemistry lecture and lab courses. She also serves as the faculty advisor to the student affiliate chapter of the American Chemical Society. In 2003, she received the Chancellor's Citation for Meritorious Teaching.

Jan resides in Hawai'i with her husband Dan, an emergency medicine physician. She has four children: Matthew and Zachary (scuba photo on page 190); Jenna, a first-year law student at Temple University in Philadelphia; and Erin, a 2006 graduate of Brown University School of Medicine and co-author of the Student Study Guide/Solutions Manual for this text. When not teaching, writing, or enjoying her family, Jan bikes, hikes, snorkels, and scuba dives in sunny Hawai'i, and time permitting, enjoys travel and Hawai'ian quilting.

Brief Contents

Contents

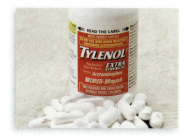

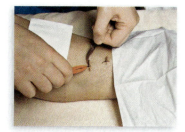

10 Nuclear Chemistry 298

11 Introduction to Organic Molecules and Functional Groups 322

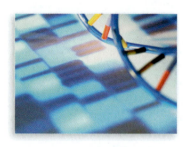

22   Nucleic Acids and Protein Synthesis   682

23   Digestion and the Conversion of Food into Energy   718

Preface

My goal in writing this text was to relate the fundamental concepts of general, organic, and biological chemistry to the world around us, and in this way illustrate how chemistry explains many aspects of everyday life. I have followed two guiding principles: use relevant and interesting applications for all basic chemical concepts, and present the material in a student-friendly fashion using bulleted lists, extensive illustrations, and step-by-step problem solving.

This text is different—by design. Since today's students rely more heavily on visual imagery to learn than ever before, this text uses less prose and more diagrams and figures to reinforce the major themes of chemistry. A key feature is the use of molecular art to illustrate and explain common phenomena we encounter every day. Each topic is broken down into small chunks of information that are more manageable and easily learned. Students are given enough detail to understand basic concepts, such as how soap cleans away dirt and why trans fats are undesirable in the diet, without being overwhelmed.

This textbook is written for students who have an interest in nursing, nutrition, environmental science, food science, and a wide variety of other health-related professions. The content of this book is designed for an introductory chemistry course with no chemistry prerequisite, and is suitable for either a two-semester sequence or a one-semester course. I have found that by introducing one new concept at a time, keeping the basic themes in focus, and breaking down complex problems into small pieces, many students in these chemistry courses acquire a new appreciation of both the human body and the larger world around them.

BUILDING THE TEXT

Writing a textbook is a multifaceted process. McGraw-Hill's 360° Development Process is an ongoing, never ending market-oriented approach to building accurate and innovative print and digital products. It is dedicated to continual large scale and incremental improvement, driven by multiple customer feedback loops and checkpoints. This is initiated during the early planning stages of new products and intensifies during the development and production stages, and then begins again upon publication, in anticipation of the next edition. This process is designed to provide a broad, comprehensive spectrum of feedback for refinement and innovation of learning tools, for both student and instructor. The 360° Development Process includes market research, content reviews, faculty and student focus groups, course- and product-specific symposia, accuracy checks, and art reviews, all guided by a carefully selected Board of Advisors.

THE LEARNING SYSTEM USED IN *GENERAL, ORGANIC, AND BIOLOGICAL CHEMISTRY*

- **Writing Style** A concise writing style allows students to focus on learning major concepts and themes of general, organic, and biological chemistry. Relevant materials from everyday life are used to illustrate concepts, and topics are broken into small chunks of information that are more easily learned.
- **Chapter Outline** The chapter outline lists the main headings of the chapter, to help students map out the organization of each chapter's content.

- **Chapter Goals, tied to end-of-chapter Key Concepts**   The Chapter Goals at the beginning of each chapter identify what students will learn, and are tied numerically to the end-of-chapter Key Concepts, which serve as bulleted summaries of the most important concepts for study.

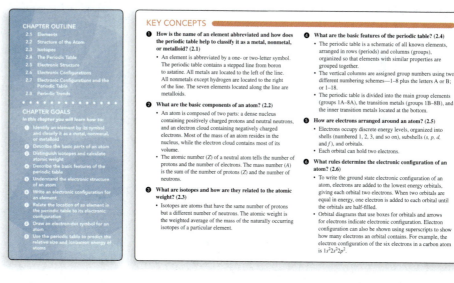

- **Macro-to-Micro Illustrations**   Because today's students are visual learners, and because visualizing molecular-level representations of macroscopic phenomena is critical to the understanding of any chemistry course, many illustrations in this text include photos or drawings of everyday objects, paired with their molecular representation, to help students visualize and understand the chemistry behind ordinary occurrences.
- **Problem Solving**   Sample Problems lead students through the thought process tied to successful problem solving by employing Analysis and Solution parts. Sample Problems are categorized sequentially by topic to match chapter organization, and are often paired with practice problems to allow students to apply what they have just learned. Students can immediately verify their answers to the follow-up problems in the appendix at the end of the book.
- **How To's**   Key processes are taught to students in a straightforward and easy-to-understand manner by using examples and multiple, detailed steps to solving problems.
- **Applications**   Common applications of chemistry to everyday life are found in margin-placed Health Notes, Consumer Notes, and Environmental Notes, as well as sections entitled "Focus on Health & Medicine," "Focus on the Environment," and "Focus on the Human Body."

OUR COMMITMENT TO SERVING TEACHERS AND LEARNERS

TO THE INSTRUCTOR   Writing a new chemistry textbook is a colossal task. Teaching chemistry for over 20 years at both a private, liberal arts college and a large state university has given me a unique perspective with which to write this text. I have found that students arrive with vastly different levels of preparation and widely different expectations for their college experience. As an instructor and now an author I have tried to channel my love and knowledge of chemistry into a form that allows this spectrum of students to understand chemical science more clearly, and then see everyday phenomena in a new light.

TO THE STUDENT   I hope that this text and its ancillary program will help you to better understand and appreciate the world of chemistry. My interactions with thousands of students in my long teaching career have profoundly affected the way I teach and write about chemistry, so please feel free to email me with any comments or questions at jgsmith@hawaii.edu.

P.A.V.E. the Way to Student Learning

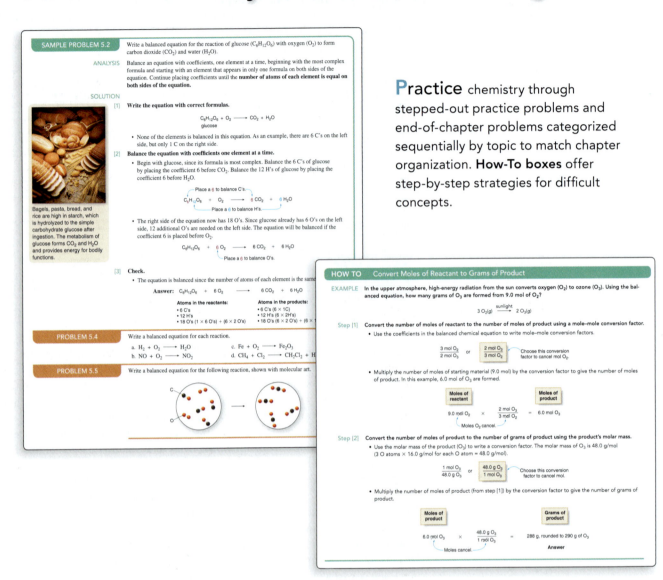

Practice chemistry through stepped-out practice problems and end-of-chapter problems categorized sequentially by topic to match chapter organization. **How-To boxes** offer step-by-step strategies for difficult concepts.

Apply chemistry through **"Focus on Health & Medicine," "Focus on the Human Body,"** and **"Focus on the Environment"** sections woven throughout the text. Chemistry applications are also woven into margin notes that cover topics on consumer, health, and environmental issues.

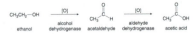

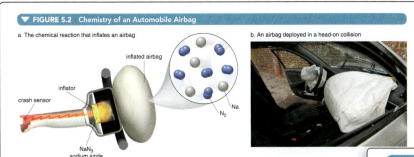

▼ **FIGURE 5.2**   Chemistry of an Automobile Airbag

a. The chemical reaction that inflates an airbag

inflated airbag

inflator

crash sensor

Na

N₂

NaN₃
sodium azide

b. An airbag deployed in a head-on collision

A severe car crash triggers an airbag to deploy when an electric sensor causes sodium azide (NaN₃) to ignite, converting it [to sodium] (Na) and nitrogen gas (N₂). The nitrogen gas causes the bag to inflate fully in 40 milliseconds, helping to protect passengers [from] injury. The sodium atoms formed in this first reaction are hazardous and subsequently converted to a safe sodium salt. It too[k years to] develop a reliable airbag system for automobiles.

Visualize chemistry through a dynamic art program that brings together macroscopic and microscopic representations of images to help students comprehend on a molecular level. Many illustrations include photos or drawings of everyday objects, paired with their molecular representation, to help students understand the chemistry behind ordinary occurrences. Many illustrations of the human body include magnifications for specific anatomic regions, as well as representations at the microscopic level, for today's visual learners.

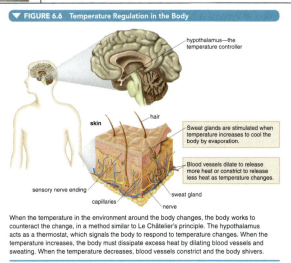

▼ **FIGURE 6.6**   Temperature Regulation in the Body

hypothalamus—the temperature controller

skin

hair

Sweat glands are stimulated when temperature increases to cool the body by evaporation.

Blood vessels dilate to release more heat or constrict to release less heat as temperature changes.

sensory nerve ending

capillaries

sweat gland

nerve

When the temperature in the environment around the body changes, the body works to counteract the change, in a method similar to Le Châtelier's principle. The hypothalamus acts as a thermostat, which signals the body to respond to temperature changes. When the temperature increases, the body must dissipate excess heat by dilating blood vessels and sweating. When the temperature decreases, blood vessels constrict and the body shivers.

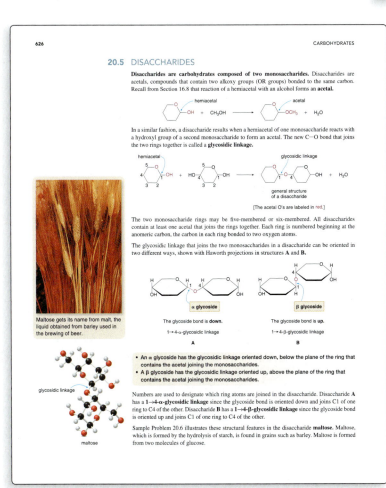

626   CARBOHYDRATES

20.5   DISACCHARIDES

Disaccharides are carbohydrates composed of two monosaccharides. Disaccharides are acetals, compounds that contain two alkoxy groups (OR groups) bonded to the same carbon. Recall from Section 16.8 that reaction of a hemiacetal with an alcohol forms an **acetal.**

hemiacetal

acetal

—OH + CH₃OH ⟶ —OCH₃ + H₂O

In a similar fashion, a disaccharide results when a hemiacetal of one monosaccharide reacts with a hydroxyl group of a second monosaccharide to form an acetal. The new C—O bond that joins the two rings together is called a **glycosidic linkage.**

hemiacetal

glycosidic linkage

general structure of a disaccharide

[The acetal O's are labeled in red.]

The two monosaccharide rings may be five-membered or six-membered. All disaccharides contain at least one acetal that joins the rings together. Each ring is numbered beginning at the anomeric carbon, the carbon in each ring bonded to two oxygen atoms.

The glycosidic linkage that joins the two monosaccharides in a disaccharide can be oriented in two different ways, shown with Haworth projections in structures **A** and **B.**

α glycoside

The glycoside bond is **down.**

1→4-α-glycosidic linkage

A

β glycoside

The glycoside bond is **up.**

1→4-β-glycosidic linkage

B

- An α glycoside has the glycosidic linkage oriented down, below the plane of the ring that contains the acetal joining the monosaccharides.
- A β glycoside has the glycosidic linkage oriented up, above the plane of the ring that contains the acetal joining the monosaccharides.

Numbers are used to designate which ring atoms are joined in the disaccharide. Disaccharide **A** has a **1→4-α-glycosidic linkage** since the glycoside bond is oriented down and joins C1 of one ring to C4 of the other. Disaccharide **B** has a **1→4-β-glycosidic linkage** since the glycoside bond is oriented up and joins C1 of one ring to C4 of the other.

Sample Problem 20.6 illustrates these structural features in the disaccharide **maltose.** Maltose, which is formed by the hydrolysis of starch, is found in grains such as barley. Maltose is formed from two molecules of glucose.

Maltose gets its name from malt, the liquid obtained from barley used in the brewing of beer.

glycosidic linkage

maltose

Engage students with a unique writing style that matches the method in which students learn. Key points of general, organic, and biological chemistry, along with attention-grabbing applications to consumer, environmental, and health-related fields, are woven together in a succinct style for today's to-the-point readers.

SUPPLEMENTS FOR THE INSTRUCTOR

Online Homework and Resources

McGraw-Hill offers online electronic homework along with a myriad of resources for both instructors and students. Instructors can create homework with easy-to-assign algorithmically-generated problems from the text and the simplicity of automatic grading and reporting. *General, Organic, & Biological Chemistry*'s end-of-chapter problems appear in the online homework system in diverse formats and with various tools.

Instructors also have access to PowerPoint lecture outlines, created by Andrea Leonard of the University of Louisiana, Lafayette, which appear as ready-made presentations that combine art and lecture notes for each chapter of the text. For instructors who prefer to create their lectures from scratch, all illustrations, photos, and tables are pre-inserted by chapter into blank Power-Point slides.

An online digital library contains photos, artwork, animations, and other media types that can be used to create customized lectures, visually enhanced tests and quizzes, compelling course websites, or attractive printed support materials. All assets are copyrighted by McGraw-Hill Higher Education, but can be used by instructors for classroom purposes. The visual resources in this collection include:

- **Art**   Full-color digital files of all illustrations in the book can be readily incorporated into lecture presentations, exams, or custom-made classroom materials. In addition, all files are pre-inserted into PowerPoint slides for ease of lecture preparation.
- **Photos**   The photo collection contains digital files of photographs from the text, which can be reproduced for multiple classroom uses.
- **Tables**   Every table that appears in the text has been saved in electronic form for use in classroom presentations and/or quizzes.
- **Animations**   Numerous full-color animations illustrating important processes are also provided. Harness the visual impact of concepts in motion by importing these files into classroom presentations or online course materials.

Instructor's Solutions Manual

This supplement, prepared by Lauren McMills of Ohio University, contains complete, worked out solutions for all problems in the text. It can be found within the Instructor's Resources for this text on the Connect website.

Computerized Test Bank Online

A comprehensive bank of test questions prepared by Kathy Thrush-Shaginaw is provided within a computerized test bank, enabling you to create paper and online tests or quizzes in an easy-to-use program that allows you to prepare and access your test or quiz anywhere, at any time. Instructors can create or edit questions, or drag-and-drop questions, to prepare tests quickly and easily. Tests may be published to their online course, or printed for paper-based assignments.

SUPPLEMENTS FOR THE STUDENT

Student Study Guide/Solutions Manual

The Student Study Guide/Solutions Manual, prepared by Erin Smith and Janice Gorzynski Smith, begins each chapter with a detailed chapter review that is organized around the chapter goals and key concepts. The Problem Solving section provides a number of examples for solving each type of problem essential to that chapter. The Self-Test section of each chapter quizzes chapter highlights, with answers provided. Finally, each chapter ends with the solutions to all in-chapter problems, as well as the solutions to all odd-numbered end-of-chapter problems.

Acknowledgments

Publishing the first edition of a modern chemistry textbook requires a team of knowledgeable and hard-working individuals who are able to translate an author's vision into a reality. I am thankful to work with such a group of dedicated publishing professionals at McGraw-Hill. Much thanks goes to Senior Sponsoring Editor Tami Hodge, whose enthusiasm, upbeat approach, and unflinching support led the conversion of a first draft manuscript to a completed text in record time. I was privileged to once again work with Senior Developmental Editor Donna Nemmers, who managed the day-to-day details of this project with a mix of humor and professionalism. Jayne Klein, Senior Project Manager, skillfully directed the production process, and Publisher Thomas Timp guided the project to assure that all the needed resources were available to see it to completion.

A special thanks goes to Michael Lange, Vice President, New Product Launches, whose experience and insights were critical in developing the early stages of this project. I am especially grateful to freelance Developmental Editor John Murdzek, whose unique blend of humor, chemical knowledge, and attention to detail were key ingredients at numerous stages in the creation of both the text and the student solutions manual. I have also greatly benefited from a team of advisors who have helped guide me through the preparation of my manuscript, as well as a panel of art reviewers who oversaw the conversion of my crude art manuscript into the beautiful figures present in the finished text.

Finally, I thank my family for their support and patience during the long process of publishing a textbook. My husband Dan, an emergency medicine physician, read the entire manuscript, took several photos that appear in the text, and served as a consultant for many medical applications. My daughter Erin co-authored the Student Study Guide/Solutions Manual with me, all of which was written while completing her residency in emergency medicine.

BOARD OF ADVISORS

David J. Gelormo, *Northampton Community College*
Andrea Leonard, *University of Louisiana, Lafayette*
Lauren E. H. McMills, *Ohio University*
Tammy Melton, *Middle Tennessee State University*

ART REVIEW PANEL

Cynthia Graham Brittain, *University of Rhode Island*
Celia Domser, *Mohawk Valley Community College*
Warren Gallagher, *University of Wisconsin, Eau Claire*
Mushtaq Khan, *Union County College*
Terrie Lacson-Lampe, *Georgia Perimeter College*
Paul Root, *Henry Ford Community College*
Heather Sklenicka, *Rochester Community and Technical College*

REVIEWERS

The following people were instrumental in reading and providing feedback on the manuscript, which helped to shape these ideas into cohesive pages:

Karen E. Atkinson, *Bunker Hill Community College*
Cynthia Graham Brittain, *University of Rhode Island*
Albert M. Bobst, *University of Cincinnati, Cincinnati*

David J. Butcher, *Western Carolina University*

Todd A. Carlson, *Grand Valley State University*

Ling Chen, *Borough of Manhattan Community College/CUNY*

Rajeev B. Dabke, *Columbus State University*

William M. Daniel, *Bakersfield College*

Cristina De Meo, *Southern Illinois University, Edwardsville*

Brahmadeo Dewprashad, *Borough of Manhattan Community College*

Celia Domser, *Mohawk Valley Community College*

John W. Francis, *Columbus State Community College*

Warren Gallagher, *University of Wisconsin, Eau Claire*

Zewdu Gebeyehu, *Columbus State University*

David J. Gelormo, *Northampton Community College*

Judy Dirbas George, *Grossmont College*

Marcia Gillette, *Indiana University, Kokomo*

Kevin A. Gratton, *Johnson County Community College*

Michael A. Hailu, *Columbus State Community College*

Amy Hanks, *Brigham Young University, Idaho*

John Haseltine, *Kennesaw State University*

Deborah Herrington, *Grand Valley State University*

Michael O. Hurst, *Georgia Southern University*

Tom Huxford, *San Diego State University*

Martina Kaledin, *Kennesaw State University*

Mushtaq Khan, *Union County College*

Myung-Hoon Kim, *Georgia Perimeter College, Dunwoody Campus*

Terrie Lacson-Lampe, *Georgia Perimeter College*

Richard H. Langley, *Stephen F. Austin State University*

Martin Lawrence, *Montana State University*

Andrea Leonard, *University of Louisiana, Lafayette*

Margaret Ruth Leslie, *Kent State University*

Marc D. Lord, *Columbus State Community College*

Julie Lowe, *Bakersfield College*

Ying Mao, *Camden County College*

Lauren E. H. McMills, *Ohio University*

Tammy Melton, *Middle Tennessee State University*

Mary Bethe Neely, *University of Colorado, Colorado Springs*

Kenneth O'Connor, *Marshall University*

Michael Y. Ogawa, *Bowling Green State University*

Beng Guat Ooi, *Middle Tennessee State University*

John A. Paparelli, *San Antonio College*

Dwight J. Patterson, *Middle Tennessee State University*

Tomislav Pintauer, *Duquesne University*

Douglas Raynie, *South Dakota State University*

Mike E. Rennekamp, *Columbus State Community College*

Jonathan Rhoad, *Missouri Western State University*

Paul Root, *Henry Ford Community College*

Raymond Sadeghi, *University of Texas, San Antonio*

Colleen Scott, *Southern Illinois University, Carbondale*

Masangu Shabangi, *Southern Illinois University, Edwardsville*

Heather M. Sklenicka, *Rochester Community and Technical College*

Denise Stiglich, *Antelope Valley College*

Susan T. Thomas, *University of Texas, San Antonio*

Lawrence Williams, *Wake Tech Community College*

Linda Arney Wilson, *Middle Tennessee State University*

Paulos Yohannes, *Georgia Perimeter College*

List of How To's

List of Applications

1

CHAPTER GOALS

In this chapter you will learn how to:

1. Describe the three states of matter
2. Classify matter as a pure substance, mixture, element, or compound
3. Report measurements using the metric units of length, mass, and volume
4. Use significant figures
5. Use scientific notation for very large and very small numbers
6. Use conversion factors to convert one unit to another
7. Convert temperature from one scale to another
8. Define density and specific gravity and use density to calculate the mass or volume of a substance

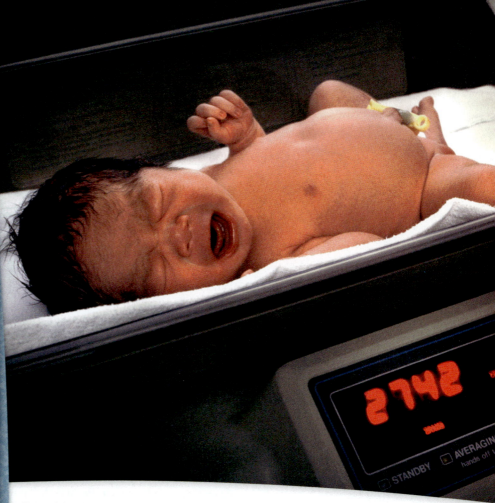

Determining the weight and length of a newborn are common measurements performed by healthcare professionals.

MATTER AND MEASUREMENT

EVERYTHING you touch, feel, or taste is composed of chemicals—that is, **matter**—so an understanding of its composition and properties is crucial to our appreciation of the world around us. Some matter—lakes, trees, sand, and soil—is naturally occurring, while other examples of matter—aspirin, CDs, nylon fabric, plastic syringes, and vaccines—are made by humans. To understand the properties of matter, as well as how one form of matter is converted to another, we must also learn about measurements. Following a recipe, pumping gasoline, and figuring out drug dosages involve manipulating numbers. Thus, Chapter 1 begins our study of chemistry by examining the key concepts of matter and measurement.

1.1 CHEMISTRY—THE SCIENCE OF EVERYDAY EXPERIENCE

What activities might occupy the day of a typical student? You may have done some or all of the following tasks: eaten some meals, drunk coffee or cola, taken a shower with soap, gone to the library to research a paper, taken notes in a class, checked email on a computer, watched some television, ridden a bike or car to a part-time job, taken an aspirin to relieve a headache, and spent some of the evening having snacks and refreshments with friends. Perhaps, without your awareness, your life was touched by chemistry in each of these activities. What, then, is this discipline we call **chemistry?**

- *Chemistry* is the study of matter—its composition, properties, and transformations.

What is **matter?**

- *Matter* is anything that has mass and takes up volume.

In other words, **chemistry studies anything that we touch, feel, see, smell, or taste,** from simple substances like water or salt, to complex substances like proteins and carbohydrates that combine to form the human body. Some matter—cotton, sand, an apple, and the cardiac drug digoxin—is **naturally occurring,** meaning it is isolated from natural sources. Other substances—nylon, Styrofoam, the plastic used in soft drink bottles, and the pain reliever ibuprofen—are **synthetic,** meaning they are produced by chemists in the laboratory (Figure 1.1).

Sometimes a chemist studies what a substance is made of, while at other times he or she might be interested in its properties. Alternatively, the focus may be how to convert one substance into another (Figure 1.2). While naturally occurring rubber exists as the sticky white liquid **latex,** the laboratory process of vulcanization converts it to the stronger, more elastic material used in tires and other products. Although the anticancer drug **taxol** was first isolated in small quantities from the bark of the Pacific yew tree, stripping the bark killed these rare and magnificent trees. Taxol, sold under the trade name of Paclitaxel, is now synthesized in the lab from a substance in the pine needles of the common English yew tree, making it readily available for many cancer patients.

▼ **FIGURE 1.1   Naturally Occurring and Synthetic Materials**

a. Naturally occurring materials   b. Synthetic materials

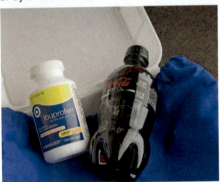

Matter occurs in nature or is synthesized in the lab. (a) Sand and apples are two examples of natural materials. Cotton fabric is woven from cotton fiber, obtained from the cotton plant. The drug digoxin, widely prescribed for decades for patients with congestive heart failure, is extracted from the leaves of the woolly foxglove plant. (b) Nylon was the first synthetic fiber made in the laboratory. It quickly replaced the natural fiber silk in parachutes and ladies' stockings. Styrofoam and PET, the plastic used for soft drink bottles, are strong yet lightweight synthetic materials used for food storage. Over-the-counter pain relievers like ibuprofen are synthetic. The starting materials for all of these useful products are obtained from petroleum.

▼ **FIGURE 1.2** Transforming Natural Materials into Useful Synthetic Products

a.

b.
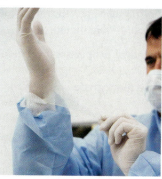

c.

d.

(a) Latex, the sticky liquid that oozes from a rubber tree when it is cut, is too soft for most applications. (b) Vulcanization converts latex to the stronger, elastic rubber used in tires and other products. (c) Taxol was first isolated by stripping the bark of the Pacific yew tree, a process that killed these ancient trees. Estimates suggest that sacrificing one 100-year-old tree provided enough taxol for only a single dose for one cancer patient. (d) Taxol, which is active against breast, ovarian, and some lung tumors, is now synthesized in the lab from a substance that occurs in the needles of the common English yew tree.

Chemistry is truly the science of everyday experience. Soaps and detergents, newspapers and CDs, lightweight exercise gear and Gore-Tex outer wear, condoms and oral contraceptives, Tylenol and penicillin—all of these items are products of chemistry. Without a doubt, advances in chemistry have transformed life in modern times.

PROBLEM 1.1	Look around you and identify five objects. Decide if they are composed of natural or synthetic materials.
PROBLEM 1.2	Imagine that your job as a healthcare professional is to take a blood sample from a patient and store it in a small container in a refrigerator until it is picked up for analysis in the hospital lab. You might have to put on gloves and a mask, use a plastic syringe with a metal needle, store the sample in a test tube or vial, and place it in a cold refrigerator. Pick five objects you might encounter during the process and decide if they are made of naturally occurring or synthetic materials.

1.2 STATES OF MATTER

Matter exists in three common states—solid, liquid, and gas.

- A *solid* has a definite volume, and maintains its shape regardless of the container in which it is placed. The particles of a solid lie close together, and are arranged in a regular three-dimensional array.
- A *liquid* has a definite volume, but takes on the shape of the container it occupies. The particles of a liquid are close together, but they can randomly move around, sliding past one another.
- A *gas* has no definite shape or volume. The particles of a gas move randomly and are separated by a distance much larger than their size. The particles of a gas expand to fill the volume and assume the shape of whatever container they are put in.

For example, water exists in its solid state as ice or snow, liquid state as liquid water, and gaseous state as steam or water vapor. Blow-up circles like those in Figure 1.3 will be used commonly in this text to indicate the composition and state of the particles that compose a substance. In this molecular art, different types of particles are shown in color-coded spheres, and the distance between the spheres signals its state—solid, liquid, or gas.

Matter is characterized by its **physical properties** and **chemical properties.**

- *Physical properties* are those that can be observed or measured without changing the composition of the material.

Common physical properties include melting point (mp), boiling point (bp), solubility, color, and odor. **A *physical change* alters a substance without changing its composition.** The most common physical changes are **changes in state.** Melting an ice cube to form liquid water, and boiling liquid water to form steam are two examples of physical changes. Water is the substance at the beginning and end of both physical changes. More details about physical changes are discussed in Chapter 7.

▼ **FIGURE 1.3   The Three States of Water—Solid, Liquid, and Gas**

a. Solid water

b. Liquid water

c. Gaseous water

- The particles of a solid are close together and highly organized. (Photo: snow-capped Mauna Kea on the Big Island of Hawaii)

- The particles of a liquid are close together but more disorganized than the solid. (Photo: Akaka Falls on the Big Island of Hawaii)

- The particles of a gas are far apart and disorganized. (Photo: steam formed by a lava flow on the Big Island of Hawaii)

Each red sphere joined to two gray spheres represents a single water particle. In proceeding from left to right, from solid to liquid to gas, the molecular art shows that the level of organization of the water particles decreases. Color-coding and the identity of the spheres within the particles will be addressed in Chapter 2.

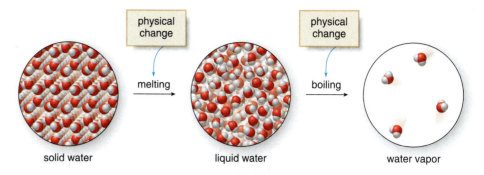

solid water   liquid water   water vapor

- *Chemical properties* are those that determine how a substance can be converted to another substance.

A *chemical change,* **or a** *chemical reaction,* **converts one material to another.** The conversion of hydrogen and oxygen to water is a chemical reaction because the composition of the material is different at the beginning and end of the process. Chemical reactions are discussed in Chapters 5 and 6.

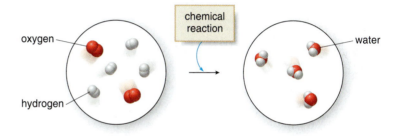

PROBLEM 1.3	Characterize each process as a physical change or a chemical change: (a) making ice cubes; (b) burning natural gas; (c) silver jewelry tarnishing; (d) a pile of snow melting; (e) baking bread.
PROBLEM 1.4	Does the molecular art represent a chemical change or a physical change? Explain your choice.

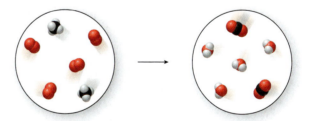

1.3 CLASSIFICATION OF MATTER

All matter can be classified as either a **pure substance** or a **mixture.**

- A *pure substance* is composed of a single component and has a constant composition, regardless of the sample size and the origin of the sample.

A pure substance, such as water or table sugar, can be characterized by its physical properties, because these properties do not change from sample to sample. **A pure substance cannot be broken down to other pure substances by any physical change.**

- A *mixture* is composed of more than one component. The composition of a mixture can vary depending on the sample.

The physical properties of a mixture may also vary from one sample to another. **A mixture can be separated into its components by physical changes.** Dissolving table sugar in water forms a mixture, whose sweetness depends on the amount of sugar added. If the water is allowed to evaporate from the mixture, pure table sugar and pure water are obtained.

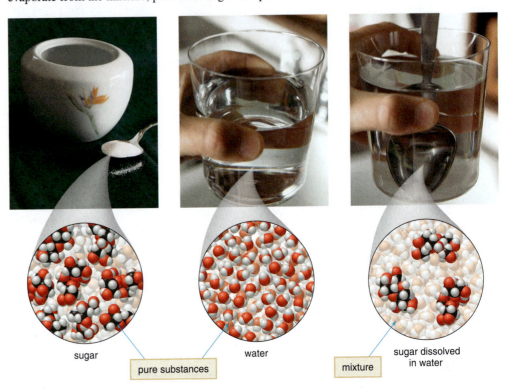

sugar   water   sugar dissolved in water

pure substances   mixture

Mixtures can be formed from solids, liquids, and gases, as shown in Figure 1.4. The compressed air breathed by a scuba diver consists mainly of the gases oxygen and nitrogen. A saline solution used in an IV bag contains solid sodium chloride (table salt) dissolved in water. Rubbing alcohol is a mixture composed of two liquids, 2-propanol and water.

A pure substance is classified as either an **element** or a **compound.**

- An *element* is a pure substance that cannot be broken down into simpler substances by a chemical reaction.
- A *compound* is a pure substance formed by chemically combining (joining together) two or more elements.

An alphabetical list of elements is located on the inside front cover of this text. The elements are commonly organized into a periodic table, also shown on the inside front cover, and discussed in much greater detail in Section 2.4.

Nitrogen gas, aluminum foil, and copper wire are all elements. Water is a compound because it is composed of the elements hydrogen and oxygen. Table salt, sodium chloride, is also a compound since it is formed from the elements sodium and chlorine (Figure 1.5). Although only 114 elements are currently known, over 20 million compounds occur naturally or have been synthesized in the laboratory. We will learn much more about elements and compounds in Chapter 2.

Figure 1.6 summarizes the categories into which matter is classified.

| PROBLEM 1.5 | Classify each item as a pure substance or a mixture: (a) blood; (b) ocean water; (c) a piece of wood; (d) a chunk of ice. |

▼ FIGURE 1.4 Three Examples of Mixtures

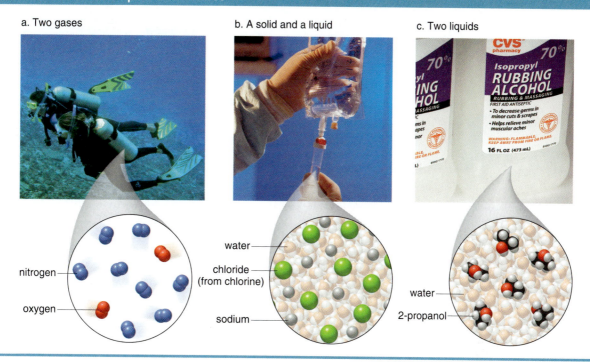

a. Two gases

nitrogen

oxygen

b. A solid and a liquid

water

chloride
(from chlorine)

sodium

c. Two liquids

water

2-propanol

▼ FIGURE 1.5 Elements and Compounds

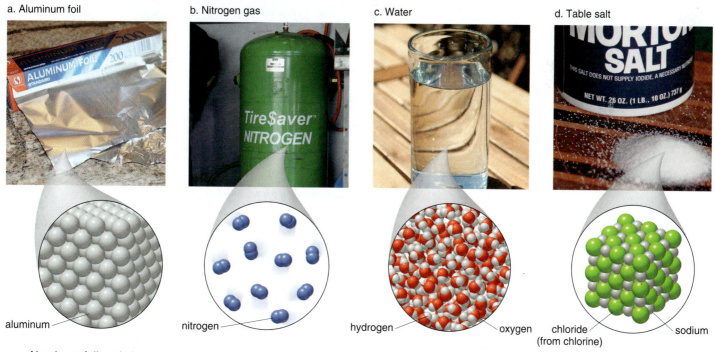

a. Aluminum foil

aluminum

b. Nitrogen gas

nitrogen

c. Water

hydrogen

oxygen

d. Table salt

chloride
(from chlorine)

sodium

- Aluminum foil and nitrogen gas are elements. Water and table salt are compounds. Color-coding of the spheres used in the molecular art indicates that water is composed of two elements—hydrogen shown as gray spheres, and oxygen shown in red. Likewise, the gray (sodium) and green (chlorine) spheres illustrate that sodium chloride is formed from two elements as well.

▼ **FIGURE 1.6   Classification of Matter**

```
                    ┌──────────────────────────────┐
                    │           Matter             │
                    │  anything with mass and volume│
                    └──────────────────────────────┘
                         │                    │
              ┌──────────────────┐   ┌──────────────────────┐
              │  Pure substance  │   │        Mixture       │
              │ a single component│  │ more than one component│
              └──────────────────┘   └──────────────────────┘
                 │            │
        ┌────────────────┐ ┌────────────────────┐
        │    Element     │ │      Compound      │
        │ can't be broken│ │  composed of two or│
        │ down into      │ │   more elements    │
        │ simpler        │ │                    │
        │ substances     │ │                    │
        └────────────────┘ └────────────────────┘
```

PROBLEM 1.6

Classify each item as an element or a compound: (a) the gas inside a helium balloon; (b) table sugar; (c) the rust on an iron nail; (d) aspirin. All elements are listed alphabetically on the inside front cover.

1.4   MEASUREMENT

Any time you check your weight on a scale, measure the ingredients of a recipe, or figure out how far it is from one location to another, you are measuring a quantity. Measurements are routine for healthcare professionals who use weight, blood pressure, pulse, and temperature to chart a patient's progress.

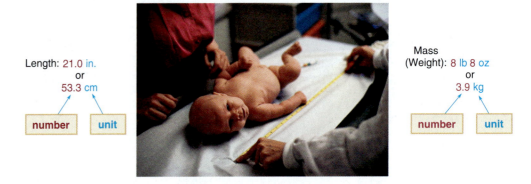

Length: 21.0 in.
or
53.3 cm

number   unit

Mass
(Weight): 8 lb 8 oz
or
3.9 kg

number   unit

- **Every measurement is composed of a *number* and a *unit*.**

Reporting the value of a measurement is meaningless without its unit. For example, if you were told to give a patient an aspirin dosage of 325, does this mean 325 ounces, pounds, grams, milligrams, or tablets? Clearly there is a huge difference among these quantities.

1.4A   THE METRIC SYSTEM

In the United States, most measurements are made with the **English system,** using units like miles (mi), gallons (gal), pounds (lb), and so forth. A disadvantage of this system is that the units are not systematically related to each other and require memorization. For example, 1 lb = 16 oz, 1 gal = 4 qt, and 1 mi = 5,280 ft.

Scientists, health professionals, and people in most other countries use the **metric system,** with units like meter (m) for length, gram (g) for mass, and liter (L) for volume. The metric system is

In 1960, the **International System of Units** was formally adopted as the uniform system of units for the sciences. **SI units,** as they are called, are based on the metric system, but the system encourages the use of some metric units over others. SI stands for the French words, *Système Internationale.*

slowly gaining popularity in the United States. Although milk is still sold in quart or gallon containers, soft drinks are now sold in one- or two-liter bottles. The weight of packaged foods is often given in both ounces and grams. Distances on many road signs are shown in miles and kilometers. Most measurements in this text will be reported using the metric system, but learning to convert English units to metric units is also a necessary skill that will be illustrated in Section 1.7.

The important features of the metric system are the following:

- Each type of measurement has a base unit—the meter (m) for length; the gram (g) for mass; the liter (L) for volume; the second (s) for time.
- All other units are related to the base unit by powers of 10.
- The prefix of the unit name indicates if the unit is larger or smaller than the base unit.

The base units of the metric system are summarized in Table 1.1, and the most common prefixes used to convert the base units to smaller or larger units are summarized in Table 1.2. **The same prefixes are used for all types of measurement.** For example, the prefix *kilo-* means 1,000 times as large. Thus,

$$1 \textbf{ kilo} \text{meter} = \textbf{1,000} \text{ meters} \quad \text{or} \quad 1 \text{ km} = 1,000 \text{ m}$$
$$1 \textbf{ kilo} \text{gram} = \textbf{1,000} \text{ grams} \quad \text{or} \quad 1 \text{ kg} = 1,000 \text{ g}$$
$$1 \textbf{ kilo} \text{liter} = \textbf{1,000} \text{ liters} \quad \text{or} \quad 1 \text{ kL} = 1,000 \text{ L}$$

The prefix *milli-* means one thousandth as large (1/1,000 or 0.001). Thus,

$$1 \textbf{ milli} \text{meter} = \textbf{0.001} \text{ meters} \quad \text{or} \quad 1 \text{ mm} = 0.001 \text{ m}$$
$$1 \textbf{ milli} \text{gram} = \textbf{0.001} \text{ grams} \quad \text{or} \quad 1 \text{ mg} = 0.001 \text{ g}$$
$$1 \textbf{ milli} \text{liter} = \textbf{0.001} \text{ liters} \quad \text{or} \quad 1 \text{ mL} = 0.001 \text{ L}$$

TABLE 1.1 The Basic Metric Units

Quantity	Metric Base Unit	Symbol
Length	Meter	m
Mass	Gram	g
Volume	Liter	L
Time	Second	s

TABLE 1.2 Common Prefixes Used for Metric Units

Prefix	Symbol	Meaning	Numerical Value[a]	Scientific Notation[b]
Mega-	M	Million	1,000,000.	10^6
Kilo-	k	Thousand	1,000.	10^3
Deci-	d	Tenth	0.1	10^{-1}
Centi-	c	Hundredth	0.01	10^{-2}
Milli-	m	Thousandth	0.001	10^{-3}
Micro-	μ	Millionth	0.000 001	10^{-6}
Nano-	n	Billionth	0.000 000 001	10^{-9}

[a]Numbers that contain five or more digits to the right of the decimal point are written with a small space separating each group of three digits.
[b]How to express numbers in scientific notation is explained in Section 1.6.

The metric symbols are all lower case except for the unit **liter** (L) and the prefix **mega-** (M). Liter is capitalized to distinguish it from the number *one*. Mega is capitalized to distinguish it from the symbol for the prefix *milli-*.

PROBLEM 1.7	What term is used for each of the following units: (a) a million liters; (b) a thousandth of a second; (c) a hundredth of a gram; (d) a tenth of a liter?

PROBLEM 1.8	What is the numerical value of each unit in terms of the base unit? (For example, 1 μL = 0.000 001 L.)

    a. 1 ng   b. 1 nm   c. 1 μs   d. 1 ML

1.4B MEASURING LENGTH

The base unit of length in the metric system is the *meter* (m). A meter, 39.4 inches in the English system, is slightly longer than a yard (36 inches). The three most common units derived from a meter are the kilometer (km), centimeter (cm), and millimeter (mm).

$$1{,}000 \text{ m} = 1 \text{ km}$$
$$1 \text{ m} = 100 \text{ cm}$$
$$1 \text{ m} = 1{,}000 \text{ mm}$$

Note how these values are related to those in Table 1.2. Since a centimeter is one *hundredth* of a meter (0.01 m), there are *100* centimeters in a meter.

PROBLEM 1.9	If a nanometer is one billionth of a meter (0.000 000 001 m), how many nanometers are there in one meter?

1.4C MEASURING MASS

Although the terms mass and weight are often used interchangeably, they really have different meanings.

- *Mass* is a measure of the amount of matter in an object.
- *Weight* is the force that matter feels due to gravity.

The mass of an object is independent of its location. The weight of an object changes slightly with its location on the earth, and drastically when the object is moved from the earth to the moon, where the gravitational pull is only one-sixth that of the earth. Although we often speak of *weighing* an object, we are really *measuring its mass*.

The basic unit of mass in the metric system is the *gram* (g), a small quantity compared to the English pound (1 lb = 454 g). The two most common units derived from a gram are the kilogram (kg) and milligram (mg).

$$1{,}000 \text{ g} = 1 \text{ kg}$$
$$1 \text{ g} = 1{,}000 \text{ mg}$$

PROBLEM 1.10	If a microgram is one millionth of a gram (0.000 001 g), how many micrograms are there in one gram?

1.4D MEASURING VOLUME

The basic unit of volume in the metric system is the *liter* (L), which is slightly larger than the English quart (1 L = 1.06 qt). One liter is defined as the volume of a cube 10 cm on an edge.

Note the difference between the units **cm** and **cm³**. The centimeter (cm) is a unit of length. A cubic centimeter (cm³ or cc) is a unit of volume.

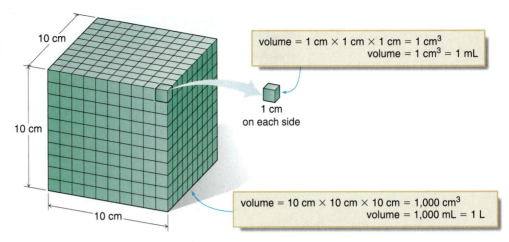

volume = 1 cm × 1 cm × 1 cm = 1 cm³
volume = 1 cm³ = 1 mL

1 cm on each side

volume = 10 cm × 10 cm × 10 cm = 1,000 cm³
volume = 1,000 mL = 1 L

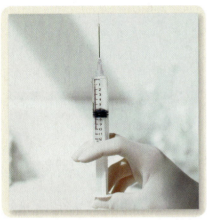

Three common units derived from a liter used in medicine and laboratory research are the deciliter (dL), milliliter (mL), and microliter (μL). **One milliliter is the same as one cubic centimeter (cm³), which is abbreviated as cc.**

$$1 \text{ L} = 10 \text{ dL}$$
$$1 \text{ L} = 1,000 \text{ mL}$$
$$1 \text{ L} = 1,000,000 \text{ μL}$$
$$1 \text{ mL} = 1 \text{ cm}^3 = 1 \text{ cc}$$

Table 1.3 summarizes common metric units of length, mass, and volume. Table 1.4 lists English units of measurement, as well as their metric equivalents.

PROBLEM 1.11

If a centiliter is one hundredth of a liter (0.01 L), how many centiliters are there in one liter?

TABLE 1.3 Summary of the Common Metric Units of Length, Mass, and Volume		
Length	**Mass**	**Volume**
1 km = 1,000 m	1 kg = 1,000 g	1 L = 10 dL
1 m = 100 cm	1 g = 1,000 mg	1 L = 1,000 mL
1 m = 1,000 mm	1 mg = 1,000 μg	1 L = 1,000,000 μL
1 cm = 10 mm		1 dL = 100 mL
		1 mL = 1 cm³ = 1 cc

1.5 SIGNIFICANT FIGURES

Numbers used in chemistry are either **exact** or **inexact**.

- **An *exact* number results from counting objects or is part of a definition.**

Our bodies have 10 fingers, 10 toes, and two kidneys. A meter is composed of 100 centimeters. These numbers are exact because there is no uncertainty associated with them.

TABLE 1.4 English Units and Their Metric Equivalents

Quantity	English Unit	Metric–English Relationship
Length	1 ft = 12 in.	2.54 cm = 1 in.
	1 yd = 3 ft	1 m = 39.4 in.
	1 mi = 5,280 ft	1 km = 0.621 mi
Mass	1 lb = 16 oz	1 kg = 2.21 lb
	1 ton = 2,000 lb	454 g = 1 lb
		28.4 g = 1 oz
Volume	1 qt = 4 cups	946 mL = 1 qt
	1 qt = 2 pints	1 L = 1.06 qt
	1 qt = 32 fl oz	29.6 mL = 1 fl oz
	1 gal = 4 qt	

Common abbreviations for English units: inch (in.), foot (ft), yard (yd), mile (mi), pound (lb), ounce (oz), gallon (gal), quart (qt), and fluid ounce (fl oz).

- An *inexact* number results from a measurement or observation and contains some uncertainty.

Whenever we measure a quantity there is a degree of uncertainty associated with the result. The last number (furthest to the right) is an estimate, and it depends on the type of measuring device we use to obtain it. For example, the length of a fish caught on a recent outing could be reported as 53 cm or 53.5 cm depending on the tape measure used.

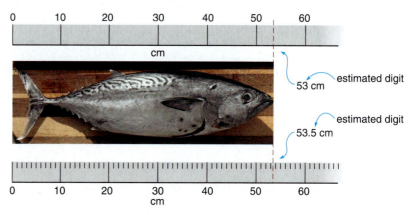

- *Significant figures* are all the digits in a measured number including *one* estimated digit.

Thus, the length 53 cm has two significant figures, and the length 53.5 cm has three significant figures.

1.5A DETERMINING THE NUMBER OF SIGNIFICANT FIGURES

How many significant figures are contained in a number?

- All nonzero digits are always significant.

65.2 g	three significant figures
1,265 m	four significant figures
25 μL	two significant figures
255.345 g	six significant figures

Whether a zero counts as a significant figure depends on its location in the number.

Rules to Determine When a Zero is a Significant Figure

Rule [1] A zero *counts* as a significant figure when it occurs:

- Between two nonzero digits

  29.05 g—four significant figures
  1.0087 mL—five significant figures

- At the end of a number with a decimal point

  25.70 cm—four significant figures
  3.7500 g—five significant figures
  620. lb—three significant figures

Rule [2] A zero does *not* count as a significant figure when it occurs:

- At the beginning of a number

  0.0245 mg—three significant figures
  0.008 mL—one significant figure

- At the end of a number that does not have a decimal point

  2,570 m—three significant figures
  1,245,500 m—five significant figures

> In reading a number with a decimal point from left to right, all digits starting with the first nonzero number are significant figures. The number 0.003 450 120 has seven significant figures, shown in red.

SAMPLE PROBLEM 1.1

How many significant figures does each number contain?

a. 34.08   b. 0.0054   c. 260.00   d. 260

ANALYSIS All nonzero digits are significant. A zero is significant only if it occurs between two nonzero digits, or at the end of a number with a decimal point.

SOLUTION Significant figures are shown in red.

a. 34.08 (four)   b. 0.0054 (two)   c. 260.00 (five)   d. 260 (two)

PROBLEM 1.12

How many significant figures does each number contain?

a. 23.45   c. 230   e. 0.202   g. 1,245,006
b. 23.057   d. 231.0   f. 0.003 60   h. 1,200,000

PROBLEM 1.13

How many significant figures does each number contain?

a. 10,040   c. 1,004.00   e. 1.0040   g. 0.001 004
b. 10,040.   d. 1.004   f. 0.1004   h. 0.010 040 0

PROBLEM 1.14

Indicate whether each zero in the following numbers is significant.

a. 0.003 04   b. 26,045   c. 1,000,034   d. 0.304 00

1.5B USING SIGNIFICANT FIGURES IN MULTIPLICATION AND DIVISION

We often must perform calculations with numbers that contain a different number of significant figures. The number of significant figures in the answer of a problem depends on the type of mathematical calculation—multiplication (and division) or addition (and subtraction).

- In multiplication and division, the answer has the same number of significant figures as the original number with the *fewest* significant figures.

Let's say you drove a car 351.2 miles in 5.5 hours, and you wanted to calculate how many miles per hour you traveled. Entering these numbers on a calculator would give the following result:

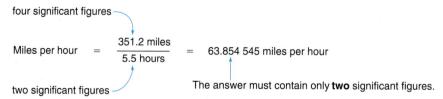

The answer to this problem can have only *two* significant figures, since one of the original numbers (5.5 hours) has only *two* significant figures. To write the answer in proper form, we must **round off the number** to give an answer with only two significant figures. Two rules are used in rounding off numbers.

- If the first number that must be dropped is 4 or less, drop it and all remaining numbers.
- If the first number that must be dropped is 5 or greater, *round the number up* by adding one to the last digit that will be retained.

In this problem:

- Since the first digit to be dropped is 8 (a 5 or greater), add 1 to the digit to its left.
- The answer 63.854 545 rounded to two digits is **64 miles per hour**.

Table 1.5 gives other examples of rounding off numbers.

TABLE 1.5 Rounding Off Numbers		
Original Number	**Rounded to**	**Rounded Number**
61.2537	Two places	61
61.2537	Three places	61.3
61.2537	Four places	61.25
61.2537	Five places	61.254

The first number to be dropped is indicated in red in each original number. When this number is 4 or fewer, it and all other digits to its right are dropped. When this number is 5 or greater, 1 is added to the digit to its left.

SAMPLE PROBLEM 1.2	Round off each number to three significant figures.

a. 1.2735   b. 0.002 536 22   c. 3,836.9

ANALYSIS   If the answer is to have *three* significant figures, look at the *fourth* number from the left. If this number is 4 or less, drop it and all remaining numbers to the right. If the fourth number from the left is 5 or greater, round the number up by adding one to the third digit.

SOLUTION   a. 1.27   b. 0.002 54   c. 3,840 (Omit the decimal point after the 0. The number 3,840. has four significant figures.)

PROBLEM 1.15 Round off each number in Sample Problem 1.2 to two significant figures.

SAMPLE PROBLEM 1.3 Carry out each calculation and give the answer using the proper number of significant figures.

a. 3.81×0.046 b. $120.085/106$

ANALYSIS Since these calculations involve multiplication and division, the answer must have the same number of significant figures as the original number with the fewest number of significant figures.

SOLUTION a. $3.81 \times 0.046 = 0.1753$

- Since 0.046 has only two significant figures, round the answer to give it two significant figures.

0.1753   Since this number is 5 (5 or greater), round the 7 to its left up by one.

Answer: 0.18

b. $120.085/106 = 1.132\ 877\ 36$

- Since 106 has three significant figures, round the answer to give it three significant figures.

1.132 877 36   Since this number is 2 (4 or less), drop it and all numbers to its right.

Answer: 1.13

PROBLEM 1.16 Carry out each calculation and give the answer using the proper number of significant figures.

a. 10.70×3.5 b. $0.206/25{,}993$ c. $1{,}300/41.2$ d. 120.5×26

1.5C USING SIGNIFICANT FIGURES IN ADDITION AND SUBTRACTION

In determining significant figures in addition and subtraction, the decimal place of the last significant digit determines the number of significant figures in the answer.

- **In addition and subtraction, the answer has the same number of decimal places as the original number with the *fewest* decimal places.**

Suppose a baby weighed 3.6 kg at birth and 10.11 kg on his first birthday. To figure out how much weight the baby gained in his first year of life, we subtract these two numbers and report the answer using the proper number of significant figures.

weight at one year = 10.11 kg   two digits after the decimal point
                                       10.11 kg
weight at birth =   3.6 kg   one digit after the decimal point
                                     $-$ 3.6 kg
                     weight gain = 6.51 kg

last significant digit

- The answer can have only **one** digit after the decimal point.
- Round 6.51 to 6.5.
- The baby gained 6.5 kg during his first year of life.

Since 3.6 kg has only one significant figure after the decimal point, the answer can have only one significant figure after the decimal point as well.

SAMPLE PROBLEM 1.4	While on a diet, a woman lost 3.52 lb the first week, 2.2 lb the second week, and 0.59 lb the third week. How much weight did she lose in all?
ANALYSIS	Add up the amount of weight loss each week to get the total weight loss. When adding, the answer has the same number of decimal places as the original number with the fewest decimal places.

SOLUTION

3.52 lb
2.2 lb ← one digit after the decimal point
0.59 lb
─────────
6.31 lb - - - - → 6.3 lb
                    round off
↑
last significant digit

- Since 2.2 lb has only one digit after the decimal point, the answer can have only one digit after the decimal point.
- Round 6.31 to 6.3.
- Total weight loss: 6.3 lb.

PROBLEM 1.17	Carry out each calculation and give the answer using the proper number of significant figures.

a. 27.8 cm + 0.246 cm

b. 102.66 mL + 0.857 mL + 24.0 mL

c. 54.6 mg – 25 mg

d. 2.35 s – 0.266 s

1.6 SCIENTIFIC NOTATION

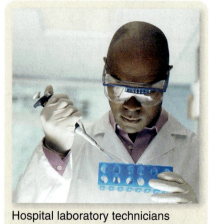

Hospital laboratory technicians determine thousands of laboratory results each day.

Healthcare professionals and scientists must often deal with very large and very small numbers. For example, the blood platelet count of a healthy adult might be 250,000 platelets per mL. At the other extreme, the level of the female sex hormone estriol during pregnancy might be 0.000 000 250 g per mL of blood plasma. Estriol is secreted by the placenta and its concentration is used as a measure of the health of the fetus.

To write numbers that contain many leading zeros (at the beginning) or trailing zeros (at the end), scientists use **scientific notation.**

- In scientific notation, a number is written as $y \times 10^x$.
- The term y, called the coefficient, is a number between 1 and 10.
- The value x is an exponent, which can be any positive or negative whole number.

First, let's recall what powers of 10 with *positive* exponents, such as 10^2 or 10^5, mean. These correspond to numbers greater than one, and the positive exponent tells how many zeros are to be written after the number one. Thus, $10^2 = 100$, a number with two zeros after the number one.

The product has two zeros.

$$10^2 = 10 \times 10 = 100$$

The exponent 2 means "multiply two 10s."

The product has five zeros.

$$10^5 = 10 \times 10 \times 10 \times 10 \times 10 = 100,000$$

The exponent 5 means "multiply five 10s."

Powers of 10 that contain *negative* exponents, such as 10^{-3}, correspond to numbers less than one. In this case the exponent tells how many places (*not* zeros) are located to the right of the decimal point.

The answer has three places to the right of the decimal point, including the number one.

$$10^{-3} = \frac{1}{10 \times 10 \times 10} = 0.001$$

The exponent −3 means "divide by three 10s."

To write a number in scientific notation, we follow a stepwise procedure.

HOW TO   Convert a Standard Number to Scientific Notation

EXAMPLE   Write each number in scientific notation: (a) 2,500; (b) 0.036.

Step [1]   **Move the decimal point to give a number between 1 and 10.**

a. 2500.

Move the decimal point three places to the left to give the number 2.5.

b. 0.036

Move the decimal point two places to the right to give the number 3.6.

Step [2]   **Multiply the result by 10^x, where x is the number of places the decimal point was moved.**

- If the decimal point is moved to the **left**, x is **positive**.
- If the decimal point is moved to the **right**, x is **negative**.

a. Since the decimal point was moved three places to the **left**, the exponent is +3, and the coefficient is multiplied by 10^3.

   Answer: $2,500 = 2.5 \times 10^3$

b. Since the decimal point was moved two places to the **right**, the exponent is –2, and the coefficient is multiplied by 10^{-2}.

   Answer: $0.036 = 3.6 \times 10^{-2}$

Notice that the number of significant figures in the coefficient in scientific notation must equal the number of significant figures in the original number. Thus, the coefficients for both 2,500 and 0.036 need two significant figures and no more. Table 1.6 shows how several numbers are written in scientific notation.

$2,500 = 2.5 \times 10^3$ ***not*** 2.50×10^3 (three significant figures)

two significant figures   ***not*** 2.500×10^3 (four significant figures)

TABLE 1.6   Numbers in Standard Form and Scientific Notation

Number	Scientific Notation
53,400	5.34×10^4
0.005 44	5.44×10^{-3}
3,500,000,000	3.5×10^9
0.000 000 000 123	1.23×10^{-10}
1,000.03	1.00003×10^3

SAMPLE PROBLEM 1.5

Write the recommended daily dietary intake of each nutrient in scientific notation: (a) sodium, 2,400 mg; (b) vitamin B_{12}, 0.000 006 g.

ANALYSIS   Move the decimal point to give a number between 1 and 10. Multiply the number by 10^x, where x is the number of places the decimal point was moved. The exponent x is (+) when the decimal point moves to the left and (–) when it moves to the right.

SOLUTION

a.

$2400. = 2.4 \times 10^3$ the number of places the decimal point was moved to the left

Move the decimal point three places to the left.

- Write the coefficient as 2.4 (two significant figures), since 2,400 contains two significant figures.

b.

$0.000\ 006 = 6 \times 10^{-6}$ the number of places the decimal point was moved to the right

Move the decimal point six places to the right.

- Write the coefficient as 6 (one significant figure), since 0.000 006 contains one significant figure.

PROBLEM 1.18	Lab results for a routine check-up showed an individual's iron level in the blood to be 0.000 098 g per deciliter, placing it in the normal range. Convert this number to scientific notation.

PROBLEM 1.19	Write each number in scientific notation. a. 93,200 c. 6,780,000 e. 4,520,000,000,000 b. 0.000 725 d. 0.000 030 f. 0.000 000 000 028

To convert a number in scientific notation to a standard number, reverse the procedure, as shown in Sample Problem 1.6. It is often necessary to add leading or trailing zeros to write the number.

- **When the exponent x is positive, move the decimal point x places to the *right*.**

$$2.800 \times 10^2 \qquad\qquad 2.800 \quad \text{---} \rightarrow \quad 280.0$$

Move the decimal point to the right two places.

- **When the exponent x is negative, move the decimal point x places to the *left*.**

$$2.80 \times 10^{-2} \qquad\qquad 002.80 \quad \text{---} \rightarrow \quad 0.0280$$

Move the decimal point to the left two places.

SAMPLE PROBLEM 1.6	As we will learn in Chapter 4, the element hydrogen is composed of two hydrogen atoms, separated by a distance of 7.4×10^{-11} m. Convert this value to a standard number.
ANALYSIS	The exponent in 10^x tells how many places to move the decimal point in the coefficient to generate a standard number. The decimal point goes to the right when x is positive and to the left when x is negative.
SOLUTION	$7.4 \times 10^{-11} \qquad\qquad 000\,000\,000\,07.4 \quad \text{---} \rightarrow \quad 0.000\,000\,000\,074$ m Move the decimal point to the left 11 places. **Answer:** The answer, 0.000 000 000 074, has two significant figures, just like 7.4×10^{-11}.

PROBLEM 1.20	There are 6.02×10^{21} "particles" called molecules (Chapter 4) of aspirin in 1.8 g. Write this number in standard form.

PROBLEM 1.21	Convert each number to its standard form. a. 6.5×10^3 c. 3.780×10^{-2} e. 2.221×10^6 b. 3.26×10^{-5} d. 1.04×10^8 f. 4.5×10^{-10}

1.7 PROBLEM SOLVING USING THE FACTOR–LABEL METHOD

Often a measurement is recorded in one unit, and then it must be converted to another unit. For example, a patient may weigh 130 lb, but we may need to know her weight in kilograms to calculate a drug dosage. The recommended daily dietary intake of potassium is 3,500 mg, but we may need to know how many grams this corresponds to.

1.7A CONVERSION FACTORS

To convert one unit to another we use one or more **conversion factors**.

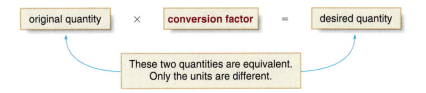

- A *conversion factor* is a term that converts a quantity in one unit to a quantity in another unit.

A conversion factor is formed by taking an equality, such as 2.21 lb = 1 kg, and writing it as a fraction. We can always write a conversion factor in two different ways.

$$\frac{2.21 \text{ lb}}{1 \text{ kg}} \quad or \quad \frac{1 \text{ kg}}{2.21 \text{ lb}} \quad \begin{array}{l}\text{numerator}\\[6pt]\text{denominator}\end{array}$$

conversion factors for pounds and kilograms

Refer to Tables 1.3 and 1.4 for metric and English units needed in problem solving. Common metric and English units are also listed on the inside back cover.

With pounds and kilograms, either of these values can be written above the division line of the fraction (the numerator) or below the division line (the denominator). The way the conversion factor is written will depend on the problem. Since the values above and below the division line are *equivalent,* a conversion factor always equals one.

SAMPLE PROBLEM 1.7

Write two conversion factors for each pair of units: (a) kilograms and grams; (b) quarts and liters.

ANALYSIS Use the equalities in Tables 1.3 and 1.4 to write a fraction that shows the relationship between the two units.

SOLUTION
a. Conversion factors for kilograms and grams:

$$\frac{1000 \text{ g}}{1 \text{ kg}} \quad or \quad \frac{1 \text{ kg}}{1000 \text{ g}}$$

b. Conversion factors for quarts and liters:

$$\frac{1.06 \text{ qt}}{1 \text{ L}} \quad or \quad \frac{1 \text{ L}}{1.06 \text{ qt}}$$

PROBLEM 1.22

Write two conversion factors for each pair of units.

a. miles and kilometers
b. meters and millimeters
c. grams and pounds
d. milligrams and micrograms

1.7B SOLVING A PROBLEM USING ONE CONVERSION FACTOR

Using conversion factors to convert a quantity in one unit to a quantity in another unit is called the **factor–label method.** In this method, **units are treated like numbers.** As a result, if a unit appears in the numerator in one term and the denominator in another term, the units *cancel*. **The goal in setting up a problem is to make sure *all unwanted units cancel.***

Let's say we want to convert 130 lb to kilograms.

$$\underset{\text{original quantity}}{130 \text{ lb}} \quad \times \quad \boxed{\textbf{conversion factor}} \quad = \quad \underset{\text{desired quantity}}{? \quad \text{kg}}$$

Two possible conversion factors: $\dfrac{2.21 \text{ lb}}{1 \text{ kg}}$ or $\dfrac{1 \text{ kg}}{2.21 \text{ lb}}$

To solve this problem we must use a conversion factor that satisfies two criteria.

- **The conversion factor must relate the two quantities in question—pounds and kilograms.**
- **The conversion factor must cancel out the unwanted unit—pounds.**

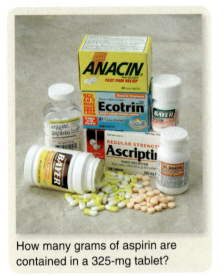

How many grams of aspirin are contained in a 325-mg tablet?

This means choosing the conversion factor with the unwanted unit—pounds—*in the denominator* to cancel out pounds in the original quantity. This leaves kilograms as the only remaining unit, and the problem is solved.

conversion factor

$$130 \text{ lb} \times \frac{1 \text{ kg}}{2.21 \text{ lb}} = 59 \text{ kg} \quad \textbf{answer in kilograms}$$

Pounds (lb) must be the denominator to cancel
the unwanted unit (lb) in the original quantity.

We must use the correct number of significant figures in reporting an answer to each problem. In this case, the value 1 kg is *defined* as 2.21 lb; in other words, 1 kg contains the exact number "1" with *no* uncertainty, so it does not limit the number of digits in the answer. Since 130 lb has two significant figures, the answer is rounded to two significant figures (59 kg).

As problems with units get more complicated, keep in mind the following general steps that are useful for solving any problem using the factor–label method.

HOW TO   Solve a Problem Using Conversion Factors

EXAMPLE   **How many grams of aspirin are contained in a 325-mg tablet?**

Step [1]   **Identify the original quantity and the desired quantity, including units.**

- In this problem the original quantity is reported in milligrams and the desired quantity is in grams.

325 mg   ? g
original quantity   desired quantity

Step [2]   **Write out the conversion factor(s) needed to solve the problem.**

- We need a conversion factor that relates milligrams and grams (Table 1.3). Since the unwanted unit is in milligrams, **choose the conversion factor that contains milligrams in the denominator so that the *units* cancel.**

Two possible conversion factors: $\dfrac{1000 \text{ mg}}{1 \text{ g}}$ or $\dfrac{1 \text{ g}}{1000 \text{ mg}}$ —— Choose this factor to cancel the unwanted unit, mg.

- Sometimes one conversion factor is all that is needed in a problem. At other times (Section 1.7C) more than one conversion factor is needed.
- If the desired answer has a single unit (grams in this case), **the conversion factor must contain the desired unit in the numerator and the unwanted unit in the denominator.**

Step [3]   **Set up and solve the problem.**

- Multiply the original quantity by the conversion factor to obtain the desired quantity.

conversion factor

$$\underset{\text{original quantity}}{325 \text{ mg}} \times \frac{1 \text{ g}}{1000 \text{ mg}} = \underset{\text{desired quantity}}{0.325 \text{ g of aspirin}}$$

The number of mg (unwanted unit) cancels.

Step [4]   **Write the answer using the correct number of significant figures and check it by estimation.**

- Use the number of significant figures in each inexact (measured) number to determine the number of significant figures in the answer. In this case the answer is limited to three significant figures by the original quantity (325 mg).
- Estimate the answer using a variety of methods. In this case we knew our answer had to be less than one, since it is obtained by dividing 325 by a number larger than itself.

PROBLEM 1.23	The distance between Honolulu, HI, and Los Angeles, CA, is 4,120 km. How many frequent flyer miles will you earn by traveling between the two cities?
PROBLEM 1.24	Convert 25 mL to μL and write the answer in scientific notation.
PROBLEM 1.25	Carry out each of the following conversions. a. 25 L to dL b. 40.0 oz to g c. 32 in. to cm d. 10 cm to mm

1.7C SOLVING A PROBLEM USING TWO OR MORE CONVERSION FACTORS

Some problems require the use of more than one conversion factor to obtain the desired units in the answer. The same four-step procedure is followed no matter how many conversion factors are needed. Keep in mind:

> • Always arrange the factors so that the denominator in one term cancels the numerator in the preceding term.

Sample Problem 1.8 illustrates how to solve a problem with two conversion factors.

SAMPLE PROBLEM 1.8	An individual donated 1.0 pint of blood at the local blood bank. How many liters of blood does this correspond to?

ANALYSIS AND SOLUTION

[1] **Identify the original quantity and the desired quantity.**

$$\underset{\text{original quantity}}{1.0\ \text{pt}} \qquad \underset{\text{desired quantity}}{?\ \text{L}}$$

[2] **Write out the conversion factors.**

- We have no conversion factor that directly relates pints to liters. We do, however, know conversions for pints to quarts, and quarts to liters.

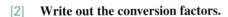

pint–quart conversion

$$\frac{2\ \text{pt}}{1\ \text{qt}} \quad \text{or} \quad \boxed{\frac{1\ \text{qt}}{2\ \text{pt}}}$$

quart–liter conversion

$$\frac{1.06\ \text{qt}}{1\ \text{L}} \quad \text{or} \quad \boxed{\frac{1\ \text{L}}{1.06\ \text{qt}}}$$

Choose the conversion factors with the unwanted units—pt and qt—in the denominator.

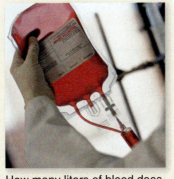

How many liters of blood does this pint of blood contain?

[3] **Solve the problem.**

- To set up the problem so that unwanted units cancel, arrange each term so that the units in the numerator of one term cancel the units of the denominator of the adjacent term. In this problem we need to cancel both pints and quarts to get liters.
- The single desired unit, liters, must be in the **numerator** of one term.

Liters do not cancel.

$$1.0\ \text{pt} \ \times\ \frac{1\ \text{qt}}{2\ \text{pt}} \ \times\ \frac{1\ \text{L}}{1.06\ \text{qt}} \ =\ 0.47\ \text{L}$$

Pints cancel.   Quarts cancel.

[4] **Check.**

- Since there are two pints in a quart and a quart is about the same size as a liter, one pint should be about half a liter. The answer, 0.47, is just about 0.5.
- Write the answer with two significant figures since one term, 1.0 pt, has two significant figures.

PROBLEM 1.26 Carry out each of the following conversions.

  a. 6,250 ft to km   b. 3 cups to L   c. 4.5 ft to cm

PROBLEM 1.27 On a recent road trip, your average speed was 65 miles per hour. What was your average speed in (a) mi/s; (b) m/s?

1.8 FOCUS ON HEALTH & MEDICINE
PROBLEM SOLVING USING CLINICAL CONVERSION FACTORS

Sometimes conversion factors don't have to be looked up in a table; they are stated in the problem. If a drug is sold as a 250-mg tablet, this fact becomes a conversion factor relating milligrams to tablets.

$$\frac{250 \text{ mg}}{1 \text{ tablet}} \quad \text{or} \quad \frac{1 \text{ tablet}}{250 \text{ mg}}$$

mg–tablet conversion factors

The active ingredient in Children's Tylenol is acetaminophen.

Alternatively, a drug could be sold as a liquid solution with a specific concentration. For example, Children's Tylenol contains 80 mg of the active ingredient acetaminophen in 2.5 mL. This fact becomes a conversion factor relating milligrams to milliliters.

$$\frac{80 \text{ mg}}{2.5 \text{ mL}} \quad \text{or} \quad \frac{2.5 \text{ mL}}{80 \text{ mg}}$$

mg of acetaminophen–mL conversion factors

Sample Problems 1.9 and 1.10 illustrate how these conversion factors are used in determining drug dosages.

SAMPLE PROBLEM 1.9 A patient is prescribed 1.25 g of amoxicillin, which is available in 250-mg tablets. How many tablets are needed?

ANALYSIS AND SOLUTION

[1] **Identify the original quantity and the desired quantity.**

  • We must convert the number of grams of amoxicillin needed to the number of tablets that must be administered.

     1.25 g   ? tablets
    original quantity   desired quantity

[2] **Write out the conversion factors.**

  • We have no conversion factor that directly relates grams to tablets. We do know, however, how to relate grams to milligrams, and milligrams to tablets.

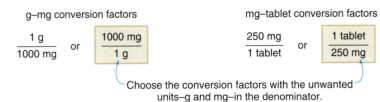

   g–mg conversion factors   mg–tablet conversion factors

$$\frac{1 \text{ g}}{1000 \text{ mg}} \quad \text{or} \quad \boxed{\frac{1000 \text{ mg}}{1 \text{ g}}} \qquad\qquad \frac{250 \text{ mg}}{1 \text{ tablet}} \quad \text{or} \quad \boxed{\frac{1 \text{ tablet}}{250 \text{ mg}}}$$

Choose the conversion factors with the unwanted units–g and mg–in the denominator.

[3]   **Solve the problem.**

- Arrange each term so that the units in the numerator of one term cancel the units in the denominator of the adjacent term. In this problem we need to cancel both grams and milligrams to get tablets.
- The single desired unit, tablets, must be located in the **numerator** of one term.

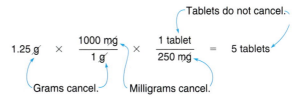

Tablets do not cancel.

$$1.25 \text{ g} \quad \times \quad \frac{1000 \text{ mg}}{1 \text{ g}} \quad \times \quad \frac{1 \text{ tablet}}{250 \text{ mg}} \quad = \quad 5 \text{ tablets}$$

Grams cancel.   Milligrams cancel.

[4]   **Check.**

- The answer of 5 tablets of amoxicillin (not 0.5 or 50) is reasonable. Since the dose in a single tablet (250 mg) is a fraction of a gram, and the required dose is more than a gram, the answer must be greater than one.

SAMPLE PROBLEM 1.10

A dose of 240 mg of acetaminophen is prescribed for a 20-kg child. How many mL of Children's Tylenol (80. mg of acetaminophen per 2.5 mL) are needed?

ANALYSIS AND SOLUTION

[1]   **Identify the original quantity and the desired quantity.**

- We must convert the number of milligrams of acetaminophen needed to the number of mL that must be administered.

$$\begin{array}{cc} 240 \text{ mg} & ? \text{ mL} \\ \text{original quantity} & \text{desired quantity} \end{array}$$

[2]   **Write out the conversion factors.**

mg of acetaminophen–mL conversion factors

$$\frac{80. \text{ mg}}{2.5 \text{ mL}} \quad \text{or} \quad \boxed{\frac{2.5 \text{ mL}}{80. \text{ mg}}}$$

Choose the conversion factor to cancel mg.

[3]   **Solve the problem.**

- Arrange the terms so that the units in the numerator of one term cancel the units of the denominator of the adjacent term. In this problem we need to cancel milligrams to obtain milliliters.
- In this problem we are given a fact we don't need to use—the child weighs 20 kg. We can ignore this quantity in carrying out the calculation.

$$240 \text{ mg} \quad \times \quad \frac{2.5 \text{ mL}}{80. \text{ mg}} \quad = \quad 7.5 \text{ mL of Children's Tylenol}$$

Milligrams cancel.

[4]   **Check.**

- The answer of 7.5 mL (not 0.75 or 75) is reasonable. Since the required dose is larger than the dose in 2.5 mL, the answer must be larger than 2.5 mL.

PROBLEM 1.28

If one teaspoon contains 5.0 mL, how many teaspoons of Children's Tylenol must be administered in Sample Problem 1.10?

| PROBLEM 1.29 | A patient is prescribed 0.100 mg of a drug that is available in 25-µg tablets. How many tablets are needed? |
| PROBLEM 1.30 | How many milliliters of Children's Motrin (100 mg of ibuprofen per 5 mL) are needed to give a child a dose of 160 mg? |

1.9 TEMPERATURE

Temperature is a measure of how hot or cold an object is. Three temperature scales are used: **Fahrenheit** (most common in the United States), **Celsius** (most commonly used by scientists and countries other than the United States), and **Kelvin** (Figure 1.7).

The Fahrenheit and Celsius scales are both divided into **degrees.** On the Fahrenheit scale, water freezes at 32 °F and boils at 212 °F. On the Celsius scale, water freezes at 0 °C and boils at 100 °C. To convert temperature values from one scale to another, we use two equations, where °C is the Celsius temperature and °F is the Fahrenheit temperature.

To convert from Celsius to Fahrenheit:

$$°F = 1.8(°C) + 32$$

To convert from Fahrenheit to Celsius:

$$°C = \frac{°F - 32}{1.8}$$

The Kelvin scale is divided into **kelvins** (K), not degrees. The only difference between the Kelvin scale and the Celsius scale is the zero point. A temperature of –273 °C corresponds to 0 K. The zero point on the Kelvin scale is called **absolute zero,** the lowest temperature possible. To convert temperature values from Celsius to Kelvin, or vice versa, use two equations.

To convert from Celsius to Kelvin:

$$K = °C + 273$$

To convert from Kelvin to Celsius:

$$°C = K - 273$$

▼ **FIGURE 1.7   Fahrenheit, Celsius, and Kelvin Temperature Scales Compared**

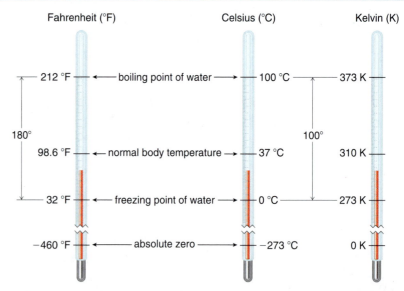

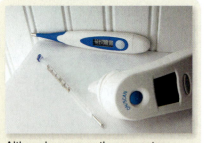

Although mercury thermometers were used in hospitals to measure temperature for many years, temperature is now more commonly recorded with a digital thermometer. Tympanic thermometers, which use an infrared sensing device placed in the ear, are also routinely used.

Since the freezing point and boiling point of water span 180° on the Fahrenheit scale, but only 100° on the Celsius scale, a Fahrenheit degree and a Celsius degree differ in size. The Kelvin scale is divided into kelvins (K), not degrees. Since the freezing point and boiling point of water span 100 kelvins, one kelvin is the same size as one Celsius degree.

SAMPLE PROBLEM 1.11   An infant had a temperature of 104 °F. Convert this temperature to both °C and K.

ANALYSIS   First convert the Fahrenheit temperature to degrees Celsius using the equation °C = (°F – 32)/1.8. Then convert the Celsius temperature to kelvins by adding 273.

SOLUTION

[1] **Convert °F to °C:**

$$°C = \frac{°F - 32}{1.8}$$

$$= \frac{104 - 32}{1.8} = 40. \, °C$$

[2] **Convert °C to K:**

$$K = °C + 273$$

$$= 40. + 273 = 313 \, K$$

PROBLEM 1.31   When the human body is exposed to extreme cold, hypothermia can result and the body's temperature can drop to 28.5 °C. Convert this temperature to °F and K.

PROBLEM 1.32   Convert each temperature to the requested temperature scale.

a. 20 °C to °F   c. 298 K to °F

b. 150 °F to °C   d. 75 °C to K

1.10 DENSITY AND SPECIFIC GRAVITY

Two additional quantities used to characterize substances are **density** and **specific gravity.**

1.10A DENSITY

***Density* is a physical property that relates the mass of a substance to its volume.** Density is reported in grams per milliliter (g/mL) or grams per cubic centimeter (g/cc).

$$\text{density} = \frac{\text{mass (g)}}{\text{volume (mL or cc)}}$$

The density of a substance depends on temperature. For most substances, the solid state is more dense than the liquid state, and as the temperature increases, the density decreases. This phenomenon occurs because the volume of a sample of a substance generally increases with temperature but the mass is always constant.

Water is an exception to this generalization. Solid water, ice, is *less* dense than liquid water, and from 0 °C to 4 °C, the density of water *increases*. Above 4 °C, water behaves like other liquids and its density decreases. Thus, water's maximum density of 1.000 g/mL occurs at 4 °C. Some representative densities are reported in Table 1.7.

TABLE 1.7 Representative Densities at 25 °C

Substance	Density [g/(mL or cc)]	Substance	Density [g/(mL or cc)]
Oxygen (0 °C)	0.001 43	Urine	1.003–1.030
Gasoline	0.66	Blood plasma	1.03
Ice (0 °C)	0.92	Table sugar	1.59
Water (4 °C)	1.00	Bone	1.80

The density (not the mass) of a substance determines whether it floats or sinks in a liquid.

- **A less dense substance floats on a more dense liquid.**

Ice floats on water because it is less dense. When petroleum leaks from an oil tanker or gasoline is spilled when fueling a boat, it floats on water because it is less dense. In contrast, a cannonball or torpedo sinks because it is more dense than water.

Knowing the density of a liquid allows us to convert the volume of a substance to its mass, or the mass of a substance to its volume.

Although a can of a diet soft drink floats in water because it is less dense, a can of a regular soft drink that contains sugar is more dense than water so it sinks.

To convert volume (mL) to mass (g):

$$\text{mL} \times \underbrace{\frac{\text{g}}{\text{mL}}}_{\text{density}} = \text{g}$$

Milliliters cancel.

To convert mass (g) to volume (mL):

$$\text{g} \times \underbrace{\frac{\text{mL}}{\text{g}}}_{\substack{\text{inverse of} \\ \text{the density}}} = \text{mL}$$

Grams cancel.

For example, one laboratory synthesis of aspirin uses the liquid acetic acid, which has a density of 1.05 g/mL. If we need 5.0 g for a synthesis, we could use density to convert this mass to a volume that could then be easily measured out using a syringe or pipette.

$$5.0 \text{ g acetic acid} \times \frac{1 \text{ mL}}{1.05 \text{ g}} = 4.8 \text{ mL of acetic acid}$$

Grams cancel.

SAMPLE PROBLEM 1.12

Calculate the mass in grams of 15.0 mL of a saline solution that has a density 1.05 g/mL.

ANALYSIS Use density (g/mL) to interconvert the mass and volume of a liquid.

SOLUTION

$$15.0 \text{ mL} \times \underbrace{\frac{1.05 \text{ g}}{1 \text{ mL}}}_{\text{density}} = 15.8 \text{ g of saline solution}$$

Milliliters cancel.

The answer, 15.8 g, is rounded to three significant figures to match the number of significant figures in both factors in the problem.

PROBLEM 1.33

Calculate the mass in grams of 10.0 mL of diethyl ether, an anesthetic that has a density of 0.713 g/mL.

PROBLEM 1.34

(a) Calculate the volume in milliliters of 100. g of coconut oil, which has a density of 0.92 g/mL. (b) How many liters does this correspond to?

PROBLEM 1.35

Ten milliliters of either hexane (density = 0.65 g/mL) or chloroform (density = 1.49 g/mL) was added to a beaker that contains 10 mL of water, forming two layers with water on top. What liquid was added to the beaker?

1.10B SPECIFIC GRAVITY

Specific gravity **is a quantity that compares the density of a substance with the density of water at the same temperature.**

$$\text{specific gravity} = \frac{\text{density of a substance (g/mL)}}{\text{density of water (g/mL)}}$$

Unlike most other quantities, specific gravity is a quantity without units, since the units in the numerator (g/mL) cancel the units in the denominator (g/mL). Since the density of water is 1.00 g/mL at and around room temperature, **the specific gravity of a substance equals its density, but it contains no units.** For example, if the density of a liquid is 1.5 g/mL at 20 °C, its specific gravity is 1.5.

The specific gravity of urine samples is often measured in a hospital lab. Normal urine has a density in the range of 1.003–1.030 g/mL (Table 1.7), so it has a specific gravity in the range of 1.003–1.030. Consistently high or low values can indicate an imbalance in metabolism. For example, the specific gravity of urine samples from patients with poorly controlled diabetes is abnormally high, because a large amount of glucose is excreted in the urine.

PROBLEM 1.36	(a) If the density of a liquid is 0.80 g/mL, what is its specific gravity? (b) If the specific gravity of a substance is 2.3, what is its density?

CHAPTER HIGHLIGHTS

KEY TERMS

Celsius scale (1.9)
Chemical properties (1.2)
Chemistry (1.1)
Compound (1.3)
Conversion factor (1.7)
Cubic centimeter (1.4)
Density (1.10)
Element (1.3)
English system of measurement (1.4)
Exact number (1.5)
Factor–label method (1.7)

Fahrenheit scale (1.9)
Gas (1.2)
Gram (1.4)
Inexact number (1.5)
Kelvin scale (1.9)
Liquid (1.2)
Liter (1.4)
Mass (1.4)
Matter (1.1)
Meter (1.4)
Metric system (1.4)

Mixture (1.3)
Physical properties (1.2)
Pure substance (1.3)
Scientific notation (1.6)
SI units (1.4)
Significant figures (1.5)
Solid (1.2)
Specific gravity (1.10)
States of matter (1.2)
Temperature (1.9)
Weight (1.4)

KEY CONCEPTS

❶ Describe the three states of matter. (1.1, 1.2)
- Matter is anything that has mass and takes up volume. Matter has three common states:
  - The solid state is composed of highly organized particles that lie close together. A solid has a definite shape and volume.
  - The liquid state is composed of particles that lie close together but are less organized than the solid state. A liquid has a definite volume but not a definite shape.
  - The gas state is composed of highly disorganized particles that lie far apart. A gas has no definite shape or volume.

❷ How is matter classified? (1.3)
- Matter is classified in one of two categories:
  - A pure substance is composed of a single component with a constant composition. A pure substance is either an element, which cannot be broken down into simpler substances by a chemical reaction, or a compound, which is formed by combining two or more elements.
  - A mixture is composed of more than one component and its composition can vary depending on the sample.

3 **What are the key features of the metric system of measurement? (1.4)**

- The metric system is a system of measurement in which each type of measurement has a base unit and all other units are related to the base unit by a prefix that indicates if the unit is larger or smaller than the base unit.
- The base units are meter (m) for length, gram (g) for mass, liter (L) for volume, and second (s) for time.

4 **What are significant figures and how are they used in calculations? (1.5)**

- Significant figures are all digits in a measured number, including one estimated digit. All nonzero digits are significant. A zero is significant only if it occurs between two nonzero digits, or at the end of a number with a decimal point. A trailing zero in a number without a decimal point is not considered significant.
- In multiplying and dividing with significant figures, the answer has the same number of significant figures as the original number with the fewest significant figures.
- In adding or subtracting with significant figures, the answer has the same number of decimal places as the original number with the fewest decimal places.

5 **What is scientific notation? (1.6)**

- Scientific notation is a method of writing a number as $y \times 10^x$, where y is a number between 1 and 10, and x is a positive or negative exponent.
- To convert a standard number to a number in scientific notation, move the decimal point to give a number between 1 and 10. Multiply the result by 10^x, where x is the number of places the decimal point was moved. When the decimal point is moved to the left, x is positive. When the decimal point is moved to the right, x is negative.

6 **How are conversion factors used to convert one unit to another? (1.7, 1.8)**

- A conversion factor is a term that converts a quantity in one unit to a quantity in another unit. To use conversion factors to solve a problem, set up the problem with any unwanted unit in the numerator of one term and the denominator of another term, so that unwanted units cancel.

7 **What is temperature and how are the three temperature scales related? (1.9)**

- Temperature is a measure of how hot or cold an object is. The Fahrenheit and Celsius temperature scales are divided into degrees. Both the size of the degree and the zero point of these scales differ. The Kelvin scale is divided into kelvins, and one kelvin is the same size as one degree Celsius.

8 **What are density and specific gravity? (1.10)**

- Density is a physical property reported in g/mL or g/cc that relates the mass of an object to its volume. A less dense substance floats on top of a more dense liquid.
- Specific gravity is a unitless quantity that relates the density of a substance to the density of water at the same temperature. Since the density of water is 1.00 g/mL at common temperatures, the specific gravity of a substance equals its density, but it contains no units.

PROMBLEMS

Selected in-chapter and end-of-chapter problems have brief answers provided in Appendix B.

Matter

1.37   What is the difference between an element and a compound?

1.38   What is the difference between a compound and a mixture?

1.39   Describe solids, liquids, and gases in terms of (a) volume (how they fill a container); (b) shape; (c) level of organization of the particles that comprise them; (d) how close the particles that comprise them lie.

1.40   How do physical properties and chemical properties differ?

1.41   Classify each process as a chemical or physical change.
   a. dissolving calcium chloride in water
   b. burning gasoline to power a car
   c. heating wax so that it melts

1.42   Classify each process as a chemical or physical change.
   a. the condensation of water on the outside of a cold glass
   b. mixing a teaspoon of instant coffee with hot water
   c. baking a cake

1.43   When a chunk of dry ice (solid carbon dioxide) is placed out in the air, the solid gradually disappears and a gas is formed above the solid. Does the molecular art drawn below indicate that a chemical or physical change has occurred? Explain your choice.

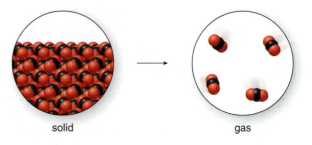

solid   gas

1.44   The inexpensive preparation of nitrogen-containing fertilizers begins with mixing together two elements, hydrogen and nitrogen, at high temperature and pressure in the presence of a metal. Does the molecular art depicted below indicate that a chemical or physical change occurs under these conditions? Explain your choice.

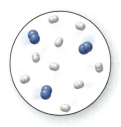 

Measurement

1.45   What is the difference between an exact number and an inexact number? Give an example of each type of number.

1.46   Label each quantity as an exact or inexact number.
   a. A recipe requires 10 cloves of garlic and two tablespoons of oil.
   b. A dog had five puppies whose combined weight was 10 lb.
   c. The four bicycles in the family have been ridden for a total of 250 mi.
   d. A child fell and had a 4 cm laceration that required 12 stitches.

1.47   Which quantity in each pair is larger?
   a. 5 mL or 5 dL   c. 5 cm or 5 mm
   b. 10 mg or 10 µg   d. 10 Ms or 10 ms

1.48   Which quantity in each pair is larger?
   a. 10 km or 10 m   c. 10 g or 10 µg
   b. 10 L or 10 mL   d. 10 cm or 10 mm

Significant Figures

1.49   How many significant figures does each number contain?
   a. 16.00   c. 0.001 60   e. 1.06   g. 1.060×10^{10}
   b. 160   d. 1,600,000   f. 0.1600   h. 1.6×10^{-6}

1.50   How many significant figures does each number contain?
   a. 160.   c. 0.000 16   e. 1,600.   g. 1.600×10^{-10}
   b. 160.0   d. 1.60   f. 1.060   h. 1.6×10^{6}

1.51   Round each number to three significant figures.
   a. 25,401   c. 0.001 265 982   e. 195.371
   b. 1,248,486   d. 0.123 456   f. 196.814

1.52   Round each number in Problem 1.51 to four significant figures.

1.53   Carry out each calculation and report the answer using the proper number of significant figures.
   a. 53.6×0.41 c. 65.2/12   e. 694.2×0.2
   b. 25.825 − 3.86   d. 41.0 + 9.135 f. 1,045 − 1.26

1.54   Carry out each calculation and report the answer using the proper number of significant figures.
   a. $49,682 \times 0.80$ c. 1,000/2.34   e. 25,000/0.4356
   b. 66.815 + 2.82   d. 21 − 0.88   f. 21.5381 + 26.55

Scientific Notation

1.55   Write each quantity in scientific notation.
   a. 1,234 g   c. 5,244,000 L   e. 44,000 km
   b. 0.000 016 2 m   d. 0.005 62 g

1.56   Write each quantity in scientific notation.
   a. 0.001 25 m   c. 54,235.6 m   e. 4,440 s
   b. 8,100,000,000 lb d. 0.000 001 899 L

1.57   Convert each number to its standard form.
   a. 3.4×10^{8} c. 3×10^{2}
   b. 5.822×10^{-5} d. 6.86×10^{-8}

1.58   Convert each number to its standard form.
   a. 4.02×10^{10} c. 6.86×10^{9}
   b. 2.46×10^{-3} d. 1.00×10^{-7}

1.59   Which number in each pair is larger?
   a. 4.44×10^{3} or 4.8×10^{2} c. 1.3×10^{8} or 52,300,000
   b. 5.6×10^{-6} or 5.6×10^{-5} d. 9.8×10^{-4} or 0.000 089

1.60   Rank the numbers in each group from smallest to largest.
   a. 5.06×10^{6}, 7×10^{4}, and 2.5×10^{8}
   b. 6.3×10^{-2}, 2.5×10^{-4}, and 8.6×10^{-6}

1.61   Write the recommended daily intake of each nutrient in scientific notation.
   a. 0.000 400 g of folate   c. 0.000 080 g of vitamin K
   b. 0.002 g of copper   d. 3,400 mg of chloride

1.62   A blood vessel is 0.40 µm in diameter. (a) Convert this quantity to meters and write the answer in scientific notation. (b) Convert this quantity to inches and write the answer in scientific notation.

1.63   A picosecond is one trillionth of a second (0.000 000 000 001 s). (a) Write this number in scientific notation. (b) How many picoseconds are there in one second? Write this answer in scientific notation.

1.64   Red light has a wavelength of 683 nm. Convert this quantity to meters and write the answer in scientific notation.

Problem Solving and Unit Conversions

1.65   Carry out each of the following conversions.
   a. 300 g to mg   d. 300 g to oz
   b. 2 L to µL   e. 2 ft to m
   c. 5.0 cm to m   f. 3.5 yd to m

1.66   Carry out each of the following conversions.
   a. 25 µL to mL   d. 300 mL to qt
   b. 35 kg to g   e. 3 cups to L
   c. 2.36 mL to L   f. 2.5 tons to kg

1.67   Carry out each of the following conversions.
   a. What is the mass in kilograms of an individual who weighs 234 lb?
   b. What is the height in centimeters of a child who is 50. in. tall?
   c. A patient required 3.0 pt of blood during surgery. How many liters does this correspond to?
   d. A patient had a body temperature of 37.7 °C. What is his body temperature in °F?

1.68   Carry out each of the following conversions.
   a. What is the mass in pounds of an individual who weighs 53.2 kg?
   b. What is the height in inches of a child who is 90. cm tall?
   c. How many mL are contained in the 5.0 qt of blood in the human body?
   d. A patient had a body temperature of 103.5 °F. What is his body temperature in °C?

1.69   (a) How many milliliters are contained in 1 qt of milk? (b) How many fluid ounces are contained in 1 L of soda?

1.70   Which gasoline is less expensive: gas that sells for $3.00 per gallon or gas that sells for $0.89 per liter?

1.71   The average mass of a human liver is 1.5 kg. Convert this quantity to (a) grams; (b) pounds; (c) ounces.

1.72   The length of a femur (thigh bone) of a patient is 18.2 in. Convert this quantity to (a) meters; (b) centimeters.

Temperature

1.73   Carry out each of the following temperature conversions.
   a. An over-the-counter pain reliever melts at 53 °C. Convert this temperature to °F and K.
   b. A cake is baked at 350 °F. Convert this temperature to °C and K.

1.74   Methane, the main component of the natural gas used for cooking and heating homes, melts at −183 °C and boils at −162 °C. Convert each temperature to °F and K.

1.75   Which temperature in each pair is higher?
   a. −10 °C or 10 °F   b. −50 °C or −50 °F

1.76   Rank the temperatures in each group from lowest to highest.
   a. 0 °F, 0 °C, 0 K   b. 100 K, 100 °C, 100 °F

Density and Specific Gravity

1.77   What is the difference between density and specific gravity?

1.78   If you have an equal mass of two different substances (**A** and **B**), but the density of **A** is twice the density of **B**, what can be said about the volumes of **A** and **B**?

1.79   If a urine sample has a mass of 122 g and a volume of 121 mL, what is its density in g/mL?

1.80   The density of sucrose, table sugar, is 1.56 g/cc. What volume (in cubic centimeters) does 20.0 g of sucrose occupy?

1.81   Isooctane is a high-octane component of gasoline. If the density of isooctane is 0.692 g/mL, how much does 220 mL weigh?

1.82   A volume of saline solution weighed 25.6 g at 4 °C. An equal volume of water at the same temperature weighed 24.5 g. What is the density of the saline solution?

1.83   If milk has a density of 1.03 g/mL, what is the mass of 1 qt, reported in kilograms?

1.84   If gasoline has a density of 0.66 g/mL, how many kilograms does 1 gal weigh?

1.85   Which is the upper layer when each of the following liquids is added to water?
   a. heptane (density = 0.684 g/mL)
   b. olive oil (density = 0.92 g/mL)
   c. chloroform (density = 1.49 g/mL)
   d. carbon tetrachloride (density = 1.59 g/mL)

1.86   Which of the following solids float on top of water and which sink?
   a. aluminum (density = 1.70 g/cc)
   b. lead (density = 11.34 g/cc)
   c. Styrofoam (density = 0.100 g/cc)
   d. maple wood (density = 0.74 g/cc)

1.87   (a) What is the specific gravity of mercury, the liquid used in thermometers, if it has a density of 13.6 g/mL? (b) What is the density of ethanol if it has a specific gravity of 0.789?

1.88   Why is specific gravity a unitless quantity?

General Questions

1.89   What are the advantages of using the metric system of measurement over the English system of measurement?

1.90   When you convert pounds to grams, how do you decide which unit of the conversion factor is located in the numerator?

1.91   Rank the quantities in each group from smallest to largest.
   a. 100 μL, 100 dL, and 100 mL
   b. 1 dL, 10 mL, and 1,000 μL
   c. 10 g, 100 mg, and 0.1 kg
   d. 1 km, 100 m, and 1,000 cm

1.92   What is the difference between mass and weight?

Applications

1.93   A lab test showed an individual's cholesterol level to be 186 mg/dL. (a) Convert this quantity to g/dL. (b) Convert this quantity to mg/L.

1.94 Hemoglobin is a protein that transports oxygen from the lungs to the rest of the body. Lab results indicated a patient had a hemoglobin concentration in the blood of 15.5 g/dL, which is in the normal range. (a) Convert the number of grams to milligrams and write the answer in scientific notation. (b) Convert the number of grams to micrograms and write the answer in scientific notation.

1.95 A woman was told to take a dose of 1.5 g of calcium daily. How many 500-mg tablets should she take?

1.96 The recommended daily calcium intake for a woman over 50 years of age is 1,200 mg. If one cup of milk has 306 mg of calcium, how many cups of milk provide this amount of calcium? (b) How many milliliters of milk does this correspond to?

1.97 A medium banana contains 451 mg of the nutrient potassium. How many bananas would you have to eat in one day to obtain the recommended daily intake of 3.5 g of potassium?

1.98 A single 1-oz serving of tortilla chips contains 250 mg of sodium. If an individual ate the entire 13-oz bag, how many grams of sodium would he ingest? If the recommended daily intake of sodium is 2.4 g, does this provide more or less than the recommended daily value, and by how much?

1.99 A bottle of liquid medication contains 300 mL and costs $10.00. (a) If the usual dose is 20. mL, how much does each dose cost? (b) If the usual dose is two tablespoons (1 tablespoon = 15 mL), how much does each dose cost?

1.100 The average nicotine content of a Camel cigarette is 1.93 mg. (a) Convert this quantity to both grams and micrograms. (b) Nicotine patches, which are used to help quit smoking, release nicotine into the body by absorption through the skin. The patches come with different amounts of nicotine. A smoker begins with the amount of nicotine that matches his typical daily intake. The maximum amount of nicotine in one brand of patch supplies a smoker with 21 mg of nicotine per day. If an individual smoked one pack of 20 Camel cigarettes each day, would a smoker get more or less nicotine per day using this patch?

1.101 A chemist synthesized 0.510 kg of aspirin in the lab. If the normal dose of aspirin is two 325-mg tablets, how many doses did she prepare?

1.102 Maalox is the trade name for an antacid and antigas medication used for relief of heartburn, bloating, and acid indigestion. Each 5-mL portion of Maalox contains 400 mg of aluminum hydroxide, 400 mg of magnesium hydroxide, and 40 mg of simethicone. If the recommended dose is two teaspoons four times a day, how many grams of each substance would an individual take in a 24-hour period. (1 teaspoon = 5 mL.)

1.103 Children's Chewable Tylenol contains 80 mg of acetaminophen per tablet. If the recommended dosage is 10 mg/kg, how many tablets are needed for a 42-lb child?

1.104 A patient is prescribed 2.0 g of a medication to be taken four times a day. If the medicine is available in 500-mg tablets, how many tablets are needed in a 24-hour period?

1.105 Children's Liquid Motrin contains 100. mg of the pain reliever ibuprofen per 5 mL. If the dose for a 45-lb child is 1.5 teaspoons, how many grams of ibuprofen would the child receive? (1 teaspoon = 5 mL.)

1.106 Often the specific amount of a drug to be administered must be calculated from a given dose in mg per kilogram of body weight. This assures that individuals who have very different body mass get the proper dose. If the proper dosage of a drug is 2 mg/kg of body weight, how many milligrams would a 110-lb individual need?

1.107 If the proper dose of a medication is 10 µg/kg of body weight, how many milligrams would a 200-lb individual need?

1.108 If a 180-lb patient is prescribed 20 mg of the cholesterol-lowering drug Lipitor daily, what dosage is the patient receiving in mg/kg of his body weight?

2

CHAPTER GOALS

In this chapter you will learn how to:

1 Identify an element by its symbol and classify it as a metal, nonmetal, or metalloid

2 Describe the basic parts of an atom

3 Distinguish isotopes and calculate atomic weight

4 Describe the basic features of the periodic table

5 Understand the electronic structure of an atom

6 Write an electronic configuration for an element

7 Relate the location of an element in the periodic table to its electronic configuration

8 Draw an electron-dot symbol for an atom

9 Use the periodic table to predict the relative size and ionization energy of atoms

Both the naturally occurring diamond used in jewelry and the synthetic carbon fibers used in high-end, lightweight bicycles are composed of the element **carbon.**

ATOMS AND THE PERIODIC TABLE

EXAMINE the ingredients listed on a box of crackers. They may include flour, added vitamins, sugar for sweetness, a natural or synthetic coloring agent, baking soda, salt for flavor, and BHT as a preservative. No matter how simple or complex each of these substances is, it is composed of the basic building block, the **atom.** The word *atom* comes from the Greek word *atomos* meaning *unable to cut.* In Chapter 2, we examine the structure and properties of atoms, the building blocks that comprise all forms of matter.

2.1 ELEMENTS

You were first introduced to elements in Section 1.3.

Elements are named for people, places, and things. For example, *carbon* (C) comes from the Latin word *carbo,* meaning *coal* or *charcoal; neptunium* (Np) was named for the planet Neptune; *einsteinium* (Es) was named for scientist Albert Einstein; and *californium* (Cf) was named for the state of California.

- An *element* is a pure substance that cannot be broken down into simpler substances by a chemical reaction.

Of the 114 elements currently known, 90 are naturally occurring and the remaining 24 have been prepared by scientists in the laboratory. Some elements, like oxygen in the air we breathe and aluminum in a soft drink can, are familiar to you, while others, like samarium and seaborgium, are probably not. An alphabetical list of all elements appears on the inside front cover.

Each element is identified by a one- or two-letter symbol. The element carbon is symbolized by the single letter **C,** while the element chlorine is symbolized by **Cl.** When two letters are used in the element symbol, the first is upper case while the second is lower case. Thus, **Co** refers to the element cobalt, but CO is carbon monoxide, which is composed of the elements carbon (C) and oxygen (O). Table 2.1 lists common elements and their symbols.

While most element symbols are derived from the first one or two letters of the element name, 11 elements have symbols derived from the Latin names for them. Table 2.2 lists these elements and their symbols.

ENVIRONMENTAL NOTE

Carbon monoxide (CO), formed in small amounts during the combustion of fossil fuels like gasoline, is a toxic component of the smoggy air in many large cities. We will learn about carbon monoxide in Section 12.8.

PROBLEM 2.1

Give the symbol for each element.

a. calcium, a nutrient needed for strong teeth and bones
b. radon, a radioactive gas produced in the soil
c. nitrogen, the main component of the earth's atmosphere
d. gold, a precious metal used in coins and jewelry

PROBLEM 2.2

An alloy is a mixture of two or more elements that has metallic properties. Give the element symbol for the components of each alloy: (a) brass (copper and zinc); (b) bronze (copper and tin); (c) pewter (tin, antimony, and lead).

PROBLEM 2.3

Give the name corresponding to each element symbol: (a) Ne; (b) S; (c) I; (d) Si; (e) B; (f) Hg.

TABLE 2.1 Common Elements and Their Symbols

Element	Symbol	Element	Symbol
Bromine	Br	Magnesium	Mg
Calcium	Ca	Manganese	Mn
Carbon	C	Molybdenum	Mo
Chlorine	Cl	Nitrogen	N
Chromium	Cr	Oxygen	O
Cobalt	Co	Phosphorus	P
Copper	Cu	Potassium	K
Fluorine	F	Sodium	Na
Hydrogen	H	Sulfur	S
Iodine	I	Zinc	Zn
Lead	Pb		

TABLE 2.2 Element Symbols with Latin Origins

Element	Symbol
Antimony	Sb (stibium)
Copper	Cu (cuprum)
Gold	Au (aurum)
Iron	Fe (ferrum)
Lead	Pb (plumbum)
Mercury	Hg (hydrargyrum)
Potassium	K (kalium)
Silver	Ag (argentum)
Sodium	Na (natrium)
Tin	Sn (stannum)
Tungsten	W (wolfram)

2.1A   ELEMENTS AND THE PERIODIC TABLE

How the periodic table is organized is discussed in Section 2.7. A periodic table appears on the inside front cover for easy reference.

Long ago it was realized that groups of elements have similar properties, and that these elements could be arranged in a schematic way called the **periodic table** (Figure 2.1). The position of an element in the periodic table tells us much about its chemical properties.

The elements in the periodic table are divided into three groups—**metals, nonmetals,** and **metalloids.** The solid line that begins with boron (B) and angles in steps down to astatine (At) marks the three regions corresponding to these groups. All metals are located to the *left* of the line. All nonmetals except hydrogen are located to the *right*. Metalloids are located along the steps.

- *Metals* are shiny solids that are good conductors of heat and electricity. All metals are solids at room temperature except for mercury, which is a liquid.
- *Nonmetals* do not have a shiny appearance, and they are generally poor conductors of heat and electricity. Nonmetals like sulfur and carbon are solids at room temperature; bromine is a liquid; and nitrogen, oxygen, and nine other elements are gases.
- *Metalloids* have properties intermediate between metals and nonmetals. Only seven elements are categorized as metalloids: boron (B), silicon (Si), germanium (Ge), arsenic (As), antimony (Sb), tellurium (Te), and astatine (At).

PROBLEM 2.4

Locate each element in the periodic table and classify it as a metal, nonmetal, or metalloid.

a. titanium	c. krypton	e. arsenic	g. selenium
b. chlorine	d. palladium	f. cesium	h. osmium

2.1B   FOCUS ON THE HUMAN BODY
THE ELEMENTS OF LIFE

Because living organisms selectively take up elements from their surroundings, the abundance of elements in the human body is very different from the distribution of elements in the earth's crust. **Four nonmetals—oxygen, carbon, hydrogen, and nitrogen—comprise 96% of the mass of the human body, and are called the** *building-block elements* (Figure 2.2). Hydrogen and oxygen are the elements that form water, the most prevalent substance in the body. Carbon, hydrogen, and oxygen are found in the four main types of biological molecules—proteins, carbohydrates, lipids, and nucleic acids. Proteins and nucleic acids contain the element nitrogen as well. These biological molecules are discussed in Chapters 19–22.

Seven other elements, called the **major minerals** or **macronutrients,** are also present in the body in much smaller amounts (0.1–2% by mass). Sodium, potassium, and chlorine are present in body fluids. Magnesium and sulfur occur in proteins, and calcium and phosphorus are present in teeth and bones. Phosphorus is also contained in all nucleic acids, such as the DNA that transfers genetic information from one generation to another. At least 100 mg of each macronutrient is needed in the daily diet.

Many other elements occur in very small amounts in the body, but are essential to good health. These **trace elements** or **micronutrients** are required in the daily diet in small quantities, usually less than 15 mg. Each trace element has a specialized function that is important for proper cellular function. For example, iron is needed for hemoglobin, the protein that carries oxygen in red blood cells, and myoglobin, the protein that stores oxygen in muscle. Zinc is needed for the proper functioning of many enzymes in the liver and kidneys, and iodine is needed for proper thyroid function. Although most of the trace elements are metals, nonmetals like fluorine and selenium are micronutrients as well.

PROBLEM 2.5

Classify each micronutrient in Figure 2.2 as a metal, nonmetal, or metalloid.

▼ **FIGURE 2.1** The Periodic Table of the Elements

1A 1																	8A 18	
1 **H** 1.0079	2A 2											3A 13	4A 14	5A 15	6A 16	7A 17	**2** **He** 4.0026	1
3 **Li** 6.941	**4** **Be** 9.0122											**5** **B** 10.811	**6** **C** 12.011	**7** **N** 14.0067	**8** **O** 15.9994	**9** **F** 18.9984	**10** **Ne** 20.1797	2
11 **Na** 22.9898	**12** **Mg** 24.3050	3B 3	4B 4	5B 5	6B 6	7B 7	←—— 8B ——→ 8 9 10			1B 11	2B 12	**13** **Al** 26.9815	**14** **Si** 28.0855	**15** **P** 30.9738	**16** **S** 32.066	**17** **Cl** 35.453	**18** **Ar** 39.948	3
19 **K** 39.0983	**20** **Ca** 40.078	**21** **Sc** 44.9559	**22** **Ti** 47.88	**23** **V** 50.9415	**24** **Cr** 51.9961	**25** **Mn** 54.9380	**26** **Fe** 55.845	**27** **Co** 58.9332	**28** **Ni** 58.693	**29** **Cu** 63.546	**30** **Zn** 65.41	**31** **Ga** 69.723	**32** **Ge** 72.64	**33** **As** 74.9216	**34** **Se** 78.96	**35** **Br** 79.904	**36** **Kr** 83.80	4
37 **Rb** 85.4678	**38** **Sr** 87.62	**39** **Y** 88.9059	**40** **Zr** 91.224	**41** **Nb** 92.9064	**42** **Mo** 95.94	**43** **Tc** (98)	**44** **Ru** 101.07	**45** **Rh** 102.9055	**46** **Pd** 106.42	**47** **Ag** 107.8682	**48** **Cd** 112.411	**49** **In** 114.82	**50** **Sn** 118.710	**51** **Sb** 121.760	**52** **Te** 127.60	**53** **I** 126.9045	**54** **Xe** 131.29	5
55 **Cs** 132.9054	**56** **Ba** 137.327	**57** **La** 138.9055	**72** **Hf** 178.49	**73** **Ta** 180.9479	**74** **W** 183.84	**75** **Re** 186.207	**76** **Os** 190.2	**77** **Ir** 192.22	**78** **Pt** 195.08	**79** **Au** 196.9665	**80** **Hg** 200.59	**81** **Tl** 204.3833	**82** **Pb** 207.2	**83** **Bi** 208.9804	**84** **Po** (209)	**85** **At** (210)	**86** **Rn** (222)	6
87 **Fr** (223)	**88** **Ra** (226)	**89** **Ac** (227)	**104** **Rf** (267)	**105** **Db** (268)	**106** **Sg** (271)	**107** **Bh** (272)	**108** **Hs** (270)	**109** **Mt** (276)	**110** **Ds** (281)	**111** **Rg** (280)	**112** – (285)		**114** – (289)		**116** – (293)			7

6	**58** **Ce** 140.115	**59** **Pr** 140.9076	**60** **Nd** 144.24	**61** **Pm** (145)	**62** **Sm** 150.36	**63** **Eu** 151.964	**64** **Gd** 157.25	**65** **Tb** 158.9253	**66** **Dy** 162.50	**67** **Ho** 164.9303	**68** **Er** 167.26	**69** **Tm** 168.9342	**70** **Yb** 173.04	**71** **Lu** 174.967	6
7	**90** **Th** 232.0381	**91** **Pa** 231.03588	**92** **U** 238.0289	**93** **Np** (237)	**94** **Pu** (244)	**95** **Am** (243)	**96** **Cm** (247)	**97** **Bk** (247)	**98** **Cf** (251)	**99** **Es** (252)	**100** **Fm** (257)	**101** **Md** (258)	**102** **No** (259)	**103** **Lr** (262)	7

metal — silver (Ag), sodium (Na), mercury (Hg)

metalloid — silicon (Si), arsenic (As), boron (B)

nonmetal — carbon (C), sulfur (S), bromine (Br)

- **Metals** like silver, sodium, and mercury are shiny substances that conduct heat and electricity.

- **Metalloids** like silicon, arsenic, and boron have properties intermediate between metals and nonmetals.

- **Nonmetals** like carbon, sulfur, and bromine are poor conductors of heat and electricity.

▼ **FIGURE 2.2** **The Elements of Life**

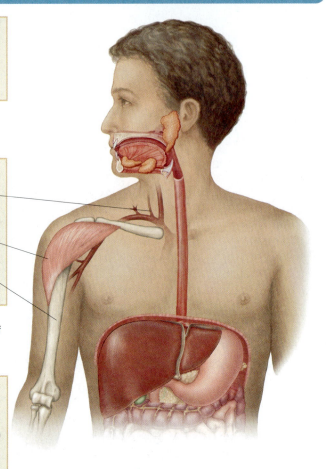

Building-Block Elements

Oxygen (O)
Carbon (C)
Hydrogen (H)
Nitrogen (N)

These four elements compose almost 96% of the mass of the human body.

Major Minerals

Potassium (K), sodium (Na), and chlorine (Cl) are present in body fluids.

Magnesium (Mg) and sulfur (S) are found in proteins.

Calcium (Ca) and phosphorus (P) are present in teeth and bones.

Each major mineral is present in 0.1–2% by mass. At least 100 mg of each mineral is needed in the daily diet.

Trace Elements

Arsenic (As)	Iron (Fe)
Boron (B)	Manganese (Mn)
Chromium (Cr)	Molybdenum (Mo)
Cobalt (Co)	Nickel (Ni)
Copper (Cu)	Selenium (Se)
Fluorine (F)	Silicon (Si)
Iodine (I)	Zinc (Zn)

Each trace element is present in less than 0.1% by mass. A small quantity (15 mg or less) of each element is needed in the daily diet.

2.1C COMPOUNDS

In Section 1.3 we learned that a *compound* **is a pure substance formed by chemically combining two or more elements together.** Element symbols are used to write chemical formulas for compounds.

- A *chemical formula* uses element symbols to show the identity of the elements forming a compound and subscripts to show the ratio of atoms (the building blocks of matter) contained in the compound.

For example, table salt is formed from sodium (Na) and chlorine (Cl) in a ratio of 1:1, so its formula is NaCl. Water, on the other hand, is formed from two hydrogen atoms for each oxygen atom, so its formula is H_2O. The subscript "1" is understood when no subscript is written. Other examples of chemical formulas are shown below.

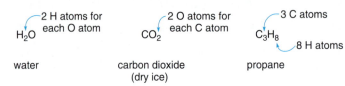

water   carbon dioxide (dry ice)   propane

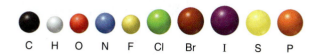

▼ **FIGURE 2.3** **Common Element Colors Used in Molecular Art**

C   H   O   N   F   Cl   Br   I   S   P

As we learned in Section 1.2, molecular art will often be used to illustrate the composition and state of elements and compounds. Color-coded spheres, shown in Figure 2.3, are used to identify the common elements that form compounds.

For example, a red sphere is used for the element oxygen and gray is used for the element hydrogen, so H_2O is represented as a red sphere joined to two gray spheres. Sometimes the spheres will be connected by "sticks" to generate a **ball-and-stick** representation for a compound. At other times, the spheres will be drawn close together to form a **space-filling** representation. No matter how the spheres are depicted, H_2O always consists of one red sphere for the oxygen atom and two gray spheres for the two hydrogen atoms.

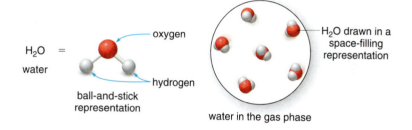

H_2O = water

ball-and-stick representation

oxygen

hydrogen

H_2O drawn in a space-filling representation

water in the gas phase

PROBLEM 2.6

Identify the elements in each chemical formula, and give the number of atoms of each element.

a. NaCN (sodium cyanide)   c. C_2H_6 (ethane)   e. CO (carbon monoxide)

b. H_2S (hydrogen sulfide)   d. SnF_2 (stannous fluoride)   f. $C_3H_8O_3$ (glycerol)

PROBLEM 2.7

Identify the elements used in each example of molecular art.

a.   b.   c.

2.2 STRUCTURE OF THE ATOM

All matter is composed of the same basic building blocks called *atoms*. An atom is much too small to be seen even by the most powerful light microscopes. The period at the end of this sentence holds about 1×10^8 atoms, and a human cheek cell contains about 1×10^{16} atoms. An atom is composed of three subatomic particles.

- A proton, symbolized by p, has a positive (+) charge.
- An electron, symbolized by e⁻, has a negative (–) charge.
- A neutron, symbolized by n, has no charge.

Protons and neutrons have approximately the same, exceedingly small mass, as shown in Table 2.3. The mass of an electron is much less, 1/1,836 the mass of a proton. These subatomic particles are not evenly distributed in the volume of an atom. There are two main components of an atom.

TABLE 2.3   Summary: The Properties of the Three Subatomic Particles

Subatomic Particle	Charge	Mass (g)	Mass (amu)
Proton	+1	1.6726×10^{-24}	1
Neutron	0	1.6749×10^{-24}	1
Electron	−1	9.1093×10^{-28}	Negligible

- The *nucleus* is a dense core that contains the protons and neutrons. Most of the mass of an atom resides in the nucleus.
- The *electron cloud* is composed of electrons that move rapidly in the almost empty space surrounding the nucleus. The electron cloud comprises most of the volume of an atom.

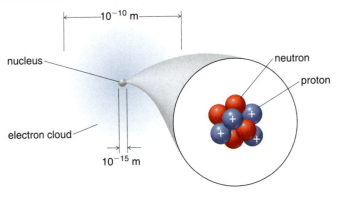

main components of an atom

While the diameter of an atom is about 10^{-10} m, the diameter of a nucleus is only about 10^{-15} m. For a macroscopic analogy, if the nucleus were the size of a baseball, an atom would be the size of Yankee Stadium!

The charged particles of an atom can either attract or repel each other.

- Opposite charges attract while like charges repel each other.

Thus, two electrons or two protons repel each other, while a proton and an electron attract each other.

Positive charges repel.   Negative charges repel.   Opposite charges attract.

Since the mass of an individual atom is so small (on the order of 10^{-24} g), chemists use a standard mass unit, the **atomic mass unit,** which defines the mass of individual atoms relative to a standard mass.

- One atomic mass unit (amu) equals one-twelfth the mass of a carbon atom that has six protons and six neutrons; 1 amu = 1.661×10^{-24} g.

Using this scale, one proton has a mass of 1.0073 amu, a value typically rounded to 1 amu. One neutron has a mass of 1.0087 amu, a value also typically rounded to 1 amu. The mass of an electron is so small that it is ignored.

Every atom of a given type of element always has the *same* number of protons in the nucleus, a value called the *atomic number,* symbolized by Z. Conversely, two *different* elements have *different* atomic numbers.

- The *atomic number (Z)* = the number of protons in the nucleus of an atom.

Thus, the element hydrogen has one proton in its nucleus, so its atomic number is one. Lithium has three protons in its nucleus, so its atomic number is three. The periodic table is arranged in order of increasing atomic number beginning at the upper left-hand corner. The atomic number appears just above the element symbol for each entry in the table.

Since a neutral atom has no overall charge:

- *Z* = the number of protons in the nucleus = the number of electrons.

Thus, the atomic number tells us *both* the number of protons in the nucleus and the number of electrons in the electron cloud of a neutral atom.

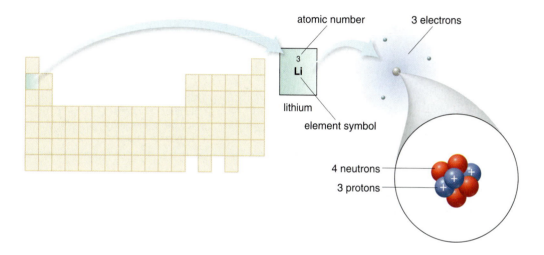

SAMPLE PROBLEM 2.1	Identify the element that has an atomic number of 19, and give the number of protons and electrons in the neutral atom.
ANALYSIS	The atomic number is unique to an element and tells the number of protons in the nucleus and the number of electrons in the electron cloud of a neutral atom.
SOLUTION	According to the periodic table, the element potassium has atomic number 19. A neutral potassium atom has 19 protons and 19 electrons.
PROBLEM 2.8	Identify the element with each atomic number, and give the number of protons and electrons in the neutral atom: (a) 2; (b) 11; (c) 20; (d) 47; (e) 78.

Both protons and neutrons contribute to the mass of an atom. The **mass number,** symbolized by *A,* is the sum of the number of protons and neutrons.

- **Mass number (A) = the number of protons (Z) + the number of neutrons.**

For example, a fluorine atom with nine protons and 10 neutrons in the nucleus has a mass number of 19. Figure 2.4 lists the atomic number, mass number, and number of subatomic particles in the four building-block elements—hydrogen, carbon, nitrogen, and oxygen—found in a wide variety of compounds including caffeine (chemical formula $C_8H_{10}N_4O_2$), the bitter-tasting mild stimulant in coffee, tea, and cola beverages.

▼ FIGURE 2.4 Atomic Composition of the Four Building-Block Elements

caffeine
($C_8H_{10}N_4O_2$)

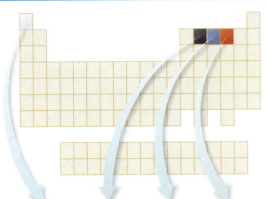

Element	Hydrogen	Carbon	Nitrogen	Oxygen
Atomic number	1	6	7	8
Mass number	1	12	14	16
Number of protons	1	6	7	8
Number of electrons	1	6	7	8
Number of neutrons	0	6	7	8

SAMPLE PROBLEM 2.2

How many protons, neutrons, and electrons are contained in an atom of argon, which has an atomic number of 18 and a mass number of 40?

ANALYSIS

- In a neutral atom, the atomic number (Z) = the number of protons = the number of electrons.
- The mass number (A) = the number of protons + the number of neutrons.

SOLUTION

The atomic number of 18 means that argon has 18 protons and 18 electrons. To find the number of neutrons, subtract the atomic number (Z) from the mass number (A).

$$\begin{aligned} \text{number of neutrons} &= \text{mass number} - \text{atomic number} \\ &= \quad\quad 40 \quad\quad - \quad\quad 18 \\ &= \quad\quad 22 \text{ neutrons} \end{aligned}$$

PROBLEM 2.9

How many protons, neutrons, and electrons are contained in each atom with the given atomic number and mass number?

a. $Z = 17, A = 35$ b. $Z = 14, A = 28$ c. $Z = 92, A = 238$

PROBLEM 2.10

What element has an atomic number of 53 and contains 74 neutrons? How many electrons does this atom contain? What is its mass number?

PROBLEM 2.11

What is the mass number of an atom that contains

a. 42 protons, 42 electrons, and 53 neutrons?   b. 24 protons, 24 electrons, and 28 neutrons?

2.3 ISOTOPES

Two atoms of the same element always have the same number of protons, but the number of neutrons can vary.

- *Isotopes* are atoms of the same element having a different number of neutrons.

2.3A   ISOTOPES, ATOMIC NUMBER, AND MASS NUMBER

Most elements in nature exist as a mixture of isotopes. For example, all atoms of the element chlorine contain 17 protons in the nucleus, but some of these atoms have 18 neutrons in the nucleus and some have 20 neutrons. Thus, chlorine has two isotopes with different mass numbers, 35 and 37. These isotopes are often referred to as chlorine-35 (or Cl-35) and chlorine-37 (or Cl-37).

Isotopes are also written using the element symbol with the atomic number written as a subscript and the mass number written as a superscript, both to the left.

> Two isotopes of the element chlorine

$$\text{mass number} \searrow \quad ^{35}_{17}\text{Cl} \qquad ^{37}_{17}\text{Cl}$$
$$\text{atomic number} \nearrow$$

chlorine-35   chlorine-37

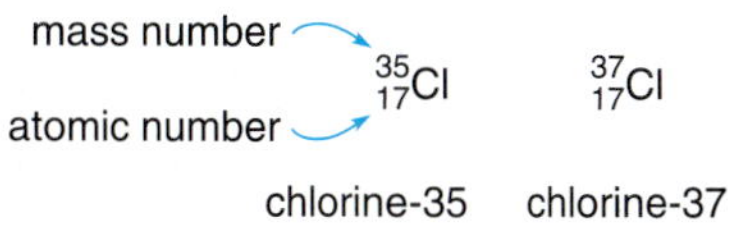

The element hydrogen has three isotopes. Most hydrogen atoms have one proton and no neutrons, giving them a mass number of one. About 1% of hydrogen atoms have one proton and one neutron, giving them a mass number of two. This isotope is called **deuterium,** and it is often symbolized as **D.** An even smaller number of hydrogen atoms contain one proton and two neutrons, giving them a mass number of three. This isotope is called **tritium,** symbolized as **T.**

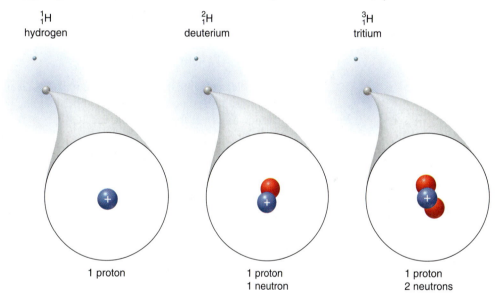

$^1_1$H
hydrogen

$^2_1$H
deuterium

$^3_1$H
tritium

1 proton

1 proton
1 neutron

1 proton
2 neutrons

SAMPLE PROBLEM 2.3	For each atom give the following information: [1] the atomic number; [2] the mass number; [3] the number of protons; [4] the number of neutrons; [5] the number of electrons.

a. $^{118}_{50}$Sn   b. $^{195}_{78}$Pt

ANALYSIS

- The superscript gives the mass number and the subscript gives the atomic number for each element.
- The atomic number = the number of protons = the number of electrons.
- The mass number = the number of protons + the number of neutrons.

SOLUTION

	Atomic Number	Mass Number	Number of Protons	Number of Neutrons	Number of Electrons
a. $^{118}_{50}$Sn	50	118	50	118 − 50 = 68	50
b. $^{195}_{78}$Pt	78	195	78	195 − 78 = 117	78

PROBLEM 2.12

For each atom give the following information: [1] the atomic number; [2] the mass number; [3] the number of protons; [4] the number of neutrons; [5] the number of electrons.

a. $^{13}_{6}C$ b. $^{121}_{51}Sb$

SAMPLE PROBLEM 2.4

Determine the number of neutrons in each isotope: (a) carbon-14; (b) ^{81}Br.

ANALYSIS
- The identity of the element tells us the atomic number.
- The number of neutrons = mass number (A) – atomic number (Z).

SOLUTION
a. Carbon's atomic number (Z) is 6. Carbon-14 has a mass number (A) of 14.

$$\text{number of neutrons} = A - Z$$
$$= 14 - 6 = 8 \text{ neutrons}$$

b. Bromine's atomic number is 35 and the mass number of the given isotope is 81.

$$\text{number of neutrons} = A - Z$$
$$= 81 - 35 = 46 \text{ neutrons}$$

PROBLEM 2.13

Magnesium has three isotopes that contain 12, 13, and 14 neutrons. For each isotope give the following information: (a) the number of protons; (b) the number of electrons; (c) the atomic number; (d) the mass number. Write the element symbol of each isotope using a superscript and subscript for mass number and atomic number, respectively.

2.3B ATOMIC WEIGHT

Some elements like fluorine occur naturally as a single isotope. More commonly, an element is a mixture of isotopes, and it is useful to know the average mass, called the **atomic weight** (or **atomic mass**), of the atoms in a sample.

- The *atomic weight* is the weighted average of the mass of the naturally occurring isotopes of a particular element reported in atomic mass units.

The atomic weights of the elements appear in the alphabetical list of elements on the inside front cover. The atomic weight is also given under the element symbol in the periodic table on the inside front cover.

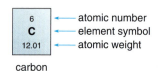

atomic number
element symbol
atomic weight

carbon

To determine the atomic weight of an element, two quantities must be known: the mass of each isotope in atomic mass units, and the frequency with which each isotope occurs.

HOW TO   Determine the Atomic Weight of an Element

EXAMPLE   **What is the atomic weight of the element chlorine?**

Step [1]   **List each isotope, along with its mass in atomic mass units (amu) and the percentage that each isotope occurs in nature.**

- Chlorine has two isotopes—Cl-35 and Cl-37.

- To solve the problem, the masses and abundances of the isotopes must be known.

	Mass (amu)	Isotopic Abundance
Cl-35	34.97	75.78% = 0.7578
Cl-37	36.97	24.22% = 0.2422

- The mass of any isotope is very close to the mass number of the isotope.
- To convert a percent to a decimal, divide by 100%, which is the same as moving the decimal point two places to the left; thus,

$$75.78\% = 0.7578$$

Step [2] **Multiply the isotopic abundance by the mass of each isotope, and add up the products. The sum is the atomic weight for the element.**

Mass due to Cl-35: 0.7578×34.97 amu $= 26.5003$ amu

Mass due to Cl-37: 0.2422×36.97 amu $= \underline{8.9541}$ amu

Atomic weight $= 35.4544$ amu rounded to 35.45 amu

Answer

SAMPLE PROBLEM 2.5

Calculate the atomic weight of copper, which has two isotopes with the following properties: Cu-63 (62.93 amu, 69.17% natural occurrence) and Cu-65 (64.93 amu, 30.83% natural occurrence).

ANALYSIS Multiply the isotopic abundance by the mass of each isotope, and add up the products to give the atomic weight for the element.

SOLUTION Mass due to Cu-63: 0.6917×62.93 amu $= 43.5287$ amu

Mass due to Cu-65: 0.3083×64.93 amu $= \underline{20.0179}$ amu

Atomic weight $= 63.5466$ amu rounded to 63.55 amu

Answer

PROBLEM 2.14

Calculate the atomic weight of each element given the mass and natural occurrence of each isotope.

a. Magnesium	Mass (amu)	Isotopic Abundance	b. Vanadium	Mass (amu)	Isotopic Abundance
Mg-24	23.99	78.99%	V-50	49.95	0.250%
Mg-25	24.99	10.00%	V-51	50.94	99.750%
Mg-26	25.98	11.01%			

2.3C FOCUS ON HEALTH & MEDICINE
ISOTOPES IN MEDICINE

Generally the chemical properties of isotopes are identical. Sometimes, however, one isotope of an element is radioactive—that is, it emits particles or energy as some form of radiation. Radioactive isotopes have both diagnostic and therapeutic uses in medicine.

As an example, iodine-131 is used in at least two different ways for thyroid disease. Iodine is a micronutrient needed by the body to synthesize the thyroid hormone thyroxine, which contains four iodine atoms. To evaluate the thyroid gland, a patient can be given sodium iodide (NaI) that contains radioactive iodine-131. Iodine-131 is taken up in the thyroid gland and as it emits radiation, it produces an image in a thyroid scan, which is then used to determine the condition of the thyroid gland, as shown in Figure 2.5.

Other applications of radioactive isotopes in medicine are discussed in Chapter 10.

Higher doses of iodine-131 can also be used to treat thyroid disease. Since the radioactive isotope is taken up by the thyroid gland, the radiation it emits can kill overactive or cancerous cells in the thyroid.

▼ **FIGURE 2.5** Iodine-131 in Medicine

a.

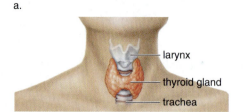

larynx

thyroid gland

trachea

b.

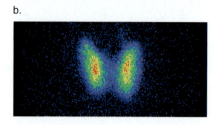

c.

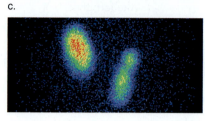

The thyroid gland is a butterfly-shaped gland in the neck, shown in (a). Uptake of radioactive iodine-131 can reveal the presence of a healthy thyroid as in (b), or an unsymmetrical thyroid gland with dense areas of iodine uptake as in (c), which may be indicative of cancer or other thyroid disease.

2.4 THE PERIODIC TABLE

Every beginning chemistry text has a periodic table in a prominent location—often the inside front cover—because it is a valuable list of all known elements organized so that groups of elements with similar characteristics are arranged together. The periodic table evolved over many years, and it resulted from the careful observations and experiments of many brilliant scientists in the nineteenth century. Most prominent was Russian chemist Dmitri Mendeleev, whose arrangement in 1869 of the 60 known elements into groups having similar properties in order of increasing atomic number became the precursor of the modern periodic table (inside front cover and Figure 2.6).

2.4A BASIC FEATURES OF THE PERIODIC TABLE

The periodic table is arranged into seven horizontal rows and 18 vertical columns. The particular row and column tell us much about the properties of an element.

- A row in the periodic table is called a *period.* Elements in the same row are similar in size.
- A column in the periodic table is called a *group.* Elements in the same group have similar electronic and chemical properties.

The rows in the periodic table are numbered 1–7. The number of elements in each row varies. The first period has just two elements, hydrogen and helium. The second and third rows have eight elements each, and the fourth and fifth rows have 18 elements. Also note that two groups of fourteen elements appear at the bottom of the periodic table. The **lanthanides,** beginning with the element cerium ($Z = 58$), immediately follow the element lanthanum (La). The **actinides,** beginning with thorium ($Z = 90$), immediately follow the element actinium (Ac).

Each column in the periodic table is assigned a **group number.** Groups are numbered in two ways. In one system, the 18 columns of the periodic table are assigned the numbers 1–18, beginning with the column farthest to the left. An older but still widely used system numbers the groups 1–8, followed by the letter A or B.

- The *main group elements* consist of the two columns on the far left and the six columns on the far right of the table. These groups are numbered 1A–8A.
- The *transition metal elements* are contained in the 10 short columns in the middle of the table, numbered 1B–8B.
- The *inner transition elements* consist of the lanthanides and actinides, and they are not assigned group numbers.

The periodic table in Figure 2.6 has both systems of numbering groups. For example, the element carbon (C) is located in the second row (period 2) of the periodic table. Its group number is 4A (or 14).

FIGURE 2.6 Basic Features of the Periodic Table

Period; Group number

Period	1A 1	2A 2	3B 3	4B 4	5B 5	6B 6	7B 7	8B 8	8B 9	8B 10	1B 11	2B 12	3A 13	4A 14	5A 15	6A 16	7A 17	8A 18	Period
1	1 **H** 1.0079																	2 **He** 4.0026	1
2	3 **Li** 6.941	4 **Be** 9.0122											5 **B** 10.811	6 **C** 12.011	7 **N** 14.0067	8 **O** 15.9994	9 **F** 18.9984	10 **Ne** 20.1797	2
3	11 **Na** 22.9898	12 **Mg** 24.3050											13 **Al** 26.9815	14 **Si** 28.0855	15 **P** 30.9738	16 **S** 32.066	17 **Cl** 35.453	18 **Ar** 39.948	3
4	19 **K** 39.0983	20 **Ca** 40.078	21 **Sc** 44.9559	22 **Ti** 47.88	23 **V** 50.9415	24 **Cr** 51.9961	25 **Mn** 54.9380	26 **Fe** 55.845	27 **Co** 58.9332	28 **Ni** 58.693	29 **Cu** 63.546	30 **Zn** 65.41	31 **Ga** 69.723	32 **Ge** 72.64	33 **As** 74.9216	34 **Se** 78.96	35 **Br** 79.904	36 **Kr** 83.80	4
5	37 **Rb** 85.4678	38 **Sr** 87.62	39 **Y** 88.9059	40 **Zr** 91.224	41 **Nb** 92.9064	42 **Mo** 95.94	43 **Tc** (98)	44 **Ru** 101.07	45 **Rh** 102.9055	46 **Pd** 106.42	47 **Ag** 107.8682	48 **Cd** 112.411	49 **In** 114.82	50 **Sn** 118.710	51 **Sb** 121.760	52 **Te** 127.60	53 **I** 126.9045	54 **Xe** 131.29	5
6	55 **Cs** 132.9054	56 **Ba** 137.327	57 **La** 138.9055	72 **Hf** 178.49	73 **Ta** 180.9479	74 **W** 183.84	75 **Re** 186.207	76 **Os** 190.2	77 **Ir** 192.22	78 **Pt** 195.08	79 **Au** 196.9665	80 **Hg** 200.59	81 **Tl** 204.3833	82 **Pb** 207.2	83 **Bi** 208.9804	84 **Po** (209)	85 **At** (210)	86 **Rn** (222)	6
7	87 **Fr** (223)	88 **Ra** (226)	89 **Ac** (227)	104 **Rf** (267)	105 **Db** (268)	106 **Sg** (271)	107 **Bh** (272)	108 **Hs** (270)	109 **Mt** (276)	110 **Ds** (281)	111 **Rg** (280)	112 **–** (285)		114 **–** (289)		116 **–** (293)			7

	Period															Period
Lanthanides	6	58 **Ce** 140.115	59 **Pr** 140.9076	60 **Nd** 144.24	61 **Pm** (145)	62 **Sm** 150.36	63 **Eu** 151.964	64 **Gd** 157.25	65 **Tb** 158.9253	66 **Dy** 162.50	67 **Ho** 164.9303	68 **Er** 167.26	69 **Tm** 168.9342	70 **Yb** 173.04	71 **Lu** 174.967	6
Actinides	7	90 **Th** 232.0381	91 **Pa** 231.03588	92 **U** 238.0289	93 **Np** (237)	94 **Pu** (244)	95 **Am** (243)	96 **Cm** (247)	97 **Bk** (247)	98 **Cf** (251)	99 **Es** (252)	100 **Fm** (257)	101 **Md** (258)	102 **No** (259)	103 **Lr** (262)	7

Main group elements; Transition metal elements; Inner transition metal elements

- Each element of the periodic table is part of a horizontal row and a vertical column.
- The periodic table consists of seven rows, labeled periods 1–7, and 18 columns that are assigned a group number. Two different numbering systems are indicated.
- Elements are divided into three categories: main group elements (groups 1A–8A, shown in light blue), transition metals (groups 1B–8B, shown in tan), and inner transition metals (shown in light green).

SAMPLE PROBLEM 2.6 Give the period and group number for each element: (a) magnesium; (b) manganese.

ANALYSIS Use the element symbol to locate an element in the periodic table. Count down the rows of elements to determine the period. The group number is located at the top of each column.

SOLUTION

a. Magnesium (Mg) is located in the third row (period 3), and has group number 2A (or 2).

b. Manganese (Mn) is located in the fourth row (period 4), and has group number 7B (or 7).

PROBLEM 2.15 Give the period and group number for each element: (a) oxygen; (b) calcium; (c) phosphorus; (d) platinum; (e) iodine.

2.4B CHARACTERISTICS OF GROUPS 1A, 2A, 7A, AND 8A

Four columns of main group elements illustrate an important fact about the periodic table.

- **Elements that comprise a particular group have similar chemical properties.**

Alkali Metals (Group 1A) and Alkaline Earth Elements (Group 2A)

The alkali metals and the alkaline earth elements are located on the far left side of the periodic table.

Although hydrogen is also located in group 1A, it is *not* an alkali metal.

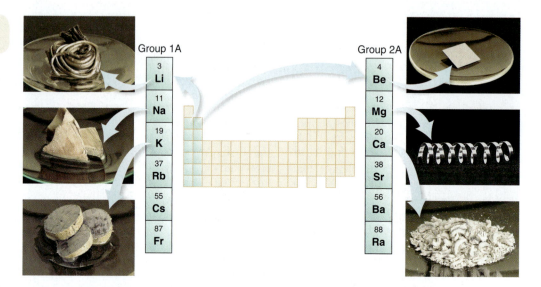

The **alkali metals,** located in group 1A (group 1), include lithium (Li), sodium (Na), potassium (K), rubidium (Rb), cesium (Cs), and francium (Fr). Alkali metals share the following characteristics:

- **They are soft and shiny and have low melting points.**
- **They are good conductors of heat and electricity.**
- **They react readily with water to form basic solutions.**

The **alkaline earth elements,** located in group 2A (group 2), include beryllium (Be), magnesium (Mg), calcium (Ca), strontium (Sr), barium (Ba), and radium (Ra). Alkaline earth metals are also shiny solids but less reactive than the alkali metals.

None of the metals in groups 1A or 2A exist in nature as pure elements; rather, they are always combined with other elements to form compounds. Examples of compounds from group 1A elements include sodium chloride (NaCl), table salt; potassium iodide (KI), an essential nutrient added to make iodized salt; and lithium carbonate (Li_2CO_3), a drug used to treat bipolar disorder. Examples of compounds from group 2A elements include magnesium sulfate ($MgSO_4$), an anticonvulsant used to prevent seizures in pregnant women; and barium sulfate ($BaSO_4$), which is used to improve the quality of X-ray images of the gastrointestinal tract.

Halogens (Group 7A) and Noble Gases (Group 8A)

The halogens and noble gases are located on the far right side of the periodic table.

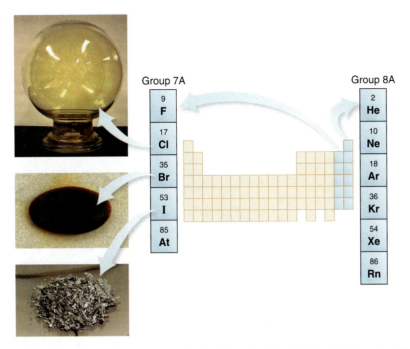

Group 7A

9			Group 8A

Group 7A: 9 F, 17 Cl, 35 Br, 53 I, 85 At

Group 8A: 2 He, 10 Ne, 18 Ar, 36 Kr, 54 Xe, 86 Rn

The **halogens,** located in group 7A (group 17), include fluorine (F), chlorine (Cl), bromine (Br), iodine (I), and the rare radioactive element astatine (At). In their elemental form, halogens contain two atoms joined together—F_2, Cl_2, Br_2, and I_2. Fluorine and chlorine are gases at room temperature, bromine is a liquid, and iodine is a solid. Halogens are very reactive and combine with many other elements to form compounds. In Chapter 14, we will learn about carbon compounds that contain halogen atoms.

The **noble gases,** located in group 8A (group 18), include helium (He), neon (Ne), argon (Ar), krypton (Kr), xenon (Xe), and radon (Rn). Unlike other elements, the noble gases are especially stable as atoms, and so they rarely combine with other elements to form compounds.

The noble gas **radon** has received attention in recent years. Radon is a radioactive gas, and generally its concentration in the air is low and therefore its presence harmless. In some types of soil, however, radon levels can be high and radon detectors are recommended for the basement of homes to monitor radon levels. High radon levels are linked to an increased risk of lung cancer.

PROBLEM 2.16	Identify the element fitting each description.

a. an alkali metal in period 4

b. a second-row element in group 7A

c. a noble gas in the third period

d. a main group element in period 5 and group 2A

e. a transition metal in group 12, period 4

f. a transition metal in group 5, period 5

PROBLEM 2.17	Identify each highlighted element in the periodic table and give its [1] element name and symbol; [2] group number; [3] period; [4] classification (main group element, transition metal, or inner transition metal).

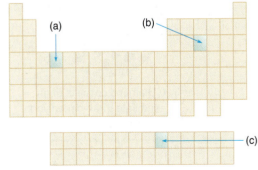

2.4C   THE UNUSUAL NATURE OF CARBON

Carbon, a second-row element in group 4A of the periodic table, is different from most other elements in that it has three elemental forms (Figure 2.7). The two most common forms of carbon are diamond and graphite. **Diamond** is hard because it contains a dense three-dimensional network of carbon atoms in six-membered rings. **Graphite,** on the other hand, is a slippery black substance used as a lubricant. It contains parallel sheets of carbon atoms in flat six-membered rings.

Buckminsterfullerene, also referred to as a bucky ball, is a third form that contains 60 carbon atoms joined together in a sphere of 20 hexagons and 12 pentagons in a pattern that resembles a soccer ball. A component of soot, this form of carbon was not discovered until 1985. Its unusual name stems from its shape, which resembles the geodesic dome invented by R. Buckminster Fuller.

Carbon's ability to join with itself and other elements gives it versatility not seen with any other element in the periodic table. In the unscientific but eloquent description by writer Bill Bryson in *A Short History of Nearly Everything,* carbon is described as "the party animal of the atomic world, latching on to many other atoms (including itself) and holding tight, forming molecular conga lines of hearty robustness—the very trick of nature necessary to build proteins and DNA." As a result, millions of compounds that contain the element carbon are known. The chemistry of these compounds is discussed at length in Chapters 11–24.

2.5   ELECTRONIC STRUCTURE

Why do elements in a group of the periodic table have similar chemical properties? **The chemistry of an element is determined by the number of *electrons* in an atom.** To understand the properties of an element, therefore, we must learn more about the electrons that surround the nucleus.

▼ **FIGURE 2.7   Three Elemental Forms of Carbon**

a. Diamond

b. Graphite

c. Buckminsterfullerene

• Diamond consists of an intricate three-dimensional network of carbon atoms.

• Graphite contains parallel sheets of carbon atoms.

• Buckminsterfullerene contains a sphere with 60 carbon atoms.

The modern description of the electronic structure of an atom is based on the following principles.

- Electrons do not move freely in space; rather, an electron is confined to a specific region, giving it a particular energy.
- Electrons occupy discrete energy levels. The energy of electrons is *quantized;* that is, the energy is restricted to specific values.

The electrons that surround a nucleus are confined to regions called the **principal energy levels,** or **shells.**

- The shells are numbered, $n = 1, 2, 3, 4$, and so forth, beginning closest to the nucleus.
- Electrons closer to the nucleus are held more tightly and are lower in energy.
- Electrons farther from the nucleus are held less tightly and are higher in energy.

The number of electrons that can occupy a given shell is determined by the value of n. **The farther a shell is from the nucleus, the larger its volume becomes, and the more electrons it can hold.** Thus, the first shell can hold only two electrons, the second holds eight, the third 18, and so forth.

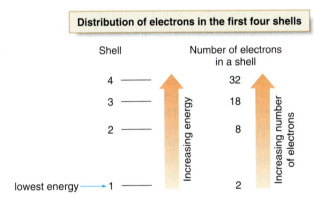

Shells are divided into **subshells,** identified by the letters s, p, d, and f. The subshells consist of **orbitals.**

- An *orbital* is a region of space where the probability of finding an electron is high. Each orbital can hold *two* electrons.

The two electrons in an orbital must have opposite spins. If one electron has a clockwise spin, the second electron in the orbital must have a counterclockwise spin.

A particular type of subshell contains a specific number of orbitals. An s subshell contains only **one** s orbital. A p subshell has **three** p orbitals. A d subshell has **five** d orbitals. An f subshell has **seven** f orbitals. The number of subshells in a given shell equals the value of n. The energy of orbitals shows the following trend:

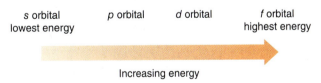

The first shell of electrons around a nucleus ($n = 1$) has only one s orbital. This orbital is called the $1s$ orbital since it is the s orbital in the first shell. Since each orbital can hold two electrons and the first shell has only one orbital, the **first shell can hold two electrons.**

shell number (principal energy level) $1s$ = the s orbital in the first shell

The second shell of electrons ($n = 2$) has two types of orbitals—one s and three p orbitals. These orbitals are called the $2s$ and $2p$ orbitals since they are located in the second shell. Since each orbital can hold two electrons and there are four orbitals, the **second shell can hold eight electrons.**

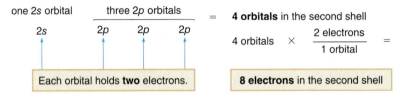

The third shell of electrons ($n = 3$) has three types of orbitals—one s, three p, and five d orbitals. These orbitals are called the $3s$, $3p$, and $3d$ orbitals since they are located in the third shell. Since each orbital can hold two electrons and the third shell has a total of nine orbitals, the **third shell can hold 18 electrons.**

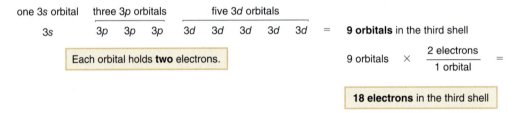

The fourth shell of electrons ($n = 4$) has four types of orbitals—one s, three p, five d, and seven f orbitals. These orbitals are called the $4s$, $4p$, $4d$, and $4f$ orbitals since they are located in the fourth shell. Since each orbital can hold two electrons and the fourth shell has a total of sixteen orbitals, the **fourth shell can hold 32 electrons.**

one 4s orbital   three 4p orbitals   five 4d orbitals   seven 4f orbitals

4s   4p 4p 4p   4d 4d 4d 4d 4d   4f 4f 4f 4f 4f 4f 4f = **16 orbitals**
                                                                                              in the fourth shell

16 orbitals $\times$ $\dfrac{2\ \text{electrons}}{1\ \text{orbital}}$ =   **32 electrons** in the fourth shell

Thus, the maximum number of electrons that can occupy a shell is determined by the number of orbitals in the shell. Table 2.4 summarizes the orbitals and electrons in the first four shells.

TABLE 2.4	Orbitals and Electrons Contained in the Principal Energy Levels ($n = 1–4$)		
Shell	Orbitals	Electrons in Each Subshell	Maximum Number of Electrons
1	1s	2	2
2	2s 2p 2p 2p	2 $3 \times 2 = 6$	8
3	3s 3p 3p 3p 3d 3d 3d 3d 3d	2 $3 \times 2 = 6$ $5 \times 2 = 10$	18
4	4s 4p 4p 4p 4d 4d 4d 4d 4d 4f 4f 4f 4f 4f 4f 4f	2 $3 \times 2 = 6$ $5 \times 2 = 10$ $7 \times 2 = 14$	32

Each type of orbital has a particular shape.

> • An *s* orbital has a sphere of electron density. It is lower in energy than other orbitals in the same shell because electrons are kept closer to the positively charged nucleus.
> • A *p* orbital has a dumbbell shape. A *p* orbital is higher in energy than an *s* orbital in the same shell because its electron density is farther from the nucleus.

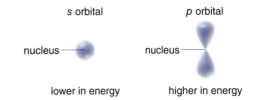

All *s* orbitals are spherical, but the orbital gets larger in size as the shell number increases. Thus, both a 1*s* orbital and a 2*s* orbital are spherical, but the 2*s* orbital is larger. The three *p* orbitals in a shell are perpendicular to each other along the *x, y,* and *z* axes.

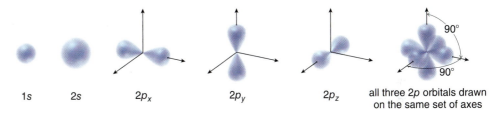

PROBLEM 2.18 How many electrons are present in each shell, subshell, or orbital?

a. a 2*p* orbital   b. the 3*d* subshell   c. a 3*d* orbital   d. the third shell

PROBLEM 2.19 What element fits each description?

a. the element with electrons that completely fill the first and second shells
b. the element with a completely filled first shell and four electrons in the second shell
c. the element with a completely filled first and second shell, and two electrons in the third shell

2.6 ELECTRONIC CONFIGURATIONS

We can now examine the **electronic configuration** of an individual atom—that is, how the electrons are arranged in an atom's orbitals. **The lowest energy arrangement of electrons is called the *ground state.*** Three rules are followed.

> ### Rules to Determine the Ground State Electronic Configuration of an Atom
>
> **Rule [1]** Electrons are placed in the lowest energy orbitals beginning with the 1*s* orbital.
> • In comparing similar types of orbitals from one shell to another (e.g., 2*s* and 3*s*), an orbital closer to the nucleus is lower in energy. Thus, the energy of a 2*s* orbital is lower than a 3*s* orbital.
> • Within a shell, orbital energies increase in the following order: *s, p, d, f.*
> These guidelines result in the following order of energies in the first three periods: 1*s*, 2*s*, 2*p*, 3*s*, 3*p*. Above the 3*p* level, however, all orbitals of one shell do *not* have to be filled before any orbital in the next higher shell gets electrons. For example, a 4*s* orbital is lower in energy than a 3*d* orbital, so it is filled first. Figure 2.8 lists the relative energy of the orbitals used by atoms in the periodic table.

▼ FIGURE 2.8   Relative Energies of Orbitals

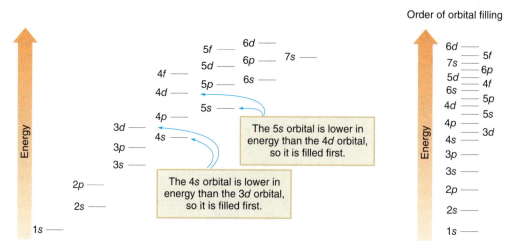

Electrons are added to orbitals in order of increasing energy. The 4s orbital is filled with electrons before the 3d orbital since it is lower in energy. The same is true for filling the 5s orbital with electrons before the 4d orbital. Likewise, the 6s orbital is filled before both the 4f and 5d orbitals, and the 7s orbital is filled before both the 5f and 6d orbitals.

Rule [2]   Each orbital holds a maximum of two electrons.

Rule [3]   When orbitals are equal in energy, one electron is added to each orbital until the orbitals are half-filled, before any orbital is completely filled.

- For example, one electron is added to each of the three *p* orbitals before filling any *p* orbital with two electrons.
- Because like charges repel each other (Section 2.2), adding electrons to *different* *p* orbitals keeps them farther away from each other, which is energetically favorable.

To illustrate how these rules are used, we can write the electronic configuration for several elements using **orbital diagrams.** An orbital diagram uses a box to represent each orbital and arrows to represent electrons. A single electron, called an **unpaired electron,** is shown with a single arrow pointing up (↑). Two electrons in an orbital have **paired spins**—that is, the spins are opposite in direction—so up and down arrows (↑↓) are used.

2.6A   FIRST-ROW ELEMENTS (PERIOD 1)

The first row of the periodic table contains only two elements—hydrogen and helium. Since the number of protons in the nucleus equals the number of electrons in a neutral atom, the **atomic number tells us how many electrons must be placed in orbitals.**

Hydrogen (H, $Z = 1$) has one electron. In the ground state, this electron is added to the lowest energy orbital, the 1s orbital. To draw an orbital diagram we use one box to represent the 1s orbital, and one up arrow to represent the electron. We can also write out the electron configuration without boxes and arrows, using a superscript with each orbital to show how many electrons it contains.

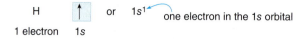

H   ↑   or $1s^1$ one electron in the 1s orbital

1 electron   1s

Helium (He, $Z = 2$) has two electrons. In the ground state, both electrons are added to the $1s$ orbital. To draw an orbital diagram we use one box to represent the $1s$ orbital, and a set of up and down arrows to represent the two electrons with paired spins. The electron configuration can also be written as $1s^2$, meaning the $1s$ orbital has two electrons. Helium has a filled first shell of electrons.

2.6B SECOND-ROW ELEMENTS (PERIOD 2)

To write orbital diagrams for the second-row elements, we must now use the four orbitals in the second shell—the $2s$ orbital and the three $2p$ orbitals. Since electrons are always added to the lowest energy orbitals first, all second-row elements have the $1s$ orbital filled with electrons, and then the remaining electrons are added to the orbitals in the second shell. Since the $2s$ orbital is lower in energy than the $2p$ orbitals, it is completely filled before adding electrons to the $2p$ orbitals.

Lithium (Li, $Z = 3$) has three electrons. In the ground state, two electrons are added to the $1s$ orbital and the remaining electron is an unpaired electron in the $2s$ orbital. Lithium's electronic configuration can also be written as $1s^2 2s^1$ to show the placement of its three electrons.

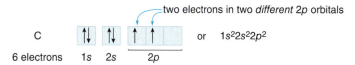

Carbon (C, $Z = 6$) has six electrons. In the ground state, two electrons are added to both the $1s$ and $2s$ orbitals. The two remaining electrons are added to two different $2p$ orbitals, giving carbon **two unpaired electrons.** These electrons spin in the same direction, so the arrows used to represent them are drawn in the same direction as well (both ↑ in this case). Carbon's electronic configuration is also written as $1s^2 2s^2 2p^2$. This method of writing an electronic configuration indicates that carbon has two electrons in its $2p$ orbitals, but it does not explicitly show that the two $2p$ electrons occupy *different* $2p$ orbitals.

two electrons in two *different* $2p$ orbitals

C   ⇅ ⇅ ↑ ↑   or $1s^2 2s^2 2p^2$

6 electrons   1s 2s   2p

Oxygen (O, $Z = 8$) has eight electrons. In the ground state, two electrons are added to both the $1s$ and $2s$ orbitals. The remaining four electrons must be distributed among the three $2p$ orbitals to give the lowest energy arrangement. This is done by pairing two electrons in one $2p$ orbital, and giving the remaining $2p$ orbitals one electron each. Oxygen has two unpaired electrons.

two electrons in two *different* $2p$ orbitals

O   ⇅ ⇅ ⇅ ↑ ↑   or $1s^2 2s^2 2p^4$

8 electrons   1s 2s   2p

Neon (Ne, $Z = 10$) has 10 electrons. In the ground state, two electrons are added to the $1s$, $2s$, and each of the three $2p$ orbitals, so that the second shell of orbitals is now completely filled with electrons.

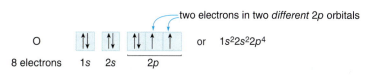

Sometimes the electronic configuration of an element is shortened by using the name of the noble gas that has a filled shell of electrons from the preceding row, and then adding the electronic configuration of all remaining electrons using orbitals and superscripts. For example, each

TABLE 2.5   Electronic Configurations of the First- and Second-Row Elements

Atomic Number	Element	Orbital Diagram			Electronic Configuration	Noble Gas Notation
		1s	2s	2p		
1	H	↑			$1s^1$	
2	He	↑↓			$1s^2$	
3	Li	↑↓	↑		$1s^2 2s^1$	[He] $2s^1$
4	Be	↑↓	↑↓		$1s^2 2s^2$	[He] $2s^2$
5	B	↑↓	↑↓	↑		$1s^2 2s^2 2p^1$	[He] $2s^2 2p^1$
6	C	↑↓	↑↓	↑ ↑		$1s^2 2s^2 2p^2$	[He] $2s^2 2p^2$
7	N	↑↓	↑↓	↑ ↑ ↑	$1s^2 2s^2 2p^3$	[He] $2s^2 2p^3$
8	O	↑↓	↑↓	↑↓ ↑ ↑	$1s^2 2s^2 2p^4$	[He] $2s^2 2p^4$
9	F	↑↓	↑↓	↑↓ ↑↓ ↑	$1s^2 2s^2 2p^5$	[He] $2s^2 2p^5$
10	Ne	↑↓	↑↓	↑↓ ↑↓ ↑↓	$1s^2 2s^2 2p^6$	[He] $2s^2 2p^6$

second-row element has a $1s^2$ configuration like the noble gas helium in the preceding row, so the electronic configuration for carbon can be shortened to $[He]2s^2 2p^2$.

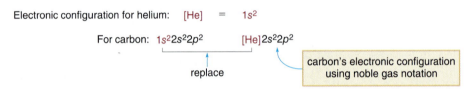

Electronic configuration for helium: [He]   = $1s^2$

For carbon: $1s^2 2s^2 2p^2$ $[He]2s^2 2p^2$

replace

carbon's electronic configuration using noble gas notation

The electronic configurations of all the first- and second-row elements are listed in Table 2.5.

PROBLEM 2.20

What element has each electronic configuration?

a. $1s^2 2s^2 2p^6 3s^2 3p^2$ c. $1s^2 2s^2 2p^6 3s^2 3p^6 4s^2 3d^1$

b. $[Ne]3s^2 3p^4$ d. $[Ar]4s^2 3d^{10}$

2.6C   OTHER ELEMENTS

Orbital diagrams and electronic configurations can be written in much the same way for every element in the periodic table. Sample Problems 2.7 and 2.8 illustrate two examples.

SAMPLE PROBLEM 2.7

Give the orbital diagram for the ground state electronic configuration of the element sulfur. Then, convert this orbital diagram to noble gas notation.

ANALYSIS

- Use the atomic number to determine the number of electrons.
- Place electrons two at a time into the lowest energy orbitals, following the order of orbital filling in Figure 2.8. When orbitals have the same energy, place electrons one at a time in the orbitals until they are half-filled.
- To convert an orbital diagram to noble gas notation, replace the electronic configuration corresponding to the noble gas in the preceding row by the symbol for the noble gas in brackets.

SOLUTION The atomic number of sulfur is 16, so 16 electrons must be placed in orbitals. Twelve electrons are added in pairs to the 1s, 2s, three 2p, and 3s orbitals. The remaining four electrons are then added to the three 3p orbitals to give two paired electrons and two unpaired electrons.

ENVIRONMENTAL NOTE

Coal that is high in **sulfur** content burns to form sulfur oxides, which in turn react with water to form sulfurous and sulfuric acids. Rain that contains these acids has destroyed acres of forests worldwide.

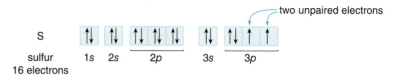

Since sulfur is in the third period, use the noble gas neon in the preceding row to write the electronic configuration in noble gas notation. **Substitute [Ne] for all of the electrons in the first and second shells.**

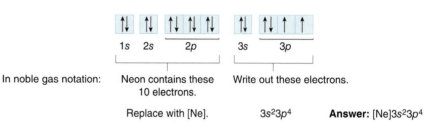

In noble gas notation: Neon contains these 10 electrons.

Replace with [Ne].

Write out these electrons.

$3s^23p^4$ **Answer:** $[Ne]3s^23p^4$

SAMPLE PROBLEM 2.8

Give the ground state electronic configuration of the element calcium. Convert the electronic configuration to noble gas notation.

ANALYSIS
- Use the atomic number to determine the number of electrons.
- Place electrons two at a time into the lowest energy orbitals, following the order of orbital filling in Figure 2.8.
- To convert the electronic configuration to noble gas notation, replace the electronic configuration corresponding to the noble gas in the preceding row by the symbol for the noble gas in brackets.

SOLUTION The atomic number of calcium is 20, so 20 electrons must be placed in orbitals. Eighteen electrons are added in pairs to the 1s, 2s, three 2p, 3s, and three 3p orbitals. Figure 2.8 shows that the 4s orbital is next highest in energy, not the 3d orbitals, so the remaining two electrons are added to the 4s orbital. Since calcium is an element in period 4, use the noble gas argon in period 3 to write the noble gas configuration.

Electronic configuration for Ca $= 1s^22s^22p^63s^23p^64s^2 = [Ar]4s^2$

(20 electrons)   The noble gas argon contains these 18 electrons.   noble gas notation

Replace with [Ar].

PROBLEM 2.21 Draw an orbital diagram for each element: (a) magnesium; (b) aluminum; (c) bromine.

PROBLEM 2.22 Give the electronic configuration for each element and then convert it to noble gas notation: (a) sodium; (b) silicon; (c) iodine.

2.7 ELECTRONIC CONFIGURATIONS AND THE PERIODIC TABLE

Having learned how electrons are arranged in the orbitals of an atom, we can now understand more about the structure of the periodic table. Considering electronic configuration, the periodic table can be divided into four regions, called **blocks**, labeled *s, p, d,* and *f,* and illustrated in

▼ **FIGURE 2.9**   **The Blocks of Elements in the Periodic Table**

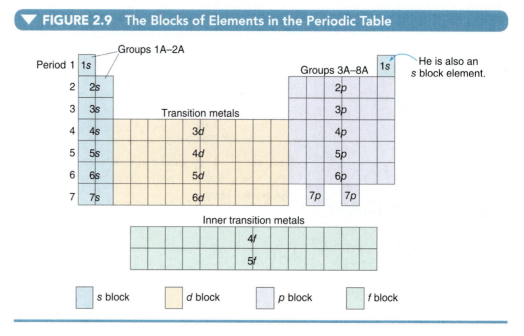

Figure 2.9. **The blocks are labeled according to the subshells that are filled with electrons *last*.**

- The *s block* consists of groups 1A and 2A and the element helium. The *s* subshell is filled last in these elements.
- The *p block* consists of groups 3A–8A (except helium). The *p* subshell is filled last in these elements.
- The *d block* consists of the 10 columns of transition metals. The *d* subshell is filled last in these elements.
- The *f block* consists of the two groups of 14 inner transition metals. The *f* subshell is filled last in these elements.

The number of electrons that can fill a given subshell determines the number of columns in a block. Since each shell contains only one *s* orbital, which can hold two electrons, the *s* block is composed of two columns, one that results from adding one electron to an *s* orbital, and one that results from adding two. Similarly, because a shell has three *p* orbitals that can hold two electrons each, there are six columns in the *p* block. The 10 columns of the *d* block result from adding up to 10 electrons to five *d* orbitals. The 14 columns of the *f* block result from adding up to 14 electrons to the seven *f* orbitals.

2.7A   VALENCE ELECTRONS

The chemical properties of an element depend on the most loosely held electrons—that is, those electrons in the outermost shell, called the **valence shell.** The period number tells the number of the valence shell.

- The electrons in the outermost shell are called the *valence electrons.*

To identify the electrons in the valence shell, always look for the shell with the *highest* number. Thus, beryllium has two valence electrons that occupy the 2*s* orbital. Chlorine has seven valence electrons since it has a total of seven electrons in the third shell, two in the 3*s* orbital and five in the 3*p* orbitals.

Be (beryllium):

$1s^2 2s^2$ 2 valence electrons

valence shell

Cl (chlorine):

$1s^2 2s^2 2p^6 3s^2 3p^5$ 7 valence electrons

valence shell

If we examine the electronic configuration of a group in the periodic table, two facts become apparent.

- Elements in the same group have the same number of valence electrons and similar electronic configurations.
- The group number (using the 1A–8A system) equals the number of valence electrons for main group elements (except helium).

As an example, the alkali metals in group 1A all have one valence electron that occupies an s orbital. Thus, a general electronic configuration for the valence electrons of an alkali metal is ns^1, where n = the period in which the element is located.

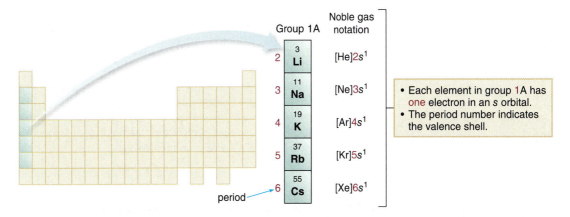

Group 1A

Noble gas notation

2	3 Li	[He]$2s^1$
3	11 Na	[Ne]$3s^1$
4	19 K	[Ar]$4s^1$
5	37 Rb	[Kr]$5s^1$
6	55 Cs	[Xe]$6s^1$

period

- Each element in group 1A has one electron in an s orbital.
- The period number indicates the valence shell.

Thus, the periodic table is organized into groups of elements with similar valence electronic configurations in the same column. The valence electronic configurations of the main group elements in the first three rows of the periodic table are given in Table 2.6.

- The chemical properties of a group are similar because these elements contain the same electronic configuration of valence electrons.

Take particular note of the electronic configuration of the noble gases in group 8A. **All of these elements have a completely filled outer shell of valence electrons.** Helium has a filled first

TABLE 2.6   Valence Electronic Configurations for the Main Group Elements in Periods 1–3

Group Number	1A	2A	3A	4A	5A	6A	7A	8A[a]
Period 1	H $1s^1$							He $1s^2$
Period 2	Li $2s^1$	Be $2s^2$	B $2s^2 2p^1$	C $2s^2 2p^2$	N $2s^2 2p^3$	O $2s^2 2p^4$	F $2s^2 2p^5$	Ne $2s^2 2p^6$
Period 3	Na $3s^1$	Mg $3s^2$	Al $3s^2 3p^1$	Si $3s^2 3p^2$	P $3s^2 3p^3$	S $3s^2 3p^4$	Cl $3s^2 3p^5$	Ar $3s^2 3p^6$
General configuration	ns^1	ns^2	$ns^2 np^1$	$ns^2 np^2$	$ns^2 np^3$	$ns^2 np^4$	$ns^2 np^5$	$ns^2 np^6$

[a]The general electronic configuration in group 8A applies to all of the noble gases except helium. Since helium is a first-row element, it has only two electrons, and these occupy the only available orbital in the first shell, the $1s$ orbital.

shell ($1s^2$ configuration). The remaining elements have a completely filled valence shell of s and p orbitals (s^2p^6). This electronic arrangement is especially stable, and as a result, these elements exist in nature as single atoms. We will learn about the consequences of having a completely filled valence shell in Chapter 3.

SAMPLE PROBLEM 2.9

Identify the total number of electrons, the number of valence electrons, and the name of the element with each electronic configuration.

  a. $1s^2 2s^2 2p^6 3s^2 3p^2$ b. $1s^2 2s^2 2p^6 3s^2 3p^6 4s^2 3d^{10} 4p^6 5s^2 4d^{10} 5p^6 6s^2 4f^{14} 5d^{10}$

ANALYSIS

To obtain the total number of electrons, add up the superscripts. This gives the atomic number and identifies the element. To determine the number of valence electrons, add up the number of electrons in the shell with the highest number.

SOLUTION

a.

valence shell

$1s^2 2s^2 2p^6 3s^2 3p^2$

4 valence electrons

Total number of electrons =
$2 + 2 + 6 + 2 + 2 = $ **14**

Answer: Silicon (Si), 14 total electrons and 4 valence electrons

b.

valence shell

$1s^2 2s^2 2p^6 3s^2 3p^6 4s^2 3d^{10} 4p^6 5s^2 4d^{10} 5p^6 6s^2 4f^{14} 5d^{10}$

2 valence electrons

Total number of electrons = **80**

Answer: Mercury (Hg), 80 total electrons and 2 valence electrons

---

PROBLEM 2.23

Identify the total number of electrons, the number of valence electrons, and the name of the element with each electronic configuration.

  a. $1s^2 2s^2 2p^6 3s^2$ c. $1s^2 2s^2 2p^6 3s^2 3p^6 4s^2 3d^{10} 4p^6 5s^2 4d^2$
  b. $1s^2 2s^2 2p^6 3s^2 3p^3$ d. $[Ar]4s^2 3d^6$

SAMPLE PROBLEM 2.10

Determine the number of valence electrons and give the electronic configuration of the valence electrons of each element: (a) nitrogen; (b) potassium.

ANALYSIS

The group number of a main group element = the number of valence electrons. Use the general electronic configurations in Table 2.6 to write the configuration of the valence electrons.

SOLUTION

  a. Nitrogen is located in group 5A so it has five valence electrons. Since nitrogen is a second-period element, its valence electronic configuration is $2s^2 2p^3$.

  b. Potassium is located in group 1A so it has one valence electron. Since potassium is a fourth-period element, its valence electronic configuration is $4s^1$.

PROBLEM 2.24

Determine the number of valence electrons and give the electronic configuration of the valence electrons of each element: (a) fluorine; (b) krypton; (c) magnesium; (d) germanium.

PROBLEM 2.25

Write the valence shell electronic configuration for the elements in periods 4, 5, and 6 of group 6A.

2.7B ELECTRON-DOT SYMBOLS

The number of valence electrons around an atom is often represented by an **electron-dot symbol.** Representative examples are shown.

	H	C	O	Cl
Number of valence electrons:	1	4	6	7
Electron-dot symbol:	H·	·Ċ·	·Ö·	·Cl:

- Each dot represents one electron.
- The dots are placed on the four sides of an element symbol.
- For one to four valence electrons, single dots are used. With more than four electrons, the dots are paired.

The location of the dots around the symbol—side, top, or bottom—does not matter. Each of the following representations for the five valence electrons of nitrogen is equivalent.

SAMPLE PROBLEM 2.11	Write an electron-dot symbol for each element: (a) sodium; (b) phosphorus.
ANALYSIS	Write the symbol for each element and use the group number to determine the number of valence electrons for a main group element. Represent each valence electron with a dot.
SOLUTION	a. The symbol for sodium is Na. Na is in group 1A and has one valence electron. Electron-dot symbol: Na·	b. The symbol for phosphorus is P. P is in group 5A and has five valence electrons. Electron-dot symbol: ·P̈·

PROBLEM 2.26	Give the electron-dot symbol for each element: (a) bromine; (b) lithium; (c) aluminum; (d) sulfur; (e) neon.

2.8 PERIODIC TRENDS

Many properties of atoms exhibit **periodic trends;** that is, they change in a regular way across a row or down a column of the periodic table. Two properties that illustrate this phenomenon are **atomic size** and **ionization energy.**

HEALTH NOTE

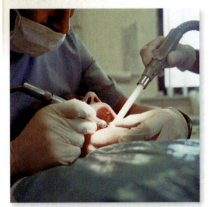

Mercury (Sample Problem 2.9) is safely used in dental amalgam to fill cavities in teeth. Mercury released into the environment, however, is converted to toxic methylmercury by microorganisms in water, so hazardous levels of this soluble mercury compound can accumulate in fish at the top of the food chain, such as sharks and swordfish.

2.8A ATOMIC SIZE

The size of an atom is measured by its atomic radius—that is, the distance from the nucleus to the outer edge of the valence shell. Two periodic trends characterize the size of atoms.

- **The size of atoms increases down a column of the periodic table, as the valence electrons are farther from the nucleus.**

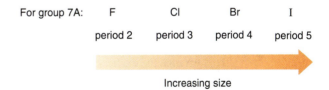

- **The size of atoms decreases across a row of the periodic table as the number of protons in the nucleus increases. An increasing number of protons pulls the electrons closer to the nucleus, so the atom gets smaller.**

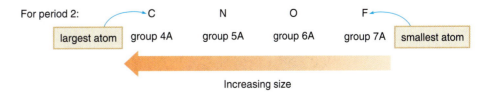

PROBLEM 2.27   Rank the atoms in each group in order of increasing size.

a. boron, carbon, neon   d. krypton, neon, xenon
b. calcium, magnesium, beryllium   e. sulfur, oxygen, silicon
c. silicon, sulfur, magnesium   f. fluorine, sulfur, aluminum

2.8B IONIZATION ENERGY

Since a negatively charged electron is attracted to a positively charged nucleus, energy is required to remove an electron from a neutral atom. The more tightly the electron is held, the greater the energy required to remove it. Removing an electron from a neutral atom forms a **cation.**

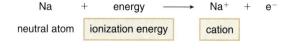

- The *ionization energy* is the energy needed to remove an electron from a neutral atom.
- A *cation* is positively charged, and has fewer electrons than the neutral atom.

Two periodic trends characterize ionization energy.

- Ionization energies decrease down a column of the periodic table as the valence electrons get farther from the positively charged nucleus.

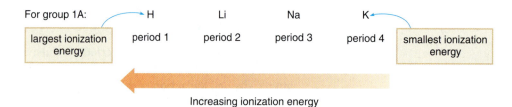

- Ionization energies generally increase across a row of the periodic table as the number of protons in the nucleus increases.

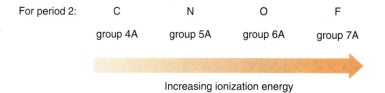

PROBLEM 2.28   Arrange the elements in each group in order of increasing ionization energy.

a. phosphorus, silicon, sulfur   d. neon, krypton, argon
b. magnesium, calcium, beryllium   e. tin, silicon, sulfur
c. carbon, fluorine, beryllium   f. calcium, aluminum, nitrogen

CHAPTER HIGHLIGHTS

KEY TERMS

Actinide (2.4)
Alkali metal (2.4)
Alkaline earth element (2.4)
Atom (2.1)
Atomic mass unit (2.2)
Atomic number (2.2)
Atomic weight (2.3)
Building-block element (2.1)
Cation (2.8)
Chemical formula (2.1)
Compound (2.1)
d Block (2.7)
Deuterium (2.3)
Electron (2.2)
Electron cloud (2.2)
Electron-dot symbol (2.7)
Electronic configuration (2.6)

Element (2.1)
f Block (2.7)
Ground state (2.6)
Group (2.4)
Group number (2.4)
Halogen (2.4)
Inner transition metal element (2.4)
Ionization energy (2.8)
Isotope (2.3)
Lanthanide (2.4)
Main group element (2.4)
Major mineral (Macronutrient, 2.1)
Mass number (2.2)
Metal (2.1)
Metalloid (2.1)
Neutron (2.2)
Noble gas (2.4)

Nonmetal (2.1)
Nucleus (2.2)
Orbital (2.5)
p Block (2.7)
Period (2.4)
Periodic table (2.1)
p Orbital (2.5)
Proton (2.2)
s Block (2.7)
Shell (2.5)
s Orbital (2.5)
Subshell (2.5)
Trace element (Micronutrient, 2.1)
Transition metal element (2.4)
Tritium (2.3)
Unpaired electron (2.6)
Valence electron (2.7)

KEY CONCEPTS

❶ How is the name of an element abbreviated and how does the periodic table help to classify it as a metal, nonmetal, or metalloid? (2.1)

- An element is abbreviated by a one- or two-letter symbol. The periodic table contains a stepped line from boron to astatine. All metals are located to the left of the line. All nonmetals except hydrogen are located to the right of the line. The seven elements located along the line are metalloids.

❷ What are the basic components of an atom? (2.2)

- An atom is composed of two parts: a dense nucleus containing positively charged protons and neutral neutrons, and an electron cloud containing negatively charged electrons. Most of the mass of an atom resides in the nucleus, while the electron cloud contains most of its volume.
- The atomic number (Z) of a neutral atom tells the number of protons and the number of electrons. The mass number (A) is the sum of the number of protons (Z) and the number of neutrons.

❸ What are isotopes and how are they related to the atomic weight? (2.3)

- Isotopes are atoms that have the same number of protons but a different number of neutrons. The atomic weight is the weighted average of the mass of the naturally occurring isotopes of a particular element.

❹ What are the basic features of the periodic table? (2.4)

- The periodic table is a schematic of all known elements, arranged in rows (periods) and columns (groups), organized so that elements with similar properties are grouped together.
- The vertical columns are assigned group numbers using two different numbering schemes—1–8 plus the letters A or B; or 1–18.
- The periodic table is divided into the main group elements (groups 1A–8A), the transition metals (groups 1B–8B), and the inner transition metals located at the bottom.

❺ How are electrons arranged around an atom? (2.5)

- Electrons occupy discrete energy levels, organized into shells (numbered 1, 2, 3, and so on), subshells (s, p, d, and f), and orbitals.
- Each orbital can hold two electrons.

❻ What rules determine the electronic configuration of an atom? (2.6)

- To write the ground state electronic configuration of an atom, electrons are added to the lowest energy orbitals, giving each orbital two electrons. When two orbitals are equal in energy, one electron is added to each orbital until the orbitals are half-filled.
- Orbital diagrams that use boxes for orbitals and arrows for electrons indicate electronic configuration. Electron configuration can also be shown using superscripts to show how many electrons an orbital contains. For example, the electron configuration of the six electrons in a carbon atom is $1s^2 2s^2 2p^2$.

❼ How is the location of an element in the periodic table related to its electronic configuration? (2.7)

- The periodic table is divided into four regions—the *s* block, *p* block, *d* block, and *f* block—based on the subshells that are filled with electrons last.
- Elements in the same group have the same number of valence electrons and similar electronic configurations.

❽ What is an electron-dot symbol? (2.7)

- An electron-dot symbol uses a dot to represent each valence electron around the symbol for an element.

❾ How are atomic size and ionization energy related to location in the periodic table? (2.8)

- The size of an atom decreases across a row and increases down a column.
- Ionization energy—the energy needed to remove an electron from an atom—increases across a row and decreases down a column.

PROBLEMS

Selected in-chapter and end-of-chapter problems have brief answers provided in Appendix B.

Elements

2.29 Give the name of the elements in each group of three element symbols.
a. Au, At, Ag　　　d. Ca, Cr, Cl
b. N, Na, Ni　　　e. P, Pb, Pt
c. S, Si, Sn　　　f. Ti, Ta, Tl

2.30 What element(s) are designated by each symbol or group of symbols?
a. CU and Cu　　　c. Ni and NI
b. Os and OS　　　d. BIN, BiN, and BIn

2.31 Does each chemical formula represent an element or a compound?
a. H_2　b. H_2O_2　c. S_8　d. Na_2CO_3　e. C_{60}

2.32 Identify the elements in each chemical formula and tell how many atoms of each are present.
a. $K_2Cr_2O_7$
b. $C_5H_8NNaO_4$ (MSG, flavor enhancer)
c. $C_{10}H_{16}N_2O_3S$ (vitamin B_7)

2.33 Identify the element that fits each description.
a. an alkali metal in period 6
b. a transition metal in period 5, group 8
c. a main group element in period 3, group 7A
d. a main group element in period 2, group 2A
e. a halogen in period 2
f. an inner transition metal with one $4f$ electron

2.34 Identify the element that fits each description.
a. an alkaline earth element in period 3
b. a noble gas in period 6
c. a main group element in period 3 that has *p* orbitals half-filled with electrons
d. a transition metal in period 4, group 11
e. an inner transition metal with its $5f$ orbitals completely filled with electrons
f. a transition metal in period 6, group 10

2.35 Give all of the terms that apply to each element:
[1] metal; [2] nonmetal; [3] metalloid; [4] alkali metal; [5] alkaline earth element; [6] halogen; [7] noble gas; [8] main group element; [9] transition metal; [10] inner transition metal.
a. sodium　　c. xenon　　e. uranium
b. silver　　d. platinum　　f. tellurium

2.36 Give all of the terms that apply to each element:
[1] metal; [2] nonmetal; [3] metalloid; [4] alkali metal; [5] alkaline earth element; [6] halogen; [7] noble gas; [8] main group element; [9] transition metal; [10] inner transition metal.
a. bromine　　c. cesium　　e. calcium
b. silicon　　d. gold　　f. chromium

Atomic Structure

2.37 Complete the following table for neutral elements.

	Element Symbol	Atomic Number	Mass Number	Number of Protons	Number of Neutrons	Number of Electrons
a.	C		12			
b.			31			15
c.					35	30
d.	Mg		24			
e.				53	74	
f.		4			5	
g.		40	91			
h.					16	16

2.38 For the given atomic number (*Z*) and mass number (*A*):
[1] identify the element; [2] give the element symbol; [3] give the number of protons, neutrons, and electrons.
a. $Z = 10, A = 20$　　d. $Z = 55, A = 133$
b. $Z = 13, A = 27$　　e. $Z = 28, A = 59$
c. $Z = 38, A = 88$　　f. $Z = 79, A = 197$

2.39 Convert the mass of a proton (1.6726×10^{-24} g) to a standard number.

2.40 Convert the mass of an electron (9.1093×10^{-28} g) to a standard number.

Periodic Table

2.41 Label each region on the periodic table.
a. noble gases
b. period 3
c. group 4A
d. *s* block elements
e. alkaline earth elements
f. *f* block elements
g. transition metals
h. group 10

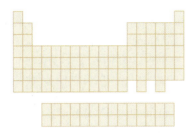

2.42 Identify each highlighted element in the periodic table and give its [1] element name and symbol; [2] group number; [3] period; [4] classification (i.e., main group element, transition metal, or inner transition metal).

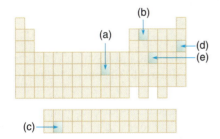

2.43 What element is located in group 1A but is not an alkali metal?

2.44 What *s* block element is not located in either group 1A or group 2A in the periodic table?

2.45 Name two elements in the periodic table that have chemical properties similar to carbon.

2.46 Name two elements in the periodic table that have chemical properties similar to calcium.

2.47 Classify each element in the fourth row of the periodic table as a metal, nonmetal, or metalloid.

2.48 To which blocks do the elements in the fifth period belong?

2.49 Which group(s) in the periodic table contain only nonmetals?

2.50 Which groups in the periodic table contain metals, nonmetals, and metalloids?

Isotopes and Atomic Weight

2.51 The most common isotope of oxygen has a mass number of 16, but two other isotopes having mass numbers of 17 and 18 are also known. For each isotope, give the following information: (a) the number of protons; (b) the number of neutrons; (c) the number of electrons in the neutral atom; (d) the group number; (e) the element symbols using superscripts and subscripts.

2.52 The three most common isotopes of tin have mass numbers 116, 118, and 120. For each isotope, give the following information: (a) the number of protons; (b) the number of neutrons; (c) the number of electrons in the neutral atom; (d) the group number; (e) the element symbols using superscripts and subscripts.

2.53 How many protons, neutrons, and electrons are contained in each element?
a. $^{27}_{13}\text{Al}$
b. $^{35}_{17}\text{Cl}$
c. $^{34}_{16}\text{S}$

2.54 Give the number of protons, neutrons, and electrons in each element: (a) silver-115; (b) Au-197; (c) Rn-222; (d) osmium-192.

2.55 Write the element symbol that fits each description, using a superscript for the mass number and a subscript for the atomic number.
a. an element that contains 53 protons and 74 neutrons
b. an element with 35 electrons and a mass number of 79
c. an element with 47 protons and 60 neutrons

2.56 Write the element symbol that fits each description. Use a superscript for the mass number and a subscript for the atomic number.
a. an element that contains 10 protons and 12 neutrons
b. an element with atomic number 24 and mass number 52
c. an element with 10 electrons and 10 neutrons

2.57 Calculate the atomic weight of silver, which has two isotopes with the following properties: Ag-107 (106.91 amu, 51.84% natural occurrence) and Ag-109 (108.90 amu, 48.16% natural occurrence).

2.58 Calculate the atomic weight of antimony, which has two isotopes with the following properties: Sb-121 (120.90 amu, 57.21% natural occurrence) and Sb-123 (122.90 amu, 42.79% natural occurrence).

2.59 Can the neutral atoms of two different elements have the same number of electrons? Explain.

2.60 Can the neutral atoms of two different elements have the same number of neutrons? Explain.

Electronic Configuration

2.61 What is the difference between a shell and a subshell of electrons?

2.62 What is the difference between a subshell and an orbital of electrons?

2.63 What is the difference between a 1*s* and 2*s* orbital?

2.64 What is the difference between a 2*s* and 2*p* orbital?

2.65 How many orbitals are contained in each of the following shells of electrons: (a) first shell ($n = 1$); (b) second shell ($n = 2$); (c) third shell ($n = 3$); (d) fourth shell ($n = 4$)?

2.66 What is the maximum number of electrons that can be contained in each shell, subshell, or orbital?
a. second shell
b. 3*s* orbital
c. 3*p* subshell
d. 4*f* orbital
e. fourth shell
f. 5*p* orbital

2.67   Why are there 10 columns of transition metal elements in the periodic table?

2.68   Why are there six columns of *p* block elements in the periodic table?

2.69   Write the electronic configuration of each element using an orbital diagram: (a) B; (b) K; (c) Se; (d) Ar; (e) Zn.

2.70   Write the electronic configuration of each element using an orbital diagram: (a) N; (b) I; (c) Ga; (d) Ti; (e) Mn.

2.71   For each element in Problem 2.69: (a) Write out the electronic configuration using a superscript with each orbital; (b) write out the electronic configuration using noble gas notation.

2.72   For each element in Problem 2.70: (a) Write out the electronic configuration using a superscript with each orbital; (b) write out the electronic configuration using noble gas notation.

2.73   How many unpaired electrons are contained in each element: (a) Al; (b) P; (c) Na?

2.74   How many unpaired electrons are contained in each element: (a) Cl; (b) Se; (c) Cs?

2.75   Give the total number of electrons, the number of valence electrons, and the identity of the element with each electronic configuration.
a. $1s^2 2s^2 2p^6 3s^2 3p^6 4s^2 3d^{10} 4p^6 5s^2$ c. $1s^2 2s^2 2p^6 3s^1$
b. $1s^2 2s^2 2p^6 3s^2 3p^4$ d. $[Ne]3s^2 3p^5$

2.76   Give the total number of electrons, the number of valence electrons, and the identity of the element with each electronic configuration.
a. $1s^2 2s^2 2p^6 3s^2 3p^6$ c. $1s^2 2s^2 2p^3$
b. $1s^2 2s^2 2p^6 3s^2 3p^6 4s^2 3d^7$ d. $[Kr]5s^2 4d^{10} 5p^2$

2.77   How do an alkali metal and an alkaline earth element in the same row differ in the electronic configuration of the valence shell electrons?

2.78   How do a halogen and a noble gas in the same row differ in the electronic configuration of the valence shell electrons?

2.79   For each element, give the following information: [1] total number of electrons; [2] group number; [3] number of valence electrons; [4] period; [5] number of the valence shell.
a. carbon   b. calcium   c. krypton

2.80   For each element, give the following information: [1] total number of electrons; [2] group number; [3] number of valence electrons; [4] period; [5] number of the valence shell.
a. oxygen   b. sodium   c. phosphorus

2.81   For each element in Problem 2.79, first write the electronic configuration of all the electrons. Then give the electronic configuration of the valence electrons only.

2.82   For each element in Problem 2.80, first write the electronic configuration of all the electrons. Then give the electronic configuration of the valence electrons only.

2.83   Give the number of valence electrons for each element in Problem 2.37.

2.84   Are the valence electrons always written last when an electronic configuration is written? Explain.

2.85   How many valence electrons does an element in each group contain: (a) 2A; (b) 4A; (c) 7A?

2.86   In what shell do the valence electrons reside for an element in period: (a) 2; (b) 3; (c) 4; (d) 5?

2.87   Can a *d* block element have valence electrons that occupy an *s* orbital? Explain.

2.88   Can an *f* block element have valence electrons that occupy an *s* orbital? Explain.

2.89   Give the number of valence electrons in each element. Write out the electronic configuration for the valence electrons.
a. sulfur   b. chlorine   c. barium   d. titanium   e. tin

2.90   Give the number of valence electrons in each element. Write out the electronic configuration for the valence electrons.
a. neon   c. aluminum   e. zirconium
b. rubidium   d. manganese

2.91   Write an electron-dot symbol for each element: (a) beryllium; (b) silicon; (c) iodine; (d) magnesium; (e) argon.

2.92   Write an electron-dot symbol for each element: (a) K; (b) B; (c) F; (d) Ca; (e) Se.

Periodic Trends

2.93   Which element in each pair is larger?
a. bromine and iodine   c. silicon and potassium
b. carbon and nitrogen   d. chlorine and selenium

2.94   Which element in each pair has its valence electrons farther from the nucleus?
a. sodium and magnesium   c. neon and krypton
b. carbon and fluorine   d. argon and bromine

2.95   For each pair of elements in Problem 2.93, label the element with the higher ionization energy.

2.96   For each pair of elements in Problem 2.94, label the element from which it is easier to remove an electron.

2.97   Rank the following elements in order of increasing size: sulfur, silicon, oxygen, magnesium, and fluorine.

2.98   Rank the following elements in order of increasing size: aluminum, nitrogen, potassium, oxygen, and phosphorus.

2.99   Rank the following elements in order of increasing ionization energy: nitrogen, fluorine, magnesium, sodium, and phosphorus.

2.100 Rank the following elements in order of decreasing ionization energy: calcium, silicon, oxygen, magnesium, and carbon.

Applications

2.101 Sesame seeds, sunflower seeds, and peanuts are good dietary sources of the trace element copper. Copper is needed for the synthesis of neurotransmitters, compounds that transmit nerve signals from one nerve cell to another. Copper is also needed for the synthesis of collagen, a protein found in bone, tendons, teeth, and blood vessels. What is unusual about the electronic configuration of the trace element copper: $1s^22s^22p^63s^23p^64s^13d^{10}$?

2.102 Platinum is a precious metal used in a wide variety of products. Besides fine jewelry, platinum is also the catalyst found in the catalytic converters of automobile exhaust systems, and platinum-containing drugs like cisplatin are used to treat some lung and ovarian cancers. Answer the following questions about the element platinum.
   a. What is its element symbol?
   b. What group number and period are assigned to platinum?
   c. What is its atomic number?
   d. Is platinum classified as a main group element, transition metal, or inner transition metal?
   e. In what block does platinum reside?
   f. What is unusual about the electronic configuration for platinum: $[Xe]6s^14f^{14}5d^9$?

2.103 Answer the following questions about the macronutrients sodium, potassium, and chlorine.
   a. Is each element classified as a metal, nonmetal, or metalloid?
   b. In which block does each element reside?
   c. Which element has the smallest atomic radius?
   d. Which element has the largest atomic radius?
   e. Which element has the largest ionization energy?
   f. Which element has the smallest ionization energy?
   g. How many valence electrons does each element possess?

2.104 Answer the following questions about the macronutrients calcium, magnesium, and sulfur.
   a. Is each element classified as a metal, nonmetal, or metalloid?
   b. In which block does each element reside?
   c. Which element has the smallest atomic radius?
   d. Which element has the largest atomic radius?
   e. Which element has the largest ionization energy?
   f. Which element has the smallest ionization energy?
   g. How many valence electrons does each element possess?

2.105 Carbon-11 is an unnatural isotope used in positron emission tomography (PET) scans. PET scans are used to monitor brain activity and diagnose dementia. How does carbon-11 compare to carbon-12 in terms of the number of protons, neutrons, and electrons? Write the element symbol of carbon-11 using superscripts and subscripts.

2.106 Strontium-90 is a radioactive element formed in nuclear reactors. When an unusually high level of strontium is released into the air, such as occurred during the Chernobyl nuclear disaster in 1986, the strontium can be incorporated into the bones of exposed individuals. High levels of strontium can cause bone cancer and leukemia. Why does Sr-90 cause this particular health problem? (Hint: What macronutrient has similar chemical properties to strontium?)

3

• • • • • • • • • • • • • •

CHAPTER GOALS

In this chapter you will learn how to:

1 Describe the basic features of ionic and covalent bonds

2 Use the periodic table to determine whether an atom forms a cation or an anion, and determine its charge using the group number

3 Describe the octet rule

4 Write the formula for an ionic compound

5 Name ionic compounds

6 Describe the properties of ionic compounds

7 Recognize the structures of common polyatomic ions and name compounds that contain them

8 List useful consumer products and drugs that are composed of ionic compounds

Zinc oxide is an ionic compound widely used in sunblocks to protect the skin from harmful ultraviolet radiation.

IONIC COMPOUNDS

ALTHOUGH much of the discussion in Chapter 2 focused on atoms, individual atoms are rarely encountered in nature. Instead, atoms are far more commonly joined together to form compounds. There are two types of chemical compounds, **ionic** and **covalent. Ionic compounds** are composed of positively and negatively charged ions held together by strong electrostatic forces—the electrical attraction between oppositely charged ions. Examples of ionic compounds include the sodium chloride (NaCl) in table salt and the calcium carbonate ($CaCO_3$) in snail shells. **Covalent compounds** are composed of individual molecules, discrete groups of atoms that share electrons. Covalent compounds include water (H_2O) and methane (CH_4), the main component of natural gas. Chapters 3 and 4 focus on the structure and properties of ionic and covalent compounds, respectively.

3.1 INTRODUCTION TO BONDING

It is rare in nature to encounter individual atoms. Instead, anywhere from two to hundreds or thousands of atoms tend to join together to form compounds. The oxygen we breathe, for instance, consists of two oxygen atoms joined together, whereas the hemoglobin that transports it to our tissues consists of thousands of carbon, hydrogen, oxygen, nitrogen, and sulfur atoms joined together. We say **two atoms are *bonded* together.**

- *Bonding* is the joining of two atoms in a stable arrangement.

Bonding is a favorable process because it always forms a compound that is more stable than the atoms from which it is made. Only the noble gases in group 8A of the periodic table are particularly stable as individual atoms; that is, the **noble gases do *not* readily react to form bonds.** Since chemical reactivity is based on electronic configuration, the electronic configuration of the noble gases must be especially stable to begin with. As a result, one overriding principle explains the process of bonding.

- In bonding, elements gain, lose, or share electrons to attain the electronic configuration of the noble gas closest to them in the periodic table.

Bonding involves only the valence electrons of an atom. There are two different kinds of bonding: **ionic** and **covalent.**

- *Ionic bonds* result from the transfer of electrons from one element to another.
- *Covalent bonds* result from the sharing of electrons between two atoms.

The position of an element in the periodic table determines the type of bonds it makes. **Ionic bonds form between a metal on the left side of the periodic table and a nonmetal on the right side.** For example, when the metal sodium (Na) bonds to the nonmetal chlorine (Cl_2), the ionic compound sodium chloride (NaCl) forms. Since ionic compounds are composed of *ions*—**charged species in which the number of protons and electrons in an atom is *not* equal**—we begin our discussion of ionic compounds with how ions are formed in Section 3.2.

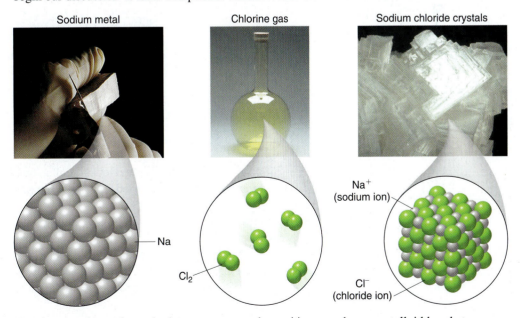

Sodium metal   Chlorine gas   Sodium chloride crystals

Na Cl_2 Na^+ (sodium ion) Cl^- (chloride ion)

Covalent bonds are formed when two nonmetals combine, or when a metalloid bonds to a nonmetal. **A *molecule* is a discrete group of atoms that share electrons.** For example, when two hydrogen atoms bond they form the molecule H_2, and two electrons are shared. Covalent bonds and molecules are discussed in Chapter 4.

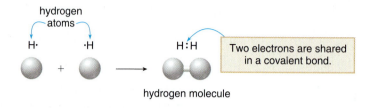

hydrogen molecule

| SAMPLE PROBLEM 3.1 | Predict whether the bonds in the following compounds are ionic or covalent: (a) NaI (sodium iodide); (b) H_2O_2 (hydrogen peroxide). |

ANALYSIS   The position of the elements in the periodic table determines the type of bonds they form. When a metal and nonmetal combine, the bond is ionic. When two nonmetals combine, or a metalloid bonds to a nonmetal, the bond is covalent.

SOLUTION
a. Since Na is a metal on the left side and I is a nonmetal on the right side of the periodic table, the bonds in NaI are ionic.

b. Since H_2O_2 contains only the nonmetals hydrogen and oxygen, the bonds must be covalent.

PROBLEM 3.1   Predict whether the bonds in the following species are ionic or covalent.

a. CO   b. CaF_2 c. MgO   d. Cl_2 e. HF   f. C_2H_6

PROBLEM 3.2   Label each of the following as a compound, element, or molecule. In some cases, more than one term applies.

a. CO_2 b. H_2O c. NaF   d. $MgBr_2$ e. F_2 f. CaO

PROBLEM 3.3   Vitamin C is important in the formation of collagen, a protein that holds together the connective tissue of skin, muscle, and blood vessels. Vitamin C has the chemical formula $C_6H_8O_6$. Even if you know nothing about how the atoms in vitamin C are arranged, what type of bonds are likely to be present in vitamin C?

3.2 IONS

Ionic compounds consist of oppositely charged **ions** that have a strong electrostatic attraction for each other.

3.2A CATIONS AND ANIONS

There are two types of ions called **cations** and **anions.**

- *Cations* are positively charged ions. A cation has fewer electrons than protons.
- *Anions* are negatively charged ions. An anion has more electrons than protons.

The nature and magnitude of the charge on an ion depend on the position of an element in the periodic table. In forming an ion, an atom of a main group element loses or gains electrons to obtain the electronic configuration of the noble gas closest to it in the periodic table. This gives the ion an especially stable electronic arrangement with a **completely filled shell of electrons;** that is, the electrons completely fill the shell farthest from the nucleus.

For example, sodium (group 1A) has an atomic number of 11, giving it 11 protons and 11 electrons in the neutral atom. This gives sodium one *more* electron than neon, the noble gas closest to it in the periodic table. In losing one electron, sodium forms a cation with a +1 charge, which still has 11 protons, but now has only 10 electrons in its electron cloud.

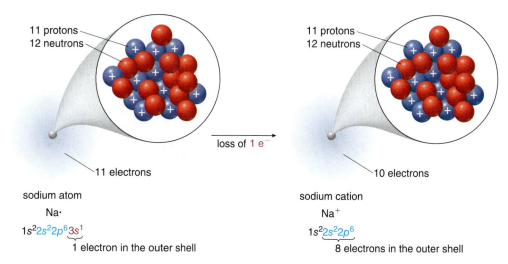

11 protons
12 neutrons

11 electrons

sodium atom
Na·
$1s^2 2s^2 2p^6 3s^1$
1 electron in the outer shell

loss of 1 e⁻

11 protons
12 neutrons

10 electrons

sodium cation
Na⁺
$1s^2 2s^2 2p^6$
8 electrons in the outer shell

What does this mean in terms of valence electrons? A neutral sodium atom, with an electronic configuration of $1s^2 2s^2 2p^6 3s^1$, has a single valence electron. Loss of this valence electron forms a **sodium cation,** symbolized as **Na⁺,** which has the especially stable electronic configuration of the noble gas neon, $1s^2 2s^2 2p^6$. The sodium cation now has eight electrons that fill the $2s$ and three $2p$ orbitals.

Magnesium (group 2A) has 12 protons and 12 electrons in the neutral atom. This gives magnesium two *more* electrons than neon, the noble gas closest to it in the periodic table. In losing two electrons, magnesium forms a cation with a +2 charge, which still has 12 protons, but now has only 10 electrons in its electron cloud.

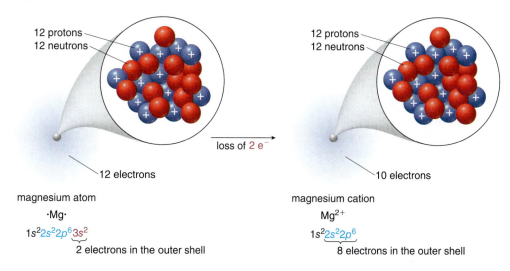

12 protons
12 neutrons

12 electrons

magnesium atom
·Mg·
$1s^2 2s^2 2p^6 3s^2$
2 electrons in the outer shell

loss of 2 e⁻

12 protons
12 neutrons

10 electrons

magnesium cation
Mg²⁺
$1s^2 2s^2 2p^6$
8 electrons in the outer shell

In terms of valence electrons, a neutral magnesium atom, with an electronic configuration of $1s^2 2s^2 2p^6 3s^2$, has two valence electrons. Loss of these valence electrons forms a **magnesium cation,** symbolized as **Mg²⁺,** which has the especially stable electronic configuration of the noble gas neon, $1s^2 2s^2 2p^6$. The magnesium cation now has eight electrons that fill the $2s$ and three $2p$ orbitals.

Sodium and magnesium are examples of metals.

Some metals—notably tin and lead—can lose *four* electrons to form cations.

- Metals are found on the left side of the periodic table.
- Metals form *cations.*
- By losing one, two, or three electrons, an atom forms a cation with a completely filled outer shell of electrons.

A neutral chlorine atom (group 7A), on the other hand, has 17 protons and 17 electrons. This gives it one *fewer* electron than argon, the noble gas closest to it in the periodic table. By gaining one electron, chlorine forms an anion with a −1 charge because it still has 17 protons, but now has 18 electrons in its electron cloud.

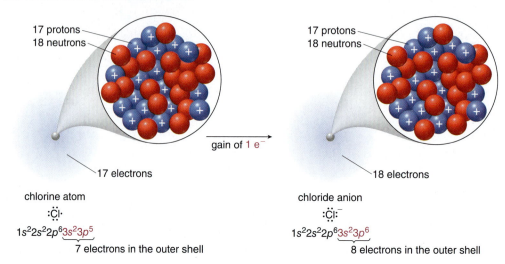

17 protons
18 neutrons

gain of 1 e⁻

17 electrons

chlorine atom

:C̈l·

$1s^2 2s^2 2p^6 \underbrace{3s^2 3p^5}$
7 electrons in the outer shell

17 protons
18 neutrons

18 electrons

chloride anion

:C̈l:⁻

$1s^2 2s^2 2p^6 \underbrace{3s^2 3p^6}$
8 electrons in the outer shell

In terms of valence electrons, a neutral chlorine atom, with an electronic configuration of $1s^2 2s^2 2p^6 3s^2 3p^5$, has seven valence electrons. Gain of one electron forms a **chloride anion,** symbolized as **Cl⁻,** which has the especially stable electronic configuration of the noble gas argon, $1s^2 2s^2 2p^6 3s^2 3p^6$. The chloride anion now has eight valence electrons that fill the 3s and three 3p orbitals.

Chlorine is an example of a nonmetal.

- Nonmetals are found on the right side of the periodic table.
- Nonmetals form *anions.*
- By gaining one, two, or sometimes three electrons, an atom forms an anion with a completely filled outer shell of electrons.

Ions are written with the element symbol followed by a superscript to indicate the charge. The number "1" is omitted in ions that have a +1 or −1 charge, as in Na⁺ or Cl⁻. When the charge is "2" or greater, it is written as 2+ or 2−, as in Mg²⁺ or O²⁻.

Each of these ions formed from a main group element has the s and three p orbitals filled with **eight electrons.** This results in the **octet rule.**

- A main group element is especially stable when it possesses an *octet* of electrons in its outer shell.

SAMPLE PROBLEM 3.2

Write the ion symbol for an atom with: (a) nine protons and 10 electrons; (b) three protons and two electrons.

ANALYSIS

Since the number of protons equals the atomic number (Section 2.2), this quantity identifies the element. The charge is determined by comparing the number of protons and electrons. If the number of electrons is greater than the number of protons, the charge is negative (an anion). If the number of protons is greater than the number of electrons, the charge is positive (a cation).

SOLUTION

a. An element with nine protons has an atomic number of nine, identifying it as fluorine (F). Since there is one more electron than proton (10 vs. 9), the charge is −1.

Answer: F⁻

b. An element with three protons has an atomic number of three, identifying it as lithium (Li). Since there is one more proton than electron (3 vs. 2), the charge is +1.

Answer: Li⁺

PROBLEM 3.4

Write the ion symbol for an atom with the given number of protons and electrons.

a. 19 protons and 18 electrons

c. 35 protons and 36 electrons

b. seven protons and 10 electrons

d. 23 protons and 21 electrons

SAMPLE PROBLEM 3.3

How many protons and electrons are present in each ion: (a) Ca^{2+}; (b) O^{2-}?

ANALYSIS

Use the identity of the element to determine the number of protons. The charge tells how many more or fewer electrons there are compared to the number of protons. A positive charge means more protons than electrons, while a negative charge means more electrons than protons.

SOLUTION

a. Ca^{2+}: The element calcium (Ca) has an atomic number of 20, so it has 20 protons. Since the charge is +2, there are two more protons than electrons, giving the ion 18 electrons.

b. O^{2-}: The element oxygen (O) has an atomic number of eight, so it has eight protons. Since the charge is –2, there are two more electrons than protons, giving the ion 10 electrons.

PROBLEM 3.5

How many protons and electrons are present in each ion?

a. Ni^{2+} b. Se^{2-} c. Zn^{2+} d. Fe^{3+}

3.2B RELATING GROUP NUMBER TO IONIC CHARGE FOR MAIN GROUP ELEMENTS

Because elements with similar electronic configurations are grouped together in the periodic table, **elements in the same group form ions of similar charge.** The group number of a main group element can be used to determine the charge on an ion derived from that element.

- **Metals form cations. For metals in groups 1A, 2A, and 3A, the group number = the charge on the cation.**

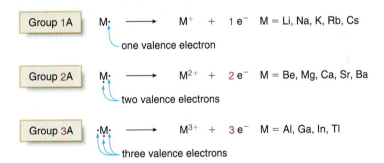

Group **1A** elements (Li, Na, K, Rb, and Cs) have **one** valence electron. Loss of this electron forms a cation with a **+1** charge. Group **2A** elements (Be, Mg, Ca, Sr, and Ba) have **two** valence electrons. Loss of both electrons forms a cation with a **+2** charge. Group **3A** elements (Al, Ga, In, and Tl) form cations, too, but only aluminum is commonly found in ionic compounds. It has **three** valence electrons, so loss of three electrons from aluminum forms a cation with a **+3** charge.

All of the cations derived from group 1A–3A elements have an octet of outer shell electrons except for Li^+ and Be^{2+}. Li^+ and Be^{2+} have a $1s^2$ electronic configuration like helium, the noble gas to which they are closest in the periodic table. Thus, these cations are especially stable because they have a filled outer shell of electrons but they do *not* have an octet of electrons.

- **Nonmetals form anions. For nonmetals in groups 6A and 7A, the anion charge = 8 – (the group number).**

Group 6A $\cdot\ddot{X}\cdot$ +   2 e$^-$ $\longrightarrow$ $:\ddot{X}:^{2-}$ X = O, S, Se

six valence
electrons

anion charge = 8 − (group number)
= 8 − 6 = 2

Group 7A $:\ddot{X}\cdot$ +   1 e$^-$ $\longrightarrow$ $:\ddot{X}:^{-}$ X = F, Cl, Br, I

seven valence
electrons

anion charge = 8 − (group number)
= 8 − 7 = 1

Group **6**A elements have **six** valence electrons. Gain of **two** electrons forms an anion with a **−2** charge (anion charge = 8 − 6). Group **7**A elements have **seven** valence electrons. Gain of one electron forms an anion with a **−1** charge (anion charge = 8 − 7).

Table 3.1 summarizes the ionic charges of the main group elements. The periodic table in Figure 3.1 gives the common ions formed by the main group elements.

TABLE 3.1	Ionic Charges of the Main Group Elements		
Group Number	Number of Valence Electrons	Number of Electrons Gained or Lost	General Structure of the Ion
1A (1)	1	1 e$^-$ lost	M$^+$
2A (2)	2	2 e$^-$ lost	M^{2+}
3A (3)	3	3 e$^-$ lost	M^{3+}
6A (16)	6	2 e$^-$ gained	M^{2-}
7A (17)	7	1 e$^-$ gained	M$^-$

▼ FIGURE 3.1   Common Ions Formed by Main Group Elements

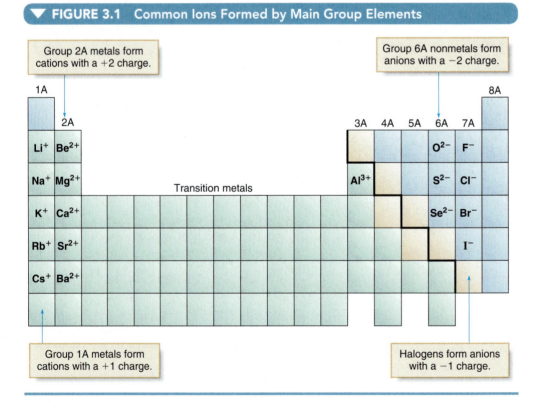

SAMPLE PROBLEM 3.4	Use the group number to determine the charge on an ion derived from each element: (a) barium; (b) sulfur.
ANALYSIS	Locate the element in the periodic table. A metal in groups 1A, 2A, or 3A forms a cation equal in charge to the group number. A nonmetal in groups 6A and 7A forms an anion whose charge equals 8 – (the group number).
SOLUTION	a. Barium (Ba) is located in group 2A, so it forms a cation with a +2 charge; Ba^{2+}.
	b. Sulfur (S) is located in group 6A, so it forms an anion with a negative charge of $8 - 6 = 2$; S^{2-}.

PROBLEM 3.6	Use the group number to determine the charge on an ion derived from each element.
	a. magnesium b. iodine c. selenium d. rubidium

PROBLEM 3.7	Nitrogen rarely forms ionic compounds. How many electrons would nitrogen have to gain to form an anion with an electronic configuration of its nearest noble gas?

PROBLEM 3.8	What noble gas has the same electronic configuration as each ion derived from the elements in Problem 3.6?

3.2C METALS WITH VARIABLE CHARGE

The transition metals form cations like other metals, but the magnitude of the charge on the cation is harder to predict. Some transition metals, and a few main group metals as well, form more than one type of cation. For example, iron forms two different cations, Fe^{2+} and Fe^{3+}. Fe^{2+} is formed by losing two valence electrons from the $4s$ orbital. Fe^{3+} is formed by loss of three electrons, two from a $4s$ orbital and one from a $3d$ orbital.

Because transition metal cations generally have additional d electrons that the nearest noble gas does not, the octet rule is not usually followed. Figure 3.2 illustrates the common cations formed from transition metals, as well as some main group elements that form more than one cation.

▼ **FIGURE 3.2 Common Cations Derived from Transition Metals and Main Group Elements, Some of Which Have Variable Charge**

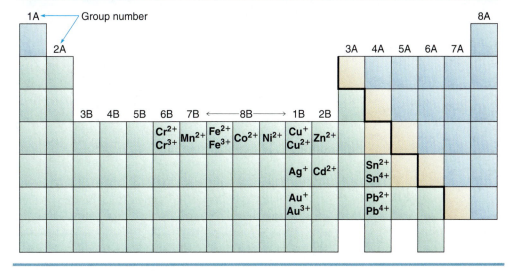

PROBLEM 3.9

How many electrons and protons are contained in each cation?

a. Au^+ b. Au^{3+} c. Sn^{2+} d. Sn^{4+}

3.2D FOCUS ON THE HUMAN BODY
IMPORTANT IONS IN THE BODY

Many different ions are required for proper cellular and organ function (Figure 3.3). The major cations in the body are Na^+, K^+, Ca^{2+}, and Mg^{2+}. K^+ and Mg^{2+} are present in high concentrations inside cells, while Na^+ and Ca^{2+} are present in a higher concentration outside of cells, in the extracellular fluids. Na^+ is the major cation present in blood and extracellular bodily fluids and its concentration is carefully regulated through a number of mechanisms to maintain blood volume and blood pressure within acceptable ranges that permit organ function. Ca^{2+} is found mainly in solid body parts such as teeth and bones, but it is also needed for proper nerve conduction and muscle contraction, as is Mg^{2+}.

In addition to these four cations, Fe^{2+} and Cl^- are also important ions. Fe^{2+} is essential for oxygen transport by red blood cells. Cl^- is present in red blood cells, gastric juices, and other body fluids. Along with Na^+, it plays a major role in regulating the fluid balance in the body.

▼ **FIGURE 3.3   Common Ions in the Human Body**

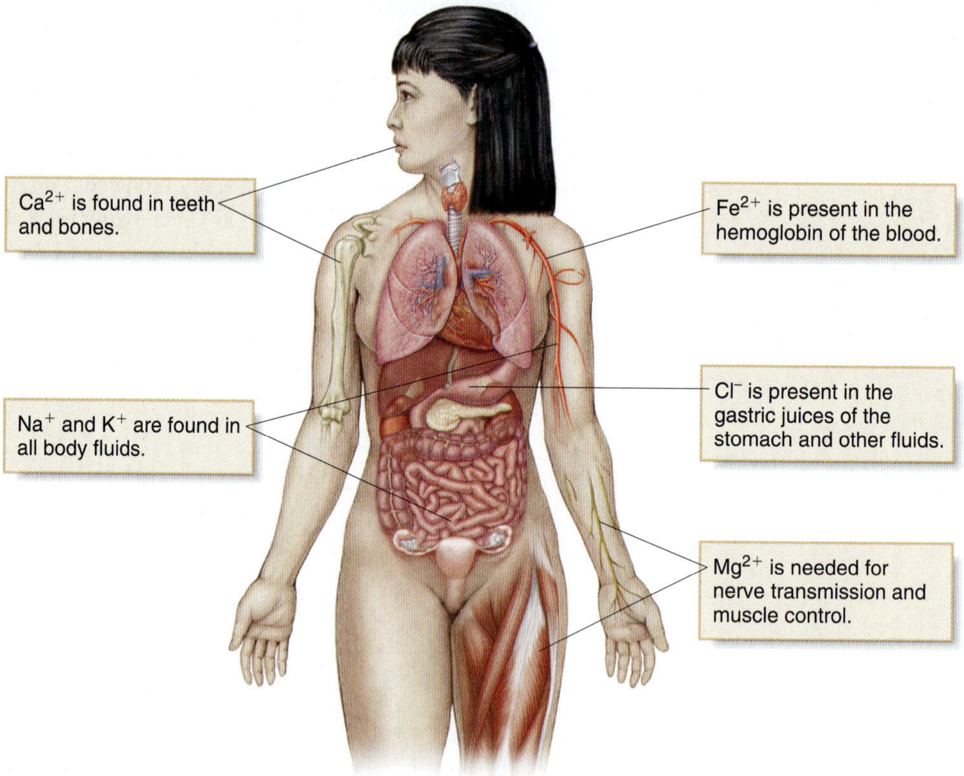

Ca^{2+} is found in teeth and bones.

Fe^{2+} is present in the hemoglobin of the blood.

Cl^- is present in the gastric juices of the stomach and other fluids.

Na^+ and K^+ are found in all body fluids.

Mg^{2+} is needed for nerve transmission and muscle control.

Na^+, K^+, Ca^{2+}, Mg^{2+}, Fe^{2+}, and Cl^- are all common ions present throughout the organs of the human body.

All of these foods are high in sodium.

TABLE 3.2 Na⁺ Content in Common Foods

Foods High in Na⁺		Foods Low in Na⁺	
Food	Na⁺ (mg)	Food	Na⁺ (mg)
Potato chips (30)	276	Banana (1)	1
Hot dog (1)	504	Orange juice (1 cup)	2
Ham, smoked (3 oz)	908	Oatmeal, cooked (1 cup)	2
Chicken soup, canned (1 cup)	1,106	Cereal, shredded wheat (3.5 oz)	3
Tomato sauce, canned (1 cup)	1,402	Raisins, dried (3.5 oz)	27
Parmesan cheese (1 cup)	1,861	Salmon (3 oz)	55

Although Na^+ is an essential mineral needed in the daily diet, the average American consumes three to five times the recommended daily allowance (RDA) of 2,400 mg. Excess sodium intake is linked to high blood pressure and heart disease. Dietary Na^+ comes from salt, NaCl, added during cooking or at the table. Na^+ is also added during the preparation of processed foods and canned products. For example, one 3.5-oz serving of fresh asparagus has only 1 mg of Na^+, but the same serving size of canned asparagus contains 236 mg of Na^+. Potato chips, snack foods, ketchup, processed meats, and many cheeses are particularly high in Na^+. Table 3.2 lists the Na^+ content of some common foods.

PROBLEM 3.10

Horseshoe crabs utilize a copper-containing protein called hemocyanin to transport oxygen. When oxygen binds to the protein it converts Cu^+ to Cu^{2+}, and the blood becomes blue in color. How many protons and electrons do each of these copper cations contain?

PROBLEM 3.11

Mn^{2+} is an essential nutrient needed for blood clotting and the formation of the protein collagen. (a) How many protons and electrons are found in a neutral manganese atom? (b) How many electrons and protons are found in the cation Mn^{2+}? (c) Write the electronic configuration of the element manganese and suggest which electrons are lost to form the Mn^{2+} cation.

3.3 IONIC COMPOUNDS

When a metal on the left side of the periodic table transfers one or more electrons to a nonmetal on the right side, **ionic bonds** are formed.

- **Ionic compounds are composed of cations and anions.**

The ions in an ionic compound are arranged to maximize the attractive force between the oppositely charged species. For example, sodium chloride, NaCl, is composed of sodium cations (Na^+) and chloride anions (Cl^-), packed together in a regular arrangement in a crystal lattice. Each Na^+ cation is surrounded by six Cl^- anions, and each Cl^- anion is surrounded by six Na^+ cations. In this way, the positively charged cations are located closer to the charged particles to which they are attracted—anions—and farther from the particles from which they are repelled—cations.

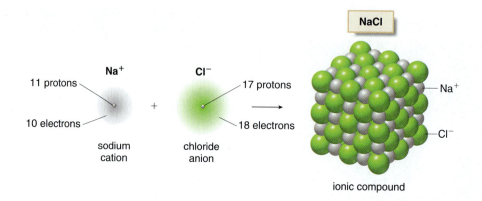

- **The sum of the charges in an ionic compound must always be zero overall.**

The formula for an ionic compound shows the ratio of ions that combine to give zero charge. Since the sodium cation has a +1 charge and the chloride anion has a −1 charge, there must be one Na^+ cation for each Cl^- anion; thus, the formula is **NaCl.**

When cations and anions having charges of different magnitude combine, the number of cations per anion is not equal. Consider an ionic compound formed from calcium (Ca) and fluorine (F). Since calcium is located in group 2A, it loses two valence electrons to form Ca^{2+}. Since fluorine is located in group 7A, it gains one electron to form F^- like other halogens. When Ca^{2+} combines with the fluorine anion F^-, there must be two F^- anions for each Ca^{2+} cation to have an overall charge of zero.

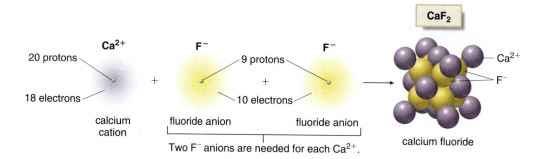

In writing a formula for an ionic compound, we use subscripts when the number of ions needed to achieve zero charge is greater than one. Since two F^- anions are needed for each calcium cation, the formula is **CaF_2.**

| PROBLEM 3.12 | Which pairs of elements will form ionic compounds? |

a. lithium and bromine   c. calcium and magnesium

b. chlorine and oxygen   d. barium and chlorine

3.3A FORMULAS FOR IONIC COMPOUNDS

Writing a formula for an ionic compound from two elements is a useful skill that can be practiced by following a series of steps.

HOW TO   Write a Formula for an Ionic Compound

Step [1]   **Identify which element is the cation and which is the anion.**

- **Metals form *cations* and nonmetals form *anions.***
- Use the group number of a main group element to determine the charge.

An ionic compound derived from calcium and oxygen has the metal calcium as the cation and the nonmetal oxygen as the anion. Calcium (group 2A) loses two electrons to form Ca^{2+}. Oxygen (group 6A) gains two electrons to form O^{2-}.

Step [2]   **Determine how many of each ion type is needed for an overall charge of zero.**

- When the cation and anion have the *same* charge only *one* of each is needed.

- When the cation and anion have different charges, use the ion charges to determine the number of ions of each needed.

An ionic compound from calcium and chlorine has two ions of unequal charges, Ca^{2+} and Cl^-. **The charges on the ions tell us how many of the *oppositely* charged ions are needed to balance charge.**

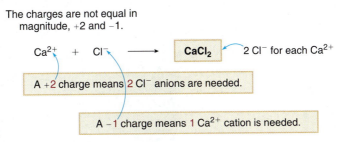

Step [3]   **To write the formula, place the cation first and then the anion, and omit charges.**

- Use subscripts to show the number of each ion needed to have zero overall charge. When no subscript is written it is assumed to be "1."

As shown in step [2], the formula for the ionic compound formed from one calcium cation (Ca^{2+}) and one oxygen anion (O^{2-}) is CaO. The formula for the ionic compound formed from one calcium cation (Ca^{2+}) and two chlorine anions (Cl^-) is $CaCl_2$.

SAMPLE PROBLEM 3.5

When sterling silver tarnishes it forms an ionic compound derived from silver and sulfur. Write the formula for this ionic compound.

ANALYSIS
- Identify the cation and the anion, and use the periodic table to determine the charges.
- When ions of equal charge combine, one of each ion is needed. When ions of unequal charge combine, use the ionic charges to determine the relative number of each ion.
- Write the formula with the cation first and then the anion, omitting charges, and using subscripts to indicate the number of each ion.

SOLUTION
Silver is a metal, so it forms the cation. Sulfur is a nonmetal, so it forms the anion. The charge on silver is +1 (Ag^+), as shown in Figure 3.2. Sulfur (group 6A) is a main group element with a −2 charge (S^{2-}). Since the charges are unequal, use their magnitudes to determine the relative number of each ion to give an overall charge of zero.

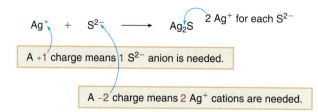

$$Ag^+ \quad + \quad S^{2-} \longrightarrow Ag_2S \quad \text{2 Ag}^+ \text{ for each S}^{2-}$$

A +1 charge means 1 S²⁻ anion is needed.

A −2 charge means 2 Ag⁺ cations are needed.

Answer: Since two Ag^+ cations are needed for each S^{2-} anion, the formula is **Ag_2S.**

Write the formula for the ionic compound formed from each pair of elements.

a. sodium and bromine

b. barium and oxygen

c. magnesium and iodine

d. lithium and oxygen

The tarnish on sterling silver is composed of an ionic compound formed from silver and sulfur (Sample Problem 3.5).

3.3B FOCUS ON HEALTH & MEDICINE
IONIC COMPOUNDS IN CONSUMER PRODUCTS

Simple ionic compounds are added to food or consumer products to prevent disease or maintain good health. For example, **potassium iodide** (KI) is an essential nutrient added to table salt. Iodine is needed to synthesize thyroid hormones. A deficiency of iodine in the diet can lead to insufficient thyroid hormone production. In an attempt to compensate, the thyroid gland may become enlarged, producing a swollen thyroid referred to as a goiter. **Sodium fluoride** (NaF) is added to toothpaste to strengthen tooth enamel and help prevent tooth decay.

HEALTH NOTE

Potassium is a critical cation for normal heart and skeletal muscle function and nerve impulse conduction. Drinking electrolyte replacement beverages like Gatorade or Powerade can replenish K⁺ lost in sweat.

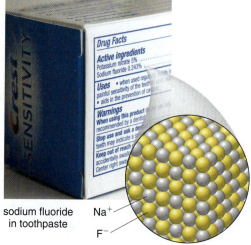

K^+ / I^- potassium iodide in table salt

sodium fluoride in toothpaste Na^+ / F^-

Potassium chloride (KCl), sold under trade names such as K–Dur, Klor–Con, and Micro–K, is an ionic compound used for patients whose potassium levels are low. Potassium chloride can be given as tablets, an oral suspension, or intravenously. Adequate potassium levels are needed for proper fluid balance and organ function. Although potassium is readily obtained from many different food sources (e.g., potatoes, beans, melon, bananas, and spinach), levels can become low when too much potassium is lost in sweat and urine or through the use of certain medications.

Zinc oxide, an ionic compound formed from zinc and oxygen, is a common component of sunblocks, as mentioned in the chapter opener. The zinc oxide crystals reflect sunlight away from the skin, and in this way, protect it from sun exposure. What is the ionic formula for zinc oxide?

3.4 NAMING IONIC COMPOUNDS

Now that we have learned how to write the formulas of some simple ionic compounds, we must learn how to name them. Assigning an unambiguous name to each compound is called chemical **nomenclature.** To name ionic compounds, we must first learn how to name the cations and anions that compose them.

3.4A NAMING CATIONS

Cations of main group metals are given the name of the element from which they are formed.

$$Na^+ \qquad K^+ \qquad Ca^{2+} \qquad Mg^{2+}$$
$$\text{sodium} \qquad \text{potassium} \qquad \text{calcium} \qquad \text{magnesium}$$

It is common to add the word "ion" after the name of the metal cation to distinguish it from the neutral metal itself. For example, when the concentration of sodium in a blood sample is determined, what is really measured is the concentration of sodium *ions* (Na^+).

When a metal is able to form two different cations, a method is needed to distinguish these cations. Two systems are used, the systematic method and the common method. The systematic method (Method [1]) will largely be followed in this text. Since many ions are still identified by older names, however, the common method (Method [2]) is also given.

- **Method [1]:** Follow the name of the cation by a Roman numeral in parentheses to indicate its charge.
- **Method [2]:** Use the suffix *-ous* for the cation with the smaller charge, and the suffix *-ic* for the cation with the higher charge. These suffixes are often added to the Latin names of the elements.

For example, the element iron (Fe) forms two cations, Fe^{2+} and Fe^{3+}, which are named in the following way:

	Systematic Name	Common Name
Fe^{2+}	iron(II)	fer**rous**
Fe^{3+}	iron(III)	fer**ric**

Table 3.3 lists the systematic and common names for several cations.

TABLE 3.3 Systematic and Common Names for Some Metal Ions

Element	Ion Symbol	Systematic Name	Common Name
Copper	Cu^+	Copper(I)	Cuprous
	Cu^{2+}	Copper(II)	Cupric
Chromium	Cr^{2+}	Chromium(II)	Chromous
	Cr^{3+}	Chromium(III)	Chromic
Iron	Fe^{2+}	Iron(II)	Ferrous
	Fe^{3+}	Iron(III)	Ferric
Tin	Sn^{2+}	Tin(II)	Stannous
	Sn^{4+}	Tin(IV)	Stannic

3.4B NAMING ANIONS

Anions are named by replacing the ending of the element name by the suffix -ide. For example:

$$Cl \quad ---\rightarrow \quad Cl^- \qquad \text{[Change -ine to -ide.]}$$
$$\text{chlor}ine \qquad\qquad\qquad \text{chlor}ide$$

$$O \quad ---\rightarrow \quad O^{2-} \qquad \text{[Change -ygen to -ide.]}$$
$$\text{ox}ygen \qquad\qquad\qquad \text{ox}ide$$

Table 3.4 lists the names of common anions derived from nonmetal elements.

TABLE 3.4	Names of Common Anions	
Element	**Ion Symbol**	**Name**
Bromine	Br^-	Bromide
Chlorine	Cl^-	Chloride
Fluorine	F^-	Fluoride
Iodine	I^-	Iodide
Oxygen	O^{2-}	Oxide
Sulfur	S^{2-}	Sulfide

PROBLEM 3.15

Give the name of each ion.

a. S^{2-} b. Cu^+ c. Cs^+ d. Al^{3+} e. Sn^{4+}

PROBLEM 3.16

Give the symbol for each ion.

a. stannic   b. iodide   c. manganese ion   d. lead(II)   e. selenide

PROBLEM 3.17

Under certain reaction conditions, an anion (H^-) can be formed from the hydrogen atom. Given the general way that anions are named, suggest a name for this anion.

3.4C NAMING IONIC COMPOUNDS WITH CATIONS FROM MAIN GROUP METALS

To name an ionic compound with a main group metal cation whose charge never varies, **name the cation and then the anion.** Do *not* specify the charge on the cation. Do *not* specify how many ions of each type are needed to balance charge.

$$Na^+ \qquad F^- \qquad ---\rightarrow \qquad NaF$$
$$\text{sodium} \quad \text{fluoride} \qquad\qquad\qquad \text{sodium fluoride}$$

$$Mg^{2+} \qquad Cl^- \qquad ---\rightarrow \qquad MgCl_2$$
$$\text{magnesium} \quad \text{chloride} \qquad\qquad\qquad \text{magnesium chloride}$$

Thus, $BaCl_2$ is named barium chloride (*not* barium *di*chloride). The number of ions of each type is inferred in the name because the net charge must be zero.

SAMPLE PROBLEM 3.6	Name each ionic compound: (a) Na_2S; (b) $AlBr_3$.
ANALYSIS	Name the cation and then the anion.
SOLUTION	a. Na_2S: The cation is sodium and the anion is sulfide (derived from sulfur); thus, the name is sodium sulfide. b. $AlBr_3$: The cation is aluminum and the anion is bromide (derived from bromine); thus, the name is aluminum bromide.

PROBLEM 3.18	Name each ionic compound.

<table>
<tr><td>a. NaF</td><td>c. $SrBr_2$</td><td>e. TiO_2</td><td>g. CaI_2</td></tr>
<tr><td>b. MgO</td><td>d. Li_2O</td><td>f. $AlCl_3$</td><td>h. $CoCl_2$</td></tr>
</table>

3.4D NAMING IONIC COMPOUNDS CONTAINING METALS WITH VARIABLE CHARGE

To name an ionic compound that contains a metal with variable charge, we must specify the charge on the cation. The formula of the ionic compound—that is, how many cations there are per anion—allows us to determine the charge on the cation.

HOW TO   Name an Ionic Compound That Contains a Metal with Variable Charge

EXAMPLE   **Give the name for $CuCl_2$.**

Step [1]   **Determine the charge on the cation.**
- Since there are two Cl^- anions, each of which has a −1 charge, the copper cation must have a +2 charge to make the overall charge zero.

$$CuCl_2 \quad 2\ Cl^- \text{ anions} \quad ---\rightarrow \quad \text{The total negative charge is } -2.$$

Cu must have a +2 charge to balance the −2 charge of the anions.

$$Cu^{2+}$$

Step [2]   **Name the cation and anion.**
- Name the cation using its element name followed by a Roman numeral to indicate its charge. In the common system, use the suffix *-ous* or *-ic* to indicate charge.
- Name the anion by changing the ending of the element name to the suffix *-ide.*

$$Cu^{2+} \quad ---\rightarrow \quad \text{copper(II)} \quad \text{or} \quad \text{cupric}$$
$$Cl^- \quad ---\rightarrow \quad \text{chloride}$$

Step [3]   **Write the name of the cation first, then the anion.**
- **Answer:** Copper(II) chloride or cupric chloride.

Sample Problem 3.7 illustrates the difference in naming ionic compounds derived from metals that have fixed or variable charge.

SAMPLE PROBLEM 3.7

SnF_2 and Al_2O_3 are both ingredients in commercial toothpastes. SnF_2 contains fluoride, which strengthens tooth enamel. Al_2O_3 is an abrasive that helps to scrub the teeth clean when they are brushed. Give names for (a) SnF_2; (b) Al_2O_3.

ANALYSIS

First determine if the cation has a fixed or variable charge. To name an ionic compound that contains a cation that always has the same charge, name the cation and then the anion (using the suffix *-ide*). When the metal has a variable charge, use the overall anion charge to determine the charge on the cation. Then name the cation (using a Roman numeral or the suffix *-ous* or *-ic*), followed by the anion.

SOLUTION

a. SnF_2: Sn cations have variable charge so the overall anion charge determines the cation charge.

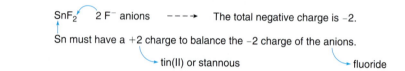

Sn must have a +2 charge to balance the −2 charge of the anions.

Answer: tin(II) fluoride or stannous fluoride

HEALTH NOTE

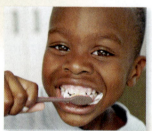

Some toothpastes contain the ionic compounds SnF_2 as a source of fluoride and Al_2O_3 as an abrasive.

b. Al_2O_3: Al has a fixed charge of +3. To name the compound, name the cation as the element (aluminum), and the anion by changing the ending of the element name to the suffix *-ide* (oxygen → oxide).

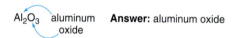

Answer: aluminum oxide

PROBLEM 3.19

Name each ionic compound.

a. $CrCl_3$ c. SnF_4 e. $FeBr_2$

b. PbS d. PbO_2 f. $AuCl_3$

PROBLEM 3.20

Several copper salts are brightly colored. Give the name for each of the following ionic copper compounds.

a. Cu_2O (brown)   b. CuO (black)   c. $CuCl$ (green)   d. $CuCl_2$ (blue)

PROBLEM 3.21

When iron rusts it forms Fe_2O_3. Name this product of air oxidation.

3.4E   WRITING A FORMULA FROM THE NAME OF AN IONIC COMPOUND

Thus far, we have focused on assigning a name to a formula for an ionic compound. Writing a formula from a name is also a useful skill.

HOW TO   Derive a Formula from the Name of an Ionic Compound

EXAMPLE   **Write the formula for tin(IV) oxide.**

Step [1]   **Identify the cation and the anion and determine their charges.**

- The name of the cation appears first, followed by the anion.
- For metals with variable charge, the Roman numeral gives the charge on the cation.

In this example, tin is the cation. The Roman numeral tells us that its charge is +4, making the cation Sn^{4+}. Oxide is the name of the oxygen anion, O^{2-} (Table 3.4).

Step [2]   **Balance charges.**

- Use the charge on the cation to determine the number of ions of the anion needed to balance charge.

$$Sn^{4+} \quad O^{2-} \quad \text{Two } -2 \text{ anions are needed}$$
$$\text{for each } +4 \text{ cation.}$$
cation   anion

Step [3]   **Write the formula with the cation first, and use subscripts to show the number of each ion needed to have zero overall charge.**

Answer: SnO_2

PROBLEM 3.22

Write the formula for each ionic compound.

a. calcium bromide   c. ferric bromide   e. chromium(II) chloride

b. copper(I) iodide   d. magnesium sulfide   f. sodium oxide

3.5 PHYSICAL PROPERTIES OF IONIC COMPOUNDS

Ionic compounds are crystalline solids composed of ions packed to maximize the interaction of the positive charge of the cations and negative charge of the anions. The relative size and charge of the ions determine the way they are packed in the crystal lattice. Ionic solids are held together by extremely strong interactions of the oppositely charged ions. How is this reflected in the melting point and boiling point of an ionic compound?

When a compound melts to form a liquid, energy is needed to overcome some of the attractive forces of the ordered solid, to form the less ordered liquid phase. Since an ionic compound is held together by very strong electrostatic interactions, it takes a great deal of energy to separate the ions from each other. As a result, **ionic compounds have very high melting points.** For example, the melting point of NaCl is 801 °C.

A great deal of energy is needed to overcome the attractive forces present in the liquid phase, too, to form ions that are far apart and very disorganized in the gas phase, so **ionic compounds have extremely high boiling points.** The boiling point of liquid NaCl is 1413 °C.

A great many ionic compounds are soluble in water. When an ionic compound dissolves in water, the ions are separated, and each anion and cation is surrounded by water molecules, as shown in Figure 3.4. The interaction of the water solvent with the ions provides the energy needed to overcome the strong ion–ion attractions of the crystalline lattice. We will learn much more about solubility in Chapter 8.

An **aqueous solution** contains a substance dissolved in liquid water.

When an ionic compound dissolves in water, the resulting aqueous solution conducts an electric current. This distinguishes ionic compounds from other compounds discussed in Chapter 4, some of which dissolve in water but do not form ions and therefore do not conduct electricity.

PROBLEM 3.23

List five physical properties of ionic compounds.

3.6 POLYATOMIC IONS

Sometimes ions are composed of more than one element. The ion bears a charge because the total number of electrons it contains is different from the total number of protons in the nuclei of all of the atoms.

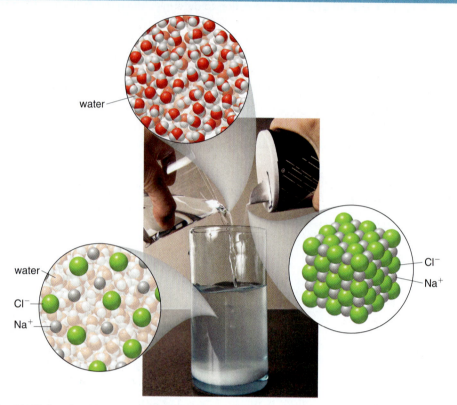

FIGURE 3.4   Dissolving NaCl in Water

water

water
Cl⁻
Na⁺

Cl⁻
Na⁺

When NaCl dissolves in water, each Na^+ ion and each Cl^- ion is surrounded by water molecules. The interactions of these ions with water molecules provide the energy needed to break apart the ions of the crystal lattice.

- A *polyatomic ion* is a cation or anion that contains more than one atom.

The atoms in the polyatomic ion are held together by covalent bonds, but since the ion bears a charge, it bonds to other ions by ionic bonding. For example, calcium sulfate, $CaSO_4$, is composed of a calcium cation, Ca^{2+}, and the polyatomic anion sulfate, SO_4^{2-}. $CaSO_4$ is used to make plaster casts for broken bones.

We will encounter only two polyatomic cations: **H_3O^+, the hydronium ion,** which will play a key role in the acid–base chemistry discussed in Chapter 9, and **NH_4^+, the ammonium ion.**

In contrast, there are several common polyatomic anions, most of which contain a nonmetal like carbon, sulfur, or phosphorus, usually bonded to one or more oxygen atoms. Common examples include **carbonate (CO_3^{2-}), sulfate (SO_4^{2-}), and phosphate (PO_4^{3-}).** Table 3.5 lists the most common polyatomic anions.

The names of most polyatomic anions end in the suffix *-ate.* Exceptions to this generalization include hydroxide (^-OH) and cyanide (^-CN). Two other aspects of nomenclature are worthy of note.

- The suffix *-ite* is used for an anion that has one fewer oxygen atom than a similar anion named with the *-ate* ending. Thus, SO_4^{2-} is sul*fate,* but SO_3^{2-} is sul*fite.*
- When two anions differ in the presence of a hydrogen, the word *hydrogen* or the prefix *bi-* is added to the name of the anion. Thus, SO_4^{2-} is sulfate, but HSO_4^- is *hydrogen* sulfate or *bi*sulfate.

Spam, a canned meat widely consumed in Alaska, Hawaii, and other parts of the United States, contains the preservative sodium nitrite, $NaNO_2$. Sodium nitrite inhibits the growth of *Clostridium botulinum,* a bacterium responsible for a lethal form of food poisoning.

TABLE 3.5   Names of Common Polyatomic Anions

Nonmetal	Formula	Name
Carbon	CO_3^{2-}	**Carbonate**
	HCO_3^-	Hydrogen carbonate or bicarbonate
	$CH_3CO_2^-$	Acetate
	^-CN	Cyanide
Nitrogen	NO_3^-	**Nitrate**
	NO_2^-	Nitrite
Oxygen	^-OH	**Hydroxide**
Phosphorus	PO_4^{3-}	**Phosphate**
	HPO_4^{2-}	Hydrogen phosphate
	$H_2PO_4^-$	Dihydrogen phosphate
Sulfur	SO_4^{2-}	**Sulfate**
	HSO_4^-	Hydrogen sulfate or bisulfate
	SO_3^{2-}	Sulfite
	HSO_3^-	Hydrogen sulfite or bisulfite

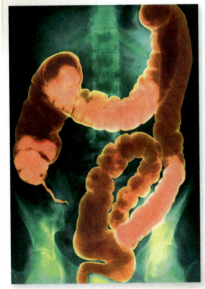

Barium sulfate is used to visualize the digestive system during an X-ray procedure.

3.6A   WRITING FORMULAS FOR IONIC COMPOUNDS WITH POLYATOMIC IONS

Writing the formula for an ionic compound with a polyatomic ion is no different than writing a formula for an ion with a single charged atom, so we follow the procedure outlined in Section 3.3A. When the cation and anion have the *same* charge, only *one* of each ion is needed for an overall charge of zero.

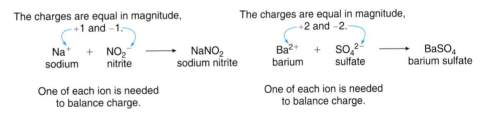

In a compound formed from ions of unequal charge, such as magnesium (Mg^{2+}) and hydroxide (^-OH), **the charges on the ions tell us how many of the** *oppositely* **charged ions are needed to balance the charge.**

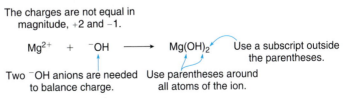

Parentheses are used around the polyatomic ion, and a subscript indicates how many of each are needed to balance charge. The formula is written as $Mg(OH)_2$ *not* MgO_2H_2.

SAMPLE PROBLEM 3.8

A dietary supplement used to prevent and treat calcium deficiencies consists of an ionic compound formed from calcium and phosphate. What is its formula?

ANALYSIS

- Identify the cation and anion and determine the charges.
- When ions of equal charge combine, one of each is needed. When ions of unequal charge combine, use the ionic charges to determine the relative number of each ion.
- Write the formula with the cation first and then the anion, omitting charges. Use parentheses around polyatomic ions when more than one appears in the formula, and use subscripts to indicate the number of each ion.

SOLUTION

The cation (Ca^{2+}) and anion (PO_4^{3-}) have different charges so the magnitude of the ionic charges determines the number of each ion giving an overall charge of zero.

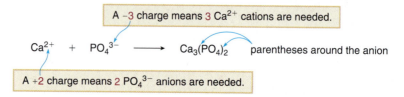

A −3 charge means 3 Ca^{2+} cations are needed.

$$Ca^{2+} \;+\; PO_4^{3-} \longrightarrow Ca_3(PO_4)_2 \quad \text{parentheses around the anion}$$

A +2 charge means 2 PO_4^{3-} anions are needed.

Answer: Since three Ca^{2+} cations are needed for two PO_4^{3-} anions, the formula is $Ca_3(PO_4)_2$.

PROBLEM 3.24

Write the formula for the compound formed when the sulfate anion (SO_4^{2-}) combines with a cation from each of the following elements: (a) magnesium; (b) sodium; (c) nickel; (d) aluminum; (e) lithium.

PROBLEM 3.25

Write the formula of the ionic compound formed from each pair of cations and anions.

a. sodium and bicarbonate   c. ammonium and sulfate   e. calcium and bisulfate
b. potassium and nitrate   d. magnesium and phosphate   f. barium and hydroxide

PROBLEM 3.26

Write the formula for the compound formed when K^+ combines with each anion.

a. ^-OH c. SO_4^{2-} e. PO_4^{3-}
b. NO_2^- d. HSO_3^- f. ^-CN

3.6B NAMING IONIC COMPOUNDS WITH POLYATOMIC IONS

Naming ionic compounds derived from polyatomic anions follows the same procedures outlined in Sections 3.4C and 3.4D. There is no easy trick for remembering the names and structures of the anions listed in Table 3.5. The names of the anions in boldface type are especially common and should be committed to memory.

SAMPLE PROBLEM 3.9

Name each ionic compound: (a) $NaHCO_3$, the active ingredient in baking soda; (b) $Al_2(SO_4)_3$, an ingredient once used in antiperspirants, but no longer considered effective.

ANALYSIS

First determine if the cation has a fixed or variable charge. To name an ionic compound that contains a cation that always has the same charge, name the cation and then the anion. When the metal has a variable charge, use the overall anion charge to determine the charge on the cation. Then name the cation (using a Roman numeral or the suffix *-ous* or *-ic*), followed by the anion.

SOLUTION

a. $NaHCO_3$: Sodium cations have a fixed charge of +1. The anion HCO_3^- is called bicarbonate or hydrogen carbonate.

  Answer: sodium bicarbonate or sodium hydrogen carbonate

b. $Al_2(SO_4)_3$: Aluminum cations have a fixed charge of +3. The anion SO_4^{2-} is called sulfate.

  Answer: aluminum sulfate

Name each compound.

a. Na_2CO_3 c. $Mg(NO_3)_2$ e. $Fe(HSO_3)_3$

b. $Ca(OH)_2$ d. $Mn(CH_3CO_2)_2$ f. $Mg_3(PO_4)_2$

3.6C FOCUS ON HEALTH & MEDICINE
USEFUL IONIC COMPOUNDS

Ionic compounds are the active ingredients in several over-the-counter drugs. Examples include **calcium carbonate ($CaCO_3$),** the antacid in Tums; **magnesium hydroxide [$Mg(OH)_2$],** one of the active components in the antacids Maalox and milk of magnesia; and **iron(II) sulfate ($FeSO_4$),** an iron supplement used to treat anemia.

The shells of oysters and other mollusks are composed largely of calcium carbonate, $CaCO_3$.

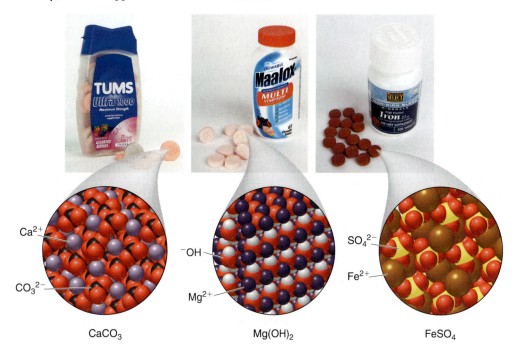

Ca²⁺ → Ca^{2+}
CO₃²⁻ → CO_3^{2-}
CaCO₃ → $CaCO_3$

⁻OH → ^-OH
Mg²⁺ → Mg^{2+}
Mg(OH)₂ → $Mg(OH)_2$

SO₄²⁻ → SO_4^{2-}
Fe²⁺ → Fe^{2+}
FeSO₄ → $FeSO_4$

Some ionic compounds are given as intravenous drugs. Bicarbonate (HCO_3^-) is an important polyatomic anion that controls the acid–base balance in the blood. When the blood becomes too acidic, sodium bicarbonate ($NaHCO_3$) is administered intravenously to decrease the acidity. Magnesium sulfate ($MgSO_4$), an over-the-counter laxative, is also given intravenously to prevent seizures caused by extremely high blood pressure associated with some pregnancies.

3.6D FOCUS ON HEALTH & MEDICINE
TREATING OSTEOPOROSIS

Although much of the body is composed of compounds held together by covalent bonds, about 70% of bone is composed largely of a complex ionic solid with the formula $Ca_{10}(PO_4)_6(OH)_2$ called **hydroxyapatite.** Throughout an individual's life, hydroxyapatite is constantly broken down and rebuilt. In postmenopausal women, however, the rate of bone loss often becomes greater than bone synthesis, and bones get brittle and easily broken. This condition is called **osteoporosis.**

In recent years, some prescription drugs have proven effective in combating osteoporosis. Sodium alendronate, trade name Fosamax, increases bone density by decreasing the rate of bone loss. Fosamax is an ionic compound with the formula $Na(C_4H_{12}NO_7P_2)$. This compound contains a sodium cation, Na^+, and a polyatomic anion, $(C_4H_{12}NO_7P_2)^-$.

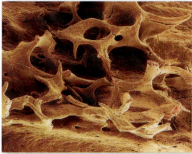

(top) Normal bone; (bottom) brittle bone due to osteoporosis. Osteoporosis results in a decrease in bone density, making bones brittle and easily fractured.

PROBLEM 3.28	Using the charges on the ions that compose hydroxyapatite, show that it has zero overall charge.

CHAPTER HIGHLIGHTS

KEY TERMS

Ammonium ion (3.6)
Anion (3.2)
Bonding (3.1)
Carbonate (3.6)
Cation (3.2)
Covalent bond (3.1)

Hydronium ion (3.6)
Hydroxide (3.6)
Ion (3.1)
Ionic bond (3.1)
Molecule (3.1)

Nomenclature (3.4)
Octet rule (3.2)
Phosphate (3.6)
Polyatomic ion (3.6)
Sulfate (3.6)

KEY CONCEPTS

❶ What are the basic features of ionic and covalent bonds? (3.1)

- Both ionic and covalent bonding follows one general rule: Elements gain, lose, or share electrons to attain the electronic configuration of the noble gas closest to them in the periodic table.
- Ionic bonds result from the transfer of electrons from one element to another. Ionic bonds form between a metal and a nonmetal. Ionic compounds consist of oppositely charged ions that feel a strong electrostatic attraction for each other.
- Covalent bonds result from the sharing of electrons between two atoms. Covalent bonds occur between two nonmetals, or when a metalloid combines with a nonmetal. Covalent bonding forms discrete molecules.

❷ How can the periodic table be used to determine whether an atom forms a cation or an anion, and its resulting ionic charge? (3.2)

- Metals form cations and nonmetals form anions.
- By gaining or losing one, two, or three electrons, an atom forms an ion with a completely filled outer shell of electrons.
- The charge on main group ions can be predicted from the position in the periodic table. For metals in groups 1A, 2A, and 3A, the group number = the charge on the cation. For nonmetals in groups 6A and 7A, the anion charge = 8 − (the group number).

❸ What is the octet rule? (3.2)

- Main group elements are especially stable when they possess an octet of electrons. Main group elements gain or lose one, two, or three electrons to form ions with eight outer shell electrons.

❹ What determines the formula of an ionic compound? (3.3)

- Cations and anions always form ionic compounds that have zero overall charge.
- Ionic compounds are written with the cation first, and then the anion, with subscripts to show how many of each are needed to have zero net charge.

❺ How are ionic compounds named? (3.4)

- Ionic compounds are always named with the name of the cation first.
- With cations having a fixed charge, the cation has the same name as its neutral element. The name of the anion usually ends in the suffix -ide if it is derived from a single atom or -ate (or -ite) if it is polyatomic.
- When the metal has a variable charge, use the overall anion charge to determine the charge on the cation. Then name the cation using a Roman numeral or the suffix -ous (for the ion with the smaller charge) or -ic (for the ion with the larger charge).

❻ Describe the properties of ionic compounds. (3.5)

- Ionic compounds are crystalline solids with the ions arranged to maximize the interactions of the oppositely charged ions.
- Ionic compounds have high melting points and boiling points.
- Most ionic compounds are soluble in water and their aqueous solutions conduct an electric current.

7 What are polyatomic ions and how are they named? (3.6)

- Polyatomic ions are charged species that are composed of more than one element.
- The names for polyatomic cations end in the suffix -*onium*.
- Many polyatomic anions have names that end in the suffix -*ate*. The suffix -*ite* is used for an anion that has one fewer oxygen atom than a similar anion named with the -*ate* ending. When two anions differ in the presence of a hydrogen, the word *hydrogen* or the prefix *bi-* is added to the name of the anion.

8 List useful consumer products and drugs that are composed of ionic compounds.

- Useful ionic compounds that contain alkali metal cations and halogen anions include KI (iodine supplement), NaF (source of fluoride in toothpaste), and KCl (potassium supplement). (3.3)
- Other products contain SnF_2 (fluoride source in toothpaste), Al_2O_3 (abrasive in toothpaste), and ZnO (sunblock agent). (3.4)
- Useful ionic compounds with polyatomic anions include $CaCO_3$ (antacid and calcium supplement), magnesium hydroxide (antacid), and $FeSO_4$ (iron supplement). (3.6)

PROBLEMS

Selected in-chapter and end-of-chapter problems have brief answers provided in Appendix B.

Ionic and Covalent Bonding

3.29 Which formulas represent ionic compounds and which represent covalent compounds?
a. CO_2 b. H_2SO_4 c. KF   d. CH_5N

3.30 Which formulas represent ionic compounds and which represent covalent compounds?
a. C_3H_8 b. ClBr   c. CuO   d. CH_4O

3.31 Which pairs of elements are likely to form ionic bonds and which pairs are likely to form covalent bonds?
a. potassium and oxygen   c. two bromine atoms
b. sulfur and carbon   d. carbon and oxygen

3.32 Which pairs of elements are likely to form ionic bonds and which pairs are likely to form covalent bonds?
a. carbon and hydrogen   c. hydrogen and oxygen
b. sodium and sulfur   d. magnesium and bromine

3.33 Why do ionic bonds form between a metal and a nonmetal?

3.34 Is it proper to speak of sodium chloride molecules? Explain.

Ions

3.35 Write the ion symbol for an atom with the given number of protons and electrons.
a. four protons and two electrons
b. 22 protons and 20 electrons
c. 16 protons and 18 electrons
d. 13 protons and 10 electrons
e. 17 protons and 18 electrons
f. 20 protons and 18 electrons

3.36 How many protons and electrons are present in each ion?
a. K^+ b. S^{2-} c. Mn^{2+} d. Fe^{2+} e. Cs^+ f. I^-

3.37 What element fits each description?
a. a period 2 element that forms a +2 cation
b. an ion from group 7A with 18 electrons
c. a cation from group 1A with 36 electrons

3.38 What element fits each description?
a. a period 3 element that forms an ion with a −1 charge
b. an ion from group 2A with 36 electrons
c. an ion from group 6A with 18 electrons

3.39 Why do elements in group 6A gain electrons to form anions?

3.40 Why do elements in group 2A lose electrons to form cations?

3.41 Give the ion symbol for each ion.
a. sodium ion   c. manganese ion   e. stannic
b. selenide   d. gold(III)

3.42 Give the ion symbol for each ion.
a. barium ion   c. oxide   e. lead(IV)
b. iron(II)   d. ferrous

3.43 What noble gas has the same electronic configuration as each ion?
a. O^{2-} b. Mg^{2+} c. Al^{3+} d. S^{2-} e. F^- f. Be^{2+}

3.44 Give two cations and two anions that have the same electronic configuration as each noble gas: (a) neon; (b) argon.

3.45 How many electrons must be gained or lost by each element to achieve a noble gas configuration of electrons?
a. lithium   b. iodine   c. sulfur   d. strontium

3.46 How many electrons must be gained or lost by each element to achieve a noble gas configuration of electrons?
a. cesium   b. barium   c. selenium   d. aluminum

3.47 Which ions are likely to form? For those ions that are not likely to form, explain why this is so.
a. S^- b. S^{2-} c. S^{3-} d. Na^+ e. Na^{2+} f. Na^-

3.48 Which ions are likely to form? For those ions that are not likely to form, explain why this is so.
a. Mg^+ b. Mg^{2+} c. Mg^{3+} d. Cl^+ e. Cl^- f. Cl^{2-}

3.49 For each of the general electron-dot formulas for elements, give the following information: [1] the number of valence electrons; [2] the group number of the element; [3] how many electrons would be gained or lost to achieve a noble gas configuration; [4] the charge on the resulting ion; [5] an example of the element.
a. X·   b. ·Q·   c. ·Ż·   d. :Ä·

3.50 Label each of the following elements or regions in the periodic table.

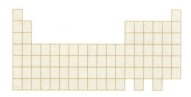

a. a group that forms cations with a +2 charge
b. a group that forms anions with a −2 charge
c. a group that forms cations with a +1 charge
d. a group that forms anions with a −1 charge
e. elements that form ions with the same electronic configuration as Ne
f. elements that form ions with the same electronic configuration as He

3.51 Give the formula for each polyatomic ion.
a. sulfate   c. hydrogen carbonate
b. ammonium   d. cyanide

3.52 Give the formula for each polyatomic ion.
a. acetate   c. dihydrogen phosphate
b. bisulfite   d. hydronium

3.53 How many protons and electrons are contained in each polyatomic ion?
a. ^-OH b. H_3O^+ c. PO_4^{3-}

3.54 How many protons and electrons are contained in each polyatomic ion?
a. NH_4^+ b. ^-CN c. CO_3^{2-}

3.55 Why don't elements in group 4A readily form ions?

3.56 Do all isotopes of an element form the same type of ions? Explain.

3.57 Why isn't the octet rule followed by transition metals when they form cations?

3.58 Why don't the elements in group 8A form ions?

Ionic Compounds

3.59 How does the compound NaF illustrate the octet rule?

3.60 How does the compound LiF "violate" the octet rule?

3.61 Write the formula for the ionic compound formed from each pair of elements.
a. calcium and sulfur   d. nickel and chlorine
b. aluminum and bromine   e. sodium and selenium
c. lithium and iodine

3.62 Write the formula for the ionic compound formed from each pair of elements.
a. barium and bromine
b. aluminum and sulfur
c. manganese and chlorine
d. zinc and sulfur
e. magnesium and fluorine

3.63 Write the formula for the ionic compound formed from each cation and anion.
a. lithium and nitrite
b. calcium and acetate
c. sodium and bisulfite
d. manganese and phosphate
e. magnesium and hydrogen sulfite

3.64 Write the formula for the ionic compound formed from each cation and anion.
a. potassium and bicarbonate
b. magnesium and nitrate
c. lithium and carbonate
d. potassium and cyanide
e. ammonium and phosphate

3.65 Complete the following table by filling in the formula of the ionic compound derived from the cations on the left and each of the anions across the top.

	Br^-	^-OH	HCO_3^-	SO_3^{2-}	PO_4^{3-}
Na^+					
Co^{2+}					
Al^{3+}					

3.66 Complete the following table by filling in the formula of the ionic compound derived from the cations on the left and each of the anions across the top.

	I^-	^-CN	NO_3^-	SO_4^{2-}	HPO_4^{2-}
K^+					
Mg^{2+}					
Cr^{3+}					

3.67 Write the formula for the ionic compound formed from the bisulfate anion (HSO_4^-) and each cation: (a) K^+; (b) Ba^{2+}; (c) Al^{3+}; (d) Zn^{2+}.

3.68 Write the formula for the ionic compound formed from the sulfite anion (SO_3^{2-}) and each cation: (a) K^+; (b) Ba^{2+}; (c) Al^{3+}; (d) Zn^{2+}.

3.69 Write the formula for the ionic compound formed from the barium cation (Ba^{2+}) and each anion: (a) ^-CN; (b) PO_4^{3-}; (c) HPO_4^{2-}; (d) $H_2PO_4^-$.

3.70 Write the formula for the ionic compound formed from the iron(III) cation (Fe^{3+}) and each anion: (a) ^-CN; (b) PO_4^{3-}; (c) HPO_4^{2-}; (d) $H_2PO_4^-$.

Naming Ionic Compounds

3.71 Name each ionic compound.
   a. Na_2O d. $AgCl$ f. $RbBr$
   b. BaS e. $CoBr_2$ g. $PbBr_2$
   c. PbS_2

3.72 Name each ionic compound.
   a. KF d. SnO f. Li_2S
   b. $ZnCl_2$ e. $AuBr_3$ g. $SnBr_4$
   c. Cu_2S

3.73 Name each ionic compound.
   a. $FeCl_2$ b. $FeBr_3$ c. FeS d. Fe_2S_3

3.74 Name each ionic compound.
   a. $CrCl_2$ b. $CrBr_3$ c. CrO d. Cr_2O_3

3.75 Why is a Roman numeral needed in the name for $CuBr_2$ but not $CaBr_2$? Name both compounds.

3.76 Why is a Roman numeral needed in the name for PbO but not ZnO? Name both compounds.

3.77 Write formulas to illustrate the difference between each pair of compounds.
   a. sodium sulfide and sodium sulfate
   b. magnesium oxide and magnesium hydroxide
   c. magnesium sulfate and magnesium bisulfate

3.78 Write formulas to illustrate the difference between each pair of compounds.
   a. lithium sulfite and lithium sulfide
   b. sodium carbonate and sodium hydrogen carbonate
   c. calcium phosphate and calcium dihydrogen phosphate

3.79 Name each ionic compound.
   a. NH_4Cl c. $Cu(NO_3)_2$ e. $Fe(NO_3)_2$
   b. $PbSO_4$ d. $Ca(HCO_3)_2$

3.80 Name each ionic compound.
   a. $(NH_4)_2SO_4$ c. $Cr(CH_3CO_2)_3$ e. $Ni_3(PO_4)_2$
   b. NaH_2PO_4 d. $Sn(HPO_4)_2$

3.81 Write a formula from each name.
   a. magnesium carbonate
   b. nickel sulfate
   c. copper(II) hydroxide
   d. potassium hydrogen phosphate
   e. gold(III) nitrate
   f. lithium phosphate
   g. aluminum bicarbonate
   h. chromous cyanide

3.82 Write a formula from each name.
   a. copper(I) sulfite
   b. aluminum nitrate
   c. tin(II) acetate
   d. lead(IV) carbonate
   e. zinc hydrogen phosphate
   f. manganese dihydrogen phosphate
   g. ammonium cyanide
   h. iron(II) nitrate

3.83 Write the formula for the ionic compound formed from Pb^{4+} and each anion. Then name each compound.
   a. ^-OH c. HCO_3^- e. PO_4^{3-}
   b. SO_4^{2-} d. NO_3^- f. $CH_3CO_2^-$

3.84 Write the formula for the ionic compound formed from Fe^{3+} and each anion. Then name each compound.
   a. ^-OH c. HPO_4^{2-} e. PO_4^{3-}
   b. CO_3^{2-} d. NO_2^- f. $CH_3CO_2^-$

Properties of Ionic Compounds

3.85 Label each statement as "true" or "false." Correct any false statement to make it true.
   a. Ionic compounds have high melting points.
   b. Ionic compounds can be solid, liquid, or gas at room temperature.
   c. Most ionic compounds are insoluble in water.
   d. An ionic solid like sodium chloride consists of discrete pairs of sodium cations and chloride anions.

3.86 Label each statement as "true" or "false." Correct any false statement to make it true.
   a. Ionic compounds have high boiling points.
   b. The ions in a crystal lattice are arranged randomly and the overall charge is zero.
   c. When an ionic compound dissolves in water, the solution conducts electricity.
   d. In an ionic crystal, ions having like charges are arranged close to each other.

3.87 Why do ionic solids have high melting points?

3.88 Would you expect the gases in the atmosphere to be composed of ionic compounds or covalent molecules? Explain your choice.

3.89 Which compound has the highest melting point: $NaCl$, CH_4, or H_2SO_4?

3.90 Which compound or element has the lowest boiling point: Cl_2, KI, or LiF?

Applications

3.91 Zinc is an essential nutrient needed by many enzymes to maintain proper cellular function. Zinc is obtained in many dietary sources, including oysters, beans, nuts, whole grains, and sunflower seeds. (a) How many protons and electrons are found in a neutral zinc atom? (b) How many electrons and protons are found in the Zn^{2+} cation? (c) Write the electronic configuration of the element zinc, and suggest which electrons are lost to form the Zn^{2+} cation.

3.92 Wilson's disease is an inherited defect in copper metabolism in which copper accumulates in tissues, causing neurological problems and liver disease. The disease can be treated with compounds that bind to copper and thus remove it from the tissues. (a) How many protons and electrons are found in a neutral copper atom? (b) How many electrons and protons are found in the Cu^+ cation? (c) How many electrons and protons are found in the Cu^{2+} cation? (d) Zinc acetate inhibits copper absorption and so it is used to treat Wilson's disease. What is the structure of zinc acetate?

3.93 Na^+, K^+, Ca^{2+}, and Mg^{2+} are the four major cations in the body. For each cation, give the following information: (a) the number of protons; (b) the number of electrons; (c) the noble gas that has the same electronic configuration; (d) its role in the body.

3.94   Unlike many ionic compounds, calcium carbonate is insoluble in water. What information contained in this chapter suggested that calcium carbonate is water insoluble?

3.95   Write the formula for silver nitrate, an antiseptic and germ killing agent.

3.96   Ammonium carbonate is the active ingredient in smelling salts. Write its formula.

3.97 $CaSO_3$ is used to preserve cider and fruit juices. Name this ionic compound.

3.98   Many ionic compounds are used as paint pigments. Name each of the following pigments.
a. CdS (yellow)   c. Cr_2O_3 (white)
b. TiO_2 (white)   d. $Mn_3(PO_4)_2$ (purple)

3.99   Ammonium nitrate is the most common source of the element nitrogen in fertilizers. When it is mixed with water, the solution gets cold, so it is used in instant cold packs. When mixed with diesel fuel it forms an explosive mixture that can be used as a bomb. Write the structure of ammonium nitrate.

3.100 Write the formula for sodium phosphate, a key ingredient in many commercial detergents.

CHALLENGE QUESTIONS

3.101 Energy bars contain ionic compounds that serve as a source of the trace elements that the body needs each day for proper cellular function. Answer the following questions about some of the ingredients in one commercial product.
a. Write the formulas for magnesium oxide and potassium iodide.
b. The ingredient $CaHPO_4$ is called dicalcium phosphate on the label. What name would you give to this ionic compound?
c. Give two different names for the ingredient $FePO_4$.
d. Sodium selenite is one ingredient. Selenite is a polyatomic anion that contains a selenium atom in place of the sulfur atom in sulfite. With this in mind, suggest a structure for sodium selenite.
e. Another ingredient is listed as chromium chloride. What is wrong with this name?

3.102 Some polyatomic anions contain a metal as part of the anion. For example, the anion dichromate has the structure $Cr_2O_7^{2-}$ and the anion permanganate has the structure MnO_4^-. Write the formula of the ionic compound formed from each of these anions and a potassium cation. Name each compound.

4

- - - - - - - - - - - - - - - - -

CHAPTER GOALS

In this chapter you will learn how to:

1. Recognize the bonding characteristics of covalent compounds

2. Draw Lewis structures for covalent compounds

3. Draw resonance structures for some ions and molecules

4. Name covalent compounds that contain two types of elements

5. Predict the shape around an atom in a molecule

6. Use electronegativity to determine whether a bond is polar or nonpolar

7. Determine whether a molecule is polar or nonpolar

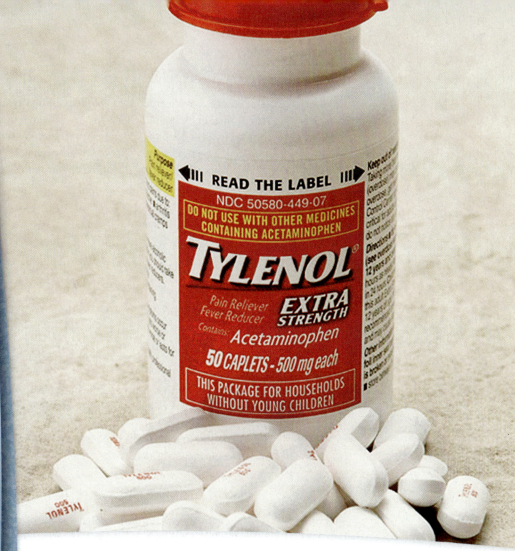

Like many drugs, **acetaminophen,** the active ingredient in the pain reliever Tylenol, is composed of covalent bonds.

COVALENT COMPOUNDS

ALTHOUGH all of Chapter 3 was devoted to ionic compounds, in truth, most compounds that we come in contact with in our daily lives are **covalent compounds,** formed by sharing electrons between atoms. The air we breathe is composed largely of the gases nitrogen and oxygen, two covalent molecules. The body is mostly water, which contains two covalent hydrogen–oxygen bonds. Most of the drugs routinely used—aspirin, acetaminophen, ibuprofen, and all antibiotics—are covalent compounds. Virtually all products of the chemical industry—polyethylene, nylon, synthetic dyes, gasoline, and pesticides, to name a few—are covalent compounds. In Chapter 4, we learn about the important features of covalent compounds.

4.1 INTRODUCTION TO COVALENT BONDING

In Section 3.1 we learned that **covalent bonds result from the** *sharing* **of electrons between two atoms.** For example, when two hydrogen atoms with one electron each (H·) combine, they form a covalent bond that contains two electrons. The two negatively charged electrons are now attracted to both positively charged hydrogen nuclei, forming the hydrogen **molecule,** H_2. This is an especially stable arrangement, since the shared electrons give each hydrogen atom the noble gas configuration of helium.

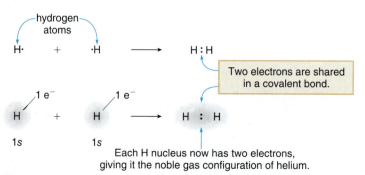

- A covalent bond is a two-electron bond in which the bonding atoms share the electrons.
- A *molecule* is a discrete group of atoms held together by covalent bonds.

We use a **solid line between two element symbols to represent a two-electron bond.** Thus, the H_2 molecule can be drawn as:

$$H_2 \;=\; H:H \;=\; H{-}H \quad \text{two-electron bond}$$

Hydrogen is called a **diatomic molecule** because it contains just two atoms. In addition to hydrogen, six other elements exist as diatomic molecules: nitrogen (N_2), oxygen (O_2), fluorine (F_2), chlorine (Cl_2), bromine (Br_2), and iodine (I_2).

Hydrogen fluoride, HF, is an example of a diatomic molecule formed between two different atoms, hydrogen and fluorine. Hydrogen has one valence electron and fluorine has seven. H and F each donate one electron to form a single two-electron bond.

$$H\cdot \;+\; \cdot\ddot{\underset{..}{F}}: \;\longrightarrow\; H:\ddot{\underset{..}{F}}: \;\text{ or }\; H{-}\ddot{\underset{..}{F}}: \quad \text{three lone pairs}$$

two electrons around H — eight electrons around F

The resulting molecule gives both H and F a filled valence shell: H is surrounded by two electrons, giving it the noble gas configuration of helium, and F is surrounded by eight electrons, giving it the noble gas configuration of neon. The F atom shares two electrons in one covalent bond, and it also contains three pairs of electrons that it does not share with hydrogen. These unshared electron pairs are called **nonbonded electron pairs** or **lone pairs.**

Nonbonded electron pair = lone pair.

- In covalent bonding, atoms share electrons to attain the electronic configuration of the noble gas closest to them in the periodic table.

As a result, hydrogen shares two electrons. Other main group elements are especially stable when they possess an *octet* **of electrons in their outer shell.**

PROBLEM 4.1

Use electron-dot symbols to show how a hydrogen atom and a chlorine atom form the diatomic molecule HCl. Explain how each atom has the electronic configuration of the noble gas closest to it in the periodic table.

PROBLEM 4.2	Use electron-dot symbols to show how two chlorine atoms form the diatomic molecule Cl_2. Explain how each atom has the electronic configuration of the noble gas closest to it in the periodic table.

4.1A   COVALENT BONDING AND THE PERIODIC TABLE

When do two atoms form covalent bonds rather than ionic bonds? **Covalent bonds are formed when two nonmetals combine.** Nonmetals do not easily lose electrons, and as a result, one non-metal does not readily transfer an electron to another nonmetal. **Covalent bonds are also formed when a metalloid bonds to a nonmetal. Covalent bonding is preferred with elements in the middle of the periodic table** that would otherwise have to gain or lose several electrons to form an ion with a complete outer shell of electrons.

Methane (CH_4), ammonia (NH_3), and water (H_2O) are three examples of covalent molecules in which each main group element is surrounded by eight electrons. Methane, the main component of natural gas, contains four covalent carbon–hydrogen bonds, each having two electrons. The nitrogen atom in NH_3, an agricultural fertilizer, is surrounded by an octet since it has three bonds and one lone pair. The oxygen atom in H_2O is also surrounded by an octet since it has two bonds and two lone pairs.

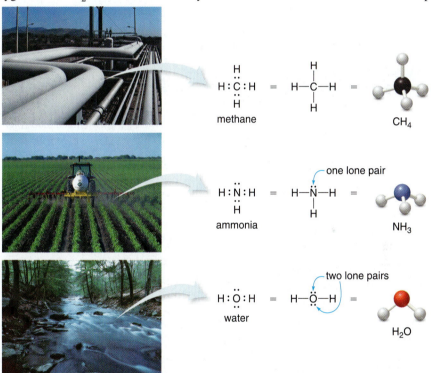

These electron-dot structures for molecules are called Lewis structures. Lewis structures show the location of all valence electrons in a molecule, both the shared electrons in bonds, and the nonbonded electron pairs. In Section 4.2, we will learn a general procedure for drawing Lewis structures.

How many covalent bonds will a particular atom typically form? As you might expect, it depends on the location of the atom in the periodic table. In the first row, hydrogen forms one covalent bond with its one valence electron. Other main group elements generally have no more than eight electrons around them. For neutral molecules, two consequences result.

- Atoms with one, two, or three valence electrons generally form one, two, or three bonds, respectively.
- Atoms with four or more valence electrons form enough bonds to give an octet. Thus, for atoms with four or more valence electrons:

$$\text{Predicted number of bonds} = 8 - \text{number of valence electrons}$$

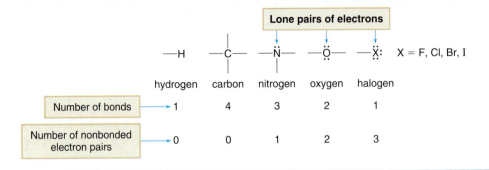

▼ FIGURE 4.1 Bonding Patterns for Common Main Group Elements

These guidelines are used in Figure 4.1 to summarize the usual number of covalent bonds formed by some common atoms. Except for hydrogen, these common elements generally follow one rule in bonding:

$$\boxed{\text{Number of bonds}} + \boxed{\text{Number of lone pairs}} = \boxed{4}$$

SAMPLE PROBLEM 4.1   Without referring to Figure 4.1, how many covalent bonds are predicted for each atom: (a) B; (b) N?

ANALYSIS   Atoms with one, two, or three valence electrons form one, two, or three bonds, respectively. Atoms with four or more valence electrons form enough bonds to give an octet.

SOLUTION   a. B has three valence electrons. Thus, it is expected to form three bonds.

b. N has five valence electrons. Since it contains more than four valence electrons, it is expected to form 8 − 5 = 3 bonds.

PROBLEM 4.3   How many covalent bonds are predicted for each atom: (a) F; (b) Si; (c) Br; (d) O; (e) P; (f) S?

PROBLEM 4.4   Fill in the lone pairs on each atom to give every main group element except hydrogen an octet.

a. $H-\overset{\overset{\displaystyle H}{|}}{\underset{\underset{\displaystyle H}{|}}{C}}-Cl$ b. $H-\overset{\overset{\displaystyle H}{|}}{\underset{\underset{\displaystyle H}{|}}{N}}-O-H$ c. $H-\overset{\overset{\displaystyle H}{|}}{\underset{\underset{\displaystyle H}{|}}{C}}-O-H$ d. $Br-\overset{\overset{\displaystyle H}{|}}{\underset{\underset{\displaystyle H}{|}}{C}}-Br$

PROBLEM 4.5   A nonmetal like oxygen forms both ionic and covalent bonds, depending on the identity of the element to which it bonds. What type of bonding is observed in CaO and CO_2? Explain why two different types of bonding are observed.

4.1B   FOCUS ON THE HUMAN BODY
COVALENT MOLECULES AND THE
CARDIOVASCULAR SYSTEM

Living organisms are a sea of covalent molecules. The major component in the body, water, is a covalent molecule. In addition, the proteins that compose muscle, the carbohydrates that are metabolized for energy, stored fat, and DNA, the carrier of genetic information, are all covalent molecules. Some of these molecules are very large, and are composed of hundreds or thousands of covalent bonds.

FIGURE 4.2 Covalent Molecules and the Human Heart

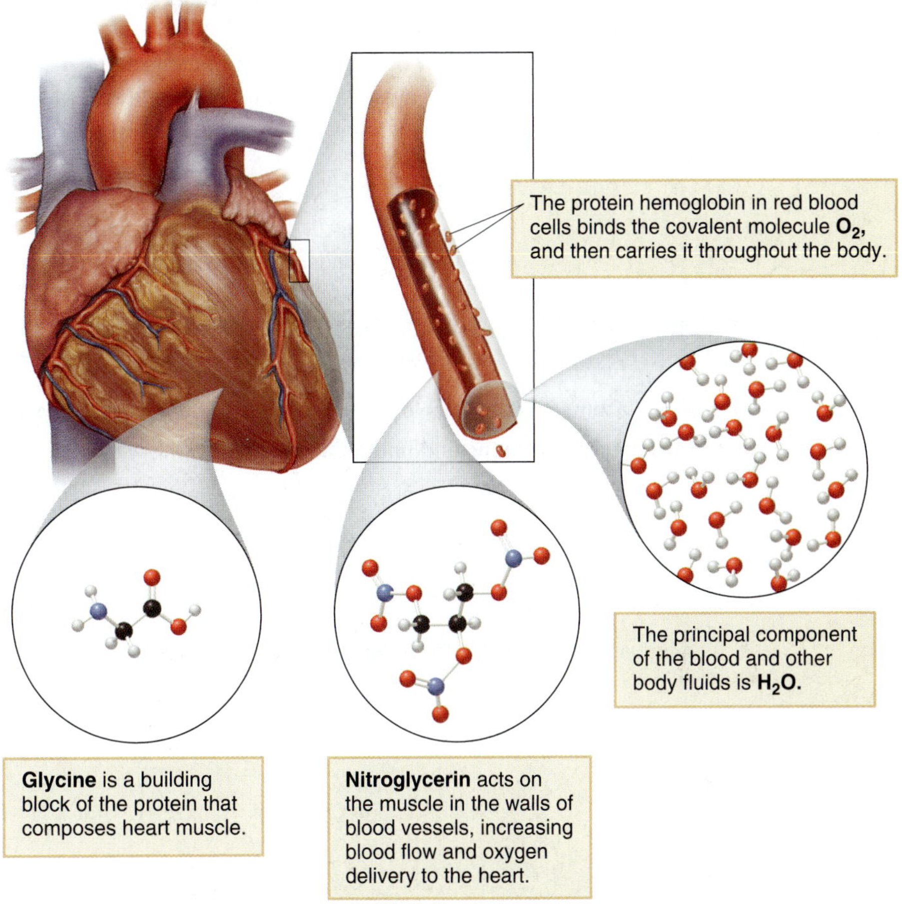

The protein hemoglobin in red blood cells binds the covalent molecule O_2, and then carries it throughout the body.

The principal component of the blood and other body fluids is H_2O.

Glycine is a building block of the protein that composes heart muscle.

Nitroglycerin acts on the muscle in the walls of blood vessels, increasing blood flow and oxygen delivery to the heart.

Some covalent compounds related to the chemistry of the heart include **water,** the most prevalent covalent compound in the body; **oxygen,** which is carried by the protein **hemoglobin** to the tissues; **glycine,** a building block of the proteins that compose heart muscle; and **nitroglycerin,** a drug used to treat some forms of heart disease.

Figure 4.2 contains a schematic of a blood vessel inside the heart, and it illustrates a few covalent molecules—water, hemoglobin, oxygen, glycine, and nitroglycerin—that play a role in the cardiovascular system. Blood is composed of water and red blood cells that contain the protein hemoglobin. Hemoglobin is a large covalent compound that complexes oxygen molecules, and carries oxygen to tissues throughout the body. Heart muscle is composed of complex covalent protein molecules, which are synthesized from smaller molecules. The three-dimensional structure of one of those molecules, glycine, is pictured. Finally, covalent compounds are used to treat heart disease. For example, nitroglycerin, a drug used when blood vessels have become narrow, increases blood flow and thereby oxygen delivery to the heart.

4.2 LEWIS STRUCTURES

A molecular formula shows the number and identity of all of the atoms in a compound, but it does not tell us what atoms are bonded to each other. Thus, the formula NH_3 for ammonia shows that ammonia contains one nitrogen atom and three hydrogen atoms, but it does not tell us that ammonia has three covalent nitrogen–hydrogen bonds and that the N atom has a lone pair. **A Lewis structure, in contrast, shows the connectivity between the atoms, as well as where all the bonding and nonbonding valence electrons reside.**

4.2A   DRAWING LEWIS STRUCTURES

There are three general rules for drawing Lewis structures.

> 1. Draw only the *valence electrons.*
> 2. Give every main group element (except hydrogen) an *octet* of electrons.
> 3. Give each hydrogen two electrons.

In Section 4.1, Lewis structures were drawn for several covalent molecules. While drawing a Lewis structure for a diatomic molecule with one bond is straightforward, drawing Lewis structures for compounds with three or more atoms is easier if you follow a general procedure.

HOW TO   Draw a Lewis Structure

Step [1]   Arrange the atoms next to each other that you think are bonded together.

- **Always place hydrogens and halogens on the periphery** since these atoms form only one bond.

For CH₄:
```
       H                    H
                            |
   H   C   H     not    H   C   H   H
                                ↑
       H                This H cannot form two bonds.
```

- As a first approximation, **use the common bonding patterns** in Figure 4.1 to arrange the atoms.

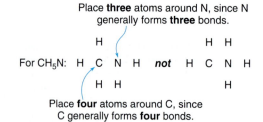

Place **three** atoms around N, since N generally forms **three** bonds.

For CH₅N: H C N H   **not**   H C N H

Place **four** atoms around C, since C generally forms **four** bonds.

In truth, sometimes atom arrangement is not obvious. For this reason, atom arrangement will be specified for you in some problems.

Step [2]   Count the valence electrons.

- Use the group number of a main group element to give the number of valence electrons.
- This sum gives the total number of electrons that must be used in drawing the Lewis structure.

Step [3]   Arrange the electrons around the atoms.

- **Place one bond between every two atoms,** giving two electrons to each H and no more than eight to all other main group atoms.
- Use all remaining electrons to **fill octets with lone pairs,** beginning with atoms on the periphery.
- If all valence electrons are used and an atom does not have an octet, proceed to Step [4].

Step [4]   Use multiple bonds to fill octets when needed.

- **Convert one lone pair to one bonding pair of electrons for each two electrons needed to complete an octet.** This forms double or triple bonds in some molecules, as shown in Section 4.2B. While a single covalent bond contains two electrons, a double bond consists of four electrons and a triple bond consists of six electrons.

Sample Problems 4.2 and 4.3 illustrate how to draw Lewis structures in two molecules that contain only single bonds.

SAMPLE PROBLEM 4.2

Draw a Lewis structure for chloromethane, CH_3Cl, a compound produced by giant kelp and a component of volcanic emissions.

ANALYSIS AND SOLUTION

[1] **Arrange the atoms.**

```
    H
H   C   Cl
    H
```

- Place C in the center and 3 H's and 1 Cl on the periphery.
- In this arrangement, C is surrounded by four atoms, its usual number.

[2] **Count the electrons.**

$$1\,C \times 4\,e^- = 4\,e^-$$
$$3\,H \times 1\,e^- = 3\,e^-$$
$$1\,Cl \times 7\,e^- = 7\,e^-$$
$$\overline{\textbf{14 e}^-\textbf{ total}}$$

[3] **Add the bonds and lone pairs.**

```
    H                                       H  ⌐eight electrons around C              H  ⌐eight electrons
                                            |                                        |     around Cl
H   C   Cl   - - - - - - - - - →        H — C — Cl    - - - - - - - - →          H — C — Cl:
                                            |                                        |    ¨
    H            Add a bond between         H         Add three lone pairs           H
                 the C and each atom.                 to Cl to form an octet.
```

two electrons around H⌐   14 e⁻ used altogether

First add four single bonds, three C—H bonds and one C—Cl bond. This uses eight valence electrons, and gives carbon an octet (four two-electron bonds) and each hydrogen two electrons. Next, give Cl an octet by adding three lone pairs. This uses all 14 valence electrons. To check if a Lewis structure is valid, we must answer YES to three questions.

- Have all the electrons been used?
- Is each H surrounded by two electrons?
- Is every other main group element surrounded by eight electrons?

Since the answer to all three questions is YES, we have drawn a valid Lewis structure for CH_3Cl.

The covalent molecule CH_3Cl is one of many gases released into the air from an erupting volcano.

SAMPLE PROBLEM 4.3

Draw a Lewis structure for methanol, a compound with molecular formula CH_4O. Methanol is a toxic compound that can cause blindness and even death when ingested in small quantities (Section 14.4).

ANALYSIS AND SOLUTION

[1] **Arrange the atoms.**

```
        H
H   C   O   H
        H
```

- four atoms around C
- two atoms around O

[2] **Count the electrons.**

$$1\,C \times 4\,e^- = 4\,e^-$$
$$1\,O \times 6\,e^- = 6\,e^-$$
$$4\,H \times 1\,e^- = 4\,e^-$$
$$\overline{\textbf{14 e}^-\textbf{ total}}$$

[3] **Add the bonds and lone pairs.**

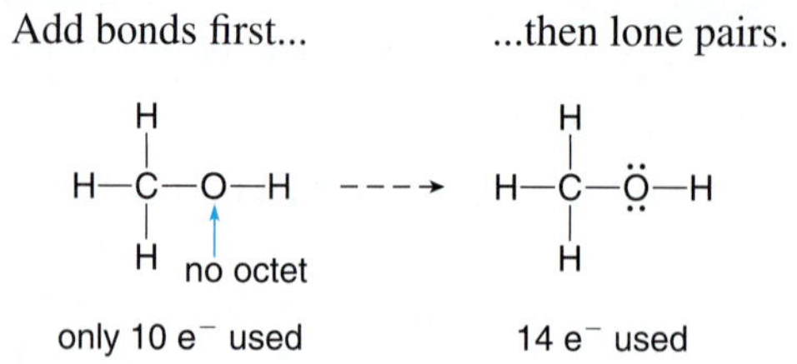

Add bonds first...   ...then lone pairs.

only 10 e⁻ used   14 e⁻ used

In step [3], placing bonds between all atoms uses only 10 electrons, and the O atom, with only four electrons, does not yet have a complete octet. To complete the structure, give the O atom two lone pairs. This uses all 14 electrons, giving every H two electrons and every main group element eight. We have now drawn a valid Lewis structure.

PROBLEM 4.6	Draw a Lewis structure for each covalent molecule. a. HBr b. CH_3F c. H_2O_2 d. N_2H_4 e. C_2H_6 f. CH_2Cl_2
PROBLEM 4.7	Draw a Lewis structure for dimethyl ether (C_2H_6O) with the given arrangement of atoms. H H H C O C H H H

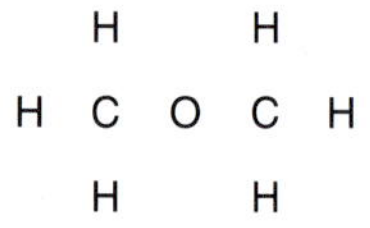

4.2B MULTIPLE BONDS

Sometimes it is not possible to give every main group element (except hydrogen) an octet of electrons by placing only single bonds in a molecule. For example, in drawing a Lewis structure for N_2, each N has five valence electrons, so there are 10 electrons to place. If there is only one N—N bond, adding lone pairs gives one or both N's fewer than eight electrons.

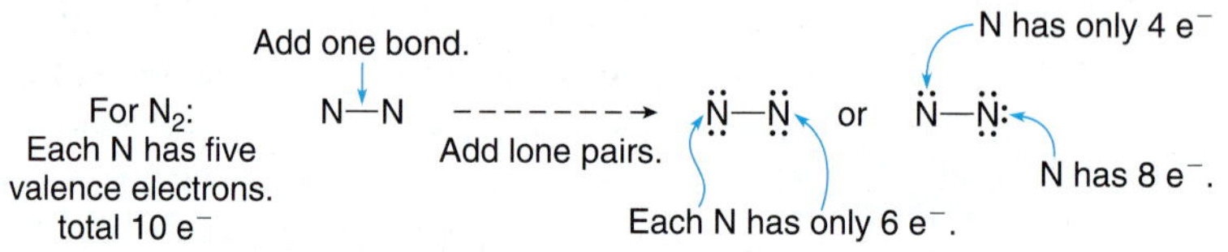

In this case, we must convert a lone pair to a bonding pair of electrons to form a multiple bond. Since we have four fewer electrons than needed, **we must convert *two* lone pairs to *two* bonding pairs of electrons and form a *triple* bond.**

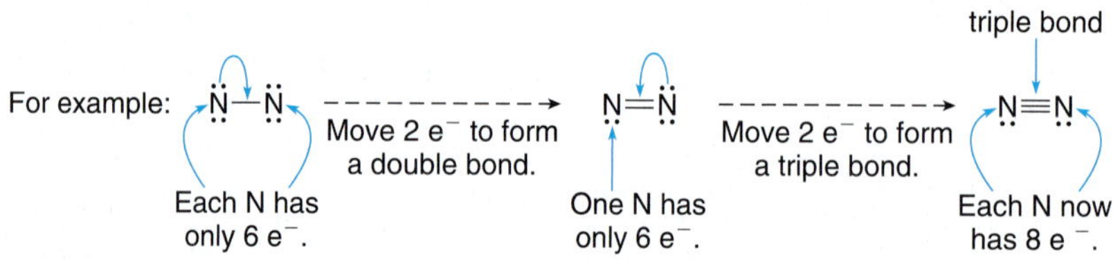

- A triple bond contains six electrons in three two-electron bonds.

Sample Problem 4.4 illustrates another example of a Lewis structure that contains a double bond.

- A double bond contains four electrons in two two-electron bonds.

| SAMPLE PROBLEM 4.4 | Draw a Lewis structure for ethylene, a compound of molecular formula C_2H_4 in which each carbon is bonded to two hydrogens. |

ANALYSIS AND SOLUTION   Follow steps [1]–[3] to draw a Lewis structure.

[1]   **Arrange the atoms.**

$$\text{H C C H}$$
$$\quad\ \text{H H}$$

• Each C gets 2 H's.

[2]   **Count the electrons.**

$$2\,C \times 4\,e^- = 8\,e^-$$
$$4\,H \times 1\,e^- = 4\,e^-$$
$$\overline{\qquad\qquad}$$
$$\textbf{12 } e^- \textbf{ total}$$

[3]   **Add the bonds and lone pairs.**

Add bonds first...   ...then lone pairs.

$$\underset{\text{H}\ \ \text{H}}{\text{H}-\text{C}-\text{C}-\text{H}} \ \ ---\rightarrow \ \ \underset{\text{H}\ \ \text{H}}{\text{H}-\text{C}-\overset{..}{\text{C}}-\text{H}}$$
no octet

After placing five bonds between the atoms and adding the two remaining electrons as a lone pair, one C still has no octet.

[4]   **To give both C's an octet, change *one* lone pair into *one* bonding pair of electrons between the two C atoms, forming a *double* bond.**

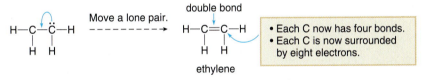

double bond

Move a lone pair.

$$\underset{\text{H}\ \ \text{H}}{\text{H}-\text{C}-\overset{..}{\text{C}}-\text{H}} \ \ ---------\rightarrow \ \ \underset{\text{H}\ \ \text{H}}{\text{H}-\text{C}=\text{C}-\text{H}}$$

ethylene

• Each C now has four bonds.
• Each C is now surrounded by eight electrons.

This uses all 12 electrons, each C has an octet, and each H has two electrons. The Lewis structure is valid. **Ethylene contains a carbon–carbon double bond.**

• After placing all electrons in bonds and lone pairs, use a lone pair to form a multiple bond if an atom does not have an octet.

PROBLEM 4.8	Draw a valid Lewis structure for each compound, using the given arrangement of atoms. a. HCN H C N b. CH_2O H C O c. C_2H_3Cl H C C Cl hydrogen cyanide formaldehyde H vinyl chloride H H
PROBLEM 4.9	The Lewis structure for acetylene (C_2H_2) is drawn as $H-C\equiv C-H$. Explain why it is possible to answer YES to the three questions posed in Sample Problem 4.2 for this Lewis structure.

PROBLEM 4.10	Formic acid (CH_2O_2) is responsible for the sting of some types of ants. Draw a Lewis structure for formic acid with the given arrangement of atoms.
	
	                        O
	
	                H C O H

4.3   EXCEPTIONS TO THE OCTET RULE

Most of the common elements in covalent compounds—carbon, nitrogen, oxygen, and the halogens—generally follow the octet rule. **Hydrogen is a notable exception, because it accommodates only two electrons in bonding.** Additional exceptions include elements such as **boron** in group 3A, and elements in the third row and later in the periodic table, particularly **phosphorus** and **sulfur.**

4.3A   ELEMENTS IN GROUP 3A

Elements in group 3A of the periodic table, such as boron, do not have enough valence electrons to form an octet in a neutral molecule. A Lewis structure for BF_3 illustrates that the boron atom has only six electrons around it. There is nothing we can do about this! There simply aren't enough electrons to form an octet.

:F̈:
|
:F̈—B—F̈:   only six electrons around B

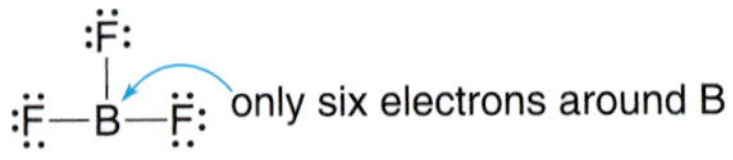

4.3B   ELEMENTS IN THE THIRD ROW

Another exception to the octet rule occurs with some elements located in the third row and later in the periodic table. These elements have empty *d* orbitals available to accept electrons, and thus they may have ***more than eight*** electrons around them. The two most common elements in this category are phosphorus and sulfur, which can have 10 or even 12 electrons around them.

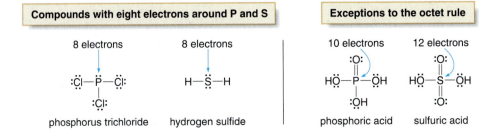

Compounds with eight electrons around P and S		Exceptions to the octet rule	
8 electrons	8 electrons	10 electrons	12 electrons
:C̈l—P̈—C̈l: | :C̈l:	H—S̈—H	:O: || HÖ—P—ÖH | :ÖH	:O: || HÖ—S—ÖH | :O:
phosphorus trichloride	hydrogen sulfide	phosphoric acid	sulfuric acid

While PCl_3 and H_2S contain phosphorus and sulfur atoms that follow the octet rule, H_3PO_4 (phosphoric acid) and H_2SO_4 (sulfuric acid) do not. The phosphorus atom in H_3PO_4 is surrounded by 10 electrons, and the sulfur atom in H_2SO_4 is surrounded by 12 electrons.

PROBLEM 4.11

Draw a Lewis structure for BBr_3, and explain why it does not follow the octet rule.

PROBLEM 4.12

Glyphosate is the most widely used weed killer and the active ingredient in Roundup and other lawn products. (a) Add lone pairs around all N and O atoms to complete octets. (b) How many electrons surround phosphorus in the given structure? (c) Which atoms in glyphosate do not follow the octet rule?

O H   H O
|| | | ||
H—O—P—C—N—C—C—O—H
| | | |
H—O H H H

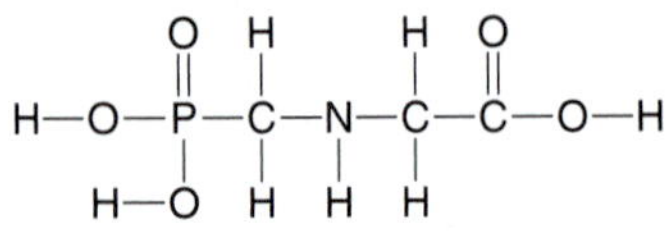

glyphosate

4.4 RESONANCE

We sometimes must draw Lewis structures for ions that contain covalent bonds—that is, polyatomic ions. In this case, the charge on the ion must be taken into account when counting the number of valence electrons that must be placed around the atoms. In counting valence electrons:

- **Add one electron for each negative charge.**
- **Subtract one electron for each positive charge.**

For example, in drawing a Lewis structure for the cyanide anion, ^-CN, there are 10 valence electrons—four from carbon, five from nitrogen, and one additional electron from the negative charge. In order to give each atom an octet, the two atoms must be joined by a triple bond, and both carbon and nitrogen have a lone pair.

Number of valence electrons   Lewis structure

1 e⁻   ⁻CN
4 e⁻   5 e⁻

$$\left[:C\equiv N: \right]^-$$

10 valence electrons   Each atom has an octet.

4.4A DRAWING RESONANCE STRUCTURES

Sometimes two or more valid Lewis structures are possible for a given arrangement of atoms. Sample Problem 4.5 illustrates that two Lewis structures are possible for the bicarbonate anion (HCO_3^-).

SAMPLE PROBLEM 4.5   Draw a Lewis structure for HCO_3^- with the following arrangement of atoms:

                                    O

                          H   O   C   O

ANALYSIS AND SOLUTION   Follow steps [1]–[3] to draw a Lewis structure.

[1]   **Arrange the atoms.**

                                    O

                          H   O   C   O

[2]   **Count the electrons.**

$$
\begin{array}{ccccc}
1\,C & \times & 4\,e^- & = & 4\,e^- \\
3\,O & \times & 6\,e^- & = & 18\,e^- \\
1\,H & \times & 1\,e^- & = & 1\,e^- \\
1\,(-) & \times & 1\,e^- & = & 1\,e^- \\
\end{array}
$$
                                    _____

                                    24 e⁻ total

[3]   **Add the bonds and lone pairs.**

Add bonds first...   ...then lone pairs.

          O   :Ö:
          | |
  H—O—C—O   H—Ö—C—Ö:

                          no octet

After placing four bonds and adding the remaining 16 electrons as lone pairs, the carbon atom does not have an octet.

[4]   **Convert one lone pair on O into one bonding pair to form a double bond. There are two ways to do this.**

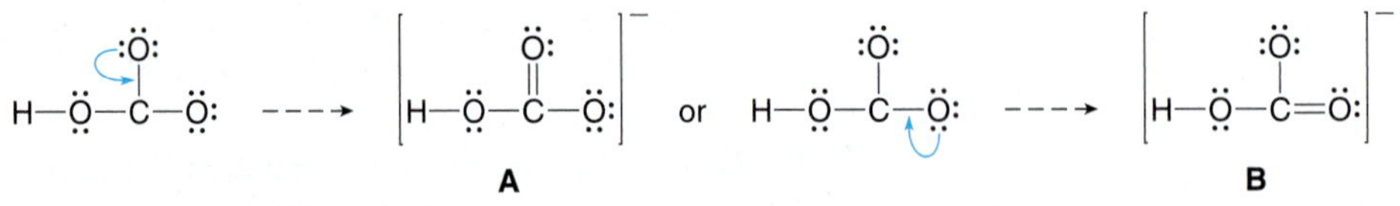

Thus, there are two different Lewis structures, **A** and **B,** for the bicarbonate anion.

The two different Lewis structures (**A** and **B**) for HCO_3^- are called **resonance structures.**

> • Resonance structures are two Lewis structures having the same arrangement of atoms but a different arrangement of electrons.

Two resonance structures differ in the location of multiple bonds and the position of lone pairs. In Lewis structures **A** and **B,** the location of one C=O and one lone pair is different. We often use a **double-headed arrow** (⟷) to show that two Lewis structures are resonance structures.

The position of the double bond is different.

The position of a lone pair is different.

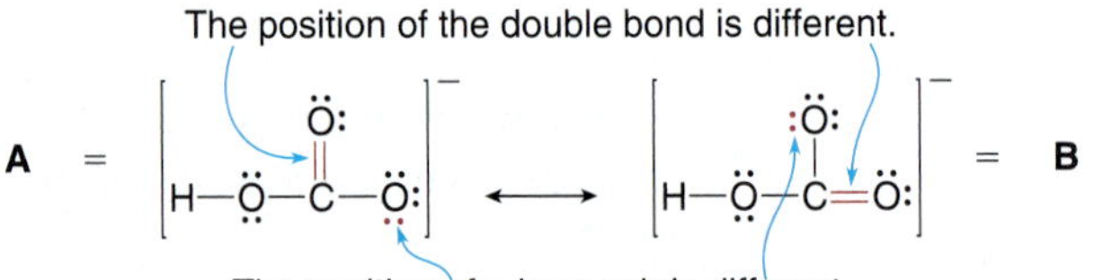

Which structure, **A** or **B,** is an accurate representation for HCO_3^-? The answer is ***neither of them.*** The true structure is a composite of both resonance forms and is called a **hybrid.** Experimentally it is shown that the carbon–oxygen bonds that appear as a double bond in one resonance structure and a single bond in the other, are really somewhere in between a C=O and a C—O. **Resonance stabilizes a molecule** by spreading out lone pairs and electron pairs in multiple bonds over a larger region of space. We say a molecule or ion that has two or more resonance structures is **resonance-stabilized.**

PROBLEM 4.13

Draw a second resonance structure for each ion.

a. 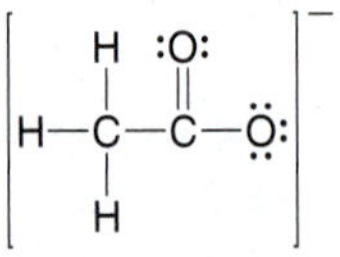 and b. 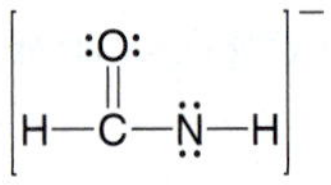 structures shown

PROBLEM 4.14

Draw resonance structures for each polyatomic anion.

a. NO_2^- (two resonance structures, central N atom)

b. HCO_2^- (two resonance structures, central C atom)

4.4B FOCUS ON THE ENVIRONMENT
OZONE

In addition to polyatomic ions, resonance structures can be drawn for neutral molecules as well. For example, the molecule **ozone, O_3,** can be drawn as two resonance structures that differ in the placement of a double bond and a lone pair.

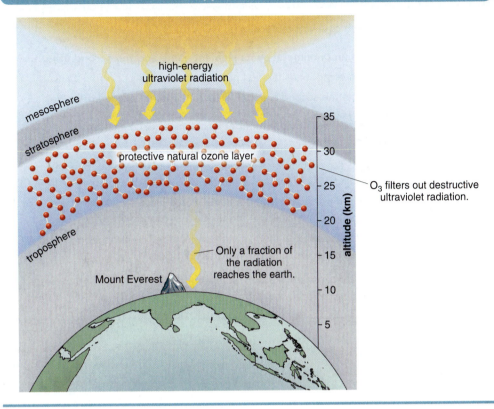

▼ **FIGURE 4.3** Ozone in the Upper Atmosphere

$$O_3: \quad :\ddot{O}=\ddot{O}-\ddot{O}: \quad \longleftrightarrow \quad :\ddot{O}-\ddot{O}=\ddot{O}:$$

Bonds and lone pairs that differ in the two resonance structures are drawn in red.

Ozone is formed in the upper atmosphere (the stratosphere) by the reaction of oxygen molecules (O_2) with oxygen atoms (O). Stratospheric ozone is vital to life: it acts as a shield, protecting the earth's surface from destructive ultraviolet radiation (Figure 4.3). A decrease in ozone concentration in this protective layer would have some immediate consequences, including an increase in the incidence of skin cancer and eye cataracts. We will learn about the interaction of ozone with covalent molecules that contain carbon–chlorine bonds in Chapter 14.

PROBLEM 4.15	When fossil fuels containing sulfur are burned in power plants to generate electricity, large amounts of sulfur dioxide (SO_2) are formed and released into the atmosphere, where some of it eventually forms the acid in acid rain. If the structure of SO_2 consists of a central sulfur atom bonded to both oxygen atoms, draw two resonance structures for sulfur dioxide.

4.5 NAMING COVALENT COMPOUNDS

Although some covalent compounds are always referred to by their common names—H_2O (water) and NH_3 (ammonia)—these names tell us nothing about the atoms that the molecule contains. Other covalent compounds with two elements are named to indicate the identity and number of elements they contain.

HOW TO   Name a Covalent Molecule

EXAMPLE   **Name each covalent molecule: (a) NO_2; (b) N_2O_4.**

Step [1]   **Name the first nonmetal by its element name and the second using the suffix -ide.**

- In both compounds the first nonmetal is nitrogen.
- To name the second element, change the name oxygen to **ox**ide.

Step [2]   **Add prefixes to show the number of atoms of each element.**

- Use a prefix from Table 4.1 for each element.
- Usually, the prefix *mono-* is omitted when only one atom of an element is present. An exception to this rule is the molecule CO, named as carbon monoxide, to distinguish it from CO_2, carbon dioxide.
- When the prefix and element name would place two vowels next to each other, omit the first vowel. For example, mono- + oxide = monoxide (*not* monooxide).

a. NO_2 contains one N atom, so the prefix *mono-* is understood. Since NO_2 contains two O atoms, use the prefix *di-* → dioxide. Thus, NO_2 is **nitrogen dioxide.**

b. N_2O_4 contains two N atoms, so use the prefix *di-* → dinitrogen. Since N_2O_4 contains four O atoms, use the prefix *tetra-* and omit the *a* → tetroxide (*not* tetraoxide). Thus, N_2O_4 is **dinitrogen tetroxide.**

PROBLEM 4.16   Name each compound: (a) CS_2; (b) SO_2; (c) PCl_5; (d) BF_3.

To write a formula from a name, write the element symbols in the order of the elements in the name. Then use the prefixes to determine the subscripts of the formula, as shown in Sample Problem 4.6.

SAMPLE PROBLEM 4.6   Give the formula for each compound: (a) silicon tetrafluoride; (b) diphosphorus pentoxide.

ANALYSIS
- Determine the symbols for the elements in the order given in the name.
- Use the prefixes to write the subscripts.

SOLUTION

a. silicon   tetrafluoride
        ↓   ↓   ↓
        Si   4 F atoms

Answer: SiF_4

b. diphosphorus pentoxide
        ↓   ↓   ↓
        2 P atoms   5 O atoms

Answer: P_2O_5

PROBLEM 4.17   Give the formula for each compound: (a) silicon dioxide; (b) phosphorus trichloride; (c) sulfur trioxide; (d) dinitrogen trioxide.

4.6   MOLECULAR SHAPE

We can now use Lewis structures to determine the shape around a particular atom in a molecule. Consider the H_2O molecule. The Lewis structure tells us only which atoms are connected to each other, but it implies nothing about the geometry. What does the overall molecule look like? Is H_2O a bent or linear molecule?

What is the bond angle?

H—Ö—H

TABLE 4.1 Common Prefixes in Nomenclature

Number of Atoms	Prefix
1	Mono
2	Di
3	Tri
4	Tetra
5	Penta
6	Hexa
7	Hepta
8	Octa
9	Nona
10	Deca

To determine geometry: [1] Draw a valid Lewis structure. [2] Count groups around a given atom.

To determine the shape around a given atom, we must first determine how many groups surround the atom. **A group is either an atom or a lone pair of electrons.** Then we use the **valence shell electron pair repulsion (VSEPR) theory** to determine the shape. VSEPR is based on the fact that electron pairs repel each other; thus:

- The most stable arrangement keeps these groups as far away from each other as possible.

In general, an atom has three possible arrangements of the groups that surround it.

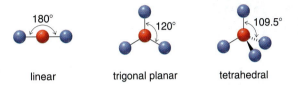

linear   trigonal planar   tetrahedral

- An atom surrounded by two groups is linear and has a bond angle of 180°.
- An atom surrounded by three groups is trigonal planar and has bond angles of 120°.
- An atom surrounded by four groups is tetrahedral and has bond angles of 109.5°.

Let's examine several molecules. In each case we use the number of groups around a given atom in a Lewis structure to predict its geometry.

4.6A TWO GROUPS AROUND AN ATOM

Any atom surrounded by only two groups is linear and has a bond angle of 180°. Two examples illustrating this geometry are CO_2 (carbon dioxide) and **HCN** (hydrogen cyanide). To determine the shape around the central atom in both molecules, we draw the Lewis structure and count the number of groups—atoms and lone pairs—that surround the central atom.

The Lewis structure for CO_2 contains a central carbon atom surrounded by two oxygen atoms. To give every atom an octet and the usual number of bonds requires two carbon–oxygen double bonds. The carbon atom is surrounded by two oxygen atoms and no lone pairs; that is, it is surrounded by two groups, making the molecule linear and the O—C—O bond angle 180°.

four bonds

$:\ddot{O}=C=\ddot{O}:$ $:\ddot{O}\overset{180°}{=}C=\ddot{O}:$ =   linear molecule

two bonds   two bonds   two atoms around C
                              two groups

Carbon dioxide illustrates another important feature of VSEPR theory: *ignore multiple bonds in predicting geometry.* **Count only atoms and lone pairs.**

Similarly, the Lewis structure for HCN contains a central carbon atom surrounded by one hydrogen and one nitrogen. To give carbon and nitrogen an octet and the usual number of bonds requires a carbon–nitrogen triple bond. The carbon atom is surrounded by two atoms and no lone pairs; that is, it is surrounded by two groups, making the molecule linear and the H—C—N bond angle 180°.

four bonds

$H-C\equiv N:$ $H\overset{180°}{-}C\equiv N:$ =   linear molecule

three bonds   two atoms around C
                    two groups

HCN, an extremely toxic gas, is produced by some naturally occurring molecules. For example, cassava, a woody shrub grown as a root crop in South America and Africa, contains the compound linamarin. Linamarin is not toxic itself, but it forms HCN in the presence of water and

Trigonal = three-sided.

some enzymes. Cassava is safe to eat when the root has been peeled and boiled, so that the linamarin is removed during processing. If the root is eaten without processing, illness and even death can result from high levels of HCN formed from linamarin.

4.6B THREE GROUPS AROUND AN ATOM

Any atom surrounded by three groups is trigonal planar and has bond angles of 120°. Two examples illustrating this geometry are **BF₃** (boron trifluoride) and **H₂C═O** (formaldehyde).

three atoms around B
three groups   trigonal planar molecule   three atoms around C
three groups   trigonal planar molecule

In BF_3, the boron atom is surrounded by three fluorines and no lone pairs; that is, the boron is surrounded by three groups. A similar situation occurs with the carbon atom in $H_2C{=}O$. The carbon atom is surrounded by three atoms (two H's and one O) and no lone pairs—that is, three groups. To keep the three groups as far from each other as possible, they are arranged in a trigonal planar fashion, with bond angles of 120°.

4.6C FOUR GROUPS AROUND AN ATOM

Any atom surrounded by four groups is tetrahedral and has bond angles of (approximately) 109.5°. For example, the simple organic compound methane, **CH₄,** has a central carbon atom with four bonds to hydrogen, each pointing to the corners of a tetrahedron.

tetrahedral carbon

How can we represent the three-dimensional geometry of a tetrahedron on a two-dimensional piece of paper? **Place two of the bonds in the plane of the paper, one bond in front, and one bond behind,** using the following conventions:

- A solid line is used for bonds in the plane.
- A wedge is used for a bond in front of the plane.
- A dashed line is used for a bond behind the plane.

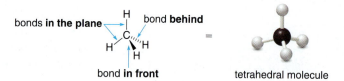

bonds **in the plane**   bond **behind**   bond **in front**   tetrahedral molecule

Up to now, each of the groups around the central atom has been another atom. **A group can also be a lone pair of electrons.** NH_3 and H_2O represent two examples of molecules with atoms surrounded by four groups, some of which are lone pairs.

The Lewis structure for ammonia, **NH₃,** has an N atom surrounded by three hydrogen atoms and one lone pair of electrons—four groups. To keep four groups as far apart as possible, the three H atoms and the one lone pair around N point to the corners of a tetrahedron. The H—N—H bond angle of 107° is close to the theoretical tetrahedral bond angle of 109.5°. This shape is referred to as a **trigonal pyramid,** since one of the groups around the N is a nonbonded electron pair, not another atom.

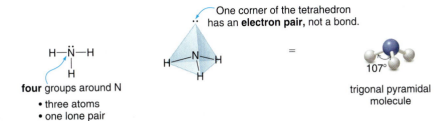

The Lewis structure for water, H_2O, has an O atom surrounded by two hydrogen atoms and two lone pairs of electrons—four groups. In H_2O, the two H atoms and the two lone pairs around O point to the corners of a tetrahedron. The H—O—H bond angle of 105° is close to the theoretical tetrahedral bond angle of 109.5°. Water has a **bent** shape, because two of the groups around oxygen are lone pairs of electrons.

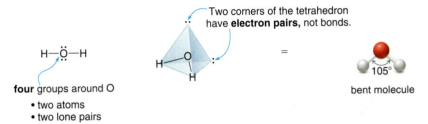

Common molecular shapes are summarized in Table 4.2. The three-dimensional shape determines the polarity of a molecule, which is discussed in Section 4.8. All of Chapter 15 is devoted to the importance of the three-dimensional shapes of molecules.

TABLE 4.2 Common Molecular Shapes Around Atoms

Total Number of Groups	Number of Atoms	Number of Lone Pairs	Shape Around an Atom (A)	Approximate Bond Angle (°)	Example
2	2	0	A linear	180	CO_2, HC≡CH
3	3	0	A trigonal planar	120	BF_3, H_2C=O
4	4	0	A tetrahedral	109.5	CH_4
4	3	1	A trigonal pyramidal	~109.5[a]	NH_3
4	2	2	A bent	~109.5[a]	H_2O

[a]The symbol "~" means approximately.

SAMPLE PROBLEM 4.7

Using the given Lewis structure, determine the shape around the second-row elements in each compound.

a. $H-C\equiv C-H$

acetylene

b. $\left[H-\overset{\displaystyle H}{\underset{\displaystyle H}{N}}-H \right]^{+}$

ammonium ion

ANALYSIS

To predict the shape around an atom, we need a valid Lewis structure, which is given in this problem. Then count groups around the atom to determine molecular shape using the information in Table 4.2.

SOLUTION

a. Each C in $H-C\equiv C-H$ is surrounded by two atoms (one C and one H) and no lone pairs—that is, two groups. An atom surrounded by two groups is linear with a 180° bond angle.

180°

$H-C\equiv C-H$

180°

b. The N atom in NH_4^+ is surrounded by four H atoms—that is, four groups. An atom surrounded by four groups is tetrahedral, with 109.5° bond angles.

109.5°

tetrahedral

PROBLEM 4.18

What is the shape around each carbon atom in ethylene, $H_2C=CH_2$?

PROBLEM 4.19

What is the shape around the indicated atom in each molecule? Don't forget to draw in all needed lone pairs before determining molecular shape.

a. H_2S b. CH_2Cl_2 c. NCl_3 d. BBr_3

4.7 ELECTRONEGATIVITY AND BOND POLARITY

When two atoms share electrons in a covalent bond, are the electrons in the bond attracted to both nuclei to the same extent? That depends on the **electronegativity** of the atoms in the bond.

- *Electronegativity* is a measure of an atom's attraction for electrons in a bond. Electronegativity tells us how much a particular atom "*wants*" electrons.

The electronegativity of an atom is assigned a value from 0 to 4; the *higher* the value, the *more electronegative* an atom is, and the *more it is attracted* to the electrons in a bond. The electronegativity values for main group elements are shown in Figure 4.4. The noble gases are not assigned values, since they do not typically form bonds.

Electronegativity values exhibit periodic trends.

- Electronegativity *increases* across a row of the periodic table as the nuclear charge increases (excluding the noble gases).
- Electronegativity *decreases* down a column of the periodic table as the atomic radius increases, pushing the valence electrons farther from the nucleus.

Thus, nonmetals have high electronegativity values compared to metals, because nonmetals have a strong tendency to hold on to and attract electrons. As a result, the most electronegative elements—fluorine and oxygen—are located at the **upper right-hand corner** of the periodic table, and the least electronegative elements are located in the lower left-hand corner.

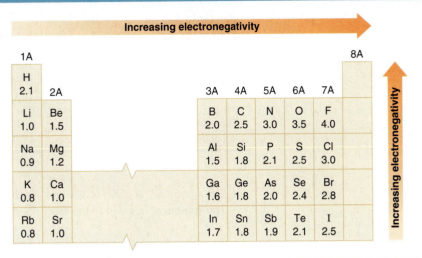

FIGURE 4.4   Electronegativity Values for Main Group Elements

---

PROBLEM 4.20

Using the trends in the periodic table, rank the following atoms in order of increasing electronegativity.

a. Li, Na, H   b. O, C, Be   c. Cl, I, F   d. B, O, N

Electronegativity values are used as a guideline to indicate whether the electrons in a bond are *equally* shared or *unequally* shared between two atoms. For example, whenever two *identical* atoms are bonded together, each atom attracts the electrons in the bond to the same extent. The electrons are equally shared, and the bond is said to be **nonpolar.** Thus, a **carbon–carbon bond is nonpolar,** as is the fluorine–fluorine bond in F_2. The same is true whenever two different atoms having *similar* electronegativities are bonded together. **C—H bonds are considered to be nonpolar,** because the electronegativity difference between C (2.5) and H (2.1) is small.

The small electronegativity difference
between C and H is ignored.

In contrast, bonding between atoms of *different* electronegativity results in the *unequal* sharing of electrons. For example, in a C—O bond, the electrons are pulled away from C (2.5) towards the element of higher electronegativity, O (3.5). **The bond is *polar,* or *polar covalent.*** The bond is said to have a **dipole**—that is, **a separation of charge.**

A C–O bond is a *polar* bond.

The direction of polarity in a bond is often indicated by an arrow, with the head of the arrow pointing towards the more electronegative element. The tail of the arrow, with a perpendicular line drawn through it, is drawn at the less electronegative element. Alternatively, the lower case Greek letter delta (δ) with a positive or negative charge is used, resulting in the symbols δ^+ and δ^- to indicate this unequal sharing of electron density.

TABLE 4.3   Electronegativity Difference and Bond Type

Electronegativity Difference	Bond Type	Electron Sharing
Less than 0.5 units	Nonpolar	Electrons are equally shared.
0.5–1.9 units	Polar covalent	Electrons are unequally shared; they are pulled towards the more electronegative element.
Greater than 1.9 units	Ionic	Electrons are transferred from the less electronegative element to the more electronegative element.

- The symbol δ^+ is given to the less electronegative atom.
- The symbol δ^- is given to the more electronegative atom.

Students often wonder how large an electronegativity difference must be to consider a bond polar. That's hard to say. We will set an arbitrary value for this difference and use it as an approximation. **Usually, a polar bond will be one in which the electronegativity difference between two atoms is 0.5 units or greater.**

As the electronegativity difference between the two atoms in a bond increases, the shared electrons are pulled more and more towards the more electronegative element. When the electronegativity difference is larger than 1.9 units, the electrons are essentially transferred from the less electronegative element to the more electronegative element and the bond is considered ionic. Table 4.3 summarizes the relationship between the electronegativity difference of the atoms in a bond and the type of bond formed.

SAMPLE PROBLEM 4.8

Use electronegativity values to classify each bond as nonpolar, polar covalent, or ionic: (a) Cl_2; (b) HCl; (c) NaCl.

ANALYSIS   Calculate the electronegativity difference between the two atoms and use the following rules: less than 0.5 (nonpolar); 0.5–1.9 (polar covalent); and greater than 1.9 (ionic).

SOLUTION

	Electronegativity Difference	**Bond Type**
a. Cl_2	3.0 (Cl) – 3.0 (Cl) = 0	Nonpolar
b. HCl	3.0 (Cl) – 2.1 (H) = 0.9	Polar covalent
c. NaCl	3.0 (Cl) – 0.9 (Na) = 2.1	Ionic

PROBLEM 4.21

Use electronegativity values to classify the bond(s) in each compound as nonpolar, polar covalent, or ionic.

a. HF   b. MgO   c. F_2 d. ClF   e. H_2O f. NH_3

PROBLEM 4.22

Show the direction of the dipole in each bond. Label the atoms with δ^+ and δ^-.

a. H—F   b. —B—C—   c. —C—Li   d. —C—Cl

4.8   POLARITY OF MOLECULES

Thus far, we have been concerned with the polarity of a single bond. Is an entire covalent molecule polar or nonpolar? That depends on two factors: the polarity of the individual bonds and the overall shape. When a molecule contains zero or one polar bond, the following can be said:

- A molecule with no polar bonds is a nonpolar molecule.
- A molecule with one polar bond is a polar molecule.

Thus, CH_4 is a nonpolar molecule because all of the C—H bonds are nonpolar. In contrast, CH_3Cl contains only one polar bond, so it is a polar molecule. The dipole is in the same direction as the dipole of the only polar bond.

CH_4
no polar bonds
nonpolar molecule

polar bond   net dipole of the molecule

CH_3Cl
one polar bond
polar molecule

With covalent compounds that have more than one polar bond, the shape of the molecule determines the overall polarity.

- If the individual bond dipoles do not cancel, the molecule is polar.
- If the individual bond dipoles cancel, the molecule is nonpolar.

To determine the polarity of a molecule that has two or more polar bonds:

1. Identify all polar bonds based on electronegativity differences.
2. Determine the shape around individual atoms by counting groups.
3. Decide if individual dipoles cancel or reinforce.

Figure 4.5 illustrates several examples of polar and nonpolar molecules that contain polar bonds. The net dipole is the sum of all the bond dipoles in a molecule.

▼ FIGURE 4.5   Examples of Polar and Nonpolar Molecules

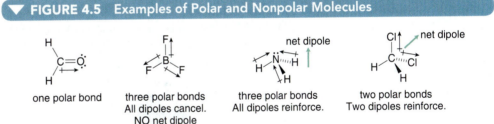

one polar bond	three polar bonds All dipoles cancel. NO net dipole	net dipole three polar bonds All dipoles reinforce.	net dipole two polar bonds Two dipoles reinforce.
polar molecule	**nonpolar** molecule	**polar** molecule	**polar** molecule

SAMPLE PROBLEM 4.9

Determine whether each molecule is polar or nonpolar: (a) H_2O; (b) CO_2.

ANALYSIS

To determine the overall polarity of a molecule: identify the polar bonds; determine the shape around individual atoms; decide if the individual bond dipoles cancel or reinforce.

SOLUTION

a. **H_2O:** Each O—H bond is polar because the electronegativity difference between O (3.5) and H (2.1) is 1.4. Since the O atom of H_2O has two atoms and two lone pairs around it, H_2O is a bent molecule around the O atom. The two dipoles reinforce (both point *up*), so **H_2O has a net dipole; that is, H_2O is a polar molecule.**

The two individual dipoles reinforce.

net dipole

Do NOT draw H_2O as:

H—Ö—H

The net dipole bisects the H—O—H bond angle.
The bent representation shows that the dipoles reinforce.

Note: **We must know the geometry to determine if two dipoles cancel or reinforce.** For example, do **not** draw H_2O as a linear molecule, because you might think that the two dipoles cancel, when in reality, they reinforce.

b. **CO_2:** Each C—O bond is polar because the electronegativity difference between O (3.5) and C (2.5) is 1.0. The Lewis structure of CO_2 (Section 4.6A) shows that the C atom is surrounded by two groups (two O atoms), making it linear. In this case, the two dipoles are equal and opposite in direction so they cancel. Thus, CO_2 is a **nonpolar molecule** with **no net dipole.**

> **The two dipoles cancel.**
>
> $$\overset{\longleftarrow +\quad + \longrightarrow}{:\ddot{O}=C=\ddot{O}:}$$
> $$\quad \delta^- \quad \delta^+ \quad \delta^-$$
>
> NO net dipole

PROBLEM 4.23

Label the polar bonds in each molecule, and then decide if the molecule is polar or nonpolar.

a. HCl   b. C_2H_6 c. CH_2F_2 d. HCN   e. CCl_4

4.9 FOCUS ON HEALTH & MEDICINE
COVALENT DRUGS AND MEDICAL PRODUCTS

Most drugs and products used in medicine are made up of covalent molecules. Some are simple molecules containing only a few atoms, while others are very complex. The principles learned in this chapter apply to all molecules regardless of size. Two examples are shown.

Hydrogen peroxide, H_2O_2, is a simple covalent molecule used to disinfect wounds. We now know a great deal about the structure of H_2O_2. The Lewis structure for H_2O_2 contains an O—O bond and each O atom has two lone pairs to give it an octet of electrons. Since each O atom is surrounded by two atoms and two lone pairs, it has a bent structure. While the O—O bond is nonpolar, both O—H bonds are polar, since the electronegativity difference between oxygen (3.5) and hydrogen (2.1) is large (1.4).

$$H-\ddot{O}-\ddot{O}-H \quad = \quad$$

hydrogen peroxide

δ^- δ^-
109.5°
δ^+ δ^+

Acetaminophen, a pain-reliever and antipyretic (an agent that reduces fever), is more complex, but we still know much about its structure. Each O atom has two lone pairs and the N atom has one lone pair. Acetaminophen has six polar bonds, labeled in red. The number of groups around each atom determines its shape. One tetrahedral C surrounded by four groups and two trigonal planar C's are labeled.

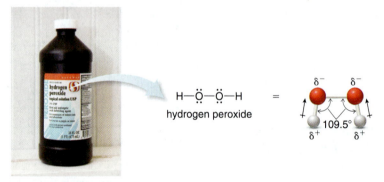

$C_8H_9NO_2$
acetaminophen

120°   trigonal planar C's
109.5°
tetrahedral C

SAMPLE PROBLEM 4.10

Glycolic acid is a starting material used to manufacture dissolving sutures. (a) Place lone pairs where needed in the Lewis structure. (b) Give the shape around each atom in red. (c) Label all polar bonds.

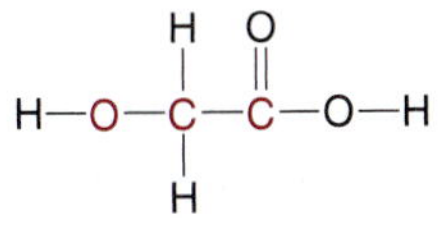

glycolic acid

ANALYSIS AND SOLUTION

a. Each O atom needs two lone pairs to make an octet.

b. Count groups to determine shape.

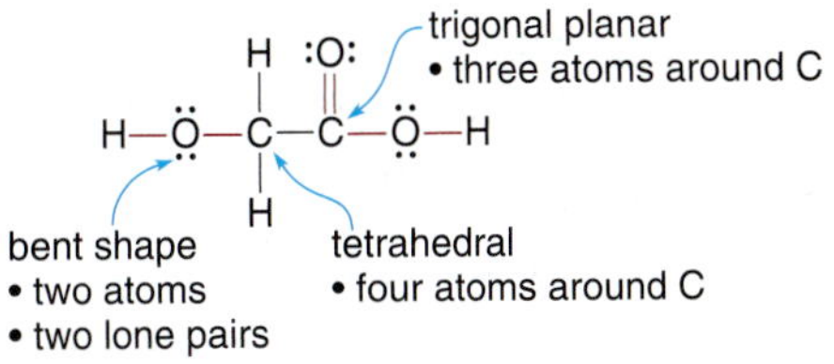

- trigonal planar
  • three atoms around C
- bent shape
  • two atoms
  • two lone pairs
- tetrahedral
  • four atoms around C

c. Since O is much more electronegative (3.5) than hydrogen (2.1) and carbon (2.5), all C—O and O—H bonds are polar and are labeled in red.

PROBLEM 4.24

In each compound: [1] determine the geometry around each carbon; [2] label each bond as polar or nonpolar.

a. 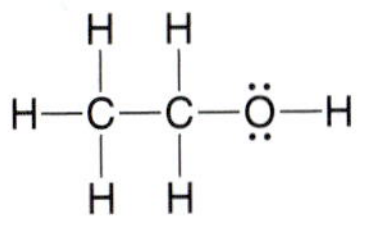 ethanol

b. 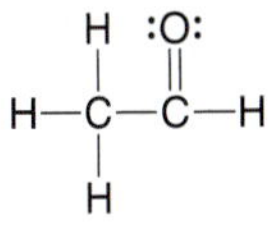 acetaldehyde

Ethanol, the "alcohol" in alcoholic beverages, is the world's most widely abused drug. It is metabolized to acetaldehyde, a toxic compound that produces some of the ill effects of ingesting too much ethanol. Ethanol is discussed in greater detail in Chapter 14.

CHAPTER HIGHLIGHTS

KEY TERMS

Covalent bond (4.1)
Diatomic molecule (4.1)
Dipole (4.7)
Double bond (4.2)
Double-headed arrow (4.4)
Electronegativity (4.7)
Hybrid (4.4)

Lewis structure (4.1)
Lone pair (4.1)
Molecular formula (4.2)
Molecule (4.1)
Multiple bond (4.2)
Nonbonded electron pair (4.1)

Nonpolar bond (4.7)
Polar bond (4.7)
Resonance structure (4.4)
Triple bond (4.2)
Valence shell electron pair repulsion
   theory (4.6)

KEY CONCEPTS

❶ **What are the characteristic bonding features of covalent compounds? (4.1)**

- Covalent bonds result from the sharing of electrons between two atoms, forming molecules. Atoms share electrons to attain the electronic configuration of the noble gas nearest

them in the periodic table. For many main group elements, this results in an octet of electrons.

- Covalent bonds are formed when two nonmetals combine or when a metalloid bonds to a nonmetal. Covalent bonds are preferred with elements in the middle of the periodic table that would have to gain or lose too many electrons to form an ion.

- Except for hydrogen, the common elements—C, N, O, and the halogens—follow one rule: the number of bonds + the number of lone pairs = four.

❷ What are Lewis structures and how are they drawn? (4.2)

- Lewis structures are electron-dot representations of molecules. Two-electron bonds are drawn with a solid line and nonbonded electrons are drawn with dots (:).
- Lewis structures contain only valence electrons. Each H gets two electrons and main group elements generally get eight.
- After placing all electrons in bonds and lone pairs in a Lewis structure, it may be necessary to use lone pairs to form double or triple bonds if an atom does not have an octet.

❸ What are resonance structures? (4.4)

- Resonance structures are two Lewis structures having the same arrangement of atoms but a different arrangement of electrons.
- The hybrid is a composite of all resonance structures that spreads out electron pairs in multiple bonds and lone pairs.

❹ How are covalent compounds with two elements named? (4.5)

- Name the first nonmetal by its element name and the second using the suffix -ide. Add prefixes to indicate the number of atoms of each element.

❺ How is the molecular shape around an atom determined? (4.6)

- To determine the shape around an atom, count groups—atoms and lone pairs—and keep the groups as far away from each other as possible.
- Two groups = linear, 180° bond angle; three groups = trigonal planar, 120° bond angle; four groups = tetrahedral, 109.5° bond angle.

❻ How does electronegativity determine bond polarity? (4.7)

- Electronegativity is a measure of an atom's attraction for electrons in a bond.
- When two atoms have the same electronegativity value, or the difference is less than 0.5 units, the electrons are equally shared and the bond is nonpolar.
- When two atoms have very different electronegativity values—a difference of 0.5–1.9 units—the electrons are unequally shared and the bond is polar.

❼ When is a molecule polar or nonpolar? (4.8)

- A polar molecule has either one polar bond, or two or more bond dipoles that do not cancel.
- A nonpolar molecule has either all nonpolar bonds, or two or more bond dipoles that cancel.

PROBLEMS

Selected in-chapter and end-of-chapter problems have brief answers provided in Appendix B.

Covalent Bonding

4.25 For each pair of compounds, classify the bonding as ionic or covalent and explain your choice.
   a. LiCl and HCl   b. KBr and HBr

4.26 For each pair of compounds, classify the bonding as ionic or covalent and explain your choice.
   a. BeH_2 and $BeCl_2$ b. Na_3N and NH_3

4.27 How many bonds and lone pairs are typically observed with each element: (a) C; (b) Se; (c) I; (d) P?

4.28 How many bonds and lone pairs are typically observed with each element: (a) O; (b) Si; (c) Ge; (d) B?

4.29 Fill in the lone pairs needed to give the main group elements (except hydrogen) an octet. Acrylonitrile is a starting material used to manufacture synthetic Orlon and Acrilan fibers. Cysteine is an amino acid used to synthesize proteins.

a. acrylonitrile   b. cysteine

4.30 Fill in the lone pairs needed to give the main group elements (except hydrogen) an octet. Glycerol is a product of the metabolism of fats. Acrylamide is used to make polyacrylamide, which is used in some cosmetics and food packaging.

a. glycerol   b. acrylamide

Lewis Structures

4.31 Draw a valid Lewis structure for each molecule.
   a. HI   b. CH_2F_2 c. H_2Se d. CO   e. C_2Cl_6

4.32 Draw a valid Lewis structure for each molecule.
   a. CH_3Br b. PH_3 c. HBr   d. SiF_4 e. C_2HCl

4.33 Draw a valid Lewis structure for each compound using the given arrangement of atoms.

a. CH_5N

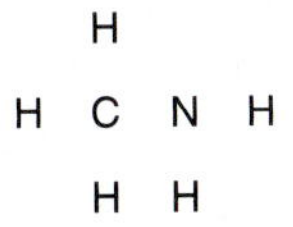

```
      H
H  C  N  H
   H  H
```

b. HNO_2

```
H  O  N  O
```

c. C_3H_4

```
      H
H  C  C  C  H
      H
```

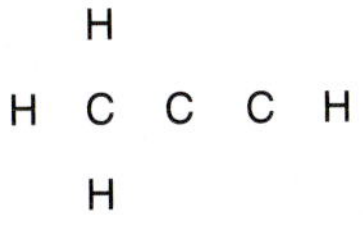

4.34 Draw a valid Lewis structure for each compound using the given arrangement of atoms.

a. C_2H_3N

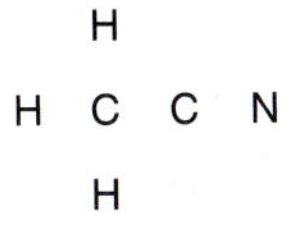

```
      H
H  C  C  N
      H
```

b. HNO_3

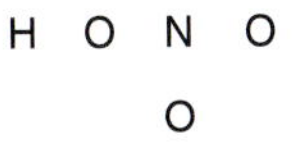

```
H  O  N  O
         O
```

c. C_3H_6

```
      H
H  C  C  C  H
   H  H  H
```

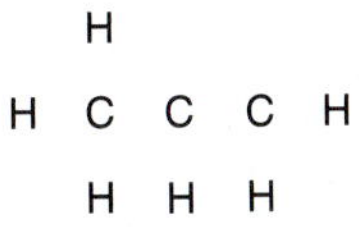

4.35 Draw a valid Lewis structure for tetrafluoroethylene, C_2F_4, the industrial starting material used to prepare Teflon. Teflon is most widely used as a nonstick surface on pots and pans, but it has also found application in tape used by plumbers to seal joints, nail polish, and coatings on eye glasses. Assume that each carbon is bonded to two fluorine atoms.

4.36 Draw a valid Lewis structure for phosgene, CCl_2O, which contains a central carbon atom. Phosgene is an extremely toxic gas used as a chemical weapon during World War I. It is now an important industrial starting material for the synthesis of Lexan, a lightweight transparent material used in bike helmets, goggles, and catcher's masks.

4.37 Draw a valid Lewis structure for each ion: (a) NH_2^-; (b) H_3O^+.

4.38 Draw a valid Lewis structure for each ion: (a) OCl^-; (b) CH_3O^-.

4.39 Keeping in mind that some elements violate the octet rule, draw a Lewis structure for each compound: (a) BCl_3; (b) SO_3.

4.40 Keeping in mind that some elements violate the octet rule, draw a Lewis structure for each compound: (a) BeH_2; (b) PCl_5.

Resonance Structures

4.41 What is the difference between a resonance structure and a resonance hybrid?

4.42 Briefly explain why having two resonance structures for a molecule stabilizes it.

4.43 Draw a second resonance structure for the following anion:

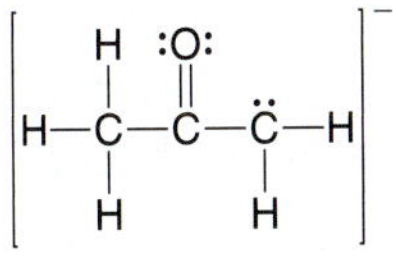

4.44 Draw a second resonance structure for nitromethane, a compound used in drag racing fuels and in the manufacture of pharmaceuticals, pesticides, and fibers.

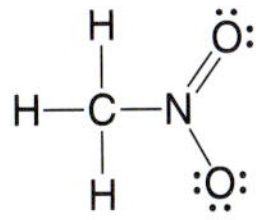

nitromethane

4.45 Label each pair of compounds as resonance structures or not resonance structures.

a. 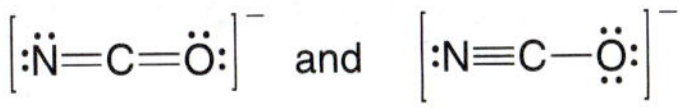 $[:\ddot{N}=C=\ddot{O}:]^-$ and $[:N\equiv C-\ddot{O}:]^-$

b.

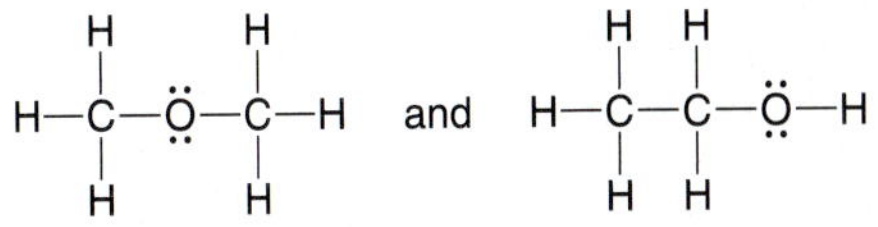

4.46 Label each pair of compounds as resonance structures or not resonance structures.

a. 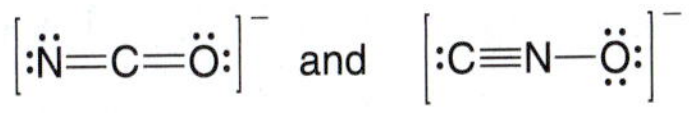 $[:\ddot{N}=C=\ddot{O}:]^-$ and $[:C\equiv N-\ddot{O}:]^-$

b.

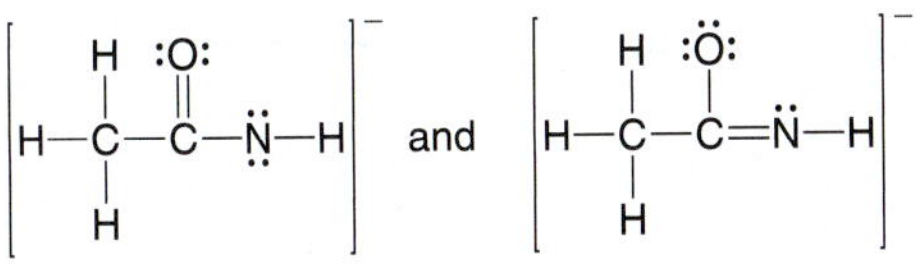

4.47 Draw three resonance structures for the carbonate anion (CO_3^{2-}) that contain a central carbon atom.

4.48 Draw three resonance structures for the nitrate anion (NO_3^-) that contain a central N atom.

Naming Covalent Compounds

4.49 Name each covalent compound.
a. PBr_3 b. SO_3 c. NCl_3 d. P_2S_5

4.50 Name each covalent compound.
a. SF_6 b. CBr_4 c. N_2O d. P_4O_{10}

4.51 Write a formula that corresponds to each name.
     a. selenium dioxide
     b. carbon tetrachloride
     c. dinitrogen pentoxide

4.52 Write a formula that corresponds to each name.
     a. silicon tetrafluoride
     b. nitrogen oxide
     c. phosphorus triiodide

4.53 What is the common name for dihydrogen oxide?

4.54 What is the systematic name for H_2S, the compound we commonly call hydrogen sulfide?

Molecular Shape

4.55 Add lone pairs where needed to give octets and then determine the shape around each indicated atom.

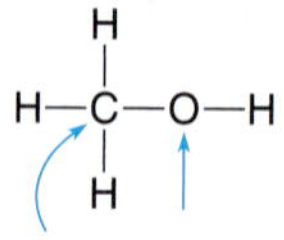

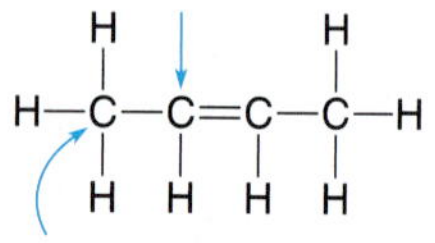

     b.  NF_3

4.56 Add lone pairs where needed to give octets and then determine the shape around each indicated atom.

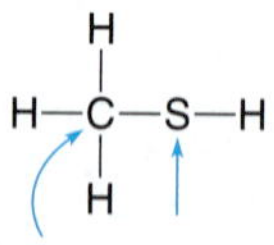

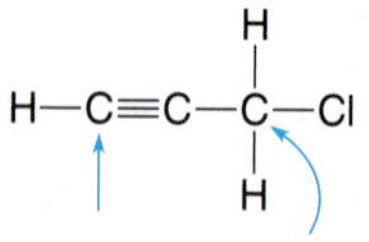

     b. 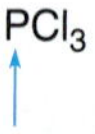 PCl_3

4.57 Add lone pairs where needed to give octets and then determine the shape around each indicated atom.

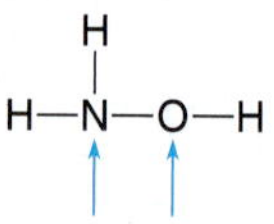

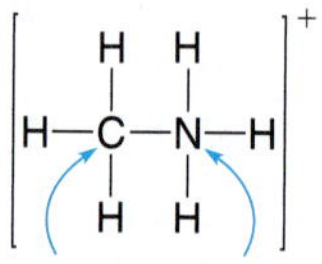

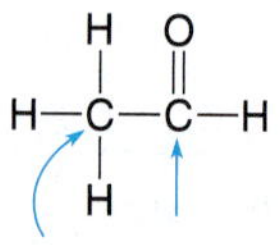

4.58 Add lone pairs where needed to give octets and then determine the shape around each indicated atom.

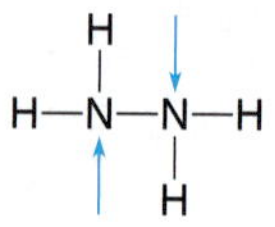

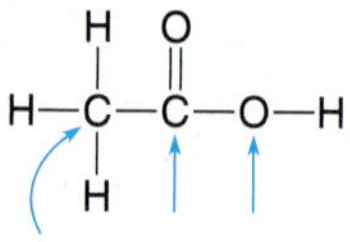

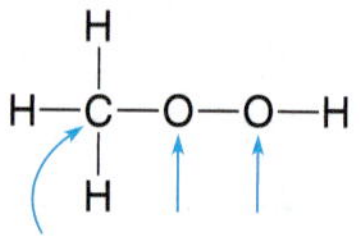

4.59 Give the molecular shape around the boron atom in BCl_3 and the nitrogen atom in NCl_3 and explain why they are different.

4.60 Give the molecular shape for the oxygen atom in H_2O and H_3O^+ and explain why they are different.

4.61 Predict the bond angles around the indicated atoms in each compound. Don't forget to draw in lone pairs where needed to give octets.

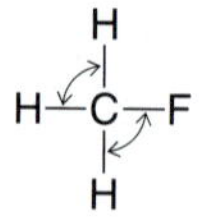

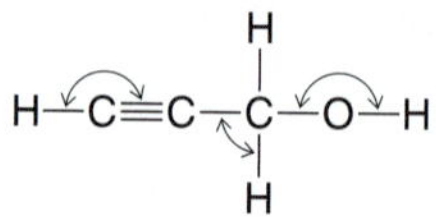

4.62 Predict the bond angles around the indicated atoms in each compound. Don't forget to draw in lone pairs where needed to give octets.

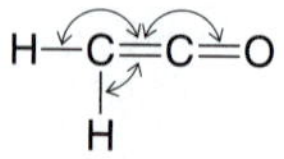

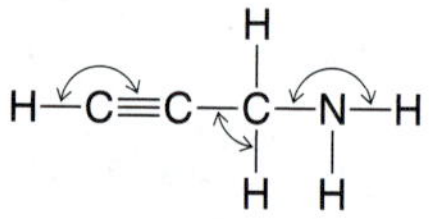

4.63 Draw Lewis structures for CCl_4 and C_2Cl_4. Give the molecular shape around each carbon atom. Explain why the carbon atoms in the two molecules have different shapes.

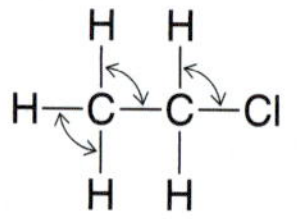

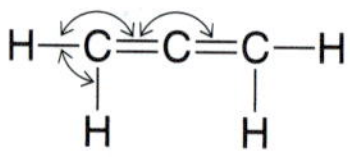

4.64 Draw a Lewis structure for N_2H_4 and explain why the shape around each N atom should be described as trigonal pyramidal.

Electronegativity and Polarity

4.65 Rank the atoms in each group in order of increasing electronegativity.
     a. Se, O, S   c. Cl, S, F
     b. P, Na, Cl   d. O, P, N

4.66 Rank the atoms in each group in order of increasing electronegativity.
     a. Si, P, S   c. Se, Cl, Br
     b. Be, Mg, Ca   d. Li, Be, Na

4.67 What is the difference between a polar bond and a nonpolar bond? Give an example of each.

4.68 What is the difference between a polar covalent bond and an ionic bond? Give an example of each.

4.69 Using electronegativity values, classify the bond formed between each pair of elements as polar covalent or ionic.
     a. hydrogen and bromine   c. sodium and sulfur
     b. nitrogen and carbon   d. lithium and oxygen

4.70 Using electronegativity values, classify the bond formed between each pair of elements as polar covalent or ionic.
a. nitrogen and oxygen   c. sulfur and chlorine
b. oxygen and hydrogen   d. sodium and chlorine

4.71 Label the bond formed between carbon and each of the following elements as nonpolar, polar, or ionic.
a. carbon   c. lithium   e. hydrogen
b. oxygen   d. chlorine

4.72 Label the bond formed between fluorine and each of the following elements as nonpolar, polar, or ionic.
a. hydrogen   c. carbon   e. sulfur
b. fluorine   d. lithium

4.73 Which bond in each pair is more polar—that is, has the larger electronegativity difference between atoms?
a. C—O or C—N
b. C—F or C—Cl
c. Si—C or P—H

4.74 Which bond in each pair is more polar—that is, has the larger electronegativity difference between atoms?
a. Si—O or Si—S
b. H—F or H—Br
c. C—B or C—Li

4.75 Label each bond in Problem 4.73 with δ^+ and δ^- to show the direction of polarity.

4.76 Label each bond in Problem 4.74 with δ^+ and δ^- to show the direction of polarity.

4.77 Explain why the carbon atom in CH_3NH_2 bears a partial positive charge (δ^+), but the carbon atom in CH_3MgBr bears a partial negative charge (δ^-).

4.78 Explain why the carbon atom in CH_3Cl bears a partial positive charge (δ^+), but the carbon atom in CH_3Li bears a partial negative charge (δ^-).

4.79 Can a compound be polar if it contains all nonpolar bonds? Explain.

4.80 Can a compound be nonpolar if it contains some polar bonds? Explain.

4.81 Can a compound be nonpolar if it contains one polar bond? Explain.

4.82 Is a compound that contains polar bonds always polar? Explain.

4.83 Label the polar bonds and then decide if each molecule is polar or nonpolar.

4.84 Label the polar bonds and then decide if each molecule is polar or nonpolar.

4.85 Explain why $CHCl_3$ is a polar molecule but CCl_4 is not.

4.86 Explain why H_2O is a polar molecule but H_2S is not.

General Questions

4.87 Answer the following questions about the molecule Cl_2O.
a. How many valence electrons does Cl_2O contain?
b. Draw a valid Lewis structure.
c. Label all polar bonds.
d. What is the shape around the O atom?
e. Is Cl_2O a polar molecule? Explain.

4.88 Answer the following questions about the molecule OCS.
a. How many valence electrons does OCS contain?
b. Draw a valid Lewis structure.
c. Label all polar bonds.
d. What is the shape around the C atom?
e. Is OCS a polar molecule? Explain.

Applications

4.89 Glycine is a building block used to make proteins, such as those in heart muscle (Figure 4.2).

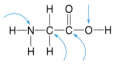

glycine

a. Add lone pairs where needed, and then count the total number of valence electrons in glycine.
b. Determine the shape around the four indicated atoms.
c. Label all of the polar bonds.
d. Is glycine a polar or nonpolar molecule? Explain.

4.90 Lactic acid gives sour milk its distinctive taste. Lactic acid is also an ingredient in several skin care products that purportedly smooth fine lines and improve skin texture.

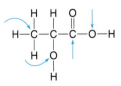

lactic acid

a. Add lone pairs where needed, and then count the total number of valence electrons in lactic acid.
b. Determine the shape around the four indicated atoms.
c. Label all of the polar bonds.
d. Is lactic acid a polar or nonpolar molecule? Explain.

4.91 Serotonin ($C_{10}H_{12}N_2O$) is a neurotransmitter that is important in mood, sleep, perception, and temperature regulation. Fill in all lone pairs and double bonds to give every atom its usual bonding pattern.

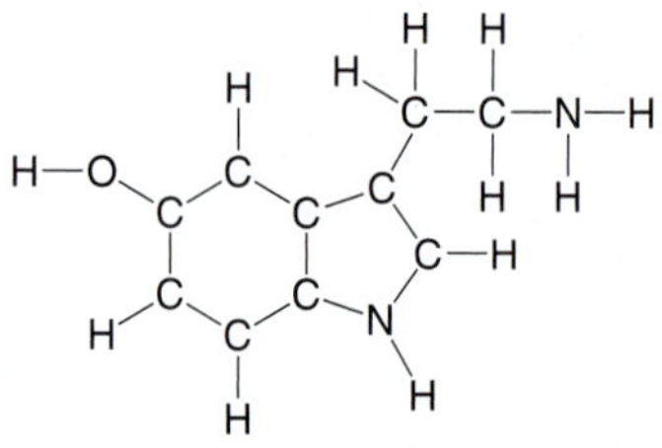

serotonin

4.92 Phenylephrine ($C_9H_{13}NO_2$) is the decongestant in Sudafed PE. Phenylephrine replaced the decongestant pseudoephedrine, which was readily converted to the illegal stimulant methamphetamine. Fill in all lone pairs and double bonds to give every atom its usual bonding pattern.

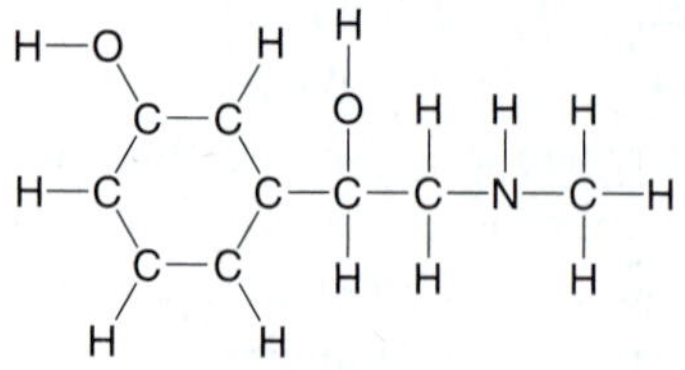

phenylephrine

CHALLENGE QUESTIONS

4.93 Cyclopropane is a stable compound that contains three carbon atoms in a three-membered ring.

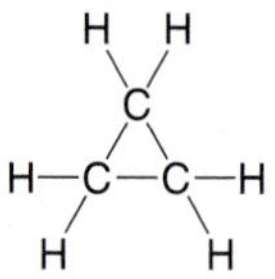

cyclopropane

a. What is the predicted shape around each carbon atom in the ring, given the number of groups around carbon?
b. What is the predicted C—C—C bond angle, given the shape and size of the ring?
c. Explain why cyclopropane is less stable than similar three-carbon compounds that do not contain a ring.

4.94 Although carbon has four bonds in stable molecules, sometimes reactive carbon intermediates that contain carbon atoms without four bonds are formed for very short time periods. Examples of these unstable intermediates include the methyl carbocation $(CH_3)^+$ and the methyl carbanion $(CH_3)^-$. Draw Lewis structures for both unstable ions and predict the shape around carbon.

5

CHAPTER GOALS

In this chapter you will learn how to:

1. Write and balance chemical equations
2. Define a mole and use Avogadro's number in calculations
3. Calculate formula weight and molar mass
4. Relate the mass of a substance to its number of moles
5. Carry out mole and mass calculations in chemical equations
6. Calculate percent yield
7. Define oxidation and reduction and recognize the components of a redox reaction
8. Give examples of common or useful redox reactions

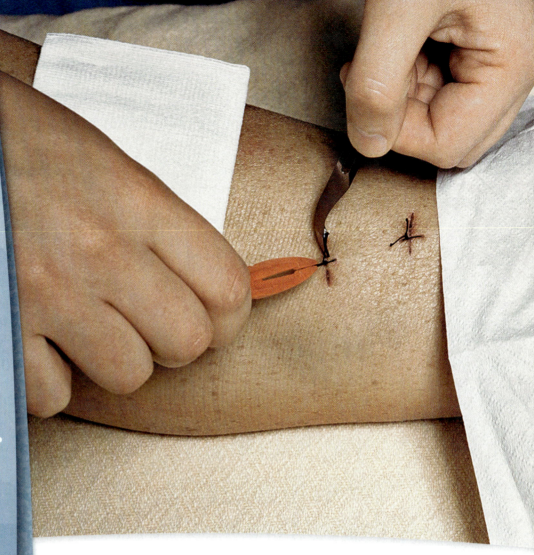

Thread for suturing wounds is made from **nylon,** one of the countless products synthesized by the chemical industry using chemical reactions.

CHEMICAL REACTIONS

HAVING learned about atoms, ionic compounds, and covalent molecules in Chapters 2–4, we now turn our attention to chemical reactions. Reactions are at the heart of chemistry. An understanding of chemical processes has made possible the conversion of natural substances into new compounds with different and sometimes superior properties. Aspirin, ibuprofen, and nylon are all products of chemical reactions utilizing substances derived from petroleum. Chemical reactions are not limited to industrial processes. The metabolism of food involves a series of reactions that both forms new compounds and also provides energy for the body's maintenance and growth. Burning gasoline, baking a cake, and photosynthesis involve chemical reactions. In Chapter 5 we learn the basic principles about chemical reactions.

5.1 INTRODUCTION TO CHEMICAL REACTIONS

Now that we have learned about compounds and the atoms that compose them, we can better understand the difference between the physical and chemical changes that were first discussed in Section 1.2.

- A physical change alters the physical state of a substance without changing its composition.

Changes in state—such as melting and boiling—are familiar examples of physical changes. When ice (solid water) melts to form liquid water, the highly organized water molecules in the solid phase become more disorganized in the liquid phase, but the chemical **bonds do *not*** change. Each water molecule (H_2O) is composed of two O—H bonds in both the solid and liquid phases.

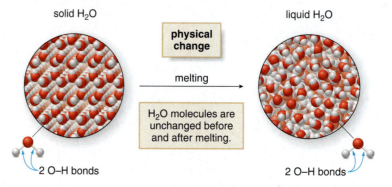

- A chemical change—chemical reaction—converts one substance into another.

Chemical reactions involve breaking bonds in the starting materials, called *reactants*, and forming new bonds in the *products*. The combustion of methane (CH_4), the main constituent of natural gas, in the presence of oxygen (O_2) to form carbon dioxide (CO_2) and water (H_2O) is an example of a chemical reaction. The carbon–hydrogen bonds in methane and the oxygen–oxygen bond in elemental oxygen are broken, and new carbon–oxygen and hydrogen–oxygen bonds are formed in the products.

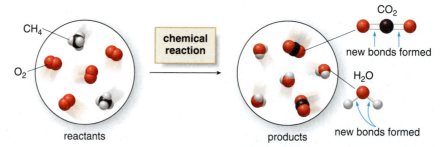

A chemical reaction may be accompanied by a visible change: two colorless reactants can form a colored product; a gas may be given off; two liquid reactants may yield a solid product. Sometimes heat is produced so that a reaction flask feels hot. A reaction having a characteristic visible change occurs when hydrogen peroxide (H_2O_2) is used to clean a bloody wound. An enzyme in the blood called catalase converts the H_2O_2 to water (H_2O) and oxygen (O_2), and bubbles of oxygen appear as a foam, as shown in Figure 5.1.

- A *chemical equation* is an expression that uses chemical formulas and other symbols to illustrate what reactants constitute the starting materials in a reaction and what products are formed.

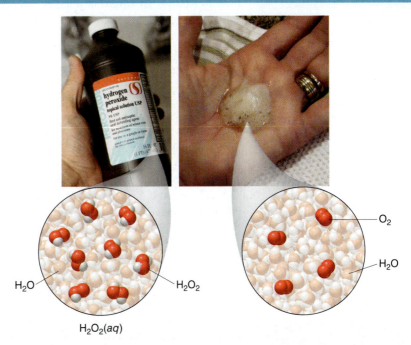

▼ FIGURE 5.1 Treating Wounds with Hydrogen Peroxide—A Visible Chemical Reaction

$H_2O_2(aq)$

The enzyme catalase in red blood converts hydrogen peroxide (H_2O_2) to water and oxygen gas, which appears as a visible white foam on the bloody surface. Hydrogen peroxide does not foam when it comes in contact with skin because skin cells do not contain the catalase needed for the reaction to occur.

Chemical equations are written with the **reactants on the left** and the **products on the right,** separated by a horizontal arrow—a **reaction arrow**—that points from the reactants to the products. In the combustion of methane, methane (CH_4) and oxygen (O_2) are the reactants on the left side of the arrow, and carbon dioxide (CO_2) and water (H_2O) are the products on the right side.

coefficient　　　　　coefficient

Chemical equation　　CH_4 + 2 O_2 ⟶ CO_2 + 2 H_2O

reactants　　　　　products

The numbers written in front of any formula are called **coefficients. Coefficients show the number of molecules of a given element or compound that react or are formed.** When no number precedes a formula, the coefficient is assumed to be "1." In the combustion of methane, the coefficients tell us that one molecule of CH_4 reacts with two molecules of O_2 to form one molecule of CO_2 and two molecules of H_2O.

When a formula contains a subscript, **multiply its coefficient by the subscript** to give the total number of atoms of a given type in that formula.

$$2\ O_2 = 4\ O\ \text{atoms}$$
$$2\ H_2O = 4\ H\ \text{atoms} + 2\ O\ \text{atoms}$$

Coefficients are used because all chemical reactions follow a fundamental principle of nature, the **law of conservation of mass,** which states:

- **Atoms cannot be created or destroyed in a chemical reaction.**

TABLE 5.1 Symbols Used in Chemical Equations

Symbol	Meaning
$\longrightarrow$	Reaction arrow
Δ	Heat
(s)	Solid
(l)	Liquid
(g)	Gas
(aq)	Aqueous solution

Although bonds are broken and formed in reactions, the number of atoms of each element in the reactants must be the same as the number of atoms of each type in the products. **Coefficients are used to *balance* an equation,** making the number of atoms of each element the same on both sides of the equation.

$$CH_4 + 2 O_2 \longrightarrow CO_2 + 2 H_2O$$

Atoms in the reactants:
- 1 C atom
- 4 H atoms
- 4 O atoms

Atoms in the products:
- 1 C atom
- 4 H atoms
- 4 O atoms

Two other important features are worthy of note. If heat is needed for a reaction to occur, the Greek letter delta (Δ) may be written over the arrow. The physical states of the reactants and products are sometimes indicated next to each formula—solid (*s*), liquid (*l*), or gas (*g*). If an aqueous solution is used—that is, if a reactant is dissolved in water—the symbol (*aq*) is used next to the reactant. When these features are added, the equation for the combustion of methane becomes:

Combustion of methane

$$CH_4(g) + 2 O_2(g) \xrightarrow{\Delta} CO_2(g) + 2 H_2O(g)$$

The symbols used for chemical equations are summarized in Table 5.1.

SAMPLE PROBLEM 5.1

Label the reactants and products, and indicate how many atoms of each type of element are present on each side of the equation.

$$C_2H_6O(l) + 3 O_2(g) \longrightarrow 2 CO_2(g) + 3 H_2O(g)$$

ANALYSIS Reactants are on the left side of the arrow and products are on the right side in a chemical equation. When a formula contains a subscript, multiply its coefficient by the subscript to give the total number of atoms of a given type in the formula.

SOLUTION In this equation, the reactants are C_2H_6O and O_2, while the products are CO_2 and H_2O. If no coefficient is written, it is assumed to be "1." To determine the number of each type of atom when a formula has both a coefficient and a subscript, multiply the coefficient by the subscript.

1 C_2H_6O	= 2 C's + 6 H's + 1 O	
3 O_2	= 6 O's	Multiply the coefficient 3 by the subscript 2.
2 CO_2	= 2 C's + 4 O's	Multiply the coefficient 2 by each subscript; 2×1 C = 2 C's; 2×2 O's = 4 O's.
3 H_2O	= 6 H's + 3 O's	Multiply the coefficient 3 by each subscript; 3×2 H's = 6 H's; 3×1 O = 3 O's.

Add up the atoms on each side to determine the total number for each type of element.

$$C_2H_6O(l) + 3 O_2(g) \longrightarrow 2 CO_2(g) + 3 H_2O(g)$$

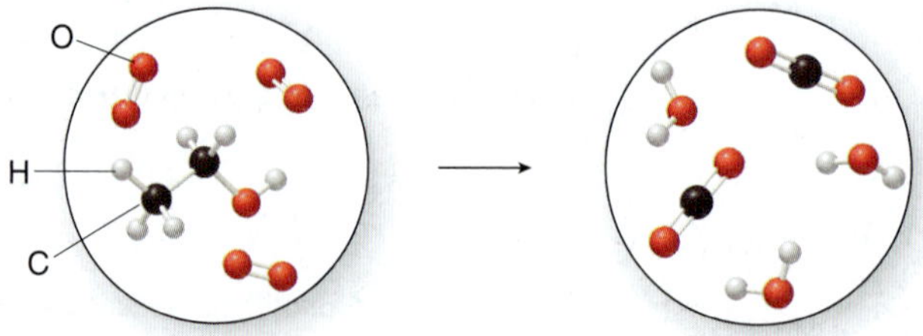

Atoms in the reactants:
- 2 C's   6 H's   7 O's

Atoms in the products:
- 2 C's   6 H's   7 O's

PROBLEM 5.1

Label the reactants and products, and indicate how many atoms of each type of element are present on each side of the following equations.

  a. $2 H_2O_2(aq) \longrightarrow 2 H_2O(l) + O_2(g)$
  b. $2 C_8H_{18} + 25 O_2 \longrightarrow 16 CO_2 + 18 H_2O$
  c. $2 Na_3PO_4(aq) + 3 MgCl_2(aq) \longrightarrow Mg_3(PO_4)_2(s) + 6 NaCl(aq)$

PROBLEM 5.2

One term in a balanced chemical equation contained the coefficient 3 in front of the formula $Al_2(SO_4)_3$. How many atoms of each type of element does this represent?

PROBLEM 5.3

Write a chemical equation from each of the following descriptions of reactions.

  a. One molecule of gaseous methane (CH_4) is heated with four molecules of gaseous chlorine (Cl_2), forming one molecule of liquid carbon tetrachloride (CCl_4) and four molecules of gaseous hydrogen chloride (HCl).
  b. One molecule of liquid methyl acetate ($C_3H_6O_2$) reacts with two molecules of hydrogen gas (H_2) to form one molecule each of liquid ethanol (C_2H_6O) and methanol (CH_4O).

ENVIRONMENTAL NOTE

The reaction of propane with oxygen forms carbon dioxide, water, and a great deal of energy that can be used for cooking, heating homes and water, drying clothes, and powering generators and vehicles. The combustion of propane and other fossil fuels adds a tremendous amount of CO_2 to the atmosphere each year, with clear environmental consequences (Section 12.8).

5.2 BALANCING CHEMICAL EQUATIONS

In order to carry out a reaction in the laboratory, we must know how much of each reactant we must combine to give the desired product. For example, if we wanted to synthesize aspirin ($C_9H_8O_4$) from a given amount of salicylic acid ($C_7H_6O_3$), say 100 g, we would have to determine how much acetic acid would be needed to carry out the reaction. A calculation of this sort begins with a **balanced chemical equation.**

salicylic acid $C_7H_6O_3$ +   acetic acid $C_2H_4O_2$ $\longrightarrow$ aspirin $C_9H_8O_4$ + H_2O

In this example, the equation is balanced as written and the coefficient in front of each formula is "1." Thus, one molecule of salicylic acid reacts with one molecule of acetic acid to form one molecule of aspirin and one molecule of water. More often, however, an equation must be balanced by adding coefficients in front of some formulas so that the **number of atoms of each element is equal on both sides of the equation.**

HOW TO   Balance a Chemical Equation

EXAMPLE   Write a balanced chemical equation for the reaction of propane (C_3H_8) with oxygen (O_2) to form carbon dioxide (CO_2) and water (H_2O).

Step [1]   **Write the equation with the correct formulas.**

  • Write the reactants on the left side and the products on the right side of the reaction arrow, and check if the equation is balanced without adding any coefficients.

$$C_3H_8 + O_2 \longrightarrow CO_2 + H_2O$$

Continued

How To, continued...

- This equation is not balanced as written since none of the elements—carbon, hydrogen, and oxygen—has the same number of atoms on both sides of the equation. For example, there are 3 C's on the left and only 1 C on the right.
- **The subscripts in a formula can *never* be changed to balance an equation.** Changing a subscript changes the identity of the compound. For example, changing CO_2 to CO would balance oxygen (there would be 2 O's on both sides of the equation), but that would change CO_2 (carbon dioxide) into CO (carbon monoxide).

Step [2]   **Balance the equation with coefficients one element at a time.**

- Begin with the most complex formula, and start with an element that appears in only one formula on both sides of the equation. In this example, begin with either the C's or H's in C_3H_8. Since there are 3 C's on the left, place the coefficient 3 before CO_2 on the right.

$$C_3H_8 \ + \ O_2 \ \longrightarrow \ 3\,CO_2 \ + \ H_2O$$

3 C's on the left   Place a 3 to balance C's.

- To balance the 8 H's in C_3H_8, place the coefficient 4 before H_2O on the right.

$$C_3H_8 \ + \ O_2 \ \longrightarrow \ 3\,CO_2 \ + \ 4\,H_2O$$

8 H's on the left   Place a 4 to balance H's.
(4×2 H's in H_2O = 8 H's)

- The only element not balanced is oxygen, and at this point there are a total of 10 O's on the right—six from three CO_2 molecules and four from four H_2O molecules. To balance the 10 O's on the right, place the coefficient 5 before O_2 on the left.

$$C_3H_8 \ + \ 5\,O_2 \ \longrightarrow \ 3\,CO_2 \ + \ 4\,H_2O$$

Place a 5 to balance O's.   10 O's on the right

Step [3]   **Check to make sure that the smallest set of whole numbers is used.**

$$C_3H_8 \ + \ 5\,O_2 \ \longrightarrow \ 3\,CO_2 \ + \ 4\,H_2O$$

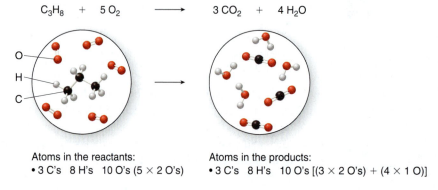

Atoms in the reactants:
- 3 C's   8 H's   10 O's (5×2 O's)

Atoms in the products:
- 3 C's   8 H's   10 O's [(3×2 O's) + (4×1 O)]

- This equation is balanced because the same number of C's, O's, and H's is present on both sides of the equation.
- Sometimes an equation is balanced but the lowest set of whole numbers is not used as coefficients. Say, for example, that balancing yielded the following equation:

$$2\,C_3H_8 + 10\,O_2 \longrightarrow 6\,CO_2 + 8\,H_2O$$

- This equation has the same number of C's, O's, and H's on both sides, but *each coefficient must be divided by two* to give the lowest set of whole numbers for the balanced equation, as drawn in the first equation in step [3].

Sample Problems 5.2–5.4 illustrate additional examples of balancing chemical equations. Sample Problem 5.3 gives an example that uses fractional coefficients in balancing. Sample Problem 5.4 illustrates how to balance an equation that contains several polyatomic anions.

SAMPLE PROBLEM 5.2

Write a balanced equation for the reaction of glucose ($C_6H_{12}O_6$) with oxygen (O_2) to form carbon dioxide (CO_2) and water (H_2O).

ANALYSIS

Balance an equation with coefficients, one element at a time, beginning with the most complex formula and starting with an element that appears in only one formula on both sides of the equation. Continue placing coefficients until the **number of atoms of each element is equal on both sides of the equation.**

SOLUTION

Bagels, pasta, bread, and rice are high in starch, which is hydrolyzed to the simple carbohydrate glucose after ingestion. The metabolism of glucose forms CO_2 and H_2O and provides energy for bodily functions.

[1] **Write the equation with correct formulas.**

$$C_6H_{12}O_6 + O_2 \longrightarrow CO_2 + H_2O$$
glucose

- None of the elements is balanced in this equation. As an example, there are 6 C's on the left side, but only 1 C on the right side.

[2] **Balance the equation with coefficients one element at a time.**

- Begin with glucose, since its formula is most complex. Balance the 6 C's of glucose by placing the coefficient 6 before CO_2. Balance the 12 H's of glucose by placing the coefficient 6 before H_2O.

Place a **6** to balance C's.

$$C_6H_{12}O_6 + O_2 \longrightarrow 6\,CO_2 + 6\,H_2O$$

Place a **6** to balance H's.

- The right side of the equation now has 18 O's. Since glucose already has 6 O's on the left side, 12 additional O's are needed on the left side. The equation will be balanced if the coefficient 6 is placed before O_2.

$$C_6H_{12}O_6 + 6\,O_2 \longrightarrow 6\,CO_2 + 6\,H_2O$$

Place a **6** to balance O's.

[3] **Check.**

- The equation is balanced since the number of atoms of each element is the same on both sides.

Answer: $C_6H_{12}O_6 + 6\,O_2 \longrightarrow 6\,CO_2 + 6\,H_2O$

Atoms in the reactants:	Atoms in the products:
• 6 C's	• 6 C's (6 × 1 C)
• 12 H's	• 12 H's (6 × 2 H's)
• 18 O's (1 × 6 O's) + (6 × 2 O's)	• 18 O's (6 × 2 O's) + (6 × 1 O)

PROBLEM 5.4

Write a balanced equation for each reaction.

a. $H_2 + O_2 \longrightarrow H_2O$

b. $NO + O_2 \longrightarrow NO_2$

c. $Fe + O_2 \longrightarrow Fe_2O_3$

d. $CH_4 + Cl_2 \longrightarrow CH_2Cl_2 + HCl$

PROBLEM 5.5

Write a balanced equation for the following reaction, shown with molecular art.

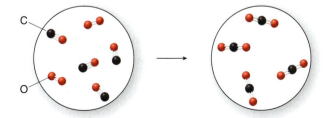

▼ **FIGURE 5.2**   Chemistry of an Automobile Airbag

a. The chemical reaction that inflates an airbag

inflated airbag

inflator

crash sensor

Na

N₂

NaN_3
sodium azide

b. An airbag deployed in a head-on collision

A severe car crash triggers an airbag to deploy when an electric sensor causes sodium azide (NaN_3) to ignite, converting it to sodium (Na) and nitrogen gas (N_2). The nitrogen gas causes the bag to inflate fully in 40 milliseconds, helping to protect passengers from serious injury. The sodium atoms formed in this first reaction are hazardous and subsequently converted to a safe sodium salt. It took 30 years to develop a reliable airbag system for automobiles.

SAMPLE PROBLEM 5.3

The airbag in an automobile inflates when ionic sodium azide (NaN_3), which is composed of Na^+ cations and the polyatomic anion, N_3^- (azide), rapidly decomposes to sodium (Na) and gaseous N_2 (Figure 5.2). Write a balanced equation for this reaction.

ANALYSIS   Balance an equation with coefficients, one element at a time, beginning with the most complex formula and starting with an element that appears in only one formula on both sides of the equation. Continue placing coefficients until the **number of atoms of each element is equal on both sides of the equation.**

SOLUTION

[1]   **Write the equation with correct formulas.**

$$NaN_3 \longrightarrow Na + N_2$$
sodium azide

- The N atoms are not balanced since there are 3 N's on the left side and only 2 N's on the right.

[2]   **Balance the equation with coefficients.**

- To balance the N atoms, we can use a **fractional coefficient** on the right side. A coefficient of 3/2 before N_2 is the equivalent of 3 N atoms.

$$NaN_3 \longrightarrow Na + \frac{3}{2} N_2$$

3 N atoms

$$\frac{3}{2} \times \frac{2 \text{ N atoms}}{1 \text{ N}_2 \text{ molecule}} = 3 \text{ N atoms}$$

[3]   **Check and simplify.**

- The equation is balanced since the number of atoms of each element is the same on both sides.

$$NaN_3 \longrightarrow Na + \frac{3}{2} N_2$$

Atoms in the reactants:
- 1 Na
- 3 N's

Atoms in the products:
- 1 Na
- 3 N's [(3/2) × 2 N's]

• Since a properly balanced equation uses the lowest set of *whole* numbers, **multiply each coefficient by 2** to convert the fraction 3/2 to the whole number 3.

$$\text{Answer:} \quad 2\,NaN_3 \longrightarrow 2\,Na + 3\,N_2$$

$$\underbrace{}_{(2 \times 1)} \quad \underbrace{}_{(2 \times 1)} \quad \underbrace{}_{\left(2 \times \frac{3}{2}\right)}$$

PROBLEM 5.6

Write a balanced equation for the reaction of ethane (C_2H_6) with O_2 to form CO_2 and H_2O.

PROBLEM 5.7

The Haber process is an important industrial reaction that converts N_2 and H_2 to ammonia (NH_3), an agricultural fertilizer and starting material for the synthesis of nitrate fertilizers. Write a balanced equation for the Haber process.

SAMPLE PROBLEM 5.4

Balance the following equation.

$$Ca_3(PO_4)_2 + H_2SO_4 \longrightarrow CaSO_4 + H_3PO_4$$
$$\text{calcium phosphate} \quad \text{sulfuric acid} \qquad \text{calcium sulfate} \quad \text{phosphoric acid}$$

ANALYSIS Balance an equation with coefficients, one element at a time, beginning with the most complex formula and starting with an element that appears in only one formula on both sides of the equation. Continue placing coefficients until the **number of atoms of each element is equal on both sides of the equation.**

SOLUTION

[1] **Write the equation with correct formulas.**

• The correct formula for each compound is given in the problem statement. When the reactants and products contain polyatomic ions, PO_4^{3-} and SO_4^{2-} in this case, **balance each ion as a *unit*,** rather than balancing the individual atoms. Thus, phosphate is not balanced in the equation as written, because the left side has two PO_4^{3-} anions while the right side has only one.

[2] **Balance the equation with coefficients.**

• Begin with $Ca_3(PO_4)_2$. Balance the 3 Ca's by placing the coefficient 3 before $CaSO_4$. Balance the 2 PO_4^{3-} anions by placing the coefficient 2 before H_3PO_4.

Place a 3 to balance Ca's.
$$Ca_3(PO_4)_2 + H_2SO_4 \longrightarrow 3\,CaSO_4 + 2\,H_3PO_4$$
Place a 2 to balance PO_4^{3-}.

• Two components are still not balanced—H atoms and sulfate anions (SO_4^{2-}). Both can be balanced by placing the coefficient 3 before H_2SO_4 on the left.

$$\overset{\text{6 H's}}{} \qquad\qquad\qquad \overset{\text{6 H's}}{}$$
$$Ca_3(PO_4)_2 + 3\,H_2SO_4 \longrightarrow 3\,CaSO_4 + 2\,H_3PO_4$$
3 SO_4^{2-} in both

Place a 3 to balance H and SO_4^{2-}.

[3] **Check.**

• The equation is balanced since the number of atoms and polyatomic anions is the same on both sides.

$$\text{Answer:} \quad Ca_3(PO_4)_2 + 3\,H_2SO_4 \longrightarrow 3\,CaSO_4 + 2\,H_3PO_4$$

Atoms or ions in the reactants:		Atoms or ions in the products:	
• 3 Ca's	• 6 H's	• 3 Ca's	• 6 H's
• 2 PO_4^{3-}	• 3 SO_4^{2-}	• 2 PO_4^{3-}	• 3 SO_4^{2-}

Ammonium hydrogen phosphate [$(NH_4)_2HPO_4$], the major phosphorus fertilizer, is formed from phosphoric acid, H_3PO_4, which is synthesized industrially by the chemical reaction in Sample Problem 5.4.

Balance each chemical equation.

a. $Al + H_2SO_4 \longrightarrow Al_2(SO_4)_3 + H_2$

b. $Na_2SO_3 + H_3PO_4 \longrightarrow H_2SO_3 + Na_3PO_4$

5.3 THE MOLE AND AVOGADRO'S NUMBER

Although the chemical equations in Section 5.2 were discussed in terms of individual atoms and molecules, atoms are exceedingly small. It is more convenient to talk about larger quantities of atoms, and for this reason, scientists use the **mole.** A mole defines a quantity, much like a dozen items means 12, and a case of soda means 24 cans. The only difference is that a mole is much larger.

- A *mole* is a quantity that contains 6.02×10^{23} items—usually atoms, molecules, or ions.

The definition of a mole is based on the number of atoms contained in exactly 12 g of the carbon-12 isotope. This number is called **Avogadro's number,** after the Italian scientist Amadeo Avogadro, who first proposed the concept of a mole in the nineteenth century. One mole, abbreviated as **mol,** always contains an Avogadro's number of particles.

1 mole of C atoms =	6.02×10^{23} C atoms
1 mole of CO_2 molecules =	6.02×10^{23} CO_2 molecules
1 mole of H_2O molecules =	6.02×10^{23} H_2O molecules
1 mole of vitamin C molecules =	6.02×10^{23} vitamin C molecules

How many items are contained in one mole of (a) baseballs; (b) bicycles; (c) Cheerios; (d) CH_4 molecules?

We can use Avogadro's number as a conversion factor to relate the number of moles of a substance to the number of atoms or molecules it contains.

Two possible conversion factors: $\dfrac{1 \text{ mol}}{6.02 \times 10^{23} \text{ atoms}}$ or $\dfrac{6.02 \times 10^{23} \text{ atoms}}{1 \text{ mol}}$

These conversion factors allow us to determine how many atoms or molecules are contained in a given number of moles. To carry out calculations that contain numbers written in scientific notation, we must first learn how to multiply and divide numbers written in this form.

- **To multiply two numbers in scientific notation, multiply the coefficients together and add the exponents in the powers of 10.**

Add exponents.
(5 + 2)

(3.0×10^5) $\times$ (2.0×10^2) = 6.0×10^7

Multiply coefficients.
(3.0×2.0)

- **To divide two numbers in scientific notation, divide the coefficients and subtract the exponents in the powers of 10.**

Divide coefficients.
$(6.0/2.0)$
$\dfrac{6.0 \times 10^2}{2.0 \times 10^{20}}$ Subtract exponents. = 3.0×10^{-18}
(2 − 20)

Each sample contains one mole of the substance—water (H_2O molecules), salt (NaCl, one mole of Na^+ and one mole of Cl^-), and aspirin ($C_9H_8O_4$ molecules). Pictured is a mole of aspirin *molecules,* not a mole of aspirin *tablets,* which is a quantity too large to easily represent. If a mole of aspirin tablets were arranged next to one another to cover a football field and then stacked on top of each other, they would occupy a volume 100 yards long, 53.3 yards wide, and over 20,000,000,000 miles high!

For a number written in scientific notation as $y \times 10^x$, y is the coefficient and x is the exponent in the power of 10 (Section 1.6).

Sample Problems 5.5 and 5.6 illustrate how to interconvert moles and molecules. In both problems we follow the stepwise procedure for problem solving using conversion factors outlined in Section 1.7B.

SAMPLE PROBLEM 5.5

How many molecules are contained in 5.0 moles of carbon dioxide (CO_2)?

ANALYSIS AND SOLUTION

[1] **Identify the original quantity and the desired quantity.**

$$5.0 \text{ mol of } CO_2 \qquad ? \text{ number of molecules of } CO_2$$

original quantity   desired quantity

[2] **Write out the conversion factors.**

- Choose the conversion factor that places the unwanted unit, mol, in the denominator so that the units cancel.

$$\frac{1 \text{ mol}}{6.02 \times 10^{23} \text{ molecules}} \quad \text{or} \quad \boxed{\frac{6.02 \times 10^{23} \text{ molecules}}{1 \text{ mol}}}$$

Choose this conversion factor to cancel mol.

[3] **Set up and solve the problem.**

- Multiply the original quantity by the conversion factor to obtain the desired quantity.

Convert to a number between 1 and 10.

$$5.0 \text{ mol} \quad \times \quad \frac{6.02 \times 10^{23} \text{ molecules}}{1 \text{ mol}} \quad = \quad 30. \times 10^{23} \text{ molecules}$$

Moles cancel.

$$= \quad 3.0 \times 10^{24} \text{ molecules of } CO_2$$

Answer

- Multiplication first gives an answer that is not written in scientific notation since the coefficient (30.) is greater than 10. Moving the decimal point one place to the *left* and *increasing* the exponent by one gives the answer written in the proper form.

PROBLEM 5.10

How many carbon atoms are contained in each of the following number of moles: (a) 2.00 mol; (b) 6.00 mol; (c) 0.500 mol; (d) 25.0 mol?

PROBLEM 5.11

How many molecules are contained in each of the following number of moles?

a. 2.5 mol of penicillin molecules

c. 0.40 mol of sugar molecules

b. 0.25 mol of NH_3 molecules

d. 55.3 mol of acetaminophen molecules

SAMPLE PROBLEM 5.6

How many moles of aspirin contain 8.62×10^{25} molecules?

ANALYSIS AND SOLUTION

[1] **Identify the original quantity and the desired quantity.**

$$8.62 \times 10^{25} \text{ molecules of aspirin} \qquad ? \text{ mole of aspirin}$$

original quantity   desired quantity

[2] **Write out the conversion factors.**

- Choose the conversion factor that places the unwanted unit, number of molecules, in the denominator so that the units cancel.

$$\frac{6.02 \times 10^{23} \text{ molecules}}{1 \text{ mol}} \quad \text{or} \quad \frac{1 \text{ mol}}{6.02 \times 10^{23} \text{ molecules}}$$

Choose this conversion factor to cancel molecules.

[3]   **Set up and solve the problem.**

- Multiply the original quantity by the conversion factor to obtain the desired quantity.
- To divide numbers using scientific notation, divide the coefficients (8.62/6.02) and subtract the exponents (25 − 23).

$$8.62 \times 10^{25} \text{ molecules} \quad \times \quad \frac{1 \text{ mol}}{6.02 \times 10^{23} \text{ molecules}} \quad = \quad 1.43 \times 10^{2} \text{ mol}$$

Molecules cancel.   =   143 mol of aspirin

Answer

PROBLEM 5.12

How many moles of water contain each of the following number of molecules?

a. 6.02×10^{25} molecules   b. 3.01×10^{22} molecules   c. 9.0×10^{24} molecules

---

5.4   MASS TO MOLE CONVERSIONS

In Section 2.3, we learned that the *atomic weight* **is the average mass of an element,** reported in atomic mass units (amu), and that the atomic weight of each element appears just below its chemical symbol in the periodic table. Thus, carbon has an atomic weight of 12.01 amu. We use atomic weights to calculate the mass of a compound.

- **The *formula weight* is the sum of the atomic weights of all the atoms in a compound, reported in atomic mass units (amu).**

The term "formula weight" is used for both ionic and covalent compounds. Often the term **"molecular weight"** is used in place of formula weight for covalent compounds, since they are composed of molecules, not ions. The formula weight of ionic sodium chloride (NaCl) is 58.44 amu, which is determined by adding up the atomic weights of Na (22.99 amu) and Cl (35.45 amu). The stepwise procedure for calculating the formula weight of compounds whose chemical formulas contain subscripts is shown in the accompanying *How To*.

Formula weight of NaCl:

Atomic weight of 1 Na   = 22.99 amu
Atomic weight of 1 Cl   = 35.45 amu
Formula weight of NaCl = 58.44 amu

HOW TO   Calculate the Formula Weight of a Compound

EXAMPLE   Calculate the formula weight for iron(II) sulfate, $FeSO_4$, an iron supplement used to treat anemia.

Step [1]   Write the correct formula and determine the number of atoms of each element from the subscripts.

- $FeSO_4$ contains 1 Fe atom, 1 S atom, and 4 O atoms.

Step [2]   Multiply the number of atoms of each element by the atomic weight and add the results.

1 Fe atom $\times$ 55.85 amu   =   55.85 amu
1 S atom $\times$ 32.07 amu   =   32.07 amu
4 O atoms $\times$ 16.00 amu   =   64.00 amu
Formula weight of $FeSO_4$ = 151.92 amu

| PROBLEM 5.13 | Calculate the formula weight of each ionic compound. |

a. $CaCO_3$, a common calcium supplement

b. KI, the essential nutrient added to NaCl to make iodized salt

| PROBLEM 5.14 | Calculate the molecular weight of each covalent compound. |

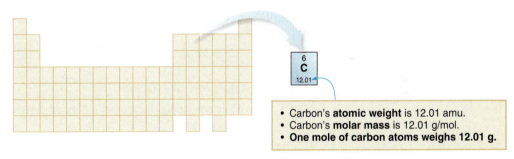

a. ethanol
(alcohol in alcoholic beverages)

b. phenol
(antiseptic)

c. halothane
(general anesthetic)

5.4A MOLAR MASS

When reactions are carried out in the laboratory, single atoms and molecules are much too small to measure out. Instead, substances are weighed on a balance and amounts are typically reported in grams, not atomic mass units. To determine how many atoms or molecules are contained in a given mass, we use its **molar mass.**

- The *molar mass* is the mass of one mole of any substance, reported in grams per mole.

The value of the molar mass of an element in the periodic table (in grams per mole) is the same as the value of its atomic weight (in amu). Thus, the molar mass of carbon is 12.01 g/mol, since its atomic weight is 12.01 amu; that is, one mole of carbon atoms weighs 12.01 g.

When a consumer product contains a great many lightweight small objects—for example, Cheerios—it is typically sold by weight, not by the number of objects. We buy Cheerios in an 8.9-oz box, not a box that contains 2,554 Cheerios.

| 6 |
| C |
| 12.01 |

- Carbon's **atomic weight** is 12.01 amu.
- Carbon's **molar mass** is 12.01 g/mol.
- **One mole of carbon atoms weighs 12.01 g.**

- The value of the molar mass of a compound in grams equals the value of its formula weight in amu.

Since the formula weight of NaCl is 58.44 amu, its molar mass is 58.44 g/mol. One mole of NaCl weighs 58.44 g. Since we know how to calculate a compound's formula weight, we also know how to calculate its molar mass, as shown in Sample Problem 5.7.

| SAMPLE PROBLEM 5.7 | What is the molar mass of nicotine ($C_{10}H_{14}N_2$), the toxic and addictive stimulant in tobacco? |

ANALYSIS Determine the number of atoms of each element from the subscripts in the chemical formula, multiply the number of atoms of each element by the atomic weight, and add up the results.

SOLUTION

10 C atoms $\times$ 12.01 amu $= 120.1$ amu
14 H atoms $\times$ 1.01 amu $= 14.14$ amu
2 N atoms $\times$ 14.01 amu $= \underline{28.02 \text{ amu}}$

Formula weight of nicotine $= 162.26$ amu rounded to 162.3 amu

Answer: Since the formula weight of nicotine is 162.3 amu, the molar mass of nicotine is 162.3 g/mol.

PROBLEM 5.15

What is the molar mass of $CaCO_3$ and KI, whose formula weights were calculated in Problem 5.13?

PROBLEM 5.16

What is the molar mass of each compound?

a. Li_2CO_3 (lithium carbonate), a drug used to treat bipolar disorder

b. C_2H_5Cl (ethyl chloride), a local anesthetic

c. $C_{13}H_{21}NO_3$ (albuterol), a drug used to treat asthma

5.4B   RELATING GRAMS TO MOLES

The molar mass is a very useful quantity because it relates the number of *moles* to the number of *grams* of a substance. In this way, the molar mass can be used as a conversion factor. For example, since the molar mass of H_2O is 18.0 g/mol, two conversion factors can be written.

$$\frac{18.0 \text{ g } H_2O}{1 \text{ mol}} \quad \text{or} \quad \frac{1 \text{ mol}}{18.0 \text{ g } H_2O}$$

Using these conversion factors, we can convert a given number of moles of water to grams, or a specific number of grams of water to moles. To solve problems of this sort, we once again return to the stepwise procedure for problem solving using conversion factors that was outlined in Section 1.7B.

SAMPLE PROBLEM 5.8

Converting moles to mass: How many grams does 0.25 moles of water weigh?

ANALYSIS AND SOLUTION

[1]   **Identify the original quantity and the desired quantity.**

$$\begin{array}{cc} 0.25 \text{ mol of } H_2O & ? \text{ g of } H_2O \\ \text{original quantity} & \text{desired quantity} \end{array}$$

[2]   **Write out the conversion factors.**

- Choose the conversion factor that places the unwanted unit, moles, in the denominator so that the units cancel.

$$\frac{1 \text{ mol}}{18.0 \text{ g } H_2O} \quad \text{or} \quad \boxed{\frac{18.0 \text{ g } H_2O}{1 \text{ mol}}} \quad \begin{array}{l} \text{Choose this conversion} \\ \text{factor to cancel mol.} \end{array}$$

[3]   **Set up and solve the problem.**

- Multiply the original quantity by the conversion factor to obtain the desired quantity.

$$0.25 \text{ mol} \quad \times \quad \frac{18.0 \text{ g } H_2O}{1 \text{ mol}} \quad = \quad 4.5 \text{ g of } H_2O$$

Moles cancel.   **Answer**

PROBLEM 5.17

Calculate the number of grams contained in each of the following number of moles.

a. 0.500 mol of NaCl   c. 3.60 mol of C_2H_4 (ethylene)

b. 2.00 mol of KI   d. 0.820 mol of CH_4O (methanol)

| SAMPLE PROBLEM 5.9 | **Converting mass to moles:** How many moles are present in 100. g of aspirin ($C_9H_8O_4$, molar mass 180.2 g/mol)? |

ANALYSIS AND SOLUTION

[1] **Identify the original quantity and the desired quantity.**

100. g of aspirin   ? mol of aspirin
original quantity   desired quantity

[2] **Write out the conversion factors.**

- Choose the conversion factor that places the unwanted unit, grams, in the denominator so that the units cancel.

$$\frac{180.2 \text{ g aspirin}}{1 \text{ mol}} \quad \text{or} \quad \boxed{\frac{1 \text{ mol}}{180.2 \text{ g aspirin}}} \quad \text{Choose this conversion factor to cancel g.}$$

[3] **Set up and solve the problem.**

- Multiply the original quantity by the conversion factor to obtain the desired quantity.

$$100. \cancel{g} \times \frac{1 \text{ mol}}{180.2 \cancel{g} \text{ aspirin}} = 0.555 \text{ mol of aspirin}$$

Grams cancel.   **Answer**

| PROBLEM 5.18 | How many moles are contained in each of the following? |

a. 100. g of NaCl   c. 0.250 g of aspirin ($C_9H_8O_4$)

b. 25.5 g of CH_4 d. 25.0 g of H_2O

5.4C   RELATING GRAMS TO NUMBER OF ATOMS OR MOLECULES

Since the molar mass of a substance gives the number of grams in a mole and a mole contains 6.02×10^{23} molecules (or atoms), **we can use molar mass to show the relationship between grams and *number of molecules* (or atoms).** For example, since the molar mass of aspirin is 180.2 g/mol, the following relationships exist:

$$\frac{180.2 \text{ g aspirin}}{1 \text{ mol}} = \frac{180.2 \text{ g aspirin}}{6.02 \times 10^{23} \text{ molecules}}$$

1 mol = 6.02×10^{23} molecules

| SAMPLE PROBLEM 5.10 | **Converting mass to number of molecules:** How many molecules are contained in a 325-mg tablet of aspirin ($C_9H_8O_4$, molar mass 180.2 g/mol)? |

ANALYSIS AND SOLUTION

[1] **Identify the original quantity and the desired quantity.**

325 mg of aspirin   ? molecules of aspirin
original quantity   desired quantity

[2] **Write out the conversion factors.**

- We have no conversion factor that directly relates milligrams to number of molecules. We do know, however, how to relate milligrams to grams, and grams to number of molecules. In other words, we need two conversion factors to solve this problem.

- Choose the conversion factors that place the unwanted units, grams and milligrams, in the denominator so that the units cancel.

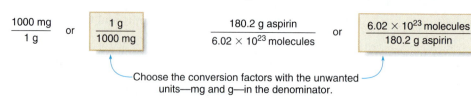

g–mg conversion factors

$$\frac{1000 \text{ mg}}{1 \text{ g}} \quad \text{or} \quad \frac{1 \text{ g}}{1000 \text{ mg}}$$

gram–number of molecules conversion factors

$$\frac{180.2 \text{ g aspirin}}{6.02 \times 10^{23} \text{ molecules}} \quad \text{or} \quad \frac{6.02 \times 10^{23} \text{ molecules}}{180.2 \text{ g aspirin}}$$

Choose the conversion factors with the unwanted units—mg and g—in the denominator.

[3]   **Set up and solve the problem.**

- Arrange each term so that the units in the numerator of one term cancel the units in the denominator of the adjacent term. The single desired unit, number of molecules, must be located in the numerator of one term.

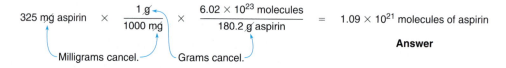

$$325 \text{ mg aspirin} \times \frac{1 \text{ g}}{1000 \text{ mg}} \times \frac{6.02 \times 10^{23} \text{ molecules}}{180.2 \text{ g aspirin}} = 1.09 \times 10^{21} \text{ molecules of aspirin}$$

Answer

Milligrams cancel.   Grams cancel.

PROBLEM 5.19	How many molecules are contained in two 500.-mg tablets of penicillin ($C_{16}H_{18}N_2O_4S$, molar mass 334.4 g/mol)?
PROBLEM 5.20	What is the mass in grams of 2.0×10^{19} molecules of ibuprofen ($C_{13}H_{18}O_2$)?

5.5   MOLE CALCULATIONS IN CHEMICAL EQUATIONS

Having learned about moles and molar mass, we can now return to balanced chemical equations. As we learned in Section 5.2, the coefficients in a balanced chemical equation tell us the number of *molecules* of each compound that react or are formed in a given reaction.

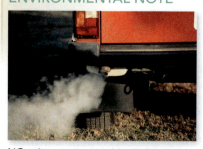
- **A balanced chemical equation also tells us the number of *moles* of each reactant that combine and the number of *moles* of each product formed.**

$$1 \text{ N}_2(g) \quad + \quad 1 \text{ O}_2(g) \quad \xrightarrow{\Delta} \quad 2 \text{ NO}(g)$$

one molecule of N_2 one molecule of O_2 two molecules of NO
one mole of N_2 one mole of O_2 two moles of NO

[The coefficient "1" has been written for emphasis.]

For example, the balanced chemical equation for the high temperature reaction of N_2 and O_2 to form nitrogen monoxide, NO, shows that one *molecule* of N_2 combines with one *molecule* of O_2 to form two *molecules* of NO. It also shows that one *mole* of N_2 combines with one *mole* of O_2 to form two *moles* of NO.

Coefficients are used to form mole ratios, which can serve as conversion factors. These ratios tell us the relative number of moles of reactants that combine in a reaction, as well as the relative number of moles of product formed from a given reactant, as shown in Sample Problem 5.11.

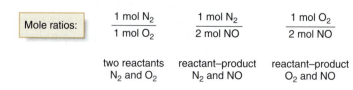

Mole ratios:

$$\frac{1 \text{ mol N}_2}{1 \text{ mol O}_2} \qquad \frac{1 \text{ mol N}_2}{2 \text{ mol NO}} \qquad \frac{1 \text{ mol O}_2}{2 \text{ mol NO}}$$

two reactants   reactant–product   reactant–product
N_2 and O_2 N_2 and NO O_2 and NO

• Use the mole ratio from the coefficients in the balanced equation to convert the number of moles of one compound (A) into the number of moles of another compound (B).

Moles of **A** $\longrightarrow$ Moles of **B**

mole–mole
conversion factor

SAMPLE PROBLEM 5.11

HEALTH NOTE

Meters that measure CO levels in homes are sold commercially. CO, a colorless, odorless gas, is a minor product formed whenever fossil fuels and wood are burned. In poorly ventilated rooms, such as those found in modern, well-insulated homes, CO levels can reach unhealthy levels.

Carbon monoxide (CO) is a poisonous gas that combines with hemoglobin in the blood, thus reducing the amount of oxygen that can be delivered to tissues. Under certain conditions, CO is formed when ethane (C_2H_6) in natural gas is burned in the presence of oxygen. Using the balanced equation, how many moles of CO are produced from 3.5 mol of C_2H_6?

$$2\,C_2H_6(g) \;+\; 5\,O_2(g) \;\xrightarrow{\Delta}\; 4\,CO(g) \;+\; 6\,H_2O(g)$$

ANALYSIS AND SOLUTION

[1] **Identify the original quantity and the desired quantity.**

3.5 mol of C_2H_6 ? mol of CO
original quantity   desired quantity

[2] **Write out the conversion factors.**

• Use the coefficients in the balanced equation to write mole–mole conversion factors for the two compounds, C_2H_6 and CO. Choose the conversion factor that places the unwanted unit, moles of C_2H_6, in the denominator so that the units cancel.

$$\frac{2\text{ mol }C_2H_6}{4\text{ mol CO}} \quad or \quad \boxed{\frac{4\text{ mol CO}}{2\text{ mol }C_2H_6}}$$

Choose this conversion factor to cancel mol C_2H_6.

[3] **Set up and solve the problem.**

• Multiply the original quantity by the conversion factor to obtain the desired quantity.

$$3.5\text{ mol }C_2H_6 \quad \times \quad \frac{4\text{ mol CO}}{2\text{ mol }C_2H_6} \quad = \quad 7.0\text{ mol CO}$$

Moles C_2H_6 cancel.   **Answer**

PROBLEM 5.21

Use the balanced equation for the reaction of N_2 and O_2 to form NO at the beginning of Section 5.5 to answer each question.

a. How many moles of NO are formed from 3.3 moles of N_2?

b. How many moles of NO are formed from 0.50 moles of O_2?

c. How many moles of O_2 are needed to completely react with 1.2 moles of N_2?

PROBLEM 5.22

Use the balanced equation in Sample Problem 5.11 to answer each question.

a. How many moles of O_2 are needed to react completely with 3.0 moles of C_2H_6?

b. How many moles of H_2O are formed from 0.50 moles of C_2H_6?

c. How many moles of C_2H_6 are needed to form 3.0 moles of CO?

5.6   MASS CALCULATIONS IN CHEMICAL EQUATIONS

Since a mole represents an enormously large number of very small molecules, there is no way to directly count the number of moles or molecules used in a chemical reaction. Instead, we utilize a balance to measure the number of grams of a compound used and the number of grams of product formed. The number of grams of a substance and the number of moles it contains are related by the molar mass (Section 5.4).

5.6A   CONVERTING MOLES OF REACTANT TO GRAMS OF PRODUCT

To determine how many grams of product are expected from a given number of moles of reactant, two operations are necessary. First, we must determine how many moles of product to expect using the coefficients of the balanced chemical equation (Section 5.5). Then, we convert the number of moles of product to the number of grams using the molar mass (Section 5.4). Each step needs a conversion factor. The stepwise procedure is outlined in the accompanying *How To*, and then illustrated with an example in Sample Problem 5.12.

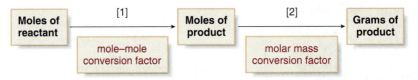

HOW TO   Convert Moles of Reactant to Grams of Product

EXAMPLE   In the upper atmosphere, high-energy radiation from the sun converts oxygen (O_2) to ozone (O_3). Using the balanced equation, how many grams of O_3 are formed from 9.0 mol of O_2?

$$3\ O_2(g) \xrightarrow{\text{sunlight}} 2\ O_3(g)$$

Step [1]   **Convert the number of moles of reactant to the number of moles of product using a mole–mole conversion factor.**

- Use the coefficients in the balanced chemical equation to write mole–mole conversion factors.

$$\frac{3\ \text{mol}\ O_2}{2\ \text{mol}\ O_3} \quad \text{or} \quad \boxed{\frac{2\ \text{mol}\ O_3}{3\ \text{mol}\ O_2}}$$ Choose this conversion factor to cancel mol O_2.

- Multiply the number of moles of starting material (9.0 mol) by the conversion factor to give the number of moles of product. In this example, 6.0 mol of O_3 are formed.

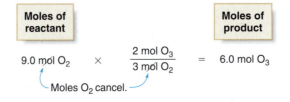

$$9.0\ \cancel{\text{mol}\ O_2} \times \frac{2\ \text{mol}\ O_3}{3\ \cancel{\text{mol}\ O_2}} = 6.0\ \text{mol}\ O_3$$

Moles O_2 cancel.

Step [2]   **Convert the number of moles of product to the number of grams of product using the product's molar mass.**

- Use the molar mass of the product (O_3) to write a conversion factor. The molar mass of O_3 is 48.0 g/mol (3 O atoms $\times$ 16.0 g/mol for each O atom = 48.0 g/mol).

$$\frac{1\ \text{mol}\ O_3}{48.0\ \text{g}\ O_3} \quad \text{or} \quad \boxed{\frac{48.0\ \text{g}\ O_3}{1\ \text{mol}\ O_3}}$$ Choose this conversion factor to cancel mol.

- Multiply the number of moles of product (from step [1]) by the conversion factor to give the number of grams of product.

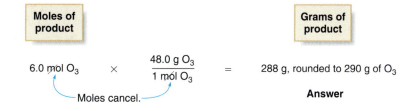

$$6.0 \text{ mol } O_3 \quad \times \quad \frac{48.0 \text{ g } O_3}{1 \text{ mol } O_3} \quad = \quad 288 \text{ g, rounded to } 290 \text{ g of } O_3$$

Moles cancel.

Answer

It is also possible to combine the multiplication operations from steps [1] and [2] into a single operation using both conversion factors. This converts the moles of starting material to grams of product all at once. Both the one-step and stepwise approaches give the same overall result.

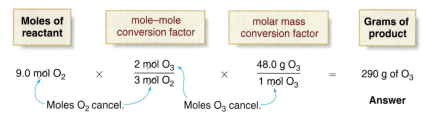

$$9.0 \text{ mol } O_2 \quad \times \quad \frac{2 \text{ mol } O_3}{3 \text{ mol } O_2} \quad \times \quad \frac{48.0 \text{ g } O_3}{1 \text{ mol } O_3} \quad = \quad 290 \text{ g of } O_3$$

Moles O_2 cancel. Moles O_3 cancel.

Answer

SAMPLE PROBLEM 5.12

Wine is produced by the fermentation of grapes. In fermentation, the carbohydrate glucose ($C_6H_{12}O_6$) is converted to ethanol and carbon dioxide according to the given balanced equation. How many grams of ethanol (C_2H_6O, molar mass 46.1 g/mol) are produced from 5.00 mol of glucose?

$$C_6H_{12}O_6(aq) \longrightarrow 2\, C_2H_6O(aq) + 2\, CO_2(g)$$

glucose   ethanol

ANALYSIS AND SOLUTION

[1] **Convert the number of moles of reactant to the number of moles of product using a mole–mole conversion factor.**

- Use the coefficients in the balanced chemical equation to write mole–mole conversion factors for the two compounds—one mole of glucose ($C_6H_{12}O_6$) forms two moles of ethanol (C_2H_6O).
- Multiply the number of moles of reactant (glucose) by the conversion factor to give the number of moles of product (ethanol).

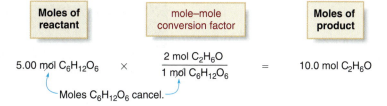

$$5.00 \text{ mol } C_6H_{12}O_6 \quad \times \quad \frac{2 \text{ mol } C_2H_6O}{1 \text{ mol } C_6H_{12}O_6} \quad = \quad 10.0 \text{ mol } C_2H_6O$$

Moles $C_6H_{12}O_6$ cancel.

[2] **Convert the number of moles of product to the number of grams of product using the product's molar mass.**

- Use the molar mass of the product (C_2H_6O, molar mass 46.1 g/mol) to write a conversion factor.
- Multiply the number of moles of product (from step [1]) by the conversion factor to give the number of grams of product.

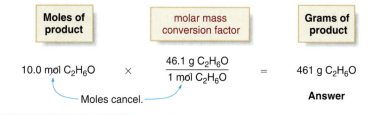

$$10.0 \text{ mol } C_2H_6O \quad \times \quad \frac{46.1 \text{ g } C_2H_6O}{1 \text{ mol } C_2H_6O} \quad = \quad 461 \text{ g } C_2H_6O$$

Moles cancel.

Answer

PROBLEM 5.23

Using the balanced equation for fermentation written in Sample Problem 5.12, answer the following questions.

   a. How many grams of ethanol are formed from 0.55 mol of glucose?

   b. How many grams of CO_2 are formed from 0.25 mol of glucose?

   c. How many grams of glucose are needed to form 1.0 mol of ethanol?

PROBLEM 5.24

Using the balanced equation for the combustion of ethanol, answer the following questions.

$$C_2H_6O(l) + 3\,O_2(g) \longrightarrow 2\,CO_2(g) + 3\,H_2O(g)$$
$$\text{ethanol}$$

   a. How many grams of CO_2 are formed from 0.50 mol of ethanol?

   b. How many grams of H_2O are formed from 2.4 mol of ethanol?

   c. How many grams of O_2 are needed to react with 0.25 mol of ethanol?

5.6B CONVERTING GRAMS OF REACTANT TO GRAMS OF PRODUCT

The **coefficients in chemical equations tell us the ratio of the number of *molecules* or *moles* that are involved in a chemical reaction.** The coefficients do *not*, however, tell us directly about the number of grams. That's because the molar mass—the number of grams in one mole—of a substance depends on the identity of the elements that compose it. One mole of H_2O molecules weighs 18.0 g, one mole of NaCl weighs 58.4 g, and one mole of sugar molecules weighs 342.3 g (Figure 5.3).

In the laboratory, we measure out the number of grams of a reactant on a balance. This does not tell us directly the number of grams of a particular product that will form, because in all likelihood, the molar masses of the reactant and product are different. To carry out this type of calculation—grams of one compound (reactant) to grams of another compound (product)—three operations are necessary.

▼ **FIGURE 5.3   One Mole of Water, Table Salt, and Table Sugar**

one mole of table sugar
C₁₂H₂₂O₁₁
342.3 g/mol

one mole of table salt
NaCl
58.4 g/mol

one mole of water molecules
H₂O
18.0 g/mol

One mole of each substance has the same number of units—6.02×10^{23} H_2O molecules, 6.02×10^{23} Na^+ and Cl^- ions, and 6.02×10^{23} sugar molecules. The molar mass of each substance is *different,* however, because they are each composed of *different* elements.

First, we must determine how many moles of reactant are contained in the given number of grams using the molar mass. Then, we can determine the number of moles of product expected using the coefficients of the balanced chemical equation. Finally, we convert the number of moles of product to the number of grams of product using its molar mass. Now there are three steps and three conversion factors. The stepwise procedure is outlined in the accompanying *How To*, and then illustrated with an example in Sample Problem 5.13.

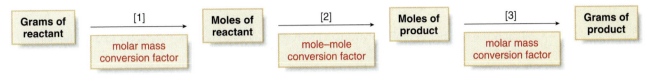

HOW TO   Convert Grams of Reactant to Grams of Product

EXAMPLE   Ethanol (C_2H_6O, molar mass 46.1 g/mol) is synthesized by reacting ethylene (C_2H_4, molar mass 28.1 g/mol) with water. How many grams of ethanol are formed from 14 g of ethylene?

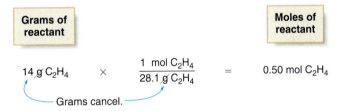

ethylene   ethanol

Step [1]   **Convert the number of grams of reactant to the number of moles of reactant using the reactant's molar mass.**

• Use the molar mass of the reactant (C_2H_4) to write a conversion factor.

$$\frac{28.1 \text{ g } C_2H_4}{1 \text{ mol } C_2H_4} \quad \text{or} \quad \frac{1 \text{ mol } C_2H_4}{28.1 \text{ g } C_2H_4}$$ Choose this conversion factor to cancel g.

• Multiply the number of grams of reactant by the conversion factor to give the number of moles of reactant.

Grams of reactant		**Moles of reactant**

$$14 \text{ g } C_2H_4 \quad \times \quad \frac{1 \text{ mol } C_2H_4}{28.1 \text{ g } C_2H_4} \quad = \quad 0.50 \text{ mol } C_2H_4$$

Grams cancel.

Step [2]   **Convert the number of moles of reactant to the number of moles of product using a mole–mole conversion factor.**

• Use the coefficients in the balanced chemical equation to write mole–mole conversion factors.

$$\frac{1 \text{ mol } C_2H_4}{1 \text{ mol } C_2H_6O} \quad \text{or} \quad \frac{1 \text{ mol } C_2H_6O}{1 \text{ mol } C_2H_4}$$ Choose this conversion factor to cancel mol C_2H_4.

• Multiply the number of moles of reactant by the conversion factor to give the number of moles of product. In this example, 0.50 mol of C_2H_6O is formed.

Moles of reactant		**Moles of product**

$$0.50 \text{ mol } C_2H_4 \quad \times \quad \frac{1 \text{ mol } C_2H_6O}{1 \text{ mol } C_2H_4} \quad = \quad 0.50 \text{ mol } C_2H_6O$$

Moles C_2H_4 cancel.

Continued

How To, continued...

Step [3]   **Convert the number of moles of product to the number of grams of product using the product's molar mass.**

- Use the molar mass of the product (C_2H_6O) to write a conversion factor.

$$\frac{1 \text{ mol } C_2H_6O}{46.1 \text{ g } C_2H_6O} \quad or \quad \boxed{\frac{46.1 \text{ g } C_2H_6O}{1 \text{ mol } C_2H_6O}} \quad \text{Choose this conversion factor to cancel mol } C_2H_6O.$$

- Multiply the number of moles of product (from step [2]) by the conversion factor to give the number of grams of product.

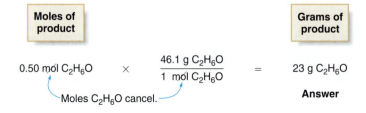

Moles of product				Grams of product
0.50 mol C_2H_6O	×	$\frac{46.1 \text{ g } C_2H_6O}{1 \text{ mol } C_2H_6O}$	=	23 g C_2H_6O
Moles C_2H_6O cancel.				**Answer**

It is also possible to combine the multiplication operations from steps [1], [2], and [3] into a single operation using all three conversion factors. This converts grams of starting material to grams of product all at once. Both the one-step and stepwise approaches give the same overall result.

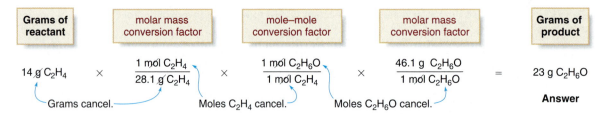

Grams of reactant		molar mass conversion factor		mole–mole conversion factor		molar mass conversion factor		Grams of product
14 g C_2H_4	×	$\frac{1 \text{ mol } C_2H_4}{28.1 \text{ g } C_2H_4}$	×	$\frac{1 \text{ mol } C_2H_6O}{1 \text{ mol } C_2H_4}$	×	$\frac{46.1 \text{ g } C_2H_6O}{1 \text{ mol } C_2H_6O}$	=	23 g C_2H_6O
Grams cancel.			Moles C_2H_4 cancel.		Moles C_2H_6O cancel.			**Answer**

SAMPLE PROBLEM 5.13   How many grams of aspirin are formed from 10.0 g of salicylic acid using the given balanced equation?

$$C_7H_6O_3(s) + C_2H_4O_2(l) \longrightarrow C_9H_8O_4(s) + H_2O(l)$$

salicylic acid   acetic acid   aspirin

ANALYSIS AND SOLUTION

[1]   **Convert the number of grams of reactant to the number of moles of reactant using the reactant's molar mass.**

- Use the molar mass of the reactant ($C_7H_6O_3$, molar mass 138.1 g/mol) to write a conversion factor. Multiply the number of grams of reactant by the conversion factor to give the number of moles of reactant.

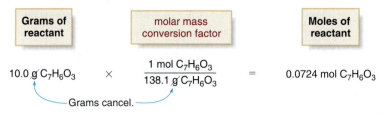

Grams of reactant		molar mass conversion factor		Moles of reactant
10.0 g $C_7H_6O_3$	×	$\frac{1 \text{ mol } C_7H_6O_3}{138.1 \text{ g } C_7H_6O_3}$	=	0.0724 mol $C_7H_6O_3$
		Grams cancel.		

[2]   **Convert the number of moles of reactant to the number of moles of product using a mole–mole conversion factor.**

Ethanol is used as a gasoline additive. Although some of the ethanol used for this purpose comes from corn and other grains, much of it is still produced by the reaction of ethylene with water. Ethanol produced from grains is a renewable resource, whereas ethanol produced from ethylene is not, because ethylene is made from crude oil. Thus, running your car on gasohol (gasoline mixed with ethanol) reduces our reliance on fossil fuels only if the ethanol is produced from renewable sources such as grains or sugarcane.

- Use the coefficients in the balanced chemical equation to write mole–mole conversion factors for the two compounds—one mole of salicylic acid ($C_7H_6O_3$) forms one mole of aspirin ($C_9H_8O_4$).
- Multiply the number of moles of reactant (salicylic acid) by the conversion factor to give the number of moles of product (aspirin).

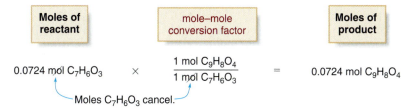

Moles of reactant		mole–mole conversion factor		Moles of product
0.0724 mol $C_7H_6O_3$	×	$\dfrac{1 \text{ mol } C_9H_8O_4}{1 \text{ mol } C_7H_6O_3}$	=	0.0724 mol $C_9H_8O_4$

Moles $C_7H_6O_3$ cancel.

[3] **Convert the number of moles of product to the number of grams of product using the product's molar mass.**

- Use the molar mass of the product ($C_9H_8O_4$, molar mass 180.2 g/mol) to write a conversion factor. Multiply the number of moles of product (from step [2]) by the conversion factor to give the number of grams of product.

Moles of product		molar mass conversion factor		Grams of product
0.0724 mol $C_9H_8O_4$	×	$\dfrac{180.2 \text{ g } C_9H_8O_4}{1 \text{ mol } C_9H_8O_4}$	=	13.0 g $C_9H_8O_4$

Moles cancel.   **Answer**

PROBLEM 5.25

Use the balanced equation in Sample Problem 5.13 for the conversion of salicylic acid and acetic acid to aspirin to answer the following questions.

a. How many grams of aspirin are formed from 55.5 g of salicylic acid?
b. How many grams of acetic acid are needed to react with 55.5 g of salicylic acid?
c. How many grams of water are formed from 55.5 g of salicylic acid?

PROBLEM 5.26

Use the balanced equation, $N_2 + O_2 \longrightarrow 2 \text{ NO}$, to answer the following questions.

a. How many grams of NO are formed from 10.0 g of N_2?
b. How many grams of NO are formed from 10.0 g of O_2?
c. How many grams of O_2 are needed to react completely with 10.0 g of N_2?

The actual yield is an experimental value determined by weighing the product obtained on a balance. The theoretical yield, on the other hand, is calculated from the coefficients of the balanced equation.

5.7 PERCENT YIELD

In determining the number of moles or grams of product in Sections 5.5 and 5.6, we assumed that each reaction gives the maximum amount of product from a given amount of reactant. This value is called the **theoretical yield** of a reaction.

- The *theoretical yield* is the amount of product expected from a given amount of reactant based on the coefficients in the balanced chemical equation.

Usually, however, the amount of product formed is *less* than the maximum amount predicted. Sometimes other undesired reactions, called **side reactions,** occur between the reactants. At other times the desired product is formed, but it goes on to react further to form another product. Moreover, each time material is weighed and transferred to a reaction vessel, or anytime a separation or purification step is carried out, material is inadvertently lost.

• The *actual yield* is the amount of product isolated from a reaction.

PROBLEM 5.27   Is it possible for the actual yield to be greater than the theoretical yield? Explain your answer.

5.7A CALCULATING PERCENT YIELD

The theoretical and actual yields are typically reported in grams. The amount of product actually formed in a particular reaction is reported as a **percent yield.**

$$\boxed{\text{Percent yield}} = \frac{\text{actual yield (g)}}{\text{theoretical yield (g)}} \times 100\%$$

For example, if a reaction forms 25.0 g of product and the theoretical yield is 40.0 g, the percent yield is calculated as follows.

$$\textbf{Percent yield} = \frac{25.0 \text{ g}}{40.0 \text{ g}} \times 100\% = 62.5\%$$

SAMPLE PROBLEM 5.14   Consider the reaction of ethylene (C_2H_4) and water to form ethanol (C_2H_6O), which was mentioned in Section 5.6B. If the theoretical yield of ethanol is 23 g in a reaction, what is the percent yield of ethanol if only 15 g of ethanol are actually formed?

ANALYSIS   Use the formula, percent yield = (actual yield/theoretical yield) × 100% to calculate the percent yield.

SOLUTION

$$\text{Percent yield} = \frac{\text{actual yield (g)}}{\text{theoretical yield (g)}} \times 100\%$$

$$= \frac{15 \text{ g}}{23 \text{ g}} \times 100\% = 65\%$$

Answer

SAMPLE PROBLEM 5.15   When charcoal is burned, the carbon (C) it contains reacts with oxygen (O_2) to form carbon dioxide (CO_2). (a) What is the theoretical yield of CO_2 in grams from 0.50 mol of C? (b) What is the percent yield if 10.0 g of CO_2 are formed?

$$C(s) + O_2(g) \longrightarrow CO_2(g)$$

ANALYSIS   To answer part (a), convert the number of moles of reactant to the number of moles of product, as in Section 5.6A. Then, convert the number of moles of product to the number of grams of product using the product's molar mass. This is the theoretical yield. To answer part (b), use the formula, percent yield = (actual yield/theoretical yield) × 100%.

SOLUTION   a. Calculate the theoretical yield using the procedure outlined in Sample Problem 5.12.

[1]   **Convert the number of moles of reactant to the number of moles of product using a mole–mole conversion factor.**

• Use the coefficients in the balanced equation to write a mole–mole conversion factor for the two compounds—one mole of carbon forms one mole of CO_2. Multiply the number of moles of reactant (carbon) by the conversion factor to give the number of moles of product (CO_2).

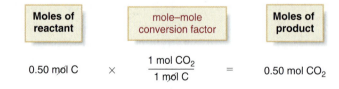

[2] **Convert the number of moles of product to the number of grams of product—the theoretical yield—using the product's molar mass.**

- Use the molar mass of the product (CO_2, molar mass 44.0 g/mol) to write a conversion factor. Multiply the number of moles of product (from step [1]) by the conversion factor to give the number of grams of product.

Moles of product		molar mass conversion factor		Grams of product
0.50 mol CO_2	$\times$	$\dfrac{44.0 \text{ g } CO_2}{1 \text{ mol } CO_2}$	$=$	22 g CO_2

Theoretical yield
Answer: part (a)

b. Use the theoretical yield from part (a) and the given actual yield to calculate the percent yield.

$$\text{Percent yield} = \frac{\text{actual yield (g)}}{\text{theoretical yield (g)}} \times 100\%$$

$$= \frac{10.0 \text{ g}}{22 \text{ g}} \times 100\% = 45\%$$

Answer: part (b)

Charcoal is composed of a network of covalently bonded carbon atoms. When charcoal is burned in a grill or coal is burned in a furnace, carbon atoms combine with O_2 from the air to form carbon dioxide (CO_2).

PROBLEM 5.28

Using the chemical equation in Sample Problem 5.15, answer each question. (a) What is the theoretical yield of CO_2 from 3.50 mol of charcoal? (b) What is the percent yield if the reaction gives 53.5 g of CO_2?

PROBLEM 5.29

Consider the conversion of oxygen (O_2) to ozone (O_3) described in Section 5.6A; that is, $3 \text{ O}_2 \longrightarrow 2 \text{ O}_3$. (a) What is the theoretical yield of O_3 in grams from 8.0 mol of O_2? (b) What is the percent yield if the reaction actually gives 155 g of O_3?

PROBLEM 5.30

Suppose the theoretical yield in a reaction is 10.5 g and the percent yield is 75.5%. What is the actual yield of product obtained?

5.7B CALCULATING PERCENT YIELD FROM GRAMS OF REACTANT

Since we weigh the amount of reactants and products on a balance, we must be able to calculate a percent yield from the number of grams of reactant used and the number of grams of product formed. This latter quantity is the **actual yield,** and it is always experimentally determined; that is, it is *not* a calculated value. We can carry out this lengthy calculation by putting together the steps in Sample Problems 5.13 and 5.14. Sample Problem 5.16 illustrates the steps used to calculate a percent yield when the number of grams of reactant used and the actual yield of product formed are both known.

SAMPLE PROBLEM 5.16

Acetaminophen, the active ingredient in Tylenol, can be prepared by the chemical reaction given below. What is the percent yield when 60.0 g of 4-aminophenol reacts with acetyl chloride to form 70.0 g of acetaminophen?

4-aminophenol
molar mass 109.1 g/mol

acetyl chloride

acetaminophen
molar mass 151.2 g/mol

ANALYSIS To find the theoretical yield, convert grams of reactant to grams of product, following the three steps in Sample Problem 5.13. Then, use the formula, percent yield = (actual yield/theoretical yield) × 100%.

SOLUTION

[1] **Convert the number of grams of reactant to the number of moles of reactant using the reactant's molar mass.**

- Use the molar mass of the reactant, 4-aminophenol, to write a conversion factor. Multiply the number of grams of reactant by the conversion factor to give the number of moles of reactant.

Grams of reactant		molar mass conversion factor		Moles of reactant
60.0 g 4-aminophenol	×	1 mol 4-aminophenol / 109.1 g 4-aminophenol	=	0.550 mol 4-aminophenol

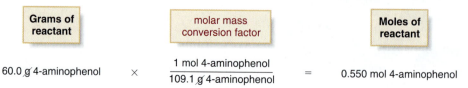

[2] **Convert the number of moles of reactant to the number of moles of product using a mole–mole conversion factor.**

- Use the coefficients in the balanced chemical equation to write a mole–mole conversion factor—one mole of 4-aminophenol forms one mole of acetaminophen. Multiply the number of moles of reactant by the conversion factor to give the number of moles of product.

Moles of reactant		mole–mole conversion factor		Moles of product
0.550 mol 4-aminophenol	×	1 mol acetaminophen / 1 mol 4-aminophenol	=	0.550 mol acetaminophen

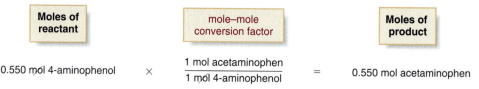

[3] **Convert the number of moles of product to the number of grams of product using the product's molar mass.**

- Use the molar mass of the product to write a conversion factor. Multiply the number of moles of product (from step [2]) by the conversion factor to give the number of grams of product.

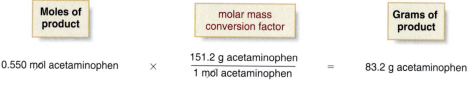

Moles of product		molar mass conversion factor		Grams of product
0.550 mol acetaminophen	×	151.2 g acetaminophen / 1 mol acetaminophen	=	83.2 g acetaminophen
				Theoretical yield

[4] **Use the theoretical yield and the given actual yield to calculate the percent yield.**

$$\text{Percent yield} = \frac{\text{actual yield (g)}}{\text{theoretical yield (g)}} \times 100\%$$

$$= \frac{70.0\text{ g}}{83.2\text{ g}} \times 100\% = 84.1\%$$

Answer

PROBLEM 5.31 Using the equation in Sample Problem 5.16, answer each question. (a) What is the theoretical yield of acetaminophen from 80.0 g of 4-aminophenol? (b) What is the percent yield if the reaction gives 65.5 g of acetaminophen?

PROBLEM 5.32	Consider the conversion of oxygen (O_2) to ozone (O_3) described in Section 5.6A; that is, $3\ O_2 \longrightarrow 2\ O_3$. (a) What is the theoretical yield of O_3 in grams from 324 g of O_2? (b) What is the percent yield if the reaction actually gives 122 g of O_3?

5.7C FOCUS ON HEALTH & MEDICINE
THE IMPORTANCE OF PERCENT YIELD IN THE PHARMACEUTICAL INDUSTRY

Although some drugs, like the cardiac drug digoxin (used to treat congestive heart failure, Section 1.1), are isolated directly from a natural source, most widely used drugs are synthesized in the laboratory. All common pain relievers—aspirin, acetaminophen, and ibuprofen—are synthetic. The same is true for the bronchodilator albuterol (trade name Proventil or Ventolin), the antidepressant fluoxetine (trade name Prozac), and the cholesterol-lowering medication atorvastatin (trade name Lipitor), whose three-dimensional structures are shown in Figure 5.4.

Once it has been determined that a drug is safe and effective, a pharmaceutical company must be able to prepare large quantities of the material cost-efficiently. This means that cheap and readily available starting materials must be used. It also means that the reactions used to synthesize a drug must proceed in high yield. Rarely is a drug prepared in a single step, and typically, five or more steps may be required in a synthesis.

- To determine the overall percent yield in a synthesis that has more than one step, multiply the percent yield for each step.

▼ **FIGURE 5.4** Three Widely Used Synthetic Drugs—Albuterol, Fluoxetine, and Atorvastatin

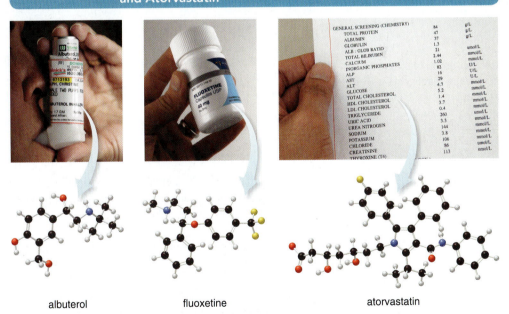

albuterol   fluoxetine   atorvastatin

Most commonly prescribed drugs are synthesized in the laboratory. Albuterol (Proventil, Ventolin) is a bronchodilator—that is, it widens airways—and so it is used to treat asthma. Fluoxetine (Prozac) is one of the most common antidepressants currently on the market, used by over 40 million individuals since 1986. Atorvastatin (Lipitor) lowers cholesterol levels, and in this way decreases the risk of heart attack and stroke.

For example, if a synthesis has five steps and each step has a 90.% yield (0.90 written as a decimal), the overall yield is

$$0.90 \times 0.90 \times 0.90 \times 0.90 \times 0.90 = 0.59 = 59\%$$

yield for each step,   **overall yield**
written as a decimal   **for five steps**

Thus, even if all steps proceed in high yield, the overall yield is considerably lower—59% in this example. If only one step has a lower yield, say 50.%, the overall yield drops even more, from 59% to 33%.

$$0.50 \times 0.90 \times 0.90 \times 0.90 \times 0.90 = 0.33 = 33\%$$

one low-yield reaction   **overall yield**
                                                for five steps

Moreover, many drugs are synthesized by routes that require 10 or more steps, resulting in a low overall yield. Thus, pharmaceutical companies are faced with the task of developing drugs that have the desired physiological effects, which are prepared by reactions that give high yields of the desired compounds.

PROBLEM 5.33

The synthetic antiviral drug Tamiflu, currently the most effective agent against avian influenza, is prepared by a 10-step synthesis. What is the overall yield of Tamiflu in each of the following 10-step syntheses?

a. Each step proceeds in 90.% yield.

b. Each step proceeds in 80.% yield.

c. One step proceeds in 50.% yield, while the rest occur in 90.% yield.

d. The following yields are recorded: 20.% (one reaction), 50.% (two reactions), 80.% (all remaining reactions).

5.8 OXIDATION AND REDUCTION

Another group of reactions—acid–base reactions—is discussed in Chapter 9.

Thus far we have examined features that are common to all types of chemical reactions. We conclude with an examination of one class of reactions that involves electron transfer—oxidation–reduction reactions.

5.8A GENERAL FEATURES OF OXIDATION–REDUCTION REACTIONS

A common type of chemical reaction involves the transfer of electrons from one element to another. When iron rusts, methane and wood burn, and a battery generates electricity, one element gains electrons and another loses them. These reactions involve **oxidation** and **reduction.**

• **Oxidation is the loss of electrons from an atom.**
• **Reduction is the gain of electrons by an atom.**

Oxidation and reduction are opposite processes, and both occur together in a single reaction called an **oxidation–reduction** or **redox reaction.** A redox reaction always has two components—one that is oxidized and one that is reduced.

• **A redox reaction involves the transfer of electrons from one element to another.**

An example of an oxidation–reduction reaction occurs when Zn metal reacts with Cu^{2+} cations, as shown in Figure 5.5.

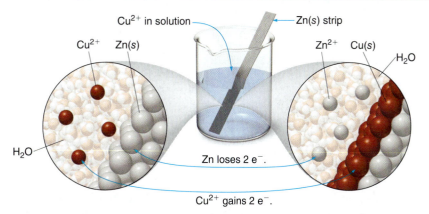

▼ FIGURE 5.5   A Redox Reaction—The Transfer of Electrons from Zn to Cu^{2+}

A redox reaction occurs when a strip of Zn metal is placed in a solution of Cu^{2+} ions. In this reaction, Zn loses two electrons to form Zn^{2+}, which goes into solution. Cu^{2+} gains two electrons to form Cu metal, which precipitates out of solution, forming a coating on the zinc strip.

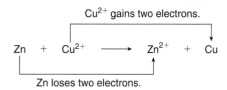

- Zn loses two electrons to form Zn^{2+}, so Zn is oxidized.
- Cu^{2+} gains two electrons to form Cu metal, so Cu^{2+} is reduced.

Each of these processes can be written as individual reactions, called **half reactions,** to emphasize which electrons are gained and lost.

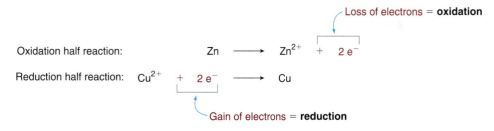

- A compound that gains electrons (is reduced) while causing another compound to be oxidized is called an *oxidizing agent.*
- A compound that loses electrons (is oxidized) while causing another compound to be reduced is called a *reducing agent.*

In this example, Zn loses electrons to Cu^{2+}. We can think of Zn as a **reducing agent** since it causes Cu^{2+} to gain electrons and become reduced. We can think of Cu^{2+} as an **oxidizing agent** since it causes Zn to lose electrons and become oxidized.

To draw the products of an oxidation–reduction reaction, we must decide which element or ion gains electrons and which element or ion loses electrons. Use the following guidelines.

- When considering neutral atoms, metals lose electrons and nonmetals gain electrons.
- When considering ions, cations tend to gain electrons and anions tend to lose electrons.

▼ FIGURE 5.6 Examples of Oxidation and Reduction Reactions

▼ FIGURE 5.6   Examples of Oxidation and Reduction Reactions

Oxidation Reactions

$$Na \longrightarrow Na^+ + e^-$$
$$Mg \longrightarrow Mg^{2+} + 2\,e^-$$
$$2\,Cl^- \longrightarrow Cl_2 + 2\,e^-$$
$$2\,O^{2-} \longrightarrow O_2 + 4\,e^-$$

Electrons are **lost.**

Reduction Reactions

$$Cl_2 + 2\,e^- \longrightarrow 2\,Cl^-$$
$$O_2 + 4\,e^- \longrightarrow 2\,O^{2-}$$
$$Cu^{2+} + 2\,e^- \longrightarrow Cu$$
$$Ag^+ + e^- \longrightarrow Ag$$

Electrons are **gained.**

Thus, the metals sodium (Na) and magnesium (Mg) readily lose electrons to form the cations Na^+ and Mg^{2+}, respectively; that is, they are oxidized. The nonmetals O_2 and Cl_2 readily gain electrons to form $2\,O^{2-}$ and $2\,Cl^-$, respectively; that is, they are reduced. A positively charged ion like Cu^{2+} is reduced to Cu by gaining two electrons, while two negatively charged Cl^- anions are oxidized to Cl_2 by losing two electrons. These reactions and additional examples are shown in Figure 5.6.

SAMPLE PROBLEM 5.17

Identify the species that is oxidized and the species that is reduced in the following reaction. Write out half reactions to show how many electrons are gained or lost by each species.

$$Mg(s) + 2\,H^+(aq) \longrightarrow Mg^{2+}(aq) + H_2(g)$$

ANALYSIS

Metals and anions tend to lose electrons and thus undergo oxidation. Nonmetals and cations tend to gain electrons and thus undergo reduction.

SOLUTION

The metal Mg is oxidized to Mg^{2+}, thus losing two electrons. Two H^+ cations gain a total of two electrons, and so are reduced to the nonmetal H_2.

$$Mg(s) \longrightarrow Mg^{2+}(aq) + 2\,e^- \qquad\qquad 2\,H^+(aq) + 2\,e^- \longrightarrow H_2(g)$$

Mg is oxidized.

H^+ is reduced.   Two electrons are needed to balance charge.

We need enough electrons so that **the total charge is the same on both sides of the equation.** Since 2 H^+ cations have a +2 overall charge, this means that 2 e^- must be gained so that the total charge on both sides of the equation is zero.

PROBLEM 5.34

Identify the species that is oxidized and the species that is reduced in each reaction. Write out half reactions to show how many electrons are gained or lost by each species.

a. $Zn(s) + 2\,H^+(aq) \longrightarrow Zn^{2+}(aq) + H_2(g)$ c. $2\,I^- + Br_2 \longrightarrow I_2 + 2\,Br^-$

b. $Fe^{3+}(aq) + Al(s) \longrightarrow Al^{3+}(aq) + Fe(s)$ d. $2\,AgBr \longrightarrow 2\,Ag + Br_2$

PROBLEM 5.35

Classify each reactant in Problem 5.34 as an oxidizing agent or a reducing agent.

5.8B   EXAMPLES OF OXIDATION–REDUCTION REACTIONS

Many common processes involve oxidation and reduction. For example, common antiseptics like iodine (I_2) and hydrogen peroxide (H_2O_2) are oxidizing agents that clean wounds by oxidizing, thereby killing bacteria that might cause infection.

When iron (Fe) rusts, it is oxidized by the oxygen in air to form iron(III) oxide, Fe_2O_3. In this redox reaction, neutral iron atoms are oxidized to Fe^{3+} cations, and elemental O_2 is reduced to O^{2-} anions.

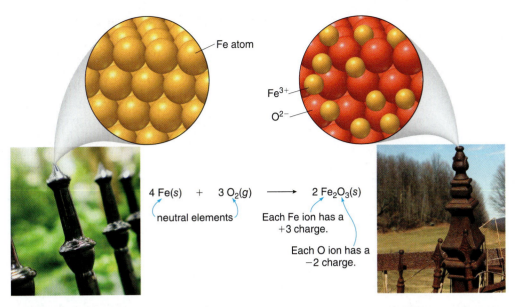

$$4 \, Fe(s) \; + \; 3 \, O_2(g) \longrightarrow 2 \, Fe_2O_3(s)$$

neutral elements

Each Fe ion has a +3 charge.

Each O ion has a −2 charge.

Batteries consist of a metal and a cation that undergo a redox reaction. When the electrons are transferred from the metal to the cation, an electric current results, which can supply power for a lightbulb, radio, computer, or watch. For example, an alkaline battery usually contains zinc powder and Mn^{4+} cations, together with sodium or potassium hydroxide (NaOH or KOH), as shown in Figure 5.7.

$$Zn \; + \; 2 \, MnO_2 \longrightarrow ZnO \; + \; Mn_2O_3$$

Mn^{4+} $\qquad$ Zn^{2+} $\quad$ Mn^{3+}

In this redox reaction, neutral Zn atoms are oxidized to Zn^{2+} cations. Mn^{4+} cations are reduced to Mn^{3+} cations. The oxygen anions (O^{2-}) just balance the charge of the metal cations and are neither oxidized nor reduced.

$$Zn \longrightarrow Zn^{2+} \; + \; 2 \, e^{-} \qquad\qquad 2 \, Mn^{4+} \; + \; 2 \, e^{-} \longrightarrow 2 \, Mn^{3+}$$

Zn is oxidized. $\qquad\qquad\qquad\qquad\qquad$ Mn^{4+} is reduced.

▼ **FIGURE 5.7   A Flashlight Battery—An Example of a Redox Reaction**

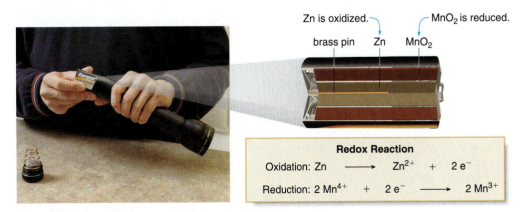

Zn is oxidized. $\qquad$ MnO_2 is reduced.

brass pin $\quad$ Zn $\quad$ MnO_2

Redox Reaction		
Oxidation: Zn $\longrightarrow$ Zn^{2+} + $2 \, e^{-}$		
Reduction: $2 \, Mn^{4+}$ + $2 \, e^{-}$ $\longrightarrow$ $2 \, Mn^{3+}$		

Alkaline batteries consist of zinc powder (Zn) and manganese dioxide (MnO_2), along with a paste of sodium hydroxide (NaOH) or potassium hydroxide (KOH). When electrical contact is made, Zn atoms lose electrons, which flow towards the Mn^{4+} cations in MnO_2. The resulting electric current can be used to power a lightbulb, radio, or other electrical device.

In some reactions it is much less apparent which reactant is oxidized and which is reduced. For example, in the combustion of methane (CH_4) with oxygen to form CO_2 and H_2O, there are no metals or cations that obviously lose or gain electrons, yet this is a redox reaction. In these instances, it is often best to count oxygen and hydrogen atoms.

- Oxidation results in the gain of oxygen atoms or the loss of hydrogen atoms.
- Reduction results in the loss of oxygen atoms or the gain of hydrogen atoms.

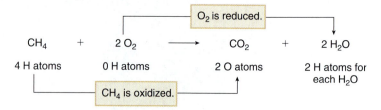

CH_4 is oxidized since it gains two oxygen atoms to form CO_2. O_2 is reduced since it gains two hydrogen atoms to form H_2O.

PROBLEM 5.36

The following redox reaction occurs in mercury batteries for watches. Identify the species that is oxidized and the species that is reduced, and write out two half reactions to show how many electrons are gained or lost.

$$Zn + HgO \longrightarrow ZnO + Hg$$

PROBLEM 5.37

Identify the species that is oxidized and the species that is reduced in the following redox reaction. Explain your choices.

$$C_2H_4O_2 + 2 H_2 \longrightarrow C_2H_6O + H_2O$$

5.9 FOCUS ON HEALTH & MEDICINE
PACEMAKERS

A pacemaker is a small electrical device implanted in an individual's chest and used to maintain an adequate heart rate (Figure 5.8). When a pacemaker detects that the heart is beating too slowly, it sends an electrical signal to the heart so that the heart muscle beats faster. A pacemaker contains a small, long-lasting battery that generates an electrical impulse by a redox reaction.

Most pacemakers used today contain a lithium–iodine battery. Each neutral lithium atom is oxidized to Li^+ by losing one electron. Each I_2 molecule is reduced by gaining two electrons and forming $2 I^-$. Since the balanced equation contains two Li atoms for each I_2 molecule, the number of electrons lost by Li atoms equals the number of electrons gained by I_2.

- I_2 gains 2 e^-, forming 2 I^-.
- I_2 is **reduced.**

$$2 Li + I_2 \longrightarrow 2 LiI$$

- Each Li atom loses 1 e^-, forming Li^+.
- Li metal is **oxidized.**

The lithium–iodine battery has a much longer battery life (over 10 years) than earlier batteries, greatly improving the quality of life for the many individuals with pacemakers.

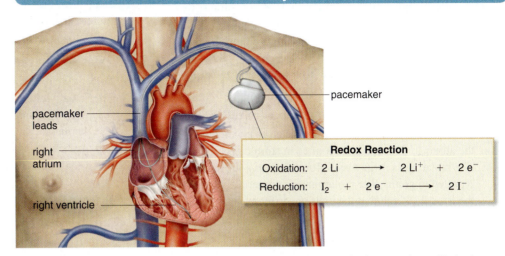

▼ **FIGURE 5.8   The Lithium–Iodine Battery in a Pacemaker**

Redox Reaction

Oxidation: $2\,Li \longrightarrow 2\,Li^+ + 2\,e^-$

Reduction: $I_2 + 2\,e^- \longrightarrow 2\,I^-$

A pacemaker generates a small electrical impulse that triggers the heart to beat. Today's pacemakers sense when the heart beats normally and provide an electrical signal only when the heart rate slows. Such devices are called "demand" pacemakers, and they quickly replaced earlier "fixed" rate models that continuously produced impulses to set the heart rate at a fixed value.

PROBLEM 5.38

Early pacemakers generated an electrical impulse by the following reaction. What species is the oxidizing agent and what species is the reducing agent in this reaction?

$$Zn + Hg^{2+} \longrightarrow Zn^{2+} + Hg$$

CHAPTER HIGHLIGHTS

KEY TERMS

Actual yield (5.7)
Avogadro's number (5.3)
Balanced chemical equation (5.2)
Chemical equation (5.1)
Formula weight (5.4)
Half reaction (5.8)
Law of conservation of mass (5.1)

Molar mass (5.4)
Mole (5.3)
Molecular weight (5.4)
Oxidation (5.8)
Oxidizing agent (5.8)
Percent yield (5.7)

Product (5.1)
Reactant (5.1)
Redox reaction (5.8)
Reducing agent (5.8)
Reduction (5.8)
Theoretical yield (5.7)

KEY CONCEPTS

❶ What do the terms in a chemical equation mean and how is an equation balanced? (5.1, 5.2)

- A chemical equation contains the reactants on the left side of an arrow and the products on the right. The coefficients tell how many molecules or moles of a substance react or are formed.

- A chemical equation is balanced by placing coefficients in front of chemical formulas one at a time, beginning with the most complex formula, so that the number of atoms of each element is the same on both sides. You must *not* balance the chemical equation by changing the subscripts in the chemical formulas of the reactants or products.

❷ Define the terms mole and Avogadro's number. (5.3)

- A mole is a quantity that contains 6.02×10^{23} atoms, molecules, or ions.
- Avogadro's number is the number of particles in a mole—6.02×10^{23}.
- The number of molecules in a given number of moles is calculated using Avogadro's number.

❸ How are formula weight and molar mass calculated? (5.4)

- The formula weight is the sum of the atomic weights of all the atoms in a compound, reported in atomic mass units.
- The molar mass is the mass of one mole of a substance, reported in grams. The molar mass is numerically equal to the formula weight but the units are different (g/mol not amu).

❹ How are the mass of a substance and its number of moles related? (5.4)

- The molar mass is used as a conversion factor to determine how many grams are contained in a given number of moles of a substance. Similarly, the molar mass is used to determine how many moles of a substance are contained in a given number of grams.

❺ How can a balanced equation and molar mass be used to calculate the number of moles and mass of a reaction product? (5.5, 5.6)

- The coefficients in a balanced chemical equation tell us the number of moles of each reactant that combine and the number of moles of each product formed. Coefficients are used to form mole ratios that serve as conversion factors relating the number of moles of reactants and products.

- When the mass of a substance in a reaction must be calculated, first its number of moles is determined using mole ratios, and then the molar mass is used to convert moles to grams.

❻ What is percent yield? (5.7)

- Percent yield = (actual yield/theoretical yield) $\times$ 100%.
- The actual yield is the amount of product formed in a reaction, determined by weighing a product on a balance. The theoretical yield is a quantity calculated from a balanced chemical equation, using mole ratios and molar masses. The theoretical yield is the maximum amount of product that can form in a chemical reaction from the amount of reactants used.

❼ What are oxidation and reduction reactions? (5.8)

- Oxidation–reduction or redox reactions are electron transfer reactions.
- Oxidation results in the loss of electrons. Metals and anions tend to undergo oxidation. In some reactions, oxidation results in the gain of O atoms or the loss of H atoms.
- Reduction results in the gain of electrons. Nonmetals and cations tend to undergo reduction. In some reactions, reduction results in the loss of O atoms or the gain of H atoms.

❽ Give some examples of common or useful redox reactions. (5.8, 5.9)

- Common examples of redox reactions include the rusting of iron and the combustion of methane. The electric current generated in batteries used for flashlights and pacemakers results from redox reactions.

PROBLEMS

Selected in-chapter and end-of-chapter problems have brief answers provided in Appendix B.

Chemical Equations

5.39 What is the difference between a coefficient in a chemical equation and a subscript in a chemical formula?

5.40 Why is it not possible to change the subscripts of a chemical formula to balance an equation?

5.41 What is the difference between a chemical equation and a chemical reaction?

5.42 What do the symbols Δ and (*aq*) mean in a chemical equation?

5.43 How many atoms of each element are drawn on each side of the following equations? Label the equations as balanced or not balanced.

    a. $2\,HCl(aq) + Ca(s) \longrightarrow CaCl_2(aq) + H_2(g)$
    b. $TiCl_4 + 2\,H_2O \longrightarrow TiO_2 + HCl$
    c. $Al(OH)_3 + H_3PO_4 \longrightarrow AlPO_4 + 3\,H_2O$

5.44 How many atoms of each element are drawn on each side of the following equations? Label the equations as balanced or not balanced.

    a. $3\,NO_2 + H_2O \longrightarrow HNO_3 + 2\,NO$
    b. $2\,H_2S + 3\,O_2 \longrightarrow H_2O + 2\,SO_2$
    c. $Ca(OH)_2 + 2\,HNO_3 \longrightarrow 2\,H_2O + Ca(NO_3)_2$

5.45 Balance each equation.

    a. $Ni(s) + HCl(aq) \longrightarrow NiCl_2(aq) + H_2(g)$
    b. $CH_4(g) + Cl_2(g) \longrightarrow CCl_4(g) + HCl(g)$
    c. $KClO_3 \longrightarrow KCl + O_2$
    d. $Al_2O_3 + HCl \longrightarrow AlCl_3 + H_2O$
    e. $Al(OH)_3 + H_2SO_4 \longrightarrow Al_2(SO_4)_3 + H_2O$

5.46 Balance each equation.

    a. $Mg(s) + HBr(aq) \longrightarrow MgBr_2(s) + H_2(g)$
    b. $CO(g) + O_2(g) \longrightarrow CO_2(g)$
    c. $PbS(s) + O_2(g) \longrightarrow PbO(s) + SO_2(g)$
    d. $H_2SO_4 + NaOH \longrightarrow Na_2SO_4 + H_2O$
    e. $H_3PO_4 + Ca(OH)_2 \longrightarrow Al_2(SO_4)_3 + H_2O$

5.47 Hydrocarbons are compounds that contain only C and H atoms. When a hydrocarbon reacts with O_2, CO_2 and H_2O are formed. Write a balanced equation for the combustion of each of the following hydrocarbons, all of which are high-octane components of gasoline.
   a. C_6H_6 (benzene)
   b. C_7H_8 (toluene)
   c. C_8H_{18} (isooctane)

5.48 MTBE ($C_5H_{12}O$) is a high-octane gasoline additive with a sweet, nauseating odor. Because small amounts of MTBE have contaminated the drinking water in some towns, it is now banned as a fuel additive in some states. MTBE reacts with O_2 to form CO_2 and H_2O. Write a balanced equation for the combustion of MTBE.

5.49 Some coal is high in sulfur (S) content, and when it burns, it forms sulfuric acid (H_2SO_4), a major component of acid rain, by a series of reactions. Balance the equation for the overall conversion drawn below.

$$S(s) + O_2(g) + H_2O(l) \longrightarrow H_2SO_4(l)$$

5.50 Balance the equation for the formation of magnesium hydroxide [$Mg(OH)_2$], one of the active ingredients in milk of magnesia.

$$MgCl_2 + NaOH \longrightarrow Mg(OH)_2 + NaCl$$

5.51 Consider the reaction, $O_3 + CO \longrightarrow O_2 + CO_2$. Molecular art is used to show the starting materials for this reaction. Fill in the molecules of the products using the balanced equation and following the law of conservation of mass.

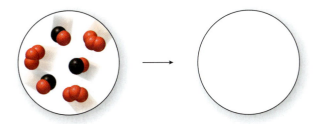

5.52 Consider the reaction, $2\ NO + 2\ CO \longrightarrow N_2 + 2\ CO_2$. Molecular art is used to show the starting materials for this reaction. Fill in the molecules of the products using the balanced equation and following the law of conservation of mass.

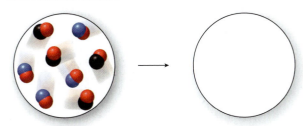

Formula Weight and Molar Mass

5.53 What is the difference between formula weight and molecular weight?

5.54 What is the difference between formula weight and molar mass?

5.55 Calculate the formula weight and molar mass of each compound.
   a. $NaNO_2$ (sodium nitrite), a preservative in hot dogs, ham, and other cured meats
   b. C_2H_4 (ethylene), the industrial starting material for the plastic polyethylene
   c. $Al_2(SO_4)_3$ (aluminum sulfate), once used as a common antiperspirant

5.56 Calculate the formula weight and molar mass of each compound.
   a. $MgSO_4$ (magnesium sulfate), a laxative
   b. C_2H_5Cl (chloroethane), a local anesthetic
   c. $Ca_3(PO_4)_2$ (calcium phosphate), a calcium supplement

5.57 Calculate the formula weight and molar mass of each biologically active compound.
   a. $C_6H_8O_6$ (vitamin C)
   b. $C_9H_{13}NO_2$ (phenylephrine), a decongestant in Sudafed PE
   c. $C_{16}H_{16}ClNO_2S$ (Plavix), a drug used to treat coronary artery disease

5.58 Calculate the formula weight and molar mass of each biologically active compound.
   a. $C_{29}H_{50}O_2$ (vitamin E)
   b. $C_6H_{13}NO_5$ (glucosamine), an over-the-counter arthritis medication
   c. $C_{17}H_{18}F_3NO$ (Prozac), a common antidepressant

5.59 L-Dopa is a drug used to treat Parkinson's disease.

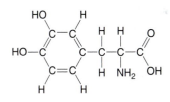

L-dopa

   a. What is the molecular formula of L-dopa?
   b. What is the formula weight of L-dopa?
   c. What is the molar mass of L-dopa?

5.60 Niacin, vitamin B_3, is found in soybeans, which contain it naturally, and cereals, which are fortified with it.

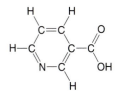

niacin

   a. What is the molecular formula of niacin?
   b. What is the formula weight of niacin?
   c. What is the molar mass of niacin?

Moles, Mass, and Avogadro's Number

5.61 Which quantity has the greater mass?
   a. 1 mol of Fe atoms or 1 mol of Sn atoms
   b. 1 mol of C atoms or 6.02×10^{23} N atoms
   c. 1 mol of N atoms or 1 mol of N_2 molecules
   d. 1 mol of CO_2 molecules or 3.01×10^{23} N_2O molecules

5.62   Which quantity has the greater mass?
       a. 1 mol of Si atoms or 1 mol of Ar atoms
       b. 1 mol of He atoms or 6.02×10^{23} H atoms
       c. 1 mol of Cl atoms or 1 mol of Cl_2 molecules
       d. 1 mol of C_2H_4 molecules or 3.01×10^{23} C_2H_4
            molecules

5.63   How many grams are contained in 5.00 mol of each
       compound?
       a. HCl   b. Na_2SO_4 c. C_2H_2 d. $Al(OH)_3$

5.64   How many grams are contained in 0.50 mol of each
       compound?
       a. NaOH   b. $CaSO_4$ c. C_3H_6 d. $Mg(OH)_2$

5.65   How many moles are contained in each number of grams
       of table sugar ($C_{12}H_{22}O_{11}$, molar mass 342.3 g/mol)?
       a. 0.500 g   b. 5.00 g   c. 25.0 g   d. 0.0250 g

5.66   How many moles are contained in each number of
       grams of fructose ($C_6H_{12}O_6$, molar mass 180.2 g/mol), a
       carbohydrate that is about twice as sweet as table sugar?
       "Lite" food products use half as much fructose as table
       sugar to achieve the same sweet taste, but with fewer
       calories.
       a. 0.500 g   b. 5.00 g   c. 25.0 g   d. 0.0250 g

5.67   Which has the greater mass: 0.050 mol of aspirin or
       10.0 g of aspirin ($C_9H_8O_4$)?

5.68   What is the mass in grams of 2.02×10^{20} molecules
       of the pain reliever ibuprofen ($C_{13}H_{18}O_2$, molar mass
       206.3 g/mol)?

5.69   How many molecules of butane (C_4H_{10}) are contained in
       the following number of moles: (a) 2.00 mol; (b) 0.250 mol;
       (c) 26.5 mol; (d) 222 mol; (e) 5.00×10^5 mol?

5.70   How many moles of pentane (C_5H_{12}) are contained in the
       following number of molecules?
       a. 5.00×10^{19} molecules   c. 8.32×10^{21} molecules
       b. 6.51×10^{28} molecules   d. 3.10×10^{20} molecules

5.71   What is the mass in grams of each quantity of lactic
       acid ($C_3H_6O_3$, molar mass 90.1 g/mol), the compound
       responsible for the aching feeling of tired muscles during
       vigorous exercise?
       a. 3.60 mol   c. 7.3×10^{24} molecules
       b. 0.580 mol   d. 6.56×10^{22} molecules

5.72   What is the mass in grams of each quantity of vitamin D
       (molar mass 384.7 g/mol), which is needed for forming
       and maintaining healthy bones?
       a. 3.6 mol   c. 7.3×10^{24} molecules
       b. 0.58 mol   d. 6.56×10^{22} molecules

5.73   Spinach, cabbage, and broccoli are excellent sources of
       vitamin K (molar mass 450.7 g/mol), which is needed in
       adequate amounts for blood to clot. The recommended
       daily intake of vitamin K is 120 μg. How many molecules
       of vitamin K does this correspond to?

5.74   How many molecules of amoxicillin ($C_{16}H_{19}N_3O_5S$, molar
       mass 365.4 g/mol) are contained in a 250-mg tablet?

Mass and Mole Calculations in Chemical Equations

5.75   Using the balanced equation for the combustion of
       acetylene, answer the following questions.

$$2\ H\!-\!C\!\equiv\!C\!-\!H\ +\ 5\ O_2\ \longrightarrow\ 4\ CO_2\ +\ 2\ H_2O$$
       acetylene

       a. How many moles of O_2 are needed to react completely
            with 5.00 mol of C_2H_2?
       b. How many moles of CO_2 are formed from 6.0 mol
            of C_2H_2?
       c. How many moles of H_2O are formed from 0.50 mol
            of C_2H_2?
       d. How many moles of C_2H_2 are needed to form 0.80
            mol of CO_2?

5.76   Sodium metal (Na) reacts violently when added to water
       according to the following balanced equation.

$$2\ Na(s)\ +\ 2\ H_2O(l)\ \longrightarrow\ 2\ NaOH(aq)\ +\ H_2(g)$$

       a. How many moles of H_2O are needed to react
            completely with 3.0 mol of Na?
       b. How many moles of H_2 are formed from 0.38 mol
            of Na?
       c. How many moles of H_2 are formed from 3.64 mol
            of H_2O?

5.77   Using the balanced equation for the combustion of
       acetylene in Problem 5.75, answer the following questions.
       a. How many grams of CO_2 are formed from 2.5 mol
            of C_2H_2?
       b. How many grams of CO_2 are formed from 0.50 mol
            of C_2H_2?
       c. How many grams of H_2O are formed from 0.25 mol
            of C_2H_2?
       d. How many grams of O_2 are needed to react with
            3.0 mol of C_2H_2?

5.78   Using the balanced equation for the reaction of Na with
       H_2O in Problem 5.76, answer the following questions.
       a. How many grams of NaOH are formed from 3.0 mol
            of Na?
       b. How many grams of H_2 are formed from 0.30 mol
            of Na?
       c. How many grams of H_2O are needed to react
            completely with 0.20 mol of Na?

5.79   Under certain conditions, the combustion of charcoal
       (C) in the presence of O_2 forms carbon monoxide (CO)
       according to the given balanced equation.

$$2\ C(s)\ +\ O_2(g)\ \longrightarrow\ 2\ CO(g)$$

       a. How many grams of CO are formed from 24.0 g
            of charcoal?
       b. How many grams of CO are formed from 0.16 g of O_2?

5.80   Iron, like most metals, does not occur naturally as the
       pure metal. Rather, it must be produced from iron ore,
       which contains iron(III) oxide, according to the given
       balanced equation.

$$Fe_2O_3(s) + 3\,CO(g) \longrightarrow 2\,Fe(s) + 3\,CO_2(g)$$

a. How many grams of Fe are formed from 10.0 g of Fe_2O_3?

b. How many grams of Fe are formed from 25.0 g of Fe_2O_3?

Theoretical Yield and Percent Yield

5.81 What is the difference between the theoretical yield and the actual yield?

5.82 What is the difference between the actual yield and the percent yield?

5.83 What is the percent yield of **B** in a reaction that uses 10.0 g of starting material **A**, has a theoretical yield of 12.0 g of **B**, and an actual yield of 9.0 g of **B?**

5.84 What is the percent yield of **B** in a reaction that uses 25.0 g of starting material **A**, has a theoretical yield of 20.0 g of **B**, and an actual yield of 17.0 g of **B?**

5.85 The reaction of methane (CH_4) with Cl_2 forms chloroform ($CHCl_3$) and HCl. Although $CHCl_3$ is a general anesthetic, it is no longer used for this purpose since it is also carcinogenic. The molar masses for all substances are given under the balanced equation.

$$\begin{array}{cccc} CH_4(g) & + \ 3\,Cl_2(g) \longrightarrow & CHCl_3(l) & + \ 3\,HCl(g) \\ 16.0\ \text{g/mol} & 70.9\ \text{g/mol} & 119.4\ \text{g/mol} & 36.5\ \text{g/mol} \end{array}$$

a. What is the theoretical yield of $CHCl_3$ in grams from 3.20 g of CH_4?

b. What is the percent yield if 15.0 g of $CHCl_3$ are actually formed in this reaction?

5.86 Methanol (CH_4O), which is used as a fuel in high-performance racing cars, burns in the presence of O_2 to form CO_2 and H_2O. The molar masses for all substances are given under the balanced equation.

$$\begin{array}{cccc} 2\,CH_4O(l) & + \ 3\,O_2(g) \longrightarrow & 2\,CO_2(g) & + \ 4\,H_2O(g) \\ 32.0\ \text{g/mol} & 32.0\ \text{g/mol} & 44.0\ \text{g/mol} & 18.0\ \text{g/mol} \end{array}$$

a. What is the theoretical yield of CO_2 from 48.0 g of methanol?

b. What is the percent yield of CO_2 if 48.0 g of CO_2 are formed?

Oxidation–Reduction Reactions

5.87 What is the difference between a substance that is oxidized and an oxidizing agent?

5.88 What is the difference between a substance that is reduced and a reducing agent?

5.89 Identify the species that is oxidized and the species that is reduced in each reaction. Write out two half reactions to show how many electrons are gained or lost by each species.

a. $Fe + Cu^{2+} \longrightarrow Fe^{2+} + Cu$

b. $Cl_2 + 2\,I^- \longrightarrow I_2 + 2\,Cl^-$

c. $2\,Na + Cl_2 \longrightarrow 2\,NaCl$

5.90 Identify the species that is oxidized and the species that is reduced in each reaction. Write out two half reactions to show how many electrons are gained or lost by each species.

a. $Mg + Fe^{2+} \longrightarrow Mg^{2+} + Fe$

b. $Cu^{2+} + Sn \longrightarrow Sn^{2+} + Cu$

c. $4\,Na + O_2 \longrightarrow 2\,Na_2O$

5.91 Zinc–silver oxide batteries are used in cameras and hearing aids. Identify the species that is oxidized and the species that is reduced in the following redox reaction. Identify the oxidizing agent and the reducing agent.

$$Zn + Ag_2O \longrightarrow ZnO + 2\,Ag$$

5.92 Rechargeable nickel–cadmium batteries are used in appliances and power tools. Identify the species that is oxidized and the species that is reduced in the following redox reaction. Identify the oxidizing agent and the reducing agent.

$$Cd + Ni^{4+} \longrightarrow Cd^{2+} + Ni^{2+}$$

5.93 The reaction of hydrogen (H_2) with acetylene (C_2H_2) forms ethane (C_2H_6). Is acetylene oxidized or reduced in this reaction? Explain your choice.

5.94 When Cl_2 is used to disinfect drinking water, Cl^- is formed. Is Cl_2 oxidized or reduced in this process?

5.95 The reaction of magnesium metal (Mg) with oxygen (O_2) forms MgO. Write a balanced equation for this redox reaction. Write two half reactions to show how many electrons are gained or lost by each species.

5.96 The reaction of aluminum metal (Al) with oxygen (O_2) forms Al_2O_3. Write a balanced equation for this redox reaction. Write two half reactions to show how many electrons are gained or lost by each species.

General Questions

5.97 Answer the following questions about the conversion of the sucrose ($C_{12}H_{22}O_{11}$) in sugarcane to ethanol (C_2H_6O) and CO_2 according to the following unbalanced equation. In this way sugarcane is used as a renewable source of ethanol, which is used as a fuel additive in gasoline.

$$\underset{\text{sucrose}}{C_{12}H_{22}O_{11}(s)} + H_2O(l) \longrightarrow \underset{\text{ethanol}}{C_2H_6O(l)} + CO_2(g)$$

a. What is the molar mass of sucrose?

b. Balance the given equation.

c. How many moles of ethanol are formed from 2 mol of sucrose?

d. How many moles of water are needed to react with 10 mol of sucrose?

e. How many grams of ethanol are formed from 0.550 mol of sucrose?

f. How many grams of ethanol are formed from 34.2 g of sucrose?

g. What is the theoretical yield of ethanol in grams from 17.1 g of sucrose?

h. If 1.25 g of ethanol are formed in the reaction in part (g), what is the percent yield of ethanol?

5.98 Answer the following questions about diethyl ether ($C_4H_{10}O$), the first widely used general anesthetic. Diethyl ether can be prepared from ethanol according to the following unbalanced equation.

$$C_2H_6O(l) \longrightarrow C_4H_{10}O(l) + H_2O(l)$$
   ethanol   diethyl ether

a. What is the molar mass of diethyl ether?
b. Balance the given equation.
c. How many moles of diethyl ether are formed from 2 mol of ethanol?
d. How many moles of water are formed from 10 mol of ethanol?
e. How many grams of diethyl ether are formed from 0.55 mol of ethanol?
f. How many grams of diethyl ether are formed from 4.60 g of ethanol?
g. What is the theoretical yield of diethyl ether in grams from 2.30 g of ethanol?
h. If 1.80 g of diethyl ether are formed in the reaction in part (g), what is the percent yield of diethyl ether?

Applications

5.99 A bottle of the pain reliever ibuprofen ($C_{13}H_{18}O_2$, molar mass 206.3 g/mol) contains 500 200.-mg tablets. (a) How many moles of ibuprofen does the bottle contain? (b) How many molecules of ibuprofen does the bottle contain?

5.100 One dose of Maalox contains 500. mg each of $Mg(OH)_2$ and $Al(OH)_3$. How many moles of each compound are contained in a single dose?

5.101 The average nicotine ($C_{10}H_{14}N_2$, molar mass 162.3 g/mol) content of a Camel cigarette is 1.93 mg. Suppose an individual smokes one pack of 20 cigarettes a day.
a. How many molecules of nicotine are smoked in a day?
b. How many moles of nicotine are smoked in a day?

5.102 How many moles of sucrose (table sugar, $C_{12}H_{22}O_{11}$, molar mass 342.3 g/mol) are contained in a 5-lb bag of sugar?

5.103 If the daily recommended intake of sodium ions is 2,400 mg, how many Na^+ ions does this correspond to?

5.104 How many molecules are contained in a glass that holds 250 g of water? How many moles does that correspond to?

5.105 DDT, a pesticide that kills disease-carrying mosquitoes, is synthesized by the given equation. DDT is now banned in the United States because it is a persistent environmental pollutant that only slowly degrades.

chlorobenzene
112.6 g/mol

DDT
$C_{14}H_9Cl_5$

a. What is the molar mass of DDT?
b. How many grams of DDT would be formed from 0.10 mol of chlorobenzene?
c. What is the theoretical yield of DDT in grams from 11.3 g of chlorobenzene?
d. If 15.0 g of DDT are formed in the reaction in part (c), what is the percent yield of DDT?

5.106 Fats, such as butter, and oils, such as corn oil, are formed from compounds called fatty acids, one of which is linolenic acid ($C_{18}H_{30}O_2$). Linolenic acid undergoes reactions with hydrogen and oxygen to form the products shown in each equation.

[1] $C_{18}H_{30}O_2 + H_2 \longrightarrow C_{18}H_{36}O_2$
linolenic acid

[2] $C_{18}H_{30}O_2 + O_2 \longrightarrow CO_2 + H_2O$
linolenic acid

a. Calculate the molar mass of linolenic acid.
b. Balance Equation [1], which shows the reaction with hydrogen.
c. Balance Equation [2], which shows the reaction with oxygen.
d. How many grams of product are formed from 10.0 g of linolenic acid in Equation [1]?

CHALLENGE QUESTIONS

5.107 TCDD, also called dioxin ($C_{12}H_4Cl_4O_2$, molar mass 322.0 g/mol), is a potent poison. The average lethal dose in humans is estimated to be 3.0×10^{-2} mg per kg of body weight. (a) How many grams constitute a lethal dose for a 70.-kg individual? (b) How many molecules of TCDD does this correspond to?

5.108 The lead–acid battery in a car consists of lead (Pb), lead(IV) oxide (PbO_2), and sulfuric acid (H_2SO_4), which undergo a redox reaction according to the given equation. Explain the oxidation and reduction reactions that occur with the lead atoms and ions in this battery.

$$Pb + PbO_2 + 2\ H_2SO_4 \longrightarrow 2\ PbSO_4 + 2\ H_2O$$

6

• • • • • • • • • • • • •

CHAPTER GOALS

In this chapter you will learn how to:

1 Define energy and become familiar with the units of energy

2 Use bond dissociation energies to predict bond strength

3 Describe energy changes in a reaction, and classify reactions as endothermic or exothermic

4 Draw energy diagrams

5 Predict the effect of concentration, temperature, and the presence of a catalyst on the rate of a reaction

6 Describe the basic features of chemical equilibrium and write an expression for an equilibrium constant

7 Use Le Châtelier's principle to predict what happens when equilibrium is disturbed

8 Use Le Châtelier's principle and reaction rates to explain the regulation of body temperature

The combustion of gasoline and the metabolism of carbohydrates during exercise are examples of oxidation reactions that release a great deal of useful energy.

ENERGY CHANGES, REACTION RATES, AND EQUILIBRIUM

IN Chapter 6 we turn our attention to two facets of chemical reactions: energy changes and reaction rates. Why do some reactions—like the combustion of fossil fuels or the metabolism of carbohydrates—release a great deal of energy that can be used for powering a car or running a marathon, while other reactions absorb energy from the environment? What factors affect how fast a reaction proceeds? To answer these and related questions, we must learn about what happens when molecules come together in a reaction, as well as what energy changes are observed when bonds are broken and formed.

6.1 ENERGY

Energy **is the capacity to do work.** Whenever you throw a ball, ride a bike, or read a newspaper, you use energy to do work. There are two types of energy.

- Potential energy is stored energy.
- Kinetic energy is the energy of motion.

A ball at the top of a hill or the water in a reservoir behind a dam are examples of potential energy. When the ball rolls down the hill or the water flows over the dam, the stored potential energy is converted to the kinetic energy of motion. Although energy can be converted from one form to another, one rule, the **law of conservation of energy,** governs the process.

- The total energy in a system does not change. Energy cannot be created or destroyed.

The energy stored in chemical bonds—both ionic and covalent—is a form of potential energy. In chemical reactions, potential energy may be released and converted to heat, the kinetic energy of the moving particles of the product. **Reactions that form products having** *lower* **potential energy than the reactants are favored.**

- A compound with lower potential energy is more stable than a compound with higher potential energy.

6.1A THE UNITS OF ENERGY

The joule, named after the nineteenth-century English physicist James Prescott Joule, is pronounced *jewel*.

Energy can be measured using two different units, **calories (cal)** and **joules (J).** A **calorie** is the amount of energy needed to raise the temperature of 1 g of water 1 °C. Joules and calories are related in the following way.

$$1 \text{ cal} = 4.184 \text{ J}$$

Since both the calorie and the joule are small units of measurement, more often energies in reactions are reported with kilocalories (kcal) and kilojoules (kJ). Recall from Table 1.2 that the prefix *kilo* means 1,000.

$$1 \text{ kcal} = 1,000 \text{ cal}$$
$$1 \text{ kJ} = 1,000 \text{ J}$$
$$1 \text{ kcal} = 4.184 \text{ kJ}$$

To convert a quantity from one unit of measurement to another, set up conversion factors and use the method first shown in Section 1.7B and illustrated in Sample Problem 6.1.

SAMPLE PROBLEM 6.1

ANALYSIS AND SOLUTION

A reaction releases 421 kJ of energy. How many kilocalories does this correspond to?

[1] **Identify the original quantity and the desired quantity.**

421 kJ   ? kcal
original quantity   desired quantity

[2] **Write out the conversion factors.**

- Choose the conversion factor that places the unwanted unit, kilojoules, in the denominator so that the units cancel.

kJ–kcal conversion factors

$$\frac{4.184 \text{ kJ}}{1 \text{ kcal}} \quad \text{or} \quad \boxed{\frac{1 \text{ kcal}}{4.184 \text{ kJ}}} \quad \text{Choose this conversion factor to cancel kJ.}$$

[3] **Set up and solve the problem.**

- Multiply the original quantity by the conversion factor to obtain the desired quantity.

$$421 \text{ kJ} \quad \times \quad \frac{1 \text{ kcal}}{4.184 \text{ kJ}} \quad = \quad 100.6 \text{ kcal, rounded to } 101 \text{ kcal}$$

Kilojoules cancel. **Answer**

PROBLEM 6.1

Carry out each of the following conversions.

a. 42 J to cal   b. 55.6 kcal to cal   c. 326 kcal to kJ   d. 25.6 kcal to J

PROBLEM 6.2

Combustion of 1 g of gasoline releases 11.5 kcal of energy. How many kilojoules of energy is released? How many joules does this correspond to?

TABLE 6.1 Caloric Value for Three Classes of Compounds

	Cal/g	cal/g
Protein	4	4,000
Carbohydrate	4	4,000
Fat	9	9,000

One nutritional Calorie (1 Cal) = 1,000 cal = 1 kcal.

6.1B FOCUS ON THE HUMAN BODY ENERGY AND NUTRITION

When we eat food, the protein, carbohydrates, and fat (lipid) in the food are metabolized to form small molecules that in turn are used to prepare new molecules that cells need for maintenance and growth. This process also generates the energy needed for the organs to function, allowing the heart to beat, the lungs to breathe, and the brain to think.

The amount of stored energy in food is measured using nutritional Calories (upper case C), where 1 Cal = 1,000 cal. Since 1,000 cal = 1 kcal, the following relationships exist.

Nutritional Calorie
1 Cal = 1 kcal
1 Cal = 1,000 cal

Upon metabolism, proteins, carbohydrates, and fat each release a predictable amount of energy, the **caloric value** of the substance. For example, one gram of protein or one gram of carbohydrate typically releases about 4 Cal/g, while fat releases 9 Cal/g (Table 6.1). If we know the amount of each of these substances contained in a food product, we can make a first approximation of the number of Calories it contains by using caloric values as conversion factors, as illustrated in Sample Problem 6.2.

When an individual eats more Calories than are needed for normal bodily maintenance, the body stores the excess as fat. The average body fat content for men and women is about 20% and 25%, respectively. This stored fat can fill the body's energy needs for two or three months. Frequent ingestion of a large excess of Calories results in a great deal of stored fat, causing an individual to be overweight.

SAMPLE PROBLEM 6.2

If a baked potato contains 3 g of protein, a trace of fat, and 23 g of carbohydrates, estimate its number of Calories.

ANALYSIS

Use the caloric value (Cal/g) of each class of molecule to form a conversion factor to convert the number of grams to Calories and add up the results.

SOLUTION

[1] **Identify the original quantity and the desired quantity.**

3 g protein
23 g carbohydrates   ? Cal
original quantities   desired quantity

[2] **Write out the conversion factors.**

- Write out conversion factors that relate the number of grams to the number of Calories for each substance. Each conversion factor must place the unwanted unit, grams, in the denominator so that the units cancel.

Cal–g conversion factor for protein   Cal–g conversion factor for carbohydrates

$$\frac{4 \text{ Cal}}{1 \text{ g protein}}$$ $$\frac{4 \text{ Cal}}{1 \text{ g carbohydrate}}$$

[3] **Set up and solve the problem.**

- Multiply the original quantity by the conversion factor for both protein and carbohydrates and add up the results to obtain the desired quantity.

Calories due to protein   Calories due to carbohydrate

Total Calories $= 3 \text{ g} \times \dfrac{4 \text{ Cal}}{1 \text{ g protein}} + 23 \text{ g carbohydrate} \times \dfrac{4 \text{ Cal}}{1 \text{ g carbohydrate}}$

Grams cancel.   Grams cancel.

$=$ 12 Cal $+$ 92 Cal

Total Calories $=$ 104 Cal, rounded to 100 Cal

Answer

PROBLEM 6.3

How many Calories are contained in one tablespoon of olive oil, which has 14 g of fat?

PROBLEM 6.4

One serving (36 crackers) of wheat crackers contains 6 g of fat, 20 g of carbohydrates, and 2 g of protein. Estimate the number of calories.

6.2 ENERGY CHANGES IN REACTIONS

When molecules come together and react, bonds are broken in the reactants and new bonds are formed in the products. **Breaking a bond requires energy.** For example, 58 kcal of energy is needed to break the chlorine–chlorine bond in a mole of chlorine molecules (Cl_2).

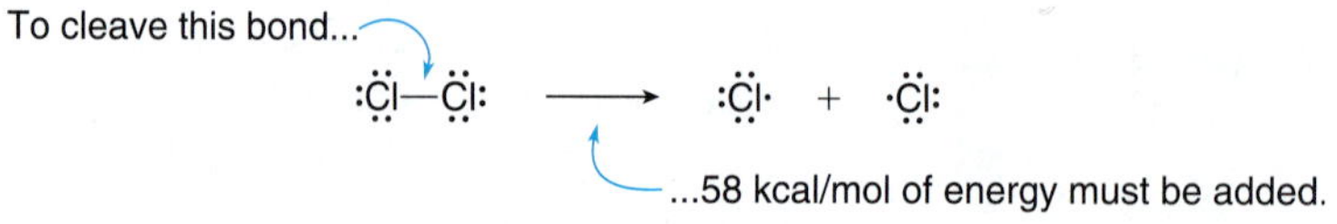

To cleave this bond...

$:\ddot{C}l\!-\!\ddot{C}l: \longrightarrow \ :\ddot{C}l\cdot \ + \ \cdot\ddot{C}l:$

...58 kcal/mol of energy must be added.

In contrast, when the chlorine–chlorine bond is formed, 58 kcal of energy is *released*. The amount of energy needed to break a bond is the same amount that is released when that bond is formed.

When this bond is formed...

$$:\ddot{Cl}\cdot \ + \ \cdot\ddot{Cl}: \ \longrightarrow \ :\ddot{Cl}-\ddot{Cl}:$$

...58 kcal/mol of energy is released.

- **Bond breaking always requires an input of energy and bond formation always releases energy.**

The energy absorbed or released in any reaction is called the **heat of reaction** or the **enthalpy change**, symbolized by ΔH. The heat of reaction is given a positive (+) or negative (–) sign depending on whether energy is absorbed or released.

- **When energy is absorbed, the reaction is said to be *endothermic* and ΔH is positive (+).**
- **When energy is released, the reaction is said to be *exothermic* and ΔH is negative (–).**

Thus, $\Delta H = +58$ kcal/mol for cleaving the Cl—Cl bond and the reaction is endothermic; on the other hand, $\Delta H = -58$ kcal/mol for forming the Cl—Cl bond and that reaction is exothermic. The heat of reaction is reported as the number of kilocalories *per mole*. The cleavage of two moles of chlorine–chlorine bonds requires twice as much energy; that is, (2 mol)(+58 kcal/mol) = +116 kcal.

6.2A BOND DISSOCIATION ENERGY

The heat of reaction (ΔH) for breaking a covalent bond by equally dividing the electrons between the two atoms in the bond is called the **bond dissociation energy.** Because bond breaking requires energy, **bond dissociation energies are always *positive* numbers,** and breaking a covalent bond into the atoms that compose it is always **endothermic.** Since **bond formation always *releases* energy,** forming a bond is **exothermic and ΔH is a *negative* number.** The H—H bond requires +104 kcal/mol to cleave and releases –104 kcal/mol when formed. Table 6.2 lists bond dissociation energies for some simple molecules.

TABLE 6.2 Bond Dissociation Energies (ΔH) for Some Common Bonds (A—B → A· + ·B)	
Bond	**ΔH (kcal/mol)**
H—H	+104
F—F	+38
Cl—Cl	+58
Br—Br	+46
I—I	+36
H—OH	+119
H—F	+136
H—Cl	+103
H—Br	+88
H—I	+71

Bond breaking is **endothermic.**
Energy must be added.

$$H-H \ \longrightarrow \ H\cdot \ + \ \cdot H \qquad \Delta H = +104 \text{ kcal/mol}$$

$$H\cdot \ + \ \cdot H \ \longrightarrow \ H-H \qquad \Delta H = -104 \text{ kcal/mol}$$

Bond making is **exothermic.**
Energy is released.

SAMPLE PROBLEM 6.3

Write the equation for the formation of HCl from H and Cl atoms. Classify the reaction as endothermic or exothermic, and give the ΔH using the values in Table 6.2.

ANALYSIS

Bond formation is exothermic and ΔH is (–). The energy released in forming a bond is (–) the bond dissociation energy.

SOLUTION

Energy is released.

$$H\cdot \ + \ \cdot\ddot{Cl}: \ \longrightarrow \ H-\ddot{Cl}: \qquad \Delta H = -103 \text{ kcal/mol}$$

Bond formation is **exothermic.**

PROBLEM 6.5

Using the values in Table 6.2, give ΔH for each reaction, and classify the reaction as endothermic or exothermic.

a. $H-\ddot{Br}: \ \longrightarrow \ H\cdot \ + \ \cdot\ddot{Br}:$ c. $H-\ddot{O}H \ \longrightarrow \ H\cdot \ + \ \cdot\ddot{O}H$

b. $H\cdot \ + \ \cdot\ddot{F}: \ \longrightarrow \ H-\ddot{F}:$

Bond dissociation energies tell us about bond strength.

- **The *stronger* the bond, the *higher* its bond dissociation energy.**

For example, since the bond dissociation energy for the H—H bond (+104 kcal/mol) is higher than the bond dissociation energy for the Cl—Cl bond (+58 kcal/mol), the H—H bond is stronger.

Bond dissociation energies exhibit periodic trends, much like atomic radius (Section 2.8A) and electronegativity (Section 4.7). In the series, HF, HCl, HBr, and HI, hydrogen is bonded to the first four elements of group 7A (the halogens). According to Table 6.2, the bond dissociation energies of these compounds *decrease* down the column from HF → HCl → HBr → HI. HI has the *weakest* of these four bonds because the valence electrons used by I to form the H—I bond are farther from the nucleus than the valence electrons in Br, Cl, or F. Similarly, HF has the *strongest* of these four bonds because the valence electrons in F are closer to the nucleus than those in Cl, Br, or I. This is a specific example of a general periodic trend.

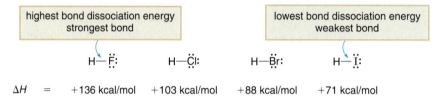

highest bond dissociation energy strongest bond		lowest bond dissociation energy weakest bond	
H—F̈:	H—C̈l:	H—B̈r:	H—Ï:
ΔH = +136 kcal/mol	+103 kcal/mol	+88 kcal/mol	+71 kcal/mol

- **In comparing bonds formed from elements in the same group of the periodic table, bond dissociation energies generally *decrease* going down the column.**

SAMPLE PROBLEM 6.4

Considering the indicated carbon–halogen bonds, which bond is predicted to have the higher bond dissociation energy? Which bond is stronger?

$$H\!-\!\overset{\overset{\displaystyle H}{|}}{\underset{\underset{\displaystyle H}{|}}{C}}\!-\!\ddot{\overset{..}{F}}\!: \qquad H\!-\!\overset{\overset{\displaystyle H}{|}}{\underset{\underset{\displaystyle H}{|}}{C}}\!-\!\ddot{\overset{..}{C}}l\!:$$

ANALYSIS

The higher the bond dissociation energy, the stronger the bond. In comparing bonds to atoms in the same group of the periodic table, bond dissociation energies and bond strength decrease down a column.

SOLUTION

Since Cl is below F in the same group of the periodic table, the C—Cl bond is predicted to have the lower bond dissociation energy, thus making it weaker. The actual values for the bond dissociation energies are given and illustrate that the prediction is indeed true.

lower bond dissociation energy
weaker bond

$$H\!-\!\overset{\overset{\displaystyle H}{|}}{\underset{\underset{\displaystyle H}{|}}{C}}\!-\!\ddot{\overset{..}{F}}\!: \qquad \Delta H = +109\ \text{kcal/mol}$$

$$H\!-\!\overset{\overset{\displaystyle H}{|}}{\underset{\underset{\displaystyle H}{|}}{C}}\!-\!\ddot{\overset{..}{C}}l\!: \qquad \Delta H = +84\ \text{kcal/mol}$$

PROBLEM 6.6

Which indicated bond in each pair of compounds has the higher bond dissociation energy? Which is the stronger bond?

a. $H\!-\!\overset{\overset{\displaystyle H}{|}}{\underset{\underset{\displaystyle H}{|}}{C}}\!-\!\ddot{\overset{..}{I}}\!:$ or $H\!-\!\overset{\overset{\displaystyle H}{|}}{\underset{\underset{\displaystyle H}{|}}{C}}\!-\!\ddot{\overset{..}{B}}r\!:$ b. $H\!-\!OH$ or $H\!-\!SH$

6.2B CALCULATIONS INVOLVING ΔH VALUES

Most reactions involve breaking and forming more than one bond. In these instances, the heat of reaction measures the difference between the energy needed to break bonds in the reactants and the energy released from the bonds formed in the products. In other words, **ΔH indicates the relative strength of bonds broken and formed in a reaction.**

- When ΔH is negative, more energy is released in forming bonds than is needed to break bonds. The bonds formed in the products are *stronger* than the bonds broken in the reactants.

For example, when methane (CH_4) burns in the presence of oxygen (O_2) to form CO_2 and H_2O, 213 kcal/mol of energy is released in the form of heat.

Heat is released.

$$CH_4(g) \ + \ 2\,O_2(g) \ \longrightarrow \ CO_2(g) \ + \ 2\,H_2O(l) \quad \Delta H = -213 \text{ kcal/mol}$$

In this reaction energy is released, ΔH is negative (–), and the reaction is exothermic. The bonds formed are stronger than the bonds broken, since more energy is released in forming the bonds in CO_2 and H_2O than is absorbed in breaking the bonds in CH_4 and O_2. **Since energy is released, the products are *lower* in energy than the reactants.**

The values for ΔH are reported in kilocalories per mole (kcal/mol). This means that the given amount of energy is released (or absorbed) for the molar quantities shown by the coefficients in the balanced chemical equation. Thus, 213 kcal of energy is released when 1 mol of CH_4 reacts with 2 mol of O_2 to form 1 mol of CO_2 and 2 mol of H_2O.

- When ΔH is positive, more energy is needed to break bonds than is released in forming bonds. The bonds broken in the reactants are *stronger* than the bonds formed in the product.

For example, in the process of photosynthesis, green plants use chlorophyll to convert CO_2 and H_2O to glucose ($C_6H_{12}O_6$, a simple carbohydrate) and O_2 and 678 kcal of energy is absorbed.

$$6\,CO_2(g) + 6\,H_2O(l) \longrightarrow C_6H_{12}O_6(aq) + 6\,O_2(g) \quad \Delta H = +678 \text{ kcal/mol}$$

In this reaction energy is absorbed, ΔH is positive (+), and the reaction is endothermic. The bonds broken are stronger than the bonds formed, since more energy is needed to break the bonds in CO_2 and H_2O than is released in forming the bonds in $C_6H_{12}O_6$ and O_2. **Since energy is absorbed, the products are *higher* in energy than the reactants.**

Table 6.3 summarizes the characteristics of energy changes in reactions.

ENVIRONMENTAL NOTE

The CH_4 produced by decomposing waste material in large landfills is burned to produce energy for heating and generating electricity.

Photosynthesis is an endothermic reaction. Energy from sunlight is absorbed in the reaction and stored in the bonds of the products.

TABLE 6.3 Endothermic and Exothermic Reactions	
Endothermic Reaction	**Exothermic Reaction**
• Heat is absorbed.	• Heat is released.
• ΔH is positive.	• ΔH is negative.
• The bonds broken in the reactants are *stronger* than the bonds formed in the products.	• The bonds formed in the products are *stronger* than the bonds broken in the reactants.
• The products are higher in energy than the reactants.	• The products are lower in energy than the reactants.

PROBLEM 6.7

Answer the following questions using the given equation and ΔH. (a) Is heat absorbed or released? (b) Which bonds are stronger, those in the reactants or those in the products? (c) Are the reactants or products lower in energy? (d) Is the reaction endothermic or exothermic?

$$2\,NH_3(g) \longrightarrow 3\,H_2(g) + N_2(g) \qquad \Delta H = +22.0\ \text{kcal/mol}$$

Once we know the ΔH for a balanced chemical reaction, we can use this information to calculate how much energy is absorbed or released for any given amount of reactant or product. A value for ΔH and the coefficients of the balanced equation are used to set up conversion factors, as shown in Sample Problem 6.5. To convert the number of grams of a reactant to the number of kilocalories released, we must use the molar mass, as shown in Sample Problem 6.6.

SAMPLE PROBLEM 6.5

The combustion of propane (C_3H_8) with O_2 according to the given balanced chemical equation releases 531 kcal/mol. How many kilocalories of energy are released when 0.750 mol of propane is burned?

$$C_3H_8(g) + 5\,O_2(g) \longrightarrow 3\,CO_2(g) + 4\,H_2O(l) \qquad \Delta H = -531\ \text{kcal/mol}$$
propane

ANALYSIS   Use the given value of ΔH to set up a conversion factor that relates the kcal of energy released to the number of moles of C_3H_8.

SOLUTION   The given ΔH value is the amount of energy released when 1 mol of C_3H_8 reacts with 5 mol of O_2. Set up a conversion factor that relates kilocalories to moles of C_3H_8, with moles in the denominator to cancel this unwanted unit.

kcal–mol
conversion factor

$$0.750\ \text{mol}\ C_3H_8 \quad \times \quad \frac{531\ \text{kcal}}{1\ \text{mol}\ C_3H_8} \quad = \quad 398\ \text{kcal of energy released}$$

Moles cancel.   **Answer**

PROBLEM 6.8

Given the ΔH and balanced equation in Sample Problem 6.5, how many kilocalories of energy are released when 1.00 mol of O_2 reacts with propane?

SAMPLE PROBLEM 6.6

Using the ΔH and balanced equation for the combustion of propane (C_3H_8) with O_2 shown in Sample Problem 6.5, how many kilocalories of energy are released when 20.0 g of propane is burned?

ANALYSIS   To relate the number of grams of propane to the number of kilocalories of energy released on combustion, two operations are needed: [1] Convert the number of grams to the number of moles using the molar mass. [2] Convert the number of moles to the number of kilocalories using ΔH (kcal/mol) and the coefficients of the balanced chemical equation.

SOLUTION

[1]   **Convert the number of grams of propane to the number of moles of propane.**

- Use the molar mass of the reactant (C_3H_8, molar mass 44.1 g/mol) to write a conversion factor. Multiply the number of grams of propane by the conversion factor to give the number of moles of propane.

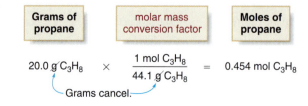

[2] **Convert the number of moles of propane to the number of kilocalories using a kcal–mole conversion factor.**

- Use the ΔH and the number of moles of propane in the balanced chemical equation to write a kcal–mole conversion factor—one mole of propane (C_3H_8) releases 531 kcal of energy. Multiply the number of moles of propane by the conversion factor to give the number of kilocalories of energy released. This process was illustrated in Sample Problem 6.5.

kcal–mol conversion factor

$$0.454 \text{ mol } C_3H_8 \quad \times \quad \frac{531 \text{ kcal}}{1 \text{ mol } C_3H_8} \quad = \quad 241 \text{ kcal of energy released}$$

Moles cancel.

Answer

PROBLEM 6.9

Answer the following questions about the fermentation of glucose ($C_6H_{12}O_6$, molar mass 180.2 g/mol) to ethanol (C_2H_6O) and CO_2.

$$C_6H_{12}O_6(s) \longrightarrow 2\ C_2H_6O(l) + 2\ CO_2(g) \qquad \Delta H = -16 \text{ kcal/mol}$$

glucose   ethanol

a. How many kilocalories of energy are released from 6.0 mol of glucose?

b. How many kilocalories of energy are released when 1.0 mol of ethanol is formed?

c. How many kilocalories of energy are released from 20.0 g of glucose?

6.3 ENERGY DIAGRAMS

On a molecular level, what happens when a reaction occurs? In order for two molecules to react, they must collide, and in the collision, the kinetic energy they possess is used to break bonds. Not every collision between two molecules, however, leads to a reaction. Collisions must have the proper orientation and enough energy for the reaction to occur.

How does the orientation of a collision affect a reaction? Consider a general reaction between two starting materials, **A**—**B** and **C** in which the **A**—**B** bond is broken and a new **B**—**C** bond is formed.

This bond breaks.   This bond forms.

$$A\text{—}B \ + \ C \ \longrightarrow \ A \ + \ B\text{—}C$$

Since **C** forms a new bond with **B**, a reaction occurs only when **C** collides with **B**. If **C** collides with **A**, no bond breaking or bond making can occur, and a reaction does not take place. Since molecular collisions are random events, many collisions are ineffective because they do not place the reacting atoms close to each other.

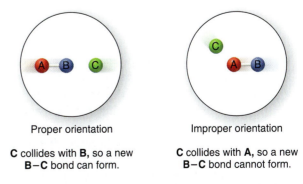

Proper orientation   Improper orientation

C collides with **B**, so a new   **C** collides with **A**, so a new
B–**C** bond can form.   **B**–**C** bond cannot form.

The energy of the reacting molecules also determines whether a particular collision will lead to a reaction. In any given sample, molecules possess a wide range of kinetic energies. Some are fast moving and thus possess more kinetic energy than others that are slow moving. Since the kinetic energy of the reacting molecules **A—B** and **C** provides the energy to break the **A—B** bond, reaction occurs only when the reactants possess sufficient energy.

- **Thus, only collisions that have sufficient energy and proper orientation lead to a reaction.**

The energy changes in a reaction are often illustrated on an **energy diagram,** which plots energy on the vertical axis, and the progress of the reaction—the **reaction coordinate**—on the horizontal axis. The reactants are written on the left side and the products on the right side, and a smooth curve that illustrates how energy changes with time connects them. Let's assume that the products, **A** and **B—C,** are lower in energy than the reactants, **A—B** and **C.**

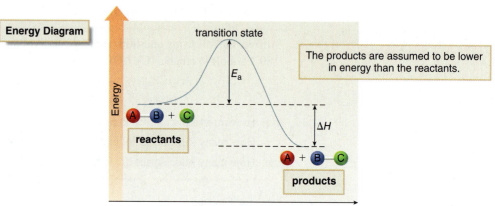

When the reactants **A—B** and **C** approach each other, their electron clouds feel some repulsion, causing an increase in energy until a maximum value is reached. This point is called the **transition state.** In the transition state, the bond between **A** and **B** is partially broken and the bond between **B** and **C** is partially formed. The transition state is located at the top of the energy hill that separates reactants from products.

At the transition state, the bond between **A** and **B** can re-form to regenerate reactants, or the bond between **B** and **C** can form to generate products. As the bond forms between **B** and **C,** the energy decreases until some stable energy minimum is reached. The products are drawn lower in energy than the reactants to reflect the initial assumption about their relative energies.

- The difference in energy between the reactants and the transition state is called the *energy of activation,* symbolized by E_a.

The energy of activation is the minimum amount of energy needed for a reaction to occur. It represents the amount of energy that the reactants must possess for the reaction to take place. The energy of activation is often called the **energy barrier** that must be crossed. The height of the energy barrier—the magnitude of the energy of activation—determines the **reaction rate, how fast the reaction occurs.**

- When the energy of activation is *high,* few molecules have enough energy to cross the energy barrier and the reaction is *slow.*
- When the energy of activation is *low,* many molecules have enough energy to cross the energy barrier and the reaction is *fast.*

The difference in energy between the reactants and products is the ΔH, which is also labeled on the energy diagram. When the products are lower in energy than the reactants, as is the case here, the bonds in the product are stronger than the bonds in the reactants. ΔH is negative (–) and the reaction is exothermic.

▼ **FIGURE 6.1** Energy Diagram for an Endothermic Reaction

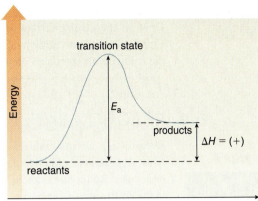

E_a is the energy difference between the reactants and the transition state. ΔH is the difference in energy between the reactants and products. Since the products are higher in energy than the reactants, ΔH is positive (+) and the reaction is endothermic.

Energy diagrams can be drawn for any reaction. In the endothermic reaction shown in Figure 6.1, the products are higher in energy than the reactants.

An energy diagram is a visual tool that helps to illustrate both the rate of a reaction (by the height of the energy barrier), and the energy difference between the reactants and the products. Keep in mind that these two quantities are independent. A large E_a does not tell us anything about the relative energies of the reactants and products.

- The size of E_a determines the rate of a reaction.
- The sign of ΔH determines whether the products or reactants are lower in energy. Reactions are favored in which ΔH is negative and the products are lower in energy, making them more stable than the reactants.

SAMPLE PROBLEM 6.7

Draw an energy diagram for a reaction with a low energy of activation and a ΔH of -10 kcal/mol. Label the axes, reactants, products, transition state, E_a, and ΔH.

ANALYSIS

A low energy of activation means a low energy barrier and a small hill that separates reactants and products. When ΔH is $(-)$, the products are lower in energy than the reactants.

SOLUTION

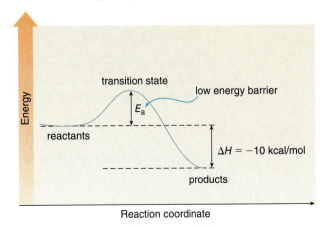

PROBLEM 6.10

Draw an energy diagram for a reaction with a high E_a and a $\Delta H = +20$ kcal/mol.

PROBLEM 6.11

Draw an energy diagram for the following reaction: $H_2O + HCl \longrightarrow H_3O^+ + Cl^-$. Assume the energy of activation is low and the products are lower in energy than the reactants. Clearly label the reactants and products on the energy diagram.

6.4 REACTION RATES

ENVIRONMENTAL NOTE

Gasoline can be safely handled in the air because its reaction with O_2 is slow unless there is a spark to provide energy to initiate the reaction.

Even though we may not realize it, the rate of chemical processes affects many facets of our lives. Aspirin is an effective pain reliever because it rapidly blocks the synthesis of pain-causing molecules. Butter turns rancid with time because its fat molecules are slowly oxidized by oxygen in the air to undesirable by-products. DDT is a persistent environmental pollutant because it does not react appreciably with water, oxygen, or any other chemical with which it comes into contact. All of these processes occur at different rates, resulting in beneficial or harmful effects.

The energy of activation, the minimum amount of energy needed for a reaction to occur, is a fundamental characteristic of a reaction. Some reactions are fast because they have low energies of activation. Other reactions are slow because the energy of activation is high. Even reactions that form products that are lower in energy than the reactants can have high energies of activation. The combustion of gasoline to form CO_2 and H_2O releases a great deal of energy, but it is a very slow reaction without a spark or flame to initiate the reaction.

6.4A HOW CONCENTRATION AND TEMPERATURE AFFECT REACTION RATE

As we learned in Section 6.3, chemical reactions occur when molecules collide. The rate of a reaction depends on the number of collisions and the effectiveness of each collision. How do changes in concentration and temperature affect the reaction rate?

- **Increasing the concentration of the reactants increases the number of collisions, so the reaction rate increases.**
- **Increasing the temperature increases the reaction rate.**

Increasing the temperature increases the reaction rate for two reasons. First, increasing the temperature increases the kinetic energy, which increases the number of collisions. Second, increasing the temperature increases the *average* kinetic energy of the reactants. Because the kinetic energy of colliding molecules is used for bond cleavage, more molecules have sufficient energy to cause bond breaking, and the reaction rate increases. As a general rule, a reaction rate *doubles* for each 10 °C the temperature is *raised*. Similarly, a reaction rate is generally *halved* for each 10 °C the temperature is *lowered*.

We frequently take advantage of the effect of temperature on reaction rate. We store food in a cold refrigerator to slow the reactions that cause food to spoil. On the other hand, we use heat to bake bread to increase the rate of the reactions that occur during baking.

PROBLEM 6.12

Consider the reaction of ozone (O_3) with nitrogen monoxide (NO), which occurs in smog. What effect would each of the following changes have on the rate of this reaction?

$$O_3(g) + NO(g) \longrightarrow O_2(g) + NO_2(g)$$

a. Increasing the concentration of O_3

b. Decreasing the concentration of NO

c. Increasing the temperature

d. Decreasing the temperature

6.4B CATALYSTS

Some reactions do not occur in a reasonable period of time unless a **catalyst** is added.

- A *catalyst* is a substance that speeds up the rate of a reaction. A catalyst is recovered unchanged in a reaction, and it does not appear in the product.

Catalysts accelerate a reaction by lowering the energy of activation (Figure 6.2). They have no effect on the energies of the reactants and products. Thus, the addition of a catalyst lowers E_a but does not affect ΔH.

Metals are often used as catalysts in reactions. For example, ethylene (CH_2=CH_2) does not react appreciably with hydrogen (H_2), but in the presence of palladium (Pd) a rapid reaction occurs and ethane (C_2H_6) is formed as the product. The metal serves as a surface that brings together both reactants, facilitating the reaction. This reaction, called **hydrogenation,** is used in the food industry to prepare margarine, peanut butter, and many other consumer products that contain vegetable oils (Section 13.7).

$$\underset{\text{ethylene}}{\overset{H}{\underset{H}{C}}=\overset{H}{\underset{H}{C}}} \quad + \quad H_2 \quad \xrightarrow[\text{catalyst}]{\text{Pd}} \quad \underset{\text{ethane}}{C_2H_6}$$

6.4C FOCUS ON THE HUMAN BODY
LACTASE, A BIOLOGICAL CATALYST

The catalysts that synthesize and break down biological molecules in living organisms are governed by the same principles as the acids and metals in organic reactions. The catalysts in living organisms, however, are protein molecules called **enzymes.**

Enzymes are discussed in greater detail in Section 21.9.

- *Enzymes* are biological catalysts held together in a very specific three-dimensional shape.

An enzyme contains a region called its **active site** that binds a reactant, which then undergoes a very specific reaction with an enhanced rate. For example, **lactase** is the enzyme that binds **lactose,** the principal carbohydrate in milk (Figure 6.3). Once bound, lactose is converted into two simpler sugars, glucose and galactose. When individuals lack adequate amounts of this enzyme, they are unable to digest lactose, and this causes abdominal cramping and diarrhea.

▼ **FIGURE 6.2** The Effect of a Catalyst on a Reaction

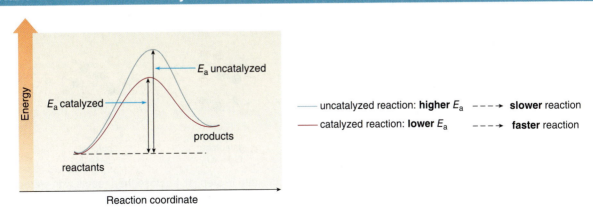

A catalyst lowers the energy of activation, thus increasing the rate of the catalyzed reaction. The energy of the reactants and products is the same in both the uncatalyzed and catalyzed reactions.

PROBLEM 6.13   The reaction of acetic acid ($C_2H_4O_2$) and ethanol (C_2H_6O) to form ethyl acetate ($C_4H_8O_2$) and water occurs only when a small amount of sulfuric acid (H_2SO_4) is added. Is H_2SO_4 a catalyst for this reaction? What effect does H_2SO_4 have on the relative energies of the reactants and products?

6.4D FOCUS ON THE ENVIRONMENT
CATALYTIC CONVERTERS

The combustion of gasoline with oxygen provides a great deal of energy, much like the oxidation reactions of methane and propane discussed in Section 6.2B, and this energy is used to power vehicles. As the number of automobiles increased in the twentieth century, the air pollution they were responsible for became a major problem, especially in congested urban areas.

▼ **FIGURE 6.3** Lactase, an Example of a Biological Catalyst

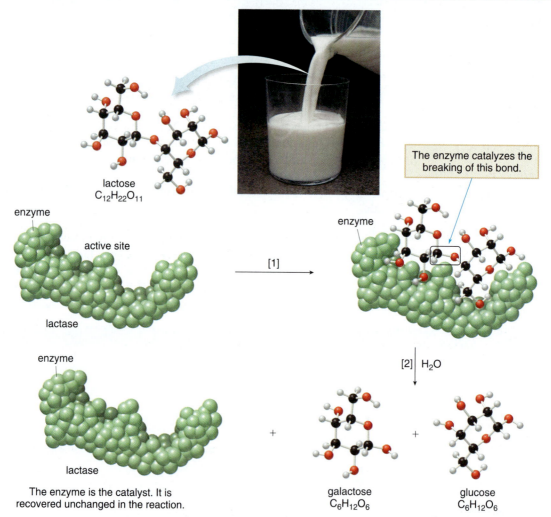

The enzyme lactase binds the carbohydrate lactose ($C_{12}H_{22}O_{11}$) in its active site in step [1]. Lactose then reacts with water to break a bond and form two simpler sugars, galactose and glucose, in step [2]. This process is the first step in digesting lactose, the principal carbohydrate in milk. Without the enzyme, individuals are unable to convert lactose to galactose and glucose, lactose cannot be metabolized, and digestive problems result.

There is a direct link between the poor air quality in large metropolitan areas like Los Angeles and an increase in respiratory diseases.

▼ **FIGURE 6.4** How a Catalytic Converter Works

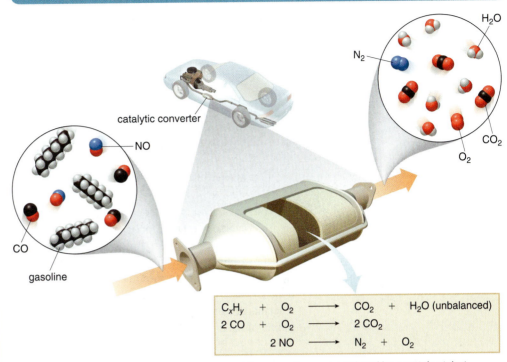

$$C_xH_y \;+\; O_2 \;\longrightarrow\; CO_2 \;+\; H_2O \;(\text{unbalanced})$$
$$2\,CO \;+\; O_2 \;\longrightarrow\; 2\,CO_2$$
$$2\,NO \;\longrightarrow\; N_2 \;+\; O_2$$

Three reactions are catalyzed by a metal catalyst, usually rhodium, platinum, or palladium.

A catalytic converter uses a metal catalyst—rhodium, platinum, or palladium—to catalyze three reactions that clean up the exhaust from an auto engine.

One problem with the auto engines of the 1970s centered on the carbon- and nitrogen-containing by-products emitted in engine exhaust. In addition to CO_2 and H_2O formed during combustion, auto exhaust also contained unreacted gasoline molecules (general formula C_xH_y), the toxic gas carbon monoxide (CO, Section 5.5), and nitrogen monoxide (NO, Section 5.5, a contributing component of acid rain). **Catalytic converters** were devised to clean up these polluting automobile emissions.

The newest catalytic converters, called three-way catalytic converters, use a metal as a surface to catalyze three reactions, as shown in Figure 6.4. Both the unreacted gasoline molecules and carbon monoxide (CO) are oxidized to CO_2 and H_2O. Nitrogen monoxide is also converted to oxygen and nitrogen. In this way, three molecules that contribute to unhealthy smog levels are removed, and the only materials in the engine exhaust are CO_2, H_2O, N_2, and O_2.

PROBLEM 6.14

Nitrogen dioxide, NO_2, also an undesired product formed during combustion, is converted to N_2 and O_2 in a catalytic converter. Write a balanced equation for this reaction.

6.5 EQUILIBRIUM

Thus far in discussing reactions we have assumed that the reactants are completely converted to products. A reaction of this sort is said to **go to completion**. Sometimes, however, a reaction is **reversible**; that is, reactants can come together and form products, and products can come together to re-form reactants.

- A *reversible reaction* can occur in either direction, from reactants to products or from products to reactants.

Consider the reversible reaction of carbon monoxide (CO) with water to form carbon dioxide (CO_2) and hydrogen. Two full-headed arrows ($\rightleftharpoons$) are used to show that the reaction can proceed from left to right and right to left as written.

The **forward** reaction proceeds to the **right.**

$$CO(g) \; + \; H_2O(g) \; \rightleftharpoons \; CO_2(g) \; + \; H_2(g)$$

The **reverse** reaction proceeds to the **left.**

- The *forward* reaction proceeds from left to right as drawn.
- The *reverse* reaction proceeds from right to left as drawn.

When CO and H_2O are mixed together they react to form CO_2 and H_2 by the forward reaction. Once CO_2 and H_2 are formed, they can react together to form CO and H_2O by the reverse reaction. The rate of the forward reaction is rapid at first, and then decreases as the concentration of reactants decreases. The rate of the reverse reaction is slow at first, but speeds up as the concentration of the products increases.

- When the rate of the forward reaction equals the rate of the reverse reaction, the net concentrations of all species do not change and the system is at *equilibrium.*

The forward and reverse reactions do not stop once equilibrium has been reached. The reactants and products continue to react. Since the rates of the forward and reverse reactions are equal, however, the **net concentrations of all reactants and products do *not* change.**

PROBLEM 6.15

Identify the forward and reverse reactions in each of the following reversible reactions.

a. $2 SO_2(g) + O_2(g) \rightleftharpoons 2 SO_3(g)$

b. $N_2(g) + O_2(g) \rightleftharpoons 2 NO(g)$

c. $C_2H_4O_2 + CH_4O \rightleftharpoons C_3H_6O_2 + H_2O$

6.5A   THE EQUILIBRIUM CONSTANT

Because the net concentrations of the reactants and products do not change at equilibrium, they are used to define an expression, the **equilibrium constant, K,** which has a characteristic value for a reaction at a given temperature. When discussing equilibrium it is not the absolute number of moles that is the important quantity, but rather it is the **concentration, the number of moles in a given volume.** Brackets, [ ], are used to symbolize concentration, which is reported in moles per liter (mol/L).

Consider the following general reaction, where **A** and **B** represent reactants, **C** and **D** represent products, and *a, b, c,* and *d* represent the coefficients in the balanced chemical equation.

$$a\,\mathbf{A} \; + \; b\,\mathbf{B} \; \rightleftharpoons \; c\,\mathbf{C} \; + \; d\,\mathbf{D}$$

The equilibrium constant K is the ratio of the concentrations of the products (**C** and **D**) multiplied together, to the concentrations of the reactants (**A** and **B**) multiplied together. Each concentration term is raised to a power equal to the coefficient in the balanced chemical equation.

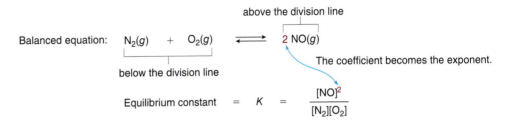

$$K = \frac{[C]^c[D]^d}{[A]^a[B]^b}$$

The expression for the equilibrium constant for any reaction can be written from a balanced equation, as shown for the reaction of N_2 and O_2 to form NO.

above the division line

Balanced equation: $N_2(g) \; + \; O_2(g) \rightleftharpoons 2\;NO(g)$

The coefficient becomes the exponent.

below the division line

Equilibrium constant $= K = \dfrac{[NO]^2}{[N_2][O_2]}$

SAMPLE PROBLEM 6.8   Write the expression for the equilibrium constant for the following balanced equation.

$$2\;CO(g) + O_2(g) \rightleftharpoons 2\;CO_2(g)$$

ANALYSIS   To write an expression for the equilibrium constant, multiply the concentration of the products together and divide this number by the product of the concentrations of the reactants. Each concentration term must be raised to a power equal to the coefficient in the balanced chemical equation.

SOLUTION   The concentration of the sole product, CO_2, is placed in the numerator and raised to the second power since this term has the coefficient "2." The denominator contains concentration terms for the two reactants, CO and O_2, multiplied together. Since the coefficient preceding CO in the balanced equation is "2," this concentration term has an exponent of "2."

$$\text{Equilibrium constant} = K = \frac{[CO_2]^2}{[CO]^2[O_2]}$$

PROBLEM 6.16   Write the expression for the equilibrium constant for each equation.

a. $PCl_3(g) + Cl_2(g) \rightleftharpoons PCl_5(g)$

b. $2\;SO_2(g) + O_2(g) \rightleftharpoons 2\;SO_3(g)$

c. $H_2(g) + Br_2(g) \rightleftharpoons 2\;HBr(g)$

d. $CH_4(g) + 3\;Cl_2(g) \rightleftharpoons CHCl_3(g) + 3\;HCl(g)$

Since K is a constant, it determines the ratio of products to reactants at equilibrium no matter how much of each substance is present at the beginning of a reaction, as shown in Figure 6.5.

6.5B   THE MAGNITUDE OF THE EQUILIBRIUM CONSTANT

The magnitude of the equilibrium constant tells us whether the products or reactants are favored once equilibrium is reached.

- When the equilibrium constant is much greater than one ($K > 1$), the concentration of the products is larger than the concentration of the reactants. We say equilibrium lies to the *right* and favors the *products*.

▼ **FIGURE 6.5** **Equilibrium Concentrations from Different Amounts of Substances**

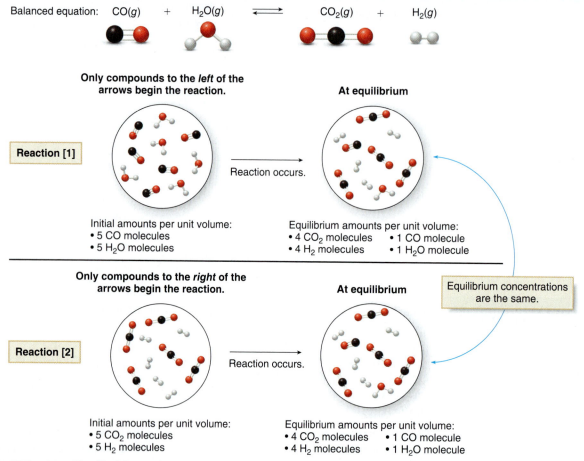

- Reaction [1] begins with only CO and H_2O molecules in equal amounts. At equilibrium there are four times as many product molecules (CO_2 and H_2) as reactant molecules (CO and H_2O).

- Reaction [2] begins with only CO_2 and H_2 molecules in equal amounts. At equilibrium, there are four times as many product molecules (CO_2 and H_2) as reactant molecules (CO and H_2O).

- Thus, it does not matter if the initial reaction mixture contains only compounds to the left of the equilibrium arrows or only compounds to the right of the equilibrium arrows. The equilibrium concentrations of the products and reactants are the same.

When K is greater than 1: $\dfrac{[\text{products}]}{[\text{reactants}]}$ The numerator is larger.

($K > 1$)   Equilibrium favors the **products**.

- When the equilibrium constant is much less than one ($K < 1$), the concentration of the reactants is larger than the concentration of the products. We say equilibrium lies to the *left* and favors the *reactants*.

When K is less than 1: $\dfrac{[\text{products}]}{[\text{reactants}]}$ The denominator is larger.

($K < 1$)   Equilibrium favors the **reactants**.

- When the equilibrium constant is around 1, anywhere in the range of 0.01–100, both reactants and products are present at equilibrium.

When K is approximately equal to 1:   [products] ← Both reactants and products are present.
$(K \approx 1)$ ————
[reactants] ←

For example, the equilibrium constant for the reaction of H_2 and O_2 to form water is much greater than one, so the product, H_2O, is highly favored at equilibrium. A reaction with such a large K essentially goes to completion, with little or no reactants left.

$$2\,H_2(g) \; + \; O_2(g) \; \rightleftharpoons \; 2\,H_2O(g) \qquad K \; = \; 2.9 \times 10^{82}$$

• The product is highly favored since $K > 1$.
• Equilibrium proceeds to the right.

In contrast, the equilibrium constant for the conversion of O_2 to O_3 is much less than one, so the reactant, O_2, is highly favored at equilibrium, and almost no product, O_3, is formed. The relationship between the equilibrium constant and the direction of equilibrium is summarized in Table 6.4.

$$3\,O_2(g) \; \rightleftharpoons \; 2\,O_3(g) \qquad K \; = \; 2.7 \times 10^{-29}$$

• The reactant is highly favored since $K < 1$.
• Equilibrium proceeds to the left.

Generally there is a relationship between the equilibrium constant K and the ΔH of a reaction.

• **The products of a reaction are favored when K is much greater than one ($K > 1$), and ΔH is negative. In other words, equilibrium favors the products when they are lower in energy than the reactants.**

There is, however, no relationship between K and the reaction rate. Some reactions with very large equilibrium constants are still very slow. Moreover, a catalyst may speed up a reaction, but it does not affect the size of K. With a catalyst, equilibrium is reached more quickly, but the relative concentrations of reactants and products do not change.

TABLE 6.4   How the Magnitude of K Relates to the Direction of Equilibrium

Value of K	Position of Equilibrium
$K > 1$	Equilibrium favors the products. Equilibrium lies to the right.
$K < 1$	Equilibrium favors the reactants. Equilibrium lies to the left.
$K \approx 1$	Both the reactants and products are present at equilibrium.

PROBLEM 6.17

Given each equilibrium constant, state whether the reactants are favored, the products are favored, or both the reactants and products are present at equilibrium.

a. 5.0×10^{-4} b. 4.4×10^5 c. 35   d. 0.35

PROBLEM 6.18

Predict whether the reactions with the equilibrium constants in Problem 6.17 are endothermic or exothermic.

PROBLEM 6.19

Consider the following reaction: $2\,H_2(g) + S_2(g) \longrightarrow 2\,H_2S(g)$ with $K = 1.1 \times 10^7$. (a) Write the expression for the equilibrium constant for this reaction. (b) Are the reactants or products favored at equilibrium? (c) Would you predict ΔH to be positive or negative? (d) Are the reactants or products lower in energy? (e) Would you predict this reaction to be fast or slow? Explain your choice.

6.5C CALCULATING THE EQUILIBRIUM CONSTANT

The equilibrium constant for a reaction is an experimentally determined value, and thus it can be calculated if the concentrations of all substances involved in a reaction are measured at equilibrium. Concentrations are reported in moles per liter (mol/L), symbolized as M.

HOW TO   Calculate the Equilibrium Constant for a Reaction

EXAMPLE   Calculate K for the reaction between the general reactants A_2 and B_2. The equilibrium concentrations are as follows: $[A_2]$ = 0.25 M; $[B_2]$ = 0.25 M; $[AB]$ = 0.50 M.

$$A_2 + B_2 \rightleftharpoons 2\ AB$$

Step [1]   Write the expression for the equilibrium constant from the balanced chemical equation, using the coefficients as exponents for the concentration terms.

$$\text{Equilibrium constant} = K = \frac{[AB]^2}{[A_2][B_2]}$$

Step [2]   Substitute the given concentrations in the equilibrium expression and calculate K.

- Since the concentration is always reported in mol/L (M), these units are omitted during the calculation.

$$K = \frac{[AB]^2}{[A_2][B_2]} = \frac{[0.50]^2}{[0.25][0.25]} = \frac{(0.50) \times (0.50)}{0.0625}$$

$$= \frac{0.25}{0.0625} = 4.0$$

Answer

SAMPLE PROBLEM 6.9

Calculate K for the reaction of N_2 and H_2 to form NH_3, with the given balanced equation and the following equilibrium concentrations: $[N_2]$ = 0.12 M; $[H_2]$ = 0.36 M; $[NH_3]$ = 1.1 M.

$$N_2(g) + 3\ H_2(g) \rightleftharpoons 2\ NH_3(g)$$

ANALYSIS   Write an expression for K using the balanced equation and substitute the equilibrium concentrations of all substances in the expression.

SOLUTION

$$K = \frac{[NH_3]^2}{[N_2][H_2]^3} = \frac{(1.1)^2}{(0.12)(0.36)^3} = \frac{1.1 \times 1.1}{0.12 \times 0.36 \times 0.36 \times 0.36}$$

$$= \frac{1.21}{0.0056} = 216 \text{ rounded to } 220$$

Answer

PROBLEM 6.20

Calculate the equilibrium constant for each reaction using the balanced chemical equations and the concentrations of the substances at equilibrium.

a. $CO(g) + H_2O(g) \rightleftharpoons CO_2(g) + H_2(g)$ $[CO]$ = 0.0236 M; $[H_2O]$ = 0.00240 M; $[CO_2]$ = 0.0164 M; $[H_2]$ = 0.0164 M

b. $2\ NO_2(g) \rightleftharpoons N_2O_4(g)$ $[NO_2]$ = 0.0760 M; $[N_2O_4]$ = 1.26 M

6.6 LE CHÂTELIER'S PRINCIPLE

What happens when a reaction is at equilibrium and something changes? For example, what happens when the temperature is increased or some additional reactant is added? **Le Châtelier's principle** is a general rule used to explain the effect of a change in reaction conditions on equilibrium. Le Châtelier's principle states:

- If a chemical system at equilibrium is disturbed or stressed, the system will react in the direction that counteracts the disturbance or relieves the stress.

Let's examine the effect of changes in concentration, temperature, and pressure on equilibrium.

6.6A CONCENTRATION CHANGES

Consider the reaction of carbon monoxide (CO) with oxygen (O_2) to form carbon dioxide (CO_2).

$$2\ CO(g) + O_2(g) \rightleftharpoons 2\ CO_2(g)$$

If the reactants and products are at equilibrium, what happens if the concentration of CO is increased? Now the equilibrium is disturbed and, as a result, the rate of the forward reaction increases to produce more CO_2. We can think of **added reactant as driving the equilibrium to the** *right*.

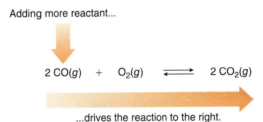

Adding more reactant...

$$2\ CO(g) + O_2(g) \rightleftharpoons 2\ CO_2(g)$$

...drives the reaction to the right.

When the system reaches equilibrium once again, the concentrations of CO_2 and CO are both higher. Since O_2 reacted with the additional CO, its new value at equilibrium is lower. Even though the concentrations of reactants and products are different, the value of K is the same.

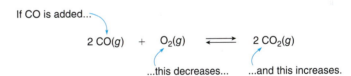

If CO is added...

$$2\ CO(g) + O_2(g) \rightleftharpoons 2\ CO_2(g)$$

...this decreases...   ...and this increases.

What happens, instead, if the concentration of CO_2 is increased? Now the equilibrium is disturbed but there is more product than there should be. As a result, the rate of the reverse reaction increases to produce more of both reactants, CO and O_2. We can think of **added product as driving the equilibrium to the** *left*.

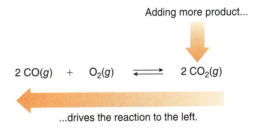

Adding more product...

$$2\ CO(g) + O_2(g) \rightleftharpoons 2\ CO_2(g)$$

...drives the reaction to the left.

When the system reaches equilibrium once again, the concentrations of CO, O_2, and CO_2 are all higher. Even though the concentrations of reactants and products are different, the value of K is the same.

The conversion of nitrogen and hydrogen to ammonia (Sample Problem 6.9) is an exceedingly important reaction since it converts molecular nitrogen from the air to a nitrogen compound (NH_3) that can be used as a fertilizer for plants. In the early twentieth century, German chemist Fritz Haber developed a catalyst that enabled this reaction to proceed quickly to afford favorable yields of NH_3. This process paved the way for large-scale agriculture, which provides food for the world's growing population.

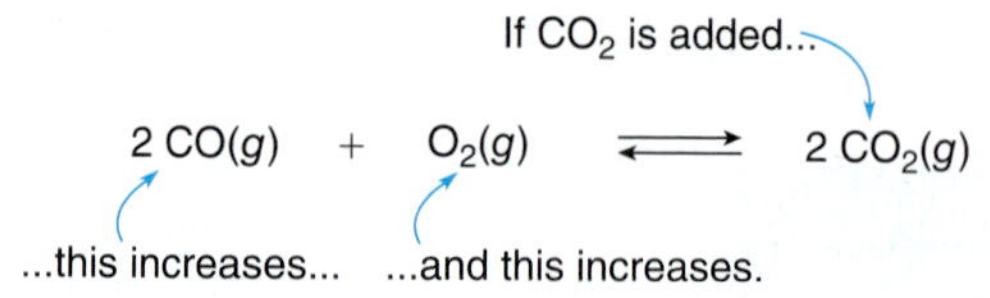

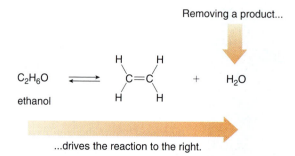

If CO₂ is added...

$$2\ CO(g)\ +\ O_2(g)\ \rightleftharpoons\ 2\ CO_2(g)$$

...this increases...   ...and this increases.

Similar arguments can be made about the effect of decreasing the concentration of a reactant or product. Sometimes, when $K < 1$ and the amount of product at equilibrium is not high, a product is removed from the reaction mixture as it is formed. For example, ethanol (C_2H_6O) can be converted to ethylene ($CH_2{=}CH_2$) and water in the presence of a small amount of acid, but equilibrium does not favor the products.

Removing a product...

C_2H_6O ⇌

$$\underset{H\qquad H}{\overset{H\qquad H}{C{=}C}}\ +\ H_2O$$

ethanol

...drives the reaction to the right.

In this case, water can be removed from the reaction as it is formed. A decrease in concentration of one product results in more of the forward reaction to form more product. **This process drives the equilibrium to the right.** If water is continuously removed, essentially all of the starting material can be converted to product.

SAMPLE PROBLEM 6.10

In which direction is the equilibrium shifted with each of the following concentration changes for the given reaction: (a) increase [SO₂]; (b) increase [SO₃]; (c) decrease [O₂]; (d) decrease [SO₃]?

$$2\ SO_2(g)\ +\ O_2(g)\ \rightleftharpoons\ 2\ SO_3(g)$$

ANALYSIS

Use Le Châtelier's principle to predict the effect of a change in concentration on equilibrium. Adding more reactant or removing product drives the equilibrium to the right. Adding more product or removing reactant drives the equilibrium to the left.

SOLUTION

a. Increasing [SO₂], a reactant, drives the equilibrium to the right to form more product.

b. Increasing [SO₃], a product, drives the equilibrium to the left to form more reactants.

c. Decreasing [O₂], a reactant, drives the equilibrium to the left to form more reactants.

d. Decreasing [SO₃], a product, drives the equilibrium to the right to form more product.

PROBLEM 6.21

In which direction is the equilibrium shifted with each of the following concentration changes for the given reaction: (a) increase [H₂]; (b) increase [HCl]; (c) decrease [Cl₂]; (d) decrease [HCl]?

$$H_2(g)\ +\ Cl_2(g)\ \rightleftharpoons\ 2\ HCl(g)$$

6.6B TEMPERATURE CHANGES

In order to predict what effect a change of temperature has on a reaction, we must know if a reaction is endothermic or exothermic.

- When temperature is increased, the reaction that removes heat is favored.
- When temperature is decreased, the reaction that adds heat is favored.

For example, the reaction of N_2 and O_2 to form NO is an endothermic reaction ($\Delta H = +43$ kcal/mol). **Since an endothermic reaction absorbs heat, increasing the temperature increases the rate of the *forward* reaction to form more product.** The equilibrium shifts to the right.

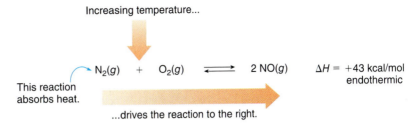

In the exothermic reaction of N_2 and H_2 to form NH_3 ($\Delta H = -22$ kcal/mol), **increasing the temperature increases the rate of the *reverse* reaction to form more reactants.** When temperature is increased, the reaction that absorbs heat, in this case the reverse reaction, predominates and the equilibrium shifts to the left.

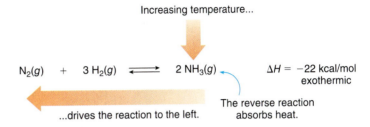

SAMPLE PROBLEM 6.11	The reaction of SO_2 with O_2 to form SO_3 is an exothermic reaction. In which direction is the equilibrium shifted when the temperature is (a) increased; (b) decreased?
ANALYSIS	When temperature is increased, the reaction that removes heat is favored. When temperature is decreased, the reaction that adds heat is favored.
SOLUTION	In an exothermic reaction, the forward reaction releases heat and the reverse reaction absorbs heat. (a) When the temperature is increased, the reverse reaction is favored because it removes heat, so the equilibrium shifts to the reactants (to the left). (b) When the temperature is decreased, the forward reaction is favored because it adds heat, so the equilibrium shifts to the products (to the right).

PROBLEM 6.22	The conversion of H_2O to H_2 and O_2 is an endothermic reaction. In which direction is the equilibrium shifted when the temperature is (a) increased; (b) decreased?
PROBLEM 6.23	The reaction of O_3 with NO to form NO_2 and O_2 is an exothermic reaction. In which direction is the equilibrium shifted when the temperature is (a) increased; (b) decreased?

6.6C PRESSURE CHANGES

When the substances involved in a reaction are gases and the number of moles of reactants and products differs, a change in pressure has an effect on equilibrium.

Pressure, the force per unit area, is discussed in greater detail in Chapter 7.

- **When pressure *increases*, equilibrium shifts in the direction that *decreases* the number of moles in order to decrease pressure.**
- **When pressure *decreases*, equilibrium shifts in the direction that *increases* the number of moles in order to increase pressure.**

In the reaction of N_2 and H_2 to form NH_3, there are four moles of reactants but only two moles of product. When the pressure of the system is increased, the equilibrium shifts to the right since there are fewer moles of product.

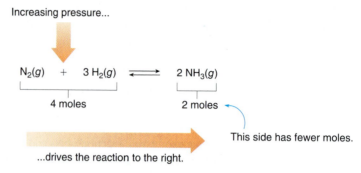

Increasing pressure...

$$N_2(g) \ + \ 3\,H_2(g) \ \rightleftharpoons \ 2\,NH_3(g)$$

4 moles   2 moles

This side has fewer moles.

...drives the reaction to the right.

In contrast, when the pressure of the system is decreased, the equilibrium shifts to the left since there are more moles of reactants.

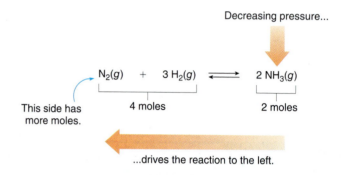

Decreasing pressure...

$$N_2(g) \ + \ 3\,H_2(g) \ \rightleftharpoons \ 2\,NH_3(g)$$

4 moles   2 moles

This side has more moles.

...drives the reaction to the left.

SAMPLE PROBLEM 6.12

In which direction is the equilibrium shifted in the following reaction when the pressure is (a) increased; (b) decreased?

$$H-C \equiv C-H(g) + 2\,H_2(g) \rightleftharpoons C_2H_6(g)$$

ANALYSIS

When pressure is increased, the equilibrium shifts in the direction that decreases the number of moles. When pressure is decreased, the equilibrium shifts in the direction that increases the number of moles.

SOLUTION

In this reaction, there are three moles of reactants and only one mole of product. (a) Increasing the pressure shifts the equilibrium to the right, to the side of fewer moles. (b) Decreasing the pressure shifts the equilibrium to the left, to the side of more moles.

PROBLEM 6.24

In which direction is the equilibrium shifted in the following reaction when the pressure is (a) increased; (b) decreased?

$$C_2H_4(g) + Cl_2(g) \rightleftharpoons C_2H_4Cl_2(g)$$

Table 6.5 summarizes the effects of changes in reaction conditions on the direction of an equilibrium.

6.7 FOCUS ON THE HUMAN BODY
BODY TEMPERATURE

The human body is an enormously complex organism that illustrates important features of energy and reaction rates. At any moment, millions of reactions occur in the body, when nutrients are metabolized and new cell materials are synthesized.

Normal body temperature, 37 °C, reflects a delicate balance between the amount of heat absorbed and released in all of the reactions and other processes. Since reaction rate increases with increasing temperature, it is crucial to maintain the right temperature for proper body function. When

TABLE 6.5 The Effects of Changes in Conditions on Equilibrium

Change	Effect on Equilibrium
Concentration • Adding reactant • Removing reactant • Adding product • Removing product	Equilibrium favors the products. Equilibrium favors the reactants. Equilibrium favors the reactants. Equilibrium favors the products.
Temperature • Increasing temperature • Decreasing temperature	In an endothermic reaction, equilibrium favors the products. In an exothermic reaction, equilibrium favors the reactants. In an endothermic reaction, equilibrium favors the reactants. In an exothermic reaction, equilibrium favors the products.
Pressure • Increasing pressure • Decreasing pressure	Equilibrium favors the side that has fewer moles. Equilibrium favors the side that has more moles.

temperature increases, reactions proceed at a faster rate. An individual must breathe more rapidly and the heart must pump harder to supply oxygen for the faster metabolic processes. When temperature decreases, reactions slow down, less heat is generated in exothermic reactions, and it becomes harder and harder to maintain an adequate body temperature.

Thermoregulation—regulating temperature—is a complex mechanism that involves the brain, the circulatory system, and the skin (Figure 6.6). Temperature sensors in the skin and body core

▼ FIGURE 6.6 Temperature Regulation in the Body

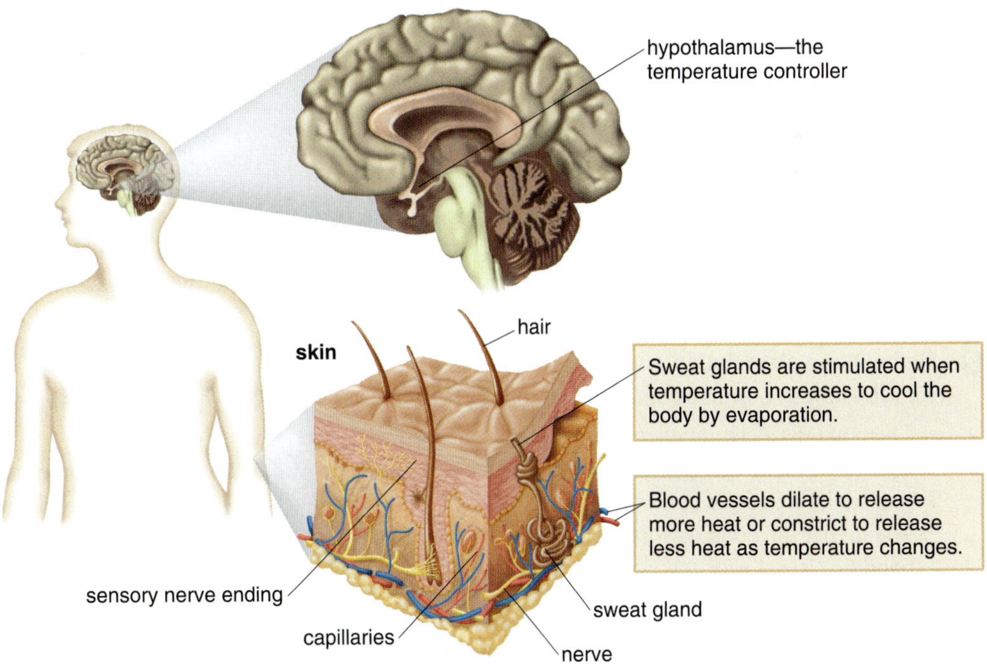

hypothalamus—the temperature controller

hair

skin

Sweat glands are stimulated when temperature increases to cool the body by evaporation.

Blood vessels dilate to release more heat or constrict to release less heat as temperature changes.

sensory nerve ending

capillaries

sweat gland

nerve

When the temperature in the environment around the body changes, the body works to counteract the change, in a method similar to Le Châtelier's principle. The hypothalamus acts as a thermostat, which signals the body to respond to temperature changes. When the temperature increases, the body must dissipate excess heat by dilating blood vessels and sweating. When the temperature decreases, blood vessels constrict and the body shivers.

signal when there is a temperature change. The hypothalamus region of the brain, in turn, responds to changes in its environment in a process that is reminiscent of Le Châtelier's principle.

When the temperature increases, the body must somehow rid itself of excess heat. Blood vessels near the surface of the skin are dilated to release more heat. Sweat glands are stimulated so the body can be cooled by the evaporation of water from the skin's surface.

When the temperature decreases, the body must generate more heat as well as slow down the loss of heat to the surroundings. Blood vessels constrict to reduce heat loss from the skin and muscles shiver to generate more heat.

An infection in the body is often accompanied by a fever; that is, the temperature in the body increases. A fever is part of the body's response to increase the rates of defensive reactions that kill bacteria. The respiratory rate and heart rate increase to supply more oxygen needed for faster reactions.

CHAPTER HIGHLIGHTS

KEY TERMS

Active site (6.4)
Bond dissociation energy (6.2)
Calorie (6.1)
Catalyst (6.4)
Endothermic reaction (6.2)
Energy (6.1)
Energy diagram (6.3)
Energy of activation (6.3)

Enthalpy (6.2)
Enzyme (6.4)
Equilibrium (6.5)
Equilibrium constant (6.5)
Exothermic reaction (6.2)
Forward reaction (6.5)
Heat of reaction (6.2)
Joule (6.1)

Kinetic energy (6.1)
Law of conservation of energy (6.1)
Le Châtelier's principle (6.6)
Potential energy (6.1)
Reaction rate (6.3)
Reverse reaction (6.5)
Reversible reaction (6.5)
Transition state (6.3)

KEY CONCEPTS

❶ What is energy and what units are used to measure energy? (6.1)

- Energy is the capacity to do work. Kinetic energy is the energy of motion, whereas potential energy is stored energy.
- Energy is measured in calories (cal) or joules (J), where 1 cal = 4.184 J.
- One nutritional calorie (Cal) = 1 kcal = 1,000 cal.

❷ Define bond dissociation energy and describe its relationship to bond strength. (6.2)

- The bond dissociation energy is the energy needed to break a covalent bond by equally dividing the electrons between the two atoms in the bond.
- The higher the bond dissociation energy, the stronger the bond.

❸ What is the heat of reaction and what is the difference between an endothermic and an exothermic reaction? (6.2)

- The heat of reaction, also called the enthalpy and symbolized by ΔH, is the energy absorbed or released in a reaction.

- In an endothermic reaction, energy is absorbed, ΔH is positive (+), and the products are higher in energy than the reactants. The bonds in the reactants are stronger than the bonds in the products.
- In an exothermic reaction, energy is released, ΔH is negative (−), and the reactants are higher in energy than the products. The bonds in the products are stronger than the bonds in the reactants.

❹ What are the important features of an energy diagram? (6.3)

- An energy diagram illustrates the energy changes that occur during the course of a reaction. Energy is plotted on the vertical axis and reaction coordinate is plotted on the horizontal axis. The transition state is located at the top of the energy barrier that separates the reactants and products.
- The energy of activation is the energy difference between the reactants and the transition state. The higher the energy of activation, the slower the reaction.
- The difference in energy between the reactants and products is the ΔH.

5 **How do temperature, concentration, and catalysts affect the rate of a reaction? (6.4)**
- The rate of a reaction depends on the number of collisions and the effectiveness of the collisions of the reacting molecules.
- Increasing the temperature and concentration increases the reaction rate.
- A catalyst speeds up the rate of a reaction without affecting the energies of the reactants and products. Enzymes are biological catalysts that increase the rate of reactions in living organisms. Catalytic converters use a catalyst to convert automobile engine exhaust to environmentally cleaner products.

6 **What are the basic features of equilibrium? (6.5)**
- At equilibrium, the rates of the forward and reverse reactions in a reversible reaction are equal and the net concentrations of all substances do not change.
- The equilibrium constant for a reaction $aA + bB \rightleftharpoons cC + dD$ is written as $K = [C]^c[D]^d/[A]^a[B]^b$.
- The magnitude of K tells the relative amount of reactants and products. When $K > 1$, the products are favored; when $K < 1$, the reactants are favored; when $K \approx 1$, both reactants and products are present at equilibrium.

7 **How does Le Châtelier's principle predict what happens when equilibrium is disturbed? (6.6)**
- Le Châtelier's principle states that a system at equilibrium reacts in such a way as to counteract any disturbance to the equilibrium. How changes in concentration, temperature, and pressure affect equilibrium are summarized in Table 6.5.
- Catalysts increase the rate at which equilibrium is reached, but do not alter the amount of any substance involved in the reaction.

8 **How can the principles that describe equilibrium and reaction rates be used to understand the regulation of body temperature? (6.7)**
- Increasing temperature increases the rates of the reactions in the body.
- When temperature is increased, the body dissipates excess heat by dilating blood vessels and sweating. When temperature is decreased, blood vessels constrict to conserve heat, and the body shivers to generate more heat.

PROBLEMS

Selected in-chapter and end-of-chapter problems have brief answers provided in Appendix B.

Energy

6.25 What is the difference between kinetic energy and potential energy? Give an example of each type.

6.26 What is the difference between the law of conservation of energy and the law of conservation of mass (Section 5.1)?

6.27 What is the difference between a calorie and a joule?

6.28 What is the difference between a calorie and a Calorie?

6.29 Riding a bicycle at 12–13 miles per hour uses 563 Calories in an hour. Convert this value to (a) calories; (b) kilocalories; (c) joules; (d) kilojoules.

6.30 Running at a rate of 6 mi/h uses 704 Calories in an hour. Convert this value to (a) calories; (b) kilocalories; (c) joules; (d) kilojoules.

6.31 Carry out each of the following conversions.
- a. 50 cal to kcal
- b. 56 cal to kJ
- c. 0.96 kJ to cal
- d. 4,230 kJ to cal

6.32 Carry out each of the following conversions.
- a. 5 kcal to cal
- b. 2,560 cal to kJ
- c. 1.22 kJ to cal
- d. 4,230 J to kcal

6.33 Estimate the number of Calories in two tablespoons of peanut butter, which contains 16 g of protein, 7 g of carbohydrates, and 16 g of fat.

6.34 Estimate the number of Calories in a serving of oatmeal that has 4 g of protein, 19 g of carbohydrates, and 2 g of fat.

6.35 A can of soda contains 120 Calories, and no protein or fat. How many grams of carbohydrates are present in each can?

6.36 Alcohol releases 29.7 kJ/g when it burns. Convert this value to the number of Calories per gram.

6.37 Which food has more Calories: 3 oz of salmon, which contains 17 g of protein and 5 g of fat, or 3 oz of chicken, which contains 20 g of protein and 3 g of fat?

6.38 Which food has more Calories: one egg, which contains 6 g of protein and 6 g of fat, or 1 cup of nonfat milk, which contains 9 g of protein and 12 g of carbohydrates?

Bond Dissociation Energy and ΔH

6.39 What is the difference between an endothermic reaction and an exothermic reaction?

6.40 What is the difference between ΔH and the bond dissociation energy?

6.41 Based on the location of the elements in the periodic table, which species in each pair has the stronger bond?
- a. Br_2 or Cl_2
- b. Cl_2 or I_2
- c. HF or HBr

6.42 Using the given bond dissociation energies, rank the indicated bonds in order of increasing strength.

$\Delta H = +104$ kcal/mol $\Delta H = +98$ kcal/mol $\Delta H = +125$ kcal/mol

6.43 Do each of the following statements describe an endothermic or exothermic reaction?
a. ΔH is a negative value.
b. The energy of the reactants is lower than the energy of the products.
c. Energy is absorbed in the reaction.
d. The bonds in the products are stronger than the bonds in the reactants.

6.44 Do each of the following statements describe an endothermic or exothermic reaction?
a. ΔH is a positive value.
b. The energy of the products is lower than the energy of the reactants.
c. Energy is released in the reaction.
d. The bonds in the reactants are stronger than the bonds in the products.

6.45 The combustion of coal with oxygen forms CO_2 and releases 94 kcal of energy.

$$C(s) + O_2(g) \longrightarrow CO_2(g) \qquad \Delta H = -94 \text{ kcal/mol}$$

a. How much energy is released when 2.5 mol of C reacts?
b. How much energy is released when 3.0 mol of O_2 reacts?
c. How much energy is released when 25.0 g of C reacts?

6.46 Ammonia (NH_3) decomposes to hydrogen and nitrogen and 22.0 kcal/mol of energy is absorbed.

$$2 NH_3(g) \longrightarrow 3 H_2(g) + N_2(g) \qquad \Delta H = +22.0 \text{ kcal/mol}$$

a. How much energy is absorbed when 1 mol of N_2 is formed?
b. How much energy is absorbed when 1 mol of NH_3 reacts?
c. How much energy is absorbed when 3.50 g of NH_3 reacts?

6.47 The metabolism of glucose with oxygen forms CO_2 and H_2O and releases 678 kcal/mol of energy.

$$C_6H_{12}O_6(aq) + 6 O_2(g) \rightleftharpoons 6 CO_2(g) + 6 H_2O(l)$$
glucose
(molar mass 180.2 g/mol)

a. Are the bonds formed in the products stronger or weaker than the bonds broken in the reactants?
b. How much energy is released when 4.00 mol of glucose is metabolized?
c. How much energy is released when 3.00 mol of O_2 reacts?
d. How much energy is released when 10.0 g of glucose reacts?

6.48 Ethanol (C_2H_6O), a gasoline additive, is formed by the reaction of ethylene ($CH_2\!=\!CH_2$) with water. The ΔH for this reaction is -9.0 kcal/mol.

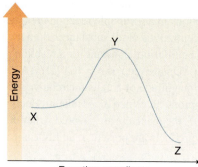

ethylene   ethanol

a. How much energy is released when 3.5 mol of ethylene reacts?
b. How much energy is released when 0.50 mol of H_2O reacts?
c. How much energy is released when 15.0 g of ethylene reacts?
d. How much energy is released when 2.5 g of ethanol is formed?

Energy Diagrams

6.49 What is the difference between the energy of activation and the transition state?

6.50 What is the difference between E_a and ΔH?

6.51 Consider the energy diagram drawn below.

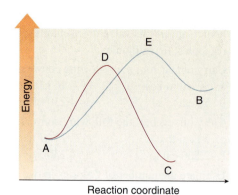

a. Which point on the graph corresponds to reactants?
b. Which point on the graph corresponds to products?
c. Which point on the graph corresponds to the transition state?
d. The difference in energy between which two points equals the energy of activation?
e. The difference in energy between which two points equals the ΔH?
f. Which point is highest in energy?
g. Which point is lowest in energy?

6.52 Compound **A** can be converted to either **B** or **C**. The energy diagrams for both processes are drawn on the graph below.

a. Label each reaction as endothermic or exothermic.

b. Which reaction is faster?

c. Which reaction generates the product lower in energy?

d. Which points on the graphs correspond to transition states?

e. Label the energy of activation for each reaction.

f. Label the ΔH for each reaction.

6.53 Draw an energy diagram that fits each description.

a. an endothermic reaction with a high E_a

b. a reaction that has a low E_a and ΔH is negative

c. a slow reaction in which the products are at a lower energy than the reactants

6.54 Draw an energy diagram that fits each description.

a. an exothermic reaction with a high E_a

b. a reaction with a low E_a and a positive value of ΔH

c. a fast reaction in which the products are at a lower energy than the reactants

6.55 Draw an energy diagram for the following reaction in which $\Delta H = -12$ kcal/mol and $E_a = 5$ kcal: $A_2 + B_2 \longrightarrow 2$ AB. Label the axes, reactants, products, transition state, E_a, and ΔH. Is the reaction endothermic or exothermic?

6.56 Draw an energy diagram for the following reaction in which $\Delta H = +13$ kcal/mol and $E_a = 21$ kcal: $A + B \longrightarrow C$. Label the axes, reactants, products, transition state, E_a, and ΔH. Are the products or reactants higher in energy?

Reaction Rates

6.57 How does collision orientation affect the rate of a reaction?

6.58 Explain why a high energy of activation causes a reaction to be slow.

6.59 State two reasons why increasing temperature increases reaction rate.

6.60 Why does decreasing concentration decrease the rate of a chemical reaction?

6.61 Which value (if any) in each pair corresponds to a faster reaction? Explain your choice.

a. $E_a = 10$ kcal or $E_a = 1$ kcal

b. $K = 10$ or $K = 100$

c. $\Delta H = -2$ kcal/mol or $\Delta H = +2$ kcal/mol

6.62 Which value (if any) in each pair corresponds to a faster reaction? Explain your choice.

a. $E_a = 0.10$ kcal or $E_a = 1$ kcal

b. $K = 10$ or $K = 0.001$

c. $\Delta H = -25$ kcal/mol or $\Delta H = -2$ kcal/mol

6.63 Which of the following affect the rate of a reaction: (a) K; (b) E_a; (c) temperature?

6.64 Which of the following affect the rate of a reaction: (a) concentration; (b) ΔH; (c) energy difference between the reactants and the transition state?

6.65 How does a catalyst affect each of the following: (a) reaction rate; (b) ΔH; (c) E_a; (d) K; (e) relative energy of the reactants and products?

6.66 What is the difference between a catalyst and an enzyme?

Equilibrium

6.67 What is the difference between the forward and reverse reactions?

6.68 What is the difference between a reversible reaction and the reverse reaction?

6.69 Given each value of the equilibrium constant, are the reactants or products favored at equilibrium?

a. $K = 5.2 \times 10^3$ c. $K = 0.002$

b. $\Delta H = -27$ kcal/mol   d. $\Delta H = +2$ kcal/mol

6.70 Given each value of the equilibrium constant, are the reactants or products favored at equilibrium?

a. $K = 5.2 \times 10^{-6}$ c. $K = 10,000$

b. $\Delta H = +16$ kcal/mol   d. $\Delta H = -21$ kcal/mol

6.71 How is the magnitude of K related to the sign of ΔH?

6.72 How are the sign and magnitude of ΔH affected by the presence of a catalyst?

6.73 Write an expression for the equilibrium constant for each reaction.

a. $2\ NO(g) + O_2(g) \rightleftharpoons 2\ NO_2(g)$

b. $CH_4(g) + 2\ Br_2(g) \rightleftharpoons CH_2Br_2(g) + 2\ HBr(g)$

6.74 Write an expression for the equilibrium constant for each reaction.

a. $HBr + H_2O \rightleftharpoons H_3O^+ + Br^-$

b. $2\ HCl(g) \rightleftharpoons H_2(g) + Cl_2(g)$

6.75 Write an expression for the equilibrium constant for each reaction.

a. $C_2H_4O_2 + 2\ H_2 \rightleftharpoons C_2H_6O + H_2O$

b. $2\ N_2O_5(g) \rightleftharpoons 4\ NO_2(g) + O_2(g)$

6.76 Write an expression for the equilibrium constant for each reaction.

a. $N_2O_4(g) \rightleftharpoons 2\ NO_2(g)$

b. $4\ NH_3(g) + 3\ O_2(g) \rightleftharpoons 2\ N_2(g) + 6\ H_2O(g)$

6.77 Use each expression for the equilibrium constant to write a chemical equation.

a. $K = \dfrac{[A_2]}{[A]^2}$

b. $K = \dfrac{[AB_3]^2}{[A_2][B_2]^3}$

6.78 Use each expression for the equilibrium constant to write a chemical equation.

a. $K = \dfrac{[AB_2]^2}{[A_2][B_2]^2}$

b. $K = \dfrac{[A_2B_3]}{[A]^2[B]^3}$

6.79 Consider the following reaction: $2 HBr(g) \rightleftharpoons H_2(g) + Br_2(g)$ with $K = 5.3 \times 10^{-20}$.
   a. Write the expression for the equilibrium constant for this reaction.
   b. Are the reactants or products favored at equilibrium?
   c. Would you predict ΔH to be positive or negative?
   d. Are the reactants or products lower in energy?
   e. Would you predict this reaction to be fast or slow? Explain your choice.

6.80 Consider the following reaction: $CO(g) + Cl_2(g) \rightleftharpoons COCl_2(g)$ with $K = 6.6 \times 10^{11}$.
   a. Write the expression for the equilibrium constant for this reaction.
   b. Are the reactants or products favored at equilibrium?
   c. Would you predict ΔH to be positive or negative?
   d. Are the reactants or products lower in energy?
   e. Would you predict this reaction to be fast or slow? Explain your choice.

6.81 Consider the following reaction.

$$CO(g) + H_2O(g) \rightleftharpoons CO_2(g) + H_2(g)$$

   a. Write the expression for the equilibrium constant.
   b. Calculate K using the following concentrations of each substance at equilibrium: $[CO] = 0.090$ M; $[H_2O] = 0.12$ M; $[CO_2] = 0.15$ M; $[H_2] = 0.30$ M.

6.82 Consider the following reaction.

$$H_2(g) + I_2(g) \rightleftharpoons 2 HI(g)$$

   a. Write the expression for the equilibrium constant.
   b. Calculate K using the following concentrations of each substance at equilibrium: $[H_2] = 0.95$ M; $[I_2] = 0.78$ M; $[HI] = 0.27$ M.

Le Châtelier's Principle

6.83 Consider the reaction of $N_2(g) + O_2(g) \rightleftharpoons 2 NO(g)$. What happens to the concentration of each substance when (a) $[O_2]$ is increased; (b) $[NO]$ is increased?

6.84 Consider the reaction of $H_2(g) + F_2(g) \rightleftharpoons 2 HF(g)$. What happens to the concentration of each substance when (a) $[H_2]$ is decreased; (b) $[HF]$ is increased?

6.85 Consider the endothermic conversion of oxygen to ozone: $3 O_2(g) \rightleftharpoons 2 O_3(g)$. What effect does each of the following changes have on the direction of equilibrium?
   a. decrease $[O_3]$ d. decrease temperature
   b. decrease $[O_2]$ e. add a catalyst
   c. increase $[O_3]$ f. increase pressure

6.86 Consider the exothermic reaction: $H_2(g) + I_2(g) \rightleftharpoons 2 HI(g)$. What effect does each of the following changes have on the direction of equilibrium?
   a. decrease $[HI]$ d. increase temperature
   b. increase $[H_2]$ e. decrease temperature
   c. decrease $[I_2]$ f. increase pressure

6.87 Consider the exothermic reaction: $C_2H_4(g) + Cl_2(g) \rightleftharpoons C_2H_4Cl_2(g)$. What effect does each of the following changes have on the direction of equilibrium?
   a. increase $[C_2H_4]$ d. decrease pressure
   b. decrease $[Cl_2]$ e. increase temperature
   c. decrease $[C_2H_4Cl_2]$ f. decrease temperature

6.88 Consider the endothermic reaction: $2 NH_3(g) \rightleftharpoons 3 H_2(g) + N_2(g)$. What effect does each of the following changes have on the direction of equilibrium?
   a. increase $[NH_3]$ d. increase temperature
   b. decrease $[N_2]$ e. decrease temperature
   c. increase $[H_2]$ f. increase pressure

General Problems

6.89 Consider the gas-phase reaction of ethylene ($CH_2{=}CH_2$) with hydrogen to form ethane (C_2H_6), which occurs in the presence of a palladium catalyst (Section 6.4B).
   a. Write the expression for the equilibrium constant for this reaction.
   b. If $\Delta H = -28$ kcal/mol, are the products or reactants higher in energy?
   c. Which is likely to be true about the equilibrium constant for the reaction: $K > 1$ or $K < 1$?
   d. How much energy is released when 20.0 g of ethylene reacts?
   e. What happens to the rate of the reaction if the concentration of ethylene is increased?
   f. What happens to the equilibrium when each of the following changes occurs: [1] an increase in $[H_2]$; [2] a decrease in $[C_2H_6]$; [3] an increase in temperature; [4] an increase in pressure; [5] removal of the palladium catalyst?

6.90 Methanol (CH_4O), which is used as a fuel in race cars, burns in oxygen (O_2) to form CO_2 and H_2O.
   a. Write a balanced equation for this reaction.
   b. Write the expression for the equilibrium constant for this reaction.
   c. If $\Delta H = -174$ kcal/mol, are the products or reactants higher in energy?
   d. How much energy is released when 10.0 g of methanol is burned?
   e. Although this reaction is exothermic, the reaction is very slow unless a spark or flame initiates the reaction. Explain how this can be possible.

Applications

6.91 What is the role of lactase and why is it important in the human body?

6.92 How does a catalytic converter clean up automobile emissions?

6.93 A patient receives 2,000 mL of a glucose solution that contains 5 g of glucose in 100 mL. How many Calories does the glucose, a simple carbohydrate, contain?

6.94 The reaction of salicylic acid with acetic acid yields aspirin and water according to the given balanced equation. Since the equilibrium constant for this reaction is close to one, both reactants and products are present at equilibrium. If the reaction has a small negative value of ΔH, suggest ways that this equilibrium could be driven to the right to favor products.

$$C_7H_6O_3 \;+\; C_2H_4O_2 \;\rightleftharpoons\; C_9H_8O_4 \;+\; H_2O$$

salicylic acid   acetic acid   aspirin

6.95 Walking at a brisk pace burns off about 280 Cal/h. How long would you have to walk to burn off the Calories obtained from eating a cheeseburger that contained 32 g of protein, 29 g of fat, and 34 g of carbohydrates?

6.96 How many kilocalories does a runner expend when he runs for 4.5 h and uses 710 Cal/h? How many pieces of pizza that each contain 12 g of protein, 11 g of fat, and 30 g of carbohydrates could be eaten after the race to replenish these Calories?

6.97 The amount of energy released when a fuel burns is called its heat content. The heat content of fuels is often reported in kcal/g not kcal/mol so that fuels with different molar masses can be compared on a mass basis. The heat content of propane (C_3H_8), used as the fuel in gas grills, is 531 kcal/mol, while the heat content of butane (C_4H_{10}), used in lighters, is 688 kcal/mol. Show that the heat content of these two fuels is similar when converted to kcal/g.

6.98 One mole of ethanol (C_2H_6O) releases 327 kcal when burned, whereas one mole of hydrogen (H_2) releases only 68.4 kcal/mol. How many kcal/g are released when each of these fuels is burned? On a per gram basis, which substance would be a better source of energy?

CHALLENGE QUESTIONS

6.99 Let's assume that a gallon of gasoline contains pure octane (C_8H_{18}) and has a density of 0.700 g/mL. When octane is burned, it releases 1,303 kcal/mol of energy. How many kilocalories of energy are released from burning one gallon of gasoline?

6.100 Determine what bonds are broken and formed in the following reaction: $H_2 + Cl_2 \longrightarrow 2\ HCl$. How could you use the bond dissociation energies in Table 6.2 to calculate the ΔH for this reaction?

7

CHAPTER GOALS

In this chapter you will learn how to:

1. Measure pressure and convert one unit of pressure to another

2. Describe the relationship between the pressure, volume, and temperature of a gas using gas laws

3. Describe the relationship between the volume and number of moles of a gas

4. Write the equation for the ideal gas law and use it in calculations

5. Use Dalton's law to determine the partial pressure and total pressure of a gas mixture

6. Determine what types of intermolecular forces a compound possesses, and how these forces determine a compound's boiling point and melting point

7. Describe the properties of a liquid, including vapor pressure, viscosity, and surface tension

8. Describe the features of different types of solids

9. Describe the energy changes that accompany changes of state

Scuba divers must carefully plan the depth and duration of their dives to avoid "the bends," a dangerous condition caused by the formation of nitrogen gas bubbles in the bloodstream.

GASES, LIQUIDS, AND SOLIDS

IN Chapter 7 we study the properties of gases, liquids, and solids. Why is air pulled into the lungs when we expand our rib cage and diaphragm? Why does a lid pop off a container of food when it is heated in the microwave? Why does sweating cool down the body? To answer questions of this sort, we must understand the molecular properties of the three states of matter, as well as the energy changes involved when one state is converted to another.

7.1 INTRODUCTION

As we first learned in Section 1.2, matter exists in three common states—**gas, liquid,** and **solid.**

- A gas consists of particles that are far apart and move rapidly and independently from each other.
- A liquid consists of particles that are much closer together but are still somewhat disorganized since they can move about. The particles in a liquid are close enough that they exert a force of attraction on each other.
- A solid consists of particles—atoms, molecules, or ions—that are close to each other and are often highly organized. The particles in a solid have little freedom of motion and are held together by attractive forces.

As shown in Figure 7.1, air is composed largely of N_2 and O_2 molecules, along with small amounts of argon (Ar), carbon dioxide (CO_2), and water molecules that move about rapidly. Liquid water is composed of H_2O molecules that have no particular organization. Sand is a solid composed of SiO_2, which contains a network of covalent silicon–oxygen bonds.

Whether a substance exists as a gas, liquid, or solid depends on the balance between the kinetic energy of its particles and the strength of the interactions between the particles. In a gas, the kinetic energy of motion is high and the particles are far apart from each other. As a result, the attractive forces between the molecules are negligible and gas molecules move freely. In a liquid, attractive forces hold the molecules much more closely together, so the distance between molecules and the kinetic energy is much less than the gas. In a solid, the attractive forces between molecules are even stronger, so the distance between individual particles is small and there is little freedom of motion. The properties of gases, liquids, and solids are summarized in Table 7.1.

▼ **FIGURE 7.1   The Three States of Matter—Solid, Liquid, and Gas**

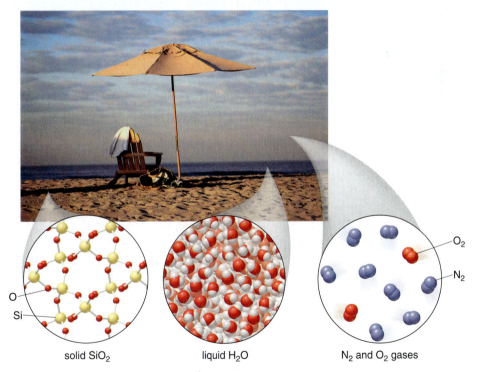

solid SiO_2 liquid H_2O N_2 and O_2 gases

Most sand is composed of silicon dioxide (SiO_2), which forms a three-dimensional network of covalent bonds. Liquid water is composed of H_2O molecules, which can move past each other but are held close together by a force of attraction (Section 7.7). Air contains primarily N_2 and O_2 molecules that move rapidly with no force of attraction for each other.

TABLE 7.1 Properties of Gases, Liquids, and Solids

Property	Gas	Liquid	Solid
Shape and Volume	Expands to fill its container	A fixed volume that takes the shape of the container it occupies	A definite shape and volume
Arrangement of Particles	Randomly arranged, disorganized, and far apart	Randomly arranged but close	Fixed arrangement of very close particles
Density	Low (< 0.01 g/mL)	High (~1 g/mL)[a]	High (1–10 g/mL)
Particle Movement	Very fast	Moderate	Slow
Interaction Between Particles	None	Strong	Very strong

[a]The symbol "~" means approximately.

PROBLEM 7.1 How do gaseous, liquid, and solid methanol (CH_4O) compare in each of the following features: (a) density; (b) the space between the molecules; (c) the attractive force between the molecules?

PROBLEM 7.2 Why is a gas much more easily compressed into a smaller volume than a liquid or solid?

7.2 GASES AND PRESSURE

Anyone who has ridden a bike against the wind knows that even though we can't see the gas molecules of the air, we can feel them as we move through them.

Simple gases in the atmosphere—oxygen (O_2), carbon dioxide (CO_2), and ozone (O_3)—are vital to life. Oxygen, which constitutes 21% of the earth's atmosphere, is needed for metabolic processes that convert carbohydrates to energy. Green plants use carbon dioxide, a minor component of the atmosphere, to store the energy of the sun in the bonds of carbohydrate molecules during photosynthesis. Ozone forms a protective shield in the upper atmosphere to filter out harmful radiation from the sun, thus keeping it from the surface of the earth.

7.2A PROPERTIES OF GASES

Helium, a noble gas composed of He atoms, and oxygen, a gas composed of diatomic O_2 molecules, behave differently in chemical reactions. Many of their properties, however, and the properties of all gases, can be explained by the **kinetic-molecular theory of gases,** a set of principles based on the following assumptions:

- A gas consists of particles—atoms or molecules—that move randomly and rapidly.
- The size of gas particles is small compared to the space between the particles.
- Because the space between gas particles is large, gas particles exert no attractive forces on each other.
- The kinetic energy of gas particles increases with increasing temperature.
- When gas particles collide with each other, they rebound and travel in new directions. When gas particles collide with the walls of a container, they exert a pressure.

Because gas particles move rapidly, two gases mix together quickly. Moreover, when a gas is added to a container, the particles rapidly move to fill the entire container.

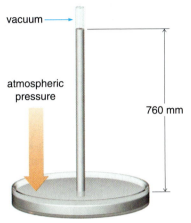

vacuum

atmospheric
pressure

760 mm

mercury-filled dish

A barometer measures atmospheric pressure. Air pressure on the Hg in the dish pushes Hg up a sealed glass tube to a height that equals the atmospheric pressure.

The term *torr* is named for Italian physicist Evangelista Torricelli, who invented the barometer in the 1600s.

7.2B GAS PRESSURE

When many gas molecules strike a surface, they exert a measurable pressure. **Pressure (P) is the force (F) exerted per unit area (A).**

$$\text{Pressure} = \frac{\text{Force}}{\text{Area}} = \frac{F}{A}$$

All of the gases in the atmosphere collectively exert **atmospheric pressure** on the surface of the earth. The value of the atmospheric pressure varies with location, decreasing with increasing altitude. Atmospheric pressure also varies slightly from day to day, depending on the weather.

Atmospheric pressure is measured with a **barometer** (Figure 7.2). A barometer consists of a column of mercury (Hg) sealed at one end and inverted in a dish of mercury. The downward pressure exerted by the mercury in the column equals the atmospheric pressure on the mercury in the dish. Thus, the height of the mercury in the column measures the atmospheric pressure. Atmospheric pressure at sea level corresponds to a column of mercury 760. mm in height.

Many different units are used for pressure. The two most common units are the **atmosphere (atm),** and **millimeters of mercury (mm Hg),** where **1 atm = 760. mm Hg.** One millimeter of mercury is also called one **torr.** In the United States, the common pressure unit is **pounds per square inch (psi),** where **1 atm = 14.7 psi.** Pressure can also be measured in pascals (Pa), where 1 mm Hg = 133.32 Pa.

$$
\begin{aligned}
1 \text{ atm} &= 760. \text{ mm Hg} \\
&= 760. \text{ torr} \\
&= 14.7 \text{ psi} \\
&= 101{,}325 \text{ Pa}
\end{aligned}
$$

To convert a value from one pressure unit to another, set up conversion factors and use the method first shown in Section 1.7B and illustrated in Sample Problem 7.1.

SAMPLE PROBLEM 7.1

A scuba diver typically begins a dive with a compressed air tank at 3,000. psi. Convert this value to (a) atmospheres; (b) mm Hg.

ANALYSIS

To solve each part, set up conversion factors that relate the two units under consideration. Use conversion factors that place the unwanted unit, psi, in the denominator to cancel.

In part (a), the conversion factor must relate psi and atm:

psi–atm conversion factor

$$\frac{1 \text{ atm}}{14.7 \text{ psi}}$$

unwanted unit

In part (b), the conversion factor must relate psi and mm Hg:

psi–mm Hg conversion factor

$$\frac{760. \text{ mm Hg}}{14.7 \text{ psi}}$$

unwanted unit

SOLUTION

a. Convert the original unit (3000. psi) to the desired unit (atm) using the conversion factor:

$$3000. \text{ psi} \times \frac{1 \text{ atm}}{14.7 \text{ psi}} = 204 \text{ atm}$$

Answer

Psi cancels.

b. Convert the original unit (3000. psi) to the desired unit (mm Hg) using the conversion factor:

$$3000. \text{ psi} \times \frac{760. \text{ mm Hg}}{14.7 \text{ psi}} = 155{,}000 \text{ mm Hg}$$

Answer

Psi cancels.

PROBLEM 7.3

Typical atmospheric pressure in Denver is 630 mm Hg. Convert this value to (a) atmospheres; (b) psi.

PROBLEM 7.4 The tires on a road bike are inflated to 90 psi. Convert this value to (a) atmospheres; (b) mm Hg.

PROBLEM 7.5 Convert each pressure unit to the indicated unit.

a. 3.0 atm to mm Hg   b. 720 mm Hg to psi   c. 424 mm Hg to atm

7.2C   FOCUS ON HEALTH & MEDICINE
BLOOD PRESSURE

Taking a patient's blood pressure is an important part of most physical examinations. Blood pressure measures the pressure in an artery of the upper arm using a device called a **sphyg-momanometer.** A blood pressure reading consists of two numbers such as 120/80, where both values represent pressures in mm Hg. The higher number is the systolic pressure and refers to the maximum pressure in the artery right after the heart contracts. The lower number is the diastolic pressure and represents the minimum pressure when the heart muscle relaxes. A desirable systolic pressure is in the range of 100–120 mm Hg. A desirable diastolic pressure is in the range of 60–80 mm Hg. Figure 7.3 illustrates how a sphygmomanometer records pressure in a blood vessel.

When a patient's systolic pressure is routinely 140 mm Hg or greater or diastolic pressure is 90 mm Hg or greater, an individual is said to have **hypertension**—that is, high blood pressure.

▼ **FIGURE 7.3   Measuring Blood Pressure**

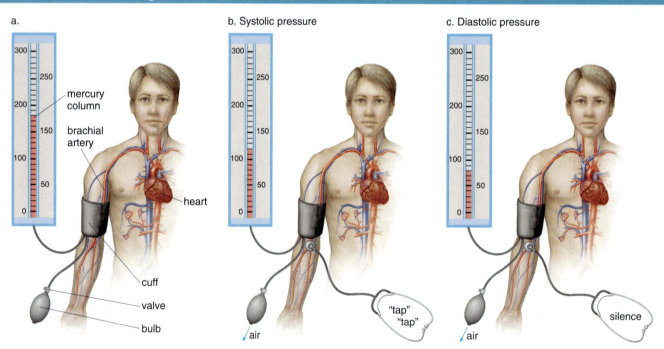

(a) To measure blood pressure, a cuff is inflated around the upper arm and a stethoscope is used to listen to the sound of blood flowing through the brachial artery. When the pressure in the cuff is high, it constricts the artery, so that no blood can flow to the lower arm. (b) Slowly the pressure in the cuff is decreased, and when it gets to the point that blood begins to spurt into the artery, a tapping sound is heard in the stethoscope. This value corresponds to the systolic blood pressure. (c) When the pressure in the cuff is further decreased, so that blood once again flows freely in the artery, the tapping sound disappears and the diastolic pressure is recorded.

Consistently high blood pressure leads to increased risk of stroke and heart attacks. Many forms of hypertension can be controlled with medications such as propranolol (trade name Inderal, Problem 15.12).

PROBLEM 7.6	Convert both values in the blood pressure reading 120/80 to atmospheres.
PROBLEM 7.7	Suppose blood pressure readings were reported in cm Hg rather than mm Hg. If this were the case, how would the pressure 140/90 be reported?

7.3 GAS LAWS THAT RELATE PRESSURE, VOLUME, AND TEMPERATURE

Four variables are important in discussing the behavior of gases—pressure (P), volume (V), temperature (T), and number of moles (n). The relationship of these variables is described by equations called **gas laws** that explain and predict the behavior of all gases as conditions change. Three gas laws illustrate the interrelationship of pressure, volume, and temperature.

- Boyle's law relates pressure and volume.
- Charles's law relates volume and temperature.
- Gay–Lussac's law relates pressure and temperature.

7.3A BOYLE'S LAW—HOW THE PRESSURE AND VOLUME OF A GAS ARE RELATED

Boyle's law describes how the volume of a gas changes as the pressure is changed.

- Boyle's law: For a fixed amount of gas at constant temperature, the pressure and volume of a gas are inversely related.

When two quantities are *inversely* related, one quantity *increases* as the other *decreases*. **The product of the two quantities, however, is a *constant*, symbolized by *k*.**

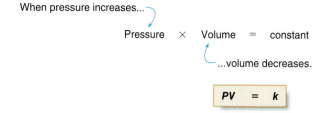

$$PV = k$$

Thus, if the volume of a cylinder of gas is halved, the pressure of the gas inside the cylinder doubles. **The same number of gas particles occupies half the volume and exerts two times the pressure.**

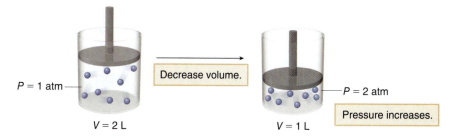

If we know the pressure and volume under an initial set of conditions (P_1 and V_1), we can calculate the pressure or volume under a different set of conditions (P_2 and V_2), since the product of pressure and volume is a constant.

$$P_1V_1 \quad = \quad P_2V_2$$

initial conditions　　new conditions

HOW TO　Use Boyle's Law to Calculate a New Gas Volume or Pressure

EXAMPLE **If a 4.0-L container of helium gas has a pressure of 10.0 atm, what pressure does the gas exert if the volume is increased to 6.0 L?**

Step [1]　**Identify the known quantities and the desired quantity.**

- To solve an equation using Boyle's law, we must know three quantities and solve for one quantity. In this case P_1, V_1, and V_2 are known and the final pressure, P_2, must be determined.

$$P_1 = 10.0 \text{ atm}$$
$$V_1 = 4.0 \text{ L} \qquad V_2 = 6.0 \text{ L} \qquad\qquad P_2 = ?$$

known quantities　　　　　　desired quantity

Step [2]　**Write the equation and rearrange it to isolate the desired quantity on one side.**

- Rearrange the equation for Boyle's law so that the unknown quantity, P_2, is present alone on one side.

$$P_1V_1 \quad = \quad P_2V_2 \qquad \text{Solve for } P_2 \text{ by dividing both sides by } V_2.$$

$$\frac{P_1V_1}{V_2} \quad = \quad P_2$$

Step [3]　**Solve the problem.**

- Substitute the known quantities into the equation and solve for P_2. Identical units must be used for two similar quantities (liters in this case) so that the units cancel.

$$P_2 \quad = \quad \frac{P_1V_1}{V_2} \quad = \quad \frac{(10.0 \text{ atm})(4.0 \text{ L})}{6.0 \text{ L}} \quad = \quad 6.7 \text{ atm}$$

Liters cancel.　**Answer**

- In this example, the volume increased so the pressure decreased.

SAMPLE PROBLEM 7.2

A tank of compressed air for scuba diving contains 8.5 L of gas at 204 atm pressure. What volume of air does this gas occupy at 1.0 atm?

ANALYSIS　Boyle's law can be used to solve this problem since an initial pressure and volume (P_1 and V_1) and a final pressure (P_2) are known, and a final volume (V_2) must be determined.

SOLUTION

[1]　**Identify the known quantities and the desired quantity.**

$$P_1 = 204 \text{ atm} \qquad P_2 = 1.0 \text{ atm}$$
$$V_1 = 8.5 \text{ L} \qquad\qquad\qquad V_2 = ?$$

known quantities　　　　desired quantity

[2]　**Write the equation and rearrange it to isolate the desired quantity, V_2, on one side.**

$$P_1V_1 \quad = \quad P_2V_2 \qquad \text{Solve for } V_2 \text{ by dividing both sides by } P_2.$$

$$\frac{P_1V_1}{P_2} \quad = \quad V_2$$

[3]   **Solve the problem.**

• Substitute the three known quantities into the equation and solve for V_2.

$$V_2 = \frac{P_1 V_1}{P_2} = \frac{(204 \text{ atm})(8.5 \text{ L})}{1.0 \text{ atm}} = 1{,}734 \text{ rounded to } 1{,}700 \text{ L}$$

Atm cancels.

Answer

• Thus, the volume increased because the pressure decreased.

PROBLEM 7.8	A sample of helium gas has a volume of 2.0 L at a pressure of 4.0 atm. What is the volume of gas at each of the following pressures? a. 5.0 atm b. 2.5 atm c. 10.0 atm d. 380 mm Hg
PROBLEM 7.9	A sample of nitrogen gas has a volume of 15.0 mL at a pressure of 0.50 atm. What is the pressure exerted by the gas if the volume is changed to each of the following values? a. 30.0 mL b. 5.0 mL c. 100. mL d. 1.0 L

Boyle's law explains how air is brought into or expelled from the lungs as the rib cage and diaphragm expand and contract when we breathe (Figure 7.4).

▼ **FIGURE 7.4   Focus on the Human Body: Boyle's Law and Breathing**

a.

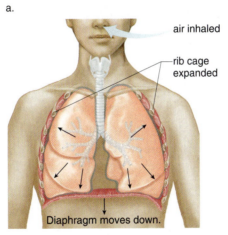

air inhaled

rib cage expanded

Diaphragm moves down.

b.

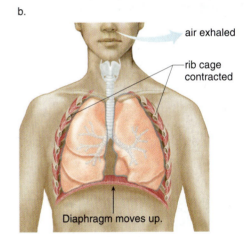

air exhaled

rib cage contracted

Diaphragm moves up.

a. When an individual inhales, the rib cage expands and the diaphragm is lowered, thus increasing the volume of the lungs. According to Boyle's law, increasing the volume of the lungs decreases the pressure inside the lungs. The decrease in pressure draws air into the lungs.

b. When an individual exhales, the rib cage contracts and the diaphragm is raised, thus decreasing the volume of the lungs. Since the volume is now decreased, the pressure inside the lungs increases, causing air to be expelled into the surroundings.

7.3B   CHARLES'S LAW—HOW THE VOLUME AND TEMPERATURE OF A GAS ARE RELATED

All gases expand when they are heated and contract when they are cooled. Charles's law describes how the volume of a gas changes as the Kelvin temperature is changed.

• **Charles's law: For a fixed amount of gas at constant pressure, the volume of a gas is proportional to its Kelvin temperature.**

A hot air balloon illustrates Charles's law. Heating the air inside the balloon causes it to expand and fill the balloon. When the air inside the balloon becomes less dense than the surrounding air, the balloon rises.

Equations for converting one temperature unit to another are given in Section 1.9.

Volume and temperature are *proportional;* that is, as one quantity *increases,* the other *increases* as well. Thus, **dividing volume by temperature is a constant (k).**

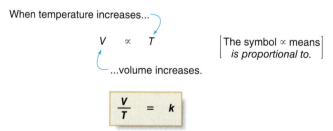

When temperature increases...

$$V \propto T$$

...volume increases.

$$\boxed{\frac{V}{T} = k}$$

$\begin{bmatrix} \text{The symbol} \propto \text{means} \\ \text{is proportional to.} \end{bmatrix}$

Increasing the temperature increases the kinetic energy of the gas particles, and they move faster and spread out, thus occupying a larger volume. Note that **Kelvin temperature** must be used in calculations involving gas laws. Any temperature reported in °C or °F must be converted to kelvins (K) prior to carrying out the calculation.

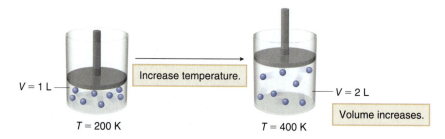

$V = 1$ L

Increase temperature.

$V = 2$ L

Volume increases.

$T = 200$ K $T = 400$ K

Since dividing the volume of a gas by the temperature gives a constant, knowing the volume and temperature under an initial set of conditions (V_1 and T_1) means we can calculate the volume or temperature under another set of conditions (V_2 and T_2) when either volume or temperature is changed.

$$\frac{V_1}{T_1} = \frac{V_2}{T_2}$$

initial conditions   new conditions

To solve a problem of this sort, we follow the same three steps listed in the *How To* outlined in Section 7.3A, except we use the equation for Charles's law in step [2] in place of the equation for Boyle's law. This procedure is illustrated in Sample Problem 7.3.

SAMPLE PROBLEM 7.3   A balloon that contains 0.50 L of air at 25 °C is cooled to −196 °C. What volume does the balloon now occupy?

ANALYSIS   Since this question deals with volume and temperature, Charles's law is used to determine a final volume because three quantities are known—the initial volume and temperature (V_1 and T_1), and the final temperature (T_2).

SOLUTION

[1] **Identify the known quantities and the desired quantity.**

$$V_1 = 0.50 \text{ L}$$
$$T_1 = 25 \text{ °C} \qquad T_2 = -196 \text{ °C} \qquad\qquad V_2 = ?$$
known quantities   desired quantity

- Both temperatures must be converted to Kelvin temperatures using the equation **K = °C + 273.**
- $T_1 = 25 \text{ °C} + 273 = 298$ K
- $T_2 = -196 \text{ °C} + 273 = 77$ K

[2] **Write the equation and rearrange it to isolate the desired quantity, V_2, on one side.**

- Use Charles's law.

$$\frac{V_1}{T_1} = \frac{V_2}{T_2} \qquad \text{Solve for } V_2 \text{ by multiplying both sides by } T_2.$$

$$\frac{V_1 T_2}{T_1} = V_2$$

[3] **Solve the problem.**

- Substitute the three known quantities into the equation and solve for V_2.

$$V_2 = \frac{V_1 T_2}{T_1} = \frac{(0.50 \text{ L})(77 \text{ K})}{298 \text{ K}} = 0.13 \text{ L}$$

Answer

Kelvins cancel.

- Since the temperature has decreased, the volume of gas must decrease as well.

PROBLEM 7.10	A volume of 0.50 L of air at 37 °C is expelled from the lungs into cold surroundings at 0.0 °C. What volume does the expelled air occupy at this temperature?
PROBLEM 7.11	(a) A volume (25.0 L) of gas at 45 K is heated to 450 K. What volume does the gas now occupy? (b) A volume (50.0 mL) of gas at 400. °C is cooled to 50. °C. What volume does the gas now occupy?
PROBLEM 7.12	Calculate the Kelvin temperature to which 10.0 L of a gas at 27 °C would have to be heated to change the volume to 12.0 L.

Charles's law can be used to explain how wind currents form at the beach (Figure 7.5). The air above land heats up faster than the air above water. As the temperature of the air above the land increases, the volume that it occupies increases; that is, the air expands, and as a result, its density decreases. This warmer, less dense air then rises, and the cooler denser air above the water moves toward the land as wind, filling the space left vacant by the warm, rising air.

▼ **FIGURE 7.5   Focus on the Environment: How Charles's Law Explains Wind Currents**

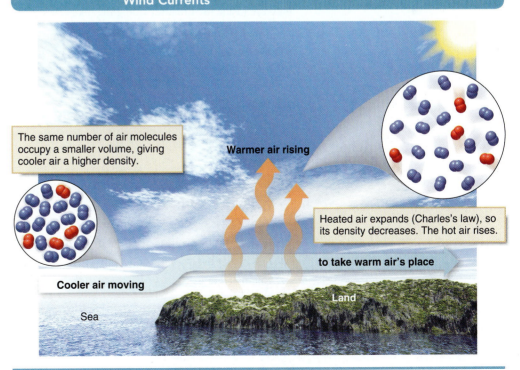

The same number of air molecules occupy a smaller volume, giving cooler air a higher density.

Warmer air rising

Heated air expands (Charles's law), so its density decreases. The hot air rises.

to take warm air's place

Cooler air moving

Land

Sea

7.3C GAY–LUSSAC'S LAW—HOW THE PRESSURE AND TEMPERATURE OF A GAS ARE RELATED

Gay–Lussac's law describes how the pressure of a gas changes as the Kelvin temperature is changed.

> • Gay–Lussac's law: For a fixed amount of gas at constant volume, the pressure of a gas is proportional to its Kelvin temperature.

Pressure and temperature are *proportional*; that is, as one quantity *increases*, the other *increases*. Thus, **dividing the pressure by the temperature is a constant (k).**

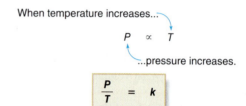

When temperature increases...

$$P \propto T$$

...pressure increases.

$$\frac{P}{T} = k$$

Increasing the temperature increases the kinetic energy of the gas particles, and if the volume is kept constant, the pressure exerted by the particles increases. Since dividing the pressure of a gas by the temperature gives a constant, knowing the pressure and Kelvin temperature under an initial set of conditions (P_1 and T_1) means we can calculate the pressure or temperature under another set of conditions (P_2 and T_2) when either pressure or temperature is changed.

$$\frac{P_1}{T_1} = \frac{P_2}{T_2}$$

initial conditions   new conditions

We solve this type of problem by following the same three steps in the *How To* in Section 7.3A, using the equation for Gay–Lussac's law in step [2].

SAMPLE PROBLEM 7.4

The tire on a bicycle stored in a cool garage at 18 °C had a pressure of 80. psi. What is the pressure inside the tire after riding the bike at 43 °C?

ANALYSIS

Since this question deals with pressure and temperature, Gay–Lussac's law is used to determine a final pressure because three quantities are known—the initial pressure and temperature (P_1 and T_1), and the final temperature (T_2).

SOLUTION

[1] **Identify the known quantities and the desired quantity.**

$$P_1 = 80.\ \text{psi}$$
$$T_1 = 18\ °C \qquad T_2 = 43\ °C \qquad\qquad\qquad P_2 = ?$$

known quantities   desired quantity

- Both temperatures must be converted to Kelvin temperatures.
- $T_1 = °C + 273 = 18\ °C + 273 = 291\ K$
- $T_2 = °C + 273 = 43\ °C + 273 = 316\ K$

[2] **Write the equation and rearrange it to isolate the desired quantity, P_2, on one side.**

- Use Gay–Lussac's law. Since the initial pressure is reported in psi, the final pressure will be calculated in psi.

$$\frac{P_1}{T_1} = \frac{P_2}{T_2} \qquad \text{Solve for } P_2 \text{ by multiplying both sides by } T_2.$$

$$\frac{P_1 T_2}{T_1} = P_2$$

[3] **Solve the problem.**

- Substitute the three known quantities into the equation and solve for P_2.

$$P_2 = \frac{P_1 T_2}{T_1} = \frac{(80.\,\text{psi})(316\,\cancel{K})}{291\,\cancel{K}} = 87\,\text{psi}$$

Kelvins cancel. **Answer**

- Since the temperature has increased, the pressure of the gas must increase as well.

PROBLEM 7.13

A pressure cooker is used to cook food in a closed pot. By heating the contents of a pressure cooker at constant volume, the pressure increases. If the steam inside the pressure cooker is initially at 100. °C and 1.00 atm, what is the final temperature of the steam if the pressure is increased to 1.05 atm?

PROBLEM 7.14

The temperature of a 0.50-L gas sample at 25 °C and 1.00 atm is changed to each of the following temperatures. What is the final pressure of the system?

a. 310. K   b. 150. K   c. 50. °C   d. 200. °C

PROBLEM 7.15

Use Gay–Lussac's law to answer the question posed at the beginning of the chapter: Why does a lid pop off a container of food when it is heated in a microwave?

CONSUMER NOTE

Food cooks faster in a pressure cooker because the reactions involved in cooking occur at a faster rate at a higher temperature.

7.3D THE COMBINED GAS LAW

All three gas laws—Boyle's, Charles's, and Gay–Lussac's laws—can be combined in a single equation, the **combined gas law,** that relates pressure, volume, and temperature.

$$\frac{P_1 V_1}{T_1} = \frac{P_2 V_2}{T_2}$$

initial conditions   new conditions

The combined gas law contains six terms that relate the pressure, volume, and temperature of an initial and final state of a gas. It can be used to calculate one quantity when the other five are known, as long as the amount of gas is constant. The combined gas law is used for determining the effect of changing two factors—such as pressure and temperature—on the third factor, volume.

We solve this type of problem by following the same three steps in the *How To* in Section 7.3A, using the equation for the combined gas law in step [2]. Sample Problem 7.5 shows how this is done. Table 7.2 summarizes the equations for the gas laws presented in Section 7.3.

SAMPLE PROBLEM 7.5

A weather balloon contains 222 L of helium at 20 °C and 760 mm Hg. What is the volume of the balloon when it ascends to an altitude where the temperature is –40 °C and 540 mm Hg?

ANALYSIS Since this question deals with pressure, volume, and temperature, the combined gas law is used to determine a final volume (V_2) because five quantities are known—the initial pressure, volume, and temperature (P_1, V_1, and T_1), and the final pressure and temperature (P_2 and T_2).

SOLUTION

[1] **Identify the known quantities and the desired quantity.**

$P_1 = 760$ mm Hg $P_2 = 540$ mm Hg
$T_1 = 20$ °C $T_2 = -40$ °C
$V_1 = 222$ L $V_2 = ?$

known quantities   desired quantity

- Both temperatures must be converted to Kelvin temperatures.
- $T_1 = °C + 273 = 20\,°C + 273 = 293$ K
- $T_2 = °C + 273 = -40\,°C + 273 = 233$ K

[2] **Write the equation and rearrange it to isolate the desired quantity, V_2, on one side.**

Use the combined gas law.

$$\frac{P_1V_1}{T_1} = \frac{P_2V_2}{T_2} \qquad \text{Solve for } V_2 \text{ by multiplying both sides by } \frac{T_2}{P_2}.$$

$$\frac{P_1V_1T_2}{T_1P_2} = V_2$$

[3] **Solve the problem.**

- Substitute the five known quantities into the equation and solve for V_2.

$$V_2 = \frac{P_1V_1T_2}{T_1P_2} = \frac{(760 \text{ mm Hg})(222 \text{ L})(233 \text{ K})}{(293 \text{ K})(540 \text{ mm Hg})} = 248.5 \text{ L rounded to 250 L}$$

Answer

Kelvins and mm Hg cancel.

TABLE 7.2   Summary of the Gas Laws That Relate Pressure, Volume, and Temperature

Law	Equation	Relationship
Boyle's law	$P_1V_1 = P_2V_2$	As P increases, V decreases for constant T and n.
Charles's law	$\dfrac{V_1}{T_1} = \dfrac{V_2}{T_2}$	As T increases, V increases for constant P and n.
Gay–Lussac's law	$\dfrac{P_1}{T_1} = \dfrac{P_2}{T_2}$	As T increases, P increases for constant V and n.
Combined gas law	$\dfrac{P_1V_1}{T_1} = \dfrac{P_2V_2}{T_2}$	The combined gas law shows the relationship of P, V, and T when two quantities are changed.

PROBLEM 7.16

The pressure inside a 1.0-L balloon at 25 °C was 750 mm Hg. What is the pressure inside the balloon when it is cooled to −40 °C and expands to 2.0 L in volume?

7.4 AVOGADRO'S LAW—HOW VOLUME AND MOLES ARE RELATED

Each equation in Section 7.3 was written for a constant amount of gas; that is, the number of moles (n) did not change. **Avogadro's law** describes the relationship between the number of moles of a gas and its volume.

- **Avogadro's law: When the pressure and temperature are held constant, the volume of a gas is proportional to the number of moles present.**

As the number of moles of a gas *increases*, its volume *increases* as well. Thus, dividing the volume by the number of moles is a constant (k). **The value of k is the same regardless of the identity of the gas.**

When the number of moles increases...

$$V \propto n$$

...the volume increases.

$$\frac{V}{n} = k$$

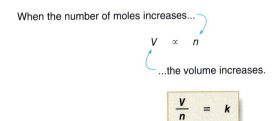

Thus, if the pressure and temperature of a system are held constant, **increasing the number of moles increases the volume of a gas.**

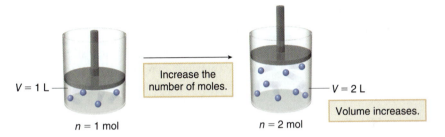

Since dividing the volume of a gas by the number of moles is a constant, knowing the volume and number of moles initially (V_1 and n_1) means we can calculate a new volume or number of moles (V_2 and n_2) when one of these quantities is changed.

$$\frac{V_1}{n_1} = \frac{V_2}{n_2}$$

initial conditions   new conditions

To solve a problem of this sort, we follow the same three steps listed in the *How To* outlined in Section 7.3A, using Avogadro's law in step [2].

SAMPLE PROBLEM 7.6

The lungs of an average male hold 0.25 mol of air in a volume of 5.8 L. How many moles of air do the lungs of an average female hold if the volume is 4.6 L?

ANALYSIS This question deals with volume and number of moles, so Avogadro's law is used to determine a final number of moles when three quantities are known—the initial volume and number of moles (V_1 and n_1), and the final volume (V_2).

SOLUTION

[1] **Identify the known quantities and the desired quantity.**

$V_1 = 5.8$ L $V_2 = 4.6$ L
$n_1 = 0.25$ mol $n_2 = ?$

known quantities   desired quantity

[2] **Write the equation and rearrange it to isolate the desired quantity, n_2, on one side.**

- Use Avogadro's law. To solve for n_2, we must invert the numerator and denominator on *both* sides of the equation, and then multiply by V_2.

$$\frac{V_1}{n_1} = \frac{V_2}{n_2} \xrightarrow[\substack{\text{Switch } V \text{ and } n \\ \text{on both sides.}}]{} \qquad \frac{n_1}{V_1} = \frac{n_2}{V_2} \qquad \text{Solve for } n_2 \text{ by multiplying both sides by } V_2.$$

$$\frac{n_1 V_2}{V_1} = n_2$$

[3] **Solve the problem.**

- Substitute the three known quantities into the equation and solve for n_2.

$$n_2 = \frac{n_1 V_2}{V_1} = \frac{(0.25 \text{ mol})(4.6\,\cancel{L})}{(5.8\,\cancel{L})} = 0.20 \text{ mol}$$

Liters cancel. **Answer**

PROBLEM 7.17   A balloon that contains 0.30 mol of helium in a volume of 6.4 L develops a leak so that its volume decreases to 3.85 L at constant temperature and pressure. How many moles of helium does the balloon now contain?

PROBLEM 7.18   A sample of nitrogen gas contains 5.0 mol in a volume of 3.5 L. Calculate the new volume of the container if the pressure and temperature are kept constant but the number of moles of nitrogen is changed to each of the following values: (a) 2.5 mol; (b) 3.65 mol; (c) 21.5 mol.

Avogadro's law allows us to compare the amounts of any two gases by comparing their volumes. Often amounts of gas are compared at a set of **standard conditions of temperature and pressure,** abbreviated as **STP.**

- STP conditions are: 1 atm (760 mm Hg) for pressure
                        273 K (0 °C) for temperature
- At STP, one mole of any gas has the same volume, 22.4 L, called the *standard molar volume.*

Under STP conditions, one mole of nitrogen gas and one mole of helium gas each contain 6.02×10^{23} molecules of gas and occupy a volume of 22.4 L at 0 °C and 1 atm pressure. Since the molar masses of nitrogen and helium are different (28.0 g for N_2 compared to 4.0 g for He), one mole of each substance has a *different* mass.

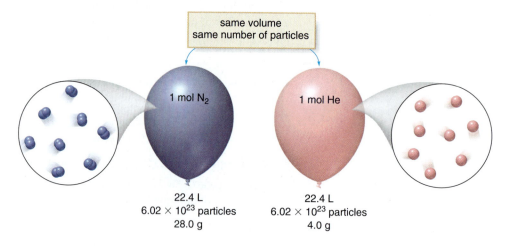

same volume
same number of particles

1 mol N_2

1 mol He

22.4 L
6.02×10^{23} particles
28.0 g

22.4 L
6.02×10^{23} particles
4.0 g

The standard molar volume can be used to set up conversion factors that relate the volume and number of moles of a gas at STP, as shown in the following stepwise procedure.

HOW TO   Convert Moles of Gas to Volume at STP

EXAMPLE   How many moles are contained in 2.0 L of N_2 at standard temperature and pressure?

Step [1]   **Identify the known quantities and the desired quantity.**

$$\underset{\text{original quantity}}{2.0 \text{ L of } N_2} \qquad \underset{\text{desired quantity}}{? \text{ moles of } N_2}$$

Step [2]   **Write out the conversion factors.**

- Set up conversion factors that relate the number of moles of a gas to volume at STP. Choose the conversion factor that places the unwanted unit, liters, in the denominator so that the units cancel.

$$\frac{22.4 \text{ L}}{1 \text{ mol}} \quad \text{or} \quad \boxed{\frac{1 \text{ mol}}{22.4 \text{ L}}}$$

Choose this conversion factor to cancel L.

Step [3]   **Solve the problem.**

- Multiply the original quantity by the conversion factor to obtain the desired quantity.

$$2.0 \, \cancel{L} \times \frac{1 \text{ mol}}{22.4 \, \cancel{L}} = 0.089 \text{ mol of } N_2$$

Answer

Liters cancel.

By using the molar mass of a gas, we can determine the volume of a gas from a given number of grams, as shown in Sample Problem 7.7.

SAMPLE PROBLEM 7.7

Burning 1 mol of propane in a gas grill adds 132.0 g of carbon dioxide (CO_2) to the atmosphere. What volume of CO_2 does this correspond to at STP?

ANALYSIS   To solve this problem, we must convert the number of grams of CO_2 to moles using the molar mass. The number of moles of CO_2 can then be converted to its volume using a mole–volume conversion factor (1 mol/22.4 L).

SOLUTION

[1]   **Identify the known quantities and the desired quantity.**

$$\underset{\text{known quantity}}{132.0 \text{ g } CO_2} \qquad \underset{\text{desired quantity}}{? \text{ L } CO_2}$$

[2]   **Convert the number of grams of CO_2 to the number of moles of CO_2 using the molar mass.**

molar mass conversion factor

$$132.0 \, \cancel{g} \, CO_2 \times \frac{1 \text{ mol } CO_2}{44.0 \, \cancel{g} \, CO_2} = 3.00 \text{ mol } CO_2$$

Grams cancel.

[3]   **Convert the number of moles of CO_2 to the volume of CO_2 using a mole–volume conversion factor.**

mole–volume conversion factor

$$3.00 \, \cancel{mol} \, CO_2 \times \frac{22.4 \text{ L}}{1 \, \cancel{mol}} = 67.2 \text{ L } CO_2$$

Answer

Moles cancel.

PROBLEM 7.19   How many liters does each of the following quantities of O_2 occupy at STP: (a) 4.5 mol; (b) 0.35 mol; (c) 18.0 g?

PROBLEM 7.20   How many moles are contained in the following volumes of air at STP: (a) 1.5 L; (b) 8.5 L; (c) 25 mL?

7.5   THE IDEAL GAS LAW

All four properties of gases—pressure, volume, temperature, and number of moles—can be combined into a single equation called the **ideal gas law.** The product of pressure and volume divided by the product of moles and Kelvin temperature is a constant, called the **universal gas constant** and symbolized by **R.**

$$\frac{PV}{nT} = R \quad \text{universal gas constant}$$

More often the equation is rearranged and written in the following way:

$$PV = nRT$$
Ideal gas law

For atm: $R = 0.0821 \dfrac{\text{L} \cdot \text{atm}}{\text{mol} \cdot \text{K}}$

For mm Hg: $R = 62.4 \dfrac{\text{L} \cdot \text{mm Hg}}{\text{mol} \cdot \text{K}}$

The value of the universal gas constant R depends on its units. The two most common values of R are given using atmospheres or mm Hg for pressure, liters for volume, and kelvins for temperature. **Be careful to use the correct value of R for the pressure units in the problem you are solving.**

The ideal gas law can be used to find any value—P, V, n, or T—as long as three of the quantities are known. Solving a problem using the ideal gas law is shown in the stepwise *How To* procedure and in Sample Problem 7.8. Although the ideal gas law gives exact answers only for a perfectly "ideal" gas, it gives a good approximation for most real gases, such as the oxygen and carbon dioxide in breathing, as well (Figure 7.6).

HOW TO   Carry Out Calculations with the Ideal Gas Law

EXAMPLE   How many moles of gas are contained in a typical human breath that takes in 0.50 L of air at 1.0 atm pressure and 37 °C?

Step [1]   Identify the known quantities and the desired quantity.

$P = 1.0$ atm
$V = 0.50$ L
$T = 37$ °C $n = ?$ mol
known quantities   desired quantity

Step [2]   Convert all values to proper units and choose the value of R that contains these units.
- Convert °C to K. K = °C + 273 = 37 °C + 273 = 310. K
- Use the value of R in atm since the pressure is given in atm; that is, $R = 0.0821$ L · atm/mol · K.

Step [3]   Write the equation and rearrange it to isolate the desired quantity on one side.
- Use the ideal gas law and solve for n by dividing both sides by RT.

$$PV = nRT \quad \text{Solve for } n \text{ by dividing both sides by } RT.$$

$$\frac{PV}{RT} = n$$

How To, continued...

Step [4]   **Solve the problem.**

- Substitute the known quantities into the equation and solve for *n*.

$$n = \frac{PV}{RT} = \frac{(1.0 \text{ atm})(0.50 \text{ L})}{\left(0.0821 \dfrac{\text{L} \cdot \text{atm}}{\text{mol} \cdot \text{K}}\right)(310. \text{ K})} = 0.0196 \text{ rounded to } 0.020 \text{ mol}$$

Answer

▼ **FIGURE 7.6   Focus on the Human Body: The Lungs**

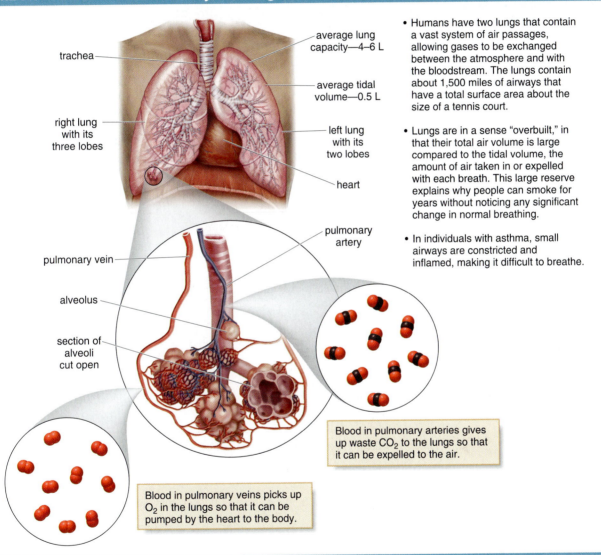

trachea

right lung with its three lobes

pulmonary vein

alveolus

section of alveoli cut open

average lung capacity—4–6 L

average tidal volume—0.5 L

left lung with its two lobes

heart

pulmonary artery

- Humans have two lungs that contain a vast system of air passages, allowing gases to be exchanged between the atmosphere and with the bloodstream. The lungs contain about 1,500 miles of airways that have a total surface area about the size of a tennis court.

- Lungs are in a sense "overbuilt," in that their total air volume is large compared to the tidal volume, the amount of air taken in or expelled with each breath. This large reserve explains why people can smoke for years without noticing any significant change in normal breathing.

- In individuals with asthma, small airways are constricted and inflamed, making it difficult to breathe.

Blood in pulmonary arteries gives up waste CO_2 to the lungs so that it can be expelled to the air.

Blood in pulmonary veins picks up O_2 in the lungs so that it can be pumped by the heart to the body.

SAMPLE PROBLEM 7.8   If a person exhales 25.0 g of CO_2 in an hour, what volume does this amount occupy at 1.00 atm and 37 °C?

ANALYSIS   Use the ideal gas law to calculate *V*, since *P* and *T* are known and *n* can be determined by using the molar mass of CO_2 (44.0 g/mol).

SOLUTION

[1] **Identify the known quantities and the desired quantity.**

$P = 1.00$ atm

$T = 37$ °C   25.0 g CO_2 $V = ?$ L

known quantities   desired quantity

[2] **Convert all values to proper units and choose the value of R that contains these units.**

- Convert °C to K. $K = °C + 273 = 37 °C + 273 = 310.$ K
- Use the value of R with atm since the pressure is given in atm; that is, $R = 0.0821$ L · atm/mol · K.
- Convert the number of grams of CO_2 to the number of moles of CO_2 using the molar mass (44.0 g/mol).

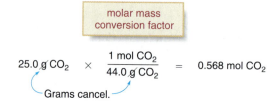

molar mass conversion factor

$$25.0 \; \cancel{g} \; CO_2 \quad \times \quad \frac{1 \; mol \; CO_2}{44.0 \; \cancel{g} \; CO_2} \quad = \quad 0.568 \; mol \; CO_2$$

Grams cancel.

[3] **Write the equation and rearrange it to isolate the desired quantity, V, on one side.**

- Use the ideal gas law and solve for V by dividing both sides by P.

$$PV = nRT \qquad \text{Solve for } V \text{ by dividing both sides by } P.$$

$$V = \frac{nRT}{P}$$

[4] **Solve the problem.**

- Substitute the three known quantities into the equation and solve for V.

$$V = \frac{nRT}{P} = \frac{(0.568 \; \cancel{mol})\left(0.0821 \; \dfrac{L \cdot atm}{\cancel{mol} \cdot \cancel{K}}\right)(310. \; \cancel{K})}{1.0 \; \cancel{atm}} = 14.5 \; L$$

Answer

PROBLEM 7.21

How many moles of oxygen (O_2) are contained in a 5.0-L cylinder that has a pressure of 175 atm and a temperature of 20. °C?

PROBLEM 7.22

Determine the pressure of N_2 under each of the following conditions.

a. 0.45 mol at 25 °C in 10.0 L   b. 10.0 g at 20. °C in 5.0 L

PROBLEM 7.23

Determine the volume of 8.50 g of He gas at 25 °C and 750 mm Hg.

Since the partial pressure of O_2 is low at very high altitudes, most mountain climbers use supplemental O_2 tanks above about 24,000 ft.

7.6 DALTON'S LAW AND PARTIAL PRESSURES

Since gas particles are very far apart compared to the size of an individual particle, gas particles behave independently. As a result, the identity of the components of a gas mixture does not matter, and **a mixture of gases behaves like a pure gas.** Each component of a gas mixture is said to exert a pressure called its **partial pressure. Dalton's law** describes the relationship between the partial pressures of the components and the total pressure of a gas mixture.

- **Dalton's law:** The *total pressure* (P_{total}) of a gas mixture is the sum of the partial pressures of its component gases.

Thus, if a mixture has three gases (**A, B,** and **C**) with partial pressures P_A, P_B, and P_C, respectively, the total pressure of the system (P_{total}) is the sum of the three partial pressures. The partial pressure of a component of a mixture is the same pressure that the gas would exert if it were a pure gas.

$$P_{total} = P_A + P_B + P_C$$

total pressure   partial pressures of **A, B,** and **C**

SAMPLE PROBLEM 7.9

A sample of exhaled air from the lungs contains four gases with the following partial pressures: N_2 (563 mm Hg), O_2 (118 mm Hg), CO_2 (30 mm Hg), and H_2O (50 mm Hg). What is the total pressure of the sample?

ANALYSIS   Using Dalton's law, the total pressure is the sum of the partial pressures.

SOLUTION   Adding up the four partial pressures gives the total:

563 + 118 + 30 + 50 = 761 mm Hg (total pressure)

PROBLEM 7.24

CO_2 was added to a cylinder containing 2.5 atm of O_2 to give a total pressure of 4.0 atm of gas. What is the partial pressure of O_2 and CO_2 in the final mixture?

We can also calculate the partial pressure of each gas in a mixture if two quantities are known—[1] the total pressure and [2] the percent of each component—as shown in Sample Problem 7.10.

SAMPLE PROBLEM 7.10

Air is a mixture of 21% O_2, 78% N_2, and 1% argon by volume. What is the partial pressure of each gas at sea level, where the total pressure is 760 mm Hg?

ANALYSIS   Convert each percent to a decimal by moving the decimal point two places to the left. Multiply each decimal by the total pressure to obtain the partial pressure for each component.

SOLUTION

                                                      Partial pressure

Fraction O_2: 21% = 0.21   0.21 × 760 mm Hg = 160 mm Hg (O_2)

Fraction N_2: 78% = 0.78   0.78 × 760 mm Hg = 590 mm Hg (N_2)

Fraction Ar: 1% = 0.01   0.01 × 760 mm Hg =   8 mm Hg (Ar)

                                                                758 rounded to 760 mm Hg

---

PROBLEM 7.25

A sample of natural gas at 750 mm Hg contains 85% methane, 10% ethane, and 5% propane. What are the partial pressures of each gas in this mixture?

HEALTH NOTE

The high pressures of a hyperbaric chamber can be used to treat patients fighting infections and scuba divers suffering from the bends.

The composition of the atmosphere does not change with location, even though the total atmospheric pressure decreases with increasing altitude. At high altitudes, therefore, the partial pressure of oxygen is much lower than it is at sea level, making breathing difficult. This is why mountain climbers use supplemental oxygen at altitudes above 8,000 meters.

In contrast, a hyperbaric chamber is a device that maintains air pressure two to three times higher than normal. Hyperbaric chambers have many uses. At this higher pressure the partial pressure of O_2 is higher. For burn patients, the higher pressure of O_2 increases the amount of O_2 in the blood, where it can be used by the body for reactions that fight infections.

When a scuba diver surfaces too quickly, the N_2 dissolved in the blood can form microscopic bubbles that cause pain in joints and can occlude small blood vessels, causing organ injury. This condition, called the bends, is treated by placing a diver in a hyperbaric chamber, where the elevated pressure decreases the size of the N_2 bubbles, which are then eliminated as N_2 gas from the lungs as the pressure is slowly decreased.

PROBLEM 7.26	Air contains 21% O_2 and 78% N_2. What are the partial pressures of N_2 and O_2 in a hyperbaric chamber that contains air at 2.5 atm?

7.7 INTERMOLECULAR FORCES, BOILING POINT, AND MELTING POINT

Unlike gases, the behavior of liquids and solids cannot be described by a set of laws that can be applied regardless of the identity of the substance. Since liquids and solids are composed of particles that are much closer together, **a force of attraction exists between them.**

Ionic compounds are composed of extensive arrays of oppositely charged ions that are held together by strong electrostatic interactions. In covalent compounds, the nature and strength of the attraction between individual molecules depend on the identity of the atoms.

- Intermolecular forces are the attractive forces that exist *between* molecules.

There are three different types of intermolecular forces in covalent molecules, presented in order of *increasing strength:*

- London dispersion forces
- Dipole–dipole interactions
- Hydrogen bonding

The strength of the intermolecular forces determines whether a compound has a high or low melting point and boiling point, and thus if the compound is a solid, liquid, or gas at a given temperature.

7.7A LONDON DISPERSION FORCES

London dispersion forces can also be called **van der Waals forces.**

- London dispersion forces are very weak interactions due to the momentary changes in electron density in a molecule.

For example, although a nonpolar methane molecule (CH_4) has no net dipole, at any one instant its electron density may not be completely symmetrical. If more electron density is present in one region of the molecule, less electron density must be present some place else, and this creates a *temporary* dipole. A temporary dipole in one CH_4 molecule induces a temporary dipole in another CH_4 molecule, with the partial positive and negative charges arranged close to each other. **The weak interaction between these temporary dipoles constitutes London dispersion forces.**

Although any single interaction is weak, a large number of London dispersion forces creates a strong force. For example, geckos stick to walls and ceilings by London dispersion forces between the surfaces and the 500,000 tiny hairs on each foot.

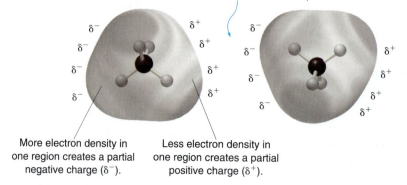

London dispersion force between two CH_4 molecules

More electron density in one region creates a partial negative charge (δ^-).

Less electron density in one region creates a partial positive charge (δ^+).

All **covalent compounds exhibit London dispersion forces.** These intermolecular forces are the only intermolecular forces present in nonpolar compounds. The strength of these forces is related to the size of the molecule.

• The larger the molecule, the larger the attractive force between two molecules, and the stronger the intermolecular forces.

PROBLEM 7.27	Which of the following compounds exhibit London dispersion forces: (a) NH_3; (b) H_2O; (c) HCl; (d) ethane (C_2H_6)?

7.7B DIPOLE–DIPOLE INTERACTIONS

How to determine whether a molecule is polar is shown in Section 4.8.

• Dipole–dipole interactions are the attractive forces between the permanent dipoles of two polar molecules.

For example, the carbon–oxygen bond in formaldehyde, $H_2C=O$, is polar because oxygen is more electronegative than carbon. This polar bond gives formaldehyde a permanent dipole, making it a polar molecule. The dipoles in adjacent formaldehyde molecules can align so that the partial positive and partial negative charges are close to each other. These attractive forces due to permanent dipoles are much stronger than London dispersion forces.

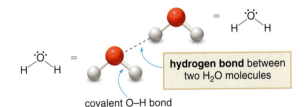

formaldehyde

PROBLEM 7.28	Draw the individual dipoles of two H—Cl molecules and show how the dipoles are aligned in a dipole–dipole interaction.

7.7C HYDROGEN BONDING

• Hydrogen bonding occurs when a hydrogen atom bonded to O, N, or F, is electrostatically attracted to an O, N, or F atom in another molecule.

Hydrogen bonding is only possible between two molecules that contain a hydrogen atom bonded to a very electronegative atom—that is, oxygen, nitrogen, or fluorine. For example, two H_2O molecules can hydrogen bond to each other: a hydrogen atom is covalently bonded to oxygen in one water molecule, and hydrogen bonded to an oxygen atom in another water molecule. **Hydrogen bonds are the strongest of the three types of intermolecular forces.** Table 7.3 summarizes the three types of intermolecular forces.

Hydrogen bonding is important in many biological molecules, including proteins and DNA. DNA, which is contained in the chromosomes of the nucleus of a cell, is responsible for the storage of all genetic information. DNA is composed of two long strands of atoms that are held together by hydrogen bonding as shown in Figure 7.7. A detailed discussion of DNA appears in Chapter 22.

TABLE 7.3   Summary of the Types of Intermolecular Forces

Type of Force	Relative Strength	Exhibited by	Example
London dispersion	Weak	All molecules	CH_4, H_2CO, H_2O
Dipole–dipole	Moderate	Molecules with a net dipole	H_2CO, H_2O
Hydrogen bonding	Strong	Molecules with an O—H, N—H, or H—F bond	H_2O

▼ **FIGURE 7.7   Hydrogen Bonding and DNA**

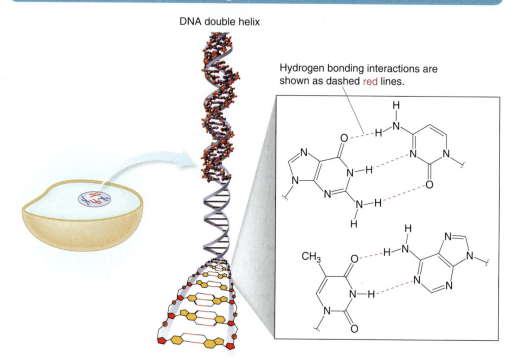

DNA double helix

Hydrogen bonding interactions are shown as dashed red lines.

DNA is composed of two long strands of atoms that wind around each other in an arrangement called a double helix. The two strands are held together by an extensive network of hydrogen bonds. In each hydrogen bond, an H atom of an N—H bond on one chain is intermolecularly hydrogen bonded to an oxygen or nitrogen atom on an adjacent chain. Five hydrogen bonds are indicated.

SAMPLE PROBLEM 7.11

What types of intermolecular forces are present in each compound: (a) HCl; (b) C_2H_6 (ethane); (c) NH_3?

ANALYSIS

• London dispersion forces are present in all covalent compounds.
• Dipole–dipole interactions are present only in polar compounds with a permanent dipole.
• Hydrogen bonding occurs only in compounds that contain an O—H, N—H, or H—F bond.

SOLUTION

a.

polar bond

$\overset{\delta+}{H}—\overset{\delta-}{Cl}$

• HCl has London forces like all covalent compounds.
• HCl has a polar bond, so it exhibits dipole–dipole interactions.
• HCl has no H atom on an O, N, or F, so it has no intermolecular hydrogen bonding.

b.

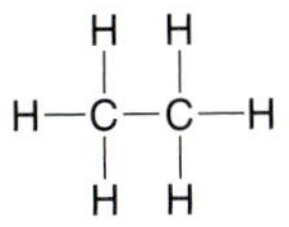

H—C—C—H (structure with H atoms)

nonpolar molecule

- C_2H_6 is a nonpolar molecule since it has only nonpolar C—C and C—H bonds. Thus, it exhibits only London forces.

c.

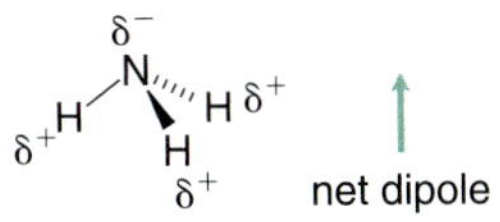

δ^- N structure with H δ^+, δ^+ H, H δ^+ net dipole

- NH_3 has London forces like all covalent compounds.
- NH_3 has a net dipole from its three polar bonds (Section 4.8), so it exhibits dipole–dipole interactions.
- NH_3 has a H atom bonded to N, so it exhibits intermolecular hydrogen bonding.

PROBLEM 7.29

What types of intermolecular forces are present in each molecule?

a. Cl_2 b. HCN   c. HF   d. CH_3Cl e. H_2

PROBLEM 7.30

Which of the compounds in each pair has stronger intermolecular forces?

a. CO_2 or H_2O b. CO_2 or HBr   c. HBr or H_2O d. CH_4 or C_2H_6

7.7D BOILING POINT AND MELTING POINT

The **boiling point (bp)** of a compound is the temperature at which a liquid is converted to the gas phase, while the **melting point (mp)** is the temperature at which a solid is converted to the liquid phase. The strength of the intermolecular forces determines the boiling point and melting point of compounds.

- **The *stronger* the intermolecular forces, the *higher* the boiling point and melting point.**

In boiling, energy must be supplied to overcome the attractive forces of the liquid state and separate the molecules to the gas phase. Similarly, in melting, energy must be supplied to overcome the highly ordered solid state and convert it to the less ordered liquid phase. A stronger force of attraction between molecules means that more energy must be supplied to overcome those intermolecular forces, increasing the boiling point and melting point.

In comparing compounds of similar size, the following trend is observed:

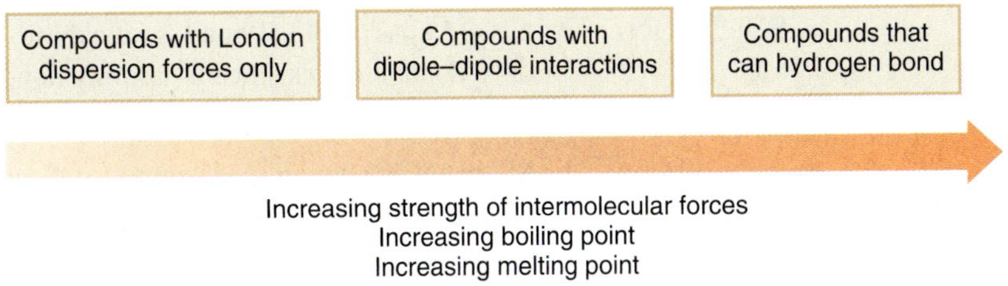

Compounds with London dispersion forces only	Compounds with dipole–dipole interactions	Compounds that can hydrogen bond

Increasing strength of intermolecular forces
Increasing boiling point
Increasing melting point

Methane (CH_4, 16.0 g/mol) and water (18.0 g/mol) have similar molar masses, so you might expect them to have similar melting points and boiling points. Methane, however, is a nonpolar molecule that exhibits only London dispersion forces, whereas water is a polar molecule that can form intermolecular hydrogen bonds. As a result, the melting point and boiling point of water are *much higher* than those of methane. In fact, the hydrogen bonds in water are so strong that it is a liquid at room temperature, whereas methane is a gas.

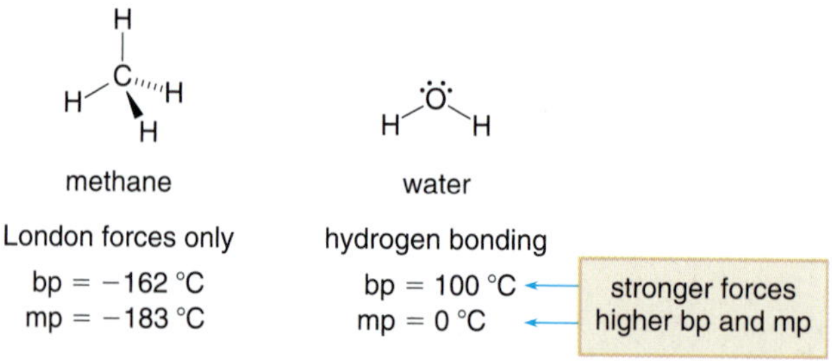

methane

water

London forces only

hydrogen bonding

bp = −162 °C

bp = 100 °C ← stronger forces

mp = −183 °C

mp = 0 °C ← higher bp and mp

In comparing two compounds with similar types of intermolecular forces, the higher molecular weight compound generally has more surface area and therefore a larger force of attraction, giving it the higher boiling point and melting point. Thus, propane (C_3H_8) and butane (C_4H_{10}) have only nonpolar bonds and London forces, but butane is larger and therefore has the higher boiling point and melting point.

propane

butane

larger molecule
stronger forces
higher bp and mp

bp = −42 °C

bp = −0.5 °C

mp = −190 °C

mp = −138 °C

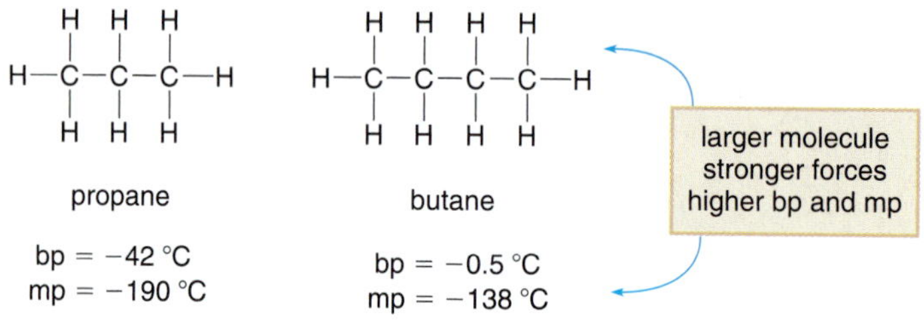

SAMPLE PROBLEM 7.12

(a) Which compound, **A** or **B,** has the higher boiling point? (b) Which compound, **C** or **D,** has the higher melting point?

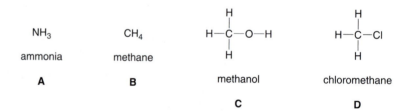

NH_3

ammonia

A

CH_4

methane

B

methanol

C

chloromethane

D

ANALYSIS   Determine the types of intermolecular forces in each compound. The compound with the stronger forces has the higher boiling point or melting point.

SOLUTION   a. NH_3 (**A**) has an N—H bond, so it exhibits intermolecular hydrogen bonding. CH_4 (**B**) has only London forces since it has only nonpolar C—H bonds. NH_3 has stronger forces and the higher boiling point.

b. Methanol (**C**) has an O—H bond, so it can intermolecularly hydrogen bond. Chloromethane (**D**) has a polar C—Cl bond, so it has dipole–dipole interactions, but it cannot hydrogen bond. **C** has stronger forces, so **C** has the higher melting point.

PROBLEM 7.31

Which compound in each pair has the higher boiling point?

a. CH_4 or C_2H_6 b. C_2H_6 or CH_3OH c. HBr or HCl   d. C_2H_6 or CH_3Br

---

PROBLEM 7.32

Which compound in each pair of compounds in Problem 7.31 has the higher melting point?

PROBLEM 7.33

Explain why CO_2 is a gas at room temperature but H_2O is a liquid.

7.8 THE LIQUID STATE

Since liquid molecules are much closer together than gas molecules, many properties of a liquid are determined by the strength of its intermolecular forces. The molecules in a liquid are still much more mobile than those of a solid, though, making liquids fluid and giving them no definite shape. Some liquid molecules move fast enough that they escape the liquid phase altogether and become gas molecules that are very far apart from each other.

7.8A VAPOR PRESSURE

When a liquid is placed in an open container, liquid molecules near the surface that have enough kinetic energy to overcome the intermolecular forces, escape to the gas phase. This process, **evaporation,** will continue until all of the liquid has become gas. A puddle of water formed after a rainstorm evaporates as all of the liquid water is converted to gas molecules called water **vapor. Evaporation is an** *endothermic* **process**—it absorbs heat from the surroundings. This explains why the skin is cooled as sweat evaporates.

evaporation

In a closed container, some liquid molecules evaporate from the surface and enter the gas phase. As more molecules accumulate in the gas phase, some molecules re-enter the liquid phase in the process of **condensation. Condensation is an** *exothermic* **process**—it gives off heat to the surroundings. At equilibrium, the rate of evaporation and the rate of condensation are equal.

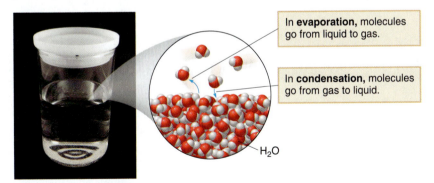

In **evaporation,** molecules go from liquid to gas.

In **condensation,** molecules go from gas to liquid.

liquid in a closed container

The gas laws we have already learned can describe the behavior of the gas molecules above a liquid. In particular, these gas molecules exert pressure, called **vapor pressure.**

- **Vapor pressure is the pressure exerted by gas molecules in equilibrium with the liquid phase.**

The vapor pressure exerted by a particular liquid depends on the identity of the liquid and the temperature. As the temperature is increased, the kinetic energy of the molecules increases and more molecules escape into the gas phase.

- Vapor pressure increases with increasing temperature.

When the temperature is high enough that the vapor pressure above the liquid equals the atmospheric pressure, even molecules below the surface of the liquid have enough kinetic energy to enter the gas phase and the liquid boils. **The *normal boiling point* of a liquid is the temperature at which its vapor pressure equals 760 mm Hg.**

The boiling point depends on the atmospheric pressure. At the lower atmospheric pressure of higher altitudes, a liquid has a lower boiling point because the vapor pressure above the liquid equals the atmospheric pressure at a lower temperature. In Denver, Colorado (elevation 1,609 m or 5,280 ft), where atmospheric pressure is typically 630 mm Hg, water boils at 93 °C.

How is the strength of the intermolecular forces related to the vapor pressure? The stronger the intermolecular forces, the less readily a compound escapes from the liquid to the gas phase. Thus,

- The stronger the intermolecular forces, the lower the vapor pressure at a given temperature.

Compounds with strong intermolecular forces have high boiling points and low vapor pressures at a given temperature. In Section 7.7 we learned that water has a *higher* boiling point than methane (CH_4) because water can hydrogen bond while methane cannot. At any given temperature, water has a *lower* vapor pressure because its molecules are held more tightly and remain in the liquid phase.

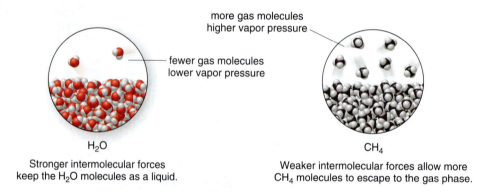

more gas molecules
higher vapor pressure

fewer gas molecules
lower vapor pressure

H_2O

Stronger intermolecular forces
keep the H_2O molecules as a liquid.

CH_4

Weaker intermolecular forces allow more
CH_4 molecules to escape to the gas phase.

SAMPLE PROBLEM 7.13

Which compound, H_2O or H_2S, has the higher boiling point? Which compound has the higher vapor pressure at a given temperature?

ANALYSIS

Determine which compound has the stronger intermolecular forces. Stronger forces mean a higher boiling point and a lower vapor pressure.

SOLUTION

Since H_2O contains an O atom bonded to H, H_2O exhibits hydrogen bonding, the strongest of the intermolecular forces. H_2S has no hydrogen bonding since it has no hydrogen bonded to O, N, or F. The stronger intermolecular forces in H_2O give it a higher boiling point. The weaker intermolecular forces in H_2S mean that the molecules escape into the gas phase more readily, giving H_2S the higher vapor pressure.

PROBLEM 7.34

Which molecule in each pair has the higher vapor pressure at a given temperature?

a. CH_4 or NH_3 b. CH_4 or C_2H_6 c. C_2H_6 or CH_3OH

PROBLEM 7.35

Explain why you feel cool when you get out of a swimming pool, even when the air temperature is quite warm. Then explain why the water feels warmer when you get back into the swimming pool.

PROBLEM 7.36 Your friend thinks that boiling water at his beach house is hotter than boiling water on Mount Haleakala in Hawaii (elevation 3,049 m or 10,003 ft). Comment on his statement.

7.8B VISCOSITY AND SURFACE TENSION

Viscosity and surface tension are two more properties of liquids that can be explained at least in part by the strength of the intermolecular forces.

Viscosity **is a measure of a fluid's resistance to flow freely.** A viscous liquid is one that feels "thicker." Compounds with stronger intermolecular forces tend to be more viscous than compounds with weaker forces. Thus water is more viscous than gasoline, which is composed of nonpolar molecules with weak intermolecular forces. The size of molecules also plays a role. Large molecules do not slide past each other as freely, so substances composed of large molecules tend to be more viscous. Olive oil, for example, is more viscous than water because olive oil is composed of compounds with three long floppy chains that contain more than 50 atoms in each chain.

Surface tension **is a measure of the resistance of a liquid to spread out.** Molecules in the interior of a liquid are surrounded by intermolecular forces on all sides, making them more stable than surface molecules that only experience intermolecular forces from neighbors on the side and below [Figure 7.8(a)]. This makes surface molecules less stable. **The** *stronger* **the intermolecular forces, the stronger surface molecules are pulled down toward the interior of a liquid and the** *higher* **the surface tension.** Because water has strong intermolecular hydrogen bonding, its surface tension is high. This explains why water striders can walk across the surface [Figure 7.8(b)], and why a paper clip can "float" on water.

▼ **FIGURE 7.8** Surface Tension

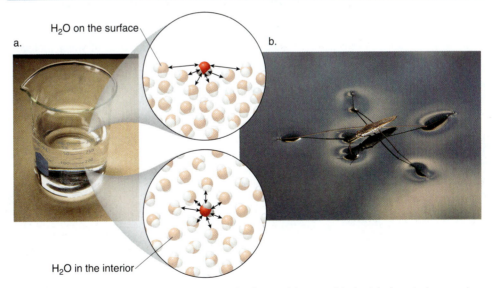

a.

b.

H_2O on the surface

H_2O in the interior

(a) Interior molecules experience intermolecular forces (shown with double-headed arrows) evenly in all directions, whereas surface molecules experience uneven interactions and are pulled downward towards the interior.
(b) Water's high surface tension allows a water strider to walk on the surface.

PROBLEM 7.37   Explain why benzene is less viscous than water, but ethylene glycol is more viscous than water.

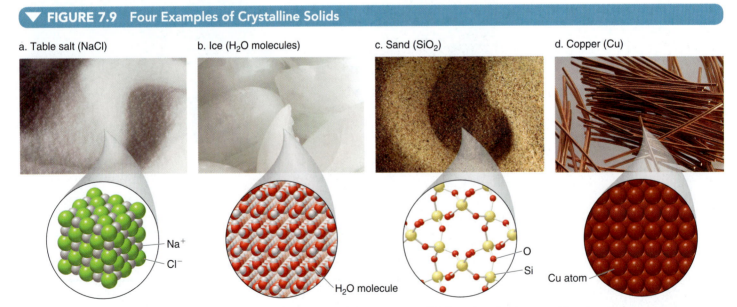

benzene   ethylene glycol

PROBLEM 7.38   Would you predict the surface tension of gasoline, composed of molecules containing only carbon and hydrogen atoms, to be higher or lower than the surface tension of water?

7.9 THE SOLID STATE

When a liquid is cooled so that the intermolecular forces are stronger than the kinetic energy of the particles, a solid is formed. Solids can be either **crystalline** or **amorphous.**

- A crystalline solid has a regular arrangement of particles—atoms, molecules, or ions—with a repeating structure.
- An amorphous solid has no regular arrangement of its closely packed particles.

There are four different types of crystalline solids—**ionic, molecular, network,** and **metallic**—as shown in Figure 7.9.

An **ionic solid** is composed of oppositely charged ions. For example, sodium chloride, NaCl, is composed of Na^+ cations and Cl^- anions, arranged so that each Na^+ cation is surrounded by six Cl^- anions and each Cl^- anion is surrounded by six Na^+ cations.

A **molecular solid** is composed of individual molecules arranged regularly. Ice, for example, contains a hexagonal arrangement of water molecules that are extensively hydrogen bonded to

▼ **FIGURE 7.9   Four Examples of Crystalline Solids**

a. Table salt (NaCl)   b. Ice (H_2O molecules)   c. Sand (SiO_2)   d. Copper (Cu)

Na^+
Cl^-
H_2O molecule
O
Si
Cu atom

(a) Sodium chloride, NaCl, an ionic solid; (b) ice, a molecular solid composed of H_2O molecules arranged in a hexagonal pattern; (c) silicon dioxide, SiO_2, a network solid that comprises sand; (d) metallic copper.

A polycarbonate helmet, polyethylene water bottle, and rubber tires are all examples of amorphous solids.

each other. In fact, the crystalline structure of water accounts for one of its unique properties. Water is one of the few substances whose solid phase is *less* dense than its liquid phase. As a result, solid ice floats on liquid water. A sheet of ice freezes on top of a lake, allowing plant and animal life to survive beneath the surface.

A **network solid** is composed of a vast number of atoms covalently bonded together, forming sheets or three-dimensional arrays. Quartz sand, SiO_2, contains an infinite network of silicon and oxygen atoms. Each silicon atom is bonded to four oxygen atoms and each oxygen is bonded to two silicon atoms, so that there are twice as many oxygen atoms as silicon. Other examples include diamond and graphite, two elemental forms of carbon, whose structures appeared in Section 2.4.

Since metals are atoms that readily give up their electrons, a **metallic solid** such as copper or silver can be thought of as a lattice of metal cations surrounded by a cloud of electrons that move freely. Because of their loosely held, delocalized electrons, metals conduct electricity and heat.

In contrast to crystalline solids, **amorphous solids have no regular arrangement of particles.** Amorphous solids can be formed when a liquid cools too quickly to allow the regular crystalline pattern to form. Substances composed of very large strands of covalent molecules also tend to form amorphous solids, because the chains can become folded and intertwined, making an organized regular arrangement impossible. Examples of amorphous solids include rubber, glass, and most plastics.

PROBLEM 7.39	Which type of crystalline solid is formed by each substance: (a) $CaCl_2$; (b) Fe (iron); (c) sugar ($C_{12}H_{22}O_{11}$); (d) $NH_3(s)$?

7.10 ENERGY AND PHASE CHANGES

In Section 7.7 we learned how the strength of intermolecular forces in a liquid and solid affect a compound's boiling point and melting point. Let's now look in more detail at the energy changes that occur during phase changes.

7.10A CONVERTING A SOLID TO A LIQUID

Converting a solid to a liquid is called *melting.* Melting is an *endothermic* process. Energy must be absorbed to overcome some of the attractive intermolecular forces that hold the organized solid molecules together to form the more random liquid phase. The amount of energy needed to melt 1 g of a substance is called its **heat of fusion.**

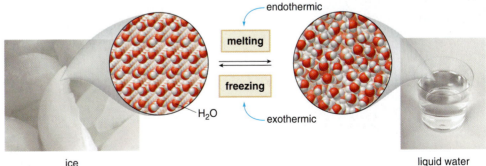

ice   liquid water

Freezing is the opposite of melting; that is, *freezing* **converts a liquid to a solid.** Freezing is an *exothermic* process because energy is released as the faster moving liquid molecules form an organized solid in which particles have little freedom of motion. For a given mass of a particular substance, the amount of energy released in freezing is the same as the amount of energy absorbed during melting.

Heats of fusion are reported in calories per gram (cal/g). A heat of fusion can be used as a conversion factor to determine how much energy is absorbed when a particular amount of a substance melts, as shown in Sample Problem 7.14.

SAMPLE PROBLEM 7.14

How much energy in calories is absorbed when 50.0 g of ice cubes melt? The heat of fusion of H_2O is 79.7 cal/g.

ANALYSIS

Use the heat of fusion as a conversion factor to determine the amount of energy absorbed in melting.

SOLUTION

[1]   **Identify the original quantity and the desired quantity.**

$$50.0 \text{ g} \qquad\qquad ? \text{ calories}$$
$$\text{original quantity} \qquad \text{desired quantity}$$

[2]   **Write out the conversion factors.**

- Use the heat of fusion as a conversion factor to convert grams to calories.

g–cal conversion factors

$$\frac{1 \text{ g}}{79.7 \text{ cal}} \quad \text{or} \quad \boxed{\frac{79.7 \text{ cal}}{1 \text{ g}}}$$

Choose this conversion factor to cancel the unwanted unit, g.

[3]   **Solve the problem.**

$$50.0 \text{ g} \times \frac{79.7 \text{ cal}}{1 \text{ g}} = 3,985 \text{ cal rounded to } 3,990 \text{ cal}$$

Grams cancel.

Answer

PROBLEM 7.40

Use the heat of fusion of water from Sample Problem 7.14 to answer each question.

a. How much energy in calories is released when 50.0 g of water freezes?
b. How much energy in calories is absorbed when 35.0 g of water melts?
c. How much energy in kilocalories is absorbed when 35.0 g of water melts?
d. How much energy in calories is absorbed when 1.00 mol of water melts?

7.10B   CONVERTING A LIQUID TO A GAS

Converting a liquid to a gas is called *vaporization*. Vaporization is an *endothermic* process. Energy must be absorbed to overcome the attractive intermolecular forces of the liquid phase to form gas molecules. The amount of energy needed to vaporize 1 g of a substance is called its **heat of vaporization.**

When an ice cube is added to a liquid at room temperature, the ice cube melts. The energy needed for melting is "pulled" from the warmer liquid molecules and the liquid cools down.

liquid water

H_2O

endothermic

vaporization

condensation

exothermic

steam

Condensation is the opposite of vaporization; that is, ***condensation* converts a gas to a liquid.** Condensation is an *exothermic* process because energy is released as the faster moving gas molecules form the more organized liquid phase. For a given mass of a particular substance, the amount of energy released in condensation equals the amount of energy absorbed during vaporization.

Heats of vaporization are reported in calories per gram (cal/g). A high heat of vaporization means that a substance absorbs a great deal of energy as it is converted from a liquid to a gas. **Water has a high heat of vaporization.** As a result, the evaporation of sweat from the skin is a very effective cooling mechanism for the body. The heat of vaporization can be used as a conversion factor to determine how much energy is absorbed when a particular amount of a substance vaporizes, as shown in Sample Problem 7.15.

SAMPLE PROBLEM 7.15

How much heat in kilocalories is absorbed when 22.0 g of 2-propanol, rubbing alcohol, evaporates after being rubbed on the skin? The heat of vaporization of 2-propanol is 159 cal/g.

ANALYSIS

Use the heat of vaporization to convert grams to an energy unit, calories. Calories must also be converted to kilocalories using a cal–kcal conversion factor.

SOLUTION

[1] **Identify the original quantity and the desired quantity.**

22.0 g   ? kilocalories

original quantity   desired quantity

[2] **Write out the conversion factors.**

- We have no conversion factor that directly relates grams and kilocalories. We do know, however, how to relate grams to calories using the heat of vaporization, and calories to kilocalories.

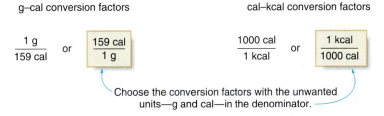

g–cal conversion factors

$$\frac{1\ g}{159\ cal} \quad \text{or} \quad \boxed{\frac{159\ cal}{1\ g}}$$

cal–kcal conversion factors

$$\frac{1000\ cal}{1\ kcal} \quad \text{or} \quad \boxed{\frac{1\ kcal}{1000\ cal}}$$

Choose the conversion factors with the unwanted units—g and cal—in the denominator.

[3] **Solve the problem.**

$$22.0\ \cancel{g} \quad \times \quad \frac{159\ \cancel{cal}}{1\ \cancel{g}} \quad \times \quad \frac{1\ kcal}{1000\ \cancel{cal}} \quad = \quad 3.50\ kcal$$

Grams cancel.   Calories cancel.

Answer

PROBLEM 7.41

Answer the following questions about water, which has a heat of vaporization of 540 cal/g.

a. How much energy in calories is absorbed when 42 g of water is vaporized?

b. How much energy in calories is released when 42 g of water is condensed?

c. How much energy in kilocalories is absorbed when 1.00 mol of water is vaporized?

d. How much energy in kilocalories is absorbed when 3.5 mol of water is vaporized?

7.10C CONVERTING A SOLID TO A GAS

Occasionally a solid phase forms a gas phase without passing through the liquid state. This process is called **sublimation.** The reverse process, conversion of a gas directly to a solid, is called **deposition.** Carbon dioxide is called *dry ice* because solid carbon dioxide (CO_2) sublimes to gaseous CO_2 without forming liquid CO_2.

Freeze-drying removes water from foods by the process of sublimation. These products can be stored almost indefinitely, since bacteria cannot grow in them without water.

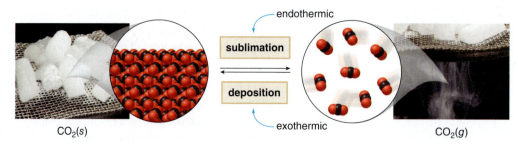

$CO_2(s)$ endothermic   sublimation   deposition   exothermic $CO_2(g)$

Carbon dioxide is a good example of a solid that undergoes this process at atmospheric pressure. At reduced pressure other substances sublime. For example, freeze-dried foods are prepared by subliming water from a food product at low pressure.

PROBLEM 7.42   Label each process as endothermic or exothermic and explain your reasoning: (a) sublimation; (b) deposition.

CHAPTER HIGHLIGHTS

KEY TERMS

Amorphous solid (7.9)
Atmosphere (7.2)
Avogadro's law (7.4)
Barometer (7.2)
Boiling point (bp, 7.7)
Boyle's law (7.3)
Combined gas law (7.3)
Condensation (7.8, 7.10)
Crystalline solid (7.9)
Dalton's law (7.6)
Deposition (7.10)
Dipole–dipole interactions (7.7)
Evaporation (7.8)
Freezing (7.10)

Gas laws (7.3)
Gay–Lussac's law (7.3)
Heat of fusion (7.10)
Heat of vaporization (7.10)
Hydrogen bonding (7.7)
Ideal gas law (7.5)
Intermolecular forces (7.7)
Ionic solid (7.9)
Kinetic-molecular theory (7.2)
London dispersion forces (7.7)
Melting (7.10)
Melting point (mp, 7.7)
Metallic solid (7.9)
Millimeters mercury (7.2)

Molecular solid (7.9)
Network solid (7.9)
Normal boiling point (7.8)
Partial pressure (7.6)
Pressure (7.2)
Standard molar volume (7.4)
STP (7.4)
Sublimation (7.10)
Surface tension (7.8)
Universal gas constant (7.5)
Vapor (7.8)
Vapor pressure (7.8)
Vaporization (7.10)
Viscosity (7.8)

KEY CONCEPTS

❶ What is pressure and what units are used to measure it? (7.2)

- Pressure is the force per unit area. The pressure of a gas is the force exerted when gas particles strike a surface. Pressure is measured by a barometer and recorded in atmospheres (atm), millimeters of mercury (mm Hg), or pounds per square inch (psi).
- 1 atm = 760 mm Hg = 14.7 psi.

❷ What are gas laws and how are they used to describe the relationship between the pressure, volume, and temperature of a gas? (7.3)

- Because gas particles are far apart and behave independently, a set of gas laws describes the behavior of all gases regardless of their identity. Three gas laws—

Boyle's law, Charles's law, and Gay–Lussac's law—describe the relationship between the pressure, volume, and temperature of a gas. These gas laws are summarized in "Key Equations—The Gas Laws" on page 223.

- For a constant amount of gas, the following relationships exist.
  - The pressure and volume of a gas are inversely related, so increasing the pressure decreases the volume at constant temperature.
  - The volume of a gas is proportional to its Kelvin temperature, so increasing the temperature increases the volume at constant pressure.
  - The pressure of a gas is proportional to its Kelvin temperature, so increasing the temperature increases the pressure at constant volume.

❸ Describe the relationship between the volume and number of moles of a gas. (7.4)

- Avogadro's law states that when temperature and pressure are held constant, the volume of a gas is proportional to its number of moles.
- One mole of any gas has the same volume, the standard molar volume of 22.4 L, at 1 atm and 273 K (STP).

❹ What is the ideal gas law? (7.5)

- The ideal gas law is an equation that relates the pressure (P), volume (V), temperature (T), and number of moles (n) of a gas; $PV = nRT$, where R is the universal gas constant. The ideal gas law can be used to calculate any one of the four variables, as long as the other three variables are known.

❺ What is Dalton's law and how is it used to relate partial pressures and the total pressure of a gas mixture? (7.6)

- Dalton's law states that the total pressure of a gas mixture is the sum of the partial pressures of its component gases. The partial pressure is the pressure exerted by each component of a mixture.

❻ What types of intermolecular forces exist and how do they determine a compound's boiling point and melting point? (7.7)

- Intermolecular forces are the forces of attraction between molecules. Three types of intermolecular forces exist in covalent compounds. London dispersion forces are due to momentary changes in electron density in a molecule. Dipole–dipole interactions are due to permanent dipoles. Hydrogen bonding, the strongest intermolecular force, results when a H atom bonded to an O, N, or F, is attracted to an O, N, or F atom in another molecule.
- The stronger the intermolecular forces, the higher the boiling point and melting point of a compound.

❼ Describe three features of the liquid state—vapor pressure, viscosity, and surface tension. (7.8)

- Vapor pressure is the pressure exerted by gas molecules in equilibrium with the liquid phase. Vapor pressure increases with increasing temperature. The higher the vapor pressure at a given temperature, the lower the boiling point of a compound.
- Viscosity measures a liquid's resistance to flow. More viscous compounds tend to have stronger intermolecular forces or they have high molecular weights.
- Surface tension measures a liquid's resistance to spreading out. The stronger the intermolecular forces, the higher the surface tension.

❽ Describe the features of different types of solids. (7.9)

- Solids can be amorphous or crystalline. An amorphous solid has no regular arrangement of particles. A crystalline solid has a regular arrangement of particles in a repeating pattern. There are four types of crystalline solids. Ionic solids are composed of ions. Molecular solids are composed of individual molecules. Network solids are composed of vast repeating arrays of covalently bonded atoms in a regular three-dimensional arrangement. Metallic solids are composed of metal cations with a cloud of delocalized electrons.

❾ Describe the energy changes that accompany changes of state. (7.10)

- A phase change converts one state to another. Energy is absorbed when a more organized state is converted to a less organized state. Thus, energy is absorbed when a solid melts to form a liquid, or when a liquid vaporizes to form a gas.
- Energy is released when a less organized state is converted to a more organized state. Thus, energy is released when a gas condenses to form a liquid, or a liquid freezes to form a solid.
- The heat of fusion is the energy needed to melt 1 g of a substance, while the heat of vaporization is the energy needed to vaporize 1 g of a substance.

KEY EQUATIONS—THE GAS LAWS

Name	Equation	Variables Related	Constant Terms
Boyle's law	$P_1V_1 = P_2V_2$	P, V	T, n
Charles's law	$\dfrac{V_1}{T_1} = \dfrac{V_2}{T_2}$	V, T	P, n
Gay–Lussac's law	$\dfrac{P_1}{T_1} = \dfrac{P_2}{T_2}$	P, T	V, n
Combined gas law	$\dfrac{P_1V_1}{T_1} = \dfrac{P_2V_2}{T_2}$	P, V, T	n
Avogadro's law	$\dfrac{V_1}{n_1} = \dfrac{V_2}{n_2}$	V, n	P, T
Ideal gas law	$PV = nRT$	P, V, T, n	R

PROBLEMS

Selected in-chapter and end-of-chapter problems have brief answers provided in Appendix B.

Pressure

7.43 What is the relationship between the units mm Hg and atm?

7.44 What is the relationship between the units mm Hg and psi?

7.45 The highest atmospheric pressure ever measured is 814.3 mm Hg, recorded in Mongolia in December, 2001. Convert this value to atmospheres.

7.46 The lowest atmospheric pressure ever measured is 652.5 mm Hg, recorded during Typhoon Tip on October 12, 1979. Convert this value to atmospheres.

7.47 Convert each quantity to the indicated unit.
a. 2.8 atm to psi   c. 20.0 atm to torr
b. 520 mm Hg to atm   d. 100. mm Hg to Pa

7.48 The compressed air tank of a scuba diver reads 3,200 psi at the beginning of a dive and 825 psi at the end of a dive. Convert each of these values to atm and mm Hg.

Boyle's Law

7.49 Assuming a fixed amount of gas at constant temperature, complete the following table.

	P_1	V_1	P_2	V_2
a.	2.0 atm	3.0 L	8.0 atm	?
b.	55 mm Hg	0.35 L	18 mm Hg	?
c.	705 mm Hg	215 mL	?	1.52 L

7.50 Assuming a fixed amount of gas at constant temperature, complete the following table.

	P_1	V_1	P_2	V_2
a.	2.5 atm	1.5 L	3.8 atm	?
b.	2.0 atm	350 mL	750 mm Hg	?
c.	75 mm Hg	9.1 mL	?	890 mL

7.51 If a scuba diver releases a 10.-mL air bubble below the surface where the pressure is 3.5 atm, what is the volume of the bubble when it rises to the surface and the pressure is 1.0 atm?

7.52 If someone takes a breath and the lungs expand from 4.5 L to 5.6 L in volume, and the initial pressure was 756 mm Hg, what is the pressure inside the lungs before any additional air is pulled in?

Charles's Law

7.53 Assuming a fixed amount of gas at constant pressure, complete the following table.

	V_1	T_1	V_2	T_2
a.	5.0 L	310 K	?	250 K
b.	150 mL	45 K	?	45 °C
c.	60.0 L	0.0 °C	180 L	?

7.54 Assuming a fixed amount of gas at constant pressure, complete the following table.

	V_1	T_1	V_2	T_2
a.	10.0 mL	210 K	?	450 K
b.	255 mL	55 °C	?	150 K
c.	13 L	−150 °C	52 L	?

7.55 If a balloon containing 2.2 L of gas at 25 °C is cooled to −78 °C, what is its new volume?

7.56 How hot must the air in a balloon be heated if initially it has a volume of 750. L at 20 °C and the final volume must be 1,000. L?

Gay–Lussac's Law

7.57 Assuming a fixed amount of gas at constant volume, complete the following table.

	P_1	T_1	P_2	T_2
a.	3.25 atm	298 K	?	398 K
b.	550 mm Hg	273 K	?	−100. °C
c.	0.50 atm	250 °C	955 mm Hg	?

7.58 Assuming a fixed amount of gas at constant volume, complete the following table.

	P_1	T_1	P_2	T_2
a.	1.74 atm	120 °C	?	20. °C
b.	220 mm Hg	150 °C	?	300. K
c.	0.75 atm	198 °C	220 mm Hg	?

7.59 An autoclave is a pressurized container used to sterilize medical equipment by heating it to a high temperature under pressure. If an autoclave containing steam at 100. °C and 1.0 atm pressure is then heated to 150. °C, what is the pressure inside it?

7.60 If a plastic container at 1.0 °C and 750. mm Hg is heated in a microwave oven to 80. °C, what is the pressure inside the container?

Combined Gas Law

7.61 Assuming a fixed amount of gas, complete the following table.

	P_1	V_1	T_1	P_2	V_2	T_2
a.	0.90 atm	4.0 L	265 K	?	3.0 L	310 K
b.	1.2 atm	75 L	5.0 °C	700. mm Hg	?	50 °C
c.	200. mm Hg	125 mL	298 K	100. mm Hg	0.62 L	?

7.62 Assuming a fixed amount of gas, complete the following table.

	P_1	V_1	T_1	P_2	V_2	T_2
a.	0.55 atm	1.1 L	340 K	?	3.0 L	298 K
b.	735 mm Hg	1.2 L	298 K	1.1 atm	?	0.0 °C
c.	7.5 atm	230 mL	−120 °C	15 atm	0.45 L	?

7.63 If a compressed air cylinder for scuba diving contains 6.0 L of gas at 18 °C and 200. atm pressure, what volume does the gas occupy at 1.0 atm and 25 °C?

7.64 What happens to the pressure of a sample with each of the following changes?
a. Double the volume and halve the Kelvin temperature.
b. Double the volume and double the Kelvin temperature.
c. Halve the volume and double the Kelvin temperature.

Avogadro's Law

7.65 What is the difference between STP and standard molar volume?

7.66 Given the same number of moles of two gases at STP conditions, how do the volumes of two gases compare? How do the masses of the two gas samples compare?

7.67 How many moles of helium are contained in each volume at STP: (a) 5.0 L; (b) 11.2 L; (c) 50.0 mL?

7.68 How many moles of argon are contained in each volume at STP: (a) 4.0 L; (b) 31.2 L; (c) 120 mL?

7.69 Calculate the volume of each substance at STP.
a. 4.2 mol Ar   b. 3.5 g CO_2 c. 2.1 g N_2

7.70 Calculate the volume of each substance at STP.
a. 4.2 mol N_2 b. 6.5 g He   c. 22.0 g CH_4

7.71 What volume does 3.01×10^{21} molecules of N_2 occupy at STP?

7.72 What volume does 1.50×10^{24} molecules of CO_2 occupy at STP?

Ideal Gas Law

7.73 How many moles of gas are contained in a human breath that occupies 0.45 L and has a pressure of 747 mm Hg at 37 °C?

7.74 How many moles of gas are contained in a compressed air tank for scuba diving that has a volume of 7.0 L and a pressure of 210 atm at 25 °C?

7.75 How many moles of air are present in the lungs if they occupy a volume of 5.0 L at 37 °C and 740 mm Hg? How many molecules of air does this correspond to?

7.76 If a cylinder contains 10.0 g of CO_2 in 10.0 L at 325 K, what is the pressure?

7.77 Which sample contains more moles: 2.0 L of O_2 at 273 K and 500 mm Hg, or 1.5 L of N_2 at 298 K and 650 mm Hg? Which sample weighs more?

7.78 An unknown amount of gas occupies 30.0 L at 2.1 atm and 298 K. How many moles does the sample contain? What is the mass if the gas is helium? What is the mass if the gas is argon?

Dalton's Law and Partial Pressure

7.79 Air pressure on the top of Mauna Loa, a 13,000-ft mountain in Hawaii, is 460 mm Hg. What are the partial pressures of O_2 and N_2, which compose 21% and 78% of the atmosphere, respectively?

7.80 If air contains 21% O_2, what is the partial pressure of O_2 in a cylinder of compressed air at 175 atm?

7.81 The partial pressure of N_2 in the air is 593 mm Hg at 1 atm. What is the partial pressure of N_2 in a bubble of air a scuba diver breathes when he is 66 ft below the surface of the water where the pressure is 3 atm?

7.82 If N_2 is added to a balloon that contains O_2 (partial pressure 450 mm Hg) and CO_2 (partial pressure 150 mm Hg) to give a total pressure of 850 mm Hg, what is the partial pressure of each gas in the final mixture?

Intermolecular Forces

7.83 What is the difference between dipole–dipole interactions and London dispersion forces?

7.84 What is the difference between dipole–dipole interactions and hydrogen bonding?

7.85 Why is H_2O a liquid at room temperature, but H_2S, which has a higher molecular weight and a larger surface area, is a gas at room temperature?

7.86 Why is Cl_2 a gas, Br_2 a liquid, and I_2 a solid at room temperature?

7.87 What types of intermolecular forces are exhibited by each compound? Chloroethane is a local anesthetic and cyclopropane is a general anesthetic.

a. chloroethane   b. cyclopropane

7.88 What types of intermolecular forces are exhibited by each compound? Acetaldehyde is formed when ethanol, the alcohol in alcoholic beverages, is metabolized, and acetic acid gives vinegar its biting odor and taste.

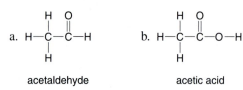

a. acetaldehyde   b. acetic acid

7.89 Which molecules are capable of intermolecular hydrogen bonding?

a. $H-C\equiv C-H$ b. CO_2 c. Br_2 d. $H-\underset{\underset{H}{|}}{\overset{\overset{H}{|}}{C}}-N-H$

7.90 Which molecules are capable of intermolecular hydrogen bonding?

a. N_2 b. $H-\underset{\underset{H}{|}}{\overset{\overset{H}{|}}{C}}-F$ c. HI d. $H-\underset{\underset{H}{|}}{\overset{\overset{H}{|}}{C}}-O-H$

7.91 Can two molecules of formaldehyde ($H_2C=O$) intermolecularly hydrogen bond to each other? Explain why or why not.

7.92 Why is the melting point of NaCl (801 °C) much higher than the melting point of water (0 °C)?

7.93 Ethylene and methanol have approximately the same molar mass.

ethylene   methanol

a. What types of intermolecular forces are present in each compound?
b. Which compound has the higher boiling point?
c. Which compound has the higher vapor pressure at a given temperature?

7.94 Ethanol and dimethyl ether have the same molecular formula.

ethanol   dimethyl ether

a. What types of intermolecular forces are present in each compound?
b. Which compound has the higher boiling point?
c. Which compound has the higher vapor pressure at a given temperature?

Liquids and Solids

7.95 What is the difference between vapor pressure and partial pressure?

7.96 What is the difference between viscosity and surface tension?

7.97 Given the following vapor pressures at 20 °C, arrange the compounds in order of increasing boiling point: butane, 1,650 mm Hg; acetaldehyde, 740 mm Hg, Freon-113, 284 mm Hg.

7.98 Using the given boiling points, predict which compound has the higher vapor pressure at a given temperature.
a. ethanol (C_2H_6O, bp 78 °C) or 1-propanol (C_3H_8O, bp 97 °C)
b. hexane (C_6H_{14}, bp 69 °C) or octane (C_8H_{18}, bp 125 °C)

7.99 Explain why glycerol is *more* viscous than water, but acetone is *less* viscous than water. Glycerol is a component of skin lotions and creams. Acetone is the main ingredient in nail polish remover.

glycerol   acetone

7.100 Explain the following observation. When a needle is carefully placed on the surface of water, it floats, yet when its tip is pushed below the surface, it sinks to the bottom.

7.101 What is the difference between an ionic solid and a metallic solid?

7.102 What is the difference between a molecular solid and a network solid?

7.103 Classify each solid as amorphous, ionic, molecular, network, or metallic.
a. KI
b. CO_2
c. bronze, an alloy of Cu and Sn
d. diamond
e. the plastic polyethylene

7.104 Classify each solid as amorphous, ionic, molecular, network, or metallic.
a. $CaCO_3$
b. CH_3COOH (acetic acid)
c. Ag
d. graphite
e. the plastic polypropylene

Energy and Phase Changes

7.105 What is the difference between evaporation and condensation?

7.106 What is the difference between vaporization and condensation?

7.107 What is the difference between sublimation and deposition?

7.108 What is the difference between melting and freezing?

7.109 Indicate whether heat is absorbed or released in each process.
a. melting 100 g of ice
b. freezing 25 g of water
c. condensing 20 g of steam
d. vaporizing 30 g of water

7.110 What is the difference between the heat of fusion and the heat of vaporization?

7.111 Which process requires more energy, melting 250 g of ice or vaporizing 50.0 g of water? The heat of fusion of water is 79.7 cal/g and the heat of vaporization is 540 cal/g.

7.112 How much energy in kilocalories is needed to vaporize 255 g of water? The heat of vaporization of water is 540 cal/g.

General Problems

7.113 Explain the difference between Charles's law and Gay–Lussac's law, both of which deal with the temperature of gases.

7.114 Explain the difference between Charles's law and Avogadro's law, both of which deal with the volume of gases.

7.115 Explain the difference between Boyle's law and Gay–Lussac's law, both of which deal with the pressure of gases.

7.116 What is the difference between the combined gas law and the ideal gas law?

7.117 A balloon is filled with helium at sea level. What happens to the volume of the balloon in each instance? Explain each answer.
a. The balloon floats to a higher altitude.
b. The balloon is placed in a bath of liquid nitrogen at −196 °C.
c. The balloon is placed inside a hyperbaric chamber at a pressure of 2.5 atm.
d. The balloon is heated inside a microwave.

7.118 Suppose you have a fixed amount of gas in a container with a movable piston, as drawn. Re-draw the container and piston to illustrate what it looks like after each of the following changes takes place.

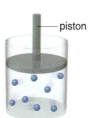

a. The temperature is held constant and the pressure is doubled.
b. The pressure is held constant and the Kelvin temperature is doubled.
c. The pressure is halved and the Kelvin temperature is halved.

Applications

7.119 What is the difference between the systolic and diastolic blood pressure?

7.120 What is hypertension and what are some of its complications?

7.121 If you pack a bag of potato chips for a snack on a plane ride, the bag appears to have inflated when you take it out to open. Explain why this occurs. If the initial volume of air in the bag was 250 mL at 760 mm Hg, and the plane is pressurized at 650 mm Hg, what is the final volume of the bag?

7.122 Why does a bubble at the bottom of a glass of a soft drink get larger as it rises to the surface?

7.123 Explain why cooling a full glass water bottle to −10 °C causes the bottle to crack.

7.124 What happens to the density of a gas if the temperature is increased but the pressure is held constant? Use this information to explain how wind currents arise.

7.125 A common laboratory test for a patient is to measure blood gases—that is, the partial pressures of O_2 and CO_2 in oxygenated blood. Normal values are 100 mm Hg for O_2 and 40 mm Hg for CO_2. A high or low level of one or both readings has some underlying cause. Offer an explanation for each of the following situations.
a. If a patient comes in agitated and hyperventilating—breathing very rapidly—the partial pressure of O_2 is normal but the partial pressure of CO_2 is 22 mm Hg.
b. A patient with chronic lung disease has a partial pressure of O_2 of 60 mm Hg and a partial pressure of CO_2 of 60 mm Hg.

7.126 If a scuba diver inhales 0.50 L of air at a depth of 100. ft and 4.0 atm pressure, what volume does this air occupy at the surface of the water, assuming air pressure is 1.0 atm? When a scuba diver must make a rapid ascent to the surface, he is told to exhale slowly as he ascends. How does your result support this recommendation?

CHALLENGE QUESTIONS

7.127 A gas (4.0 g) occupies 11.2 L at 2 atm and 273 K. What is the molar mass of the gas? What is the identity of the gas?

7.128 As we learned in Chapter 5, an automobile airbag inflates when NaN_3 is converted to Na and N_2 according to the equation, $2\ NaN_3 \longrightarrow 2\ Na + 3\ N_2$. What volume of N_2 would be produced if 100. g of NaN_3 completely reacted at STP?

8

- - - - - - - - - - - - - - - - -

CHAPTER GOALS

In this chapter you will learn how to:

1. Describe the fundamental properties of a solution
2. Predict whether a substance is soluble in water or a nonpolar solvent
3. Predict the effect of temperature and pressure on solubility
4. Calculate the concentration of a solution
5. Prepare a dilute solution from a more concentrated solution
6. Describe the effect of dissolved particles on the boiling point and melting point of a solution
7. Describe the process of osmosis and how it relates to biological membranes and dialysis

A sports drink is a solution of dissolved ions and carbohydrates, used to provide energy and hydration during strenuous exercise.

SOLUTIONS

IN Chapter 8 we study **solutions**—homogeneous mixtures of two or more substances. Why are table salt (NaCl) and sugar (sucrose) soluble in water but vegetable oil and gasoline are not? How does a healthcare professional take a drug as supplied by the manufacturer and prepare a dilute solution to administer a proper dose to a patient? An understanding of solubility and concentration is needed to explain each of these phenomena.

8.1 INTRODUCTION

Thus far we have concentrated primarily on **pure substances**—elements, covalent compounds, and ionic compounds. Most matter with which we come into contact, however, is a **mixture** composed of two or more pure substances. The air we breathe is composed of nitrogen and oxygen, together with small amounts of argon, water vapor, carbon dioxide, and other gases. Seawater is composed largely of sodium chloride and water. A mixture may be **heterogeneous** or **homogeneous.**

- A *heterogeneous* mixture does not have a uniform composition throughout a sample.
- A *homogeneous* mixture has a uniform composition throughout a sample.

A pepperoni pizza is an example of a heterogeneous mixture, while a sports drink is a homogeneous mixture. Homogeneous mixtures are either **solutions** or **colloids.**

- A *solution* is a homogeneous mixture that contains small particles. Liquid solutions are transparent.
- A *colloid* is a homogeneous mixture with larger particles, often having an opaque appearance.

A cup of hot coffee, vinegar, and gasoline are solutions, whereas milk and whipped cream are colloids. Any phase of matter can form a solution (Figure 8.1). Air is a solution of gases. An intravenous saline solution contains solid sodium chloride (NaCl) in liquid water. A dental filling contains liquid mercury (Hg) in solid silver.

When two substances form a solution, the substance present in the lesser amount is called the **solute,** and the substance present in the larger amount is the **solvent.** A solution with water as the solvent is called an **aqueous solution.**

Although a solution can be separated into its pure components, one component of a solution cannot be filtered away from the other component. For a particular solute and solvent, solutions

▼ **FIGURE 8.1 Three Different Types of Solutions**

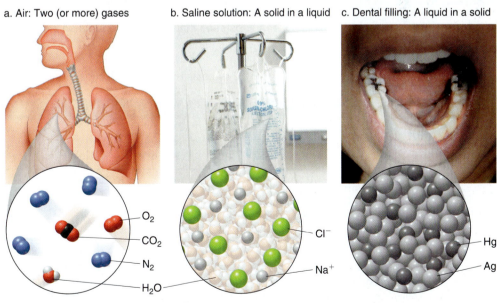

a. Air: Two (or more) gases   b. Saline solution: A solid in a liquid   c. Dental filling: A liquid in a solid

a. Air is a solution of gases, primarily N_2 and O_2. The lungs contain significant amounts of H_2O and CO_2 as well.

b. An IV saline solution contains solid sodium chloride (NaCl) dissolved in liquid water.

c. A dental filling contains a liquid, mercury (Hg), dissolved in solid silver (Ag).

having different compositions are possible. For example, 1.0 g of NaCl can be mixed with 50.0 g of water or 10.0 g of NaCl can be mixed with 50.0 g of water.

An aqueous solution that contains ions conducts electricity, whereas one that contains only neutral molecules does not. Thus, an aqueous solution of sodium chloride, NaCl, contains Na^+ cations and Cl^- anions and conducts electricity. An aqueous solution of hydrogen peroxide, H_2O_2, contains only neutral H_2O_2 molecules in H_2O, so it does not conduct electricity.

- A substance that conducts an electric current in water is called an *electrolyte.* NaCl is an electrolyte.
- A substance that does not conduct an electric current in water is called a *nonelectrolyte.* H_2O_2 is a nonelectrolyte.

Figure 8.2 summarizes the classification of matter.

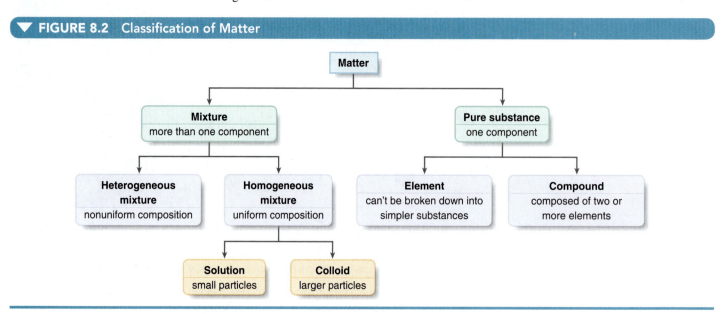

▼ **FIGURE 8.2** **Classification of Matter**

PROBLEM 8.1	Classify each substance as a heterogeneous mixture, solution, or colloid: (a) Cherry Garcia ice cream (cherry ice cream + chocolate bits + cherries); (b) mayonnaise; (c) seltzer water; (d) nail polish remover; (e) brass (an alloy of Cu and Zn).
PROBLEM 8.2	Classify each solution as an electrolyte or nonelectrolyte: (a) KCl in H_2O; (b) sucrose ($C_{12}H_{22}O_{11}$) in H_2O; (c) KI in H_2O.

8.2 SOLUBILITY—GENERAL FEATURES

Solubility is the amount of solute that dissolves in a given amount of solvent, usually reported in grams of solute per 100 mL of solution (g/100 mL). A solution that has less than the maximum number of grams of solute is said to be **unsaturated.** A solution that has the maximum number of grams of solute that can dissolve is said to be **saturated.** If we added more solute to a saturated solution, the additional solute would remain undissolved in the flask.

8.2A BASIC PRINCIPLES

What determines if a compound dissolves in a particular solvent? Whether a compound is soluble in a given solvent depends on the strength of the interactions between the compound and the solvent. As a result, compounds are soluble in solvents to which they are strongly attracted. Solubility is often summed up in three words: **"Like dissolves like."**

- Most ionic and polar covalent compounds are soluble in water, a polar solvent.
- Nonpolar compounds are soluble in nonpolar solvents.

Water-soluble compounds are ionic or are small polar molecules that can hydrogen bond with the water solvent. For example, solid sodium chloride (NaCl) is held together by very strong electrostatic interactions of the oppositely charged ions. When it is mixed with water, the Na^+ and Cl^- ions are separated from each other and surrounded by polar water molecules (Figure 8.3). Each Na^+ is surrounded by water molecules arranged with their O atoms (which bear a partial negative charge) in close proximity to the positive charge of the cation. Each

▼ **FIGURE 8.3   Dissolving Sodium Chloride in Water**

When ionic NaCl dissolves in water, the Na^+ and Cl^- interactions of the crystal are replaced by new interactions of Na^+ and Cl^- ions with the solvent. Each ion is surrounded by a loose shell of water molecules arranged so that oppositely charged species are close to each other.

Cl⁻ is surrounded by water molecules arranged with their H atoms (which bear a partial positive charge) in close proximity to the negative charge of the anion.

- **The attraction of an ion with a dipole in a molecule is called an ion–dipole interaction.**

The ion–dipole interactions between Na^+, Cl^-, and water provide the energy needed to break apart the ions from the crystal lattice. The water molecules form a loose shell of solvent around each ion. The process of surrounding particles of a solute with solvent molecules is called **solvation.**

Small neutral molecules that can hydrogen bond with water are also soluble. Thus, ethanol (C_2H_5OH), which is present in alcoholic beverages, dissolves in water because hydrogen bonding occurs between the OH group in ethanol and the OH group of water.

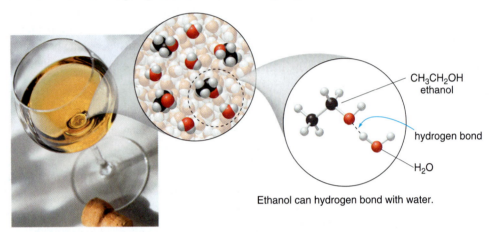

CH₃CH₂OH
ethanol

hydrogen bond

H₂O

Ethanol can hydrogen bond with water.

The basic principles of solubility explain why some vitamins are fat soluble while others are water soluble, as discussed in Chapter 11.

Water solubility for neutral molecules occurs only with small polar molecules or those with many O or N atoms that can hydrogen bond to water. Thus, stearic acid ($C_{18}H_{36}O_2$), a component of animal fats, is *insoluble* in water because its nonpolar part (C—C and C—H bonds) is large compared to its polar part (C—O and O—H bonds). On the other hand, glucose ($C_6H_{12}O_6$), a simple carbohydrate, is *soluble* in water because it has many OH groups and thus many opportunities for hydrogen bonding with water.

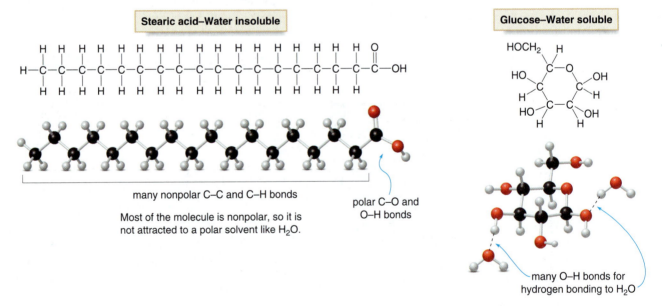

Stearic acid–Water insoluble

many nonpolar C–C and C–H bonds

Most of the molecule is nonpolar, so it is not attracted to a polar solvent like H_2O.

polar C–O and O–H bonds

Glucose–Water soluble

many O–H bonds for hydrogen bonding to H_2O

Nonpolar compounds are soluble in nonpolar solvents. As a result, octane (C_8H_{18}), a component of gasoline, dissolves in the nonpolar solvent carbon tetrachloride (CCl_4), as shown in Figure 8.4. Animal fat and vegetable oils, which are composed largely of nonpolar C—C and C—H bonds, are soluble in CCl_4, but are insoluble in a polar solvent like water. These solubility properties explain why "oil and water don't mix."

FIGURE 8.4 Solubility—A Nonpolar Compound in a Nonpolar Solvent

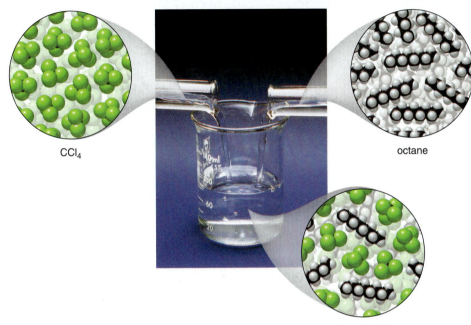

CCl$_4$

octane

octane in CCl$_4$

Octane (C$_8$H$_{18}$) dissolves in CCl$_4$ because both are nonpolar liquids that exhibit only London dispersion forces.

Dissolving a solute in a solvent is a physical process that is accompanied by an energy change. Breaking up the particles of the solute requires energy, and forming new attractive forces between the solute and the solvent releases energy.

- When solvation releases more energy than that required to separate particles, the overall process is exothermic (heat is released).
- When the separation of particles requires more energy than is released during solvation, the process is endothermic (heat is absorbed).

These energy changes are used to an advantage in commercially available hot packs and cold packs. A hot pack, sometimes used for pain relief of sore muscles, contains calcium chloride (CaCl$_2$) or magnesium sulfate (MgSO$_4$) and water. Breaking the seal that separates them allows the salt to dissolve in the water, releasing heat, and the pouch gets warm. In contrast, ammonium nitrate (NH$_4$NO$_3$) absorbs heat on mixing with water, so this salt is found in instant cold packs used to reduce swelling.

SAMPLE PROBLEM 8.1	Predict the water solubility of each compound: (a) KCl; (b) methanol (CH$_3$OH); (c) hexane (C$_6$H$_{14}$).
ANALYSIS	Use the general solubility rule—"like dissolves like." Generally, ionic and small polar compounds that can hydrogen bond are soluble in water. Nonpolar compounds are soluble in nonpolar solvents.
SOLUTION	a. KCl is an ionic compound, so it dissolves in water, a polar solvent.
	b. CH$_3$OH is a small polar molecule that contains an OH group. As a result, it can hydrogen bond to water, making it soluble.
	c. Hexane (C$_6$H$_{14}$) has only nonpolar C—C and C—H bonds, making it a nonpolar molecule that is therefore water insoluble.

PROBLEM 8.3   Which compounds are water soluble?

$$\text{a. NaNO}_3 \qquad \text{b. CH}_4 \qquad \text{c. HO}-\overset{\displaystyle H}{\underset{\displaystyle H}{C}}-\overset{\displaystyle H}{\underset{\displaystyle H}{C}}-\text{OH} \qquad \text{d. KBr} \qquad \text{e. NH}_2\text{OH}$$

PROBLEM 8.4   Explain why table sugar, which has molecular formula $C_{12}H_{22}O_{11}$ and eight OH groups, is water soluble. Would you expect table sugar to dissolve in CCl_4? Explain.

PROBLEM 8.5   Which pairs of compounds will form a solution?

  a. Benzene (C_6H_6) and hexane (C_6H_{14})   c. NaCl and hexane (C_6H_{14})

  b. Na_2SO_4 and H_2O d. H_2O and CCl_4

8.2B   IONIC COMPOUNDS—ADDITIONAL PRINCIPLES

Although ionic compounds are generally water soluble, some are not. If the attraction between the ions in a crystalline solid is stronger than the forces of attraction between the ions and water, the ionic compound does not dissolve. The identity of the cation and anion in the ionic compound determines its water solubility. Two rules can be used to predict water solubility.

General Rules for the Solubility of Ionic Compounds

Rule [1]   A compound is soluble if it contains one of the following cations:
   • Group 1A cations: Li^+, Na^+, K^+, Rb^+, Cs^+
   • Ammonium, NH_4^+

Rule [2]   A compound is soluble if it contains one of the following anions:
   • Halide: Cl^-, Br^-, I^-, except for salts with Ag^+, Hg_2^{2+}, and Pb^{2+}
   • Nitrate, NO_3^-
   • Acetate, $CH_3CO_2^-$
   • Sulfate, SO_4^{2-}, except for salts with Ba^{2+}, Hg_2^{2+}, and Pb^{2+}

Thus, Na_2CO_3 is water soluble because it contains a Na^+ cation (rule [1]), but $CaCO_3$ is water insoluble because it contains none of the ions listed in rules [1] and [2]. In dealing with the solubility of ionic compounds in this text, we will assume the compound to be water soluble unless specifically asked to consider the solubility rules just mentioned.

SAMPLE PROBLEM 8.2   Use the solubility rules to predict whether the following ionic compounds are soluble in water: (a) Na_3PO_4; (b) $Mg_3(PO_4)_2$; (c) KOH.

ANALYSIS   Identify the cation and anion and use the solubility rules to predict if the ionic compound is water soluble.

SOLUTION
   a. Na_3PO_4 contains a Na^+ cation, and all Na^+ salts are soluble regardless of the anion.

   b. $Mg_3(PO_4)_2$ contains none of the cations or anions listed under the solubility rules, so it is water insoluble.

   c. KOH contains a K^+ cation, and all K^+ salts are soluble regardless of the anion.

PROBLEM 8.6 Use the solubility rules to predict whether the following ionic compounds are soluble in water: (a) Li_2CO_3; (b) $MgCO_3$; (c) KBr; (d) $PbSO_4$; (e) $CaCl_2$; (f) $MgCl_2$.

PROBLEM 8.7 Use the solubility rules to predict whether the following ionic compounds are soluble in water: (a) $AgCl$; (b) $AgNO_3$; (c) $Ca(NO_3)_2$; (d) $Ca(OH)_2$.

PROBLEM 8.8 Use the solubility rules for ionic compounds to explain why milk of magnesia, which contains $Mg(OH)_2$ and water, is a heterogeneous mixture rather than a solution.

8.3 SOLUBILITY—EFFECTS OF TEMPERATURE AND PRESSURE

Both temperature and pressure can affect solubility.

8.3A TEMPERATURE EFFECTS

For most ionic and molecular solids, solubility generally increases as temperature increases. Thus, sugar is much more soluble in a cup of hot coffee than in a glass of iced tea. If a solid is dissolved in a solvent at high temperature and then the solution is slowly cooled, the solubility of the solute decreases and it precipitates from the solution. Sometimes, however, if cooling is very slow, the solution becomes **supersaturated** with solute; that is, the solution contains more than the predicted maximum amount of solute at a given temperature. Such a solution is unstable, and when it is disturbed, the solute precipitates rapidly.

In contrast, **the solubility of gases *decreases* with increasing temperature.** Because increasing temperature increases the kinetic energy, more molecules escape into the gas phase and fewer remain in solution. Increasing temperature decreases the solubility of oxygen in lakes and streams. In cases where industrial plants operating near lakes or streams have raised water temperature, marine life dies from lack of sufficient oxygen in solution.

PROBLEM 8.9 Why does a soft drink become "flat" faster when it is left open at room temperature compared to when it is left open in the refrigerator?

8.3B PRESSURE EFFECTS

Pressure changes do not affect the solubility of liquids and solids, but pressure affects the solubility of gases a great deal. **Henry's law** describes the effect of pressure on gas solubility.

- **Henry's law: The solubility of a gas in a liquid is proportional to the partial pressure of the gas above the liquid.**

Thus, the **higher the pressure, the higher the solubility of a gas in a solvent.** A practical demonstration of Henry's law occurs whenever we open a carbonated soft drink. Soft drinks containing dissolved CO_2 are sealed under greater than 1 atm pressure. When a can is opened, the pressure above the liquid decreases to 1 atm, so the solubility of the CO_2 in the soda decreases as well and some of the dissolved CO_2 fizzes out of solution (Figure 8.5).

As we learned in Section 7.6, increasing gas solubility affects scuba divers because more N_2 is dissolved in the blood under the higher pressures experienced under water. Divers must ascend slowly to avoid forming bubbles of N_2 in joints and small blood vessels. If a diver ascends slowly, the external pressure around the diver slowly decreases and by Henry's law, the solubility of the gas in the diver's blood slowly decreases as well.

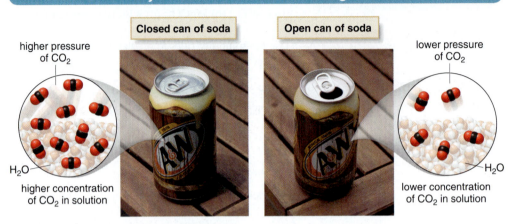

▼ FIGURE 8.5 Henry's Law and Carbonated Beverages

The air pressure in a closed can of soda is approximately 2 atm. When the can is opened, the pressure above the liquid in the can decreases to 1 atm, so the CO_2 concentration in the soda decreases as well, and the gas fizzes from the soda.

PROBLEM 8.10

Predict the effect each change has on the solubility of [1] $Na_2CO_3(s)$; [2] $N_2(g)$.

  a. increasing the temperature   c. increasing the pressure

  b. decreasing the temperature   d. decreasing the pressure

8.4 CONCENTRATION UNITS—PERCENT CONCENTRATION

In using a solution in the laboratory or in administering the proper dose of a liquid medication, we must know its ***concentration*—how much solute is dissolved in a given amount of solution.** Concentration can be measured in several different ways that use mass, volume, or moles. Two useful measures of concentration are reported as percentages—that is, the number of grams or milliliters of solute per 100 mL of solution.

8.4A WEIGHT/VOLUME PERCENT

One of the most common measures of concentration is **weight/volume percent concentration, (w/v)%—that is, the number of grams of solute dissolved in 100 mL of solution.** Mathematically, weight/volume percent is calculated by dividing the number of grams of solute in a given number of milliliters of solution, and multiplying by 100%.

Weight/volume percent concentration
$$(w/v)\% = \frac{\text{mass of solute (g)}}{\text{volume of solution (mL)}} \times 100\%$$

For example, vinegar contains 5 g of acetic acid dissolved in 100 mL of solution, so the acetic acid concentration is 5% (w/v).

$$(w/v)\% = \frac{5 \text{ g acetic acid}}{100 \text{ mL vinegar solution}} \times 100\% = 5\% \text{ (w/v) acetic acid}$$

Note that the volume used to calculate concentration is the *final* volume of the solution, not the volume of solvent added to make the solution. A special flask called a **volumetric flask** is used to make a solution of a given concentration (Figure 8.6). The solute is placed in the flask and then enough solvent is added to dissolve the solute by mixing. Next, additional solvent is added until it reaches a calibrated line that measures the final volume of the solution.

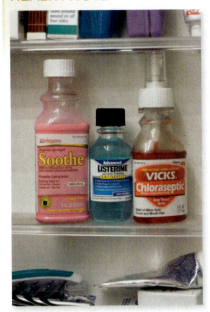

Mouthwash, sore throat spray, and many other over-the-counter medications contain ingredients whose concentrations are reported in (w/v)%.

▼ **FIGURE 8.6** **Making a Solution with a Particular Concentration**

a. Add the solute.

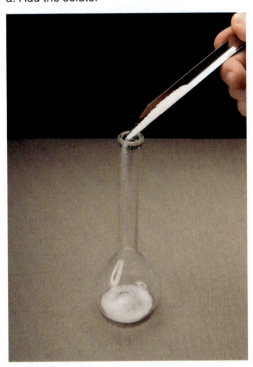

b. Add the solvent.

To make a solution of a given concentration, (a) add a measured number of grams of solute to a volumetric flask; (b) then add solvent to dissolve the solid, bringing the level of the solvent to the calibrated mark on the neck of the flask.

SAMPLE PROBLEM 8.3	Chloraseptic sore throat spray contains 0.35 g of the antiseptic phenol dissolved in 25 mL of solution. What is the weight/volume percent concentration of phenol?
ANALYSIS	Use the formula (w/v)% = (grams of solute)/(mL of solution) × 100%.
SOLUTION	

$$(w/v)\% = \frac{0.35 \text{ g phenol}}{25 \text{ mL solution}} \times 100\% = 1.4\% \text{ (w/v) phenol}$$

Answer

PROBLEM 8.11	Pepto-Bismol, an over-the-counter medication used for upset stomach and diarrhea, contains 525 mg of bismuth subsalicylate in each 15-mL tablespoon. What is the weight/volume percent concentration of bismuth subsalicylate?
PROBLEM 8.12	A commercial mouthwash contains 4.3 g of ethanol and 0.021 g of antiseptic in each 30.-mL portion. Calculate the weight/volume percent concentration of each component.

8.4B VOLUME/VOLUME PERCENT

When the solute in a solution is a liquid, its concentration is often reported using **volume/volume percent concentration, (v/v)%—that is, the number of milliliters of solute dissolved in 100 mL of solution.** Mathematically, volume/volume percent is calculated by dividing the number of milliliters of solute in a given number of milliliters of solution, and multiplying by 100%.

Volume/volume percent concentration $(v/v)\% = \dfrac{\text{volume of solute (mL)}}{\text{volume of solution (mL)}} \times 100\%$

For example, a bottle of rubbing alcohol that contains 70 mL of 2-propanol in 100 mL of solution has a 70% (v/v) concentration of 2-propanol.

$$(v/v)\% = \frac{70 \text{ mL 2-propanol}}{100 \text{ mL rubbing alcohol}} \times 100\% = 70\% \text{ (v/v) 2-propanol}$$

SAMPLE PROBLEM 8.4	A 750-mL bottle of wine contains 101 mL of ethanol. What is the volume/volume percent concentration of ethanol?
ANALYSIS	Use the formula (v/v)% = (mL of solute)/(mL of solution) × 100%.
SOLUTION	$$(v/v)\% = \frac{101 \text{ mL ethanol}}{750 \text{ mL wine}} \times 100\% = 14\% \text{ (v/v) ethanol}$$ **Answer**

PROBLEM 8.13	A 250-mL bottle of mouthwash contains 21 mL of ethanol. What is the volume/volume percent concentration of ethanol?

8.4C USING A PERCENT CONCENTRATION AS A CONVERSION FACTOR

The alcohol (ethanol) content of wine, beer, and other alcoholic beverages is reported using volume/volume percent concentration. Wines typically contain 10–13% (v/v) ethanol, whereas beer usually contains 3–5%.

Percent concentration can be used as a conversion factor to relate the amount of solute (either grams or milliliters) to the amount of solution. For example, ketamine, an anesthetic especially useful for children, is supplied as a 5.0% (w/v) solution, meaning that 5.0 g of ketamine are present in 100 mL of solution. Two conversion factors derived from the percent concentration can be written.

5.0% (w/v) ketamine
weight/volume
percent concentration

$$\frac{5.0 \text{ g ketamine}}{100 \text{ mL solution}} \quad \text{or} \quad \frac{100 \text{ mL solution}}{5.0 \text{ g ketamine}}$$

We can use these conversion factors to determine the amount of solute contained in a given volume of solution (Sample Problem 8.5), or to determine how much solution contains a given number of grams of solute (Sample Problem 8.6). Each of these types of problems is solved using conversion factors in the stepwise procedure first outlined in Section 1.7B.

SAMPLE PROBLEM 8.5	A saline solution used in intravenous drips for patients who cannot take oral fluids contains 0.92% (w/v) NaCl in water. How many grams of NaCl are contained in 250 mL of this solution?

ANALYSIS AND SOLUTION

[1] **Identify the known quantities and the desired quantity.**

0.92% (w/v) NaCl solution

250 mL	? g NaCl
known quantities	desired quantity

[2] **Write out the conversion factors.**

- Set up conversion factors that relate grams of NaCl to the volume of the solution using the weight/volume percent concentration. Choose the conversion factor so that the unwanted unit, mL solution, cancels.

$$\frac{100 \text{ mL solution}}{0.92 \text{ g NaCl}} \quad \text{or} \quad \boxed{\frac{0.92 \text{ g NaCl}}{100 \text{ mL solution}}}$$

Choose this conversion factor to cancel mL.

[3]   **Solve the problem.**

- Multiply the original quantity by the conversion factor to obtain the desired quantity.

$$250 \; \cancel{mL} \times \frac{0.92 \text{ g NaCl}}{100 \; \cancel{mL} \text{ solution}} = 2.3 \text{ g NaCl}$$
$$\textbf{Answer}$$

SAMPLE PROBLEM 8.6

What volume of a 5.0% (w/v) solution of ketamine contains 75 mg?

ANALYSIS AND SOLUTION

[1]   **Identify the known quantities and the desired quantity.**

5.0% (w/v) ketamine solution
75 mg	? mL ketamine
known quantities	desired quantity

[2]   **Write out the conversion factors.**

- Use the weight/volume percent concentration to set up conversion factors that relate grams of ketamine to mL of solution. Since percent concentration is expressed in grams, a mg–g conversion factor is needed as well. Choose the conversion factors that place the unwanted units, mg and g, in the denominator to cancel.

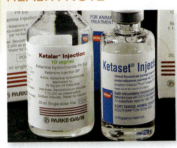

HEALTH NOTE

Ketamine is a widely used anesthetic in both human and veterinary medicine. It has been illegally used as a recreational drug because it can produce hallucinations.

mg–g conversion factors   g–mL solution conversion factors

$$\frac{1000 \text{ mg}}{1 \text{ g}} \quad \text{or} \quad \boxed{\frac{1 \text{ g}}{1000 \text{ mg}}} \qquad\qquad \frac{5.0 \text{ g ketamine}}{100 \text{ mL solution}} \quad \text{or} \quad \boxed{\frac{100 \text{ mL solution}}{5.0 \text{ g ketamine}}}$$

Choose the conversion factors with the unwanted units—g and mg—in the denominator.

[3]   **Solve the problem.**

- Multiply the original quantity by the conversion factors to obtain the desired quantity.

$$75 \; \cancel{mg} \text{ ketamine} \times \frac{1 \; \cancel{g}}{1000 \; \cancel{mg}} \times \frac{100 \text{ mL solution}}{5.0 \; \cancel{g} \text{ ketamine}} = 1.5 \text{ mL solution}$$
$$\textbf{Answer}$$

PROBLEM 8.14

How many mL of ethanol are contained in a 30.-mL portion of a mouthwash that has 8.0% (v/v) of ethanol?

PROBLEM 8.15

A drink sold in a health food store contains 0.50% (w/v) of vitamin C. What volume would you have to ingest to obtain 1,000. mg of vitamin C?

PROBLEM 8.16

A cough medicine contains 0.20% (w/v) dextromethorphan, a cough suppressant, and 2.0% (w/v) guaifenisin, an expectorant. How many milligrams of each drug would you obtain from 3.0 tsp of cough syrup? (1 tsp = 5 mL)

8.4D   PARTS PER MILLION

When a solution contains a very small concentration of solute, concentration is often expressed in **parts per million (ppm).** Whereas percent concentration is the number of "parts"—grams or milliliters—in 100 parts (100 mL) of solution, parts per million is the number of "parts" in 1,000,000 parts of solution. The "parts" may be expressed in either mass or volume units as long as the *same* unit is used for both the numerator and denominator.

ENVIRONMENTAL NOTE

Seabirds such as osprey that feed on fish contaminated with the pesticide DDT accumulate an average of 25 parts per million of DDT in their fatty tissues. When DDT concentration is high, mother osprey produce eggs with very thin shells that are easily crushed, so fewer osprey chicks hatch.

Parts per million

$$\text{ppm} = \frac{\text{mass of solute (g)}}{\text{mass of solution (g)}} \times 10^6$$

or

$$\text{ppm} = \frac{\text{volume of solute (mL)}}{\text{volume of solution (mL)}} \times 10^6$$

A sample of seawater that contains 1.3 g of magnesium ions in 10^6 g of solution contains 1.3 ppm of magnesium.

$$\text{ppm} = \frac{1.3 \text{ g magnesium}}{10^6 \text{ g seawater}} \times 10^6 = 1.3 \text{ ppm magnesium}$$

Parts per million is used as a concentration unit for very dilute solutions. When water is the solvent, the density of the solution is close to the density of pure water, which is 1.0 g/mL at room temperature. In this case, the numerical value of the denominator is the same no matter if the unit is grams or milliliters. Thus, an aqueous solution that contains 2 ppm of MTBE, a gasoline additive and environmental pollutant, can be written in the following ways:

$$\frac{2 \text{ g MTBE}}{10^6 \text{ g solution}} \times 10^6 = \frac{2 \text{ g MTBE}}{10^6 \text{ mL solution}} \times 10^6 = 2 \text{ ppm MTBE}$$

10^6 mL has a mass of 10^6 g.

SAMPLE PROBLEM 8.7

What is the concentration in parts per million of DDT in the tissues of a seabird that contains 50. mg of DDT in 1,900 g of tissue? DDT, a nonbiodegradable pesticide that is a persistent environmental pollutant, has been banned from use in the United States since 1973.

ANALYSIS Use the formula ppm = (g of solute)/(g of solution) $\times 10^6$.

SOLUTION

[1] **Convert milligrams of DDT to grams of DDT so that both the solute and solution have the same unit.**

$$50. \text{ mg DDT} \times \frac{1 \text{ g}}{1000 \text{ mg}} = 0.050 \text{ g DDT}$$

[2] **Use the formula to calculate parts per million.**

$$\frac{0.050 \text{ g DDT}}{1900 \text{ g tissue}} \times 10^6 = 26 \text{ ppm DDT}$$

Answer

PROBLEM 8.17

What is the concentration in parts per million of DDT in each of the following?

a. 0.042 mg in 1,400 g plankton
b. 5×10^{-4} g in 1.0 kg minnow tissue
c. 2.0 mg in 1.0 kg needlefish tissue
d. 225 µg in 1.0 kg breast milk

8.5 CONCENTRATION UNITS—MOLARITY

The most common measure of concentration in the laboratory is *molarity*—**the number of moles of solute per liter of solution,** abbreviated as M.

$$\text{Molarity} \ = \ M \ = \ \frac{\text{moles of solute (mol)}}{\text{liter of solution (L)}}$$

A solution that is formed from 1.00 mol (58.4 g) of NaCl in enough water to give 1.00 L of solution has a molarity of 1.00 M. A solution that is formed from 2.50 mol (146 g) of NaCl in enough water to give 2.50 L of solution is also a 1.00 M solution. Both solutions contain the *same number of moles per unit volume.*

$$M \ = \ \frac{\text{moles of solute (mol)}}{V \ (L)} \ = \ \frac{1.00 \ \text{mol NaCl}}{1.00 \ \text{L solution}} \ = \ 1.00 \ M$$

$$M \ = \ \frac{\text{moles of solute (mol)}}{V \ (L)} \ = \ \frac{2.50 \ \text{mol NaCl}}{2.50 \ \text{L solution}} \ = \ 1.00 \ M$$

same concentration
same number of moles per unit volume (*V*)

Since quantities in the laboratory are weighed on a balance, we must learn how to determine molarity beginning with a particular number of grams of a substance, as shown in the accompanying stepwise procedure.

HOW TO Calculate Molarity from a Given Number of Grams of Solute

EXAMPLE Calculate the molarity of a solution made from 20.0 g of NaOH in 250 mL of solution.

Step [1] **Identify the known quantities and the desired quantity.**

20.0 g NaOH

250 mL solution   ? M (mol/L)

known quantities   desired quantity

Step [2] **Convert the number of grams of solute to the number of moles. Convert the volume of the solution to liters, if necessary.**

- Use the molar mass to convert grams of NaOH to moles of NaOH (molar mass 40.0 g/mol).

molar mass
conversion factor

$$20.0 \ \text{g NaOH} \ \times \ \frac{1 \ \text{mol}}{40.0 \ \text{g NaOH}} \ = \ 0.500 \ \text{mol NaOH}$$

Grams cancel.

- Convert milliliters of solution to liters of solution using a mL–L conversion factor.

mL–L
conversion factor

$$250 \ \text{mL solution} \ \times \ \frac{1 \ \text{L}}{1000 \ \text{mL}} \ = \ 0.25 \ \text{L solution}$$

Milliliters cancel.

Step [3] **Divide the number of moles of solute by the number of liters of solution to obtain the molarity.**

$$M \ = \ \frac{\text{moles of solute (mol)}}{V \ (L)} \ = \ \frac{0.500 \ \text{mol NaOH}}{0.25 \ \text{L solution}} \ = \ 2.0 \ M$$

molarity   **Answer**

SAMPLE PROBLEM 8.8

What is the molarity of an intravenous glucose solution prepared from 108 g of glucose in 2.0 L of solution?

ANALYSIS AND SOLUTION

[1] **Identify the known quantities and the desired quantity.**

108 g glucose

2.0 L solution   ? M (mol/L)

known quantities   desired quantity

[2] **Convert the number of grams of glucose to the number of moles using the molar mass (180.2 g/mol).**

$$108 \text{ g glucose} \times \frac{1 \text{ mol}}{180.2 \text{ g}} = 0.599 \text{ mol glucose}$$

Grams cancel.

- Since the volume of the solution is given in liters, no conversion is necessary for volume.

[3] **Divide the number of moles of solute by the number of liters of solution to obtain the molarity.**

$$M = \frac{\text{moles of solute (mol)}}{V \text{ (L)}} = \frac{0.599 \text{ mol glucose}}{2.0 \text{ L solution}} = 0.30 \text{ M}$$

molarity   **Answer**

PROBLEM 8.18

Calculate the molarity of each aqueous solution with the given amount of NaCl (molar mass 58.4 g/mol) and final volume.

a. 1.0 mol in 0.50 L   c. 0.050 mol in 5.0 mL   e. 24.4 g in 350 mL

b. 2.0 mol in 250 mL   d. 12.0 g in 2.0 L   f. 60.0 g in 750 mL

PROBLEM 8.19

Which solution has the higher concentration, one prepared from 10.0 g of NaOH in a final volume of 150 mL, or one prepared from 15.0 g of NaOH in a final volume of 250 mL of solution?

Molarity is a conversion factor that relates the number of moles of solute to the volume of solution it occupies. Thus, if we know the molarity and volume of a solution, we can calculate the number of moles it contains. If we know the molarity and number of moles, we can calculate the volume in liters.

To calculate the moles of solute...   ...rearrange the equation for molarity (M):

$$\frac{\text{moles of solute (mol)}}{V \text{ (L)}} = M$$

$$\boxed{\text{moles of solute (mol)} = M \times V \text{ (L)}}$$

To calculate the volume of solution...   ...rearrange the equation for molarity (M):

$$\frac{\text{moles of solute (mol)}}{V \text{ (L)}} = M$$

$$\boxed{V \text{ (L)} = \frac{\text{moles of solute (mol)}}{M}}$$

SAMPLE PROBLEM 8.9

What volume in milliliters of a 0.30 M solution of glucose contains 0.025 mol of glucose?

ANALYSIS

Use the equation, $V = $ (moles of solute)/M, to find the volume in liters, and then convert the liters to milliliters.

SOLUTION

[1] **Identify the known quantities and the desired quantity.**

0.30 M   ? V (L) solution

0.025 mol glucose

known quantities   desired quantity

[2] **Divide the number of moles by molarity to obtain the volume in liters.**

$$V\text{ (L)} = \frac{\text{moles of solute (mol)}}{M}$$

$$= \frac{0.025 \text{ mol glucose}}{0.30 \text{ mol/L}} = 0.083 \text{ L solution}$$

[3] **Use a mL–L conversion factor to convert liters to milliliters.**

mL–L conversion factor

$$0.083 \text{ L solution} \quad \times \quad \frac{1000 \text{ mL}}{1 \text{ L}} \quad = \quad 83 \text{ mL glucose solution}$$

Liters cancel.   **Answer**

PROBLEM 8.20

How many milliliters of a 1.5 M glucose solution contain each of the following number of moles?

a. 0.15 mol   b. 0.020 mol   c. 0.0030 mol   d. 3.0 mol

PROBLEM 8.21

How many moles of NaCl are contained in each volume of aqueous NaCl solution?

a. 2.0 L of a 2.0 M solution   c. 25 mL of a 2.0 M solution

b. 2.5 L of a 0.25 M solution   d. 250 mL of a 0.25 M solution

---

Since the number of grams and moles of a substance is related by the molar mass, we can convert a given volume of solution to the number of grams of solute it contains by carrying out the step-wise calculation shown in Sample Problem 8.10.

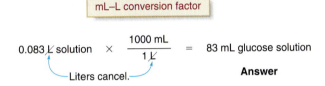

Volume of solution —[1]→ **Moles of solute** —[2]→ **Grams of solute**

M (mol/L) conversion factor   molar mass conversion factor

SAMPLE PROBLEM 8.10

How many grams of aspirin are contained in 50.0 mL of a 0.050 M solution?

ANALYSIS   Use the molarity to convert the volume of the solution to moles of solute. Then use the molar mass to convert moles to grams.

SOLUTION

[1] **Identify the known quantities and the desired quantity.**

0.050 M   ? g aspirin

50.0 mL solution

known quantities   desired quantity

[2] **Determine the number of moles of aspirin using the molarity.**

volume   molarity   mL–L conversion factor

$$50.0 \text{ mL solution} \quad \times \quad \frac{0.050 \text{ mol aspirin}}{1 \text{ L}} \quad \times \quad \frac{1 \text{ L}}{1000 \text{ mL}} \quad = \quad 0.0025 \text{ mol aspirin}$$

[3] Convert the number of moles of aspirin to grams using the molar mass (180.2 g/mol).

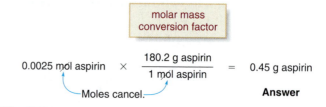

$$0.0025 \text{ mol aspirin} \quad \times \quad \frac{180.2 \text{ g aspirin}}{1 \text{ mol aspirin}} \quad = \quad 0.45 \text{ g aspirin}$$

Moles cancel.   **Answer**

PROBLEM 8.22   How many grams of NaCl are contained in each of the following volumes of a 1.25 M solution?

a. 0.10 L   b. 2.0 L   c. 0.55 L   d. 50. mL

PROBLEM 8.23   How many milliliters of a 0.25 M sucrose solution contain each of the following number of grams? The molar mass of sucrose ($C_{12}H_{22}O_{11}$) is 342.3 g/mol.

a. 0.500 g   b. 2.0 g   c. 1.25 g   d. 50.0 mg

8.6 DILUTION

Sometimes a solution has a higher concentration than is needed. *Dilution* **is the addition of solvent to decrease the concentration of solute.** For example, a stock solution of a drug is often supplied in a concentrated form to take up less space on a pharmacy shelf, and then it is diluted so that it can be administered in a reasonable volume and lower concentration that allows for more accurate dosing.

A key fact to keep in mind is that the **amount of solute is** *constant.* Only the volume of the solution is changed by adding solvent.

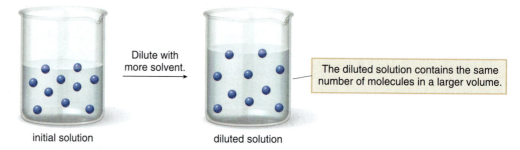

Dilute with more solvent.

The diluted solution contains the same number of molecules in a larger volume.

initial solution   diluted solution

In using molarity as a measure of concentration in Section 8.5, we learned that the number of moles of solute can be calculated from the molarity and volume of a solution.

$$\text{moles of solute} \quad = \quad \text{molarity} \quad \times \quad \text{volume}$$

$$\text{mol} \; = \; MV$$

Thus, if we have initial values for the molarity and volume (M_1 and V_1), we can calculate a new value for the molarity or volume (M_2 or V_2), since the product of the molarity and volume equals the number of moles, a constant.

$$M_1V_1 \qquad = \qquad M_2V_2$$

initial values   final values

Although molarity is the most common concentration measure in the laboratory, the same facts hold in diluting solutions reported in other concentration units—percent concentration and parts per million—as well. In general, therefore, if we have initial values for the concentration and volume (C_1 and V_1), we can calculate a new value for the concentration or volume (C_2 or V_2), since the product of the concentration and volume is a constant.

$$C_1V_1 \quad = \quad C_2V_2$$

initial values   final values

SAMPLE PROBLEM 8.11

What is the concentration of a solution formed by diluting 5.0 mL of a 3.2 M glucose solution to 40.0 mL?

ANALYSIS Since we know an initial molarity and volume (M_1 and V_1) and a final volume (V_2), we can calculate a new molarity (M_2) using the equation $M_1V_1 = M_2V_2$.

SOLUTION

[1] **Identify the known quantities and the desired quantity.**

$$M_1 = 3.2 \text{ M}$$
$$V_1 = 5.0 \text{ mL} \qquad V_2 = 40.0 \text{ mL} \qquad\qquad M_2 = ?$$

known quantities   desired quantity

[2] **Write the equation and rearrange it to isolate the desired quantity, M_2, on one side.**

$$M_1V_1 \quad = \quad M_2V_2 \qquad \text{Solve for } M_2 \text{ by dividing both sides by } V_2.$$

$$\frac{M_1V_1}{V_2} \quad = \quad M_2$$

[3] **Solve the problem.**

- Substitute the three known quantities into the equation and solve for M_2.

$$M_2 \quad = \quad \frac{M_1V_1}{V_2} \quad = \quad \frac{(3.2 \text{ M})(5.0 \text{ mL})}{(40.0 \text{ mL})} \quad = \quad 0.40 \text{ M glucose solution}$$

Answer

SAMPLE PROBLEM 8.12

Dopamine is a potent drug administered intravenously to increase blood pressure in seriously ill patients. How many milliliters of a 4.0% (w/v) solution must be used to prepare 250 mL of a 0.080% (w/v) solution?

ANALYSIS Since we know an initial concentration (C_1), a final concentration (C_2), and a final volume (V_2), we can calculate the volume (V_1) of the initial solution that must be used with the equation, $C_1V_1 = C_2V_2$.

SOLUTION

[1] **Identify the known quantities and the desired quantity.**

$$C_1 = 4.0\% \text{ (w/v)} \qquad C_2 = 0.080\% \text{ (w/v)}$$
$$V_2 = 250 \text{ mL} \qquad\qquad V_1 = ?$$

known quantities   desired quantity

[2] **Write the equation and rearrange it to isolate the desired quantity, V_1, on one side.**

$$C_1V_1 \quad = \quad C_2V_2 \qquad \text{Solve for } V_1 \text{ by dividing both sides by } C_1.$$

$$V_1 \quad = \quad \frac{C_2V_2}{C_1}$$

[3] **Solve the problem.**

- Substitute the three known quantities into the equation and solve for V_1.

$$V_1 \quad = \quad \frac{(0.080\%)(250 \text{ mL})}{4.0\%} \quad = \quad 5.0 \text{ mL dopamine solution}$$

Answer

PROBLEM 8.24	What is the concentration of a solution formed by diluting 25.0 mL of a 3.8 M glucose solution to 275 mL?

PROBLEM 8.25	How many milliliters of a 6.0 M NaOH solution would be needed to prepare each solution?

a. 525 mL of a 2.5 M solution   c. 450 mL of a 0.10 M solution
b. 750 mL of a 4.0 M solution   d. 25 mL of a 3.5 M solution

PROBLEM 8.26	Ketamine, an anesthetic, is supplied in a solution of 100. mg/mL. If 2.0 mL of this solution is diluted to a volume of 10.0 mL, how much of the diluted solution should be administered to supply a dose of 75 mg?

8.7 COLLIGATIVE PROPERTIES

Although many properties of a solution are similar to those of a pure solvent, the boiling point and melting point of a solution differ from the boiling point and melting point of the solvent used to make it.

- *Colligative properties* are properties of a solution that depend on the concentration of the solute but not its identity.

Thus, the *number* of dissolved particles of solute affects the properties of the solution, but the identity of the solute does not. In this section we examine how a dissolved solute increases the boiling point and decreases the melting point of a solution. In Section 8.8, we look at **osmosis,** a process that involves the diffusion of solvent across a semipermeable membrane.

8.7A BOILING POINT ELEVATION

A solute in a solution can be **volatile** or **nonvolatile.**

- A volatile solute readily escapes into the vapor phase.
- A nonvolatile solute does not readily escape into the vapor phase, and thus it has a negligible vapor pressure at a given temperature.

Figure 8.7 compares the vapor pressure above a pure liquid (water) with the vapor pressure above a solution made by dissolving a nonvolatile solute in water. The vapor pressure of a solution

▼ **FIGURE 8.7** *Vapor Pressure Above a Liquid Solution*

| H₂O | pure liquid | solution | solute | H₂O |

When a nonvolatile solute is added to a solvent, there are fewer molecules of solvent in the gas phase, so the vapor pressure of the solution above the solvent is lower.

composed of a nonvolatile solute and a liquid solvent consists solely of gas molecules derived from the solvent. Since there are fewer solvent molecules in the solution than there are in the pure liquid, there are fewer molecules in the gas phase as well. **As a result, the vapor pressure above the solution is *lower* than the vapor pressure of the pure solvent.**

What effect does this lower vapor pressure have on the boiling point of the solution? The boiling point is the temperature at which the vapor pressure equals the atmospheric pressure. A lower vapor pressure means that the solution must be heated to a higher temperature to get the vapor pressure to equal the atmospheric pressure. This results in **boiling point elevation.**

- A liquid solution that contains a nonvolatile solute has a higher boiling point than the solvent alone.

The amount that the boiling point increases depends only on the number of dissolved particles. For example,

- One mole of any nonvolatile solute raises the boiling point of one kilogram of water the same amount, 0.51 °C.

Thus, one mole of glucose molecules raises the boiling point of 1 kg of water by 0.51 °C, to 100.51 °C. Since NaCl contains two particles—Na^+ cations and Cl^- anions—per mole, one mole of NaCl raises the boiling point of 1 kg of water by 2×0.51 °C or 1.02 °C, to 101.02 °C, rounded to 101.0 °C.

SAMPLE PROBLEM 8.13	What is the boiling point of a solution that contains 0.45 mol of KCl in 1.00 kg of water?
ANALYSIS	Determine the number of "particles" contained in the solute. Use 0.51 °C/mol as a conversion factor to relate the temperature change to the number of moles of solute particles.
SOLUTION	Each KCl provides two "particles," K^+ and Cl^-.

$$\text{temperature increase} = \frac{0.51 \text{ °C}}{\text{mol particles}} \times 0.45 \text{ mol KCl} \times \frac{2 \text{ mol particles}}{\text{mol KCl}} = 0.46 \text{ °C}$$

The boiling point of the solution is 100.0 °C + 0.46 °C = 100.46 °C, rounded to 100.5 °C.

PROBLEM 8.27	What is the boiling point of a solution prepared from the given quantity of solute in 1.00 kg of water?

    a. 2.0 mol of sucrose molecules   c. 2.0 mol of $CaCl_2$

    b. 2.0 mol of KNO_3 d. 20.0 g of NaCl

8.7B FREEZING POINT DEPRESSION

In a similar manner, a dissolved solute lowers the freezing point of a solvent. The presence of solute molecules makes it harder for solvent molecules to form an organized crystalline solid, thus lowering the temperature at which the liquid phase becomes solid. This results in **freezing point depression.**

- A liquid solution that contains a nonvolatile solute has a lower freezing point than the solvent alone.

The amount of freezing point depression depends only on the number of dissolved particles. For example,

- One mole of any nonvolatile solute lowers the freezing point of one kilogram of water the same amount, 1.86 °C.

Airplane wings are de-iced with a solution that contains ethylene glycol, which lowers the freezing point, so the ice melts.

Thus, one mole of glucose molecules lowers the freezing point of 1 kg of water to –1.86 °C. Since NaCl contains two particles—Na^+ cations and Cl^- anions—per mole, one mole of NaCl lowers the freezing point of 1 kg of water by 2 × (–1.86 °C) or –3.72 °C.

Several practical applications exploit freezing point depression. Antifreeze, which contains non-volatile ethylene glycol, is added to automobile radiators to lower the freezing point of the water in the cooling system, so it does not freeze in cold climates. Fish that inhabit cold environments produce large amounts of glycerol, $C_3H_8O_3$, which lowers the freezing point of their blood, thus allowing it to remain fluid in very cold water.

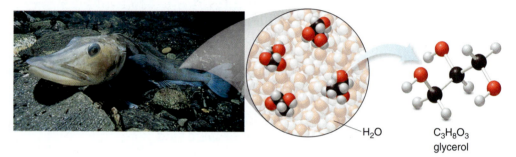

H_2O $C_3H_8O_3$
              glycerol

SAMPLE PROBLEM 8.14	What is the melting point of a solution that contains 3.00 mol of $CaCl_2$ dissolved in 1.00 kg of water? $CaCl_2$ is used in rock salt to melt ice and snow on highways and sidewalks in the winter.
ANALYSIS	Determine the number of "particles" contained in the solute. Use 1.86 °C/mol as a conversion factor to relate the temperature change to the number of moles of solute particles.
SOLUTION	Each $CaCl_2$ provides three "particles," Ca^+ and 2 Cl^-.

$$\text{temperature decrease} = \frac{1.86\ °C}{\text{mol particles}} \times 3.00\ \text{mol } CaCl_2 \times \frac{3\ \text{mol particles}}{\text{mol } CaCl_2} = 16.7\ °C$$

The melting point of the solution is 0.0 °C + –16.7 °C = –16.7 °C.

PROBLEM 8.28	What is the melting point of a solution prepared from the given quantity of solute in 1.00 kg of water?

a. 2.0 mol of sucrose molecules   c. 2.0 mol of $CaCl_2$
b. 2.0 mol of KNO_3 d. 20.0 g of NaCl

PROBLEM 8.29	What is the melting point of a solution that is formed when 250 g of ethylene glycol ($C_2H_6O_2$) is dissolved in 1.00 kg of water?

8.8   OSMOSIS AND DIALYSIS

The membrane that surrounds living cells is an example of a **semipermeable membrane**—a membrane that allows water and small molecules to pass across, but ions and large molecules cannot.

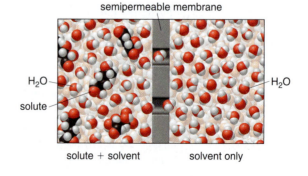

semipermeable membrane

H_2O

solute

H_2O

solute + solvent   solvent only

- *Osmosis* is the passage of water and small molecules across a semipermeable membrane from a solution of low solute concentration to a solution of higher solute concentration.

8.8A   OSMOTIC PRESSURE

What happens when water and an aqueous glucose solution are separated by a semipermeable membrane? Water flows back and forth across the membrane, but more water flows from the side that has pure solvent towards the side that has dissolved glucose. This decreases the volume of pure solvent on one side of the membrane and increases the volume of the glucose solution on the other side.

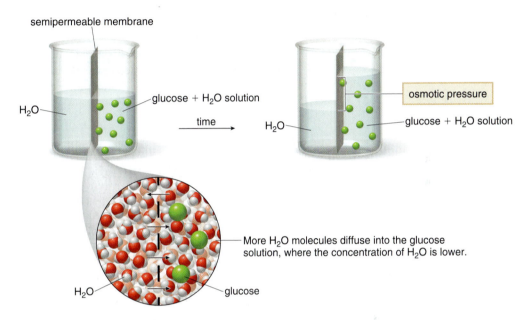

The increased weight of the glucose solution creates increased pressure on one side of the membrane. When the increased pressure gets to a certain point, it prevents more water movement to further dilute the glucose solution. Water continues to diffuse back and forth across the membrane, but the level of the two liquids does not change any further.

- *Osmotic pressure* is the pressure that prevents the flow of additional solvent into a solution on one side of a semipermeable membrane.

Osmotic pressure is a colligative property, so it depends only on the number of particles in a solution. **The greater the number of dissolved particles, the greater the osmotic pressure.** A 0.1 M NaCl solution has twice the osmotic pressure as a 0.1 M glucose solution, since each NaCl is composed of two particles, Na^+ cations and Cl^- anions.

If, instead of having pure water on one side of the membrane, there were two solutions of different concentrations, water would flow from the side of the *less* concentrated solution to dilute the *more* concentrated solution.

SAMPLE PROBLEM 8.15

A 0.1 M glucose solution is separated from a 0.2 M glucose solution by a semipermeable membrane. (a) Which solution exerts the greater osmotic pressure? (b) In which direction will water flow between the two solutions? (c) Describe the level of the two solutions when equilibrium is reached.

ANALYSIS   The solvent (water) flows from the less concentrated solution to the more concentrated solution.

SOLUTION
a. The greater the number of dissolved particles, the higher the osmotic pressure, so the 0.2 M glucose solution exerts the greater pressure.

b. Water will flow from the less concentrated solution (0.1 M) to the more concentrated solution (0.2 M).

c. Since water flows into the 0.2 M solution, its height will increase, and the height of the 0.1 M glucose solution will decrease.

PROBLEM 8.30

Which solution in each pair exerts the greater osmotic pressure?

a. 1.0% sugar solution or 5.0% sugar solution

b. 3.0 M NaCl solution or a 4.0 M NaCl solution

c. 1.0 M glucose solution or a 0.75 M NaCl solution

PROBLEM 8.31

Describe the process that occurs when a 1.0 M NaCl solution is separated from a 1.5 M NaCl solution by a semipermeable membrane in terms of each of the following: (a) the identity of the substances that flow across the membrane; (b) the direction of flow before and after equilibrium is achieved; (c) the height of the solutions after equilibrium is achieved.

8.8B FOCUS ON THE HUMAN BODY
OSMOSIS AND BIOLOGICAL MEMBRANES

Since cell membranes are semipermeable and biological fluids contain dissolved ions and molecules, osmosis is an ongoing phenomenon in living cells. Fluids on both sides of a cell membrane must have the same osmotic pressure to avoid pressure build up inside or outside the cell. Any intravenous solution given to a patient, therefore, must have the same osmotic pressure as the fluids in the body.

- **Two solutions with the same osmotic pressure are said to be *isotonic*.**

Isotonic solutions used in hospitals include 0.92% (w/v) NaCl solution (or 0.15 M NaCl solution) and 5.0% (w/v) glucose solution. Although these solutions do not contain exactly the same ions or molecules present in body fluids, they exert the same osmotic pressure. ***Remember:* With a colligative property the concentration of particles is important, but not the identity of those particles.**

If a red blood cell is placed in an isotonic NaCl solution, called physiological saline solution, the red blood cells retain their same size and shape because the osmotic pressure inside and outside the cell is the same (Figure 8.8a). What happens if a red blood cell is placed in a solution having a different osmotic pressure?

- **A *hypotonic* solution has a lower osmotic pressure than body fluids.**
- **A *hypertonic* solution has a higher osmotic pressure than body fluids.**

In a hypotonic solution, the concentration of particles outside the cell is lower than the concentration of particles inside the cell. In other words, the concentration of water outside the cell is *higher* than the concentration of water inside the cell, so water diffuses inside (Figure 8.8b). As a result, the cell swells and eventually bursts. This swelling and rupture of red blood cells is called **hemolysis.**

In a hypertonic solution, the concentration of particles outside the cell is higher than the concentration of particles inside the cell. In other words, the concentration of water inside the cell is *higher* than the concentration of water outside the cell, so water diffuses out of the cell (Figure 8.8c). As a result, the cell shrinks. This process is called **crenation.**

FIGURE 8.8   The Effect of Osmotic Pressure Differences on Red Blood Cells

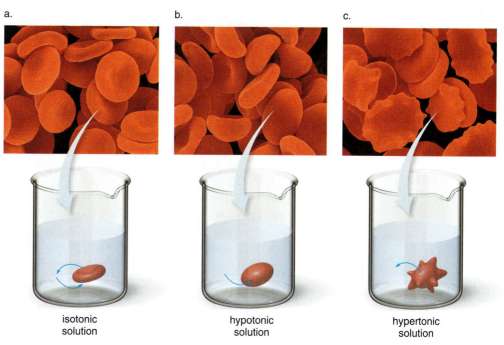

isotonic
solution

hypotonic
solution

hypertonic
solution

(a) In an isotonic solution, the movement of water into and out of the red blood cell occurs to an equal extent and the red blood cell keeps its normal volume. (b) In a hypotonic solution, more water moves into the cell than diffuses out, so the cell swells and eventually it can rupture (hemolysis). (c) In a hypertonic solution, more water moves out of the cell than diffuses in, so the cell shrivels (crenation).

PROBLEM 8.32

What happens to a red blood cell when it is placed in each of the following solutions: (a) 3% (w/v) glucose solution; (b) 0.15 M KCl solution; (c) 0.15 M Na_2CO_3 solution?

8.8C   FOCUS ON HEALTH & MEDICINE
DIALYSIS

Dialysis is also a process that involves the selective passage of substances across a semipermeable membrane, called a dialyzing membrane. In dialysis, however, water, small molecules, and ions can travel across the membrane; only large biological molecules like proteins and starch cannot.

In the human body, blood is filtered through the kidneys by the process of dialysis (Figure 8.9). Each kidney contains over a million nephrons, tubelike structures with filtration membranes. These membranes filter small molecules—glucose, amino acids, urea, ions, and water—from the blood. Useful materials are then reabsorbed, but urea and other waste products are eliminated in urine.

When an individual's kidneys are incapable of removing waste products from the blood, **hemodialysis** is used (Figure 8.10). A patient's blood flows through a long tube connected to a cellophane membrane suspended in an isotonic solution that contains NaCl, KCl, $NaHCO_3$, and glucose. Small molecules like urea cross the membrane into the solution, thus removing them from the blood. Red blood cells and large molecules are not removed from the blood because they are too big to cross the dialyzing membrane.

▼ **FIGURE 8.9** Dialysis of Body Fluids by the Kidneys

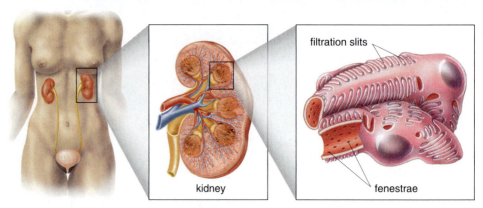

Body fluids are dialyzed by passage through the kidneys, which contain more than a million nephrons that filter out small molecules and ions from the blood. Useful materials are then reabsorbed while urea and other waste products are eliminated in urine.

▼ **FIGURE 8.10** Hemodialysis

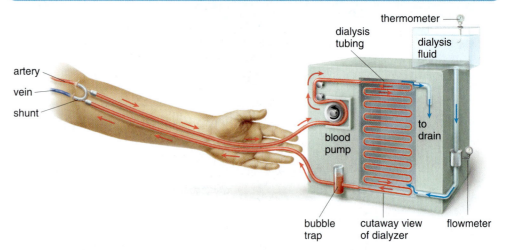

When a patient's kidneys no longer function properly, periodic dialysis treatments are used to remove waste products from the blood. Blood is passed through a dialyzer, which contains a membrane that allows small molecules to pass through, thus acting as an artificial kidney. Each treatment takes several hours. Patients usually require two to three treatments per week.

CHAPTER HIGHLIGHTS

KEY TERMS

Aqueous solution (8.1)
Boiling point elevation (8.7)
Colligative properties (8.7)
Colloid (8.1)
Concentration (8.4)
Dialysis (8.8)
Dilution (8.6)
Electrolyte (8.1)
Freezing point depression (8.7)
Henry's law (8.3)
Heterogeneous mixture (8.1)
Homogeneous mixture (8.1)

Hypertonic solution (8.8)
Hypotonic solution (8.8)
Ion–dipole interaction (8.2)
Isotonic solution (8.8)
Molarity (8.5)
Nonelectrolyte (8.1)
Nonvolatile (8.7)
Osmosis (8.8)
Osmotic pressure (8.8)
Parts per million (8.4)
Saturated solution (8.2)

Semipermeable membrane (8.8)
Solubility (8.2)
Solute (8.1)
Solution (8.1)
Solvation (8.2)
Solvent (8.1)
Supersaturated solution (8.3)
Unsaturated solution (8.2)
Volatile (8.7)
Volume/volume percent concentration (8.4)
Weight/volume percent concentration (8.4)

KEY CONCEPTS

❶ What are the fundamental features of a solution? (8.1)

- A solution is a homogeneous mixture that contains small dissolved particles. Any phase of matter can form solutions. The substance present in the lesser amount is called the solute, and the substance present in the larger amount is the solvent.
- A solution conducts electricity if it contains dissolved ions, but does not conduct electricity if it contains atoms or neutral molecules.

❷ What determines whether a substance is soluble in water or a nonpolar solvent? (8.2)

- One rule summarizes solubility: "Like dissolves like."
- Most ionic compounds are soluble in water. If the attractive forces between the ions and water are stronger than the attraction between the ions in the crystal, an ionic compound dissolves in water.
- Small polar compounds that can hydrogen bond are soluble in water.
- Nonpolar compounds are soluble in nonpolar solvents. Compounds with many nonpolar C—C and C—H bonds are soluble in nonpolar solvents.

❸ What effect do temperature and pressure have on solubility? (8.3)

- The solubility of solids in a liquid solvent generally increases with increasing temperature. The solubility of gases decreases with increasing temperature.
- Increasing pressure increases the solubility of a gas in a solvent. Pressure changes do not affect the solubility of liquids and solids.

❹ How is the concentration of a solution expressed? (8.4, 8.5)

- Concentration is a measure of how much solute is dissolved in a given amount of solution, and can be measured using mass, volume, or moles.

- Weight/volume (w/v) percent concentration is the number of grams of solute dissolved in 100 mL of solution.
- Volume/volume (v/v) percent concentration is the number of milliliters of solute dissolved in 100 mL of solution.
- Parts per million (ppm) is the number of parts of solute in 1,000,000 parts of solution, where the units for both the solute and the solution are the same.
- Molarity (M) is the number of moles of solute per liter of solution.

❺ How are dilutions performed? (8.6)

- Dilution is the addition of solvent to decrease the concentration of a solute. Since the number of moles of solute is constant in carrying out a dilution, a new molarity or volume (M_2 and V_2) can be calculated from a given molarity and volume (M_1 and V_1) using the equation $M_1 V_1 = M_2 V_2$, as long as three of the four quantities are known.

❻ How do dissolved particles affect the boiling point and melting point of a solution? (8.7)

- A nonvolatile solute lowers the vapor pressure above a solution, thus increasing its boiling point.
- A nonvolatile solute makes it harder for solvent molecules to form a crystalline solid, thus decreasing its melting point.

❼ What is osmosis? (8.8)

- Osmosis is the passage of water and small molecules across a semipermeable membrane. Solvent always moves from the less concentrated solution to the more concentrated solution, until the osmotic pressure prevents additional flow of solvent.
- Since living cells contain and are surrounded by biological solutions separated by a semipermeable membrane, the osmotic pressure must be the same on both sides of the membrane. Dialysis is similar to osmosis in that it involves the selective passage of several substances—water, small molecules, and ions—across a dialyzing membrane.

KEY EQUATIONS—CONCENTRATION

Weight/volume percent concentration

$$(w/v)\% = \frac{\text{mass of solute (g)}}{\text{volume of solution (mL)}} \times 100\%$$

Volume/volume percent concentration

$$(v/v)\% = \frac{\text{volume of solute (mL)}}{\text{volume of solution (mL)}} \times 100\%$$

Parts per million

$$ppm = \frac{\text{parts of solute (g or mL)}}{\text{parts of solution (g or mL)}} \times 10^6$$

Molarity

$$M = \frac{\text{moles of solute (mol)}}{\text{liter of solution (L)}}$$

PROBLEMS

Selected in-chapter and end-of-chapter problems have brief answers provided in Appendix B.

Mixtures and Solutions

8.33 What is the difference between a solution and a colloid?

8.34 What is the difference between a homogeneous mixture and a solution?

8.35 Classify each of the following as a heterogeneous mixture, a solution, or a colloid.
   a. bronze (an alloy of Sn and Cu)
   b. diet soda
   c. orange juice with pulp
   d. household ammonia
   e. gasoline
   f. fog

8.36 Classify each of the following as a heterogeneous mixture, a solution, or a colloid.
   a. soft drink   c. wine   e. bleach
   b. cream   d. lava rock   f. apple juice

Solubility

8.37 What is the difference between a solute and a solvent?

8.38 What is the difference between an unsaturated solution and a supersaturated solution?

8.39 If the solubility of KCl in 100 mL of H_2O is 34 g at 20 °C and 43 g at 50 °C, label each of the following solutions as unsaturated, saturated, or supersaturated. If more solid is added than can dissolve in the solvent, assume that undissolved solid remains at the bottom of the flask.
   a. adding 30 g to 100 mL of H_2O at 20 °C
   b. adding 65 g to 100 mL of H_2O at 50 °C
   c. adding 20 g to 50 mL of H_2O at 20 °C
   d. adding 42 g to 100 mL of H_2O at 50 °C and slowly cooling to 20 °C to give a clear solution with no precipitate

8.40 If the solubility of sucrose in 100 mL of H_2O is 204 g at 20 °C and 260 g at 50 °C, label each of the following solutions as unsaturated, saturated, or supersaturated.

If more solid is added than can dissolve in the solvent, assume that undissolved solid remains at the bottom of the flask.
   a. adding 200 g to 100 mL of H_2O at 20 °C
   b. adding 245 g to 100 mL of H_2O at 50 °C
   c. adding 110 g to 50 mL of H_2O at 20 °C
   d. adding 220 g to 100 mL of H_2O at 50 °C and slowly cooling to 20 °C to give a clear solution with no precipitate

8.41 Which compounds are soluble in water?
   a. LiCl
   b. (structure of methylbenzene / toluene)
   c. (structure: $H-C(H)(H)-C(=O)-C(H)(H)-H$)
   d. Na_3PO_4

8.42 Which compounds are soluble in water?
   a. C_5H_{12}
   b. $CaCl_2$
   c. (structure: $H-C(H)(H)-N(H)-H$)
   d. CH_3Br

8.43 Explain the statement, "Oil and water don't mix."

8.44 Explain why a bottle of salad dressing that contains oil and vinegar has two layers.

8.45 Predict the solubility of solid I_2 in water and in CCl_4. Explain your choices.

8.46 Glycine is a covalent compound that contains two charged atoms. Explain why glycine, an amino acid used to make proteins, is soluble in water.

glycine

8.47 Explain why cholesterol, a compound with molecular formula $C_{27}H_{46}O$ and one OH group, is soluble in CCl_4 but insoluble in water.

8.48 Which of the following pairs of compounds form a solution?
a. KCl and CCl_4
b. 1-propanol (C_3H_8O) and H_2O
c. cyclodecanone ($C_{10}H_{18}O$) and H_2O
d. pentane (C_5H_{12}) and hexane (C_6H_{14})

8.49 How is the solubility of solid NaCl in water affected by each of the following changes?
a. increasing the temperature from 25 °C to 50 °C
b. decreasing the temperature from 25 °C to 0 °C
c. increasing the pressure from 1 atm to 2 atm
d. decreasing the pressure from 5 atm to 1 atm

8.50 How is the solubility of helium gas in water affected by each of the following changes?
a. increasing the temperature from 25 °C to 50 °C
b. decreasing the temperature from 25 °C to 0 °C
c. increasing the pressure from 1 atm to 2 atm
d. decreasing the pressure from 5 atm to 1 atm

8.51 Explain the effect of a decrease in temperature on the solubility of each type of solute in a liquid solvent: (a) gas; (b) solid.

8.52 Explain the effect of a decrease in pressure on the solubility of each type of solute in a liquid solvent: (a) gas; (b) solid.

8.53 Explain why many ionic compounds are soluble in water.

8.54 Explain why some ionic compounds are insoluble in water.

8.55 Use the solubility rules listed in Section 8.2B to predict whether each of the following ionic compounds is soluble in water.
a. K_2SO_4 e. $Fe(NO_3)_3$
b. $MgSO_4$ f. $PbCl_2$
c. $ZnCO_3$ g. CsCl
d. KI   h. $Ni(HCO_3)_2$

8.56 Use the solubility rules listed in Section 8.2B to predict whether each of the following ionic compounds is soluble in water.
a. $Al(NO_3)_3$ e. $CuCO_3$
b. $NaHCO_3$ f. $(NH_4)_2SO_4$
c. $Cr(OH)_2$ g. $Fe(OH)_3$
d. LiOH   h. $(NH_4)_3PO_4$

Concentration

8.57 What is the difference between weight/volume percent concentration and molarity?

8.58 What is the difference between volume/volume percent concentration and parts per million?

8.59 Write two conversion factors for each concentration.
a. 5% (w/v)   b. 6.0 M   c. 10 ppm

8.60 Write two conversion factors for each concentration.
a. 15% (v/v)   b. 12.0 M   c. 15 ppm

8.61 What is the weight/volume percent concentration using the given amount of solute and total volume of solution?
a. 10.0 g of LiCl in 750 mL of solution
b. 25 g of $NaNO_3$ in 150 mL of solution
c. 40.0 g of NaOH in 500. mL of solution

8.62 What is the weight/volume percent concentration using the given amount of solute and total volume of solution?
a. 5.5 g of LiCl in 550 mL of solution
b. 12.5 g of $NaNO_3$ in 250 mL of solution
c. 20.0 g of NaOH in 400. mL of solution

8.63 What is the volume/volume percent concentration of a solution prepared from 25 mL of ethyl acetate in 150 mL of solution?

8.64 What is the volume/volume percent concentration of a solution prepared from 75 mL of acetone in 250 mL of solution?

8.65 What is the molarity of a solution prepared using the given amount of solute and total volume of solution?
a. 3.5 mol of KCl in 1.50 L of solution
b. 0.44 mol of $NaNO_3$ in 855 mL of solution
c. 25.0 g of NaCl in 650 mL of solution
d. 10.0 g of $NaHCO_3$ in 3.3 L of solution

8.66 What is the molarity of a solution prepared using the given amount of solute and total volume of solution?
a. 2.4 mol of NaOH in 1.50 L of solution
b. 0.48 mol of KNO_3 in 750 mL of solution
c. 25.0 g of KCl in 650 mL of solution
d. 10.0 g of Na_2CO_3 in 3.8 L of solution

8.67 How would you use a 250-mL volumetric flask to prepare each of the following solutions?
a. 4.8% (w/v) acetic acid in water
b. 22% (v/v) ethyl acetate in water
c. 2.5 M NaCl solution

8.68 How would you use a 250-mL volumetric flask to prepare each of the following solutions?
a. 2.0% (w/v) KCl in water
b. 34% (v/v) ethanol in water
c. 4.0 M NaCl solution

8.69 How many moles of solute are contained in each solution?
a. 150 mL of a 0.25 M $NaNO_3$ solution
b. 45 mL of a 2.0 M HNO_3 solution
c. 2.5 L of a 1.5 M HCl solution

8.70 How many moles of solute are contained in each solution?
a. 250 mL of a 0.55 M $NaNO_3$ solution
b. 145 mL of a 4.0 M HNO_3 solution
c. 6.5 L of a 2.5 M HCl solution

8.71 How many grams of solute are contained in each solution in Problem 8.69?

8.72 How many grams of solute are contained in each solution in Problem 8.70?

8.73 How many mL of ethanol are contained in a 750-mL bottle of wine that contains 11.0% (v/v) of ethanol?

8.74 What is the molarity of a 20.0% (v/v) aqueous ethanol solution? The density of ethanol (C_2H_6O, molar mass 46.1 g/mol) is 0.790 g/mL.

8.75 A 1.89-L bottle of vinegar contains 5.0% (w/v) of acetic acid ($C_2H_4O_2$, molar mass 60.1 g/mol) in water.
   a. How many grams of acetic acid are present in the container?
   b. How many moles of acetic acid are present in the container?
   c. Convert the weight/volume percent concentration to molarity.

8.76 What is the molarity of a 15% (w/v) glucose solution?

8.77 The maximum safe level of each compound in drinking water is given below. Convert each value to parts per million.
   a. chloroform ($CHCl_3$, a solvent), 80 μg/kg
   b. glyphosate (a pesticide), 700 μg/kg

8.78 The maximum safe level of each metal in drinking water is given below. Convert each value to parts per million.
   a. copper, 1,300 μg/kg
   b. arsenic, 10 μg/kg
   c. chromium, 100 μg/kg

Dilution

8.79 How are the concepts of concentration and dilution related?

8.80 Explain why it is impossible to prepare 200 mL of a 5.0 M NaOH solution by diluting a 2.5 M NaOH solution.

8.81 What is the weight/volume percent concentration of a 30.0% (w/v) solution of vitamin C after each of the following dilutions?
   a. 100. mL diluted to 200. mL
   b. 100. mL diluted to 500. mL
   c. 250 mL diluted to 1.5 L
   d. 0.35 L diluted to 750 mL

8.82 One gram (1.00 g) of vitamin B_3 (niacin) is dissolved in water to give 10.0 mL of solution. (a) What is the weight/volume percent concentration of this solution? (b) What is the concentration of a solution formed by diluting 1.0 mL of this solution to each of the following volumes: [1] 10.0 mL; [2] 2.5 mL; [3] 50.0 mL; [4] 120 mL?

8.83 What is the concentration of a solution formed by diluting 125 mL of 12.0 M HCl solution to 850 mL?

8.84 What is the concentration of a solution formed by diluting 250 mL of 6.0 M NaOH solution to 0.45 L?

8.85 How many milliliters of a 2.5 M NaCl solution would be needed to prepare each solution?
   a. 25 mL of a 1.0 M solution
   b. 1.5 L of a 0.75 M solution
   c. 15 mL of a 0.25 M solution
   d. 250 mL of a 0.025 M solution

8.86 How many milliliters of a 5.0 M sucrose solution would be needed to prepare each solution?
   a. 45 mL of a 4.0 M solution
   b. 150 mL of a 0.5 M solution
   c. 1.2 L of a 0.025 M solution
   d. 750 mL of a 1.0 M solution

Colligative Properties

8.87 What is the difference between a volatile solute and a nonvolatile solute?

8.88 Does pure water have osmotic pressure? Explain why or why not.

8.89 What is the boiling point of a solution that contains each of the following quantities of solute in 1.00 kg of water?
   a. 3.0 mol of fructose molecules
   b. 1.2 mol of KI
   c. 1.5 mol of Na_3PO_4

8.90 What is the freezing point of each solution in Problem 8.89?

8.91 If 150 g of ethylene glycol is added to 1,000. g of water, what is the freezing point?

8.92 How many grams of ethylene glycol must be added to 1,000. g of water to form a solution that has a freezing point of –10. °C?

8.93 In comparing a 1.0 M NaCl solution and a 1.0 M glucose solution, which solution has the higher: (a) boiling point; (b) melting point; (c) osmotic pressure; (d) vapor pressure at a given temperature?

8.94 In comparing a 1.0 M NaCl solution and a 1.0 M $CaCl_2$ solution, which solution has the higher: (a) boiling point; (b) melting point; (c) osmotic pressure; (d) vapor pressure at a given temperature?

8.95 Which solution in each pair has the higher melting point?
   a. 0.10 M NaOH or 0.10 M glucose
   b. 0.20 M NaCl or 0.15 M $CaCl_2$
   c. 0.10 M Na_2SO_4 or 0.10 M Na_3PO_4
   d. 0.10 M glucose or 0.20 M glucose

8.96 Which solution in each pair in Problem 8.95 has the higher boiling point?

Osmosis

8.97 What is the difference between osmosis and osmotic pressure?

8.98 What is the difference between osmosis and dialysis?

8.99 What is the difference between a hypotonic solution and an isotonic solution?

8.100 What is the difference between a hypertonic solution and a hypotonic solution?

8.101 A flask contains two compartments (**A** and **B**) with equal volumes of solution separated by a semipermeable membrane. Describe the final level of the liquids when **A** and **B** contain each of the following solutions.

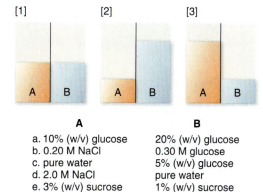

A	**B**
a. 1% (w/v) glucose solution	pure water
b. 0.10 M glucose solution	0.20 M glucose solution
c. 0.10 M NaCl solution	0.10 M NaI solution
d. 0.10 M CaCl$_2$ solution	0.10 M NaCl solution
e. 0.20 M glucose solution	0.10 M NaCl solution

8.102 A flask contains two compartments (**A** and **B**) with equal volumes of solution separated by a semipermeable membrane. Which diagram represents the final level of the liquids when **A** and **B** contain each of the following solutions?

[1]   [2]   [3]

A	**B**
a. 10% (w/v) glucose	20% (w/v) glucose
b. 0.20 M NaCl	0.30 M glucose
c. pure water	5% (w/v) glucose
d. 2.0 M NaCl	pure water
e. 3% (w/v) sucrose	1% (w/v) sucrose

Applications

8.103 Explain why opening a warm can of soda causes a louder "whoosh" and more fizzing than opening a cold can of soda.

8.104 Explain why more sugar dissolves in a cup of hot coffee than a glass of iced coffee.

8.105 If the concentration of glucose in the blood is 90 mg/100 mL, what is the weight/volume percent concentration of glucose? What is the molarity of glucose (molar mass 180.2 g/mol) in the blood?

8.106 If the human body contains 5.0 L of blood, how many grams of glucose are present in the blood if the concentration is 90. mg/100. mL?

8.107 Mannitol, a carbohydrate, is supplied as a 25% (w/v) solution. This hypertonic solution is given to patients who have sustained a head injury with associated brain swelling. (a) What volume should be given to provide a dose of 70. g? (b) How does the hypertonic mannitol benefit brain swelling?

8.108 A patient receives 750 mL of a 10.% (w/v) aqueous glucose solution. (a) How many grams of glucose does the patient receive? (b) How many moles of glucose (molar mass 180.2 g/mol) does the patient receive?

8.109 Explain why a cucumber placed in a concentrated salt solution shrivels.

8.110 Explain why a raisin placed in water swells.

8.111 Explain why the solution contained in a dialyzer used in hemodialysis contains NaCl, KCl, and glucose dissolved in water.

8.112 Explain why pure water is not used in the solution contained in a dialyzer during hemodialysis.

8.113 A sports drink contains 15 g of soluble complex carbohydrates in 8.0 oz (1 oz = 29.6 mL). What weight/volume percent concentration does this represent?

8.114 A sports drink contains 25 mg of magnesium in an 8.0-oz portion (1 oz = 29.6 mL). How many parts per million does this represent? Assume that the mass of 1.0 mL of the solution is 1.0 g.

8.115 Each day, the stomach produces 2.0 L of gastric juice that contains 0.10 M HCl. How many grams of HCl does this correspond to?

8.116 Describe what happens when a red blood cell is placed in pure water.

8.117 An individual is legally intoxicated with a blood alcohol level of 0.08% (w/v) of ethanol. How many milligrams of ethanol are contained in 5.0 L of blood with this level?

8.118 A bottle of vodka labeled "80 proof" contains 40.% (v/v) ethanol in water. How many mL of ethanol are contained in 250 mL of vodka?

CHALLENGE QUESTIONS

8.119 The therapeutic concentration—the concentration needed to be effective—of acetaminophen (C$_8$H$_9$NO$_2$, molar mass 151.2 g/mol) is 10–20 µg/mL. Assume that the density of blood is 1.0 g/mL.
   a. If the concentration of acetaminophen in the blood was measured at 15 ppm, is this concentration in the therapeutic range?
   b. How many moles of acetaminophen are present at this concentration in 5.0 L of blood?

8.120 Very dilute solutions can be measured in parts per billion—that is, the number of parts in 1,000,000,000 parts of solution. To be effective, the concentration of digoxin, a drug used to treat congestive heart failure, must be 0.5–2.0 ng/mL. Convert both values to parts per billion (ppb).

9

CHAPTER GOALS

In this chapter you will learn how to:

① Identify acids and bases and describe their characteristics

② Write equations for acid–base reactions

③ Relate acid strength to the direction of equilibrium of an acid–base reaction

④ Define the acid dissociation constant and relate its magnitude to acid strength

⑤ Define the ion–product of water and use it to calculate hydronium or hydroxide ion concentration

⑥ Calculate pH

⑦ Draw the products of common acid–base reactions

⑧ Determine whether a salt solution is acidic, basic, or neutral

⑨ Use a titration to determine the concentration of an acid or a base

⑩ Describe the basic features of a buffer

⑪ Understand the importance of buffers in maintaining pH in the body

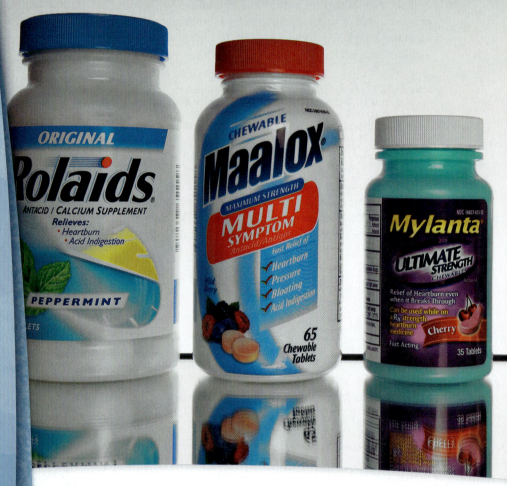

Many over-the-counter medications are acids or bases. The pain reliever aspirin—acetylsalicylic *acid*—is an acid, and the antacids Maalox, Mylanta, and Rolaids all contain a base as their active ingredient.

ACIDS AND BASES

CHEMICAL terms such as *anion* and *cation* may be unfamiliar to most nonscientists, but **acid** has found a place in everyday language. Commercials advertise the latest remedy for the heartburn caused by excess stomach *acid*. The nightly news may report the latest environmental impact of *acid* rain. Wine lovers often know that wine sours because its alcohol has turned to *acid*. *Acid* comes from the Latin word *acidus,* meaning sour, because when tasting compounds was a routine method of identification, these compounds were found to be sour. Acids commonly react with **bases,** and many products, including antacid tablets, glass cleaners, and drain cleaners, all contain bases. In Chapter 9 we learn about the characteristics of acids and bases and the reactions they undergo.

9.1 INTRODUCTION TO ACIDS AND BASES

The earliest definition of acids and bases was suggested by Swedish chemist Svante Arrhenius in the late nineteenth century.

- An *acid* contains a hydrogen atom and dissolves in water to form a hydrogen ion, H^+.
- A *base* contains hydroxide and dissolves in water to form ^-OH.

By the Arrhenius definition, hydrogen chloride (HCl) is an acid because it forms aqueous H^+ and Cl^- when it dissolves in water. Sodium hydroxide (NaOH) is a base because it contains ^-OH and forms solvated Na^+ and ^-OH ions when it dissolves in water.

$$HCl(g) \longrightarrow H^+(aq) + Cl^-(aq)$$

H^+ is formed from HCl.

acid

$$NaOH(s) \longrightarrow Na^+(aq) + {}^-OH(aq)$$

base

^-OH is formed from NaOH.

While the Arrhenius definition correctly predicts the behavior of many acids and bases, this definition is limited and sometimes inaccurate. We now know, for example, that the hydrogen ion, H^+, does *not* exist in water. H^+ is a naked proton with no electrons, and this concentrated positive charge reacts rapidly with a lone pair on H_2O to form the **hydronium ion, H_3O^+.** Although $H^+(aq)$ will sometimes be written in an equation for emphasis, $H_3O^+(aq)$ is actually the reacting species.

$$H^+(aq) + H_2O(l) \longrightarrow H_3O^+(aq)$$

hydrogen ion   hydronium ion

actually present in aqueous solution

does not really exist in aqueous solution

$H^+(aq)$ and $H_3O^+(aq)$ are sometimes used interchangeably by chemists. Keep in mind, however, that $H^+(aq)$ does not really exist in aqueous solution.

Moreover, several compounds contain no hydroxide anions, yet they still exhibit the characteristic properties of a base. Examples include the neutral molecule ammonia (NH_3) and the salt sodium carbonate (Na_2CO_3). As a result, a more general definition of acids and bases, proposed by Johannes **Brønsted** and Thomas **Lowry** in the early twentieth century, is widely used today.

In the Brønsted–Lowry definition, acids and bases are classified according to whether they can donate or accept a **proton**—a positively charged hydrogen ion, H^+.

- A Brønsted–Lowry acid is a *proton donor.*
- A Brønsted–Lowry base is a *proton acceptor.*

Consider what happens when HCl is dissolved in water.

This proton is donated. H_2O accepts a proton.

$$HCl(g) + H_2O(l) \longrightarrow H_3O^+(aq) + Cl^-(aq)$$

Brønsted–Lowry acid   **Brønsted–Lowry base**

- HCl is a Brønsted–Lowry *acid* because it *donates* a proton to the solvent water.
- H_2O is a Brønsted–Lowry *base* because it *accepts* a proton from HCl.

Before we learn more about the details of this process, we must first learn about the characteristics of Brønsted–Lowry acids and bases.

9.1A BRØNSTED–LOWRY ACIDS

A Brønsted–Lowry acid must contain a hydrogen atom. HCl is a Brønsted–Lowry acid because it *donates* a proton (H^+) to water when it dissolves, forming the hydronium ion (H_3O^+) and chloride (Cl^-).

This proton is donated to H_2O.

$$HCl(g) \; + \; H_2O(l) \longrightarrow H_3O^+(aq) \; + \; Cl^-(aq)$$

Brønsted–Lowry acid

Although hydrogen chloride, HCl, is a covalent molecule and a gas at room temperature, when it dissolves in water it **ionizes,** forming two ions, H_3O^+ and Cl^-. An aqueous solution of hydrogen chloride is called **hydrochloric acid.**

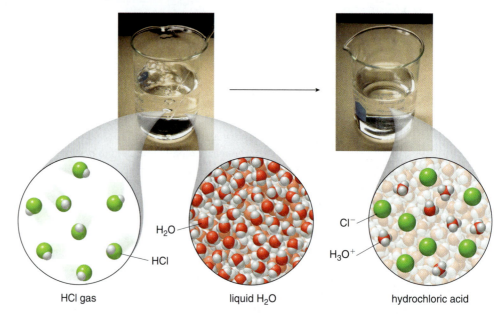

HCl gas   liquid H_2O hydrochloric acid

Because a Brønsted–Lowry acid contains a hydrogen atom, a general Brønsted–Lowry acid is often written as **HA. A** can be a single atom such as Cl or Br. Thus, HCl and HBr are Brønsted–Lowry acids. **A** can also be a polyatomic ion. Sulfuric acid (H_2SO_4) and nitric acid (HNO_3) are Brønsted–Lowry acids, as well. Carboxylic acids are a group of Brønsted–Lowry acids that contain the atoms COOH arranged so that the carbon atom is doubly bonded to one O atom and singly bonded to another. Acetic acid, CH_3COOH, is a simple carboxylic acid. Although carboxylic acids may contain several hydrogen atoms, the **H atom of the OH group is the acidic proton that is donated.**

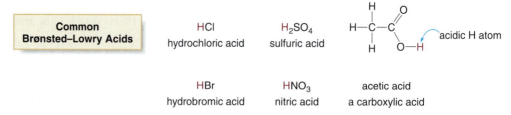

A Brønsted–Lowry acid may contain one or more protons that can be donated.

- A monoprotic acid contains *one* acidic proton. HCl is a monoprotic acid.
- A diprotic acid contains *two* acidic protons. H_2SO_4 is a diprotic acid.
- A triprotic acid contains *three* acidic protons. H_3PO_4 is a triprotic acid.

▼ FIGURE 9.1 Examples of Brønsted–Lowry Acids in Food Products

a.   b.   c.

acetic acid   citric acid   carbonic acid
CH_3COOH $C_6H_8O_7$ H_2CO_3

a. Acetic acid is the sour-tasting component of vinegar. The air oxidation of ethanol to acetic acid is the process that makes "bad" wine taste sour.

b. Citric acid imparts a sour taste to oranges, lemons, and other citrus fruits.

c. Carbonated beverages contain carbonic acid, H_2CO_3.

Although a Brønsted–Lowry acid must contain a hydrogen atom, **it may be a neutral molecule or contain a net positive or negative charge.** Thus, H_3O^+, HCl, and HSO_4^- are all Brønsted–Lowry acids even though their net charges are +1, 0, and –1, respectively. Vinegar, citrus fruits, and carbonated soft drinks all contain Brønsted–Lowry acids, as shown in Figure 9.1.

SAMPLE PROBLEM 9.1	Which of the following species can be Brønsted–Lowry acids: (a) HF; (b) HSO_3^-; (c) Cl_2?
ANALYSIS	A Brønsted–Lowry acid must contain a hydrogen atom, but it may be neutral or contain a net positive or negative charge.
SOLUTION	a. HF is a Brønsted–Lowry acid since it contains a H.
	b. HSO_3^- is a Brønsted–Lowry acid since it contains a H.
	c. Cl_2 is not a Brønsted–Lowry acid because it does not contain a H.

PROBLEM 9.1	Which of the following species can be Brønsted–Lowry acids: (a) HI; (b) SO_4^{2-}; (c) $H_2PO_4^-$; (d) Cl^-?

9.1B BRØNSTED–LOWRY BASES

A Brønsted–Lowry base is a proton acceptor and as such, it must be able to form a bond to a proton. Because a proton has no electrons, **a base must contain a lone pair of electrons** that can be donated to form a new bond. Thus, ammonia (NH_3) is a Brønsted–Lowry base because it contains a nitrogen atom with a lone pair of electrons. When NH_3 is dissolved in water, its N atom accepts a proton from H_2O, forming an ammonium cation (NH_4^+) and hydroxide (^-OH).

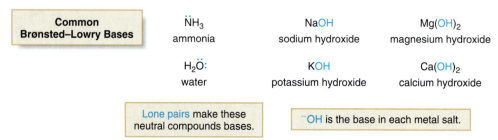

This electron pair forms a new bond to a H from H_2O.

$$H-\overset{\overset{\displaystyle H}{|}}{\underset{\underset{\displaystyle H}{|}}{N}}-H \ + \ H_2O(l) \ \longrightarrow \ \left[H-\overset{\overset{\displaystyle H}{|}}{\underset{\underset{\displaystyle H}{|}}{N}}-H\right]^+ \ + \ {}^-OH(aq)$$

Brønsted–Lowry base

A general Brønsted–Lowry base is often written as **B:** to emphasize that the base must contain a lone pair of electrons to bond to a proton. A base may be neutral or, more commonly, have a net negative charge. Ammonia (NH_3) and water (H_2O) are both Brønsted–Lowry bases because each contains an atom with a lone pair of electrons. Hydroxide (^-OH), which contains an oxygen atom with three lone pairs of electrons, is the most common Brønsted–Lowry base. The source of hydroxide anions can be a variety of metal salts, including NaOH, KOH, $Mg(OH)_2$, and $Ca(OH)_2$.

Common Brønsted–Lowry Bases	$\overset{..}{N}H_3$ ammonia	NaOH sodium hydroxide	$Mg(OH)_2$ magnesium hydroxide
	$H_2\overset{..}{O}:$ water	KOH potassium hydroxide	$Ca(OH)_2$ calcium hydroxide

Lone pairs make these neutral compounds bases.

^-OH is the base in each metal salt.

Many consumer products contain Brønsted–Lowry bases, as shown in Figure 9.2.

▼ **FIGURE 9.2   Examples of Brønsted–Lowry Bases in Consumer Products**

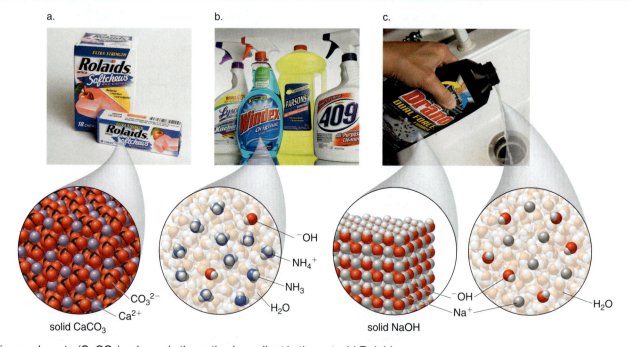

a.

b.

c.

^-OH
NH_4^+
NH_3
CO_3^{2-}
Ca^{2+}
H_2O
solid $CaCO_3$

^-OH
Na^+
H_2O
solid NaOH

a. Calcium carbonate ($CaCO_3$), a base, is the active ingredient in the antacid Rolaids.

b. Windex and other household cleaners contain ammonia (NH_3) dissolved in water, forming NH_4^+ cations and ^-OH anions.

c. Drain cleaners contain pellets of solid sodium hydroxide (NaOH), which form Na^+ cations and ^-OH anions when mixed with water.

SAMPLE PROBLEM 9.2	Which of the following species can be Brønsted–Lowry bases: (a) LiOH; (b) Cl^-; (c) CH_4?
ANALYSIS	A Brønsted–Lowry base must contain a lone pair of electrons, but it may be neutral or have a net negative charge.
SOLUTION	a. LiOH is a base since it contains hydroxide, ^-OH, which has three lone pairs on its O atom.
	b. Cl^- is a base since it has four lone pairs.
	c. CH_4 is not a base since it has no lone pairs.

PROBLEM 9.2	Which of the following species can be Brønsted–Lowry bases: (a) $Al(OH)_3$; (b) Br^-; (c) NH_4^+; (d) ^-CN?

SAMPLE PROBLEM 9.3	Classify each reactant as a Brønsted–Lowry acid or base.
	a. $HF(g) + H_2O(l) \longrightarrow F^-(aq) + H_3O^+(aq)$
	b. $SO_4^{2-}(aq) + H_2O(l) \longrightarrow HSO_4^-(aq) + {}^-OH(aq)$
ANALYSIS	In each equation, the Brønsted–Lowry acid is the species that loses a proton and the Brønsted–Lowry base is the species that gains a proton.
SOLUTION	a. HF is the acid since it loses a proton (H^+) to form F^-, and H_2O is the base since it gains a proton to form H_3O^+.

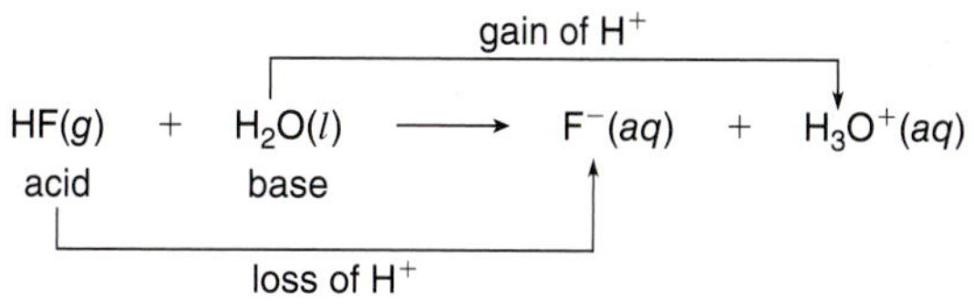

b. H_2O is the acid since it loses a proton (H^+) to form ^-OH, and SO_4^{2-} is the base since it gains a proton to form HSO_4^-.

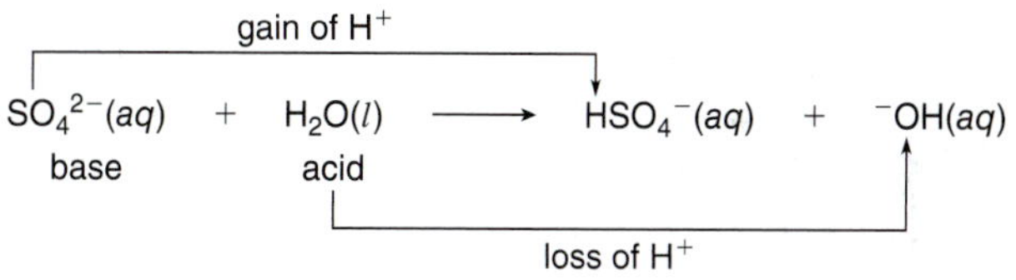

PROBLEM 9.3	Classify each reactant as a Brønsted–Lowry acid or base.
	a. $HCl(g) + NH_3(g) \longrightarrow Cl^-(aq) + NH_4^+(aq)$
	b. $CH_3COOH(l) + H_2O(l) \longrightarrow CH_3COO^-(aq) + H_3O^+(aq)$
	c. $^-OH(aq) + HSO_4^-(aq) \longrightarrow H_2O(l) + SO_4^{2-}(aq)$

9.2 PROTON TRANSFER—THE REACTION OF A BRØNSTED–LOWRY ACID WITH A BRØNSTED–LOWRY BASE

When a Brønsted–Lowry acid reacts with a Brønsted–Lowry base, a proton is *transferred* from the acid to the base. **The Brønsted–Lowry acid donates a proton to the Brønsted–Lowry base, which accepts it.**

Consider, for example, the reaction of the general acid H—A with the general base B:. **In an acid–base reaction, one bond is broken and one bond is formed.** The electron pair of the base B: forms a new bond to the proton of the acid, forming H—B$^+$. The acid H—A loses a proton, leaving the electron pair in the H—A bond on A, forming A:$^-$.

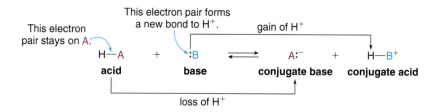

- The product formed by loss of a proton from an acid is called its *conjugate base.*
- The product formed by gain of a proton by a base is called its *conjugate acid.*

Thus, the conjugate base of the acid HA is A:$^-$. The conjugate acid of the base B: is HB$^+$.

- Two species that differ by the presence of a proton are called a conjugate acid–base pair.

Thus, in an acid–base reaction, the acid and the base on the left side of the equation (HA and B:) form two products that are also an acid and a base (HB$^+$ and A:$^-$). Equilibrium arrows ($\rightleftharpoons$) are often used to separate reactants and products because the reaction can proceed in either the forward or the reverse directions. In some reactions, the products are greatly favored, as discussed in Section 9.3.

When HBr is dissolved in water, for example, the acid HBr loses a proton to form its conjugate base Br$^-$, and the base H_2O gains a proton to form H_3O^+.

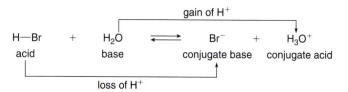

Thus, HBr and Br$^-$ are a conjugate acid–base pair since these two species differ by the presence of a proton (H$^+$). H_2O and H_3O^+ are also a conjugate acid–base pair because these two species differ by the presence of a proton as well.

The net charge must be the same on both sides of the equation. In this example, the two reactants are neutral (zero net charge), and the sum of the −1 and +1 charges in the products is also zero.

Take particular note of what happens to the charges in each conjugate acid–base pair. **When a species gains a proton (H$^+$), it gains a +1 charge.** Thus, if a reactant is neutral to begin with, it ends up with a +1 charge. **When a species loses a proton (H$^+$), it effectively gains a −1 charge** since the product has one fewer proton (+1 charge) than it started with. Thus, if a reactant is neutral to begin with, it ends up with a −1 charge.

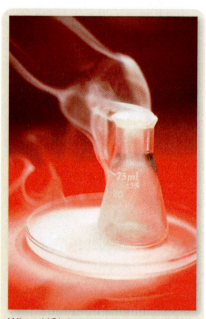

When HCl donates a proton to NH_3 in the absence of water, NH_4^+ and Cl$^-$ are formed, which combine to form solid ammonium chloride, NH_4Cl.

The reaction of ammonia (NH_3) with HCl is also a Brønsted–Lowry acid–base reaction. In this example, NH_3 is the base since it gains a proton to form its conjugate acid, NH_4^+. HCl is the acid since it donates a proton, forming its conjugate base, Cl$^-$.

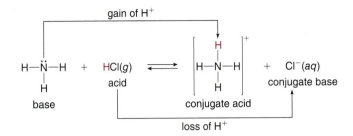

- A Brønsted–Lowry acid–base reaction is a *proton transfer reaction* since it always results in the transfer of a proton from an acid to a base.

The ability to identify and draw a conjugate acid or base from a given starting material is a necessary skill, illustrated in Sample Problems 9.4 and 9.5.

SAMPLE PROBLEM 9.4

Draw the conjugate acid of each base: (a) F^-; (b) NO_3^-.

ANALYSIS

Conjugate acid–base pairs differ by the presence of a proton. To draw a conjugate acid from a base, *add* a proton, H^+. This adds +1 to the charge of the base to give the charge on the conjugate acid.

SOLUTION

a. $F^- + H^+$ gives HF as the conjugate acid. HF has no charge since a proton with a +1 charge is added to an anion with a –1 charge.

b. $NO_3^- + H^+$ gives HNO_3 (nitric acid) as the conjugate acid. HNO_3 has no charge since a proton with a +1 charge is added to an anion with a –1 charge.

PROBLEM 9.4

Draw the conjugate acid of each species: (a) H_2O; (b) I^-; (c) HCO_3^-.

SAMPLE PROBLEM 9.5

Draw the conjugate base of each acid: (a) H_2O; (b) HCO_3^-.

ANALYSIS

Conjugate acid–base pairs differ by the presence of a proton. To draw a conjugate base from an acid, *remove* a proton, H^+. This adds –1 to the charge of the acid to give the charge on the conjugate base.

SOLUTION

a. Remove H^+ from H_2O to form ^-OH, the conjugate base. ^-OH has a –1 charge since –1 is added to a molecule that was neutral to begin with.

b. Remove H^+ from HCO_3^- to form CO_3^{2-}, the conjugate base. CO_3^{2-} has a –2 charge since –1 is added to an anion that had a –1 charge to begin with.

PROBLEM 9.5

Draw the conjugate base of each species: (a) H_2S; (b) HCN; (c) HSO_4^-.

A compound that contains both a hydrogen atom and a lone pair of electrons can be either an acid or a base, depending on the particular reaction. Such a compound is said to be **amphoteric.** For example, when H_2O acts as a base it gains a proton, forming H_3O^+. Thus, H_2O and H_3O^+ are a conjugate acid–base pair. When H_2O acts as an acid it loses a proton, forming ^-OH. H_2O and ^-OH are also a conjugate acid–base pair.

H_2O as a base	$H\!-\!\overset{\cdot\cdot}{\underset{\cdot\cdot}{O}}\!-\!H$ base	$\xrightarrow{\text{add } H^+}$	$\left[H\!-\!\overset{\overset{H}{\mid}}{\underset{}{O}}\!-\!H \right]^+$ conjugate acid
H_2O as an acid	$H\!-\!\overset{\cdot\cdot}{\underset{\cdot\cdot}{O}}\!-\!H$ acid	$\xrightarrow{\text{remove } H^+}$	$\left[H\!-\!\overset{\cdot\cdot}{\underset{\cdot\cdot}{O}}\!: \right]^-$ conjugate base

SAMPLE PROBLEM 9.6

Label the acid and the base and the conjugate acid and the conjugate base in the following reaction.

$$NH_4^+(aq) + {}^-OH(aq) \rightleftharpoons NH_3(g) + H_2O(l)$$

ANALYSIS

The Brønsted–Lowry acid loses a proton to form its conjugate base. The Brønsted–Lowry base gains a proton to form its conjugate acid.

SOLUTION

NH_4^+ is the acid since it loses a proton to form NH_3, its conjugate base. ^-OH is the base since it gains a proton to form its conjugate acid, H_2O.

$$NH_4^+(aq) \quad + \quad {}^-OH(aq) \rightleftharpoons NH_3(g) \quad + \quad H_2O(l)$$
acid   base   conjugate base   conjugate acid

gain of H^+

loss of H^+

PROBLEM 9.6

Label the acid and the base and the conjugate acid and the conjugate base in each reaction.

a. $H_2O(l) + HI(g) \rightleftharpoons I^-(aq) + H_3O^+(aq)$
b. $CH_3COOH(l) + NH_3(g) \rightleftharpoons CH_3COO^-(aq) + NH_4^+(aq)$
c. $Br^-(aq) + HNO_3(aq) \rightleftharpoons HBr(aq) + NO_3^-(aq)$

PROBLEM 9.7

Ammonia, NH_3, is amphoteric. (a) Draw the conjugate acid of NH_3. (b) Draw the conjugate base of NH_3.

PROBLEM 9.8

HSO_3^- can act as an acid or a base. (a) Draw the conjugate base of HSO_3^-. (b) Draw the conjugate acid of HSO_3^-.

9.3 ACID AND BASE STRENGTH

Although all Brønsted–Lowry acids contain protons, some acids readily donate protons while others do not. Similarly, some Brønsted–Lowry bases accept a proton much more readily than others. How readily proton transfer occurs is determined by the strength of the acid and base.

9.3A RELATING ACID AND BASE STRENGTH

When a covalent acid dissolves in water, proton transfer forms H_3O^+ and an anion. This process is called **dissociation**. Acids differ in their tendency to donate a proton; that is, acids differ in the extent to which they dissociate in water.

- A strong acid readily donates a proton. When a strong acid dissolves in water, essentially 100% of the acid dissociates into ions.
- A weak acid less readily donates a proton. When a weak acid dissolves in water, only a small fraction of the acid dissociates into ions.

Common strong acids include **HI, HBr, HCl, H$_2$SO$_4$, and HNO$_3$** (Table 9.1). When each acid is dissolved in water, 100% of the acid dissociates, forming H_3O^+ and the conjugate base.

- Use a single reaction arrow.
- The product is greatly favored at equilibrium.

$$HCl(g) + H_2O(l) \longrightarrow H_3O^+(aq) + Cl^-(aq)$$
strong acid   conjugate base

$$H_2SO_4(l) + H_2O(l) \longrightarrow H_3O^+(aq) + HSO_4^-(aq)$$
strong acid   conjugate base

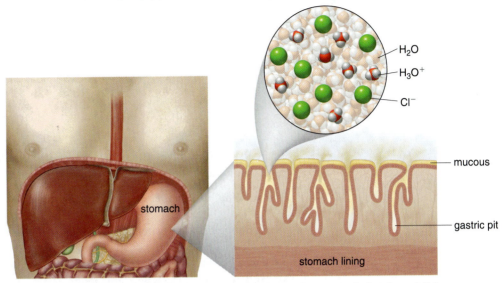

▼ **FIGURE 9.3   Focus on the Human Body: Hydrochloric Acid in the Stomach**

The thick mucous layer protects the stomach lining.

Although HCl is a corrosive acid secreted in the stomach, a thick layer of mucous covering the stomach wall protects it from damage by the strong acid.

HCl, hydrochloric acid, is secreted by the stomach to digest food (Figure 9.3), and **H_2SO_4,** sulfuric acid, is an important industrial starting material in the synthesis of phosphate fertilizers. A single reaction arrow (⟶) is drawn to show that essentially all of the reactants are converted to products.

TABLE 9.1   Relative Strength of Acids and Their Conjugate Bases

Acid		Conjugate Base		
Strong Acids				
Hydroiodic acid	HI	I^-	Iodide ion	
Hydrobromic acid	HBr	Br^-	Bromide ion	
Hydrochloric acid	HCl	Cl^-	Chloride ion	
Sulfuric acid	H_2SO_4	HSO_4^-	Hydrogen sulfate ion	
Nitric acid	HNO_3	NO_3^-	Nitrate ion	
Hydronium ion	H_3O^+	H_2O	Water	
Weak Acids				
Phosphoric acid	H_3PO_4	$H_2PO_4^-$	Dihydrogen phosphate ion	
Hydrofluoric acid	HF	F^-	Fluoride ion	
Acetic acid	CH_3COOH	CH_3COO^-	Acetate ion	
Carbonic acid	H_2CO_3	HCO_3^-	Bicarbonate ion	
Ammonium ion	NH_4^+	NH_3	Ammonia	
Hydrocyanic acid	HCN	^-CN	Cyanide ion	
Water	H_2O	^-OH	Hydroxide ion	

Increasing acid strength ↑

Increasing base strength ↓

Acetic acid, **CH₃COOH,** is a weak acid. When acetic acid dissolves in water, only a small fraction of acetic acid molecules donate a proton to water to form H_3O^+ and the conjugate base, CH_3COO^-. The major species at equilibrium is the undissociated acid, CH_3COOH. Equilibrium arrows that are unequal in length ($\longleftarrow\rightleftharpoons$) are used to show that the equilibrium lies to the left. Other weak acids and their conjugate bases are listed in Table 9.1.

> • Use unequal reaction arrows.
> • The reactants are favored at equilibrium.

$$CH_3COOH(l) \; + \; H_2O(l) \; \rightleftharpoons \; H_3O^+(aq) \; + \; CH_3COO^-(aq)$$
weak acid   conjugate base

Figure 9.4 illustrates the difference between an aqueous solution of a strong acid that is completely dissociated and a weak acid that contains much undissociated acid.

Bases also differ in their ability to accept a proton.

> • A strong base readily accepts a proton. When a strong base dissolves in water, essentially 100% of the base dissociates into ions.
> • A weak base less readily accepts a proton. When a weak base dissolves in water, only a small fraction of the base forms ions.

The most common strong base is hydroxide, **⁻OH,** used as a variety of metal salts, including NaOH and KOH. Solid NaOH dissolves in water to form solvated Na^+ cations and ^-OH anions. In contrast, when **NH₃,** a weak base, dissolves in water, only a small fraction of NH_3 molecules react to form NH_4^+ and ^-OH. The major species at equilibrium is the undissociated molecule, NH_3. Figure 9.5 illustrates the difference between aqueous solutions of strong and weak bases. Table 9.1 lists common bases.

▼ **FIGURE 9.4   A Strong and Weak Acid Dissolved in Water**

hydrochloric acid

Cl⁻

H₃O⁺

A strong acid is
completely dissociated.

vinegar

CH₃COO⁻

H₃O⁺

CH₃COOH

A weak acid contains mostly
undissociated acid, CH₃COOH.

• The strong acid HCl completely dissociates into H_3O^+ and Cl^- in water.

• Vinegar contains CH_3COOH dissolved in H_2O. The weak acid CH_3COOH is only slightly dissociated into H_3O^+ and CH_3COO^-, so mostly CH_3COOH is present at equilibrium.

▼ FIGURE 9.5   A Strong and Weak Base Dissolved in Water

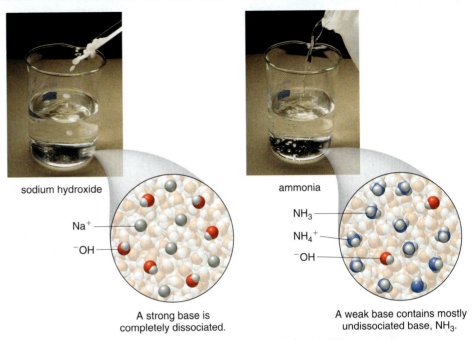

sodium hydroxide

Na$^+$
$^-$OH

A strong base is
completely dissociated.

ammonia

NH$_3$
NH$_4^+$
$^-$OH

A weak base contains mostly
undissociated base, NH$_3$.

- The strong base NaOH completely dissociates into Na$^+$ and $^-$OH in water.

- The weak base NH$_3$ is only slightly dissociated into NH$_4^+$ and $^-$OH, so mostly NH$_3$ is present at equilibrium.

- Use a single reaction arrow.
- The products are greatly favored at equilibrium.

NaOH(s) + H$_2$O(l) $\longrightarrow$ Na$^+$(aq) + $^-$OH(aq)
strong base

NH$_3$(g) + H$_2$O(l) $\rightleftharpoons$ NH$_4^+$(aq) + $^-$OH(aq)
weak base

- Use unequal reaction arrows.
- The reactants are favored at equilibrium.

An inverse relationship exists between acid and base strength.

- A strong acid readily donates a proton, forming a *weak* conjugate base.
- A strong base readily accepts a proton, forming a *weak* conjugate acid.

Why does this inverse relationship exist? Since a strong acid readily donates a proton, it forms a conjugate base that has little ability to accept a proton. Since a strong base readily accepts a proton, it forms a conjugate acid that tightly holds onto its proton, making it a weak acid.

Thus, a *strong* acid like HCl forms a *weak* conjugate base (Cl$^-$), and a *strong* base like $^-$OH forms a *weak* conjugate acid (H$_2$O). The entries in Table 9.1 are arranged in order of *decreasing* acid strength. This means that Table 9.1 is also arranged in order of *increasing* strength of the resulting conjugate bases. Knowing the relative strength of two acids makes it possible to predict the relative strength of their conjugate bases.

SAMPLE PROBLEM 9.7 Using Table 9.1: (a) Is H_3PO_4 or HF the stronger acid? (b) Draw the conjugate base of each acid and predict which base is stronger.

ANALYSIS The stronger the acid, the weaker the conjugate base.

SOLUTION

a. H_3PO_4 is located above HF in Table 9.1, making it the stronger acid.

b. To draw each conjugate base, remove a proton (H^+). Since each acid is neutral, both conjugate bases have a –1 charge. Since HF is the weaker acid, F^- is the stronger conjugate base.

$$\underset{\text{stronger acid}}{H_3PO_4} \xrightarrow{\text{lose } H^+} H_2PO_4^- \quad \boxed{\text{weaker base}}$$

$$\underset{\text{weaker acid}}{HF} \xrightarrow{\text{lose } H^+} F^- \quad \boxed{\text{stronger base}}$$

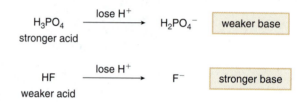

PROBLEM 9.9 Label the stronger acid in each pair. Which acid has the stronger conjugate base?

a. H_2SO_4 or H_3PO_4 b. HF or HCl c. H_2CO_3 or NH_4^+ d. HCN or HF

PROBLEM 9.10 If HCOOH is a stronger acid than CH_3COOH, which compound forms the stronger conjugate base?

PROBLEM 9.11 (a) Draw the conjugate acids of NO_2^- and NO_3^-. (b) If NO_2^- is the stronger base, which acid is stronger?

9.3B USING ACID STRENGTH TO PREDICT THE DIRECTION OF EQUILIBRIUM

A Brønsted–Lowry acid–base reaction represents an equilibrium. Since an acid donates a proton to a base, forming a conjugate acid and conjugate base, there are always two acids and two bases in the reaction mixture. Which pair of acids and bases is favored at equilibrium? **The position of the equilibrium depends upon the strength of the acids and bases.**

- **The stronger acid reacts with the stronger base to form the weaker acid and weaker base.**

Since a strong acid readily donates a proton and a strong base readily accepts one, these two species react to form a weaker conjugate acid and base that do not donate or accept a proton as readily. Thus, **when the stronger acid and base are the reactants on the left side, the reaction readily occurs and the reaction proceeds to the *right*.**

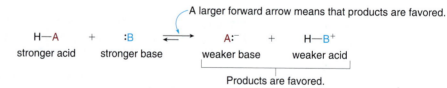

On the other hand, **if an acid–base reaction would form the stronger acid and base, equilibrium favors the reactants and little product forms.**

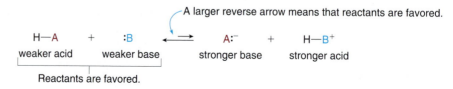

Predicting the direction of equilibrium using the information in Table 9.1 is illustrated in the accompanying stepwise *How To* procedure.

HOW TO   Predict the Direction of Equilibrium in an Acid–Base Reaction

EXAMPLE   **Are the reactants or products favored in the following acid–base reaction?**

$$HCN(g) + {}^-OH(aq) \rightleftharpoons {}^-CN(aq) + H_2O(l)$$

Step [1]   **Identify the acid in the reactants and the conjugate acid in the products.**

gain of H$^+$

$$HCN(g) \quad + \quad {}^-OH(aq) \rightleftharpoons {}^-CN(aq) \quad + \quad H_2O(l)$$
acid   base   conjugate base   conjugate acid

loss of H$^+$

- HCN is the acid since it donates a proton.
- H$_2$O is the conjugate acid formed from the hydroxide base.

Step [2]   **Determine the relative strength of the acid and the conjugate acid.**
- According to Table 9.1, HCN is a stronger acid than H$_2$O.

Step [3]   **Equilibrium favors the formation of the weaker acid.**
- Since the stronger acid HCN is a reactant, the reaction *proceeds to the right* as written, to form the weaker acid, H$_2$O. Unequal equilibrium arrows should be drawn with the larger arrow pointing towards the product side on the right.

$$HCN(g) \quad + \quad {}^-OH(aq) \rightleftharpoons {}^-CN(aq) \quad + \quad H_2O(l)$$
stronger acid   weaker acid

Products are favored.

PROBLEM 9.12

Are the reactants or products favored at equilibrium in each reaction?

a. $HF(g) + {}^-OH(aq) \rightleftharpoons F^-(aq) + H_2O(l)$

b. $NH_4^+(aq) + Cl^-(aq) \rightleftharpoons NH_3(g) + HCl(aq)$

c. $HCO_3^-(aq) + H_3O^+(aq) \rightleftharpoons H_2CO_3(aq) + H_2O(l)$

PROBLEM 9.13

If lactic acid is similar in strength to acetic acid (Table 9.1), predict whether reactants or products are favored in each reaction.

$C_3H_6O_3$
lactic acid

a. $C_3H_6O_3(aq) + H_2O(l) \rightleftharpoons C_3H_5O_3^-(aq) + H_3O^+(aq)$
   lactic acid

b. $C_3H_6O_3(aq) + HCO_3^-(aq) \rightleftharpoons C_3H_5O_3^-(aq) + H_2CO_3(aq)$
   lactic acid

HEALTH NOTE

Lactic acid accumulates in tissues during vigorous exercise, making muscles feel tired and sore. The formation of lactic acid is discussed in greater detail in Section 24.4.

9.4 EQUILIBRIUM AND ACID DISSOCIATION CONSTANTS

Like all equilibria, we can write an expression for the equilibrium constant for the acid–base reaction that takes place when an acid HA dissolves in water. The equilibrium constant (K) shows the ratio of the concentrations of the products to the concentrations of the reactants.

Reaction

$$HA(aq) + H_2O(l) \rightleftharpoons H_3O^+(aq) + A{:}^-$$

Equilibrium constant

$$K = \frac{[H_3O^+][A{:}^-]}{[HA][H_2O]}$$

concentrations of the products

concentrations of the reactants

Water serves as both the base and the solvent. Since its concentration is essentially constant, the equation can be rearranged by multiplying both sides by [H₂O]. This forms a new constant called the **acid dissociation constant, K_a.**

$$K_a = K[H_2O] = \frac{[H_3O^+][A{:}^-]}{[HA]}$$

acid dissociation constant

How is K_a related to acid strength? The stronger the acid, the higher the concentration of the products of an acid–base reaction, and the larger the numerator in the expression for K_a. As a result:

- **The stronger the acid, the larger the value of K_a.**

The strong acids listed in Table 9.1 all have K_a values much greater than 1. Weak acids have K_a values less than 1. The K_a's for several weak acids are listed in Table 9.2.

HEALTH NOTE

Citrus fruits (oranges, grapefruit, and lemons) are well known sources of vitamin C (ascorbic acid, Sample Problem 9.9), but guava, kiwifruit, and rose hips are excellent sources, too.

TABLE 9.2   Acid Dissociation Constants (K_a) for Common Weak Acids

Acid	Structure	K_a
Hydrogen sulfate ion	HSO_4^-	1.2×10^{-2}
Phosphoric acid	H_3PO_4	7.5×10^{-3}
Hydrofluoric acid	HF	7.2×10^{-4}
Acetic acid	CH_3COOH	1.8×10^{-5}
Carbonic acid	H_2CO_3	4.3×10^{-7}
Dihydrogen phosphate ion	$H_2PO_4^-$	6.2×10^{-8}
Ammonium ion	NH_4^+	5.6×10^{-10}
Hydrocyanic acid	HCN	4.9×10^{-10}
Bicarbonate ion	HCO_3^-	5.6×10^{-11}
Hydrogen phosphate ion	HPO_4^{2-}	2.2×10^{-13}

Increasing acidity

SAMPLE PROBLEM 9.8

Which acid in each pair is stronger: (a) HCN or HSO_4^-; (b) CH_3COOH or NH_4^+?

ANALYSIS   Use Table 9.2 to find the K_a for each acid. The acid with the larger K_a is the stronger acid.

SOLUTION

a.

HCN	HSO_4^-
$K_a = 4.9 \times 10^{-10}$	$K_a = 1.2 \times 10^{-2}$
	larger K_a
	stronger acid

b.

CH_3COOH	NH_4^+
1.8×10^{-5}	5.6×10^{-10}
larger K_a	
stronger acid	

PROBLEM 9.14

Rank the acids in each group in order of increasing acid strength.

a. H_3PO_4, $H_2PO_4^-$, HPO_4^{2-} b. HCN, HF, CH_3COOH

PROBLEM 9.15

(a) Which compound is the stronger acid, H_3PO_4 or CH_3COOH? (b) Draw the conjugate base of each compound and predict which base is stronger.

Because K_a values tell us the relative strength of two acids, we can use K_a's to predict the direction of equilibrium in an acid–base reaction, as shown in Sample Problem 9.9.

- **Equilibrium favors the formation of the weaker acid—that is, the acid with the *smaller* K_a value.**

SAMPLE PROBLEM 9.9

Ascorbic acid, vitamin C, is needed for the formation of collagen, a common protein in connective tissues in muscles and blood vessels. If vitamin C has a K_a of 7.9×10^{-5}, are the reactants or products favored in the following acid–base reaction?

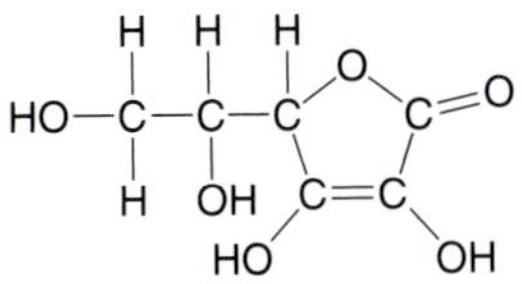

$C_6H_8O_6(aq)$ + $NH_3(aq)$ ⇌ $C_6H_7O_6^-(aq)$ + $NH_4^+(aq)$
vitamin C   conjugate base
     of vitamin C

vitamin C
ascorbic acid
$C_6H_8O_6$

ANALYSIS

To determine the direction of equilibrium, we must identify the acid in the reactants and the conjugate acid in the products. Then compare their K_a's. **Equilibrium favors the formation of the acid with the smaller K_a value.**

SOLUTION

Vitamin C is the acid and NH_3 is the base on the reactant side. NH_3 gains a proton to form its conjugate acid, NH_4^+, which has a K_a of 5.6×10^{-10} (Table 9.2). The conjugate acid, therefore, has a *smaller* K_a than vitamin C (7.9×10^{-5}), making it the *weaker acid*. Thus, the products are favored at equilibrium.

products favored

$C_6H_8O_6(aq)$ + $NH_3(aq)$ ⇌ $C_6H_7O_6^-(aq)$ + $NH_4^+(aq)$
vitamin C   conjugate acid

$K_a = 7.9 \times 10^{-5}$ $K_a = 5.6 \times 10^{-10}$
larger K_a smaller K_a
stronger acid   **weaker acid**

PROBLEM 9.16

Use the acid dissociation constants in Table 9.2 to determine whether the reactants or products are favored in the following reaction.

$HCO_3^-(aq) + NH_3(aq) \rightleftharpoons CO_3^{2-}(aq) + NH_4^+(aq)$

PROBLEM 9.17

Consider the weak acids, HCN and H_2CO_3.

a. Which acid has the larger K_a?

b. Which acid is stronger?

c. Which acid has the stronger conjugate base?

d. Which acid has the weaker conjugate base?

e. When each acid is dissolved in water, for which acid does the equilibrium lie further to the right?

9.5 DISSOCIATION OF WATER

In Section 9.2 we learned that water can behave as *both* a Brønsted–Lowry acid and a Brønsted–Lowry base. As a result, two molecules of water can react together in an acid–base reaction.

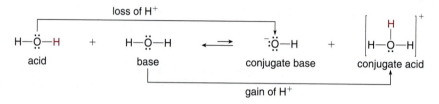

- One molecule of H_2O donates a proton (H^+), forming its conjugate base ^-OH.
- One molecule of H_2O accepts a proton, forming its conjugate acid H_3O^+.

Equilibrium favors the starting materials in this reaction, since the reactant acid, H_2O, is much weaker than the conjugate acid, H_3O^+. Thus, pure water contains an exceedingly low concentration of ions, H_3O^+ and ^-OH.

As usual, an expression for the equilibrium constant can be written that shows the ratio of the concentrations of the products, H_3O^+ and ^-OH, to the concentration of the reactants, two molecules of H_2O. Since the concentration of water is essentially constant, this equation can be rearranged by multiplying by $[H_2O]^2$ to afford a new equilibrium constant, K_w, **the ion–product constant** for water.

$$K = \frac{[H_3O^+][^-OH]}{[H_2O][H_2O]} = \frac{[H_3O^+][^-OH]}{[H_2O]^2}$$ — Multiply both sides by $[H_2O]^2$.

$$\underset{\substack{\textbf{ion–product} \\ \textbf{constant}}}{K_w} = K[H_2O]^2 = [H_3O^+][^-OH]$$

$$\boxed{K_w = [H_3O^+][^-OH]}$$

Since one H_3O^+ ion and one ^-OH ion are formed in each reaction, the concentration of H_3O^+ and ^-OH are equal in pure water. Experimentally it can be shown that the $[H_3O^+] = [^-OH] = 1.0 \times 10^{-7}$ M at 25 °C. Thus,

$$K_w = [H_3O^+][^-OH]$$
$$K_w = (1.0 \times 10^{-7}) \times (1.0 \times 10^{-7})$$
$$K_w = 1.0 \times 10^{-14}$$

- The product, $[H_3O^+][^-OH]$, is a constant, 1.0×10^{-14}, for all aqueous solutions at 25 °C.

Thus, **the value of K_w applies to any aqueous solution,** not just pure water. If we know the concentration of one ion, H_3O^+ or ^-OH, we can find the concentration of the other by rearranging the expression for K_w.

To calculate $[^-OH]$ when $[H_3O^+]$ is known:

$$K_w = [H_3O^+][^-OH]$$

$$[^-OH] = \frac{K_w}{[H_3O^+]}$$

$$[^-OH] = \frac{1.0 \times 10^{-14}}{[H_3O^+]}$$

To calculate $[H_3O^+]$ when $[^-OH]$ is known:

$$K_w = [H_3O^+][^-OH]$$

$$[H_3O^+] = \frac{K_w}{[^-OH]}$$

$$[H_3O^+] = \frac{1.0 \times 10^{-14}}{[^-OH]}$$

How to write numbers in scientific notation was presented in Section 1.6. Multiplying and dividing numbers written in scientific notation was described in Section 5.3.

Thus, if the concentration of H_3O^+ in a cup of coffee is 1.0×10^{-5} M, we can use this value to calculate [$^-$OH].

$$[^-OH] = \frac{K_w}{[H_3O^+]} = \frac{1.0 \times 10^{-14}}{1.0 \times 10^{-5}}$$

$$[^-OH] = 1.0 \times 10^{-9} \text{ M}$$

hydroxide ion concentration
in a cup of coffee

In a cup of coffee, therefore, the concentration of H_3O^+ ions is greater than the concentration of $^-$OH ions, but the product of these concentrations, 1.0×10^{-14}, is a constant, K_w.

Pure water and any solution that has an equal concentration of H_3O^+ and $^-$OH ions (1.0×10^{-7}) is said to be *neutral*. Other solutions are classified as **acidic** or **basic,** depending on which ion is present in a higher concentration.

Coffee is an *acidic* solution since the concentration of H_3O^+ is *greater* than the concentration of $^-$OH.

- In an *acidic* solution, [H_3O^+] > [$^-$OH]; thus, [H_3O^+] > 10^{-7} M.
- In a *basic* solution, [$^-$OH] > [H_3O^+]; thus, [$^-$OH] > 10^{-7} M.

In an acidic solution, the concentration of the acid H_3O^+ is greater than the concentration of the base $^-$OH. In a basic solution, the concentration of the base $^-$OH is greater than the concentration of the acid H_3O^+. Table 9.3 summarizes information about neutral, acidic, and basic solutions.

TABLE 9.3   Neutral, Acidic, and Basic Solutions

Type	[H_3O^+] and [$^-$OH]	[H_3O^+]	[$^-$OH]
Neutral	[H_3O^+] = [$^-$OH]	10^{-7} M	10^{-7} M
Acidic	[H_3O^+] > [$^-$OH]	> 10^{-7} M	< 10^{-7} M
Basic	[H_3O^+] < [$^-$OH]	< 10^{-7} M	> 10^{-7} M

SAMPLE PROBLEM 9.10

If [H_3O^+] in blood is 4.0×10^{-8} M, what is the value of [$^-$OH]? Is blood acidic, basic, or neutral?

ANALYSIS   Use the equation [$^-$OH] = K_w/[H_3O^+] to calculate the hydroxide ion concentration.

SOLUTION   Substitute the given value of [H_3O^+] in the equation to find [$^-$OH].

$$[^-OH] = \frac{K_w}{[H_3O^+]} = \frac{1.0 \times 10^{-14}}{4.0 \times 10^{-8}} = 2.5 \times 10^{-7} \text{ M}$$

hydroxide ion concentration
in the blood

Since [$^-$OH] > [H_3O^+], blood is a basic solution.

PROBLEM 9.18

Calculate the value of [$^-$OH] from the given [H_3O^+] in each solution and label the solution as acidic or basic: (a) [H_3O^+] = 10^{-3} M; (b) [H_3O^+] = 10^{-11} M; (c) [H_3O^+] = 2.8×10^{-10} M; (d) [H_3O^+] = 5.6×10^{-4} M.

PROBLEM 9.19

Calculate the value of [H_3O^+] from the given [$^-$OH] in each solution and label the solution as acidic or basic: (a) [$^-$OH] = 10^{-6} M; (b) [$^-$OH] = 10^{-9} M; (c) [$^-$OH] = 5.2×10^{-11} M; (d) [$^-$OH] = 7.3×10^{-4} M.

Since a strong acid like HCl is completely dissociated in aqueous solution, the concentration of the acid tells us the concentration of hydronium ions present. Thus, a 0.1 M HCl solution completely dissociates, so the concentration of H_3O^+ is 0.1 M. This value can then be used to calculate the hydroxide ion concentration. Similarly, a strong base like NaOH completely dissociates, so the concentration of the base gives the concentration of hydroxide ions present. Thus, the concentration of $^-$OH in a 0.1 M NaOH solution is 0.1 M.

In 0.1 M HCl solution: $[H_3O^+]$ = 0.1 M = 1×10^{-1} M
        strong acid

In 0.1 M NaOH solution: $[^-OH]$ = 0.1 M = 1×10^{-1} M
        strong base

SAMPLE PROBLEM 9.11

Calculate the value of $[H_3O^+]$ and $[^-OH]$ in a 0.01 M NaOH solution.

ANALYSIS

Since NaOH is a strong base that completely dissociates to form Na^+ and ^-OH, the concentration of NaOH gives the concentration of ^-OH ions. The $[^-OH]$ can then be used to calculate $[H_3O^+]$ from the expression for K_w.

SOLUTION

The value of $[^-OH]$ in a 0.01 M NaOH solution is 0.01 M = 1×10^{-2} M.

$$[H_3O^+] = \frac{K_w}{[^-OH]} = \frac{1 \times 10^{-14}}{1 \times 10^{-2}} = 1 \times 10^{-12} \text{ M}$$

concentration of ^-OH concentration of H_3O^+

PROBLEM 9.20

Calculate the value of $[H_3O^+]$ and $[^-OH]$ in each solution: (a) 0.001 M NaOH; (b) 0.001 M HCl; (c) 1.5 M HCl; (d) 0.30 M NaOH.

9.6 THE pH SCALE

Knowing the hydronium ion concentration is necessary in many different instances. The blood must have an H_3O^+ concentration in a very narrow range for an individual's good health. Plants thrive in soil that is not too acidic or too basic. The H_3O^+ concentration in a swimming pool must be measured and adjusted to keep the water clean and free from bacteria and algae.

9.6A CALCULATING pH

Since values for the hydronium ion concentration are very small, with negative powers of ten, the **pH scale** is used to more conveniently report $[H_3O^+]$. The pH of a solution is a number generally between 0 and 14, defined in terms of the *logarithm* (log) of the H_3O^+ concentration.

$$\boxed{pH = -\log [H_3O^+]}$$

A logarithm is an exponent of a power of ten.

The log is the exponent.

$\log(10^5)$ = 5 $\log(10^{-10})$ = -10 $\log(0.001) = \log(10^{-3})$ = -3

                 The log is the exponent.   Convert to scientific notation.

In calculating pH, first consider an H_3O^+ concentration that has a coefficient of *one* when the number is written in scientific notation. For example, the value of $[H_3O^+]$ in apple juice is about 1×10^{-4}, or 10^{-4} written without the coefficient. The pH of this solution is calculated as follows:

$$pH = -\log [H_3O^+] = -\log(10^{-4})$$
$$= -(-4) \qquad = 4$$

pH of apple juice

Apple juice has a pH of about 4, so it is an acidic solution.

Since pH is defined as the *negative* logarithm of $[H_3O^+]$ and these concentrations have *negative* exponents (10^{-x}), pH values are *positive* numbers.

Whether a solution is acidic, neutral, or basic can now be defined in terms of its pH.

- Acidic solution: pH < 7 $\longrightarrow$ $[H_3O^+]$ > 1 × 10^{-7}
- Neutral solution: pH = 7 $\longrightarrow$ $[H_3O^+]$ = 1 × 10^{-7}
- Basic solution: pH > 7 $\longrightarrow$ $[H_3O^+]$ < 1 × 10^{-7}

Note the relationship between $[H_3O^+]$ and pH.

- The *lower* the pH, the *higher* the concentration of H_3O^+.

Since pH is measured on a *logarithmic* scale, **a small difference in pH translates to a large change in H_3O^+ concentration.** For example, a difference of one pH unit means a ten-fold difference in H_3O^+ concentration. A difference of three pH units means a thousand-fold difference in H_3O^+ concentration.

A difference of one pH unit...

pH = 2 $[H_3O^+]$ = 1 × 10^{-2}
pH = 3 $[H_3O^+]$ = 1 × 10^{-3}

...means the $[H_3O^+]$ differs by a factor of 10.

A difference of three pH units...

pH = 2 $[H_3O^+]$ = 1 × 10^{-2}
pH = 5 $[H_3O^+]$ = 1 × 10^{-5}

...means the $[H_3O^+]$ differs by a factor of 1,000.

The pH of a solution can be measured using a pH meter as shown in Figure 9.6. Approximate pH values are determined using pH paper or indicators that turn different colors depending on the pH of the solution. The pH of various substances is shown in Figure 9.7.

Converting a given H_3O^+ concentration to a pH value is shown in Sample Problem 9.12. The reverse process, converting a pH value to an H_3O^+ concentration, is shown in Sample Problem 9.13.

▼ **FIGURE 9.6   Measuring pH**

a.

b.

c.
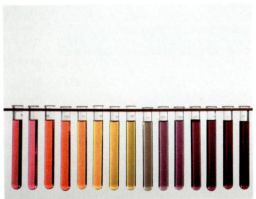

a. A pH meter is a small electronic device that measures pH when an electrode is dipped into a solution.

b. Paper strips called pH paper change color corresponding to a particular pH, when a drop of an aqueous solution is applied to them.

c. An acid–base indicator can be used to give an approximate pH. The indicator is a dye that changes color depending on the pH of the solution.

▼ **FIGURE 9.7   The pH of Some Common Substances**

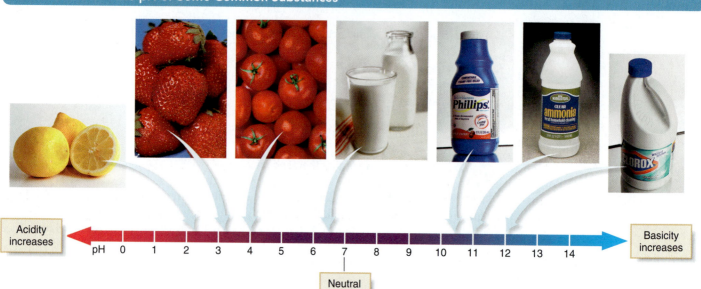

Acidity increases ⟵ pH 0 1 2 3 4 5 6 7 8 9 10 11 12 13 14 ⟶ Basicity increases

Neutral

The pH of many fruits is less than 7, making them acidic. Many cleaning agents, such as household ammonia and bleach, are basic (pH > 7).

SAMPLE PROBLEM 9.12

What is the pH of a urine sample that has an H_3O^+ concentration of 1×10^{-5} M? Classify the solution as acidic, basic, or neutral.

ANALYSIS   Use the formula pH = $-\log [H_3O^+]$. When the coefficient of a number written in scientific notation is one, the pH equals the value x in 10^{-x}.

SOLUTION

$$pH = -\log [H_3O^+] = -\log(10^{-5})$$
$$= -(-5) = 5 \quad \text{pH of urine sample}$$

Answer

The urine sample is acidic since the pH < 7.

PROBLEM 9.21

Convert each H_3O^+ concentration to a pH value.

  a. 1×10^{-6} M   b. 1×10^{-12} M   c. 0.000 01 M   d. 0.000 000 000 01 M

SAMPLE PROBLEM 9.13

What is the H_3O^+ concentration in lemon juice that has a pH of about 2? Classify the solution as acidic, basic, or neutral.

ANALYSIS   To find $[H_3O^+]$ from a pH, which is logarithm, we must determine what number corresponds to the given logarithm. When the pH is a whole number x, the value of x becomes the exponent in the expression $1 \times 10^{-x} = [H_3O^+]$.

SOLUTION   If the pH of lemon juice is 2, $[H_3O^+] = 1 \times 10^{-2}$ M. Since the pH is less than 7, the lemon juice is acidic.

PROBLEM 9.22

What H_3O^+ concentration corresponds to each pH value: (a) 13; (b) 7; (c) 3?

PROBLEM 9.23

Label each solution in Problem 9.22 as acidic, basic, or neutral.

9.6B CALCULATING pH USING A CALCULATOR

Determining logarithms and anti-logarithms using an electronic calculator is shown in Appendix A.

To calculate the pH of a solution in which the hydronium ion concentration has a coefficient in scientific notation that is *not* equal to one—as in 2.0×10^{-3}—you need a calculator that has a log function. How the keys are labeled and the order of the steps depends on your particular calculator.

Enter this number on your calculator.

$$pH = -\log[H_3O^+] = -\log(2.0 \times 10^{-3})$$
$$= -(-2.70) = 2.70$$

Similarly, when a reported pH is *not* a whole number—as in the pH = 8.50 for a sample of seawater—you need a calculator to calculate an *antilogarithm*—that is, the number that has a logarithm of 8.50. To make sure your calculation is correct, note that since the pH of seawater is between 8 and 9, the H_3O^+ concentration must be between 10^{-8} and 10^{-9}.

$$[H_3O^+] = antilog(-pH) = antilog(-8.50)$$
$$= 3.2 \times 10^{-9} \text{ M}$$

Care must be taken in keeping track of significant figures when using logarithms.

Because seawater contains dissolved salts, its pH is 8.50, making it slightly basic, not neutral like pure water.

- **A logarithm has the same number of digits to the right of the decimal point as are contained in the coefficient of the original number.**

$$[H_3O^+] = 3.2 \times 10^{-9} \text{ M} \qquad pH = 8.50 \quad \text{two digits after the decimal point}$$
two significant figures

SAMPLE PROBLEM 9.14

What is the pH of wine that has an H_3O^+ concentration of 3.2×10^{-4} M?

ANALYSIS Use a calculator to determine the logarithm of a number that contains a coefficient in scientific notation that is not a whole number; $pH = -\log[H_3O^+]$.

SOLUTION The order of the steps in using an electronic calculator, as well as the labels on the calculator buttons, vary. In some cases it is possible to calculate the pH by following three steps: enter the number (H_3O^+ concentration); press the *log* button; and press the change sign key. Consult your calculator manual if these steps do not give the desired value. Because the coefficient in the original number had two significant figures, the pH must have two digits to the right of the decimal point.

two significant figures

$$pH = -\log[H_3O^+] = -\log(3.2 \times 10^{-4})$$
$$= -(-3.49) = 3.49 \quad \text{two digits to the right of the decimal point}$$

PROBLEM 9.24

What H_3O^+ concentration corresponds to each pH value: (a) 10.2; (b) 7.8; (c) 4.3?

SAMPLE PROBLEM 9.15

What is the H_3O^+ concentration in sweat that has a pH of 5.8?

ANALYSIS Use a calculator to determine the antilogarithm of the negative of the pH; $[H_3O^+] = antilog(-pH)$.

SOLUTION The order of the steps in using an electronic calculator, as well as the labels on the calculator buttons, vary. In some cases it is possible to calculate $[H_3O^+]$ by the following steps: enter the pH value; press the change sign key; and press the *2nd + log* buttons. Since the pH has only one number to the right of the decimal point, the H_3O^+ concentration must have only one significant figure in its coefficient.

one digit to the right of the decimal point

$$[H_3O^+] = antilog(-pH) = antilog(-5.8)$$
$$[H_3O^+] = 2 \times 10^{-6} \text{ M}$$
one significant figure

PROBLEM 9.25

Convert each H_3O^+ concentration to a pH value.

   a. 1.8×10^{-6} M   b. 9.21×10^{-12} M   c. 0.000 088 M   d. 0.000 000 000 076 2 M

9.6C FOCUS ON THE HUMAN BODY
THE pH OF BODY FLUIDS

The human body contains fluids that vary in pH as shown in Figure 9.8. While saliva is slightly acidic, the gastric juice in the stomach has the lowest pH found in the body. The strongly acidic environment of the stomach aids in the digestion of food. It also kills many types of bacteria that might be inadvertently consumed along with food and drink. When food leaves the stomach, it passes to the basic environment of the small intestines. Bases in the small intestines react with acid from the stomach.

The pH of some body fluids must occupy a very narrow range. For example, a healthy individual has a blood pH in the range of 7.35–7.45. Maintaining this pH is accomplished by a complex mechanism described in Section 9.11. The pH of other fluids can be more variable. Urine has a pH anywhere from 4.6–8.0, depending on an individual's recent diet and exercise.

PROBLEM 9.26

Label each organ or fluid in Figure 9.8 as being acidic, basic, or neutral.

9.7 COMMON ACID–BASE REACTIONS

Although we have already seen a variety of acid–base reactions in Sections 9.2–9.4, two common reactions deserve additional attention—reaction of acids with hydroxide bases (^-OH), and reaction of acids with bicarbonate (HCO_3^-) or carbonate (CO_3^{2-}).

▼ **FIGURE 9.8** Variation in pH Values in the Human Body

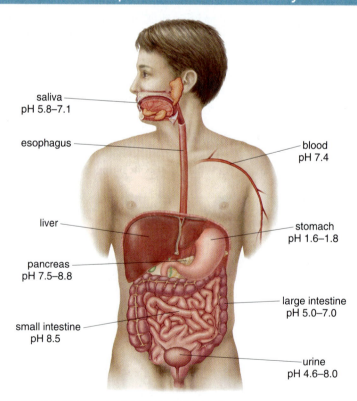

saliva
pH 5.8–7.1

esophagus

blood
pH 7.4

liver

stomach
pH 1.6–1.8

pancreas
pH 7.5–8.8

large intestine
pH 5.0–7.0

small intestine
pH 8.5

urine
pH 4.6–8.0

9.7A   REACTION OF ACIDS WITH HYDROXIDE BASES

The reaction of a Brønsted–Lowry acid (HA) with the metal salt of a hydroxide base (MOH) is an example of a *neutralization* reaction—an acid–base reaction that produces a salt and water as products.

$$HA(aq) + MOH(aq) \longrightarrow H{-}OH(l) + MA(aq)$$
$$\text{acid} \qquad \text{base} \qquad\qquad \text{water} \qquad \text{salt}$$

- The acid HA donates a proton (H^+) to the $^-$OH base to form H_2O.
- The anion A^- from the acid combines with the cation M^+ from the base to form the salt MA.

For example, hydrochloric acid, HCl, reacts with sodium hydroxide, NaOH, to form water and sodium chloride, NaCl. A single reaction arrow is drawn between reactants and products because equilibrium greatly favors the products.

$$HCl(aq) + NaOH(aq) \longrightarrow H{-}OH(l) + NaCl(aq)$$
$$\text{acid} \qquad \text{base} \qquad\qquad \text{water} \qquad \text{salt}$$

The important reacting species in this reaction are H^+ from the acid HCl and $^-$OH from the base NaOH. To more clearly see the acid–base reaction, we can write an equation that contains only the species that are actually involved in the reaction. Such an equation is called a **net ionic equation.**

- A *net ionic equation* contains only the species involved in a reaction.

To write a net ionic equation for an acid–base reaction, we first write the acid, base, and salt as individual ions in solution. This process is simplified if we use H^+ (not H_3O^+) as the reacting species of the acid, since it is the H^+ ion that is transferred to the base. The reaction of HCl with NaOH using individual ions is then drawn as:

$$H^+(aq) + Cl^-(aq) + Na^+(aq) + \ ^-OH(aq) \longrightarrow H{-}OH(l) + Na^+(aq) + Cl^-(aq)$$

Writing the equation in this manner shows that the Na^+ and Cl^- ions are unchanged in the reaction. Ions that appear on both sides of an equation but undergo no change in a reaction are called **spectator ions.** Removing the spectator ions from the equation gives the net ionic equation.

$$H^+(aq) \ + \ Cl^-(aq) \ + \ Na^+(aq) \ + \ ^-OH(aq) \longrightarrow H{-}OH(l) \ + \ Na^+(aq) \ + \ Cl^-(aq)$$

Omit the spectator ions.   Omit the spectator ions.

Net ionic equation $H^+(aq) \ + \ ^-OH(aq) \longrightarrow H{-}OH(l)$

- Whenever a strong acid and strong base react, the net ionic equation is always the same—H^+ reacts with $^-$OH to form H_2O.

To draw the products of these neutralization reactions, keep in mind that **two products are always formed—water and a metal salt.** Balancing an acid–base equation can be done with the stepwise procedure for balancing a general reaction outlined in Section 5.2. The coefficients in a balanced chemical equation illustrate that one H^+ ion is always needed to react with each $^-$OH anion.

HOW TO   Draw a Balanced Equation for a Neutralization Reaction Between HA and MOH

EXAMPLE Write a balanced equation for the reaction of Mg(OH)$_2$, an active ingredient in the antacid product Maalox, with the hydrochloric acid (HCl) in the stomach.

Step [1] Identify the acid and base in the reactants and draw H$_2$O as one product.

- HCl is the acid and Mg(OH)$_2$ is the base. H$^+$ from the acid reacts with $^-$OH from the base to form H$_2$O.

$$\underset{\text{acid}}{HCl(aq)} + \underset{\text{base}}{Mg(OH)_2(aq)} \longrightarrow H_2O(l) + \text{salt}$$

Step [2] Determine the structure of the salt formed as product.

- The salt is formed from the elements of the acid and base that are *not* used to form H$_2$O. The anion of the salt comes from the acid and the cation of the salt comes from the base.
- In this case, Cl$^-$ (from HCl) and Mg^{2+} [from Mg(OH)$_2$] combine to form the salt MgCl$_2$.

Step [3] Balance the equation.

- Follow the procedure in Section 5.2 to balance an equation. The balanced equation shows that *two* moles of HCl are needed for *each* mole of Mg(OH)$_2$, since each mole of Mg(OH)$_2$ contains two moles of $^-$OH.

Place a 2 to balance H and O.

$$2\,HCl(aq) + Mg(OH)_2(aq) \longrightarrow 2\,H_2O(l) + MgCl_2$$

Place a 2 to balance Cl.

PROBLEM 9.27 Write a balanced equation for each acid–base reaction.

a. HNO$_3$(aq) + NaOH(aq) $\longrightarrow$

b. H$_2$SO$_4$(aq) + KOH(aq) $\longrightarrow$

PROBLEM 9.28 Write the net ionic equation for each reaction in Problem 9.27.

9.7B   REACTION OF ACIDS WITH BICARBONATE AND CARBONATE

Acids react with the bases bicarbonate (HCO$_3^-$) and carbonate (CO$_3^{2-}$). A bicarbonate base reacts with *one* proton to form carbonic acid, H$_2$CO$_3$. A carbonate base reacts with *two* protons. The carbonic acid formed in these reactions is unstable and decomposes to form CO$_2$ and H$_2$O. Thus, when an acid reacts with either base, bubbles of CO$_2$ gas are given off.

$$\underset{\text{1 H}^+ \text{ needed}}{H^+(aq)} + \underset{\text{bicarbonate}}{HCO_3^-(aq)} \longrightarrow [H_2CO_3(aq)] \longrightarrow H_2O(l) + CO_2(g)$$

$$\underset{\text{2 H}^+ \text{ needed}}{2\,H^+(aq)} + \underset{\text{carbonate}}{CO_3^{2-}(aq)} \longrightarrow [H_2CO_3(aq)] \longrightarrow H_2O(l) + CO_2(g)$$

bubbles of CO$_2$

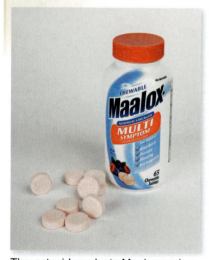

The antacid products Maalox and Mylanta both contain two bases—Mg(OH)$_2$ and Al(OH)$_3$—that react with excess stomach acid. A combination of bases is used so that the constipating effect of the aluminum salt is counteracted by the laxative effect of the magnesium salt.

HEALTH NOTE

Sodium bicarbonate (NaHCO$_3$), an ingredient in the over-the-counter antacid Alka-Seltzer, is the metal salt of a bicarbonate base that reacts with excess stomach acid, releasing CO$_2$. Like the neutralization reactions in Section 9.7A, a salt, NaCl, is formed in which the cation (Na$^+$) comes from the base and the anion (Cl$^-$) comes from the acid.

$$\underset{\text{acid}}{HCl(aq)} + \underset{\text{base}}{NaHCO_3(aq)} \longrightarrow \underset{\text{salt}}{NaCl(aq)} + H_2CO_3(aq)$$
$$\longrightarrow H_2O(l) + CO_2(g)$$

Calcium carbonate (CaCO$_3$), a calcium supplement and antacid in Tums, also reacts with excess stomach acid with release of CO$_2$. Since each carbonate ion reacts with two protons, the balanced equation shows a 2:1 ratio of HCl to CaCO$_3$.

$$\underset{\text{acid}}{2\,HCl(aq)} + \underset{\text{base}}{CaCO_3(aq)} \longrightarrow \underset{\text{salt}}{CaCl_2(aq)} + H_2CO_3(aq)$$
$$\longrightarrow H_2O(l) + CO_2(g)$$

SAMPLE PROBLEM 9.16

Write a balanced equation for the reaction of H_2SO_4 with $NaHCO_3$.

ANALYSIS

The acid and base react to form a salt and carbonic acid (H_2CO_3), which decomposes to CO_2 and H_2O.

SOLUTION

H_2SO_4 is the acid and $NaHCO_3$ is the base. H^+ from the acid reacts with HCO_3^- from the base to give H_2CO_3, which decomposes to H_2O and CO_2. A salt (Na_2SO_4) is also formed from the cation of the base (Na^+) and the anion of the acid (SO_4^{2-}).

Unbalanced equation: $H_2SO_4(aq)$ + $NaHCO_3(aq)$ $\longrightarrow$ $Na_2SO_4(aq)$ + $H_2O(l)$ + $CO_2(g)$
                              acid   base   salt $\underbrace{\qquad}_{\text{from } H_2CO_3}$

To balance the equation, place coefficients so the number of atoms on both sides of the arrow is the same.

Place a **2** to balance Na...

$H_2SO_4(aq)$ +   **2** $NaHCO_3(aq)$ $\longrightarrow$ $Na_2SO_4(aq)$ +   **2** $H_2O(l)$ +   **2** $CO_2(g)$
      acid   base   salt

...then place **2**'s to balance C, H, and O.

PROBLEM 9.29

The acid in acid rain is generally sulfuric acid (H_2SO_4). When this rainwater falls on statues composed of marble ($CaCO_3$), the H_2SO_4 slowly dissolves the $CaCO_3$. Write a balanced equation for this acid–base reaction.

PROBLEM 9.30

Write a balanced equation for the reaction of nitric acid (HNO_3) with each base: (a) $NaHCO_3$; (b) $MgCO_3$.

9.8   THE ACIDITY AND BASICITY OF SALT SOLUTIONS

Thus far we have discussed what occurs when an acid or a base dissolves in water. What happens to the pH of water when a salt is dissolved? A salt is the product of a neutralization reaction (Section 9.7A). Does this mean that a salt dissolved in water forms a neutral solution with a pH of 7?

- **A salt can form an acidic, basic, or neutral solution depending on whether its cation and anion are derived from a strong or weak acid and base.**

Let's first examine what acid and base are used to form a salt M^+A^-.

- **The cation M^+ comes from the base.**
- **The anion A^- comes from the acid HA.**

Thus, we consider NaCl to be a salt formed when the strong base NaOH and the strong acid HCl react together. On the other hand, $NaHCO_3$ is formed from the strong base NaOH and the weak acid H_2CO_3, and NH_4Cl is formed from the weak base NH_3 and the strong acid HCl.

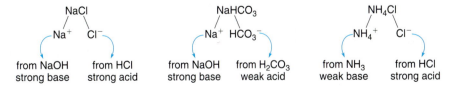

NaCl		NaHCO₃		NH₄Cl	
Na^+	Cl^-	Na^+	HCO_3^-	NH_4^+	Cl^-
from NaOH strong base	from HCl strong acid	from NaOH strong base	from H_2CO_3 weak acid	from NH_3 weak base	from HCl strong acid

A salt derived from a strong base and a strong acid forms a neutral solution (pH = 7). When one ion of a salt is derived from a weak acid or base, one principle is followed: **the ion derived from the *stronger* acid or base determines whether the solution is acidic or basic.**

- A salt derived from a *strong base* and a *weak acid* forms a *basic* solution (pH > 7).
- A salt derived from a *weak base* and a *strong acid* forms an *acidic* solution (pH < 7).

For example, when $NaHCO_3$ dissolves in water, it forms $Na^+(aq)$ and $HCO_3^-(aq)$. Alkali metal cations like Na^+ do not react with H_2O, but HCO_3^- reacts with H_2O to form ^-OH. Thus, **$NaHCO_3$, a salt derived from a *strong* base and a *weak* acid, forms a *basic* solution.**

$$HCO_3^-(aq) + H_2O(l) \rightleftharpoons H_2CO_3(aq) + \boxed{^-OH(aq)}$$

Hydroxide makes the solution basic, so the pH > 7.

When NH_4Cl dissolves in water it forms $NH_4^+(aq)$ and $Cl^-(aq)$. Halide anions like Cl^- do not react with H_2O, but NH_4^+ reacts with H_2O to form H_3O^+. Thus, **NH_4Cl, a salt derived from a *strong* acid and a *weak* base, forms an *acidic* solution.**

$$NH_4^+(aq) + H_2O(l) \rightleftharpoons NH_3(aq) + \boxed{H_3O^+(aq)}$$

H_3O^+ makes the solution acidic, so the pH < 7.

When NaCl dissolves in water it forms $Na^+(aq)$ and $Cl^-(aq)$. Neither ion reacts with water. Since no acid–base reaction occurs, the solution remains neutral. Thus, **NaCl, a salt derived from a strong acid and strong base, forms a neutral solution.** The acidity and basicity of salt solutions is summarized in Table 9.4.

TABLE 9.4 The Acidity and Basicity of Salt Solutions

Cation Derived from	Anion Derived from	Solution	pH	Examples
Strong base	Strong acid	neutral	7	$NaCl$, KBr, $NaNO_3$
Strong base	Weak acid	basic	> 7	$NaHCO_3$, KCN, CaF_2
Weak base	Strong acid	acidic	< 7	NH_4Cl, NH_4NO_3

SAMPLE PROBLEM 9.17

Determine whether each salt forms an acidic, basic, or neutral solution when dissolved in water: (a) NaF; (b) KNO_3; (c) NH_4Br.

ANALYSIS

Determine what type of acid and base (strong or weak) are used to form the salt. When the ions in the salt come from a strong acid and strong base, the solution is neutral. When the ions come from acids and bases of different strength, the ion derived from the stronger reactant determines the acidity.

SOLUTION

a.
    NaF
   Na⁺ F⁻
from NaOH   from HF
strong base weak acid
   basic solution

b.
    KNO_3
   K⁺ NO_3^-
from KOH   from HNO_3
strong base **strong acid**
   neutral solution

c.
    NH_4Br
   NH_4^+ Br^-
from NH_3 from HBr
weak base **strong acid**
   acidic solution

PROBLEM 9.31

Determine whether each salt forms an acidic, basic, or neutral solution when dissolved in water: (a) KI; (b) K_2CO_3; (c) $Ca(NO_3)_2$; (d) NH_4I; (e) $BaCl_2$; (f) Na_3PO_4.

PROBLEM 9.32

Which of the following salts forms an aqueous solution that has a pH > 7: (a) LiCl; (b) K_2CO_3; (c) NH_4Br; (d) $MgCO_3$?

9.9 TITRATION

Sometimes it is necessary to know the exact concentration of acid or base in a solution. To determine the molarity of a solution, we carry out a **titration.** A titration uses a *buret,* a calibrated tube with a stopcock at the bottom that allows a solution of known molarity to be added in small quantities to a solution of unknown molarity. The procedure for determining the total acid concentration of a solution of HCl is illustrated in Figure 9.9.

How does a titration tell us the concentration of an HCl solution? A titration is based on the acid–base reaction that occurs between the acid in the flask (HCl) and the base that is added (NaOH). **When the number of moles of base added equals the number of moles of acid in the flask, the acid is *neutralized*, forming a salt and water.**

At the end point, the number of moles of
the acid (H^+) and the base (^-OH) are equal.

$$HCl(aq) \; + \; NaOH(aq) \; \rightleftharpoons \; NaCl(aq) \; + \; H_2O(aq)$$

To determine an unknown molarity from titration data requires three operations.

▼ FIGURE 9.9   Titration of an Acid with a Base of Known Concentration

a.

b.

c.

Steps in determining the molarity of a solution of HCl:

a. Add a measured volume of HCl solution to a flask. Add an acid–base indicator, often phenolphthalein, which is colorless in acid but turns bright pink in base.

b. Fill a buret with an NaOH solution of known molarity and slowly add it to the HCl solution.

c. Add NaOH solution until the *end point* is reached, the point at which the indicator changes color. At the end point, the **number of moles of NaOH added equals the number of moles of HCl** in the flask. In other words, all of the HCl has reacted with NaOH and the solution is no longer acidic. Read the volume of NaOH solution added from the buret. Using the known volume and molarity of the NaOH solution and the known volume of HCl solution, the molarity of the HCl solution can be calculated.

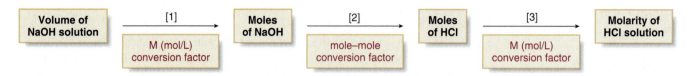

First, we determine the number of moles of base added using its known molarity and volume. Then we use coefficients in the balanced acid–base equation to tell us the number of moles of acid that react with the base. Finally, we determine the molarity of the acid from the calculated number of moles and the known volume of the acid.

HOW TO   Determine the Molarity of an Acid Solution from a Titration

EXAMPLE   **What is the molarity of an HCl solution if 22.5 mL of a 0.100 M NaOH solution are needed to titrate a 25.0 mL sample of the acid?**

Step [1]   **Determine the number of moles of base used to neutralize the acid.**
- Convert milliliters to liters of base using a mL–L conversion factor. Use the molarity (M) and volume (V) of the base to calculate the number of moles (mol = MV).

$$22.5 \ \text{mL NaOH} \quad \times \quad \frac{1 \ \text{L}}{1000 \ \text{mL}} \quad \times \quad \frac{0.100 \ \text{mol NaOH}}{1 \ \text{L}} \quad = \quad 0.00225 \ \text{mol NaOH}$$

Step [2]   **Determine the number of moles of acid that react from the balanced chemical equation.**
- In this reaction, one mole of HCl reacts with one mole of NaOH, so the number of moles of NaOH equals the number of moles of HCl at the end point.

$$\underset{0.00225 \ \text{mol}}{HCl(aq)} \ + \ \underset{0.00225 \ \text{mol}}{NaOH(aq)} \ \longrightarrow \ NaCl(aq) \ + \ H_2O(l)$$

Step [3]   **Determine the molarity of the acid from the number of moles and known volume.**
- Convert milliliters to liters of acid using a mL–L conversion factor. Use the number of moles and known volume of the acid to calculate the molarity (M = mol/L).

$$M = \frac{\text{mol}}{\text{L}} = \frac{0.00225 \ \text{mol HCl}}{25.0 \ \text{mL solution}} \times \frac{1000 \ \text{mL}}{1 \ \text{L}} = 0.0900 \ \text{M HCl}$$

Sample Problem 9.18 illustrates a calculation for a titration in which the ratio of acid to base in the balanced acid–base equation is something other than 1:1.

SAMPLE PROBLEM 9.18

Acid rain is rainwater with a lower-than-normal pH, caused by the presence of dissolved acids such as H_2SO_4. What is the molarity of H_2SO_4 in rainwater if 5.22 mL of a 0.20 M NaOH solution are needed to titrate 125 mL of the sample? The balanced equation for this acid–base reaction is given.

$$H_2SO_4(aq) + 2 \ NaOH(aq) \longrightarrow Na_2SO_4(aq) + 2 \ H_2O(l)$$

ANALYSIS AND SOLUTION

[1] **Determine the number of moles of base used to neutralize the acid.**

- Use the molarity (M) and volume (V) of the base to calculate the number of moles (mol = MV).

$$5.22 \text{ mL NaOH} \times \frac{1 \text{ L}}{1000 \text{ mL}} \times \frac{0.20 \text{ mol NaOH}}{1 \text{ L}} = 0.0010 \text{ mol NaOH}$$

[2] **Determine the number of moles of acid that react from the balanced chemical equation.**

- Since each H_2SO_4 molecule contains two protons, *one* mole of the acid H_2SO_4 reacts with *two* moles of the base NaOH in the neutralization reaction. The coefficients in the balanced equation form a mole ratio to calculate the number of moles of acid that react.

$$0.0010 \text{ mol NaOH} \times \frac{1 \text{ mol H}_2\text{SO}_4}{2 \text{ mol NaOH}} = 0.000\,50 \text{ mol H}_2\text{SO}_4$$

[3] **Determine the molarity of the acid from the number of moles and known volume.**

$$\underset{\text{molarity}}{M} = \frac{\text{mol}}{\text{L}} = \frac{0.000\,50 \text{ mol H}_2\text{SO}_4}{125 \text{ mL solution}} \times \frac{1000 \text{ mL}}{1 \text{ L}} = \underset{\textbf{Answer}}{0.0040 \text{ M H}_2\text{SO}_4}$$

PROBLEM 9.33 What is the molarity of an HCl solution if 25.5 mL of a 0.24 M NaOH solution are needed to neutralize 15.0 mL of the sample?

PROBLEM 9.34 How many milliliters of 2.0 M NaOH are needed to neutralize 5.0 mL of a 6.0 M H_2SO_4 solution?

9.10 BUFFERS

A *buffer* is a solution whose pH changes very little when acid or base is added. Most buffers are solutions composed of approximately equal amounts of a weak acid and the salt of its conjugate base.

- **The weak acid of the buffer reacts with added base, ⁻OH.**
- **The conjugate base of the buffer reacts with added acid, H_3O^+.**

9.10A GENERAL CHARACTERISTICS OF A BUFFER

The effect of a buffer can be illustrated by comparing the pH change that occurs when a small amount of strong acid or strong base is added to water, compared to the pH change that occurs when the same amount of strong acid or strong base is added to a buffer, as shown in Figure 9.10. When 0.020 mol of HCl is added to 1.0 L of water, the pH changes from 7 to 1.7, and when 0.020 mol of NaOH is added to 1.0 L of water, the pH changes from 7 to 12.3. In this example, addition of a small quantity of a strong acid or strong base to neutral water changes the pH by over 5 pH units.

In contrast, a buffer prepared from 0.50 M acetic acid (CH_3COOH) and 0.50 M sodium acetate ($NaCH_3COO$) has a pH of 4.74. Addition of the same quantity of acid, 0.020 mol HCl, changes the pH to 4.70, and addition of the same quantity of base, 0.020 mol of NaOH, changes the pH to 4.77. In this example, the change of pH in the presence of the buffer is no more than 0.04 pH units!

Why is a buffer able to absorb acid or base with very little pH change? Let's use as an example a buffer that contains equal concentrations of acetic acid (CH_3COOH), and the sodium salt of its conjugate base, sodium acetate ($NaCH_3COO$). CH_3COOH is a weak acid, so when it dissolves in water, only a small fraction dissociates to form its conjugate base CH_3COO^-. In the buffer solution, however, the sodium acetate provides an equal amount of the conjugate base.

▼ **FIGURE 9.10   The Effect of a Buffer on pH Changes**

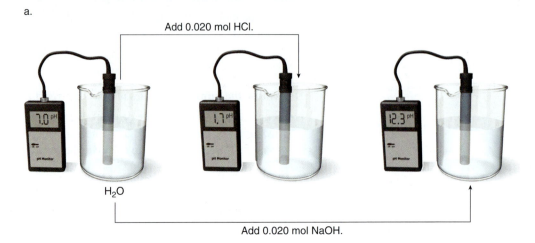

a. The pH of pure water changes drastically when a small amount of strong acid or strong base is added.

b. The pH of a buffer changes very little when the same amount of strong acid or strong base is added.

$$CH_3COOH(aq) \quad + \quad H_2O(l) \quad \rightleftharpoons \quad H_3O^+(aq) \quad + \quad CH_3COO^-(aq)$$

approximately equal amounts

In Section 9.4, we learned how to write the expression for the acid dissociation constant K_a for this reaction. Rearranging this expression to solve for $[H_3O^+]$ then illustrates why a buffer does not change pH much when acid or base is added.

Expression for K_a: $\quad K_a \quad = \quad \dfrac{[H_3O^+][CH_3COO^-]}{[CH_3COOH]}$

Rearranging the expression: $\quad [H_3O^+] \quad = \quad K_a \quad \times \quad \dfrac{[CH_3COOH]}{[CH_3COO^-]}$

If this ratio does not change much, then $[H_3O^+]$ does not change much.

The H_3O^+ concentration depends on two terms—K_a, which is a constant, and the ratio of the concentrations of the weak acid and its conjugate base. If these concentrations do not change much, then the concentration of H_3O^+ and therefore the pH do not change much.

Suppose a small amount of strong acid is added to the buffer. Added H_3O^+ reacts with CH_3COO^- to form CH_3COOH, so that $[CH_3COO^-]$ decreases slightly and $[CH_3COOH]$ increases slightly. However, the ratio of these two concentrations is not altered significantly, so the $[H_3O^+]$ and therefore the pH change only slightly.

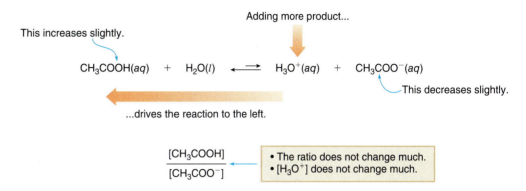

On the other hand, if a small amount of strong base is added to the buffer, ^-OH reacts with CH_3COOH to form CH_3COO^-, so that $[CH_3COOH]$ decreases slightly and $[CH_3COO^-]$ increases slightly. However, the ratio of these two concentrations is not altered significantly, so the $[H_3O^+]$ and therefore the pH change only slightly.

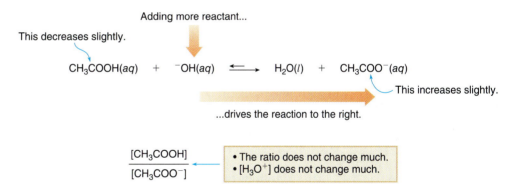

For a buffer to be effective, the amount of added acid or base must be small compared to the amount of buffer present. When a large amount of acid or base is added to a buffer, the concentrations of the weak acid and its conjugate base change a great deal, so the H_3O^+ concentration changes a great deal as well. Some common buffers are listed in Table 9.5.

TABLE 9.5	Common Buffers		
Buffer	**Weak Acid**	**Conjugate Base**	**K_a**
Acetic acid/acetate	CH_3COOH	CH_3COO^-	1.8×10^{-5}
Bicarbonate/carbonate	HCO_3^-	CO_3^{2-}	5.6×10^{-11}
Dihydrogen phosphate/ hydrogen phosphate	$H_2PO_4^-$	HPO_4^{2-}	6.2×10^{-8}
Hydrogen phosphate/ phosphate	HPO_4^{2-}	PO_4^{3-}	2.2×10^{-13}

PROBLEM 9.35

Determine whether a solution containing each of the following substances is a buffer. Explain your reasoning.

a. HBr and NaBr   b. HF and KF   c. CH_3COOH alone

PROBLEM 9.36

Consider a buffer prepared from the weak acid HCO_3^- and its conjugate base CO_3^{2-}.

$$HCO_3^-(aq) + H_2O(l) \rightleftharpoons CO_3^{2-}(aq) + H_3O^+(aq)$$

a. Explain why both HCO_3^- and CO_3^{2-} are needed to prepare the buffer.

b. What happens to the concentrations of HCO_3^- and CO_3^{2-} when a small amount of acid is added to the buffer?

c. What happens to the concentrations of HCO_3^- and CO_3^{2-} when a small amount of base is added to the buffer?

9.10B CALCULATING THE pH OF A BUFFER

The effective pH range of a buffer depends on its K_a. The pH of a buffer can be calculated from the K_a of the weak acid (HA), and the concentrations of the weak acid (HA) and conjugate base (A:$^-$) used to prepare it, as shown in Sample Problem 9.19.

Acid dissociation constant K_a for the general acid HA:

$$K_a = \frac{[H_3O^+][A:^-]}{[HA]}$$

Rearranging the expression to solve for $[H_3O^+]$:

$$[H_3O^+] = K_a \times \frac{[HA]}{[A:^-]}$$

determines the buffer pH

SAMPLE PROBLEM 9.19

What is the pH of a buffer that contains 0.20 M CH_3COOH and 0.20 M $NaCH_3COO$?

ANALYSIS

Use the K_a of the weak acid of the buffer in Table 9.5 and the expression $[H_3O^+] = K_a([HA]/[A:^-])$ to calculate $[H_3O^+]$. Calculate the pH using the expression, pH = –log $[H_3O^+]$.

SOLUTION

[1] **Substitute the given concentrations of CH_3COOH and CH_3COO^- for [HA] and [A:$^-$], respectively. K_a for CH_3COOH is 1.8×10^{-5}.**

$$[H_3O^+] = K_a \times \frac{[CH_3COOH]}{[CH_3COO^-]} = (1.8 \times 10^{-5}) \times \frac{[0.20\ M]}{[0.20\ M]}$$

$$[H_3O^+] = 1.8 \times 10^{-5}\ M$$

[2] **Use an electronic calculator to convert the H_3O^+ concentration to pH, as in Sample Problem 9.14.**

$$pH = -log\ [H_3O^+] = -log(1.8 \times 10^{-5})$$

$$pH = 4.74$$

PROBLEM 9.37

Calculate the pH of a dihydrogen phosphate/hydrogen phosphate buffer prepared with each of the following concentrations. What do you conclude about the pH of a buffer when equal concentrations of the weak acid and conjugate base are used to prepare it?

a. 0.10 M NaH_2PO_4 and 0.10 M Na_2HPO_4

b. 1.0 M NaH_2PO_4 and 1.0 M Na_2HPO_4

c. 0.50 M NaH_2PO_4 and 0.50 M Na_2HPO_4

PROBLEM 9.38   What is the pH of a buffer that contains 0.20 M CH_3COOH and 0.15 M $NaCH_3COO$?

9.10C FOCUS ON THE ENVIRONMENT
ACID RAIN AND A NATURALLY BUFFERED LAKE

Unpolluted rainwater is not a neutral solution with a pH of 7; rather, because it contains dissolved carbon dioxide, it is slightly acidic with a pH of about 5.6.

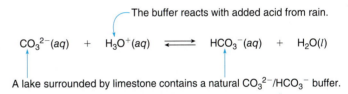

$$CO_2(g) \; + \; 2\,H_2O(l) \; \rightleftharpoons \; H_3O^+(aq) \; + \; HCO_3^-(aq)$$

carbon dioxide
from the air

A low concentration of H_3O^+ gives rainwater a pH < 7.

Rainwater that contains dissolved H_2SO_4 (or HNO_3) from burning fossil fuels has a pH lower than 5.6. In some parts of the United States, rainwater often has a pH range of 4–5, and readings as low as pH = 1.8 have been recorded. When the rain in a region consistently has a lower-than-normal pH, this acid rain can have a devastating effect on plant and animal life.

The pH of some lakes changes drastically as the result of acid rain, whereas the pH of other lakes does not. In fact, the ability of some lakes to absorb acid rain without much pH change is entirely due to buffers (Figure 9.11). Lakes that are surrounded by limestone-rich soil are in contact with solid calcium carbonate, $CaCO_3$. As a result, the lake contains a natural carbonate/bicarbonate buffer. When acid precipitation falls on the lake, the dissolved carbonate (CO_3^{2-}) reacts with the acid to form bicarbonate (HCO_3^-).

The buffer reacts with added acid from rain.

$$CO_3^{2-}(aq) \; + \; H_3O^+(aq) \; \rightleftharpoons \; HCO_3^-(aq) \; + \; H_2O(l)$$

A lake surrounded by limestone contains a natural CO_3^{2-}/HCO_3^- buffer.

▼ **FIGURE 9.11   Acid Rain and a Naturally Buffered Lake**

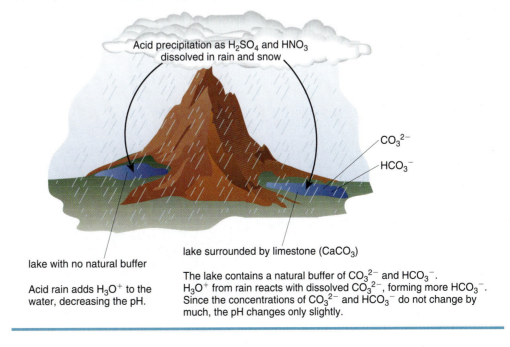

lake with no natural buffer

Acid rain adds H_3O^+ to the water, decreasing the pH.

lake surrounded by limestone ($CaCO_3$)

The lake contains a natural buffer of CO_3^{2-} and HCO_3^-. H_3O^+ from rain reacts with dissolved CO_3^{2-}, forming more HCO_3^-. Since the concentrations of CO_3^{2-} and HCO_3^- do not change by much, the pH changes only slightly.

The carbonate/bicarbonate buffer thus allows the lake to resist large pH changes when acid is added. In some areas acidic lakes have been treated with limestone, thus adding calcium carbonate to neutralize the acid and restore the natural pH. This procedure is expensive and temporary because with time and more acid rain, the pH of the lakes decreases again.

9.11 FOCUS ON THE HUMAN BODY
BUFFERS IN THE BLOOD

The normal blood pH of a healthy individual is in the range of 7.35 to 7.45. A pH above or below this range is generally indicative of an imbalance in respiratory or metabolic processes. The body is able to maintain a very stable pH because the blood and other tissues are buffered. The principal buffer in the blood is carbonic acid/bicarbonate (H_2CO_3/HCO_3^-).

In examining the carbonic acid/bicarbonate buffer system in the blood, two equilibria are important. First of all, carbonic acid (H_2CO_3) is in equilibrium with CO_2 dissolved in the bloodstream (Section 9.7). Second, since carbonic acid is a weak acid, it is also dissociated in water to form its conjugate base, bicarbonate (HCO_3^-). Bicarbonate is also generated in the kidneys.

$$CO_2(g) \; + \; H_2O(l) \; \rightleftharpoons \; \underset{\text{carbonic acid}}{H_2CO_3(aq)} \; \overset{H_2O}{\rightleftharpoons} \; H_3O^+(aq) \; + \; \underset{\text{bicarbonate}}{HCO_3^-(aq)}$$

principal buffer in the blood

CO_2 is constantly produced by metabolic processes in the body and then transported to the lungs to be eliminated. Thus, the amount of CO_2 dissolved in the blood is directly related to the H_3O^+ concentration and therefore the pH of the blood. If the pH of the blood is lower than 7.35, the blood is more acidic than normal, and the condition is called **acidosis.** If the pH of the blood is higher than 7.45, the blood is more basic than normal, and the condition is called **alkalosis.**

Le Châtelier's principle explains the effect of increasing or decreasing the level of dissolved CO_2 on the pH of the blood. A higher-than-normal CO_2 concentration shifts the equilibrium to the right, increasing the H_3O^+ concentration and lowering the pH. **Respiratory acidosis** results when the body fails to eliminate adequate amounts of CO_2 through the lungs. This may occur in patients with advanced lung disease or respiratory failure.

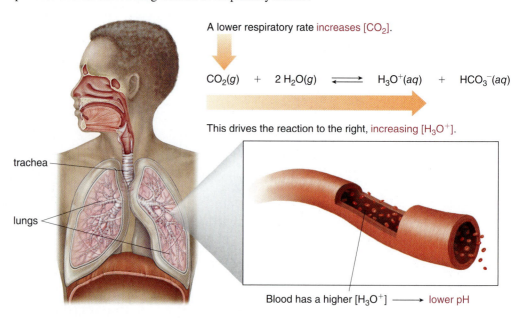

A lower respiratory rate increases [CO_2].

$$CO_2(g) \; + \; 2\,H_2O(g) \; \rightleftharpoons \; H_3O^+(aq) \; + \; HCO_3^-(aq)$$

This drives the reaction to the right, increasing [H_3O^+].

trachea

lungs

Blood has a higher [H_3O^+] ⟶ lower pH

A lower-than-normal CO_2 concentration shifts the equilibria to the left, decreasing the H_3O^+ concentration and raising the pH. **Respiratory alkalosis** is caused by hyperventilation, very rapid breathing that occurs when an individual experiences excitement or panic.

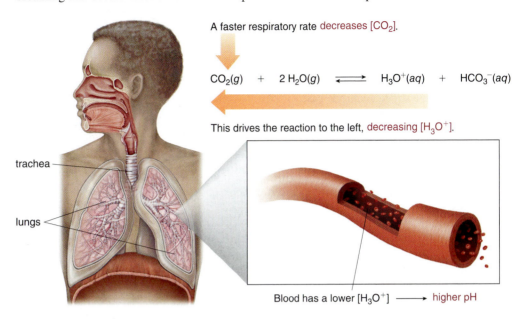

A faster respiratory rate decreases [CO₂].

$$CO_2(g) \quad + \quad 2\,H_2O(g) \quad \rightleftharpoons \quad H_3O^+(aq) \quad + \quad HCO_3^-(aq)$$

This drives the reaction to the left, decreasing [H₃O⁺].

trachea

lungs

Blood has a lower [H₃O⁺] ⟶ higher pH

The pH of the blood may also be altered when the metabolic processes of the body are not in balance. **Metabolic acidosis** results when excessive amounts of acid are produced and the blood pH falls. This may be observed in patients with severe infections (sepsis) when a large amount of lactic acid accumulates. It may also occur in poorly controlled diabetes when keto acid levels rise (Chapter 24). **Metabolic alkalosis** may occur when recurrent vomiting decreases the amount of acid in the stomach, thus causing a rise in pH.

CHAPTER HIGHLIGHTS

KEY TERMS

Acid (9.1)
Acid dissociation constant (9.4)
Acidic solution (9.5)
Amphoteric (9.2)
Base (9.1)
Basic solution (9.5)
Brønsted–Lowry acid (9.1)
Brønsted–Lowry base (9.1)

Buffer (9.10)
Conjugate acid (9.2)
Conjugate acid–base pair (9.2)
Conjugate base (9.2)
Diprotic acid (9.1)
Dissociation (9.3)
Ion–product constant (9.5)
Monoprotic acid (9.1)

Net ionic equation (9.7)
Neutral solution (9.5)
Neutralization reaction (9.7)
pH scale (9.6)
Proton transfer reaction (9.2)
Spectator ion (9.7)
Titration (9.9)
Triprotic acid (9.1)

KEY CONCEPTS

❶ Describe the basic features of acids and bases. (9.1)

- The Brønsted–Lowry definition is most commonly used to describe acids and bases.
- A Brønsted–Lowry acid is a proton donor, often symbolized by HA. A Brønsted–Lowry acid must contain one or more hydrogen atoms.
- A Brønsted–Lowry base is a proton acceptor, often symbolized by B:. To form a bond to a proton, a Brønsted–Lowry base must contain a lone pair of electrons.

❷ What are the basic features of an acid–base reaction? (9.2)

- In a Brønsted–Lowry acid–base reaction, a proton is transferred from the acid (HA) to the base (B:). In this reaction, the acid loses a proton to form its conjugate base (A^-) and the base gains a proton to form its conjugate acid (HB^+).

❸ How is acid strength related to the direction of equilibrium in an acid–base reaction? (9.3)

• A strong acid readily donates a proton, and when dissolved in water, 100% of the acid dissociates into ions. A strong base readily accepts a proton, and when dissolved in water, 100% of the base dissociates into ions.

• An inverse relationship exists between acid and base strength. A strong acid forms a weak conjugate base, whereas a weak acid forms a strong conjugate base.

• In an acid–base reaction, the stronger acid reacts with the stronger base to form the weaker acid and the weaker base.

❹ What is the acid dissociation constant and how is it related to acid strength? (9.4)

• For a general acid HA, the acid dissociation constant K_a is defined by the equation:

$$K_a = \frac{[H_3O^+][A:^-]}{[HA]}$$

• The stronger the acid, the larger the K_a. Equilibrium in an acid–base reaction favors formation of the acid with the smaller K_a value.

❺ What is the ion–product of water and how is it used to calculate hydronium or hydroxide ion concentration? (9.5)

• The ion–product of water, K_w, is a constant for all aqueous solutions; $K_w = [H_3O^+][^-OH] = 1.0 \times 10^{-14}$ at 25 °C. If either $[H_3O^+]$ or $[^-OH]$ is known, the other value can be calculated from K_w.

❻ What is pH? (9.6)

• The pH of a solution measures the concentration of H_3O^+; pH = $-\log [H_3O^+]$.
• A pH = 7 means $[H_3O^+] = [^-OH]$ and the solution is neutral.
• A pH < 7 means $[H_3O^+] > [^-OH]$ and the solution is acidic.
• A pH > 7 means $[H_3O^+] < [^-OH]$ and the solution is basic.

❼ Draw the products of some common acid–base reactions. (9.7)

• In a Brønsted–Lowry acid–base reaction with hydroxide bases (MOH), the acid HA donates a proton to ^-OH to form

H_2O. The anion from the acid HA combines with the cation M^+ of the base to form the salt MA. This reaction is called a neutralization reaction.

• In acid–base reactions with bicarbonate (HCO_3^-) or carbonate (CO_3^{2-}) bases, carbonic acid (H_2CO_3) is formed, which decomposes to form H_2O and CO_2.

❽ What happens to the pH of an aqueous solution when a salt is dissolved? (9.8)

• A salt can form an acidic, basic, or neutral solution depending on whether its cation and anion are derived from strong or weak acids and bases. A salt derived from a strong acid and strong base forms a neutral solution with pH = 7. When one ion of a salt is derived from a weak acid or base, the ion derived from the stronger acid or base determines whether the solution is acidic or basic.

❾ How is a titration used to determine the concentration of an acid or base? (9.9)

• A titration is a procedure that uses a base (or acid) of known volume and molarity to react with a known volume of acid (or base) of unknown molarity. The volume and molarity of the base are used to calculate the number of moles of base that react, and from this value, the molarity of the acid can be determined.

❿ What is a buffer? (9.10)

• A buffer is a solution whose pH changes very little when acid or base is added. Most buffers are composed of approximately equal amounts of a weak acid and the salt of its conjugate base.

⓫ What is the principal buffer present in the blood? (9.11)

• The principal buffer in the blood is carbonic acid/bicarbonate. Since carbonic acid (H_2CO_3) is in equilibrium with dissolved CO_2, the amount of CO_2 in the blood affects its pH, which is normally maintained in the range of 7.35–7.45. When the CO_2 concentration in the blood is higher than normal, the acid–base equilibrium shifts to form more H_3O^+ and the pH decreases. When the CO_2 concentration in the blood is lower than normal, the acid–base equilibrium shifts to consume $[H_3O^+]$, so $[H_3O^+]$ decreases, and the pH increases.

PROBLEMS

Selected in-chapter and end-of-chapter problems have brief answers provided in Appendix B.

Acids and Bases

9.39 Explain the difference between the Arrhenius definition of acids and bases and the Brønsted–Lowry definition of acids and bases.

9.40 Explain why NH_3 is a Brønsted–Lowry base but not an Arrhenius base.

9.41 Which of the following species can be Brønsted–Lowry acids?
  a. HBr   c. $AlCl_3$ e. NO_2^-
  b. Br_2 d. HCOOH   f. HNO_2

9.42 Which of the following species can be Brønsted–Lowry acids?
  a. H_2O c. HOCl   e. CH_3CH_2COOH
  b. I^- d. $FeBr_3$ f. CO_2

9.43 Which of the following species can be Brønsted–Lowry bases?
  a. ^-OH c. C_2H_6 e. ^-OCl
  b. Ca^{2+} d. PO_4^{3-} f. $MgCO_3$

9.44 Which of the following species can be Brønsted–Lowry bases?
a. Cl^- c. H_2O e. $Ca(OH)_2$
b. BH_3 d. Na^+ f. $HCOO^-$

9.45 Why is NH_3 a Brønsted–Lowry base but CH_4 is not?

9.46 Why is HCl a Brønsted–Lowry acid but NaCl is not?

9.47 Draw the conjugate acid of each base.

a. HS^- b. CO_3^{2-} c. NO_2^- d.
$$H\!-\!\underset{\underset{H}{|}}{\overset{\overset{H}{|}}{C}}\!-\!\overset{..}{\underset{\underset{H}{|}}{N}}\!-\!H$$

9.48 Draw the conjugate acid of each base.

a. Br^- b. HPO_4^{2-} c. CH_3COO^- d.
$$H\!-\!\underset{\underset{H}{|}}{\overset{\overset{H}{|}}{C}}\!-\!\overset{..}{\underset{..}{O}}\!-\!H$$

9.49 Draw the conjugate base of each acid.
a. HNO_2 b. NH_4^+ c. H_2O_2

9.50 Draw the conjugate base of each acid.
a. H_3O^+ b. H_2Se c. HSO_4^-

9.51 Label the conjugate acid–base pairs in each equation.
a. $HI(g) + NH_3(g) \rightleftharpoons NH_4^+(aq) + I^-(aq)$
b. $HCOOH(l) + H_2O(l) \rightleftharpoons H_3O^+(aq) + HCOO^-(aq)$
c. $HSO_4^-(aq) + H_2O(l) \rightleftharpoons H_2SO_4(aq) + {}^-OH(aq)$

9.52 Label the conjugate acid–base pairs in each equation.
a. $Cl^-(aq) + HSO_4^-(aq) \rightleftharpoons HCl(aq) + SO_4^{2-}(aq)$
b. $HPO_4^{2-}(aq) + {}^-OH(aq) \rightleftharpoons PO_4^{3-}(aq) + H_2O(l)$
c. $NH_3(g) + HF(g) \rightleftharpoons NH_4^+(aq) + F^-(aq)$

9.53 Like H_2O, HCO_3^- is amphoteric. (a) Draw the conjugate acid of HCO_3^-. (b) Draw the conjugate base of HCO_3^-.

9.54 Like H_2O, $H_2PO_4^-$ is amphoteric. (a) Draw the conjugate acid of $H_2PO_4^-$. (b) Draw the conjugate base of $H_2PO_4^-$.

9.55 Write the equation for the acid–base reaction that takes place when nitric acid (HNO_3) dissolves in H_2O.

9.56 Write the equation for the acid–base reaction that takes place when formic acid (HCOOH) dissolves in H_2O.

Acid and Base Strength

9.57 How do the following two processes differ: dissolving a strong acid in water compared to dissolving a weak acid in water?

9.58 How do the following two processes differ: dissolving a strong base in water compared to dissolving a weak base in water?

9.59 Which diagram represents an aqueous solution of HF and which represents HCl? Explain your choice.

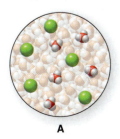
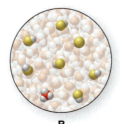

A   B

9.60 Which diagram represents what happens when HCN dissolves in water? Explain your choice.

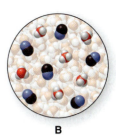

A   B

9.61 Use the data in Tables 9.1 and 9.2 to label the stronger acid in each pair.
a. H_2O or CH_3COOH
b. H_3PO_4 or HCO_3^-
c. H_2SO_4 or HSO_4^-

9.62 Use the data in Tables 9.1 and 9.2 to label the stronger acid in each pair.
a. HPO_4^{2-} or HCN
b. HSO_4^- or NH_4^+
c. H_2O or HF

9.63 Which acid in each pair in Problem 9.61 has the stronger conjugate base?

9.64 Which acid in each pair in Problem 9.62 has the stronger conjugate base?

9.65 Explain why a strong acid has a weak conjugate base.

9.66 Explain why a strong base has a weak conjugate acid.

9.67 Which acid, **A** or **B**, is stronger in each part?
a. **A** dissociates to a greater extent in water.
b. **A** has a smaller K_a.
c. The conjugate base of **A** is stronger than the conjugate base of **B**.

9.68 Which acid, **A** or **B**, is stronger in each part?
a. **B** dissociates to a greater extent in water.
b. **A** has a larger K_a.
c. The conjugate base of **B** is stronger than the conjugate base of **A**.

Equilibrium and Acid Dissociation Constants

9.69 What is the difference between the equilibrium constant K and the acid dissociation constant K_a?

9.70 What is the difference between the acid dissociation constant K_a and the ion–product constant K_w for water?

9.71 For each pair of acids: [1] Label the stronger acid. [2] Draw the conjugate bases. [3] Label the stronger conjugate base.

a. HSO_4^- or $H_2PO_4^-$
    bisulfate   dihydrogen phosphate
    $K_a = 1.2 \times 10^{-2}$ $K_a = 6.2 \times 10^{-8}$

b. CH_3CH_2COOH or CH_3COOH
    propanoic acid   acetic acid
    $K_a = 1.3 \times 10^{-5}$ $K_a = 1.8 \times 10^{-5}$

9.72 For each pair of acids: [1] Label the stronger acid. [2] Draw the conjugate bases. [3] Label the stronger conjugate base.

a. H_3PO_4 or   HCOOH
    phosphoric acid   formic acid
    $K_a = 7.5 \times 10^{-3}$ $K_a = 1.8 \times 10^{-4}$

b.   HCOOH   or C_6H_5COOH
    formic acid   benzoic acid
    $K_a = 1.8 \times 10^{-4}$ $K_a = 6.5 \times 10^{-5}$

9.73 Label the acid in the reactants and the conjugate acid in the products in each reaction. Use the data in Tables 9.1 and 9.2 to determine whether the reactants or products are favored at equilibrium. Explain your reasoning.
a. $H_3PO_4(aq) + {}^-CN(aq) \rightleftharpoons H_2PO_4{}^-(aq) + HCN(aq)$
b. $Br^-(aq) + HSO_4{}^-(aq) \rightleftharpoons SO_4{}^{2-}(aq) + HBr(aq)$
c. $CH_3COO^-(aq) + H_2CO_3(aq) \rightleftharpoons$
                    $CH_3COOH(aq) + HCO_3{}^-(aq)$

9.74 Label the acid in the reactants and the conjugate acid in the products in each reaction. Use the data in Tables 9.1 and 9.2 to determine whether the reactants or products are favored at equilibrium. Explain your reasoning.
a. $HF(g) + NH_3(g) \rightleftharpoons NH_4{}^+(aq) + F^-(aq)$
b. $Br^-(aq) + H_2O(l) \rightleftharpoons HBr(aq) + {}^-OH(aq)$
c. $HCN(aq) + HCO_3{}^-(aq) \rightleftharpoons H_2CO_3(aq) + {}^-CN(aq)$

Water and the pH Scale

9.75 Calculate the value of [$^-$OH] from the given [H_3O^+] and label the solution as acidic or basic.
a. 10^{-8} M   c. 3.0×10^{-4} M
b. 10^{-10} M   d. 2.5×10^{-11} M

9.76 Calculate the value of [$^-$OH] from the given [H_3O^+] and label the solution as acidic or basic.
a. 10^{-1} M   c. 2.6×10^{-7} M
b. 10^{-13} M   d. 1.2×10^{-12} M

9.77 Calculate the value of [H_3O^+] from the given [$^-$OH] and label the solution as acidic or basic.
a. 10^{-2} M   c. 6.2×10^{-7} M
b. 4.0×10^{-8} M   d. 8.5×10^{-13} M

9.78 Calculate the value of [H_3O^+] from the given [$^-$OH] and label the solution as acidic or basic.
a. 10^{-12} M   c. 6.0×10^{-4} M
b. 5.0×10^{-10} M   d. 8.9×10^{-11} M

9.79 Calculate the pH from each H_3O^+ concentration calculated in Problem 9.77.

9.80 Calculate the pH from each H_3O^+ concentration calculated in Problem 9.78.

9.81 Calculate the H_3O^+ concentration from each pH: (a) 12; (b) 1; (c) 1.80; (d) 8.90.

9.82 Calculate the H_3O^+ concentration from each pH: (a) 4; (b) 8; (c) 2.60; (d) 11.30.

9.83 If a urine sample has a pH of 5.90, calculate the concentrations of H_3O^+ and $^-$OH in the sample.

9.84 If pancreatic fluids have a pH of 8.2, calculate the concentrations of H_3O^+ and $^-$OH in the pancreas.

9.85 What are the concentrations of H_3O^+ and $^-$OH in tomatoes that have a pH of 4.10?

9.86 What are the concentrations of H_3O^+ and $^-$OH in a cola beverage that has a pH of 3.15?

9.87 Calculate the pH of each aqueous solution: (a) 0.0025 M HCl; (b) 0.015 M KOH.

9.88 Calculate the pH of each aqueous solution: (a) 0.015 M HNO_3; (b) 0.0025 M NaOH.

9.89 Why is the pH of 0.10 M HCl lower than the pH of 0.10 M CH_3COOH solution (1.0 vs. 2.88)?

9.90 Why is the pH of 0.0050 M CH_3COOH solution higher than the pH of 0.0050 M HCl solution (3.5 vs. 2.3)?

Acid–Base Reactions

9.91 Write a balanced equation for each reaction.
a. $HBr(aq) + KOH(aq) \longrightarrow$
b. $HNO_3(aq) + Ca(OH)_2(aq) \longrightarrow$
c. $HCl(aq) + NaHCO_3(aq) \longrightarrow$
d. $H_2SO_4(aq) + Mg(OH)_2(aq) \longrightarrow$

9.92 Write a balanced equation for each reaction.
a. $HNO_3(aq) + LiOH(aq) \longrightarrow$
b. $H_2SO_4(aq) + NaOH(aq) \longrightarrow$
c. $K_2CO_3(aq) + HCl(aq) \longrightarrow$
d. $HI(aq) + NaHCO_3(aq) \longrightarrow$

9.93 Marble statues, which are composed of calcium carbonate ($CaCO_3$), are slowly eaten away by the nitric acid (HNO_3) in acid rain. Write a balanced equation for the reaction of $CaCO_3$ with HNO_3.

9.94 Some liquid antacids contain suspensions of aluminum hydroxide [$Al(OH)_3$]. Write a balanced equation for the reaction of $Al(OH)_3$ with the HCl in stomach acid.

Salt Solutions

9.95 Determine whether each salt forms an acidic, basic, or neutral solution when it dissolves in water.
a. NaI   c. NH_4NO_3 e. $MgBr_2$
b. LiF   d. $KHCO_3$ f. NaH_2PO_4

9.96 Determine whether each salt forms an acidic, basic, or neutral solution when it dissolves in water.
a. NaBr   c. KCH_3COO e. $CaBr_2$
b. NaCN   d. CsF   f. K_3PO_4

Titration

9.97 What is the molarity of an HCl solution if 35.5 mL of 0.10 M NaOH are needed to neutralize 25.0 mL of the sample?

9.98 What is the molarity of an HCl solution if 17.2 mL of 0.15 M NaOH are needed to neutralize 5.00 mL of the sample?

9.99 What is the molarity of an acetic acid (CH_3COOH) solution if 15.5 mL of 0.20 M NaOH are needed to neutralize 25.0 mL of the sample?

9.100 What is the molarity of an H_2SO_4 solution if 18.5 mL of 0.18 M NaOH are needed to neutralize 25.0 mL of the sample?

9.101 How many milliliters of 1.0 M NaOH solution are needed to neutralize 10.0 mL of 2.5 M CH_3COOH solution?

9.102 How many milliliters of 2.0 M NaOH solution are needed to neutralize 8.0 mL of 3.5 M H_2SO_4 solution?

Buffers

9.103 Why is a buffer most effective at minimizing pH changes when the concentrations of the weak acid and its conjugate base are equal?

9.104 Although most buffers are prepared from a weak acid and its conjugate base, explain why a buffer can also be prepared from a weak base such as NH_3 and its conjugate acid NH_4^+.

9.105 Can a buffer be prepared from equal amounts of NaCN and HCN? Explain why or why not.

9.106 Can a buffer be prepared from equal amounts of HNO_3 and KNO_3? Explain why or why not.

9.107 Consider a buffer prepared from the weak acid HNO_2 and its conjugate base NO_2^-.

$$HNO_2(aq) + H_2O(l) \rightleftarrows NO_2^-(aq) + H_3O^+(aq)$$

a. Explain why both HNO_2 and NO_2^- are needed to prepare the buffer.
b. What happens to the concentrations of HNO_2 and NO_2^- when a small amount of acid is added to the buffer?
c. What happens to the concentrations of HNO_2 and NO_2^- when a small amount of base is added to the buffer?

9.108 Consider a buffer prepared from the weak acid HF and its conjugate base F^-.

$$HF(aq) + H_2O(l) \rightleftarrows F^-(aq) + H_3O^+(aq)$$

a. Explain why both HF and F^- are needed to prepare the buffer.
b. What happens to the concentrations of HF and F^- when a small amount of acid is added to the buffer?
c. What happens to the concentrations of HF and F^- when a small amount of base is added to the buffer?

9.109 Using the K_a values in Table 9.5, calculate the pH of a buffer that contains the given concentrations of a weak acid and its conjugate base.
a. 0.10 M Na_2HPO_4 and 0.10 M Na_3PO_4
b. 0.22 M $NaHCO_3$ and 0.22 M Na_2CO_3

9.110 Using the K_a values in Table 9.5, calculate the pH of a buffer that contains the given concentrations of a weak acid and its conjugate base.
a. 0.55 M CH_3COOH and 0.55 M $NaCH_3COO$
b. 0.15 M NaH_2PO_4 and 0.15 M Na_2HPO_4

9.111 Calculate the pH of an acetic acid/acetate buffer in which the concentration of acetic acid is always 0.20 M, but the concentration of sodium acetate ($NaCH_3COO$) corresponds to each of the following values: (a) 0.20 M; (b) 0.40 M; (c) 0.10 M.

9.112 Calculate the pH of a bicarbonate/carbonate buffer in which the concentration of sodium bicarbonate ($NaHCO_3$) is always 0.20 M, but the concentration of sodium carbonate (Na_2CO_3) corresponds to each of the following values: (a) 0.20 M; (b) 0.40 M; (c) 0.10 M.

General Questions

9.113 What is the difference between the concentration of an acid and the strength of an acid?

9.114 Do equal volumes of a 1.0 M HCl solution and a 1.0 M CH_3COOH solution require the same volume of NaOH to reach the end point in a titration?

Applications

9.115 Why is the pH of unpolluted rainwater lower than the pH of pure water?

9.116 Why is the pH of acid rain lower than the pH of rainwater?

9.117 The optimum pH of a swimming pool is 7.50. Calculate the value of $[H_3O^+]$ and $[^-OH]$ at this pH.

9.118 A sample of rainwater has a pH of 4.18. (a) Calculate the H_3O^+ concentration in the sample. (b) Suggest a reason why this pH differs from the pH of unpolluted rainwater (5.6).

9.119 When an individual hyperventilates, he is told to blow into a paper bag held over his mouth. What effect should this process have on the CO_2 concentration and pH of the blood?

9.120 What is the difference between respiratory acidosis and respiratory alkalosis?

9.121 How is CO_2 concentration related to the pH of the blood?

9.122 Explain why a lake on a bed of limestone is naturally buffered against the effects of acid rain.

CHALLENGE QUESTIONS

9.123 Calcium hypochlorite $[Ca(OCl)_2]$ is used to chlorinate swimming pools. $Ca(OCl)_2$ acts as a source of the weak acid hypochlorous acid, HOCl, a disinfectant that kills bacteria. Write the acid–base reaction that occurs when ^-OCl dissolves in water and explain why this reaction makes a swimming pool more basic.

9.124 Most buffer solutions are prepared using a weak acid and a salt of its conjugate base. Explain how the following combination can also form a buffer solution: 0.20 M H_3PO_4 and 0.10 M NaOH.

10

CHAPTER GOALS

In this chapter you will learn how to:

1. Describe the different types of
   radiation emitted by a radioactive
   nucleus

2. Write equations for nuclear reactions

3. Define half-life

4. Recognize the units used for
   measuring radioactivity

5. Give examples of common
   radioisotopes used in medical
   diagnosis and treatment

6. Describe the general features of
   nuclear fission and nuclear fusion

7. Describe the features of medical
   imaging techniques that do not use
   radioactivity

CT scans and MRIs are diagnostic medical imaging techniques that utilize forms of electromagnetic radiation. In addition, radioactive isotopes are commonly used to diagnose and treat
thyroid disease, as well as leukemia and many other forms of cancer.

NUCLEAR CHEMISTRY

THUS far our study of reactions has concentrated on processes that involve the
valence electrons of atoms. In these reactions, bonds that join atoms are broken
and new bonds between atoms are formed, but the identity of the atoms does not
change. In Chapter 10, we turn our attention to **nuclear reactions,** processes that
involve changes in the nucleus of atoms. While certainly much less common than
chemical reactions that occur with electrons, nuclear reactions form a useful group
of processes with a wide range of applications. Nuclear medicine labs in hospitals
use radioactive isotopes to diagnose disease, visualize organs, and treat tumors.
Generating energy in nuclear power plants, dating archaeological objects using the
isotope carbon-14, and designing a simple and reliable smoke detector all utilize the
concepts of nuclear chemistry discussed here in Chapter 10.

10.1 INTRODUCTION

Although most reactions involve valence electrons, a small but important group of reactions, **nuclear reactions,** involves the subatomic particles of the nucleus. To understand nuclear reactions we must first review facts presented in Chapter 2 regarding isotopes and the characteristics of the nucleus.

10.1A ISOTOPES

The nucleus of an atom is composed of protons and neutrons.

- **The atomic number (Z) = the number of protons in the nucleus.**
- **The mass number (A) = the number of protons and neutrons in the nucleus.**

Atoms of the same type of element have the same atomic number, but the number of neutrons may vary.

- **Isotopes are atoms of the same element having a different number of neutrons.**

As a result, isotopes have the same atomic number (Z) but different mass numbers (A). Carbon, for example, has three naturally occurring isotopes. Each isotope has six protons in the nucleus (i.e., $Z = 6$), but the number of neutrons may be six, seven, or eight. Thus, the mass numbers (A) of these isotopes are 12, 13, and 14, respectively. As we learned in Chapter 2, we can refer to these isotopes as carbon-12, carbon-13, and carbon-14. Isotopes are also written with the mass number to the upper left of the element symbol and the atomic number to the lower left.

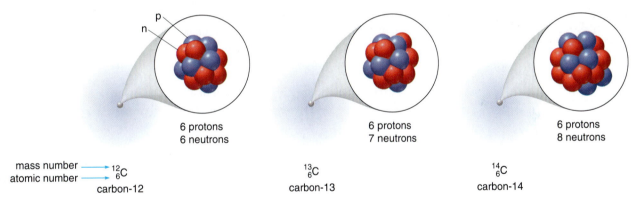

6 protons 6 neutrons	6 protons 7 neutrons	6 protons 8 neutrons

mass number ⟶ $^{12}_{6}\text{C}$

atomic number ⟶

carbon-12 $^{13}_{6}\text{C}$ $^{14}_{6}\text{C}$

carbon-13   carbon-14

Many isotopes are stable, but a larger number are not.

- **A *radioactive isotope*, called a *radioisotope*, is unstable and spontaneously emits energy to form a more stable nucleus.**

Radioactivity **is the nuclear radiation emitted by a radioactive isotope.** Of the known isotopes of all the elements, 264 are stable and 300 are naturally occurring but unstable. An even larger number of radioactive isotopes, called **artificial isotopes,** have been produced in the laboratory. Both carbon-12 and carbon-13 are stable isotopes and occur in higher natural abundance than carbon-14, a radioactive isotope.

SAMPLE PROBLEM 10.1

Iodine-123 and iodine-131 are radioactive isotopes used for the diagnosis or treatment of thyroid disease. Complete the following table for both isotopes.

	Atomic number	Mass number	Number of protons	Number of neutrons	Isotope symbol
Iodine-123					
Iodine-131					

ANALYSIS

- The atomic number = the number of protons.
- The mass number = the number of protons + the number of neutrons.
- Isotopes are written with the mass number to the upper left of the element symbol and the atomic number to the lower left.

SOLUTION

	Atomic number	Mass number	Number of protons	Number of neutrons	Isotope symbol
Iodine-123	53	123	53	$123 - 53 = 70$	$^{123}_{53}\text{I}$
Iodine-131	53	131	53	$131 - 53 = 78$	$^{131}_{53}\text{I}$

PROBLEM 10.1

Complete the following table for two isotopes of cobalt. Cobalt-60 is commonly used in cancer therapy.

	Atomic number	Mass number	Number of protons	Number of neutrons	Isotope symbol
Cobalt-59					
Cobalt-60					

PROBLEM 10.2

Each of the following radioisotopes is used in medicine. For each isotope give its: [1] atomic number; [2] mass number; [3] number of protons; [4] number of neutrons.

a. $^{85}_{38}\text{Sr}$ b. $^{67}_{31}\text{Ga}$ c.   selenium-75

     used in bone scans   used in abdominal scans   used in pancreas scans

10.1B TYPES OF RADIATION

Different forms of radiation are emitted when a radioactive nucleus is converted to a more stable nucleus, including **alpha particles, beta particles, positrons,** and **gamma radiation.**

- An *alpha particle* is a high-energy particle that contains two protons and two neutrons.

$$\text{alpha particle:} \quad \alpha \quad \text{or} \quad {}^{4}_{2}\text{He}$$

An alpha particle, symbolized by the Greek letter **alpha (α)** or the element symbol for helium, has a +2 charge and a mass number of 4.

- A *beta particle* is a high-energy electron.

$$\text{beta particle:} \quad \beta \quad \text{or} \quad {}^{0}_{-1}\text{e}$$

An electron has a –1 charge and a negligible mass compared to a proton. A beta particle, symbolized by the **Greek letter beta (β),** is also drawn with the symbol for an electron, **e,** with a mass number of 0 in the upper left corner and a charge of –1 in the lower left corner. A β particle is formed when a neutron (n) is converted to a proton (p) and an electron.

$$\overset{\text{mass = 1 amu}}{{}^{1}_{0}\text{n} \longrightarrow {}^{1}_{1}\text{p} + {}^{0}_{-1}\text{e}}$$

neutron   proton β particle

- A *positron* is called an *antiparticle* of a β particle, since their charges are different but their masses are the same.

Thus, a **positron** has a negligible mass like a β particle, but is opposite in charge, +1. A positron, symbolized as β⁺, is also drawn with the symbol for an electron, **e,** with a mass number of 0 in the upper left corner and a charge of +1 in the lower left corner. A positron, which can be thought of as a "positive electron," is formed when a proton is converted to a neutron.

$$\text{Symbol:} \quad {}^{0}_{+1}e \quad \text{or} \quad \beta^{+} \qquad\qquad \text{Formation:} \quad {}^{1}_{1}p \longrightarrow {}^{1}_{0}n + {}^{0}_{+1}e$$

positron $\qquad\qquad\qquad$ proton $\qquad$ neutron positron

- *Gamma rays* are high-energy radiation released from a radioactive nucleus.

Gamma rays, symbolized by the **Greek letter gamma (γ),** are a form of energy and thus they have no mass or charge. Table 10.1 summarizes the properties of some of the different types of radiation.

gamma ray: γ

TABLE 10.1 Types of Radiation

Type of Radiation	Symbol	Charge	Mass
Alpha particle	α or ${}^{4}_{2}He$	+2	4
Beta particle	β or ${}^{0}_{-1}e$	−1	0
Positron	β^{+} or ${}^{0}_{+1}e$	+1	0
Gamma ray	γ	0	0

PROBLEM 10.3   What is the difference between an α particle and a helium atom?

PROBLEM 10.4   What is the difference between an electron and a positron?

PROBLEM 10.5   Identify Q in each of the following symbols.

a. ${}^{0}_{-1}Q$ b. ${}^{4}_{2}Q$ c. ${}^{0}_{+1}Q$

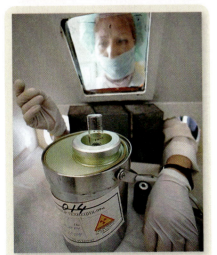

A lab worker must use protective equipment when working with radioactive substances.

10.1C FOCUS ON HEALTH & MEDICINE THE EFFECTS OF RADIOACTIVITY

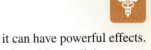

Radioactivity cannot be seen, smelled, tasted, heard, or felt, and yet it can have powerful effects. Because it is high in energy, nuclear radiation penetrates the surface of an object or living organism, where it can damage or kill cells. The cells that are most sensitive to radiation are those that undergo rapid cell division, such as those in bone marrow, reproductive organs, skin, and the intestinal tract. Since cancer cells also rapidly divide, they are also particularly sensitive to radiation, a fact that makes radiation an effective method of cancer treatment (Section 10.5).

Alpha (α) particles, β particles, and γ rays differ in the extent to which they can penetrate a surface. Alpha particles are the heaviest of the radioactive particles, and as a result they move the slowest and penetrate the least. Individuals who work with radioisotopes that emit α particles wear lab coats and gloves that provide a layer of sufficient protection. Beta particles move much faster since they have negligible mass, and they can penetrate into body tissue. Lab workers and health professionals must wear heavy lab coats and gloves when working with substances that

give off β particles. Gamma rays travel the fastest and readily penetrate body tissue. Working with substances that emit γ rays is extremely hazardous, and a thick lead shield is required to halt their penetration.

That γ rays kill cells is used to an advantage in the food industry. To decrease the incidence of harmful bacteria in foods, certain fruits and vegetables are irradiated with γ rays that kill any bacteria contained in them. Foods do not come into contact with radioisotopes and the food is not radioactive after radiation. Gamma rays merely penetrate the food and destroy any live organism, and often as a result, the food product has a considerably longer shelf life.

10.2   NUCLEAR REACTIONS

Radioactive decay **is the process by which an unstable radioactive nucleus emits radiation, forming a nucleus of new composition.** A nuclear equation can be written for this process, which contains the original nucleus, the new nucleus, and the radiation emitted. Unlike a chemical equation that balances atoms, in a nuclear equation the mass numbers and the atomic numbers of the nuclei must be balanced.

- The sum of the mass numbers (*A*) must be equal on both sides of a nuclear equation.
- The sum of the atomic numbers (*Z*) must be equal on both sides of a nuclear equation.

10.2A   ALPHA EMISSION

Alpha emission **is the decay of a nucleus by emitting an α particle.** For example, uranium-238 decays to thorium-234 by loss of an α particle.

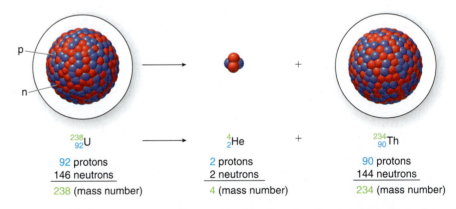

$$^{238}_{92}U \longrightarrow \,^{4}_{2}He + \,^{234}_{90}Th$$

92 protons	2 protons	90 protons
146 neutrons	2 neutrons	144 neutrons
238 (mass number)	4 (mass number)	234 (mass number)

Since an α particle has two protons, **the new nucleus has** *two fewer protons* **than the original nucleus.** Because it has a *different* number of protons, the new nucleus represents a *different* element. Uranium-238 has 92 protons, so loss of two forms the element thorium with 90 protons. The thorium nucleus has a mass number that is four fewer than the original—234—because it has been formed by loss of an α particle with a mass number of four.

As a result, the sum of the mass numbers is equal on both sides of the equation—238 = 4 + 234. The sum of the atomic numbers is also equal on both sides of the equation—92 = 2 + 90.

HOW TO   Balance an Equation for a Nuclear Reaction

EXAMPLE   Write a balanced nuclear equation showing how americium-241, a radioactive atom used in smoke detectors, decays to form an α particle.

Step [1]   **Write an incomplete equation with the original nucleus on the left and the particle emitted on the right.**

- Include the mass number and atomic number (from the periodic table) in the equation.

$$^{241}_{95}\text{Am} \longrightarrow {}^{4}_{2}\text{He} + \text{?}$$

Step [2]   **Calculate the mass number and atomic number of the newly formed nucleus on the right.**

- Mass number: Subtract the mass of an α particle (4) to obtain the mass of the new nucleus; $241 - 4 = 237$.
- Atomic number: Subtract the two protons of an α particle to obtain the atomic number of the new nucleus; $95 - 2 = 93$.

Step [3]   **Use the atomic number to identify the new nucleus and complete the equation.**

- From the periodic table, the element with an atomic number of 93 is neptunium, Np.
- Write the mass number and the atomic number with the element symbol to complete the equation.

$$
\begin{array}{c}
241 = 4 + 237 \\
95 = 2 + 93
\end{array}
\qquad
^{241}_{95}\text{Am} \longrightarrow {}^{4}_{2}\text{He} + {}^{237}_{93}\text{Np}
$$

PROBLEM 10.6   Radon, a radioactive gas formed in the soil, can cause lung cancers when inhaled in high concentrations for a long period of time. Write a balanced nuclear equation for the decay of radon-222, which emits an α particle.

PROBLEM 10.7   Radon (Problem 10.6) is formed in the soil as a product of radioactive decay that produces an α particle. Write a balanced nuclear equation for the formation of radon-222 and an α particle.

PROBLEM 10.8   Write a balanced equation showing how each nucleus decays to form an α particle: (a) polonium-218; (b) thorium-230; (c) Es-252.

10.2B   BETA EMISSION

Beta emission **is the decay of a nucleus by emitting a β particle.** For example, carbon-14 decays to nitrogen-14 by loss of a β particle. The decay of carbon-14 is used to date archaeological specimens (Section 10.3)

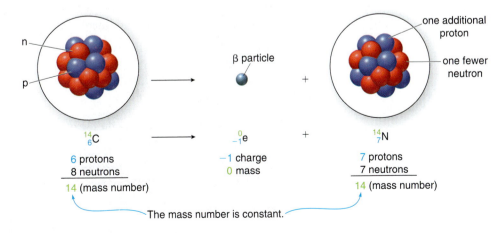

$$^{14}_{6}\text{C} \longrightarrow {}^{0}_{-1}\text{e} + {}^{14}_{7}\text{N}$$

$^{14}_{6}$C	$^{0}_{-1}$e	$^{14}_{7}$N
6 protons	−1 charge	7 protons
8 neutrons	0 mass	7 neutrons
14 (mass number)		14 (mass number)

The mass number is constant.

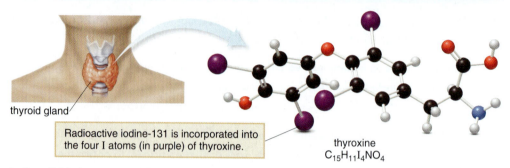

▼ FIGURE 10.1   The Use of Iodine-131 to Treat Hyperthyroidism

thyroid gland

Radioactive iodine-131 is incorporated into the four I atoms (in purple) of thyroxine.

thyroxine
$C_{15}H_{11}I_4NO_4$

Iodine-131 is incorporated into the thyroid hormone thyroxine. Beta radiation emitted by the radioactive isotope destroys nearby thyroid cells, thus decreasing the activity of the thyroid gland and bringing the disease under control.

In β emission, one neutron of the original nucleus decays to a β particle and a proton. As a result, the **new nucleus has** *one more proton* **and** *one fewer neutron* **than the original nucleus.** In this example, a carbon atom with six protons decays to a nitrogen atom with seven protons. Since the total number of particles in the nucleus does not change, the **mass number is constant.**

The subscripts that represent the atomic numbers are balanced because the β particle has a charge of –1. Seven protons on the right side plus a –1 charge for the β particle gives a total "charge" of +6, the atomic number of carbon on the left. The mass numbers are also balanced since a β particle has zero mass, and both the original nucleus and the new nucleus contain 14 subatomic particles (protons + neutrons).

Radioactive elements that emit β radiation are widely used in medicine. Since β radiation is composed of high-energy, rapidly moving electrons that penetrate tissue in a small, localized region, radioactive elements situated in close contact with tumor cells kill them. Although both healthy and diseased cells are destroyed by this internal radiation therapy, rapidly dividing tumor cells are more sensitive to its effects and therefore their growth and replication are affected the most.

Iodine-131, a radioactive element that emits β radiation, is used to treat hyperthyroidism, a condition resulting from an overactive thyroid gland (Figure 10.1). When iodine-131 is administered, it is incorporated into thyroxine, an iodine-containing hormone that is concentrated in the thyroid gland. The β radiation emitted by the iodine-131 kills some of the thyroid tissue, so that the gland is no longer overactive.

SAMPLE PROBLEM 10.2

Write a balanced nuclear equation for the β emission of phosphorus-32, a radioisotope used to treat leukemia and other blood disorders.

ANALYSIS

Balance the atomic numbers and mass numbers on both sides of a nuclear equation. With β emission, treat the β particle as an electron with zero mass in balancing mass numbers, and a –1 charge when balancing the atomic numbers.

SOLUTION

[1] **Write an incomplete equation with the original nucleus on the left and the particle emitted on the right.**

- Use the identity of the element to determine the atomic number; phosphorus has an atomic number of 15.

$$^{32}_{15}P \longrightarrow \, ^{0}_{-1}e \, + \, ?$$

[2]   **Calculate the mass number and the atomic number of the newly formed nucleus on the right.**

- Mass number: Since a β particle has no mass, the masses of the new particle and the original particle are the same, 32.
- Atomic number: Since β emission converts a neutron into a proton, the new nucleus has one more proton than the original nucleus; $15 = -1 + ?$. Thus the new nucleus has an atomic number of 16.

[3]   **Use the atomic number to identify the new nucleus and complete the equation.**

- From the periodic table, the element with an atomic number of 16 is sulfur, S.
- Write the mass number and the atomic number with the element symbol to complete the equation.

$$^{32}_{15}\text{P} \longrightarrow \, ^{0}_{-1}\text{e} \, + \, ^{32}_{16}\text{S}$$

PROBLEM 10.9   Write a balanced nuclear equation for the β emission iodine-131.

PROBLEM 10.10   Write a balanced nuclear equation for the β emission of each of the following isotopes.

a. $^{20}_{9}\text{F}$ b. $^{92}_{38}\text{Sr}$ c. chromium-55

10.2C   POSITRON EMISSION

Positron emission is the decay of a nucleus by emitting a positron (β⁺). For example, carbon-11, an artificial radioactive isotope of carbon, decays to boron-11 by loss of a β⁺ particle. Positron emitters are used in a relatively new diagnostic technique, positron emission tomography (PET), described in Section 10.5.

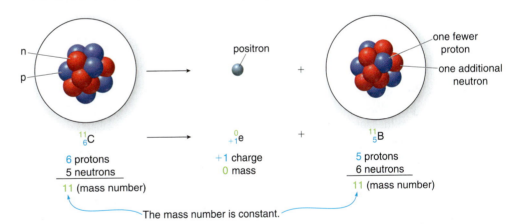

In positron emission, one proton of the original nucleus decays to a β⁺ particle and a neutron. As a result, the **new nucleus has *one fewer proton* and *one more neutron* than the original nucleus.** In this example, a carbon atom with six protons decays to a boron atom with five protons. Since the total number of particles in the nucleus does not change, the **mass number is constant.**

SAMPLE PROBLEM 10.3   Write a balanced nuclear equation for the positron emission of fluorine-18, a radioisotope used for imaging in PET scans.

ANALYSIS   Balance the atomic numbers and mass numbers on both sides of a nuclear equation. With β⁺ emission, treat the positron as a particle with zero mass when balancing mass numbers, and a +1 charge when balancing the atomic numbers.

SOLUTION

[1] **Write an incomplete equation with the original nucleus on the left and the particle emitted on the right.**

- Use the identity of the element to determine the atomic number; fluorine has an atomic number of 9.

$$^{18}_{9}\text{F} \longrightarrow \, ^{0}_{+1}\text{e} \; + \; ?$$

[2] **Calculate the mass number and the atomic number of the newly formed nucleus on the right.**

- Mass number: Since a β^+ particle has no mass, the masses of the new particle and the original particle are the same, 18.
- Atomic number: Since β^+ emission converts a proton into a neutron, the new nucleus has one fewer proton than the original nucleus; $9 - 1 = 8$. Thus, the new nucleus has an atomic number of 8.

[3] **Use the atomic number to identify the new nucleus and complete the equation.**

- From the periodic table, the element with an atomic number of 8 is oxygen, O.
- Write the mass number and the atomic number with the element symbol to complete the equation.

$$^{18}_{9}\text{F} \longrightarrow \, ^{0}_{+1}\text{e} \; + \; ^{18}_{8}\text{O}$$

PROBLEM 10.11

Write a balanced nuclear equation for the positron emission of each of the following nuclei: (a) arsenic-74; (b) oxygen-15.

10.2D GAMMA EMISSION

Gamma emission **is the decay of a nucleus by emitting γ radiation.** Since γ rays are simply a form of energy, their emission causes **no change in the atomic number or mass number** of a radioactive nucleus. Gamma emission sometimes occurs alone. For example, one form of technetium-99, written as technetium-99m, is an energetic form of the technetium nucleus that decays with emission of γ rays to technetium-99, a more stable but still radioactive element.

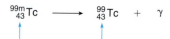

$$^{99m}_{43}\text{Tc} \longrightarrow \, ^{99}_{43}\text{Tc} \; + \; \gamma$$

The mass number and atomic number are the same.

> The *m* in technetium-99m stands for *metastable.* This designation is meant to indicate that the isotope decays to a more stable form of the same isotope.

Technetium-99m is a widely used radioisotope in medical imaging. Because it emits high-energy γ rays but decays in a short period of time, it is used to image the brain, thyroid, lungs, liver, skeleton, and many other organs. It has also been used to detect ulcers in the gastrointestinal system, and combined with other compounds, it is used to map the circulatory system and gauge damage after a heart attack.

More commonly, γ emission accompanies α or β emission. For example, cobalt-60 decays with both β and γ emission. Because a β particle is formed, decay generates an element with the *same* mass but a *different* number of protons, and thus a new element, nickel-60.

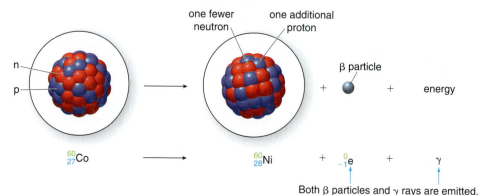

Both β particles and γ rays are emitted.

▼ FIGURE 10.2 Focus on Health & Medicine: External Radiation Treatment for Tumors

a.

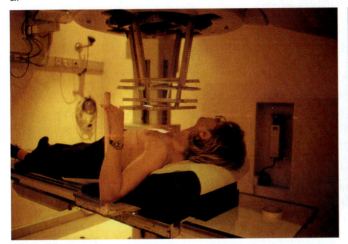

b. c.

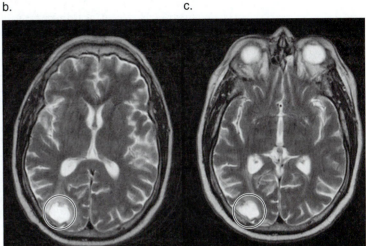

a. Gamma radiation from the decay of cobalt-60 is used to treat a variety of tumors, especially those that cannot be surgically removed.

b. A tumor (bright area in circle) before radiation treatment

c. A tumor (bright area in circle) that has decreased in size after six months of radiation treatment

Cobalt-60 is used in external radiation treatment for cancer. Radiation generated by cobalt-60 decay is focused on a specific site in the body that contains cancerous cells (Figure 10.2). By directing the radiation on the tumor, damage to surrounding healthy tissues is minimized.

PROBLEM 10.12	Write a nuclear equation for the decay of iridium-192 with β and γ emission. Iridium implants have been used to treat breast cancer. After the correct dose is administered, the iridium source is removed.

PROBLEM 10.13	Complete each nuclear equation. a. $^{11}_{5}\text{B} \longrightarrow ? + \gamma$ b. $^{40}_{19}\text{K} \longrightarrow ? + {}^{0}_{-1}\text{e} + \gamma$

10.3 HALF-LIFE

How fast do radioactive isotopes decay? It depends on the isotope.

- The *half-life* ($t_{1/2}$) of a radioactive isotope is the time it takes for one-half of the sample to decay.

10.3A GENERAL FEATURES

Suppose we have a sample that contains 16 g of phosphorus-32, a radioactive isotope that decays to sulfur-32 by β emission (Sample Problem 10.2). Phosphorus-32 has a half-life of approximately 14 days. Thus, after 14 days, the sample contains only half the amount of P-32—8.0 g. After another 14 days (a total of two half-lives), the 8.0 g of P-32 is again halved to 4.0 g. After another 14 days (a total of three half-lives), the 4.0 g of P-32 is halved to 2.0 g, and so on. Every 14 days, half of the P-32 decays.

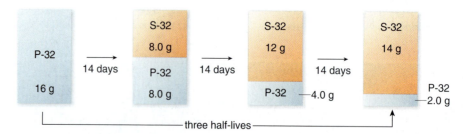

Many naturally occurring isotopes have long half-lives. Examples include carbon-14 (5,730 years) and uranium-235 (7.0×10^8 years). Radioisotopes that are used for diagnosis and imaging in medicine have short half-lives so they do not linger in the body. Examples include technetium-99m (6.0 hours) and iodine-131 (8.0 days). The half-lives of several elements are given in Table 10.2.

The half-life of a radioactive isotope is a property of a given isotope and is independent of the amount of sample, temperature, and pressure. Thus, if the half-life and amount of a sample are known, it is possible to predict how much of the radioactive isotope will remain after a period of time.

TABLE 10.2   Half-Lives of Some Common Radioisotopes

Radioisotope	Symbol	Half-Life	Use
Carbon-14	$^{14}_{6}\text{C}$	5,730 years	Archaeological dating
Cobalt-60	$^{60}_{27}\text{Co}$	5.3 years	Cancer therapy
Iodine-131	$^{131}_{53}\text{I}$	8.0 days	Thyroid therapy
Potassium-40	$^{40}_{19}\text{K}$	1.3×10^9 years	Geological dating
Phosphorus-32	$^{32}_{15}\text{P}$	14.3 days	Leukemia treatment
Technetium-99m	$^{99m}_{43}\text{Tc}$	6.0 hours	Organ imaging
Uranium-235	$^{235}_{92}\text{U}$	7.0×10^8 years	Nuclear reactors

HOW TO   Use a Half-Life to Determine the Amount of Radioisotope Present

EXAMPLE   If the half-life of iodine-131 is 8.0 days, how much of a 100. mg sample of iodine-131 remains after 32 days?

Step [1]   **Determine how many half-lives occur in the given amount of time.**

- Use the half-life of iodine-131 as a conversion factor to convert the number of days to the number of half-lives.

$$32 \text{ days} \times \frac{1 \text{ half-life}}{8.0 \text{ days}} = 4.0 \text{ half-lives}$$

Step [2]   **For each half-life, multiply the initial mass by one-half to obtain the final mass.**

- Since 32 days corresponds to *four* half-lives, multiply the initial mass by ½ *four* times to obtain the final mass. After four half-lives, 6.25 mg of iodine-131 remains.

$$100. \text{ mg} \times \frac{1}{2} \times \frac{1}{2} \times \frac{1}{2} \times \frac{1}{2} = 6.25 \text{ mg of iodine-131 remains.}$$

initial mass   The mass is halved four times.

PROBLEM 10.14   How much phosphorus-32 remains from a 1.00 g sample after each of the following number of half-lives: (a) 2; (b) 4; (c) 8; (d) 20?

PROBLEM 10.15   If a 160. mg sample of technetium-99m is used for a diagnostic procedure, how much Tc-99m remains after each interval: (a) 6.0 h; (b) 18.0 h; (c) 24.0 h; (d) 2 days?

10.3B ARCHAEOLOGICAL DATING

Archaeologists use the half-life of carbon-14 to determine the age of carbon-containing material derived from plants or animals. The technique, **radiocarbon dating,** is based on the fact that the ratio of radioactive carbon-14 to stable carbon-12 is a constant value in a living organism that is constantly taking in CO_2 and other carbon-containing nutrients from its surroundings. Once the organism dies, however, the radioactive isotope (C-14) decays (Section 10.2B) without being replenished, thus decreasing its concentration, while the stable isotope of carbon (C-12) remains at a constant value. By comparing the ratio of C-14 to C-12 in an artifact to the ratio of C-14 to C-12 in organisms today, the age of the artifact can be determined. Radiocarbon dating can be used to give the approximate age of wood, cloth, bone, charcoal, and many other substances that contain carbon.

The half-life of carbon-14 is 5,730 years, so half of the C-14 has decayed after about 6,000 years. Thus, a 6,000-year-old object has a ratio of C-14 to C-12 that has decreased by a factor of two, a 12,000-year-old object has a ratio of C-14 to C-12 that has decreased by a factor of four, and so forth.

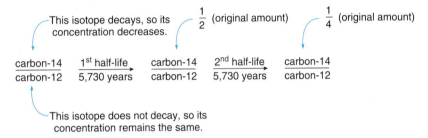

Using this technique, archaeologists have determined the age of the paintings on cave walls in Algeria to be about 8,000 years old (Figure 10.3). Because the amount of carbon-14 decreases with time, artifacts older than about 20,000 years have too little carbon-14 to accurately estimate their age.

PROBLEM 10.16	Estimate the age of an artifact that has 1/8 of the amount of C-14 (relative to C-12) compared to living organisms.

▼ **FIGURE 10.3   Radiocarbon Dating**

Radiocarbon dating has been used to estimate the age of this Algerian cave painting at about 8,000 years.

10.4 DETECTING AND MEASURING RADIOACTIVITY

We all receive a miniscule daily dose of radiation from cosmic rays and radioactive substances in the soil. Additional radiation exposure comes from television sets, dental X-rays, and other man-made sources. Moreover, we are still exposed to nuclear fallout, residual radiation resulting from the testing of nuclear weapons in the atmosphere decades ago.

Although this background radiation is unavoidable and minute, higher levels can be harmful and life-threatening because radiation is composed of high-energy particles and waves that damage cells and disrupt key biological processes, often causing cell death. How can radiation be detected and measured when it can't be directly observed by any of the senses?

A Geiger counter is a device used to detect radiation.

A **Geiger counter** is a small portable device used for measuring radioactivity. It consists of a tube filled with argon gas that is ionized when it comes into contact with nuclear radiation. This in turn generates an electric current that produces a clicking sound or registers on a meter. Geiger counters are used to locate a radiation source or a site that has become contaminated by radioactivity.

Individuals who work with radioactivity wear protective clothing (Section 10.1) as well as radiation badges. A radiation badge contains photographic film that fogs when it comes into contact with radioactivity. These badges are regularly monitored to assure that these individuals are not exposed to unhealthy levels of harmful radiation.

Individuals who work with radioactivity wear badges to monitor radiation levels.

10.4A MEASURING THE RADIOACTIVITY IN A SAMPLE

The amount of radioactivity in a sample is measured by the number of nuclei that decay per unit time—disintegrations per second. The most common unit is the **curie** (Ci), and smaller units derived from it, the **millicurie** (mCi) and the **microcurie** (μCi). One curie equals 3.7×10^{10} disintegrations/second, which corresponds to the decay rate of 1 g of the element radium.

$$1 \text{ Ci} = 3.7 \times 10^{10} \text{ disintegrations/second}$$
$$1 \text{ Ci} = 1{,}000 \text{ mCi}$$
$$1 \text{ Ci} = 1{,}000{,}000 \text{ μCi}$$

The **becquerel** (Bq), an SI unit, is also used to measure radioactivity; 1 Bq = 1 disintegration/second. Since each nuclear decay corresponds to one becquerel, $1 \text{ Ci} = 3.7 \times 10^{10}$ Bq. Radioactivity units are summarized in Table 10.3.

Often a dose of radiation is measured in the number of millicuries that must be administered. For example, a diagnostic test for thyroid activity uses sodium iodide that contains iodine-131—that is, $Na^{131}I$. The radioisotope is purchased with a known amount of radioactivity per milliliter, such as 3.5 mCi/mL. By knowing the amount of radioactivity a patient must be given, as well as the concentration of radioactivity in the sample, one can calculate the volume of radioactive isotope that must be administered (Sample Problem 10.4).

TABLE 10.3   Units Used to Measure Radioactivity
$1 \text{ Ci} = 3.7 \times 10^{10}$ disintegrations/s
$1 \text{ Ci} = 3.7 \times 10^{10}$ Bq
$1 \text{ Ci} = 1{,}000$ mCi
$1 \text{ Ci} = 1{,}000{,}000$ μCi

SAMPLE PROBLEM 10.4

A patient must be given a 4.5-mCi dose of iodine-131, which is available as a solution that contains 3.5 mCi/mL. What volume of solution must be administered?

ANALYSIS

Use the amount of radioactivity (mCi/mL) as a conversion factor to convert the dose of radioactivity from millicuries to a volume in milliliters.

SOLUTION

The dose of radioactivity is known in millicuries, and the amount of radioactivity per unit volume (3.5 mCi/mL) is also known. Use 3.5 mCi/mL as a millicurie–milliliter conversion factor.

The curie is named for Polish chemist Marie Skłodowska Curie who discovered the radioactive elements polonium and radium, and received Nobel Prizes for both Chemistry and Physics in the early twentieth century.

$$\text{4.5 mCi dose} \times \frac{1 \text{ mL}}{3.5 \text{ mCi}} = \text{1.3 mL dose}$$

mCi–mL conversion factor

Millicuries cancel.

Answer

PROBLEM 10.17   To treat a thyroid tumor, a patient must be given a 110-mCi dose of iodine-131, supplied in a vial containing 25 mCi/mL. What volume of solution must be administered?

10.4B MEASURING HUMAN EXPOSURE TO RADIOACTIVITY

Several units are used to measure the amount of radiation *absorbed* by an organism.

> • The **rad**—radiation absorbed dose—is the amount of radiation absorbed by one gram of a substance. The amount of energy absorbed varies with both the nature of the substance and the type of radiation.
> • The **rem**—radiation equivalent for man—is the amount of radiation that also factors in its energy and potential to damage tissue. Using rem as a measure of radiation, 1 rem of any type of radiation produces the same amount of tissue damage.

Other units to measure absorbed radiation include the **gray** (1 Gy = 100 rad) and the **sievert** (1 Sv = 100 rem).

Although background radiation varies with location, the average radiation dose per year for an individual is estimated at 0.27 rem. Generally, no detectable biological effects are noticed when the dose of radiation is less than 25 rem. A single dose of 25–100 rem causes a temporary decrease in white blood cell count. The symptoms of radiation sickness—nausea, vomiting, fatigue, and prolonged decrease in white blood cell count—are visible at a dose of more than 100 rem.

Death results at still higher doses of radiation. The **LD$_{50}$—the lethal dose that kills 50% of a population**—is 500 rem in humans, and exposure to 600 rem of radiation is fatal for an entire population.

PROBLEM 10.18   The unit millirem (1 rem = 1,000 mrem) is often used to measure the amount of radiation absorbed. (a) The average yearly dose of radiation from radon gas is 200 mrem. How many rem does this correspond to? (b) If a thyroid scan exposes a patient to 0.014 rem of radiation, how many mrem does this correspond to? (c) Which represents the larger dose?

10.5 FOCUS ON HEALTH & MEDICINE
MEDICAL USES OF RADIOISOTOPES

Radioactive isotopes are used for both diagnostic and therapeutic procedures in medicine. In a diagnostic test to measure the function of an organ or to locate a tumor, low doses of radioactivity are generally given. When the purpose of using radiation is therapeutic, such as to kill diseased cells or cancerous tissue, a much higher dose of radiation is required.

10.5A RADIOISOTOPES USED IN DIAGNOSIS

Radioisotopes are routinely used to determine if an organ is functioning properly or to detect the presence of a tumor. The isotope is ingested or injected and the radiation it emits can be used to produce a scan. Sometimes the isotope is an atom or ion that is not part of a larger molecule. Examples include iodine-131, which is administered as the salt sodium iodide (Na^{131}I), and xenon-133, which is a gas containing radioactive xenon atoms. At other times the radioactive atom is bonded to a larger molecule that targets a specific organ. An organ that has increased or decreased uptake of the radioactive element can indicate disease, the presence of a tumor, or other conditions.

A HIDA scan (hepatobiliary iminodiacetic acid scan) uses a technetium-99m-labeled molecule to evaluate the functioning of the gall bladder and bile ducts (Figure 10.4). After injection, the

▼ **FIGURE 10.4   HIDA Scan Using Technetium-99m**

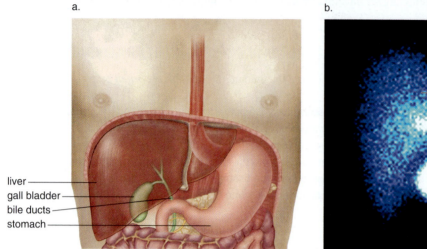

a. Schematic showing the location of the liver, gall bladder, and bile ducts

b. A scan using technetium-99m showing bright areas for the liver, gall bladder, and bile ducts, indicating normal function

technetium-99m travels through the bloodstream and into the liver, gall bladder, and bile ducts, where, in a healthy individual, the organs are all clearly visible on a scan. When the gall bladder is inflamed or the bile ducts are obstructed by gallstones, uptake of the radioisotope does not occur and these organs are not visualized because they do not contain the radioisotope.

Red blood cells tagged with technetium-99m are used to identify the site of internal bleeding in an individual. Bone scans performed with technetium-99m can show the location of metastatic cancer, so that specific sites can be targeted for radiation therapy (Figure 10.5).

Thallium-201 is used in stress tests to diagnose coronary artery disease. Thallium injected into a vein crosses cell membranes into normal heart muscle. Little radioactive thallium is found in areas of the heart that have a poor blood supply. This technique is used to identify individuals who may need bypass surgery or other interventions because of blocked coronary arteries.

PROBLEM 10.19   The half-life of thallium-201 is three days. What fraction of thallium-201 is still present in an individual after nine days?

10.5B   RADIOISOTOPES USED IN TREATMENT

The high-energy radiation emitted by radioisotopes can be used to kill rapidly dividing tumor cells. Two techniques are used. Sometimes the radiation source is external to the body. For example, a beam of radiation produced by decaying cobalt-60 can be focused at a tumor. Such a radiation source must have a much longer half-life—5.3 years in this case—than radioisotopes that are ingested for diagnostic purposes. With this method some destruction of healthy tissue often occurs, and a patient may experience some signs of radiation sickness, including vomiting, fatigue, and hair loss.

A more selective approach to cancer treatment involves using a radioactive isotope internally at the site of the tumor within the body. Using iodine-131 to treat hyperthyroidism has already been discussed (Section 10.1). Other examples include using radioactive "seeds" or wire that can be implanted close to a tumor. Iodine-125 seeds are used to treat prostate cancer and iridium-192 wire is used to treat some cancers of the breast.

Figure 10.6 illustrates radioisotopes that are used for diagnosis or treatment.

▼ **FIGURE 10.5** Bone Scan Using Technetium-99m

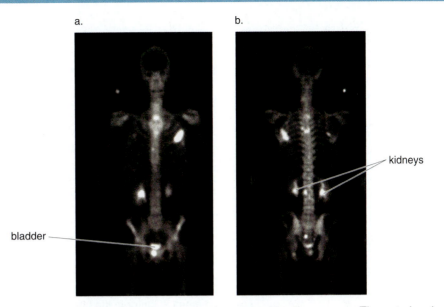

The bone scan of a patient whose lung cancer has spread to other organs. The anterior view [from the front in (a)] shows the spread of disease to the ribs, while the posterior view [from the back in (b)] shows spread of disease to the ribs and spine. The bright areas in the mid-torso and lower pelvis are due to a collection of radioisotope in the kidneys and bladder, before it is eliminated in the urine.

▼ **FIGURE 10.6** Common Radioisotopes Used in Medicine

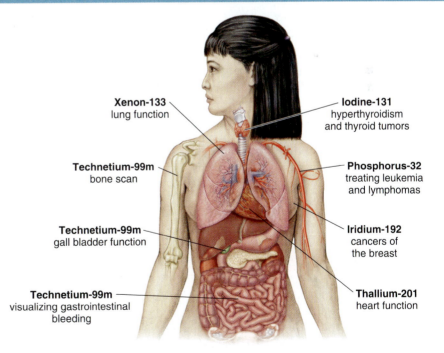

10.5C POSITRON EMISSION TOMOGRAPHY—PET SCANS

Positron emission tomography (PET) scans use radioisotopes that emit positrons when the nucleus decays. Once formed, a positron combines with an electron to form two γ rays, which create a scan of an organ.

▼ **FIGURE 10.7 PET Scans**

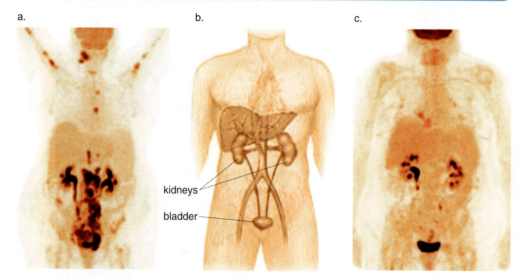

a.   b.   c.

kidneys

bladder

a. The PET scan shows cancer of the lymph nodes in the neck and abdomen, as well as scattered areas of tumor in the bone marrow of the arms and spine before treatment.

b. The schematic of selected organs in the torso and pelvis.

c. The PET scan shows significant clearing of disease after chemotherapy by the decrease in intensity of the radioisotope. The dark regions in the kidneys (in the torso) and bladder (in the lower pelvis) are due to the concentration of the radioisotope before elimination in the urine.

$$^{0}_{+1}e \; + \; ^{0}_{-1}e \; \longrightarrow \; 2\,\gamma$$

positron electron   gamma rays

Carbon-11, oxygen-15, nitrogen-13, and fluorine-18 are common radioactive isotopes used in PET scans. For example, a carbon-11 or fluorine-18 isotope can be incorporated in a glucose molecule. When this radioactive molecule is taken internally, its concentration becomes highest in areas in the body that continually use glucose. A healthy brain shows a high level of radioactivity from labeled glucose. When an individual suffers a stroke or has Alzheimer's disease, brain activity is significantly decreased and radioactivity levels are decreased.

PET scans are also used to detect tumors and coronary artery disease, and determine whether cancer has spread to other organs of the body. A PET scan is also a noninvasive method of monitoring whether cancer treatment has been successful (Figure 10.7).

PROBLEM 10.20   Write a nuclear equation for the emission of a positron from nitrogen-13.

10.6 NUCLEAR FISSION AND NUCLEAR FUSION

The nuclear reactions used in nuclear power plants occur by a process called *nuclear fission,* whereas the nuclear reactions that take place in the sun occur by a process called *nuclear fusion.*

- *Nuclear fission* is the splitting apart of a heavy nucleus into lighter nuclei and neutrons.
- *Nuclear fusion* is the joining together of two light nuclei to form a larger nucleus.

10.6A   NUCLEAR FISSION

When uranium-235 is bombarded by a neutron, it undergoes **nuclear fission** and splits apart into two lighter nuclei. Several different fission products have been identified. One common nuclear reaction is the fission of uranium-235 into krypton-91 and barium-142.

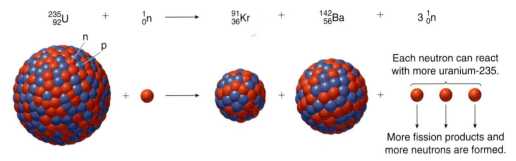

$$^{235}_{92}\text{U} \quad + \quad ^{1}_{0}\text{n} \quad \longrightarrow \quad ^{91}_{36}\text{Kr} \quad + \quad ^{142}_{56}\text{Ba} \quad + \quad 3\,^{1}_{0}\text{n}$$

Each neutron can react with more uranium-235.

More fission products and more neutrons are formed.

Three high-energy neutrons are also produced in the reaction as well as a great deal of energy. Whereas burning 1 g of methane in natural gas releases 13 kcal of energy, fission of 1 g of uranium-235 releases 3.4×10^8 kcal. Each neutron produced during fission can go on to bombard three other uranium-235 nuclei to produce more nuclei and more neutrons. Such a process is called a **chain reaction.**

In order to sustain a chain reaction there must be a sufficient amount of uranium-235. When that amount—the **critical mass**—is present, the chain reaction occurs over and over again and an atomic explosion occurs. When less than the critical mass of uranium-235 is present, there is a more controlled production of energy, as is the case in a nuclear power plant.

A nuclear power plant utilizes the tremendous amount of energy produced by fission of the uranium-235 nucleus to heat water to steam, which powers a generator to produce electricity (Figure 10.8). While nuclear energy accounts for a small but significant fraction of the electricity needs in the United States, most of the electricity generated in some European countries comes from nuclear power.

▼ **FIGURE 10.8**   A Nuclear Power Plant

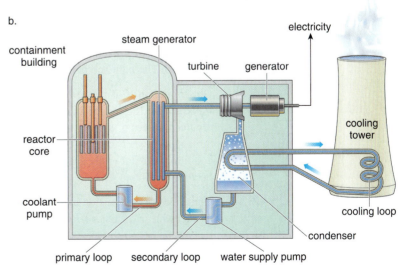

a. Nuclear power plant with steam rising from a cooling tower

b. Fission occurs in a nuclear reactor core that is housed in a containment facility. Water surrounding the reactor is heated by the energy released during fission, and this energy drives a turbine, which produces electricity. Once the steam has been used to drive the turbine, it is cooled and re-circulated around the core of the reactor. To prevent the loss of any radioactive material to the environment, the water that surrounds the reactor core never leaves the containment building.

Two problems that surround nuclear power generation are the possibility of radiation leaks and the disposal of nuclear waste. Plants are designed and monitored to contain the radioactive materials within the nuclear reactor. The reactor core itself is located in a containment facility with thick walls, so that should a leak occur, the radiation should in principle be kept within the building. The nuclear reactor in Chernobyl, Russia, was built without a containment facility and in 1986 it exploded, releasing high levels of radioactivity to the immediate environment and sending a cloud of reactivity over much of Europe.

The products of nuclear fission are radioactive nuclei with long half-lives, often hundreds or even thousands of years. As a result, nuclear fission generates radioactive waste that must be stored in a secure facility so that it does not pose a hazard to the immediate surroundings. Burying waste far underground is currently considered the best option, but this issue is still unresolved.

PROBLEM 10.21

Write a nuclear equation for each process.

a. Fission of uranium-235 by neutron bombardment forms strontium-90, an isotope of xenon, and three neutrons.

b. Fission of uranium-235 by neutron bombardment forms antimony-133, three neutrons, and one other isotope.

10.6B   NUCLEAR FUSION

Nuclear fusion occurs when two light nuclei join together to form a larger nucleus. For example, fusion of a deuterium nucleus with a tritium nucleus forms helium and a neutron. Recall from Section 2.3 that deuterium is an isotope of hydrogen that contains one proton and one neutron in its nucleus, while tritium is an isotope of hydrogen that contains one proton and two neutrons in its nucleus.

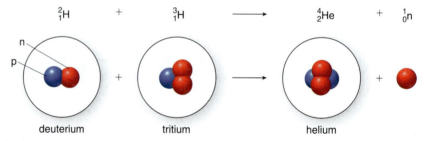

$$_{1}^{2}\text{H} \quad + \quad _{1}^{3}\text{H} \quad \longrightarrow \quad _{2}^{4}\text{He} \quad + \quad _{0}^{1}\text{n}$$

deuterium   tritium   helium

Like fission, fusion also releases a great deal of energy—namely, 5.3×10^{8} kcal/mol of helium produced. The light and heat of the sun and other stars result from nuclear fusion.

One limitation of using fusion to provide energy for mankind is the extreme experimental conditions needed to produce it. Because it takes a considerable amount of energy to overcome the repulsive forces of the like charges of two nuclei, fusion can only be accomplished at high temperatures (greater than 100,000,000 °C) and pressures (greater than 100,000 atm). Since these conditions are not easily achieved, using controlled nuclear fusion as an energy source has yet to become a reality.

Controlled nuclear fusion has the potential of providing cheap and clean power. It is not plagued by the nuclear waste issues of fission reactors, and the needed reactants are readily available.

PROBLEM 10.22

Nuclear fusion in the stars occurs by a series of reactions. Identify **X, Y,** and **Z** in the following nuclear reactions that ultimately convert hydrogen into helium.

a. $_{1}^{1}\text{H} + \textbf{X} \longrightarrow \, _{1}^{2}\text{H} + \, _{+1}^{0}\text{e}$

b. $_{1}^{1}\text{H} + \, _{1}^{2}\text{H} \longrightarrow \textbf{Y}$

c. $_{1}^{1}\text{H} + \, _{2}^{3}\text{He} \longrightarrow \, _{2}^{4}\text{He} + \textbf{Z}$

10.7 FOCUS ON HEALTH & MEDICINE
MEDICAL IMAGING WITHOUT RADIOACTIVITY

X-rays, CT scans, and **MRIs** are also techniques that provide an image of an organ or extremity that is used for diagnosis of a medical condition. Unlike PET scans and other procedures discussed thus far, however, **these procedures are *not* based on nuclear reactions and they do *not* utilize radioactivity.** In each technique, an energy source is directed towards a specific region in the body, and a scan is produced that is analyzed by a trained medical professional.

X-rays are a high-energy type of radiation called electromagnetic radiation. Tissues of different density interact differently with an X-ray beam, and so a map of bone and internal organs is created on an X-ray film. Dense bone is clearly visible in an X-ray, making it a good diagnostic technique for finding fractures (Figure 10.9a). Although X-rays are a form of high-energy radiation, they are lower in energy than the γ rays produced in nuclear reactions. Nonetheless, X-rays still cause adverse biological effects on the cells with which they come in contact, and the exposure of both the patient and X-ray technician must be limited.

CT (computed tomography) scans, which also use X-rays, provide high resolution images of "slices" of the body. Historically, CT images have shown a slice of tissue perpendicular to the long axis of the body. Modern CT scanners can now provide a three-dimensional view of the body's organs. CT scans of the head are used to diagnose bleeding and tumors in the brain (Figure 10.9b).

MRI (magnetic resonance imaging) uses low-energy radio waves to visualize internal organs. Unlike methods that use high-energy radiation, MRIs do not damage cells. An MRI is a good diagnostic method for visualizing soft tissue (Figure 10.9c), and thus it complements X-ray techniques.

▼ **FIGURE 10.9   Imaging the Human Body**

a.

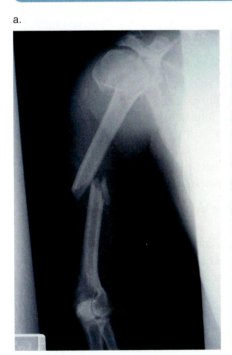

b.

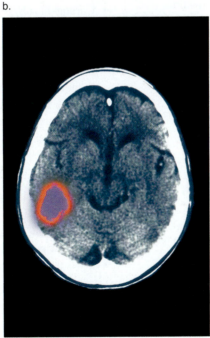

c.

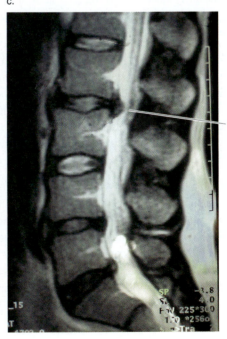

herniated disc

a. X-ray of a broken humerus in a patient's arm

b. A color-enhanced CT scan of the head showing the site of a stroke

c. MRI of the spinal cord showing spinal compression from a herniated disc

CHAPTER HIGHLIGHTS

KEY TERMS

Alpha (α) particle (10.1)
Becquerel (10.4)
Beta (β) particle (10.1)
Chain reaction (10.6)
Critical mass (10.6)
Curie (10.4)
Gamma (γ) ray (10.1)
Geiger counter (10.4)

Gray (10.4)
Half-life (10.3)
LD_{50} (10.4)
Nuclear fission (10.6)
Nuclear fusion (10.6)
Nuclear reaction (10.1)
Positron (10.1)
Rad (10.4)

Radioactive decay (10.2)
Radioactive isotope (10.1)
Radioactivity (10.1)
Radiocarbon dating (10.3)
Rem (10.4)
Sievert (10.4)
X-ray (10.7)

KEY CONCEPTS

❶ Describe the different types of radiation emitted by a radioactive nucleus. (10.1)

- A radioactive nucleus can emit α particles, β particles, positrons, or γ rays.
- An α particle is a high-energy nucleus that contains two protons and two neutrons.
- A β particle is a high-energy electron.
- A positron is an antiparticle of a β particle. A positron has a +1 charge and negligible mass.
- A γ ray is high-energy radiation with no mass or charge.

❷ How are equations for nuclear reactions written? (10.2)

- In an equation for a nuclear reaction, the sum of the mass numbers (A) must be equal on both sides of the equation. The sum of the atomic numbers (Z) must be equal on both sides of the equation as well.

❸ What is the half-life of a radioactive isotope? (10.3)

- The half-life ($t_{1/2}$) is the time it takes for one-half of a radioactive sample to decay. Knowing the half-life and the amount of a radioactive substance, one can calculate how much sample remains after a period of time.
- The half-life of radioactive C-14 can be used to date archaeological artifacts.

❹ What units are used to measure radioactivity? (10.4)

- Radiation in a sample is measured by the number of disintegrations per second, most often using the curie (Ci); $1 \text{ Ci} = 3.7 \times 10^{10}$ disintegrations/s. The becquerel (Bq) is also used; $1 \text{ Bq} = 1$ disintegration/s; $1 \text{ Ci} = 3.7 \times 10^{10}$ Bq.
- The exposure of a substance to radioactivity is measured with the rad (radiation absorbed dose) or the rem (radiation equivalent for man).

❺ Give examples of common radioisotopes used in medicine. (10.5)

- Iodine-131 is used to diagnose and treat thyroid disease.
- Technetium-99m is used to evaluate the functioning of the gall bladder and bile ducts, and in bone scans to evaluate the spread of cancer.
- Red blood cells tagged with technetium-99m are used to find the site of a gastrointestinal bleed.
- Thallium-201 is used to diagnose coronary artery disease.
- Cobalt-60 is used as an external source of radiation for cancer treatment.
- Iodine-125 and iridium-192 are used in internal radiation treatment of prostate cancer and breast cancer, respectively.
- Carbon-11, oxygen-15, nitrogen-13, and fluorine-18 are used in positron emission tomography.

❻ What are nuclear fission and nuclear fusion? (10.6)

- Nuclear fission is the splitting apart of a heavy nucleus into lighter nuclei and neutrons.
- Nuclear fusion is the joining together of two light nuclei to form a larger nucleus.
- Both nuclear fission and nuclear fusion release a great deal of energy. Nuclear fission is used in nuclear power plants to generate electricity. Nuclear fusion occurs in stars.

❼ What medical imaging techniques do not use radioactivity? (10.7)

- X-rays and CT scans both use X-rays, a high-energy form of electromagnetic radiation.
- MRIs use low-energy radio waves to image soft tissue.

PROBLEMS

Selected in-chapter and end-of-chapter problems have brief answers provided in Appendix B.

Isotopes and Radiation

10.23 Compare fluorine-18 and fluorine-19 with regard to each of the following: (a) atomic number; (b) number of protons; (c) number of neutrons; (d) mass number. Give the isotope symbol for each isotope. F-19 is a stable nucleus and F-18 is used in PET scans.

10.24 Compare nitrogen-13 and nitrogen-14 with regard to each of the following: (a) atomic number; (b) number of protons; (c) number of neutrons; (d) mass number. Give the isotope symbol for each isotope. N-14 is a stable nucleus and N-13 is used in PET scans.

10.25 Complete the table of isotopes, each of which has found use in medicine.

	Atomic number	Mass number	Number of protons	Number of neutrons	Isotope symbol
a. Chromium-51					
b.		46	103		
c.			19	23	
d.		133	54		

10.26 Complete the table of isotopes, each of which has found use in medicine.

	Atomic number	Mass number	Number of protons	Number of neutrons	Isotope symbol
a. Sodium-24					
b.		89		51	
c.			59	26	
d. Samarium-153					

10.27 How much does the mass and charge of a nucleus change when each type of radiation is emitted: (a) α particle; (b) β particle; (c) γ ray; (d) positron?

10.28 Compare α particles, β particles, and γ rays with regard to each of the following: (a) speed the radiation travels; (b) penetrating power; (c) protective equipment that must be worn when handling.

10.29 What is the mass and charge of radiation that has each of the following symbols: (a) α; (b) n; (c) γ; (d) β?

10.30 What is the mass and charge of radiation that has each of the following symbols?
a. $_{-1}^{0}e$ b. $_{+1}^{0}e$ c. $_{2}^{4}He$ d. β^{+}

Nuclear Reactions

10.31 Complete the nuclear equation by drawing the nucleus of the missing atom. Give the symbol for each atom and type of radiation. (The blue spheres represent protons and the red spheres represent neutrons.)

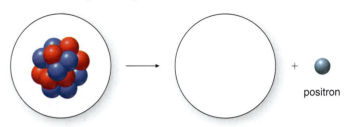

10.32 Complete the nuclear equation by drawing the nucleus of the missing atom. Give the symbol for each atom and type of radiation. (The blue spheres represent protons and the red spheres represent neutrons.)

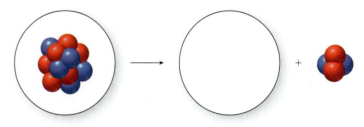

10.33 Complete each nuclear equation.
a. $_{26}^{59}Fe \longrightarrow ? + _{-1}^{0}e$ c. $_{80}^{178}Hg \longrightarrow ? + _{+1}^{0}e$
b. $_{78}^{190}Pt \longrightarrow ? + _{2}^{4}He$

10.34 Complete each nuclear equation.
a. $_{37}^{77}Rb \longrightarrow ? + _{+1}^{0}e$ c. $_{29}^{66}Cu \longrightarrow ? + _{-1}^{0}e$
b. $_{102}^{251}No \longrightarrow ? + _{2}^{4}He$

10.35 Complete each nuclear equation.
a. $_{39}^{90}Y \longrightarrow _{40}^{90}Zr + ?$ c. $_{83}^{210}Bi \longrightarrow ? + _{2}^{4}He$
b. $? \longrightarrow _{59}^{135}Pr + _{+1}^{0}e$

10.36 Complete each nuclear equation.
a. $? \longrightarrow _{39}^{90}Y + _{-1}^{0}e$ c. $_{84}^{214}Po \longrightarrow ? + _{2}^{4}He$
b. $_{15}^{29}P \longrightarrow _{14}^{29}Si + ?$

10.37 Bismuth-214 can decay to form either polonium-214 or thallium-210, depending on what type of radiation is emitted. Write a balanced nuclear equation for each process.

10.38 Lead-210 can be formed by the decay of either thallium-210 or polonium-214, depending on what type of radiation is emitted. Write a balanced nuclear equation for each process.

10.39 Write a balanced nuclear equation for each reaction.
a. decay of thorium-232 by α emission
b. decay of sodium-25 by β emission
c. decay of xenon-118 by positron emission
d. decay of curium-243 by α emission

10.40   Write a balanced nuclear equation for each reaction.
        a. decay of sulfur-35 by β emission
        b. decay of thorium-225 by α emission
        c. decay of rhodium-93 by positron emission
        d. decay of silver-114 by β emission

Half-Life

10.41   If the amount of a radioactive element decreases from 2.4 g to 0.30 g in 12 days, what is its half-life?

10.42   If the amount of a radioactive element decreases from 0.36 g to 90. mg in 22 min, what is its half-life?

10.43   Radioactive iodine-131 ($t_{1/2}$ = 8.0 days) decays to form xenon-131 by emission of a β particle. How much of each isotope is present after each time interval if 64 mg of iodine-131 was present initially: (a) 8.0 days; (b) 16 days; (c) 24 days; (d) 32 days?

10.44   Radioactive phosphorus-32 decays to form sulfur-32 by emission of a β particle. Estimating the half-life to be 14 days, how much of each isotope is present after each time interval if 124 mg of phosphorus-32 was present initially: (a) 14 days; (b) 28 days; (c) 42 days; (d) 56 days?

10.45   If the half-life of an isotope is 24 hours, has all the isotope decayed in 48 hours?

10.46   Explain how the half-life of carbon-14 is used to date objects.

10.47   Why can't radiocarbon dating be used to determine the age of an artifact that is over 50,000 years old?

10.48   Why can't radiocarbon dating be used to estimate the age of rocks?

10.49   A patient is injected with a sample of technetium-99m ($t_{1/2}$ = 6.0 h), which has an activity of 20 mCi. What activity is observed after each interval: (a) 6 h; (b) 12 h; (c) 24 h?

10.50   A sample of iodine-131 ($t_{1/2}$ = 8.0 days) has an activity of 200. mCi. What activity is observed after each interval: (a) 8.0 days; (b) 24 days; (c) 48 days?

Measuring Radioactivity

10.51   If a radioactive sample had an activity of 5.0 mCi, how many disintegrations per second does this correspond to?

10.52   Why is the average amount of background radiation generally higher at higher elevations?

10.53   A patient must be administered a 28-mCi dose of technetium-99m, which is supplied in a vial containing a solution with an activity of 12 mCi/mL. What volume of solution must be given?

10.54   A radioactive isotope used for imaging is supplied in an 8.0-mL vial containing a solution with an activity of 108 mCi. What volume must be given to a patient who needs a 12-mCi dose?

10.55   Radioactive sodium-24, administered as $^{24}NaCl$, is given to treat leukemia. If a patient must receive 190 μCi/kg

and the isotope is supplied as a solution that contains 5.0 mCi/mL, what volume is needed for a 68-kg patient?

10.56   Radioactive phosphorus-32, administered as sodium phosphate ($Na_3{}^{32}PO_4$), is used to treat chronic leukemia. The activity of an intravenous solution is 670 μCi/mL. What volume of solution must be used to supply a dose of 15 mCi?

10.57   The units chosen to report radiation amounts give us different information. What is measured using the curie compared to the rad?

10.58   The units chosen to report radiation amounts give us different information. What is measured using the millicurie compared to the rem?

10.59   The initial responders to the Chernobyl nuclear disaster were exposed to 20 Sv of radiation. Convert this value to rem. Did these individuals receive a fatal dose of radiation?

10.60   Many individuals who fought fires at the Chernobyl nuclear disaster site were exposed to 0.25 Sv of radiation. Convert this value to rem. Did these individuals receive a fatal dose of radiation? Would you expect any of these individuals to have shown ill health effects?

Nuclear Fission and Nuclear Fusion

10.61   What is the difference between nuclear fission and nuclear fusion?

10.62   What is the difference between the nuclear fission process that takes place in a nuclear reactor and the nuclear fission that occurs in an atomic bomb?

10.63   For which process does each statement apply—nuclear fission, nuclear fusion, both fission and fusion?
        a. The reaction occurs in the sun.
        b. A neutron is used to bombard a nucleus.
        c. A large amount of energy is released.
        d. Very high temperatures are required.

10.64   For which process does each statement apply—nuclear fission, nuclear fusion, both fission and fusion?
        a. The reaction splits a nucleus into lighter nuclei.
        b. The reaction joins two lighter nuclei into a heavier nucleus.
        c. The reaction is used to generate energy in a nuclear power plant.
        d. The reaction generates radioactive waste with a long half-life.

10.65   Complete each nuclear fission equation.
        a. $^{235}_{92}U + {}^{1}_{0}n \longrightarrow ? + {}^{97}_{42}Mo + 2\,{}^{1}_{0}n$
        b. $^{235}_{92}U + {}^{1}_{0}n \longrightarrow ? + {}^{140}_{56}Ba + 3\,{}^{1}_{0}n$

10.66   Complete each nuclear fission equation.
        a. $^{235}_{92}U + {}^{1}_{0}n \longrightarrow ? + {}^{139}_{57}La + 2\,{}^{1}_{0}n$
        b. $^{235}_{92}U + {}^{1}_{0}n \longrightarrow ? + {}^{140}_{58}Ce + 2\,{}^{1}_{0}n + 6\,{}^{0}_{-1}e$

10.67   The fusion of two deuterium nuclei (hydrogen-2) forms a hydrogen nucleus (hydrogen-1) as one product. What other product is formed?

10.68 Fill in the missing product in the following nuclear fusion reaction.

$$\ce{^3_2He} + \ce{^3_2He} \longrightarrow \; ? \; + 2\,\ce{^1_1H}$$

10.69 Discuss two problems that surround the generation of electricity from a nuclear power plant.

10.70 Why are there as yet no nuclear power plants that use nuclear fusion to generate electricity?

10.71 All nuclei with atomic numbers around 100 or larger do not exist naturally; rather they have been synthesized by fusing two lighter-weight nuclei together. Complete the following nuclear equation by giving the name, atomic number, and mass number of the element made by this reaction.

$$\ce{^{209}_{83}Bi} + \ce{^{58}_{26}Fe} \longrightarrow \; ? \; + \ce{^1_0n}$$

10.72 Complete the following nuclear equation, and give the name, atomic number, and mass number of the element made by this reaction.

$$\ce{^{235}_{92}U} + \ce{^{14}_7N} \longrightarrow \; ? \; + 5\,\ce{^1_0n}$$

General Questions

10.73 Arsenic-74 is a radioisotope used for locating brain tumors.
   a. Write a balanced nuclear equation for the positron emission of arsenic-74.
   b. If $t_{1/2}$ for As-74 is 18 days, how much of a 120-mg sample remains after 90 days?
   c. If the radioactivity of a 2.0-mL vial of arsenic-74 is 10.0 mCi, what volume must be administered to give a 7.5-mCi dose?

10.74 Sodium-24 is a radioisotope used for examining circulation.
   a. Write a balanced nuclear equation for the β decay of sodium-24.
   b. If $t_{1/2}$ for Na-24 is 15 h, how much of an 84-mg sample remains after 2.5 days?
   c. If the radioactivity of a 5.0-mL vial of sodium-24 is 10.0 mCi, what volume must be administered to give a 6.5-mCi dose?

10.75 Answer the following questions about radioactive iridium-192.
   a. Write a balanced nuclear equation for the decay of iridium-192, which emits both a β particle and a γ ray.
   b. If $t_{1/2}$ for Ir-192 is 74 days, estimate how much of a 120-mg sample remains after five months.
   c. If a sample of Ir-192 had an initial activity of 36 Ci, estimate how much activity remained in the sample after 10 months.

10.76 Answer the following questions about radioactive samarium-153.
   a. Write a balanced nuclear equation for the decay of samarium-153, which emits both a β particle and a γ ray.
   b. If $t_{1/2}$ for Sm-153 is 46 h, estimate how much of a 160-mg sample remains after four days.
   c. If a sample of Sm-153 had an initial activity of 48 Ci, estimate how much activity remained in the sample after six days.

Applications

10.77 Explain how each isotope is used in medicine.
   a. iodine-131   b. iridium-192   c. thallium-201

10.78 Explain how each isotope is used in medicine.
   a. iodine-125   b. technetium-99m   c. cobalt-60

10.79 How does the half-life of each of the following isotopes of iodine affect the manner in which it is administered to a patient: (a) iodine-125, $t_{1/2} = 60$ days; (b) iodine-131, $t_{1/2} = 8$ days?

10.80 Explain why food is irradiated with γ rays.

10.81 A mammogram is an X-ray of the breast. Why does an X-ray technician leave the room or go behind a shield when a mammogram is performed on a patient?

10.82 Why is a lead apron placed over a patient's body when dental X-rays are taken?

10.83 One of the radioactive isotopes that contaminated the area around Chernobyl after the nuclear accident in 1986 was iodine-131. Suggest a reason why individuals in the affected region were given doses of NaI that contained the stable iodine-127 isotope.

10.84 The element strontium has similar properties to calcium. Suggest a reason why exposure to strontium-90, a product of nuclear testing in the atmosphere, is especially hazardous for children.

CHALLENGE QUESTIONS

10.85 An article states that the fission of 1.0 g of uranium-235 releases 3.4×10^8 kcal, the same amount of energy as burning one ton (2,000 lb) of coal. If this report is accurate, how much energy is released when 1.0 g of coal is burned?

10.86 Radioactive isotopes with high atomic numbers often decay to form isotopes that are themselves radioactive, and once formed, decay to form new isotopes. Sometimes a series of such decays occurs over many steps until a stable nucleus is formed. The following series of decays occurs: Polonium-218 decays with emission of an α particle to form **X**, which emits a β particle to form **Y**, which emits an α particle to form **Z**. Identify **X**, **Y**, and **Z**.

11

CHAPTER GOALS

In this chapter you will learn how to:

1. Recognize the characteristic features of organic compounds
2. Predict the shape around atoms in organic molecules
3. Use shorthand methods to draw organic molecules
4. Recognize the common functional groups and understand their importance
5. Distinguish organic compounds from ionic inorganic compounds
6. Determine whether an organic compound is polar or nonpolar
7. Determine solubility properties of organic compounds
8. Determine whether a vitamin is fat soluble or water soluble

Vitamin A, a key component of the vision receptors in the eye, is synthesized in the body from β-carotene, the orange pigment found in carrots.

INTRODUCTION TO ORGANIC MOLECULES AND FUNCTIONAL GROUPS

CONSIDER for a moment the activities that occupied your past 24 hours. You likely showered with soap, drank a caffeinated beverage, ate at least one form of starch, took some medication, read a newspaper, listened to a CD, and traveled in a vehicle that had rubber tires and was powered by fossil fuels. If you did any *one* of these, your life was touched by organic chemistry. In Chapter 11, we learn about the characteristic features of organic molecules.

11.1 INTRODUCTION TO ORGANIC CHEMISTRY

What is organic chemistry?

- **Organic chemistry is the study of compounds that contain the element carbon.**

While it may seem odd that an entire discipline is devoted to the study of a single element in the periodic table, millions of organic compounds are known, far more than the inorganic compounds discussed in Chapters 1–10. **These organic chemicals affect virtually every facet of our lives, and for this reason, it is important and useful to know something about them.**

Clothes, foods, medicines, gasoline, refrigerants, and soaps are composed almost solely of organic compounds. Some, like cotton, wool, or silk are naturally occurring; that is, they can be isolated directly from natural sources. Others, such as nylon and polyester are synthetic, meaning they are produced by chemists in the laboratory. By studying the principles and concepts of organic chemistry, you can learn more about compounds such as these and how they affect the world around you. Figure 11.1 illustrates some common products of organic chemistry used in medicine.

▼ **FIGURE 11.1 Some Common Products of Organic Chemistry Used in Medicine**

a. Oral contraceptives

b. Plastic syringes

c. Antibiotics

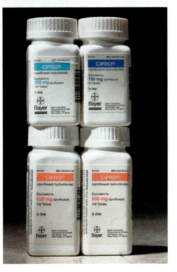

d. Synthetic heart valves

Organic chemistry has given us contraceptives, plastics, antibiotics, synthetic heart valves, and a myriad of other materials. Our lives would be vastly different today without these products of organic chemistry.

PROBLEM 11.1	Which molecular formulas represent organic compounds and which represent inorganic compounds?

a. C_6H_{12} c. KI   e. CH_4O

b. H_2O d. $MgSO_4$ f. NaOH

11.2 CHARACTERISTIC FEATURES OF ORGANIC COMPOUNDS

Perhaps the best way to appreciate the variety of organic molecules is to look at a few. Simple organic compounds that contain just one or two carbon atoms, respectively, are methane and ethanol.

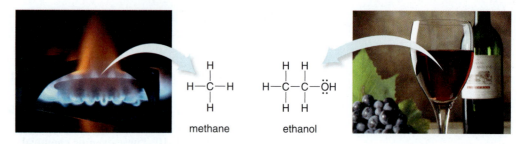

methane   ethanol

Methane, the main component of natural gas, burns in the presence of oxygen. The natural gas we use today was formed by the decomposition of organic material millions of years ago. **Ethanol,** the alcohol present in wine and other alcoholic beverages, is formed by the fermentation of sugar. Ethanol can also be made in the lab by a totally different process. **Ethanol produced in the lab is identical to the ethanol produced by fermentation.**

Two more complex organic molecules are capsaicin and caffeine. **Capsaicin,** the compound responsible for the characteristic spiciness of hot peppers, is the active ingredient in several topical creams for pain relief. **Caffeine** is the bitter-tasting stimulant found in coffee, tea, cola beverages, and chocolate.

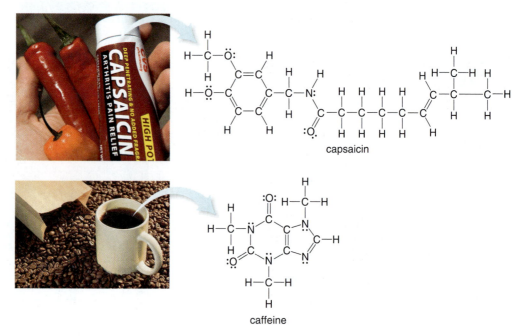

capsaicin

caffeine

What are the common features of these organic compounds?

[1] All organic compounds contain carbon atoms and most contain hydrogen atoms. Carbon always forms four covalent bonds, and hydrogen forms one covalent bond.

Carbon is located in group 4A of the periodic table, so a carbon atom has four valence electrons available for bonding (Section 4.1). Since hydrogen has a single valence electron, methane (CH_4) consists of four single bonds, each formed from one electron from a hydrogen atom and one electron from carbon.

Remember that each solid line represents one two-electron bond.

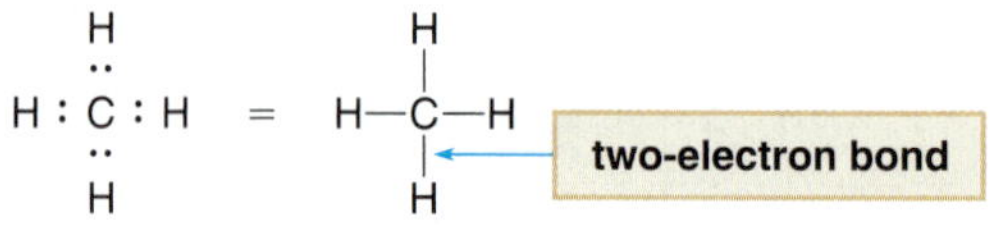

$$H:C:H = H-C-H$$ two-electron bond

methane

[2] Carbon forms single, double, and triple bonds to other carbon atoms.

When a compound contains two or more carbon atoms, the type of bonding is determined by the number of atoms around carbon. Consider the three compounds drawn below:

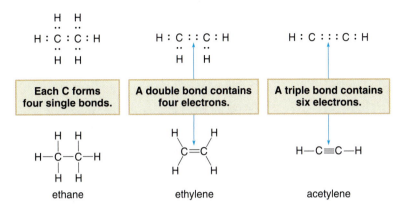

- **A C atom surrounded by four atoms forms four single bonds.** In ethane (C_2H_6), each carbon atom is bonded to three hydrogen atoms and one carbon atom. All bonds are single bonds.
- **A C atom surrounded by three atoms forms one double bond.** In ethylene (C_2H_4), each carbon atom is surrounded by three atoms (two hydrogens and one carbon); thus, each C forms a single bond to each hydrogen atom and a double bond to carbon.
- **A C atom surrounded by two atoms generally forms one triple bond.** In acetylene (C_2H_2), each carbon atom is surrounded by two atoms (one hydrogen and one carbon); thus, each C forms a single bond to hydrogen and a triple bond to carbon.

[3] Some compounds have chains of atoms and some compounds have rings.

For example, three carbon atoms can bond in a row to form propane, or form a ring called cyclo-propane. Propane is the fuel burned in gas grills, and cyclopropane is an anesthetic.

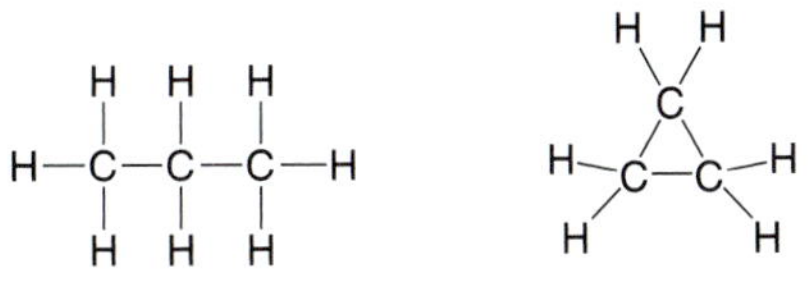

propane
C_3H_8

cyclopropane
C_3H_6

[4] Organic compounds may also contain elements other than carbon and hydrogen. Any atom that is not carbon or hydrogen is called a *heteroatom*.

The most common heteroatoms are nitrogen, oxygen, and the halogens (F, Cl, Br, and I).

- **Each heteroatom forms a characteristic number of bonds, determined by its location in the periodic table.**
- **The common heteroatoms also have nonbonding, lone pairs of electrons, so that each atom is surrounded by eight electrons.**

The number of bonds formed by common elements was first discussed in Section 4.1.

Thus, nitrogen forms three bonds and has one lone pair of electrons, while oxygen forms two bonds and has two additional lone pairs. The halogens form one bond and have three additional lone pairs. Common bonding patterns for atoms in organic compounds are summarized in Table 11.1. Except for hydrogen, these common elements in organic compounds follow one rule in bonding:

Number of bonds + Number of lone pairs = 4

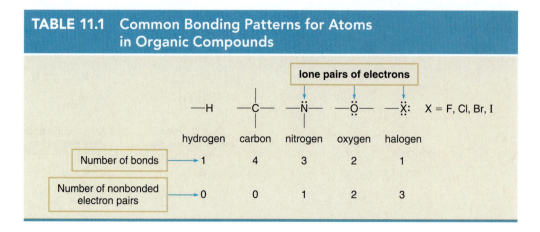

TABLE 11.1 Common Bonding Patterns for Atoms in Organic Compounds

	hydrogen	carbon	nitrogen	oxygen	halogen
Number of bonds	1	4	3	2	1
Number of nonbonded electron pairs	0	0	1	2	3

Oxygen and nitrogen form both single and multiple bonds to carbon. **The most common multiple bond between carbon and a heteroatom is a carbon–oxygen double bond (C=O).** The bonding patterns remain the same even when an atom is part of a multiple bond, as shown with methanol (CH_3OH) and formaldehyde ($H_2C=O$, a preservative).

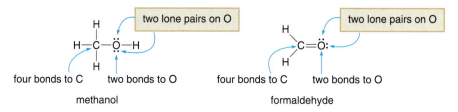

These features explain why there are so many organic compounds: **Carbon forms four strong bonds with itself and other elements. Carbon atoms combine together to form rings and chains.**

SAMPLE PROBLEM 11.1

Draw in all H's and lone pairs in each compound.

    a. C—C—Cl   b. C—C=O   c. C—C≡N

ANALYSIS

Each C and heteroatom must be surrounded by eight electrons. Use the common bonding patterns in Table 11.1 to fill in the needed H's and lone pairs. C needs four bonds; Cl needs one bond and three lone pairs; O needs two bonds and two lone pairs; N needs three bonds and one lone pair.

SOLUTION

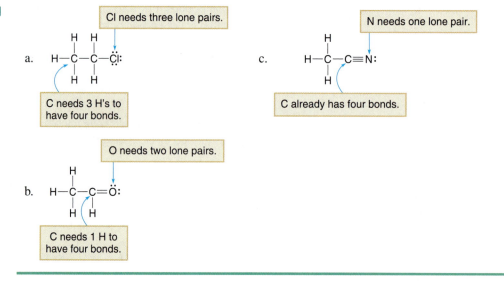

PROBLEM 11.2

Fill in all H's and lone pairs in each compound.

a. 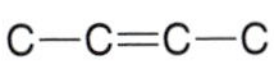 C—C=C—C b. C—C—C c. C—C—C d. 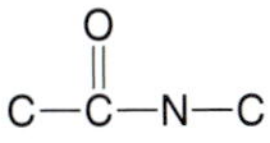 C—C—N—C e.

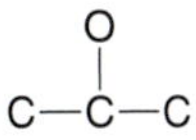

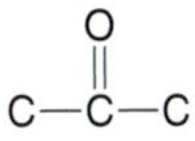

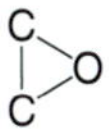

11.3 SHAPES OF ORGANIC MOLECULES

Because acetylene produces a very hot flame on burning, it is often used in welding torches.

The shape around atoms in organic molecules is determined by counting groups using the principles of VSEPR theory (Section 4.6). Recall that a group is either another atom or a lone pair of electrons. **The most stable arrangement keeps these groups as far away from each other as possible.**

• An atom surrounded by two groups is linear and has a bond angle of 180°.

Each carbon in acetylene (HC≡CH) is surrounded by two atoms and no lone pairs. Thus, each H—C—C bond angle is 180°, making all four atoms of acetylene linear.

180°
H—C≡C—H   =
180°

two atoms around each C   acetylene
ball-and-stick model

• An atom surrounded by three groups is trigonal planar and has a bond angle of 120°.

Each carbon in ethylene ($CH_2=CH_2$) is surrounded by three atoms and no lone pairs. Thus, each H—C—C bond angle is 120°, and all six atoms of ethylene lie in one plane.

Ethylene is an important starting material in the preparation of the plastic polyethylene.

three atoms around each C   ethylene

• An atom surrounded by four groups places these four groups at the corners of a tetrahedron, giving bond angles of approximately 109.5°.

The carbon atom in methane (CH_4) is bonded to four hydrogen atoms, pointing to the corners of a tetrahedron.

109.5°

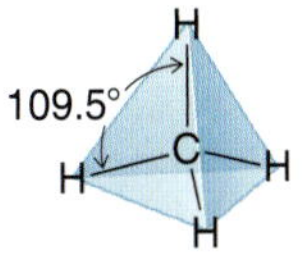

tetrahedral carbon

To draw the three-dimensional tetrahedron on a two-dimensional page, place two of the bonds in the plane of the page, one bond in front, and one bond behind. Then, use the drawing conventions first presented in Section 4.6:

• A solid line is used for bonds in the plane.
• A wedge is used for a bond in front of the plane.
• A dashed line is used for a bond behind the plane.

Ethane is a constituent of natural gas.

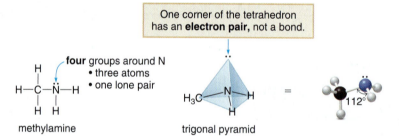

All carbons having four single bonds are tetrahedral. For example, each carbon of ethane is tetrahedral and can be drawn with two bonds in the plane, one in front on a wedge, and one behind on a dashed line.

Don't forget about the lone pairs of electrons on nitrogen and oxygen when determining the shape around these atoms. The N atom in methylamine (CH_3NH_2) is surrounded by three atoms and one lone pair—four groups. To keep these four groups as far apart as possible, each of these groups occupies the corner of a tetrahedron. In this way, the N atom of methylamine resembles the N atom of ammonia (NH_3) discussed in Section 4.6. The H—N—C bond angle of 112° is close to the theoretical tetrahedral bond angle of 109.5°. This molecular shape is referred to as a **trigonal pyramid,** because one of the groups around the N is a lone pair, not another atom.

Similarly, the O atom in methanol (CH_3OH) is surrounded by two atoms and two lone pairs—four groups. To keep these groups as far apart as possible, each of these groups occupies the corner of a tetrahedron. In this way the O atom of methanol resembles the O atom of water (H_2O) discussed in Section 4.6. The C—O—H bond angle of 109° is close to the theoretical tetrahedral bond angle of 109.5°. Methanol has a bent molecular shape around O because two of the groups around oxygen are lone pairs. Table 11.2 summarizes the possible shapes around atoms in organic compounds.

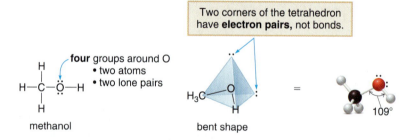

SAMPLE PROBLEM 11.2   Determine the shape around each atom in dimethyl ether.

dimethyl ether

ANALYSIS First, draw in all lone pairs on the heteroatom. A heteroatom needs eight electrons around it, so add two lone pairs to O. Then count groups to determine the molecular shape.

SOLUTION

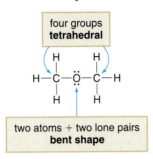

- Each C is surrounded by four atoms (3 H's and 1 O), making each C tetrahedral.
- The O atom is surrounded by 2 C's and two lone pairs (four groups), giving it a bent shape.
- This results in the following three-dimensional shape:

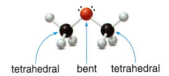

tetrahedral   bent   tetrahedral

PROBLEM 11.3

Determine the shape around the indicated atoms in each compound. Don't forget to add electron pairs to heteroatoms before determining the shape.

(1) (2)

a. 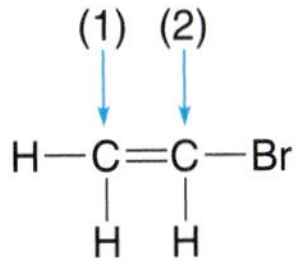 $H-C=C-Br$ (with H, H below)

c. $H-C-O-H$ (with O double-bonded above)
   (1) (2)

(1)   (3)

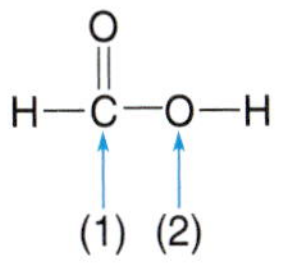

b. 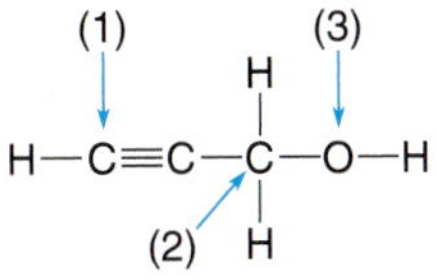 $H-C\equiv C-C-O-H$ (with H above and H below at (2))
   (2)

d. (benzene ring) $H-C \cdots C=C \cdots C-C-C\equiv N$
   (1)   (2)

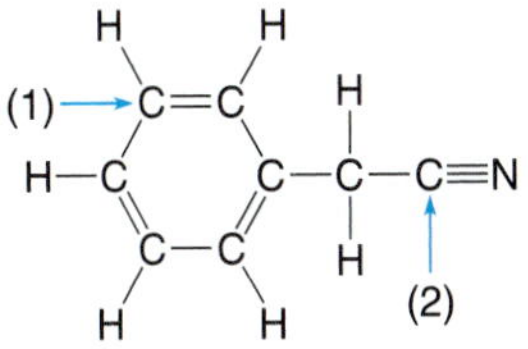

TABLE 11.2	Molecular Shape Around Atoms in Organic Compounds			
Total Number of Groups	**Number of Atoms**	**Number of Lone Pairs**	**Shape Around an Atom (A)**	**Approximate Bond Angle (°)**
2	2	0	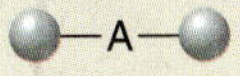 linear	180
3	3	0	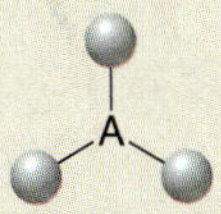 trigonal planar	120
4	4	0	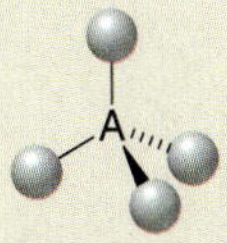 tetrahedral	109.5
4	3	1	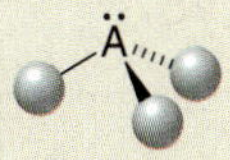 trigonal pyramidal	~109.5[a]
4	2	2	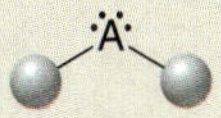 bent	~109.5[a]

[a]The symbol "~" means approximately.

PROBLEM 11.4   Predict the indicated bond angles in each compound.

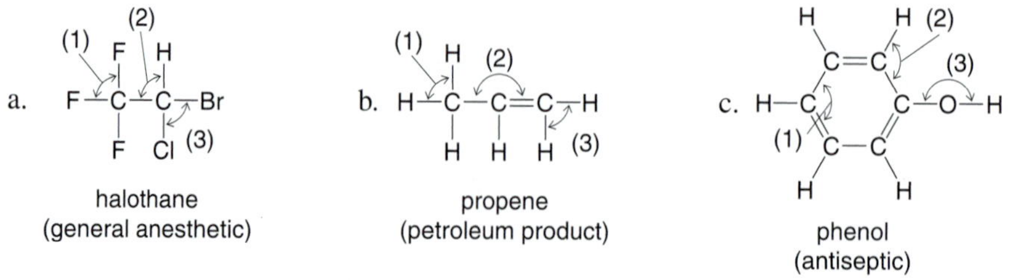

a. halothane (general anesthetic)

b. propene (petroleum product)

c. phenol (antiseptic)

PROBLEM 11.5   Draw the structure of each compound using solid lines, wedges, and dashes to indicate the three-dimensional shape around each carbon.

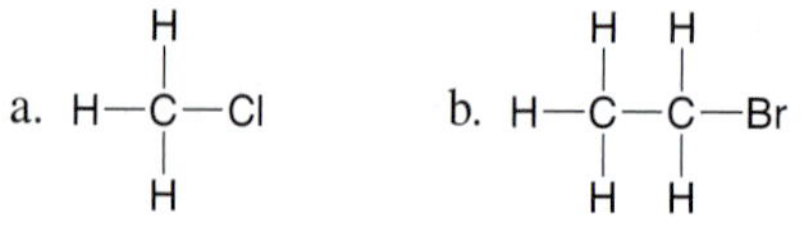

a. H—C—Cl   b. H—C—C—Br

Don't let the size or complexity of an organic molecule intimidate you! Using the principles described in Section 11.3, you can predict the shape around any atom in any molecule no matter how complex it might be.

SAMPLE PROBLEM 11.3   L-Dopa is a drug used to treat Parkinson's disease. Give the shape around each indicated atom.

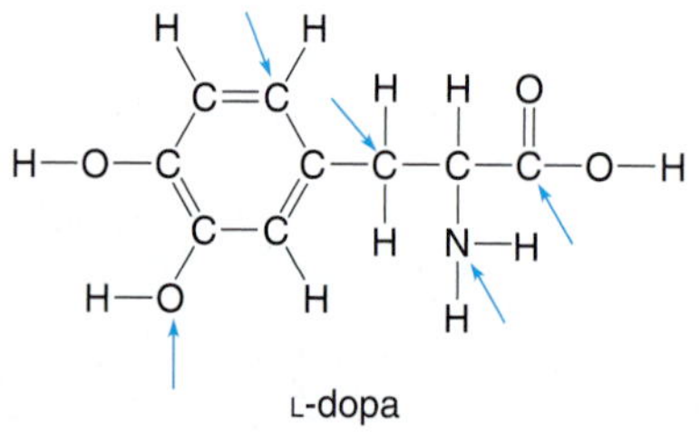

L-dopa

ANALYSIS   First, add lone pairs of electrons to each heteroatom so that it is surrounded by eight electrons. Each O gets two lone pairs and N gets one. Then count groups to predict shape.

SOLUTION   Adding electron pairs around O and N:

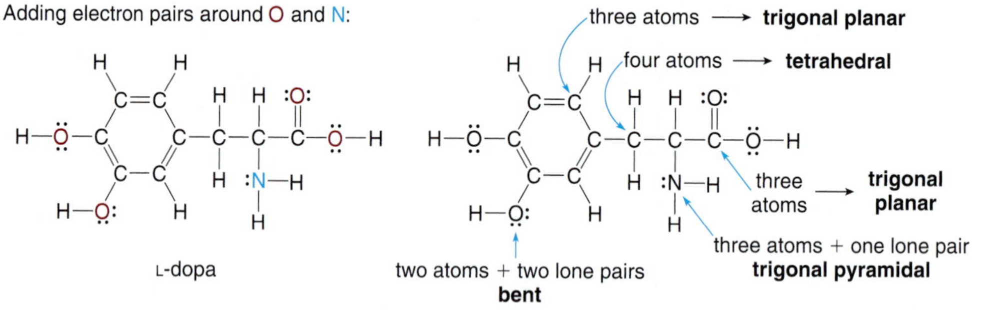

three atoms ⟶ **trigonal planar**

four atoms ⟶ **tetrahedral**

three atoms ⟶ **trigonal planar**

three atoms + one lone pair **trigonal pyramidal**

L-dopa   two atoms + two lone pairs **bent**

Pay special attention to the O and N atoms, and note the importance of adding the lone pairs before determining shape. O atoms are surrounded by *four* groups—two atoms *and* two electron pairs—giving them a bent shape. O atoms are *never* surrounded by two groups, making them linear.

Enanthotoxin is a poisonous compound isolated from a common variety of hemlock grown in England. Predict the shape around the indicated atoms in enanthotoxin.

Hemlock is the source of the poison enanthotoxin.

enanthotoxin

11.4 DRAWING ORGANIC MOLECULES

Because organic molecules often contain many atoms, we need shorthand methods to simplify their structures. The two main types of shorthand representations used for organic compounds are **condensed structures** and **skeletal structures.**

11.4A CONDENSED STRUCTURES

Condensed structures are most often used for a compound having a chain of atoms bonded together, rather than a ring. The following conventions are used.

- **All of the atoms are drawn in, but the two-electron bond lines are generally omitted.**
- **Lone pairs on heteroatoms are omitted.**

To interpret a condensed formula, it is usually best to start at the *left side* of the molecule and remember that the *carbon atoms must have four bonds.*

- **A carbon bonded to 3 H's becomes CH_3.**
- **A carbon bonded to 2 H's becomes CH_2.**
- **A carbon bonded to 1 H becomes CH.**

$$= CH_3CH_2CH_2CH_3$$

$$= CH_3CHCH_2CH_3$$

Keep the vertical bond line to show where the CH_3 group is bonded.

Sometimes these structures are further simplified by using parentheses around like groups. Two CH$_2$ groups bonded together become (CH$_2$)$_2$. Two CH$_3$ groups bonded to the same carbon become (CH$_3$)$_2$C.

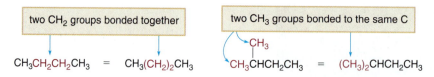

SAMPLE PROBLEM 11.4

Convert each compound into a condensed structure.

a. 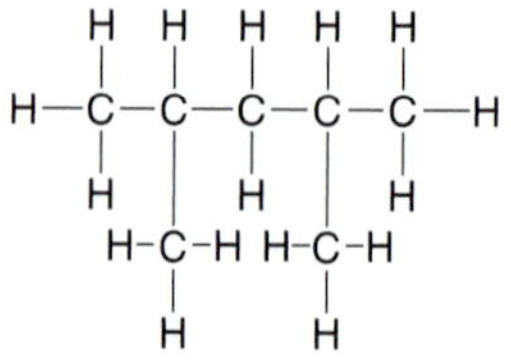 H—C—C—C—C—C—H (with H's and two H—C—H branches)

b. 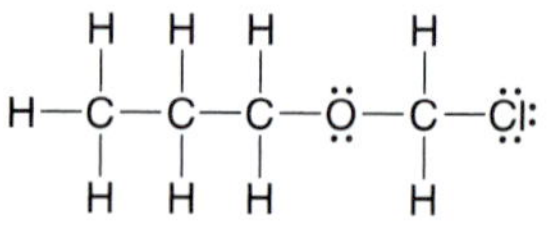 H—C—C—C—Ö—C—Cl: (with H's)

ANALYSIS

Start at the left and proceed to the right, making sure that each carbon has four bonds. Omit lone pairs on the heteroatoms O and Cl. When like groups are bonded together or bonded to the same atom, use parentheses to further simplify the structure.

SOLUTION

	Condensed structure	Further simplified

a. 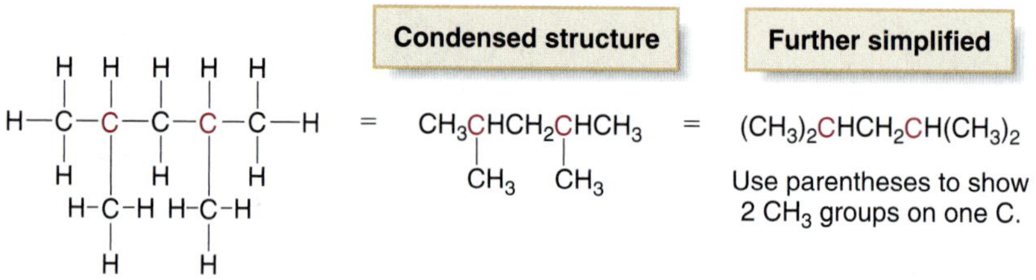 structure = CH$_3$CHCH$_2$CHCH$_3$ with CH$_3$ CH$_3$ = (CH$_3$)$_2$CHCH$_2$CH(CH$_3$)$_2$

Use parentheses to show 2 CH$_3$ groups on one C.

b. 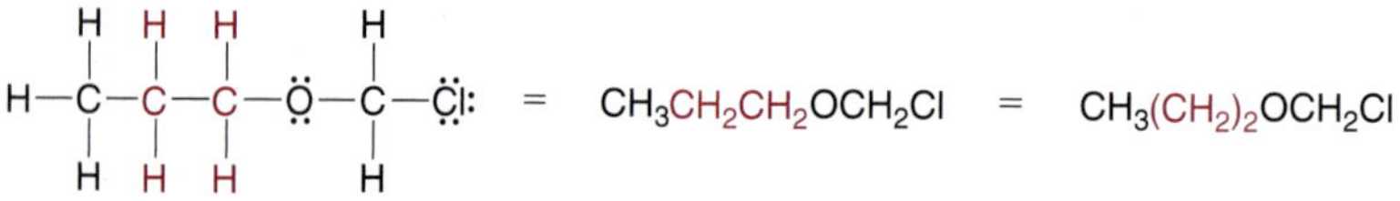 structure = CH$_3$CH$_2$CH$_2$OCH$_2$Cl = CH$_3$(CH$_2$)$_2$OCH$_2$Cl

PROBLEM 11.7

Convert each compound to a condensed formula.

a. 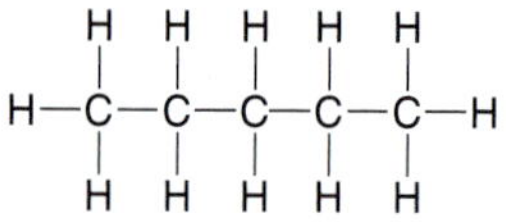 H—C—C—C—C—C—H (with H's)

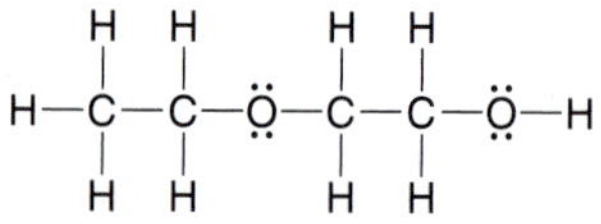

b. 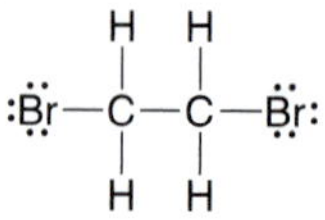 :Br—C—C—Br: (with H's)

c. H—C—C—Ö—C—C—Ö—H (with H's)

d. 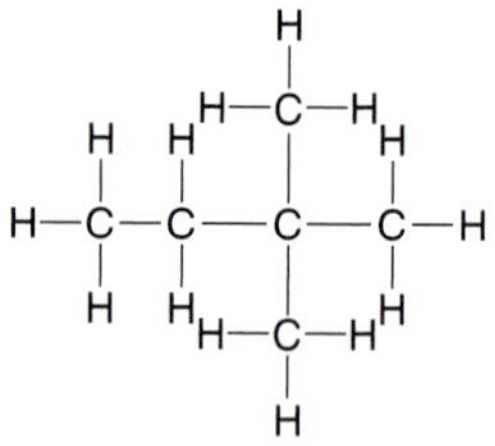 H—C—C—C—C—H (with CH$_2$ branches)

PROBLEM 11.8

Convert each condensed formula to a complete structure.

a. CH$_3$(CH$_2$)$_8$CH$_3$
b. CH$_3$(CH$_2$)$_4$OH
c. CH$_3$CCl$_3$
d. CH$_3$(CH$_2$)$_4$CH(CH$_3$)$_2$
e. (CH$_3$)$_2$CHCH$_2$NH$_2$

11.4B SKELETAL STRUCTURES

Skeletal structures are used for organic compounds containing both rings and chains of atoms. Three important rules are used in drawing them.

- Assume there is a carbon atom at the junction of any two lines or at the end of any line.
- Assume there are enough hydrogens around each carbon to give it four bonds.
- Draw in all heteroatoms and the hydrogens directly bonded to them.

Rings are drawn as polygons with a carbon atom "understood" at each vertex, as shown for cyclohexane and cyclopentanol. All carbons and hydrogens in these molecules are understood, except for H's bonded to heteroatoms.

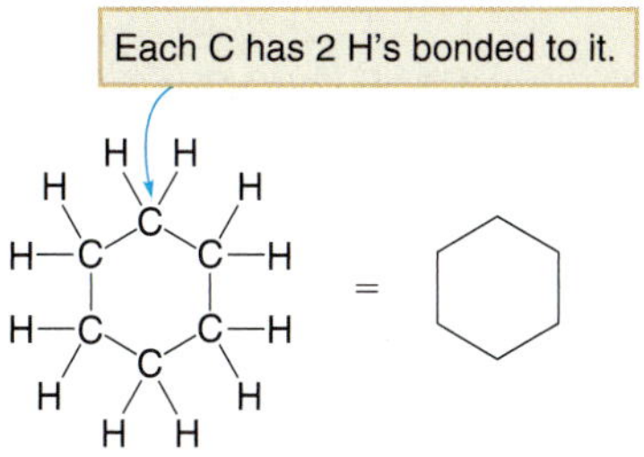

Each C has 2 H's bonded to it.

Draw in the H on O.

cyclohexane   skeletal structure 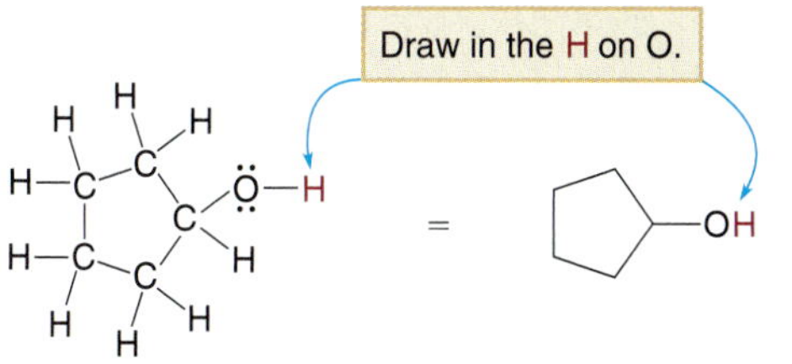   cyclopentanol   skeletal structure

---

SAMPLE PROBLEM 11.5

Convert each skeletal structure to a complete structure with all C's and H's drawn in.

a. ▢   b. 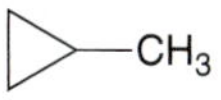 ▷—CH₃

ANALYSIS To draw each complete structure, place a C atom at the corner of each polygon and add H's to give carbon four bonds.

SOLUTION

a. 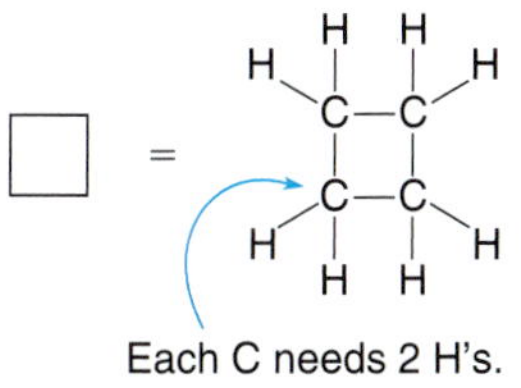 ▢ = [structure]   b. 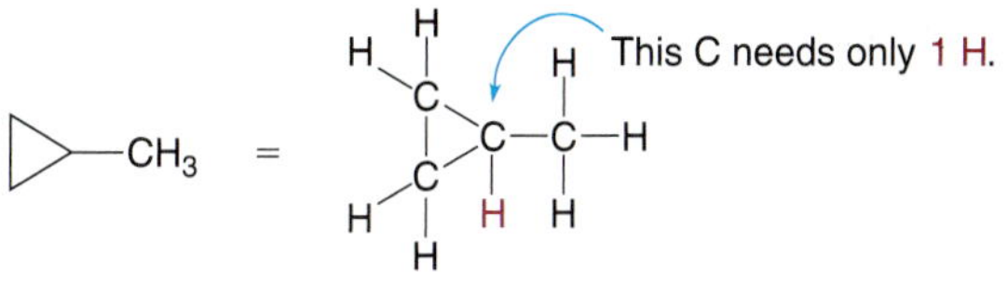 ▷—CH₃ = [structure] This C needs only 1 H.

Each C needs 2 H's.

PROBLEM 11.9

Convert each skeletal structure to a complete structure with all C's and H's drawn in. Add lone pairs on all heteroatoms.

a. 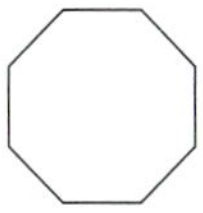 [octagon]   b. 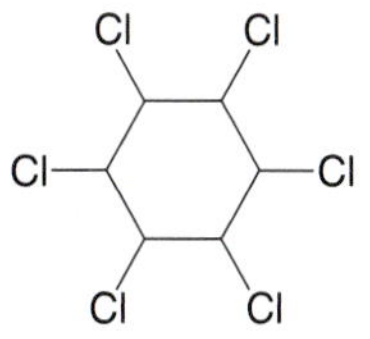 Cl—[ring with Cl's]—Cl   c. CH_3—[ring with OH]—$CH(CH_3)_2$

cyclooctane   lindane   menthol
                       (an insecticide)   (isolated from peppermint)

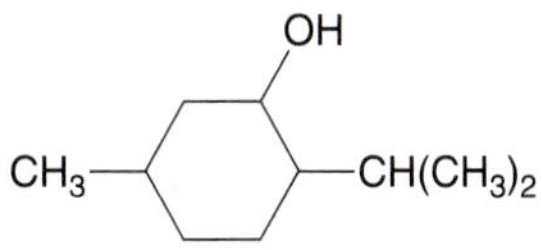

SAMPLE PROBLEM 11.6

Shorthand structures for more complex organic compounds are drawn with the same principles used for simpler structures. Answer the following questions about estradiol, a female sex hormone that controls the development of secondary sex characteristics.

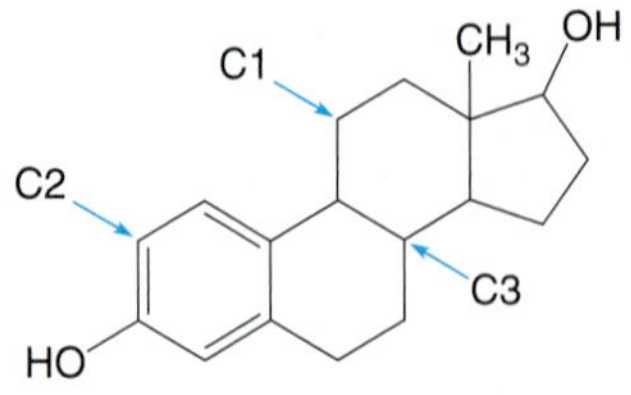

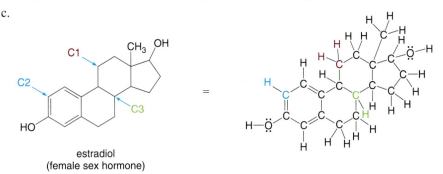

estradiol
(female sex hormone)

a. How many C atoms are present in estradiol?

b. How many H atoms are bonded to C1? C2? C3?

c. Draw a complete structure for estradiol, including all lone pairs on O atoms.

ANALYSIS   Place a C atom at each corner of all polygons. Add enough H's to give each C four bonds. Each O needs two lone pairs to have eight electrons around it.

SOLUTION
a. Since each corner of a polygon represents a C, estradiol has a total of 18 C's.

b. C1, labeled in red (below), needs 2 H's to give it four bonds. C2, labeled in blue, needs 1 H to give it four bonds. C3, labeled in green, needs 1 H to give it four bonds.

c.

estradiol
(female sex hormone)

PROBLEM 11.10

How many H's are bonded to each indicated carbon (C1–C5) in the following drugs.

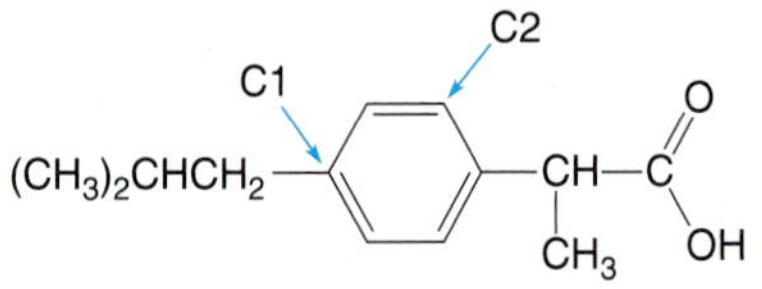

ibuprofen
(anti-inflammatory agent)

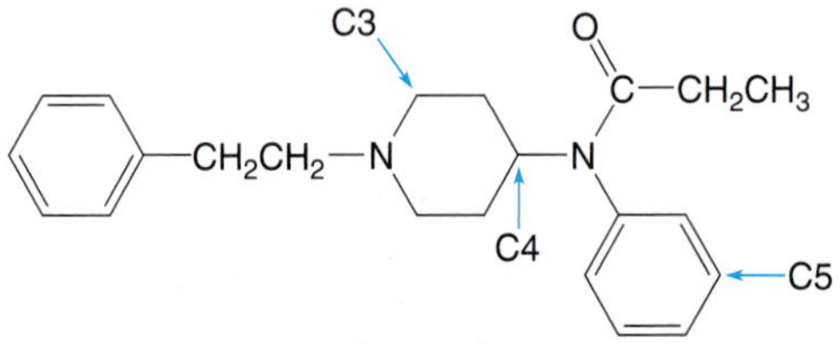

fentanyl
(narcotic pain reliever)

11.5   FUNCTIONAL GROUPS

In addition to strong C—C and C—H bonds, organic molecules may have other structural features as well. Although over 20 million organic compounds are currently known, only a limited number of common structural features, called **functional groups,** are found in these molecules.

- A *functional group* is an atom or a group of atoms with characteristic chemical and physical properties.
- A functional group contains a heteroatom, a multiple bond, or sometimes both a heteroatom *and* a multiple bond.

A functional group determines a molecule's shape, properties, and the type of reactions it undergoes. A functional group behaves the same whether it is bonded to a carbon backbone having as few as two or as many as 20 carbons. For this reason, we often abbreviate the carbon and hydrogen portion of the molecule by a capital letter **R,** and draw the **R** bonded to a particular functional group.

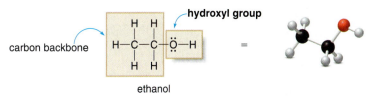

Ethanol (CH_3CH_2OH), for example, has two carbons and five hydrogens in its carbon backbone, as well as an OH group, a functional group called a **hydroxyl** group. The hydroxyl group determines the physical properties of ethanol as well as the type of reactions it undergoes. Moreover, any organic molecule containing a hydroxyl group has properties similar to ethanol. Compounds that contain a hydroxyl group are called **alcohols.**

ethanol

The most common functional groups can be subdivided into three types.

- Hydrocarbons
- Compounds containing a single bond to a heteroatom
- Compounds containing a C=O group

11.5A   HYDROCARBONS

Hydrocarbons **are compounds that contain only the elements of carbon and hydrogen,** as shown in Table 11.3.

- *Alkanes* have only C—C single bonds and no functional group. Ethane, **CH_3CH_3,** is a simple alkane.
- *Alkenes* have a C—C double bond as their functional group. Ethylene, **CH_2=CH_2,** is a simple alkene.
- *Alkynes* have a C—C triple bond as their functional group. Acetylene, **HC≡CH,** is a simple alkyne.
- *Aromatic hydrocarbons* contain a benzene ring, a six-membered ring with three double bonds.

TABLE 11.3   Hydrocarbons

Type of Compound	General Structure	Example	Functional Group
Alkane	R—H	CH_3CH_3	—
Alkene	C=C	C=C	Carbon–carbon double bond
Alkyne	—C≡C—	H—C≡C—H	Carbon–carbon triple bond
Aromatic compound	(benzene ring)	(benzene ring)	Benzene ring

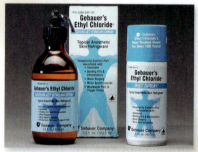
All hydrocarbons other than alkanes contain multiple bonds. Alkanes, which have no functional groups and therefore no reactive sites, are notoriously unreactive except under very drastic conditions. For example, **polyethylene** is a synthetic plastic and high molecular weight alkane, consisting of long chains of $-CH_2-$ groups bonded together, hundreds or even thousands of atoms long. Because it has no reactive sites, it is a very stable compound that does not readily degrade and thus persists for years in landfills.

polyethylene

The chain continues in both directions.

11.5B   COMPOUNDS CONTAINING A SINGLE BOND TO A HETEROATOM

Several types of functional groups contain a carbon atom singly bonded to a heteroatom. Common examples include alkyl halides, alcohols, ethers, and amines, as shown in Table 11.4.

Several simple compounds in this category are widely used. Bromomethane (CH_3Br) is an alkyl halide that has been used to control the spread of the destructive Asian long-horned beetle to the United States. Chloroethane (CH_3CH_2Cl, commonly called ethyl chloride) is an alkyl halide used as a local anesthetic. When chloroethane is sprayed on a wound it quickly evaporates, causing a cooling sensation that numbs the site of injury.

Molecules containing these functional groups may be simple or very complex. It doesn't matter what else is present in other parts of the molecule. **Always dissect it into small pieces to identify the functional groups.** For example, diethyl ether, the first general anesthetic, is an ether because it has an O atom bonded to two C's. Tetrahydrocannabinol (THC), the active component in marijuana, is also an ether because it contains an O atom bonded to two carbon atoms. In this case the O atom is also part of a ring.

ether

$CH_3CH_2-\ddot{O}-CH_2CH_3$

diethyl ether

tetrahydrocannabinol (THC)

ether

Identify the functional groups in each compound. Some compounds contain more than one functional group.

a.
$$\begin{array}{c} OH \\ | \\ CH_3CHCH_3 \end{array}$$

2-propanol
(disinfectant in rubbing alcohol)

b. ⬡—$CH=CH_2$

styrene
(starting material used to synthesize Styrofoam)

c. $H_2NCH_2CH_2CH_2CH_2NH_2$

putrescine
(putrid odor of rotting fish)

TABLE 11.4 Compounds Containing a Carbon–Heteroatom Single Bond

Type of Compound	General Structure	Example	3-D Structure	Functional Group
Alkyl halide	R—$\ddot{\text{X}}$: (X = F, Cl, Br, I)	CH_3—$\ddot{\text{Br}}$:	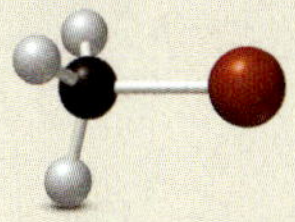	–X
Alcohol	R—$\ddot{\text{O}}$H	CH_3—$\ddot{\text{O}}$H	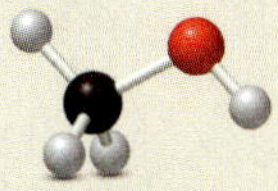	–OH hydroxyl group
Ether	R—$\ddot{\text{O}}$—R	CH_3—$\ddot{\text{O}}$—CH_3	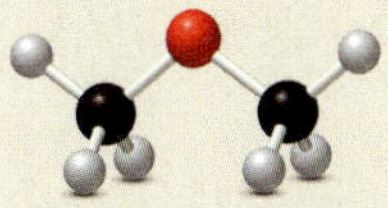	–OR
Amine	R—$\ddot{\text{N}}H_2$ or $R_2\ddot{\text{N}}H$ or $R_3\ddot{\text{N}}$	CH_3—$\ddot{\text{N}}H_2$	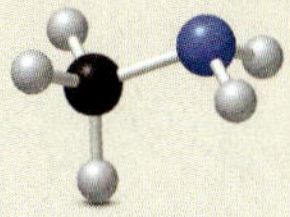	–NH_2 amino group

11.5C COMPOUNDS CONTAINING A C=O GROUP

Many different kinds of compounds contain a carbon–oxygen double bond (**C=O, carbonyl group**), as shown in Table 11.5. Carbonyl compounds include aldehydes, ketones, carboxylic acids, esters, and amides. The type of atom bonded to the carbonyl carbon—hydrogen, carbon, or a heteroatom—determines the specific class of carbonyl compound.

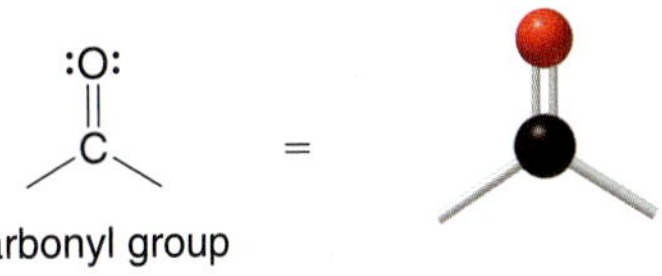

carbonyl group

Take special note of the condensed structures used to draw aldehydes, carboxylic acids, and esters.

- **An aldehyde has a hydrogen atom bonded directly to the carbonyl carbon.**

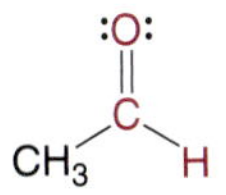

acetaldehyde = CH_3CHO | C is bonded to both H and O. |
condensed structure

- **A carboxylic acid contains an OH group bonded directly to the carbonyl carbon.**

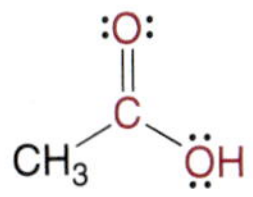

acetic acid = CH_3COOH or CH_3CO_2H | C is bonded to both O atoms. |
condensed structures

- **An ester contains an OR group bonded directly to the carbonyl carbon.**

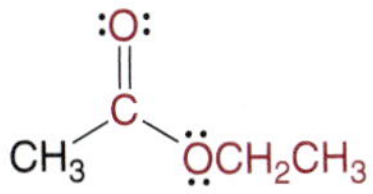

ethyl acetate = $CH_3COOCH_2CH_3$ or $CH_3CO_2CH_2CH_3$
condensed structures | C is bonded to both O atoms. |

TABLE 11.5   Compounds Containing a C=O Group

Type of Compound	General Structure	Example	3-D Structure	Functional Group
Aldehyde	$\overset{\displaystyle :O:}{\underset{\displaystyle R \quad H}{\parallel C}}$	$\overset{\displaystyle :O:}{\underset{\displaystyle CH_3 \quad H}{\parallel C}}$		$\overset{\displaystyle :O:}{\underset{\displaystyle H}{\parallel C}}$
Ketone	$\overset{\displaystyle :O:}{\underset{\displaystyle R \quad R}{\parallel C}}$	$\overset{\displaystyle :O:}{\underset{\displaystyle CH_3 \quad CH_3}{\parallel C}}$		$\overset{\displaystyle :O:}{\underset{\displaystyle }{\parallel C}}$
Carboxylic acid	$\overset{\displaystyle :O:}{\underset{\displaystyle R \quad \ddot{O}H}{\parallel C}}$	$\overset{\displaystyle :O:}{\underset{\displaystyle CH_3 \quad \ddot{O}H}{\parallel C}}$		$\overset{\displaystyle :O:}{\underset{\displaystyle \ddot{O}H}{\parallel C}}$ carboxyl group
Ester	$\overset{\displaystyle :O:}{\underset{\displaystyle R \quad \ddot{O}R}{\parallel C}}$	$\overset{\displaystyle :O:}{\underset{\displaystyle CH_3 \quad \ddot{O}CH_3}{\parallel C}}$		$\overset{\displaystyle :O:}{\underset{\displaystyle \ddot{O}R}{\parallel C}}$
Amide	$\overset{\displaystyle :O:}{\underset{\displaystyle R \quad \underset{\displaystyle H\,(or\,R)}{\overset{\displaystyle H\,(or\,R)}{N}}}{\parallel C}}$	$\overset{\displaystyle :O:}{\underset{\displaystyle CH_3 \quad \ddot{N}H_2}{\parallel C}}$		$\overset{\displaystyle :O:}{\underset{\displaystyle \ddot{N}}{\parallel C}}$

Carbonyl compounds occur widely in nature, and many are responsible for the characteristic odor or flavor of the fruits shown in Figure 11.2.

SAMPLE PROBLEM 11.7   Identify the functional group in each compound.

a. $\underset{\displaystyle H}{\overset{\displaystyle H}{C}} = \underset{\displaystyle CH_3}{\overset{\displaystyle CH_3}{C}}$

A

b. ⬡—OH

B

c. (cyclopentane ring)=O

C

ANALYSIS   Concentrate on the multiple bonds and heteroatoms and refer to Tables 11.3, 11.4, and 11.5.

SOLUTION

a. $\underset{\displaystyle H}{\overset{\displaystyle H}{C}} = \underset{\displaystyle CH_3}{\overset{\displaystyle CH_3}{C}}$

A

b. ⬡—[OH]

B

c. (cyclopentane ring)=O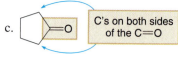

C

A is a hydrocarbon with a carbon–carbon double bond, making it an alkene.

B has a carbon atom bonded to a hydroxyl group (OH), making it an alcohol.

C contains a C=O. Since the carbonyl carbon is bonded to two other carbons in the ring, **C** is a ketone.

▼ **FIGURE 11.2**   **What We Eat—Naturally Occurring Carbonyl Compounds in Fruit**

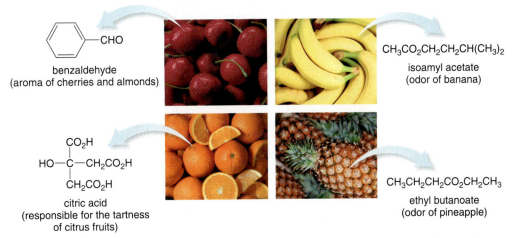

benzaldehyde
(aroma of cherries and almonds)

$CH_3CO_2CH_2CH_2CH(CH_3)_2$
isoamyl acetate
(odor of banana)

citric acid
(responsible for the tartness
of citrus fruits)

$CH_3CH_2CH_2CO_2CH_2CH_3$
ethyl butanoate
(odor of pineapple)

In addition to water, fruits (and other foods) contain a variety of organic compounds that give them taste, odor, texture, and nutritional value. Although one organic compound might be the major component in a food's flavor or odor, what we taste and smell is often a complex mixture of compounds. When even one of these components is missing, the odor and flavor can be somewhat different. This explains why artificial flavors and colors mimic, but don't really replace, the natural flavor.

PROBLEM 11.12

For each compound: [1] Identify the functional group; [2] draw out the complete compound, including lone pairs on heteroatoms.

a. ⬡—CHO

c. ⬡=O

e. (cyclopentane)—C(=O)—NHCH₃

b. $CH_3CH_2CH_2CO_2H$

d. $CH_3CH_2CO_2CH_2CH_3$

PROBLEM 11.13

Identify the type of carbonyl group in each of the compounds in Figure 11.2.

Many useful organic compounds contain complex structures with two or more functional groups. Sample Problem 11.8 illustrates an example of identifying several functional groups in a single molecule.

SAMPLE PROBLEM 11.8

Tamiflu is an antiviral drug effective against avian influenza. Identify all of the functional groups in Tamiflu.

$(CH_3CH_2)_2CH$

$CO_2CH_2CH_3$

HN

NH_2

CH_3

Tamiflu

ANALYSIS   To identify functional groups, look for multiple bonds and heteroatoms. With functional groups that contain O atoms, look at what is bonded to the O's to decide if the group is an alcohol, ether, or other group. With carbonyl-containing groups, look at what is bonded to the carbonyl carbon.

SOLUTION   Re-draw Tamiflu to further clarify the functional groups:

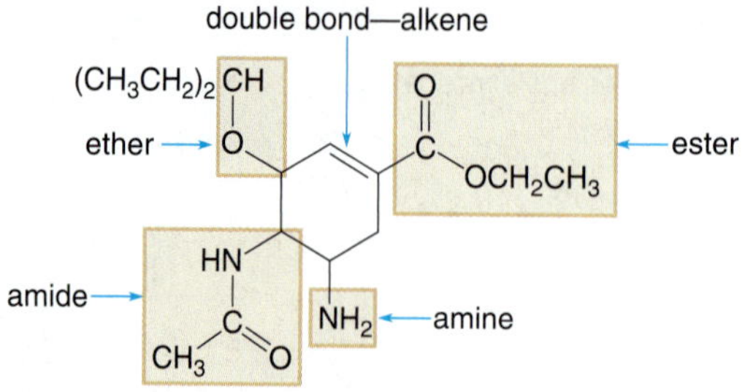

- Tamiflu contains a total of five functional groups: a carbon–carbon double bond (an alkene), ether, ester, amide, and amine.
- Note the difference between an amine and an amide. **An amine contains a N atom but no C=O. An amide contains a N atom directly bonded to a C=O.**

PROBLEM 11.14   Besides an ether group, tetrahydrocannabinol (Section 11.5B) contains three additional functional groups. Identify all of the functional groups in THC.

PROBLEM 11.15   Identify all of the functional groups in each drug. Atenolol is an antihypertensive agent; that is, it is used to treat high blood pressure. Mestranol is a synthetic estrogen used in oral contraceptives.

a.

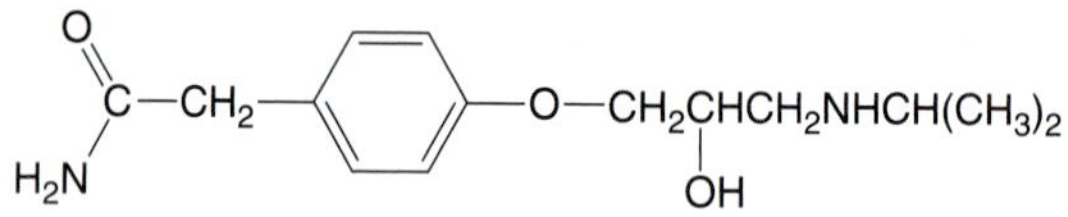

atenolol

b.

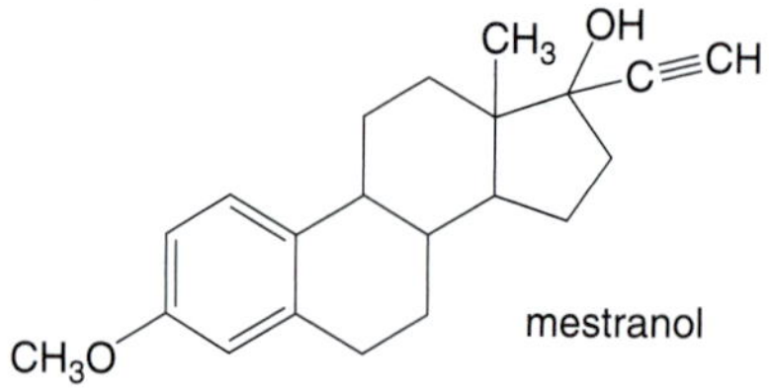

mestranol

11.6 PROPERTIES OF ORGANIC COMPOUNDS

Because organic compounds are composed of covalent bonds, their properties differ a great deal from those of ionic inorganic compounds.

The intermolecular forces in covalent compounds were discussed in Section 7.7.

- Organic compounds exist as discrete molecules with much weaker intermolecular forces—the forces that exist *between* molecules—than those seen in ionic compounds, which are held together by very strong interactions of oppositely charged ions.

As a result, organic compounds resemble other covalent compounds in that they have much lower melting points and boiling points than ionic compounds. While ionic compounds are generally solids at room temperature, many organic compounds are liquids and some are even gases. Table 11.6 compares these and other properties of a typical organic compound (butane, $CH_3CH_2CH_2CH_3$) and a typical ionic inorganic compound (sodium chloride, NaCl).

Review Sections 4.6 and 4.7 on electronegativity, bond polarity, and polar molecules. Recall that **electronegativity** measures how much a particular atom "wants" the electron density in a bond.

11.6A POLARITY

Are organic compounds polar or nonpolar? That depends. Like other covalent compounds, the polarity of an organic compound is determined by two factors: the polarity of the individual

PROPERTIES OF ORGANIC COMPOUNDS

TABLE 11.6 Comparing the Properties of an Organic Compound (CH₃CH₂CH₂CH₃) and an Inorganic Compound (NaCl)

Property	CH₃CH₂CH₂CH₃ (An Organic Compound)	NaCl (An Inorganic Compound)
Bonding	 CH₃CH₂CH₂CH₃ butane discrete molecules with covalent bonding	 Na⁺ Cl⁻ ionic bonding
Physical state	Gas at room temperature	Solid at room temperature
Boiling point	Low (−0.5 °C)	High (1413 °C)
Melting point	Low (−138 °C)	High (801 °C)
Solubility in water	Insoluble	Soluble
Solubility in organic solvents	Soluble	Insoluble
Flammability	Flammable	Nonflammable

bonds and the overall shape of the molecule. **The polarity of an individual bond depends on the electronegativity of the atoms that form it.**

- A covalent bond is *nonpolar* when two atoms of identical or similar electronegativity are joined together. Thus, C—C and C—H bonds are nonpolar bonds.
- A covalent bond is *polar* when atoms of different electronegativity are joined together. Thus, bonds between carbon and common heteroatoms (N, O, and the halogens) are polar bonds.

Since the common heteroatoms are *more* electronegative than carbon, these atoms pull electron density away from carbon, making it partially positive and the heteroatom partially negative.

$$\overset{\delta^+}{C}-\overset{\delta^-}{\ddot{O}}- \qquad \overset{\delta^+}{C}-\overset{\delta^-}{\ddot{N}}- \qquad \overset{\delta^+}{C}-\overset{\delta^-}{\ddot{X}}:$$

X = halogen

- The symbol δ⁺ is given to the less electronegative atom (usually C or H).
- The symbol δ⁻ is given to the more electronegative atom (usually N, O, or a halogen).

PROBLEM 11.16   Indicate the polar bonds in each compound. Label the bonds with δ^+ and δ^-.

a.

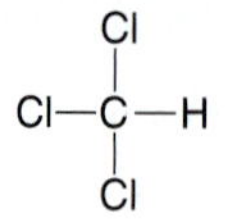

$$\text{Cl}-\overset{\overset{\displaystyle Cl}{|}}{\underset{\underset{\displaystyle Cl}{|}}{C}}-H$$

chloroform
(once used as an anesthetic)

c.

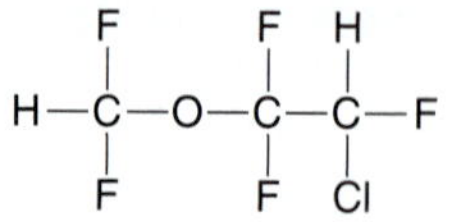

$$H-\overset{\overset{\displaystyle F}{|}}{\underset{\underset{\displaystyle F}{|}}{C}}-O-\overset{\overset{\displaystyle F}{|}}{\underset{\underset{\displaystyle F}{|}}{C}}-\overset{\overset{\displaystyle H}{|}}{\underset{\underset{\displaystyle Cl}{|}}{C}}-F$$

enflurane
(anesthetic)

b.

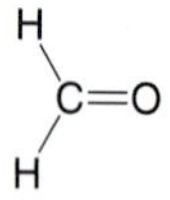

$$\overset{\displaystyle H}{\underset{\displaystyle H}{C}}=O$$

formaldehyde
(preservative)

d.

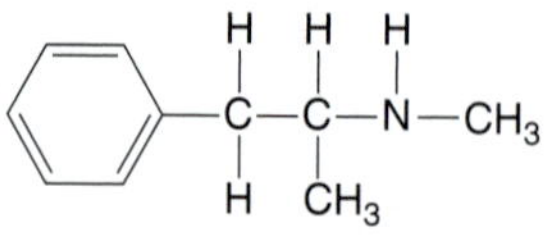

methamphetamine
(illegal, addictive stimulant)

Hydrocarbons contain only nonpolar C—C and C—H bonds, so **hydrocarbons are nonpolar molecules.** In contrast, when an organic compound contains a *single* polar bond, as in the alkyl halide CH_3CH_2Cl, the molecule contains a net dipole, and is therefore a polar molecule.

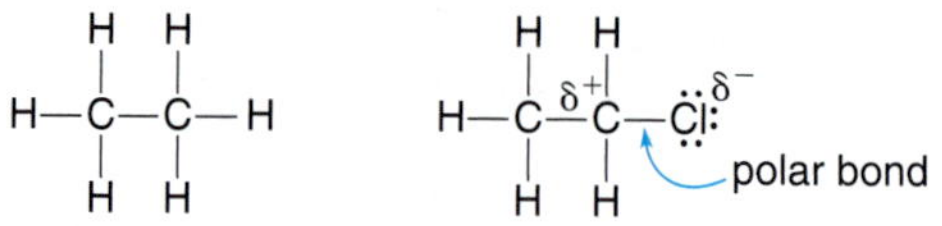

ethane

no polar bonds
nonpolar molecule

chloroethane

one polar bond
polar molecule

With organic compounds that have more than one polar bond, **the shape of the molecule determines the overall polarity.**

- If the individual bond dipoles cancel in a molecule, the molecule is nonpolar.
- If the individual bond dipoles do not cancel, the molecule is polar.

A bond dipole can be indicated by an arrow with the head of the arrow pointing towards the more electronegative element, and a perpendicular line drawn through the tail of the arrow ($\longmapsto$). Thus, dichloroacetylene ($ClC{\equiv}CCl$) is a linear molecule with two polar C—Cl bonds, but the individual bond dipoles are equal in magnitude and opposite in direction. As a result, dichloroacetylene is a *nonpolar* molecule. In contrast, methanol, CH_3OH, has two polar bonds (C—O and O—H), and a bent shape around oxygen (Section 11.3). The individual bond dipoles do *not* cancel, so CH_3OH is a *polar* molecule.

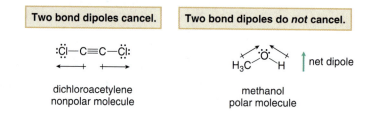

Two bond dipoles cancel.	Two bond dipoles do *not* cancel.
dichloroacetylene nonpolar molecule	methanol polar molecule

SAMPLE PROBLEM 11.9

Explain why dichloromethane (CH_2Cl_2), a solvent used in paint remover, is a polar molecule.

ANALYSIS

CH_2Cl_2 has two polar C—Cl bonds. We must determine the shape of the molecule around the carbon atom and check whether the two bond dipoles cancel or reinforce.

SOLUTION

Since the C atom in CH_2Cl_2 is surrounded by four atoms, it is tetrahedral. The two bond dipoles do not cancel, and **CH_2Cl_2 is a polar molecule.**

Draw CH_2Cl_2 as a tetrahedron to see that the bond dipoles do not cancel.

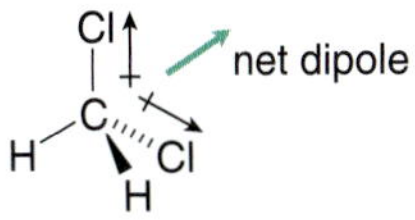

Do *not* draw CH_2Cl_2 as:

net dipole

polar molecule

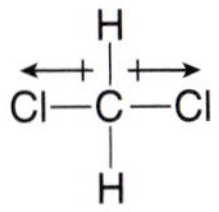

In this drawing it appears as if the bond dipoles cancel, but in reality they do *not*.

PROBLEM 11.17

Label each molecule as polar or nonpolar and explain your choice.

a. △

b. 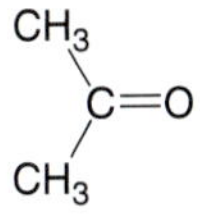 $\begin{matrix} CH_3 \\ CH_3 \end{matrix} C{=}O$

c. CCl_4

d. $CH_3CH_2CH_2NH_2$

PROBLEM 11.18

Explain why dimethyl ether (CH_3OCH_3) is a polar molecule.

11.6B SOLUBILITY

Review the general principles of solubility in Section 8.2.

Understanding the solubility properties of organic compounds allows us to understand many interesting phenomena. One rule governs solubility: **"Like dissolves like."** Three facts can then be used to explain the solubility of organic compounds.

- Most organic compounds are soluble in organic solvents.
- Hydrocarbons and other nonpolar organic compounds are not soluble in water.
- Polar organic compounds are water soluble only if they are small and contain a nitrogen or oxygen atom that can hydrogen bond with water.

Three compounds—hexane, ethanol, and cholesterol—illustrate these principles. All of these compounds are organic and therefore soluble in an organic solvent like diethyl ether ($CH_3CH_2OCH_2CH_3$). Their water solubility depends on their polarity and size.

Since hexane ($CH_3CH_2CH_2CH_2CH_2CH_3$) is a nonpolar hydrocarbon, it is insoluble in water. Ethanol (CH_3CH_2OH), on the other hand, is water soluble because it is small and contains a polar OH group. The ethanol in gin dissolves in the water of tonic water, resulting in the alcoholic beverage called a "gin and tonic."

Cholesterol has 27 carbon atoms and only one OH group. Its nonpolar hydrocarbon part is too large to dissolve in water, making it water insoluble. As a result, cholesterol can't dissolve in the aqueous medium of the blood. When it must be transported through the body in the bloodstream, cholesterol is bound to other substances that are water soluble.

CH₃CH₂—OH
ethanol

H₂O soluble

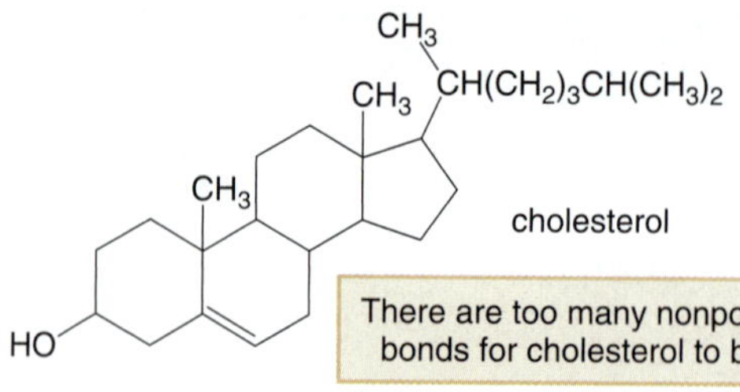

cholesterol

There are too many nonpolar C—C and C—H
bonds for cholesterol to be soluble in water.

HO

H₂O insoluble

Table 11.7 summarizes the solubility characteristics of hexane, ethanol, and cholesterol.

**TABLE 11.7   The Solubility Characteristics of Three Representative
Organic Molecules**

Compound	Solubility in an Organic Solvent	Solubility in Water
CH₃CH₂CH₂CH₂CH₂CH₃ Hexane	Soluble	Insoluble
CH₃CH₂OH Ethanol	Soluble	Soluble
Cholesterol	Soluble	Insoluble

PROBLEM 11.19

Predict the water solubility of each compound.

a.   CH₃(CH₂)₆CH₃

octane
(gasoline component)

b.

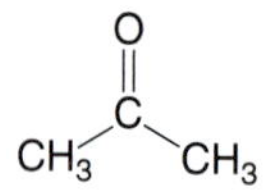

$$\underset{CH_3}{}\overset{O}{\underset{}{C}}\underset{CH_3}{}$$

acetone
(solvent)

c.   CH₃(CH₂)₁₆CO₂H

stearic acid
(a fatty acid)

11.6C   FOCUS ON THE ENVIRONMENT
ENVIRONMENTAL POLLUTANTS

MTBE (*tert*-butyl methyl ether) and 4,4'–dichlorobiphenyl (a PCB) demonstrate that solubility properties can help to determine the fate of organic pollutants in the environment.

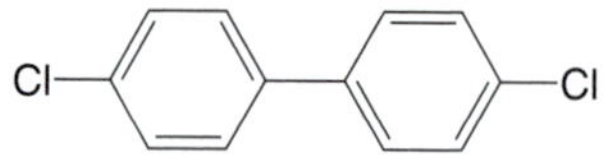

CH₃—O—C(CH₃)₃   Cl—⟨　⟩—⟨　⟩—Cl

MTBE   4,4'-dichlorobiphenyl
tert-butyl methyl ether   (a PCB)

MTBE is a small polar organic molecule that has been used as a high-octane additive in unleaded gasoline. The use of MTBE, however, has now also caused environmental concern. Although MTBE is not toxic or carcinogenic, it has a distinctive, nauseating odor, and **it is water soluble.** Small amounts of MTBE have contaminated the drinking water in several California communities, making it unfit for consumption. These solubility properties of MTBE have now led to its ban in several states.

4,4'-Dichlorobiphenyl is one of a group of compounds called **PCB**s, which have been used in the manufacture of polystyrene coffee cups and other products. They have been released into the environment during production, use, storage, and disposal, making them one of the most widespread organic pollutants. **PCBs are insoluble in H₂O, but very soluble in organic media,** so they are soluble in fatty tissue, including that found in all types of fish and birds around the world. Although not acutely toxic, ingesting large quantities of fish contaminated with PCBs has been shown to retard growth and memory retention in children.

<table>
<tr><td>**PROBLEM 11.20**</td><td>DDT is a persistent environmental pollutant once widely used as an insecticide. Using solubility principles, explain why shore birds that populate areas sprayed with DDT are found to have a high concentration of DDT in their tissues.</td></tr>
</table>

DDT

11.7 FOCUS ON HEALTH & MEDICINE
VITAMINS

Vitamins **are organic compounds needed in small amounts for normal cell function.** Our bodies cannot synthesize these compounds, so they must be obtained in our diet. Most vitamins are identified by a letter, such as A, C, D, E, and K. There are several different B vitamins, though, so a subscript is added to distinguish them—for example, B_1, B_2, and B_{12}.

Vitamins are classified on the basis of their solubility properties.

- A *fat-soluble vitamin* dissolves in an organic solvent but is insoluble in water. Fat-soluble vitamins have many nonpolar C—C and C—H bonds and few polar functional groups.
- A *water-soluble vitamin* dissolves in water. Water-soluble vitamins have many polar bonds.

The name **vitamin** was first used in 1912 by the Polish chemist Casimir Funk, who called them *vitamines,* since he thought that they all contained an *amine* functional group. Later the word was shortened to vitamin, because some are amines but others, like vitamins A and C, are not.

Whether a vitamin is fat soluble or water soluble can be determined by applying the solubility principles discussed in Section 11.6. Vitamins A and C illustrate the differences between fat-soluble and water-soluble vitamins.

11.7A   VITAMIN A

Vitamin A, or **retinol,** is an essential component of the vision receptors in the eyes. It also helps to maintain the health of mucous membranes and the skin, so many anti-aging creams contain vitamin A. A deficiency of this vitamin leads to a loss of night vision.

vitamin A

Vitamin A contains 20 carbons and a single OH group, making it **water insoluble.** Because it is organic, it is **soluble in any organic medium.** To understand the consequences of these solubility characteristics, we must learn about the chemical environment of the body.

About 70% of the body is composed of water. Fluids such as blood, gastric juices in the stomach, and urine are largely water with dissolved ions such as Na^+ and K^+. Vitamin A is insoluble in these fluids. There are also fat cells composed of organic compounds having C—C and C—H bonds. Vitamin A is soluble in this organic environment, and thus it is readily stored in these fat cells, particularly in the liver.

In addition to obtaining vitamin A directly from the diet, β-carotene, the orange pigment found in many plants including carrots, is readily converted to vitamin A in our bodies.

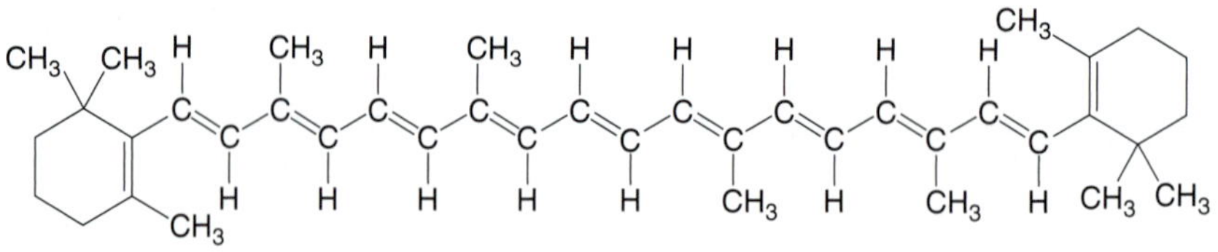

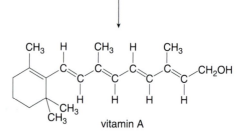

β–carotene
(orange pigment in carrots)

vitamin A

HEALTH NOTE

Vitamin C can be obtained from eating oranges and other citrus fruits. It can also be obtained by taking tablets that contain vitamin C synthesized in the laboratory. The vitamin C from both sources is exactly the same, so that "natural" vitamin C in oranges and "synthetic" vitamin C in tablets are identical.

Eating too many carrots does not result in storing an excess of vitamin A. If you consume more β-carotene in carrots than you need, the body stores this precursor until it needs more vitamin A. Some of the stored β-carotene reaches the surface tissues of the skin, giving it an orange color. This phenomenon is harmless and reversible. When stored β-carotene is converted to vitamin A and is no longer in excess, these tissues will return to their normal hue.

Ingesting a moderate excess of vitamin A doesn't cause any harm, but a large excess causes headaches, loss of appetite, and even death. Early Arctic explorers who ate polar bear livers that contain an unusually large amount of vitamin A, are thought to have died from consuming too much vitamin A.

11.7B VITAMIN C

Although most animal species can synthesize vitamin C, humans, the Indian fruit bat, and the bulbul bird must obtain this vitamin from dietary sources. Citrus fruits, strawberries, tomatoes, and sweet potatoes are all excellent sources of vitamin C.

Vitamin C, or **ascorbic acid,** is important in the formation of collagen, a protein that forms the connective tissues of skin, muscle, and blood vessels. A deficiency of vitamin C causes scurvy, a common disease of sailors in the 1600s who had no access to fresh fruits on long voyages.

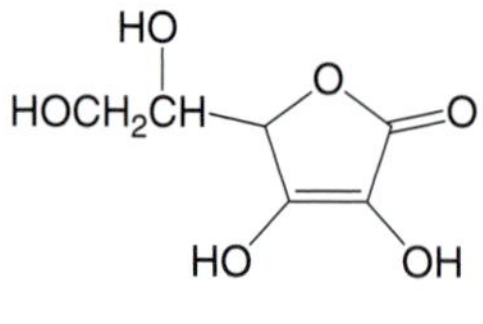

vitamin C
(ascorbic acid)

Vitamin C has six carbon atoms and six oxygen atoms, making it **water soluble.** Vitamin C thus dissolves in urine. Although it has been acclaimed as a deterrent for all kinds of diseases, from the common cold to cancer, the benefits of taking large amounts of vitamin C are not really known, since an excess of the minimum daily requirement is excreted unchanged in the urine.

The solubility characteristics of vitamins A and C are shown in Figure 11.3.

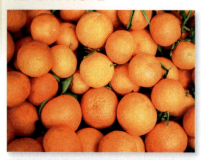

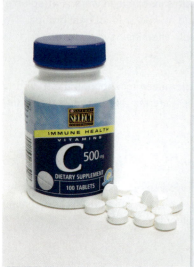

▼ **FIGURE 11.3   Solubility Characteristics of Vitamins A and C**

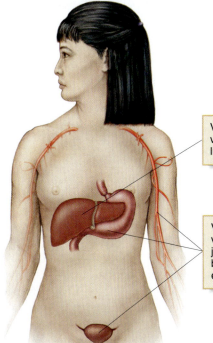

Vitamin A and other fat-soluble vitamins are readily stored in the liver and fatty tissue.

Vitamin C and other water-soluble vitamins are soluble in the gastric juices of the stomach and the blood. These vitamins are readily excreted in urine.

PROBLEM 11.21

Predict the solubility of vitamin B_3 in water.

vitamin B_3
(niacin)

PROBLEM 11.22

Predict the solubility of vitamin K_1 in water and in organic solvents.

CH_3

CH_2 C $(CH_2)_3CH(CH_2)_3CH(CH_2)_3CH(CH_3)_2$

H CH_3 CH_3

CH_3

vitamin K_1
(phylloquinone)

CHAPTER HIGHLIGHTS

KEY TERMS

Alcohol (11.5)
Aldehyde (11.5)
Alkane (11.5)
Alkene (11.5)
Alkyl halide (11.5)
Alkyne (11.5)
Amide (11.5)
Amine (11.5)
Amino group (11.5)
Aromatic compound (11.5)

Carbonyl group (11.5)
Carboxyl group (11.5)
Carboxylic acid (11.5)
Condensed structure (11.4)
Ester (11.5)
Ether (11.5)
Fat-soluble vitamin (11.7)
Functional group (11.5)
Heteroatom (11.2)
Hydrocarbon (11.5)

Hydroxyl group (11.5)
Ketone (11.5)
"Like dissolves like" (11.6)
Nonpolar molecule (11.6)
Organic chemistry (11.1)
Polar molecule (11.6)
Skeletal structure (11.4)
Vitamin (11.7)
Water-soluble vitamin (11.7)

KEY CONCEPTS

❶ What are the characteristic features of organic compounds? (11.2)

- Organic compounds contain carbon atoms and most contain hydrogen atoms. Carbon forms four bonds.
- Carbon forms single, double, and triple bonds to itself and other atoms.
- Carbon atoms can bond to form chains or rings.
- Organic compounds often contain heteroatoms, commonly N, O, and the halogens.

❷ How can we predict the shape around an atom in an organic molecule? (11.3)

- The shape around an atom is determined by counting groups, and then arranging them to keep them as far away from each other as possible.

Number of groups	Number of atoms	Number of lone pairs	Shape	Bond angle	Example
2	2	0	Linear	180°	$HC\equiv CH$
3	3	0	Trigonal planar	120°	$CH_2=CH_2$
4	4	0	Tetrahedral	109.5°	CH_3CH_3
4	3	1	Trigonal pyramidal	~109.5°	CH_3NH_2 (N atom)
4	2	2	Bent	~109.5°	CH_3OH (O atom)

❸ What shorthand methods are used to draw organic molecules? (11.4)

- In condensed structures, atoms are drawn in but the two-electron bonds are generally omitted. Lone pairs are omitted as well. Parentheses are used around like groups bonded together or to the same atom.
- Three assumptions are used in drawing skeletal structures: [1] There is a carbon at the intersection of two lines or at the end of any line. [2] Each carbon has enough hydrogens to give it four bonds. [3] Heteroatoms and the hydrogens bonded to them are drawn in.

❹ What is a functional group and why are functional groups important? (11.5)

- A functional group is an atom or a group of atoms with characteristic chemical and physical properties.
- A functional group determines all of the properties of a molecule—its shape, physical properties, and the type of reactions it undergoes.

❺ How do organic compounds differ from ionic inorganic compounds? (11.6)

- The main difference between organic compounds and ionic inorganic compounds is their bonding. Organic compounds are composed of discrete molecules with covalent bonds. Ionic inorganic compounds are composed of ions held together by the strong attraction of oppositely charged ions. Other properties that are consequences of these bonding differences are summarized in Table 11.6.

❻ How can we determine if an organic molecule is polar or nonpolar? (11.6A)

- A nonpolar molecule has either no polar bonds or two or more bond dipoles that cancel.
- A polar molecule has either one polar bond, or two or more bond dipoles that do not cancel.

❼ What are the solubility properties of organic compounds? (11.6B)

- Most organic molecules are soluble in organic solvents.
- Hydrocarbons and other nonpolar organic molecules are not soluble in water.
- Polar organic molecules are water soluble only if they are small and contain a nitrogen or oxygen atom that can hydrogen bond with water.

❽ What are vitamins and when is a vitamin fat soluble or water soluble? (11.7)

- A vitamin is an organic compound that cannot be synthesized by the body but is needed in small amounts for cell function.
- A fat-soluble vitamin has many nonpolar C—C and C—H bonds and few polar bonds, making it insoluble in water.
- A water-soluble vitamin has many polar bonds, so it dissolves in water.

PROBLEMS

Selected in-chapter and end-of-chapter problems have brief answers provided in Appendix B.

General Characteristics of Organic Molecules

11.23 Which molecular formulas represent organic compounds and which represent inorganic compounds: (a) H_2SO_4; (b) Br_2; (c) C_5H_{12}?

11.24 Which molecular formulas represent organic compounds and which represent inorganic compounds: (a) LiBr; (b) HCl; (c) CH_5N?

11.25 Complete each structure by filling in all H's and lone pairs.

a. C—C=C—C≡C

b. C=C—C—O

c. C≡C—C—C
        |
        Cl

d. C=C—C—C—N
          ‖
          O

11.26 Complete each structure by filling in all H's and lone pairs.

a.
  O O
  ‖ ‖
C—C—C=O

glyceraldehyde
(simple carbohydrate)

b.
  O O
  ‖ ‖
C—C—C=O

lactic acid
(product of carbohydrate metabolism)

c.
        O
        ‖
N—C—C—O
    |
    C

alanine
(amino acid in proteins)

d.
      O
      ‖
O—C—C—C—O

dihydroxyacetone
(ingredient in artificial tanning agents)

Predicting Shape and Bond Angles

11.27 Determine the shape around the indicated atoms in each compound. Don't forget to add lone pairs to heteroatoms before determining the shape around them.

a.
     (1)   O (2)
          ‖
     H—C—O—H

formic acid
(responsible for the sting of some ants)

c.
    (1) (2) (3)
    CH_3—O—$C(CH_3)_3$

MTBE
(gasoline additive)

b.
    (1)   (2)
    H—C=C—C≡N
       | |
       H H

acrylonitrile
(used to prepare Orlon fibers)

11.28 Ethambutol is a drug used to treat tuberculosis. Determine the shape around the indicated atoms in ethambutol. Don't forget to add lone pairs to heteroatoms before determining the shape around them.

            CH_2CH_3 CH_2CH_3
              | |
H—O—CH_2—C—N—CH_2CH_2—N—C—CH_2—O—H
            | | | |
            H H   H H

ethambutol

11.29 Predict the indicated bond angles in each compound.

a.
         (1)
  CH_3—C≡C—Cl
           (2)

b.
        H (2)
        |
  CH_2=C—Cl
       (1)

c.
         H
      (1) | (2)
  CH_3—C—Cl
         |
         H

11.30 Predict the indicated bond angles in each compound. Don't forget to add lone pairs to heteroatoms before determining the shape around them.

a.
                      O
                      ‖
$CH_3(CH_2)_7$—C=C—$(CH_2)_7$—C—O—H
             | |
             H H

oleic acid
(fatty acid in vegetable oil)

b.
           H
           |
  H—N—C—$(CH_2)_5NH_2$
      | |
      H H

1,6-diaminohexane
(used to synthesize nylon)

11.31 "Ecstasy" is a widely used illegal stimulant. Determine the shape around each indicated atom.

                             (2) (3)
                                    H
(1)—O ... CH_2CHNCH_3
                                  CH_3 (4)

MDMA
"Ecstasy"

11.32 Fenfluramine is one component of the appetite suppressant Fen–Phen, which was withdrawn from the market in 1997 after it was shown to damage the heart valves in some patients. Determine the shape around each indicated atom.

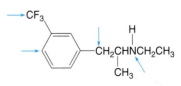

fenfluramine

11.33 Explain why each C—C—C bond angle in benzene (Table 11.3) is 120°.

11.34 Hexachloroethane (Cl_3C—CCl_3), an alkyl chloride with a camphor-like odor, is a derivative of ethane (CH_3CH_3), formed by replacing each H by a Cl atom. Draw the structure of hexachloroethane using solid lines, wedges, and dashes to indicate the three-dimensional shape around each atom.

Drawing Organic Molecules

11.35 Convert each compound to a condensed structure.

a.

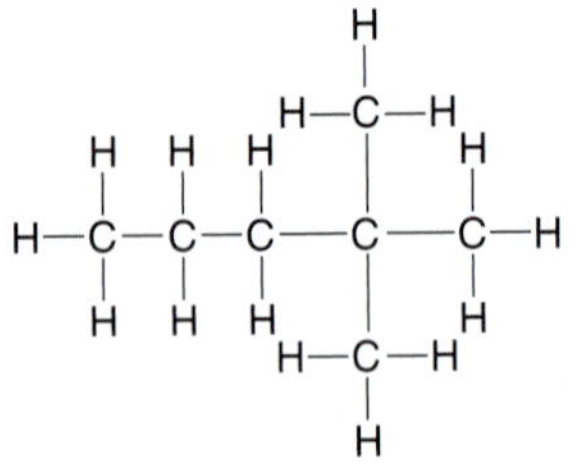

b.

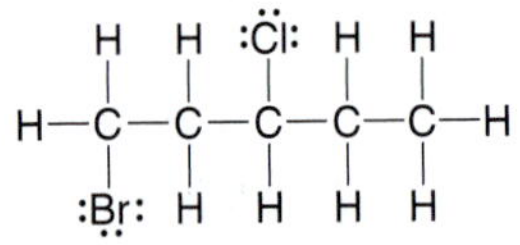

c.

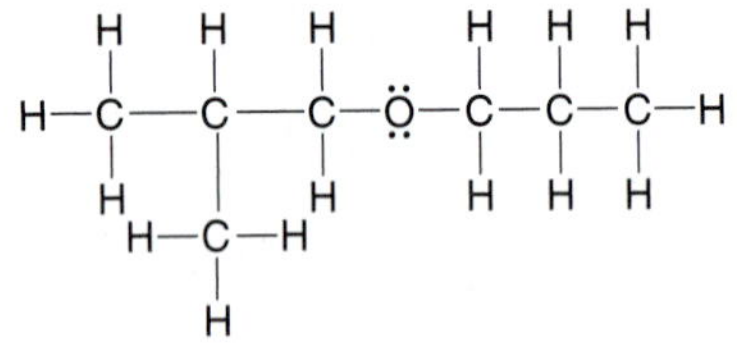

d.

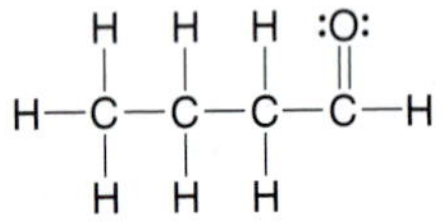

11.36 Convert each compound to a condensed structure.

a.

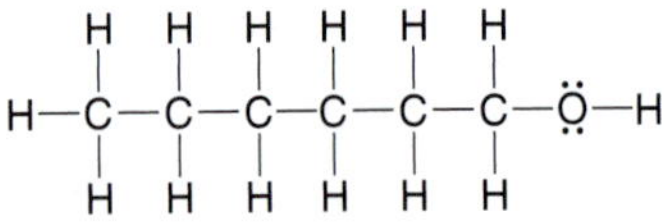

b.

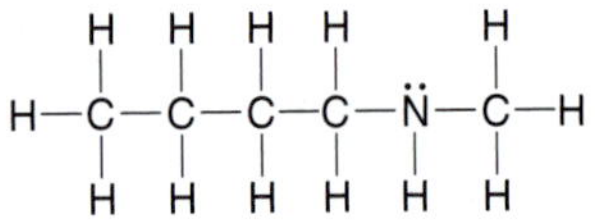

c.

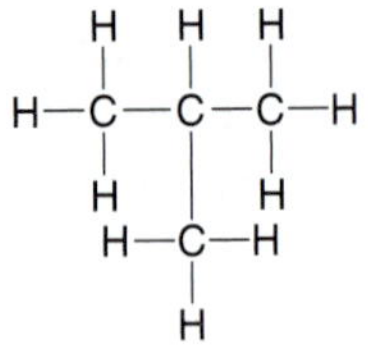

d.

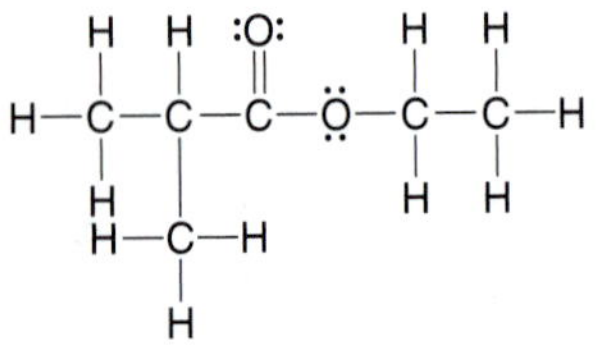

11.37 Convert each compound to a skeletal structure.

a.

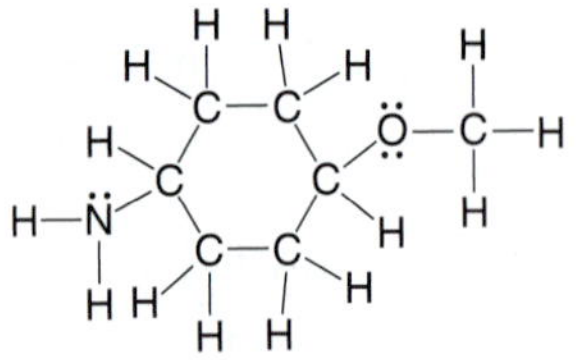

b.

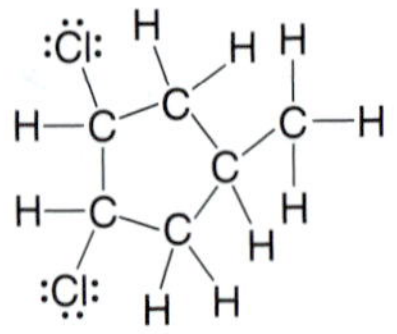

11.38 Convert each compound to a skeletal structure.

a.

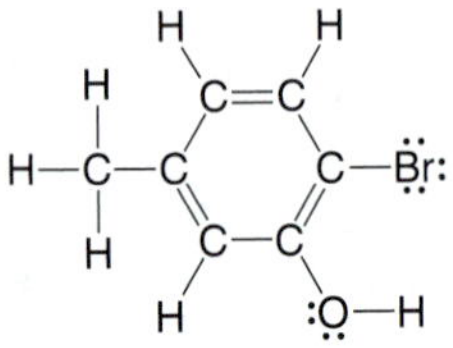

b.

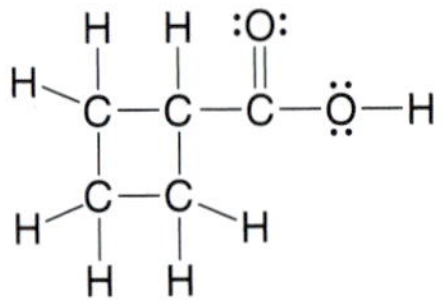

11.39 Convert each shorthand structure to a complete structure with all atoms and lone pairs drawn in.

a. $(CH_3)_2CH(CH_2)_6CH_3$

b. $(CH_3)_3COH$

c. $CH_3CO_2(CH_2)_3CH_3$

d.

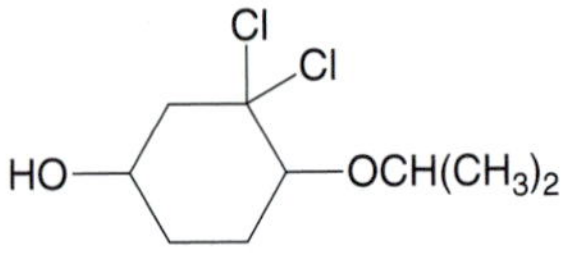

11.40 Convert each shorthand structure to a complete structure with all atoms and lone pairs drawn in.

a. $(CH_3)_2CHO(CH_2)_4CH_3$

b. $(CH_3)_3C(CH_2)_3CBr_3$

c. $(CH_3)_2CHCONH_2$

d.

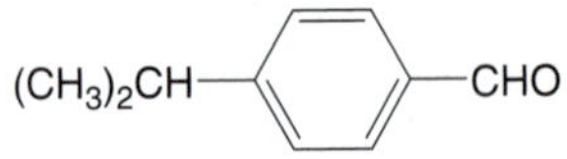

11.41 What is wrong in each of the following shorthand structures?

a. $(CH_3)_3CHCH_2CH_3$

b. $CH_3(CH_2)_4CH_3OH$

c. $(CH_3)_2C=CH_3$

d.

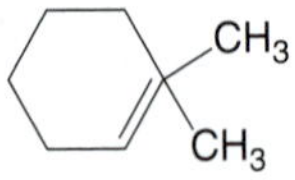

e.

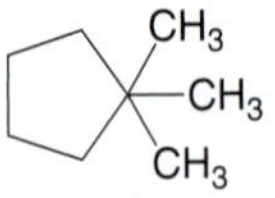

11.42 What is wrong in each of the following shorthand structures?

a. $HC\equiv CCH_2C=CH_2$

b. $CH_3C\equiv C(CH_3)_2$

c. $HC\equiv CCH_2C(CH_3)_2$

d. 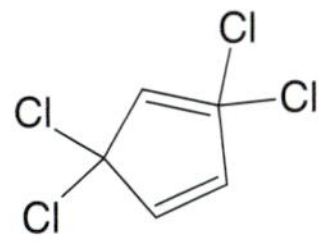 [structure]

e. 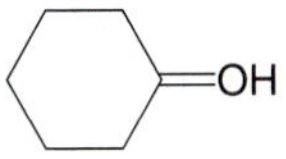 [structure]

11.43 Convert the shorthand structure of each biologically active compound to a complete structure with all atoms and lone pairs drawn in. Tyrosine is one of the 20 amino acids used for protein synthesis. Glucose is the most common simple carbohydrate.

a. 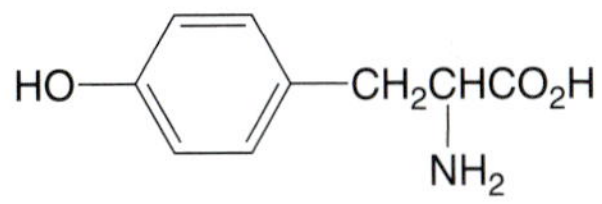 HO—[ring]—CH_2CHCO_2H with NH_2

tyrosine

b. 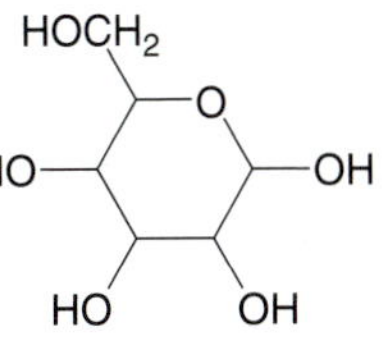 glucose structure

glucose

11.44 Testosterone is synthesized in the testes and controls the development of the secondary sex characteristics in males. Draw a complete structure of testosterone with all atoms and lone pairs.

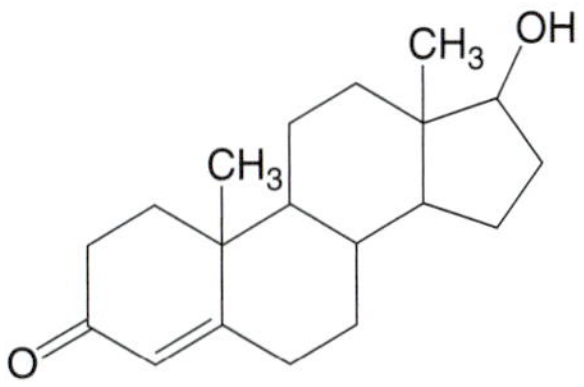

testosterone

11.45 Albuterol (trade names: Proventil and Ventolin) is a bronchodilator, a drug that widens airways, thus making it an effective treatment for individuals suffering from asthma. Draw a complete structure for albuterol with all atoms and lone pairs.

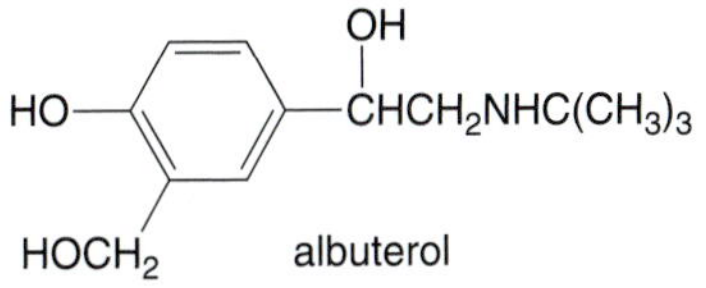

albuterol

11.46 Naproxen is the anti-inflammatory agent in Aleve and Naprosyn. Draw a complete structure for naproxen including all atoms and lone pairs.

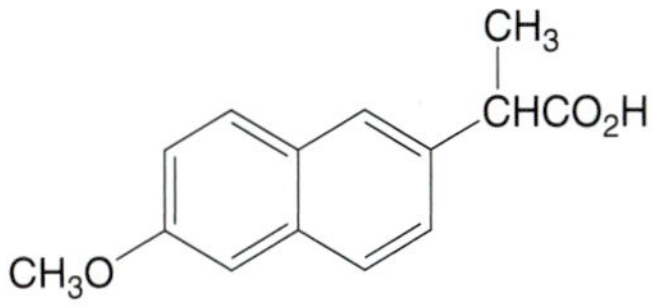

naproxen

Functional Groups

11.47 Identify the functional groups in each molecule.

a. 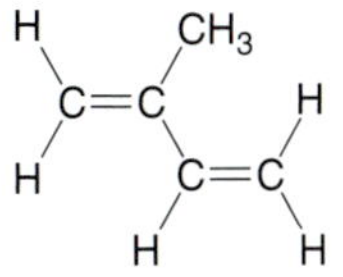 isoprene (emitted by plants)

b. 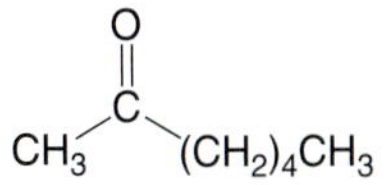 2-heptanone (from blue cheese)

c. 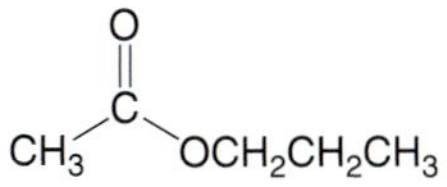 propyl acetate (from pears)

11.48 Identify the functional groups in each molecule.

a. 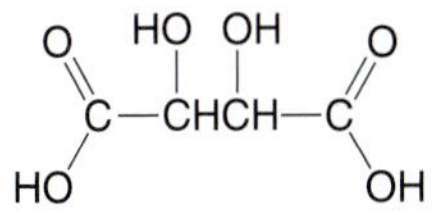 tartaric acid (from grapes)

b. 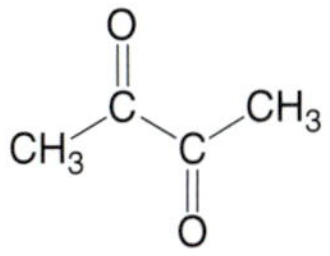 butanedione (component of butter flavor)

c. 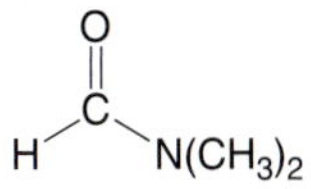 dimethylformamide (solvent)

11.49 Identify the functional groups in each molecule.

a. $HOCH_2CH_2OH$ ethylene glycol (used in antifreeze)

b. 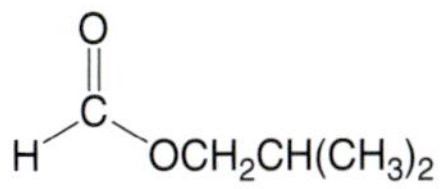 isobutyl formate (odor of raspberries)

c. 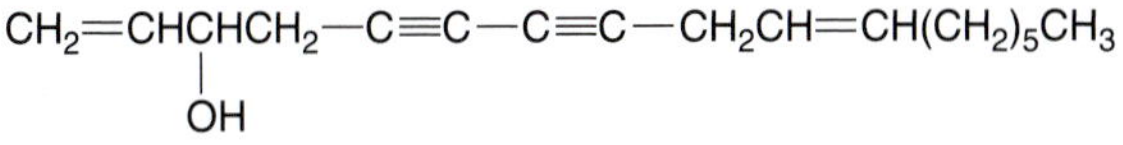 $CH_2=CHCHCH_2-C\equiv C-C\equiv C-CH_2CH=CH(CH_2)_5CH_3$ with OH — carotatoxin (neurotoxin from carrots)

d. 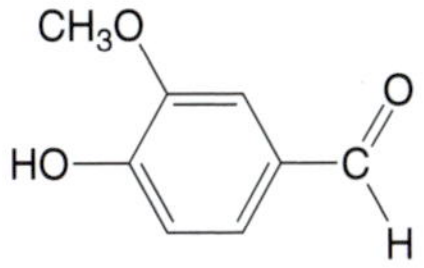 vanillin (isolated from vanilla beans)

11.50 Identify the functional groups in each drug.

a.

pseudoephedrine
(nasal decongestant)

b.

acetaminophen
(analgesic in Tylenol)

c.

ketoprofen
(analgesic)

11.51 Describe how the functional groups in an alcohol and ether differ. Give an example of an alcohol with two carbons and an ether with two carbons.

11.52 Describe how the functional groups in an amine and an amide differ. Give an example of an amine with two carbons and an amide with two carbons.

11.53 Describe how the functional groups in a carboxylic acid and an ester differ. Give an example of a carboxylic acid with three carbons and an ester with three carbons.

11.54 Describe how the functional groups in an aldehyde and a ketone differ. Give an example of an aldehyde with three carbons and a ketone with three carbons.

11.55 Draw an organic compound that fits each of the following criteria.
   a. a hydrocarbon having molecular formula C_3H_4 that contains a triple bond
   b. an alcohol containing three carbons
   c. an aldehyde containing three carbons
   d. a ketone having molecular formula C_4H_8O
   e. an ester containing a CH_3 group bonded to the ester O atom

11.56 Draw an organic compound that fits each of the following criteria.
   a. an amine containing two CH_3 groups bonded to the N atom
   b. an alkene that has the double bond in a ring
   c. an ether having two different R groups bonded to the ether oxygen
   d. an amide that has molecular formula C_3H_7NO
   e. a compound that contains an amino group bonded to a benzene ring

11.57 Draw the structure of an alkane, an alkene, and an alkyne, each having five carbon atoms. What is the molecular formula for each compound?

11.58 Some functional groups can occur as part of a ring. Although these functional groups might look quite different compared to the functional group in a molecule that has no ring, their properties and reactions are the same. Identify the functional group in each cyclic molecule.

a.

c.

b.

d.

11.59 Aspartame (trade name NutraSweet) is a widely used substitute sweetener, 180 times sweeter than table sugar. Identify the functional groups in aspartame.

aspartame

11.60 Trenbolone is a synthetic anabolic steroid used by weight lifters and body builders to increase muscle mass. Identify the functional groups in trenbolone.

trenbolone

Properties of Organic Compounds

11.61 You are given two unlabeled bottles of solids, one containing sodium chloride (NaCl) and one containing cholesterol ($C_{27}H_{46}O$, Section 11.6B). You are also given two labeled bottles of liquids, one containing water and one containing dichloromethane (CH_2Cl_2). How can you determine which solid is sodium chloride and which solid is cholesterol?

11.62 State how potassium iodide (KI) and pentane ($CH_3CH_2CH_2CH_2CH_3$) differ with regards to each of the following properties: (a) type of bonding; (b) solubility in water; (c) solubility in an organic solvent; (d) melting point; (e) boiling point.

11.63 Why do we need to know the shape of a molecule with polar bonds before deciding if the molecule is polar or nonpolar?

11.64 1,1-dichloroethylene ($CH_2=CCl_2$) is a starting material used to prepare the plastic packaging material Saran wrap. Your friend thinks that 1,1-dichloroethylene is a nonpolar molecule because the dipoles in the two C—Cl bonds cancel. Draw out the proper three-dimensional structure of 1,1-dichloroethylene, and then explain why your friend is correct or incorrect.

11.65 Indicate the polar bonds in each molecule. Label the bonds with δ^+ and δ^-.

a. $CH_3CH_2OCH_2CH_3$
   diethyl ether
   (general anesthetic)

b.
   trichlorofluoromethane
   (responsible for atmospheric ozone destruction)

c. $HOCH_2CHCH_2OH$
   glycerol
   (used in cosmetics)

11.66 Indicate the polar bonds in each molecule. Label the bonds with δ^+ and δ^-.

a.
   vinyl chloride

b.
   cyclohexanol

c.
   anisole

11.67 Classify each molecule as polar or nonpolar.

a.

b.

c. CH_2Br_2

11.68 Classify each molecule as polar or nonpolar.

a. $CH_3OCH_2CH_3$ b. $(CH_3CH_2)_2NH$ c.

11.69 Sucrose (table sugar) and 1-dodecanol [$(CH_3(CH_2)_{11}OH$] both contain 12 carbon atoms and hydroxyl groups. Explain why sucrose is water soluble, but 1-dodecanol is not water soluble.

   sucrose

11.70 Acetic acid (CH_3CO_2H) and palmitic acid [$CH_3(CH_2)_{14}CO_2H$] are both carboxylic acids. Acetic acid is the water-soluble carboxylic acid in vinegar, while palmitic acid is a water-insoluble fatty acid. Give a reason for this solubility difference.

11.71 Spermaceti wax [$CH_3(CH_2)_{14}CO_2(CH_2)_{15}CH_3$], a compound isolated from sperm whales, was once a common ingredient in cosmetics. Its use is now banned to help protect whales. (a) Identify the functional group in spermaceti wax. (b) Predict its solubility properties in water and organic solvents and explain your reasoning.

11.72 Beeswax has the structure $CH_3(CH_2)_{14}CO_2(CH_2)_{29}CH_3$. Consider its solubility in three different solvents—water, hexane [$CH_3(CH_2)_4CH_3$], and ethanol (CH_3CH_2OH). In which solvent is beeswax most soluble? In which solvent is beeswax least soluble? Explain your choices.

Vitamins

11.73 Explain why regularly taking a large excess of a fat-soluble vitamin can lead to severe health problems, whereas taking a large excess of a water-soluble vitamin often causes no major health problems.

11.74 You can obtain the minimum daily requirement of vitamin E by eating leafy greens or by taking vitamin E tablets. As a source of vitamin E, which of these methods is better for an individual?

11.75 Vitamin E is a natural antioxidant found in fish oil, almonds, hazelnuts, and leafy greens. Predict the solubility of vitamin E in water and in organic solvents.

   vitamin E

11.76 Vitamin B_6 is obtained by eating a diet that includes meat, fish, wheat germ, and bananas. A deficiency of vitamin B_6 can cause convulsions and anemia. Predict the solubility of vitamin B_6 in water and in organic solvents.

   vitamin B_6
   pyridoxine

General Questions

11.77 Can a hydrocarbon have each of the following molecular formulas? Explain why or why not in each case. (a) C_3H_8; (b) C_3H_9; (c) C_3H_6.

11.78 Can an oxygen-containing organic compound have molecular formula C_2H_6O? Explain your reasoning.

11.79   (a) Draw a condensed structure for an amine having molecular formula $C_4H_{11}N$. (b) Label all polar bonds using the symbols δ^+ and δ^-. (c) Determine the geometry around the N atom. (d) Is the compound you drew polar or nonpolar?

11.80   (a) Draw a condensed structure for an alcohol having molecular formula $C_4H_{10}O$. (b) Label all polar bonds using the symbols δ^+ and δ^-. (c) Determine the geometry around the O atom. (d) Is the compound you drew polar or nonpolar?

11.81   Answer the following questions about aldosterone, a compound that helps to control the absorption of Na^+ and Cl^- ions in the kidneys, and as a result, affects water retention.

aldosterone

a. Identify the functional groups.
b. Draw in all lone pairs on O atoms.
c. How many C's does aldosterone contain?
d. How many H's are present at each C labeled with an asterisk (*)?
e. Give the shape around each atom indicated with an arrow.
f. Label all polar bonds.

11.82   Answer the following questions about levonorgestrel (trade name: Plan B). Levonorgestrel interferes with

ovulation, the release of an egg from an ovary, so it prevents pregnancy if taken within a few days of unprotected sex.

levonorgestrel

a. Identify the functional groups.
b. How many C's does levonorgestrel contain?
c. How many H's are present at each C labeled with an asterisk (*)?
d. Give the shape around each atom indicated with an arrow.
e. Label all polar bonds.
f. Is levonorgestrel soluble in an organic solvent?
g. Is levonorgestrel soluble in water?

Applications

11.83   Cabbage leaves are coated with a hydrocarbon of molecular formula $C_{29}H_{60}$. What purpose might this hydrocarbon coating serve?

11.84   Skin moisturizers come in two types. (a) One type of moisturizer is composed mainly of hydrocarbon material. Suggest a reason as to how a moisturizer of this sort helps to keep the skin from drying out. (b) A second type of moisturizer is composed mainly of propylene glycol [$CH_3CH(OH)CH_2OH$]. Suggest a reason as to how a moisturizer of this sort helps to keep the skin from drying out.

CHALLENGE QUESTIONS

11.85   THC is the active component in marijuana (Section 11.5), and ethanol (CH_3CH_2OH) is the alcohol in alcoholic beverages. Using the solubility principles in Section 11.6, predict the water solubility of each of these compounds. Then, explain why drug screenings are able to detect the presence of THC but not ethanol weeks after these substances have been introduced into the body.

11.86   Cocaine is a widely abused, addicting drug. Cocaine is usually obtained as its hydrochloride salt (cocaine hydrochloride, an ionic salt). This salt can be converted to crack (a neutral molecule) by treatment with base.

cocaine (crack)
neutral organic molecule

cocaine hydrochloride
a salt

a. Identify the functional groups in cocaine.
b. Given what you have learned about ionic and covalent bonding, which of the two compounds—crack or cocaine hydrochloride—has a higher boiling point?
c. Which compound is more soluble in water?
d. Can you use the relative solubility to explain why crack is usually smoked but cocaine hydrochloride is injected directly into the bloodstream?

12

CHAPTER GOALS

In this chapter you will learn how to:

1. Identify and draw acyclic alkanes and cycloalkanes
2. Identify constitutional isomers
3. Name alkanes using the IUPAC system of nomenclature
4. Predict the physical properties of alkanes
5. Write equations for the complete and incomplete combustion of alkanes

Petroleum is a complex mixture of compounds that includes hydrocarbons such as hexane and decane, two members of the family of organic compounds called **alkanes.**

ALKANES

IN Chapter 12, we apply the principles of bonding, shape, and physical properties discussed in Chapter 11 to our first family of organic compounds, the **alkanes.** Since alkanes have no functional group, their chemical reactions are limited, and much of Chapter 12 is devoted to learning how to name and draw them. Alkanes do, however, undergo one important reaction, combustion, a process that powers automobiles and heats homes. The combustion of alkanes has also increased the carbon dioxide concentration in the atmosphere over the last 50 years, which may have long-term and far-reaching environmental consequences.

12.1 INTRODUCTION

Alkanes **are hydrocarbons having only C—C and C—H single bonds.** The carbons of an alkane can be joined together to form chains or rings of atoms.

The prefix *a-* means *not,* so an acyclic alkane is *not* cyclic.

- **Alkanes that contain chains of carbon atoms but no rings are called** *acyclic alkanes.* An acyclic alkane has the molecular formula $C_nH_{2n + 2}$, where *n* is the number of carbons it contains. Acyclic alkanes are also called **saturated hydrocarbons** because they have the maximum number of hydrogen atoms per carbon.
- *Cycloalkanes* **contain carbons joined in one or more rings.** Since a cycloalkane has two fewer H's than an acyclic alkane with the same number of carbons, its general formula is C_nH_{2n}.

Undecane and cyclohexane are examples of two naturally occurring alkanes. Undecane is an acyclic alkane with molecular formula $C_{11}H_{24}$. Undecane is a *pheromone,* **a chemical substance used for communication** in a specific animal species, most commonly an insect population. Secretion of undecane by a cockroach causes other members of the species to aggregate. Cyclohexane, a cycloalkane with molecular formula C_6H_{12}, is one component of the mango, the most widely consumed fruit in the world.

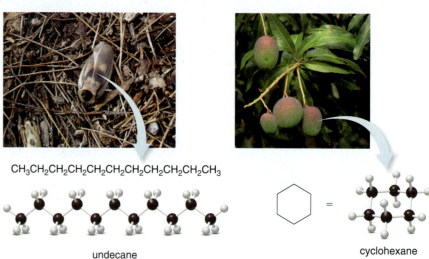

$CH_3CH_2CH_2CH_2CH_2CH_2CH_2CH_2CH_2CH_2CH_3$

undecane   cyclohexane

PROBLEM 12.1

How many hydrogen atoms are present in each compound?

a. an acyclic alkane with three carbons   c. a cycloalkane with nine carbons

b. a cycloalkane with four carbons   d. an acyclic alkane with seven carbons

PROBLEM 12.2

Which formulas represent acyclic alkanes and which represent cycloalkanes?

a. C_5H_{12} b. C_4H_8 c. $C_{12}H_{24}$ d. $C_{10}H_{22}$

12.2 SIMPLE ALKANES

We begin our study of alkanes with acyclic alkanes, and then move on to cycloalkanes in Section 12.5.

12.2A ACYCLIC ALKANES HAVING FEWER THAN FIVE CARBONS

The structures for the two simplest acyclic alkanes were given in Chapter 11.

- **Methane, CH_4,** has a single carbon atom surrounded by four hydrogens to give it four bonds.
- **Ethane, CH_3CH_3,** has two carbon atoms joined together by a single bond. Each carbon is also bonded to three hydrogens to give it four bonds total.

To draw the structure of an alkane, join the carbon atoms together with single bonds, and add enough H's to give each C four bonds.

Since each carbon in an alkane is surrounded by four atoms, each carbon is **tetrahedral,** and all bond angles are 109.5°.

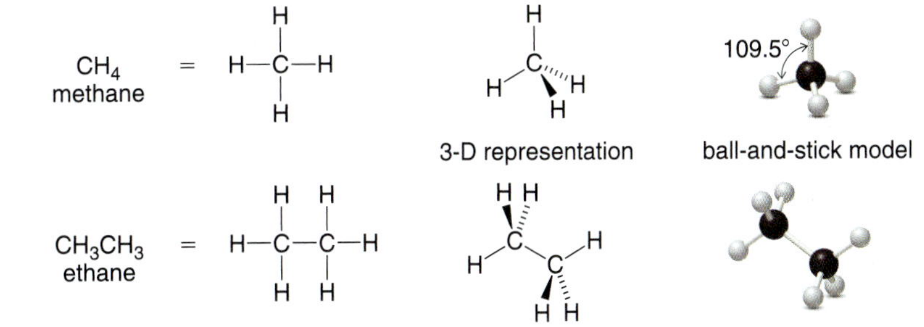

3-D representation   ball-and-stick model

CH_4
methane

CH_3CH_3
ethane

To draw a three-carbon alkane, draw three carbons joined together with single bonds and add enough hydrogens to give each carbon four bonds. This forms **propane, $CH_3CH_2CH_3$.**

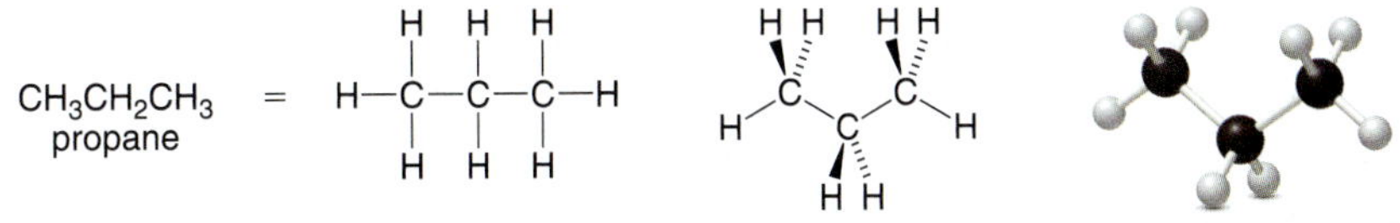

$CH_3CH_2CH_3$
propane

3-D representation   ball-and-stick model

The carbon skeleton in propane and other alkanes can be drawn in a variety of different ways and still represent the same molecule. For example, the three carbons of propane can be drawn in a horizontal row or with a bend. *These representations are equivalent.* **If you follow the carbon chain from one end to the other, you move across the *same* three carbon atoms in both representations.**

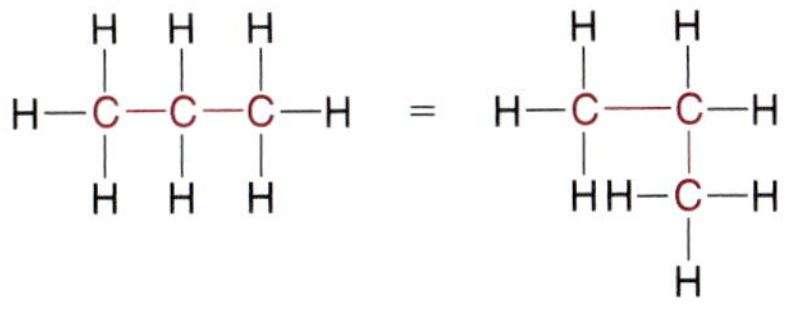

3 C's in a row   3 C's with a bend

- **The bends in a carbon chain don't matter when it comes to identifying different compounds.**

There are two different ways to arrange four carbons, giving two compounds with molecular formula C_4H_{10}.

- **Butane, $CH_3CH_2CH_2CH_3$,** has four carbon atoms in a row. Butane is a **straight-chain alkane,** an alkane that has all of its carbons in one continuous chain.
- **Isobutane, $(CH_3)_3CH$,** has three carbon atoms in a row and one carbon bonded to the middle carbon. Isobutane is a **branched-chain alkane,** an alkane that contains one or more carbon branches bonded to a carbon chain.

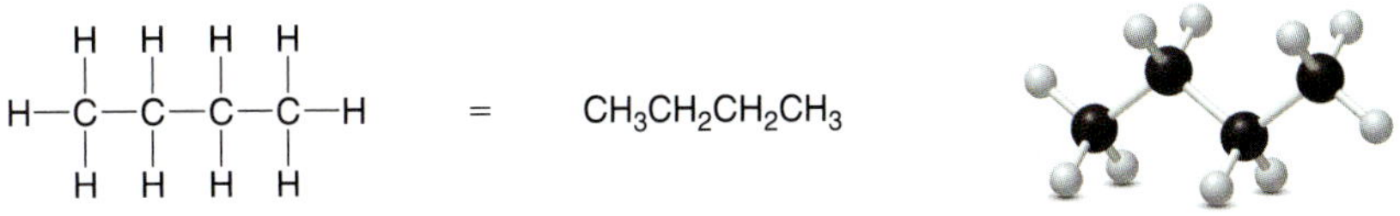

butane   4 C's in a row   **straight-chain alkane**

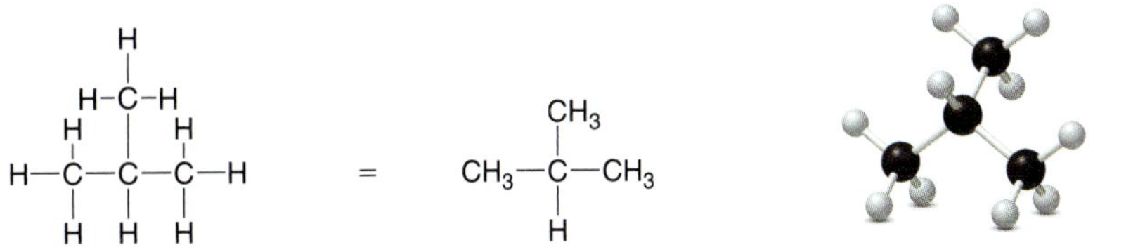

isobutane   3 C's with a one-carbon branch   **branched-chain alkane**

Butane and isobutane are *isomers,* **two different compounds with the same molecular formula.** They belong to one of the two major classes of isomers called **constitutional isomers.**

- *Constitutional isomers* differ in the way the atoms are connected to each other.

Constitutional isomers like butane and isobutane belong to the same family of compounds: they are both **alkanes.** This is not always the case. For example, there are two different arrangements of atoms for a compound of molecular formula C_2H_6O.

$$H-\underset{\underset{H}{|}}{\overset{\overset{H}{|}}{C}}-\underset{\underset{H}{|}}{\overset{\overset{H}{|}}{C}}-O-H \qquad\qquad H-\underset{\underset{H}{|}}{\overset{\overset{H}{|}}{C}}-O-\underset{\underset{H}{|}}{\overset{\overset{H}{|}}{C}}-H$$

$$CH_3CH_2OH \qquad\qquad CH_3OCH_3$$

ethanol   dimethyl ether

Ethanol (CH_3CH_2OH) and dimethyl ether (CH_3OCH_3) are constitutional isomers with different functional groups: CH_3CH_2OH is an **alcohol** and CH_3OCH_3 is an **ether.**

SAMPLE PROBLEM 12.1

Are the compounds in each pair constitutional isomers or are they not isomers of each other?

a. $CH_3CH_2CH_3$ and $\underset{H}{\overset{H}{C}}=\underset{CH_3}{\overset{H}{C}}$

b. $CH_3CH_2CH_2NH_2$ and $CH_3-\underset{\underset{CH_3}{|}}{N}-CH_3$

ANALYSIS

First compare molecular formulas; two compounds are isomers only if they have the same molecular formula. Then, check how the atoms are connected to each other. Constitutional isomers have atoms bonded to different atoms.

SOLUTION

a. $CH_3CH_2CH_3$ and $\underset{H}{\overset{H}{C}}=\underset{CH_3}{\overset{H}{C}}$

molecular formula C_3H_8 molecular formula C_3H_6

The two compounds have the same number of C's but a different number of H's, so they have different molecular formulas. Thus, they are *not* isomers of each other.

b. $CH_3CH_2CH_2NH_2$ and $CH_3-\underset{\underset{CH_3}{|}}{N}-CH_3$

Both compounds have molecular formula C_3H_9N. Since one compound has C–C bonds and the other does not, the atoms are connected differently. These compounds are *constitutional isomers.*

PROBLEM 12.3

Are the compounds in each pair isomers or are they not isomers of each other?

a. $CH_3CH_2CH_2CH_3$ and $CH_3CH_2CH_3$

c. [hexagon] and [pentagon]—CH_3

b. $CH_3CH_2CH_2OH$ and $CH_3OCH_2CH_3$

PROBLEM 12.4

Label each representation as butane or isobutane.

a. $H-\underset{\underset{H}{|}}{\overset{\overset{H}{|}}{C}}-\underset{\underset{H}{|}}{\overset{\overset{H}{|}}{C}}-\underset{\underset{CH_3}{|}}{\overset{\overset{H}{|}}{C}}-H$

b. $HC(CH_3)_3$

c. $H-\underset{\underset{H-\underset{\underset{H}{|}}{\overset{\overset{|}{C}}{}}-H}{}}{\overset{\overset{H}{|}}{C}}-\underset{\underset{H-\underset{\underset{H}{|}}{\overset{\overset{|}{C}}{}}-H}{}}{\overset{\overset{H}{|}}{C}}-H$

d. $H-\underset{\underset{H}{|}}{\overset{\overset{CH_3}{|}}{C}}-\underset{\underset{H}{|}}{\overset{\overset{CH_3}{|}}{C}}-H$

12.2B   ACYCLIC ALKANES HAVING FIVE OR MORE CARBONS

As the number of carbon atoms in an alkane increases, so does the number of isomers. There are three constitutional isomers for the five-carbon alkane, each having molecular formula C_5H_{12}: **pentane**, **isopentane** (or 2-methylbutane), and **neopentane** (or 2,2-dimethylpropane).

$CH_3CH_2CH_2CH_2CH_3$

pentane

$CH_3—\overset{\displaystyle CH_3}{\underset{\displaystyle H}{C}}—CH_2CH_3$

isopentane
(2-methylbutane)

$CH_3—\overset{\displaystyle CH_3}{\underset{\displaystyle CH_3}{C}}—CH_3$

neopentane
(2,2-dimethylpropane)

With alkanes having five or more carbons, the names of the straight-chain isomers are derived from Greek roots: **pent**ane for **five** carbons, **hex**ane for **six**, and so on. Table 12.1 lists the names and structures for the straight-chain alkanes having up to 10 carbons. The suffix *-ane* identifies a molecule as an alk*ane*. The remainder of the name—meth-, eth-, prop-, and so forth—indicates the number of carbons in the long chain.

> The suffix *-ane* identifies a molecule as an alk*ane*.

TABLE 12.1   Straight-Chain Alkanes

Number of C's	Molecular Formula	Structure	Name
1	CH_4	CH_4	*meth*ane
2	C_2H_6	CH_3CH_3	*eth*ane
3	C_3H_8	$CH_3CH_2CH_3$	*prop*ane
4	C_4H_{10}	$CH_3CH_2CH_2CH_3$	*but*ane
5	C_5H_{12}	$CH_3CH_2CH_2CH_2CH_3$	*pent*ane
6	C_6H_{14}	$CH_3CH_2CH_2CH_2CH_2CH_3$	*hex*ane
7	C_7H_{16}	$CH_3CH_2CH_2CH_2CH_2CH_2CH_3$	*hept*ane
8	C_8H_{18}	$CH_3CH_2CH_2CH_2CH_2CH_2CH_2CH_3$	*oct*ane
9	C_9H_{20}	$CH_3CH_2CH_2CH_2CH_2CH_2CH_2CH_2CH_3$	*non*ane
10	$C_{10}H_{22}$	$CH_3CH_2CH_2CH_2CH_2CH_2CH_2CH_2CH_2CH_3$	*dec*ane

---

SAMPLE PROBLEM 12.2   Draw two isomers with molecular formula C_6H_{14} that have five carbon atoms in the longest chain and a one-carbon branch coming off the chain.

ANALYSIS   Since isomers are different compounds with the same molecular formula, we must add a one-carbon branch to two different carbons to form two different products. Then add enough H's to give each C four bonds.

SOLUTION

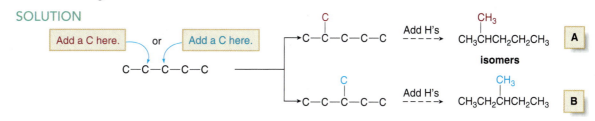

Compounds **A** and **B** are isomers because the CH_3 group is bonded to different atoms in the five-carbon chain. Note, too, that we cannot add the one-carbon branch to an *end* carbon because that creates a continuous six-carbon chain. Remember that **bends in the chain don't matter.**

Adding a C to an end C makes a six-carbon chain.

$$C-C-C-C-C \dashrightarrow C-C-C-C-\overset{\overset{C}{|}}{C} = C-C-C-C-C-C$$

equivalent structures

PROBLEM 12.5

Draw two isomers with molecular formula C_6H_{14} that have four carbon atoms in the longest chain and two one-carbon branches coming off the chain.

PROBLEM 12.6

Which of the following is *not* another representation for isopentane?

a. $CH_3CH_2-\overset{\overset{H}{|}}{\underset{\underset{CH_3}{|}}{C}}-CH_3$
b. $H-\overset{\overset{H}{|}}{\underset{\underset{H}{|}}{C}}-CH_3$ with $CH_3-\overset{|}{C}-CH_3$
c. $CH_3CH_2CH(CH_3)_2$
d. $H-\overset{\overset{H}{|}}{\underset{\underset{CH_3}{|}}{C}}-\overset{\overset{H}{|}}{\underset{\underset{H}{|}}{C}}-H$ with CH_3

12.2C CLASSIFYING CARBON ATOMS

Carbon atoms in alkanes and other organic compounds are classified by the number of other carbons directly bonded to them.

- A *primary carbon* (1° C) is bonded to *one* other C.
- A *secondary carbon* (2° C) is bonded to *two* other C's.
- A *tertiary carbon* (3° C) is bonded to *three* other C's.
- A *quaternary carbon* (4° C) is bonded to *four* other C's.

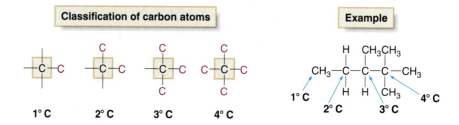

Classification of carbon atoms — 1° C, 2° C, 3° C, 4° C. **Example**

SAMPLE PROBLEM 12.3

Classify each carbon atom in the following molecule as 1°, 2°, 3°, or 4°.

$$CH_3-\overset{\overset{CH_3}{|}}{\underset{\underset{CH_3}{|}}{C}}-CH_2CH_2CH_3$$

ANALYSIS   To classify a carbon atom, count the number of C's bonded to it. Draw a complete structure with all bonds and atoms to clarify the structure if necessary.

SOLUTION

$$CH_3-\overset{\overset{CH_3}{|}}{\underset{\underset{CH_3}{|}}{C}}-CH_2CH_2CH_3 = CH_3-\overset{\overset{CH_3\ H}{|\ |}}{\underset{\underset{CH_3\ H}{|\ |}}{C}}-\overset{H}{\underset{H}{C}}-\overset{H}{\underset{H}{C}}-CH_3$$

2° C, 4° C. All C's labeled in red are 1° C's.

PROBLEM 12.7

Classify the carbon atoms in each compound as 1°, 2°, 3°, or 4°.

a. $CH_3CH_2CH_2CH_3$ b. $(CH_3)_3CH$ c. 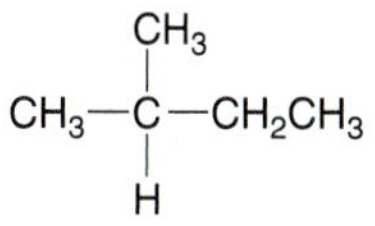 $CH_3-\overset{\overset{\displaystyle CH_3}{|}}{\underset{\underset{\displaystyle H}{|}}{C}}-CH_2CH_3$ d. 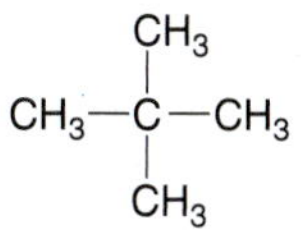 $CH_3-\overset{\overset{\displaystyle CH_3}{|}}{\underset{\underset{\displaystyle CH_3}{|}}{C}}-CH_3$

PROBLEM 12.8

Draw a condensed structure for an alkane that fits each description.

a. an alkane that contains only 1° and 2° carbons
b. an alkane of molecular formula C_5H_{12} that has one 3° carbon
c. an alkane with two 4° carbons

12.2D   BOND ROTATION AND SKELETAL STRUCTURES FOR ACYCLIC ALKANES

A ball-and-stick model of ethane with two labeled hydrogens reveals an important feature. **Rotation can occur around carbon–carbon single bonds.** This allows the hydrogens on one CH_3 group to adopt different orientations relative to the hydrogens on the other CH_3 group. This is a characteristic of *all* carbon–carbon single bonds.

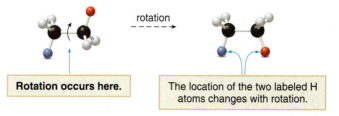

Rotation occurs here.   The location of the two labeled H atoms changes with rotation.

In a similar fashion, each carbon–carbon bond in butane ($CH_3CH_2CH_2CH_3$) undergoes rotation, a process that is rapid at room temperature. As a result, the carbon backbone of butane can be arranged in a variety of ways. The most stable arrangement avoids crowding between nearby carbons, so that the carbon skeleton is extended in a zigzag fashion.

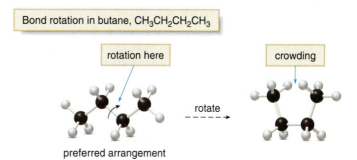

Bond rotation in butane, $CH_3CH_2CH_2CH_3$

rotation here   crowding

rotate

preferred arrangement

Sometimes zigzag skeletal structures are used to represent these extended carbon skeletons. Recall from Section 11.4 that a skeletal structure has a carbon atom at the intersection of any two lines and at the end of each line. Each carbon has enough hydrogens to give it four bonds.

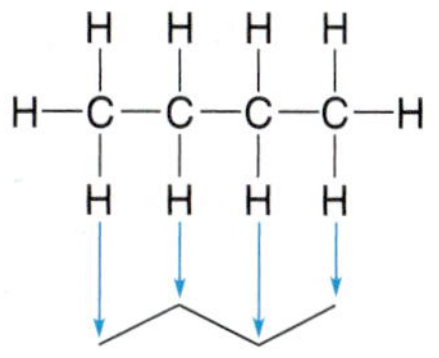

skeletal structure

PROBLEM 12.9	Draw a skeletal structure for each compound: (a) $CH_3CH_2CH_2CH_2CH_2CH_3$ (hexane); (b) $CH_3(CH_2)_5CH_3$ (heptane).
PROBLEM 12.10	Convert each skeletal structure to a complete structure with all atoms and bond lines.

a.   b.   c.

12.3 AN INTRODUCTION TO NOMENCLATURE

12.3A THE IUPAC SYSTEM OF NOMENCLATURE

How are organic compounds named? Long ago, the name of a compound was often based on the plant or animal source from which it was obtained. For example, the name allicin, the principal component of the odor of garlic, is derived from the botanical name for garlic, *Allium sativum.* Other compounds were named by their discoverer for more personal reasons. Adolf von Baeyer supposedly named barbituric acid after a woman named Barbara, although no one knows Barbara's identity—a lover, a Munich waitress, or even St. Barbara.

Garlic has been used in Chinese herbal medicine for over 4,000 years. Today it is sold as a dietary supplement because of its reported health benefits. **Allicin,** the molecule responsible for garlic's odor, is not stored in the garlic bulb, but rather is produced by the action of enzymes when the bulb is crushed or bruised.

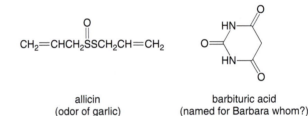

allicin
(odor of garlic)

barbituric acid
(named for Barbara whom?)

With the isolation and preparation of thousands of new organic compounds it became obvious that each organic compound must have an unambiguous name. A systematic method of naming compounds (a system of **nomenclature**) was developed by the *I*nternational *U*nion of *P*ure and *A*pplied *C*hemistry. It is referred to as the **IUPAC** system of nomenclature, and how it can be used to name alkanes is explained in Section 12.4.

12.3B FOCUS ON HEALTH & MEDICINE
NAMING NEW DRUGS

Naming organic compounds has become big business for drug companies. The IUPAC name of an organic compound can be long and complex. As a result, most drugs have three names:

- **Systematic:** The systematic name follows the accepted rules of nomenclature; this is the IUPAC name.
- **Generic:** The generic name is the official, internationally approved name for the drug.
- **Trade:** The trade name for a drug is assigned by the company that manufactures it. Trade names are often "catchy" and easy to remember. Companies hope that the public will continue to purchase a drug with an easily recalled trade name long after a cheaper generic version becomes available.

Consider the world of over-the-counter anti-inflammatory agents. The compound a chemist calls 2-[4-(2-methylpropyl)phenyl]propanoic acid has the generic name ibuprofen. It is marketed under a variety of trade names, including Motrin and Advil.

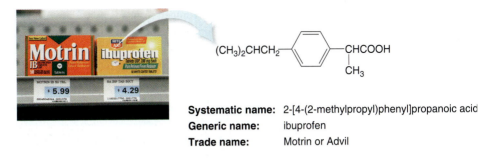

$(CH_3)_2CHCH_2$—⬡—$CHCOOH$
                          |
                          CH_3

Systematic name: 2-[4-(2-methylpropyl)phenyl]propanoic acid
Generic name: ibuprofen
Trade name: Motrin or Advil

12.4 ALKANE NOMENCLATURE

Although the names of the straight-chain alkanes having 10 carbons or fewer were already given in Table 12.1, we must also learn how to name alkanes that have carbon branches, called **substituents,** bonded to a long chain. The names of these organic molecules have three parts.

- The **parent name** indicates the number of carbons in the longest continuous carbon chain in the molecule.
- The **suffix** indicates what functional group is present.
- The **prefix** tells us the identity, location, and number of substituents attached to the carbon chain.

| Prefix | + | Parent | + | Suffix |

What is the longest carbon chain?

What and where
are the substituents?

What is the functional group?

The names of the straight-chain alkanes in Table 12.1 consist of two parts. The suffix *-ane* indicates that the compounds are alkanes. The remainder of the name is the parent name, which indicates the number of carbon atoms in the longest carbon chain. The parent name for **one carbon is *meth-*,** for **two carbons is *eth-*,** and so on. Thus, we are already familiar with two parts of the name of an organic compound.

To determine the third part of a name, the prefix, we must learn how to name the substituents that are bonded to the longest carbon chain.

12.4A NAMING SUBSTITUENTS

Carbon substituents bonded to a long carbon chain are called *alkyl groups.*

- An *alkyl group* is formed by removing one hydrogen from an alkane.

An alkyl group is a part of a molecule that is now able to bond to another atom or a functional group. **To name an alkyl group, change the *-ane* ending of the parent alkane to *-yl.*** Thus, methane (CH_4) becomes methyl (CH_3–) and ethane (CH_3CH_3) becomes ethyl (CH_3CH_2–).

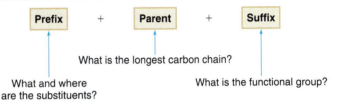

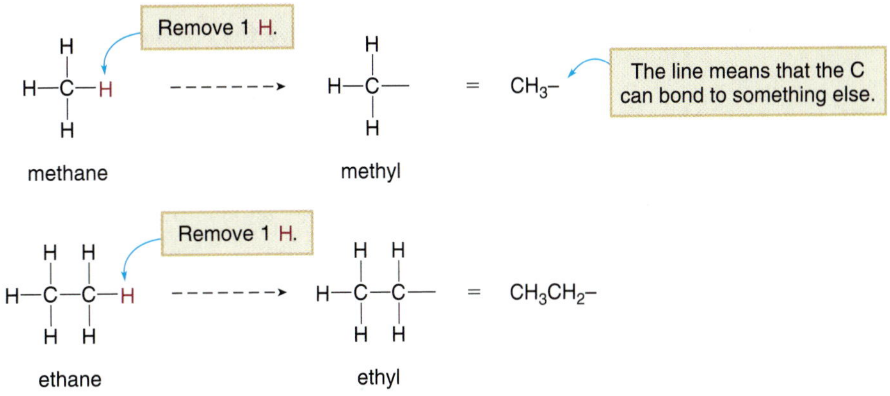

Remove 1 H.

methane   methyl

The line means that the C
can bond to something else.

Remove 1 H.

ethane   ethyl

Removing one hydrogen from an *end* carbon in any straight-chain alkane forms other alkyl groups named in a similar fashion. Thus, propane ($CH_3CH_2CH_3$) becomes propyl ($CH_3CH_2CH_2-$) and butane ($CH_3CH_2CH_2CH_3$) becomes butyl ($CH_3CH_2CH_2CH_2-$). The names of alkyl groups having six carbons or fewer are summarized in Table 12.2.

TABLE 12.2   Some Common Alkyl Groups

Number of C's	Structure	Name
1	CH_3-	methyl
2	CH_3CH_2-	ethyl
3	$CH_3CH_2CH_2-$	propyl
4	$CH_3CH_2CH_2CH_2-$	butyl
5	$CH_3CH_2CH_2CH_2CH_2-$	pentyl
6	$CH_3CH_2CH_2CH_2CH_2CH_2-$	hexyl

12.4B   NAMING AN ACYCLIC ALKANE

Four steps are needed to name an alkane.

HOW TO   Name an Alkane Using the IUPAC System

Step [1]   Find the parent carbon chain and add the suffix.

- **Find the longest continuous carbon chain,** and name the molecule using the parent name for that number of carbons, given in Table 12.1. To the name of the parent, add the suffix *-ane* for an alkane. Each functional group has its own suffix.

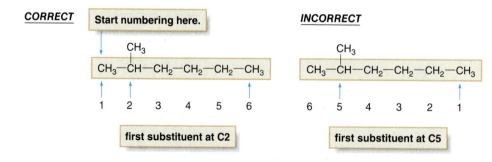

- The longest chain may not be written horizontally across the page. Remember that **it does not matter if the chain is *straight* or has *bends*.** All of the following representations are equivalent.

Step [2]   Number the atoms in the carbon chain to give the first substituent the lower number.

Step [3]   **Name and number the substituents.**

- Name the substituents as alkyl groups, and use the numbers from step [2] to designate their location.

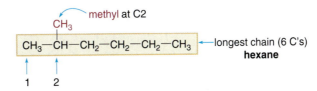

methyl at C2

CH₃—CH—CH₂—CH₂—CH₂—CH₃   ← longest chain (6 C's)
hexane

1   2

- Every carbon belongs to *either* the longest chain or a substituent, but *not both.*
- Each substituent needs its *own* number.
- If two or more identical substituents are bonded to the longest chain, use **prefixes** to indicate how many: **di-** for two groups, **tri-** for three groups, **tetra-** for four groups, and so forth. The following compound has two methyl groups so its name contains the prefix di- before methyl → *dimethyl.*

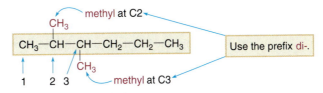

Step [4]   **Combine substituent names and numbers + parent + suffix.**

- Precede the name of the parent by the names of the substituents. **Alphabetize** the names of the substituents, ignoring any prefixes like *di-*. For example, triethyl precedes dimethyl because the *e* of **ethyl** comes before the *m* of **methyl** in the alphabet.
- Precede the name of each substituent by the number that indicates its location. There must be **one number for each substituent.**
- Separate numbers by commas and separate numbers from letters by dashes.

To help identify which carbons belong to the longest chain and which are substituents, always box in the atoms of the long chain. Every other carbon atom then becomes a substituent that needs its own name as an alkyl group.

methyl at C2

CH₃—CH—CH₂—CH₂—CH₂—CH₃

1   2   longest chain (6 C's)
hexane

Answer: 2-methylhexane

methyl at C2

CH₃—CH—CH—CH₂—CH₂—CH₃   ← **hexane**

CH₃

1   2 3   methyl at C3

- Two numbers are needed, one for each methyl group.

Answer: 2,3-dimethylhexane

SAMPLE PROBLEM 12.4   Give the IUPAC name for the following compound.

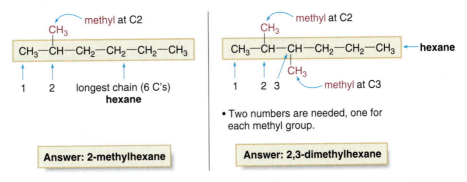

CH₃   CH₃CH₂ H

CH₃—C—CH₂CH₂—C—C—CH₂CH₃

H   H   CH₃

ANALYSIS AND SOLUTION

[1]   **Name the parent and use the suffix -ane since the molecule is an alkane.**

CH₃   CH₃CH₂ H

CH₃—C—CH₂CH₂—C—C—CH₂CH₃

H   H   CH₃

8 C's in the longest chain   - - -→   **octane**

- Box in the atoms of the longest chain to clearly show which carbons are part of the longest chain and which carbons are substituents.

[2]   **Number the chain to give the first substituent the lower number.**

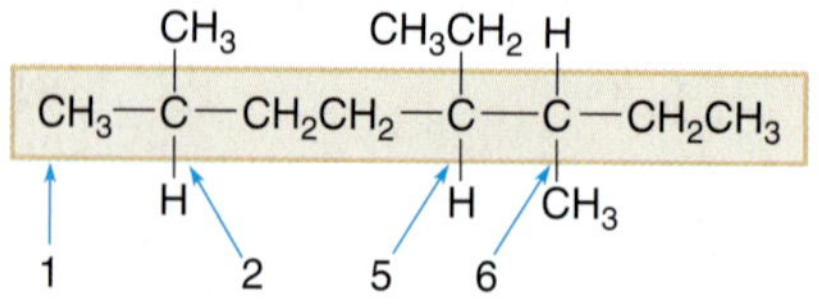

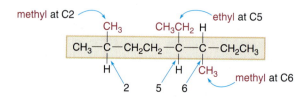

- Numbering from left to right puts the first substituent at C2.

[3]   **Name and number the substituents.**

- This compound has three substituents: two methyl groups at C2 and C6 and an ethyl group at C5.

[4]   **Combine the parts.**

- Write the name as one word and use the prefix di- before methyl since there are two methyl groups.
- Alphabetize the *e* of **e**thyl before the *m* of **m**ethyl. The prefix di- is ignored when alphabetizing.

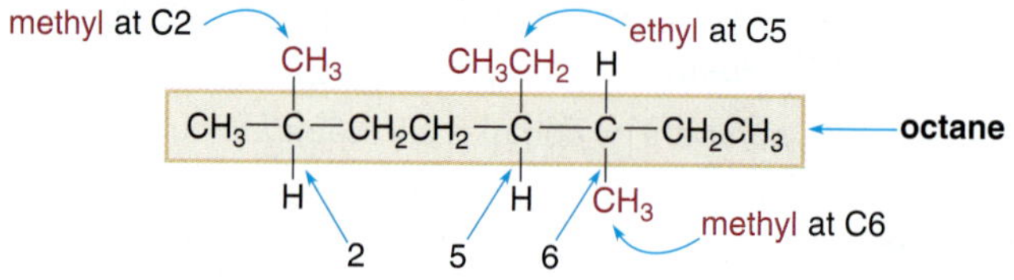

Answer: 5-ethyl-2,6-dimethyloctane

PROBLEM 12.11

Give the IUPAC name for each compound.

a. 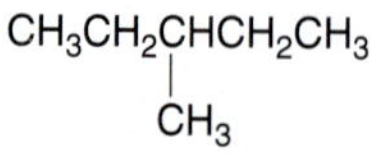 CH₃CH₂CHCH₂CH₃
              |
             CH₃

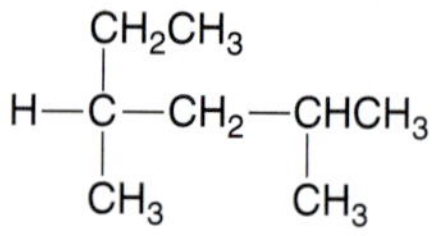

b. 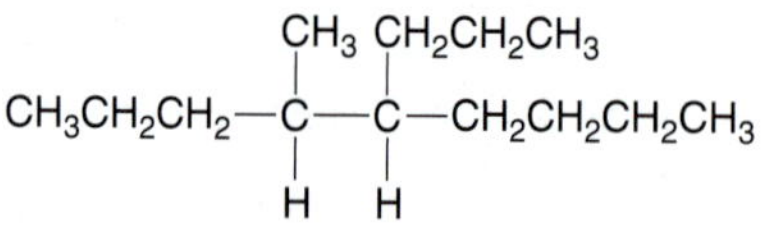 CH₃CH₂CH₂—C—C—CH₂CH₂CH₂CH₃
                     | |
      (CH₃ CH₂CH₂CH₃ above)
                     H   H

c. H—C—CH₂—CHCH₃
      | |
     CH₃   CH₃
   (CH₂CH₃ above C)

PROBLEM 12.12

Give the IUPAC name for each compound.

a. (CH₃)₂CHCH(CH₃)₂

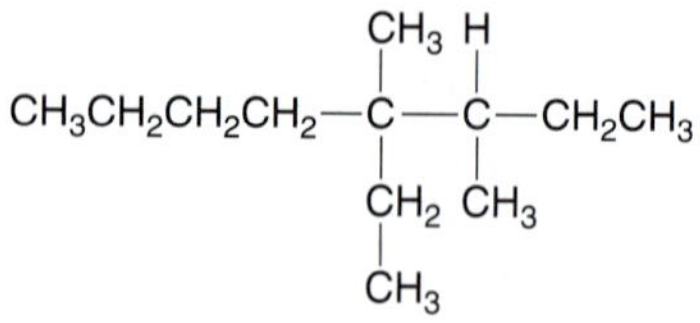

b. 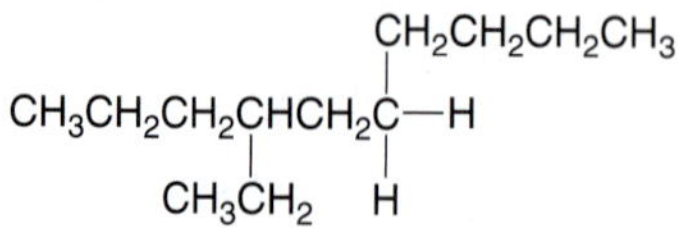 CH₃CH₂CH₂CHCH₂C—H
            | |
         CH₃CH₂   H
   (CH₂CH₂CH₂CH₃ above the C—H carbon)

c. CH₃CH₂CH₂CH₂—C—C—CH₂CH₃
                       | |
                     CH₂ CH₃
                       |
                     CH₃
   (CH₃ H above the C—C carbons)

You must also know how to derive a structure from a given name. Sample Problem 12.5 demonstrates a stepwise method.

SAMPLE PROBLEM 12.5

Give the structure with the following IUPAC name: 2,3,3-trimethylpentane.

ANALYSIS

To derive a structure from a name, first look at the end of the name to find the parent name and suffix. From the parent we know the number of C's in the longest chain, and the suffix tells us the functional group; the suffix -ane = an alkane. Then, number the carbon chain from either end and add the substituents. Finally, add enough H's to give each C four bonds.

SOLUTION

2,3,3-Trimethyl**pentane** has **pentane** (5 C's) as the longest chain and three methyl groups at carbons 2, 3, and 3.

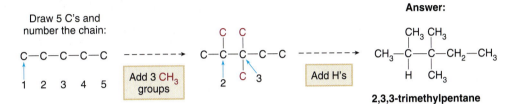

PROBLEM 12.13

Give the structure corresponding to each IUPAC name.

a. 3-methylhexane

b. 3,3-dimethylpentane

c. 3,5,5-trimethyloctane

d. 3-ethyl-4-methylhexane

PROBLEM 12.14

Give the structure corresponding to each IUPAC name.

a. 2,2-dimethylbutane

b. 6-butyl-3-methyldecane

c. 4,4,5,5-tetramethylnonane

d. 3-ethyl-5-propylnonane

12.5 CYCLOALKANES

Cycloalkanes contain carbon atoms arranged in a ring. Think of a cycloalkane as being formed by removing two H's from the end carbons of a chain, and then bonding the two carbons together.

12.5A SIMPLE CYCLOALKANES

Simple cycloalkanes are named by adding the prefix *cyclo-* to the name of the acyclic alkane having the same number of carbons. Cycloalkanes having three to six carbon atoms are shown in the accompanying figure. They are drawn using polygons in skeletal representations (Section 11.4). Each corner of the polygon has a carbon atom with two hydrogen atoms to give it four bonds.

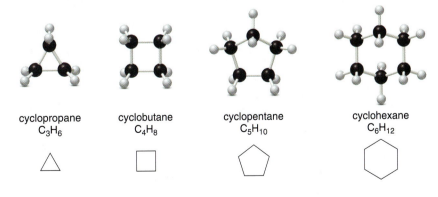

cyclopropane
C_3H_6

cyclobutane
C_4H_8

cyclopentane
C_5H_{10}

cyclohexane
C_6H_{12}

Cyclic compounds occur widely in nature and many have interesting and important biological properties. For example, **pyrethrin I,** a naturally occurring insecticide obtained from chrysanthemums, contains both a three- and five-membered ring. **Vitamin D₃,** a fat-soluble vitamin that regulates calcium and phosphorus metabolism, contains six- and five-membered rings. Many foods, particularly milk, are fortified with vitamin D₃ so that we get enough of this vital nutrient in our diet.

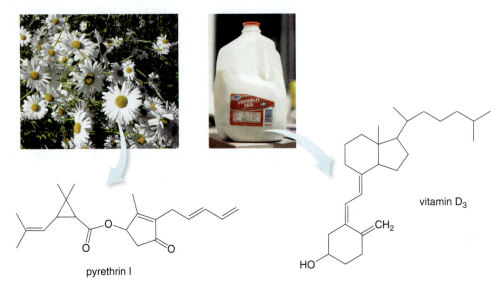

pyrethrin I

vitamin D₃

Although we draw cycloalkanes as flat polygons, in reality cycloalkanes with more than three carbons are not planar molecules. Cyclohexane, for example, adopts a puckered arrangement called the **chair** form, in which all bond angles are 109.5°.

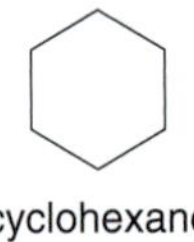

cyclohexane

skeletal structure

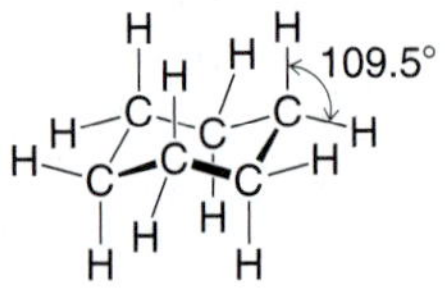

nonplanar cyclohexane
chair form

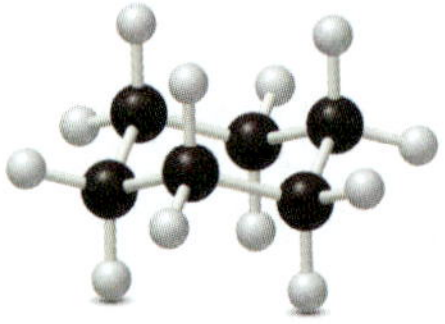

ball-and-stick model

PROBLEM 12.15   Review: Label the functional groups in pyrethrin I.

PROBLEM 12.16   Review: Explain why vitamin D₃ is a fat-soluble vitamin.

12.5B   NAMING CYCLOALKANES

Cycloalkanes are named using the rules in Section 12.4, but the prefix *cyclo-* immediately precedes the name of the parent.

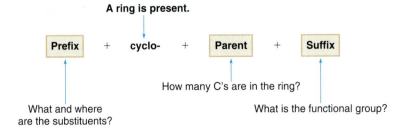

A ring is present.

Prefix + cyclo- + Parent + Suffix

What and where are the substituents?

How many C's are in the ring?

What is the functional group?

HOW TO   Name a Cycloalkane Using the IUPAC System

Step [1]   **Find the parent cycloalkane.**
  - Count the number of carbon atoms in the ring and use the parent name for that number of carbons. Add the prefix *cyclo-* and the suffix *-ane* to the parent name.

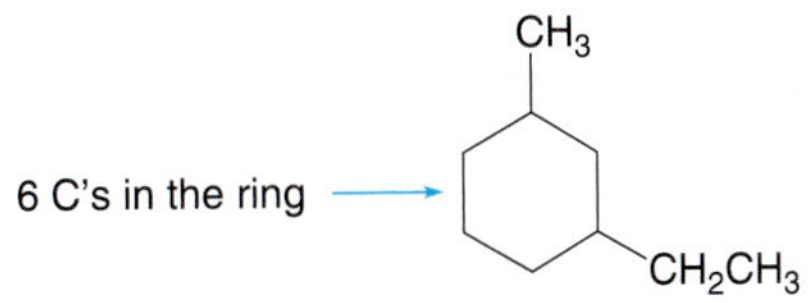

6 C's in the ring ⟶

cyclohexane

Step [2]   **Name and number the substituents.**
  - No number is needed to indicate the location of a **single** substituent.

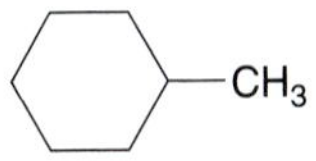

methylcyclohexane 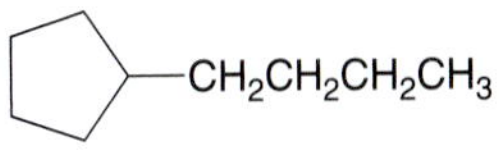   butylcyclopentane

  - For rings with more than one substituent, begin numbering at one substituent, and then give the **second substituent** the lower number. With two **different** substituents, number the ring to assign the lower number to the substituents **alphabetically.**

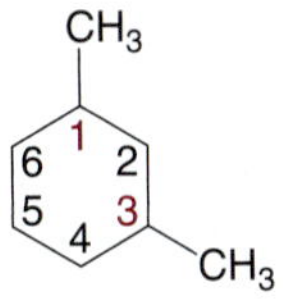

  - Place CH$_3$ groups at C1 and C3.

1,3-dimethylcyclohexane
(*not* 1,5-dimethylcyclohexane)

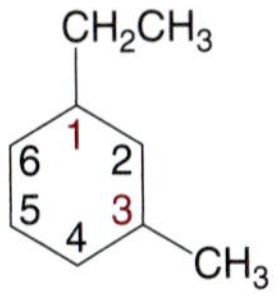

Earlier letter ----→ lower number

  - **ethyl** group at **C1**
  - **methyl** group at **C3**

1-ethyl-3-methylcyclohexane
(*not* 3-ethyl-1-methylcyclohexane)

SAMPLE PROBLEM 12.6   Give the IUPAC name for the following compound.

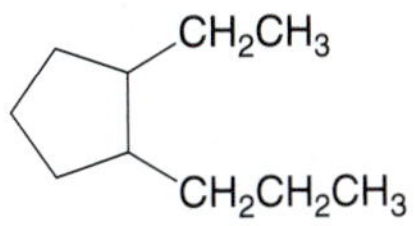

CH$_2$CH$_3$

CH$_2$CH$_2$CH$_3$

ANALYSIS AND SOLUTION

[1]   **Name the ring.** The ring has 5 C's so the molecule is named as a cyclopentane.

[2]   **Name and number the substituents.**
  - There are two substituents: CH$_3$CH$_2$– is an ethyl group and CH$_3$CH$_2$CH$_2$– is a propyl group.
  - Number to put the two groups at C1 and C2 (not C1 and C5).
  - Place the ethyl group at C1 since the *e* of **ethyl** comes before the *p* of **propyl** in the alphabet.

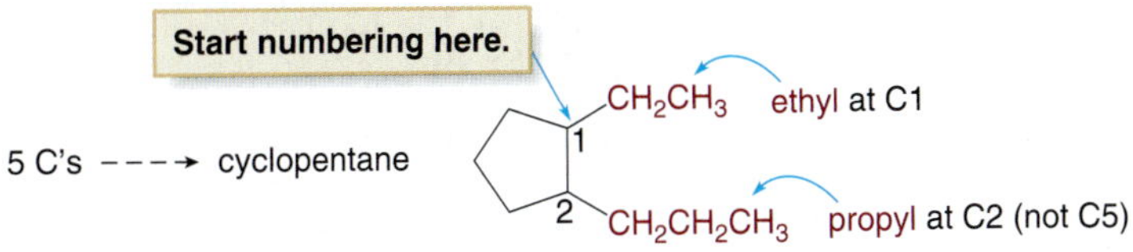

Start numbering here.

5 C's ---→ cyclopentane

CH$_2$CH$_3$ ethyl at C1

CH$_2$CH$_2$CH$_3$ propyl at C2 (not C5)

Answer: 1-ethyl-2-propylcyclopentane

PROBLEM 12.17

Give the IUPAC name for each compound.

a. [structure: cyclobutane with CH₃]   b. [structure: cyclohexane with two CH₃ groups]   c. CH₃CH₂—[cyclopentane]—CH₂CH₂CH₃   d. [structure: cyclohexane with CH₂CH₃ and CH₃]

PROBLEM 12.18

Give the structure corresponding to each IUPAC name.

a. propylcyclopentane

b. 1,2-dimethylcyclobutane

c. 1,1,2-trimethylcyclopropane

d. 4-ethyl-1,2-dimethylcyclohexane

12.6 FOCUS ON THE ENVIRONMENT
FOSSIL FUELS

Many alkanes occur in nature, primarily in natural gas and petroleum. Both of these fossil fuels serve as energy sources, formed long ago by the degradation of organic material.

Natural gas is composed largely of **methane** (60–80% depending on its source), with lesser amounts of ethane, propane, and butane. These organic compounds burn in the presence of oxygen, releasing energy for cooking and heating (Section 12.8).

Petroleum is a complex mixture of compounds, most of which are hydrocarbons containing 1–40 carbon atoms. Distilling crude petroleum, a process called **refining,** separates it into usable fractions that differ in boiling point (Figure 12.1). Most products of petroleum refining provide fuel for home heating, automobiles, diesel engines, and airplanes. Each fuel type has a different composition of hydrocarbons, as indicated below.

- **Gasoline:** C_5H_{12}–$C_{12}H_{26}$
- **Kerosene:** $C_{12}H_{26}$–$C_{16}H_{34}$
- **Diesel fuel:** $C_{15}H_{32}$–$C_{18}H_{38}$

CONSUMER NOTE

Natural gas is odorless. The smell observed in a gas leak is due to minute amounts of a sulfur additive such as methanethiol, CH_3SH, which provides an odor for easy detection and safety.

ENVIRONMENTAL NOTE

Methane is formed and used in a variety of ways. The CH_4 released from decaying vegetable matter in New York City's main landfill is used for heating homes. CH_4 generators in China convert cow manure into energy in rural farming towns.

▼ **FIGURE 12.1   Refining Crude Petroleum into Usable Fuel and Other Petroleum Products**

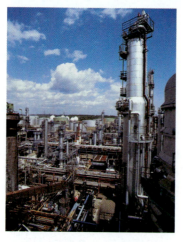

(a) **An oil refinery.** At an oil refinery, crude petroleum is separated into fractions of similar boiling point.

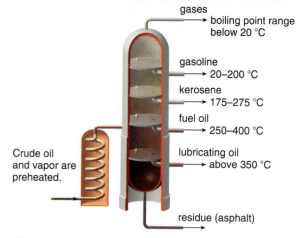

(b) **A refinery tower.** As crude petroleum is heated, the lower-boiling components come off at the top of the tower, followed by fractions of higher boiling point.

Petroleum provides more than fuel. About 3% of crude oil is used to make plastics and other synthetic compounds, including drugs, fabrics, dyes, and pesticides. These products are responsible for many of the comforts we now take for granted in industrialized countries. Imagine what life would be like without air-conditioning, refrigeration, anesthetics, and pain relievers, all products of the petroleum industry. Consider college students living without CDs and spandex!

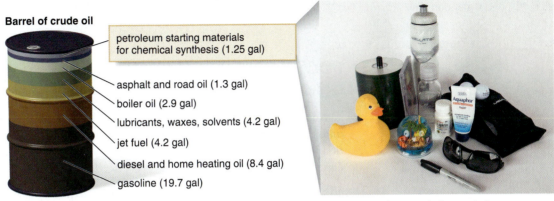

Barrel of crude oil

petroleum starting materials
for chemical synthesis (1.25 gal)

asphalt and road oil (1.3 gal)

boiler oil (2.9 gal)

lubricants, waxes, solvents (4.2 gal)

jet fuel (4.2 gal)

diesel and home heating oil (8.4 gal)

gasoline (19.7 gal)

products made from petroleum

Energy from petroleum is *nonrenewable,* and the remaining known oil reserves are limited. Given our dependence on petroleum, not only for fuel, but also for the many necessities of modern society, it becomes obvious that we must both conserve what we have and find alternate energy sources.

12.7 PHYSICAL PROPERTIES

Crude oil that leaks into the sea forms an insoluble layer on the surface.

Alkanes contain only nonpolar C—C and C—H bonds, so they exhibit only **weak intermolecular forces.** As a result, alkanes have low melting points and boiling points. Low molecular weight alkanes are gases at room temperature, and alkanes used in gasoline are all liquids.

The melting points and boiling points of alkanes increase as the number of carbons increases. Increased surface area increases the force of attraction between molecules, thus raising the boiling point and melting point. This is seen in comparing the boiling points of three straight-chain alkanes.

$$CH_3CH_2CH_2CH_3 \qquad CH_3CH_2CH_2CH_2CH_3 \qquad CH_3CH_2CH_2CH_2CH_2CH_3$$

butane   pentane   hexane

bp = −0.5 °C   bp = 36 °C   bp = 69 °C

Increasing surface area
Increasing boiling point

Because nonpolar alkanes are water insoluble and less dense than water, crude petroleum spilled into the sea from a ruptured oil tanker creates an insoluble oil slick on the surface. The insoluble hydrocarbon oil poses a special threat to birds whose feathers are coated with natural nonpolar oils for insulation. Because these oils dissolve in the crude petroleum, birds lose their layer of natural protection and many die.

PROBLEM 12.19

Rank the following products of petroleum refining in order of increasing boiling point: diesel fuel, kerosene, and gasoline.

Answer the following questions about pentane (C_5H_{12}), heptane (C_7H_{16}), and decane ($C_{10}H_{22}$).

a. Which compound has the highest boiling point?
b. Which compound has the lowest boiling point?
c. Which compound has the highest melting point?
d. Which compound has the lowest melting point?

12.8 FOCUS ON THE ENVIRONMENT
COMBUSTION

Alkanes are the only family of organic molecules that has no functional group, so **alkanes undergo few reactions.** In this chapter, we consider only one reaction of alkanes—**combustion.** Combustion is an example of an **oxidation–reduction** reaction, a class of reactions first discussed in Section 5.8.

To determine if an organic compound undergoes oxidation or reduction, we concentrate on the carbon atoms of the starting material and product, and **compare the relative number of C—H and C—O bonds.**

• *Oxidation* results in an *increase* in the number of C—O bonds or a *decrease* in the number of C—H bonds.
• *Reduction* results in a *decrease* in the number of C—O bonds or an *increase* in the number of C—H bonds.

Alkanes undergo **combustion**—that is, **they burn in the presence of oxygen to form carbon dioxide (CO_2) and water.** This is a practical example of oxidation. Every C—H and C—C bond in the starting material is converted to a C—O bond in the product. Equations are written with two different alkanes.

Review Section 5.2 on how to balance a chemical equation.

$$CH_4 \quad + \quad 2\,O_2 \quad \xrightarrow{\text{flame}} \quad CO_2 \quad + \quad 2\,H_2O \quad + \quad \text{energy}$$
methane
(natural gas)

$$2\,(CH_3)_3CCH_2CH(CH_3)_2 \quad + \quad 25\,O_2 \quad \xrightarrow{\text{flame}} \quad 16\,CO_2 \quad + \quad 18\,H_2O \quad + \quad \text{energy}$$
isooctane
(high-octane component
of gasoline)

Note that the products, **CO_2 + H_2O,** are the same regardless of the identity of the starting material. Combustion of alkanes in the form of natural gas, gasoline, or heating oil releases energy for heating homes, powering vehicles, and cooking food.

A spark or a flame is needed to initiate combustion. Gasoline, which is composed largely of alkanes, can be safely handled and stored in the air, but the presence of a spark or open flame causes immediate and violent combustion.

The combustion of alkanes and other hydrocarbons obtained from fossil fuels adds a tremendous amount of CO_2 to the atmosphere each year. Quantitatively, data show over a 20% increase in the atmospheric concentration of CO_2 in the last 50 years (from 315 parts per million in 1958 to 386 parts per million in 2007; Figure 12.2). Although the composition of the atmosphere has changed over the lifetime of the earth, this is likely the first time that the actions of mankind have altered that composition significantly and so quickly.

An increased CO_2 concentration in the atmosphere may have long-range and far-reaching effects. CO_2 is a **greenhouse gas** because it absorbs thermal energy that normally radiates from the

ENVIRONMENTAL NOTE

Driving an automobile 10,000 miles at 25 miles per gallon releases 10,000 lb of CO_2 into the atmosphere.

The 2007 Nobel Peace Prize was awarded to former Vice President Al Gore and the Intergovernmental Panel on Climate Change for their roles in focusing attention on the potentially disastrous effects of rapid climate change caused by human activity.

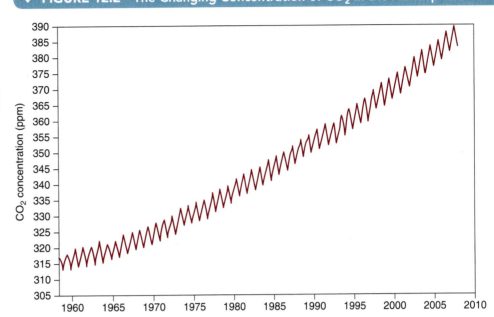

▼ **FIGURE 12.2** **The Changing Concentration of CO_2 in the Atmosphere**

The graph clearly shows the increasing level of CO_2 in the atmosphere (1958–2007). Two data points are recorded each year. The sawtooth nature of the graph is due to seasonal variation of CO_2 level with the seasonal variation in photosynthesis. (Data recorded at Mauna Loa, Hawaii.)

earth's surface, and redirects it back to the surface. Higher levels of CO_2 may therefore contribute to an increase in the average temperature of the earth's atmosphere. This **global warming,** as it has been termed, has many consequences—the melting of polar ice caps, the rise in sea level, and drastic global climate changes to name a few.

PROBLEM 12.21

Write a balanced equation for each combustion reaction.

a. $CH_3CH_2CH_3$ + O_2 $\xrightarrow{\text{flame}}$
propane

b. $CH_3CH_2CH_2CH_3$ + O_2 $\xrightarrow{\text{flame}}$
butane

When there is not enough oxygen available to completely burn a hydrocarbon, **incomplete combustion** may occur and **carbon monoxide** (CO) is formed instead of carbon dioxide (CO_2).

Incomplete combustion $2 CH_4$ + $3 O_2$ $\xrightarrow{\text{flame}}$ $2 CO$ + $4 H_2O$ + energy
methane   carbon monoxide

HEALTH NOTE

Meters that measure CO levels can be purchased, as described in Section 5.5. When wood is burned in a poorly ventilated fireplace situated in a well-insulated room, the CO concentration can reach an unhealthy level.

Carbon monoxide is a poisonous gas that binds to hemoglobin in the blood, thus reducing the amount of oxygen that can be transported through the bloodstream to cells. CO can be formed whenever hydrocarbons burn. When an automobile engine burns gasoline, unwanted carbon monoxide can be produced. Yearly car inspections measure CO and other pollutant levels and are designed to prevent cars from emitting potentially hazardous substances into ambient air. Carbon monoxide is also formed when cigarettes burn, so heavy smokers have an unhealthy concentration of CO in their bloodstream.

PROBLEM 12.22

Write a balanced equation for the incomplete combustion of ethane (CH_3CH_3) to form carbon monoxide as one product.

CHAPTER HIGHLIGHTS

KEY TERMS

Acyclic alkane (12.1)
Alkane (12.1)
Alkyl group (12.4)
Branched-chain alkane (12.2)
Combustion (12.8)
Constitutional isomer (12.2)
Cycloalkane (12.1)
Greenhouse gas (12.8)

Incomplete combustion (12.8)
Isomer (12.2)
IUPAC nomenclature (12.3)
Oxidation (12.8)
Parent name (12.4)
Petroleum (12.6)
Pheromone (12.1)
Primary carbon, 1° C (12.2)

Quaternary carbon, 4° C (12.2)
Reduction (12.8)
Refining (12.6)
Saturated hydrocarbon (12.1)
Secondary carbon, 2° C (12.2)
Straight-chain alkane (12.2)
Tertiary carbon, 3° C (12.2)

KEY CONCEPTS

❶ **What are the characteristics of an alkane? (12.1, 12.2)**

- Alkanes are hydrocarbons having only nonpolar C—C and C—H single bonds.
- There are two types of alkanes: Acyclic alkanes (C_nH_{2n+2}) have no rings. Cycloalkanes (C_nH_{2n}) have one or more rings.
- A carbon is classified as 1°, 2°, 3°, or 4° by the number of carbons bonded to it. A 1° C is bonded to one carbon; a 2° C is bonded to two carbons, and so forth.

❷ **What are constitutional isomers? (12.2)**

- Isomers are different compounds with the same molecular formula.
- Constitutional isomers differ in the way the atoms are connected to each other. $CH_3CH_2CH_2CH_3$ and $HC(CH_3)_3$ are constitutional isomers because they have molecular formula C_4H_{10} but one compound has a chain of four carbons in a row and the other does not.

❸ **How are alkanes named? (12.4, 12.5)**

- Alkanes are named using the IUPAC system of nomenclature. A name has three parts: the parent indicates the number of carbons in the longest chain or the ring; the suffix indicates the functional group (-ane = alkane); the

prefix tells the number and location of substituents coming off the chain or ring.
- Alkyl groups are formed by removing one hydrogen from an alkane. Alkyl groups are named by changing the -ane ending of the parent alkane to the suffix -yl.

❹ **Characterize the physical properties of alkanes. (12.7)**

- Alkanes are nonpolar, so they have weak intermolecular forces, low melting points, and low boiling points.
- The melting points and boiling points of alkanes increase as the number of carbons increases due to increased surface area.
- Alkanes are insoluble in water.

❺ **What are the products of the combustion and incomplete combustion of an alkane? (12.8)**

- Alkanes burn in the presence of air. Combustion forms CO_2 and H_2O as products. Incomplete combustion forms CO and H_2O.
- Alkane combustion has increased the level of CO_2, a greenhouse gas, in the atmosphere.
- Incomplete combustion forms carbon monoxide, a toxin and air pollutant.

PROBLEMS

Selected in-chapter and end-of-chapter problems have brief answers provided in Appendix B.

Alkanes and Isomers

12.23 The waxy coating that covers tobacco leaves contains a straight-chain alkane having 31 carbons. How many hydrogens does this alkane contain?

12.24 The largest known cycloalkane with a single ring has 288 carbons. What is its molecular formula?

12.25 The sex attractant of the female tiger moth is an alkane of molecular formula $C_{18}H_{38}$. Is this molecule an acyclic alkane or a cycloalkane?

12.26 In addition to other compounds, beeswax contains an acyclic alkane that has 64 hydrogens. How many carbon atoms does this alkane contain?

12.27 Paraffin wax is a mixture of straight- and branched-chain alkanes having 26–30 carbons. How many hydrogens does each of these alkanes contain?

12.28 An alkane has 20 hydrogen atoms. How many carbon atoms would it contain if it were (a) a straight-chain alkane; (b) a branched-chain alkane; (c) a cycloalkane?

12.29 Explain why a cycloalkane has two fewer hydrogen atoms than an acyclic alkane with the same number of carbon atoms.

12.30 What is the difference between a branched-chain alkane and a cycloalkane?

12.31 Classify each carbon as 1°, 2°, 3°, or 4°.

a. $CH_3(CH_2)_3CH_3$

c. $(CH_3)_3CC(CH_3)_3$

b. $CH_3CH_2CHCHCH_3$
      | (CH₃) | (CH₃)

d. (cyclopentane with CH₃ and two CH₃ groups)

12.32 Give the structure of an alkane that fits each description.
a. an alkane that contains only 1° and 4° carbons
b. a cycloalkane that contains only 2° carbons
c. an alkane of molecular formula C_6H_{14} that contains a 4° carbon
d. a cycloalkane that contains 2° and 3° carbons

12.33 Draw the structure of an alkane with molecular formula C_7H_{16} that (a) contains one 4° carbon; (b) contains only 1° and 2° carbons.

12.34 Is it possible to have an alkane that contains only 1° carbons? Explain why or why not with an example.

12.35 Label each pair of compounds as constitutional isomers or identical molecules.

a. $CH_3CHCHCH_3$ (with CH_2CH_3 and CH_2CH_3) and $CH_3CH_2CHCH_2CH(CH_3)_2$ (with CH_3)

b. $CH_3CH_2CHCHCH_2CH_3$ (with CH_3 and CH_3) and $CH_3CHCHCH_3$ (with CH_2CH_3 and CH_2CH_3)

c. (cyclopropane with two CH_3) and (cyclopropane with CH_2CH_3)

d. (cyclohexane with two CH_3) and (cyclohexane with two CH_3)

12.36 Label each pair of compounds as constitutional isomers, identical molecules, or not isomers of each other.

a. $CH_3CH_2CH_2CH_3$ and $CH_3CH_2CH_2CH_2CH_3$

b. (CH_2—CH_2 / CH_3CH_2 CH_2CH_3) and $CH_3(CH_2)_4CH_3$

c. $CH_3CH_2CH_2CH_2CH_3$ and (cyclopentane)

d. CH_3—C(=O)—OCH_3 and CH_3CH_2—C(=O)—OH

12.37 How is each of the following compounds related to $(CH_3)_2CHCH_2CH_2CH_3$? Choose from constitutional isomers or identical molecules.

a. $CH_3CHCHCH_3$ (with two CH_3)

b. $CH_3CHCH_2CH_2$ (with CH_3 and CH_3)

c. CH_3CHCH_3 (with $CH_2CH_2CH_3$)

d. $CH_3CHCH_2CH_3$ (with CH_2CH_3)

e. (branched structure)

12.38 Consider compounds **A–E** and label each pair of compounds as identical molecules, constitutional isomers, or not isomers of each other: (a) **A** and **B**; (b) **A** and **C**; (c) **B** and **D**; (d) **D** and **E**; (e) **B** and **E**.

A (zigzag chain)

B $CH_3(CH_2)_3CHCH_3$ (with CH_3)

C H—C(H)(CH₃)—$CH_2CH_2CH_2$—C(H)(CH₃)—H

D $CH_3CH_2CH_2CHCH_2CH_3$ (with CH_3)

E $CH_3CH_2CHCH_2CH_2CH_3$ (with CH_3)

12.39 Draw structures that fit the following descriptions:
a. two cycloalkanes that are constitutional isomers with molecular formula C_7H_{14}
b. an ether and an alcohol that are constitutional isomers with molecular formula $C_5H_{12}O$
c. two constitutional isomers of molecular formula C_3H_7Cl

12.40 Draw the five constitutional isomers having molecular formula C_6H_{14}.

12.41 Draw all constitutional isomers having molecular formula C_8H_{18} that contain seven carbons in the longest chain and a single CH_3 group bonded to the chain.

12.42 Draw the five constitutional isomers that have molecular formula C_5H_{10} and contain one ring.

12.43 Draw three constitutional isomers with molecular formula C_3H_6O. Draw the structure of one alcohol, one ketone, and one cyclic ether.

12.44 Draw one constitutional isomer of each compound.

a. CH_3CH_2—C(=O)—CH_3

b. $CH_3CH_2CHCH_3$ (with NH_2)

c. (cyclohexane with OH)

Alkane Nomenclature

12.45 Give the IUPAC name for each of the five constitutional isomers of molecular formula C_6H_{14} in Problem 12.40.

12.46 Give the IUPAC name for each of the five cyclic isomers of molecular formula C_5H_{10} in Problem 12.42.

12.47 Give the IUPAC name for each compound.

a. 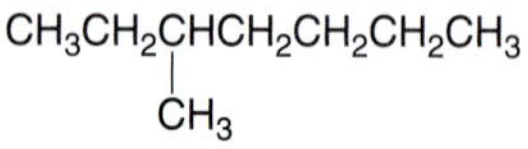 CH₃CH₂CHCH₂CH₂CH₂CH₃
      |
      CH₃

b. 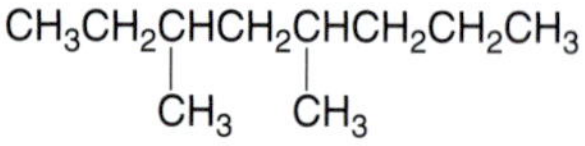 CH₃CH₂CHCH₂CHCH₂CH₂CH₃
      | |
      CH₃ CH₃

c. CH₃CH₂CH₂C(CH₂CH₃)₃

d. 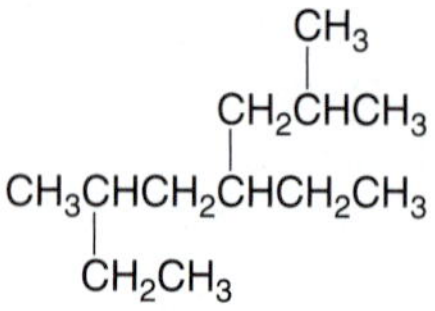
             CH₃
             |
         CH₂CHCH₃
         |
   CH₃CHCH₂CHCH₂CH₃
        |
        CH₂CH₃

e. (CH₃CH₂)₂CHCH₂CH₂CH₂CH(CH₃)₂

f. 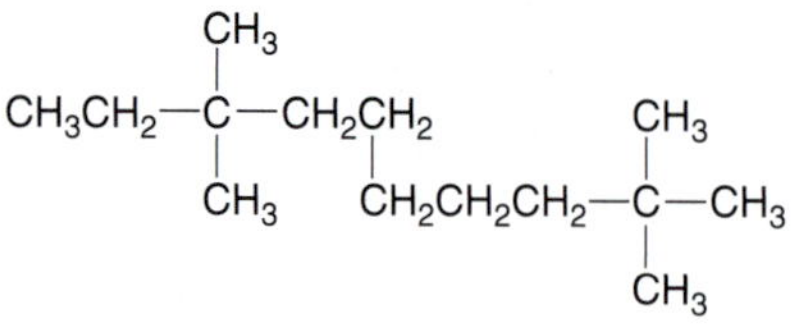
           CH₃
           |
CH₃CH₂—C—CH₂CH₂   CH₃
       | |
       CH₃   CH₂CH₂CH₂—C—CH₃
                            |
                           CH₃

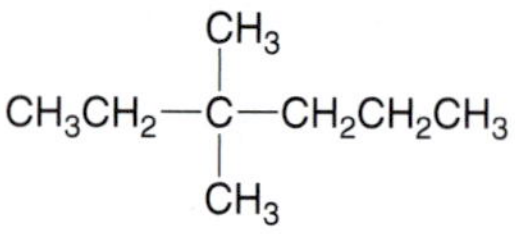

12.48 Give the IUPAC name for each compound.

a. 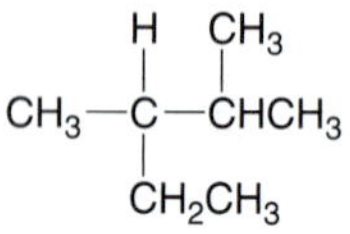
         H CH₃
         | |
   CH₃—C—CHCH₃
         |
       CH₂CH₃

b.
          CH₃
          |
  CH₃CH₂—C—CH₂CH₂CH₃
          |
         CH₃

c.
          CH₂CH₂CH₂CH₃
          |
  CH₃CH₂CHCH₂CH₂CH₂CH₃

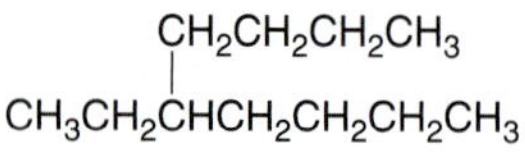

d. (CH₃CH₂)₂CHCH(CH₂CH₃)₂

e. 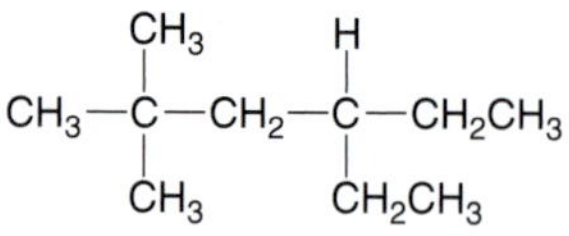
        CH₃   H
        | |
  CH₃—C—CH₂—C—CH₂CH₃
        | |
       CH₃   CH₂CH₃

f. 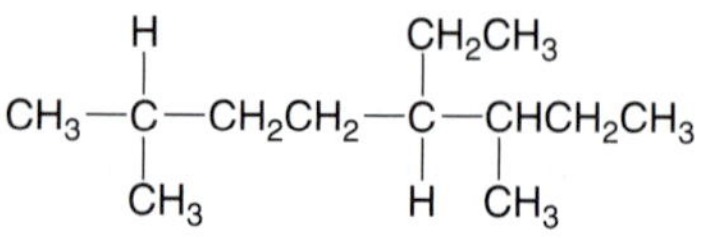
        H   CH₂CH₃
        | |
  CH₃—C—CH₂CH₂—C—CHCH₂CH₃
       | | |
      CH₃   H CH₃

12.49 Give the IUPAC name for each cycloalkane.

a.

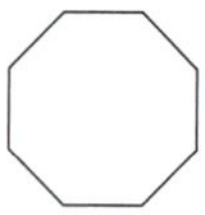

c.

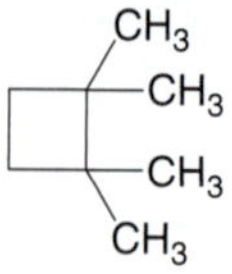

b.

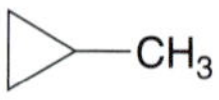

d.

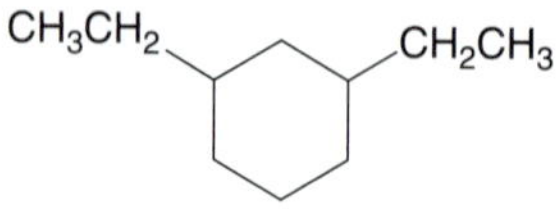

12.50 Give the IUPAC name for each cycloalkane.

a.

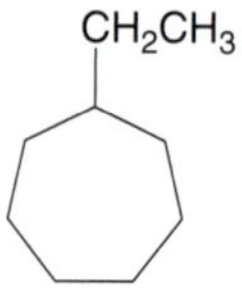

b.

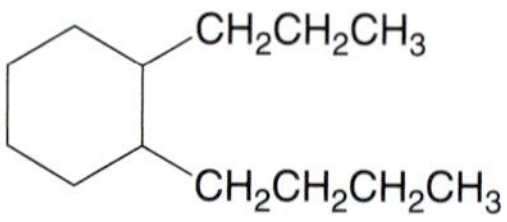

c.

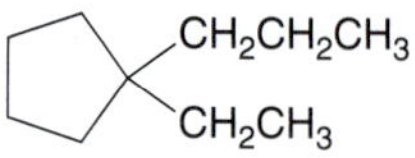

d.

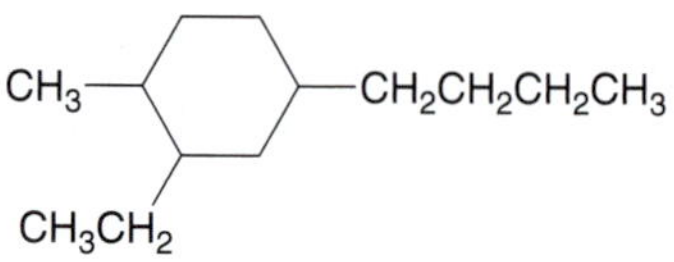

12.51 Give the structure corresponding to each IUPAC name.
a. 3-ethylhexane
b. 3-ethyl-3-methyloctane
c. 2,3,4,5-tetramethyldecane
d. cyclononane
e. 1,1,3-trimethylcyclohexane
f. 1-ethyl-2,3-dimethylcyclopentane

12.52 Give the structure corresponding to each IUPAC name.
a. 3-ethyl-3-methylhexane
b. 2,2,3,4-tetramethylhexane
c. 4-ethyl-2,2-dimethyloctane
d. 1,3,5-triethylcycloheptane
e. 3-ethyl-3,4-dimethylnonane
f. 2-ethyl-1-methyl-3-propylcyclopentane

12.53 Each of the following IUPAC names is incorrect. Explain why it is incorrect and give the correct IUPAC name.
a. 3-methylbutane
b. 1-methylcyclopentane
c. 1,3-dimethylbutane
d. 5-ethyl-2-methylhexane
e. 1,5-dimethylcyclohexane
f. 1-propyl-2-ethylcyclopentane

12.54 Each of the following IUPAC names is incorrect. Explain why it is incorrect and give the correct IUPAC name.
a. 4-methylpentane
b. 2,3,3-trimethylbutane
c. 1,3-dimethylpentane
d. 3-methyl-5-ethylhexane
e. 2-ethyl-2-methylcycloheptane
f. 1,6-diethylcyclohexane

12.55 Draw three constitutional isomers having molecular formula C_7H_{14} that contain a five-membered ring and two CH_3 groups bonded to the ring. Give the IUPAC name for each isomer.

12.56 Draw four constitutional isomers having molecular formula C_6H_{12} that contain a four-membered ring. Give the IUPAC name for each isomer.

Drawing Acyclic Alkanes and Cycloalkanes

12.57 Draw a skeletal structure for each compound.
a. octane
b. 1,2-dimethylcyclopentane
c.

12.58 Convert each compound to a skeletal structure.
a. $CH_3(CH_2)_7CH_3$
b. 1,1-diethylcyclohexane
c. $(CH_3CH_2)_2CHCH_2CH_2CH_3$

12.59 Convert each skeletal structure to a complete structure with all atoms drawn in.

a.    c.

b.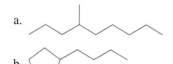

12.60 Convert each skeletal structure to a complete structure with all atoms drawn in.

a. 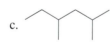   c.

b.

Physical Properties

12.61 Which compound in each pair has the higher melting point?
a. $CH_3CH_2CH_3$ or $CH_3CH_2CH_2CH_3$
b.

12.62 Which compound in each pair has the higher boiling point?
a. cyclobutane or cyclopentane
b. cyclopentane or ethylcyclopentane

12.63 Branching in an alkane chain decreases surface area. As a result, a branched-chain alkane tends to have a lower boiling point than a straight-chain alkane with the same number of carbons. Keeping this in mind, rank the alkane isomers in each group in order of increasing boiling point.
a. $CH_3CH_2CH_2CH_2CH_3$, $(CH_3)_2CHCH_2CH_3$, $(CH_3)_4C$
b. $(CH_3)_2CHCH(CH_3)_2$, $CH_3(CH_2)_4CH_3$, $CH_3CH_2CH_2CH(CH_3)_2$

12.64 Explain why the boiling points of heptane $[CH_3(CH_2)_5CH_3]$ and H_2O are similar even though heptane has a much higher molecular weight and greater surface area.

12.65 Explain why hexane is more soluble in dichloromethane (CH_2Cl_2) than in water.

12.66 Mineral oil and Vaseline are both mixtures of alkanes, but mineral oil is a liquid at room temperature and Vaseline is a solid. Which product is composed of alkanes that contain a larger number of carbon atoms? Explain your choice.

Reactions

12.67 What products are formed from the combustion of an alkane?

12.68 What products are formed from the incomplete combustion of an alkane?

12.69 Write a balanced equation for the combustion of each alkane: (a) CH_3CH_3; (b) $(CH_3)_2CHCH_2CH_3$.

12.70 Write a balanced equation for the combustion of each cycloalkane.

a. [hexagon]   b. [pentagon]

12.71 Write a balanced equation for the incomplete combustion of each alkane: (a) $CH_3CH_2CH_3$; (b) $CH_3CH_2CH_2CH_3$.

12.72 Benzene (C_6H_6) is a fuel additive sometimes used to make gasoline burn more efficiently. Write a balanced equation for the incomplete combustion of benzene.

12.73 In the body, glucose, a simple carbohydrate with molecular formula $C_6H_{12}O_6$, is oxidized to CO_2 and H_2O with the release of a great deal of energy. Thus, the metabolism of glucose provides energy for our bodies, in much the same way as the oxidation of an alkane provides energy to power an automobile. Write a balanced equation for the oxidation of glucose to form CO_2 and H_2O.

12.74 Gasohol is a mixture of 90% gasoline and 10% ethanol, CH_3CH_2OH. Ethanol is considered an environmentally friendly fuel additive because it can be made from a renewable source—sugarcane. Ethanol burns in air to form CO_2 and H_2O, and, like the combustion of alkanes, this reaction also releases a great deal of energy. Write a balanced equation for the combustion of ethanol.

Applications

12.75 The gasoline industry seasonally changes the composition of gasoline in locations where it gets very hot in the summer and very cold in the winter. Gasoline is refined to contain a larger fraction of higher molecular weight alkanes in warmer weather. In colder weather, it is refined to contain a larger fraction of lower molecular weight alkanes. What is the purpose of producing different types of gasoline for different temperatures?

12.76 Polyethylene (Section 11.5) is a high molecular weight alkane that contains hundreds or even thousands of carbon atoms, bonded together in long carbon chains. When a new home is built, the concrete foundation is often wrapped with polyethylene (sold under the trade name of Tyvek). What purpose does the polyethylene serve?

12.77 Mineral oil, a mixture of high molecular weight alkanes, is sometimes used as a laxative. Individuals who use mineral oil for this purpose are advised to avoid taking it at the same time they consume foods rich in fat-soluble vitamins such as vitamin A. Why is this advice given?

12.78 Explain why having a fire in a fireplace in a poorly ventilated room poses a health hazard.

12.79 You went running on a newly paved asphalt road, which soiled your sneakers. What solvent might be best for removing asphalt, a mixture of high molecular weight hydrocarbons, from your sneakers: (a) water; (b) nail polish remover, which is mainly acetone [$(CH_3)_2C{=}O$]; (c) paint thinner, which is largely liquid alkanes? Explain your choice.

CHALLENGE QUESTIONS

12.83 What is the molecular formula for a compound with 10 carbons and two rings?

12.84 Draw the structure of the 12 constitutional isomers that have the molecular formula C_6H_{12} and contain one ring.

12.80 A major component of animal fat is tristearin. (a) Identify the three functional groups in tristearin. (b) Label all the polar bonds. (c) Explain why tristearin will dissolve in an organic solvent like hexane, but is insoluble in water.

tristearin

General Questions

12.81 Answer the following questions about the alkane drawn below.

  a. Give the IUPAC name.
  b. Draw one constitutional isomer.
  c. Predict the solubility in water.
  d. Predict the solubility in an organic solvent.
  e. Write a balanced equation for complete combustion.
  f. Draw a skeletal structure.

12.82 Answer the following questions about the alkane drawn below.

  a. Give the IUPAC name.
  b. Draw one constitutional isomer.
  c. Predict the solubility in water.
  d. Predict the solubility in an organic solvent.
  e. Write a balanced equation for complete combustion.
  f. Draw a skeletal structure.

12.85 Cyclopentane has a higher boiling point than pentane even though both compounds have the same number of carbons (49 °C vs. 36 °C). Can you suggest a reason for this phenomenon?

12.86 Draw the structure of the seven constitutional isomers having molecular formula $C_4H_{10}O$. Each compound is either an ether or an alcohol.

13

CHAPTER GOALS

In this chapter you will learn how to:

1. Identify the three major types of unsaturated hydrocarbons—alkenes, alkynes, and aromatic compounds

2. Name alkenes, alkynes, and substituted benzenes

3. Recognize the difference between constitutional isomers and stereoisomers, as well as identify cis and trans isomers

4. Identify saturated and unsaturated fatty acids and predict their relative melting points

5. Draw the products of addition reactions of alkenes

6. Draw the products of reactions that follow Markovnikov's rule

7. Explain what products are formed when a vegetable oil is partially hydrogenated

8. Draw the structure of polymers formed from alkene monomers

9. Draw the products of substitution reactions of benzene

Vegetable oils obtained from olives, peanuts, and corn are formed from unsaturated organic molecules such as oleic acid and linoleic acid.

UNSATURATED HYDROCARBONS

IN Chapter 13 we continue our study of hydrocarbons by examining three families of compounds that contain carbon–carbon multiple bonds. **Alkenes** contain a double bond and **alkynes** contain a triple bond. **Aromatic hydrocarbons** contain a benzene ring, a six-membered ring with three double bonds. These compounds differ from the alkanes of Chapter 12 because they each have a functional group, making them much more reactive. Thousands of biologically active molecules contain these functional groups, and many useful synthetic products result from their reactions.

13.1 ALKENES AND ALKYNES

Alkenes and alkynes are two families of organic molecules that contain multiple bonds.

- Alkenes are compounds that contain a carbon–carbon double bond.

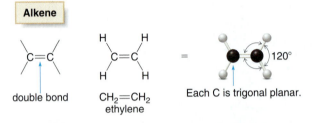

A ripe banana speeds up the ripening of green tomatoes because the banana gives off ethylene, a plant growth hormone.

The general molecular formula of an alkene is C_nH_{2n}, so an alkene has **two** fewer hydrogens than an acyclic alkane, which has a general molecular formula of C_nH_{2n+2}. Ethylene (C_2H_4) is the simplest alkene. Since each carbon of ethylene is surrounded by three atoms, each carbon is **trigonal planar.** All six atoms of ethylene lie in the same plane, and all bond angles are **120°.**

Ethylene is a hormone that regulates plant growth and fruit ripening. This allows individuals to enjoy bananas, strawberries, and tomatoes that were grown in faraway countries. Fruit can be picked green, and then sprayed with ethylene when ripening is desired upon arrival at its destination.

- Alkynes are compounds that contain a carbon–carbon triple bond.

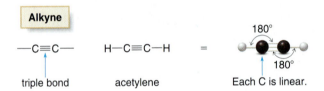

The general molecular formula for an alkyne is C_nH_{2n-2}, so an alkyne has **four** fewer hydrogens than an acyclic alkane. Acetylene (C_2H_2) is the simplest alkyne. Each carbon of acetylene is surrounded by two atoms, making each carbon **linear** with bond angles of **180°.**

Because alkenes and alkynes are composed of nonpolar carbon–carbon and carbon–hydrogen bonds, their physical properties are similar to other hydrocarbons. Like alkanes:

- Alkenes and alkynes have low melting points and boiling points and are insoluble in water.

Recall from Chapter 12 that acyclic alkanes are called saturated hydrocarbons, because they contain the maximum number of hydrogen atoms per carbon. In contrast, **alkenes and alkynes are called *unsaturated* hydrocarbons.**

- Unsaturated hydrocarbons are compounds that contain fewer than the maximum number of hydrogen atoms per carbon.

The multiple bond of an alkene or alkyne is always drawn in a condensed structure. To translate a condensed structure to a complete structure with all bond lines drawn in, make sure that each carbon of a double bond has three atoms around it, and each carbon of a triple bond has two atoms around it, as shown in Sample Problem 13.1. Keep in mind that in every structure, each carbon always has four bonds.

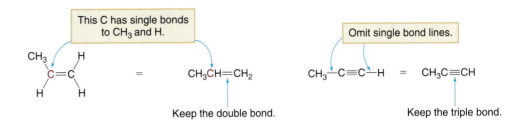

$CH_3CH=CH_2$

Keep the double bond.

$CH_3C\equiv CH$

Keep the triple bond.

SAMPLE PROBLEM 13.1

Draw a complete structure for each alkene or alkyne.

a. $CH_2=CHCH_2CH_3$

b. $CH_3C\equiv CCH_2CH_3$

ANALYSIS

First, draw the multiple bond in each structure. Draw an alkene so that each C of the double bond has three atoms around it. Draw an alkyne so that each C of the triple bond has two atoms around it. All other C's have four single bonds.

SOLUTION

a. The C labeled in red, has single bonds to 2 H's. The C labeled in blue has single bonds to 1 H and a CH_2CH_3 group.

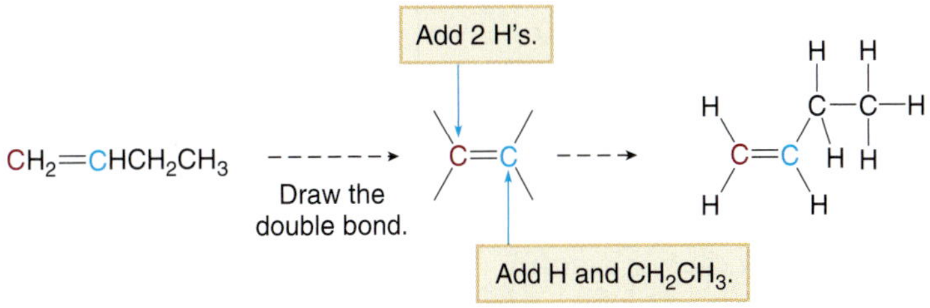

Add 2 H's.

$CH_2=CHCH_2CH_3$ — — — → Draw the double bond. — — → Add H and CH_2CH_3.

b. The C labeled in red has a single bond to a CH_3 group. The C labeled in blue has a single bond to a CH_2CH_3 group.

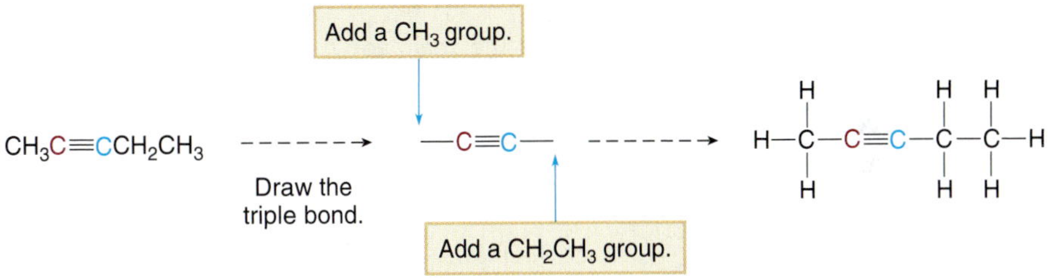

Add a CH_3 group.

$CH_3C\equiv CCH_2CH_3$ — — — → Draw the triple bond. — — — → Add a CH_2CH_3 group.

PROBLEM 13.1

Convert each condensed structure to a complete structure with all atoms and bond lines drawn in.

a. $CH_2=CHCH_2OH$

b. $(CH_3)_2C=CH(CH_2)_2CH_3$

c. $(CH_3)_2CHC\equiv CCH_2C(CH_3)_3$

PROBLEM 13.2

Myrcene is an alkene obtained from the bayberry plant. Draw a complete structure for myrcene with all atoms and bond lines.

$$CH_2$$
$$\|$$
$$(CH_3)_2C=CHCH_2CH_2CCH=CH_2$$

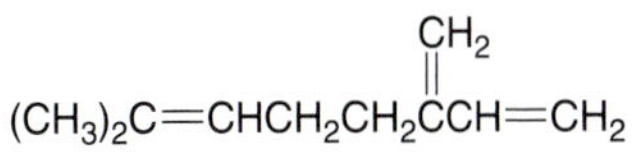

myrcene

PROBLEM 13.3

Determine whether each molecular formula corresponds to a saturated hydrocarbon, an alkene, or an alkyne.

a. C_3H_6

b. C_5H_{12}

c. C_8H_{14}

d. C_6H_{12}

PROBLEM 13.4

Give the molecular formula for each of the following compounds.

a. an alkene that has four carbons

b. a saturated hydrocarbon that has six carbons

c. an alkyne that has seven carbons

d. an alkene that has five carbons

13.2 NOMENCLATURE OF ALKENES AND ALKYNES

Whenever we encounter a new functional group, we must learn how to use the IUPAC system to name it. There are two new facts to learn: how to name the suffix that identifies the functional group, and how to number the carbon skeleton. In the IUPAC system:

- An alkene is identified by the suffix -ene.
- An alkyne is identified by the suffix -yne.

HOW TO   Name an Alkene or an Alkyne

EXAMPLE   **Give the IUPAC name of each alkene and alkyne.**

$$CH_3$$
a. 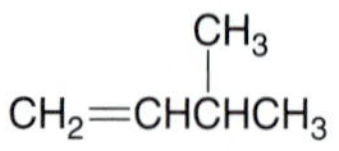 $CH_2{=}CHCHCH_3$

$$CH_2CH_3$$
b. 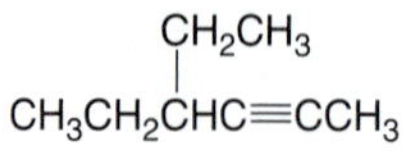 $CH_3CH_2CHC{\equiv}CCH_3$

Step [1]   Find the longest chain that contains *both* carbon atoms of the double or triple bond.

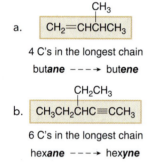

a.
$$CH_3$$
$CH_2{=}CHCHCH_3$

4 C's in the longest chain

but**ane** ---→ but**ene**

- Since the compound is an alkene, change the *-ane* ending of the parent alkane to *-ene.*

b.
$$CH_2CH_3$$
$CH_3CH_2CHC{\equiv}CCH_3$

6 C's in the longest chain

hex**ane** ---→ hex**yne**

- Since the compound is an alkyne, change the *-ane* ending of the parent alkane to *-yne.*

Step [2]   Number the carbon chain from the end that gives the multiple bond the lower number.

For each compound, number the chain and name the compound using the *first* number assigned to the multiple bond.

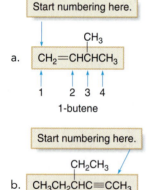

Start numbering here.

a.
$$CH_3$$
$CH_2{=}CHCHCH_3$
$\quad$ 1$\quad$ 2 3 4
1-butene

- Numbering the chain from left to right puts the double bond at C1 (not C3). The alkene is named using the *first* number assigned to the double bond, making it 1-butene.

Start numbering here.

b. 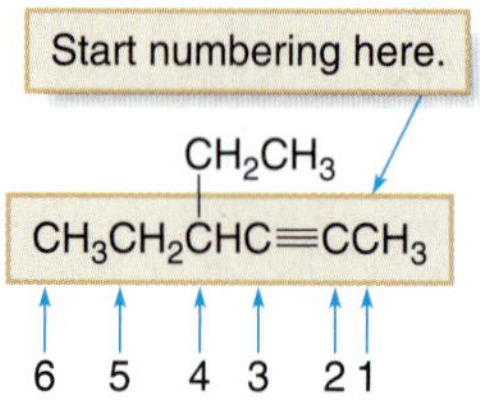
$$CH_2CH_3$$
$CH_3CH_2CHC{\equiv}CCH_3$
$\quad$ 6$\quad$ 5$\quad$ 4 3$\quad$ 2 1
2-hexyne

- Numbering the chain from right to left puts the triple bond at C2 (not C4). The alkyne is named using the *first* number assigned to the triple bond, making it 2-hexyne.

Step [3]   Number and name the substituents, and write the name.

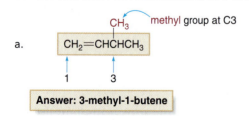

a.
CH_3 methyl group at C3
$CH_2{=}CHCHCH_3$
$\quad$ 1$\qquad$ 3

Answer: 3-methyl-1-butene

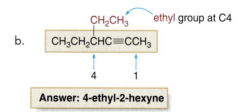

b.
CH_2CH_3 ethyl group at C4
$CH_3CH_2CHC{\equiv}CCH_3$
$\qquad$ 4$\quad$ 1

Answer: 4-ethyl-2-hexyne

Compounds with two double bonds are called **dienes.** They are named by changing the *-ane* ending of the parent alkane to the suffix *-adiene.* Each double bond gets its own number.

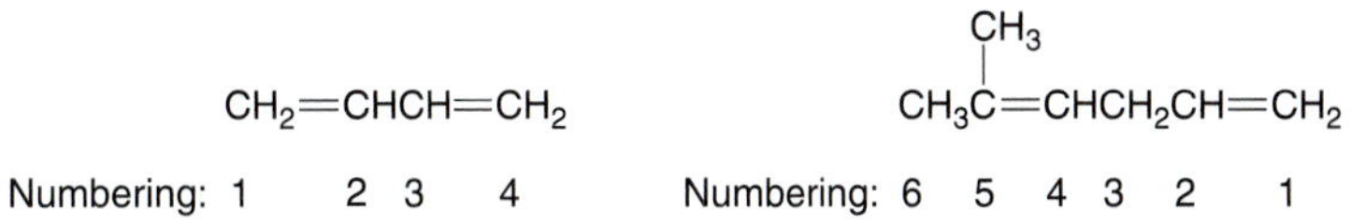

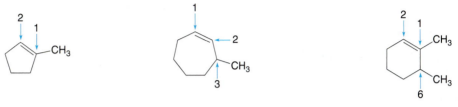

CH2=CHCH=CH2
Numbering: 1 2 3 4
CH3
CH3C=CHCH2CH=CH2
Numbering: 6 5 4 3 2 1

Name: **1,3-butadiene** Name: **5-methyl-1,4-hexadiene**

In naming cycloalkenes, the double bond is located between C1 and C2, and the "1" is usually omitted in the name. The ring is numbered to give the first substituent the lower number.

2
1
CH3

1-methylcyclopentene

1
2
CH3
3

3-methylcycloheptene

[Number clockwise beginning at the C=C and place the CH_3 at C3.]

2
1
CH3
CH3
6

1,6-dimethylcyclohexene

[Number counterclockwise beginning at the C=C and place the first CH_3 at C1.]

A few simple alkenes and alkynes have names that do not follow the IUPAC system. The simplest alkene, $CH_2{=}CH_2$, is called *ethene* in the IUPAC system, but it is commonly called **ethylene.** The simplest alkyne, HC≡CH, is called *ethyne* in the IUPAC system, but it is commonly named **acetylene.** We will use these common names since they are more widely used than their systematic IUPAC names.

SAMPLE PROBLEM 13.2

Give the IUPAC name for the following compound.

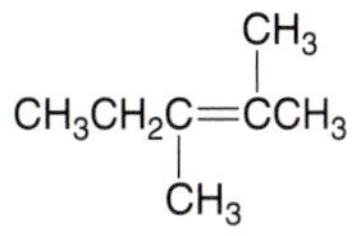

CH3
CH3CH2C=CCH3
CH3

ANALYSIS AND SOLUTION

[1] **Find the longest chain containing both carbon atoms of the multiple bond.**

CH3
CH3CH2C=CCH3
CH3

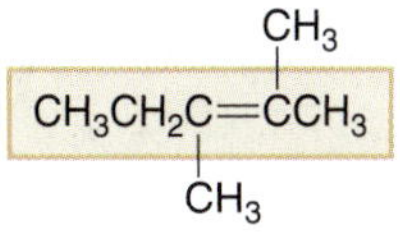

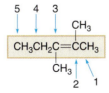

5 C's in the longest chain ---→ **pentene**

[2] **Number the chain to give the double bond the lower number.**

5 4 3
CH3
CH3CH2C=CCH3
CH3
2 1

- Numbering from right to left is preferred since the double bond begins at C2 (not C3). The molecule is named as a **2-pentene.**

[3] **Name and number the substituents and write the complete name.**

- The alkene has two methyl groups located at C2 and C3. Use the prefix di- before methyl → 2,3-**di**methyl.

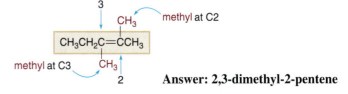

3
CH3
methyl at C2
CH3CH2C=CCH3
methyl at C3
CH3
2

Answer: 2,3-dimethyl-2-pentene

PROBLEM 13.5

Give the IUPAC name for each compound.

a. 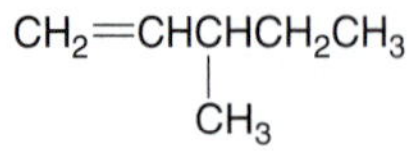 $CH_2\!\!=\!\!CHCHCH_2CH_3$
            $|$
          CH_3

c. $CH_3CH_2CH\!\!=\!\!CHCH\!\!=\!\!CHCH_3$

d.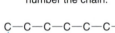

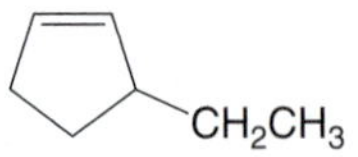

b. $(CH_3CH_2)_2C\!\!=\!\!CHCH_2CH_2CH_3$

PROBLEM 13.6

Give the IUPAC name for each compound.

a. $CH_3CH_2CH_2CH_2CH_2C\!\!\equiv\!\!CCH(CH_3)_2$

b. 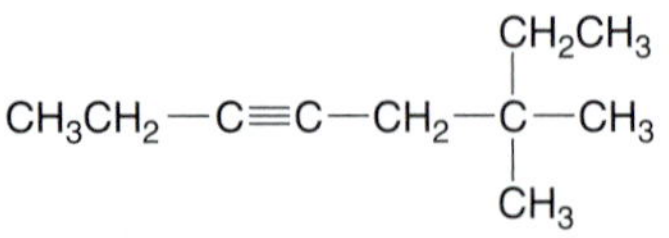 $CH_3CH_2\!-\!C\!\equiv\!C\!-\!CH_2\!-\!\underset{\underset{\displaystyle CH_3}{|}}{\overset{\overset{\displaystyle CH_2CH_3}{|}}{C}}\!-\!CH_3$

SAMPLE PROBLEM 13.3

Draw the structure corresponding to each IUPAC name: (a) 5,5-dimethyl-3-heptyne; (b) 1,3-dimethylcyclohexene.

ANALYSIS

First identify the parent name to find the longest carbon chain or ring, and then use the suffix to determine the functional group; the suffix -ene = an alkene and -yne = an alkyne. Then number the carbon chain or ring and place the functional group at the indicated carbon. Add the substituents and enough hydrogens to give each carbon four bonds.

SOLUTION

a. 5,5-Dimethyl-3-heptyne has 7 C's as its longest chain (hept-) and a triple bond that begins at C3 (3-heptyne). Two methyl groups are bonded to C5.

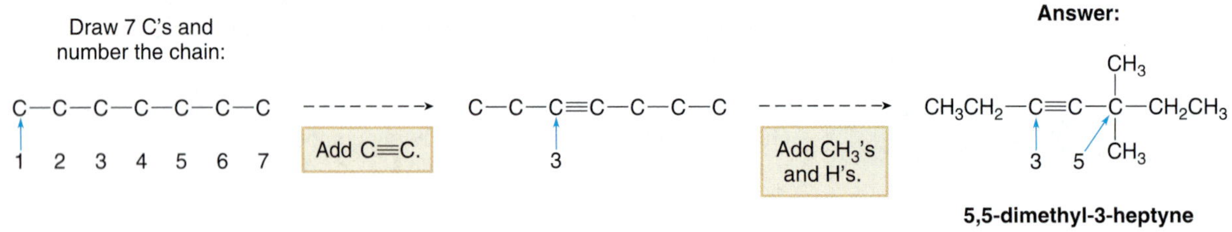

Draw 7 C's and number the chain:

Answer:

C—C—C—C—C—C—C - - - - → C—C—C≡C—C—C—C - - - - → $CH_3CH_2\!-\!C\!\equiv\!C\!-\!\overset{\overset{\displaystyle CH_3}{|}}{\underset{\underset{\displaystyle CH_3}{|}}{C}}\!-\!CH_2CH_3$
1 2 3 4 5 6 7 | Add C≡C. | 3 | Add CH₃'s and H's. | 3 5

5,5-dimethyl-3-heptyne

b. 1,3-Dimethylcyclohexene contains a six-membered ring with a double bond (cyclohexene). Number the ring to put the double bond between C1 and C2, and then add two methyl groups at C1 and C3.

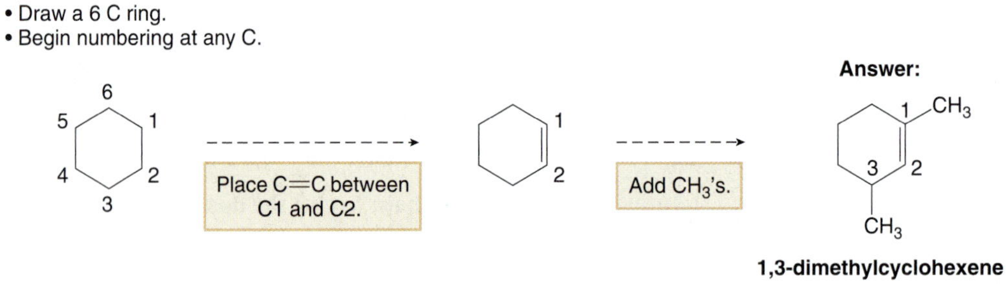

• Draw a 6 C ring.
• Begin numbering at any C.

Answer:

(ring numbered 1–6) - - - - → | Place C=C between C1 and C2. | - - - - → (ring with double bond) | Add CH₃'s. | - - - - → (ring with two CH₃ groups)

1,3-dimethylcyclohexene

PROBLEM 13.7

Give the structure corresponding to each of the following names.

a. 4-methyl-1-hexene
b. 5-ethyl-2-methyl-2-heptene

c. 2,5-dimethyl-3-hexyne
d. 1-propylcyclobutene

PROBLEM 13.8

Give the structure corresponding to each of the following names.

a. 2-methyl-2-octene
b. 5-ethyl-2-methyl-1-nonene

c. 1,3-cyclohexadiene
d. 4-ethyl-1-decyne

13.3 CIS–TRANS ISOMERS

As we learned in Section 12.2 on alkanes, constitutional isomers are possible for alkenes of a given molecular formula. For example, there are three constitutional isomers for an alkene of molecular formula C_4H_8—1-butene, 2-butene, and 2-methylpropene.

$$CH_2\!\!=\!\!CHCH_2CH_3 \qquad CH_3CH\!\!=\!\!CHCH_3 \qquad \begin{matrix} CH_3 \\ | \\ CH_2\!\!=\!\!CCH_3 \end{matrix}$$

1-butene   2-butene   2-methylpropene

These compounds are constitutional isomers because the atoms are bonded to each other in different ways. 1-Butene and 2-butene both have a four-carbon chain, but the location of the double bond is different, and 2-methylpropene has a longest chain containing only three carbon atoms.

13.3A STEREOISOMERS—A NEW CLASS OF ISOMER

2-Butene illustrates another important aspect about alkenes. In contrast to the free rotation observed around carbon–carbon single bonds (Section 12.2D), there is **restricted rotation** around the carbon atoms of a double bond. As a result, the groups on one side of the double bond *cannot* rotate to the other side.

With 2-butene, there are two ways to arrange the atoms on the double bond. The two CH_3 groups can be on the *same side* of the double bond or they can be on *opposite sides* of the double bond. These molecules are *different* compounds with the same molecular formula; that is, they are **isomers.**

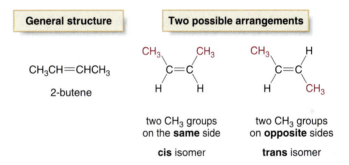

- When the two CH_3 groups are on the *same side* of the double bond, the compound is called the *cis* isomer.
- When the two CH_3 groups are on *opposite sides* of the double bond, the compound is called the *trans* isomer.

To give these compounds unique IUPAC names, we use the prefixes *cis* and *trans* before the remainder of the name to indicate the relative location of the two alkyl groups on the double bond. Thus, one isomer is called ***cis*-2-butene,** and the other isomer is called ***trans*-2-butene.**

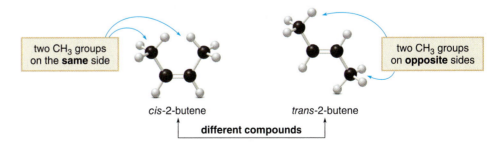

The cis and trans isomers of 2-butene are a specific example of a general class of isomer that occurs at carbon–carbon double bonds. **Whenever the two groups on *each* end of a C=C are *different from each other*, two isomers are possible.**

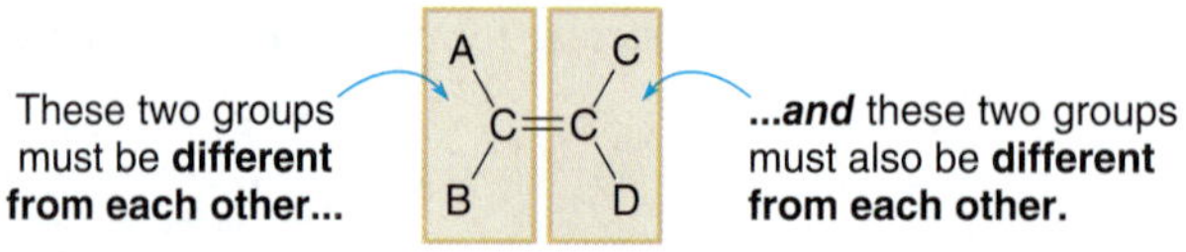

These two groups must be **different from each other...**   ...*and* these two groups must also be **different from each other.**

When the two groups on one end of the double bond are identical, there is still restricted rotation, but no cis and trans isomers are possible. With 1-butene, $CH_2=CHCH_2CH_3$, one end of the double bond has two hydrogens, so the ethyl group (CH_2CH_3) is always cis to a hydrogen, no matter how the molecule is drawn.

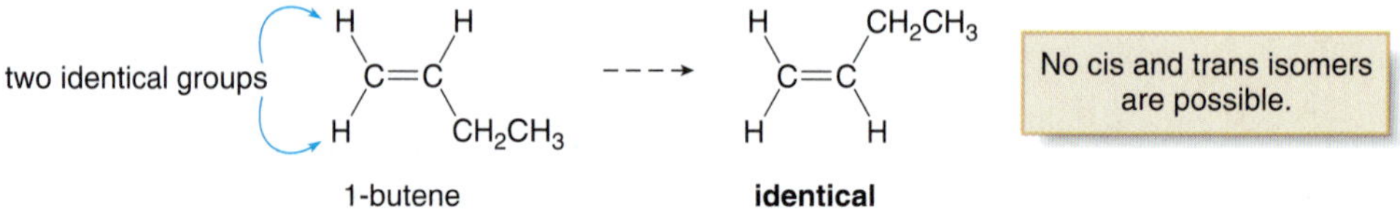

two identical groups

1-butene   identical

No cis and trans isomers are possible.

SAMPLE PROBLEM 13.4

Draw *cis*- and *trans*-3-hexene.

ANALYSIS

First, use the parent name to draw the carbon skeleton, and place the double bond at the correct carbon; 3-hexene indicates a 6 C chain with the double bond beginning at C3. Then use the definitions of cis and trans to draw the isomers.

SOLUTION

Each C of the double bond is bonded to a CH_3CH_2 group and a hydrogen. A cis isomer has the CH_3CH_2 groups bonded to the same side of the double bond. A trans isomer has the two CH_3CH_2 groups bonded to the opposite sides of the double bond.

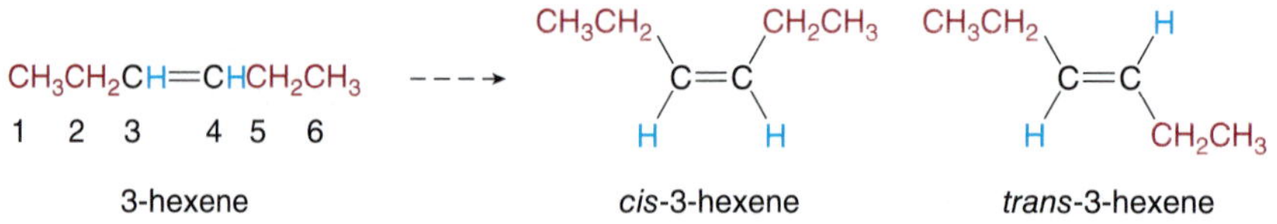

$CH_3CH_2CH=CHCH_2CH_3$
1 2 3   4 5 6

3-hexene   *cis*-3-hexene   *trans*-3-hexene

PROBLEM 13.9

Draw a condensed structure for each compound: (a) *cis*-2-octene; (b) *trans*-3-heptene; (c) *trans*-4-methyl-2-pentene.

PROBLEM 13.10

For which compounds are cis and trans isomers possible?

a. $CH_3CH_2CH=CHCH_3$ b. $CH_2=CHCH_2CH_2CH_3$ c. $CH_3C=CHCH_2CH_2CH_3$ (with CH_3 on the C)

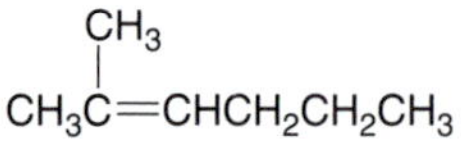

PROBLEM 13.11

Bombykol is secreted by the female silkworm moth (*Bombyx mori*) to attract mates. Bombykol contains two double bonds, and each double bond must have a particular three-dimensional arrangement of groups around it to be biologically active. Label the double bonds of bombykol as cis or trans.

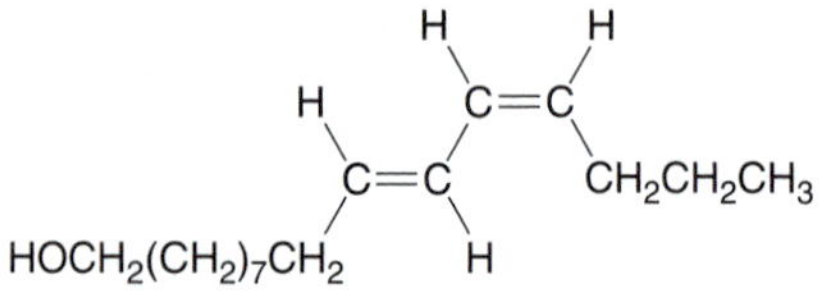

$HOCH_2(CH_2)_7CH_2$ $CH_2CH_2CH_3$

bombykol

ENVIRONMENTAL NOTE

The female silkworm moth, *Bombyx mori,* secretes the sex pheromone bombykol. Pheromones like bombykol have been used to control insect populations. In one method, the pheromone is placed in a trap containing a poison or sticky substance, and the male is lured to the trap by the pheromone. Alternatively, the pheromone can be released into infested areas, confusing males who are then unable to find females.

▼ **FIGURE 13.1** Comparing Three Isomers: 1-Butene, *cis*-2-Butene, and *trans*-2-Butene

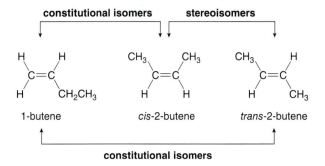

Cis and trans compounds are isomers, but they are *not* constitutional isomers. Each carbon atom of *cis*- and *trans*-2-butene is bonded to the same atoms. The only difference is the three-dimensional arrangement of the groups around the double bond. Isomers of this sort are called **stereoisomers.**

• *Stereoisomers* are isomers that differ *only* in the 3-D arrangement of atoms.

Thus, *cis*-2-butene and *trans*-2-butene are stereoisomers, but each of these compounds is a constitutional isomer of 1-butene, as shown in Figure 13.1.

We have now learned the two major classes of isomers.

We will learn more about stereoisomers in Chapter 15.

• **Constitutional isomers differ in the way the atoms are bonded to each other.**
• **Stereoisomers differ only in the three-dimensional arrangement of atoms.**

PROBLEM 13.12

Label each pair of alkenes as constitutional isomers or stereoisomers.

a. $CH_3CH=CHCH_2CH_3$ and $CH_2=CHCH_2CH_2CH_3$

b. (structures) and (structures)

c. (structures) and (structures)

13.3B FOCUS ON HEALTH & MEDICINE
SATURATED AND UNSATURATED FATTY ACIDS

Naturally occurring animal fats and vegetable oils are formed from fatty acids. **Fatty acids are carboxylic acids (RCOOH) with long carbon chains of 12–20 carbon atoms.** Because a fatty acid has many nonpolar C—C and C—H bonds and few polar bonds, fatty acids are insoluble in water. There are two types of fatty acids.

• *Saturated* fatty acids have no double bonds in their long hydrocarbon chains.
• *Unsaturated* fatty acids have one or more double bonds in their long hydrocarbon chains.

TABLE 13.1　Common Saturated and Unsaturated Fatty Acids

Name	Structure	Mp (°C)
Stearic acid (0 C=C)	$CH_3CH_2CH_2CH_2CH_2CH_2CH_2CH_2CH_2CH_2CH_2CH_2CH_2CH_2CH_2CH_2CH_2COOH$	71
Oleic acid (1 C=C)	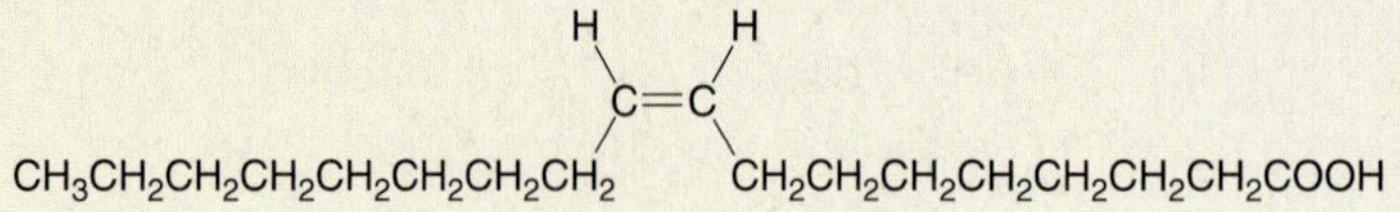	16
Linoleic acid (2 C=C)	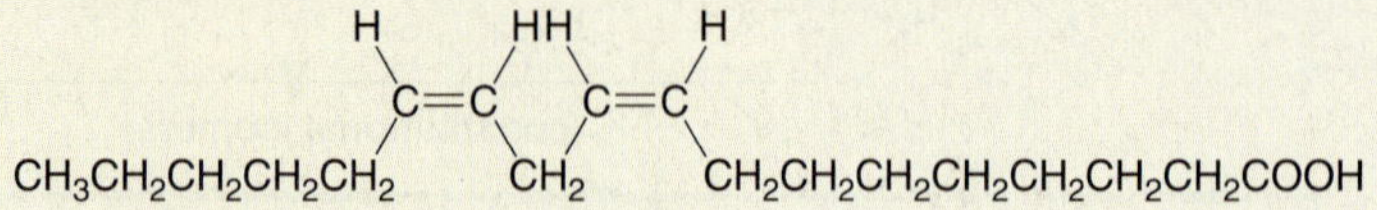	−5
Linolenic acid (3 C=C)	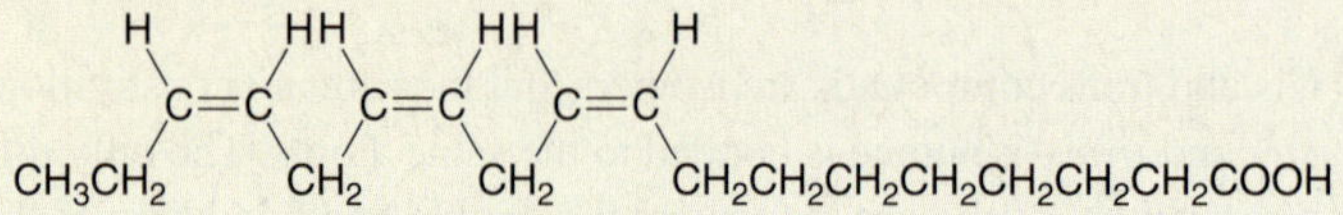	−11

Table 13.1 lists the structure and melting point of four fatty acids containing 18 carbon atoms. Stearic acid is one of the two most common saturated fatty acids, while oleic and linoleic acids are the most common unsaturated ones.

Linoleic and linolenic acids are **essential fatty acids,** meaning they cannot be synthesized in the human body and must therefore be obtained in the diet. A common source of these essential fatty acids is whole milk. Babies fed a diet of nonfat milk in their early months do not thrive because they do not obtain enough of these essential fatty acids.

One structural feature of unsaturated fatty acids is especially noteworthy.

- **Generally, double bonds in naturally occurring fatty acids are cis.**

The presence of cis double bonds affects the melting point of these fatty acids greatly.

- **As the number of double bonds in the fatty acid *increases,* the melting point *decreases.***

The cis double bonds introduce kinks in the long hydrocarbon chain, as shown in Figure 13.2. This makes it difficult for the molecules to pack closely together in a solid. **The larger the number of cis double bonds, the more kinks in the hydrocarbon chain, and the lower the melting point.**

Fats and oils are organic molecules synthesized in plant and animal cells from fatty acids. Fats and oils have different physical properties.

- **Fats are solids at room temperature. Fats are generally formed from fatty acids having few double bonds.**
- **Oils are liquids at room temperature. Oils are generally formed from fatty acids having a larger number of double bonds.**

Saturated fats are typically obtained from animal sources, while unsaturated oils are common in vegetable sources. Thus, butter and lard are formed from saturated fatty acids, while olive oil and safflower oil are formed from unsaturated fatty acids. An exception to this generalization is coconut oil, which is composed largely of saturated fatty acids.

Considerable evidence suggests that an elevated cholesterol level is linked to increased risk of heart disease. Saturated fats stimulate cholesterol synthesis in the liver, resulting in an increase in cholesterol concentration in the blood. We will learn more about fats and oils in Section 13.7 and Chapter 19.

▼ **FIGURE 13.2**   **The Three-Dimensional Structure of Four Fatty Acids**

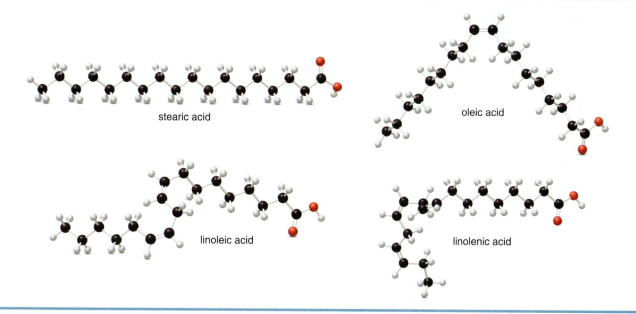

stearic acid

oleic acid

linoleic acid

linolenic acid

PROBLEM 13.13

Draw out a structure of arachidonic acid, indicating the arrangement of groups around the four double bonds.

$$CH_3(CH_2)_4CH{=}CHCH_2CH{=}CHCH_2CH{=}CHCH_2CH{=}CHCH_2(CH_2)_2CO_2H$$

arachidonic acid

PROBLEM 13.14

You have two fatty acids, one with a melting point of 63 °C, and one with a melting point of 1 °C. Which structure corresponds to each melting point?

$$CH_3(CH_2)_{14}COOH \quad CH_3(CH_2)_5CH{=}CH(CH_2)_7COOH$$

palmitic acid   palmitoleic acid

13.4   INTERESTING ALKENES IN FOOD AND MEDICINE

In addition to the plant hormone ethylene, $CH_2{=}CH_2$ (Section 13.1), many useful compounds contain one or more carbon–carbon double bonds.

More examples of antioxidants are given in Section 13.12.

Lycopene, the red pigment in tomatoes and watermelon, contains 13 double bonds. Lycopene is an **antioxidant,** a compound that prevents an unwanted oxidation reaction from occurring. Diets that contain a high intake of antioxidants like lycopene have been shown to decrease the risk of heart disease and certain forms of cancer. Unlike other antioxidants that can be destroyed when fresh fruits or vegetables are processed, even foods such as tomato paste, tomato juice, and ketchup are high in lycopene.

lycopene

Processed tomato products contain the antioxidant lycopene.

Lycopene's red color is due to the 11 double bonds that are each separated by one single bond (labeled in red). These double bonds absorb some (but not all) wavelengths of visible light. When a compound absorbs visible light, it takes on the color of those wavelengths of light it does *not* absorb. Since lycopene absorbs blue-green light, it appears red because it does not absorb the red light of the visible spectrum.

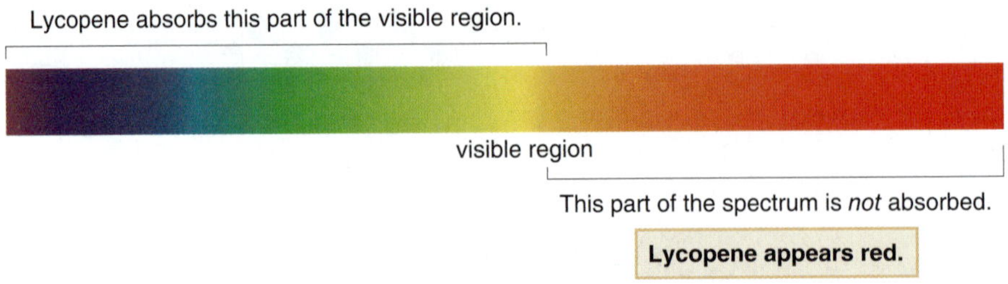

Lycopene absorbs this part of the visible region.

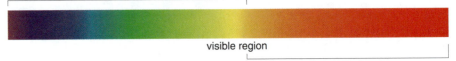

visible region

This part of the spectrum is *not* absorbed.

Lycopene appears red.

Tamoxifen is a potent anticancer drug that contains a carbon–carbon double bond in addition to other functional groups. Tamoxifen is used widely in the treatment of certain breast cancers that require the female sex hormone estrogen for growth. Tamoxifen binds to estrogen receptors, and in this way it inhibits the growth of breast cancers that are estrogen dependent. Tamoxifen is sold under the trade name of Novaldex.

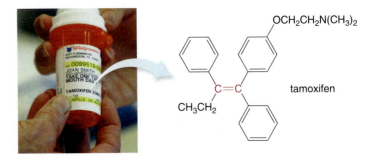

tamoxifen

PROBLEM 13.15   Identify all of the functional groups in tamoxifen.

13.5 FOCUS ON HEALTH & MEDICINE
ORAL CONTRACEPTIVES

The development of synthetic oral contraceptives in the 1960s revolutionized the ability to control fertility. Prior to that time, women ingested all sorts of substances—iron rust, gunpowder, tree bark, sheep's urine, elephant dung, and others—in the hope of preventing pregnancy.

Synthetic birth control pills are similar in structure to the female sex hormones **estradiol** and **progesterone,** but they also contain a carbon–carbon triple bond. Most oral contraceptives contain two synthetic hormones that are more potent than these natural hormones, so they can be administered in lower doses.

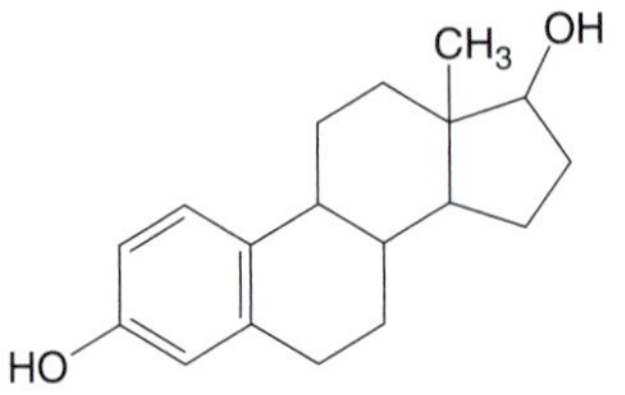

estradiol

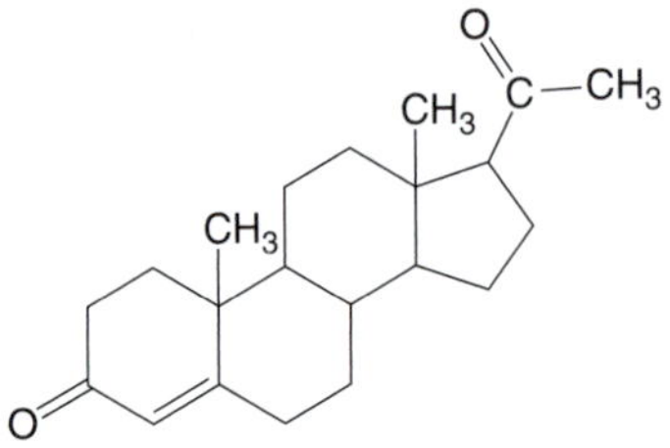

progesterone

▼ **FIGURE 13.3   How Oral Contraceptives Work**

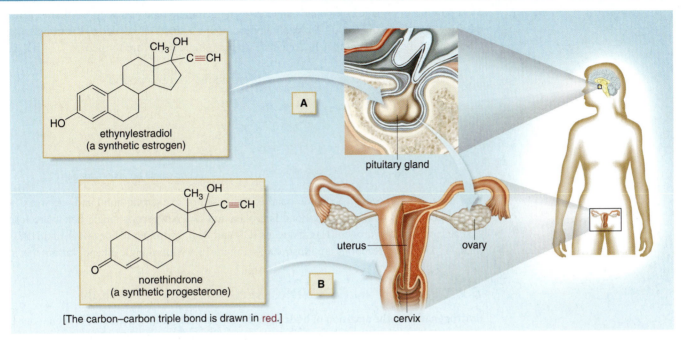

ethynylestradiol
(a synthetic estrogen)

norethindrone
(a synthetic progesterone)

[The carbon–carbon triple bond is drawn in red.]

pituitary gland

uterus

ovary

cervix

Monthly cycles of hormones from the pituitary gland cause ovulation, the release of an egg from an ovary. To prevent pregnancy, the two synthetic hormones in many oral contraceptives have different effects on the female reproductive system. **A:** The elevated level of **ethynylestradiol,** a synthetic estrogen, "fools" the pituitary gland into thinking a woman is pregnant, so ovulation does not occur. **B:** The elevated level of **norethindrone,** a synthetic progesterone, stimulates the formation of a thick layer of mucus in the cervix, making it difficult for sperm to reach the uterus.

Two common components of birth control pills are **ethynylestradiol** and **norethindrone.** Ethynylestradiol is a synthetic estrogen that resembles the structure and biological activity of estradiol. Norethindrone is a synthetic progesterone that is similar to the natural hormone progesterone. These compounds act by artificially elevating hormone levels in a woman, and this prevents pregnancy, as illustrated in Figure 13.3.

Two other synthetic hormones with triple bonds are **RU 486** and **levonorgestrel.** RU 486 blocks the effects of progesterone, and by this means, prevents implantation of a fertilized egg. RU 486 is used to induce abortions within the first few weeks of pregnancy. Levonorgestrel interferes with ovulation, and thereby prevents pregnancy if taken within a few days of unprotected sex.

RU 486
(Trade name: Mifepristone)

levonorgestrel
(Trade name: Plan B)

[The carbon–carbon triple bond is drawn in red.]

PROBLEM 13.16   Identify the functional groups in RU 486 and levonorgestrel.

13.6 REACTIONS OF ALKENES

Most families of organic compounds undergo a characteristic type of reaction. **Alkenes undergo addition reactions.** In an addition reaction, new groups X and Y are added to a starting material. One bond of the double bond is broken and two new single bonds are formed.

- Addition is a reaction in which elements are added to a compound.

$$\text{Addition} \quad \underset{\text{}}{\text{C}}=\underset{\text{}}{\text{C}} \; + \; \text{X}-\text{Y} \;\longrightarrow\; -\underset{|}{\overset{|}{\text{C}}}-\underset{|}{\overset{|}{\text{C}}}-$$

One bond is broken.   Two single bonds are formed.

Why does addition occur? A double bond is composed of one strong bond and one weak bond. In an addition reaction, the weak bond is broken and two new strong single bonds are formed. Alkenes react with hydrogen (H_2), halogens (Cl_2 and Br_2), hydrogen halides (HCl and HBr), and water (H_2O). These reactions are illustrated in Figure 13.4 with a single alkene starting material.

13.6A ADDITION OF HYDROGEN—HYDROGENATION

Hydrogenation is the addition of hydrogen (H_2) to an alkene. Two bonds are broken—one bond of the carbon–carbon double bond and the H—H bond—and two new C—H bonds are formed.

$$\text{Hydrogenation} \quad \underset{\text{alkene}}{\overset{}{\text{C}}=\text{C}} \; + \; \text{H}-\text{H} \;\xrightarrow[\text{catalyst}]{\text{metal}}\; -\underset{\underset{\text{alkane}}{\text{H} \quad \text{H}}}{\overset{|}{\text{C}}-\overset{|}{\text{C}}}- \quad \longleftarrow \text{H}_2 \text{ is added.}$$

The addition of H_2 occurs only in the presence of a **metal catalyst** such as palladium (Pd). The metal provides a surface that binds both the alkene and H_2, and this speeds up the rate of reaction. Hydrogenation of an alkene forms an **alkane** since the product has only C—C single bonds.

▼ **FIGURE 13.4 Four Addition Reactions of Alkenes**

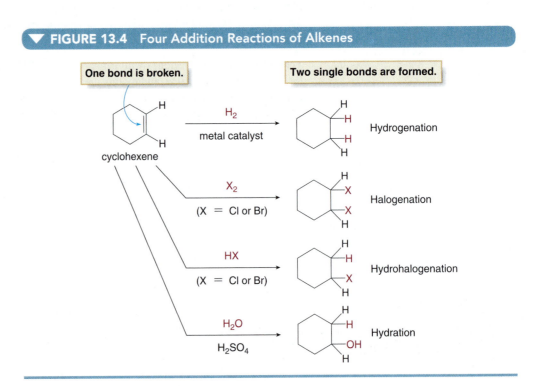

Example

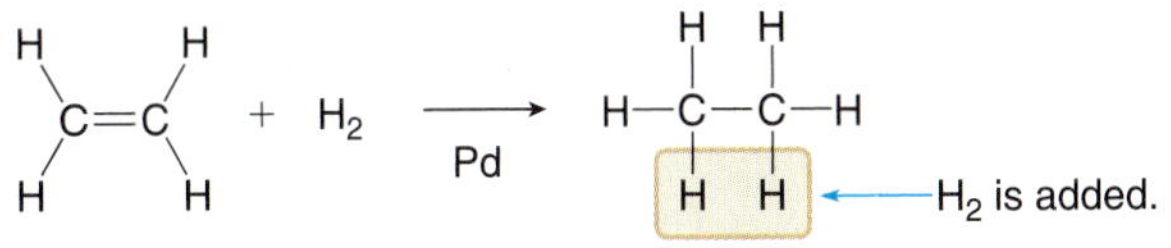

$$CH_2=CH_2 \quad + \quad H_2 \quad \xrightarrow{Pd} \quad H-\underset{|}{\overset{|}{C}}-\underset{|}{\overset{|}{C}}-H \quad \leftarrow \text{H}_2 \text{ is added.}$$

ethylene   ethane

SAMPLE PROBLEM 13.5

Draw the product of the following reaction.

$$CH_3CH_2CH_2CH=CH_2 \quad + \quad H_2 \quad \xrightarrow{Pd}$$

ANALYSIS

To draw the product of a hydrogenation reaction:

- Locate the C=C and mentally break one bond in the double bond.
- Mentally break the H—H bond of the reagent.
- Add one H atom to each C of the C=C, thereby forming two new C—H single bonds.

SOLUTION

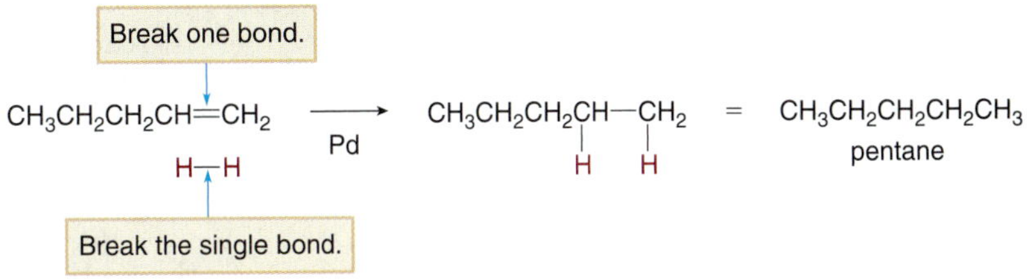

Break one bond.

$$CH_3CH_2CH_2CH=CH_2 \quad \xrightarrow{Pd} \quad CH_3CH_2CH_2\underset{|}{\overset{|}{CH}}-\underset{|}{\overset{|}{CH_2}} \quad = \quad CH_3CH_2CH_2CH_2CH_3$$

H—H

Break the single bond.

pentane

PROBLEM 13.17

What product is formed when each alkene is treated with H_2 and a Pd catalyst?

a. $CH_3CH_2CH=CHCH_2CH_3$

b. 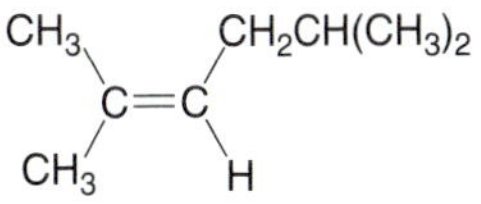 $\underset{CH_3}{\overset{CH_3}{>}}C=C\underset{H}{\overset{CH_2CH(CH_3)_2}{<}}$

c. (cyclohexene with CH_3)

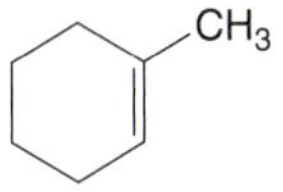

13.6B ADDITION OF HALOGEN—HALOGENATION

Halogenation is the addition of halogen (X_2) to an alkene. Two bonds are broken—one bond of the carbon–carbon double bond and the X—X bond—and two new C—X bonds are formed.

Halogenation

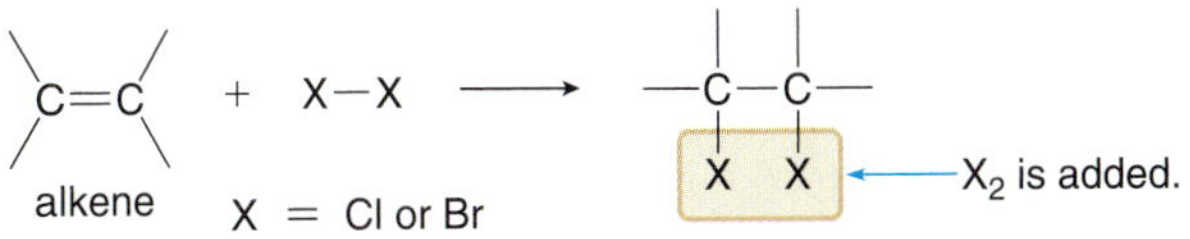

$$\overset{}{\underset{}{C}}=\overset{}{\underset{}{C} \quad + \quad X-X \quad \longrightarrow \quad -\underset{X}{\overset{}{C}}-\underset{X}{\overset{}{C}}- \quad \leftarrow X_2 \text{ is added.}}$$

alkene   X = Cl or Br

dihalide

Halogenation readily occurs with Cl_2 and Br_2. The product of halogenation is a **dihalide.**

Examples

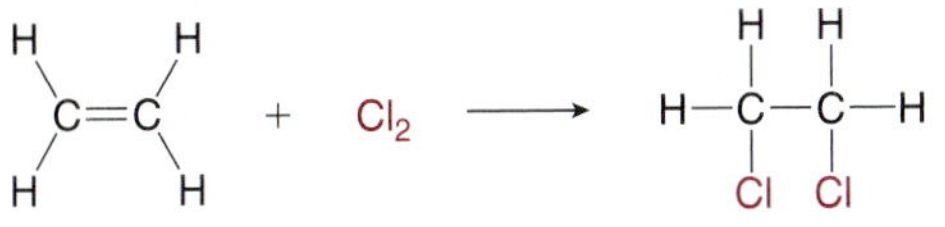

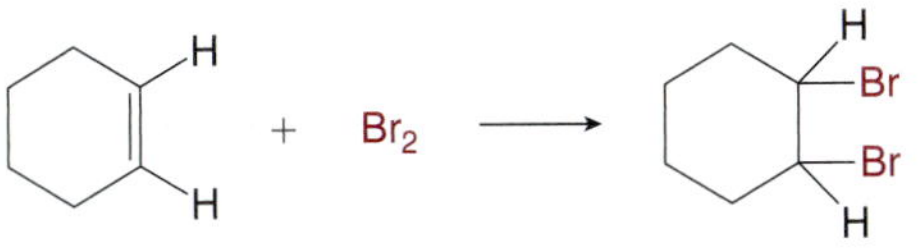

$$\underset{H}{\overset{H}{>}}C=C\underset{H}{\overset{H}{<}} \quad + \quad Cl_2 \quad \longrightarrow \quad H-\underset{Cl}{\overset{H}{C}}-\underset{Cl}{\overset{H}{C}}-H$$

(cyclohexene) $\quad + \quad Br_2 \quad \longrightarrow \quad$ (cyclohexane with Br, Br)

When an alkene is treated with red liquid bromine, the Br_2 adds to the carbon–carbon double bond to give a colorless product.

Halogenation with bromine is a simple chemical test used to determine if a double bond is present in an unknown compound. Bromine is a fuming red liquid. When bromine is added to an alkene, the bromine adds to the double bond and the red color disappears. The disappearance of the red color shows that a double bond is present.

PROBLEM 13.18

Draw the product of each reaction.

a. $CH_3CH_2CH{=}CH_2$ + Cl_2 ⟶

b. [structure: cyclohexene with two CH_3 groups on the double bond carbons] + Br_2 ⟶

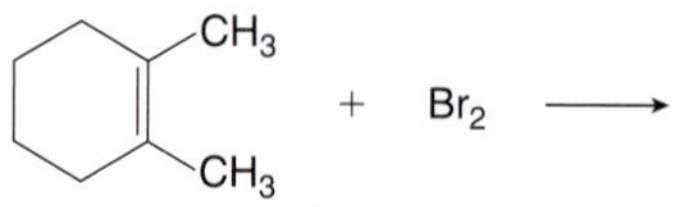

13.6C ADDITION OF HYDROGEN HALIDES— HYDROHALOGENATION

Hydrohalogenation is the addition of HX (X = Cl or Br) to an alkene. Two bonds are broken— one bond of the carbon–carbon double bond and the H—X bond—and new C—H and C—X bonds are formed.

Hydrohalogenation

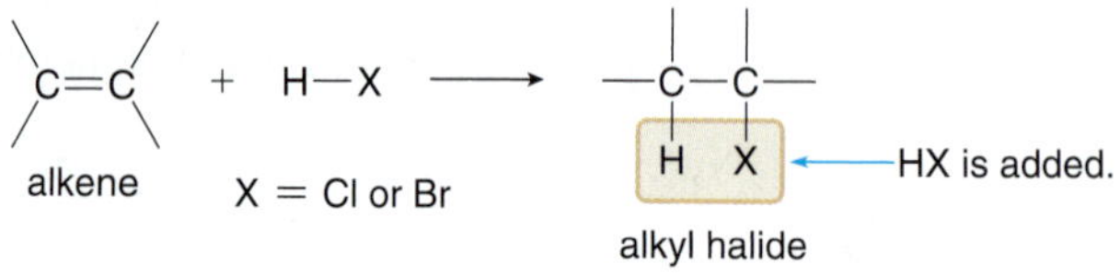

$$\underset{\text{alkene}}{\text{C}{=}\text{C}} + \underset{\substack{\text{H}-\text{X} \\ \text{X = Cl or Br}}}{} \longrightarrow \underset{\substack{\text{H X} \\ \text{alkyl halide}}}{-\text{C}-\text{C}-} \quad \longleftarrow \text{HX is added.}$$

Hydrohalogenation readily occurs with HCl and HBr. The product of hydrohalogenation is an **alkyl halide.**

Examples

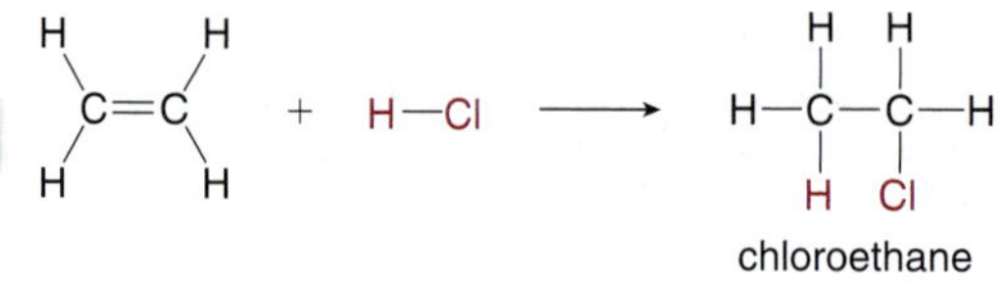

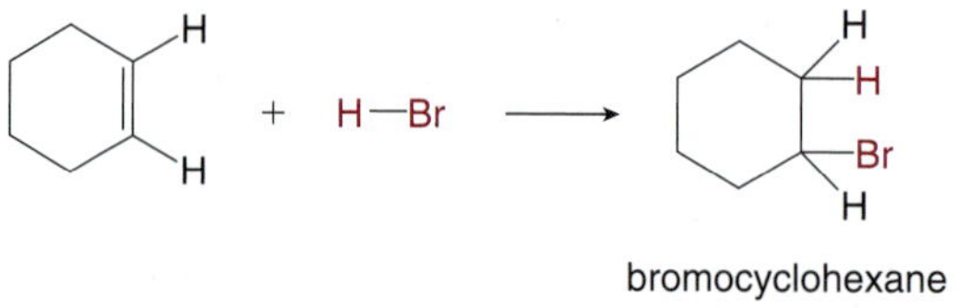

[structure: $H_2C{=}CH_2$] + H—Cl ⟶ [structure: H_3C-CH_2Cl] chloroethane

[structure: cyclohexene] + H—Br ⟶ [structure: bromocyclohexane] bromocyclohexane

There is one important difference in this addition reaction compared to the addition of H_2 and X_2. In this case, addition puts different atoms—H and X—on the two carbons of the double bond. As a result, HX can add to the double bond to give two constitutional isomers when an unsymmetrical alkene is used as starting material.

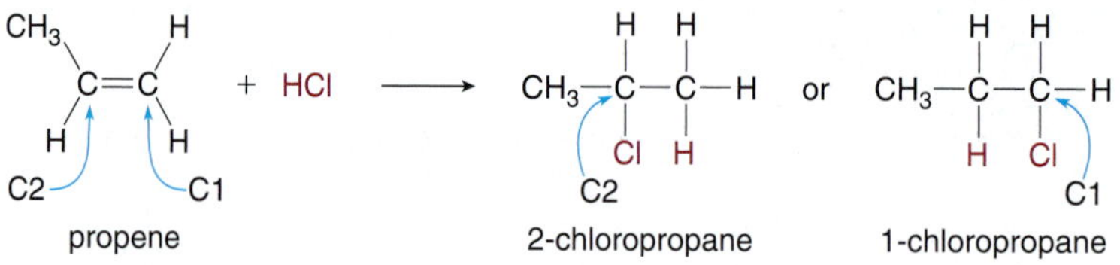

[structure: propene, C2 and C1 labeled] + HCl ⟶ [structure: $CH_3-CHCl-CH_3$] 2-chloropropane (C2) **or** [structure: $CH_3-CH_2-CH_2Cl$] 1-chloropropane (C1)

only product

For example, the addition of HCl to propene could in theory form two products. If H adds to the end carbon (labeled C1) and Cl adds to the middle carbon (C2), 2-chloropropane is formed. If Cl adds to the end carbon (C1) and H adds to the middle carbon (C2), 1-chloropropane is formed. In fact, **addition forms *only* 2-chloropropane.** This is a specific example of a general trend called **Markovnikov's rule,** named for the Russian chemist who first determined the selectivity of the addition of HX to an alkene.

- In the addition of HX to an unsymmetrical alkene, the H atom bonds to the less substituted carbon atom—that is, the carbon that has more H's to begin with.

The end carbon (C1) of propene has two hydrogens while the middle carbon (C2) has just one hydrogen. Addition puts the H of HCl on C1 since it had more hydrogens (two versus one) to begin with.

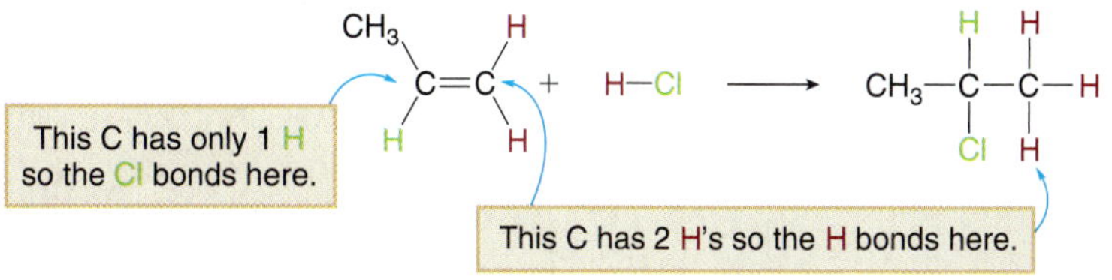

This C has only 1 H so the Cl bonds here.

This C has 2 H's so the H bonds here.

$$CH_3\text{-}CH=CH_2 + H\text{-}Cl \longrightarrow CH_3\text{-}CHCl\text{-}CH_3$$

SAMPLE PROBLEM 13.6

What product is formed when 2-methylpropene [$(CH_3)_2C=CH_2$] is treated with HBr?

ANALYSIS

Alkenes undergo addition reactions, so the elements of H and Br must be added to the double bond. Since the alkene is unsymmetrical, the H atom of HBr bonds to the carbon that has more H's to begin with.

SOLUTION

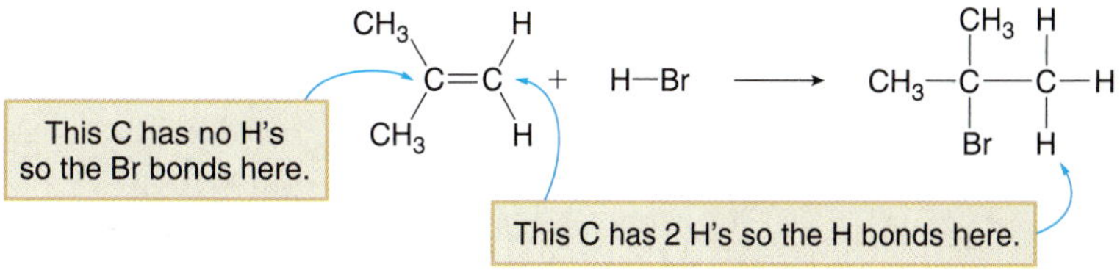

This C has no H's so the Br bonds here.

This C has 2 H's so the H bonds here.

PROBLEM 13.19

What product is formed in each of the following reactions?

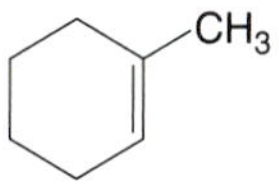

a. $CH_3CH=CHCH_3 + HBr \longrightarrow$

c. $(CH_3)_2C=CHCH_3 + HCl \longrightarrow$

b. [cyclohexene with CH₃] + HBr $\longrightarrow$

d. 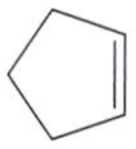 [cyclopentene] + HCl $\longrightarrow$

13.6D ADDITION OF WATER—HYDRATION

Hydration is the addition of water to an alkene. Two bonds are broken—one bond of the carbon–carbon double bond and the H—OH bond—and new C—H and C—OH bonds are formed.

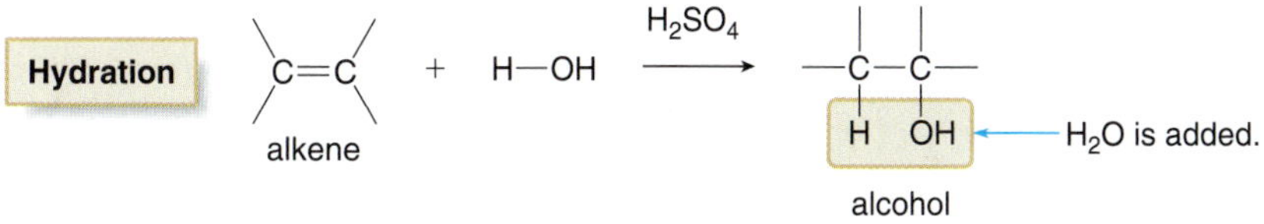

Hydration

$$\underset{\text{alkene}}{C=C} + H\text{-}OH \xrightarrow{H_2SO_4} \underset{\substack{|\ \ \ | \\ H\ \ OH \\ \text{alcohol}}}{-C-C-}$$

H_2O is added.

Hydration occurs only if a strong acid such as H_2SO_4 is added to the reaction mixture. The product of hydration is an **alcohol.** For example, hydration of ethylene forms ethanol.

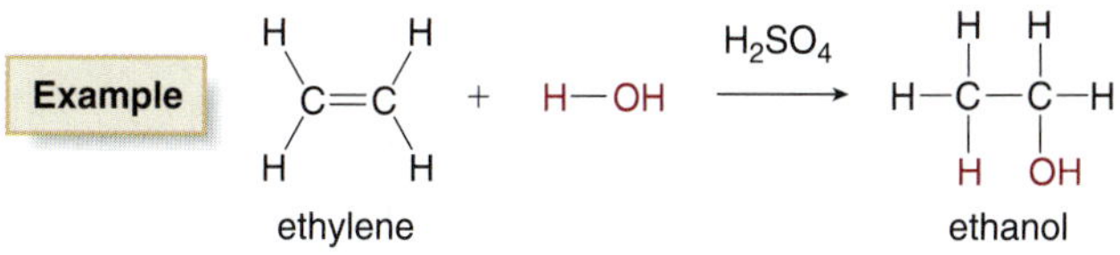

Example

$$\underset{\text{ethylene}}{H_2C=CH_2} + H\text{-}OH \xrightarrow{H_2SO_4} \underset{\text{ethanol}}{H_3C\text{-}CH_2OH}$$

Ethanol is used as a solvent in many reactions in the laboratory. **Ethanol is also used as a gasoline additive** because, like alkanes, it burns in the presence of oxygen to form CO_2 and H_2O with the release of a great deal of energy. Although ethanol can also be formed by the fermentation of carbohydrates in grains and potatoes, much of the ethanol currently used in gasoline and solvent comes from the hydration of ethylene.

Gasohol contains 10% ethanol.

The hydration of unsymmetrical alkenes follows Markovnikov's rule.

> • In the addition of H_2O to an unsymmetrical alkene, the H atom bonds to the less substituted carbon atom—that is, the carbon that has more H's to begin with.

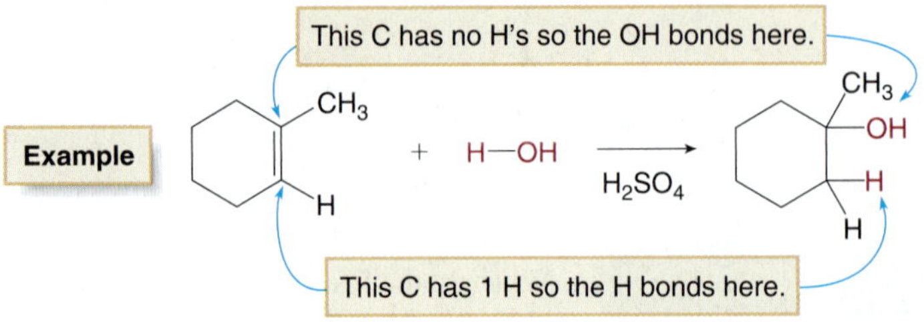

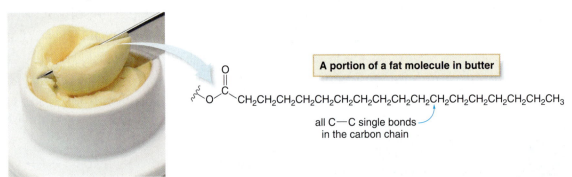

Example

This C has no H's so the OH bonds here.

This C has 1 H so the H bonds here.

PROBLEM 13.20

What alcohol is formed when each alkene is treated with H_2O in the presence of H_2SO_4?

a. $CH_3CH{=}CHCH_3$

b. $CH_3CH_2CH{=}CH_2$

c.

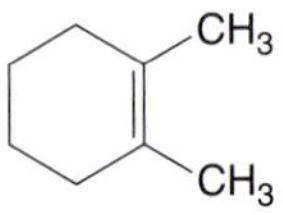

PROBLEM 13.21

What product is formed when 1-pentene ($CH_3CH_2CH_2CH{=}CH_2$) is treated with each reagent?

a. H_2, Pd   c. Br_2 e. HCl

b. Cl_2 d. HBr   f. H_2O, H_2SO_4

13.7 FOCUS ON HEALTH & MEDICINE
MARGARINE OR BUTTER?

One addition reaction of alkenes, hydrogenation, is especially important in the food industry. It lies at the heart of the debate over which product, butter or margarine, is better for the consumer.

As we learned in Section 13.3, butter is derived from saturated fatty acids like stearic acid [$CH_3(CH_2)_{16}COOH$], compounds with long carbon chains that contain only carbon–carbon single bonds. As a result, butter is a solid at room temperature.

A portion of a fat molecule in butter

$$CH_2CH_2CH_2CH_2CH_2CH_2CH_2CH_2CH_2CH_2CH_2CH_2CH_2CH_2CH_2CH_2CH_3$$

all C—C single bonds
in the carbon chain

Margarine, on the other hand, is a synthetic product that mimics the taste and texture of butter. It is prepared from vegetable oils derived from unsaturated fatty acids like linoleic acid [$CH_3(CH_2)_4CH{=}CHCH_2CH{=}CH(CH_2)_7COOH$]. Margarine is composed mainly of *partially hydrogenated* vegetable oils formed by adding hydrogen to the double bonds in the carbon chain derived from unsaturated fatty acids.

▼ FIGURE 13.5 Partial Hydrogenation of the Double Bonds in a Vegetable Oil

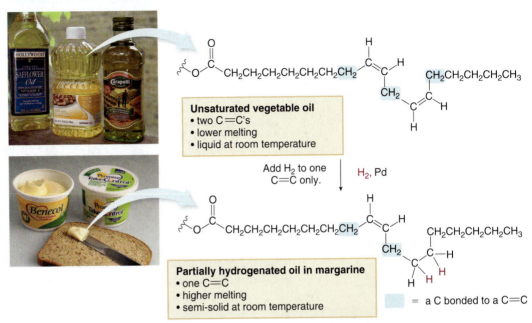

- When an oil is *partially* hydrogenated, some double bonds react with H_2, while some double bonds remain in the product. Since the product has fewer double bonds, it has a higher melting point. Thus, a liquid oil is converted to a semi-solid.

- Partial hydrogenation decreases the number of carbons bonded to double bonds, making the molecule less easily oxidized, thereby increasing its shelf life.

When an unsaturated liquid vegetable oil is treated with hydrogen, some (or all) of the double bonds add H_2, as shown in Figure 13.5. This increases the melting point of the oil, thus giving it a semi-solid consistency that more closely resembles butter. This process is called *hardening.*

As we will learn in Section 19.4, unsaturated oils with carbon–carbon double bonds are healthier than saturated fats with no double bonds. Why, then, does the food industry partially hydrogenate oils, thus reducing the number of double bonds? The reasons relate to texture and shelf life. Consumers prefer the semi-solid consistency of margarine to a liquid oil. Imagine pouring vegetable oil on a piece of toast or pancakes!

Furthermore, unsaturated oils are more susceptible than saturated fats to oxidation at the carbons adjacent to the double bond carbons. Oxidation makes the oil rancid and inedible. Hydrogenating the double bonds reduces the number of carbons bonded to double bonds (also illustrated in Figure 13.5), thus reducing the likelihood of oxidation and increasing the shelf life of the food product. This process reflects a delicate balance between providing consumers with healthier food products, while maximizing shelf life to prevent spoilage.

One other fact is worthy of note. During hydrogenation, some of the cis double bonds in vegetable oils are converted to trans double bonds, forming so-called "trans fats." The shape of the resulting fatty acid chain is very different, closely resembling the shape of a *saturated* fatty acid chain. As a result, trans fats are thought to have the same negative effects on blood cholesterol levels as saturated fats; that is, trans fats stimulate cholesterol synthesis in the liver, thus increasing blood cholesterol levels, a factor linked to increased risk of heart disease.

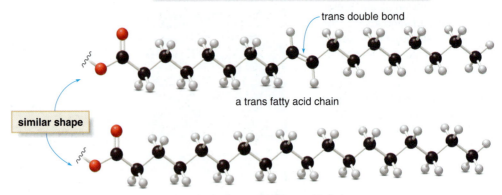

A "trans" fat and a saturated fat have a similar 3-D shape.

trans double bond

a trans fatty acid chain

similar shape

a saturated fatty acid chain

Which should you choose—butter or margarine—and if it's margarine, which of the many varieties is best for you? Margarine can be hard or soft. It can be made from olive oil, corn oil, safflower oil, or even yogurt. One fact remains clear to nutritionists. It is best to limit your intake of both butter (high in saturated fat) and margarine (high in trans fat).

PROBLEM 13.22	(a) When linolenic acid (Table 13.1) is partially hydrogenated with one equivalent of H_2 in the presence of a Pd catalyst, three constitutional isomers are formed. One of them is linoleic acid. Draw the structures of the other two possible products. (b) What product is formed when linolenic acid is completely hydrogenated with three equivalents of H_2?
PROBLEM 13.23	Draw the structure of a stereoisomer of linoleic acid (Table 13.1) that has two trans double bonds. Which compound would you predict to have the higher melting point? Explain your reasoning.

13.8   POLYMERS—THE FABRIC OF MODERN SOCIETY

Polymers **are large molecules made up of repeating units of smaller molecules—called** *monomers*—**covalently bonded together.** Polymers include the naturally occurring proteins that compose hair, tendons, and fingernails. They also include such industrially important plastics as polyethylene, poly(vinyl chloride) (PVC), and polystyrene. Since 1976, the U.S. production of synthetic polymers has exceeded its steel production.

13.8A   SYNTHETIC POLYMERS

Many synthetic polymers—that is, those synthesized in the lab—are among the most widely used organic compounds in modern society. Soft drink bottles, plastic bags, food wrap, compact discs, Teflon, and Styrofoam are all made of synthetic polymers. Figure 13.6 illustrates several consumer products and the polymers from which they are made.

To form a polymer from an alkene monomer, the weak bond that joins the two carbons of the double bond is broken and new strong carbon–carbon single bonds join the monomers together. For example, joining **ethylene monomers** together forms the polymer **polyethylene,** a plastic used in milk containers and sandwich bags.

▼ FIGURE 13.6 Polymers in Some Common Products

poly(vinyl chloride)
PVC

polyacrylonitrile

polyisobutylene

polystyrene
(Styrofoam)

HDPE (high-density polyethylene) and **LDPE** (low-density polyethylene) are common types of polyethylene prepared under different reaction conditions and having different physical properties. HDPE is opaque and rigid, and used in milk containers and water jugs. LDPE is less opaque and more flexible, and used in plastic bags and electrical insulation.

ethylene
monomers

$$CH_2{=}CH_2 \ + \ CH_2{=}CH_2 \ + \ CH_2{=}CH_2$$

polymerization

polyethylene
polymer

$$\}{-}CH_2CH_2{-}CH_2CH_2{-}CH_2CH_2{-}\{$$

New bonds are shown in red.

three monomer units joined together

- *Polymerization* is the joining together of monomers to make polymers.

Many ethylene derivatives having the general structure $CH_2{=}CHZ$ are also used as monomers for polymerization. The identity of Z affects the physical properties of the resulting polymer, making some polymers more suitable for one consumer product (e.g., plastic bags or food wrap) than another (e.g., soft drink bottles or compact discs). Polymerization of $CH_2{=}CHZ$ usually yields polymers with the Z groups on every other carbon atom in the chain.

Three monomer units joined together

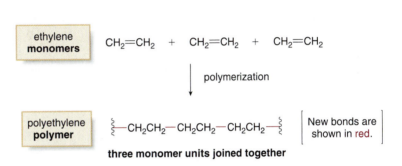

$$CH_2{=}CHZ \ + \ CH_2{=}CHZ \ + \ CH_2{=}CHZ \longrightarrow$$

Repeating unit: Many monomers are joined together.

shorthand structure

Polymer structures are often abbreviated by placing the atoms in the repeating unit in brackets, as shown. Table 13.2 lists some common monomers and polymers used in medicine.

TABLE 13.2 Common Monomers and Polymers Used in Medicine

Monomer	⟶	Polymer	Product

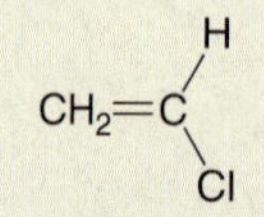

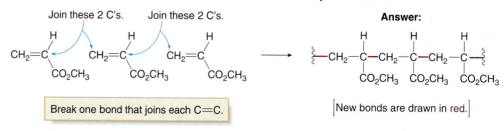

Row 1:

$CH_2\!=\!C$ with H and Cl — **vinyl chloride** ⟶ $-CH_2C\!-\!CH_2C\!-\!CH_2C-$ with Cl, Cl, Cl — **poly(vinyl chloride) PVC** — PVC blood bags and tubing

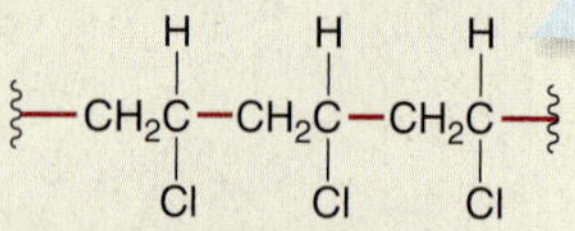

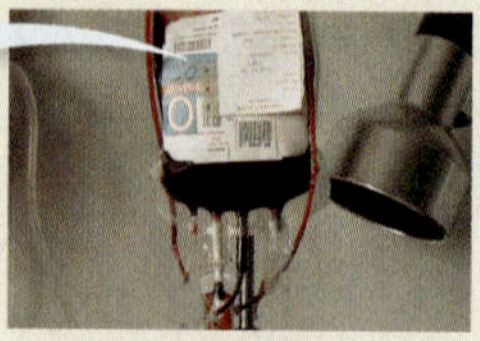

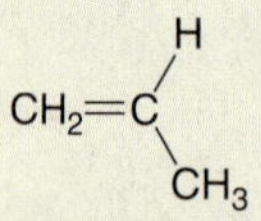

Row 2:

$CH_2\!=\!C$ with H and CH_3 — **propene** ⟶ $-CH_2C\!-\!CH_2C\!-\!CH_2C-$ with CH_3, CH_3, CH_3 — **polypropylene** — polypropylene syringes

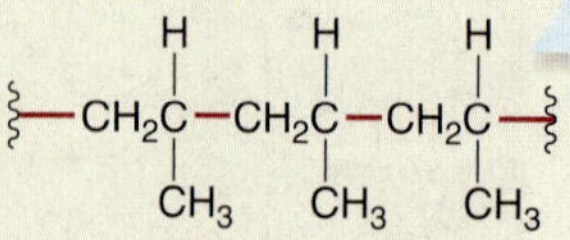

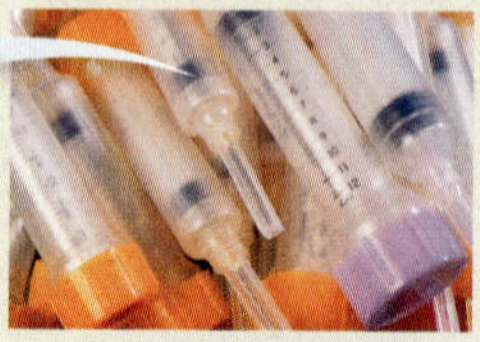

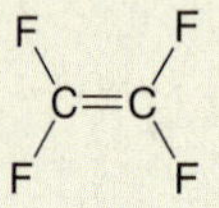

Row 3:

$C\!=\!C$ with F, F, F, F — **tetrafluoroethylene** ⟶ $-C\!-\!C\!-\!C\!-\!C\!-\!C\!-\!C-$ with F, F, F, F, F, F — **polytetrafluoroethylene Teflon** — dental floss

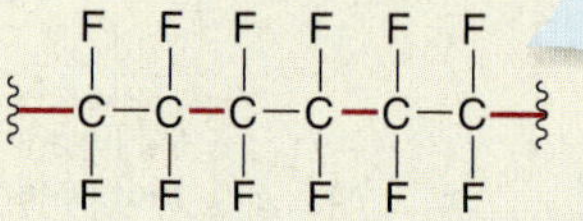

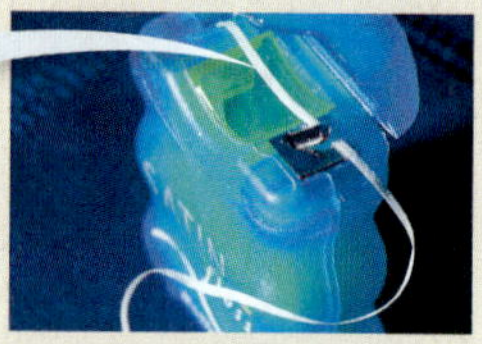

SAMPLE PROBLEM 13.7

What polymer is formed when $CH_2\!=\!CHCO_2CH_3$ (methyl acrylate) is polymerized?

ANALYSIS

Draw three or more alkene molecules and arrange the carbons of the double bonds next to each other. Break one bond of each double bond, and join the alkenes together with single bonds. With unsymmetrical alkenes, substituents are bonded to every other carbon.

SOLUTION

Join these 2 C's.   Join these 2 C's.

$CH_2\!=\!C$ (CO_2CH_3) $CH_2\!=\!C$ (CO_2CH_3) $CH_2\!=\!C$ (CO_2CH_3)   ⟶   **Answer:** $-CH_2\!-\!C\!-\!CH_2\!-\!C\!-\!CH_2\!-\!C-$ with CO_2CH_3, CO_2CH_3, CO_2CH_3

Break one bond that joins each C=C.

[New bonds are drawn in red.]

PROBLEM 13.24

What polymer is formed when each compound is polymerized?

a. 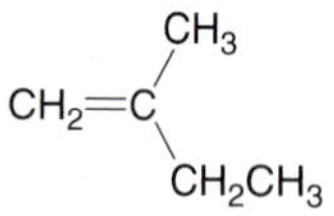 $CH_2\!=\!C$ with CH_3 and CH_2CH_3

b. $CH_2\!=\!C$ with CH_3 and CN

c. $CH_2\!=\!C$ with H and a benzene ring bearing Cl

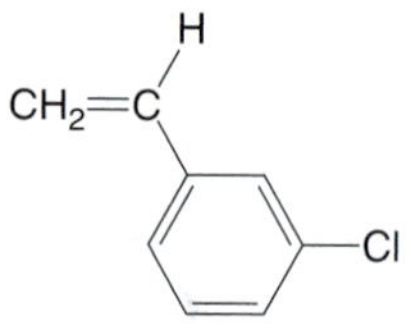

PROBLEM 13.25

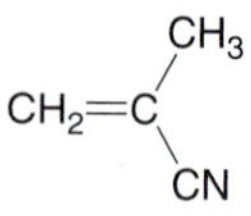

What monomer is used to form poly(vinyl acetate), a polymer used in paints and adhesives?

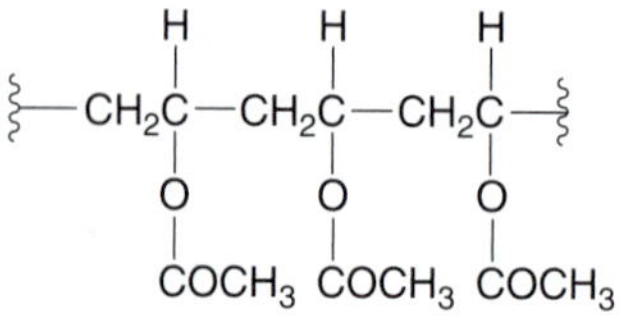

$-CH_2C\!-\!CH_2C\!-\!CH_2C-$ with O, O, O each connected to $COCH_3$, $COCH_3$, $COCH_3$

poly(vinyl acetate)

13.8B FOCUS ON THE ENVIRONMENT
POLYMER RECYCLING

The same desirable characteristics that make polymers popular materials for consumer products—durability, strength, and lack of reactivity—also contribute to environmental problems. Polymers do not degrade readily, and as a result, billions of pounds of polymers end up in landfills every year. Estimates suggest that synthetic polymers comprise 11% of solid municipal waste, 30% of which comes from packaging materials. Recycling existing polymer types to make new materials is one solution to the waste problem created by polymers.

Although thousands of different synthetic polymers have now been prepared, six compounds, called the **"Big Six,"** account for 76% of the synthetic polymers produced in the U.S. each year. Each polymer is assigned a recycling code (1–6) that indicates its ease of recycling; **the lower the number, the easier it is to recycle.** Table 13.3 lists these six most common polymers, as well as the type of products made from each recycled polymer.

Recycling begins with sorting plastics by type, shredding into small chips, and washing to remove adhesives and labels. After drying and removing any metal caps or rings, the polymer is melted and molded for reuse.

Of the Big Six, only the polyethylene terephthalate (PET) in soft drink bottles and the high-density polyethylene (HDPE) in milk jugs and juice bottles are recycled to any great extent. Since recycled polymers are often still contaminated with small amounts of adhesives and other materials, these recycled polymers are generally not used for storing food or drink products. Recycled HDPE is converted to Tyvek, an insulating wrap used in new housing construction, and recycled PET is used to make fibers for fleece clothing and carpeting. Currently about 23% of all plastics are recycled in the U.S.

TABLE 13.3 Recyclable Polymers

Recycling Code	Polymer Name	Shorthand Structure	Recycled Product
1	PET polyethylene terephthalate	$\left[CH_2CH_2-O-\overset{O}{\underset{O}{C}}-\text{(benzene ring)}-\overset{O}{\underset{O}{C}}-O \right]_n$	fleece jackets carpeting plastic bottles
2	HDPE high-density polyethylene	$\left[CH_2CH_2 \right]_n$	Tyvek insulation sports clothing
3	PVC poly(vinyl chloride)	$\left[CH_2\underset{Cl}{CH} \right]_n$	floor mats
4	LDPE low-density polyethylene	$\left[CH_2CH_2 \right]_n$	trash bags
5	PP polypropylene	$\left[CH_2\underset{CH_3}{CH} \right]_n$	furniture
6	PS polystyrene	$\left[CH_2\underset{\text{(benzene ring)}}{CH} \right]_n$	molded trays trash cans

13.9 AROMATIC COMPOUNDS

Aromatic compounds represent another example of unsaturated hydrocarbons. Aromatic compounds were originally named because many simple compounds in this family have characteristic odors. Today, the word **aromatic refers to compounds that contain a benzene ring,** or rings that react in a similar fashion to benzene.

Benzene, the simplest and most widely known aromatic compound, contains a six-membered ring and three double bonds. Since each carbon of the ring is also bonded to a hydrogen atom, the molecular formula for benzene is C_6H_6. Each carbon is surrounded by three groups, making it trigonal planar. Thus, **benzene is a planar molecule,** and all bond angles are **120°.**

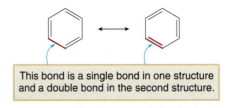

benzene
C_6H_6

planar molecule

Although benzene is drawn with a six-membered ring and three double bonds, there are two different ways to arrange the double bonds so that they alternate with single bonds around the ring.

This bond is a single bond in one structure and a double bond in the second structure.

Drawing resonance structures was first presented in Section 4.4.

Each of these representations has the same arrangement of atoms. Only the location of the electrons is different. These structures are called **resonance structures.**

- Resonance structures are Lewis structures with the same arrangement of atoms but a different arrangement of electrons.

A **double-headed arrow** indicates that the two structures are resonance structures. In actuality, neither structure represents the true structure of benzene. The true structure of benzene is really a combination of both resonance structures, called a **hybrid.** Because the electron pair in a given double bond in a benzene ring is not really located between two particular carbon atoms, we say that the three electron pairs in the double bonds are *delocalized* in the six-membered ring. This gives benzene added stability compared to other unsaturated hydrocarbons.

Sometimes, a benzene ring is drawn with a circle inside a hexagon in place of the three double bonds.

three equivalent representations of benzene

The circle is meant to show that the six electrons from the three double bonds move freely around the ring. Realize, however, that we can draw benzene using either resonance structure or a benzene ring with a circle inside. Each of these representations is equivalent.

The physical properties of aromatic hydrocarbons are similar to other hydrocarbons—they have low melting points and boiling points and are water insoluble. Their chemical properties, however, are very different. **Aromatic hydrocarbons do not undergo the addition reactions that characterize alkenes.** For example, bromine adds to ethylene to form an addition product with two new C—Br bonds, but benzene does not react under similar conditions.

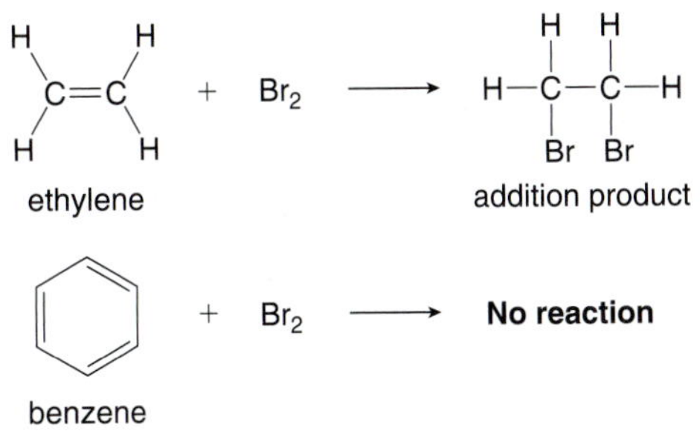

* Aromatic compounds resemble benzene—they are unsaturated compounds that do not undergo the addition reactions characteristic of alkenes.

13.10 NOMENCLATURE OF BENZENE DERIVATIVES

Many organic molecules contain a benzene ring with one or more substituents, so we must learn how to name them. Many common names are recognized by the IUPAC system, however, so this complicates the nomenclature of benzene derivatives somewhat.

13.10A MONOSUBSTITUTED BENZENES

To name a benzene ring with one substituent, **name the substituent and add the word *benzene*.** Carbon substituents are named as alkyl groups. When a halogen is a substituent, name the halogen by changing the *-ine* ending of the name of the halogen to the suffix *-o*; for example, chlor*ine* → chlor*o*.

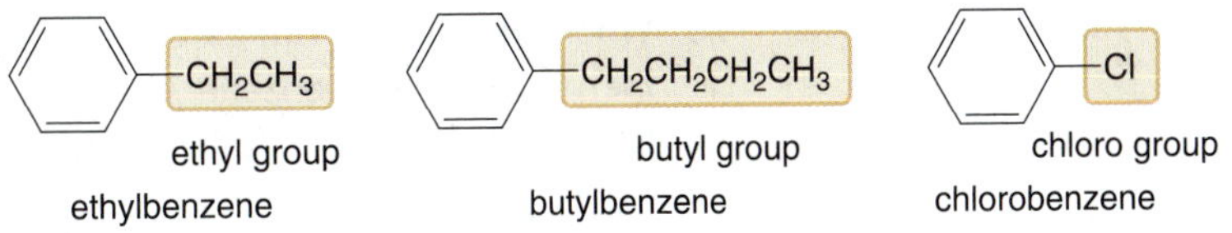

Many monosubstituted benzenes, such as those with methyl (CH_3-), hydroxyl (–OH), and amino (–NH_2) groups, have common names that you must learn, too.

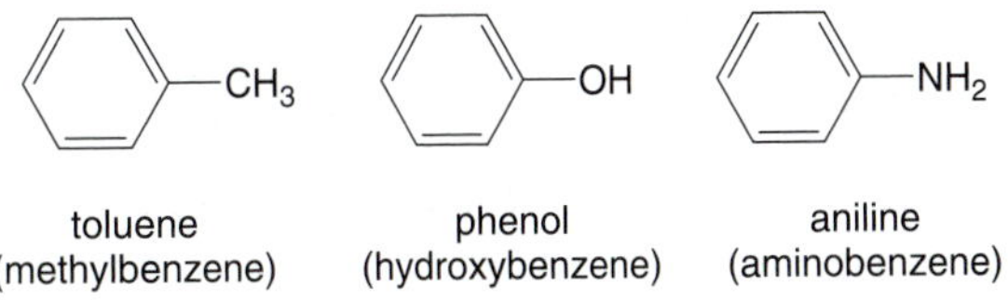

13.10B DISUBSTITUTED BENZENES

There are three different ways that two groups can be attached to a benzene ring, so a prefix—**ortho, meta,** or **para**—is used to designate the relative position of the two substituents. Ortho, meta, and para are also abbreviated as *o, m,* and *p,* respectively.

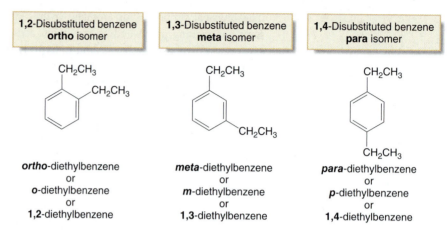

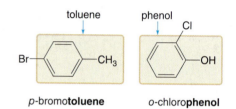

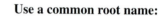

If the two groups on the benzene ring are different, **alphabetize the name of the substituents** preceding the word benzene. If one of the substituents is part of a **common root,** name the molecule as a derivative of that monosubstituted benzene.

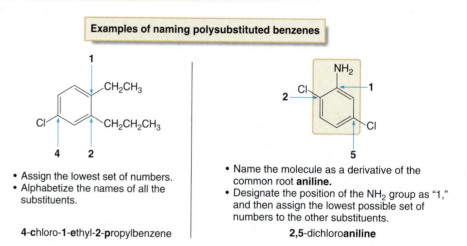

13.10C   POLYSUBSTITUTED BENZENES

For three or more substituents on a benzene ring:

> 1. Number to give the lowest possible numbers around the ring.
> 2. Alphabetize the substituent names.
> 3. When substituents are part of common roots, name the molecule as a derivative of that monosubstituted benzene. The substituent that comprises the common root is located at C1, but the "1" is omitted from the name.

13.10D   AROMATIC COMPOUNDS WITH MORE THAN ONE RING

Some aromatic compounds contain two or more benzene rings joined together. Naphthalene has two benzene rings joined together. There are two different aromatic hydrocarbons with three benzene rings, named anthracene and phenanthrene.

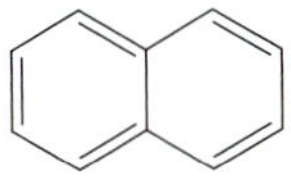

naphthalene 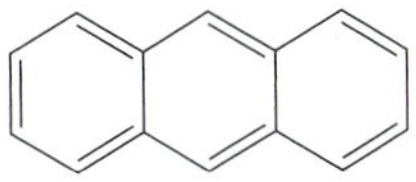   anthracene 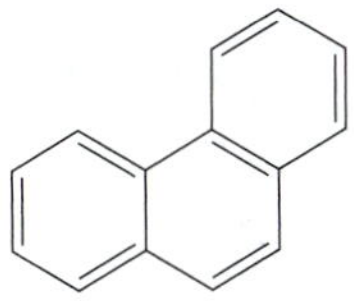   phenanthrene

SAMPLE PROBLEM 13.8

Name each of the following aromatic compounds.

a. 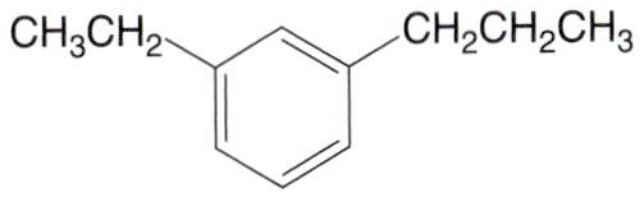 CH_3CH_2 —   — $CH_2CH_2CH_3$

b. 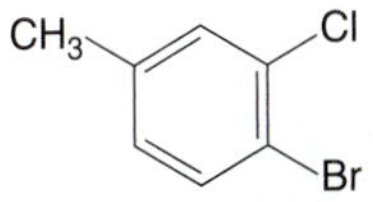 CH_3 —   Cl / Br

ANALYSIS

Name the substituents on the benzene ring. With two groups, alphabetize the substituent names and use the prefix ortho, meta, or para to indicate their location. With three substituents, alphabetize the substituent names, and number to give the lowest set of numbers.

SOLUTION

a. 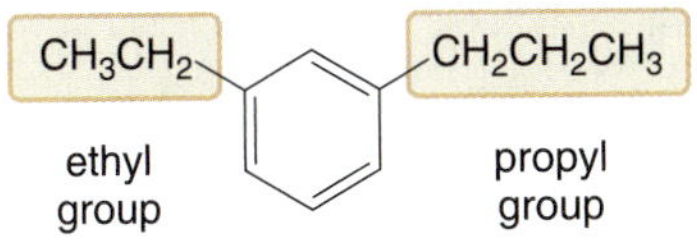 CH_3CH_2 —   — $CH_2CH_2CH_3$

    ethyl group   propyl group

- The two substituents are located 1,3- or **meta** to each other.
- Alphabetize the *e* of **e**thyl before the *p* of **p**ropyl.

Answer: *m*-ethylpropylbenzene

b. 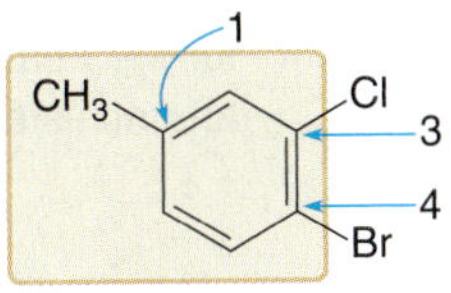 CH_3 — 1 / Cl 3 / 4 / Br

    toluene

- Since a CH_3– group is bonded to the ring, name the molecule as a derivative of toluene.
- Place the CH_3 group at the "1" position, and number to give the lowest set of numbers.

Answer: 4-bromo-3-chlorotoluene

PROBLEM 13.26

Give the IUPAC name of each compound.

a. 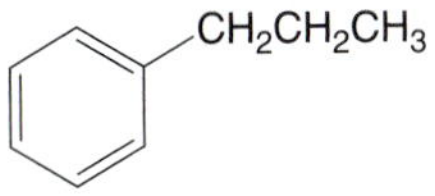 — $CH_2CH_2CH_3$

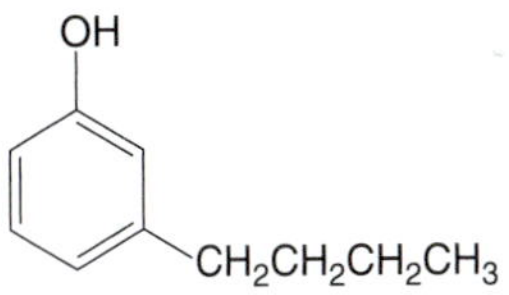

b. 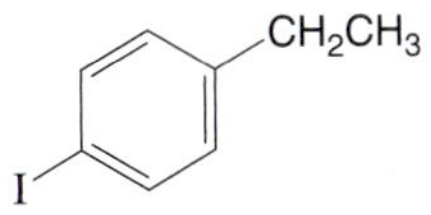 I — — CH_2CH_3

c. OH / — $CH_2CH_2CH_2CH_3$

d. 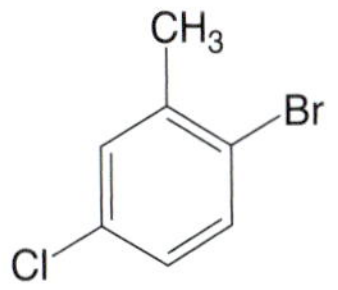 CH_3 / Br / Cl

PROBLEM 13.27

Draw the structure corresponding to each name.

a. pentylbenzene

b. *o*-dichlorobenzene

c. *m*-bromoaniline

d. 4-chloro-1,2-diethylbenzene

13.11 FOCUS ON HEALTH & MEDICINE
AROMATIC DRUGS, SUNSCREENS, AND CARCINOGENS

Many widely used drugs contain an aromatic ring. Notable examples include the analgesic acetaminophen, the local anesthetic lidocaine, and the antihistamine loratadine (Figure 13.7). These drugs are better known by their trade names Tylenol, Xylocaine, and Claritin, respectively.

▼ FIGURE 13.7   Three Drugs That Contain a Benzene Ring

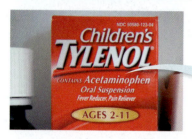

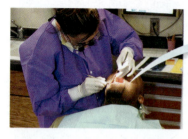

acetaminophen
(Trade name: Tylenol)

- Acetaminophen reduces fever and pain, but it is not anti-inflammatory. Thus, it is ineffective in treating conditions like arthritis, which have a significant inflammatory component.

lidocaine
(Trade name: Xylocaine)

- Lidocaine is a local anesthetic used to numb the skin when suturing an injury. It is also commonly used by dentists.

loratadine
(Trade name: Claritin)

- Loratadine is an over-the-counter antihistamine used for the treatment of seasonal allergies. Loratadine is an antihistamine that does not cause drowsiness.

Commercial sunscreens are given an **SPF** rating (sun protection factor), according to the amount of sunscreen present. The higher the number, the greater the protection.

All commercially available sunscreens contain a benzene ring. A sunscreen absorbs ultraviolet radiation and thus shields the skin for a time from its harmful effects. Two sunscreens that have been used for this purpose are *p*-aminobenzoic acid (PABA) and Padimate O.

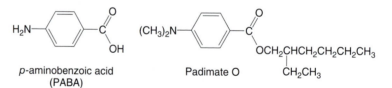

p-aminobenzoic acid (PABA)   Padimate O

Compounds containing two or more benzene rings that share carbon–carbon bonds are called **polycyclic aromatic hydrocarbons (PAHs)**. **Benzo[*a*]pyrene**, a PAH containing five rings joined together, is a widespread environmental pollutant, produced during the combustion of all types of organic material—gasoline, fuel oil, wood, garbage, and cigarettes.

tobacco plant

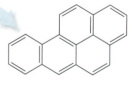

benzo[*a*]pyrene
(a polycyclic aromatic hydrocarbon)

When benzo[*a*]pyrene, a nonpolar and water-insoluble hydrocarbon, is inhaled or ingested, it is oxidized in the liver. Oxidation is a common fate of foreign substances that are not useful nutrients for the body. The oxidation products are more polar, making them much more water soluble, and more readily excreted in urine. The oxidation product from benzo[*a*]pyrene is also a potent **carcinogen;** that is, it causes cancer. The carcinogen binds to a needed protein or DNA in the cell and disrupts normal cell function, causing cancer or cell death.

PROBLEM 13.28

Which of the following compounds might be an ingredient in a commercial sunscreen? Explain why or why not.

a.

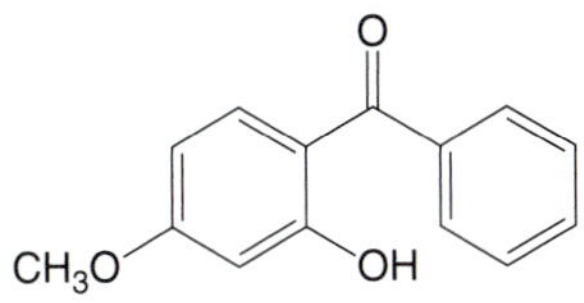

b.

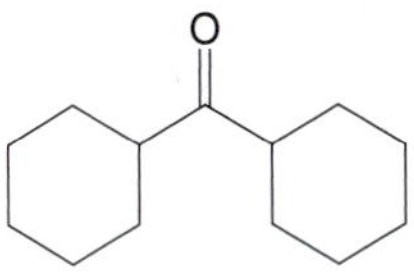

PROBLEM 13.29

What oxidation products are formed when benzo[*a*]pyrene is completely combusted (see Section 12.8)?

13.12 FOCUS ON HEALTH & MEDICINE PHENOLS AS ANTIOXIDANTS

A wide variety of phenols, compounds that contain a hydroxyl group bonded to a benzene ring, occur in nature. **Vanillin** from the vanilla bean and **eugenol** from cloves are both phenols.

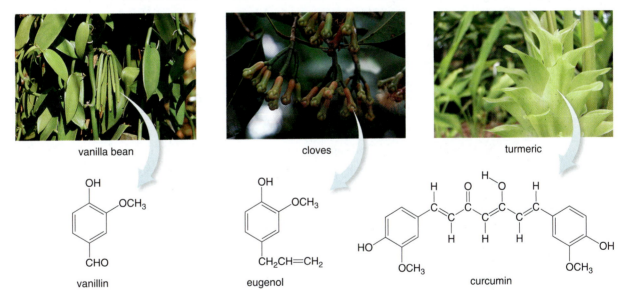

vanilla bean

cloves

turmeric

vanillin

eugenol

curcumin

Curcumin is a yellow pigment isolated from turmeric, a tropical perennial in the ginger family and a principal ingredient in curry powder. Curcumin has long been used as an anti-inflammatory agent in traditional eastern medicine. In some preliminary research carried out with mice, curcumin was shown to correct the defect that causes cystic fibrosis, a fatal genetic disease that afflicts 30,000 children and young adults in the United States.

Some phenols are antiseptics and others are irritants. The disinfectant Lysol contains 2-benzyl-4-chlorophenol. The phenol urushiol in poison ivy causes an irritating rash on contact.

Nuts are an excellent source of vitamin E.

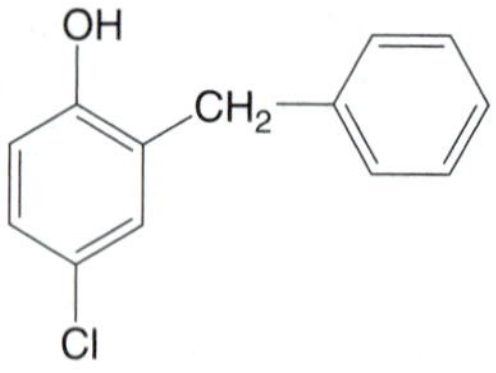

2-benzyl-4-chlorophenol
(active ingredient in Lysol)

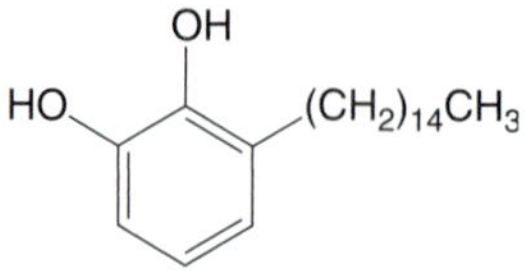

urushiol
(found in poison ivy)

Like lycopene (Section 13.4), many **phenols are antioxidants.** Two examples are naturally occurring vitamin E and synthetic BHT. **The OH group on the benzene ring is the key functional group that prevents unwanted oxidation reactions from occurring.**

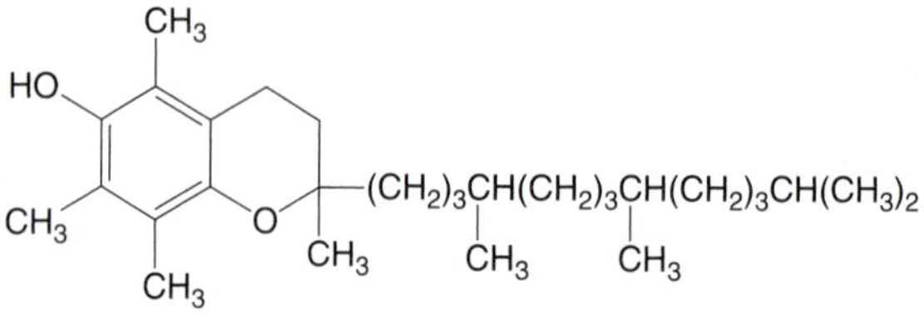

vitamin E

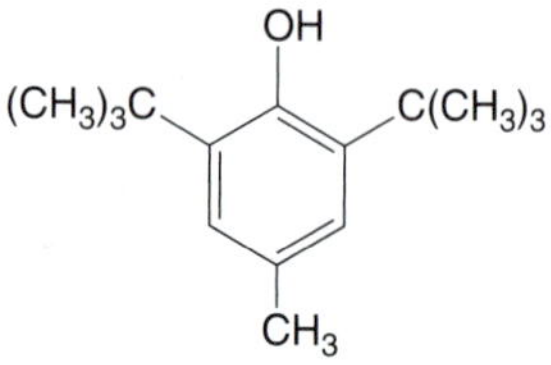

BHT
(**b**utylated **h**ydroxy **t**oluene)

Vitamin E is a natural antioxidant found in fish oil, peanut oil, wheat germ, and leafy greens. Although the molecular details of its function remain obscure, it is thought that vitamin E prevents the unwanted oxidation of unsaturated fatty acid residues in cell membranes. In this way, vitamin E helps retard the aging process.

Synthetic antioxidants such as **BHT**—**b**utylated **h**ydroxy **t**oluene—are added to packaged and prepared foods to prevent oxidation and spoilage. BHT is a common additive in breakfast cereals.

PROBLEM 13.30

Which of the following compounds might be antioxidants?

a. 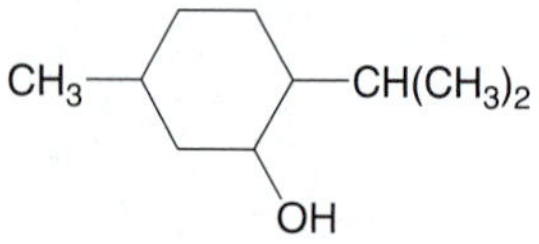
menthol

b. curcumin

c. 
p-xylene
(gasoline additive)

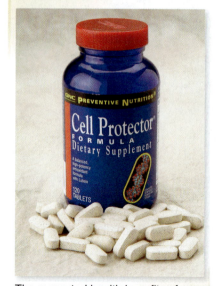

The purported health benefits of antioxidants have made them a popular component in anti-aging formulations.

13.13 REACTIONS OF AROMATIC COMPOUNDS

Like the alkenes, aromatic compounds undergo a characteristic type of reaction. In contrast to the alkenes, however, aromatic compounds undergo **substitution** *not* addition.

- Substitution is a reaction in which an atom is *replaced* by another atom or a group of atoms.

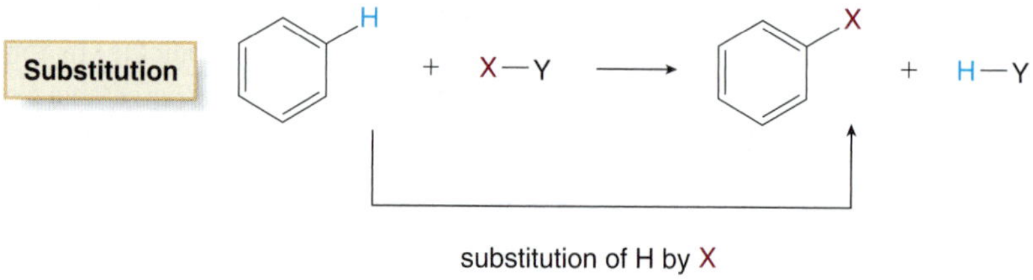

Substitution   + X—Y ⟶   + H—Y

substitution of H by X

Why does substitution occur? **Substitution of H by X keeps the stable aromatic ring intact.** Benzene does *not* undergo addition reactions like other unsaturated hydrocarbons because the product would not contain a benzene ring.

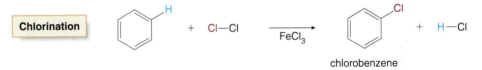

Addition does not occur.

The product is *not* aromatic.

Three specific reactions of benzene are considered: **chlorination, nitration,** and **sulfonation.** In each reaction, one hydrogen atom on the benzene ring is replaced by another atom or group: $-Cl$, $-NO_2$, or $-SO_3H$, respectively.

13.13A CHLORINATION AND THE SYNTHESIS OF THE PESTICIDE DDT

Benzene reacts with Cl_2 in the presence of an iron catalyst ($FeCl_3$) to yield chlorobenzene. Chlorobenzene is an **aryl halide,** an organic halide with the halogen bonded directly to the aromatic ring. Reaction with chlorine is called **chlorination.** In chlorination, a Cl atom substitutes for a hydrogen atom on the benzene ring.

Chlorination

$$\text{C}_6\text{H}_5\text{H} + \text{Cl}-\text{Cl} \xrightarrow{\text{FeCl}_3} \text{C}_6\text{H}_5\text{Cl} + \text{H}-\text{Cl}$$

chlorobenzene

The chlorination of benzene is the first step in the two-step synthesis of the pesticide **DDT** (*d*ichloro*d*iphenyl*t*richloroethane).

$$\text{benzene} + \text{Cl}_2 \xrightarrow{\text{FeCl}_3} \text{C}_6\text{H}_5\text{Cl} \xrightarrow[\text{H}_2\text{SO}_4]{\text{Cl}_3\text{CCHO}} \text{DDT}$$

DDT
(nonbiodegradable pesticide)

DDT is an organic molecule with valuable short-term effects that has caused long-term problems. DDT kills insects that spread diseases such as malaria and typhus, and in controlling insect populations, DDT has saved millions of lives worldwide. DDT is a weakly polar and very stable organic compound, and so it (and compounds like it) persists in the environment for years. Since DDT is soluble in organic media, it accumulates in the fatty tissues of most animals. Most adults in the U.S. have low concentrations of DDT (or a degradation product of DDT) in them. The long-term effect on humans is not known, but DDT has had a direct harmful effect on the eggshell formation of certain predator birds such as eagles and hawks.

13.13B FOCUS ON HEALTH & MEDICINE NITRATION AND SULFA DRUGS

Benzene reacts with nitric acid (HNO_3) in the presence of sulfuric acid (H_2SO_4) to yield nitrobenzene. Substitution of a **nitro group** (NO_2) for a hydrogen is called **nitration.**

Nitration

$$\text{C}_6\text{H}_5\text{H} + \text{HNO}_3 \xrightarrow{\text{H}_2\text{SO}_4} \text{C}_6\text{H}_5\text{NO}_2 + \text{H}-\text{OH}$$

nitrobenzene

Nitration is a very valuable reaction because nitro groups are readily converted to amino groups, (NH_2) with H_2 and a palladium catalyst.

ENVIRONMENTAL NOTE

"DDT is good for me-e-e!"

Time Magazine, June 30, 1947.

Winston Churchill called DDT a "miraculous" chemical in 1945 for the many lives it saved during World War II. Less than 20 years later, as prolonged use of large quantities of this nonspecific pesticide led to harmful environmental effects, Rachel Carson called DDT the "elixir of death" in her book *Silent Spring.* DDT use was banned in the U.S. in 1973, but given its low cost, it is still widely used to control insect populations in developing countries.

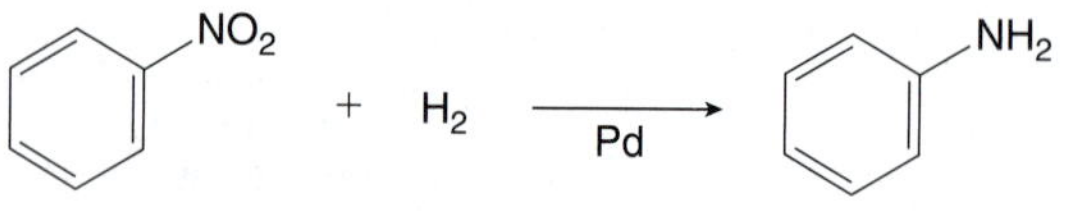

Many of the antibacterial sulfa drugs have NH_2 groups bonded to the benzene ring. Examples include sulfanilamide, one of the first antibiotics synthesized in the 1930s, and newer sulfa drugs such as sulfamethoxazole (trade name Bactrim) and sulfisoxazole (trade name Gantrisin), which are used to treat ear and urinary tract infections.

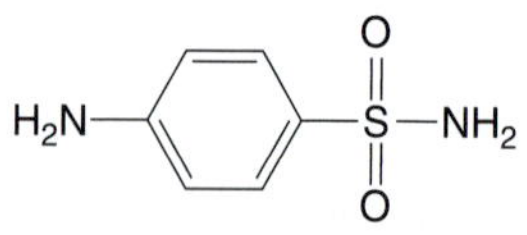

sulfanilamide

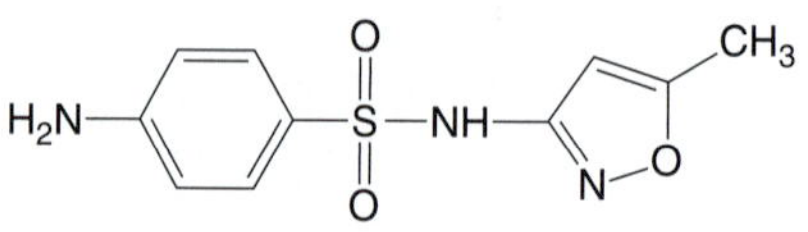

sulfamethoxazole
(Trade name: Bactrim)

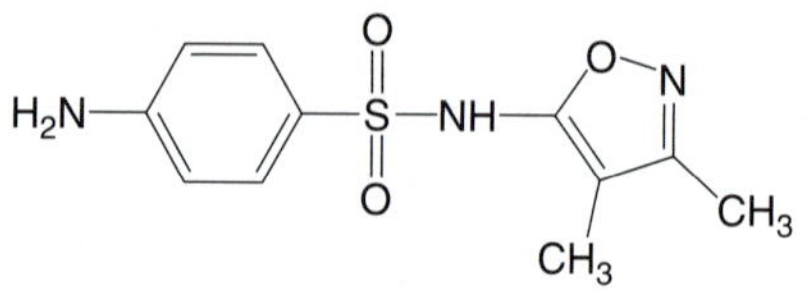

sulfisoxazole
(Trade name: Gantrisin)

In cells, sulfamethoxazole and sulfisoxazole are both metabolized to sulfanilamide, the active agent. To understand how sulfanilamide functions as an antibacterial agent we must examine **folic acid,** which microorganisms synthesize from *p*-aminobenzoic acid.

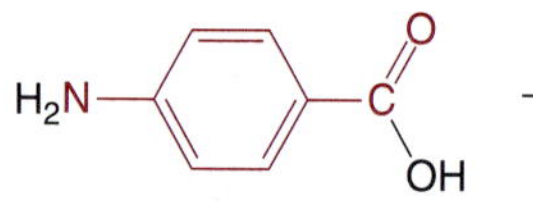

p-aminobenzoic acid
PABA

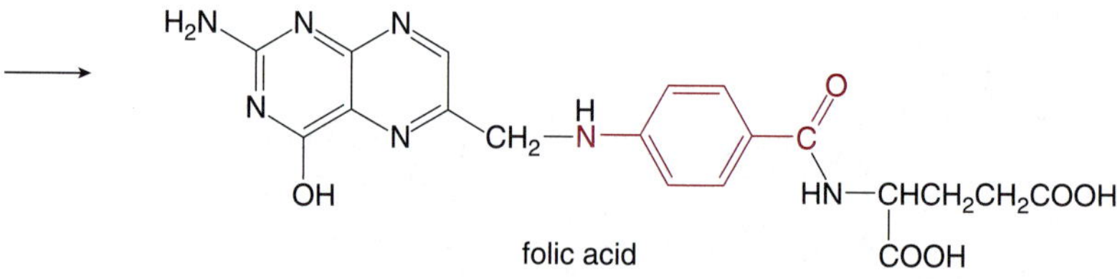

folic acid

Sulfanilamide and *p*-aminobenzoic acid are similar in size and shape and have related functional groups. Thus, when sulfanilamide is administered, bacteria attempt to use it in place of *p*-aminobenzoic acid to synthesize folic acid. Derailing folic acid synthesis means that the bacteria cannot grow and reproduce. Sulfanilamide only affects bacterial cells, though, because humans do not synthesize folic acid, and must obtain it from their diets.

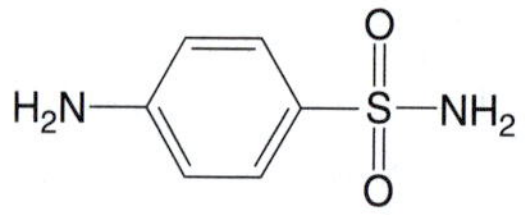

sulfanilamide

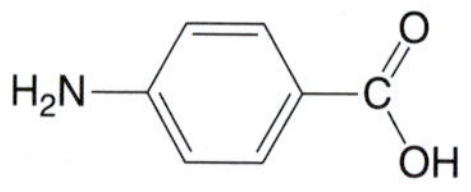

p-aminobenzoic acid

13.13C   SULFONATION AND DETERGENT SYNTHESIS

Benzene reacts with sulfur trioxide (SO_3) in the presence of sulfuric acid (H_2SO_4) to yield benzenesulfonic acid. Substitution of SO_3H for a hydrogen is called **sulfonation.**

Sulfonation

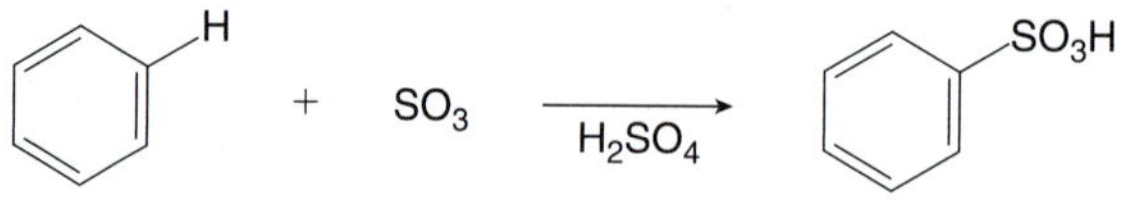

benzenesulfonic acid

Although less widely used than chlorination and nitration, sulfonation is an important step in the synthesis of detergents. Many detergents are sodium salts of sulfonic acids, and are prepared by the two-step sequence shown in Figure 13.8.

▼ **FIGURE 13.8** Synthesis of a Detergent Using Sulfonation

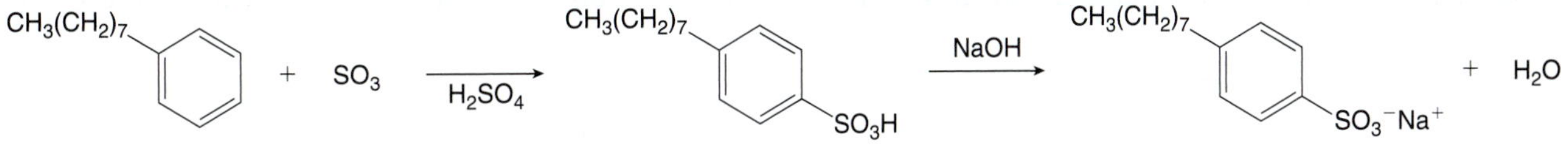

CH₃(CH₂)₇ ⬡ + SO₃ →(H₂SO₄) CH₃(CH₂)₇ ⬡ SO₃H → (NaOH) CH₃(CH₂)₇ ⬡ SO₃⁻Na⁺ + H₂O

a benzenesulfonic acid

sodium salt of
a benzenesulfonic acid

synthetic detergent

PROBLEM 13.31	What product is formed when *p*-dichlorobenzene is treated with each reagent: (a) Cl_2, $FeCl_3$; (b) HNO_3, H_2SO_4; (c) SO_3, H_2SO_4?
PROBLEM 13.32	When toluene (methylbenzene) is treated with Cl_2 and $FeCl_3$, a total of three products are possible. Draw the structure of the three products and give their IUPAC names.

CHAPTER HIGHLIGHTS

KEY TERMS

Addition reaction (13.6)
Alkene (13.1)
Alkyne (13.1)
Antioxidant (13.4)
Aromatic compound (13.9)
Aryl halide (13.13)
Chlorination (13.13)
Cis isomer (13.3)
Double-headed arrow (13.9)
Fat (13.3)
Fatty acid (13.3)
Halogenation (13.6)

Hardening (13.7)
Hybrid (13.9)
Hydration (13.6)
Hydrogenation (13.6)
Hydrohalogenation (13.6)
Markovnikov's rule (13.6)
Meta isomer (13.10)
Monomer (13.8)
Nitration (13.13)
Nitro group (13.13)
Oil (13.3)
Ortho isomer (13.10)

Para isomer (13.10)
Partial hydrogenation (13.7)
Polycyclic aromatic hydrocarbon (13.11)
Polymer (13.8)
Polymerization (13.8)
Resonance structure (13.9)
Stereoisomer (13.3)
Substitution reaction (13.13)
Sulfonation (13.13)
Trans isomer (13.3)
Unsaturated hydrocarbon (13.1)

KEY REACTIONS

[1] Addition reactions of alkenes (13.6)

  a. Hydrogenation (13.6A)

$$CH_2{=}CH_2 + H_2 \xrightarrow{Pd} \underset{H \quad H}{CH_2{-}CH_2}$$

  b. Halogenation (13.6B)

$$CH_2{=}CH_2 + X_2 \longrightarrow \underset{X \quad X}{CH_2{-}CH_2} \quad (X = Cl \text{ or } Br)$$

  c. Hydrohalogenation (13.6C)

$$CH_2{=}CH_2 + HX \longrightarrow \underset{H \quad X}{CH_2{-}CH_2} \quad (X = Cl \text{ or } Br)$$

  d. Hydration (13.6D)

$$CH_2{=}CH_2 + H_2O \xrightarrow{H_2SO_4} \underset{H \quad OH}{CH_2{-}CH_2}$$

[2] Substitution reactions of benzene (13.13)

  a. Chlorination (13.13A)

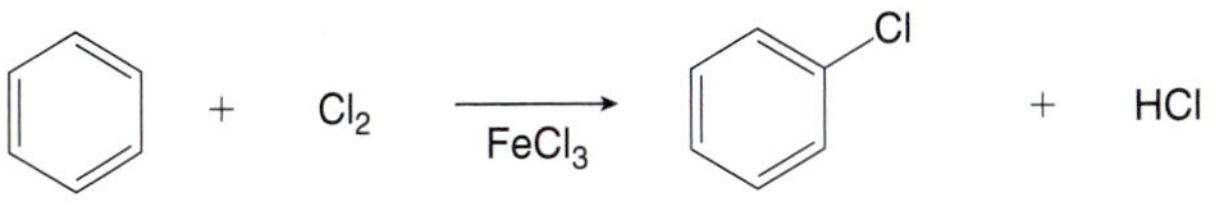

⬡ + Cl₂ →(FeCl₃) ⬡–Cl + HCl

  b. Nitration (13.13B)

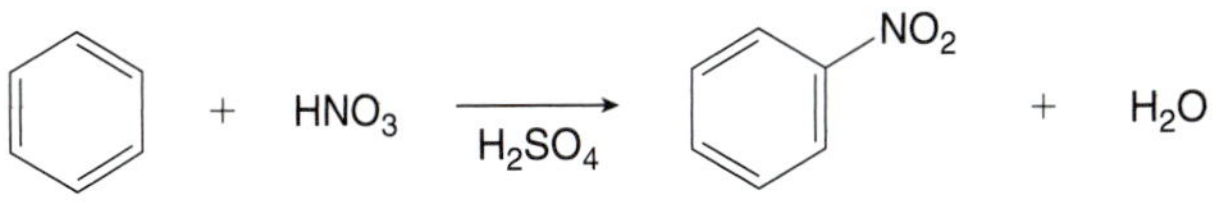

⬡ + HNO₃ →(H₂SO₄) ⬡–NO₂ + H₂O

  c. Sulfonation (13.13C)

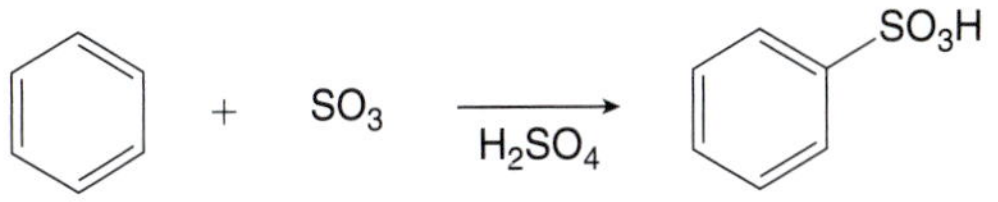

⬡ + SO₃ →(H₂SO₄) ⬡–SO₃H

KEY CONCEPTS

❶ What are the characteristics of alkenes, alkynes, and aromatic compounds?

- Alkenes are unsaturated hydrocarbons that contain a carbon–carbon double bond and have molecular formula C_nH_{2n}. Each carbon of the double bond is trigonal planar. (13.1)
- Alkynes are unsaturated hydrocarbons that contain a carbon–carbon triple bond and have molecular formula C_nH_{2n-2}. Each carbon of the triple bond is linear. (13.1)
- Benzene, molecular formula C_6H_6, is the most common aromatic hydrocarbon. Benzene is a stable hybrid of two resonance structures, each containing a six-membered ring and three double bonds. Each carbon of benzene is trigonal planar. (13.9)

❷ How are alkenes, alkynes, and substituted benzenes named?

- An alkene is identified by the suffix *-ene,* and the carbon chain is numbered to give to the C=C the lower number. (13.2)
- An alkyne is identified by the suffix *-yne,* and the carbon chain is numbered to give to the C≡C the lower number. (13.2)
- Substituted benzenes are named by naming the substituent and adding the word *benzene.* When two substituents are bonded to the ring, the prefixes ortho, meta, and para are used to show the relative positions of the two groups: 1,2-, 1,3- or 1,4-, respectively. With three substituents on a benzene ring, number to give the lowest possible numbers. (13.10)

❸ What is the difference between constitutional isomers and stereoisomers? How are cis and trans isomers different? (13.3)

- Constitutional isomers differ in the way the atoms are bonded to each other.
- Stereoisomers differ only in the three-dimensional arrangement of atoms.
- Cis and trans isomers are one type of stereoisomer. A cis alkene has two alkyl groups on the same side of the double bond. A trans alkene has two alkyl groups on opposite sides of the double bond.

❹ How do saturated and unsaturated fatty acids differ? (13.3B)

- Fatty acids are carboxylic acids (RCOOH) with long carbon chains. Saturated fatty acids have no double bonds in the carbon chain and unsaturated fatty acids have one or more double bonds in their long carbon chains.
- All double bonds in naturally occurring fatty acids are cis.
- As the number of double bonds in the fatty acid increases, the melting point decreases.

❺ What types of reactions do alkenes undergo? (13.6)

- Alkenes undergo addition reactions with reagents X—Y. One bond of the double bond and the X—Y bond break and two new single bonds (C—X and C—Y) are formed.
- Alkenes react with four different reagents—H_2 (Pd catalyst), X_2 (X = Cl or Br), HX (X = Cl or Br), and H_2O (with H_2SO_4).

❻ What is Markovnikov's rule? (13.6)

- Markovnikov's rule explains the selectivity observed when an unsymmetrical reagent HX adds to an unsymmetrical alkene like propene ($CH_3CH=CH_2$). The H of HX is added to the end of the C=C that has more H's to begin with, forming $CH_3CH(X)CH_3$.

❼ What products are formed when a vegetable oil is partially hydrogenated? (13.7)

- When an unsaturated oil is partially hydrogenated, some but not all of the cis C=C's add H_2, reducing the number of double bonds and increasing the melting point.
- Some of the cis double bonds are converted to trans double bonds, forming trans fats, whose shape and properties closely resemble those of saturated fats.

❽ What are polymers, and how are they formed from alkene monomers? (13.8)

- Polymers are large molecules made up of repeating smaller molecules called monomers covalently bonded together. When alkenes are polymerized, one bond of the double bond breaks, and two new single bonds join the alkene monomers together in long carbon chains.

❾ What types of reactions does benzene undergo? (13.13)

- To keep the stable aromatic ring intact, benzene undergoes substitution, not addition, reactions. One H atom on the ring is replaced by another atom or group of atoms. Reactions include chlorination (substitution by Cl), nitration (substitution by NO_2), and sulfonation (substitution by SO_3H).

PROBLEMS

Selected in-chapter and end-of-chapter problems have brief answers provided in Appendix B.

Alkene, Alkyne, and Benzene Structure

13.33 What is the molecular formula for a hydrocarbon with 10 carbons that is (a) completely saturated; (b) an alkene; (c) an alkyne?

13.34 Draw the structure of a hydrocarbon with molecular formula C_6H_{10} that also contains: (a) a carbon–carbon triple bond; (b) two carbon–carbon double bonds; (c) one ring and one C=C.

13.35 Draw structures for the three alkynes having molecular formula C_5H_8.

13.36 Draw the structures of the five constitutional isomers of molecular formula C_5H_{10} that contain a double bond.

13.37 Label each carbon in the following molecules as tetrahedral, trigonal planar, or linear.

a.

c.

b. $-C\equiv CH$

13.38 Falcarinol is a natural pesticide found in carrots that protects them from fungal diseases. Predict the indicated bond angles in falcarinol.

falcarinol

Nomenclature of Alkenes and Alkynes

13.39 Give the IUPAC name for each compound.
a. CH_2=$CHCH_2CH_2C(CH_3)_3$

b. $(CH_3CH_2)_2C$=$CHCHCH_2CHCH_3$
   with CH_3 and CH_3 substituents

c. CH_2=CCH_2CH_3
   with $CH_2CH_2CH_2CH_2CH_3$

d. $CH_3C\equiv CCH_2C(CH_3)_3$

e. $CH_3C\equiv C-CH_2-CH-C-CH_2CH_3$
   with CH_3, CH_3, and CH_2CH_3 substituents

f. CH_2=$CHCH_2-C-CH$=CH_2
   with CH_3 and CH_3 substituents

13.40 Give the IUPAC name for each compound.
a. CH_2=$CHCH_2CHCH_2CH_3$
   with CH_3

b. $(CH_3)_2C$=$CHCH_2CHCH_2CH_3$
   with $CH_2CH_2CH_3$

c. $CH_3C\equiv CCHCH_3$
   with $CH_2CH_2CH_3$

d. $(CH_3)_3CC\equiv CC(CH_3)_3$

e. $(CH_3CH_2CH_2CH_2)_2C$=$CHCH_3$

f. $(CH_3)_2C$=$CHCH_2CH_2CH_2CH$=$C(CH_3)_2$

13.41 Give the IUPAC name for each alkyne in Problem 13.35.

13.42 Give the IUPAC name for each alkene in Problem 13.36.

13.43 Give the IUPAC name for each cyclic compound.

a. $-CH_3$

b. with CH_2CH_3 and CH_2CH_3

c. with CH_3 and CH_3

13.44 Give the IUPAC name for each cyclic compound.

a. $-CH_2CH_2CH_3$

b.

c. with CH_3 and CH_3

13.45 Give the structure corresponding to each IUPAC name.
a. 3-methyl-1-octene
b. 1-ethylcyclobutene
c. 2-methyl-3-hexyne
d. 3,5-diethyl-2-methyl-3-heptene
e. 1,3-heptadiene
f. *cis*-7-methyl-2-octene

13.46 Give the structure corresponding to each IUPAC name.
a. 1,2-dimethylcyclopentene
b. 6-ethyl-2-octyne
c. 3,3-dimethyl-1,4-pentadiene
d. *trans*-5-methyl-2-hexene
e. 5,6-dimethyl-2-heptyne
f. 3,4,5,6-tetramethyl-1-decyne

13.47 Each of the following IUPAC names is incorrect. Explain why it is incorrect and give the correct IUPAC name.
   a. 5-methyl-4-hexene
   b. 1-methylbutene
   c. 2,3-dimethylcyclohexene
   d. 3-butyl-1-butyne

13.48 Each of the following IUPAC names is incorrect. Explain why it is incorrect and give the correct IUPAC name.
   a. *cis*-2-methyl-2-hexene
   b. 2-methyl-2,4-pentadiene
   c. 2,4-dimethylcyclohexene
   d. 1,1-dimethyl-2-cyclohexene

Isomers

13.49 Leukotriene C_4 is a key compound that causes the constriction of small airways, and in this way contributes to the asthmatic response. Label each double bond in leukotriene C_4 as cis or trans.

leukotriene C_4

13.50 Draw the complete structure of each naturally occurring compound using the proper cis or trans arrangement around the carbon–carbon double bond. Muscalure is the sex attractant of the housefly. Cinnamaldehyde is responsible for the odor of cinnamon.
   a. $CH_3(CH_2)_8CH=CH(CH_2)_{12}CH_3$

   muscalure
   (cis double bond)

   b. —CH=CHCHO

   cinnamaldehyde
   (trans double bond)

13.51 Draw the cis and trans isomers for each compound: (a) 2-nonene; (b) 2-methyl-3-heptene.

13.52 Draw the cis and trans isomers for each compound: (a) 3-heptene; (b) 4,4-dimethyl-2-hexene.

13.53 Explain the difference between a constitutional isomer and a stereoisomer.

13.54 Draw all of the possible stereoisomers for 2,4-hexadiene ($CH_3CH=CH—CH=CHCH_3$).

13.55 How are the compounds in each pair related? Choose from constitutional isomers, stereoisomers, or identical.
   a. $CH_3C\equiv CCH_3$ and $HC\equiv CCH_2CH_3$
   b.
   c.
   d.

13.56 Consider alkenes **A, B,** and **C.** How are the compounds in each pair related? Choose from constitutional isomers, stereoisomers, or identical: (a) **A** and **B**; (b) **A** and **C**; (c) **B** and **C.**

A   **B**   **C**

13.57 Why are there no cis and trans isomers when an alkene has two like groups bonded to one end of the double bond?

13.58 Why can't an alkyne have cis and trans isomers?

Reactions of Alkenes

13.59 What alkane is formed when each alkene is treated with H_2 in the presence of a Pd catalyst?
   a. $CH_2=CHCH_2CH_2CH_2CH_3$
   b. $(CH_3)_2C=CHCH_2CH_2CH_3$
   c.
   d.

13.60 What dihalide is formed when each of the alkenes in Problem 13.59 is treated with Br_2?

13.61 What alkyl halide is formed when each alkene is treated with HCl?
   a.
   b. $(CH_3)_2C=C(CH_3)_2$
   c. $CH_2=CHCH_2CH(CH_3)_2$
   d.

13.62 What alcohol is formed when each alkene in Problem 13.61 is treated with H_2O and H_2SO_4?

13.63 Draw the product formed when 1-ethylcyclohexene is treated with each reagent.

a. H_2, Pd   d. HCl
b. Cl_2 e. HBr
c. Br_2 f. H_2O, H_2SO_4

1-ethylcyclohexene

13.64 Draw the products formed in each reaction.

a. $CH_3CH{=}CHCH_3$ + HCl ⟶

b. [structure] + H_2 $\xrightarrow{Pd}$

c. [structure with CH_3, CH_3] + Cl_2 ⟶

d. $CH_3CH{=}CHCH_2CH_2CH_3$ + Br_2 ⟶

e. $(CH_3)_2C{=}CHCH_2CH_3$ + HBr ⟶

f. [structure] + H_2O $\xrightarrow{H_2SO_4}$

13.65 The hydration of 2-pentene ($CH_3CH{=}CHCH_2CH_3$) with H_2O and H_2SO_4 forms two alcohols. Draw the structure of both products and explain why more than one product is formed.

13.66 When myrcene (Problem 13.2) is treated with three equivalents of H_2O in the presence of H_2SO_4, a single addition product of molecular formula $C_{10}H_{22}O_3$ is formed. Keeping Markovnikov's rule in mind, draw the product.

13.67 What alkene is needed as a starting material to prepare each of the following alkyl halides or dihalides. In each case, also indicate what reagent is needed to react with the alkene.

a. CH_3CH_2Br

b. [structure with Cl]

c. [cyclopentane structure with Cl, Cl, CH_3]

d. $BrCH_2CHCH_2CH(CH_3)_2$ with Br

13.68 2-Bromobutane can be formed as the only product of the addition of HBr to two different alkenes. In contrast, 2-bromopentane can be formed as the *only* product of the addition of HBr to just one alkene. Draw the structures of the alkene starting materials and explain the observed results.

$CH_3CHCH_2CH_3$ with Br — 2-bromobutane

$CH_3CHCH_2CH_2CH_3$ with Br — 2-bromopentane

13.69 What reagent is needed to convert 2-methylpropene $[(CH_3)_2C{=}CH_2]$ to each compound?

a. $(CH_3)_3CCl$ d. $(CH_3)_3CBr$
b. $(CH_3)_3CH$ e. $(CH_3)_2CCH_2Br$ with Br
c. $(CH_3)_3COH$ f. $(CH_3)_2CCH_2Cl$ with Cl

13.70 What chemical test can be used to distinguish between each pair of compounds? Describe what you would visually observe (color changes or no changes in color) for each compound.

a.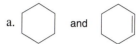

b.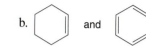

Polymers

13.71 What is the difference between a polymer and a monomer?

13.72 What is the difference between HDPE and LDPE?

13.73 Draw the structure of poly(acrylic acid), the polymer formed by polymerizing acrylic acid ($CH_2{=}CHCOOH$). Poly(acrylic acid) is used in disposable diapers because it can absorb 30 times its weight in water.

13.74 What polymer is formed when methyl α-methylacrylate $[CH_2{=}C(CH_3)CO_2CH_3]$ is polymerized? This polymer is used in Lucite and Plexiglas, transparent materials that are lighter but more impact resistant than glass.

13.75 What polymer is formed when each compound is polymerized?

a. $CH_2{=}C$ with CH_2CH_3 and H

b. $CH_2{=}C$ with Cl and CN

c. $CH_2{=}C$ with Cl and Cl

13.76 What polymer is formed when each compound is polymerized?

a. $CH_2{=}C$ with OCH_3 and H

b. $CH_2{=}C$ with Cl and CO_2CH_3

c. $CH_2{=}C$ with H and $NHCOCH_3$

13.77 What monomer is used to form the following polymer?

[polymer structure: —CH_2—C(Br)(Cl)—CH_2—C(Br)(Cl)—CH_2—C(Br)(Cl)—]

13.78 What monomer is used to form the following polymer?

[polymer structure: —CH_2—C(CH_3)(C=O, OCH_2CH_3)—CH_2—C(CH_3)(C=O, OCH_2CH_3)—CH_2—C(CH_3)(C=O, OCH_2CH_3)—]

Structure and Nomenclature of Benzene

13.79 Draw two resonance structures for chlorobenzene.

13.80 Explain why structures **A** and **B** are the same compound, even though **A** has the two Cl atoms on the same double bond and **B** has the two Cl atoms on different double bonds.

A   **B**

13.81 What is the difference between a resonance structure and an isomer?

13.82 What is the difference between a resonance structure and a hybrid?

13.83 Give the IUPAC name for each substituted benzene.

a.

b.

c.

d.

13.84 Give the IUPAC name for each substituted benzene.

a. $CH_3(CH_2)_3$—⟨benzene⟩—$(CH_2)_3CH_3$

b.

c.

d.

13.85 Draw the structure of the three constitutional isomers that have a Cl atom and an NH_2 group bonded to a benzene ring. Name each compound using the IUPAC system.

13.86 Draw the structure of TNT, the explosive that has the IUPAC name 2,4,6-trinitrotoluene.

13.87 Give the structure corresponding to each IUPAC name.
a. *p*-nitropropylbenzene   d. 2-bromo-4-chlorotoluene
b. *m*-dibutylbenzene   e. 2-chloro-6-iodoaniline
c. *o*-iodophenol

13.88 Give the structure corresponding to each IUPAC name.
a. *m*-ethylnitrobenzene   d. 1,3,5-trinitrobenzene
b. *o*-difluorobenzene   e. 2,4-dibromophenol
c. *p*-bromotoluene

13.89 Each of the following IUPAC names is incorrect. Explain why it is incorrect and give the correct IUPAC name:
(a) 5,6-dichlorophenol; (b) *m*-dibromoaniline.

13.90 Each of the following IUPAC names is incorrect. Explain why it is incorrect and give the correct IUPAC name:
(a) 1,5-dichlorobenzene; (b) 1,3-dibromotoluene.

Reactions of Aromatic Compounds

13.91 What product is formed when *p*-dimethylbenzene (also called *p*-xylene) is treated with each reagent?

p-dimethylbenzene

a. Cl_2, $FeCl_3$
b. HNO_3, H_2SO_4
c. SO_3, H_2SO_4

13.92 What reagents (**A–D**) are needed to carry out each reaction in the following sequence?

13.93 When bromobenzene is treated with HNO_3 and H_2SO_4, three different products are formed. Draw the structures of the three products formed.

13.94 You have three bottles (labeled **A, B,** and **C**) and each contains one of the *o*-, *m*-, or *p*-dibromobenzene isomers, but you don't know which isomer is in which bottle. When the dibromobenzene isomer in bottle **A** is treated with Cl_2 and $FeCl_3$, one product of molecular formula $C_6H_3Br_2Cl$ is obtained. When the dibromobenzene isomer in bottle **B** is treated with Cl_2 and $FeCl_3$, three products of molecular formula $C_6H_3Br_2Cl$ are obtained. When the dibromobenzene isomer in bottle **C** is treated with Cl_2 and $FeCl_3$, two products of molecular formula $C_6H_3Br_2Cl$ are obtained. Which isomer (ortho, meta, or para) is contained in which bottle (**A, B,** or **C**)?

Applications

13.95 The breakfast cereal Cheerios lists vitamin E as one of its ingredients. What function does vitamin E serve?

13.96 Why is BHA an ingredient in some breakfast cereals and other packaged foods?

BHA
(**b**utylated **h**ydroxy **a**nisole)

13.97 Although nonpolar compounds tend to dissolve and remain in fatty tissues, polar substances are more water soluble, and more readily excreted into an environment where they may be degraded by other organisms. Explain why methoxychlor is more biodegradable than DDT.

methoxychlor

13.98 Explain why the pesticide DDT is insoluble in water, but the herbicide 2,4-D is water soluble. 2,4-D is one component of the defoliant Agent Orange used extensively during the Vietnam War.

DDT   2,4-D

13.99 Kukui nuts contain oil that is high in linoleic acid content (Table 13.1). (a) What two constitutional isomers are formed when linoleic acid is partially hydrogenated with one equivalent of H_2? (b) What product is formed when linoleic acid is completely hydrogenated with H_2? (c) What product would be formed if, during hydrogenation, one equivalent of H_2 is added, *and* one of the cis double bonds is converted to a trans double bond?

13.100 Eleostearic acid is an unsaturated fatty acid found in tung oil, obtained from the seeds of the tung oil tree (*Aleurites fordii*), a deciduous tree native to China. Eleostearic acid is unusual in that the double bond at C9 is cis, but the other two double bonds are trans. (a) Draw the structure of eleostearic acid, showing the arrangement of groups around each double bond. (b) Draw a stereoisomer of eleostearic acid in which all of the double bonds are trans. (c) Which compound, eleostearic acid or its all-trans isomer, has the higher melting point? Explain your reasoning.

$$CH_3(CH_2)_3CH=CHCH=CHCH=CH(CH_2)_7CO_2H$$

C9

eleostearic acid

13.101 Which of the following compounds might be an antioxidant?

13.102 Which of the following compounds might be an ingredient in a commercial sunscreen?

13.103 When benzene enters the body, it is oxidized to phenol (C_6H_5OH). What is the purpose of this oxidation reaction?

13.104 Explain what a PAH is, and give an example of a PAH mentioned in this chapter, other than benzo[*a*]pyrene.

13.105 Macadamia nuts have a high concentration of unsaturated oils formed from palmitoleic acid [$CH_3(CH_2)_5CH=CH(CH_2)_7COOH$]. (a) Draw the structure of the naturally occurring fatty acid with a cis double bond. (b) Draw a stereoisomer of palmitoleic acid. (c) Draw a constitutional isomer of palmitoleic acid.

13.106 What products are formed when polyethylene is completely combusted?

General Questions

13.107 Explain the difference between a substitution reaction and an addition reaction.

13.108 What is the difference between a phenol and an alcohol? Give an example of each compound.

13.109 Why is the six-membered ring in benzene flat but the six-membered ring in cyclohexane is puckered (not flat)?

13.110 Explain why *p*-dichlorobenzene is a nonpolar molecule but *o*-dichlorobenzene is a polar molecule.

CHALLENGE QUESTIONS

13.111 Are *cis*-2-hexene and *trans*-3-hexene constitutional isomers or stereoisomers? Explain.

13.112 Some polymers are copolymers, formed from two different alkene monomers joined together. An example is Saran, the polymer used in the well known plastic food wrap. What two alkene monomers combine to form Saran?

Saran

CHAPTER GOALS

In this chapter you will learn how to:

1. Identify alcohols, ethers, alkyl halides, and thiols

2. Classify alcohols and alkyl halides as 1°, 2°, or 3°

3. Determine the properties of alcohols, ethers, alkyl halides, and thiols

4. Name alcohols, ethers, alkyl halides, and thiols

5. Draw the products of the dehydration of alcohols

6. Determine the products of alcohol oxidation

7. Convert thiols to disulfides

Ethanol, the alcohol in wine, beer, and other alcoholic beverages, is formed by the fermentation of carbohydrates in grapes, grains, and potatoes.

ORGANIC COMPOUNDS THAT CONTAIN OXYGEN, HALOGEN, OR SULFUR

CHAPTER 14 concentrates on four families of organic compounds that contain a carbon singly bonded to a heteroatom—**alcohols** (ROH), **ethers** (ROR), **alkyl halides** (RX, X = F, Cl, Br, or I), and **thiols** (RSH). The polar carbon–heteroatom bond gives these compounds different properties than the hydrocarbons of Chapters 12 and 13. Of the four families, alcohols are the most widely occurring, represented by simple compounds such as ethanol (CH_3CH_2OH) and by complex compounds such as starch and cellulose. Ethers are the most common anesthetics in use today, and alkyl halides, widely used as industrial solvents and refrigerants, are responsible for the destruction of the ozone layer. The –SH group of thiols plays an important role in protein chemistry. In Chapter 14, we learn about the properties of these four groups of organic compounds.

14.1 INTRODUCTION

Alcohols, ethers, alkyl halides, and **thiols** are four families of compounds that contain a carbon atom singly bonded to a heteroatom—oxygen, halogen, or sulfur.

Alcohols and **ethers** are organic derivatives of H_2O, formed by replacing one or two hydrogens on the oxygen by alkyl groups, respectively. *Alcohols* **contain a hydroxyl group (OH group)** **bonded to a tetrahedral carbon atom, while** *ethers* **have two alkyl groups bonded to an oxygen atom.** The oxygen atom in both compounds has two lone pairs of electrons, so it is surrounded by eight electrons.

General structures	Examples
R—Ö—H R—Ö—R	CH_3CH_2—Ö—H CH_3CH_2—Ö—CH_2CH_3
alcohol ether	ethanol diethyl ether

Alkyl halides contain a halogen atom (X = F, Cl, Br, or I) bonded to a tetrahedral carbon. Each halogen atom forms one bond to carbon and has three lone pairs of electrons. Thiols contain a **sulfhydryl group (SH group)** bonded to a tetrahedral carbon atom. A thiol is a sulfur analogue of an alcohol, formed by replacing the oxygen by sulfur. Like oxygen, the sulfur atom has two lone pairs of electrons around it.

General structures	Examples
R—Ẍ: R—S̈—H	CH_3CH_2—C̈l: CH_3CH_2—S̈—H
alkyl halide thiol	chloroethane ethanethiol
X = F, Cl, Br, I	

Mercaptomethylfuran (characteristic aroma of coffee), enflurane (a common anesthetic), and taxol (an anticancer drug) are compounds that contain these functional groups.

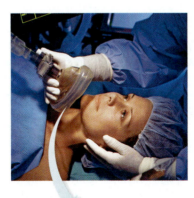

2-mercaptomethylfuran
(coffee aroma)

enflurane
(Trade name: Ethrane)

taxol
(Trade name: Paclitaxel)

[The OH, SH, halogen, and ether O atoms are labeled in red.]

Taxol, a complex anticancer agent active against ovarian, breast, and some lung tumors, is especially interesting. Initial biological studies with taxol were carried out with natural material isolated from the bark of the Pacific yew tree, but stripping the bark killed these magnificent trees. Taxol is now synthesized in the laboratory in four steps from a compound isolated from the needles of the common English yew tree, thus providing ample quantities for patients.

PROBLEM 14.1

a. Label the –OH groups, –SH groups, halogens, and ether oxygens in each compound.
b. Which –OH group in salmeterol (**C**) is *not* part of an alcohol? Explain.

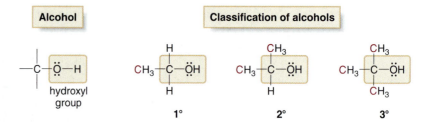

A
chondrochole A
(isolated from red seaweed,
Chondrococcus hornemanni)

B
(responsible for the
characteristic flavor of
grapefruit)

C
salmeterol
Trade name: Serevent
(dilates airways; used to treat asthma)

PROBLEM 14.2

Besides its three hydroxyl groups and one ether oxygen, taxol contains 10 other functional groups. Identify each of these functional groups.

14.2 STRUCTURE AND PROPERTIES OF ALCOHOLS

Alcohols **contain a hydroxyl group (OH** group) bonded to a tetrahedral carbon atom. Alcohols are classified as **primary (1°), secondary (2°),** or **tertiary (3°)** based on the number of carbon atoms bonded to the carbon with the OH group.

Alcohol	Classification of alcohols		

$$-\overset{|}{\underset{|}{C}}-\ddot{O}-H$$
hydroxyl group

$$CH_3-\overset{H}{\underset{H}{C}}-\ddot{O}H \quad 1° \qquad CH_3-\overset{CH_3}{\underset{H}{C}}-\ddot{O}H \quad 2° \qquad CH_3-\overset{CH_3}{\underset{CH_3}{C}}-\ddot{O}H \quad 3°$$

- A primary (1°) alcohol has an OH group on a carbon bonded to one carbon.
- A secondary (2°) alcohol has an OH group on a carbon bonded to two carbons.
- A tertiary (3°) alcohol has an OH group on a carbon bonded to three carbons.

SAMPLE PROBLEM 14.1

Classify each alcohol as 1°, 2°, or 3°.

a. ⬡—CH₂CH₂OH   b. ⬡—OH

ANALYSIS

To determine whether an alcohol is 1°, 2°, or 3°, locate the C with the OH group and count the number of C's bonded to it. A 1° alcohol has the OH group on a C bonded to one C, and so forth.

SOLUTION

Draw out the structure or add H's to the skeletal structure to clearly see how many C's are bonded to the C bearing the OH group.

a. ⬡—$\overset{H}{\underset{H}{C}}$—$\overset{H}{\underset{H}{C}}$—OH

This C is bonded to 1 C.
1° alcohol

b. ⬡$\overset{OH}{\underset{H}{}}$

This C is bonded to 2 C's in the ring.
2° alcohol

PROBLEM 14.3

Classify each alcohol as 1°, 2°, or 3°.

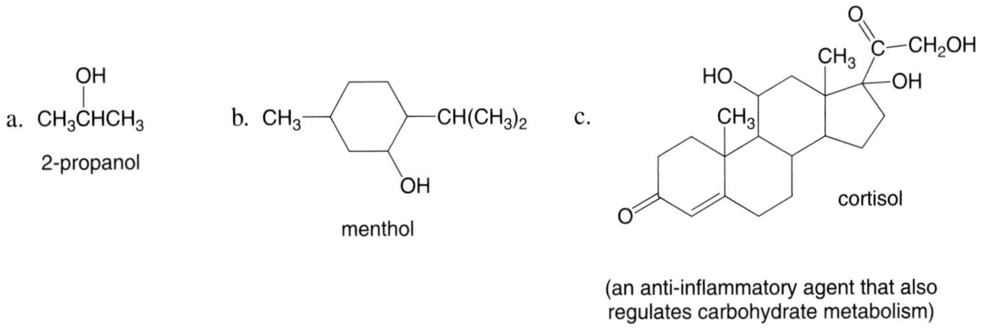

a. CH₃CHCH₃ (OH)

2-propanol

b. CH₃—[cyclohexane ring]—CH(CH₃)₂ with OH

menthol

c.

cortisol

(an anti-inflammatory agent that also regulates carbohydrate metabolism)

An alcohol contains an oxygen atom surrounded by two atoms and two lone pairs of electrons, giving the oxygen atom a **bent** shape like H₂O. The C—O—H bond angle is similar to the tetrahedral bond angle of 109.5°.

CH₃—O—H = bent shape 109.5°

methanol   bent shape

Because oxygen is much more electronegative than carbon or hydrogen, the C—O and O—H bonds are polar. Since an alcohol contains two polar bonds and a bent shape, it has a net dipole. **Alcohols are also capable of intermolecular hydrogen bonding,** since they possess a hydrogen atom bonded to an oxygen. This gives alcohols much stronger intermolecular forces than the hydrocarbons of Chapters 12 and 13.

| An alcohol has a net dipole. | Alcohols exhibit intermolecular hydrogen bonding. |

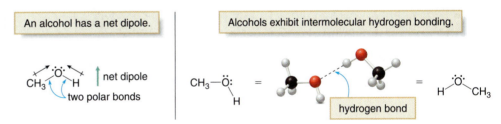

CH₃—O—H net dipole / two polar bonds

CH₃—O—H = = H—O—CH₃

hydrogen bond

As a result:

- **Alcohols have higher boiling points and melting points than hydrocarbons of comparable size and shape.**

CH₃CH₂CH₂CH₃   CH₃CH₂CH₂OH ← **stronger intermolecular forces higher boiling point and melting point**

butane   1-propanol

melting point: −138 °C   melting point: −127 °C
boiling point: −0.5 °C   boiling point: 97 °C

The general rule governing solubility—**"like dissolves like"**—explains the solubility properties of alcohols, as we learned in Section 11.6B.

- **Alcohols are soluble in organic solvents.**
- **Low molecular weight alcohols (those having less than six carbons) are soluble in water.**
- **Higher molecular weight alcohols (those having six carbons or more) are not soluble in water.**

Thus, both ethanol (CH_3CH_2OH) and 1-octanol [$CH_3(CH_2)_7OH$] are soluble in organic solvents, but ethanol is water soluble and 1-octanol is not.

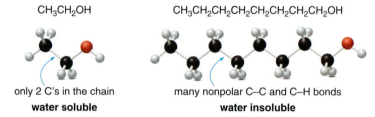

CH₃CH₂OH   CH₃CH₂CH₂CH₂CH₂CH₂CH₂CH₂OH

only 2 C's in the chain   many nonpolar C–C and C–H bonds
water soluble   **water insoluble**

PROBLEM 14.4

Which compound in each pair has the higher boiling point?

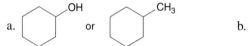

a. [cyclohexane with OH] or [cyclohexane with CH₃]

b. ($CH_3)_3C$—OH   or   ($CH_3)_4C$

PROBLEM 14.5

Label each compound as water soluble or water insoluble.

a. 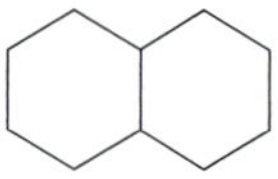 [decalin structure]   b. ($CH_3)_3C$—OH   c. CH_2=$CHCH_2CH_3$ d. 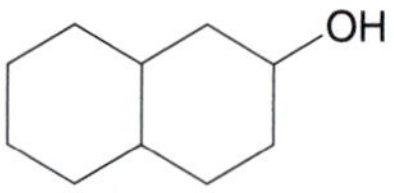 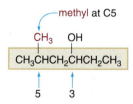 [decalin with OH]

14.3 NOMENCLATURE OF ALCOHOLS

In the IUPAC system, alcohols are identified by the suffix *-ol.*

HOW TO   Name an Alcohol Using the IUPAC System

EXAMPLE   **Give the IUPAC name of the following alcohol.**

$$CH_3CHCH_2CHCH_2CH_3$$
with CH_3 and OH substituents

Step [1]   **Find the longest carbon chain containing the carbon bonded to the OH group.**

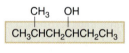

6 C's in the longest chain

hexane ----→ **hexanol**

• Change the *-e* ending of the parent alkane to the suffix *-ol.*

Step [2]   **Number the carbon chain to give the OH group the lower number, and apply all other rules of nomenclature.**

a. **Number** the chain.

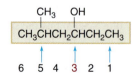

6   5   4   3   2   1

• Number the chain to put the OH group at C3, not C4.

3-hexanol

b. **Name** and **number** the substituents.

methyl at C5

CH₃   OH
CH₃CHCH₂CHCH₂CH₃

5   3

Answer: 5-methyl-3-hexanol

When an OH group is bonded to a ring, the **ring is numbered beginning with the OH group,** and the "1" is usually omitted from the name. The ring is then numbered in a clockwise or counterclockwise fashion to give the next substituent the lower number.

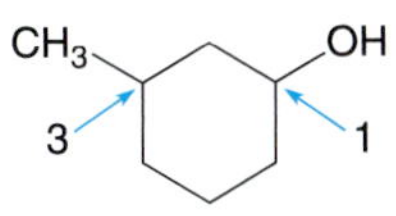

3-methylcyclohexanol

⎡ The OH group is at C1; the second substituent (CH₃) gets the lower number. ⎤

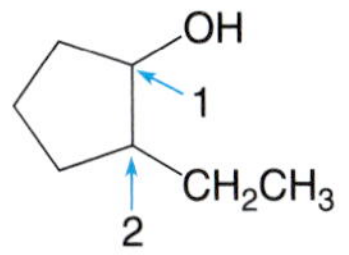

2-ethylcyclopentanol

⎡ The OH group is at C1; the second substituent (CH₃CH₂) gets the lower number. ⎤

Common names are often used for simple alcohols. To assign a common name:

- Name all the carbon atoms of the molecule as a single alkyl group.
- Add the word *alcohol,* separating the words with a space.

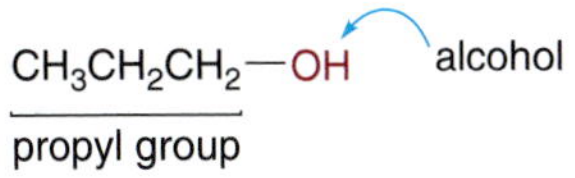

$CH_3CH_2CH_2$—OH   alcohol ---→ **propyl alcohol**

propyl group   a common name

Compounds with two hydroxyl groups are called **diols** (using the IUPAC system) or **glycols.** Compounds with three hydroxyl groups are called **triols,** and so forth. To name a diol, for example, the suffix **-diol** is added to the name of the parent alkane and numbers are used in the prefix to indicate the location of the two OH groups.

$HOCH_2CH_2OH$

IUPAC name: 1,2-ethanediol
Common name: **ethylene glycol**

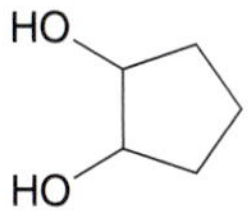

1,2-cyclopentanediol

SAMPLE PROBLEM 14.2   Give the IUPAC name of the following alcohol.

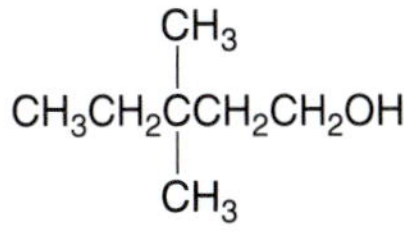

$$\underset{\underset{CH_3}{|}}{\overset{\overset{CH_3}{|}}{CH_3CH_2CCH_2CH_2OH}}$$

ANALYSIS AND SOLUTION

[1] **Find the longest carbon chain that contains the carbon bonded to the OH group.**

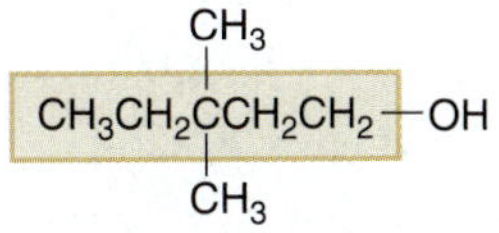

$$\underset{\underset{CH_3}{|}}{\overset{\overset{CH_3}{|}}{\boxed{CH_3CH_2CCH_2CH_2}}}\!-OH$$

5 C's in the longest chain ---→ **pentanol**

- Change the *-e* ending of the parent alkane to the suffix *-ol.*

[2] **Number the carbon chain to give the OH group the lower number, and apply all other rules of nomenclature.**

a. **Number** the chain.

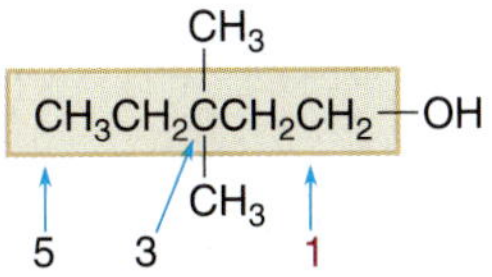

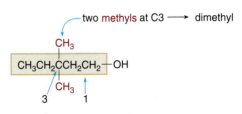

- Number the chain to put the OH group at C1, not C5.

1-pentanol

b. **Name and number** the substituents.

—two methyls at C3 ⟶ dimethyl

$$\underset{\underset{CH_3}{|}}{\overset{\overset{CH_3}{|}}{\boxed{CH_3CH_2CCH_2CH_2}}}\!-OH$$
3   1

Answer: 3,3-dimethyl-1-pentanol

PROBLEM 14.6

Give the IUPAC name for each compound.

a. CH₃CH(OH)(CH₂)₄CH₃

b. (CH₃CH₂)₂CHCH(OH)CH₂CH₃

c. [cyclohexane with CH₃ and OH]

d. CH₃CH₂CH₂CH(CH₃)CH(OH)CH₂CH(CH₂CH₃)CH₂CH₃

PROBLEM 14.7

Give the structure corresponding to each name.

a. 7,7-dimethyl-4-octanol
b. 5-methyl-4-propyl-3-heptanol
c. 2-ethyl-3-methylcyclohexanol
d. 1,3-cyclohexanediol

14.4 INTERESTING ALCOHOLS

The most well known alcohol is ethanol, CH_3CH_2OH. **Ethanol** (Figure 14.1), formed by the fermentation of carbohydrates in grains and grapes, is the alcohol present in alcoholic beverages. Fermentation requires yeast, which provides the needed enzymes for the conversion. Ethanol is likely the first organic compound synthesized by humans, since alcohol has been produced for at least 4,000 years.

Ethanol is also a common laboratory solvent, which is sometimes made unfit to ingest by adding small amounts of benzene or methanol (CH_3OH), both of which are toxic. Ethanol is used as a gasoline additive since it readily combusts to form CO_2 and H_2O with the release of energy. Other simple alcohols are listed in Figure 14.2.

▼ **FIGURE 14.1 Ethanol—The Alcohol in Alcoholic Beverages**

Fermentation

$C_6H_{12}O_6$ (glucose) $\xrightarrow{yeast}$ 2 CH_3CH_2OH (ethanol) + 2 CO_2

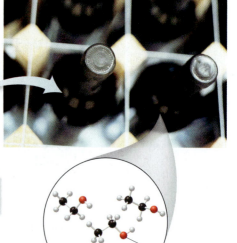

CH_3CH_2OH ethanol

Ethanol is the alcohol in red wine, obtained by the fermentation of grapes. All alcoholic beverages are mixtures of ethanol and water in various proportions. Beer has 3–8% ethanol, wines have 10–14% ethanol, and other liquors have 35–90% ethanol.

▼ **FIGURE 14.2   Some Simple Alcohols**

CH₃OH

- **Methanol (CH₃OH)** is a useful solvent and starting material for the synthesis of plastics. Methanol is extremely toxic because of the oxidation products formed when metabolized in the liver (Section 14.6A). Ingestion of as little as 15 mL (about half an ounce) causes blindness and 100 mL causes death.

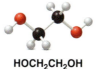

(CH₃)₂CHOH

- **2-Propanol [(CH₃)₂CHOH, isopropyl alcohol]** is the major component of rubbing alcohol. When rubbed on the skin it evaporates readily, producing a pleasant cooling sensation. 2-Propanol is used to clean skin before medical procedures and to sterilize medical instruments.

HOCH₂CH₂OH

- **Ethylene glycol (HOCH₂CH₂OH)** is the major component of antifreeze. It is sweet tasting but extremely toxic. Ethylene glycol is a starting material in the synthesis of synthetic fibers such as Dacron (Section 17.10).

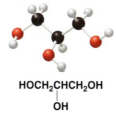

HOCH₂CHCH₂OH
       |
       OH

- **Glycerol [(HOCH₂)₂CHOH]** is a triol used in lotions, liquid soap, and shaving cream. Since it is sweet tasting and nontoxic, it is also an additive in candy and some prepared foods. Its three OH groups readily hydrogen bond to water, so it helps to retain moisture in these products.

The six-membered rings in starch and cellulose are drawn in the puckered "chair" form mentioned in Section 12.5A.

Starch and **cellulose** are two polymers that contain many OH groups and belong to the family of molecules called **carbohydrates.**

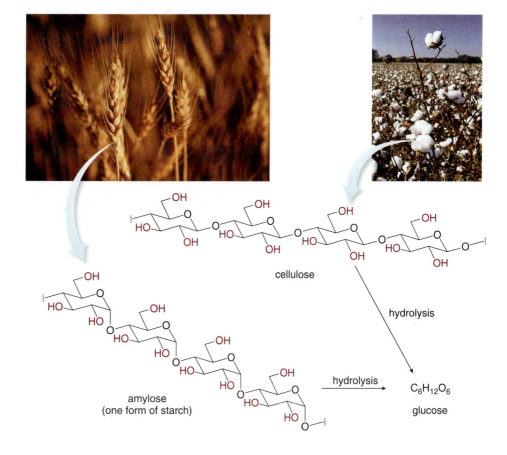

cellulose

amylose
(one form of starch)

hydrolysis → $C_6H_{12}O_6$

glucose

Starch is the main carbohydrate in the seeds and roots of plants. When humans ingest wheat, rice, and potatoes, for example, they consume starch, which is then hydrolyzed to the simple sugar **glucose. Cellulose,** nature's most abundant organic material, gives rigidity to tree trunks and plant stems. Wood, cotton, and flax are composed largely of cellulose. Complete hydrolysis of cellulose also forms glucose, but unlike starch, humans cannot metabolize cellulose to glucose; that is, humans can digest starch but not cellulose.

PROBLEM 14.8	Write a balanced equation for the combustion of ethanol with O_2 to form CO_2 and H_2O.

14.5 REACTIONS OF ALCOHOLS

Alcohols undergo two useful reactions—**dehydration** and **oxidation.**

14.5A DEHYDRATION

When an alcohol is treated with a strong acid such as H_2SO_4, the elements of water are lost and an alkene is formed as product. **Loss of H_2O from a starting material is called** *dehydration.* Dehydration takes place by breaking bonds on two adjacent atoms—the C—OH bond and an adjacent C—H bond.

Dehydration is an example of a general type of organic reaction called an **elimination reaction.**

- *Elimination* is a reaction in which elements of the starting material are "lost" and a new multiple bond is formed.

For example, dehydration of ethanol (CH_3CH_2OH) with H_2SO_4 forms ethylene ($CH_2\!=\!CH_2$), as shown. To draw the product of any dehydration, remove the elements of H and OH from two adjacent atoms and draw a carbon–carbon double bond in the product.

Sometimes more than one alkene can be formed as the product of a dehydration reaction, as is the case when 2-butanol [$CH_3CH(OH)CH_2CH_3$] is the starting material. If H and OH are removed from C1 and C2, 1-butene is formed. If H and OH are removed from C3 and C2, 2-butene is formed.

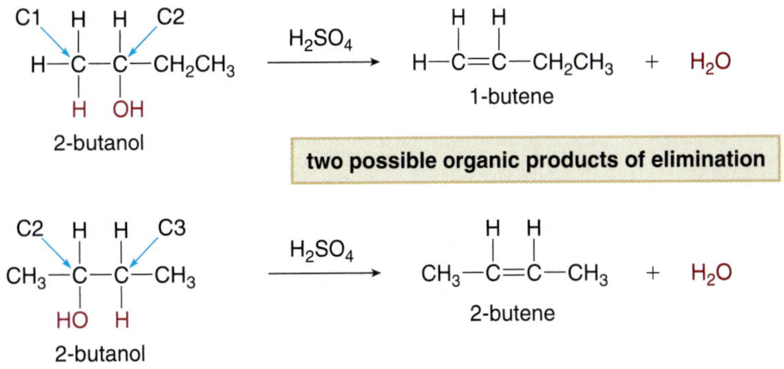

Although both products are formed, **the major product is the alkene with more alkyl groups bonded to the carbons of the double bond.** In this case, therefore, 2-butene is the major product, because its C=C has two alkyl groups (two CH_3 groups) bonded to it, whereas the C=C of 1-butene has only one alkyl group (one CH_2CH_3 group) bonded to it.

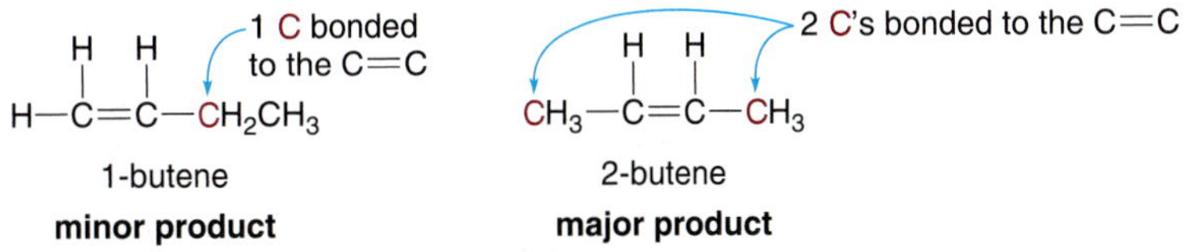

This is a specific example of a general trend called the **Zaitsev rule.**

- **The Zaitsev rule: The major product in elimination is the alkene that has more alkyl groups bonded to it.**

| SAMPLE PROBLEM 14.3 | Draw all possible products of dehydration of the following alcohol, and predict which one is the major product. |

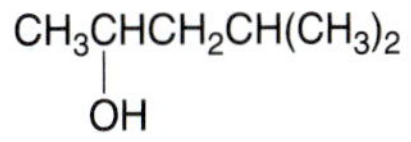

$$CH_3CHCH_2CH(CH_3)_2$$
$$|$$
$$OH$$

ANALYSIS   To draw the products of dehydration:
- First find the carbon bonded to the OH group, and then identify all carbons with H's bonded to this carbon.
- Remove the elements of H and OH from two adjacent C's, and draw a double bond between these C's in the product.
- When two different alkenes are formed, the major product has more C's bonded to the C=C.

SOLUTION   In this example, there are two different C's bonded to the C with the OH. Elimination of H and OH from C1 and C2 forms 4-methyl-1-pentene. Elimination of H and OH from C3 and C2 forms 4-methyl-2-pentene. The major product is 4-methyl-2-pentene because it has two C's bonded to the C=C, whereas 4-methyl-1-pentene has only one C bonded to the C=C.

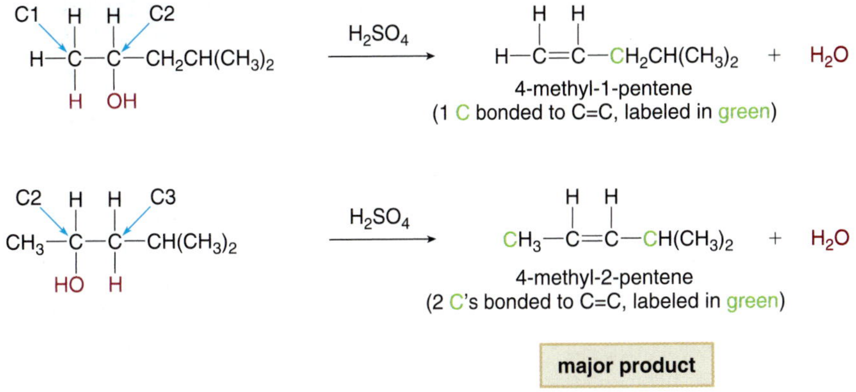

major product

PROBLEM 14.9

Draw the products formed when each alcohol is dehydrated with H_2SO_4.

a. $CH_3-\overset{\overset{\displaystyle H}{|}}{\underset{\underset{\displaystyle OH}{|}}{C}}-CH_3$

b. ⬡$-CH_2CH_2OH$

c. ⬠$-OH$

PROBLEM 14.10

What alkenes are formed when each alcohol is treated with H_2SO_4? Use the Zaitsev rule to predict the major product.

a. $CH_3\underset{\underset{\displaystyle OH}{|}}{CH}CH_2CH_2CH_3$

b. $CH_3-\overset{\overset{\displaystyle CH_3}{|}}{\underset{\underset{\displaystyle OH}{|}}{C}}-CH_2CH_3$

c. (cyclohexane ring with CH_3 and OH substituents)

14.5B OXIDATION

Recall from Section 12.8 that to determine if an organic compound has been oxidized, we compare the relative number of C—H and C—O bonds in the starting material and product.

- Oxidation results in an increase in the number of C—O bonds or a decrease in the number of C—H bonds.

Thus, an organic compound like CH_4 can be oxidized by replacing C—H bonds with C—O bonds, as shown in Figure 14.3. When an alkane like CH_4 undergoes complete oxidation, as it does during combustion, all C—H bonds are converted to C—O bonds, forming CO_2. Organic chemists use the symbol [O] to indicate oxidation.

Alcohols can be oxidized to a variety of compounds, depending on the type of alcohol and the reagent. **Oxidation occurs by replacing the C—H bonds *on the carbon bearing the OH group* by C—O bonds.** All oxidation products from alcohol starting materials contain a **C═O, a carbonyl group.**

Alcohol   **Carbonyl compound**

O—H   O
| | ||
—C—H   [O]→   C   ←A new C—O bond is formed.
| | /   \
Two bonds are broken.

A common reagent for alcohol oxidation is potassium dichromate, $K_2Cr_2O_7$, a red-orange solid. During oxidation the reagent is converted a green Cr^{3+} product. Thus, these oxidations are accompanied by a characteristic color change. Since a variety of reagents can be used, we will write oxidation with the general symbol for oxidation, [O].

▼ **FIGURE 14.3   A General Scheme for Oxidizing an Organic Compound**

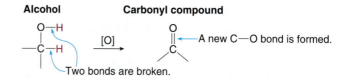

4 C—H bonds	**3 C—H bonds**	**2 C—H bonds**	**1 C—H bond**	**0 C—H bonds**

4 C—H bonds   3 C—H bonds   2 C—H bonds   1 C—H bond   0 C—H bonds
0 C—O bonds   1 C—O bond   2 C—O bonds   3 C—O bonds   4 C—O bonds

Increasing number of C—O bonds

- Primary (1°) alcohols are first oxidized to aldehydes (RCHO), which are further oxidized to carboxylic acids (RCOOH) by replacing one and then two C—H bonds by C—O bonds.

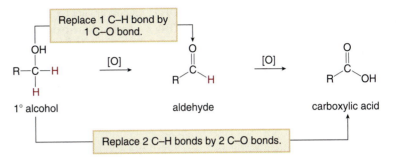

Thus, oxidation of one C—H bond of ethanol forms acetaldehyde. Since acetaldehyde still contains a hydrogen atom on the carbonyl carbon, converting this C—H bond to a C—O bond forms acetic acid, a carboxylic acid with three C—O bonds.

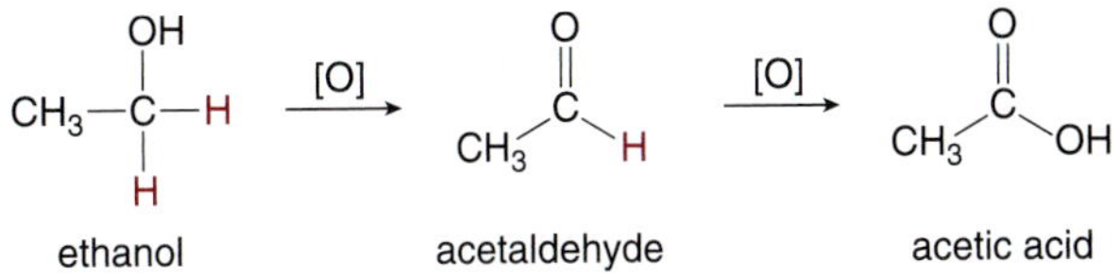

- Secondary (2°) alcohols are oxidized to ketones (R₂CO), by replacing one C—H bond by one C—O bond.

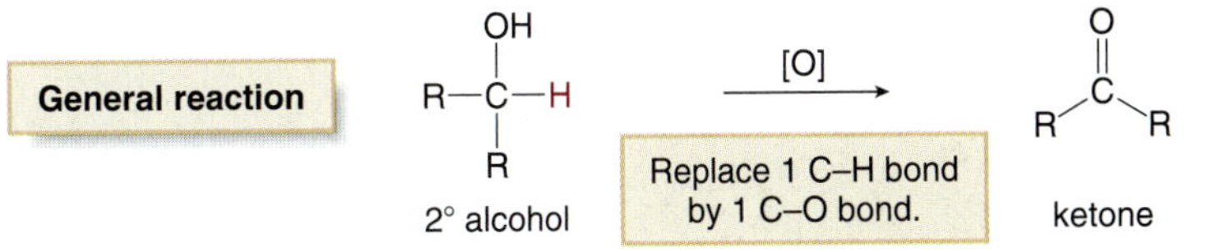

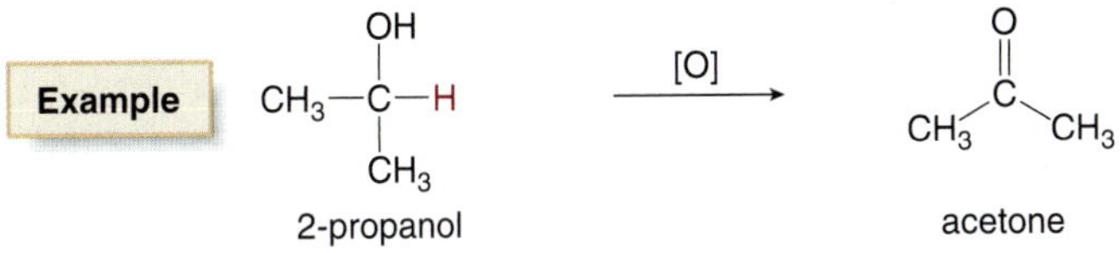

Since 2° alcohols have only one hydrogen atom bonded to the carbon with the OH group, they can be oxidized to only one type of compound, a ketone. Thus, 2-propanol is oxidized to acetone.

- Tertiary 3° alcohols have no H atoms on the carbon with the OH group, so they are not oxidized.

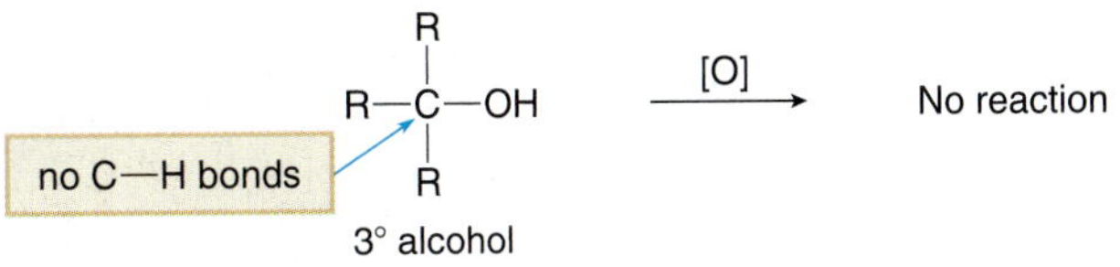

SAMPLE PROBLEM 14.4   Draw the carbonyl products formed when each alcohol is oxidized with $K_2Cr_2O_7$.

a. 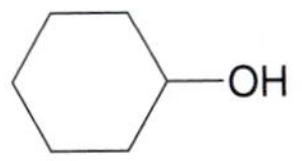 cyclohexanol —OH

b. $CH_3CH_2CH_2CH_2OH$

1-butanol

ANALYSIS   Classify the alcohol as 1°, 2°, or 3° by drawing in all of the H atoms on the C with the OH. Then concentrate on the C with the OH group and replace H atoms by bonds to O. Keep in mind:

- RCH_2OH (1° alcohols) are oxidized to RCHO, which are then oxidized to RCOOH.
- R_2CHOH (2° alcohols) are oxidized to R_2CO.
- R_3COH (3° alcohols) are not oxidized since they have no H atom on the C with the OH.

SOLUTION   a. Since cyclohexanol is a 2° alcohol with only one H atom on the C bonded to the OH group, it is oxidized to a ketone, cyclohexanone.

only 1 H atom

cyclohexanol   [O]   cyclohexanone
2° alcohol   ketone

b. 1-Butanol is a 1° alcohol with two H atoms on the C bonded to the OH group. Thus, it is first oxidized to an aldehyde and then to a carboxylic acid.

$$CH_3CH_2CH_2-\overset{\overset{\displaystyle OH}{|}}{\underset{\underset{\displaystyle H}{|}}{C}}-H \xrightarrow{[O]} CH_3CH_2CH_2-\overset{\displaystyle O}{\overset{\|}{C}}-H \xrightarrow{[O]} CH_3CH_2CH_2-\overset{\displaystyle O}{\overset{\|}{C}}-OH$$

2 H atoms

1-butanol   butanal   butanoic acid
1° alcohol   aldehyde   carboxylic acid

PROBLEM 14.11

What products are formed when each alcohol is oxidized with $K_2Cr_2O_7$? In some cases, no reaction occurs.

a. $CH_3\underset{\underset{\displaystyle OH}{|}}{CH}CH_2CH_2CH_3$ b. $(CH_3)_3CCH_2CH_2OH$ c. ⬠—OH   d. ⬡ with CH₃ and OH

14.5C   FOCUS ON THE HUMAN BODY
OXIDATION AND BLOOD ALCOHOL SCREENING

Oxidations with chromium reagents are characterized by a color change, as the **red-orange chromium reagent** is reduced to a **green Cr^{3+} product.** The first devices used to measure blood alcohol content in individuals suspected of "driving under the influence," made use of this color change. Oxidation of CH_3CH_2OH, the 1° alcohol in alcoholic beverages, with red-orange $K_2Cr_2O_7$ forms CH_3COOH and green Cr^{3+}.

$$CH_3CH_2-OH \;+\; \boxed{\begin{array}{c} \textbf{K}_2\textbf{Cr}_2\textbf{O}_7 \\ \textbf{red-orange} \end{array}} \longrightarrow CH_3-\overset{\displaystyle O}{\overset{\|}{C}}-OH \;+\; \boxed{\begin{array}{c} \textbf{Cr}^{3+} \\ \textbf{green} \end{array}}$$

ethanol   acetic acid
"alcohol"

Blood alcohol level can be determined by having an individual blow into a tube containing $K_2Cr_2O_7$ and an inert solid. The alcohol in the exhaled breath is oxidized by the chromium reagent, which turns green in the tube (Figure 14.4). The higher the concentration of CH_3CH_2OH in the breath, the more chromium reagent is reduced, and the farther the green color extends down the length of the sample tube. This value is then correlated with blood alcohol content to determine if an individual has surpassed the legal blood alcohol limit.

▼ **FIGURE 14.4**   Blood Alcohol Screening

Schematic of an alcohol testing device	Consumer product

The tube contains $K_2Cr_2O_7$.

An individual exhales → into the tube.

$K_2Cr_2O_7$ (red-orange) reacts with CH_3CH_2OH, forming Cr^{3+} (green).

The balloon inflates with exhaled air.

$K_2Cr_2O_7$

- The oxidation of CH_3CH_2OH with $K_2Cr_2O_7$ to form CH_3COOH and Cr^{3+} was the first available method for the routine testing of alcohol concentration in exhaled air. Some consumer products for alcohol screening are still based on this technology.

- A driver is considered "under the influence" in most states with a blood alcohol concentration of 0.08%.

14.6 FOCUS ON HEALTH & MEDICINE
ETHANOL, THE MOST WIDELY ABUSED DRUG

Throughout history, humans have ingested alcoholic beverages for their pleasant taste and the feeling of euphoria they impart. Although we think of alcohol as a stimulant, largely because small amounts decrease social inhibitions, the ethanol (CH_3CH_2OH) in an alcoholic beverage actually depresses the central nervous system. The chronic and excessive consumption of alcoholic beverages has become a major health and social crisis, making ethanol the most widely abused drug in the United States. One estimate suggests that there are 40 times more alcoholics than heroin addicts.

14.6A   THE METABOLISM OF ETHANOL

When ethanol is consumed, it is quickly absorbed in the stomach and small intestines and then rapidly transported in the bloodstream to other organs. Ethanol is metabolized in the liver, by a two-step oxidation sequence. The body does not use chromium reagents as oxidants. Instead, high molecular weight enzymes, alcohol dehydrogenase and aldehyde dehydrogenase, and a small molecule called a **coenzyme** carry out these oxidations.

The products of the biological oxidation of ethanol are the same as the products formed in the laboratory. When ethanol (CH_3CH_2OH, a 1° alcohol) is ingested, it is oxidized in the liver first to CH_3CHO (acetaldehyde), and then to CH_3COOH (acetic acid).

$$CH_3CH_2{-}OH \xrightarrow[\text{alcohol dehydrogenase}]{[O]} \underset{\text{acetaldehyde}}{CH_3\overset{O}{\overset{\|}{C}}H} \xrightarrow[\text{aldehyde dehydrogenase}]{[O]} \underset{\text{acetic acid}}{CH_3\overset{O}{\overset{\|}{C}}OH}$$

ethanol

If more ethanol is ingested than can be metabolized in a given time period, the concentration of acetaldehyde accumulates. This toxic compound is responsible for the feelings associated with a hangover.

Antabuse, a drug given to alcoholics to prevent them from consuming alcoholic beverages, acts by interfering with the normal oxidation of ethanol. Antabuse inhibits the oxidation of acetaldehyde to acetic acid. Since the first step in ethanol metabolism occurs but the second does not, the concentration of acetaldehyde rises, causing an individual to become violently ill.

While alcohol use is socially acceptable, alcohol-related traffic fatalities are common with irresponsible alcohol consumption. In 2004, almost 40% of all fatalities in car crashes in the United States were alcohol-related.

Antabuse

Methanol (CH_3OH, Section 14.4) is toxic because of these same oxidation reactions. Methanol is oxidized by enzymes to yield formaldehyde and formic acid. Both of these oxidation products are extremely toxic since they cannot be further metabolized by the body. As a result, the pH of the blood decreases, and blindness and death can follow.

Since the enzymes have a higher affinity for ethanol than methanol, methanol poisoning is treated by giving ethanol to an individual. With both methanol and ethanol in a patient's system, the enzymes react more readily with ethanol, allowing the methanol to be excreted unchanged without the formation of methanol's toxic oxidation products.

PROBLEM 14.12

Ethylene glycol, $HOCH_2CH_2OH$, is an extremely toxic diol because its oxidation products are also toxic. Draw the oxidation products formed during the metabolism of ethylene glycol.

14.6B HEALTH EFFECTS OF ALCOHOL CONSUMPTION

Alcohol consumption has both short-term and long-term effects. In small amounts it can cause dizziness, giddiness, and a decrease in social inhibitions. In larger amounts it decreases coordination and reaction time and causes drowsiness. Intoxicated individuals have exaggerated emotions, slurred speech, amnesia, and a lack of coordination. Coma and death can result at even higher concentrations.

Ethanol affects many organs. It dilates blood vessels, resulting in a flushed appearance. It increases urine production and stimulates the secretion of stomach acid. Chronic excessive alcohol consumption damages the heart and can lead to **cirrhosis of the liver,** an incurable and fatal disorder in which the liver is scarred and loses its ability to function.

The consumption of ethanol during pregnancy is not recommended. The liver of a fetus does not contain the necessary enzyme to metabolize ethanol. As a result, any alcohol consumed by a pregnant woman crosses the placenta and can cause a variety of abnormalities. **Fetal alcohol syndrome** can result in infants of women who chronically consume alcohol during pregnancy. Infants with fetal alcohol syndrome are mentally retarded, grow poorly, and have certain facial abnormalities. Even modest alcohol consumption during pregnancy can negatively affect the fetus. Currently the American Medical Association recommends that a woman completely abstain from alcohol during pregnancy.

14.7 STRUCTURE AND PROPERTIES OF ETHERS

Ethers are organic compounds that have two alkyl groups bonded to an oxygen atom. These two alkyl groups can be the same, or they can be different.

The oxygen atom of an ether is surrounded by two carbon atoms and two lone pairs of electrons, giving it a **bent** shape like the oxygen in H_2O. The C—O—C bond angle is similar to the tetrahedral bond angle of 109.5°.

The ether oxygen can be contained in a ring. **A ring that contains a heteroatom is called a heterocycle.** When the ether is part of a three-membered ring, the ether is called an *epoxide*.

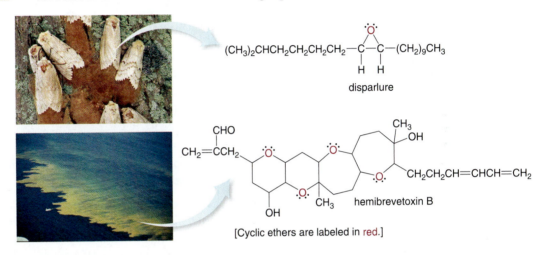

epoxide

a three-membered ring
with an ether O atom

Many compounds that contain ether heterocycles are naturally occurring. **Disparlure,** the sex pheromone of the female gypsy moth, is an epoxide used to control the spread of the gypsy moth caterpillar. This pest has periodically devastated the forests of the northeastern United States by eating the leaves of many shade and fruit-bearing trees. **Hemibrevetoxin B,** a complex neuro-toxin with four ether oxygens, is produced by algal blooms referred to as "red tides," because of the color of the ocean waters when these algae proliferate.

$(CH_3)_2CHCH_2CH_2CH_2CH_2$—C—C—$(CH_2)_9CH_3$

disparlure

hemibrevetoxin B

[Cyclic ethers are labeled in red.]

PROBLEM 14.13

What other functional groups are present in hemibrevetoxin B?

PROBLEM 14.14

Draw the structure of the three constitutional isomers of molecular formula $C_4H_{10}O$ that contain an ether.

14.7A PHYSICAL PROPERTIES

Because oxygen is more electronegative than carbon, the C—O bonds of an ether are both polar. Since an ether contains two polar bonds and a bent shape, it has a **net dipole.** In this way, ethers resemble alcohols.

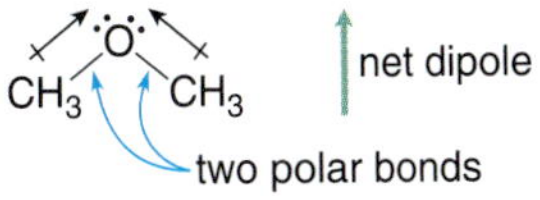

net dipole

two polar bonds

Ethers do not contain a hydrogen atom bonded to oxygen, so unlike alcohols, two ether molecules *cannot* intermolecularly hydrogen bond to each other. **This gives ethers stronger intermolecular forces than alkanes but weaker intermolecular forces than alcohols.** As a result:

- Ethers have higher melting points and boiling points than hydrocarbons of comparable size and shape.
- Ethers have lower melting points and boiling points than alcohols of comparable size and shape.

$CH_3CH_2CH_2CH_3$	$CH_3OCH_2CH_3$	$CH_3CH_2CH_2OH$
butane	ethyl methyl ether	1-propanol
boiling point −0.5 °C	boiling point 11 °C	boiling point 97 °C

Increasing boiling point →

All ethers are soluble in organic solvents. Like alcohols, **low molecular weight ethers are water soluble,** because the oxygen atom of the ether can hydrogen bond to one of the hydrogens of water. When the alkyl groups of the ether have more than a total of five carbons, the nonpolar portion of the molecule is too large, so the ether is water insoluble.

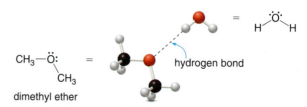

CH₃—Ö: =   hydrogen bond
       CH₃
dimethyl ether

SAMPLE PROBLEM 14.5

Rank the following compounds in order of increasing boiling point:

—OCH₃   —CH₂CH₃   —CH₂OH

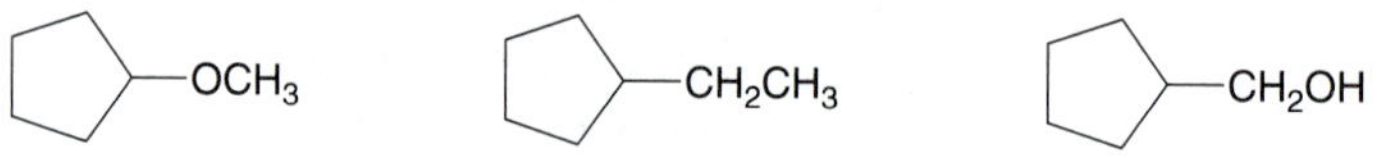

A   **B**   **C**

ANALYSIS

Look at the functional groups to determine the strength of the intermolecular forces—**the stronger the forces, the higher the boiling point.**

SOLUTION

B is an alkane with nonpolar C—C and C—H bonds, so it has the weakest intermolecular forces and therefore the lowest boiling point. **C** is an alcohol capable of intermolecular hydrogen bonding, so it has the strongest intermolecular forces and the highest boiling point. **A** is an ether, so it contains a net dipole but is incapable of intermolecular hydrogen bonding. **A,** therefore, has intermolecular forces of intermediate strength and has a boiling point between the boiling points of **B** and **C.**

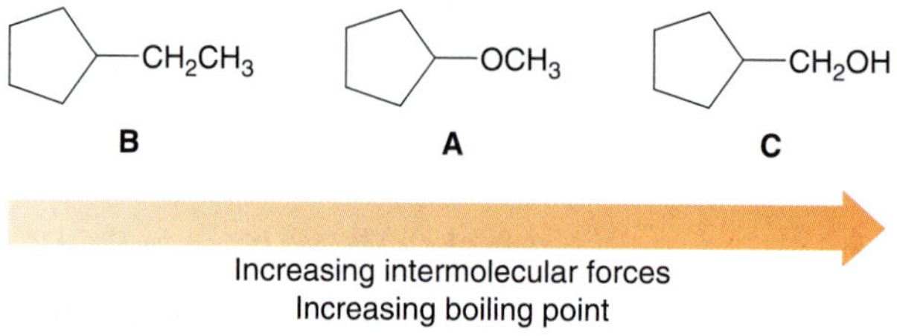

—CH₂CH₃   —OCH₃   —CH₂OH

B   **A**   **C**

Increasing intermolecular forces →
Increasing boiling point

PROBLEM 14.15

Which compound in each pair has the higher boiling point?

a. $CH_3(CH_2)_6CH_3$ or $CH_3(CH_2)_5OCH_3$

b. 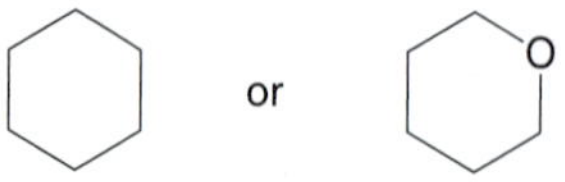 ⬡   or   ⬡O

c. $CH_3(CH_2)_6OH$ or $CH_3(CH_2)_5OCH_3$

d. $CH_3(CH_2)_5OCH_3$ or CH_3OCH_3

PROBLEM 14.16

Label each ether as water soluble or water insoluble.

a. CH_3CH_2—O—CH_3 b. 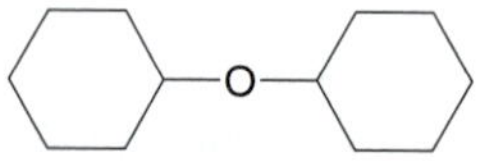 ⬡—O—⬡   c. 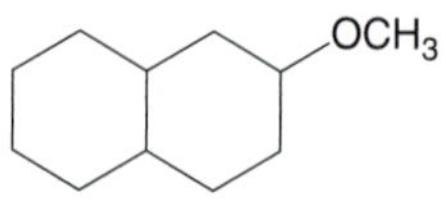 ⬡⬡—OCH₃

14.7B NAMING ETHERS

Simple ethers are usually assigned common names. To do so, **name both alkyl groups** bonded to the oxygen, arrange these names alphabetically, and add the word **ether.** For ethers with identical alkyl groups, name the alkyl group and add the prefix **di-.**

CH$_3$—O—CH$_2$CH$_3$

methyl   ethyl

ethyl methyl ether

CH$_3$CH$_2$—O—CH$_2$CH$_3$

ethyl   ethyl

diethyl ether

More complex ethers are named using the IUPAC system. One alkyl group is named as a hydrocarbon chain, and the other is named as part of a substituent bonded to that chain.

- Name the simpler alkyl group + O atom as an alkoxy substituent by changing the *-yl* ending of the alkyl group to *-oxy.*

CH$_3$O–   CH$_3$CH$_2$O–

methoxy   ethoxy

- Name the remaining alkyl group as an alkane, with the alkoxy group as a substituent bonded to this chain.

SAMPLE PROBLEM 14.6

Give the IUPAC name for the following ether.

CH$_3$CH$_2$CH$_2$CH$_2$CHCH$_2$CH$_2$CH$_3$
　　　　　　　　　　OCH$_2$CH$_3$

ANALYSIS AND SOLUTION

[1] Name the longer chain as an alkane and the shorter chain as an alkoxy group.

CH$_3$CH$_2$CH$_2$CH$_2$CHCH$_2$CH$_2$CH$_3$
　　　　　　　　　　OCH$_2$CH$_3$
ethoxy group

8 C's – – – → octane

[2] Apply other nomenclature rules to complete the name.

4　　　　　　1

CH$_3$CH$_2$CH$_2$CH$_2$CHCH$_2$CH$_2$CH$_3$
　　　　　　　　　　OCH$_2$CH$_3$

Answer: 4-ethoxyoctane

PROBLEM 14.17

Name each ether.

a. CH$_3$—O—CH$_2$CH$_2$CH$_2$CH$_3$

b. 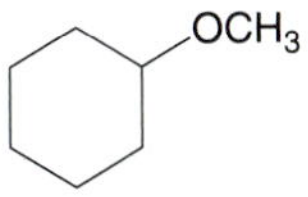 ⬡—OCH$_3$

c. CH$_3$CH$_2$CH$_2$—O—CH$_2$CH$_2$CH$_3$

PROBLEM 14.18

Draw the structure corresponding to each name.

a. dibutyl ether

b. ethyl propyl ether

c. 1-methoxypentane

d. 3-ethoxyhexane

14.8   FOCUS ON HEALTH & MEDICINE
ETHERS AS ANESTHETICS

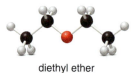

diethyl ether

A general anesthetic is a drug that interferes with nerve transmission in the brain, resulting in a loss of consciousness and the sensation of pain. The discovery that **diethyl ether** ($CH_3CH_2OCH_2CH_3$) is a general anesthetic revolutionized surgery in the nineteenth century. Early experiments performed by a dentist, Dr. William Morton, resulted in a public demonstration of diethyl ether as an anesthetic in Boston in 1846. Diethyl ether quickly supplanted the crude methods then used for a patient to tolerate the excruciating pain of surgery—making an individual unconscious by excess alcohol intake or by striking a blow to the head.

Diethyl ether is an imperfect anesthetic, but considering the alternatives in the nineteenth century, it was considered revolutionary. It is safe and easy to administer with low patient mortality, but it is highly flammable, and it causes nausea in many patients.

For these reasons, alternatives to diethyl ether are now widely used. Many of these newer general anesthetics, which cause little patient discomfort, are also ethers. These include isoflurane, enflurane, and methoxyflurane. Replacing some of the hydrogen atoms in the ether by halogens results in compounds with similar anesthetic properties but decreased flammability.

This painting by Robert Hinckley depicts a public demonstration of the use of diethyl ether as an anesthetic at the Massachusetts General Hospital in Boston, MA in the 1840s.

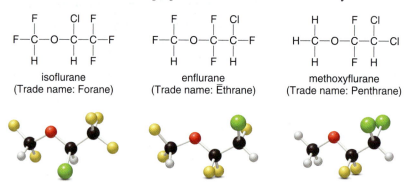

isoflurane
(Trade name: Forane)

enflurane
(Trade name: Ēthrane)

methoxyflurane
(Trade name: Penthrane)

PROBLEM 14.19

Halothane ($CF_3CHClBr$) is a general anesthetic that is not an ether. Draw a three-dimensional structure for halothane using solid lines, wedges, and dashes (see Section 11.3).

14.9   ALKYL HALIDES

Alkyl halides have the general molecular formula $C_nH_{2n+1}X$, and are formally derived from an alkane by replacing a hydrogen atom with a halogen.

Alkyl halides **are organic molecules containing a halogen atom X (X = F, Cl, Br, I) bonded to a tetrahedral carbon atom.** Alkyl halides are classified as **primary (1°), secondary (2°),** or **tertiary (3°)** depending on the number of carbons bonded to the carbon with the halogen.

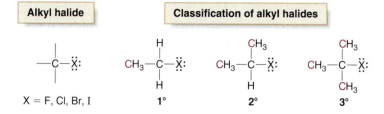

Alkyl halide	Classification of alkyl halides		
X = F, Cl, Br, I	1°	2°	3°

- A primary (1°) alkyl halide has a halogen on a carbon bonded to one carbon.
- A secondary (2°) alkyl halide has a halogen on a carbon bonded to two carbons.
- A tertiary (3°) alkyl halide has a halogen on a carbon bonded to three carbons.

PROBLEM 14.20	Classify each alkyl halide as 1°, 2°, or 3°.

a. $CH_3CH_2CH_2CH_2CH_2$—Br

b. 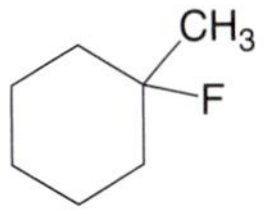 (cyclohexane ring with CH_3 and F substituents on the same carbon)

c. 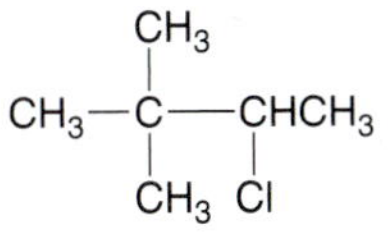 CH_3—$\underset{\underset{CH_3}{|}}{\overset{\overset{CH_3}{|}}{C}}$—$\underset{\underset{Cl}{|}}{CHCH_3}$

PROBLEM 14.21	Draw the structure of an alkyl halide that fits each description: (a) a 2° alkyl bromide of molecular formula C_4H_9Br; (b) a 3° alkyl chloride of molecular formula C_4H_9Cl; (c) a 1° alkyl fluoride of molecular formula C_3H_7F.

14.9A PHYSICAL PROPERTIES

Alkyl halides contain a polar C—X bond, but since they have all of their hydrogens bonded to carbon, they are incapable of intermolecular hydrogen bonding. Alkyl halides with one halogen are polar molecules, therefore, because they contain a net dipole. As a result, they have higher melting points and boiling points than alkanes with the same number of carbons.

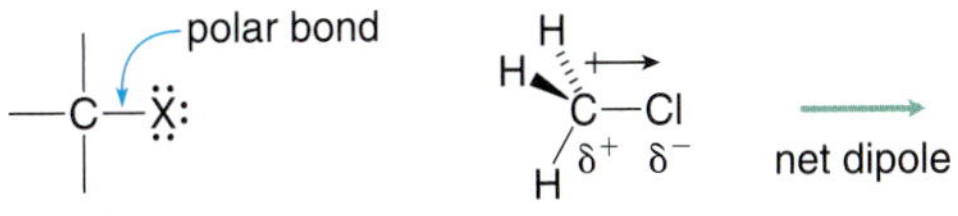

X = F, Cl, Br, I

The size of the alkyl groups and the halogen also affects the physical properties of an alkyl halide.

- **The boiling point and melting point of an alkyl halide increase with the size of the alkyl group because of increased surface area.**
- **The boiling point and melting point of an alkyl halide increase with the size of the halogen.**

Thus, $CH_3CH_2CH_2Cl$ has a higher boiling point than CH_3CH_2Cl because it has one more carbon, giving it a larger surface area. $CH_3CH_2CH_2Br$ has a higher boiling point than $CH_3CH_2CH_2Cl$ because Br is further down the column of the periodic table, making it larger than Cl.

CH_3CH_2Cl	$CH_3CH_2CH_2Cl$	$CH_3CH_2CH_2Br$
chloroethane	1-chloropropane	1-bromopropane
bp = 12 °C	bp = 47 °C	bp = 71 °C

Increasing boiling point
Increasing size of the alkyl group
Increasing size of the halogen

Since alkyl halides are incapable of hydrogen bonding, they are insoluble in water regardless of size.

PROBLEM 14.22	Rank the compounds in each group in order of increasing boiling point. a. $CH_3CH_2CH_2I$, $CH_3CH_2CH_2Cl$, $CH_3CH_2CH_2F$ b. $CH_3(CH_2)_4CH_3$, $CH_3(CH_2)_5Br$, $CH_3(CH_2)_5Cl$

14.9B NOMENCLATURE

In the IUPAC system, an alkyl halide is named as an alkane with a halogen substituent—that is, as a *halo alkane.* To name a halogen substituent, change the *-ine* ending of the name of the halogen to the suffix *-o* (e.g., chlor*ine* → chlor*o*).

HOW TO   Name an Alkyl Halide Using the IUPAC System

EXAMPLE   Give the IUPAC name of the following alkyl halide.

$$\underset{\substack{\text{CH}_3 \qquad\qquad \text{Cl}}}{\text{CH}_3\text{CH}_2\overset{|}{\text{C}}\text{HCH}_2\text{CH}_2\overset{|}{\text{C}}\text{HCH}_3}$$

Step [1]   Find the parent carbon chain containing the halogen.

$$\underset{\substack{\text{CH}_3 \qquad\qquad \text{Cl}}}{\boxed{\text{CH}_3\text{CH}_2\overset{|}{\text{C}}\text{HCH}_2\text{CH}_2\overset{|}{\text{C}}\text{HCH}_3}}$$

7 C's in the longest chain

7 C's - - - → **heptane**

• Name the parent chain as an *alkane,* with the halogen as a substituent bonded to the longest chain.

Step [2]   Apply all other rules of nomenclature.

a. **Number** the chain.

$$\underset{\substack{\text{CH}_3 \qquad\qquad \text{Cl} \\ 7\ \ 6\ \ 5\ \ 4\ \ 3\ \ 2\ \ 1}}{\boxed{\text{CH}_3\text{CH}_2\overset{|}{\text{C}}\text{HCH}_2\text{CH}_2\overset{|}{\text{C}}\text{HCH}_3}}$$

• Begin at the end nearest the first substituent, either alkyl or halogen.

b. **Name** and **number** the substituents.

methyl at C5   chloro at C2

$$\underset{\substack{7\ \ 6\ \ 5\ \ 4\ \ 3\ \ 2\ \ 1}}{\boxed{\text{CH}_3\text{CH}_2\overset{|}{\text{C}}\text{HCH}_2\text{CH}_2\overset{|}{\text{C}}\text{HCH}_3}}$$

c. Alphabetize: *c* for **c**hloro, then *m* for **m**ethyl.

ANSWER: 2-chloro-5-methylheptane

Common names for alkyl halides are used only for simple alkyl halides. To assign a common name, name the carbon atoms as an alkyl group. Then name the halogen by changing the *-ine* ending of the halogen name to the suffix *-ide;* for example, **brom*ine*** → **brom*ide.***

$$\text{CH}_3\text{CH}_2\text{—Cl} \qquad \text{chlor}\textit{ine} \longrightarrow \text{chlor}\textit{ide}$$

ethyl group

Common name: **ethyl chloride**

PROBLEM 14.23

Give the IUPAC name for each compound.

a. $\underset{\substack{\text{Br}}}{\text{CH}_3\overset{|}{\text{C}}\text{HCH}_2\text{CH}_2\text{CH}_2\text{CH}_3}$

b. $\underset{\substack{\text{Cl}}}{(\text{CH}_3)_2\text{CH}\overset{|}{\text{C}}\text{HCH}_2\text{CH}_3}$

c.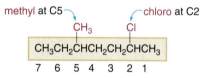

PROBLEM 14.24

Give the structure corresponding to each name.

a. 3-chloro-2-methylhexane

b. 4-ethyl-5-iodo-2,2-dimethyloctane

c. 1,1,3-tribromocyclohexane

d. propyl chloride

14.9C   INTERESTING ALKYL HALIDES

Many simple alkyl halides make excellent solvents because they are not flammable and they dissolve a wide variety of organic compounds. Two compounds in this category include $CHCl_3$ (chloroform or trichloromethane) and CCl_4 (carbon tetrachloride or tetrachloromethane). Large quantities of these solvents are produced industrially each year, but like many chlorinated organic compounds, both chloroform and carbon tetrachloride are toxic if inhaled or ingested. Two other simple alkyl halides are given in Figure 14.5.

FIGURE 14.5   Two Simple Alkyl Halides

CH₃Cl

- **Chloromethane (CH₃Cl)** is produced by giant kelp and algae and is also found in emissions from volcanoes such as Hawaii's Kilauea. Almost all of the atmospheric chloromethane results from these natural sources.

CH₂Cl₂

- **Dichloromethane (or methylene chloride, CH₂Cl₂)** is an important solvent, once used to decaffeinate coffee. Coffee is now decaffeinated using liquid CO₂ due to concerns over the possible ill effects of trace amounts of residual CH₂Cl₂ in the coffee. Subsequent studies on rats have shown, however, that no cancers occurred when animals ingested the equivalent of over 100,000 cups of decaffeinated coffee per day.

14.9D FOCUS ON THE ENVIRONMENT
ALKYL HALIDES AND THE OZONE LAYER

The 1995 Nobel Prize in Chemistry was awarded to Professors Mario Molina, Paul Crutzen, and F. Sherwood Rowland for their work in elucidating the interaction of ozone with CFCs. What began as a very fundamental research project turned out to have important implications in the practical world.

While the beneficial effects of many organic halides are undisputed, certain synthetic chlorinated organics, such as the **chlorofluorocarbons,** have caused lasting harm to the environment. **Chlorofluorocarbons (CFCs)** are simple halogen-containing compounds having the general molecular structure CF_xCl_{4-x}. CFCs are manufactured under the trade name Freons. Two examples are $CFCl_3$ (Freon 11) and CF_2Cl_2 (Freon 12).

CFCs are inert and nontoxic, and they have been used as refrigerants, solvents, and aerosol propellants. Because CFCs have low boiling points and are water insoluble, they readily escape into the upper atmosphere. There, CFCs are decomposed by sunlight. This process forms highly reactive intermediates that have been shown to destroy the ozone layer (Figure 14.6).

FIGURE 14.6   CFCs and Ozone Destruction

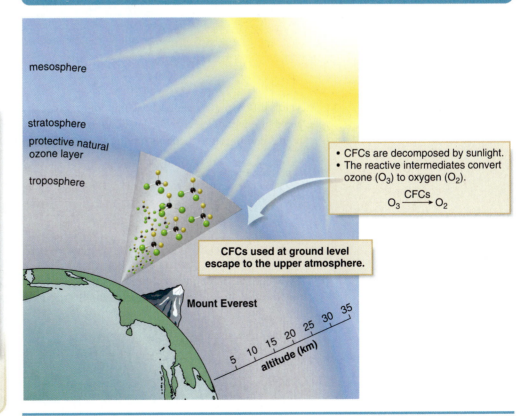

mesosphere

stratosphere

protective natural ozone layer

troposphere

- CFCs are decomposed by sunlight.
- The reactive intermediates convert ozone (O_3) to oxygen (O_2).

$$O_3 \xrightarrow{\text{CFCs}} O_2$$

CFCs used at ground level escape to the upper atmosphere.

Mount Everest

5 10 15 20 25 30 35
altitude (km)

Propane and butane are now used as propellants in spray cans in place of CFCs.

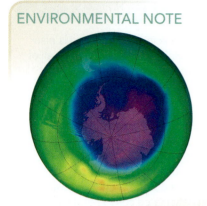
The ozone (O_3) in the upper atmosphere is vital to life: it acts as a shield, protecting the earth's surface from destructive ultraviolet radiation. A decrease in ozone concentration in this protective layer will have some immediate consequences, including an increase in the incidence of skin cancer and eye cataracts. Other long-term effects include a reduced immune response, interference with photosynthesis in plants, and harmful effects on the growth of plankton, the mainstay of the ocean food chain.

These findings led to a ban on the use of CFCs in aerosol propellants in the United States in 1978 and to the phasing out of their use in refrigeration systems. While it is now easy to second-guess the extensive use of CFCs, it is also easy to see why they were used so widely. **CFCs made refrigeration available to the general public.** Would you call your refrigerator a comfort or a necessity?

There are two newer alternatives to CFCs. **Hydrochlorofluorocarbons (HCFCs)** such as CF_3CHCl_2 contain the elements of H, Cl, and F bonded to carbon, and **hydrofluorocarbons (HFCs)** such as FCH_2CF_3 contain the elements of H and F bonded to carbon. These compounds have many properties in common with CFCs, but they are decomposed more readily before they reach the upper atmosphere, so they are less destructive to the O_3 layer.

PROBLEM 14.25   Classify each compound as a CFC, HCFC, or HFC: (a) CF_3Cl; (b) $CHFCl_2$; (c) CH_2F_2.

14.10 ORGANIC COMPOUNDS THAT CONTAIN SULFUR

Thiols **are organic compounds that contain a sulfhydryl group (SH group) bonded to a tetrahedral carbon.** Since sulfur is directly below oxygen in the periodic table, thiols can be considered sulfur analogues of alcohols. Because the sulfur atom is surrounded by two atoms and two lone pairs, thiols have a bent shape around sulfur.

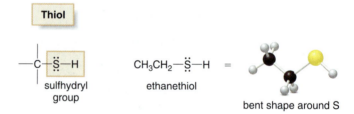

Thiols differ from alcohols in one important way. They contain no O—H bonds, so they are incapable of intermolecular hydrogen bonding. **This gives thiols lower boiling points and melting points compared to alcohols having the same size and shape.**

Hydrogen bonding can occur. stronger intermolecular forces **higher boiling point** → CH_3CH_2—OH CH_3CH_2—SH ← Hydrogen bonding is impossible. weaker intermolecular forces **lower boiling point**

ethanol
bp 78 °C

ethanethiol
bp 35 °C

To name a thiol in the IUPAC system:

- Name the parent hydrocarbon as an alkane and add the suffix *-thiol.*
- Number the carbon chain to give the SH group the lower number.

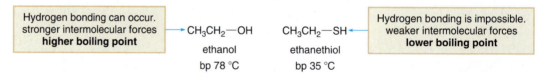

CH₃—SH   CH₃CH₂CH₂—SH   3-methyl-1-butanethiol

methanethiol   1-propanethiol

SAMPLE PROBLEM 14.7   Give the IUPAC name for the following thiol.

$$CH_3CH_2\overset{\displaystyle CH_2CH_3}{\underset{\displaystyle |}{C}}HCH_2CH_2CH_2\overset{\displaystyle SH}{\underset{\displaystyle |}{C}}HCH_3$$

ANALYSIS AND SOLUTION

[1] **Find the longest carbon chain that contains the carbon bonded to the SH group.**

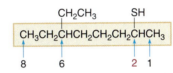

8 C's in the longest chain – – – → **octane**

- Name the alkane and add the suffix -*thiol*: **octanethiol.**

[2] **Number the carbon chain to give the SH group the lower number and apply all other rules of nomenclature.**

a. **Number** the chain.

$$\underset{8}{CH_3}CH_2\overset{CH_2CH_3}{\underset{6}{C}}HCH_2CH_2\underset{2}{C}H\underset{1}{C}H_3$$

- Number to put the SH at C2, not C7.

2-octanethiol

b. **Name** and **number** the substituents.

ethyl group at C6

$$CH_3CH_2\overset{CH_2CH_3}{\underset{6}{C}}HCH_2CH_2\underset{2}{C}HCH_3$$

Answer: 6-ethyl-2-octanethiol

PROBLEM 14.26   Give the IUPAC name for each thiol.

a. $CH_3CH_2\underset{\underset{SH}{|}}{C}HCH_2CH_3$

b. [cyclohexane ring with —SH]

c. $(CH_3CH_2)_2CHCH_2CH_2CH_2CH_2SH$

PROBLEM 14.27   Skunk odor contains *trans*-2-butene-1-thiol. Draw the structure of this thiol.

The most obvious physical property of thiols is their distinctive foul odor. 1-Propanethiol ($CH_3CH_2CH_2SH$) is partly responsible for the characteristic odor of onions. 3-Methyl-1-butanethiol [$(CH_3)_2CHCH_2CH_2SH$] is one of the main components of the defensive spray of skunks.

Thiols undergo one important reaction: **thiols are oxidized to disulfides,** compounds that contain a sulfur–sulfur bond. This is an oxidation reaction because two hydrogen atoms are removed in forming the disulfide.

new S—S bond

Oxidation $2\ CH_3CH_2$—S—H $\xrightarrow{[O]}$ CH_3CH_2—S⌒S—CH_2CH_3

thiol   disulfide

Disulfides can also be converted back to thiols with a reducing agent. The symbol for a general reducing agent is **[H]**, since hydrogen atoms are often added to a molecule during reduction.

new S—H bond

Reduction CH_3CH_2—S—S—CH_2CH_3 $\xrightarrow{[H]}$ $2\ CH_3CH_2$—S⌒H

disulfide   thiol

The chemistry of thiols and disulfides plays an important role in determining the properties and shape of some proteins, as we will learn in Chapter 21. For example, α-keratin, the protein in hair, contains many disulfide bonds. Straight hair can be made curly by cleaving the disulfide bonds in α-keratin, then rearranging and re-forming them, as shown schematically in Figure 14.7.

▼ **FIGURE 14.7** Focus on the Human Body: Making Straight Hair Curly

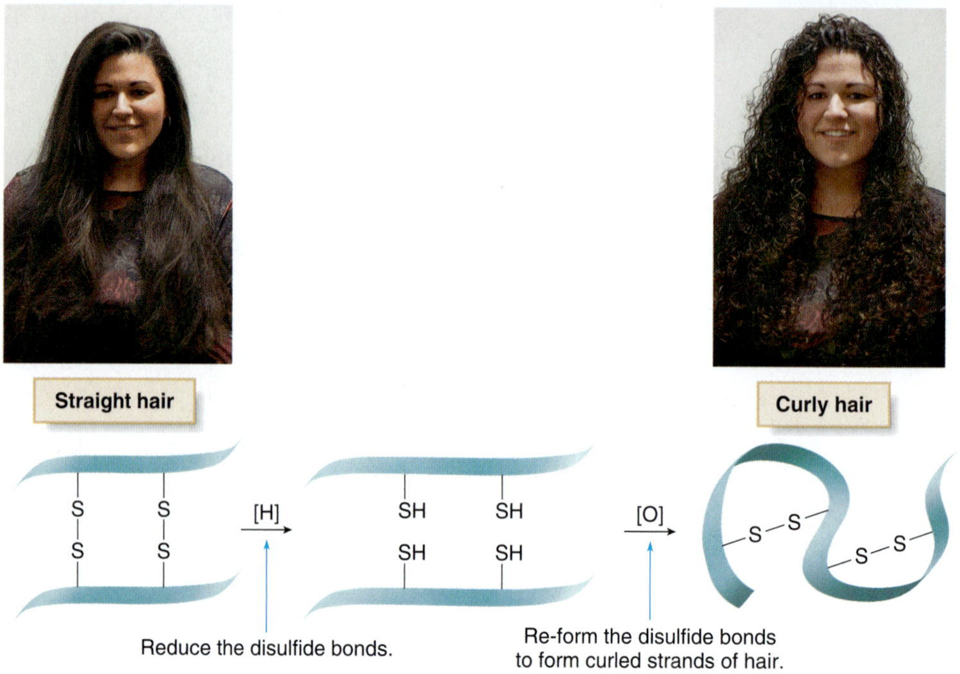

Straight hair

Curly hair

Reduce the disulfide bonds.

Re-form the disulfide bonds
to form curled strands of hair.

To make straight hair curly, the disulfide bonds holding the protein chains together are reduced. This forms free SH groups. The hair is turned around curlers and then an oxidizing agent is applied. This re-forms the disulfide bonds to the hair, now giving it a curly appearance.

First, the disulfide bonds in the straight hair are reduced to thiol groups, so the bundles of α-keratin chains are no longer held in their specific "straight" orientation. Then, the hair is wrapped around curlers and treated with an oxidizing agent that converts the thiol groups back to disulfide bonds, now with twists and turns in the keratin backbone. This makes the hair look curly and is the chemical basis for a "permanent."

PROBLEM 14.28

(a) Draw the disulfide formed when $CH_3CH_2CH_2SH$ is oxidized. (b) Draw the product formed when the following disulfide is reduced: $CH_3CH_2CH_2CH_2SSCH_2CH_3$.

CHAPTER HIGHLIGHTS

KEY TERMS

Alcohol (14.1)
Alkoxy group (14.7)
Alkyl halide (14.1)
Carbonyl group (14.5)
Chlorofluorocarbon (14.9)
Coenzyme (14.6)
Dehydration (14.5)
Diol (14.3)
Disulfide (14.10)
Elimination (14.5)

Epoxide (14.7)
Ether (14.1)
Glycol (14.3)
Heterocycle (14.7)
Hydrochlorofluorocarbon (14.9)
Hydrofluorocarbon (14.9)
Hydroxyl group (14.1)
Oxidation (14.5)
Primary (1°) alcohol (14.2)
Primary (1°) alkyl halide (14.9)

Secondary (2°) alcohol (14.2)
Secondary (2°) alkyl halide (14.9)
Sulfhydryl group (14.1)
Tertiary (3°) alcohol (14.2)
Tertiary (3°) alkyl halide (14.9)
Thiol (14.1)
Triol (14.3)
Zaitsev rule (14.5)

KEY REACTIONS

[1] Dehydration of alcohols (14.5A)

$$-\overset{|}{\underset{H}{C}}-\overset{|}{\underset{OH}{C}}- \xrightarrow{H_2SO_4} \quad C=C \quad + \quad H_2O$$

[2] Oxidation of alcohols (14.5B)

a. Primary (1°) alcohols $RCH_2OH \xrightarrow{[O]} RCHO \xrightarrow{[O]} RCOOH$
                                                    aldehyde   carboxylic acid

b. Secondary (2°) alcohols $R_2CHOH \xrightarrow{[O]} R_2CO$
                                                    ketone

c. Tertiary (3°) alcohols $R_3COH \xrightarrow{[O]}$ No reaction

[3] Oxidation of thiols and reduction of disulfides (14.10)

$$2\ R{-}S{-}H \underset{[H]}{\overset{[O]}{\rightleftharpoons}} R{-}S{-}S{-}R$$
   thiol   disulfide

KEY CONCEPTS

❶ What are the characteristics of alcohols, ethers, alkyl halides, and thiols?

- Alcohols contain a hydroxyl group (OH group) bonded to a tetrahedral carbon. Since the O atom is surrounded by two atoms and two lone pairs, alcohols have a bent shape around O. (14.2)
- Ethers have two alkyl groups bonded to an oxygen atom. Since the O atom is surrounded by two atoms and two lone pairs, ethers have a bent shape around O. (14.7)
- Alkyl halides contain a halogen atom (X = F, Cl, Br, I) singly bonded to a tetrahedral carbon. (14.9)
- Thiols contain a sulfhydryl group (SH group) bonded to a tetrahedral carbon. Since the S atom is surrounded by two atoms and two lone pairs, thiols have a bent shape around S. (14.10)

❷ How are alcohols and alkyl halides classified?

- Alcohols and alkyl halides are classified by the number of C's bonded to the C with the functional group.
- RCH_2OH = 1° alcohol; R_2CHOH = 2° alcohol; R_3COH = 3° alcohol. (14.2)
- RCH_2X = 1° alkyl halide; R_2CHX = 2° alkyl halide; R_3CX = 3° alkyl halide. (14.9)

❸ What are the properties of alcohols, ethers, alkyl halides, and thiols?

- Alcohols have a bent shape and polar C—O and O—H bonds, so they have a net dipole. Their OH bond allows for intermolecular hydrogen bonding between two alcohol

molecules or between an alcohol molecule and water. As a result, alcohols have the strongest intermolecular forces of the four families of molecules in this chapter. (14.2)
- Ethers have a bent shape and two polar C—O bonds, so they have a net dipole. (14.7)
- Alkyl halides with one halogen have one polar bond and a net dipole. (14.9)
- Thiols have a bent shape and lower boiling points than alcohols with the same number of carbons. (4.10)

❹ How are alcohols, ethers, alkyl halides, and thiols named?

- Alcohols are identified by the suffix -ol. (14.3)
- Ethers are named in two ways. Simple ethers are named by naming the alkyl groups bonded to the ether oxygen and adding the word *ether*. More complex ethers are named as *alkoxy alkanes;* that is, the simpler alkyl group is named as an alkoxy group (RO) bonded to an alkane. (14.7B)
- Alkyl halides are named as *halo alkanes;* that is, the halogen is named as a substituent (halo group) bonded to an alkane. (14.9B)
- Thiols are identified by the suffix -thiol. (14.10)

❺ What products are formed when an alcohol undergoes dehydration? (14.5A)

- Alcohols form alkenes on treatment with strong acid. The elements of H and OH are lost from two adjacent atoms and a new double bond is formed.
- Dehydration follows the Zaitsev rule: When more than one alkene can be formed, the major product of elimination is the alkene that has more alkyl groups bonded to it.

6 **What products are formed when an alcohol is oxidized? (14.5B)**

- Primary alcohols (RCH_2OH) are oxidized to aldehydes (RCHO), which are further oxidized to carboxylic acids (RCO_2H).
- Secondary alcohols (R_2CHOH) are oxidized to ketones.
- Tertiary alcohols have no C—H bond on the carbon with the OH group, so they are not oxidized.

7 **What product is formed when a thiol is oxidized? (14.10)**

- Thiols (RSH) are oxidized to disulfides (RSSR).
- Disulfides are reduced to thiols.

PROBLEMS

Selected in-chapter and end-of-chapter problems have brief answers provided in Appendix B.

Alcohols, Ethers, Alkyl Halides, and Thiols

14.29 Classify each alcohol as 1°, 2°, or 3°.

a. $CH_3CH_2CH_2OH$

b. $(CH_3CH_2)_3COH$

c.

d. $CH_3CH_2CHCHCH_3$ (with OH and CH_3 substituents)

14.30 Classify each alcohol as 1°, 2°, or 3°.

a. (cyclohexane with OH and $CH_2CH_2CH_3$)

b. $(CH_3)_3CCH_2CH_2OH$

c. $CH_3CH_2CH_2CHOH$ (with CH_3 substituent)

d. $CH_3CH_2CH_2CCH_2OH$ (with two CH_3 substituents)

14.31 Classify each alkyl halide as 1°, 2°, or 3°.

a. $CH_3(CH_2)_5CHCH_3$ (with Cl substituent)

b. $CH_3CH_2CCH_2CH_3$ (with CH_3 and Cl substituents)

c. (cyclohexane with I)

d. $CH_3CH_2—C—CH_2Br$ (with two CH_3 substituents)

14.32 Classify each halide in **A** as 1°, 2°, or 3°. **A** is a component of the red seaweed *C. hornemanni* found in Sri Lanka.

$CH_3C{=}CHCH_2CH_2CCH{=}CH_2$ (with CH_3, CH_2Br, and Cl substituents)

A

14.33 Draw the structure of a molecule that fits each description:

a. a 2° alcohol of molecular formula $C_6H_{14}O$

b. an ether with molecular formula $C_6H_{14}O$ that has a methyl group bonded to oxygen

c. a 3° alkyl halide with molecular formula $C_5H_{11}Br$

14.34 Draw the structure of a molecule that fits each description:

a. a 2° alcohol of molecular formula $C_6H_{12}O$

b. a cyclic ether with molecular formula $C_5H_{10}O$

c. a 1° alkyl halide with molecular formula $C_5H_{11}Cl$

14.35 Draw the structure of the six constitutional isomers of molecular formula $C_5H_{12}O$ that contain an ether as functional group.

14.36 Draw the structure of the four constitutional isomers of molecular formula $C_3H_6Br_2$.

Nomenclature

14.37 Give the IUPAC name for each alcohol.

a. $CH_3CHCH_2CH_2CH_3$ (with OH substituent)

b. $CH_3CHCCH_2CH_2CH_3$ (with HO, CH_3, and CH_3 substituents)

c. $(CH_3)_2CHCH_2CHCH_2CH_3$ (with $CH_2CH_2CH_2OH$ substituent)

d. (cyclohexane with HO and $CH_2CH_2CH_2CH_3$)

e. (cyclobutane with OH, CH_3, and CH_3 substituents)

f. CH_3CH_2 (cyclopentane with OH and CH_2CH_3)

14.38 Give the IUPAC name for each alcohol.

a. 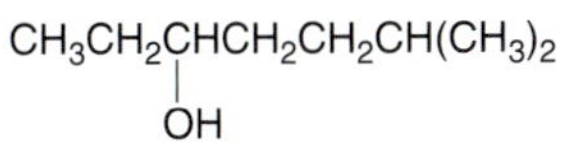

b. 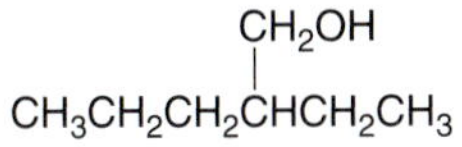

c. 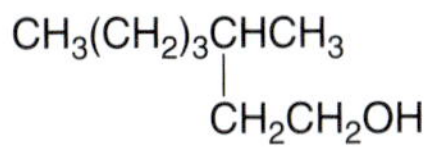

d. 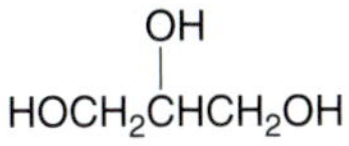

e. 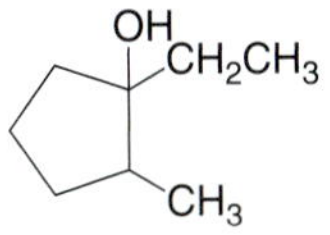

f. 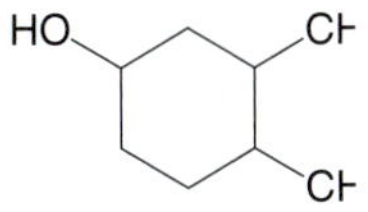

14.39 Give the structure corresponding to each IUPAC name.
a. 3-hexanol
b. propyl alcohol
c. 2-methylcyclopropanol
d. 1,2-butanediol
e. 4,4,5-trimethyl-3-heptanol
f. 3,5-dimethyl-1-heptanol

14.40 Give the structure corresponding to each IUPAC name.
a. 3-methyl-3-pentanol
b. 4-methyl-2-pentanol
c. 2,4-dimethyl-2-hexanol
d. 1,3-propanediol
e. 3,5-dimethylcyclohexanol
f. 6,6-diethyl-4-nonanol

14.41 Give an acceptable name (IUPAC or common) for each ether.

a. $CH_3CH_2OCH_2CH_2CH_2CH_3$

b. 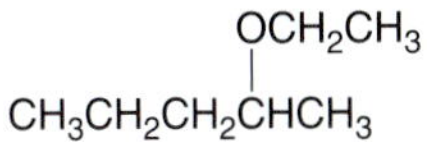

c. 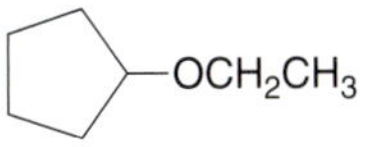

14.42 Give an acceptable name (IUPAC or common) for each ether.

a. $CH_3OCH_2CH_2CH_3$

b. 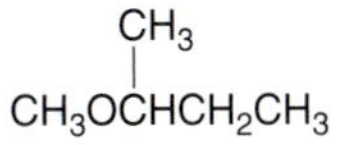

c. 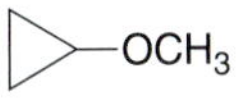

14.43 MTBE [$CH_3OC(CH_3)_3$] is a high-octane gasoline additive that has the common name *tert*-butyl methyl ether. What is its IUPAC name?

14.44 What is the IUPAC name for diethyl ether?

14.45 Draw the structures and give the IUPAC names for all constitutional isomers of molecular formula $C_7H_{16}O$ that contain an OH group and have seven carbons in the longest chain.

14.46 Draw structures for the four constitutional isomers of molecular formula $C_4H_{10}O$ that contain an OH group. Give the IUPAC name for each alcohol.

14.47 Give the IUPAC name for each alkyl halide and thiol.

a. $(CH_3CH_2CH_2)_2CHBr$

b. 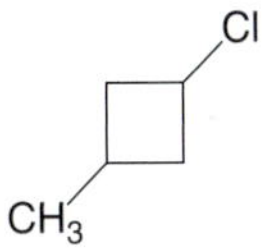

c. $CH_3CH_2CH_2CH_2CH_2SH$

14.48 Give the IUPAC name for each alkyl halide and thiol.

a. 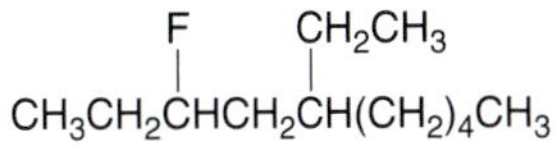

b. 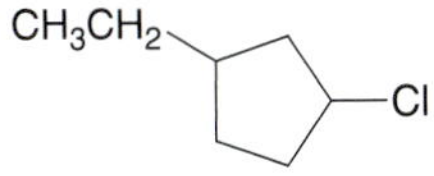

c. $(CH_3)_2CHSH$

14.49 Give the structure corresponding to each IUPAC name.
a. 2-methoxypropane
b. cyclobutyl ethyl ether
c. 1-ethoxy-2-ethylcyclohexane
d. butyl chloride
e. 2-methylcyclohexanethiol
f. 1-ethyl-2-fluorocyclobutane

14.50 Give the structure corresponding to each IUPAC name.
a. dicyclohexyl ether
b. 2-ethoxy-2-methylpropane
c. propyl iodide
d. 3-methyl-1-pentanethiol
e. 1-chloro-2-methylpropane
f. 2-bromo-3-methylheptane

Physical Properties

14.51 Which compound in each pair has the higher boiling point?
a. $CH_3CH_2CH_2Cl$ or $CH_3CH_2CH_2I$
b. $CH_3CH_2CH_2OH$ or $HOCH_2CH_2OH$
c. $CH_3CH_2CH_2OH$ or $CH_3CH_2CH_2F$

14.52 Rank the compounds in each group in order of increasing boiling point.
a. $CH_3CH_2OCH_2CH_3$, $CH_3(CH_2)_3CH_3$, $CH_3CH_2CH_2CH_2OH$
b. $CH_3CH_2CH_2CH_2OH$, $CH_3CH_2CH_2OH$, CH_3CH_2OH
c. $CH_3CH_2CH_2CH_2OH$, $CH_3CH_2CH_2CH_3$, $CH_3CH_2CH_2CH_2Cl$

14.53 Explain the following observation. Dimethyl ether [$(CH_3)_2O$] and ethanol (CH_3CH_2OH) are both water-soluble compounds. The boiling point of ethanol (78 °C), however, is much higher than the boiling point of dimethyl ether (–24 °C).

14.54 Explain why the boiling point of $CH_3CH_2CH_2CH_2OH$ (117 °C) is higher than the boiling point of $CH_3CH_2CH_2CH_2SH$ (98 °C), even though $CH_3CH_2CH_2CH_2SH$ has a higher molecular weight.

14.55 Explain why two four-carbon organic molecules have very different solubility properties: 1-butanol ($CH_3CH_2CH_2CH_2OH$) is water soluble but 1-butene ($CH_2{=}CHCH_2CH_3$) is water insoluble.

14.56 Explain why 1,6-hexanediol [$HO(CH_2)_6OH$] is much more water soluble than 1-hexanol [$CH_3(CH_2)_5OH$].

Reactions of Alcohols

14.57 Draw the products formed when each alcohol is dehydrated with H_2SO_4. Use the Zaitsev rule to predict the major product when a mixture forms.

a.

c. $CH_3CHCH_2CH_2CH_2CH_3$

b. (with OH group) —CHCH$_2$CH$_3$

d. $CH_3CH_2\overset{\underset{\displaystyle |}{OH}}{C}CH_2CH_3$
$\quad\quad\quad\quad\;\;\underset{\displaystyle CH_3}{|}$

14.58 Draw the products formed when each alcohol is dehydrated with H_2SO_4. Use the Zaitsev rule to predict the major product when a mixture forms.

a. $(CH_3)_2CHCH_2CH_2CH_2OH$

c. $CH_3\overset{\underset{\displaystyle |}{OH}}{C}H(CH_2)_5CH_3$

b. $(CH_3CH_2)_3COH$

d. (cyclopentane with CH_3 and OH)

14.59 Dehydration of 3-hexanol [$CH_3CH_2CH(OH)CH_2CH_2CH_3$] yields a mixture of two alkenes. (a) Draw the structures of both constitutional isomers formed. (b) Use the Zaitsev rule to explain why both products are formed in equal amounts.

14.60 Dehydration of 3-methylcyclohexanol yields a mixture of two alkenes. (a) Draw the structures of both constitutional isomers formed. (b) Use the Zaitsev rule to explain why both products are formed in equal amounts.

14.61 What alcohol is needed to form each alkene as a product of a dehydration reaction?

a. (two benzene rings connected by) —CH=CH—

b. (cyclooctene)

14.62 What alcohol is needed to form each alkene as the major product of a dehydration reaction?

a. (cyclohexene with two CH$_3$ groups)

b. $\underset{\displaystyle CH_3}{\overset{\displaystyle CH_3}{}}C=C\underset{\displaystyle CH_3}{\overset{\displaystyle CH_3}{}}$

14.63 Propene ($CH_3CH=CH_2$) can be formed by dehydrating two different alcohols. Draw the structures of both alcohols.

14.64 2-Methylpropene [$(CH_3)_2C=CH_2$] can be formed by dehydrating two different alcohols. Draw the structures of both alcohols.

14.65 Draw the products formed when each alcohol is oxidized with $K_2Cr_2O_7$. In some cases, no reaction occurs.

a. $CH_3(CH_2)_6CH_2OH$

c. $CH_3CH_2\overset{\underset{\displaystyle |}{}}{C}HCH_2OH$
$\quad\quad\quad\quad\;\;\underset{\displaystyle CH_3}{|}$

b. $(CH_3CH_2)_2CHOH$

d. $CH_3CH_2\overset{\underset{\displaystyle |}{}}{C}OH$
$\quad\quad\quad\;\;\overset{\displaystyle CH_3}{|}\;\underset{\displaystyle CH_3}{|}$

14.66 Draw the products formed when each alcohol is oxidized with $K_2Cr_2O_7$. In some cases, no reaction occurs.

a. (cyclopentane with OH and CH$_3$)

c. (cyclopentane)—CH$_2$OH

b. $CH_3(CH_2)_8CH_2OH$

d. $(CH_3CH_2)_3COH$

14.67 Cortisol is an anti-inflammatory agent that also regulates carbohydrate metabolism. What oxidation product is formed when cortisol is treated with $K_2Cr_2O_7$?

cortisol

14.68 Xylitol is a nontoxic compound as sweet as table sugar but with only one-third the calories, so it is often used as a sweetening agent in gum and hard candy.

$$CH_2OH$$
$$H-C-OH$$
$$HO-C-H$$
$$H-C-OH$$
$$CH_2OH$$

xylitol

a. Classify the OH groups as 1°, 2°, or 3°.
b. What product is formed when only the 1° OH groups are oxidized?
c. What product is formed when all of the OH groups are oxidized?

14.69 What alcohol starting material is needed to prepare each carbonyl compound as a product of an oxidation reaction?

a. (cycloheptanone) =O

c. (benzene ring)—C(=O)—OH

b. $CH_3-\overset{\underset{\displaystyle |}{CH_3}}{C}-\overset{\displaystyle O}{\overset{\|}{C}}-CH_3$
$\quad\quad\;\;\underset{\displaystyle CH_3}{|}$

14.70 What alcohol starting material is needed to prepare each carbonyl compound as a product of an oxidation reaction?

a. (cyclohexanone with CH$_3$) =O

c. $(CH_3CH_2)_2C=O$

b. $CH_3CH_2CH_2CH_2CH_2CO_2H$

14.71 What products are formed when 4-heptanol [$(CH_3CH_2CH_2)_2CHOH$] is treated with each reagent: (a) H_2SO_4; (b) $K_2Cr_2O_7$?

14.72 What products are formed when 2-methyl-4-heptanol [$(CH_3)_2CHCH_2CH(OH)CH_2CH_2CH_3$] is treated with each reagent: (a) H_2SO_4; (b) $K_2Cr_2O_7$?

14.73 Often there is no one-step method to convert one organic compound into another. Instead, two or more steps must be carried out. For example, 1-butene is converted to 2-butanone by the two-step reaction sequence drawn below. Fill in the reagents (**A** and **B**) needed to carry out each step.

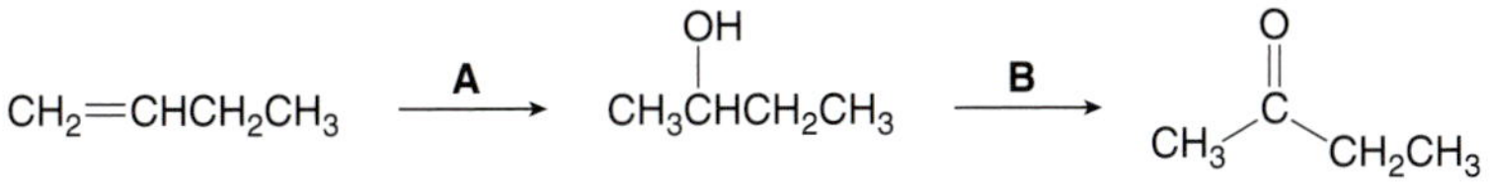

CH$_2$=CHCH$_2$CH$_3$ **A** →   CH$_3$CHCH$_2$CH$_3$ (OH)   **B** →   CH$_3$C(O)CH$_2$CH$_3$

1-butene   2-butanol   2-butanone

14.74 Fill in the reagents (**C** and **D**) needed to convert each starting material to the product drawn. This two-step reaction sequence is used to convert an alkene like cyclopentene to a ketone like cyclopentanone, as shown.

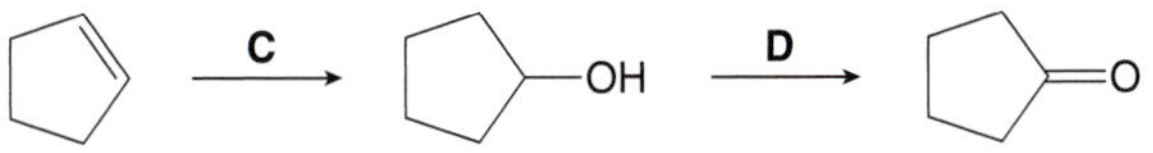

cyclopentene   **C** →   cyclopentanol   **D** →   cyclopentanone

Reactions of Thiols

14.75 What disulfide is formed when each thiol is oxidized?

a. 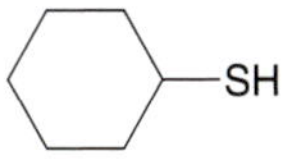   b. $CH_3(CH_2)_4SH$

14.76 As we will learn in Chapter 21, cysteine is a naturally occurring amino acid that contains a thiol group. What disulfide is formed when cysteine is oxidized?

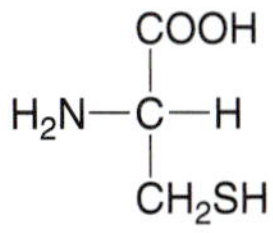

cysteine

14.77 The smell of fried onions is in part determined by the two disulfides drawn below. What thiols are formed when each disulfide is reduced?

a. $CH_3S-SCH_2CH_2CH_3$ b. CH_2=$CHCH_2S-SCH_2CH_2CH_3$

14.78 Sometimes disulfide bonds occur in a ring. As we will learn in Chapter 21, these disulfide bonds play an important role in determining the shape and properties of proteins. What *di*thiol is formed when the following disulfide is reduced?

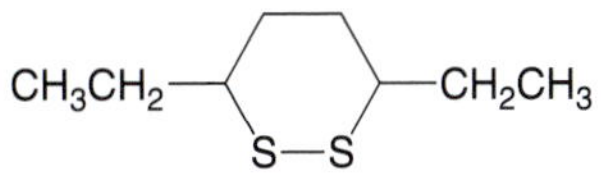

Applications

14.79 Diethyl ether [$(CH_3CH_2)_2O$] is an imperfect anesthetic in part because it readily combusts in air to form CO_2 and H_2O. Write a balanced equation for the complete combustion of diethyl ether.

14.80 Methanol (CH_3OH) is used as a fuel for some types of race cars, because it readily combusts in air to form CO_2 and H_2O. Write a balanced equation for the complete combustion of methanol.

14.81 With reference to the halogenated organic compounds called CFCs, HCFCs, and HFCs: (a) How do these compounds differ in structure? (b) How do these compounds differ in their interaction with ozone?

14.82 Give an example of a CFC and explain why the widespread use of CFCs has been detrimental to the environment.

14.83 As we learned in Chapter 13, polymers have very different properties depending (in part) on their functional groups. For example, poly(ethylene glycol) is commonly used in shampoos, where it forms viscous solutions in water. Poly(vinyl chloride), on the other hand, repels water and is used to make pipes and garden hoses. Discuss the water solubility of these polymers, and explain why there is a difference, even though both polymers have several polar bonds.

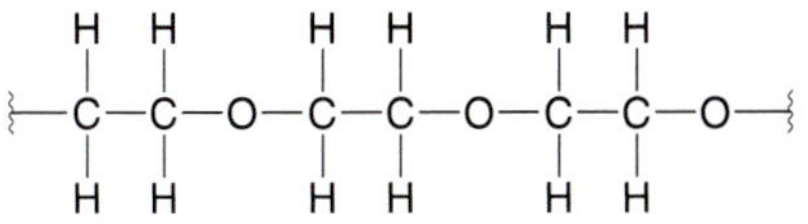

poly(ethylene glycol)
PEG

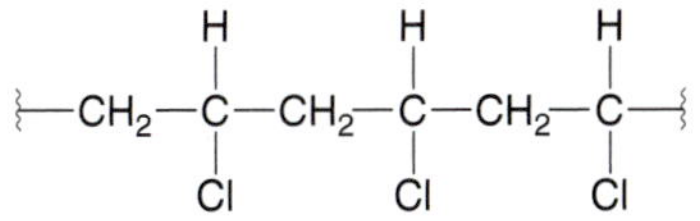

poly(vinyl chloride)
PVC

14.84 Give an example of a halogenated ether commonly used as an anesthetic. Give two reasons why halogenated ethers are now used in place of diethyl ether as anesthetics.

14.85 Write out the chemical reaction that occurs when a breathalyzer test is carried out. How does this test give an indication of blood alcohol content?

14.86 2,3-Butanedione is commonly used to give microwave popcorn a buttery flavor. What diol is needed to prepare 2,3-butanedione by an oxidation reaction?

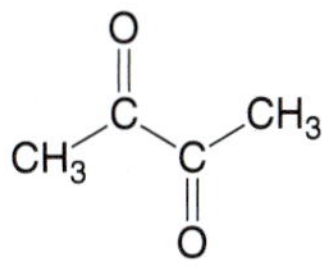

2,3-butanedione
(butter flavor)

14.87   In contrast to ethylene glycol ($HOCH_2CH_2OH$), which is extremely toxic, propylene glycol [$CH_3CH(OH)CH_2OH$] is nontoxic because it is oxidized in the body to a product produced during the metabolism of carbohydrates. What product is formed when propylene glycol is oxidized?

14.88   Lactic acid [$CH_3CH(OH)CO_2H$] gives sour milk its distinctive taste. What product is formed when lactic acid is oxidized by $K_2Cr_2O_7$?

14.89   Write out the chemical reactions that occur during the metabolism of ethanol. Explain how Antabuse is used in the treatment of alcoholism.

14.90   What chemical reactions occur when straight hair is made curly with a "permanent"?

General Questions

14.91   An unknown alcohol of molecular formula $C_4H_{10}O$ forms only one alkene when treated with H_2SO_4. No reaction is observed when the alcohol is treated with an oxidizing agent. Draw the structure of the alcohol.

14.92   You have two test tubes, one containing ($CH_3CH_2)_3COH$ and one containing $CH_3(CH_2)_6OH$, but you do not know which alcohol is in which tube. Explain how treating the alcohols with an oxidizing agent like $K_2Cr_2O_7$ can help you to determine which test tube contains which alcohol.

14.93   Explain why two ether molecules can't hydrogen bond to each other, but an ether molecule can hydrogen bond to a water molecule.

14.94   When an alcohol hydrogen bonds with water, two different hydrogen bonding interactions are possible. The oxygen atom of the alcohol (ROH) can hydrogen bond with a hydrogen atom of water, or a hydrogen atom of the alcohol can hydrogen bond to the oxygen atom of water. Illustrate both of these possibilities with CH_3CH_2OH.

14.95   Answer the following questions about alcohol **A**.

a. Give the IUPAC name.
b. Classify the alcohol as 1°, 2°, or 3°.
c. Draw the products formed when **A** is dehydrated with H_2SO_4.
d. What product is formed when **A** is oxidized with $K_2Cr_2O_7$?
e. Draw a constitutional isomer of **A** that contains an OH group.
f. Draw a constitutional isomer of **A** that contains an ether.

14.96   Answer the following questions about alcohol **B**.

$$CH_3CH_2CH_2CH_2CH_2CHCH_2OH$$
$$\text{B} \quad CH_3$$

a. Give the IUPAC name.
b. Classify the alcohol as 1°, 2°, or 3°.
c. Draw the products formed when **B** is dehydrated with H_2SO_4.
d. What product is formed when **B** is oxidized with $K_2Cr_2O_7$?
e. Draw a constitutional isomer of **B** that contains an OH group.
f. Draw a constitutional isomer of **B** that contains an ether.

CHALLENGE QUESTIONS

14.97   2-Methyl-1-butene can be formed by dehydrating two different alcohols (**A** and **B**) of molecular formula ($C_5H_{12}O$). When **A** undergoes dehydration, 2-methyl-1-butene is formed as the *only* product of dehydration. When alcohol **B** undergoes dehydration, two different alkenes are formed and 2-methyl-1-butene is formed as the *minor* product. (a) Identify the structures of **A** and **B**. (b) What is the major alkene formed when **B** undergoes dehydration?

14.98   Dehydration of alcohol **C** forms two products of molecular formula $C_{14}H_{12}$ that are isomers, but they are not constitutional isomers. Draw the structures of these two products.

15

CHAPTER GOALS

In this chapter you will learn how to:

1. Recognize whether a molecule is chiral or achiral
2. Identify chirality centers
3. Draw two enantiomers in three dimensions around the chirality center
4. Explain why some chiral drugs have very different properties from their mirror images
5. Draw Fischer projection formulas
6. Recognize the difference between enantiomers and diastereomers
7. Explain the relationship between the shape and odor of a molecule

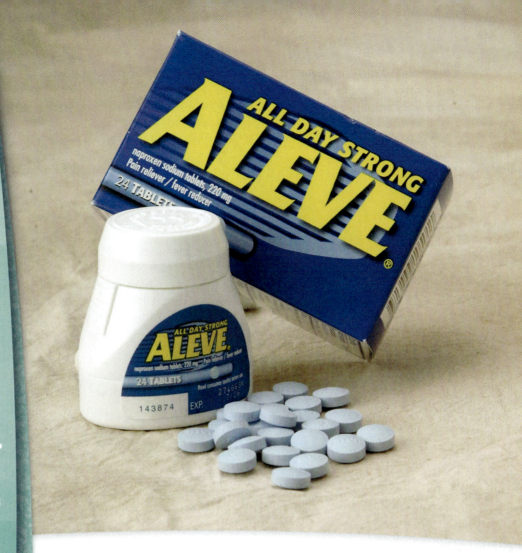

Naproxen is the active ingredient in the widely used pain relievers Naprosyn and Aleve. The three-dimensional arrangement of two atoms on a single carbon atom in naproxen determines its therapeutic properties.

THE THREE-DIMENSIONAL SHAPE OF MOLECULES

ARE you left-handed or right-handed? If you're right-handed, you've probably spent little time thinking about your hand preference. If you're left-handed, though, you probably learned at an early age that many objects—like scissors and baseball gloves—"fit" for righties, but are "backwards" for lefties. **Hands, like many objects in the world around us, are mirror images that are *not* identical.** In Chapter 15 we examine the "handedness" of molecules, and ask, "How important is the three-dimensional shape of a molecule?"

15.1 ISOMERS—A REVIEW

Chapter 15 concentrates on *stereochemistry*—**that is, the three-dimensional structure of molecules.** Since an understanding of isomers is needed in the discussion of stereochemistry, let's begin with an overview of isomers. Recall from Section 12.2:

- **Isomers are different compounds with the same molecular formula.**

There are two major classes of isomers: **constitutional isomers** (Section 12.2) and **stereoisomers** (Section 13.3).

- *Constitutional isomers* differ in the way the atoms are connected to one another.

Constitutional isomers have atoms bonded to *different* **atoms.** As a result, constitutional isomers have different IUPAC names. Constitutional isomers may have the same functional group or they may have different functional groups.

For example, 2-methylpentane and 3-methylpentane are constitutional isomers with molecular formula C_6H_{14}. 2-Methylpentane has a CH_3 group bonded to the second atom (C2) of a five-carbon chain, while 3-methylpentane has a CH_3 group bonded to the middle atom (C3) of a five-carbon chain. Both molecules are alkanes, so they both belong to the same family of organic compounds.

Constitutional isomers with the same functional group

CH_3 group at C2

$CH_3CHCH_2CH_2CH_3$ and $CH_3CH_2CHCH_2CH_3$

CH_3 group at C3

CH_3 CH_3

2-methylpentane   **3-methylpentane**
C_6H_{14} C_6H_{14}

Ethanol and dimethyl ether are constitutional isomers with molecular formula C_2H_6O. Ethanol contains an O—H bond and dimethyl ether does not. Thus, the atoms are bonded to different atoms and these constitutional isomers have *different* functional groups. Dimethyl ether is an ether and ethanol is an alcohol.

Constitutional isomers with different functional groups

CH_3CH_2—O—H CH_3—O—CH_3

ethanol   dimethyl ether
C_2H_6O C_2H_6O

- *Stereoisomers* differ *only* in the three-dimensional arrangement of atoms.

Stereoisomers always have the *same* **functional group,** since they differ only in the way the atoms are oriented in space. Two stereoisomers need a prefix like cis or trans added to their IUPAC names to distinguish them. The cis and trans isomers of 2-butene are an example of stereoisomers. Each stereoisomer of 2-butene contains the same sequence of atoms. The only difference is the three-dimensional arrangement of the two methyl groups on the double bond.

One example of stereoisomers—Isomers at a double bond

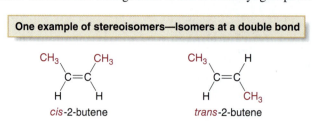

cis-2-butene   *trans*-2-butene

- The *cis* isomer has the two CH_3 groups on the same side of the double bond.
- The *trans* isomer has the two CH_3 groups on opposite sides of the double bond.

Cis and trans isomers on a carbon–carbon double bond are one example of stereoisomers. In Section 15.2, we learn about another type of stereoisomer that occurs at a tetrahedral carbon.

PROBLEM 15.1	

Classify each pair of compounds as identical molecules, constitutional isomers, or stereoisomers.

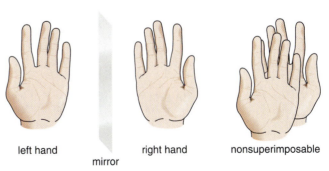

a. $CH_3CH_2\overset{\underset{|}{CH_3}}{CH}\overset{\underset{|}{CH_3}}{CH}CH_3$ and $CH_3\overset{\underset{|}{CH_3}}{CH}CH_2\overset{\underset{|}{CH_3}}{CH}CH_3$

b.

c. $CH_3CH=CHCH_2CH_3$ and $CH_3CH_2CH=CHCH_3$

d.

15.2 LOOKING GLASS CHEMISTRY—MOLECULES AND THEIR MIRROR IMAGES

To learn more about stereoisomers that occur at tetrahedral carbon atoms, we must turn our attention to molecules and their mirror images. **Everything, including molecules, has a mirror image.** What's important in chemistry is **whether a molecule is *identical* to or *different* from its mirror image.**

15.2A   WHAT IT MEANS TO BE CHIRAL OR ACHIRAL

Some molecules are like hands. **Left and right hands are mirror images of each other, but they are *not* identical.** If you try to mentally place one hand inside the other hand you can never superimpose either all the fingers, or the tops and palms. To *superimpose* an object on its mirror image means to align *all* parts of the object with its mirror image. With molecules, this means aligning all atoms and all bonds.

The dominance of right-handedness over left-handedness occurs in all races and cultures. Despite this fact, even identical twins can exhibit differences in hand preference. Pictured are Matthew (right-handed) and Zachary (left-handed), identical twin sons of the author.

left hand   mirror   right hand   nonsuperimposable

- A molecule (or object) that is *not* superimposable on its mirror image is said to be *chiral.*

The adjective *chiral* (pronounced ky-rel) comes from the Greek *cheir*, meaning *hand*. Left and right hands are **chiral**: they are mirror images that do not superimpose on each other.

Other molecules are like socks. **Two socks from a pair are mirror images that *are* superimposable.** One sock can fit inside another, aligning toes and heels, and tops and bottoms. A sock and its mirror image are *identical*.

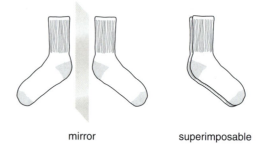

mirror   superimposable

• A molecule (or object) that *is* superimposable on its mirror image is said to be *achiral*.

PROBLEM 15.2

Classify each object as chiral or achiral: (a) nail; (b) screw; (c) glove; (d) pencil.

15.2B   THE CHIRALITY OF MOLECULES

Let's determine whether three molecules—H_2O, CH_2BrCl, and CHBrClF—are superimposable on their mirror images; that is, **are H_2O, CH_2BrCl, and CHBrClF chiral or achiral?**

To test **chirality**—that is, to test whether a molecule and its mirror image are identical or different:

1. Draw the molecule in three dimensions.
2. Draw its mirror image.
3. Try to align all bonds and atoms. To superimpose a molecule and its mirror image you can perform any rotation but **you cannot break bonds.**

Following this procedure, H_2O and CH_2BrCl are both **achiral** molecules: each molecule is superimposable on its mirror image.

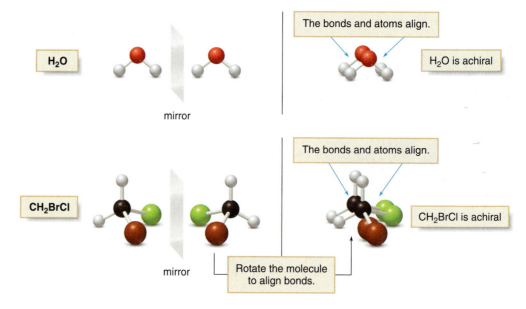

With CHBrClF, the result is different. The molecule (labeled **A**) and its mirror image (labeled **B**) are not superimposable. No matter how you rotate **A** and **B,** all of the atoms never align. **CHBrClF is thus a chiral molecule,** and **A** and **B** are different compounds.

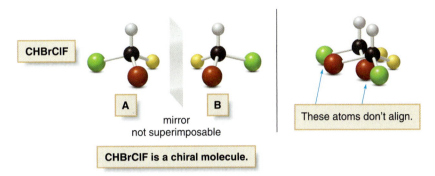

CHBrClF

A   B
mirror
not superimposable

These atoms don't align.

CHBrClF is a chiral molecule.

A and **B** are **stereoisomers** since they are isomers differing only in the three-dimensional arrangement of substituents. They represent a new type of stereoisomer that occurs at tetrahedral carbon atoms. These stereoisomers are called *enantiomers.*

- Enantiomers are mirror images that are not superimposable.

CHBrClF contains a carbon atom bonded to four different groups. **A carbon atom bonded to four different groups is called a *chirality center.***

We have now learned two related but different concepts, and it is necessary to distinguish between them.

- A molecule that is not superimposable on its mirror image is a *chiral molecule.*
- A carbon atom surrounded by four different groups is a *chirality center.*

CHBrClF is a chiral molecule because it is not superimposable on its mirror image. CHBrClF contains one chirality center, one carbon atom bonded to four different groups. Molecules can contain zero, one, or more chirality centers.

> Naming a carbon atom with four different groups is a topic that currently has no firm agreement among organic chemists. IUPAC recommends the term *chirality center,* and so this is the term used in this text. Other terms in common use are chiral center, chiral carbon, asymmetric carbon, and stereogenic center.

15.2C   CHIRALITY IN NATURE

Why must we study a topic like stereochemistry that may seem so esoteric? In truth, chirality pervades our existence. On a molecular level, many biological molecules discussed in Chapters 19–22 are chiral. On a macroscopic level, many naturally occurring objects possess handedness. Examples include helical seashells shaped like right-handed screws, and plants such as honeysuckle that wind in a left-handed helix (Figure 15.1). The human body is chiral, and hands, feet, and ears are not superimposable.

PROBLEM 15.3   Explain the difference between the terms *chiral* and *chirality center.*

15.3   CHIRALITY CENTERS

A necessary skill in studying the three-dimensional structure of molecules is the ability to locate and draw chirality centers—carbon atoms bonded to four different groups.

15.3A   LOCATING CHIRALITY CENTERS

To locate a chirality center, examine each **tetrahedral** carbon atom in a molecule, and look at the four *groups*—not the four *atoms*—bonded to it. CBrClFI has one chirality center since its carbon

▼ **FIGURE 15.1 Chirality in Nature**

a.

b.

a. Many types of snails possess a chiral, right-handed helical shell. When a right-handed shell is held in the right hand with the thumb pointing towards the wider end, the opening is on the right side.

b. Plants like honeysuckle wind in a chiral left-handed helix.

atom is bonded to four different elements—Br, Cl, F, and I. 3-Bromohexane also has one chirality center since one carbon is bonded to H, Br, CH_2CH_3, and $CH_2CH_2CH_3$. We consider all atoms in a group as a *whole unit,* not just the atom directly bonded to the carbon in question.

$$Cl-\underset{\underset{F}{|}}{\overset{\overset{Br}{|}}{C}}-I$$

chirality center

$$CH_3CH_2-\underset{\underset{Br}{|}}{\overset{\overset{H}{|}}{C}}-CH_2CH_2CH_3$$

chirality center
3-bromohexane

This C is bonded to: **H**
                     Br
                     $\mathbf{CH_2CH_3}$
                     $\mathbf{CH_2CH_2CH_3}$

two *different* alkyl groups

Keep in mind the following:

- A carbon atom bonded to two or more like groups is *never* a chirality center. CH_2 and CH_3 groups have more than one H atom bonded to the same carbon, so each of these carbons is *never* a chirality center.
- A carbon that is part of a multiple bond does not have four groups around it, so it can *never* be a chirality center.

SAMPLE PROBLEM 15.1

Locate the chirality center in each drug.

a. $$H_2N-\underset{\underset{CH_3-\underset{\underset{SH}{|}}{\overset{|}{C}}-CH_3}{|}}{\overset{\overset{\overset{O}{\diagdown}\overset{}{\underset{}{C}}\diagup^{OH}}{|}}{C}}-H$$

penicillamine

b. $$HO-\!\!\bigcirc\!\!-\underset{\underset{H}{|}}{\overset{\overset{OH}{|}}{C}}-CH_2NHC(CH_3)_3$$

$HOCH_2$

albuterol

ANALYSIS   In compounds with many C's, look at each C individually and eliminate those C's that can't be chirality centers. Thus, omit all CH_2 and CH_3 groups and all multiply bonded C's. Check all remaining C's to see if they are bonded to four different groups.

SOLUTION   a. **Penicillamine** is used to treat Wilson's disease, a genetic disorder that leads to a buildup of copper in the liver, kidneys, and the brain. The CH_3 groups and COOH group in penicillamine are not chirality centers. One C is bonded to two CH_3 groups, so it can be eliminated from consideration as well. This leaves one C with four different groups bonded to it.

b. **Albuterol** is a bronchodilator—that is, it widens air passages—so it is used to treat asthma. Omit all CH_2 and CH_3 groups and all C's that are part of a double bond. Also omit from consideration the one C bonded to three CH_3 groups. This leaves one C with four different groups bonded to it.

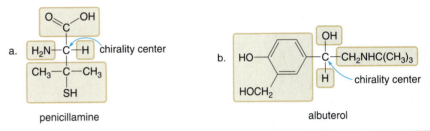

penicillamine

albuterol

PROBLEM 15.4

Label the chirality center in the given molecules. The compounds contain zero or one chirality center.

a. 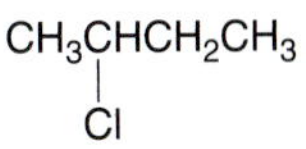 $CH_3CHCH_2CH_3$
        |
        Cl

b. $(CH_3)_3CH$

c. 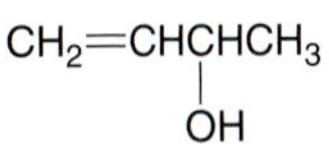 $CH_2{=}CHCHCH_3$
              |
              OH

d. 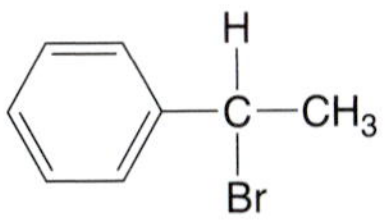 ⬡—$\overset{\text{H}}{\underset{\text{Br}}{C}}$—$CH_3$

PROBLEM 15.5

Label the chirality center in each drug.

a. 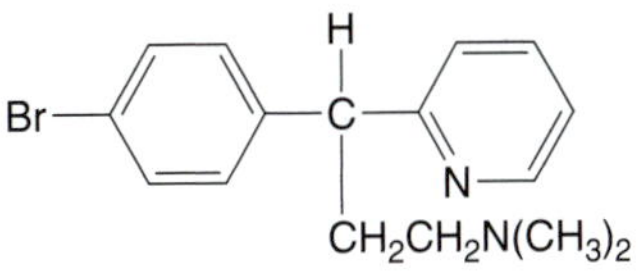 Br—⬡—$\overset{\text{H}}{\underset{CH_2CH_2N(CH_3)_2}{C}}$—⬡(N)

brompheniramine
(antihistamine)

b. 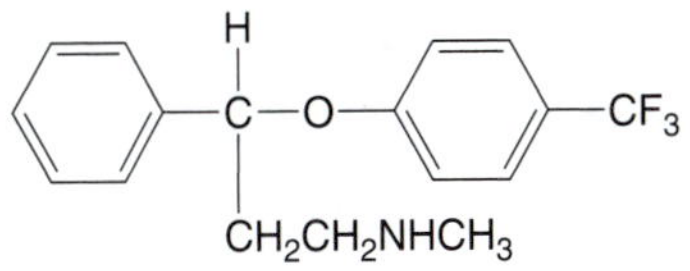 ⬡—$\overset{\text{H}}{\underset{CH_2CH_2NHCH_3}{C}}$—O—⬡—$CF_3$

fluoxetine
Trade name: Prozac
(antidepressant)

Larger organic molecules can have two, three, or even hundreds of chirality centers. Tartaric acid, isolated from grapes, and ephedrine, isolated from the herb ma huang (used to treat respiratory ailments in traditional Chinese medicine), each have two chirality centers. Once a popular drug to promote weight loss and enhance athletic performance, ephedrine use has now been linked to episodes of sudden death, heart attack, and stroke.

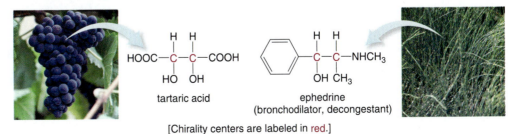

tartaric acid

ephedrine
(bronchodilator, decongestant)

[Chirality centers are labeled in red.]

PROBLEM 15.6

Label the chirality centers in each molecule. Compounds may have one or two chirality centers.

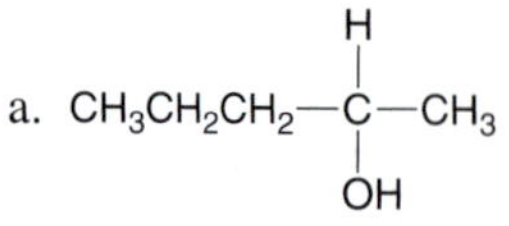

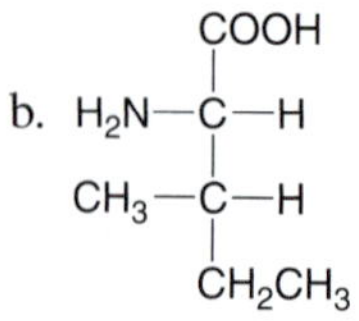

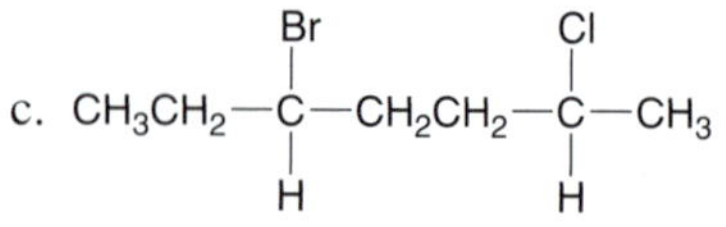

a. CH₃CH₂CH₂—C—CH₃ (with H on top, OH on bottom)

b. H₂N—C—H (with COOH on top; CH₃—C—H below; CH₂CH₃ at bottom)

c. CH₃CH₂—C—CH₂CH₂—C—CH₃ (with Br and Cl on top, H and H on bottom)

PROBLEM 15.7

Label the two chirality centers in vitamin K. Vitamin K is essential to blood clot formation. A deficiency of this vitamin may lead to excessive and often fatal bleeding.

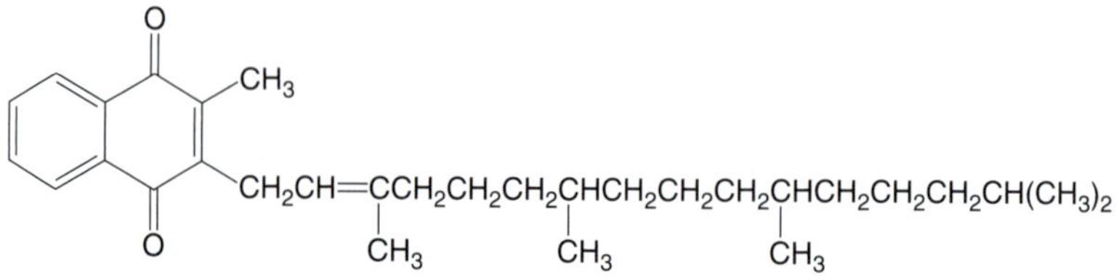

Lettuce, spinach, broccoli, and cabbage are all excellent sources of vitamin K.

CH₂CH=CCH₂CH₂CH₂CHCH₂CH₂CH₂CHCH₂CH₂CH₂CH(CH₃)₂

vitamin K

15.3B   DRAWING A PAIR OF ENANTIOMERS

How are chirality centers related to chiral molecules?

- **Any molecule with one chirality center is a chiral compound.**

Recall from Section 15.2 that a **chiral compound is not superimposable on its mirror image,** and that these mirror image isomers are called **enantiomers.** Let's learn how to draw two enantiomers using 2-butanol [CH₃CH(OH)CH₂CH₃], a compound with one chirality center, as an example.

HOW TO   Draw Two Enantiomers of a Chiral Compound

EXAMPLE   Draw two enantiomers of 2-butanol in three dimensions around the chirality center.

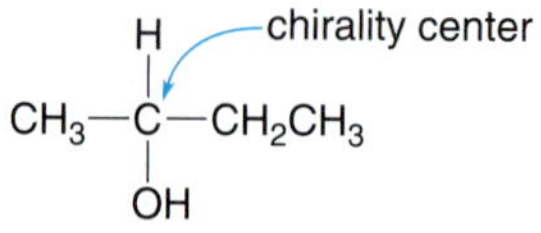

CH₃—C—CH₂CH₃ (with H on top labeled "chirality center", OH on bottom)

2-butanol

Step [1]   **Draw one enantiomer by arbitrarily placing the four different groups on any bond to the chirality center.**

- Use the convention learned in Section 11.3 for drawing a tetrahedron: **place two bonds in the plane, one in front of the plane on a wedge, and one behind the plane on a dashed line.**

- Arbitrarily place the four groups—H, OH, CH₃, and CH₂CH₃—on any bond to the chirality center, forming enantiomer **A.**

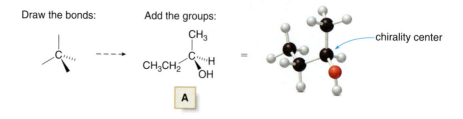

Draw the bonds: Add the groups: chirality center

A

Step [2] Draw a mirror plane and arrange the substituents in the mirror image so that they are a reflection of the groups in the first molecule.

- Place the four groups—H, OH, CH_3, and CH_2CH_3—on the chirality center in the mirror image to form enantiomer **B**.

- **A** and **B** are mirror images, but no matter how **A** and **B** are rotated their bonds do not align, so **A** and **B** are chiral compounds.

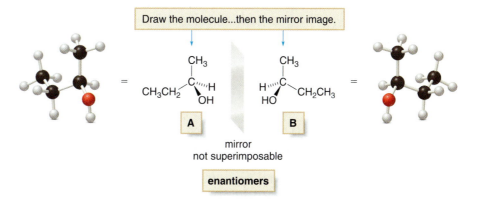

Draw the molecule...then the mirror image.

A   **B**

mirror
not superimposable

enantiomers

PROBLEM 15.8

Locate the chirality center in each compound and draw both enantiomers.

a. 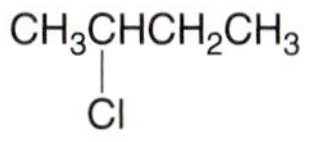 $CH_3CHCH_2CH_3$
    |
    Cl

b. 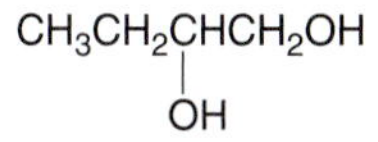 $CH_3CH_2CHCH_2OH$
    |
    OH

PROBLEM 15.9

Locate the chirality center and draw both enantiomers of fenfluramine, one component of the appetite suppressant Fen–Phen (see also Problem 11.32).

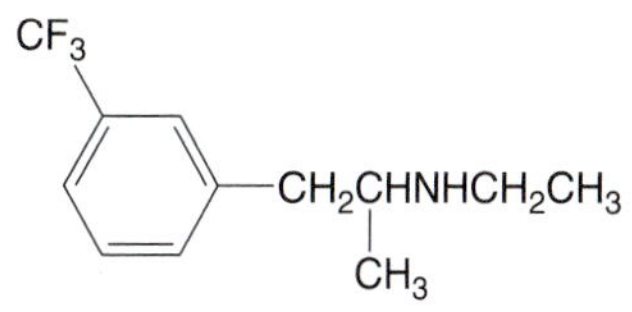

CF_3

—$CH_2CHNHCH_2CH_3$
        |
        CH_3

fenfluramine

15.4 CHIRALITY CENTERS IN CYCLIC COMPOUNDS

Chirality centers may also occur at carbon atoms that are part of a ring. To find chirality centers on ring carbons, always draw the rings as flat polygons, and look for tetrahedral carbons that are bonded to four different groups, as usual.

15.4A LOCATING CHIRALITY CENTERS ON RING CARBONS

Does methylcyclopentane have a chirality center?

Is C1 a chirality center?

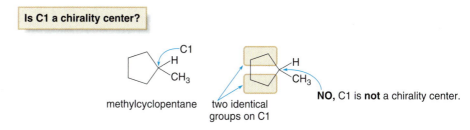

C1
H
CH_3

H
CH_3

methylcyclopentane   two identical
                      groups on C1

NO, C1 is **not** a chirality center.

All of the carbon atoms are bonded to two or three hydrogen atoms except for C1, the ring carbon bonded to the methyl group. Next, compare the ring atoms and bonds on both sides the same distance from C1, and continue until a point of difference is reached, or until both sides meet. In this case, there is no point of difference on either side, so C1 is considered to be bonded to identical alkyl groups that happen to be part of a ring. **C1 is therefore *not* a chirality center.**

With 3-methylcyclohexene, the result is different.

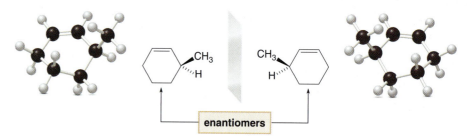

Is C3 a chirality center?

These 2 C's are different.

3-methylcyclohexene

YES, C3 is a chirality center.

All carbon atoms are bonded to two or three hydrogen atoms or are part of a double bond except for C3, the ring carbon bonded to the methyl group. In this case, the atoms equidistant from C3 are different, so C3 is considered to be bonded to *different* alkyl groups in the ring. **C3 is therefore bonded to four different groups, making it a chirality center.**

Since 3-methylcyclohexene has one tetrahedral chirality center, it is a chiral compound, and exists as a pair of enantiomers. Substituents above and below the ring are drawn with wedges and dashes.

enantiomers

15.4B FOCUS ON HEALTH & MEDICINE
THE UNFORGETTABLE LEGACY OF THALIDOMIDE

Many biologically active compounds contain one or more chirality centers on ring carbons. For example, **thalidomide,** which contains one such chirality center, was once used as a sedative and anti-nausea drug for pregnant women in Europe and Great Britain from 1959–1962.

Two enantiomers of thalidomide

chirality center

chirality center

anti-nausea drug

teratogen

Unfortunately, thalidomide was sold as a mixture of its two enantiomers, and each of these stereoisomers has a different biological activity, a property not uncommon to chiral drugs, as we will see in Section 15.5. While one enantiomer has the desired therapeutic effect, the other enantiomer was responsible for thousands of catastrophic birth defects in children born to women who took the drug during pregnancy. Thalidomide was never approved for use in the United States due to the diligence of Frances Oldham Kelsey, a medical reviewing officer for the Food and Drug Administration, who insisted that the safety data on thalidomide were inadequate.

HEALTH NOTE

Although it is a potent teratogen (a substance that causes fetal abnormalities), thalidomide exhibits several beneficial effects, and for this reason, it is now prescribed under strict control for the treatment of Hansen's disease (leprosy). Because it was once thought to be extremely contagious, individuals suffering from Hansen's disease in Hawaii were sent to Kaulapapa, a remote and inaccessible peninsula on the island of Molokai. Hansen's disease is now known to be a treatable bacterial infection completely cured by the sulfa drugs described in Section 13.13.

PROBLEM 15.10

Label the chirality centers in each compound. A molecule may have zero, one, or two chirality centers.

a. [cyclohexene ring with OH]

b. [benzene ring with CH₂CH₃]

c. [cyclobutane ring with two Cl groups]

PROBLEM 15.11

Label the two chirality centers in sertraline, the generic name for the antidepressant Zoloft.

CH_3NH— [tetrahydronaphthalene structure attached to dichlorophenyl ring, with Cl and Cl]

sertraline

15.5 FOCUS ON HEALTH & MEDICINE
CHIRAL DRUGS

A living organism is a sea of chiral molecules. Many drugs are chiral, and often they must interact with a chiral receptor to be effective. One enantiomer of a drug may be effective in treating a disease whereas its mirror image may be ineffective.

Why should such a difference in biological activity be observed? For an everyday analogy, consider what happens when you are given a right-handed glove. A glove is a chiral object, and your right and left hands are also chiral—they are mirror images that are not identical. Only your right hand can fit inside a right glove, not your left. Thus, the glove's utility depends on the hand—it either fits or it doesn't fit.

The same phenomenon is observed with drugs that must interact with a chiral receptor in a cell. One enantiomer "fits" the receptor and evokes a specific response. Its mirror image doesn't fit the same receptor, making it ineffective; or if it "fits" another receptor, it can evoke a totally different response. Figure 15.2 schematically illustrates this difference in binding between two enantiomers and a chiral receptor.

▼ **FIGURE 15.2** The Interaction of Two Enantiomers With a Chiral Receptor

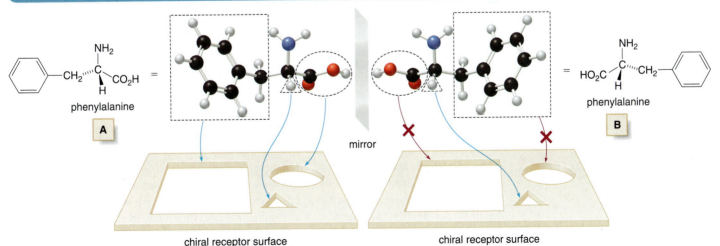

One enantiomer of the amino acid phenylalanine, labeled **A**, has three groups that can interact with the appropriate binding sites of the chiral receptor. The groups around the chirality center in the mirror image **B**, however, can never be arranged so that all three groups can bind to these same three binding sites. Thus, enantiomer **B** does not "fit" the same receptor, so it does not evoke the same response.

15.5A CHIRAL PAIN RELIEVERS

Ibuprofen is commonly used to relieve headaches, and muscle and joint pain.

Ibuprofen and naproxen are nonsteroidal anti-inflammatory drugs (NSAIDs) that illustrate how two enantiomers can have different biological activities. Both drugs are common over-the-counter remedies that relieve pain and fever, and reduce inflammation.

Ibuprofen is the generic name for the pain relievers known as Motrin and Advil. Ibuprofen has one chirality center, and thus exists as a pair of enantiomers. Only one enantiomer, labeled **A,** is an active anti-inflammatory agent. Its enantiomer **B** is inactive. **B,** however, is slowly converted to **A** in the body. Ibuprofen is sold as a mixture of both enantiomers. **An equal mixture of two enantiomers is called a** *racemic mixture* (pronounced ra-see′-mik).

ibuprofen
active anti-inflammatory agent

A

inactive enantiomer

B

One enantiomer of **naproxen,** the pain reliever in Naprosyn and Aleve, is an active anti-inflammatory agent, but its enantiomer is a harmful liver toxin. Changing the orientation of two substituents to form a mirror image can thus alter biological activity to produce an undesirable side effect in the other enantiomer.

naproxen
active anti-inflammatory agent

A

enantiomer
liver toxin

B

These examples illustrate the benefit of marketing a chiral drug as a single active enantiomer. In these instances it should be possible to use smaller doses with fewer side effects, yet many chiral drugs continue to be sold as racemic mixtures, because it is more difficult and therefore more costly to obtain a single enantiomer.

Recent rulings by the Food and Drug Administration have encouraged the development of so-called *racemic switches,* the patenting and marketing of a single enantiomer that was originally sold as a racemic mixture. To obtain a new patent on a single enantiomer, however, a company must show evidence that it provides significant benefit over the racemic mixture.

| PROBLEM 15.12 | Propranolol (trade name: Inderal) is used in the treatment of high blood pressure and heart disease. (a) Locate the chirality center in propranolol. (b) Draw both enantiomers of propranolol. Only one of these enantiomers is biologically active. (c) Draw a diagram similar to Figure 15.2 that illustrates how the two enantiomers of propranolol bind differently to a chiral receptor. |

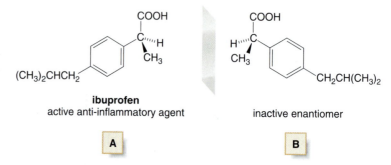

propranolol

15.5B PARKINSON'S DISEASE AND L-DOPA

L-Dopa was first isolated from the seeds of the broad bean *Vicia faba*, and since 1967 it has been the drug of choice for the treatment of Parkinson's disease. L-Dopa is a chiral molecule with one chirality center. Unlike ibuprofen, which is still marketed as a racemic mixture, L-dopa, also called levodopa, is sold as a single enantiomer. Like many chiral drugs, only one enantiomer is active against Parkinson's disease. The inactive enantiomer is also responsible for neutropenia, a decrease in certain white blood cells that help to fight infection.

The broad bean *Vicia faba* produces the drug L-dopa. To meet the large demand for L-dopa today, the commercially available drug is synthesized in the laboratory.

L-dopa
active against
Parkinson's disease

D-dopa
enantiomer responsible for
undesired side effects

Parkinson's disease, which afflicts 1.5 million individuals in the United States, results from the degeneration of neurons that produce the neurotransmitter dopamine in the brain. Dopamine affects brain processes that control movement and emotions, so proper dopamine levels are needed to maintain an individual's mental and physical health. When the level of dopamine drops, the loss of motor control symptomatic of Parkinson's disease results.

L-Dopa is an oral medication that is transported to the brain by the bloodstream. In the brain it is converted to dopamine (Figure 15.3). Dopamine itself cannot be given as a medication because it cannot pass from the bloodstream into the brain; that is, it does not cross the blood–brain barrier.

▼ **FIGURE 15.3**   L-Dopa, Dopamine, and Parkinson's Disease

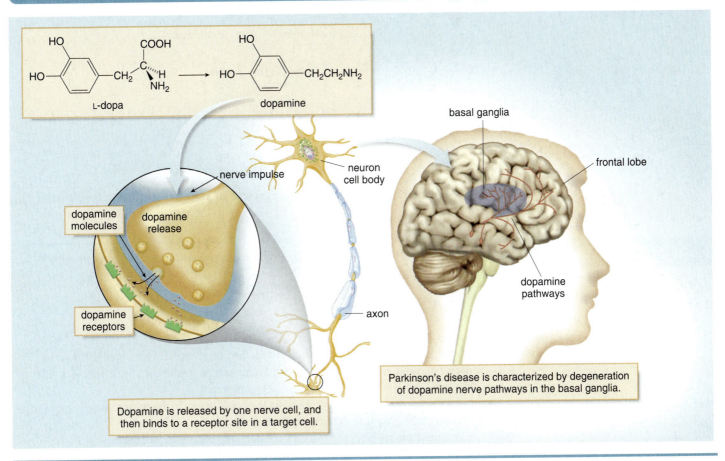

L-dopa

dopamine

basal ganglia

frontal lobe

nerve impulse

neuron
cell body

dopamine
molecules

dopamine
release

dopamine
pathways

dopamine
receptors

axon

Parkinson's disease is characterized by degeneration
of dopamine nerve pathways in the basal ganglia.

Dopamine is released by one nerve cell, and
then binds to a receptor site in a target cell.

15.6 FISCHER PROJECTIONS

The chirality centers in some organic compounds, most notably carbohydrates (Chapter 20), are often drawn using a different convention than is used for other chirality centers.

Instead of drawing a tetrahedron with two bonds in the plane, one in front of the plane, and one behind it, the **tetrahedron is tipped so that both horizontal bonds come forward (drawn on wedges) and both vertical bonds go behind (on dashed lines).** This structure is then abbreviated by a **cross formula,** also called a **Fischer projection formula.** In a Fischer projection formula, therefore,

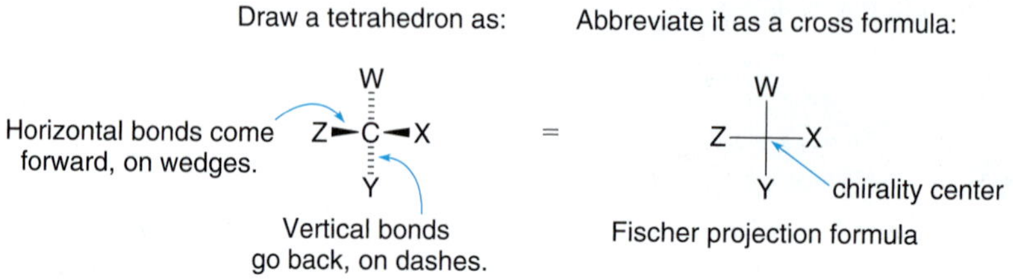

Draw a tetrahedron as: Abbreviate it as a cross formula:

Horizontal bonds come   Z—C—X   =   Z——X
forward, on wedges.
                              Vertical bonds   Fischer projection formula
                              go back, on dashes.

- A carbon atom is located at the intersection of the two lines of the cross.
- The horizontal bonds come forward, on wedges.
- The vertical bonds go back, on dashed lines.

For example, to draw the chirality center of 2-butanol [$CH_3CH(OH)CH_2CH_3$, Section 15.3B] using this convention, **draw the tetrahedron with horizontal bonds on wedges and vertical bonds on dashed lines.** Then, **replace the chirality center with a cross** to draw the Fischer projection.

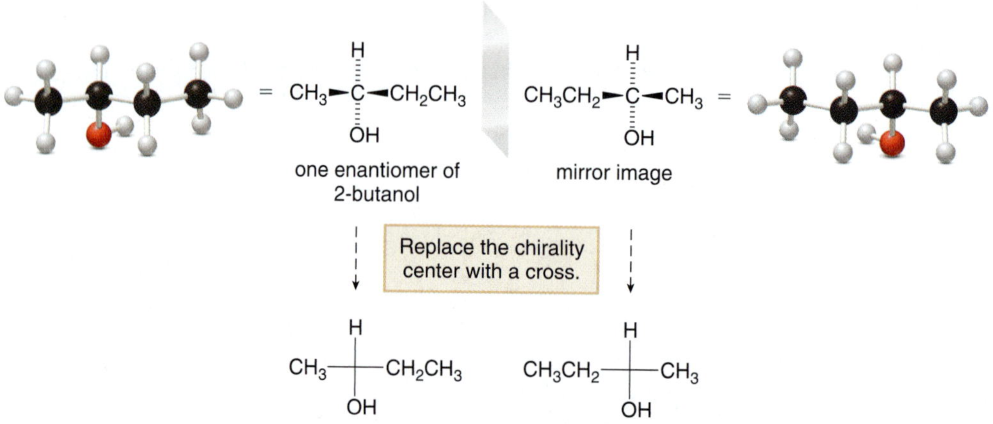

$$= CH_3—C—CH_2CH_3 \qquad CH_3CH_2—C—CH_3 =$$

one enantiomer of   mirror image
2-butanol

Replace the chirality
center with a cross.

$$CH_3——CH_2CH_3 \qquad CH_3CH_2——CH_3$$

Fischer projection formulas for both enantiomers of 2-butanol

Sample Problem 15.2 illustrates another example of drawing two enantiomers using this convention.

SAMPLE PROBLEM 15.2

Draw both enantiomers of glyceraldehyde, a simple carbohydrate, using Fischer projection formulas.

$$HOCH_2CHCHO$$
$$|$$
$$OH$$

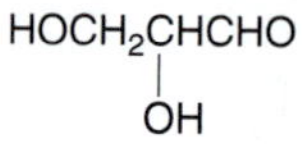

glyceraldehyde

ANALYSIS

- Draw the tetrahedron of one enantiomer with the horizontal bonds on wedges, and the vertical bonds on dashed lines. Arrange the four groups on the chirality center—H, OH, CHO, and CH_2OH—arbitrarily in the first enantiomer.
- Draw the second enantiomer by arranging the substituents in the mirror image so they are a reflection of the groups in the first molecule.
- Replace the chirality center with a cross to draw the Fischer projections.

SOLUTION

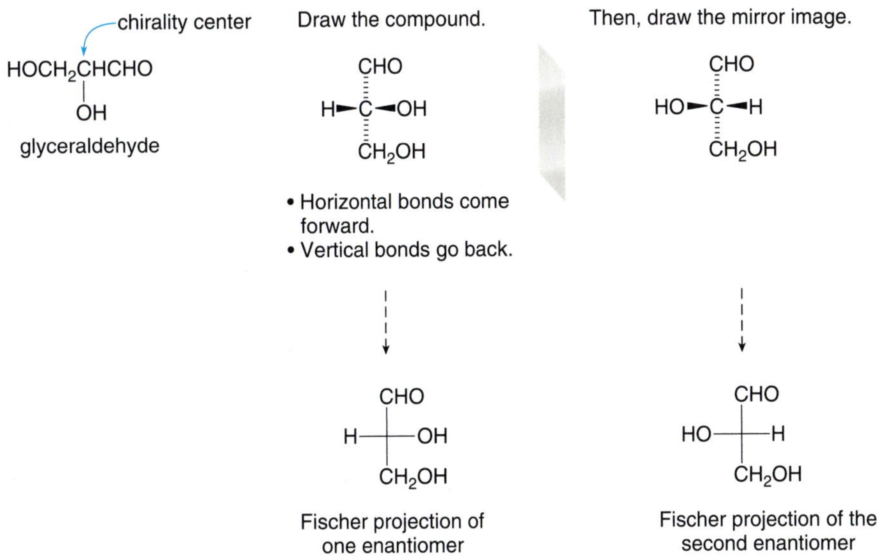

chirality center

HOCH₂CHCHO
       |
       OH

glyceraldehyde

Draw the compound.

CHO
 |
H━C━OH
 |
CH₂OH

• Horizontal bonds come forward.
• Vertical bonds go back.

⇩

CHO
 |
H———OH
 |
CH₂OH

Fischer projection of one enantiomer

Then, draw the mirror image.

CHO
 |
HO━C━H
 |
CH₂OH

⇩

CHO
 |
HO———H
 |
CH₂OH

Fischer projection of the second enantiomer

PROBLEM 15.13

Convert each molecule into a Fischer projection formula.

a.
CH₃
 |
H━C━Br
 |
Cl

b.
COOH
 |
H━C━OH
 |
CH₂Cl

PROBLEM 15.14

Convert each Fischer projection formula into a representation that uses wedges and dashes.

a.
        CH₃
         |
CH₃CH₂——+——Br
         |
        Cl

b.
        H
        |
CH₃CH₂——+——CH₃
        |
       NH₂

PROBLEM 15.15

Draw Fischer projections of both enantiomers for each compound.

a. CH₃CHCH₂CH₂CH₃
        |
        OH

b. CH₃CH₂CHCH₂Cl
           |
           Cl

15.7 COMPOUNDS WITH TWO OR MORE CHIRALITY CENTERS

Although most chiral compounds we have seen thus far have one chirality center, molecules may have two, three, or hundreds of chirality centers. Two examples—tartaric acid and ephedrine—were shown in Section 15.3.

- **A compound with one chirality center has two stereoisomers—a pair of enantiomers.**
- **A compound with two chirality centers has four possible stereoisomers.**

Let's examine the four stereoisomers of the amino acid threonine. The two chirality centers in threonine are labeled in the given structure. Threonine is one of the 20 naturally occurring amino acids needed for protein synthesis. Since humans cannot synthesize threonine, it is an essential amino acid, meaning that it must be obtained in the diet.

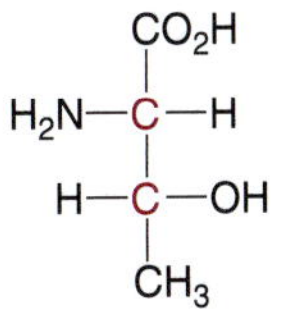

[The chirality centers are labeled in red.]

carbon skeleton of the amino acid threonine

There are four different ways that the groups around two chirality centers can be arranged, shown in structures **A–D.**

The four stereoisomers of a compound with two chirality centers

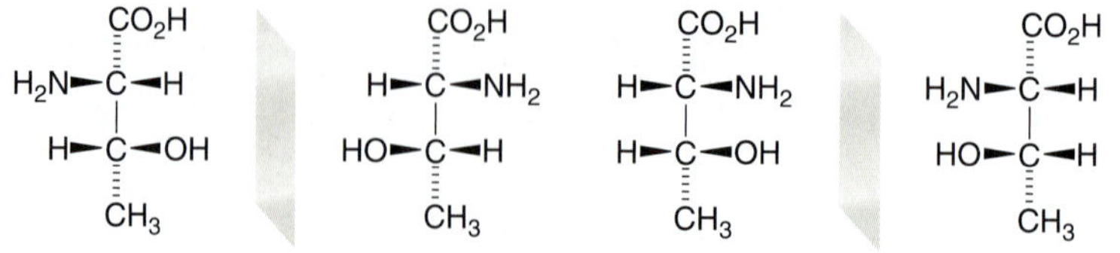

A — B (enantiomers)   C — D (enantiomers)

A and **B** are mirror images, but they are not superimposable, so they are enantiomers. **C** and **D** are also mirror images, but they are not superimposable, so they are enantiomers as well. In other words, with *two* chirality centers, there can be *two* pairs of enantiomers.

Because these structures are drawn with the horizontal bonds of the chirality centers on wedges and the vertical bonds on dashed lines, Fischer projection formulas can be drawn for each stereoisomer by replacing each chirality center by a cross.

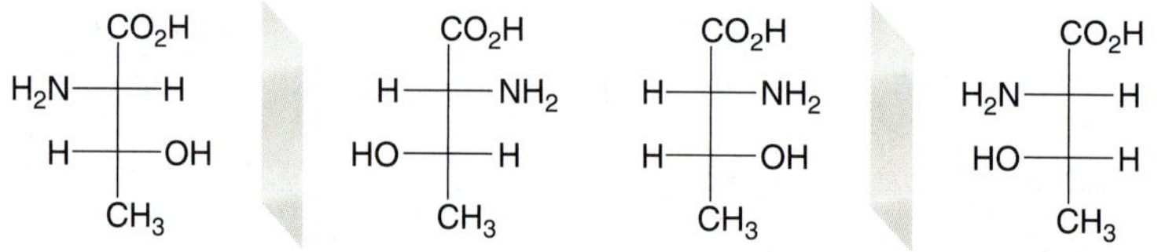

A — B (enantiomers)   C — D (enantiomers)

Thus, there are four stereoisomers for threonine: enantiomers **A** and **B,** and enantiomers **C** and **D.** What is the relationship between two stereoisomers like **A** and **C**? **A** and **C** are stereoisomers, but if you arrange a mirror plane between **A** and **C** you can see that they are *not* mirror images of each other. **A** and **C** represent a second broad class of stereoisomers, called **diastereomers.**

• *Diastereomers* are stereoisomers but they are *not* mirror images of each other.

A and **B** are diastereomers of **C** and **D,** and vice versa. Naturally occurring threonine is a single stereoisomer corresponding to structure **A.** The other stereoisomers do not occur in nature and cannot be used to synthesize biologically active proteins.

SAMPLE PROBLEM 15.3

Considering the four stereoisomers of 2,3-pentanediol (**E–H**): (a) Which compound is an enantiomer of **E**? (b) Which compound is an enantiomer of **F**? (c) What two compounds are diastereomers of **G**?

Four stereoisomers of 2,3-pentanediol:

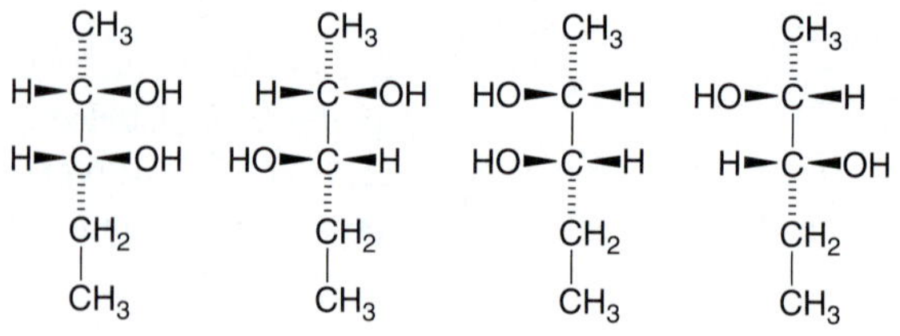

E   F   G   H

ANALYSIS

Keep in mind the definitions:

- **Enantiomers are stereoisomers that are nonsuperimposable mirror images of each other.** Look for two compounds that when placed side-by-side have the groups on both chirality centers drawn as a *reflection* of each other. A given compound has only one possible enantiomer.
- **Diastereomers are stereoisomers that are *not* mirror images of each other.**

SOLUTION

a. **E** and **G** are enantiomers because they are mirror images that do not superimpose on each other.

b. **F** and **H** are enantiomers for the same reason.

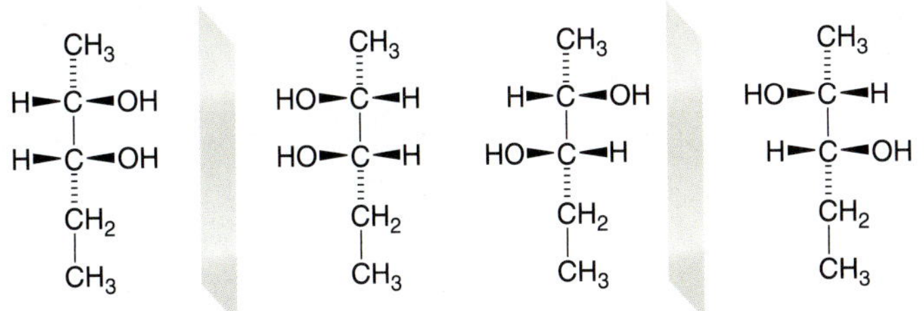

$$E \quad\quad G \quad\quad\quad F \quad\quad H$$

enantiomers   enantiomers

c. Any compound that is a stereoisomer of **G** but is not its mirror image is a diastereomer. Thus, **F** and **H** are diastereomers of **G**.

PROBLEM 15.16

Answer the following questions about the four stereoisomers of 3-bromo-2-chloropentane (**W–Z**). (a) Which compound is the enantiomer of **W**? (b) Which compound is the enantiomer of **X**? (c) What two compounds are diastereomers of **Y**? (d) What two compounds are diastereomers of **Z**?

Four stereoisomers of 3-bromo-2-chloropentane:

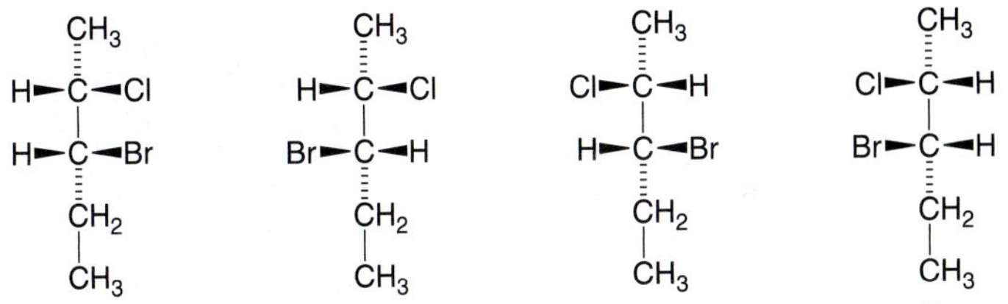

$$W \quad\quad X \quad\quad Y \quad\quad Z$$

PROBLEM 15.17

Convert each stereoisomer (**W–Z**) in Problem 15.16 into a Fischer projection.

15.8 FOCUS ON THE HUMAN BODY
THE SENSE OF SMELL

Research suggests that the odor of a particular molecule is determined more by its shape than by the presence of a particular functional group (Figure 15.4). For example, hexachloroethane (Cl_3CCCl_3) and cyclooctane have no obvious structural similarities. However, both molecules have a camphor-like odor, a fact attributed to their similar spherical shape, which is readily shown using space-filling models.

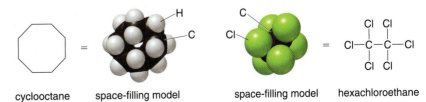

cyclooctane   space-filling model   space-filling model   hexachloroethane

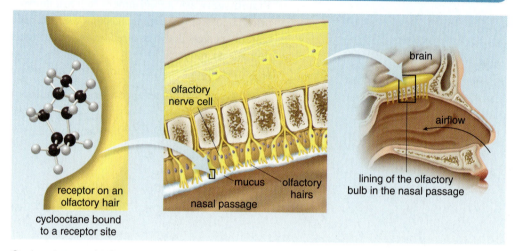

▼ **FIGURE 15.4   The Shape of Molecules and the Sense of Smell**

brain

olfactory
nerve cell

airflow

mucus   olfactory
hairs

lining of the olfactory
bulb in the nasal passage

nasal passage

receptor on an
olfactory hair

cyclooctane bound
to a receptor site

Cyclooctane and other molecules similar in shape bind to a particular olfactory receptor on the nerve cells that lie at the top of the nasal passage. Binding results in a nerve impulse that travels to the brain, which interprets impulses from particular receptors as specific odors.

Since enantiomers interact with chiral smell receptors, some enantiomers have different odors. There are a few well-characterized examples of this phenomenon in nature. For example, one enantiomer of carvone is responsible for the odor of caraway, whereas the other carvone enantiomer is responsible for the odor of spearmint.

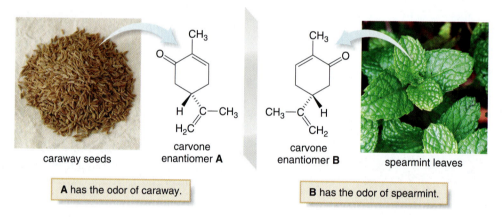

caraway seeds

carvone
enantiomer **A**

carvone
enantiomer **B**

spearmint leaves

A has the odor of caraway.

B has the odor of spearmint.

Thus, the three-dimensional structure of a molecule is important in determining its odor.

PROBLEM 15.18	Limonene is similar to carvone in that each enantiomer has a different odor; one enantiomer occurs in lemons and the other occurs in oranges. Identify the chirality center in limonene and draw both enantiomers of limonene.

$$CH_3 - \overset{\displaystyle CH_2}{\underset{\displaystyle CH_3}{C}}$$

limonene

CHAPTER HIGHLIGHTS

KEY TERMS

Achiral (15.2)
Chiral (15.2)
Chirality center (15.2)
Constitutional isomer (15.1)

Cross formula (15.6)
Diastereomer (15.7)
Enantiomer (15.2)
Fischer projection formula (15.6)

Racemic mixture (15.5)
Stereochemistry (15.1)
Stereoisomer (15.1)

KEY CONCEPTS

❶ When is a molecule chiral or achiral? (15.2)
- A chiral molecule is not superimposable on its mirror image.
- An achiral molecule is superimposable on its mirror image.
- To see whether a molecule is chiral, draw it and its mirror image and see if all of the atoms and bonds align.

❷ What is a chirality center? (15.2, 15.3)
- A chirality center is a carbon with four different groups around it.
- A molecule with one chirality center is chiral.

❸ What are enantiomers? (15.2, 15.3)
- Enantiomers are mirror images that are not superimposable on each other. To draw two enantiomers, draw one molecule in three dimensions around the chirality center. Then draw the mirror image so that the substituents are a reflection of the groups in the first molecule.

❹ Why do some chiral drugs have different properties from their mirror-image isomers? (15.5)
- When a chiral drug must interact with a chiral receptor, only one enantiomer fits the receptor properly and evokes a specific response. Ibuprofen, naproxen, and L-dopa are examples of chiral drugs in which the two enantiomers have very different properties.

❺ What is a Fischer projection? (15.6)
- A Fischer projection is a specific way of depicting a chirality center. The chirality center is located at the intersection of a cross. The horizontal lines represent bonds that come out of the plane on wedges, and the vertical lines represent bonds that go back on dashed lines.

❻ What is the difference between an enantiomer and a diastereomer? (15.7)
- Enantiomers are stereoisomers that are nonsuperimposable mirror images of each other.
- Diastereomers are stereoisomers that are *not* mirror images of each other.

❼ How is the shape of a molecule related to its odor? (15.8)
- It is thought that the odor of a molecule is determined more by its shape than the presence of a particular functional group, so compounds of similar shape have similar odors. For an odor to be perceived, a molecule must bind to an olfactory receptor, resulting in a nerve impulse that travels to the brain. Enantiomers may have different odors because they bind with chiral receptors and each enantiomer fits the chiral receptors in a different way.

PROBLEMS

Selected in-chapter and end-of-chapter problems have brief answers provided in Appendix B.

Chiral Compounds, Chirality Centers, and Enantiomers

15.19 Label each of the following objects as chiral or achiral: (a) chalk; (b) shoe; (c) baseball glove; (d) soccer ball.

15.20 Label each of the following objects as chiral or achiral: (a) boot; (b) index card; (c) scissors; (d) drinking glass.

15.21 Explain the following statement. Butane ($CH_3CH_2CH_2CH_3$) has a mirror image but it does not have an enantiomer.

15.22 Explain why the human body is chiral.

15.23 Draw a mirror image for each compound. Label each compound as chiral or achiral.

a.

halothane
(general anesthetic)

b.

valine
(naturally occurring amino acid)

c.

dichloromethane
(common solvent)

15.24 Draw a mirror image for each compound. Label each compound as chiral or achiral.

a. 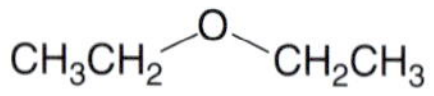 CH_3CH_2—O—CH_2CH_3

diethyl ether
(general anesthetic)

c. 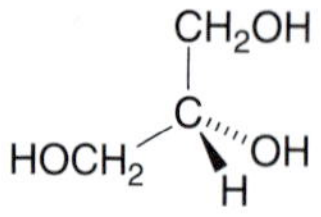 $HOCH_2$—C(—CH_2OH)(—OH)(—H)

glycerol
(triol used in lotions)

b. H—C(—$COCH_2OH$)(—OH)—CH_2OH

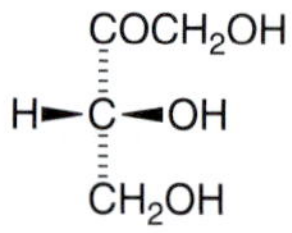

erythrulose
(a carbohydrate)

15.25 Label the chirality center (if one exists) in each compound. Compounds contain zero or one chirality center.

a. 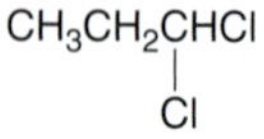 CH_3CH_2CHCl
   |
   Cl

b. 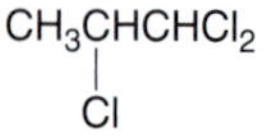 $CH_3CHCHCl_2$
   |
   Cl

c. 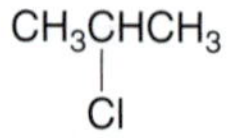 CH_3CHCH_3
   |
   Cl

d. 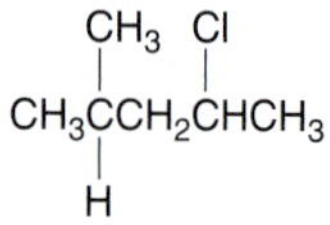 $CH_3CCH_2CHCH_3$ (with CH_3 and Cl on the first carbon, H below)

15.26 Label the chirality center (if one exists) in each compound. Compounds contain zero or one chirality center.

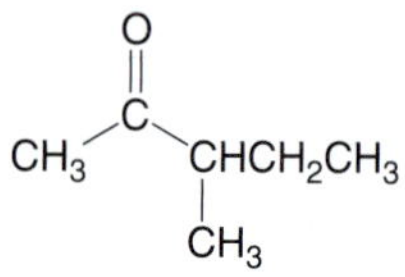

a. 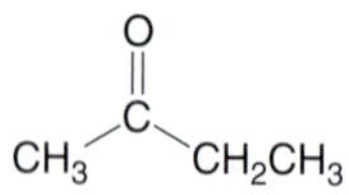 CH_3—C(=O)—CH_2CH_3

b. CH_3—C(=O)—$CHCH_2CH_3$
                    |
                    CH_3

c. CH_3CHCHO
   |
   CH_3

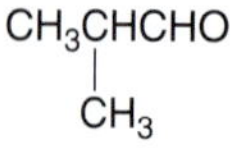

d. 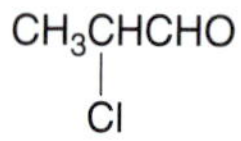 CH_3CHCHO
   |
   Cl

15.27 Label the chirality center (if one exists) in each cyclic compound. Some compounds contain no chirality centers.

a. 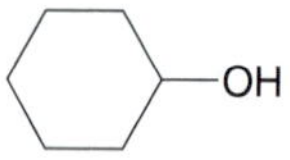 cyclohexane—OH

b. 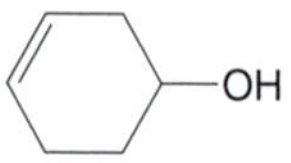 cyclohexene—OH

c. 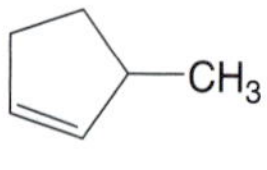 cyclopentene—CH_3

d. 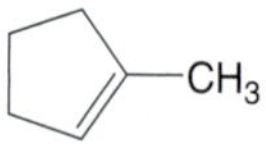 cyclopentene—CH_3

15.28 Label the chirality center (if one exists) in each cyclic compound. Some compounds contain no chirality centers.

a. 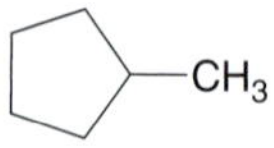 cyclopentane—CH_3

b. 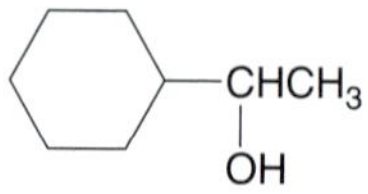 cyclohexane—$CHCH_3$
                    |
                    OH

c. 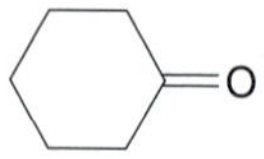 cyclohexane=O

d. 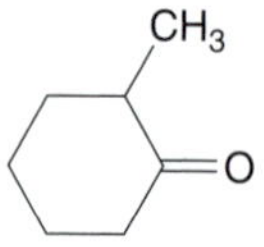 cyclohexane with CH_3 and =O

15.29 Draw the structure of a compound that fits each description:
a. an alkane of molecular formula C_7H_{16} that contains one chirality center
b. an alcohol of molecular formula $C_6H_{14}O$ that contains one chirality center

15.30 Draw the structure of a compound that fits each description:
a. an alkyl bromide of molecular formula C_4H_9Br that contains one chirality center
b. a diol of molecular formula $C_5H_{12}O_2$ that contains two chirality centers

15.31 Explain why a carbonyl carbon can never be a chirality center.

15.32 Explain why a carbon atom that is part of a triple bond can never be a chirality center.

15.33 Locate the chirality center(s) in each biologically active compound.

a.

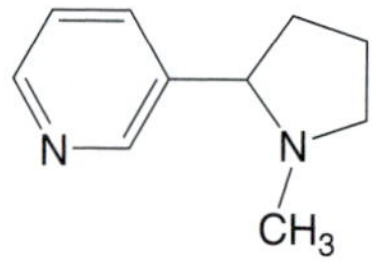

nicotine
(addictive stimulant from tobacco)

b.

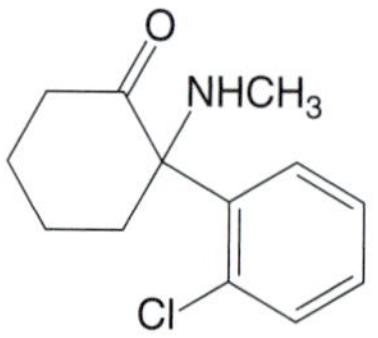

ketamine
(anesthetic)

c. 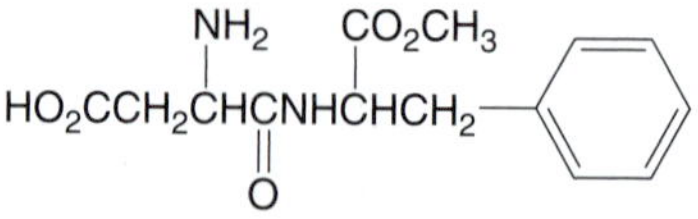 $HO_2CCH_2CHCNHCHCH_2$—phenyl (with NH_2, CO_2CH_3 groups and =O)

aspartame
Trade name: Equal
(synthetic sweetener)

15.34 Locate the chirality center(s) in each drug.

a.

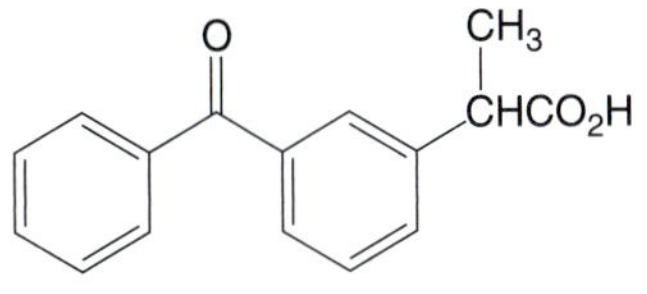

ketoprofen
(anti-inflammatory agent)

b.

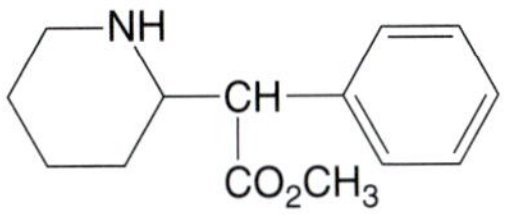

methylphenidate
Trade name: Ritalin
(used for attention deficit
hyperactivity disorder)

c.

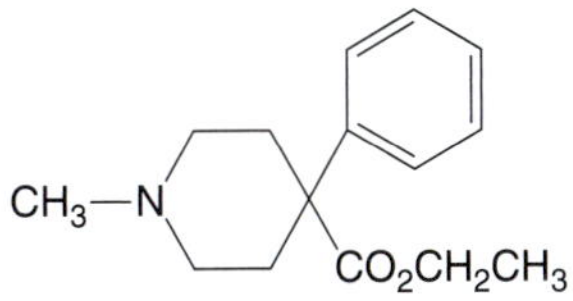

meperidine
Trade name: Demerol
(narcotic pain reliever)

15.35 Methamphetamine is an addictive stimulant sold illegally as "speed," "meth," or "crystal meth." Identify the chirality center in methamphetamine, and then draw both enantiomers in three dimensions around the chirality center.

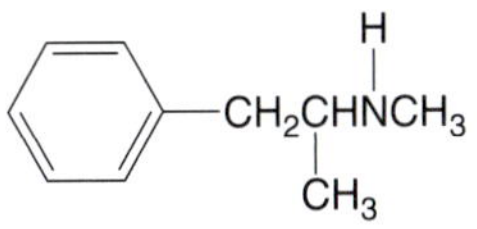

methamphetamine

15.36 Dobutamine is a heart stimulant used in stress tests to measure cardiac fitness. Identify the chirality center in dobutamine, and then draw both enantiomers in three dimensions around the chirality center.

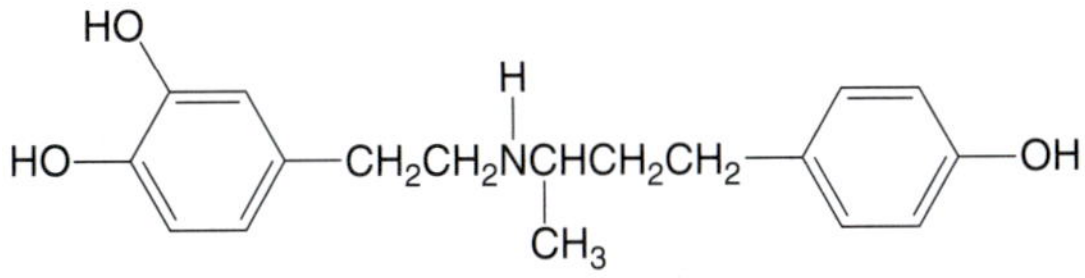

dobutamine

Isomers

15.37 How are the compounds in each pair related? Are they identical molecules or enantiomers?

a.

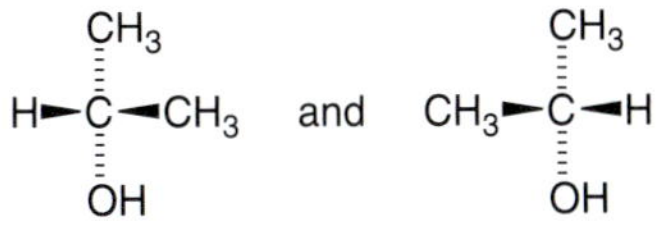

b.

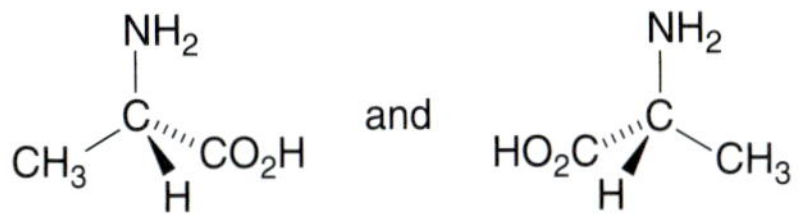

c.

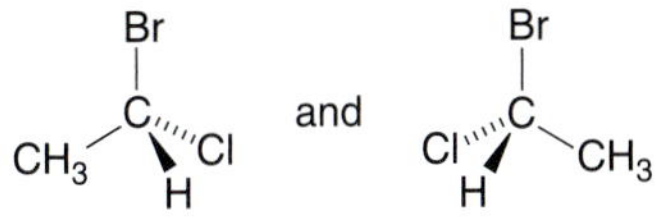

15.38 How are the compounds in each pair related? Are they identical molecules or enantiomers?

a.

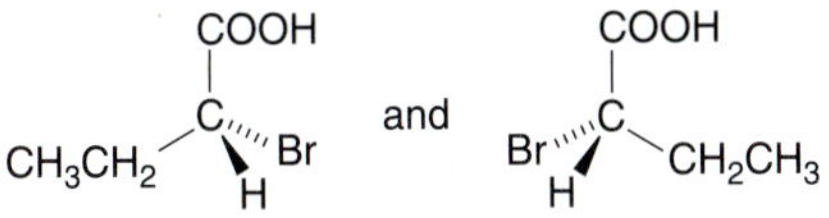

b.

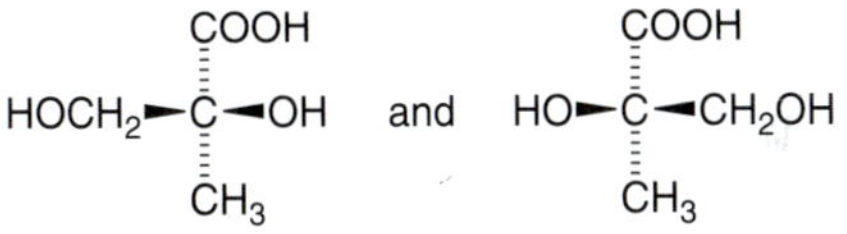

c.

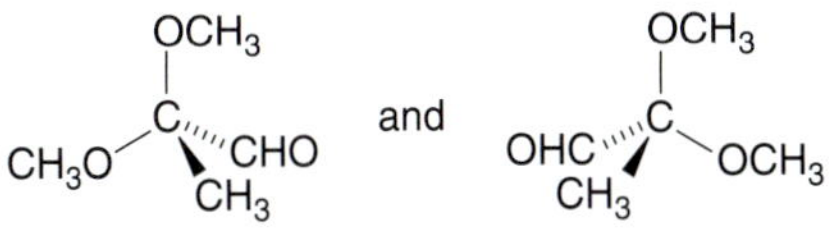

15.39 How are the molecules in each pair related? Choose from: identical molecules, constitutional isomers, enantiomers, diastereomers, or not isomers of each other.

a. $CH_3CH_2CH_2OCH_3$ and $CH_3CH_2OCH_2CH_3$

b.

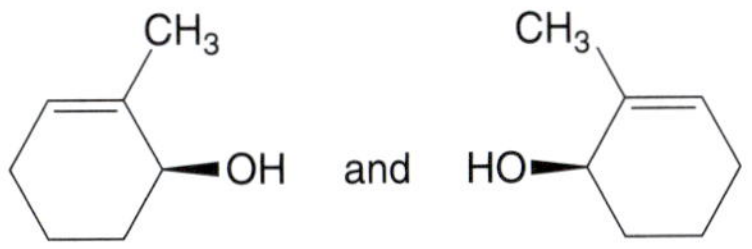

c.

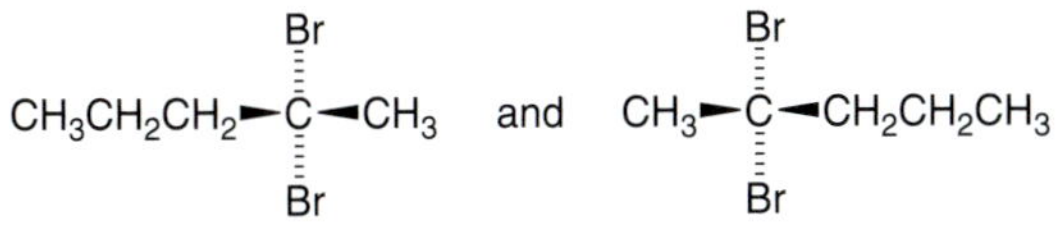

d.

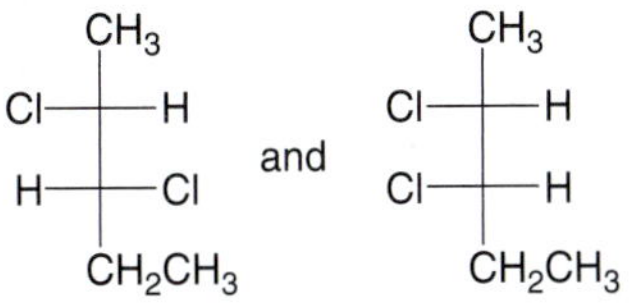

15.40 How are the molecules in each pair related? Choose from: identical molecules, constitutional isomers, enantiomers, diastereomers, or not isomers of each other.

a. 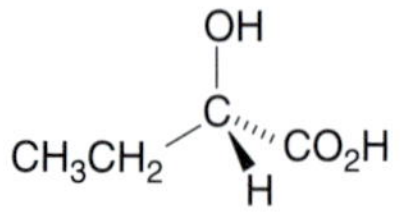   and

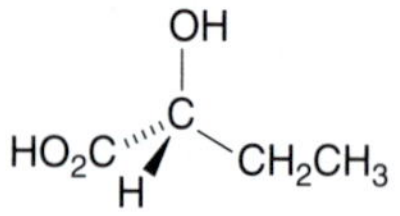

b. 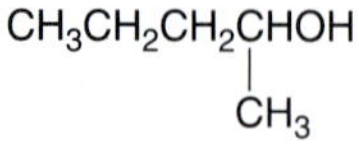 $CH_3CH_2CH_2CHOH$ and $CH_3CH_2OCH(CH_3)_2$
              |
              CH_3

c. $CH_3CH_2OCH_2CH_3$ and

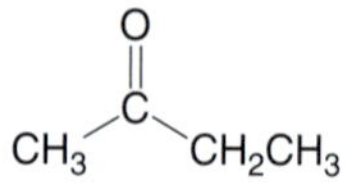

d. 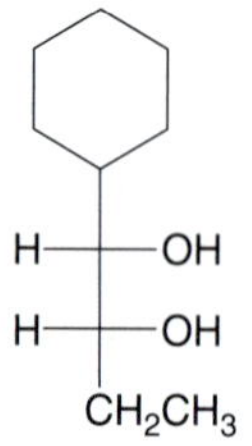   and

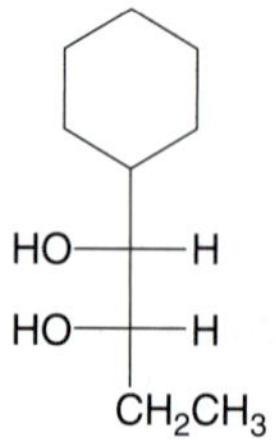

15.41 Answer each question with a compound of molecular formula $C_5H_{10}O_2$.
  a. Draw the structure of a compound that contains a COOH group and one chirality center.
  b. Draw the structure of a carboxylic acid that is a constitutional isomer of the compound drawn in part (a).
  c. Draw the structure of a constitutional isomer of the compound drawn in part (a), which contains a different functional group.

15.42 Answer each question with a compound of molecular formula $C_5H_{12}O$.
  a. Draw the structure of a compound that contains an ether group and one chirality center.
  b. Draw the structure of an ether that is a constitutional isomer of the compound drawn in part (a).
  c. Draw the structure of a constitutional isomer of the compound drawn in part (a), which contains a different functional group.

15.43 Draw the structure of the four constitutional isomers of molecular formula $C_5H_{10}O$ that contain an aldehyde (—CHO) as a functional group. Label any chirality center present in each compound. Some compounds have no chirality centers.

15.44 Although there are four constitutional isomers of molecular formula C_4H_9Cl, only one contains a chirality center and can therefore exist as a pair of enantiomers. Draw both enantiomers of this alkyl chloride in three dimensions around the chirality center.

Fischer Projections

15.45 Convert each three-dimensional representation into a Fischer projection.

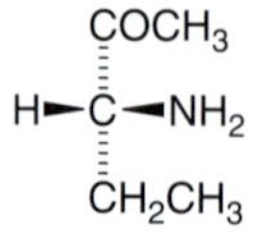

a. 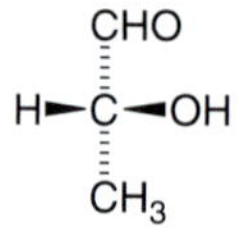   c.

b.

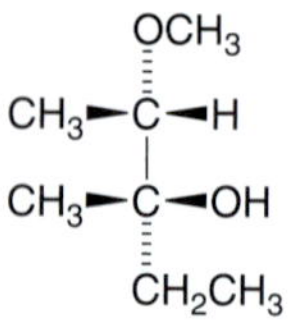

15.46 Convert each three-dimensional representation into a Fischer projection.

a. 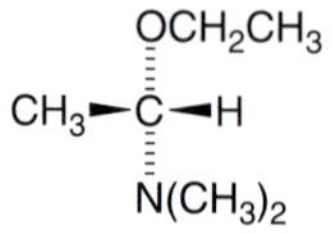   c.

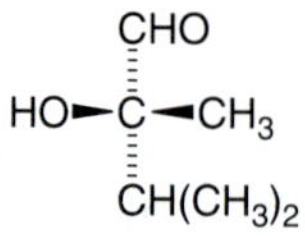

b.

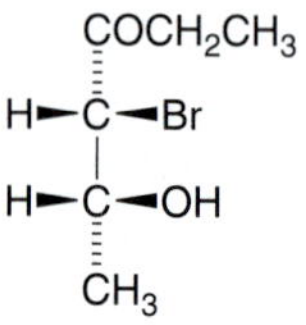

15.47 Convert each Fischer projection into a three-dimensional representation with wedges and dashed bonds.

a. 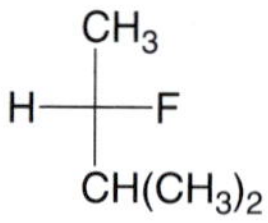   c.

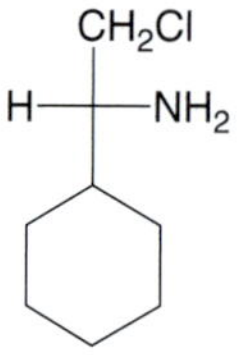

b.

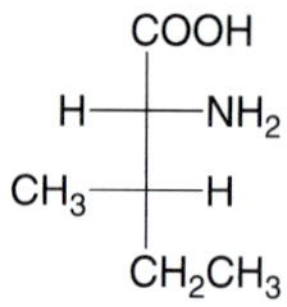

15.48 Convert each Fischer projection into a three-dimensional representation with wedges and dashed bonds.

a. 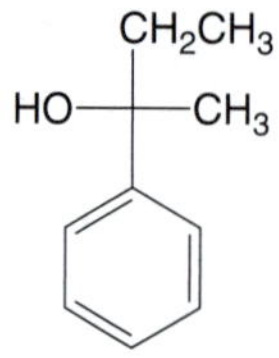   c.

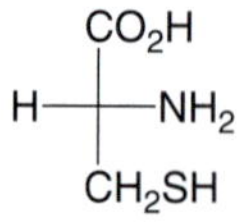

b.

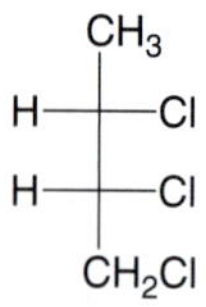

15.49 How are the Fischer projection formulas in each pair related to each other? Are they identical or are they enantiomers?

a. 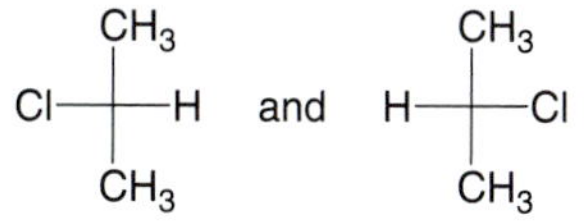 Cl—CH₃|H and H—CH₃|Cl (with CH₃ top and bottom)

b. 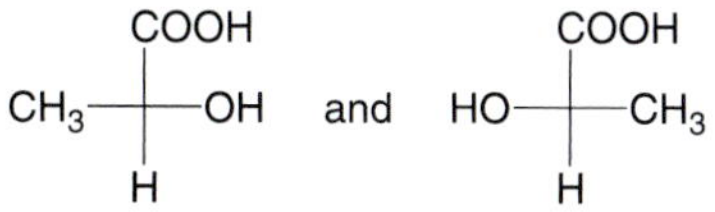 CH₃—COOH|OH|H and HO—COOH|CH₃|H

c. 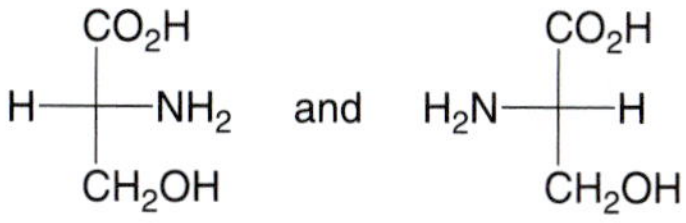 H—CO₂H|NH₂|CH₂OH and H₂N—CO₂H|H|CH₂OH

d. 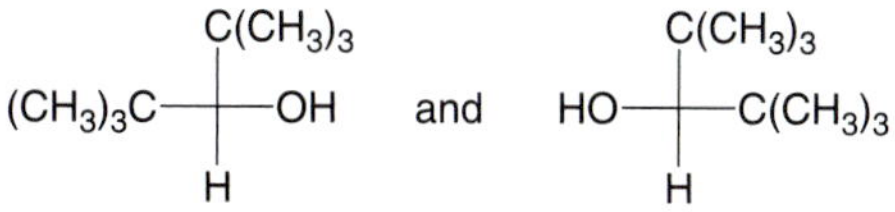 (CH₃)₃C—C(CH₃)₃|OH|H and HO—C(CH₃)₃|C(CH₃)₃|H

15.50 How are the Fischer projection formulas in each pair related to each other? Are they identical or are they enantiomers?

a. 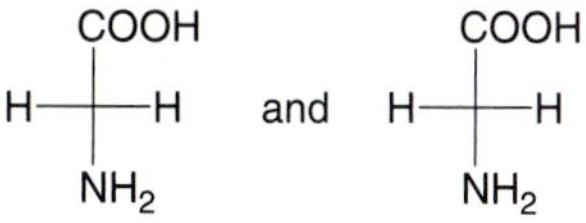 H—COOH|H|NH₂ and H—COOH|H|NH₂

b. 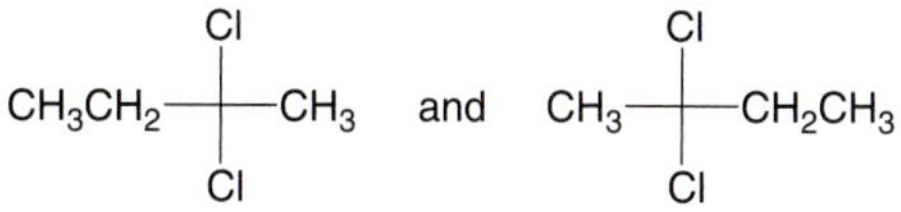 CH₃CH₂—Cl|CH₃|Cl and CH₃—Cl|CH₂CH₃|Cl

c. 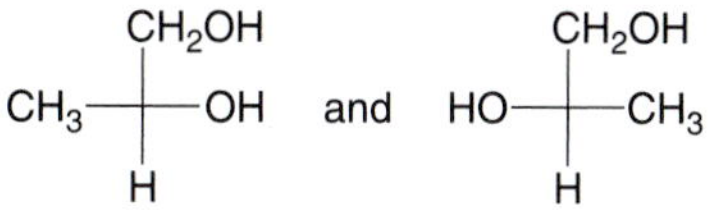 CH₃—CH₂OH|OH|H and HO—CH₂OH|CH₃|H

d. 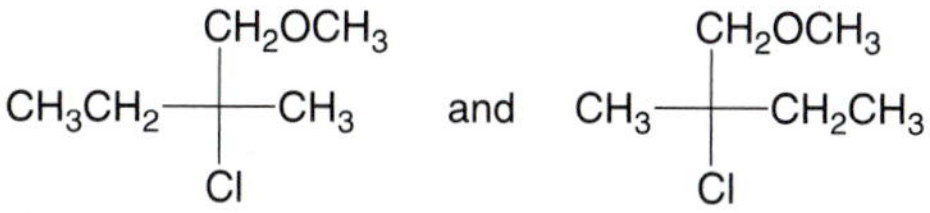 CH₃CH₂—CH₂OCH₃|CH₃|Cl and CH₃—CH₂OCH₃|CH₂CH₃|Cl

Enantiomers and Diastereomers

15.51 In what way are two enantiomers different from two diastereomers?

15.52 What are the two major types of stereoisomers? Give an example of each type.

15.53 Consider the stereoisomers (**A–D**) drawn below. These compounds are four-carbon sugars, as we will learn in Chapter 20.

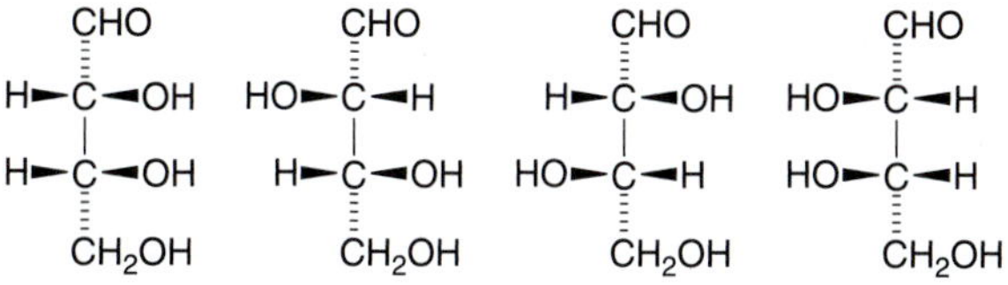

A, B, C, D — CHO / C–OH configurations / CH₂OH

a. How are the compounds in each pair related? Are the compounds enantiomers or diastereomers: [1] **A** and **B**; [2] **A** and **C**; [3] **A** and **D**; [4] **B** and **C**; [5] **B** and **D**; [6] **C** and **D**?
b. Draw a constitutional isomer of **A** that contains a ketone, not an aldehyde.

15.54 Consider the stereoisomers (**E–H**) drawn below. These compounds are five-carbon sugars, as we will learn in Chapter 20.

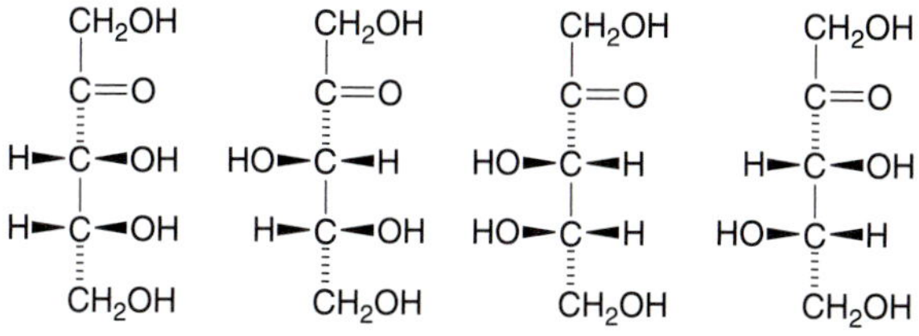

E, F, G, H

a. How are the compounds in each pair related? Are the compounds enantiomers or diastereomers: [1] **E** and **F**; [2] **E** and **G**; [3] **E** and **H**; [4] **F** and **G**; [5] **F** and **H**; [6] **G** and **H**?
b. Draw a constitutional isomer of **E** that contains an aldehyde, not a ketone.

15.55 In Chapter 13 we learned that cis and trans alkenes are stereoisomers. Are *cis*-2-butene and *trans*-2-butene enantiomers or diastereomers? Explain your choice.

15.56 Are *cis*-2-butene and *trans*-2-butene chiral or achiral compounds? Explain your choice.

Applications

15.57 Lactic acid is a product of glucose metabolism. During periods of strenuous exercise, lactic acid forms faster than it can be oxidized, resulting in the aching feeling of tired muscles.

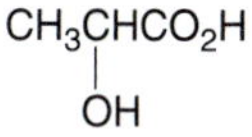

CH₃CHCO₂H | OH
lactic acid

a. Locate the chirality center in lactic acid.
b. Draw both enantiomers of lactic acid in three dimensions around the chirality center.
c. Draw Fischer projection formulas for both enantiomers of lactic acid.

15.58 Locate the chirality centers in each vitamin.

a. 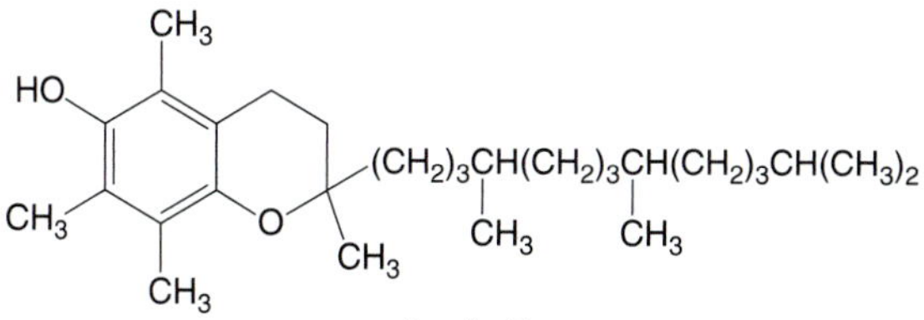 vitamin E

b. 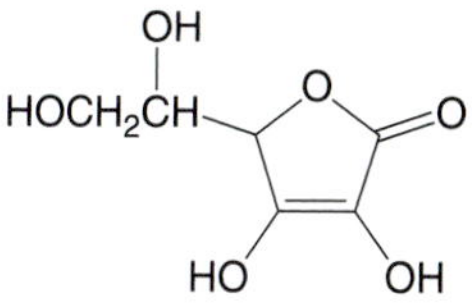 vitamin C

15.59 Hydroxydihydrocitronellal is another example of a compound that has two enantiomers that smell differently. One enantiomer smells minty, and the other smells like lily of the valley.

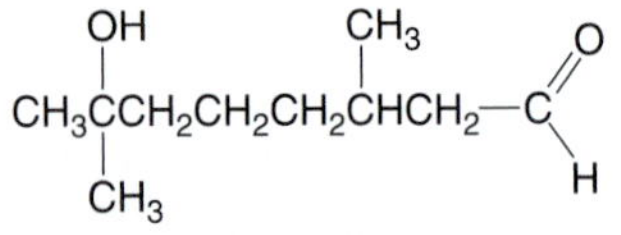

hydroxydihydrocitronellal

   a. Locate the chirality center in this compound.

   b. Draw both enantiomers in three dimensions around the chirality center.

   c. Draw Fischer projection formulas for both enantiomers.

15.60 Two enantiomers can sometimes taste very differently. L-Leucine, a naturally occurring amino acid used in protein synthesis, tastes bitter, but its enantiomer, D-leucine, tastes sweet.

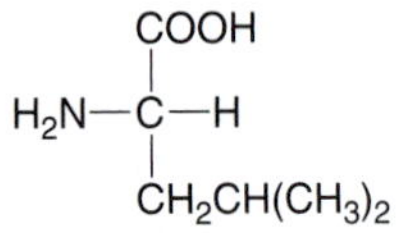

leucine

   a. Locate the chirality center in leucine.

   b. Draw both enantiomers in three dimensions around the chirality center.

   c. Draw Fischer projection formulas for both enantiomers.

15.61 Answer the following questions about the analgesic propoxyphene (trade name: Darvon).

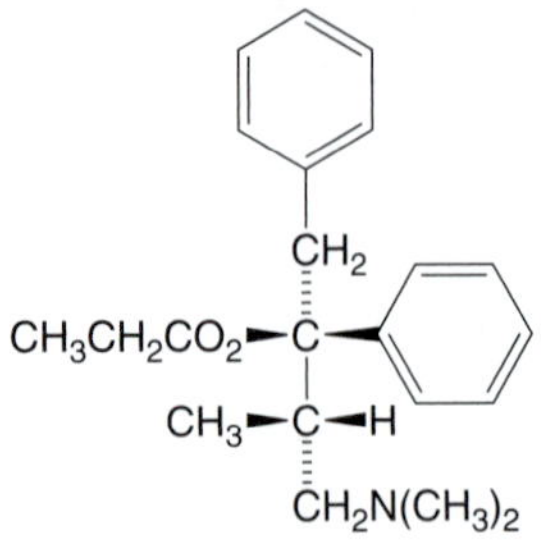

Darvon

   a. Label the chirality centers.

   b. Draw the enantiomer. Like many of the drugs discussed in Section 15.5, the enantiomer, levopropoxyphene (trade name: Novrad) has different biological properties. Levopropoxyphene is an antitussive agent, meaning it relieves coughs. (Do you notice anything about the trade names Darvon and Novrad?)

   c. Draw the structure of one diastereomer.

   d. Draw the structure of one constitutional isomer.

   e. Convert Darvon to a Fischer projection formula.

15.62 Plavix, the trade name for the generic drug clopidogrel, is used in the treatment of coronary artery disease. Like many newer drugs, Plavix is sold as a single enantiomer.

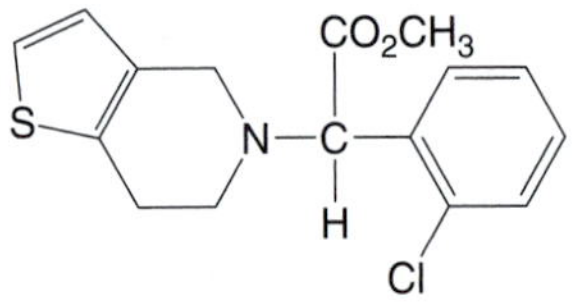

clopidogrel
Trade name: Plavix

   a. Locate the chirality center.

   b. Draw both enantiomers in three dimensions around the chirality center.

   c. Convert both enantiomers to Fischer projection formulas.

CHALLENGE QUESTIONS

15.63 Identify the nine chirality centers in sucrose, the sweet-tasting carbohydrate we use as table sugar.

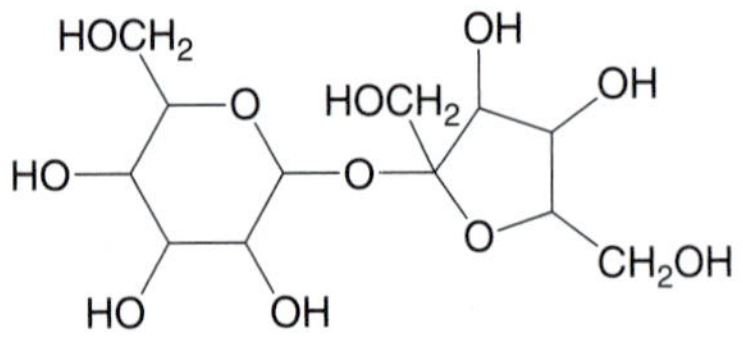

sucrose

15.64 Identify the six chirality centers in the anabolic steroid nandrolone. Nandrolone is used to increase muscle mass in body builders and athletes, although it is not permitted in competitive sports.

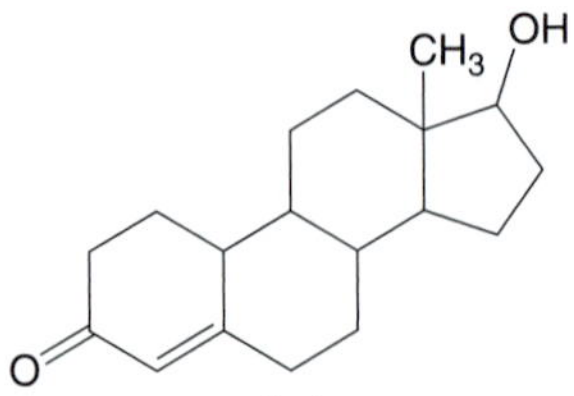

nandrolone

16

CHAPTER GOALS

In this chapter you will learn how to:

① Identify the characteristics of aldehydes and ketones

② Name aldehydes and ketones

③ Give examples of useful aldehydes and ketones

④ Draw the products of oxidation reactions of aldehydes

⑤ Draw the products of reduction reactions of aldehydes and ketones

⑥ Understand the basic reactions involved in vision

⑦ Identify and prepare hemiacetals and acetals

11-*cis*-Retinal is the light-sensitive aldehyde that plays a key role in the chemistry of vision for all vertebrates, arthropods, and mollusks.

ALDEHYDES AND KETONES

CHAPTER 16 is the first of two chapters that concentrate on compounds that contain a **carbonyl group (C=O),** perhaps the most important functional group in organic chemistry. In this chapter we examine **aldehydes** and **ketones,** compounds that contain a carbonyl carbon bonded to hydrogen or carbon atoms. Aldehydes and ketones occur widely in nature, and also serve as useful starting materials and solvents in industrial processes. All simple carbohydrates contain a carbonyl group, and more complex carbohydrates are derived from reactions discussed in this chapter.

16.1 STRUCTURE AND BONDING

Two broad classes of compounds contain a *carbonyl group.*

carbonyl group

1. Compounds that have only carbon and hydrogen atoms bonded to the carbonyl group

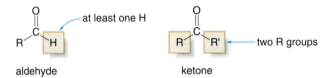

aldehyde   ketone

- An *aldehyde* has at least one H atom bonded to the carbonyl group.
- A *ketone* has two alkyl groups bonded to the carbonyl group.

2. Compounds that contain an electronegative atom bonded to the carbonyl group

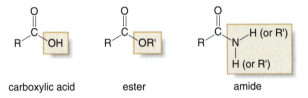

carboxylic acid   ester   amide

These compounds include **carboxylic acids, esters,** and **amides,** which contain an electronegative oxygen or nitrogen atom bonded directly to the carbonyl carbon. Esters and amides are called **carboxylic acid derivatives,** since they can be prepared from carboxylic acids. Carboxylic acids and their derivatives are discussed in Chapter 17, while aldehydes and ketones are the subject of this chapter.

Two structural features dominate the properties and chemistry of the carbonyl group.

trigonal planar   polar bond

- The carbonyl carbon atom is trigonal planar, and all bond angles are 120°. In this way the carbonyl carbon resembles the carbons of a carbon–carbon double bond.
- Since oxygen is more electronegative than carbon, a carbonyl group is polar. The carbonyl carbon is electron poor (δ^+) and the oxygen is electron rich (δ^-). In this way the carbonyl carbon is very different from the carbons of a nonpolar carbon–carbon double bond.

As mentioned in Chapter 11, the double bond of a carbonyl group is usually omitted in drawing shorthand structures. An aldehyde is often written as **RCHO.** Remember that the **H atom is bonded to the carbon atom,** not the oxygen. Likewise, a ketone is written as **RCOR,** or if both alkyl groups are the same, **R₂CO.**

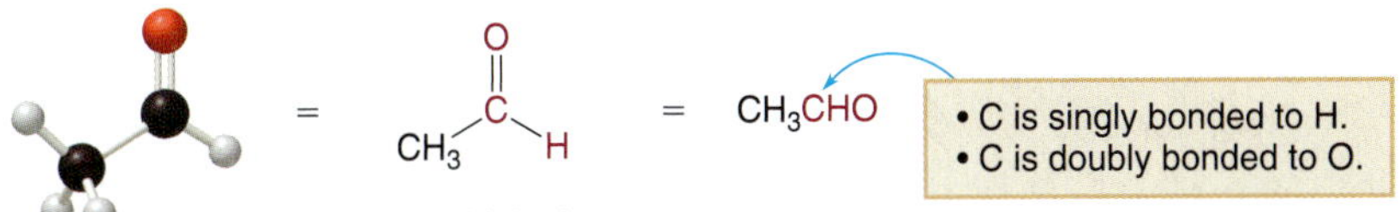

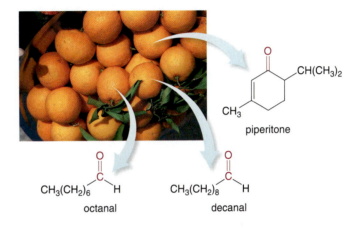

Many simple aldehydes and ketones are naturally occurring. For example, octanal, decanal, and piperitone are among the 70 organic compounds that contribute to the flavor and odor of an orange.

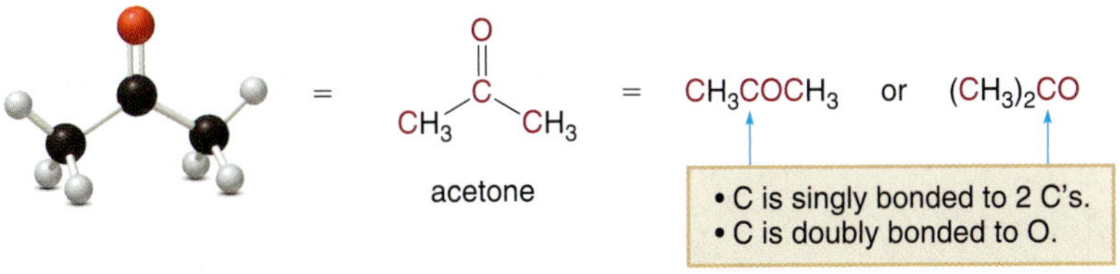

piperitone

octanal

decanal

PROBLEM 16.1	Draw out each compound to clearly show what groups are bonded to the carbonyl carbon. Label each compound as a ketone or aldehyde.

a. CH_3CH_2CHO b. $CH_3CH_2COCH_3$ c. $(CH_3)_3CCOCH_3$ d. $(CH_3CH_2)_2CHCHO$

PROBLEM 16.2	Draw the structure of the three constitutional isomers of molecular formula C_4H_8O that contain a carbonyl group. Label each compound as a ketone or aldehyde.

PROBLEM 16.3	Label each trigonal planar carbon in 11-*cis*-retinal, the chapter-opening molecule whose chemistry is discussed in Section 16.6.

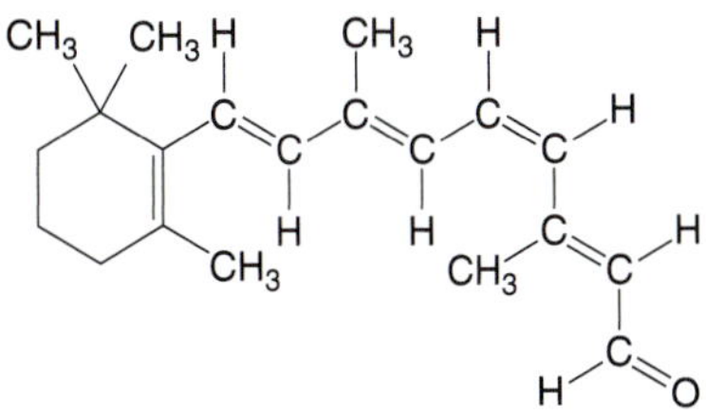

11-*cis*-retinal

16.2 NOMENCLATURE

Both IUPAC and common names are used for aldehydes and ketones.

16.2A NAMING ALDEHYDES

- **In IUPAC nomenclature, aldehydes are identified by the suffix -al.**

To name an aldehyde using the IUPAC system:

1. **Find the longest chain containing the CHO group, and change the -e ending of the parent alkane to the suffix -al.**
2. **Number the chain or ring to put the CHO group at C1, but omit this number from the name. Apply all of the other usual rules of nomenclature.**

Simple aldehydes have common names that are widely used. In fact, the common names **formaldehyde, acetaldehyde,** and **benzaldehyde** are virtually always used instead of their IUPAC names. Common names all contain the suffix **-aldehyde.**

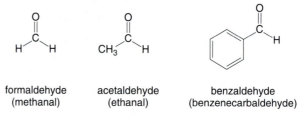

| formaldehyde
(methanal) | acetaldehyde
(ethanal) | benzaldehyde
(benzenecarbaldehyde) |

(IUPAC names are in parentheses.)

SAMPLE PROBLEM 16.1

Give the IUPAC name for each aldehyde.

a. 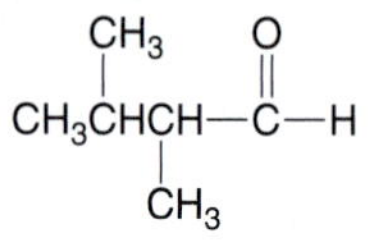
$$
\begin{array}{c}
\text{CH}_3 \quad\; \text{O} \\
| \qquad || \\
\text{CH}_3\text{CHCH}-\text{C}-\text{H} \\
| \\
\text{CH}_3
\end{array}
$$

b. 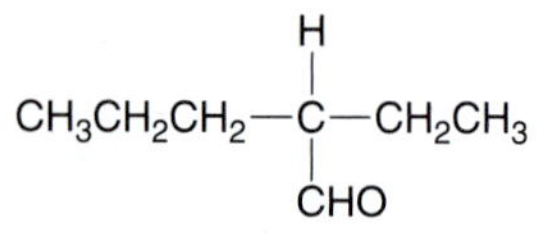
$$
\begin{array}{c}
\text{H} \\
| \\
\text{CH}_3\text{CH}_2\text{CH}_2-\text{C}-\text{CH}_2\text{CH}_3 \\
| \\
\text{CHO}
\end{array}
$$

ANALYSIS AND SOLUTION

a. [1] Find and name the longest chain containing the CHO.

$$
\boxed{\text{CH}_3\text{CHCH}-\text{C}}-\text{H}
$$
with CH₃ (above) and CH₃ (below)

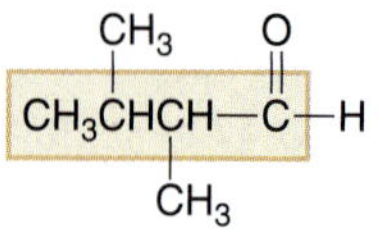

butane - - - → butan**al**
(4 C's)

[2] Number and name substituents, making sure the CHO group is at C1.

$$
\text{CH}_3\text{CHCH}-\text{C}-\text{H}
$$
3 2   1, with CH₃ above and CH₃ below

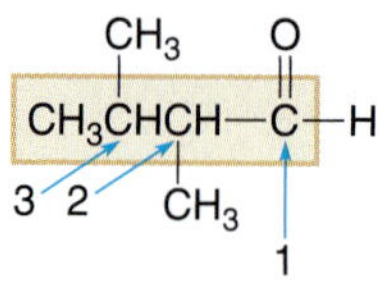

Answer: 2,3-dimethylbutanal

b. [1] Find and name the longest chain containing the CHO.

$$
\boxed{\text{CH}_3\text{CH}_2\text{CH}_2-\text{C}-\text{CH}_2\text{CH}_3}
$$
with H above and C=O (with H below) branching

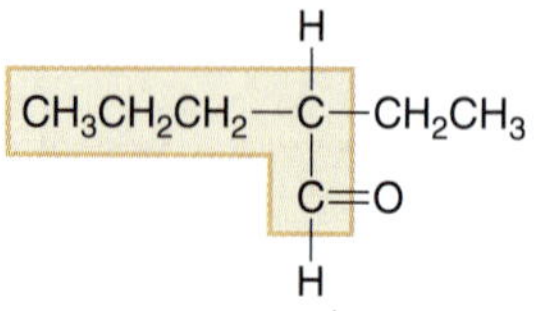

pentane - - - → pentan**al**
(5 C's)

[2] Number and name substituents, making sure the CHO group is at C1.

$$
\text{CH}_3\text{CH}_2\text{CH}_2-\text{C}-\text{CH}_2\text{CH}_3
$$
with H above, 2, and 1 → C=O with H below

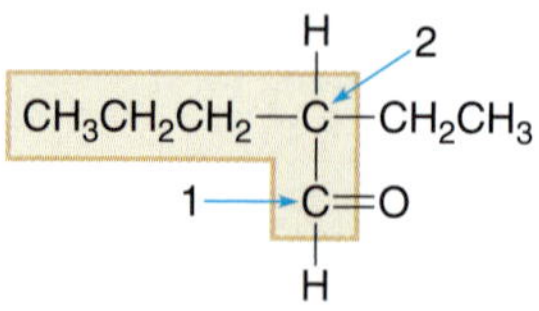

Answer: 2-ethylpentanal

PROBLEM 16.4

Give the IUPAC name for each aldehyde.

a. $(CH_3)_2CHCH_2CH_2CH_2CHO$

b. $(CH_3)_3CC(CH_3)_2CH_2CHO$

c. 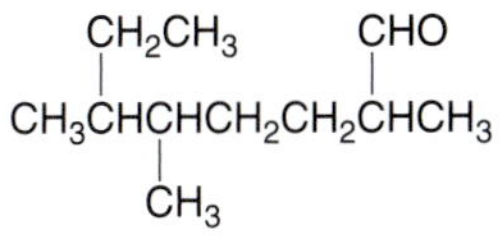 CH₃CHCHCH₂CH₂CHCH₃ with CH₂CH₃ and CHO and CH₃ substituents

$$\underset{\underset{CH_3}{|}}{CH_3CHCHCH_2CH_2CHCH_3}\quad\overset{CH_2CH_3\quad CHO}{}$$

PROBLEM 16.5

Give the structure corresponding to each IUPAC name.

a. 2-chloropropanal

b. 3,4,5-triethylheptanal

c. 3,6-diethylnonanal

d. o-ethylbenzaldehyde

16.2B NAMING KETONES

- In the IUPAC system, ketones are identified by the suffix -one.

To name an acyclic ketone using IUPAC rules:

1. Find the longest chain containing the carbonyl group, and change the -e ending of the parent alkane to the suffix -one.
2. Number the carbon chain to give the carbonyl carbon the lower number. Apply all of the other usual rules of nomenclature.

With cyclic ketones, numbering always begins at the carbonyl carbon, but the "1" is usually omitted from the name. The ring is then numbered clockwise or counterclockwise to give the first substituent the lower number.

Most common names for ketones are formed by **naming both alkyl groups** on the carbonyl carbon, **arranging them alphabetically,** and adding the word **ketone.** Using this method, the common name for 2-butanone becomes ethyl methyl ketone.

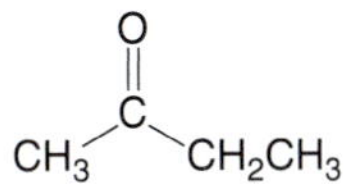

methyl group   ethyl group

CH_3 CH_2CH_3

IUPAC name: **2-butanone** 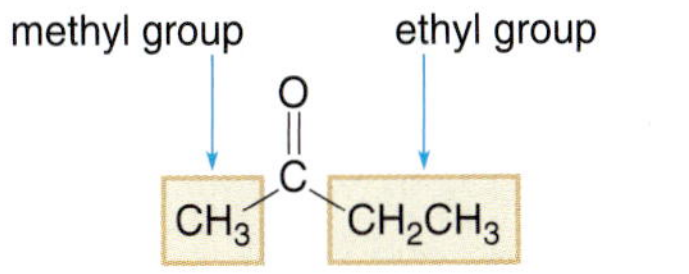   Common name: **ethyl methyl ketone**

Three widely used common names for some simple ketones do not follow this convention:

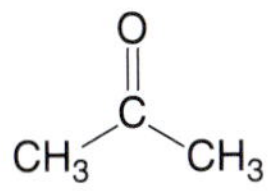

acetone 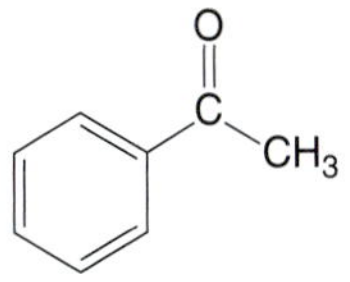   acetophenone 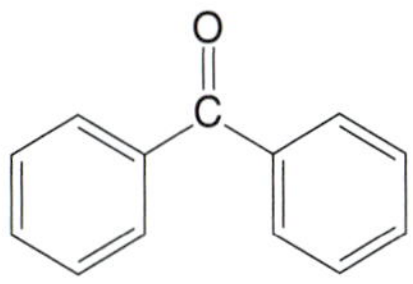   benzophenone

SAMPLE PROBLEM 16.2

Give IUPAC names for each ketone.

a. 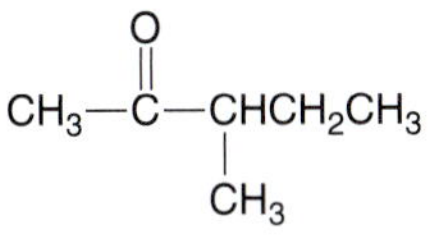 $CH_3-\overset{O}{\overset{||}{C}}-\underset{\underset{CH_3}{|}}{CH}CH_2CH_3$

b. 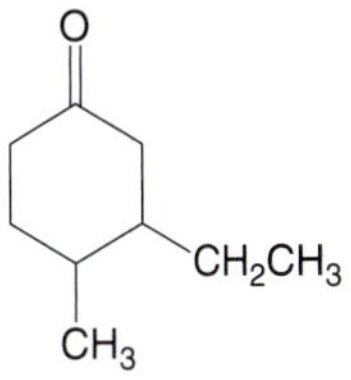 (cyclohexanone with CH₂CH₃ and CH₃ substituents)

ANALYSIS AND SOLUTION

a. **[1]** Find and name the longest chain containing the carbonyl group.

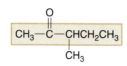

pentane ---→ pentan*one*
(5 C's)

[2] Number and name substituents, making sure the carbonyl carbon has the lowest possible number.

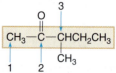

Answer: 3-methyl-2-pentanone

b. **[1]** Name the ring.

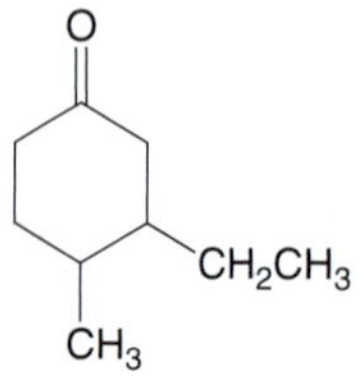

cyclohexane ---→ cyclohexan*one*
(6 C's)

[2] Number and name substituents, making sure the carbonyl carbon is at C1.

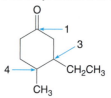

Answer: 3-ethyl-4-methylcyclohexanone

PROBLEM 16.6

Give the IUPAC name for each ketone.

a. 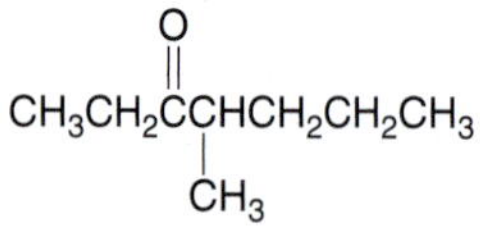 $CH_3CH_2\overset{\displaystyle O}{\overset{\|}{C}}CHCH_2CH_2CH_3$
   $\quad\quad\quad\quad\quad\underset{CH_3}{|}$

b. [structure of a cyclopentanone with a CH₃ substituent, =O]

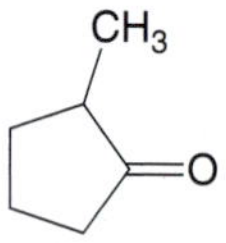

c. $(CH_3)_3C\overset{\displaystyle O}{\overset{\|}{C}}CH_2CH_2CH_2CH_3$

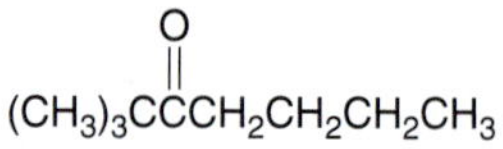

PROBLEM 16.7

Draw the structure of 2-octanone, a ketone partly responsible for the flavor of some mushrooms.

PROBLEM 16.8

Give the structure corresponding to each name.

a. butyl ethyl ketone

b. 2-methyl-3-pentanone

c. *p*-ethylacetophenone

d. 2-propylcyclobutanone

16.3 PHYSICAL PROPERTIES

Because aldehydes and ketones have a polar carbonyl group, they are **polar molecules** with stronger intermolecular forces than the hydrocarbons of Chapters 12 and 13. Since they have no O—H bond, two molecules of RCHO or RCOR are incapable of intermolecular hydrogen bonding, giving them weaker intermolecular forces than alcohols.

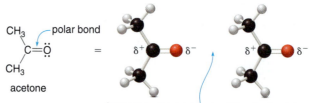

Oppositely charged ends of the dipoles interact.

As a result:

- Aldehydes and ketones have *higher* boiling points than hydrocarbons of comparable size.
- Aldehydes and ketones have *lower* boiling points than alcohols of comparable size.

$CH_3CH_2CH_2CH_2CH_3$

pentane

bp 36 °C

$CH_3CH_2CH_2CHO$

butanal

bp 76 °C

$CH_3CH_2CH_2CH_2OH$

1-butanol

bp 118 °C

$CH_3CH_2COCH_3$

2-butanone

bp 80 °C

Increasing strength of intermolecular forces
Increasing boiling point

Based on the general rule governing solubility (i.e., **"like dissolves like,"** Section 8.2), aldehydes and ketones are soluble in organic solvents. Moreover, because aldehydes and ketones contain an oxygen atom with an available lone pair, they can intermolecularly hydrogen bond to water.

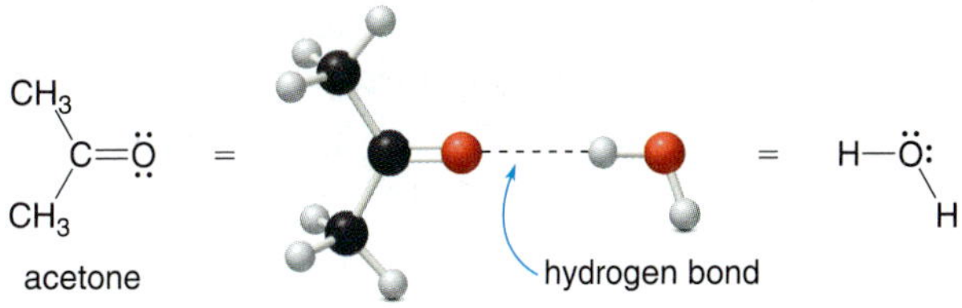

As a result:

- Low molecular weight aldehydes and ketones (those having less than six carbons) are soluble in both organic solvents and water.
- Higher molecular weight aldehydes and ketones (those having six carbons or more) are soluble in organic solvents, but insoluble in water.

PROBLEM 16.9

Which compound in each pair has the higher boiling point?

a. 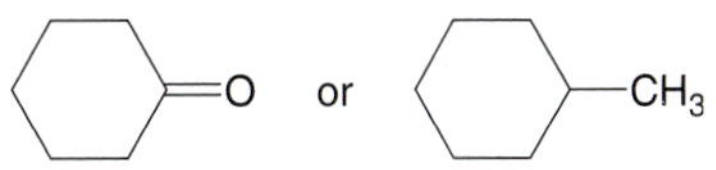 [cyclohexanone] or [methylcyclohexane]

b. $(CH_3CH_2)_2CO$ or $(CH_3CH_2)_2C=CH_2$

c. 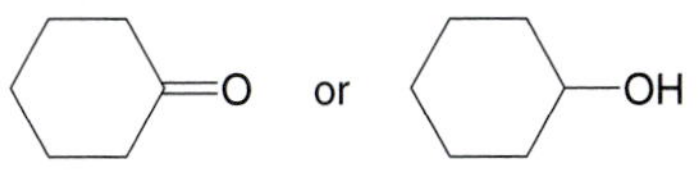 [cyclohexanone] or [cyclohexanol]

d. $CH_3(CH_2)_6CH_3$ or $CH_3(CH_2)_5CHO$

PROBLEM 16.10

Acetone and progesterone are two ketones that occur naturally in the human body. Discuss the solubility properties of both compounds in water and organic solvents.

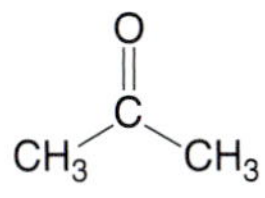

acetone

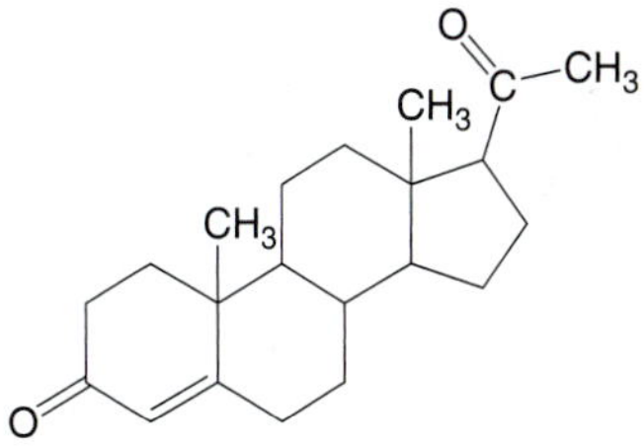

progesterone

16.4 FOCUS ON HEALTH & MEDICINE
INTERESTING ALDEHYDES AND KETONES

formaldehyde
CH_2=O

Formaldehyde (CH_2=O, the simplest aldehyde) is a starting material for the synthesis of many resins and plastics, and billions of pounds are produced annually in the United States. Formaldehyde is also sold as a 37% aqueous solution called **formalin,** a disinfectant and preservative for biological specimens. As a product of the incomplete combustion of coal and other fossil fuels, formaldehyde is partly responsible for the irritation caused by smoggy air.

acetone
$(CH_3)_2C$=O

Acetone [$(CH_3)_2C$=O, the simplest ketone] is an industrial solvent and a starting material in the synthesis of some organic polymers. Acetone is produced naturally in cells during the breakdown of fatty acids. In diabetes, a disease where normal metabolic processes are altered because of the inadequate secretion of insulin, individuals often have unusually high levels of acetone in the bloodstream. The characteristic odor of acetone can be detected on the breath of diabetic patients when their disease is poorly controlled.

Many aldehydes with characteristic odors occur in nature, as shown in Figure 16.1. It is thought that the shape of these molecules is an important factor in determining their odors, as discussed in Section 15.8.

Ketones play an important role in the tanning industry. Dihydroxyacetone is the active ingredient in commercial tanning agents that produce sunless tans. Dihydroxyacetone reacts with proteins in the skin, producing a complex colored pigment that gives the skin a brown hue. In addition, many commercial sunscreens are ketones that have the carbonyl carbon bonded to one or two benzene rings (see also Section 13.11). Examples include avobenzone, oxybenzone, and dioxybenzone.

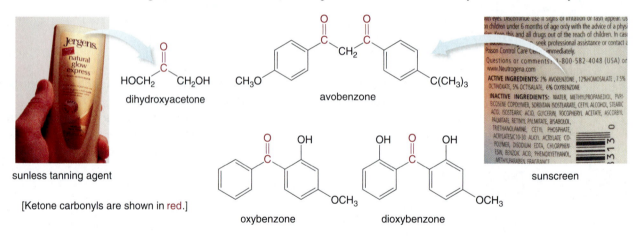

sunless tanning agent

[Ketone carbonyls are shown in red.]

dihydroxyacetone

avobenzone

oxybenzone   dioxybenzone

sunscreen

Some naturally occurring compounds do not contain a carbonyl group, but they are converted to aldehydes and ketones by enzymes in cells. One such compound is **amygdalin,** known more commonly as **laetrile.**

amygdalin
(laetrile)

$\xrightarrow{\text{enzymes}}$

glucose
(two molecules)

benzaldehyde

HCN
hydrogen
cyanide
toxic gas

Amygdalin is present in the seeds and pits of apricots, peaches, and wild cherries. In the body, amygdalin is converted to two aldehydes, glucose and benzaldehyde. Also formed as a by-product is hydrogen cyanide, **HCN,** a toxic gas. Amygdalin was once touted as an anticancer drug, and is

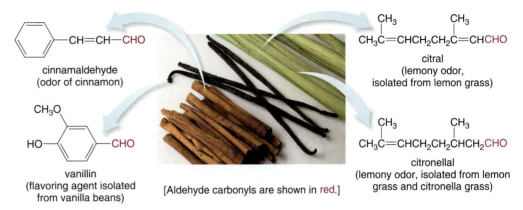

FIGURE 16.1 Some Naturally Occurring Aldehydes with Characteristic Odors

cinnamaldehyde
(odor of cinnamon)

vanillin
(flavoring agent isolated
from vanilla beans)

[Aldehyde carbonyls are shown in red.]

citral
(lemony odor,
isolated from lemon grass)

citronellal
(lemony odor, isolated from lemon
grass and citronella grass)

- **Cinnamaldehyde,** the major component of cinnamon bark, is a common flavoring agent.

- **Vanillin** is the primary component of the extract of the vanilla bean. Because natural sources cannot meet the high demand, most vanilla flavoring agents are made synthetically from starting materials derived from petroleum.

- **Citral** has the lemony odor characteristic of lemon grass. Citral is used in perfumery and as a starting material for synthesizing vitamin A.

- **Citronellal** gives the distinctive lemon odor to citronella candles, commonly used to repel mosquitoes.

still available in some countries for this purpose, although its effectiveness is unproven. It appears as if the toxic HCN produced from amygdalin indiscriminately kills cells without targeting cancer cells. Patients in some clinical trials involving amygdalin show signs of cyanide poisoning but not cancer remission.

PROBLEM 16.11	Acetone [$(CH_3)_2C=O$] is a useful solvent because it dissolves a variety of compounds well. For example, both hexane [$CH_3(CH_2)_4CH_3$] and H_2O are soluble in acetone. Explain why these solubility properties are observed.
PROBLEM 16.12	Which sunscreen—avobenzone, oxybenzone, or dioxybenzone—is probably most soluble in water, and therefore most readily washed off when an individual goes swimming? Explain your choice.

16.5 REACTIONS OF ALDEHYDES AND KETONES

16.5A GENERAL CONSIDERATIONS

Aldehydes and ketones undergo two general types of reactions.

1. Aldehydes can be oxidized to carboxylic acids.

Aldehydes are easily oxidized by oxygen in the air. When $CH_3CH_2CH_2CHO$ (butanal) stands at room temperature, it slowly develops the characteristic smell of its oxidation product, $CH_3CH_2CH_2COOH$ (butanoic acid), a compound that contributes to the distinctive odor of human sweat.

Oxidation

$$\underset{\text{aldehyde}}{\overset{\displaystyle O}{\underset{\displaystyle R}{\parallel}}\overset{}{C}H} \xrightarrow{[O]} \underset{\text{carboxylic acid}}{\overset{\displaystyle O}{\underset{\displaystyle R}{\parallel}}\overset{}{C}OH}$$

Since aldehydes contain a hydrogen atom bonded to the carbonyl carbon, they can be oxidized to carboxylic acids, as discussed in Section 16.5B.

2. Aldehydes and ketones undergo addition reactions.

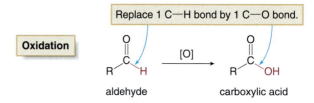

One bond is broken.   Two single bonds are formed.

Like alkenes, aldehydes and ketones contain a multiple bond (the carbonyl group) that is readily broken. As a result, aldehydes and ketones undergo addition reactions with a variety of reagents. In the addition reaction, new groups X and Y are added to the carbonyl group of the starting material. One bond of the double bond is broken and two new single bonds are formed. We examine the addition of H_2 in Section 16.6 and the addition of alcohols (ROH) in Section 16.8.

16.5B   OXIDATION OF ALDEHYDES

Since aldehydes contain a hydrogen atom bonded directly to the carbonyl carbon, they can be oxidized to carboxylic acids; that is, the aldehyde C—H bond can be converted to a C—OH bond. Since ketones have no hydrogen atom bonded to the carbonyl group, they are not oxidized under similar reaction conditions.

$K_2Cr_2O_7$, potassium dichromate, is used for oxidizing aldehydes, 1° alcohols, and 2° alcohols.

Replace 1 C—H bond by 1 C—O bond.

aldehyde   carboxylic acid

A common reagent for this oxidation is potassium dichromate, $K_2Cr_2O_7$, a red-orange solid that is converted to a green Cr^{3+} product during oxidation.

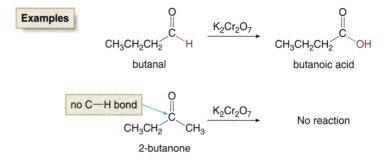

butanal   butanoic acid

2-butanone

As we learned in Section 14.5, $K_2Cr_2O_7$ oxidizes other functional groups (most notably 1° and 2° alcohols), as well. Aldehydes can be oxidized *selectively* in the presence of other functional groups using **silver(I) oxide (Ag_2O) in aqueous ammonium hydroxide (NH_4OH).** This is called **Tollens reagent.** Only aldehydes react with Tollens reagent; all other functional groups are inert. Oxidation with Tollens reagent provides a distinct color change because the Ag^+ reagent is converted to silver metal (Ag), which precipitates out of the reaction mixture as a silver mirror.

Aldehydes are said to give a *positive* Tollens test; that is, they react with Ag^+ to form RCOOH and Ag. When the reaction is carried out in a glass flask, a silver mirror is formed on its walls. Other functional groups give a *negative* Tollens test; that is, no silver mirror forms.

Only the aldehyde is oxidized.

silver mirror

SAMPLE PROBLEM 16.3

What product is formed when each carbonyl compound is treated with $K_2Cr_2O_7$?

a. 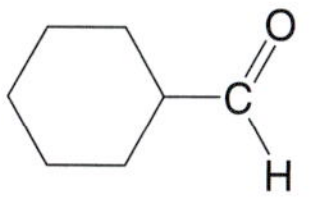

b. 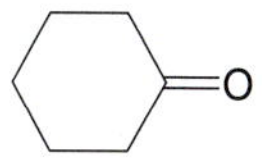

ANALYSIS Compounds that contain a C—H and C—O bond on the *same* carbon are oxidized with $K_2Cr_2O_7$. Thus:

- Aldehydes (RCHO) are oxidized to RCO_2H.
- Ketones (R_2CO) are not oxidized with $K_2Cr_2O_7$.

SOLUTION The aldehyde in part (a) is oxidized with $K_2Cr_2O_7$ to a carboxylic acid, but the ketone in part (b) is inert to oxidation.

a. 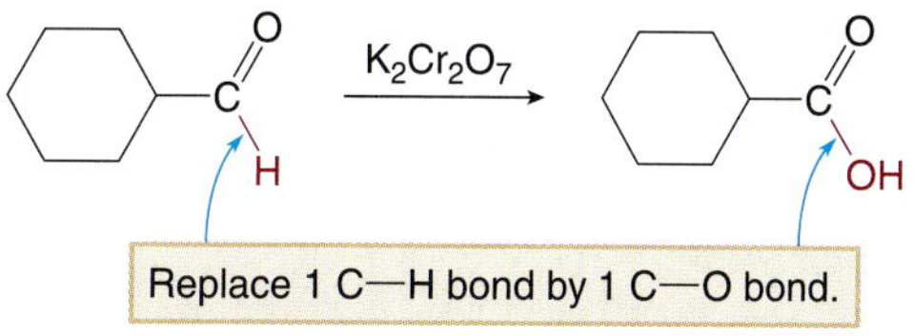
K2Cr2O7
Replace 1 C—H bond by 1 C—O bond.

b. 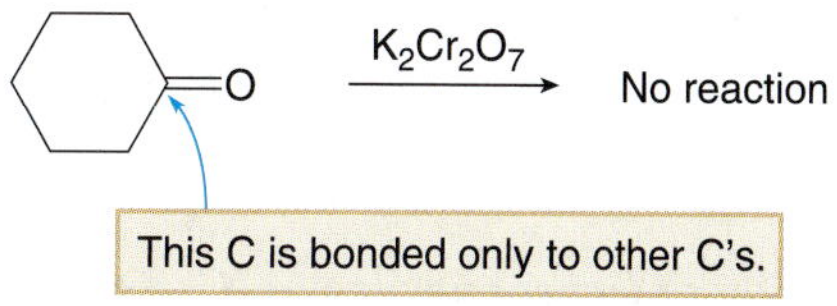
K2Cr2O7
No reaction
This C is bonded only to other C's.

PROBLEM 16.13

What product is formed when each carbonyl compound is treated with $K_2Cr_2O_7$? In some cases, no reaction occurs.

a. CH_3CH_2CHO

b. $(CH_3CH_2)_2C{=}O$

c. $CH_3C(CH_3){=}CHCH_2CH_2CH(CH_3)CH_2CHO$

SAMPLE PROBLEM 16.4

What product is formed when each compound is treated with Tollens reagent (Ag_2O, NH_4OH)?

a. 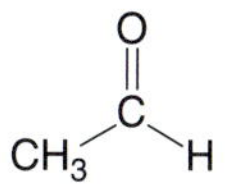

b. 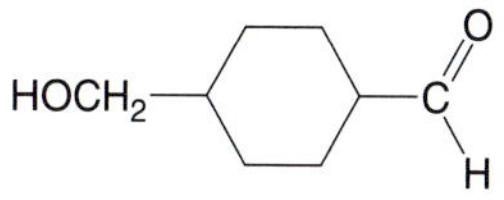

ANALYSIS **Only aldehydes (RCHO) react with Tollens reagent.** Ketones and alcohols are inert to oxidation.

SOLUTION The aldehyde in both compounds is oxidized to RCO_2H, but the 1° alcohol in part (b) does not react with Tollens reagent.

a. 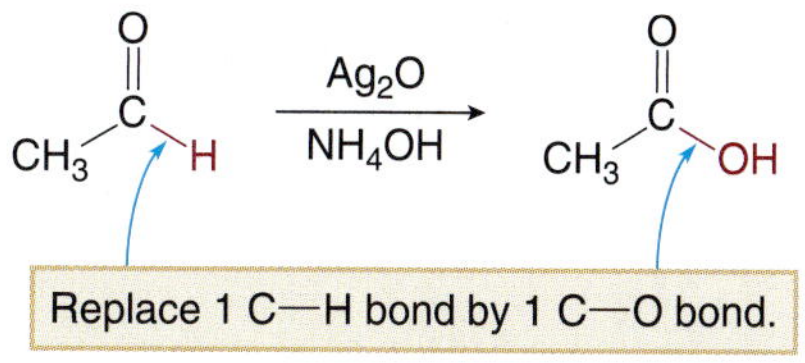
Ag2O
NH4OH
Replace 1 C—H bond by 1 C—O bond.

b. 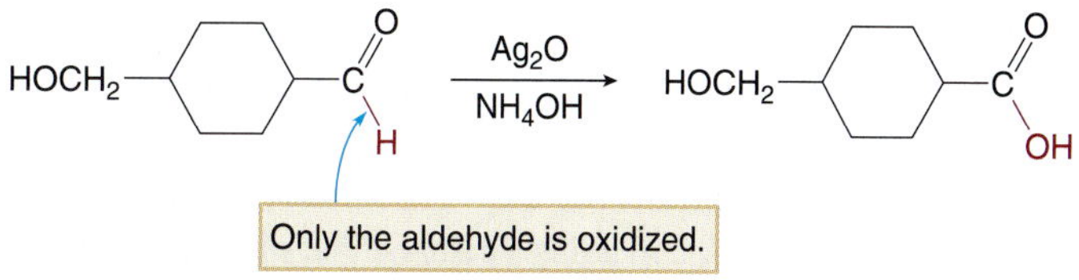
Ag2O
NH4OH
Only the aldehyde is oxidized.

What product is formed when each compound is treated with Tollens reagent (Ag_2O, NH_4OH)? In some cases, no reaction occurs.

a. $CH_3(CH_2)_6CHO$ b. 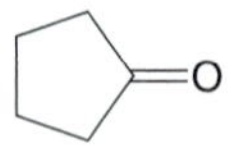 [cyclopentanone] =O   c. 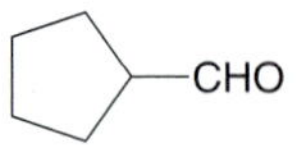 [cyclopentane]—CHO   d. 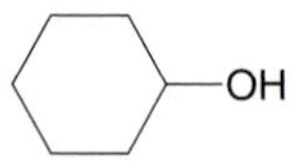 [cyclohexane]—OH

16.6 REDUCTION OF ALDEHYDES AND KETONES

Recall from Section 12.8 that to determine if an organic compound has been reduced, we compare the number of C—H and C—O bonds. **Reduction is the *opposite* of oxidation.**

- Reduction results in a *decrease* in the number of C—O bonds or an *increase* in the number of C—H bonds.

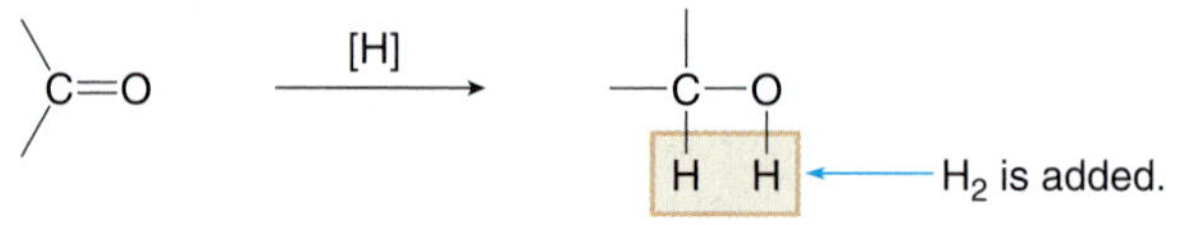

The conversion of a carbonyl group (C=O) to an alcohol (C—OH) is a reduction, since the starting material has more C—O bonds than the product (two versus one). Reduction of a carbonyl is also an **addition reaction,** since the elements of H_2 are added across the double bond, forming new C—H and O—H bonds. The symbol **[H]** is often used to represent a general reduction reaction.

16.6A SPECIFIC FEATURES OF CARBONYL REDUCTIONS

The identity of the carbonyl starting material determines the type of alcohol formed as product in a reduction reaction.

- **Aldehydes (RCHO) are reduced to 1° alcohols (RCH_2OH).**

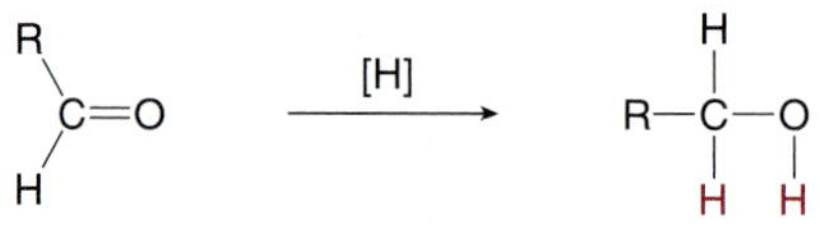

- **Ketones (RCOR) are reduced to 2° alcohols (R_2CHOH).**

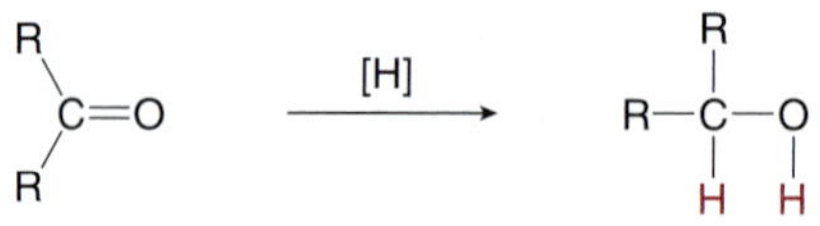

Many different reagents can be used to reduce an aldehyde or ketone to an alcohol. For example, the addition of H_2 to a carbonyl group (C=O) takes place with the same reagents used for the addition of H_2 to a C=C—namely, **H_2 gas in the presence of palladium (Pd) metal** (Section 13.6). The metal is a catalyst that provides a surface to bind both the carbonyl compound and H_2, and this speeds up the rate of reduction. The addition of hydrogen to a multiple bond is called **hydrogenation.**

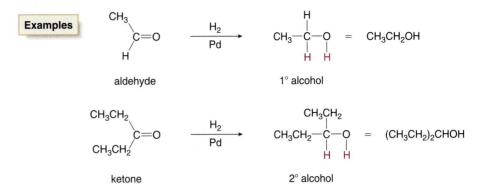

SAMPLE PROBLEM 16.5

What alcohol is formed when each aldehyde or ketone is treated with H_2 in the presence of a Pd catalyst?

a. [structure: cyclohexyl-CHO] b. [structure: cyclohexanone]

ANALYSIS

To draw the products of reduction:

- Locate the C=O and mentally break one bond in the double bond.
- Mentally break the H—H bond of the reagent.
- Add one H atom to each atom of the C=O, forming new C—H and O—H single bonds.

SOLUTION

The aldehyde (RCHO) in part (a) forms a 1° alcohol (RCH₂OH) and the ketone in part (b) forms a 2° alcohol (R₂CHOH).

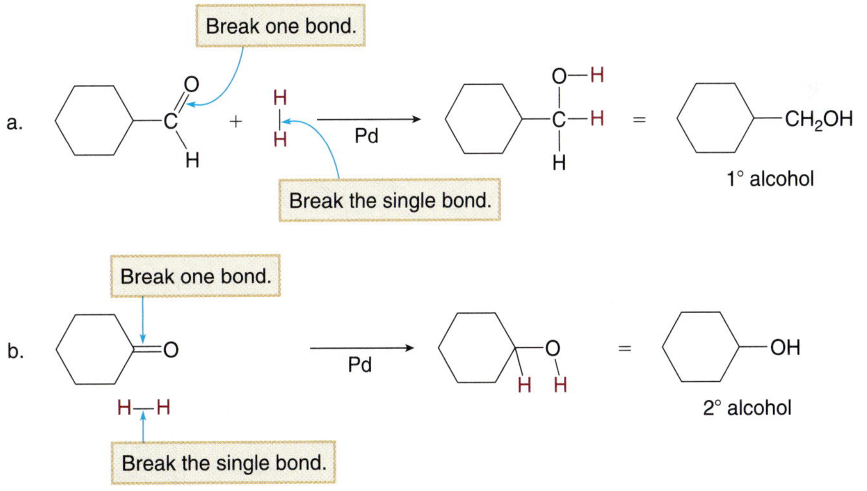

a. [reaction scheme: cyclohexyl-CHO + H—H, Pd → cyclohexyl-C(OH)(H)—H = cyclohexyl-CH₂OH]

Break one bond.
Break the single bond.
1° alcohol

b. [reaction scheme: cyclohexanone + H—H, Pd → cyclohexyl-O(H)(H) = cyclohexyl-OH]

Break one bond.
Break the single bond.
2° alcohol

PROBLEM 16.15

What alcohol is formed when each compound is treated with H_2 and a Pd catalyst?

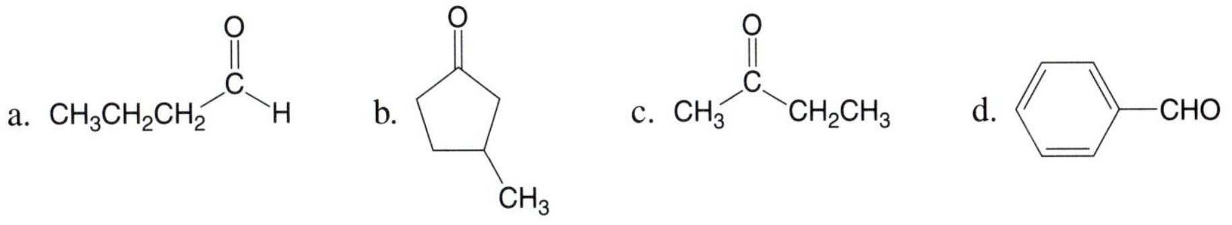

a. $CH_3CH_2CH_2$–CHO   b. [structure: 3-methylcyclopentanone]   c. CH_3–CO–CH_2CH_3 d. [structure: phenyl–CHO]

16.6B EXAMPLES OF CARBONYL REDUCTION IN ORGANIC SYNTHESIS

The reduction of aldehydes and ketones is a common reaction used in the laboratory synthesis of many useful compounds.

Chemists synthesize compounds for many different reasons. Sometimes a naturally occurring compound has useful properties but is produced by an organism in only minute amounts. Chemists then develop a method to prepare this molecule from simpler starting materials to make it more readily available and less expensive. For example, **muscone,** a strongly scented ketone isolated from musk, is an ingredient in many perfumes. Originally isolated from the male musk deer, muscone is now prepared synthetically in the lab. One step in the synthesis involves reducing a ketone to a 2° alcohol.

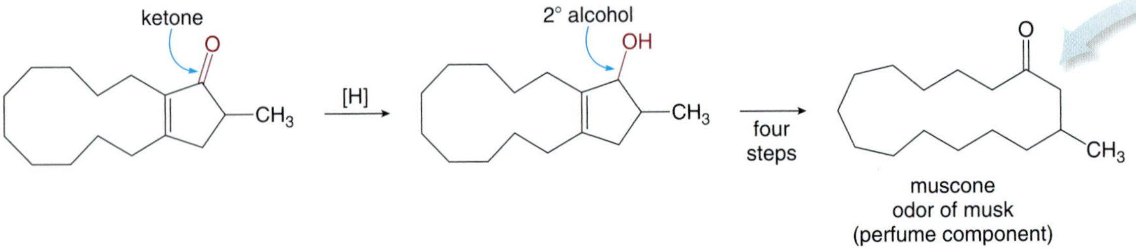

muscone
odor of musk
(perfume component)

male musk deer

Sometimes chemists prepare molecules that do not occur in nature because they have useful medicinal properties. For example, **fluoxetine** (trade name: Prozac) is a prescription antidepressant that does not occur in nature. One step in a laboratory synthesis of fluoxetine involves reduction of a ketone to a 2° alcohol. Fluoxetine is widely used because it has excellent medicinal properties, *and* because it is readily available by laboratory synthesis.

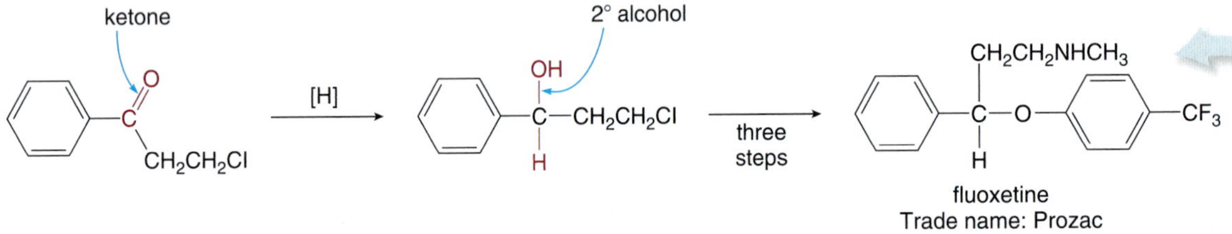

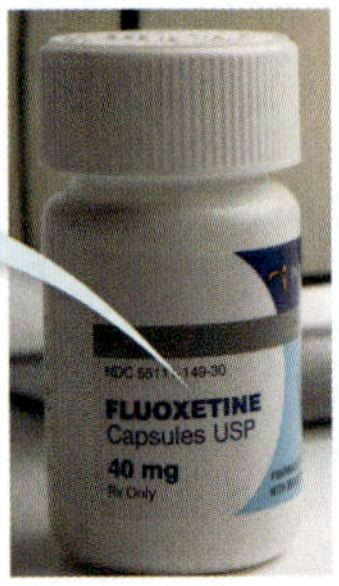

fluoxetine
Trade name: Prozac

PROBLEM 16.16

What carbonyl starting material is needed to prepare alcohol **A** by a reduction reaction. **A** can be converted to the anti-inflammatory agent ibuprofen in three steps.

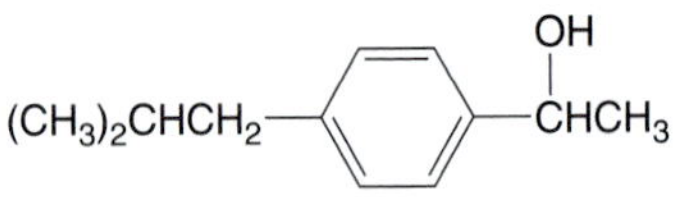

$(CH_3)_2CHCH_2$—⬡—$\overset{OH}{\underset{}{CHCH_3}}$ $(CH_3)_2CHCH_2$—⬡—$\overset{COOH}{\underset{}{CHCH_3}}$

A 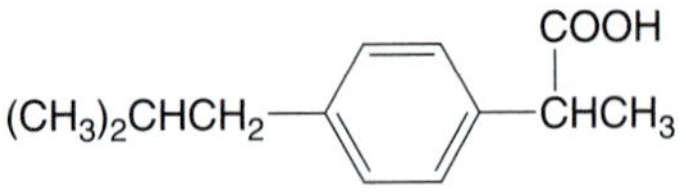   ibuprofen

16.6C FOCUS ON THE HUMAN BODY BIOLOGICAL REDUCTIONS

The reduction of carbonyl groups is common in biological systems. Biological systems do not use H_2 and Pd as a reducing agent. Instead, they use the coenzyme **NADH** (nicotinamide adenine dinucleotide, reduced form) in the presence of an enzyme. The enzyme binds both the carbonyl

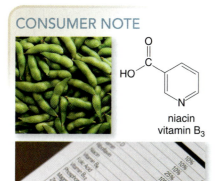
compound and NADH, holding them closely together, and this facilitates the addition of H$_2$ to the carbonyl group, forming an alcohol. The NADH itself is oxidized in the process, forming NAD$^+$. **NAD$^+$,** a biological oxidizing agent, is a coenzyme synthesized from the vitamin niacin, which can be obtained from soybeans, among other dietary sources.

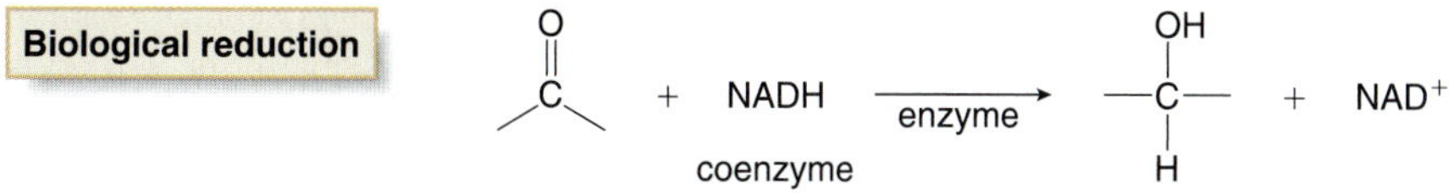

Biological reduction

$$\underset{\text{coenzyme}}{\overset{O}{\underset{}{\|}}{-}C{-}} + NADH \xrightarrow{\text{enzyme}} \underset{H}{\overset{OH}{\underset{}{|}}{-}C{-}} + NAD^+$$

For example, the reduction of pyruvic acid with NADH, catalyzed by the enzyme lactate dehydrogenase, yields lactic acid. Pyruvic acid is formed during the metabolism of the simple sugar glucose.

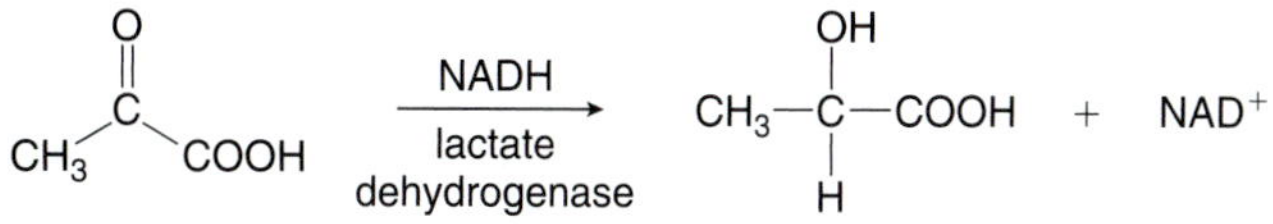

$$\underset{\text{pyruvic acid}}{CH_3{-}\overset{O}{\overset{\|}{C}}{-}COOH} \xrightarrow[\text{dehydrogenase}]{\overset{NADH}{\underset{\text{lactate}}{}}} \underset{\text{lactic acid}}{CH_3{-}\overset{OH}{\underset{H}{\overset{|}{C}}}{-}COOH} + NAD^+$$

16.7 FOCUS ON THE HUMAN BODY
THE CHEMISTRY OF VISION

The human eye consists of two types of light-sensitive cells—the rod cells, which are responsible for sight in dim light, and the cone cells, which are responsible for color vision and sight in bright light. Animals like pigeons, whose eyes have only cone cells, have color vision but see poorly in dim light, while owls, which have only rod cells, are color blind but see well in dim light.

The chemistry of vision in the rod cells centers around the aldehyde, **11-*cis*-retinal,** the chapter-opening molecule.

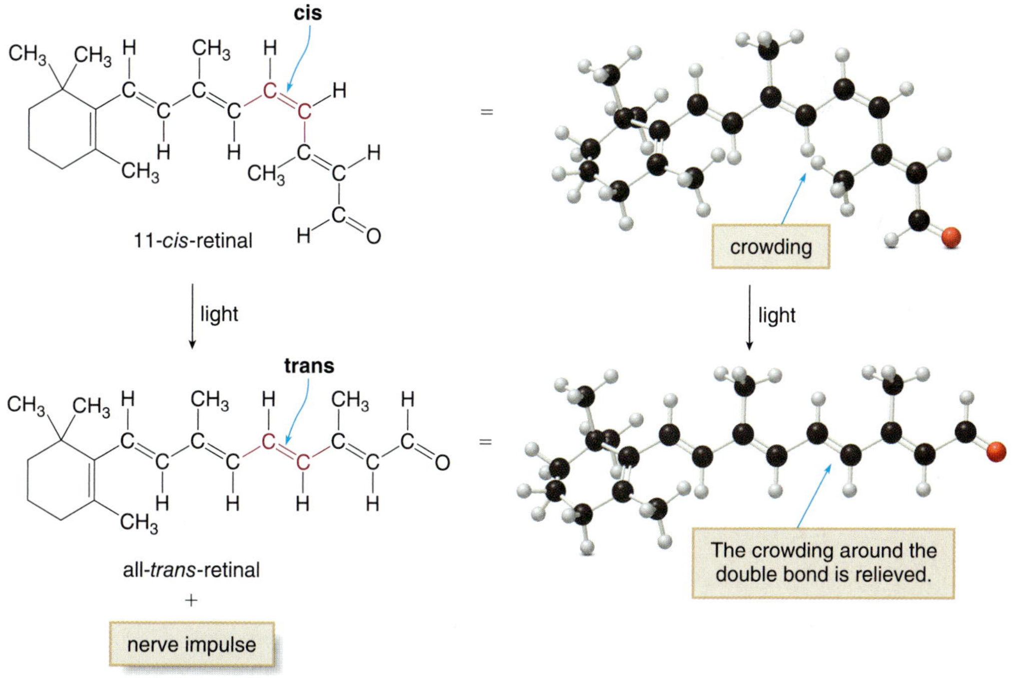

11-*cis*-retinal

↓ light

all-*trans*-retinal

+

nerve impulse

Although 11-*cis*-retinal is a stable molecule, the cis geometry around one of the double bonds causes crowding; a hydrogen atom on one double bond is close to the methyl group on an adjacent double bond. In the human retina (Figure 16.2), 11-*cis*-retinal is bonded to the protein

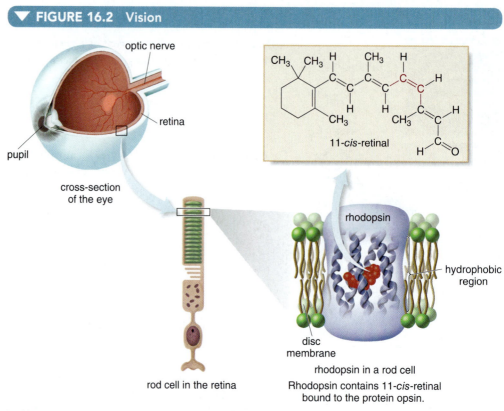

▼ **FIGURE 16.2** Vision

optic nerve

retina

pupil

cross-section
of the eye

11-*cis*-retinal

rhodopsin

hydrophobic
region

disc
membrane

rod cell in the retina

rhodopsin in a rod cell
Rhodopsin contains 11-*cis*-retinal
bound to the protein opsin.

In the rod cells of the eye, the 11-*cis*-retinal bonded to opsin absorbs light, and the crowded
11-cis double bond is isomerized to the trans isomer. This process generates a nerve impulse
that is converted to a visual image by the brain.

opsin, forming rhodopsin or visual purple. When light hits the retina, the 11-cis double bond is
isomerized to its more stable trans isomer, and all-*trans*-retinal is formed. This process sends a
nerve impulse to the brain, which is then converted into a visual image.

In order for the process to continue, the all-*trans*-retinal must be converted back to 11-*cis*-retinal.
This occurs by a series of reactions that involve biological oxidation (Section 14.5) and reduction
(Section 16.6). As shown in Figure 16.3, NADH is the coenzyme that reduces the aldehyde in all-
trans-retinal to all-*trans*-retinol, **vitamin A** (Reaction [1]). NAD$^+$ is the coenzyme that oxidizes
11-*cis*-retinol back to 11-*cis*-retinal (Reaction [3]), so the chemical cycle of vision can continue.

This scheme explains the role of vitamin A (Section 11.7) in vision. Vitamin A can be obtained
either directly in the diet or from the orange pigment β-carotene in carrots. Vitamin A, also called
all-*trans*-retinol, is converted in two steps to 11-*cis*-retinal, the aldehyde essential for vision in
the rod cells. Since the rod cells are responsible for vision in dim light, a deficiency of vitamin A
causes night blindness.

PROBLEM 16.17

How are the compounds in each pair related? Choose from stereoisomers, constitutional
isomers, or not isomers of each other.

a. all-*trans*-retinal and 11-*cis*-retinal

b. all-*trans*-retinal and vitamin A

c. vitamin A and 11-*cis*-retinol

▼ FIGURE 16.3 Vitamin A and the Chemistry of Vision

▼ FIGURE 16.3 Vitamin A and the Chemistry of Vision

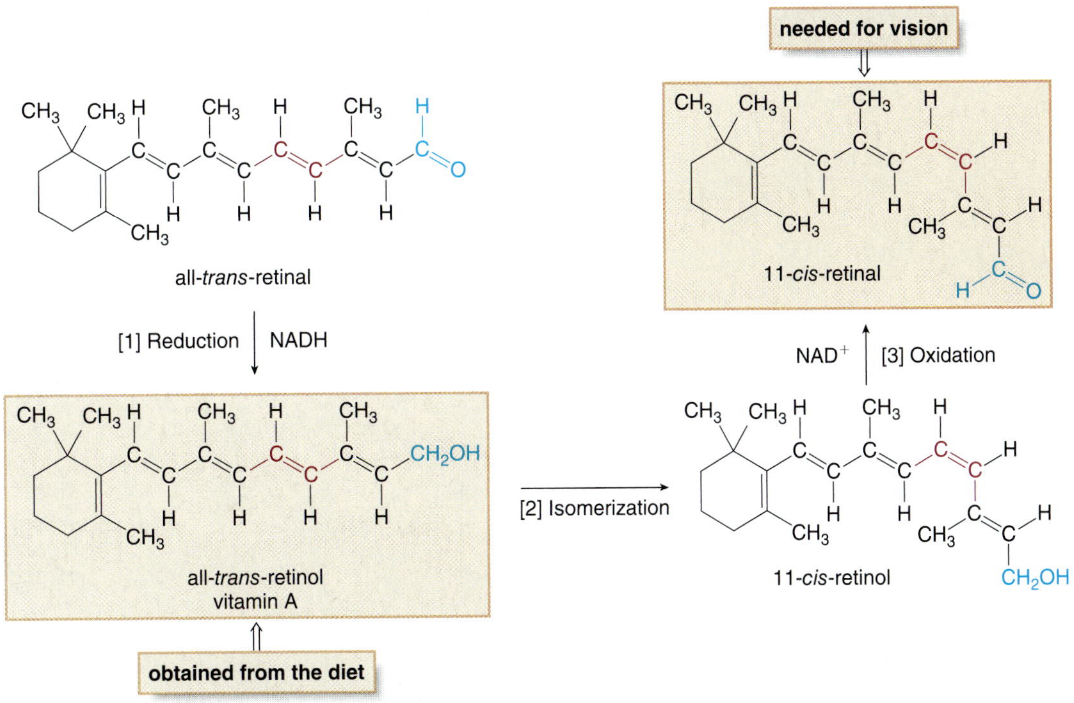

- Reaction [1] (reduction): The aldehyde in all-*trans*-retinal (labeled in blue) is reduced to a 1° alcohol using the coenzyme NADH. The product is all-*trans*-retinol.

- Reaction [2] (isomerization): The trans double bond (labeled in red) is isomerized to a cis double bond, forming 11-*cis*-retinol.

- Reaction [3] (oxidation): The 1° alcohol in 11-*cis*-retinol (labeled in blue) is oxidized with the coenzyme NAD$^+$ to an aldehyde. This re-forms 11-*cis*-retinal and the chemical cycle of vision continues.

16.8 ACETAL FORMATION

Aldehydes and ketones undergo addition reactions with alcohols (ROH) to form hemiacetals and acetals. Acetal formation is done in the presence of sulfuric acid (H_2SO_4).

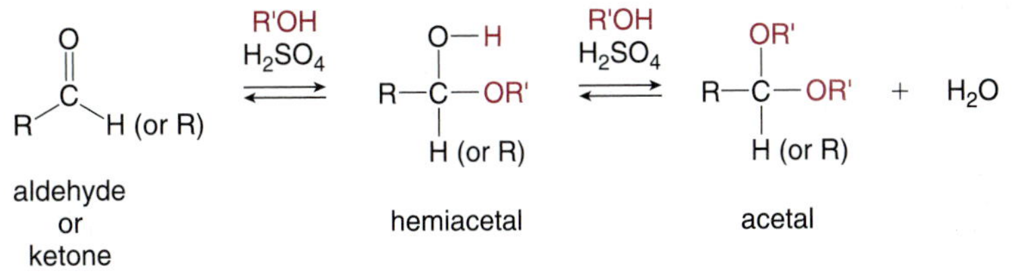

16.8A ACETALS AND HEMIACETALS

Addition of one molecule of alcohol (ROH) to an aldehyde or ketone forms a **hemiacetal.** Like other addition reactions, one bond of the C=O is broken and two new single bonds are formed. Acyclic hemiacetals are unstable. They react with a second molecule of alcohol to form **acetals.**

- A *hemiacetal* contains an OH (hydroxyl) group and an OR (alkoxy) group bonded to the *same* carbon.
- *Acetals* are compounds that contain two OR groups (alkoxy groups) bonded to the *same* carbon.

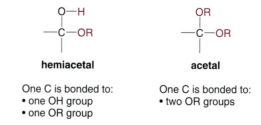

hemiacetal	**acetal**
One C is bonded to: • one OH group • one OR group	One C is bonded to: • two OR groups

Two examples of acetal formation using ethanol (CH_3CH_2OH) as the alcohol component are given.

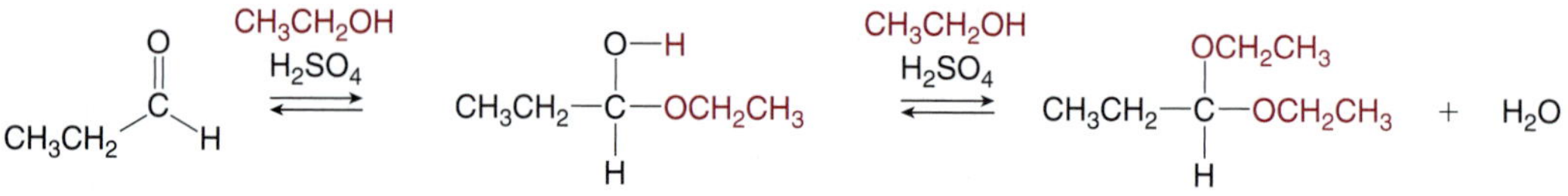

$$CH_3CH_2\overset{O}{\underset{}{\overset{\|}{C}}}H \xrightleftharpoons[H_2SO_4]{CH_3CH_2OH} CH_3CH_2-\overset{O-H}{\underset{H}{\overset{|}{C}}}-OCH_2CH_3 \xrightleftharpoons[H_2SO_4]{CH_3CH_2OH} CH_3CH_2-\overset{OCH_2CH_3}{\underset{H}{\overset{|}{C}}}-OCH_2CH_3 + H_2O$$

hemiacetal   acetal

$$CH_3\overset{O}{\underset{}{\overset{\|}{C}}}CH_3 \xrightleftharpoons[H_2SO_4]{CH_3CH_2OH} CH_3-\overset{O-H}{\underset{CH_3}{\overset{|}{C}}}-OCH_2CH_3 \xrightleftharpoons[H_2SO_4]{CH_3CH_2OH} CH_3-\overset{OCH_2CH_3}{\underset{CH_3}{\overset{|}{C}}}-OCH_2CH_3 + H_2O$$

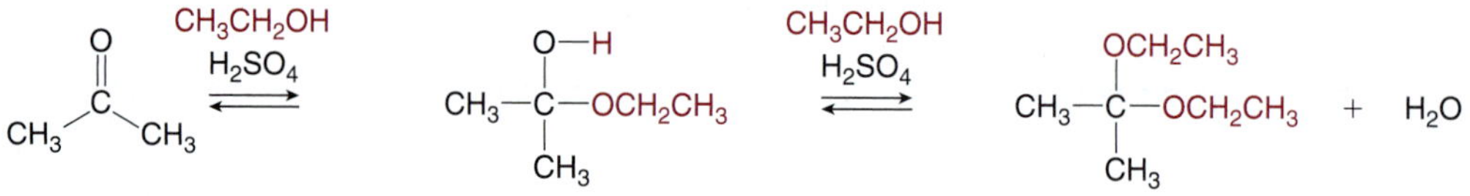

hemiacetal   acetal

SAMPLE PROBLEM 16.6

Draw the hemiacetal and acetal formed when cyclohexanone is treated with methanol (CH_3OH) in the presence of H_2SO_4.

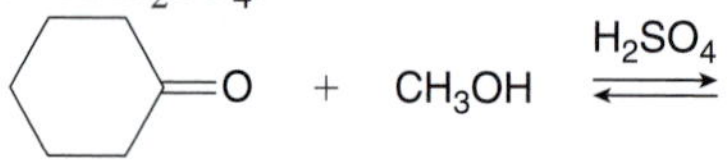

cyclohexanone   methanol (2 equivalents)

ANALYSIS

To form a hemiacetal and acetal from a carbonyl compound:

- Locate the C=O in the starting material.
- Break one C—O bond and add one equivalent of CH_3OH across the double bond, placing the OCH_3 group on the carbonyl carbon. This forms the hemiacetal.
- Replace the OH group of the hemiacetal by OCH_3 to form the acetal.

SOLUTION

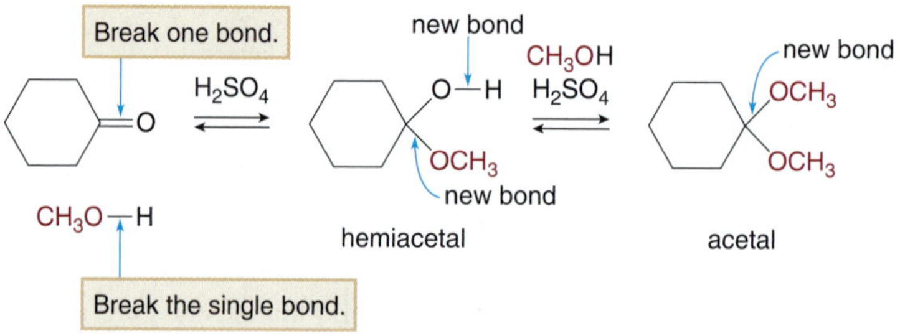

Break one bond.

new bond

CH_3OH

H_2SO_4 H_2SO_4

new bond

O—H

OCH₃ new bond

CH_3O—H

Break the single bond.

hemiacetal   acetal

PROBLEM 16.18

Draw the hemiacetal and acetal formed when each carbonyl compound is treated with two equivalents of the given alcohol in the presence of H_2SO_4.

a. 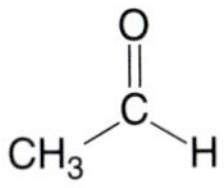 [structure: CH_3—C(=O)—H] + CH_3OH

b. $(CH_3CH_2)_2C{=}O$ + CH_3OH

c. 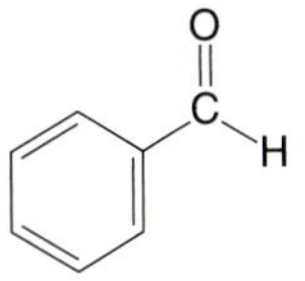 [structure: benzaldehyde, benzene ring—C(=O)—H] + CH_3CH_2OH

Acetals are *not* ethers, even though both compounds contain a C—OR bond.

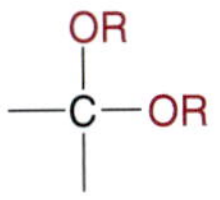

[structure: acetal, C bonded to two OR groups]

acetal

2 OR groups on one carbon

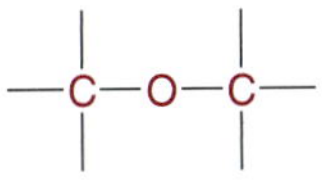

[structure: ether, —C—O—C—]

ether

2 C's on a single O atom

- An acetal contains one carbon atom bonded to *two* OR groups.
- An ether contains only *one* oxygen atom bonded to two carbons.

SAMPLE PROBLEM 16.7

Identify each compound as an ether, hemiacetal, or acetal.

a. CH_3CH_2—O—CH_2CH_3

b. 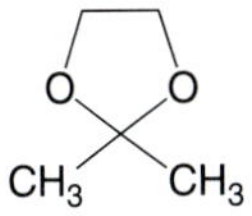 [structure: 5-membered ring with two O's, C bonded to CH_3 and CH_3]

ANALYSIS

Recall the definitions to identify the functional groups:

- An ether has the general structure ROR.
- A hemiacetal has one C bonded to OH and OR.
- An acetal has one C bonded to two OR groups.

SOLUTION

a. 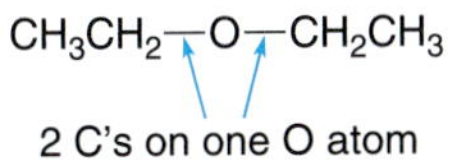 CH_3CH_2—O—CH_2CH_3

2 C's on one O atom

ether

b. 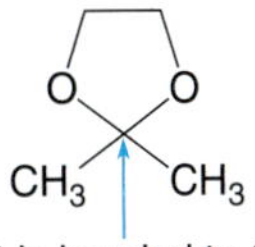 [structure: 5-membered ring with two O's, C bonded to CH_3 and CH_3]

This C is bonded to 2 O's.

acetal

Note that the acetal in part (b) is part of a ring. It contains one ring carbon bonded to two oxygens, making it an acetal.

PROBLEM 16.19

Identify each compound as an ether, hemiacetal, or acetal.

a. 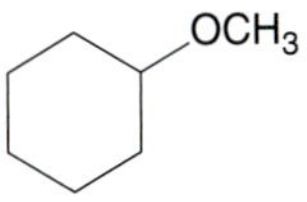 [structure: cyclohexane ring with OCH$_3$] — OCH_3

b. 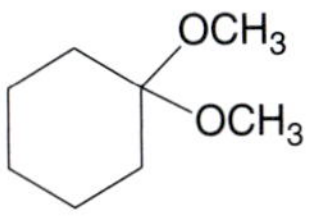 [structure: cyclohexane ring with OCH$_3$ and OCH$_3$]

c. 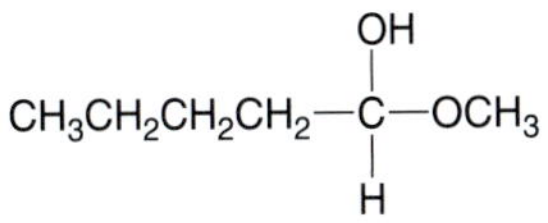 $CH_3CH_2CH_2CH_2$—C(OH)(H)—OCH_3

d. 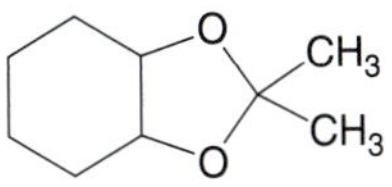 [structure: bicyclic ring with two O's, C bonded to CH_3 and CH_3]

16.8B CYCLIC HEMIACETALS

Although acyclic hemiacetals are generally unstable, **cyclic hemiacetals containing five- and six-membered rings are stable compounds** that are readily isolated.

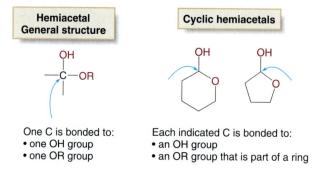

Hemiacetal General structure	Cyclic hemiacetals

One C is bonded to:
• one OH group
• one OR group

Each indicated C is bonded to:
• an OH group
• an OR group that is part of a ring

These cyclic hemiacetals are formed by an *intramolecular* reaction of a compound that contains *both* a hydroxyl group and an aldehyde or ketone.

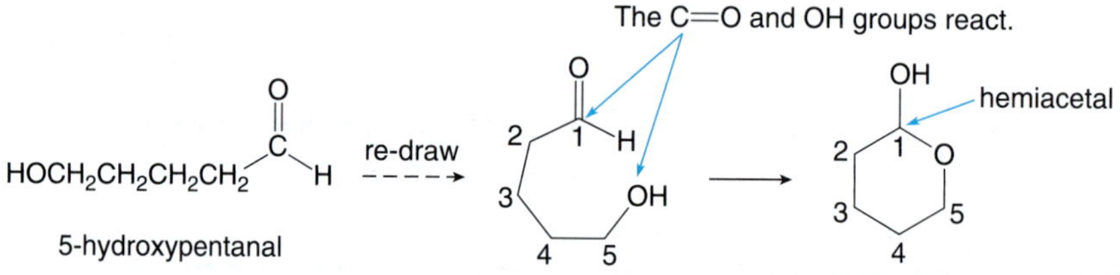

Cyclic hemiacetals are especially important in carbohydrate chemistry. **The most common simple carbohydrate, glucose, exists predominantly as a cyclic hemiacetal.** Glucose has five OH groups, but only one OH group is bonded to a carbon that is also bonded to another oxygen atom, making it a hemiacetal. All of the other OH groups in glucose are "regular" alcohols. We will learn much more about these stable cyclic hemiacetals in Chapter 20.

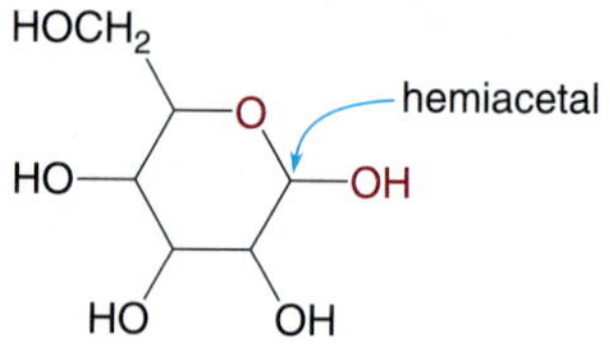

glucose
(cyclic hemiacetal form)

Cyclic hemiacetals are converted to acetals by reaction with another alcohol (ROH). To draw the structure of the resulting acetal, replace the OH group of the hemiacetal by the OR group of the alcohol.

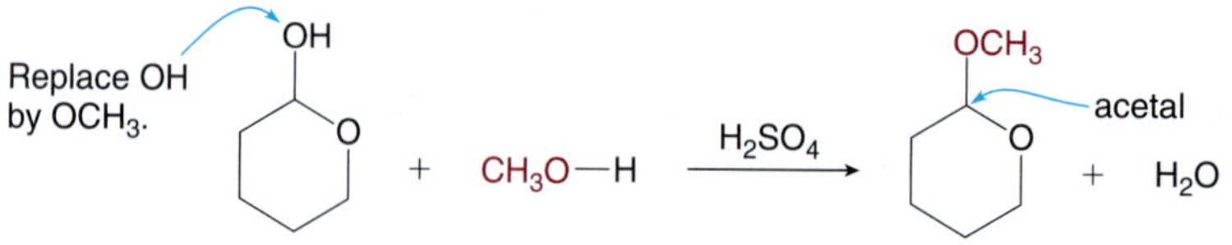

Many biologically active molecules contain acetals and cyclic hemiacetals. For example, lactose, the principal carbohydrate in milk, contains one acetal and one hemiacetal. Many individuals, mainly of Asian and African descent, lack adequate amounts of lactase, the enzyme needed to digest and absorb lactose. This condition, lactose intolerance, is associated with abdominal cramping and recurrent diarrhea.

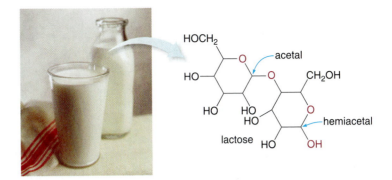

lactose

SAMPLE PROBLEM 16.8

Identify the three acetals in digoxin, a naturally occurring drug prescribed for patients with congestive heart failure and other heart ailments.

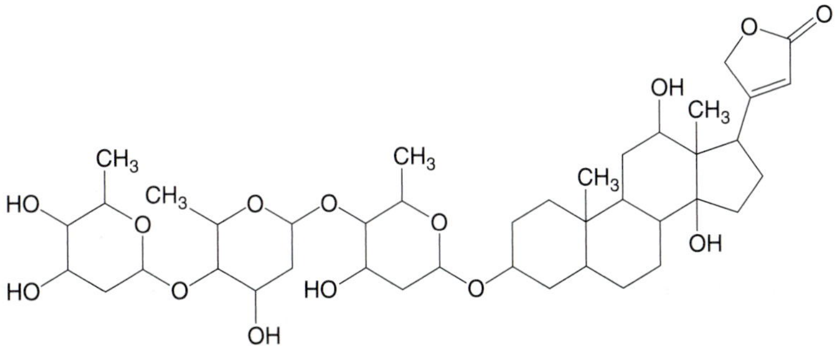

HEALTH NOTE

The leaves of the woolly foxglove plant are the source of the potent cardiac drug digoxin.

digoxin

ANALYSIS Although the structure of digoxin is complicated, look at each O atom individually and recall that an **acetal has one C bonded to two OR groups.**

SOLUTION The acetal carbons in digoxin are indicated with arrows and the oxygen atoms of the acetals are drawn in red.

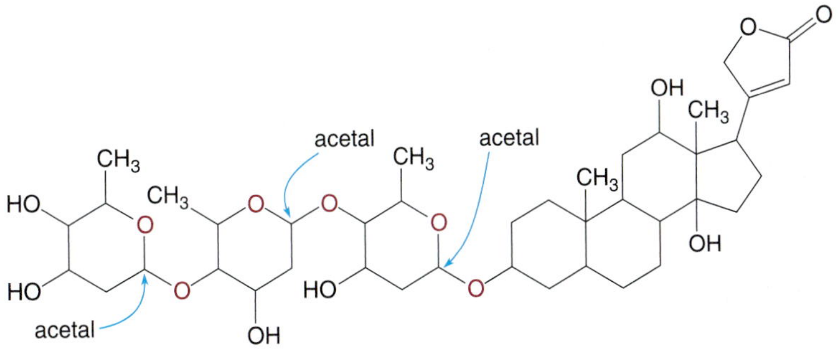

digoxin

PROBLEM 16.20

Label the hemiacetal or acetal in each compound. Glucosamine [part (a)] is available in many over-the-counter remedies for the treatment of osteoarthritis, although recent studies cast doubt on its efficacy. Salicin [part (b)] is a naturally occurring pain reliever isolated from the bark of the willow tree.

a. 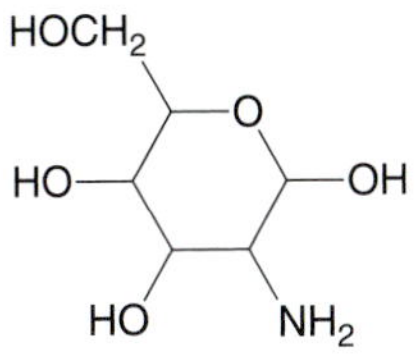 glucosamine

b. 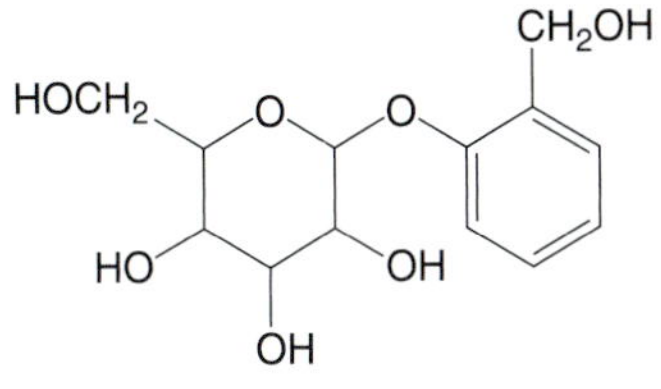 salicin

PROBLEM 16.21

What acetal is formed in each reaction?

a. 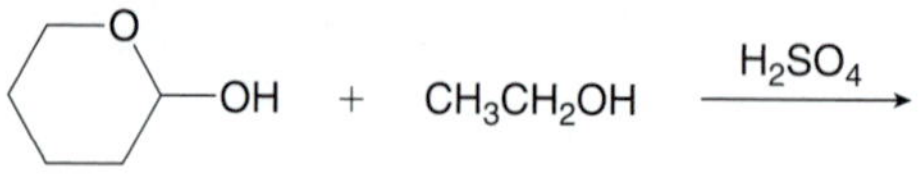 [structure: tetrahydropyran ring with —OH] + CH_3CH_2OH $\xrightarrow{H_2SO_4}$

b.  [structure: tetrahydrofuran ring with —OH] + [cyclohexane with —OH] $\xrightarrow{H_2SO_4}$

16.8C ACETAL HYDROLYSIS

Acetals are stable molecules, but they can be converted back to aldehydes and ketones by treatment with acid and water. This reaction is a *hydrolysis*, **since bonds are cleaved by reaction with water.**

Acetal hydrolysis

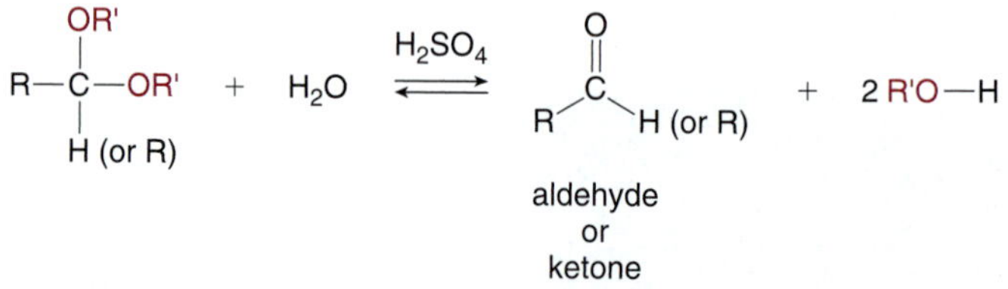

$$R-\underset{\underset{H \text{ (or R)}}{|}}{\overset{\overset{OR'}{|}}{C}}-OR' \quad + \quad H_2O \quad \underset{}{\overset{H_2SO_4}{\rightleftharpoons}} \quad \underset{R}{\overset{O}{\underset{}{C}}}\diagdown_{H \text{ (or R)}} \quad + \quad 2\ R'O-H$$

aldehyde
or
ketone

SAMPLE PROBLEM 16.9

What hydrolysis products are formed when the acetal $(CH_3CH_2)_2C(OCH_3)_2$ is treated with H_2SO_4 in H_2O?

ANALYSIS

To draw the products of hydrolysis:

- Locate the two C—OR bonds on the same carbon.
- Replace the two C—O single bonds with a carbonyl group (C=O).
- Each OR group then becomes a molecule of alcohol (ROH) product.

SOLUTION

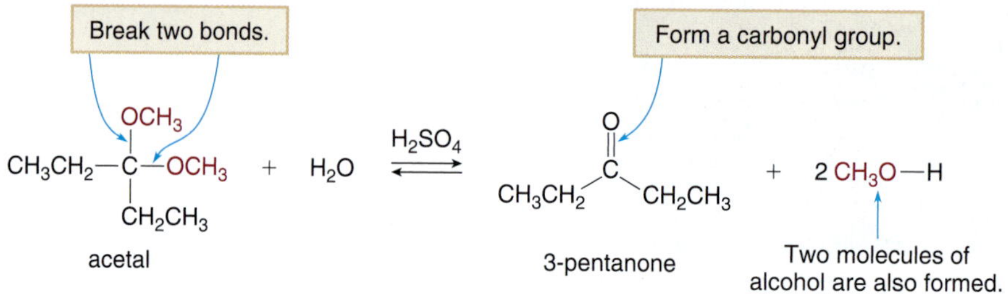

Break two bonds.		Form a carbonyl group.

$$CH_3CH_2-\underset{\underset{CH_2CH_3}{|}}{\overset{\overset{OCH_3}{|}}{C}}-OCH_3 \quad + \quad H_2O \quad \overset{H_2SO_4}{\rightleftharpoons} \quad \underset{CH_3CH_2}{\overset{O}{\underset{}{C}}}\diagdown_{CH_2CH_3} \quad + \quad 2\ CH_3O-H$$

acetal

3-pentanone

Two molecules of alcohol are also formed.

PROBLEM 16.22

What products are formed by hydrolysis of each acetal?

a. 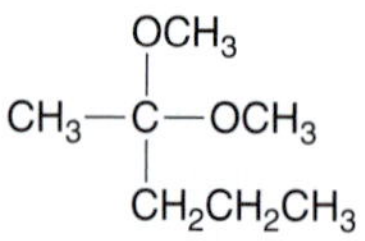 $CH_3-\underset{\underset{CH_2CH_2CH_3}{|}}{\overset{\overset{OCH_3}{|}}{C}}-OCH_3$

b. [cyclohexane ring with two substituents at one carbon: CH_3CH_2O and OCH_2CH_3]

c. [cyclohexane ring bonded to $\underset{\underset{OCH_3}{|}}{\overset{\overset{OCH_3}{|}}{C}}-H$]

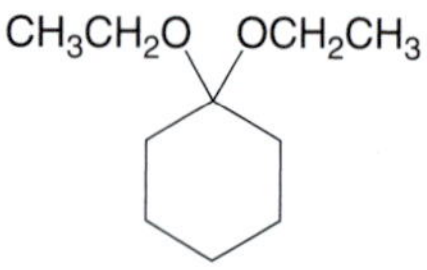

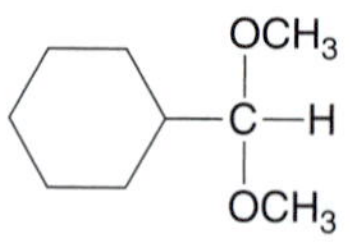

CHAPTER HIGHLIGHTS

KEY TERMS

Acetal (16.8)
Aldehyde (16.1)
Carbonyl group (16.1)
Cyclic hemiacetal (16.8)
Hemiacetal (16.8)

Hydrogenation (16.6)
Hydrolysis (16.8)
Ketone (16.1)
NAD$^+$ (16.6)

NADH (16.6)
Oxidation (16.5)
Reduction (16.6)
Tollens reagent (16.5)

KEY REACTIONS

[1]   Oxidation of aldehydes (16.5)

$$RCHO \xrightarrow[\substack{or \\ Tollens\ reagent}]{K_2Cr_2O_7} RCOOH \quad \text{carboxylic acid}$$

[2]   Reduction (16.6)

a.   Aldehydes $RCHO \xrightarrow[Pd]{H_2} RCH_2OH$ (1° alcohol)

b.   Ketones $R_2CO \xrightarrow[Pd]{H_2} R_2CHOH$ (2° alcohol)

c.   Biological reduction

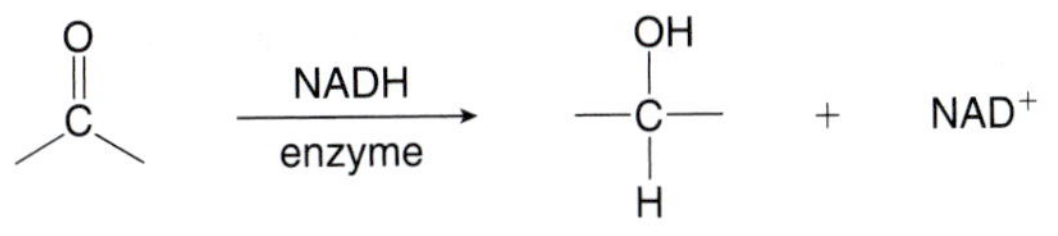

[3]   Acetals and hemiacetals (16.8)

a.   General reaction

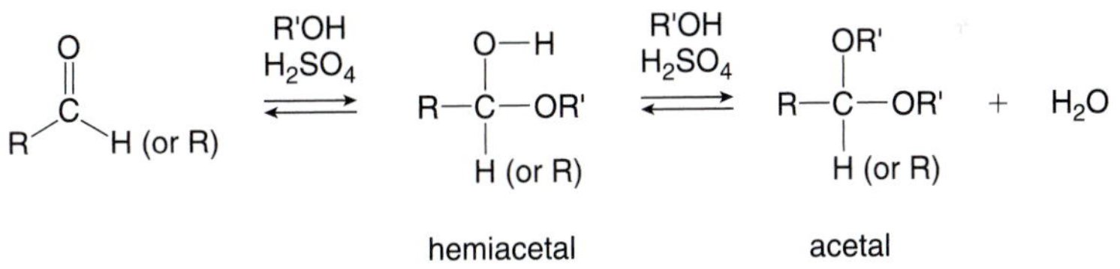

hemiacetal   acetal

b.   Cyclic hemiacetal

HOCH$_2$CH$_2$CH$_2$CH$_2$ ... 5-hydroxypentanal — re-draw — cyclic hemiacetal

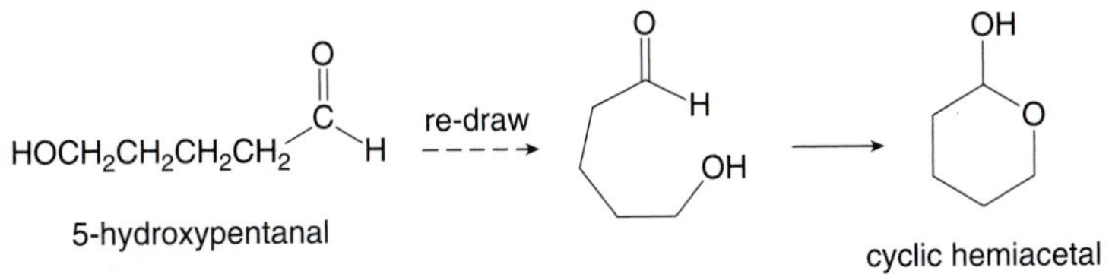

c.   Cyclic acetals

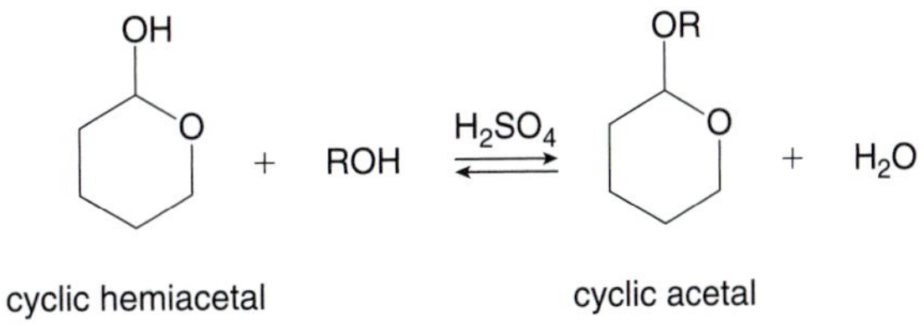

cyclic hemiacetal + ROH $\xrightleftharpoons{H_2SO_4}$ cyclic acetal + H$_2$O

d.   Acetal hydrolysis

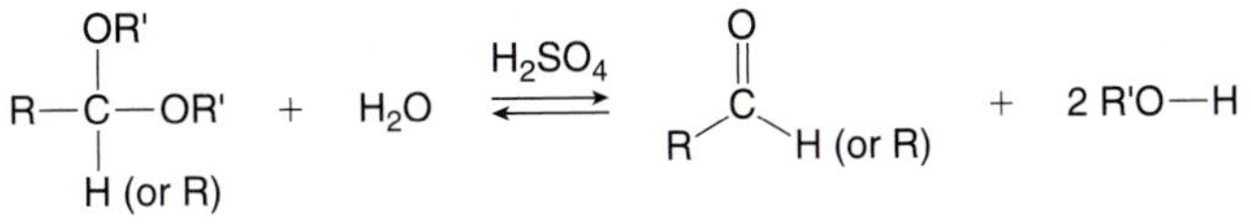

KEY CONCEPTS

❶ What are the characteristics of aldehydes and ketones?

- An aldehyde has the general structure RCHO, and contains a carbonyl group (C=O) bonded to at least one hydrogen atom. (16.1)
- A ketone has the general structure RCOR', and contains a carbonyl group (C=O) bonded to two carbon atoms. (16.1)
- The carbonyl carbon is trigonal planar, with bond angles of approximately 120°. (16.1)
- The carbonyl group is polar, giving an aldehyde or ketone stronger intermolecular forces than hydrocarbons. (16.3)
- Aldehydes and ketones have lower boiling points than alcohols, but higher boiling points than hydrocarbons of comparable size. (16.3)
- Aldehydes and ketones are soluble in organic solvents, but those having less than six C's are water soluble, too. (16.3)

❷ How are aldehydes and ketones named? (16.2)

- Aldehydes are identified by the suffix *-al,* and the carbon chain is numbered to put the carbonyl group at C1.
- Ketones are identified by the suffix *-one,* and the carbon chain is numbered to give the carbonyl group the lower number.

❸ Give examples of useful aldehydes and ketones. (16.4)

- Formaldehyde (CH_2=O) is an irritant in smoggy air, a preservative, and a disinfectant, and it is a starting material for synthesizing polymers.
- Acetone [$(CH_3)_2$C=O] is an industrial solvent and polymer starting material.
- Several aldehydes—cinnamaldehyde, vanillin, citral, and citronellal—have characteristic odors and occur naturally in fruits and plants.
- Dihydroxyacetone [$(HOCH_2)_2$C=O] is the active ingredient in artificial tanning agents, and many other ketones are useful sunscreens.
- Amygdalin, a naturally occurring carbonyl derivative, forms toxic HCN on hydrolysis.

❹ What products are formed when aldehydes are oxidized? (16.5)

- Aldehydes are oxidized to carboxylic acids (RCOOH) with $K_2Cr_2O_7$ or Tollens reagent.
- Ketones are not oxidized since they contain no H atom on the carbonyl carbon.

❺ What products are formed when aldehydes and ketones are reduced? (16.6)

- Aldehydes are reduced to 1° alcohols with H_2 and a Pd catalyst.
- Ketones are reduced to 2° alcohols with H_2 and a Pd catalyst.
- Biological reduction occurs with the coenzyme NADH.

❻ What reactions occur during vision? (16.7)

- When light hits the retina, the crowded C=C in 11-*cis*-retinal is isomerized to a more stable trans double bond, and a nerve impulse is generated. All-*trans*-retinal is converted back to the cis isomer by a three-step sequence—reduction, isomerization, and oxidation.

❼ What are hemiacetals and acetals, and how are they prepared? (16.8)

- Hemiacetals contain an OH group and an OR group bonded to the same carbon.
- Acetals contain two OR groups bonded to the same carbon.
- Treatment of an aldehyde or ketone with an alcohol (ROH) first forms an unstable hemiacetal, which reacts with more alcohol to form an acetal.
- Cyclic hemiacetals are stable compounds present in carbohydrates and some drugs.
- Acetals are converted back to aldehydes and ketones by hydrolysis with water and acid.

PROBLEMS

Selected in-chapter and end-of-chapter problems have brief answers provided in Appendix B.

Structure and Bonding

16.23 Draw the structure of a compound that fits each description:
a. an aldehyde of molecular formula $C_8H_{16}O$ that has six carbons in its longest chain
b. a ketone of molecular formula $C_6H_{12}O$ that contains five carbons in its longest chain
c. a ketone of molecular formula C_5H_8O that contains a ring
d. an aldehyde of molecular formula $C_6H_{10}O$ that has the CHO group bonded to a ring

16.24 Draw the structure of a constitutional isomer of 2-heptanone ($CH_3COCH_2CH_2CH_2CH_2CH_3$) that:
a. contains an aldehyde
b. contains a ketone
c. contains a hydroxyl group (OH)

16.25 How do a C=O and a C=C compare with regards to the following: (a) geometry; (b) polarity; (c) type of reaction?

16.26 How do an aldehyde and a ketone compare in (a) structure; (b) geometry; (c) polarity?

16.27 Can an aldehyde have molecular formula $C_5H_{12}O$? Explain why or why not.

16.28 Can a ketone have molecular formula $C_4H_{10}O$? Explain why or why not.

Nomenclature

16.29 Give an acceptable name for each aldehyde.

a. 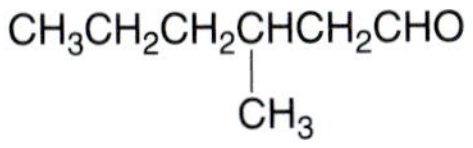 $CH_3CH_2CH_2CHCH_2CHO$
           |
           CH_3

b. 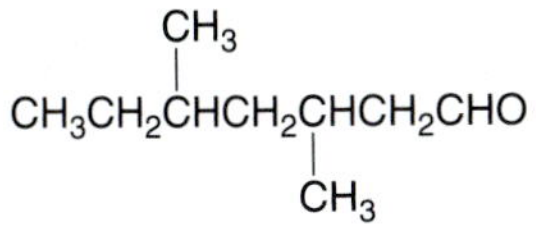 $CH_3CH_2CHCH_2CHCH_2CHO$ with CH_3 substituents

$CH_3CH_2\overset{\overset{CH_3}{|}}{C}HCH_2\overset{\overset{}{|}}{C}HCH_2CHO$
                         CH_3

c. $CH_3CH_2CH_2\overset{\overset{CH_2CHO}{|}}{C}HCH_2CH_2CH_3$

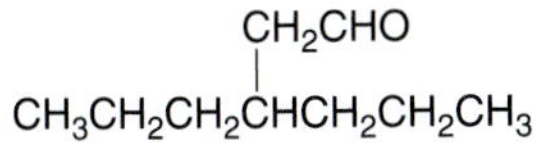

d. 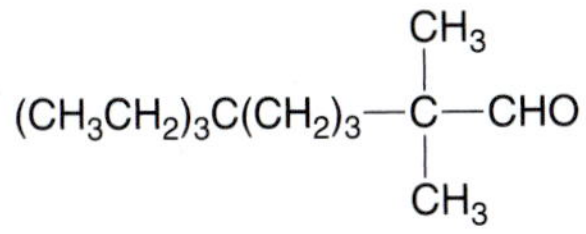 $(CH_3CH_2)_3C(CH_2)_3-\overset{\overset{CH_3}{|}}{\underset{\underset{CH_3}{|}}{C}}-CHO$

e. 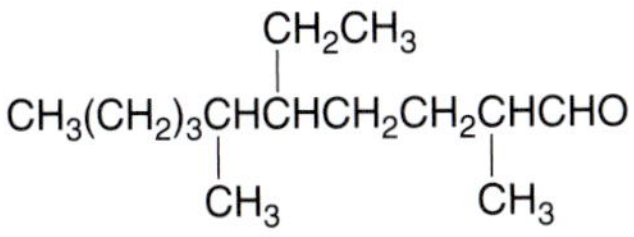 $CH_3(CH_2)_3\overset{\overset{CH_2CH_3}{|}}{C}HCH_2CH_2\overset{\overset{}{|}}{C}HCHO$ with CH_3 and CH_3

f. 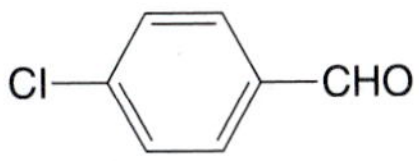 Cl—⟨benzene ring⟩—CHO

16.30 Give an acceptable name for each aldehyde.

a. $(CH_3)_3CCH_2CHO$

b. $(CH_3CH_2)_2CHCH_2CH_2CHO$

c. 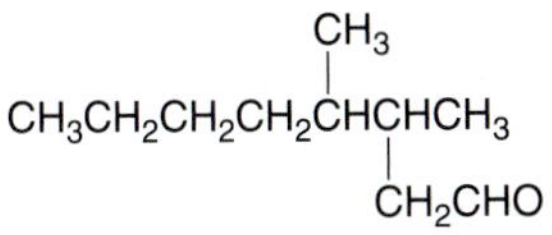 $CH_3CH_2CH_2CH_2\overset{\overset{CH_3}{|}}{C}HCHCH_3$ with CH_2CHO

d. 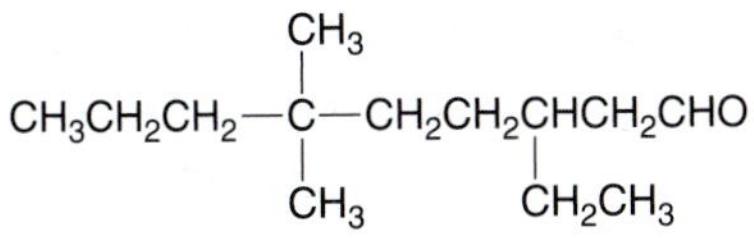 $CH_3CH_2CH_2-\overset{\overset{CH_3}{|}}{\underset{\underset{CH_3}{|}}{C}}-CH_2CH_2CHCH_2CHO$ with CH_2CH_3

e. $(CH_3CH_2CH_2CH_2)_2CHCH_2CHO$

f. 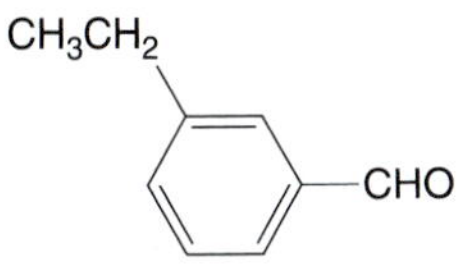 ⟨benzene ring with CH_3CH_2⟩—CHO

16.31 Draw the structure corresponding to each name.

a. 3,3-dichloropentanal   c. o-bromobenzaldehyde
b. 3,4-dimethylhexanal   d. 4-hydroxyheptanal

16.32 Draw the structure corresponding to each name.

a. 2-bromooctanal   c. 3,4-dimethoxybenzaldehyde
b. 2-propylheptanal   d. 3,4-dihydroxynonanal

16.33 Give an acceptable name for each ketone.

a. 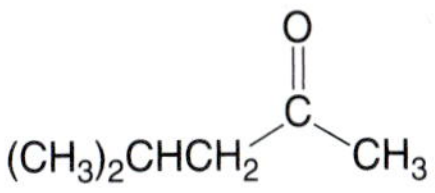 $(CH_3)_2CHCH_2\overset{\overset{O}{\|}}{C}CH_3$

b. cyclohexanone with CH_3 and CH_3 substituents (2,6-positions), ketone

c. ⟨benzene ring⟩ with $\overset{\overset{O}{\|}}{C}CH_3$ and $CH_2CH_2CH_2CH_3$

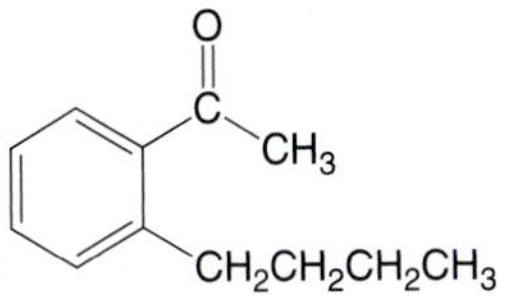

d. 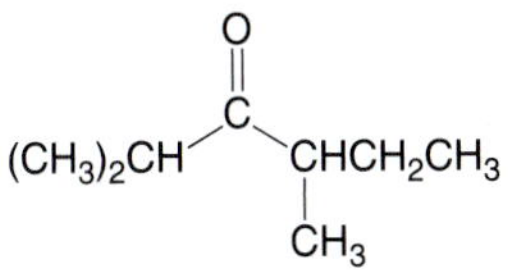 $(CH_3)_2CH\overset{\overset{O}{\|}}{C}\overset{\overset{}{|}}{C}HCH_2CH_3$ with CH_3

e. 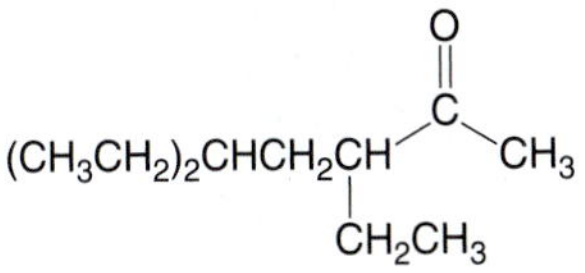 $(CH_3CH_2)_2CHCH_2\overset{}{C}H\overset{\overset{O}{\|}}{C}CH_3$ with CH_2CH_3

f. 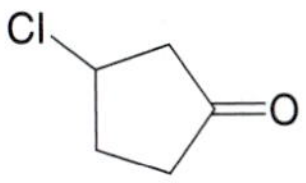 cyclopentanone with Cl substituent, O

16.34 Give an acceptable name for each ketone.

a. 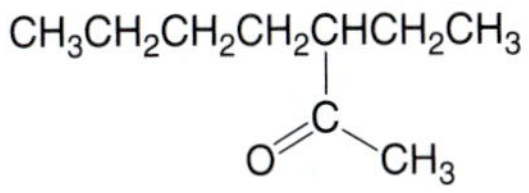 $CH_3CH_2CH_2CH_2CH_2\overset{}{C}HCH_2CH_3$ with $\overset{\overset{O}{\|}}{C}CH_3$

b. cyclobutanone with Cl, Cl substituents and O

c. $CH_3CH_2\overset{\overset{O}{\|}}{C}(CH_2)_5CH(CH_3)_2$

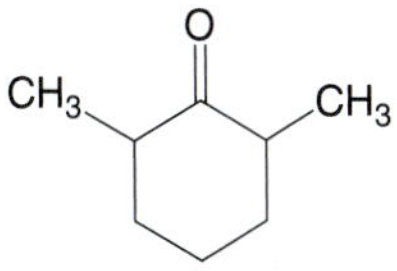

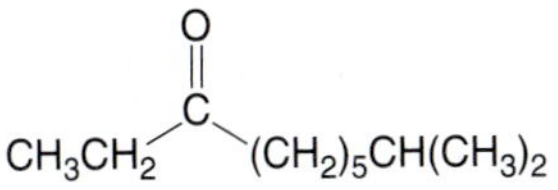

d. 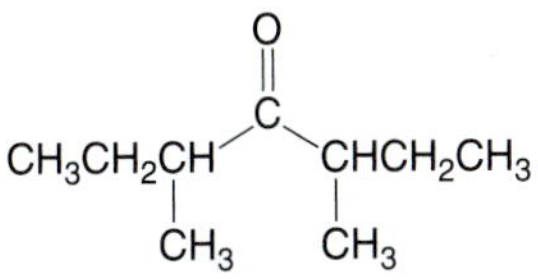 $CH_3CH_2\overset{}{C}H\overset{\overset{O}{\|}}{C}\overset{}{C}HCH_2CH_3$ with CH_3 and CH_3

e. 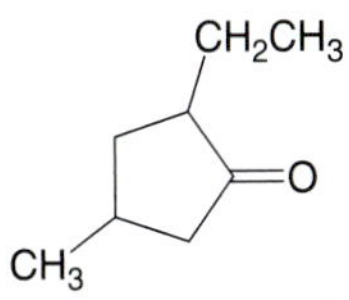 cyclopentanone with CH_2CH_3 and CH_3 substituents, O

f. 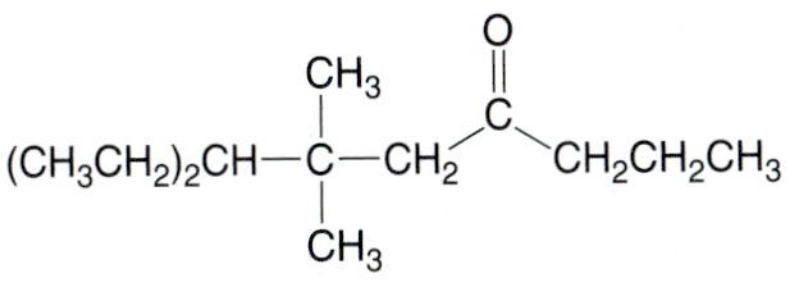 $(CH_3CH_2)_2CH-\overset{\overset{CH_3}{|}}{\underset{\underset{CH_3}{|}}{C}}-CH_2\overset{\overset{O}{\|}}{C}CH_2CH_2CH_3$

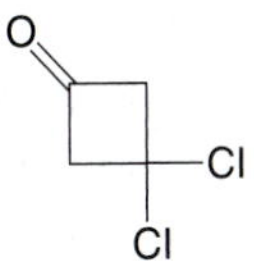

16.35 Draw the structure corresponding to each name.
   a. 3,3-dimethyl-2-hexanone
   b. methyl propyl ketone
   c. *m*-ethylacetophenone
   d. 2,4,5-triethylcyclohexanone

16.36 Draw the structure corresponding to each name.
   a. dibutyl ketone
   b. 1-chloro-3-pentanone
   c. *p*-bromoacetophenone
   d. 3-hydroxycyclopentanone

16.37 Draw the structure of the four constitutional isomers of molecular formula $C_6H_{12}O$ that contain an aldehyde and four carbons in the longest chain. Give the IUPAC name for each aldehyde.

16.38 Draw the structure of the three isomeric ketones of molecular formula $C_5H_{10}O$. Give the IUPAC name for each ketone.

16.39 Each of the following names is incorrect. Draw the structure of the compound and name it correctly.
   a. 1-pentanone   c. 3-propyl-2-butanone
   b. 4-hexanone   d. 2-methyl-1-octanal

16.40 Each of the following names is incorrect. Draw the structure of the compound and name it correctly.
   a. 6-octanone   c. 3-propyl-1-cyclopentanone
   b. 1-heptanone   d. 5-methylcyclohexanone

Physical Properties and Intermolecular Forces

16.41 Draw out the structure of benzaldehyde, including any lone pairs on oxygen. Then, show the hydrogen bonding interactions between benzaldehyde and water.

16.42 Can hydrogen bonding occur between the compounds in each pair? If hydrogen bonding is possible, draw out the compounds and indicate how that interaction would take place.
   a. two molecules of acetaldehyde (CH_3CHO)
   b. CH_3CHO and H_2O
   c. CH_3CHO and CH_3OH

16.43 Which compound in each pair has the higher boiling point?
   a. $(CH_3)_3CCH_2CH_2CH_3$ or $(CH_3)_3CCH_2CHO$
   b.
   c.

16.44 Which compound in each pair has the higher boiling point?
   a. $CH_3(CH_2)_6CHO$ or $CH_3(CH_2)_7OH$
   b. $CH_3(CH_2)_6CHO$ or $CH_3(CH_2)_2CHO$
   c.

16.45 Rank the following compounds in order of increasing melting point.

16.46 Menthone and menthol are both isolated from mint. Explain why menthol is a solid at room temperature but menthone is a liquid.
menthone   menthol

16.47 Label each compound as water soluble or water insoluble.
   a.
   b.
   c. $CH_3CH_2CH_2CH_3$

16.48 Label each compound as water soluble or water insoluble.
   a.
   b. $CH_3CH_2CH_2CHO$
   c. $CH_3CH_2CH_2OH$

16.49 Explain why 2,3-butanedione ($CH_3COCOCH_3$) is more soluble in water than acetone [$(CH_3)_2CO$] even though it has more carbon atoms. Would you expect 2,3-butanedione to be soluble in an organic solvent like diethyl ether?

16.50 Explain why the simplest ketone, acetone, has a much higher boiling point than the simplest aldehyde, formaldehyde.

Oxidation and Reduction

16.51 What product is formed when each compound is treated with $K_2Cr_2O_7$? With some compounds, no reaction occurs.
   a. $CH_3(CH_2)_4CHO$
   b.
   c.
   d. $CH_3(CH_2)_4CH_2OH$

16.52 What product is formed when each compound is treated with $K_2Cr_2O_7$? With some compounds, no reaction occurs.
   a.
   b. $CH_3(CH_2)_8CHO$
   c. $CH_3CHCH_2CH_2CH_3$ with OH group
   d.

16.53 What product is formed when each compound in Problem 16.51 is treated with Tollens reagent (Ag_2O, NH_4OH)? With some compounds, no reaction occurs.

16.54 What product is formed when each compound in Problem 16.52 is treated with Tollens reagent (Ag_2O, NH_4OH)? With some compounds, no reaction occurs.

16.55 What compound is formed when hydroxy aldehyde **A** is treated separately with each oxidizing agent?

HO— (cyclohexane ring) —CHO

a. $K_2Cr_2O_7$
b. Ag_2O, NH_4OH

A

16.56 (a) Give the structure of a compound of molecular formula C_4H_8O that reacts with Tollens reagent but not $K_2Cr_2O_7$. (b) Give the structure of a compound of molecular formula C_4H_8O that reacts with $K_2Cr_2O_7$ but not Tollens reagent.

16.57 What aldehyde is needed to prepare each carboxylic acid by an oxidation reaction?

a. $CH_3CH_2CHCH_2CO_2H$ (with CH_3 substituent)

b. CH_3—(benzene ring)—CO_2H

c. $CH_3CH_2CHCH_2CH_3$ (with CO_2H substituent)

16.58 What aldehyde is needed to prepare each carboxylic acid by an oxidation reaction?

a. (cyclopentane ring)—$CH_2CH_2CO_2H$

b. (benzene ring with Br)—CO_2H

c. $CH_3(CH_2)_8CHCO_2H$ (with Cl substituent)

16.59 What alcohol is formed when each carbonyl compound is treated with H_2 and a Pd catalyst?

a. CH_3CH_2—(benzene ring)—CHO

b. (cyclohexanone ring with CH_3 and =O)

c. (bicyclic ring with CH_3 and =O)

16.60 What alcohol is formed when each carbonyl compound is treated with H_2 and a Pd catalyst?

a. CH_3CH_2—C(=O)—$CH_2CH(CH_3)_2$

b. $CH_3(CH_2)_6CHO$

c. HO—(cyclopentane ring)—CH_2CHO

16.61 What carbonyl compound is needed to make each alcohol by a reduction reaction?

a. CH_3OH
b. $CH_3CH_2CH_2CH_2CH_2OH$
c. (cyclohexane ring with OH and CH_3)

16.62 What carbonyl compound is needed to make each alcohol by a reduction reaction?

a. $CH_3CH_2CHCH_2CH_2CH_3$ (with OH substituent)
b. $(CH_3)_2CHCH_2CH_2OH$
c. (cyclopentane ring with two CH_3 groups and OH)

16.63 Why can't 1-methylcyclohexanol be prepared by the reduction of a carbonyl compound?

(cyclohexane ring with CH_3 and OH)

1-methylcyclohexanol

16.64 Why can't $(CH_3)_3COH$ be prepared by the reduction of a carbonyl compound?

Hemiacetals and Acetals

16.65 Draw the structure of a compound of molecular formula $C_5H_{12}O_2$ that fits each description.
   a. a compound that contains an acetal
   b. a compound that contains a hemiacetal
   c. a compound that contains two ethers
   d. a compound that contains an alcohol and an ether, but is not a hemiacetal

16.66 Locate the two acetals in amygdalin (Section 16.4).

16.67 Label each functional group as an alcohol, ether, acetal, or hemiacetal.

a. $CH_3-\overset{\overset{\displaystyle OCH_3}{|}}{\underset{\underset{\displaystyle OCH_3}{|}}{C}}-H$

c. $HOCH_2\overset{\overset{\displaystyle OCH_3}{|}}{CH}CH_2CH_3$

b. $CH_3-\overset{\overset{\displaystyle OCH_2CH_3}{|}}{\underset{\underset{\displaystyle OH}{|}}{C}}-H$

d.

16.68 Label each functional group as an alcohol, ether, acetal, or hemiacetal.

a.

c.

b. $CH_3-\overset{\overset{\displaystyle OH}{|}}{\underset{\underset{\displaystyle CH_3}{|}}{C}}-OCH_2CH_2CH_3$

d.

16.69 What acetal is formed when each aldehyde or ketone is treated with two equivalents of methanol (CH_3OH) in the presence of H_2SO_4?

a.

c. $CH_3\overset{\overset{\displaystyle O}{||}}{C}CH_2CH_2CH_3$

b. $CH_2=O$

d. —CH_2CHO

16.70 What acetal is formed when each aldehyde or ketone is treated with two equivalents of ethanol (CH_3CH_2OH) in the presence of H_2SO_4?

a. $CH_3CH_2CH_2\overset{\overset{\displaystyle O}{||}}{C}CH_2CH_3$

c. $(CH_3)_2CHCH_2CH_2CHO$

b.

d. CH_3CH_2CHO

16.71 (a) What hemiacetal is formed when acetone [$(CH_3)_2CO$] is treated with one equivalent of cyclopentanol? (b) What acetal is formed when acetone is treated with two equivalents of cyclopentanol in the presence of H_2SO_4?

16.72 (a) What hemiacetal is formed when benzaldehyde (C_6H_5CHO) is treated with one equivalent of cyclohexanol? (b) What acetal is formed when benzaldehyde is treated with two equivalents of cyclohexanol in the presence of H_2SO_4?

16.73

A

a. Identify the hemiacetal carbon in compound **A**.
b. What hydroxy aldehyde would form this hemiacetal in a cyclization reaction?
c. What product is formed when **A** is treated with CH_3OH and H_2SO_4?

16.74

B

a. Identify the hemiacetal carbon in compound **B**.
b. What hydroxy aldehyde would form this hemiacetal in a cyclization reaction?
c. What product is formed when **B** is treated with CH_3CH_2OH and H_2SO_4?

16.75 What hemiacetal is formed when hydroxy aldehyde **C** is cyclized?

$HOCH_2CH_2CH_2\overset{\overset{\displaystyle}{|}}{\underset{\underset{\displaystyle CH_3}{|}}{CH}}\overset{\overset{\displaystyle O}{||}}{C}H$

C

16.76 What hemiacetal is formed when hydroxy aldehyde **D** is cyclized?

$HO\overset{\overset{\displaystyle}{|}}{\underset{\underset{\displaystyle CH_3}{|}}{CH}}CH_2CH_2CH_2\overset{\overset{\displaystyle O}{||}}{C}H$

D

16.77 What products are formed when each acetal is hydrolyzed with water and sulfuric acid?

a.

b. $H-\overset{\overset{\displaystyle OCH_3}{|}}{\underset{\underset{\displaystyle OCH_3}{|}}{C}}-CH_2CH_2CH_3$

16.78 What products are formed when each acetal is hydrolyzed with water and sulfuric acid?

a. $CH_3CH_2-\overset{\overset{\displaystyle OCH_2CH_3}{|}}{\underset{\underset{\displaystyle OCH_2CH_3}{|}}{C}}-CH_2CH_3$

b. CH_3O—

General Reactions

16.79 Benzaldehyde is the compound principally responsible for the odor of almonds. What products are formed when benzaldehyde is treated with each reagent?

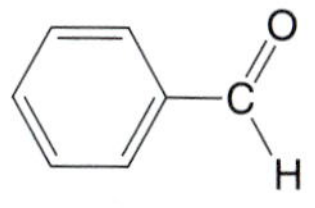

benzaldehyde

    a. H_2, Pd
    b. $K_2Cr_2O_7$
    c. Ag_2O, NH_4OH
    d. CH_3OH (2 equivalents), H_2SO_4
    e. CH_3CH_2OH (2 equivalents), H_2SO_4
    f. Product in (e), then H_2O, H_2SO_4

16.80 Anisaldehyde is a component of anise, a spice with a licorice odor used in cooking and aromatherapy. What products are formed when anisaldehyde is treated with each reagent?

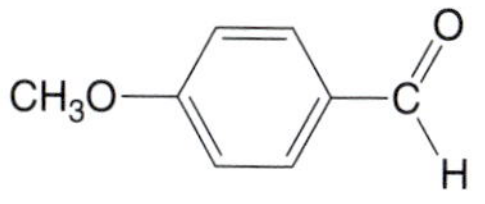

anisaldehyde

    a. H_2, Pd
    b. $K_2Cr_2O_7$
    c. Ag_2O, NH_4OH
    d. CH_3OH (2 equivalents), H_2SO_4
    e. CH_3CH_2OH (2 equivalents), H_2SO_4
    f. Product in (e), then H_2O, H_2SO_4

16.81 2-Heptanone is an alarm pheromone of the bee. What products are formed when 2-heptanone is treated with each reagent? With some reagents, no reaction occurs.

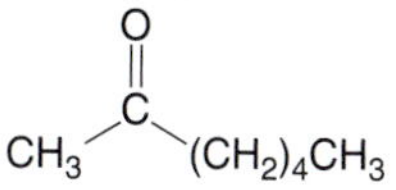

2-heptanone

    a. H_2, Pd
    b. $K_2Cr_2O_7$
    c. Ag_2O, NH_4OH
    d. CH_3OH (2 equivalents), H_2SO_4
    e. CH_3CH_2OH (2 equivalents), H_2SO_4
    f. Product in (e), then H_2O, H_2SO_4

16.82 2,6-Dimethyl-3-heptanone is a communication pheromone of the bee. What products are formed when 2,6-dimethyl-3-heptanone is treated with each reagent? With some reagents, no reaction occurs.

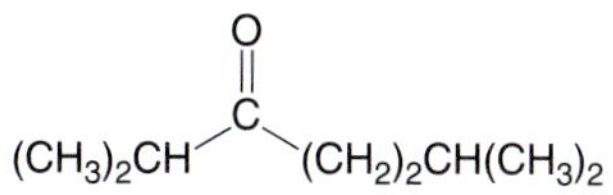

2,6-dimethyl-3-heptanone

a. H_2, Pd
b. $K_2Cr_2O_7$
c. Ag_2O, NH_4OH
d. CH_3OH (2 equivalents), H_2SO_4
e. CH_3CH_2OH (2 equivalents), H_2SO_4
f. Product in (e), then H_2O, H_2SO_4

16.83 Three constitutional isomers of molecular formula C_5H_8O can be converted to 1-pentanol ($CH_3CH_2CH_2CH_2CH_2OH$) on treatment with two equivalents of H_2 in the presence of a Pd catalyst. Draw the structures of the three possible compounds, all of which contain a carbonyl group.

16.84 Identify **A–C** in the following reaction sequence.

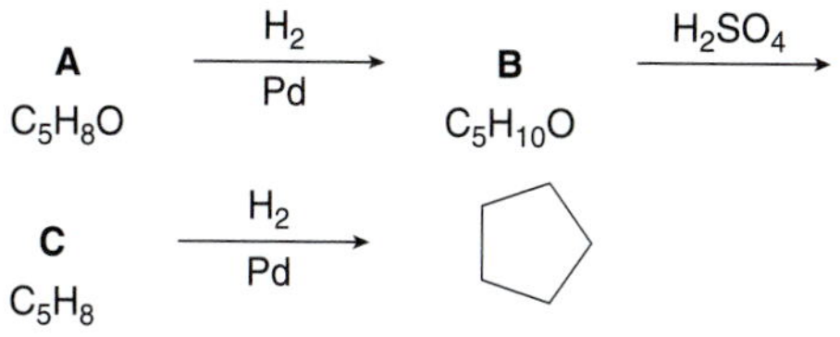

Applications

16.85 Androsterone is a male sex hormone that controls the development of secondary sex characteristics in males. Draw the organic products formed when androsterone is treated with each reagent.

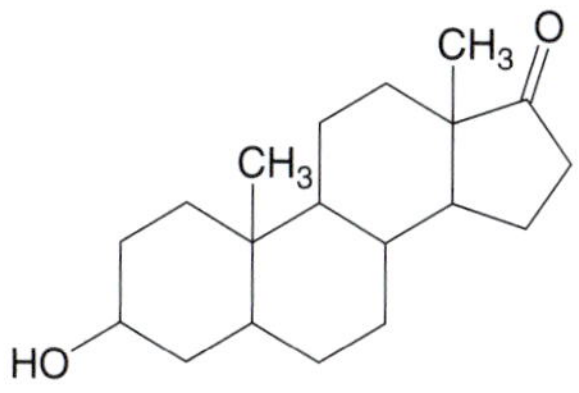

androsterone

a. H_2, Pd
b. $K_2Cr_2O_7$
c. CH_3OH (2 equivalents), H_2SO_4
d. CH_3CH_2OH (2 equivalents), H_2SO_4

16.86 Salmeterol (trade name: Serevent) is a drug that dilates airways and is used to treat asthma. It can be formed by the reduction of a dicarbonyl compound **X** with excess H_2 and a Pd catalyst.

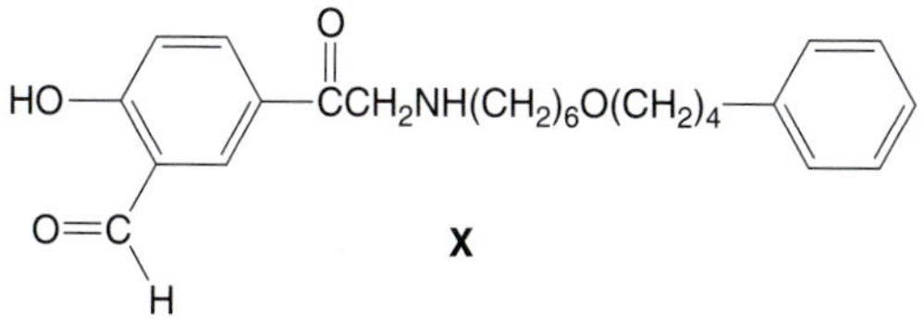

a. Draw the structure of salmeterol.
b. Label the one chirality center in salmeterol.
c. Draw both enantiomers of salmeterol in three dimensions using wedges and dashes (see Section 15.3).

16.87 When cinnamon is ingested, the cinnamaldehyde it contains is oxidized by the coenzyme NAD$^+$ and an enzyme in a similar fashion to the biological oxidation described in Section 14.6. What oxidation product is formed from cinnamaldehyde?

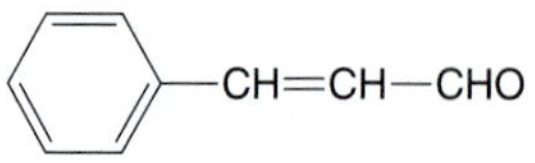

CH=CH—CHO

cinnamaldehyde

16.88 Safrole is a naturally occurring compound isolated from sassafras plants. Once used as a common food additive in root beer and other beverages, it is now banned because it is carcinogenic.

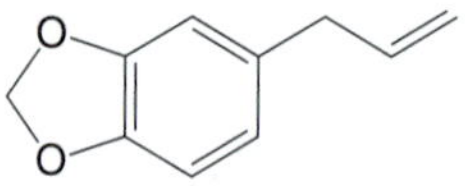

safrole

a. Label the functional groups in safrole.
b. What products are formed when safrole undergoes hydrolysis with water and sulfuric acid?

16.89 2-Deoxyribose is a building block of DNA, very large organic molecules that store all genetic information. Label each OH group in 2-deoxyribose as being part of a regular alcohol or a hemiacetal.

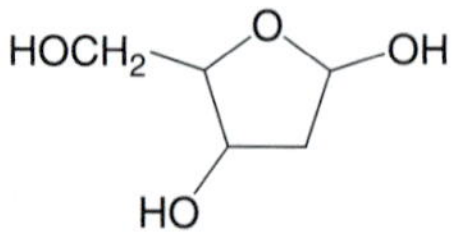

HOCH$_2$ O OH

HO

2-deoxyribose

16.90 Paraldehyde, a hypnotic and sedative once commonly used to treat seizures and induce sleep in some hospitalized patients, is formed from three molecules of acetaldehyde (CH_3CHO). Label the acetal carbons in paraldehyde.

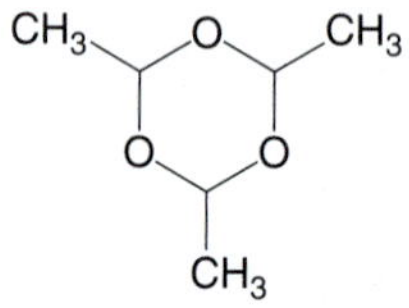

CH$_3$ O CH$_3$

O O

CH$_3$

paraldehyde

16.91 Describe the main reaction that occurs in the rod cells in the retina, which generates a nerve impulse that allows us to see. What role does the geometry of 11-*cis*-retinal play in this process?

16.92 What reactions convert all-*trans*-retinal back to 11-*cis*-retinal so that the process of vision can continue?

CHALLENGE QUESTIONS

16.93 Monensin, a complex antibiotic produced by the bacterium *Streptomyces cinamonsis,* is added to cattle feed to protect the cattle from infection. Identify each oxygen-containing functional group.

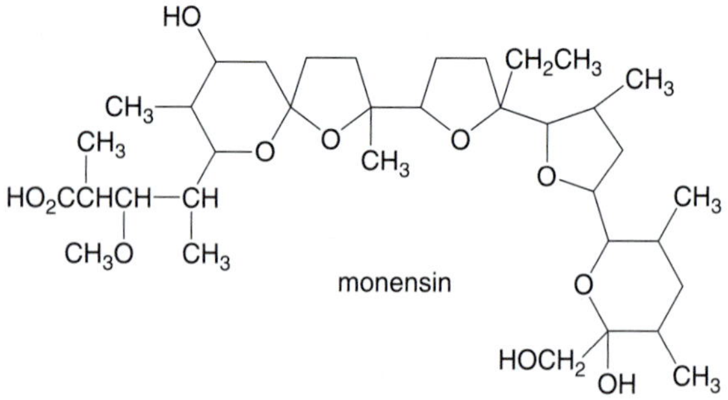

HO

CH$_3$

CH$_3$

HO$_2$CCHCH—CH

CH$_3$O CH$_3$

CH$_2$CH$_3$ CH$_3$

O CH$_3$

O

CH$_3$

monensin

O

HOCH$_2$ OH CH$_3$

16.94 Chloral hydrate is a sedative formed by the addition of water to chloral (Cl_3CCHO). Chloral hydrate is sometimes used to calm a patient prior to a surgical procedure. It has also been used for less reputable purposes. Adding it to an alcoholic beverage forms a so-called "knock-out drink," causing an individual to pass out. Although we have not studied the formation of hydrates from aldehydes, this reaction resembles the addition of alcohols to aldehydes to form hemiacetals. What is the structure of chloral hydrate?

17

CHAPTER GOALS

In this chapter you will learn how to:

1. Identify the characteristics of carboxylic acids, esters, and amides

2. Name carboxylic acids, esters, and amides

3. Give examples of useful carboxylic acids

4. Give examples of useful esters and amides

5. Draw the products of acid–base reactions of carboxylic acids

6. Explain how soap cleans away dirt

7. Discuss the acid–base chemistry of aspirin

8. Convert carboxylic acids to esters and amides

9. Draw the products of the hydrolysis of esters and amides

10. Draw the structures of polyesters and polyamides

11. Explain how penicillin works

Penicillin, an amide discovered by Scottish bacteriologist Sir Alexander Fleming in 1928, is an antibiotic produced by the mold *Penicillium notatum.*

CARBOXYLIC ACIDS, ESTERS, AND AMIDES

CHAPTER 17 is the second of two chapters that concentrates on compounds that contain a **carbonyl group (C=O).** While Chapter 16 focused on aldehydes (RCHO) and ketones (R_2CO), Chapter 17 discusses the chemistry and properties of **carboxylic acids (RCOOH), esters (RCOOR'), and amides ($RCONR'_2$).** All three families of organic compounds occur widely in nature, and many useful synthetic compounds have been prepared as well. Simple carboxylic acids like acetic acid (CH_3COOH) have a sour taste and a biting odor, and long-chain carboxylic acids form vegetable oils and animal fat. Simple esters have easily recognized fragrances, and the most abundant lipids contain three ester groups. Several useful drugs are amides, and all proteins, such as those that form hair, muscle, and connective tissue, contain many amide units.

17.1 STRUCTURE AND BONDING

Carboxylic acids, esters, and **amides** are three families of organic molecules that contain a **carbonyl group (C=O)** bonded to an element more electronegative than carbon.

- *Carboxylic acids* are organic compounds containing a carboxyl group (COOH).

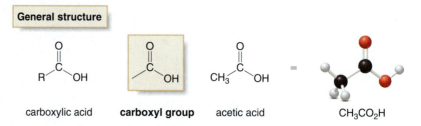

The word **carboxyl** (for a COOH group) is derived from **carb**onyl (C=O) + hyd**roxyl** (OH).

Although the structure of a carboxylic acid is often abbreviated as **RCOOH** or **RCO₂H,** keep in mind that the central carbon atom of the functional group has a double bond to one oxygen atom and a single bond to another.

- *Esters* are carbonyl compounds that contain an alkoxy group (OR') bonded to the carbonyl carbon.

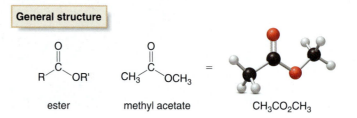

The structure of an ester is often abbreviated as RCOOR' or RCO₂R'.

- *Amides* are carbonyl compounds that contain a nitrogen atom bonded to the carbonyl carbon.

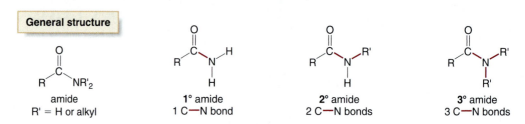

The N atom of an amide may be bonded to other hydrogen atoms or alkyl groups. Amides are classified as **1°, 2°,** or **3°** depending on the number of carbon atoms bonded directly to the *nitrogen* atom.

- A primary (1°) amide contains one C—N bond. A 1° amide has the structure RCONH₂.
- A secondary (2°) amide contains two C—N bonds. A 2° amide has the structure RCONHR'.
- A tertiary (3°) amide contains three C—N bonds. A 3° amide has the structure RCONR'₂.

Cyclic esters and amides are known. **A cyclic ester is called a *lactone*. A cyclic amide is called a *lactam*.**

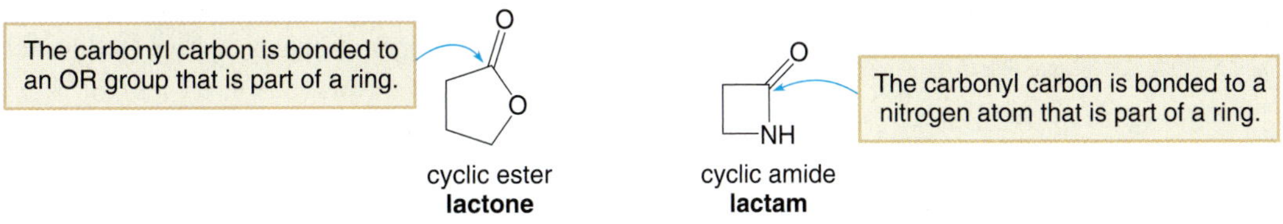

The carbonyl carbon is bonded to an OR group that is part of a ring.

cyclic ester
lactone

The carbonyl carbon is bonded to a nitrogen atom that is part of a ring.

cyclic amide
lactam

Carboxylic acids, esters, and amides are called **acyl compounds,** since they contain an **acyl group (RCO)** bonded to an electronegative atom.

acyl group

N or O atom bonded here

The two most important features of acyl compounds, like other carbonyl compounds, are the following:

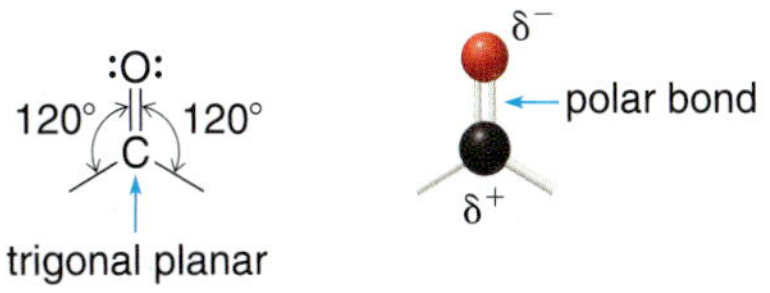

trigonal planar

polar bond

- The carbonyl carbon is trigonal planar, so all bond angles are 120°.
- The electronegative oxygen atom makes the carbonyl group polar, so the carbonyl carbon is electron poor (δ^+) and the carbonyl oxygen is electron rich (δ^-).

Oxalic acid (a carboxylic acid) in spinach, propyl acetate (an ester) in pears, and capsaicin (an amide) in hot peppers are examples of acyl compounds.

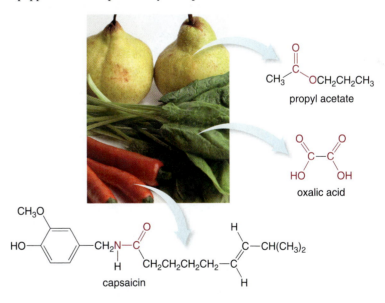

propyl acetate

oxalic acid

capsaicin

PROBLEM 17.1

Draw out each compound to clearly show what groups are bonded to the carbonyl carbon. Label each compound as a carboxylic acid, ester, or amide.

a. $CH_3CH_2CO_2CH_2CH_3$ b. $CH_3CONHCH_3$ c. $(CH_3)_3CCO_2H$ d. $(CH_3)_2CHCON(CH_3)_2$

Invirase is the trade name for sasquinavir, a drug used to treat HIV. Classify each amide in Invirase as 1°, 2°, or 3°.

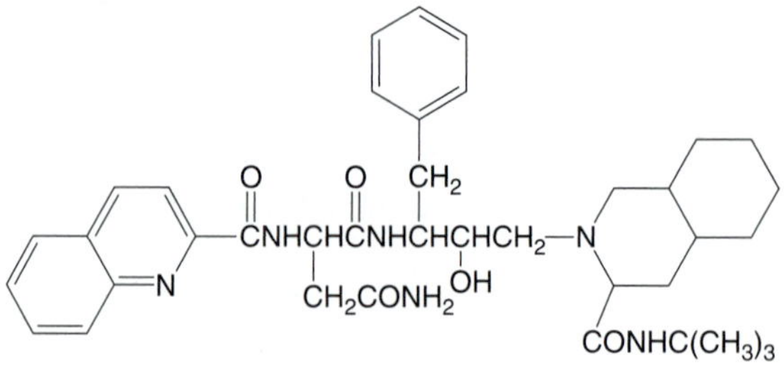

sasquinavir
(Trade name: Invirase)

17.2 NOMENCLATURE

To name carboxylic acids, esters, and amides, we must learn the suffix that identifies each functional group in the IUPAC system. Since the names of esters and amides are derived from carboxylic acids, we must first learn how to name carboxylic acids. With esters, 2° amides, and 3° amides, the name of the acyl compound has two distinct portions: an acyl group that contains the carbonyl, and one or two alkyl groups bonded to the oxygen or nitrogen atom.

17.2A NAMING A CARBOXYLIC ACID—RCOOH

- In the IUPAC system, carboxylic acids are identified by the suffix -oic acid.

To name a carboxylic acid using the IUPAC system:

1. Find the longest chain containing the COOH group, and change the -e ending of the parent alkane to the suffix -oic acid.
2. Number the carbon chain to put the COOH group at C1, but omit this number from the name. Apply all of the other usual rules of nomenclature.

Many simple carboxylic acids are often referred to by their common names. A common name uses the suffix -ic acid. Three common names are virtually always used.

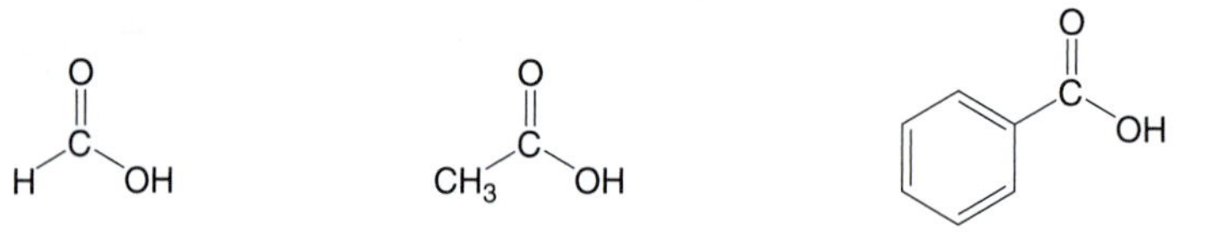

| formic acid | acetic acid | benzoic acid |
| (methanoic acid) | (ethanoic acid) | (benzenecarboxylic acid) |

[IUPAC names are given in parentheses, and are rarely used.]

Greek letters are sometimes used to designate the carbons near carbonyl groups. These terms can be used for any carbonyl group.

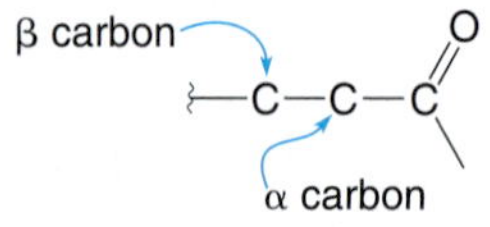

β carbon

α carbon

- The carbon adjacent to the COOH is called the α (alpha) carbon.
- The carbon bonded to the α carbon is the β (beta) carbon.

SAMPLE PROBLEM 17.1 Give the IUPAC name of the following carboxylic acid.

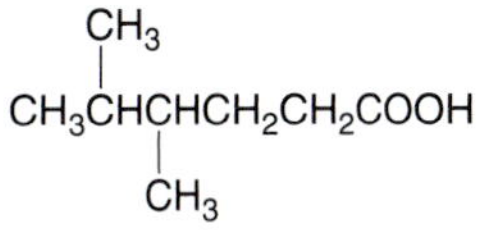

$$CH_3CHCHCH_2CH_2COOH$$

with CH_3 above and CH_3 below

ANALYSIS AND SOLUTION

[1] Find and name the longest chain containing COOH.

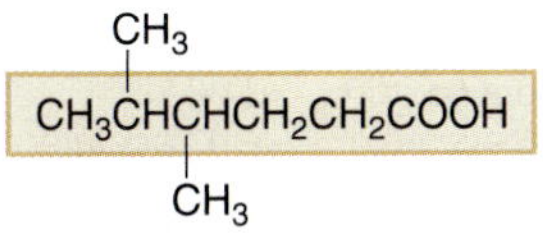

CH₃
CH₃CHCHCH₂CH₂COOH
CH₃

hexane ----→ hexan*oic acid*
(6 C's)

The COOH contributes one C to the longest chain.

[2] Number and name the substituents, making sure the COOH group is at C1.

CH₃ 4
CH₃CHCHCH₂CH₂COOH
5 CH₃ 1

two methyl substituents on C4 and C5

Answer: 4,5-dimethylhexanoic acid

PROBLEM 17.3 Give the IUPAC name for each compound.

a. 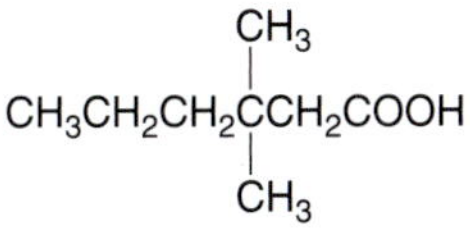 CH₃
    CH₃CH₂CH₂CCH₂COOH
    CH₃

b. 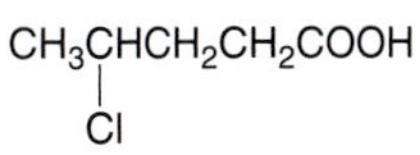 CH₃CHCH₂CH₂COOH
    Cl

c. 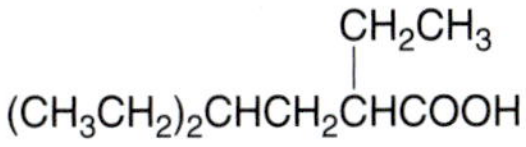 CH₂CH₃
    (CH₃CH₂)₂CHCH₂CHCOOH

PROBLEM 17.4 Give the structure corresponding to each IUPAC name.

a. 2-bromobutanoic acid

b. 2,3-dimethylpentanoic acid

c. 2-ethyl-5,5-dimethyloctanoic acid

d. 3,4,5,6-tetraethyldecanoic acid

17.2B NAMING AN ESTER—RCOOR'

The names of esters are derived from the names of the parent carboxylic acids. Keep in mind that the common names **formic acid, acetic acid,** and **benzoic acid** are used for the parent acid, so these common parent names are used for their derivatives as well.

An ester has two parts to its structure, each of which must be named separately: an **acyl group** that contains the carbonyl group (**RCO—**) and an **alkyl group** (designated as **R'**) bonded to the oxygen atom.

- **In the IUPAC system, esters are identified by the suffix -ate.**

HOW TO Name an Ester (RCO₂R') Using the IUPAC System

EXAMPLE **Give a systematic name for the ester CH₃CO₂CH₂CH₃.**

Step [1] **Name the R' group bonded to the oxygen atom as an alkyl group.**

- The name of the alkyl group, ending in the suffix -*yl*, becomes the *first* part of the ester name. In this case, a two-carbon *ethyl* group is bonded to the ester oxygen atom.

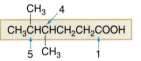

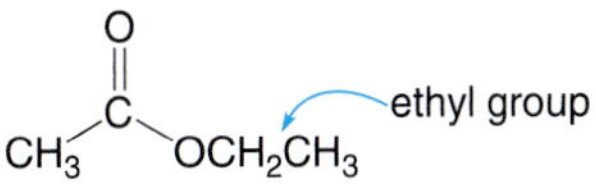

(Continued)

How To, continued...

Step [2]   **Name the acyl group (RCO–) by changing the *-ic acid* ending of the parent carboxylic acid to the suffix *-ate*.**

- The name of the acyl group becomes the *second* part of the name.
- In this case, the name of the parent two-carbon carboxylic acid, ace*tic acid*, becomes ace*tate*.

Esters are often written as RCOOR', where the alkyl group (R') is written *last*. When an ester is named, however, the R' group appears *first* in the name.

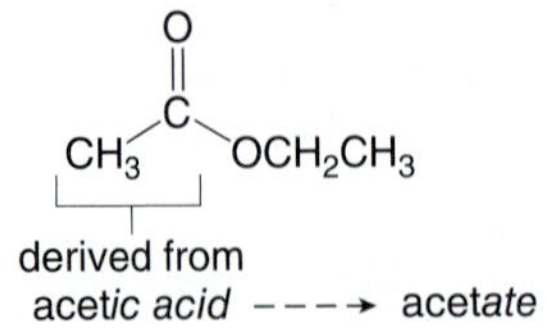

derived from
ace*tic acid* ----→ ace*tate*

**Answer:
ethyl acetate**

SAMPLE PROBLEM 17.2   Give the IUPAC name of the following ester, which occurs naturally in apples.

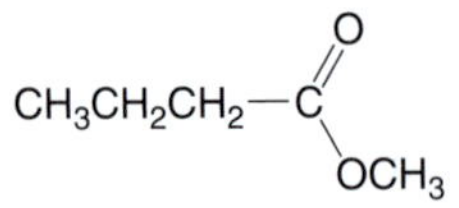

ANALYSIS AND SOLUTION

[1] Name the alkyl group on the O atom.

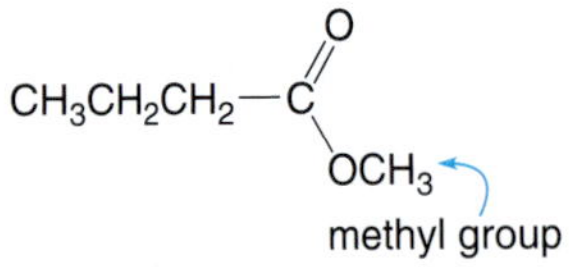

methyl group

The word *methyl* becomes the first part of the name.

[2] To name the acyl group, find and name the longest chain containing the carbonyl group, placing the C=O at C1.

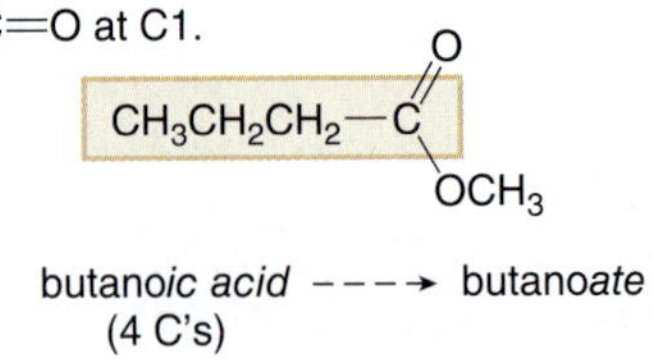

butano*ic acid* ----→ butano*ate*
(4 C's)

Answer: methyl butanoate

PROBLEM 17.5   Give an acceptable name for each ester.

a. $CH_3(CH_2)_4CO_2CH_3$

b.

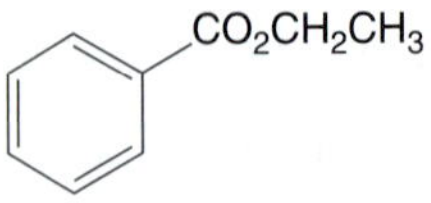

c. $CH_3CH_2CH_2CH_2COOCH_2CH_2CH_3$

PROBLEM 17.6   Give the structure corresponding to each name.

a. propyl propanoate

b. butyl acetate

c. ethyl hexanoate

d. methyl benzoate

17.2C   NAMING AN AMIDE

- **In the IUPAC system, amides are identified by the suffix *-amide*.**

All 1° amides are named by replacing the ***-oic acid*** ending (or *-ic acid* ending of a common name) with the suffix ***-amide***.

Naming
1° amides

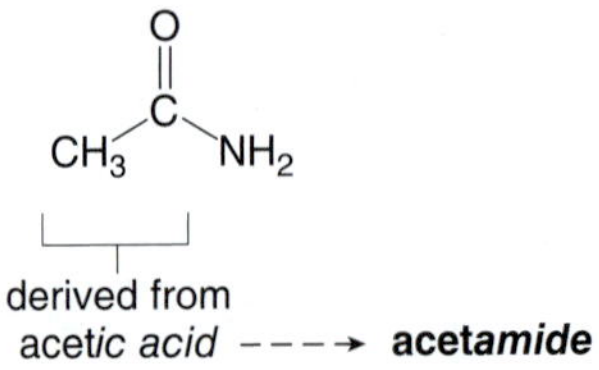

derived from
ace*tic acid* ----→ **acet*amide***

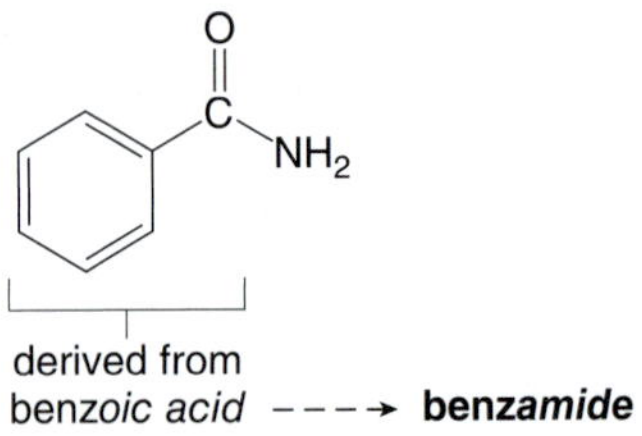

derived from
benz*oic acid* ----→ **benz*amide***

A 2° or 3° amide has two parts to its structure: an **acyl group** that contains the carbonyl group (**RCO–**) and one or two **alkyl groups** bonded to the nitrogen atom.

HOW TO Name a 2° or 3° Amide

EXAMPLE **Give a systematic name for each amide.**

a. 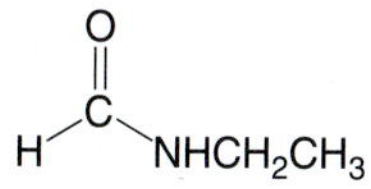
O
C
H
NHCH2CH3

b. 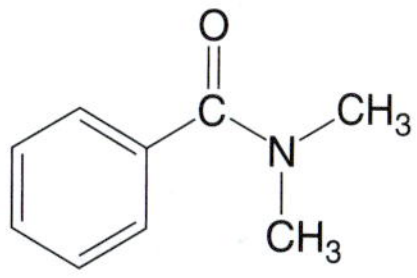
O
C
N
CH3
CH3

Step [1] **Name the alkyl group (or groups) bonded to the N atom of the amide. Use the prefix "*N*-" preceding the name of each alkyl group.**

- The name of the alkyl groups forms the first part of each amide name.
- For 3° amides, use the prefix **di-** if the two alkyl groups on N are the same. If the two alkyl groups are different, **alphabetize** their names. One "***N*-**" is needed for each alkyl group, even if both R groups are identical.

a. 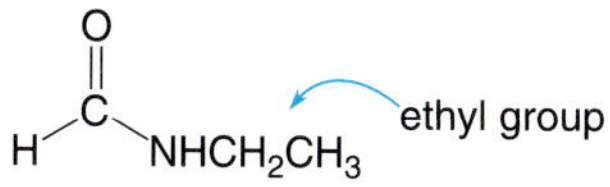
O
C
H
NHCH2CH3
ethyl group

- The compound is a 2° amide with one ethyl group → *N*-ethyl.

b. 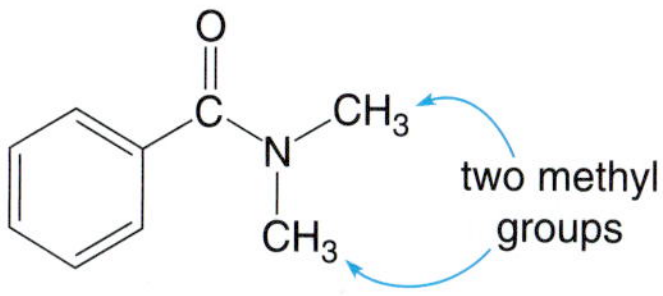
O
C
N
CH3
CH3
two methyl
groups

- The compound is a 3° amide with two methyl groups.
- Use the prefix di- and two "*N*-" to begin the name → *N,N*-dimethyl.

Step [2] **Name the acyl group (RCO–) with the suffix *-amide*.**

a. 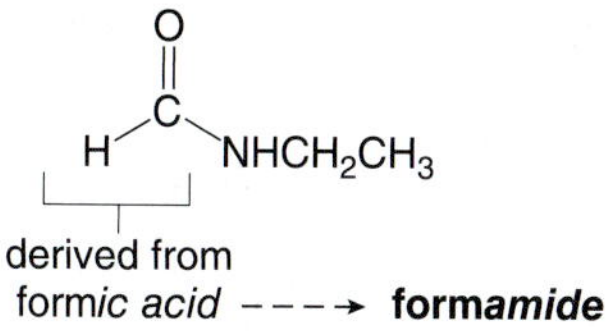
O
C
H
NHCH2CH3
derived from
formic acid
formamide

- Change the *-ic acid* or *-oic acid* suffix of the parent carboxylic acid to the suffix *-amide*.
- Put the two parts of the name together.

Answer: *N*-ethylformamide

b. 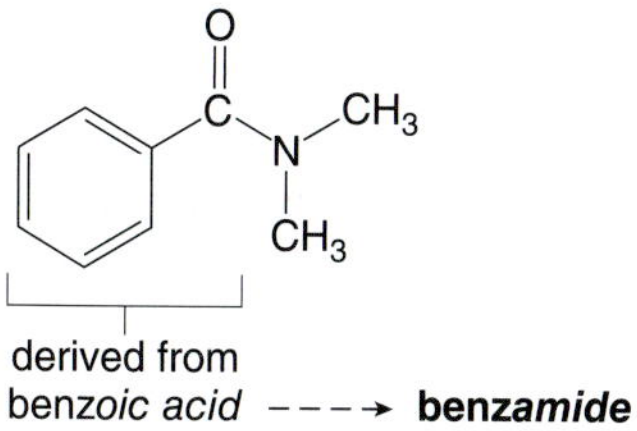
O
C
N
CH3
CH3
derived from
benzoic acid
benzamide

- Change benz*oic acid* to benz*amide*.
- Put the two parts of the name together.

Answer: *N,N*-dimethylbenzamide

PROBLEM 17.7 Give an acceptable name for each amide.

a. 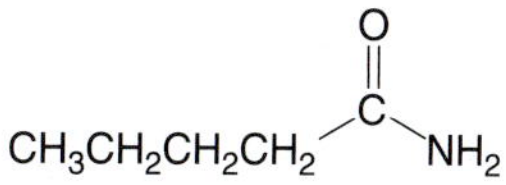
O
C
CH3CH2CH2CH2
NH2

b. 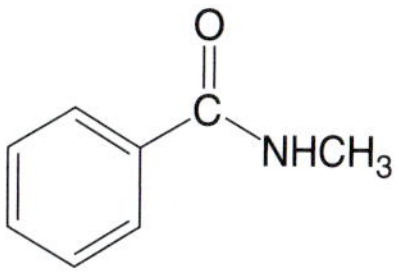
O
C
NHCH3

c. 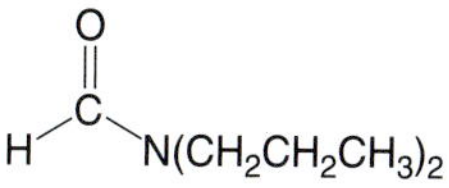
O
C
H
N(CH2CH2CH3)2

PROBLEM 17.8 Draw the structure of each amide.

a. propanamide
b. *N*-ethylhexanamide
c. *N,N*-dimethylacetamide
d. *N*-butyl-*N*-methylbutanamide

17.3 PHYSICAL PROPERTIES

Carboxylic acids, esters, and amides are polar compounds because they possess a polar carbonyl group. All C—O, O—H, C—N, and N—H bonds in these compounds are polar bonds as well, and thus contribute to the net dipole of each compound.

Carboxylic acids also exhibit intermolecular **hydrogen bonding** since they possess a hydrogen atom bonded to an electronegative oxygen atom. Carboxylic acids often exist as **dimers,** held together by *two* intermolecular hydrogen bonds, as shown in Figure 17.1. The carbonyl oxygen atom of one molecule hydrogen bonds to the hydrogen atom of another molecule. As a result:

- Carboxylic acids have stronger intermolecular forces than esters, giving them higher boiling points and melting points, when comparing compounds of similar size.

$$CH_3-\overset{\displaystyle O}{\overset{\|}{C}}\diagdown OCH_3$$
bp 58 °C

no hydrogen bonding possible

$$CH_3CH_2-\overset{\displaystyle O}{\overset{\|}{C}}\diagdown OH$$
bp 141 °C hydrogen bonding possible

stronger intermolecular forces
higher boiling point

Carboxylic acids also have stronger intermolecular forces than alcohols, even though both functional groups exhibit hydrogen bonding; two molecules of a carboxylic acid are held together by *two* intermolecular hydrogen bonds, whereas only *one* hydrogen bond is possible between two molecules of an alcohol (Section 14.2). As a result:

- Carboxylic acids have higher boiling points and melting points than alcohols of comparable size.

$$CH_3CH_2CH_2OH$$

1-propanol
bp 97 °C

$$CH_3-\overset{\displaystyle O}{\overset{\|}{C}}\diagdown OH$$

acetic acid
bp 118 °C

more hydrogen-bonding interactions possible
higher boiling point

▼ **FIGURE 17.1** Two Molecules of Acetic Acid (CH_3COOH) Held Together by Two Hydrogen Bonds

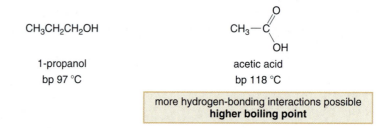

hydrogen bond

hydrogen bond

FIGURE 17.2 Intermolecular Hydrogen Bonding Between Two CH_3CONH_2 Molecules

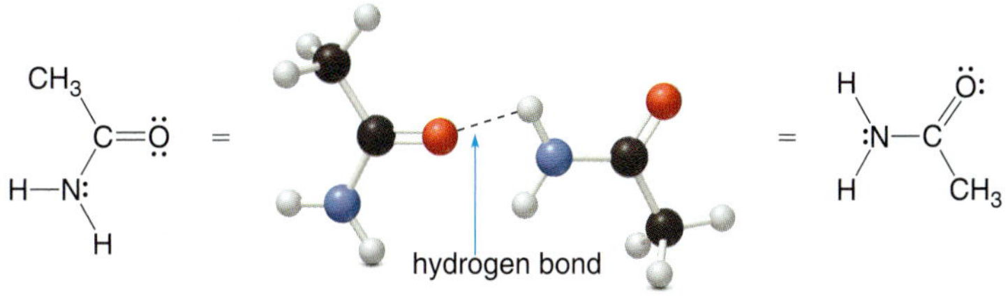

hydrogen bond

Primary (1°) and 2° amides contain N—H bonds, so hydrogen bonding is possible between two molecules of the amide, as shown in Figure 17.2. This gives 1° and 2° amides stronger intermolecular forces than 3° amides and esters, which can't intermolecularly hydrogen bond. As a result:

• **Primary (1°) and 2° amides have higher boiling points and melting points than esters and 3° amides of comparable size.**

ester
bp 58 °C

3° amide
bp 153 °C

1° amide
bp 213 °C

hydrogen bonding possible

strongest intermolecular forces
highest boiling point

Like other oxygen-containing compounds, carboxylic acids, esters, and amides having less than six carbons are soluble in water. Higher molecular weight acyl compounds are insoluble in water because the nonpolar portion of the molecule, the C—C and C—H bonds, gets larger than the polar carbonyl group.

PROBLEM 17.9

Which compound in each pair has the higher boiling point?

a. CH_3COOH or CH_3CH_2CHO

c.

b. $CH_3CH_2CH_2CONH_2$ or $CH_3CH_2CO_2CH_3$

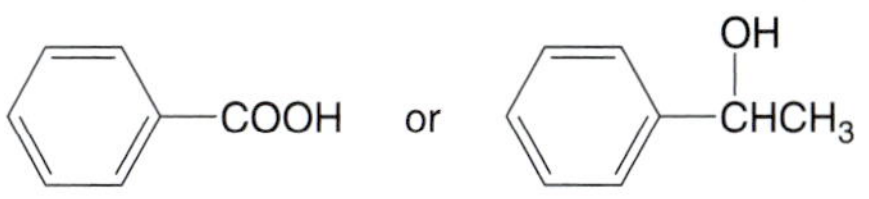

PROBLEM 17.10

Rank the following compounds in order of increasing boiling point. Which compound is the most water soluble? Which compound is the least water soluble?

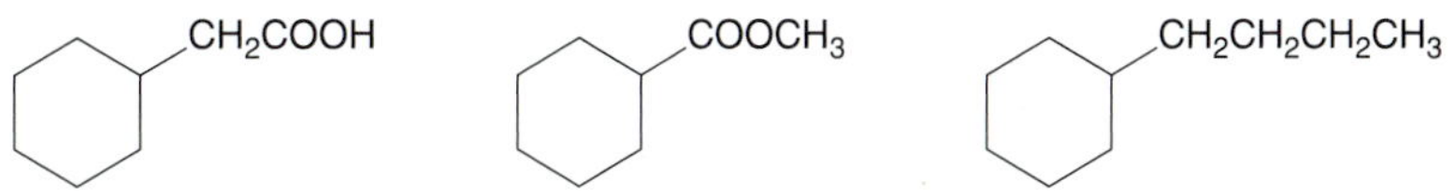

PROBLEM 17.11

Why is the boiling point of CH_3CONH_2 (221 °C) higher than the boiling point of $CH_3CON(CH_3)_2$ (166 °C), even though the latter compound has a higher molecular weight and more surface area?

17.4 INTERESTING CARBOXYLIC ACIDS IN CONSUMER PRODUCTS AND MEDICINES

Simple carboxylic acids have biting or foul odors.

Ginkgo trees have existed for over 280 million years. Extracts of the ginkgo tree have long been used in China, Japan, and India in medicine and cooking.

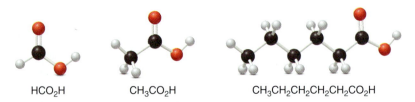

HCO_2H CH_3CO_2H $CH_3CH_2CH_2CH_2CH_2CO_2H$

- **Formic acid** (HCO_2H) is responsible for the sting of some types of ants.
- **Acetic acid** (CH_3CO_2H) is the sour-tasting component of vinegar. Air oxidation of ethanol (CH_3CH_2OH) to acetic acid makes "bad" wine taste sour.
- **Hexanoic acid** [($CH_3(CH_2)_4COOH$] has the foul odor associated with dirty socks and locker rooms. It also contributes to the unpleasant odor of ginkgo seeds. The common name for hexanoic acid is caproic acid, derived from the Latin word caper, meaning "goat."

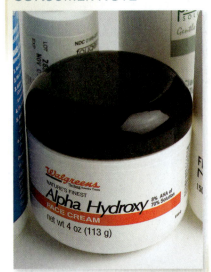

α-Hydroxy acids are used in skin care products that are promoted to remove wrinkles and age spots.

17.4A FOCUS ON HEALTH & MEDICINE SKIN CARE PRODUCTS

Several skin care products that purportedly smooth fine lines and improve skin texture contain α-hydroxy acids. **α-Hydroxy acids contain a hydroxyl group on the α (alpha) carbon to the carboxyl group.** Two common α-hydroxy acids are glycolic acid and lactic acid. Glycolic acid occurs naturally in sugarcane, and lactic acid gives sour milk its distinctive taste.

General structure

α carbon
α-hydroxy acid   glycolic acid   lactic acid

Do products that contain α-hydroxy acids make the skin look younger? α-Hydroxy acids react with the outer layer of skin cells, causing them to loosen and flake off. Underneath is a layer of healthier looking skin that has not been exposed to the sun. In this way, these skin products do not actually reverse the aging process; rather, they remove a layer of old skin that may be less pliant or contain small age spots. They can, however, give the appearance of younger skin for a time.

PROBLEM 17.12

Label the chirality center in lactic acid. Draw both enantiomers of lactic acid (Section 15.3).

PROBLEM 17.13

Which compounds are α-hydroxy acids?

a.

tartaric acid
(isolated from grapes)

b. CH_3CHCH_2—

3-hydroxybutanoic acid
(starting material for polymer synthesis)

c.

salicylic acid
(isolated from meadowsweet)

17.4B FOCUS ON HEALTH & MEDICINE
ASPIRIN AND ANTI-INFLAMMATORY AGENTS

Three common pain relievers that are also anti-inflammatory agents contain a carboxyl group—aspirin, ibuprofen, and naproxen.

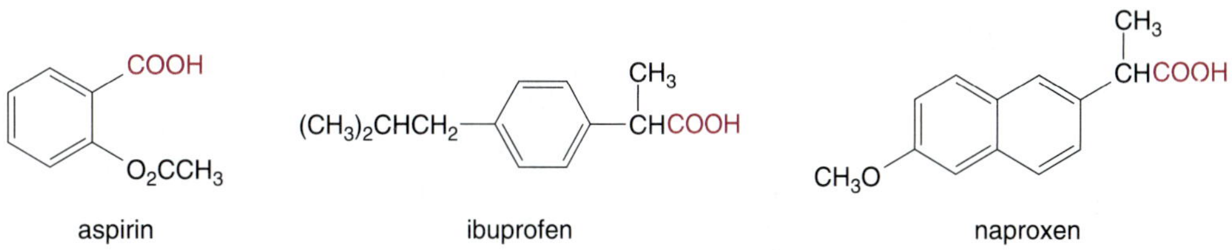

aspirin

ibuprofen

naproxen

How does aspirin relieve pain and reduce inflammation? Although aspirin is the most widely used pain reliever and anti-inflammatory agent and has been sold since the late 1800s, its mechanism of action remained unknown until the 1970s. At that time it was shown that aspirin blocks the synthesis of **prostaglandins,** carboxylic acids containing 20 carbons that are responsible for mediating pain, inflammation, and a wide variety of other biological functions.

Prostaglandins are not stored in cells. Rather, they are synthesized on an as-needed basis from arachidonic acid, an unsaturated fatty acid with four cis double bonds. **Aspirin relieves pain and decreases inflammation because it prevents the synthesis of prostaglandins, the compounds responsible for both of these physiological responses, from arachidonic acid.**

arachidonic acid

Aspirin blocks this process.

PGF$_{2\alpha}$ and compounds like it cause pain and inflammation.

PGF$_{2\alpha}$
a prostaglandin

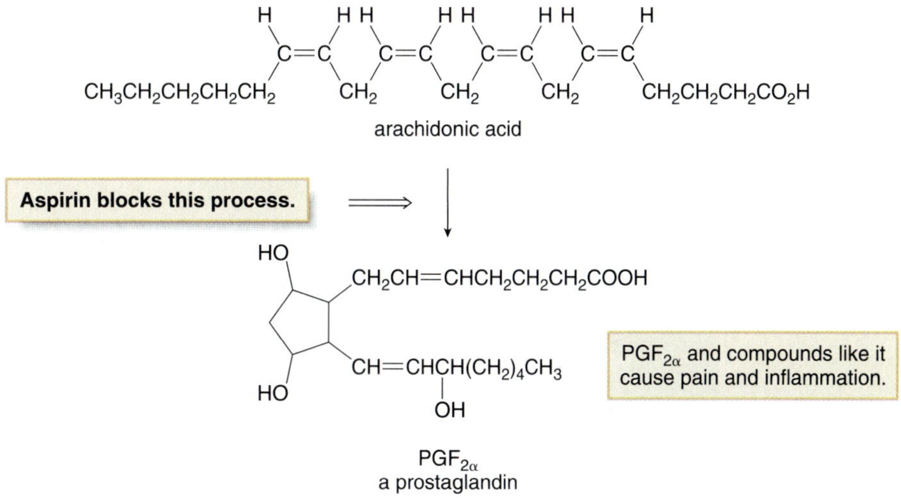

The 1982 Nobel Prize in Physiology or Medicine was awarded to John Vane, Bengt Samuelsson, and Sune Bergstrom for unraveling the details of prostaglandin synthesis, thus showing how aspirin acts.

PROBLEM 17.14

Which compound, PGF$_{2\alpha}$ (a prostaglandin) or arachidonic acid, is more water soluble? Explain your choice.

17.5   INTERESTING ESTERS AND AMIDES

Many low molecular weight esters have pleasant and very characteristic odors. These include ethyl butanoate (from mangoes), pentyl butanoate (from apricots), and methyl salicylate (oil of wintergreen).

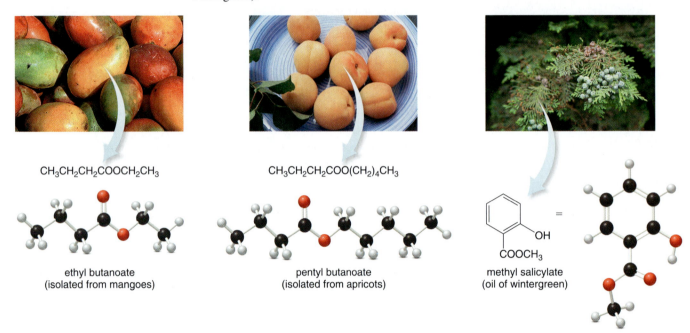

$CH_3CH_2CH_2COOCH_2CH_3$

ethyl butanoate
(isolated from mangoes)

$CH_3CH_2CH_2COO(CH_2)_4CH_3$

pentyl butanoate
(isolated from apricots)

methyl salicylate
(oil of wintergreen)

> Esters are a common functional group in many lipids, as we will learn in Section 17.9 and Chapter 19.

Melatonin, an amide synthesized by the brain's pineal gland, is thought to induce sleep. Melatonin levels in the body rise as less light falls upon the eye and drop quickly at dawn. For this reason, melatonin has become a popular supplement for travelers suffering from jet lag and individuals with mild sleep disorders.

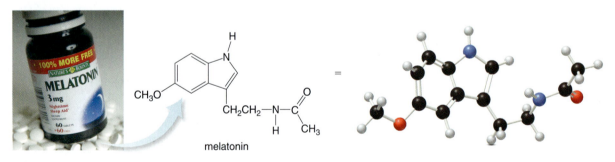

melatonin

> We will learn about proteins, naturally occurring polyamides, in Chapter 21.

Many useful drugs are esters or amides. **Benzocaine,** an ester, is the active ingredient in many over-the-counter oral topical anesthetics such as Orajel. **Acetaminophen,** an amide, is the active ingredient in Tylenol. Acetaminophen relieves pain and fever. In large doses acetaminophen causes liver damage, so dosage recommendations must be carefully followed.

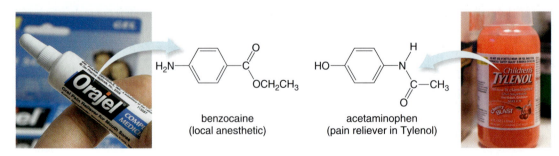

benzocaine
(local anesthetic)

acetaminophen
(pain reliever in Tylenol)

PROBLEM 17.15

Exaltolide is a synthetic lactone with a musky odor used in perfumes. Exaltolide has molecular formula $C_{14}H_{26}O_2$ and contains a 15-membered ring. Draw its structure.

PROBLEM 17.16

In addition to ethyl butanoate (Section 17.5), mangoes contain several other esters. Draw the structure of each of the following esters that has been isolated from the mango: (a) butyl formate; (b) ethyl octanoate; (c) ethyl propanoate.

17.6 THE ACIDITY OF CARBOXYLIC ACIDS

As their name implies, **carboxylic acids are acids;** that is, **they are** *proton donors.* When a carboxylic acid is dissolved in water, an acid–base equilibrium occurs: the carboxylic acid donates a proton to H_2O, forming its conjugate base, a **carboxylate anion,** and water gains a proton, forming its conjugate acid, H_3O^+.

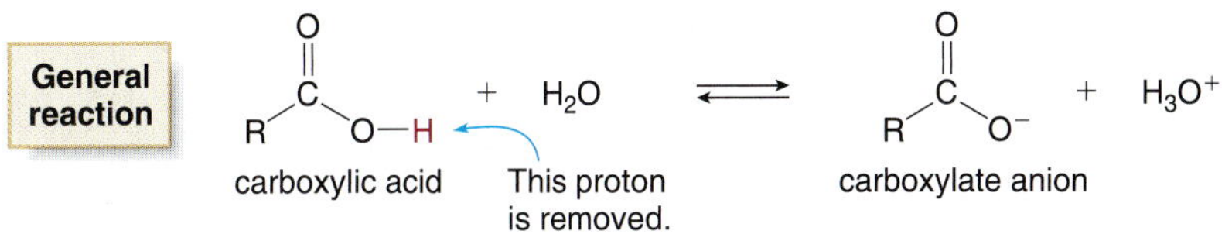

K_a values for a variety of acids are given in Section 9.4.

While carboxylic acids are more acidic than other families of organic compounds, they are weak acids compared to inorganic acids like HCl or H_2SO_4. Typical acid dissociation constants (K_a) for carboxylic acids fall in the range 10^{-4} to 10^{-5}. Thus, only a small percentage of a carboxylic acid is ionized in aqueous solution.

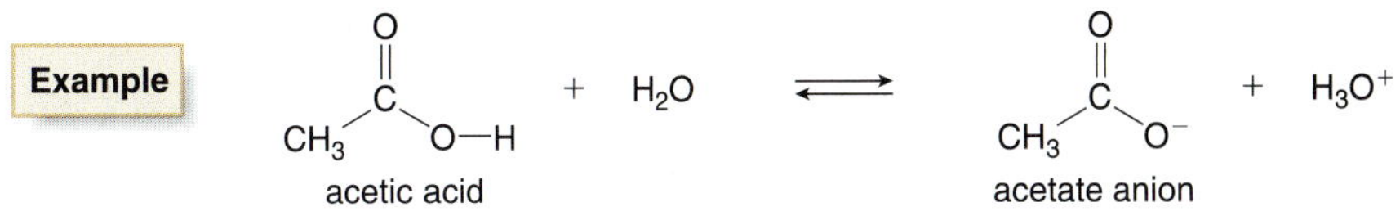

17.6A REACTION WITH BASES

Carboxylic acids react with bases such as NaOH to form water-soluble salts. In this reaction, the acid–base equilibrium is shifted to the right, and essentially all of the carboxylic acid is converted to its carboxylate anion.

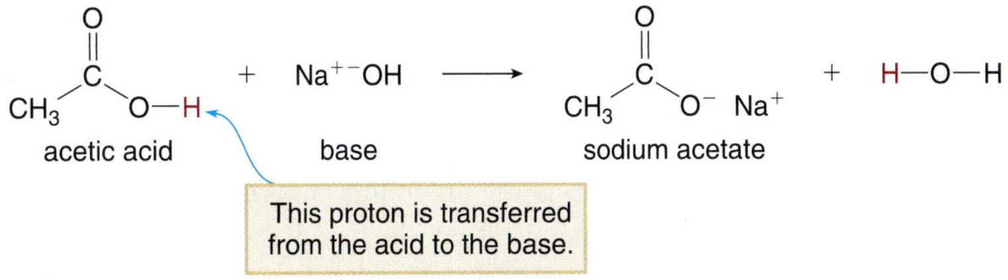

- A proton is removed from acetic acid (CH_3COOH) to form its conjugate base, the acetate anion (CH_3COO^-), which is present in solution as its sodium salt, sodium acetate.
- Hydroxide (^-OH) gains a proton to form neutral H_2O.

Similar acid–base reactions occur with other hydroxide bases (KOH), sodium bicarbonate ($NaHCO_3$), and sodium carbonate (Na_2CO_3). **In each reaction, a proton is transferred from the acid (RCOOH) to the base.**

SAMPLE PROBLEM 17.3

What products are formed when propanoic acid (CH_3CH_2COOH) reacts with potassium hydroxide (KOH)?

ANALYSIS

In any acid–base reaction with a carboxylic acid:
- Remove a proton from the carboxyl group (COOH) and form the carboxylate anion ($RCOO^-$).
- Add a proton to the base. If the base has a –1 charge to begin with, it becomes a neutral product when a proton (H^+) is added to it.
- Balance the charge of the carboxylate anion by drawing it as a salt with a metal cation.

SOLUTION

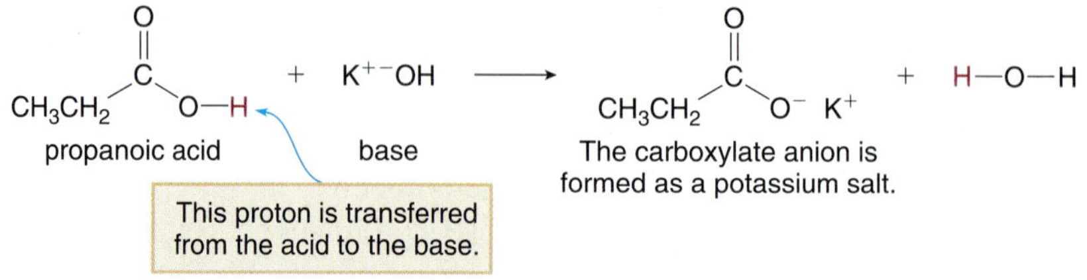

This proton is transferred from the acid to the base.

Thus, CH_3CH_2COOH loses a proton to form $CH_3CH_2COO^-$, which is present in solution as its potassium salt, $CH_3CH_2COO^- K^+$. Hydroxide (^-OH) gains a proton to form H_2O.

PROBLEM 17.17

Draw the products of each acid–base reaction.

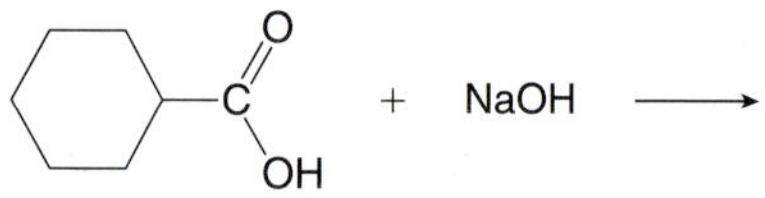

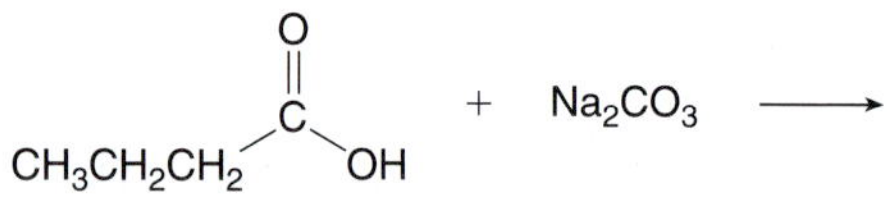

PROBLEM 17.18

What products are formed when benzoic acid (C_6H_5COOH) is treated with each base:
(a) NaOH; (b) Na_2CO_3; (c) $NaHCO_3$?

17.6B CARBOXYLATE ANIONS—SALTS OF CARBOXYLIC ACIDS

The salts of carboxylic acids that are formed by acid–base reactions are water-soluble ionic solids. Thus, a water-*insoluble* carboxylic acid like octanoic acid can be converted to its water-*soluble* sodium salt by reaction with NaOH. This reaction is important in determining the solubility of drugs that contain a carboxyl group, as we will learn in Section 17.7.

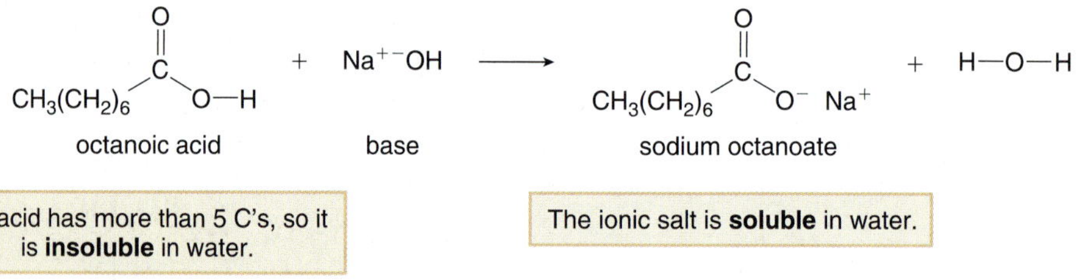

The acid has more than 5 C's, so it is **insoluble** in water.

The ionic salt is **soluble** in water.

To name the metal salts of carboxylate anions, three parts must be put together: the name of the metal cation, the parent name indicating the number of carbons in the parent carboxylic acid, and the suffix indicating that the species is a salt. The suffix is added by changing the *-ic acid* ending of the parent carboxylic acid to the suffix *-ate.*

| name of the metal cation | + | parent | + | -ate (suffix) |

SAMPLE PROBLEM 17.4

Give an acceptable name for each salt.

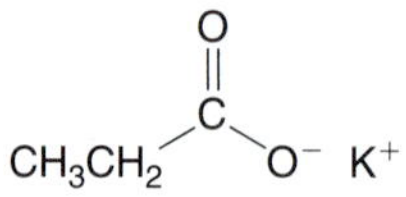

ANALYSIS

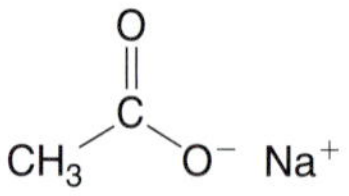

Name the carboxylate salt by putting three parts together:
- the name of the metal cation
- the parent name that indicates the number of carbons in the parent chain
- the suffix, *-ate*

SOLUTION

a.

O
‖
CH₃—C—O⁻ Na⁺

sodium cation

parent + suffix
acet-   -ate

- The first part of the name is the metal cation, sodium.
- The parent name is derived from the common name, acetic acid. Change the *-ic acid* ending to *-ate*; acetic acid → acetate.

Answer: sodium acetate

b.

O
‖
CH₃CH₂—C—O⁻ K⁺

potassium cation

parent + suffix
propano-   -ate

- The first part of the name is the metal cation, potassium.
- The parent name is derived from the IUPAC name, propanoic acid. Change the *-ic acid* ending to *-ate*; propanoic acid → propanoate.

Answer: potassium propanoate

PROBLEM 17.19

Name each salt of a carboxylic acid.

a. $CH_3CH_2CH_2CO_2{}^-Na^+$

b. ⬡—COO⁻ Li⁺

Salts of carboxylic acids are commonly used as preservatives. **Sodium benzoate,** which inhibits the growth of fungus, is a preservative used in soft drinks, and potassium sorbate is an additive that prolongs the shelf-life of baked goods and other foods. These salts do not kill bacteria or fungus. They increase the pH of the product, thus preventing further growth of microorganisms.

O
‖
⬡—C—O⁻ Na⁺

sodium benzoate

O
‖
CH₃CH=CHCH=CH—C—O⁻ K⁺

potassium sorbate

PROBLEM 17.20

Draw the structure of sodium propanoate, a common preservative.

17.6C HOW DOES SOAP CLEAN AWAY DIRT?

Soap has been used by humankind for some 2,000 years. Historical records describe its manufacture in the first century and document the presence of a soap factory in Pompeii. Prior to this time, clothes were cleaned by rubbing them on rocks in water, or by forming soapy lathers from the roots, bark, and leaves of certain plants. These plants produced natural materials called *saponins,* which act in much the same way as modern-day soaps.

Soaps are salts of carboxylic acids that have many carbon atoms in a long hydrocarbon chain. A soap molecule has two parts.

- **The ionic end is called the *polar head.***
- **The carbon chain of nonpolar C—C and C—H bonds is called the *nonpolar tail.***

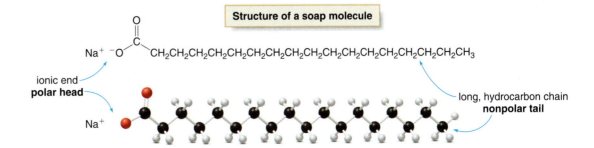

Structure of a soap molecule

O
‖
Na⁺ ⁻O—C—CH₂CH₂CH₂CH₂CH₂CH₂CH₂CH₂CH₂CH₂CH₂CH₂CH₂CH₂CH₂CH₂CH₃

ionic end
polar head

Na⁺

long, hydrocarbon chain
nonpolar tail

▼ **FIGURE 17.3   Dissolving Soap in Water**

soap molecule

$$Na^+ \ ^-O-\overset{\overset{\displaystyle O}{\|}}{C}-CH_2CH_2CH_2CH_2CH_2CH_2CH_2CH_2CH_2CH_2CH_2CH_2CH_2CH_2CH_2CH_2CH_3$$

polar head

nonpolar tail

nonpolar interior

polar exterior solvated by H_2O

soap

+

soap micelle

soap micelles in H_2O

H_2O

When soap is dissolved in H_2O, the molecules form spherical droplets with the nonpolar tails in the interior and the polar heads on the surface.

Dissolving soap in water forms **spherical droplets having the ionic heads on the surface and the nonpolar tails packed together in the interior.** These spherical droplets are called *micelles* and are illustrated in Figure 17.3. In this arrangement, the ionic heads can interact with the polar solvent water, and this brings the nonpolar, "greasy" hydrocarbon portion of the soap into solution.

How does soap dissolve grease and oil? The polar solvent water alone cannot dissolve dirt, which is composed largely of nonpolar hydrocarbons. When soap is mixed with water, however, the nonpolar hydrocarbon tails dissolve the dirt in the interior of the micelle. The polar head of the soap remains on the surface to interact with water. The nonpolar tails of the soap molecules are so well sealed off from the water by the polar head groups that the micelles are water soluble, so they can separate from the fibers of our clothes and be washed down the drain with water. In this way, soaps do a seemingly impossible task: they remove nonpolar hydrocarbon material from skin and clothes by dissolving it in the polar solvent water.

CONSUMER NOTE

All soaps are metal salts of carboxylate anions. The main difference between brands is the addition of other ingredients that do not alter their cleaning properties: dyes for color, scents for a pleasing odor, and oils for lubrication. Soaps that float have been aerated so that they are less dense than water.

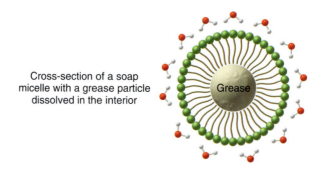

Cross-section of a soap micelle with a grease particle dissolved in the interior

Grease

PROBLEM 17.21 Draw the structure of a soap molecule that has a potassium cation and a carboxylate anion containing 16 carbons.

17.7 FOCUS ON HEALTH & MEDICINE
ASPIRIN

The modern history of aspirin dates back to 1763 when Reverend Edmund Stone reported on the analgesic effects of chewing on the bark of the willow tree. Willow bark is now known to contain salicin, which is structurally similar to aspirin.

The word *aspirin* is derived from the prefix **a-** for *acetyl* + **spir** from the Latin name *spirea* for the meadowsweet plant.

Aspirin is one of the most widely used over-the-counter drugs. Whether you purchase Anacin, Bufferin, Bayer, or a generic, the active ingredient is the same—**acetylsalicylic acid.** Aspirin is a synthetic compound; that is, it does not occur in nature, but it is similar in structure to salicin found in willow bark and salicylic acid found in meadowsweet blossoms.

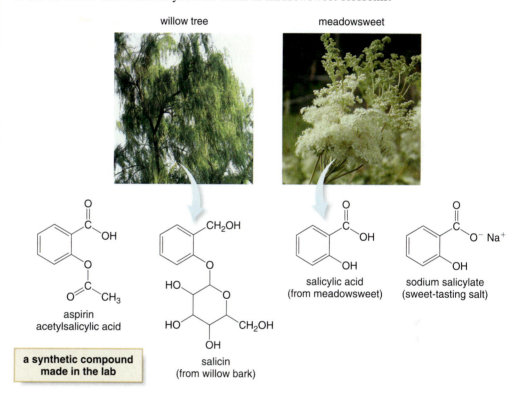

Both salicylic acid and sodium salicylate, its sodium salt, were used to relieve pain in the nineteenth century, but both had undesirable side effects. Salicylic acid irritated the mucous membranes of the mouth and stomach, and sodium salicylate was too sweet for most patients. This prompted a search for a related compound that did not have these same drawbacks. In 1899, Felix Hoffman, a German chemist at Bayer Company, developed an economic commercial synthesis of acetylsalicylic acid, and aspirin was marketed to the general public. Hoffman's work was motivated by personal reasons; his father suffered from rheumatoid arthritis and he was unable to tolerate the sweet taste of sodium salicylate.

Aspirin was first used in medicine for its analgesic (pain-relieving), antipyretic (fever-reducing), and anti-inflammatory properties. Today it is also commonly used as an antiplatelet agent; that is, it is used to prevent blood clots from forming in arteries. In this way, it is used to treat and prevent heart attacks and strokes.

Like other carboxylic acids, **aspirin is a proton donor.** When aspirin is dissolved in water, an acid–base equilibrium occurs, and the carboxyl group (COOH) donates a proton to form its conjugate base, a carboxylate anion. Because it is a weak acid, only a small percentage of the carboxylic acid is ionized in water.

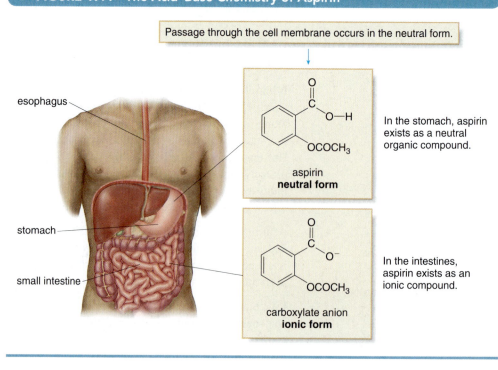

aspirin   This proton is removed.   carboxylate anion

After ingestion, aspirin travels first into the stomach and then the intestines. In the acidic environment of the stomach, aspirin remains in its neutral form. When aspirin travels into the basic environment of the small intestines, however, aspirin is converted to its carboxylate anion, as shown in Figure 17.4. In the intestines, therefore, aspirin is predominantly in its ionic form.

Why is this acid–base reaction important? **To be active, aspirin must cross a cell membrane, and to do so, it must be neutral, not ionic.** This means that aspirin crosses a cell membrane and is absorbed by the body in its neutral form in the stomach. Whether aspirin is present as its acid or its conjugate base is thus very important in determining whether it can enter a cell and act as a pain reliever and anti-inflammatory agent.

PROBLEM 17.22

Draw the acid–base reaction that occurs when aspirin is treated with NaOH.

PROBLEM 17.23

Ibuprofen is another pain reliever that is a carboxylic acid. (a) Draw the acid–base reaction that occurs when ibuprofen is treated with NaOH. (b) What form of ibuprofen (neutral or ionic) exists in the stomach? (c) What form of ibuprofen exists in the intestines?

$(CH_3)_2CHCH_2$—〈benzene ring〉—CH(CH₃)COOH

ibuprofen

▼ **FIGURE 17.4   The Acid–Base Chemistry of Aspirin**

Passage through the cell membrane occurs in the neutral form.

esophagus

stomach

small intestine

aspirin
neutral form

In the stomach, aspirin exists as a neutral organic compound.

carboxylate anion
ionic form

In the intestines, aspirin exists as an ionic compound.

17.8 THE CONVERSION OF CARBOXYLIC ACIDS TO ESTERS AND AMIDES

All acyl compounds undergo a common type of reaction—*substitution*. When an acyl compound (RCOZ) undergoes substitution, the group Z containing an electronegative atom bonded to the carbonyl carbon is *replaced* by another atom or group of atoms (Y).

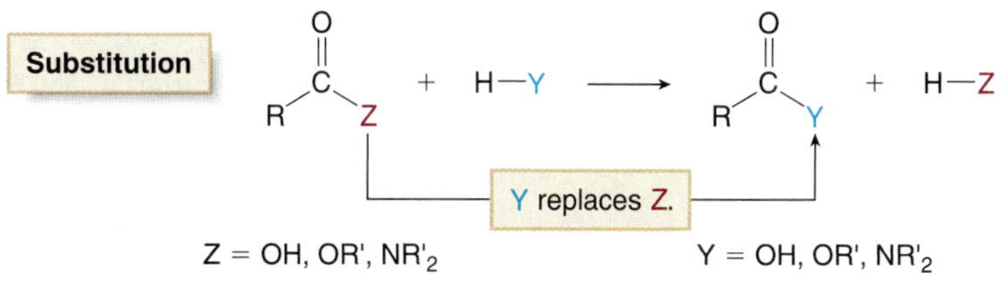

Substitution

$$\underset{R}{\overset{O}{\underset{|}{C}}}\text{—Z} \ + \ \text{H—Y} \longrightarrow \underset{R}{\overset{O}{\underset{|}{C}}}\text{—Y} \ + \ \text{H—Z}$$

Y replaces Z.

Z = OH, OR', NR'$_2$ Y = OH, OR', NR'$_2$

For example, carboxylic acids react with alcohols to form esters and with ammonia to form amides.

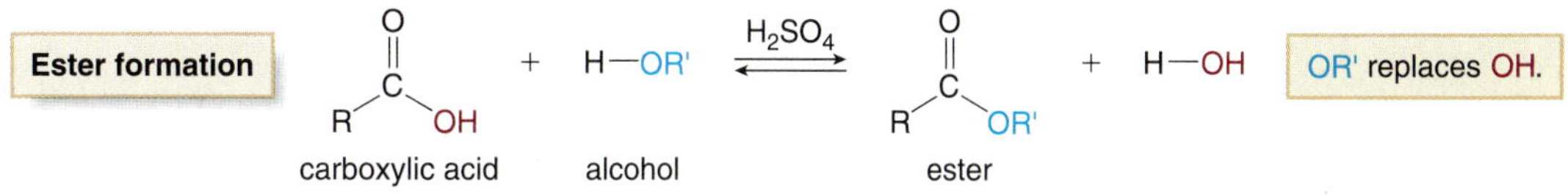

Ester formation

$$\underset{R}{\overset{O}{\underset{}{C}}}\text{—OH} \ + \ \text{H—OR'} \underset{}{\overset{H_2SO_4}{\rightleftharpoons}} \underset{R}{\overset{O}{\underset{}{C}}}\text{—OR'} \ + \ \text{H—OH}$$

carboxylic acid   alcohol   ester

OR' replaces OH.

> The Greek letter Δ (delta) is used over the arrow to indicate that a reaction requires heat (Section 5.1).

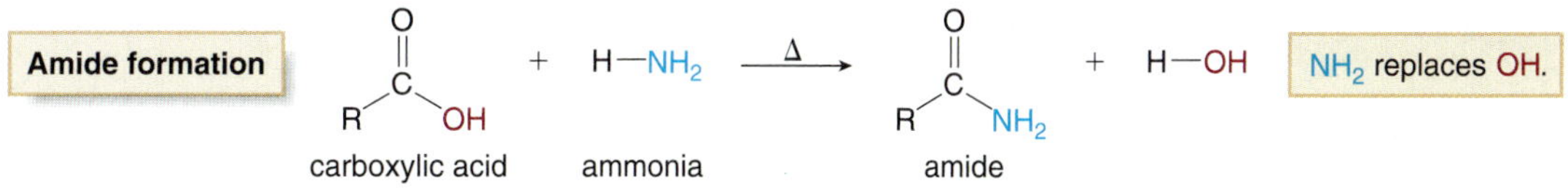

Amide formation

$$\underset{R}{\overset{O}{\underset{}{C}}}\text{—OH} \ + \ \text{H—NH}_2 \overset{\Delta}{\longrightarrow} \underset{R}{\overset{O}{\underset{}{C}}}\text{—NH}_2 \ + \ \text{H—OH}$$

carboxylic acid   ammonia   amide

NH$_2$ replaces OH.

17.8A ESTER FORMATION

Treatment of a carboxylic acid (RCOOH) with an alcohol (R'OH) in the presence of an acid catalyst forms an ester (RCOOR'). This reaction is called a **Fischer esterification.** Esterification is a substitution because the OR' group of an alcohol replaces the OH group of the starting carboxylic acid.

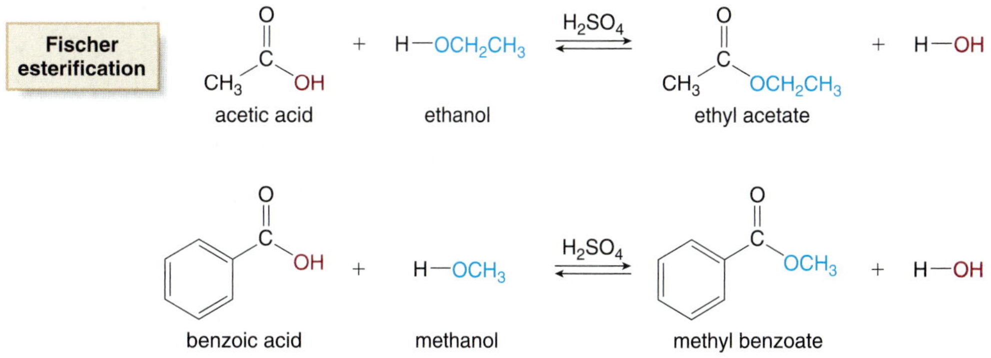

Fischer esterification

$$\underset{CH_3}{\overset{O}{\underset{}{C}}}\text{—OH} \ + \ \text{H—OCH}_2\text{CH}_3 \overset{H_2SO_4}{\rightleftharpoons} \underset{CH_3}{\overset{O}{\underset{}{C}}}\text{—OCH}_2\text{CH}_3 \ + \ \text{H—OH}$$

acetic acid   ethanol   ethyl acetate

benzoic acid   +   H—OCH$_3$ $\overset{H_2SO_4}{\rightleftharpoons}$ methyl benzoate   +   H—OH

benzoic acid   methanol   methyl benzoate

This reaction is an equilibrium process. According to Le Châtelier's principle (Section 6.6), it is driven to the right by using excess alcohol or by removing the water as it is formed. Note where the elements of H$_2$O originate: the OH group comes from the carboxylic acid, and one H atom comes from the hydrogen of the alcohol (R'OH).

SAMPLE PROBLEM 17.5

What ester is formed when propanoic acid (CH_3CH_2COOH) is treated with methanol (CH_3OH) in the presence of H_2SO_4?

ANALYSIS

To draw the products of any acyl substitution, arrange the carboxyl group of the carboxylic acid next to the functional group with which it reacts—the OH group of the alcohol in this case. Then, replace the OH group of the carboxylic acid by the OR' group of the alcohol, forming a new C—O bond at the carbonyl carbon.

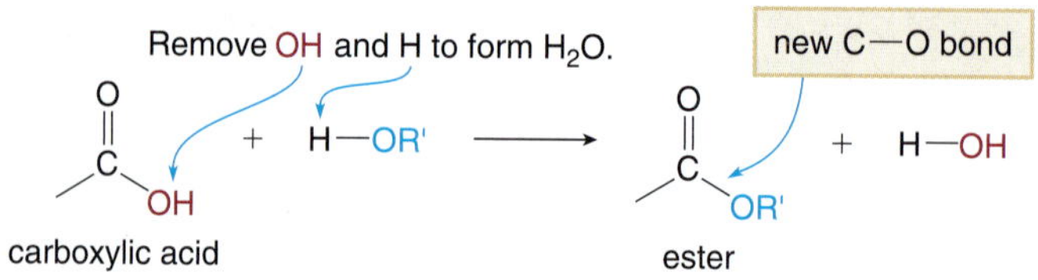

Remove OH and H to form H_2O.

new C—O bond

$$carboxylic\ acid \quad + \quad H-OR' \quad \longrightarrow \quad ester \quad + \quad H-OH$$

SOLUTION

Replace the OH group of propanoic acid by the OCH_3 group of methanol to form the ester.

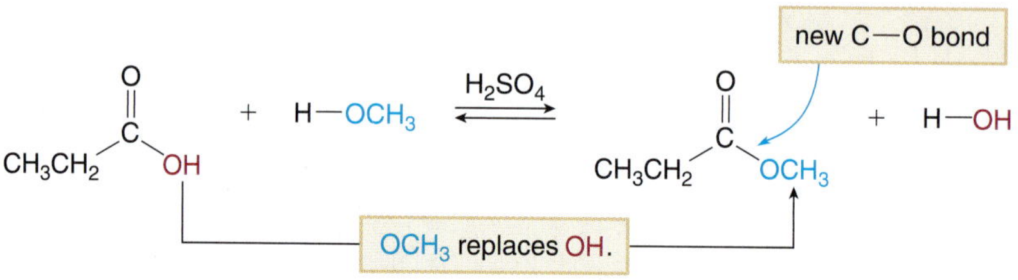

$$CH_3CH_2\text{—}COOH \quad + \quad H-OCH_3 \quad \underset{}{\overset{H_2SO_4}{\rightleftharpoons}} \quad CH_3CH_2\text{—}COOCH_3 \quad + \quad H-OH$$

new C—O bond

OCH_3 replaces OH.

PROBLEM 17.24

What ester is formed when each carboxylic acid is treated with ethanol (CH_3CH_2OH) in the presence of H_2SO_4?

a. 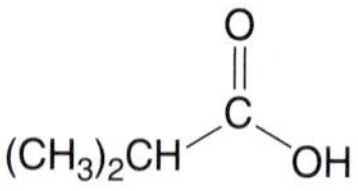 $(CH_3)_2CH\text{—}C(=O)\text{—}OH$ b. HCO_2H c. 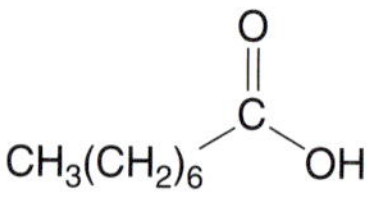 $CH_3(CH_2)_6\text{—}C(=O)\text{—}OH$ d. 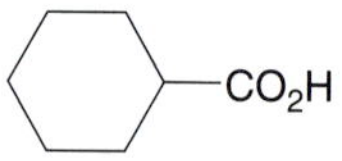 cyclohexyl—CO_2H

Fischer esterification can be used to synthesize aspirin from salicylic acid.

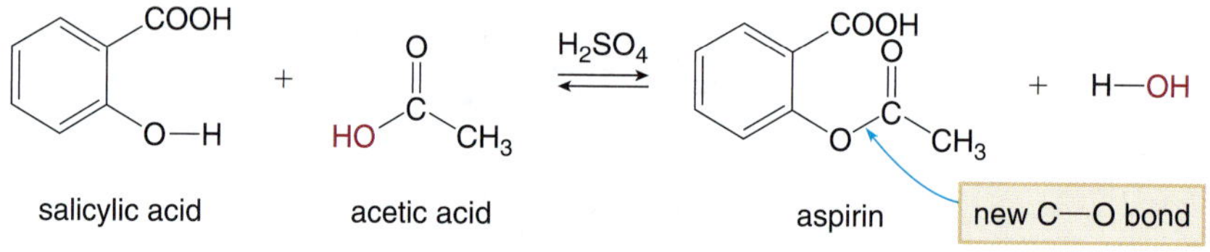

$$salicylic\ acid \quad + \quad acetic\ acid \quad \underset{}{\overset{H_2SO_4}{\rightleftharpoons}} \quad aspirin \quad + \quad H-OH$$

new C—O bond

PROBLEM 17.25

Methyl salicylate is the major component of the oil of wintergreen (Section 17.5). What carboxylic acid and alcohol are needed to synthesize methyl salicylate by Fischer esterification?

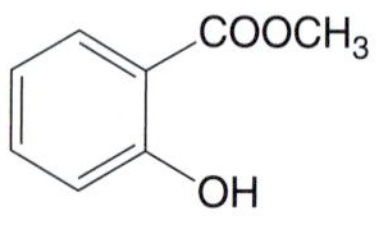

methyl salicylate

PROBLEM 17.26

Identify **A** in the following reaction. **A** was converted in one step to blattellaquinone, the sex pheromone of the female German cockroach, *Blattella germanica*.

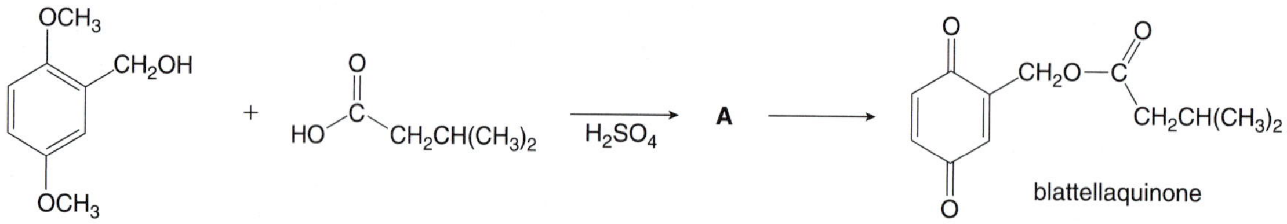

$$\text{(aromatic with } OCH_3, CH_2OH, OCH_3) \quad + \quad HO\text{—}C(=O)\text{—}CH_2CH(CH_3)_2 \quad \overset{H_2SO_4}{\longrightarrow} \quad \mathbf{A} \quad \longrightarrow \quad blattellaquinone$$

17.8B AMIDE FORMATION

Heating a carboxylic acid (RCOOH) with ammonia (NH_3) or an amine ($R'NH_2$ or R'_2NH) forms an amide. Formation of an amide from a carboxylic acid is a substitution because the OH group of the acid is replaced by a nitrogen-containing group.

Amide formation

$$\underset{\text{carboxylic acid}}{R-\overset{\overset{\displaystyle O}{\|}}{C}-OH} + \underset{R' = H \text{ or alkyl}}{H-\overset{\overset{\displaystyle |}{R'}}{\underset{\underset{\displaystyle R'}{|}}{N}}-R'} \xrightarrow{\Delta} \underset{\text{amide}}{R-\overset{\overset{\displaystyle O}{\|}}{C}-NR'_2} + H-OH$$

The identity of the nitrogen compound determines the type of amide formed.

- Reaction of RCOOH with NH_3 forms a 1° amide ($RCONH_2$).
- Reaction of RCOOH with $R'NH_2$ forms a 2° amide ($RCONHR'$).
- Reaction of RCOOH with R'_2NH forms a 3° amide ($RCONR'_2$).

Examples

$$\underset{\text{acetic acid}}{CH_3-\overset{\overset{\displaystyle O}{\|}}{C}-OH} + \underset{\text{ammonia}}{H-NH_2} \xrightarrow{\Delta} \underset{\text{1° amide}}{CH_3-\overset{\overset{\displaystyle O}{\|}}{C}-NH_2} + H-OH$$

$$\underset{\text{acetic acid}}{CH_3-\overset{\overset{\displaystyle O}{\|}}{C}-OH} + \underset{\text{methylamine}}{H-\overset{\overset{\displaystyle |}{}}{\underset{\underset{\displaystyle H}{|}}{N}}-CH_3} \xrightarrow{\Delta} \underset{\text{2° amide}}{CH_3-\overset{\overset{\displaystyle O}{\|}}{C}-NHCH_3} + H-OH$$

$$\underset{\text{acetic acid}}{CH_3-\overset{\overset{\displaystyle O}{\|}}{C}-OH} + \underset{\text{dimethylamine}}{H-\overset{\overset{\displaystyle |}{}}{\underset{\underset{\displaystyle CH_3}{|}}{N}}-CH_3} \xrightarrow{\Delta} \underset{\text{3° amide}}{CH_3-\overset{\overset{\displaystyle O}{\|}}{C}-N(CH_3)_2} + H-OH$$

Useful amides can be made by this reaction. Heating carboxylic acid **X** with diethylamine forms the 3° amide *N,N*-diethyl-*m*-toluamide, popularly known as **DEET**. DEET, the active ingredient in the most widely used insect repellents, is effective against mosquitoes, fleas, and ticks.

$$\underset{\mathbf{X}}{\text{[m-toluic acid]}} + \underset{\text{diethylamine}}{H-N(CH_2CH_3)_2} \xrightarrow{\Delta} \underset{\substack{\textit{N,N}\text{-diethyl-}m\text{-toluamide}\\ \text{(DEET)}}}{\text{[DEET structure]}} + H_2O$$

SAMPLE PROBLEM 17.6

What amide is formed when propanoic acid (CH_3CH_2COOH) is heated with ethylamine ($CH_3CH_2NH_2$)?

ANALYSIS

To draw the products of amide formation, arrange the carboxyl group of the carboxylic acid next to the H—N bond of the amine. Then, replace the OH group of the carboxylic acid by the $NHCH_2CH_3$ group of the amine, forming a new C—N bond at the carbonyl carbon.

SOLUTION   Reaction of RCOOH with an amine that has one alkyl group on the N atom (R'NH$_2$) forms a 2° amide (RCONHR').

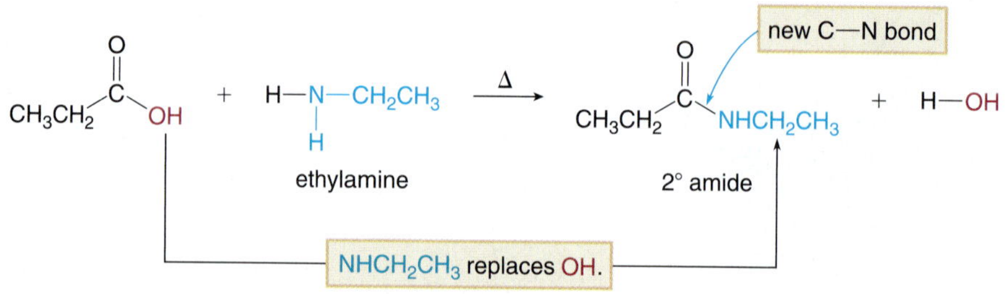

PROBLEM 17.27

What amide is formed when CH$_3$CH$_2$CH$_2$CH$_2$COOH is treated with each compound?

a. NH$_3$ b. (CH$_3$)$_2$NH   c. 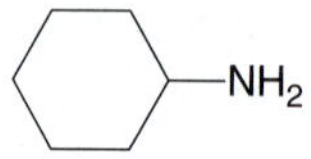 [cyclohexyl]—NH$_2$ d. CH$_3$NH$_2$

PROBLEM 17.28

What carboxylic acid and amine are needed to synthesize acetaminophen, the active pain reliever in Tylenol?

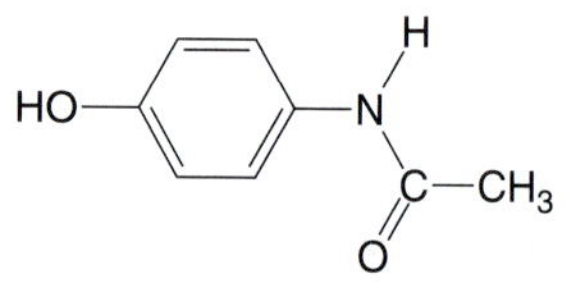

acetaminophen

17.9   HYDROLYSIS OF ESTERS AND AMIDES

Esters and amides undergo substitution at the carbonyl carbon as well. For example, **esters and amides react with water to form carboxylic acids.** This reaction is another example of **hydrolysis,** since bonds are cleaved on reaction with water.

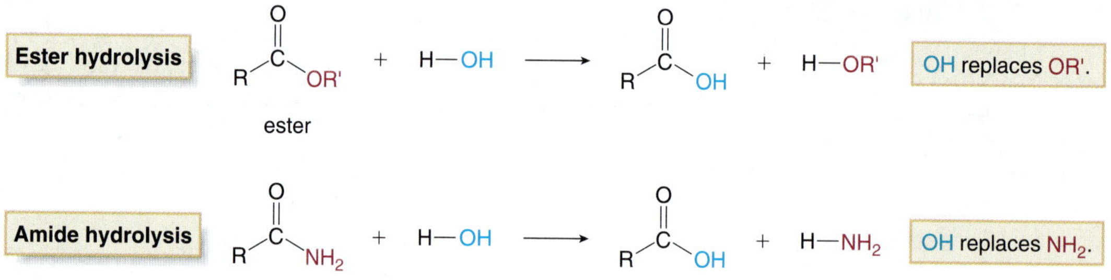

17.9A   ESTER HYDROLYSIS

Esters are hydrolyzed with water in the presence of acid or base. **Treatment of an ester (RCOOR') with water in the presence of an acid catalyst forms a carboxylic acid (RCOOH) and a molecule of alcohol (R'OH).** The hydrolysis of esters in aqueous acid is an equilibrium reaction that is driven to the right by using a large excess of water.

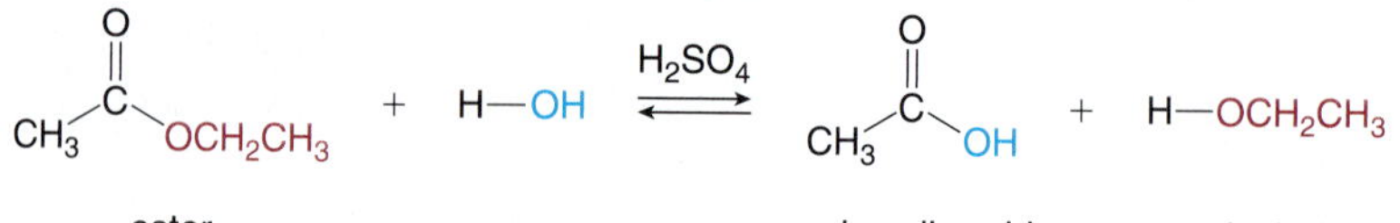

Esters are hydrolyzed in aqueous base to form carboxylate anions and a molecule of alcohol. Basic hydrolysis of an ester is called **saponification.**

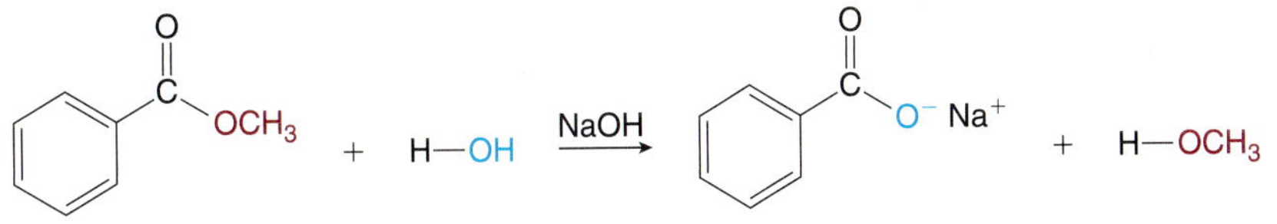

ester   carboxylate anion   alcohol

SAMPLE PROBLEM 17.7

What products are formed when ethyl propanoate ($CH_3CH_2CO_2CH_2CH_3$) is hydrolyzed with water in the presence of H_2SO_4?

ANALYSIS

To draw the products of hydrolysis in acid, replace the OR' group of the ester by an OH group from water, forming a new C—O bond at the carbonyl carbon. A molecule of alcohol (R'OH) is also formed from the alkoxy group (OR') of the ester.

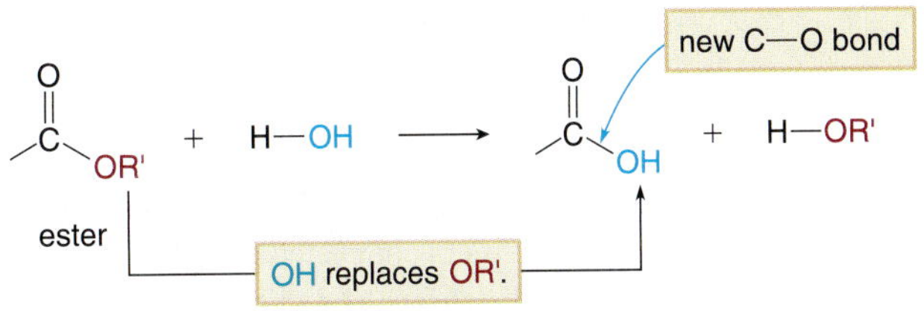

SOLUTION

Replace the OCH_2CH_3 group of ethyl propanoate by the OH group of water to form propanoic acid ($CH_3CH_2CO_2H$) and ethanol (CH_3CH_2OH).

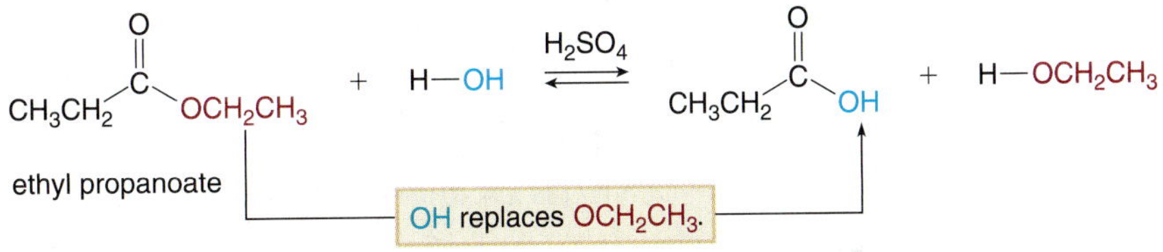

PROBLEM 17.29

What products are formed when each ester is treated with H_2O and H_2SO_4?

a. 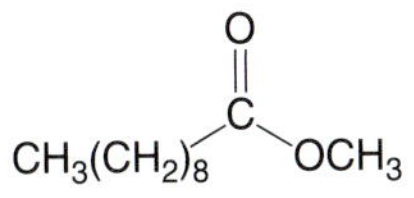 $CH_3(CH_2)_8$—CO—OCH_3

b. 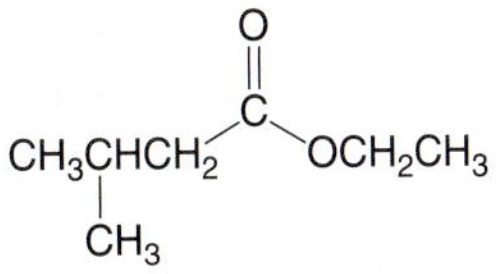 CH_3CHCH_2—CO—OCH_2CH_3 with CH_3 substituent

c. 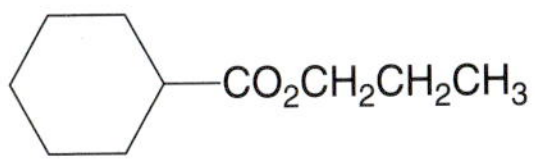 cyclohexyl—$CO_2CH_2CH_2CH_3$

PROBLEM 17.30

What products are formed when each ester in Problem 17.29 is treated with H_2O and NaOH?

PROBLEM 17.31

Aspirin cannot be sold as a liquid solution for children because it slowly undergoes hydrolysis in water. What products are formed when aspirin is hydrolyzed?

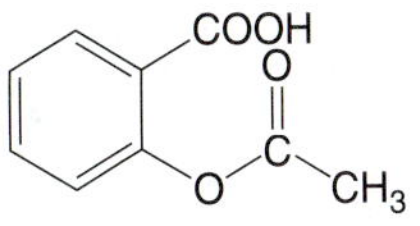

aspirin

17.9B AMIDE HYDROLYSIS

Amides are much less reactive than esters in substitution reactions. Nonetheless, under forcing conditions amides can be hydrolyzed with water in the presence of acid or base. **Treatment of an amide (RCONHR') with water in the presence of an acid catalyst (HCl) forms a carboxylic acid (RCOOH) and an amine salt (R'NH$_3^+$ Cl$^-$).**

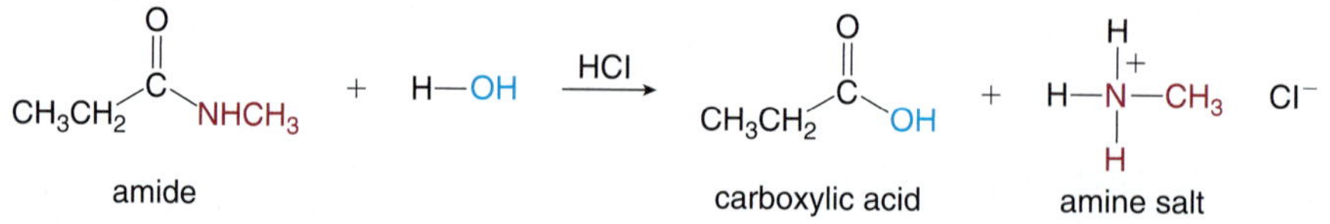

Amides are also hydrolyzed in aqueous base to form carboxylate anions and a molecule of ammonia (NH_3) or amine.

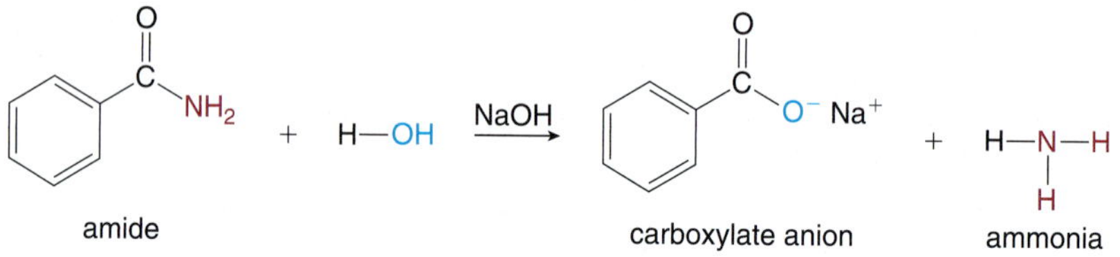

The relative lack of reactivity of the amide bond is important in the proteins in the body. Proteins are polymers connected by amide linkages, as we will learn in Chapter 21. Proteins are stable in water in the absence of acid or base, so they can perform their various functions in the cell without breaking down. The hydrolysis of the amide bonds in proteins requires a variety of specific enzymes.

SAMPLE PROBLEM 17.8

What products are formed when N-methylacetamide ($CH_3CONHCH_3$) is hydrolyzed with water in the presence of NaOH?

ANALYSIS

To draw the products of amide hydrolysis in base, replace the NHR' group of the amide by an oxygen anion (O$^-$), forming a new C—O bond at the carbonyl carbon. A molecule of amine (R'NH$_2$) is also formed from the nitrogen group (NHR') of the amide.

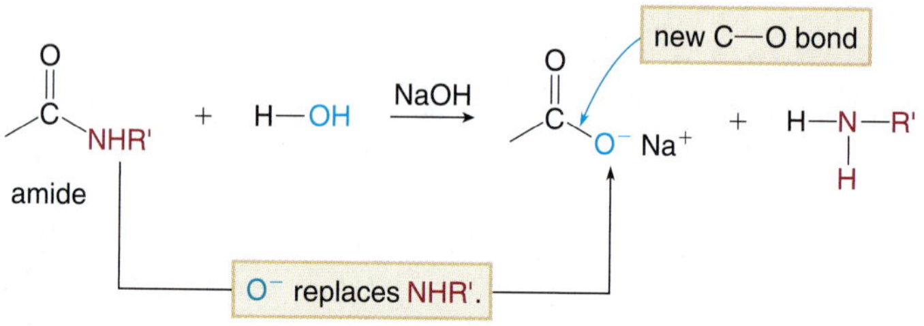

SOLUTION

Replace the NHCH$_3$ group of N-methylacetamide by a negatively charged oxygen atom (O$^-$) to form sodium acetate ($CH_3CO_2^-$ Na$^+$) and methylamine (CH_3NH_2).

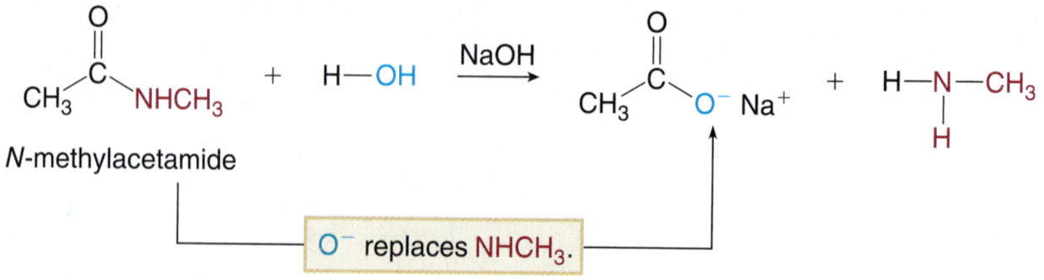

PROBLEM 17.32

What products are formed when each amide is treated with H_2O and H_2SO_4?

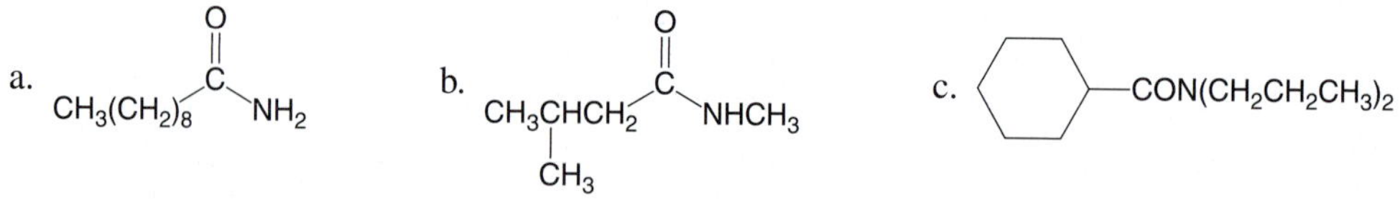

a. $CH_3(CH_2)_8$—C(=O)—NH_2

b. CH_3CHCH_2—C(=O)—$NHCH_3$ (with CH_3 branch)

c. cyclohexyl—$CON(CH_2CH_2CH_3)_2$

PROBLEM 17.33

What products are formed when each amide in Problem 17.32 is treated with H_2O and NaOH?

17.9C FOCUS ON HEALTH & MEDICINE
OLESTRA, A SYNTHETIC FAT

The most prevalent naturally occurring esters are the **triacylglycerols.** *Triacylglycerols* **contain three ester groups, each having a long carbon chain (abbreviated as R, R', and R") bonded to the carbonyl group.** Triacylglycerols are **lipids;** that is, they are water-insoluble organic compounds found in biological systems. Animal fats and vegetable oils are composed of triacylglycerols.

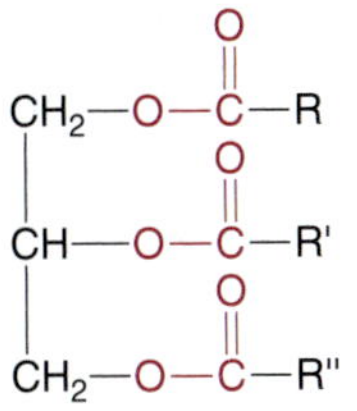

R groups have 11–19 C's.

[Three ester groups are labeled in red.]

triacylglycerol

Animals store energy in the form of triacylglycerols kept in a layer of fat cells below the surface of the skin. This fat serves to insulate the organism, as well as provide energy for its metabolic needs for long periods of time. The first step in the metabolism of a triacylglycerol is hydrolysis of the ester bonds to form glycerol and three fatty acids—long-chain carboxylic acids. **This reaction is simply ester hydrolysis.** In cells, this reaction is carried out with enzymes called **lipases.**

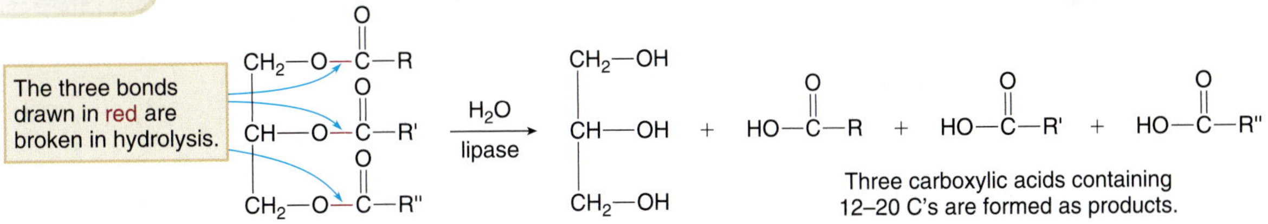

The three bonds drawn in red are broken in hydrolysis.

triacylglycerol   glycerol

Three carboxylic acids containing 12–20 C's are formed as products.

The fatty acids produced on hydrolysis are then oxidized, yielding CO_2 and H_2O, as well as a great deal of energy. Diets high in fat content can lead to a large amount of stored fat, ultimately causing an individual to be overweight. One recent attempt to reduce calories in common snack foods has been to substitute "fake fats" such as **olestra** (trade name: **Olean**) for triacylglycerols.

We will learn more about triacylglycerols in Chapter 19.

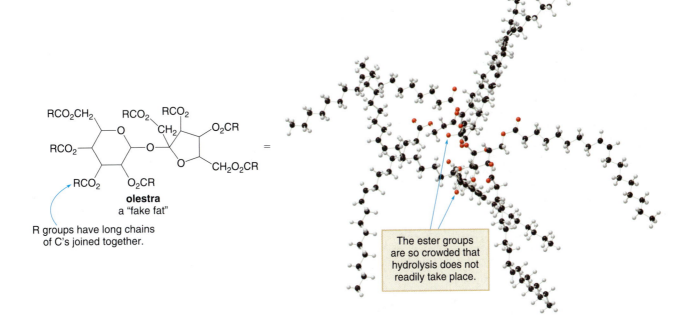

olestra
a "fake fat"

R groups have long chains of C's joined together.

The ester groups are so crowded that hydrolysis does not readily take place.

Olestra has many ester groups formed from long-chain carboxylic acids and sucrose, the sweet-tasting carbohydrate in table sugar. Olestra has many properties similar to the triacylglycerols in fats and oils. In one way, however, olestra is different. Olestra has so many ester units clustered together that they are too crowded to be hydrolyzed. As a result, olestra is *not* metabolized nor is it absorbed. Instead, it passes through the body unchanged, *providing no calories to the consumer.*

<table>
<tr><td>**PROBLEM 17.34**</td><td>What products are formed when the following triacylglycerol is hydrolyzed with water and H_2SO_4?</td></tr>
</table>

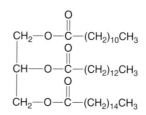

17.10 SYNTHETIC POLYMERS IN MODERN SOCIETY— POLYAMIDES AND POLYESTERS

All natural and synthetic fibers are polymers. Natural fibers are obtained from either plant or animal sources, and this determines the fundamental nature of their chemical structure. Fibers like **wool and silk obtained from animals are proteins,** and so they are joined together by many amide linkages.

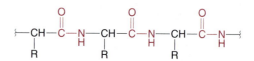

wool and silk—polymers with many amide bonds
(labeled in red)

An important practical application of organic chemistry has been the synthesis of synthetic fibers, many of which have properties that are different from and sometimes superior to their naturally occurring counterparts. Two common classes of synthetic polymers are polyamides and polyesters.

17.10A NYLON—A POLYAMIDE

The search for a synthetic fiber with properties similar to silk led to the discovery of **nylon,** a **polyamide.** There are several different kinds of nylon, but the best known is nylon 6,6.

nylon 6,6
[The amide bonds are labeled in red.]

To synthesize nylon 6,6, two monomers are needed. Each monomer has six carbons (hence its name) and two functional groups. One monomer is a dicarboxylic acid, adipic acid [$HO_2C(CH_2)_4CO_2H$], and the second monomer is a diamine, hexamethylenediamine [$H_2N(CH_2)_6NH_2$]. Heating the two monomers together at high temperature forms the amide bonds of the long polymer chain. When each amide is formed, a molecule of H_2O is lost from the starting materials. Nylon is an example of a **condensation polymer.**

Polymers were first discussed in detail in Section 13.8.

Each NH₂ and COOH react.

$$H_2N-(CH_2)_6-NH_2 \ + \ HO-\overset{\overset{\displaystyle O}{\|}}{C}-(CH_2)_4-\overset{\overset{\displaystyle O}{\|}}{C}-OH \ + \ H_2N-(CH_2)_6-NH_2 \ + \ HO-\overset{\overset{\displaystyle O}{\|}}{C}-(CH_2)_4-\overset{\overset{\displaystyle O}{\|}}{C}-OH$$

loss of H₂O   loss of H₂O   loss of H₂O

Δ

new amide bonds

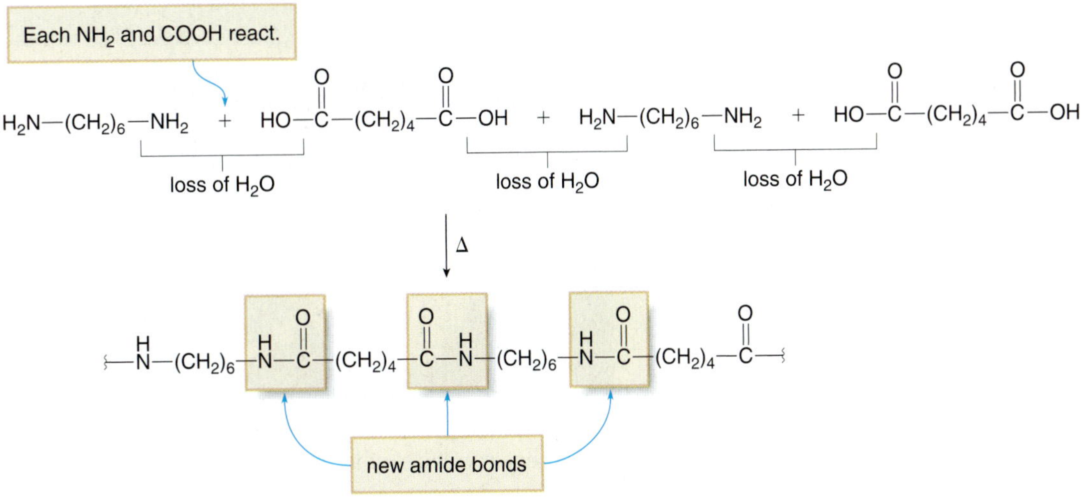

nylon 6,6

- A *condensation polymer* is a polymer formed when monomers containing two functional groups come together with loss of a small molecule such as water.

CONSUMER NOTE

Armadillo bicycle tires reinforced with Kevlar are hard to pierce with sharp objects, so a cyclist rarely gets a flat tire.

Kevlar is a polyamide formed from terephthalic acid and 1,4-diaminobenzene. The aromatic rings of the polymer backbone make the chains less flexible than those of nylon 6,6, resulting in a very strong material. Kevlar is light in weight compared to other materials that are similar in strength, so it is used in many products, such as bulletproof vests, army helmets, and the protective clothing used by fire fighters.

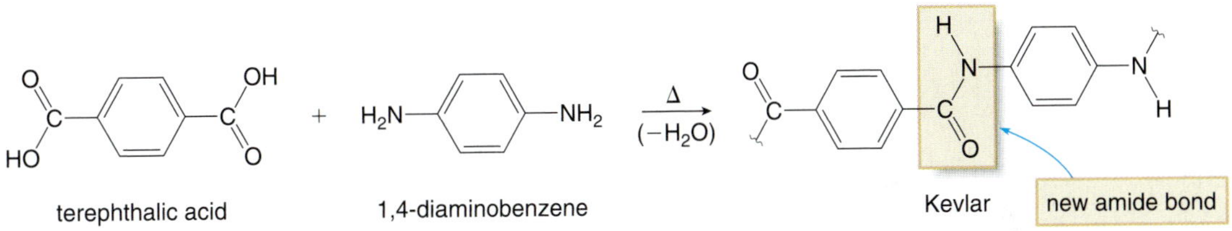

terephthalic acid   1,4-diaminobenzene

Δ
(−H₂O)

Kevlar   new amide bond

PROBLEM 17.35

What two monomers are needed to prepare nylon 6,10?

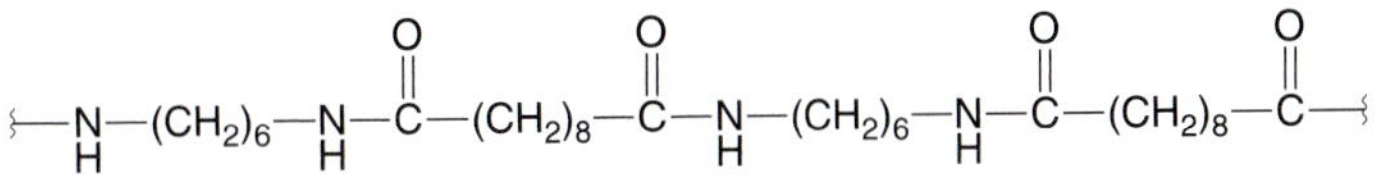

nylon 6,10

17.10B POLYESTERS

Polyesters constitute a second major class of condensation polymers. The most common polyester is polyethylene terephthalate (**PET**), which is sold under a variety of trade names depending on its use (Dacron, Terylene, and Mylar). Plastic soft drink bottles are made of PET.

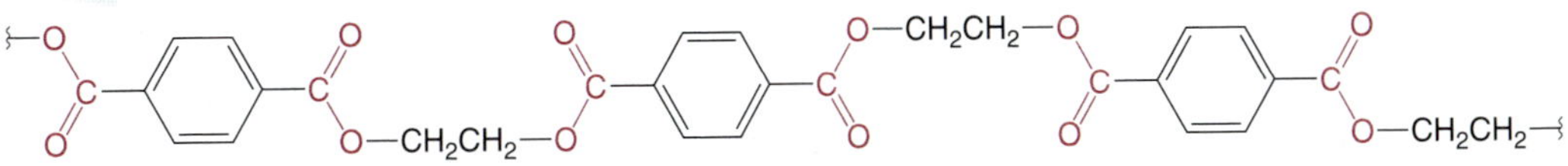

polyethylene terephthalate
PET
[Ester bonds are drawn in red.]

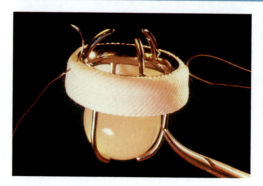

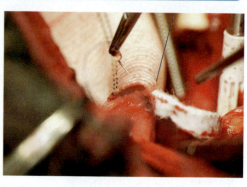

a. PET is used to make the knitted material   b. Some replacement blood vessels are
   present in artificial heart valves.   made with PET.

One method of synthesizing PET is by the acid-catalyzed reaction of a dicarboxylic acid
(terephthalic acid) with ethylene glycol ($HOCH_2CH_2OH$) in a Fischer esterification reaction.

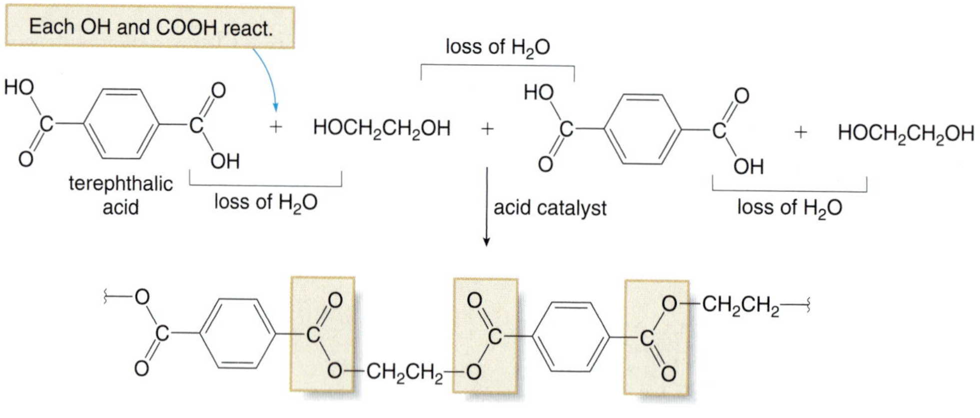

Because these polymers are easily and cheaply prepared and they form strong and chemically
stable materials, they have been used in clothing, films, tires, and plastic drink bottles. Polyester
fiber is also used in artificial heart valves and replacement blood vessels (Figure 17.5).

PROBLEM 17.36

Draw the structure of Kodel, a polyester formed from 1,4-dihydroxymethylcyclohexane and
terephthalic acid. Suggest a reason why fabrics made from Kodel are stiff and crease resistant.

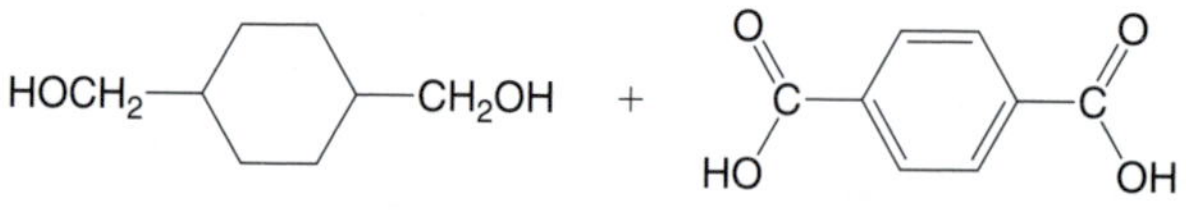

$$HOCH_2 - \bigcirc - CH_2OH \quad + \quad terephthalic\ acid$$

1,4-dihydroxymethylcyclohexane   terephthalic acid

17.10C   FOCUS ON HEALTH & MEDICINE
DISSOLVING SUTURES

Although PET is a stable material, the ester bonds in some other polyesters are quite readily
hydrolyzed to carboxylic acids and alcohols in aqueous medium. Polyesters of this sort are suited
for applications in which slow degradation is a useful feature. One example of a degradable

polymer is formed from two α-hydroxy acids—glycolic acid ($HOCH_2COOH$) and lactic acid [$CH_3CH(OH)COOH$].

Join the C=O of one acid with the OH group of the second acid.

glycolic acid   lactic acid

new C—O bonds

dissolving suture material

When glycolic acid and lactic acid are combined, the hydroxyl group of one α-hydroxy acid reacts with the carboxyl group of the second α-hydroxy acid. The resulting polymer is used by surgeons in dissolving sutures. Within weeks, the polymer is hydrolyzed to the monomers from which it was prepared, which are metabolized readily by the body. These sutures are used internally to hold tissues together while healing and scar formation occur.

17.10D FOCUS ON THE ENVIRONMENT POLYMER RECYCLING

The same desirable characteristics that make polymers popular materials for consumer products—durability, strength, and lack of reactivity—also contribute to environmental problems. Many polymers do not degrade readily, and as a result, billions of pounds of polymers end up in landfills every year. Recycling polymers is one way to help combat the waste problem they create.

The most common polymers are each assigned a recycling code (1–6); **the lower the number, the easier the polymer is to recycle.** PET materials are assigned recycling code "1" because they are the easiest to recycle. Since recycled polymers are often still contaminated with small amounts of adhesives and other materials, these recycled polymers are generally not used for storing food or drink products. Recycled PET is used to make fibers for fleece clothing and carpeting.

PET can also be converted back to the monomers from which it was made. For example, heating PET with water and acid breaks the ester bonds of the polymer chain to give ethylene glycol ($HOCH_2CH_2OH$) and terephthalic acid. These monomers then serve as starting materials for more PET.

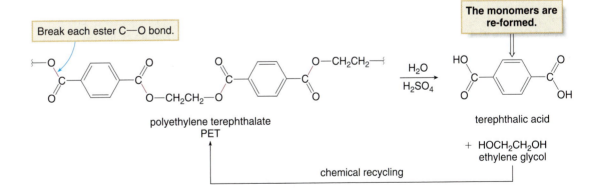

Break each ester C—O bond.

The monomers are re-formed.

polyethylene terephthalate
PET

$\xrightarrow{H_2O, H_2SO_4}$

terephthalic acid

+ $HOCH_2CH_2OH$
ethylene glycol

chemical recycling

PROBLEM 17.37   A polyester can be converted back to the monomers from which it is made much more easily than a polymer like polyethylene, $[-CH_2CH_2-]_n$. Explain why this is so.

Penicillin was used to treat injured soldiers in World War II before its structure had been conclusively determined.

17.11 FOCUS ON HEALTH & MEDICINE
PENICILLIN

In the twenty-first century it is hard to imagine that an infected cut or scrape could be life-threatening. Before antibiotics were discovered in the early twentieth century, however, that was indeed the case.

The antibiotic properties of **penicillin** were first discovered in 1928 by Sir Alexander Fleming, who noticed that a mold of the genus *Penicillium* inhibited the growth of certain bacteria. After years of experimentation, penicillin was first used to treat a female patient who had developed a streptococcal infection in 1942. By 1944, penicillin production was given high priority by the United States government, because it was needed to treat the many injured soldiers in World War II.

The penicillins are a group of related antibiotics. All penicillins contain two amide units. One amide is part of a four-membered ring called a **β-lactam.** The second amide is bonded to the α carbon of the β-lactam carbonyl group. Particular penicillins differ in the identity of the R group in the amide side chain. The first penicillin to be discovered was penicillin G. Amoxicillin is another example in common use today.

General structure Penicillin

α carbon

β-lactam

penicillin G

amoxicillin

Penicillin has no effect on mammalian cells because they are surrounded by a flexible membrane composed of a lipid bilayer (Chapter 19) and not a cell wall.

Unlike mammalian cells, bacterial cells are surrounded by a fairly rigid cell wall, which allows the bacterium to live in many different environments. **Penicillin interferes with the synthesis of the bacterial cell wall.** Penicillin kills bacteria by a substitution reaction that occurs at the amide that is part of the β-lactam. The β-lactam is more reactive than a regular amide because the bond angles in the small ring must be 90°. This feature makes the ring unstable. The β-lactam ring of the penicillin molecule reacts with an enzyme needed to synthesize the cell wall, and this deactivates the enzyme. Cell wall construction is halted, killing the bacterium. This process is discussed further in Section 21.9C.

PROBLEM 17.38	How many chirality centers does penicillin G contain?

CHAPTER HIGHLIGHTS

KEY TERMS

Acyl group (17.1)
Alpha (α) carbon (17.2)
Amide (17.1)
Beta (β) carbon (17.2)
Carboxyl group (17.1)
Carboxylate anion (17.6)
Carboxylic acid (17.1)
Condensation polymer (17.10)
Ester (17.1)

Fischer esterification (17.8)
Hydrolysis (17.9)
α-Hydroxy acid (17.4)
Lactam (17.1)
Lactone (17.1)
Lipid (17.9)
Micelle (17.6)
Nylon (17.10)
Penicillin (17.11)

Polyamide (17.10)
Polyester (17.10)
Primary (1°) amide (17.1)
Prostaglandin (17.4)
Saponification (17.9)
Secondary (2°) amide (17.1)
Soap (17.6)
Tertiary (3°) amide (17.1)
Triacylglycerol (17.9)

KEY REACTIONS

[1] Acid–base reaction of carboxylic acids (17.6)

[2] Substitution reactions of carboxylic acids (17.8)

a. Formation of esters

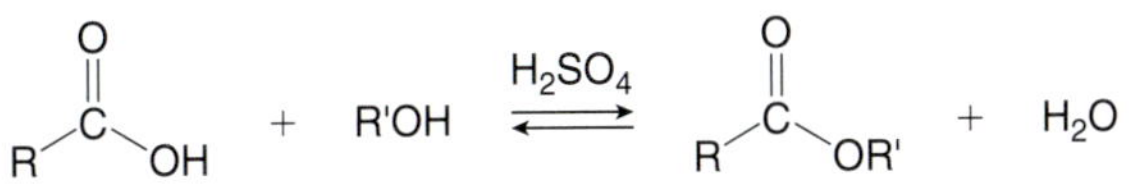

b. Formation of amides

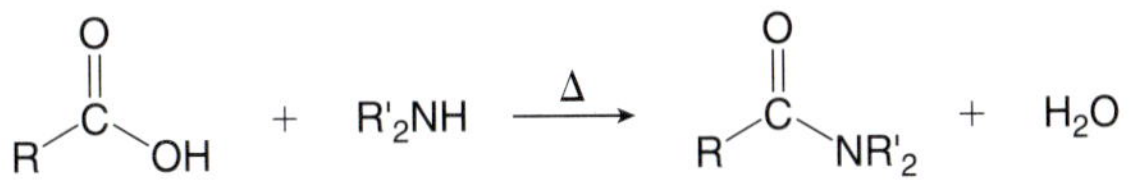

R' = H or alkyl

[3] Hydrolysis of acyl compounds (17.9)

a. Ester hydrolysis

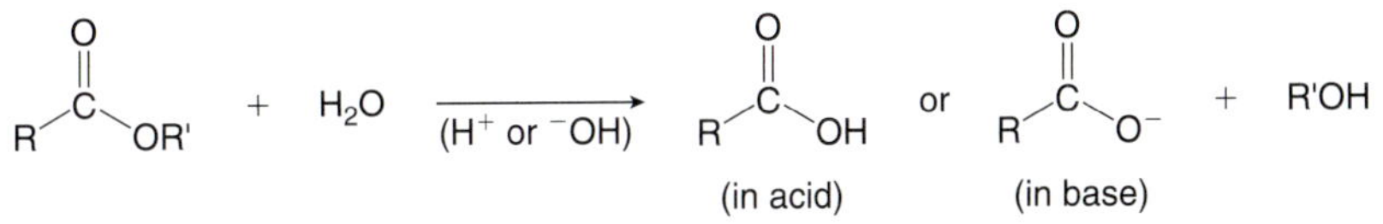

(in acid)   (in base)

b. Amide hydrolysis

(in acid)   (in base)

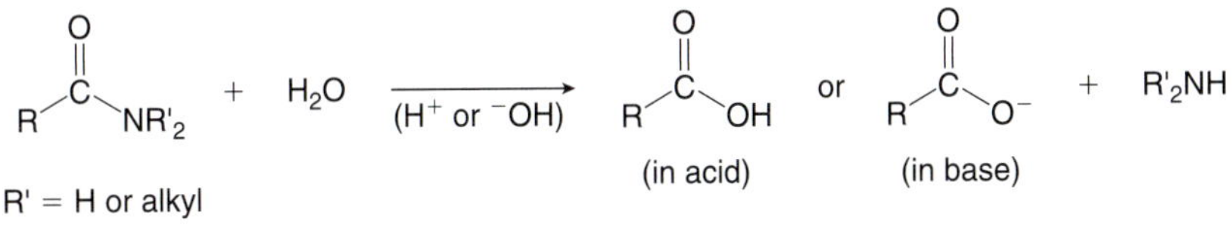

R' = H or alkyl

KEY CONCEPTS

❶ **What are the characteristics of carboxylic acids, esters, and amides?**
- Carboxylic acids have the general structure RCOOH; esters have the general structure RCOOR'; amides have the general structure RCONR'$_2$, where R' = H or alkyl. (17.1)
- The carbonyl carbon is trigonal planar with bond angles of 120°. (17.1)
- All acyl compounds have a polar C=O. RCO$_2$H, RCONH$_2$, and RCONHR' are capable of intermolecular hydrogen bonding. (17.3)

❷ **How are carboxylic acids, esters, and amides named? (17.2)**
- Carboxylic acids are identified by the suffix -oic acid.
- Esters are identified by the suffix -ate.
- Amides are identified by the suffix -amide.

❸ **Give examples of useful carboxylic acids. (17.4)**
- α-Hydroxy carboxylic acids are used in skin care products.
- Aspirin, ibuprofen, and naproxen are pain relievers and anti-inflammatory agents.
- Aspirin is an anti-inflammatory agent because it blocks the synthesis of prostaglandins from arachidonic acid.

❹ **Give examples of useful esters and amides. (17.5)**
- Some esters—ethyl butanoate, pentyl butanoate, and methyl salicylate—have characteristic odors and flavors.
- Benzocaine is the active ingredient in over-the-counter oral topical anesthetics.
- Acetaminophen is the active ingredient in Tylenol.

❺ **What products are formed when carboxylic acids are treated with base? (17.6)**
- Carboxylic acids react with bases to form carboxylate anions (RCOO$^-$).
- Carboxylate anions are water soluble and commonly used as preservatives.

❻ **How does soap clean away dirt? (17.6C)**
- Soaps are salts of carboxylic acids that have many carbon atoms in a long hydrocarbon chain. A soap molecule has an ionic head and a nonpolar hydrocarbon tail.
- Soap forms micelles in water with the polar heads on the surface and the hydrocarbon tails in the interior. Grease and dirt dissolve in the nonpolar tails, so they can be washed away with water.

❼ Discuss the acid–base chemistry of aspirin. (17.7)

- Aspirin remains in its neutral form in the stomach, and in this form it can cross a cell membrane and serve as an anti-inflammatory agent.
- A proton is removed from aspirin in the basic environment of the intestines to form an ionic carboxylate anion. This form is not absorbed.

❽ How are carboxylic acids converted to esters and amides? (17.8)

- Carboxylic acids are converted to esters by reaction with alcohols (R'OH) and acid (H_2SO_4).
- Carboxylic acids are converted to amides by heating with ammonia (NH_3) or amines (R'NH_2 or R'$_2NH$).

❾ What hydrolysis products are formed from esters and amides? (17.9)

- Esters are hydrolyzed to carboxylic acids (RCOOH) in the presence of an acid catalyst (H_2SO_4). Esters are converted to carboxylate anions ($RCOO^-$) with aqueous base (NaOH in H_2O).
- Amides are hydrolyzed to carboxylic acids (RCOOH) in the presence of an acid catalyst (HCl). Amides are converted to carboxylate anions ($RCOO^-$) with aqueous base (NaOH in H_2O).

❿ What are polyamides and polyesters and how are they formed? (17.10)

- Polyamides like nylon are polymers that contain many amide bonds. They are formed when a dicarboxylic acid is heated with a diamine.
- Polyesters like PET are polymers that contain many ester bonds. They are formed when a dicarboxylic acid is treated with a diol in the presence of acid (H_2SO_4).

⓫ How does penicillin act as an antibiotic? (17.11)

- The β-lactam of penicillin is more reactive than a regular amide and it reacts with an enzyme needed to synthesize the cell wall of a bacterium. Without a cell wall, the bacterium dies.

PROBLEMS

Selected in-chapter and end-of-chapter problems have brief answers provided in Appendix B.

Structure and Bonding

17.39 Draw the structure of a compound that fits each description:
   a. a carboxylic acid of molecular formula $C_8H_{16}O_2$ that has six carbons in its longest chain
   b. an ester of molecular formula $C_6H_{12}O_2$ that contains a methoxy group (OCH_3) bonded to the carbonyl group
   c. an ester of molecular formula $C_6H_{10}O_2$ that contains a ring
   d. a carboxylic acid of molecular formula $C_6H_{10}O_2$ that has the carboxyl group bonded to a ring

17.40 Draw the structure of a constitutional isomer of octanoic acid [$CH_3(CH_2)_6CO_2H$] that:
   a. contains a carboxyl group and no chirality center
   b. contains a carboxyl group and a chirality center
   c. contains an ester

17.41 Draw the structure of a compound of molecular formula $C_5H_{11}NO$ that contains: (a) a 1° amide; (b) a 2° amide; (c) a 3° amide.

17.42 Draw the structure of a compound of molecular formula $C_9H_{11}NO$ that contains a benzene ring and: (a) a 1° amide; (b) a 2° amide; (c) a 3° amide.

17.43 Draw the structure of four constitutional isomers of molecular formula $C_4H_8O_2$ that contain an ester.

17.44 Draw the structure of four constitutional isomers of molecular formula $C_5H_{10}O_2$ that have the general formula HCO_2R; that is, compounds that are esters of formic acid, HCO_2H.

17.45 Can a lactam be a 1° amide?

17.46 Can a lactam be a 3° amide?

17.47 What is the difference between a lactone and a lactam?

17.48 What is the difference between a lactone and an ester?

Nomenclature

17.49 Give an acceptable name for each carboxylic acid.
   a. $(CH_3)_2CHCH_2CH_2COOH$

   b. $CH_3CH_2CH_2\overset{\underset{|}{CH_2CH_3}}{CH}CHCH_2CH_2COOH$
   $\underset{|}{}$
   CH_2CH_3

   c. $CH_3CH_2CH_2CH_2CH_2\overset{}{CH}CH_2CH_3$
   $\underset{|}{}$
   CO_2H

   d. benzene ring with COOH and Cl substituents

17.50 Give an acceptable name for each carboxylic acid.

a. $(CH_3)_2CHCH_2CH_2CH_2CH_2CH_2COOH$

b. CH₃CHCHCHCH₂COOH (with CH₃, CH₃ on second and third carbons, and CH₃ below)

c. $CH_3CH_2CH_2CH_2CHCH_2CH_3$ with CO_2H substituent

d. Br—(benzene ring)—COOH

17.51 Give an acceptable name for each ester.

a. $CH_3CH_2CH_2CO_2CH_2CH_2CH_3$

b. $HCOOCH_2CH_3$

c. (benzene ring)—$CO_2(CH_2)_3CH_3$

17.52 Give an acceptable name for each ester.

a. $CH_3CO_2(CH_2)_4CH_3$

b. $HCO_2(CH_2)_5CH_3$

c. $CH_3CH_2CHCO_2CH_3$ with CH_3 substituent

17.53 Give an acceptable name for each amide.

a. $CH_3(CH_2)_4CONH_2$

b. $HCONHCH_2CH_2CH_2CH_3$

c. (benzene ring with C=O)—N with CH₂CH₃ and CH₃ groups

17.54 Give an acceptable name for each amide.

a. $CH_3(CH_2)_6CONH_2$

b. (benzene ring with C=O)—$N(CH_2CH_3)_2$

c. $(CH_3)_2CHCH_2CH_2CONH_2$

17.55 Give an acceptable name for each compound.

a. $HCONH_2$

b. (structure: CH₃—C(=O)—O⁻ Li⁺)

c. $CH_3CH_2CHCOOH$ with OH substituent

d. $CH_3CH_2CO_2CH_2CH_2CH_3$

17.56 Give an acceptable name for each compound.

a. $CH_3(CH_2)_8CO_2CH_3$

b. $(CH_3)_2CHCH_2CONH_2$

c. (structure: H—C(=O)—O⁻ K⁺)

d. $(CH_3)_3C(CH_2)_3COOH$

17.57 Draw the structure corresponding to each name.

a. 2-hydroxyheptanoic acid
b. 4-chlorononanoic acid
c. 3,4-dibromobenzoic acid
d. lithium propanoate
e. 2,2-dibromobutanoic acid
f. ethyl 2-methylpropanoate
g. potassium butanoate

17.58 Draw the structure corresponding to each name.

a. 3-methylhexanoic acid
b. 3-hydroxy-4-methylheptanoic acid
c. *p*-nitrobenzoic acid
d. sodium hexanoate
e. *m*-ethylbenzoic acid
f. propyl decanoate
g. lithium heptanoate

17.59 Draw the structure corresponding to each name.

a. propyl formate
b. butyl butanoate
c. heptyl benzoate
d. *N*-ethylhexanamide
e. *N*-ethyl-*N*-methylheptanamide

17.60 Draw the structure corresponding to each name.

a. octyl acetate
b. hexyl pentanoate
c. propyl hexanoate
d. *N*-butylbenzamide
e. *N,N*-dimethyloctanamide

17.61 What is the difference between *N*-methylbutanamide and 2-methylbutanamide? Draw structures to illustrate the difference.

17.62 What is the difference between *N*-ethylpentanamide and 3-ethylpentanamide? Draw structures to illustrate the difference.

17.63 Draw the structures for the four carboxylic acids of molecular formula $C_5H_{10}O_2$. Give the IUPAC name for each compound.

17.64 Draw the structures for the four sodium salts of the carboxylic acids in Problem 17.63, formed by treating each of these carboxylic acids with NaOH. Give the IUPAC name for each sodium salt.

Physical Properties and Intermolecular Forces

17.65 Which compound(s) can hydrogen bond to another molecule like itself? Which compounds can hydrogen bond to water? (a) HCO_2CH_3; (b) $CH_3CH_2CO_2H$.

17.66 Which compound(s) can hydrogen bond to another molecule like itself? Which compounds can hydrogen bond to water? (a) $CH_3CONHCH_3$; (b) $HCON(CH_3)_2$.

17.67 Rank the following compounds in order of increasing boiling point: $CH_3CH_2CH(CH_3)_2$, $CH_3CH_2CO_2H$, and $CH_3CH_2COCH_3$.

17.68 Rank the following compounds in order of increasing boiling point: $(CH_3)_2CHCO_2H$, $CH_3CH_2CO_2CH_3$, $(CH_3)_2CHCH(CH_3)_2$.

17.69 Explain why the boiling point of $CH_3CH_2CONH_2$ is higher than the boiling point of $CH_3CO_2CH_3$.

17.70 Which compound in each pair is more water soluble? Which compound in each pair is more soluble in an organic solvent? (a) $CH_3CH_2CH_2CH_3$ or $CH_3CH_2CH_2CH_2CO_2H$; (b) $CH_3(CH_2)_4COOH$ or $CH_3(CH_2)_4COO^- \, Na^+$.

Acidity

17.71 Draw the products of each acid–base reaction.

a. $CH_3(CH_2)_3\overset{\displaystyle O}{\overset{\|}{C}}OH$ +   KOH $\longrightarrow$

b. $(CH_3)_2CHCH_2CH_2COOH$ + Na_2CO_3 $\longrightarrow$

17.72 Draw the products of each acid–base reaction.

a. $CH_3-\!\!\!\bigcirc\!\!\!-\overset{\displaystyle O}{\overset{\|}{C}}OH$ +   NaOH $\longrightarrow$

b. Cl_3CCOOH + $NaHCO_3$ $\longrightarrow$

Formation of Esters and Amides

17.73 What ester is formed when butanoic acid $(CH_3CH_2CH_2CO_2H)$ is treated with each alcohol in the presence of H_2SO_4?

a. CH_3OH

b. $CH_3CH_2CH_2OH$

c. $\bigcirc\!\!\!-OH$

d. $\bigcirc\!\!\!-CH_2CH_2OH$

17.74 What ester is formed when each carboxylic acid is treated with 2-propanol $[(CH_3)_2CHOH]$ in the presence of H_2SO_4?

a. CH_3CH_2COOH

b. $CH_3(CH_2)_{10}\overset{\displaystyle O}{\overset{\|}{C}}OH$

c. HCO_2H

d. $CH_3O-\!\!\!\bigcirc\!\!\!-COOH$

17.75 Methylparaben is a common preservative used in cosmetics. What carboxylic acid and alcohol are needed to synthesize methylparaben by Fischer esterification?

$HO-\!\!\!\bigcirc\!\!\!-CO_2CH_3$

methylparaben

17.76 Parsol MCX is the trade name for 2-ethylhexyl 4-methoxycinnamate, a widely used sunscreen. What carboxylic acid and alcohol are needed to synthesize Parsol MCX by Fischer esterification?

$CH_3O-\!\!\!\bigcirc\!\!\!-\overset{\displaystyle H}{\underset{\displaystyle H}{C}}\!=\!C\!-\!\overset{\displaystyle O}{\overset{\|}{C}}-O-CH_2CHCH_2CH_2CH_2CH_3$ with CH_2CH_3 branch

Parsol MCX

17.77 What amide is formed when butanoic acid $(CH_3CH_2CH_2CO_2H)$ is heated with each nitrogen-containing compound?

a. NH_3

b. $CH_3CH_2NH_2$

c. $(CH_3CH_2)_2NH$

d. $CH_3NHCH_2CH_3$

17.78 What amide is formed when propylamine $(CH_3CH_2CH_2NH_2)$ is heated with each carboxylic acid?

a. CH_3CH_2COOH

b. $CH_3(CH_2)_{10}\overset{\displaystyle O}{\overset{\|}{C}}OH$

c. HCO_2H

d. $CH_3O-\!\!\!\bigcirc\!\!\!-COOH$

17.79 What carboxylic acid and alcohol are needed to synthesize each ester by Fischer esterification?

a. $\bigcirc\!\!\!-CO_2CH_3$

b. $\bigcirc\!\!\!-O-\overset{\displaystyle O}{\overset{\|}{C}}CH_3$

17.80 What carboxylic acid and alcohol are needed to synthesize each ester by Fischer esterification?

a. $(CH_3)_3CCO_2CH_3$

b. $(CH_3)_3C-O-\overset{\displaystyle O}{\overset{\|}{C}}CH_3$

17.81 What carboxylic acid and amine are needed to synthesize the pain reliever phenacetin? Phenacetin was once a component of the over-the-counter pain reliever APC (aspirin, phenacetin, caffeine), but is no longer used because of its kidney toxicity.

$CH_3CH_2O-\!\!\!\bigcirc\!\!\!-\overset{\displaystyle H}{\underset{}{N}}-\overset{\displaystyle O}{\overset{\|}{C}}CH_3$

phenacetin

17.82 What carboxylic acid and amine are needed to synthesize each amide?

a. $\bigcirc\!\!\!-CONHCH(CH_3)_2$

b. $(CH_3)_2CHCONH-\!\!\!\bigcirc$

Hydrolysis

17.83 What products are formed when each ester is hydrolyzed with water and H_2SO_4?

a. 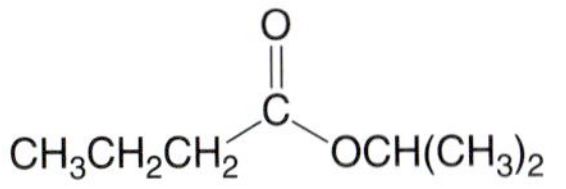
CH$_3$CH$_2$CH$_2$ ... OCH(CH$_3$)$_2$

b.

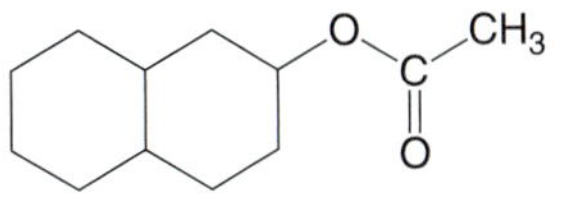

c. 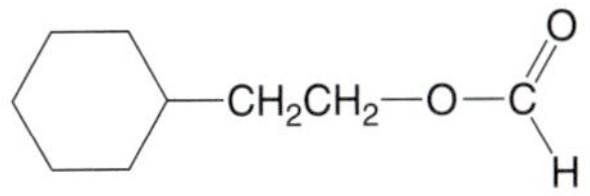
—CH$_2$CH$_2$—O—C

d. 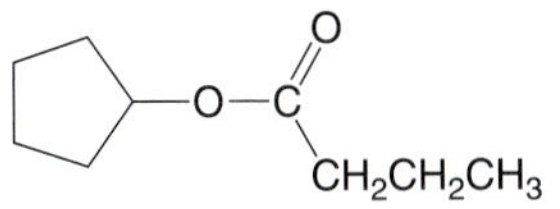
CH$_2$CH$_2$CH$_3$

17.84 What products are formed when each ester in Problem 17.83 is hydrolyzed with water and NaOH?

17.85 What products are formed when each amide is hydrolyzed with water and HCl?

a. 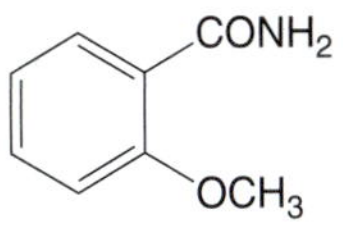
CONH$_2$
OCH$_3$

b. (CH$_3$)$_3$CCON(CH$_3$)$_2$

c.

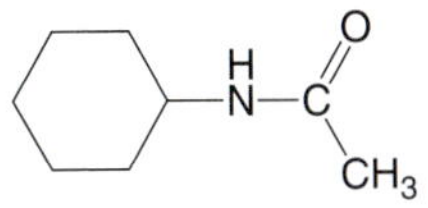

d. HO(CH$_2$)$_4$CONHCH$_3$

17.86 What products are formed when each amide in Problem 17.85 is hydrolyzed with water and NaOH?

17.87 Ethyl phenylacetate ($C_6H_5CH_2CO_2CH_2CH_3$) is a naturally occurring ester in honey. What hydrolysis products are formed when this ester is treated with water and H_2SO_4?

17.88 Benzyl acetate ($CH_3CO_2CH_2C_6H_5$) is a naturally occurring ester in peaches. What hydrolysis products are formed when this ester is treated with water and NaOH?

Polyesters and Polyamides

17.89 What dicarboxylic acid and diamine are needed to prepare the following polyamide?

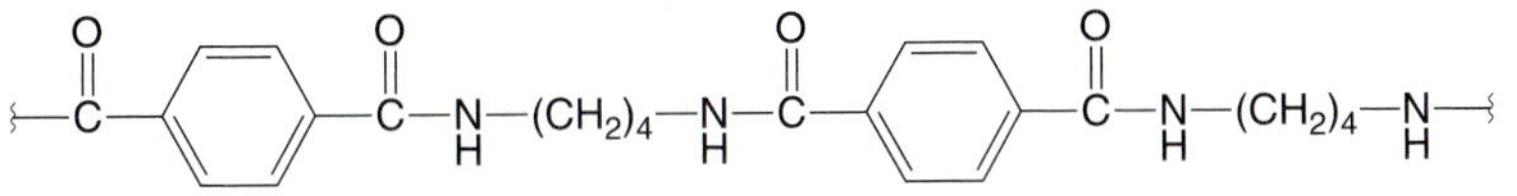

17.90 What dicarboxylic acid and diol are needed to form the following polyester?

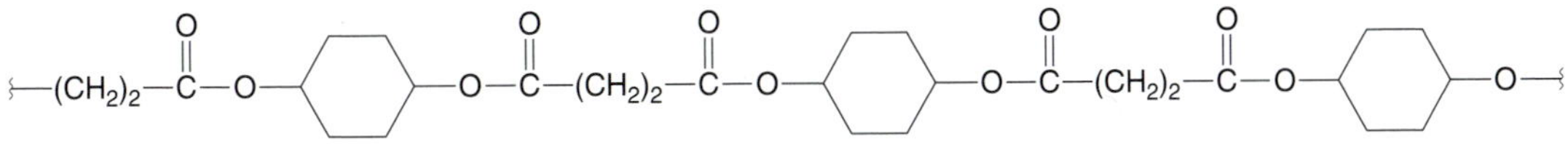

17.91 PTT (polytrimethylene terephthalate) is a polyester formed from 1,3-propanediol ($HOCH_2CH_2CH_2OH$) by the given reaction. PTT is sold under the trade name Sorona by DuPont. 1,3-Propanediol can now be made from a plant source like corn, rather than from petroleum starting materials. This makes PTT a more environmentally friendly polyester than PET. What is the structure of PTT?

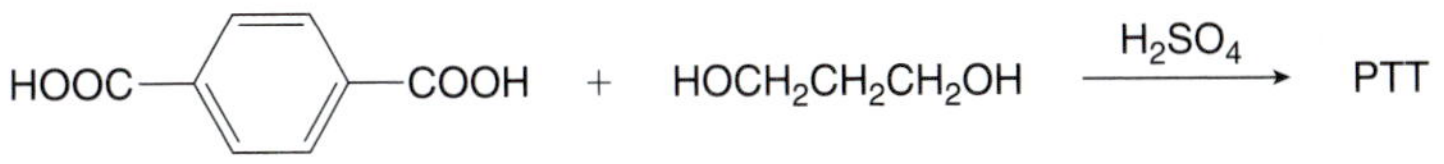

HOOC—⬡—COOH + HOCH$_2$CH$_2$CH$_2$OH $\xrightarrow{H_2SO_4}$ PTT

terephthalic acid   1,3-propanediol

17.92 What polyamide is formed when the following dicarboxylic acid and diamine are heated together?

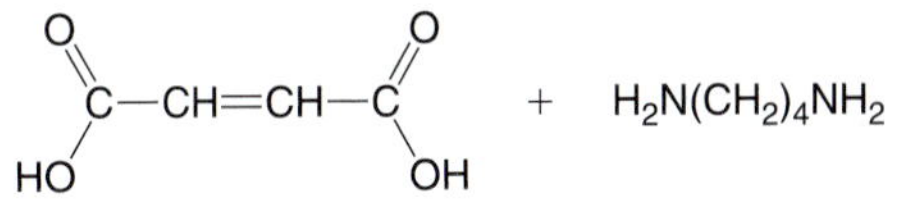

General Questions

17.93 What is the difference between saponification and esterification?

17.94 What is the difference between esterification and hydrolysis?

17.95 Draw the products formed in each reaction.

a. 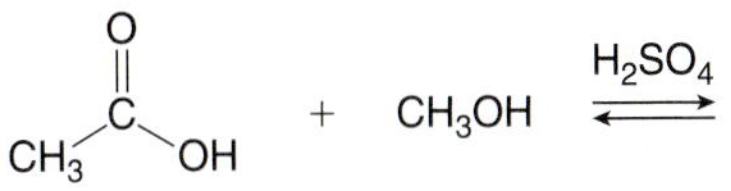
CH$_3$ ... OH + CH$_3$OH $\underset{}{\overset{H_2SO_4}{\rightleftarrows}}$

b. 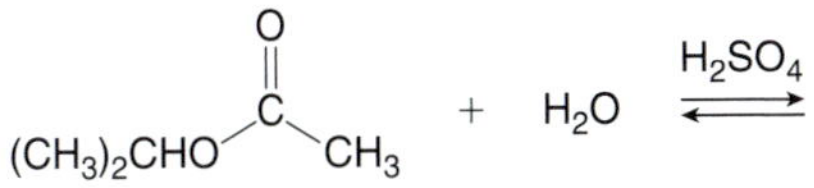
(CH$_3$)$_2$CHO ... CH$_3$ + H$_2$O $\underset{}{\overset{H_2SO_4}{\rightleftarrows}}$

c. 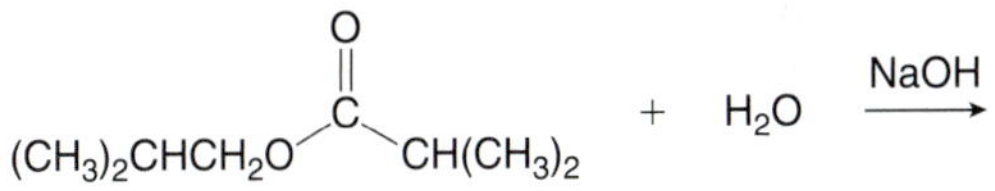
(CH$_3$)$_2$CHCH$_2$O ... CH(CH$_3$)$_2$ + H$_2$O $\xrightarrow{NaOH}$

d. 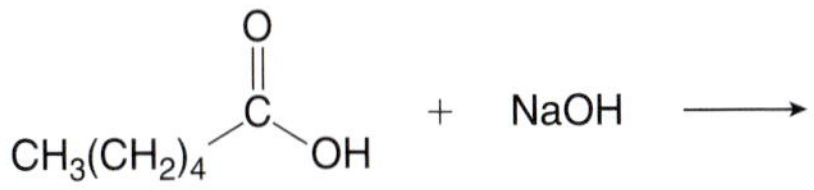
CH$_3$(CH$_2$)$_4$ ... OH + NaOH $\longrightarrow$

17.96 Draw the products formed in each reaction.

a. $(CH_3)_2CH$—C(=O)—OH + CH_3NH_2 $\xrightarrow{\Delta}$

b. $(CH_3)_2CHCH_2NH$—C(=O)—CH_3 + H_2O $\xrightarrow{HCl}$

c. $(CH_3)_2CHNH$—C(=O)—CH_2CH_3 + H_2O $\xrightarrow{NaOH}$

d. HO—C(=O)—$CH_2CH(CH_3)_2$ + NaOH $\longrightarrow$

17.97 Answer the following questions about $(CH_3)_2CHCH_2CH_2CH_2CO_2H$ (**A**).
   a. What is the IUPAC name for **A**?
   b. Draw an isomer of **A** that has the same functional group.
   c. Draw an isomer of **A** that has a different functional group.
   d. What products are formed when **A** is treated with NaOH?
   e. Predict the solubility properties of **A** in H_2O; in an organic solvent.
   f. What product is formed when **A** is treated with CH_3CH_2OH and H_2SO_4?
   g. What product is formed when **A** is heated with $CH_3CH_2NH_2$?

17.98 Answer the following questions about $(CH_3)_3C(CH_2)_6CO_2H$ (**B**).
   a. What is the IUPAC name for **B**?
   b. Draw an isomer of **B** that has the same functional group.
   c. Draw an isomer of **B** that has a different functional group.
   d. What products are formed when **B** is treated with NaOH?
   e. Predict the solubility properties of **B** in H_2O; in an organic solvent.
   f. What product is formed when **B** is treated with CH_3CH_2OH and H_2SO_4?
   g. What product is formed when **B** is heated with $CH_3CH_2NH_2$?

17.99 Identify **A** and **B** in the following two-step sequence that synthesizes benzyl benzoate from benzyl alcohol. Benzyl benzoate is a component of jasmine oil.

benzyl alcohol

benzyl benzoate

17.100 Dibutyl phthalate is a plasticizer added to make brittle polymers softer and more flexible. Dibutyl phthalate (**D**) can be prepared by the following two-step sequence. What are the structures of **C** and **D**?

Applications

17.101 Which of the following structures represent soaps? Explain your answers.
   a. $CH_3CO_2^-$ Na^+
   b. $CH_3(CH_2)_{14}CO_2^-$ Na^+
   c. $CH_3(CH_2)_{12}COOH$

17.102 Which of the following structures represent soaps? Explain your answers.
   a. $CH_3CH_2CO_2^-$ K^+
   b. $CH_3(CH_2)_{16}COOH$
   c. $CH_3(CH_2)_9CO_2^-$ Na^+

17.103 Like aspirin, naproxen contains a carboxyl group, so it undergoes similar acid–base reactions.

naproxen

   a. Draw the reaction that occurs when naproxen is treated with NaOH.
   b. What form of naproxen exists in the stomach?
   c. What form of naproxen exists in the intestines?

17.104 Sunscreens that contain an ester can undergo hydrolysis. What products are formed when octyl salicylate, a commercial sunscreen, is hydrolyzed with water?

octyl salicylate

17.105 How does aspirin act as an anti-inflammatory agent?

17.106 What structural feature is important in making penicillin a biologically active antibiotic? How does the biological activity of penicillin come about?

17.107 Explain how soap is able to dissolve nonpolar hydrocarbons in a polar solvent like H_2O.

17.108 Today, synthetic detergents like the compound drawn below are used to clean clothes, not soaps. Synthetic detergents are similar to soaps in that they contain an ionic head bonded to a large hydrocarbon group (the nonpolar tail). Explain how this detergent cleans away dirt.

$$CH_3(CH_2)_9 - \bigcirc - SO_3^- \ Na^+$$

a detergent

17.109 Draw the products formed by acidic hydrolysis of **aspartame,** the artificial sweetener used in Equal and many diet beverages. One of the hydrolysis products of this reaction is the naturally occurring amino acid phenylalanine. Infants afflicted with phenylketonuria cannot metabolize this amino acid so it accumulates, causing mental retardation. When identified early, a diet limiting the consumption of phenylalanine (and compounds like aspartame that are converted to it) can allow for a normal life.

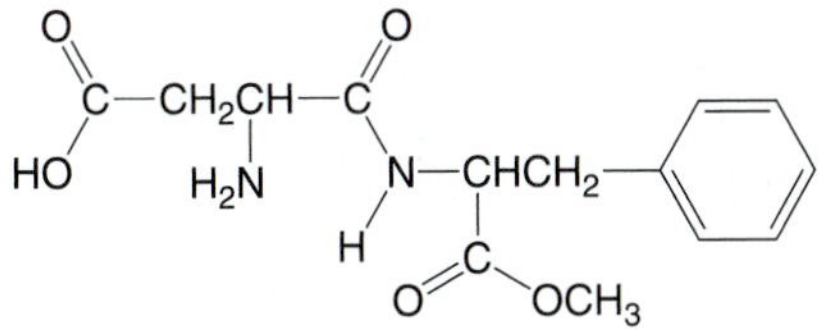

aspartame

17.110 Polyhydroxybutyrate (PHB) is a biodegradable polyester; that is, the polymer is degraded by microorganisms that naturally occur in the environment. Such biodegradable polymers are attractive alternatives to the commonly used polymers that persist in landfills for years. What is the single product formed when PHB is hydrolyzed with acid and water?

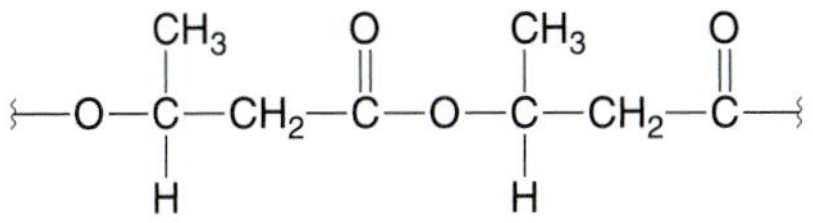

PHB

CHALLENGE QUESTIONS

17.111 Compounds that contain both a hydroxyl group (OH) and a carboxyl group (COOH) can undergo an intramolecular esterification reaction. What product is formed when each hydroxy acid undergoes an intramolecular reaction?
a. $HOCH_2CH_2CH_2CH_2CO_2H$
b. $HOCH_2CH_2CH_2CO_2H$

17.112 Lactams can be hydrolyzed with base, just like other amides. When a lactam is hydrolyzed, the ring is opened to form a product with two functional groups and no ring. What product is formed when caprolactam, the starting material for the synthesis of one type of nylon, is hydrolyzed with water and base?

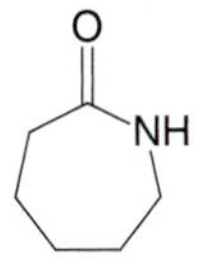

caprolactam

• • • • • • • • • • • • •

CHAPTER GOALS

In this chapter you will learn how to:

1. Identify the characteristics of amines
2. Name amines
3. Give examples of common alkaloids
4. Draw the products of acid–base reactions of amines
5. Identify and name ammonium salts
6. Discuss the function of neurotransmitters and explain how neurotransmitters and hormones differ
7. Discuss the role of dopamine and serotonin in the body
8. Give examples of important derivatives of 2-phenylethylamine
9. Discuss the chemistry of histamine, antihistamines, and anti-ulcer drugs

Tobacco products contain the addictive stimulant **nicotine,** an amine that is synthesized by tobacco plants as a defense against insect predators.

AMINES AND NEUROTRANSMITTERS

IN Chapter 18 we turn our attention to **amines,** organic compounds that contain nitrogen atoms. The amine nitrogen atom can be bonded to one, two, or three alkyl groups, and many amines have a nitrogen atom in a ring. All proteins contain amines, as do many vitamins and hormones. Many common amines, such as the caffeine in coffee and the nicotine in tobacco, are naturally occurring. Synthetic drugs used as sedatives, antihistamines, and bronchodilators also contain the amine functional group. Chapter 18 concludes with a discussion of **neurotransmitters,** important chemical messengers in the body, and many other related physiologically active amines.

18.1 STRUCTURE AND BONDING

Amines **are organic nitrogen compounds,** formed by replacing one or more hydrogen atoms of ammonia (NH_3) with alkyl groups. Amines are classified as 1°, 2°, or 3° by the number of alkyl groups bonded to the *nitrogen* atom.

R—N̈—H   R—N̈—H   R—N̈—R
   | | |
   H   R   R

1° amine   2° amine   3° amine

> **Classifying amines as 1°, 2°, or 3° is reminiscent of classifying amides** in Chapter 17, but is *different* from classifying alcohols and alkyl halides as 1°, 2°, and 3°. Compare, for example, a 2° amine and a 2° alcohol. A 2° amine (R_2NH) has *two* C—N bonds. A 2° alcohol (R_2CHOH) has only *one* C—O bond, but *two* C—C bonds on the carbon bonded to oxygen.

- A primary (1°) amine has one C—N bond and the general structure RNH_2.
- A secondary (2°) amine has two C—N bonds and the general structure R_2NH.
- A tertiary (3°) amine has three C—N bonds and the general structure R_3N.

Like ammonia, **the amine nitrogen atom has a lone pair of electrons,** which is generally omitted in condensed structures. In Section 18.7, we will also learn about quaternary ammonium ions that contain a fourth alkyl group bonded to nitrogen. In this case, the nitrogen atom has no lone pair and bears a positive charge.

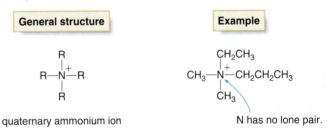

| General structure | Example |

    R CH_2CH_3
    |+ |+
R—N—R CH_3—N—$CH_2CH_2CH_3$
    | |
    R CH_3

quaternary ammonium ion   N has no lone pair.

> **Recall from Section 14.7 that a** *heterocycle* **is a ring that contains a heteroatom such as N, O, or S.**

The amine nitrogen atom can also be part of a ring to form a **heterocycle.** For example, piperidine, one compound isolated from black pepper, and coniine, the poisonous principle of hemlock, both contain a nitrogen atom in a six-membered ring.

pepper plant   hemlock

piperidine   coniine

An amine nitrogen atom is surrounded by three atoms and one nonbonded electron pair, making it trigonal pyramidal in shape, with bond angles of approximately 109.5°.

CH_3NH_2 = CH_3⟍N̈⟋H =
methylamine   H
                  109.5°

trigonal pyramidal

$(CH_3)_3N$ = CH_3⟍N̈⟋CH_3 =
trimethylamine CH_3
                  109.5°

trigonal pyramidal

SAMPLE PROBLEM 18.1

Classify each amine in the following compounds as 1°, 2°, or 3°. Putrescine is partly responsible for the foul odor of decaying fish. MDMA is the illegal stimulant commonly called "Ecstasy."

a. $H_2N(CH_2)_4NH_2$

putrescine

b. 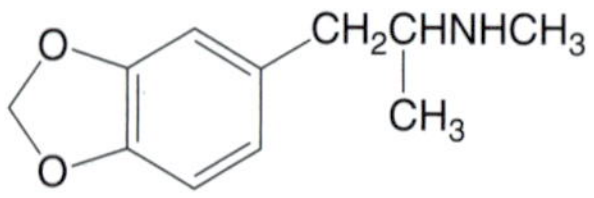 [structure of MDMA with CH₂CHNHCH₃ and CH₃ groups]

MDMA
"Ecstasy"

ANALYSIS

To determine whether an amine is 1°, 2°, or 3°, count the number of carbons bonded to the nitrogen atom. A 1° amine has one C—N bond, and so forth.

SOLUTION

Draw out the structure or add H's to the skeletal structure to clearly see how many C—N bonds the amine contains.

a. 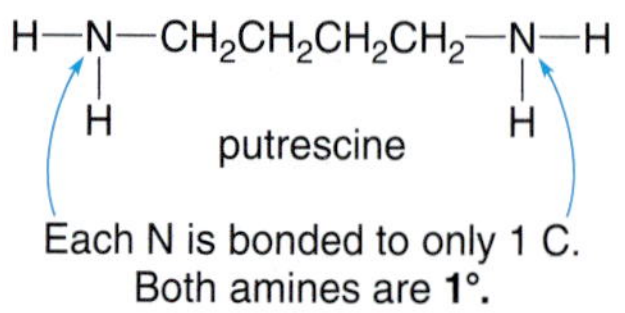 H—N—CH₂CH₂CH₂CH₂—N—H
         | |
         H   H
              putrescine

Each N is bonded to only 1 C.
Both amines are **1°**.

b. 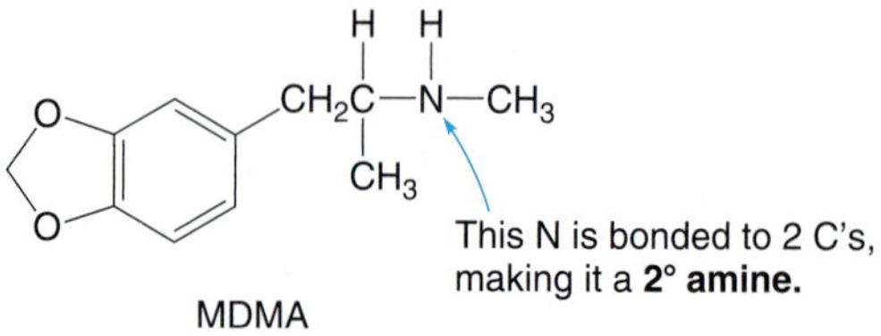 [structure of MDMA showing CH₂C—N—CH₃ with H H above N and CH₃ below]

This N is bonded to 2 C's,
making it a **2° amine**.

MDMA
"Ecstasy"

PROBLEM 18.1

Classify each amine in the following compounds as 1°, 2°, or 3°.

a. $H_2N(CH_2)_3NH(CH_2)_4NH(CH_2)_3NH_2$

spermine
(isolated from semen)

b. 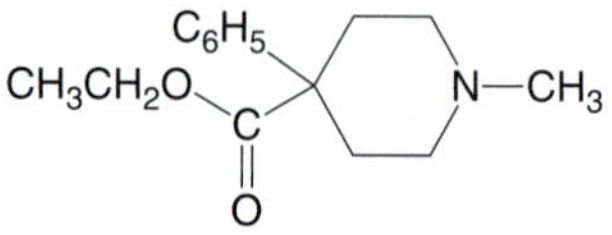 CH₃CH₂O—[structure of meperidine with C₆H₅ group, C=O, and N—CH₃]

meperidine
(Trade name: Demerol)

PROBLEM 18.2

Classify each amine in piperidine and coniine as 1°, 2°, or 3°.

Methamphetamine (Figure 18.1) is one of the many biologically active amines discussed in this chapter. Because it imparts to the user a pleasurable "high," methamphetamine is a widely abused illegal drug that has profound effects on the body and mind. Methamphetamine, known as speed, meth, or crystal meth, is highly addictive, easy to synthesize, and has adverse effects on the heart, lungs, blood vessels, and other organs. Prolonged use can lead to sleeplessness, seizures, hallucinations, paranoia, and serious heart disease.

PROBLEM 18.3

(a) What type of amine does methamphetamine contain? (b) Give the molecular shape around each indicated atom in methamphetamine.

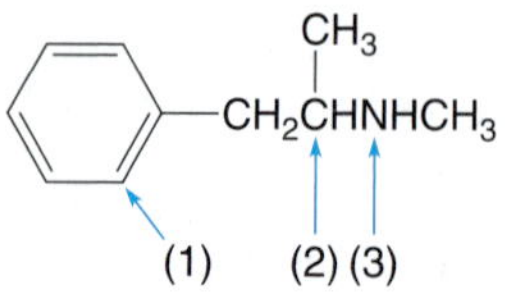

[structure of methamphetamine with benzene ring, CH₃ group, CH₂CHNHCH₃, and arrows labeled (1), (2), (3)]

(1)   (2) (3)

methamphetamine

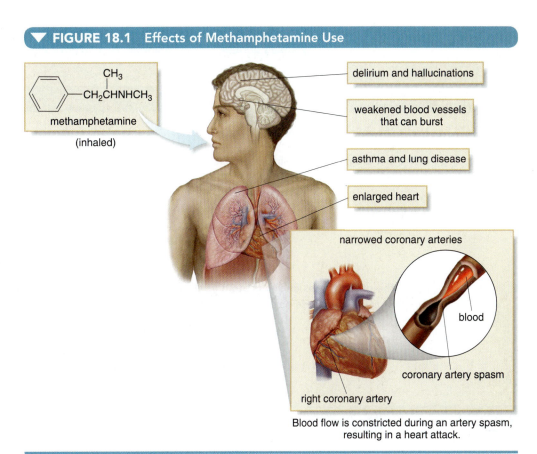

▼ **FIGURE 18.1**   Effects of Methamphetamine Use

Blood flow is constricted during an artery spasm, resulting in a heart attack.

18.2   NOMENCLATURE

18.2A   PRIMARY AMINES

Primary (1°) amines are named using either systematic or common names. Systematic names can be used for any amine. To assign the **systematic name,** find the longest carbon chain bonded to the amine nitrogen, and change the *-e* ending of the parent alkane to the suffix *-amine.* Then use the usual rules of nomenclature to number the chain and name the substituents.

Common names are used only for simple amines. To assign a **common name,** name the alkyl group bonded to the nitrogen atom and add the suffix *-amine,* forming a single word.

$$CH_3NH_2$$
Common name: methylamine
Systematic name: methanamine

$$CH_3CH_2CH_2CH_2NH_2$$
$$\quad\;\;4\quad\;3\quad\;\;2\quad\;\;1$$
Common name: butylamine
Systematic name: 1-butanamine

18.2B   SECONDARY AND TERTIARY AMINES

Secondary (2°) and 3° amines having identical alkyl groups are named by using the prefix **di-** or **tri-** with the name of the primary amine.

$$\underset{\text{triethylamine}}{CH_3CH_2-\overset{\overset{\displaystyle CH_2CH_3}{|}}{N}-CH_2CH_3}$$

$$\underset{\text{dipropylamine}}{CH_3CH_2CH_2-\overset{\overset{\displaystyle H}{|}}{N}-CH_2CH_2CH_3}$$

Secondary and tertiary amines having more than one kind of alkyl group are named as **N-substituted primary amines** using the following *How To* procedure.

HOW TO   Name 2° and 3° Amines with Different Alkyl Groups

EXAMPLE   Name the following 2° amine: $(CH_3)_2CHNHCH_3$.

Step [1]   **Name the longest alkyl chain (or largest ring) bonded to the N atom as the parent amine.**

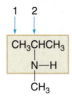

- Since the longest carbon chain has 3 C's, the parent name is *propanamine*.
- Since the N atom is bonded to the middle carbon, the name becomes *2-propanamine*.

Step [2]   **Name the other groups on the N atom as alkyl groups, alphabetize the names when there is more than one substituent, and precede each name with the prefix N-.**

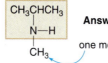

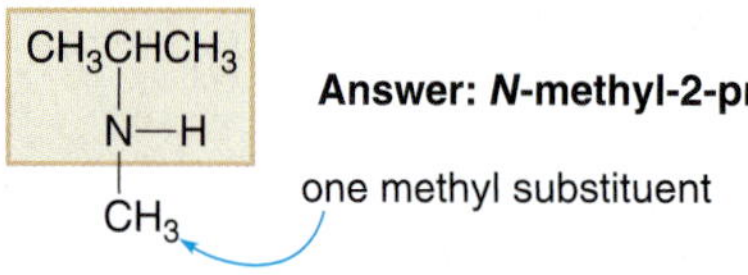

Answer: *N*-methyl-2-propanamine

one methyl substituent

SAMPLE PROBLEM 18.2

Give a systematic name for each amine.

a. $CH_3CH_2CH_2CH_2CH_2NH_2$

b. 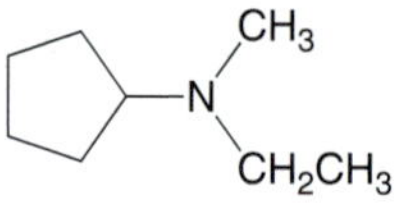 (structure: cyclopentane bonded to N with CH₃ and CH₂CH₃)

ANALYSIS AND SOLUTION

a. [1] For a 1° amine, find and name the longest chain containing the amine group.

$CH_3CH_2CH_2CH_2CH_2NH_2$

pentane - - -→ pentan*amine*
(5 C's)

[2] Number the carbon skeleton.

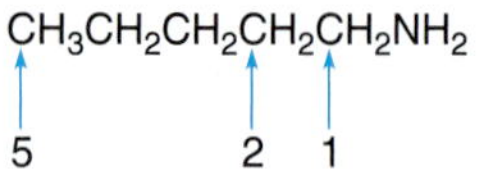

$CH_3CH_2CH_2CH_2CH_2NH_2$

5   2   1

You must use a number to show the location of the NH_2 group.
Answer: 1-pentanamine

b. For a 3° amine, one alkyl group on N is the principal R group and the others are substituents.

[1] Name the ring bonded to the N.

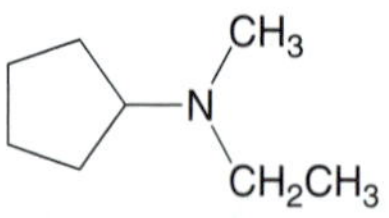

(structure: cyclopentane–N with CH₃ and CH₂CH₃)

5 C's in the ring

↓

cyclopentanamine

[2] Name the substituents.

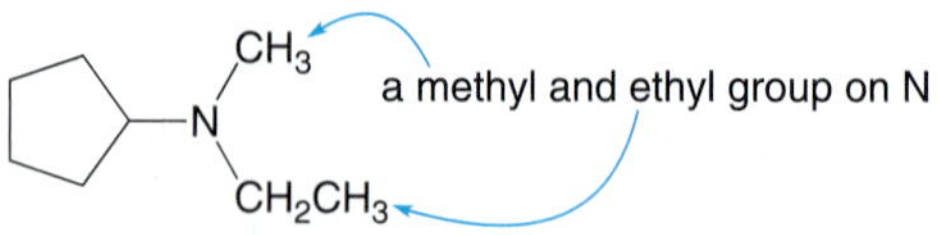

(structure with CH₃ and CH₂CH₃ — a methyl and ethyl group on N)

- 2 N's are needed, one for each alkyl group.
- Alphabetize the **e** of **e**thyl before the **m** of **m**ethyl.

Answer: *N*-ethyl-*N*-methylcyclopentanamine

PROBLEM 18.4

Give an acceptable name for each amine.

a. 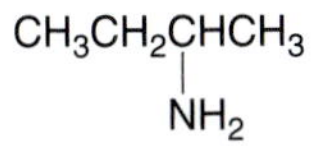 $CH_3CH_2CHCH_3$
       |
      NH_2

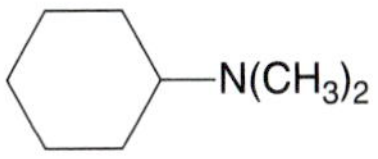

b. $CH_3CH_2CH_2NHCH_3$

c. [cyclohexyl]—$N(CH_3)_2$

d. $(CH_3CH_2CH_2CH_2)_2NH$

18.2C AROMATIC AMINES

Aromatic amines, amines having a nitrogen atom bonded directly to a benzene ring, are named as derivatives of aniline.

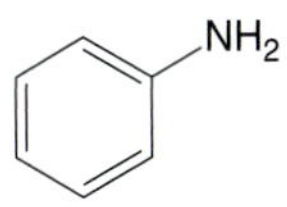

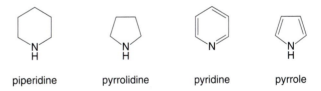

aniline 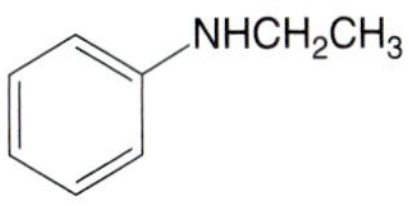   *N*-ethylaniline 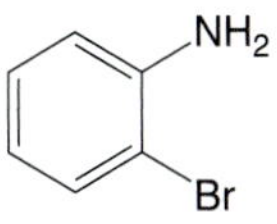   *o*-bromoaniline

PROBLEM 18.5

Draw a structure corresponding to each name: (a) *N*-methylaniline; (b) *m*-ethylaniline; (c) 3,5-diethylaniline; (d) *N,N*-diethylaniline.

18.2D MISCELLANEOUS NOMENCLATURE FACTS

An **NH_2** group named as a substituent is called an **amino group.**

There are many different **nitrogen heterocycles,** and each ring type is named differently, depending on the number of N atoms in the ring and the ring size. The structures and names of four common nitrogen heterocycles are shown.

piperidine   pyrrolidine   pyridine   pyrrole

PROBLEM 18.6

Draw a structure corresponding to each name.

a. 3-hexanamine
b. *N*-methylpentylamine
c. *p*-nitroaniline
d. *N*-methylpiperidine

e. *N,N*-dimethylethylamine
f. 2-aminocyclohexanone
g. 1-propylcyclohexanamine
h. *N*-propylaniline

18.3 PHYSICAL PROPERTIES

Many low molecular weight amines have *very* foul odors. **Trimethylamine** [$(CH_3)_3N$], formed when enzymes break down certain fish proteins, has the characteristic odor of rotting fish. **Cadaverine** ($NH_2CH_2CH_2CH_2CH_2CH_2NH_2$) is a poisonous diamine with a putrid odor also present in rotting fish, and partly responsible for the odor of semen, urine, and bad breath.

Because nitrogen is much more electronegative than carbon or hydrogen, amines contain polar C—N and N—H bonds. **Primary (1°) and 2° amines are also capable of intermolecular hydrogen bonding,** because they contain N—H bonds.

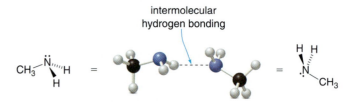

Since nitrogen is less electronegative than oxygen, however, intermolecular hydrogen bonds between N and H are *weaker* than those between O and H. As a result:

- In comparing compounds of similar size, 1° and 2° amines have higher boiling points than compounds incapable of hydrogen bonding, but lower boiling points than alcohols that have stronger intermolecular hydrogen bonds.

$CH_3CH_2OCH_2CH_3$	$CH_3CH_2CH_2CH_2NH_2$	$CH_3CH_2CH_2CH_2OH$
diethyl ether	1-butanamine	1-butanol
bp = 38 °C	bp = 78 °C	bp = 117 °C

Increasing intermolecular forces
Increasing boiling point

- Tertiary (3°) amines have lower boiling points than 1° and 2° amines of comparable size, because they have no N—H bonds.

3° amine $CH_3CH_2N(CH_3)_2$ $CH_3CH_2\overset{\overset{\text{H}}{|}}{N}CH_2CH_3$ 2° amine

no N—H bond   bp = 38 °C   bp = 56 °C   N—H bond

higher boiling point

Amines are soluble in organic solvents regardless of size. Amines with fewer than six carbons are water soluble since they can hydrogen bond with water. Larger amines are water insoluble since the nonpolar alkyl portion is too large to dissolve in the polar water solvent.

SAMPLE PROBLEM 18.3

Which compound in each pair has the higher boiling point: (a) $CH_3CH_2NHCH_3$ or $CH_3CH_2OCH_3$; (b) $(CH_3)_3N$ or $CH_3CH_2CH_2NH_2$?

ANALYSIS

Keep in mind the general rule: For compounds of comparable size, **the stronger the intermolecular forces, the higher the boiling point.** Compounds that can hydrogen bond have higher boiling points than compounds that are polar but cannot hydrogen bond. Polar compounds have higher boiling points than nonpolar compounds.

SOLUTION

a. The 2° amine ($CH_3CH_2NHCH_3$) has an N—H bond, so intermolecular hydrogen bonding is possible. The ether ($CH_3CH_2OCH_3$) has only C—H bonds, so there is no possibility of intermolecular hydrogen bonding. $CH_3CH_2NHCH_3$ has a higher boiling point because it has stronger intermolecular forces.

$CH_3CH_2\overset{\overset{}{|}}{\underset{\underset{\text{H}}{|}}{N}}CH_3$ $CH_3CH_2\text{—O—}CH_3$

a 2° amine with an N—H bond
intermolecular hydrogen bonding
higher boiling point

an ether with only C—H bonds

b. The 1° amine ($CH_3CH_2CH_2NH_2$) has N—H bonds, so intermolecular hydrogen bonding is possible. The 3° amine [$(CH_3)_3N$] has only C—H bonds, so there is no possibility of intermolecular hydrogen bonding. $CH_3CH_2CH_2NH_2$ has a higher boiling point because it has stronger intermolecular forces.

$$CH_3CH_2CH_2NH_2 \qquad CH_3-\overset{\underset{|}{CH_3}}{N}-CH_3$$

a 1° amine with N—H bonds
intermolecular hydrogen bonding
higher boiling point

a 3° amine with only C—H bonds

PROBLEM 18.7

Which compound in each pair has the higher boiling point?

a. $CH_3\overset{\overset{O}{\|}}{C}CH_2CH_3$ or $(CH_3)_2CHCH_2NH_2$ c.   ⬡—NH_2 or   ⬡—CH_3

b. $(CH_3)_2CHCH_2NH_2$ or $(CH_3)_2CHCH_2OH$

18.4 FOCUS ON HEALTH & MEDICINE
CAFFEINE AND NICOTINE

Caffeine and **nicotine** are widely used stimulants of the central nervous system that contain amine heterocycles. Caffeine and nicotine, like other amines discussed in Section 18.5, are *alkaloids,* **naturally occurring amines derived from plant sources.**

caffeine   nicotine

18.4A CAFFEINE

CONSUMER NOTE

Some chocolate bars contain as much caffeine as a 12-oz can of cola.

Caffeine (Figure 18.2) is a bitter-tasting amine found in coffee and tea. Caffeine is also present in soft drinks and chocolate bars. The pain reliever Anacin contains aspirin and caffeine, and over-the-counter remedies like NoDoz used to ward off sleep contain caffeine. Table 18.1 lists the average caffeine content of several products.

Caffeine is a mild stimulant, usually imparting a feeling of alertness after consumption. It also increases heart rate, dilates airways, and stimulates the secretion of stomach acid. These effects are observed because caffeine increases glucose production, making an individual feel energetic. Caffeine's effects are temporary, so an individual must consume it throughout the day to maintain the same "high."

In moderation, caffeine consumption poses no health risks, but large quantities can cause insomnia, anxiety, and dehydration. Since caffeine is somewhat addicting, withdrawal symptoms—headache, irritability, and mild depression—can be observed when it is eliminated from the diet. Several studies suggest that pregnant women should limit their intake of caffeine, and that excessive caffeine consumption increases the risk of miscarriage. Nursing mothers are likewise advised to limit caffeine intake, since it passes into breast milk and can adversely affect newborn babies.

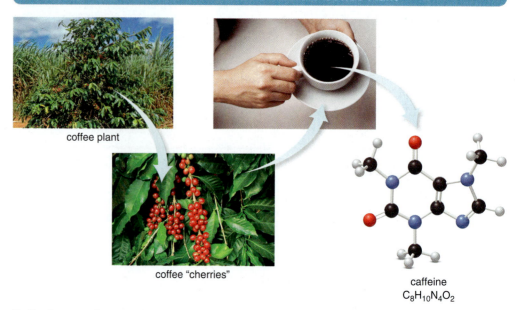

▼ FIGURE 18.2 Caffeine—An Alkaloid from the Coffee Plant

coffee plant

coffee "cherries"

caffeine
$C_8H_{10}N_4O_2$

Coffee is grown in 50 tropical countries worldwide, and ranks second only to oil in the amount traded internationally. The coffee "cherries" are picked by hand and dried to give two seeds per fruit. Raw beans are shipped to roasting facilities around the world.

TABLE 18.1 Average Caffeine Content in Common Foods and Drugs

Product	Serving Size	Caffeine (mg)
Coffee, brewed	8 oz	133
Coffee, instant	16 oz	47
Coffee, decaffeinated	8 oz	5
Tea, brewed	8 oz	53
Coca-Cola Classic	12 oz	35
Diet Coke	12 oz	47
Hershey's Chocolate Bar	1.55 oz	9
Hershey's Special Dark Chocolate	1.45 oz	31
Anacin	2 tablets	64
NoDoz	1 tablet	200

[Data taken from the web, Center for Science in the Public Interest.]

PROBLEM 18.8	Draw a complete structure of caffeine with all atoms and lone pairs.

| PROBLEM 18.9 | Decaffeinated coffee is produced by extracting the caffeine from coffee with a solvent. Initially dichloromethane (CH_2Cl_2) was used, but concerns arose over the ill health effects of trace amounts of CH_2Cl_2 remaining in the coffee. Why is caffeine soluble in CH_2Cl_2? |

Nicotine patches and nicotine-containing chewing gum have been effectively used by many individuals to quit smoking.

18.4B NICOTINE

Nicotine (Figure 18.3) is a highly toxic amine isolated from tobacco. In small doses it acts as a stimulant, but in larger doses causes depression, nausea, and even death. Nicotine is synthesized in tobacco plants as a defense against insect predators, and is used commercially as an insecticide.

Although inhaled tobacco smoke contains over 4,000 compounds, many of which are toxic or cancer causing, nicotine is the compound that makes smoking addictive. Nicotine imparts a sense of pleasure to the smoker, and once a smoker is "hooked," an individual needs to continue smoking to avoid the extreme discomforts of withdrawal—restlessness, anxiety, and craving.

Cigarette smoking causes lung disease, heart disease, and cancer, and it is recognized today as the leading cause of avoidable death in the United States. Despite this well known fact, the nicotine addiction makes it difficult for current smokers to quit. In fact, the sale of nicotine-free cigarettes many years ago was an unsuccessful enterprise because the nicotine in cigarettes is what smokers really crave when they light up.

One method employed by some to quit smoking involves the use of a nicotine patch. The patch is applied to the skin and nicotine diffuses through the skin and into the bloodstream. In this way, a smoker still derives pleasure from the nicotine in the system, but is no longer exposed to the ill effects of the other components of cigarette smoke. Slowly the concentration of nicotine in the patch is reduced, with the hope that eventually the smoker will no longer need nicotine. A similar strategy is used with chewing gum that contains nicotine.

▼ **FIGURE 18.3   Nicotine—An Alkaloid from the Tobacco Plant**

tobacco plant

dried tobacco leaves

nicotine
$C_{10}H_{14}N_2$

Tobacco is grown in temperate climates. The leaves are harvested and dried, and then converted to cigarettes and other tobacco products.

PROBLEM 18.10   What two types of nitrogen heterocycles are present in nicotine?

PROBLEM 18.11   Locate the single chirality center in nicotine. Draw both enantiomers of nicotine (see Section 15.4).

18.5   ALKALOIDS—AMINES FROM PLANT SOURCES

The word **alkaloid** is derived from the word *alkali,* since aqueous solutions of alkaloids are slightly basic.

Morphine, quinine, and **atropine** are three alkaloids with useful medicinal properties. All three compounds contain nitrogen heterocycles and have a complex structure with several functional groups.

18.5A   MORPHINE AND RELATED ALKALOIDS

The psychological and analgesic effects of the opium poppy *Papaver somniferum* have been known for probably 6,000 years, and it has been widely used as a recreational drug and pain-killing remedy for centuries. The analgesic and narcotic effects of opium are largely due to **morphine. Codeine,** an alkaloid of similar structure with a single methyl ether, is also present, but in much smaller amounts.

CONSUMER NOTE

Poppy seed tea, which contains morphine, was used as a folk remedy in parts of England until World War II. Since poppy seeds contain morphine, eating poppy seed cake or bagels can introduce enough morphine into the system to produce a positive test in a drug screen.

Morphine is especially useful in relieving severe chronic pain, so it is often prescribed for patients with terminal cancer. Morphine is also very addictive, and with time, higher doses are needed to produce the same effects. Approximately 95% of the morphine obtained from legally grown poppies is converted to codeine, a useful drug for the relief of less severe pain.

Heroin, a widely used, very addictive illicit drug, is readily prepared by converting the two hydroxyl groups of morphine to acetate esters. Heroin is less polar than morphine, making it more soluble in fat cells in the body. As a result, it is two to three times more potent than morphine in producing euphoria and pain relief.

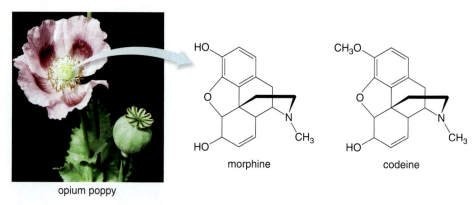

PROBLEM 18.12	Classify the amine in morphine as 1°, 2°, or 3°.

PROBLEM 18.13	Identify all the functional groups in heroin.

18.5B QUININE

Quinine is an alkaloid isolated from the bark of the cinchona tree native to the Andes Mountains. Quinine is a powerful antipyretic—that is, it reduces fever—and for centuries it was the only effective treatment for malaria. Its bitter taste gives tonic water its characteristic flavor.

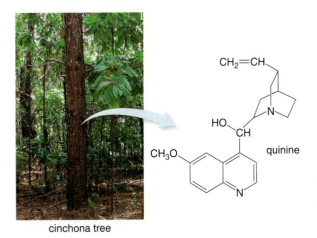

cinchona tree

PROBLEM 18.14	(a) How many trigonal planar carbon atoms does quinine contain? (b) How many tetrahedral carbon atoms does quinine contain?

18.5C ATROPINE

Atropine is an alkaloid isolated from *Atropa belladonna,* the deadly nightshade plant.

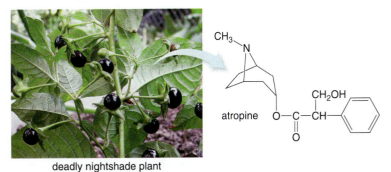

deadly nightshade plant

During the Renaissance (fourteenth to seventeenth centuries), women used the juice of the berries of the nightshade to enlarge the pupils of their eyes for cosmetic reasons. Sometimes ophthalmologists today use atropine to dilate pupils. Atropine also increases heart rate, so physicians administer atropine for this purpose. Atropine relaxes smooth muscles and interferes with nerve impulses. In higher doses atropine is toxic, leading to convulsions, coma, and death.

PROBLEM 18.15	Identify all of the functional groups in atropine, and classify the amine as 1°, 2°, or 3°.

18.6   AMINES AS BASES

Like ammonia (NH_3), **amines are bases;** that is, **they are** *proton acceptors.* When an amine is dissolved in water, an acid–base equilibrium occurs: the amine accepts a proton from H_2O, forming its conjugate acid, an **ammonium ion,** and water loses a proton, forming hydroxide, ^-OH.

General
reactions

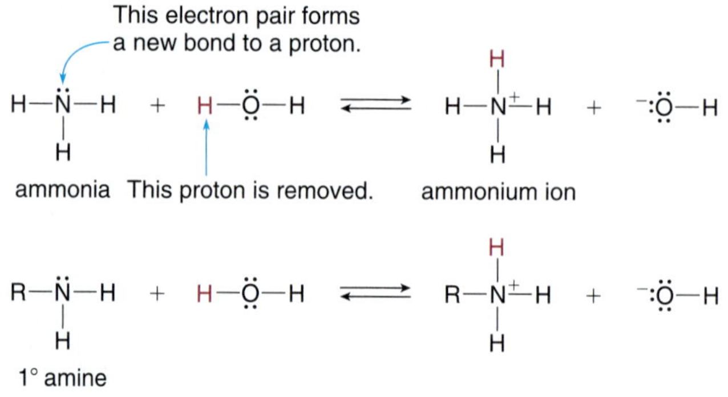

ammonia   This proton is removed.   ammonium ion

1° amine

This acid–base reaction occurs with 1°, 2°, and 3° amines. While amines are more basic than other families of organic compounds, they are weak bases compared to inorganic bases like NaOH.

Example

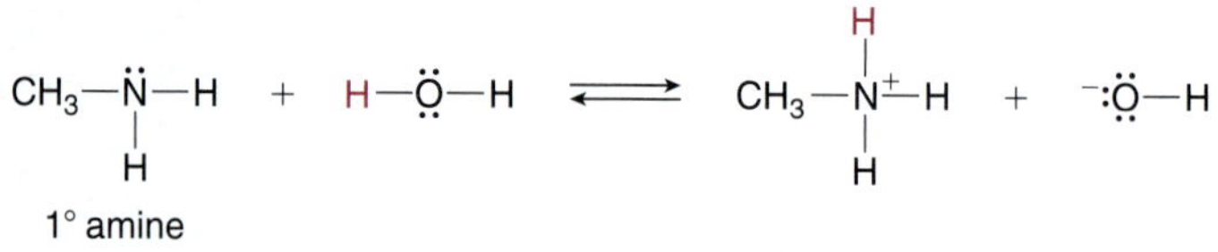

1° amine

18.6A   REACTION OF AMINES WITH ACIDS

Amines also react with acids such as HCl to form water-soluble salts. The lone pair of electrons from the amine nitrogen atom is always used to form a new bond to a proton from the acid.

Example

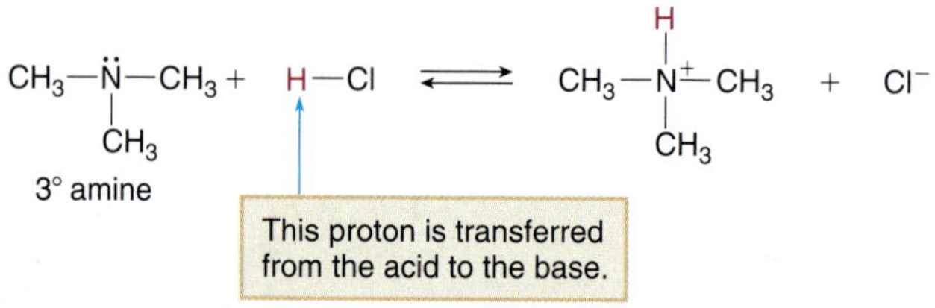

3° amine

This proton is transferred
from the acid to the base.

- The amine [$(CH_3)_3N$] gains a proton to form its conjugate acid, an ammonium cation [$(CH_3)_3NH^+$].
- A proton is removed from the acid (HCl) to form its conjugate base, the chloride anion (Cl^-).

Similar acid–base reactions occur with other inorganic acids (H_2SO_4), and with organic acids like CH_3COOH, as well. **In each reaction, the amine gains a proton and the acid loses a proton.** This fundamental reaction of amines occurs no matter how complex the amine, as shown in Sample Problem 18.4.

- In an acid–base reaction of an amine, the amine nitrogen always forms a new bond to a proton forming an ammonium ion.

SAMPLE PROBLEM 18.4

What products are formed when methamphetamine (Section 18.1) reacts with HCl?

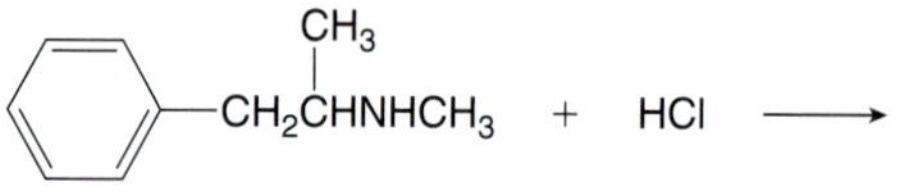

methamphetamine

ANALYSIS   In any acid–base reaction with an amine:

- Locate the N atom of the amine and add a proton to it. Since the amine nitrogen is neutral to begin with, adding a proton gives it a +1 charge.
- Remove a proton from the acid (HCl) and form its conjugate base (Cl$^-$).

SOLUTION   Transfer a proton from the acid to the base. Use the lone pair on the N atom to form the new bond to the proton of the acid.

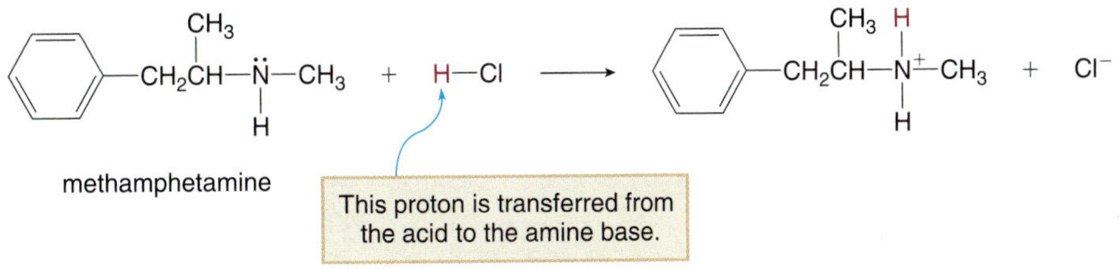

methamphetamine

This proton is transferred from the acid to the amine base.

Thus, HCl loses a proton to form Cl$^-$, and the N atom of amphetamine gains a proton to form an ammonium cation.

PROBLEM 18.16

What products are formed when each of the following amines is treated with HCl:
(a) $CH_3CH_2NH_2$; (b) $(CH_3CH_2)_2NH$; (c) $(CH_3CH_2)_3N$?

PROBLEM 18.17

Draw the products of each acid–base reaction.

a. $CH_3CH_2CH_2CH_2$—NH_2 + HCl $\longrightarrow$

c.   + H_2O $\longrightarrow$

b. $(CH_3)_2NH$ + C_6H_5COOH $\longrightarrow$

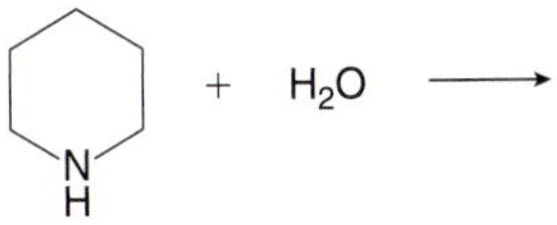

PROBLEM 18.18

What products are formed when epinephrine is treated with H_2SO_4?

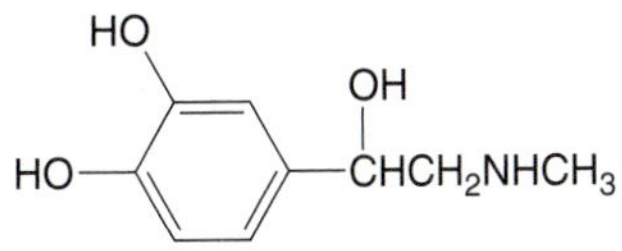

epinephrine
(adrenaline)

18.6B AMMONIUM SALTS

When an amine reacts with an acid, the product is an *ammonium salt:* **the amine forms a positively charged ammonium ion and the acid forms an anion.**

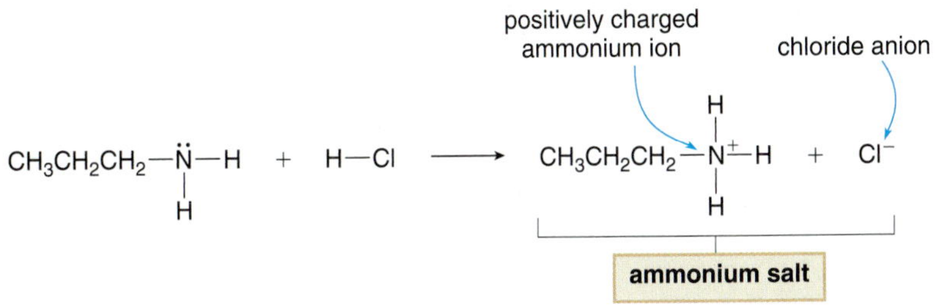

positively charged
ammonium ion   chloride anion

ammonium salt

To name an ammonium salt, change the suffix -*amine* of the parent amine from which the salt is formed to the suffix -*ammonium.* Then add the name of the anion. Two examples are shown in Sample Problem 18.5.

SAMPLE PROBLEM 18.5

Name each ammonium salt.

a. $CH_3CH_2\overset{+}{N}H_3$ Cl^- b. $(CH_3)_3\overset{+}{N}H$ CH_3COO^-

ANALYSIS

To name an ammonium salt, draw out the four groups bonded to the N atom. Remove one hydrogen from the N atom to draw the structure of the parent amine. Then put the two parts of the name together.

- Name the ammonium ion by changing the suffix -*amine* of the parent amine to the suffix -*ammonium*.
- Add the name of the anion.

SOLUTION

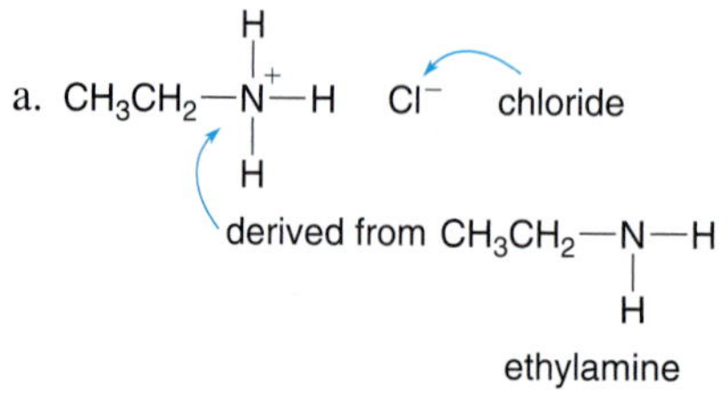

a. Change the name ethyl*amine* to ethyl*ammonium*.
- Add the name of the anion, chloride.
Answer: ethylammonium chloride

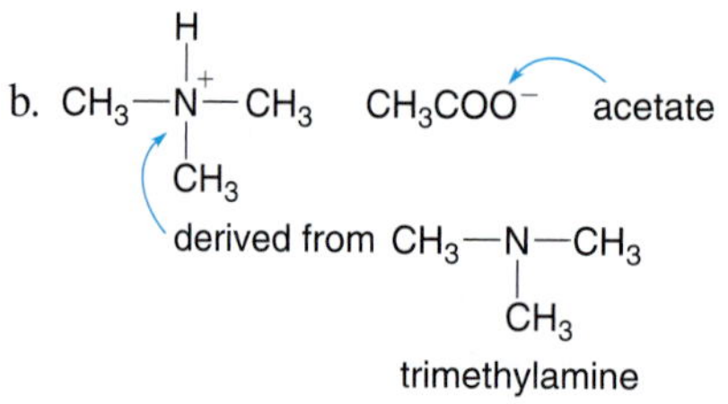

b. Change the name trimethyl*amine* to trimethyl*ammonium*.
- Add the name of the anion, acetate.
Answer: trimethylammonium acetate

PROBLEM 18.19

Name each ammonium salt.

a. $CH_3\overset{+}{N}H_3$ Cl^- b. $(CH_3CH_2CH_2)_2\overset{+}{N}H_2$ Br^- c. $(CH_3)_2\overset{+}{N}HCH_2CH_3$ CH_3COO^-

Ammonium salts are ionic compounds, and as a result:

- **Ammonium salts are water-soluble solids.**

In this way, the solubility properties of an amine can be changed by treatment with acid. For example, octylamine has eight carbons, making it water insoluble. Reaction with HCl forms octylammonium chloride. This ionic solid is now soluble in water.

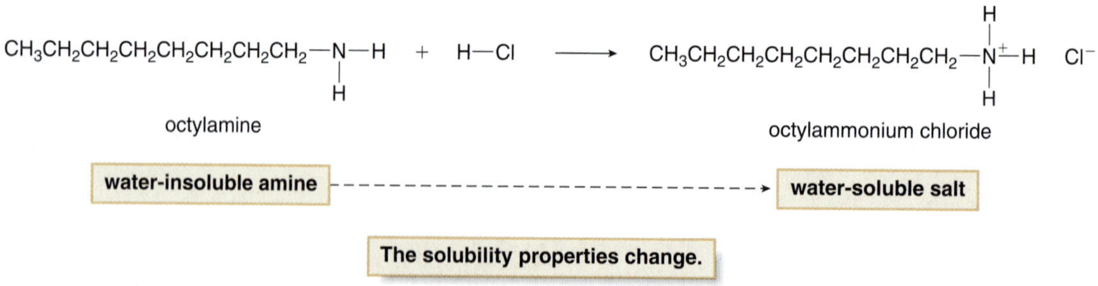

- **A water-insoluble amine can be converted to a water-soluble ammonium salt by treatment with acid.**

PROBLEM 18.20

Label each compound as water soluble or water insoluble.

a. $(CH_3CH_2)_3N$ b. $(CH_3CH_2)_3\overset{+}{N}H$ Br^- c. $CH_3CH_2NH_2$ d.

Ammonium salts can be re-converted to amines by treatment with base. Base removes a proton from the nitrogen atom of the amine, regenerating the neutral amine.

This proton is removed by the base.

$$CH_3CH_2CH_2CH_2CH_2CH_2CH_2CH_2 - \overset{H}{\underset{H}{\overset{|}{N^+}}} - H \quad Cl^- \quad + \quad Na^+ {}^-OH \longrightarrow CH_3CH_2CH_2CH_2CH_2CH_2CH_2CH_2 - \overset{H}{\underset{H}{\overset{|}{N}}} - H \quad + \quad H-OH \quad + \quad NaCl$$

octylammonium chloride
water-soluble salt

octylamine
water-insoluble amine

PROBLEM 18.21

What product is formed when each ammonium salt is treated with NaOH?

a. $(CH_3CH_2)_3\overset{+}{N}H$ Br^- b. $CH_3CH_2\overset{+}{N}H_3$ HSO_4^- c.

18.7 FOCUS ON HEALTH & MEDICINE
AMMONIUM SALTS AS USEFUL DRUGS

CONSUMER NOTE

Many antihistamines and decongestants are sold as their hydrochloride salts.

Many amines with useful medicinal properties are sold as their ammonium salts. Since the ammonium salts are more water soluble than the parent amine, they are easily transported through the body in the aqueous medium of the blood.

For example, diphenhydramine is a 3° amine that is sold as its ammonium salt under the name of Benadryl. Benadryl, formed by treating diphenhydramine with HCl, is an over-the-counter antihistamine that is used to relieve the itch and irritation of skin rashes and hives.

diphenhydramine   Benadryl
                                                   ammonium salt

Medicines sold as ammonium salts are often derived from the names of the amine and the acid used to form them. Since Benadryl is formed from diphenhydramine and HCl, it is called **diphenhydramine hydrochloride.**

Two other examples of amines sold as their ammonium salts are phenylephrine hydrochloride, the decongestant in Sudafed PE, Theraflu, and DayQuil, and methadone hydrochloride, a long-acting narcotic taken orally to treat chronic painful conditions. Methadone hydrochloride is also used as a substitute narcotic for heroin addicts.

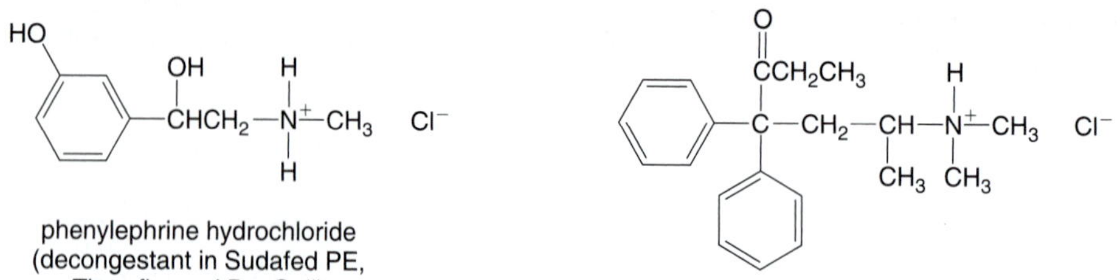

phenylephrine hydrochloride
(decongestant in Sudafed PE,
Theraflu, and DayQuil)

methadone hydrochloride
(narcotic analgesic)

Some **quaternary ammonium salts** ($R_4N^+ X^-$, Section 18.1)—that is, ammonium salts with nitrogen atoms bonded to *four* alkyl groups—are also useful agents. Bitrex is a foul-tasting non-toxic salt used on children's fingers to keep them from biting their nails or sucking their thumbs. Benzalkonium chloride is a disinfectant and antiseptic in mouthwash.

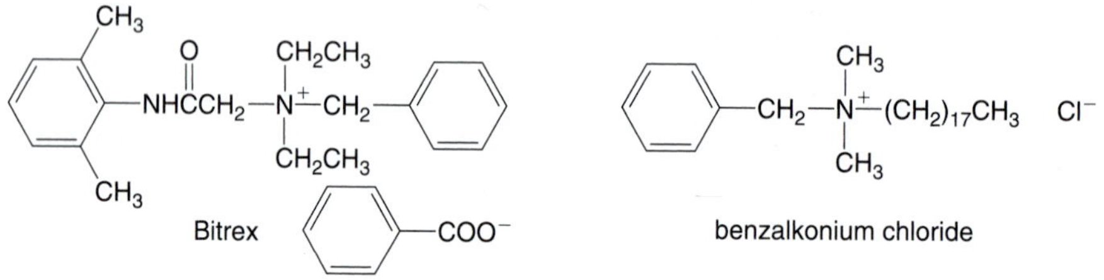

Bitrex

benzalkonium chloride

PROBLEM 18.22	Draw the structures of the amines from which phenylephrine hydrochloride and methadone hydrochloride are derived.
PROBLEM 18.23	When Benadryl is treated with a base like NaOH, the free amine, diphenhydramine, is re-formed, as we learned in Section 18.6. Explain why a similar reaction does not occur with a quaternary ammonium salt like Bitrex.

18.8   NEUROTRANSMITTERS

A *neurotransmitter* is a chemical messenger that transmits nerve impulses from one nerve cell—neuron—to another cell. As shown in Figure 18.4, a neuron consists of many short filaments called **dendrites** connected to a cell body. A long stem called an **axon** also protrudes from the cell body. The axon ends in numerous small filaments at its opposite end. These filaments are separated from the dendrites of a nearby neuron by a small gap called the **synapse.**

Neurotransmitters are stored in small packets called **vesicles** in the filaments of the axon near the synapse. When an electrical signal flows from the cell body of the neuron along the axon to the filaments, the neurotransmitter is released, diffuses across the synapse, and binds to a receptor on the dendrites of an adjacent neuron.

- The presynaptic neuron releases the neurotransmitter.
- The postsynaptic neuron contains the receptors that bind the neurotransmitter.

Once the neurotransmitter is bound to the receptor, the chemical message is delivered. The neurotransmitter is then degraded or it is returned to the presynaptic neuron to begin the process again.

Drugs may be used to affect neurotransmitters in many different ways. Some drugs prevent the release of a neurotransmitter or prevent its binding to a receptor. Other drugs increase the amount of neurotransmitter released. Some drugs affect the degradation of a neurotransmitter or prevent its reuptake by the presynaptic neuron.

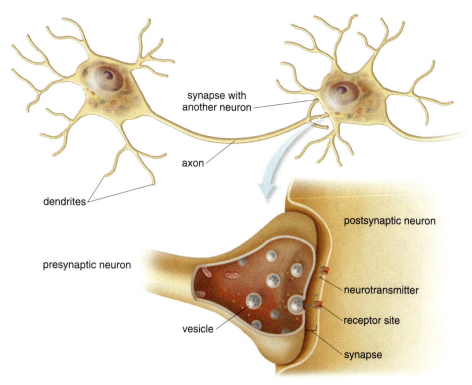

FIGURE 18.4   Neurotransmitters—Transmitting Information Between Neurons

Neurotransmitters are released from vesicles in the presynaptic neuron and diffuse through the synapse to bind to the receptor sites on the postsynaptic neuron.

All neurotransmitters contain nitrogen atoms. Three important neurotransmitters are amines—**norepinephrine, dopamine,** and **serotonin**—and one is an ammonium ion—**acetylcholine.**

norepinephrine
(noradrenaline)

dopamine

serotonin

acetylcholine

18.8A   NOREPINEPHRINE AND DOPAMINE

Norepinephrine, commonly called **noradrenaline,** and **dopamine** are two structurally related neurotransmitters that are both synthesized from the amino acid tyrosine, as shown in Figure 18.5.

Proper levels of both amines are needed for good mental health. For example, when **norepinephrine** levels are above normal, an individual feels unusually elated, and very high levels are associated with manic behavior. Norepinephrine is also converted to adrenaline when an individual experiences fear or stress, as we will learn in Section 18.9.

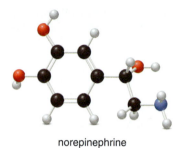

norepinephrine

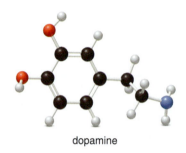

dopamine

▼ FIGURE 18.5   Synthesis of Norepinephrine and Dopamine

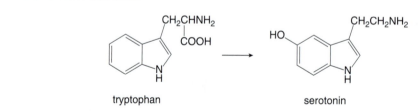

Norepinephrine and dopamine are both synthesized from the amino acid tyrosine by a multistep route. The newly introduced group at each stage is shown in red.

Dopamine affects brain processes that control movement, emotions, and pleasure. When dopamine-producing neurons die and there is too little dopamine in the brain, an individual loses control of fine motor skills and Parkinson's disease results. As we learned in Section 15.5, the level of dopamine in the brain cannot be increased simply by giving an individual dopamine, because dopamine does not cross the blood–brain barrier. L-Dopa, a precursor of dopamine shown in Figure 18.5, can travel from the bloodstream to the brain, so it is an effective treatment for Parkinson's disease. Elevated dopamine levels are also unhealthy and are associated with schizophrenia.

Dopamine plays an important role in addiction. Normal dopamine levels give an individual a pleasurable, satisfied feeling. Increased levels result in an intense "high." Drugs such as heroin, cocaine, and alcohol increase dopamine levels. When dopamine receptors are stimulated more than normal, the number and sensitivity of the receptors decreases, and an individual requires more of a drug to feel the same sense of pleasure.

PROBLEM 18.24

Indicate the chirality center in each of the compounds in Figure 18.5. Which compounds are achiral and contain no chirality centers?

18.8B   SEROTONIN

The neurotransmitter **serotonin** is synthesized from the amino acid tryptophan. Serotonin plays an important role in mood, sleep, perception, and temperature regulation. We get sleepy after eating a turkey dinner on Thanksgiving because the unusually high level of tryptophan in turkey is converted to serotonin.

serotonin

A deficiency of serotonin causes depression. Understanding the central role of serotonin in determining one's mood has led to the development of a wide variety of drugs for the treatment of depression. The most widely used antidepressants today are **selective serotonin reuptake inhibitors (SSRIs).** These drugs act by inhibiting the reuptake of serotonin by the presynaptic neuron, thus effectively increasing its concentration. Fluoxetine (trade name: Prozac) and sertraline (trade name: Zoloft) are two common antidepressants that act in this way.

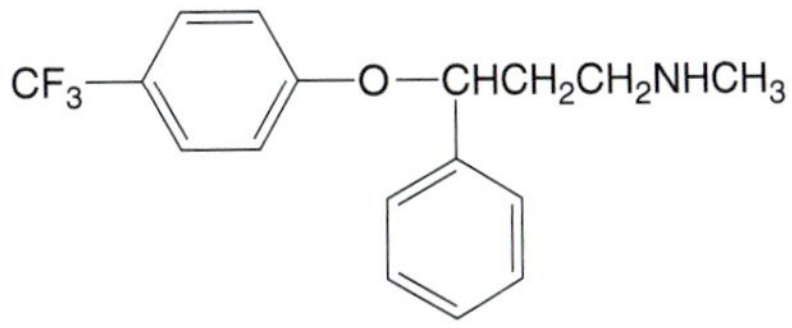

fluoxetine
(Trade name: Prozac)

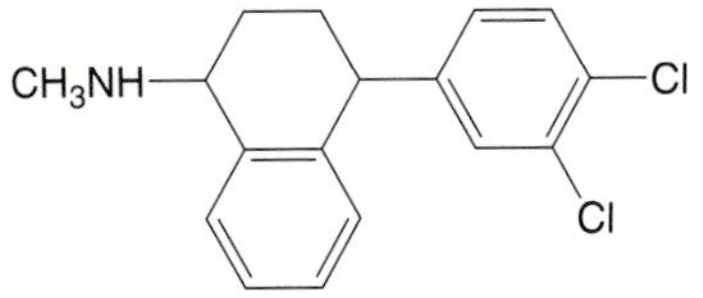

sertraline
(Trade name: Zoloft)

The discovery that migraine headaches result from improper levels of serotonin in the brain has led to the synthesis of several drugs that combat the pain, nausea, and light sensitivity that are associated with this condition. These drugs include sumatriptan (trade name: Imitrex) and rizatriptan (trade name: Maxalt).

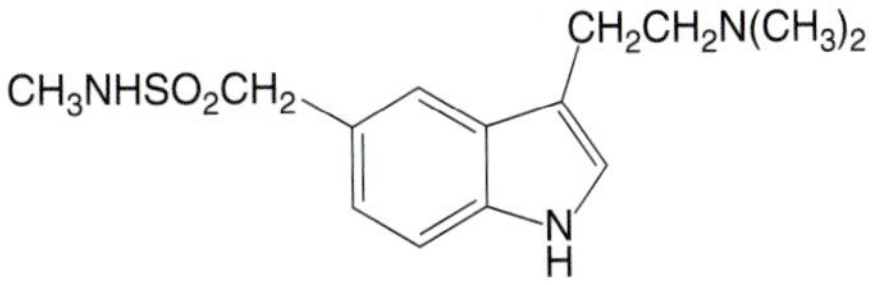

sumatriptan
(Trade name: Imitrex)

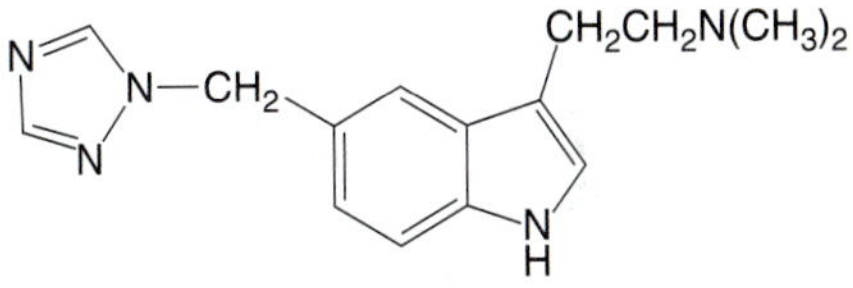

rizatriptan
(Trade name: Maxalt)

Drugs that interfere with the metabolism of serotonin have a profound effect on mental state. For example, bufotenin, isolated from *Bufo* toads from the Amazon jungle, and psilocin isolated from *Psilocybe* mushrooms, are very similar in structure to serotonin and both cause intense hallucinations.

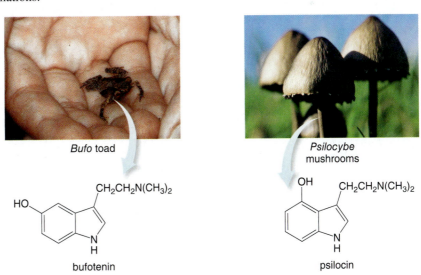

Bufo toad

Psilocybe mushrooms

bufotenin

psilocin

PROBLEM 18.25	The conversion of tryptophan to serotonin is a two-step process. What functional groups must be added or removed during this process?
PROBLEM 18.26	How are serotonin, bufotenin, and psilocin similar in structure? Classify the amines in these three compounds as 1°, 2°, or 3°.

18.8C ACETYLCHOLINE AND NICOTINE ADDICTION

Acetylcholine is a quaternary ammonium ion that serves as a neurotransmitter between neurons and muscle cells. In the brain, acetylcholine is involved in memory, mood, and other functions.

acetylcholine   nicotine

The pleasurable high that a smoker feels is due to nicotine interacting with acetylcholine receptors. At low concentrations, nicotine is a stimulant because it binds to these receptors, resulting in a feeling of well-being and alertness. Other nerve cells are also activated and this releases dopamine. The sense of pleasure resulting from the release of dopamine is what causes nicotine to be addictive. Over time receptors are inactivated and some degenerate, so tolerance to nicotine develops and larger quantities are needed to maintain the same sense of pleasure.

18.9 FOCUS ON THE HUMAN BODY
EPINEPHRINE AND RELATED COMPOUNDS

While neurotransmitters are the chemical messengers of the nervous system, **hormones** are the chemical messengers of the endocrine system. **A *hormone* is a compound produced by an endocrine gland, which then travels through the bloodstream to a target tissue or organ.**

Epinephrine, or **adrenaline** as it is commonly called, is a hormone synthesized in the adrenal glands from norepinephrine (noradrenaline).

norepinephrine   epinephrine
(noradrenaline)   (adrenaline)

When an individual senses danger or is confronted by stress, the hypothalamus region of the brain signals the adrenal glands to synthesize and release epinephrine, which enters the bloodstream and then stimulates a response in many organs (Figure 18.6). Stored carbohydrates are metabolized in the liver to form glucose, which is further metabolized to provide an energy boost. Heart rate and blood pressure increase, and lung passages are dilated. These physiological changes are commonly referred to as a "rush of adrenaline," and they prepare an individual for "fight or flight."

18.9A DERIVATIVES OF 2-PHENYLETHYLAMINE

Like epinephrine and norepinephrine, a large number of physiologically active compounds are derived from **2-phenylethylamine, $C_6H_5CH_2CH_2NH_2$.** Each of these compounds has a common structural unit: a benzene ring bonded to a two-carbon chain that is bonded to a nitrogen atom.

2-phenylethylamine   common structural feature

For example, **amphetamine** and **methamphetamine** are derivatives of 2-phenylethylamine that are powerful stimulants of the central nervous system. Both amines increase heart rate and

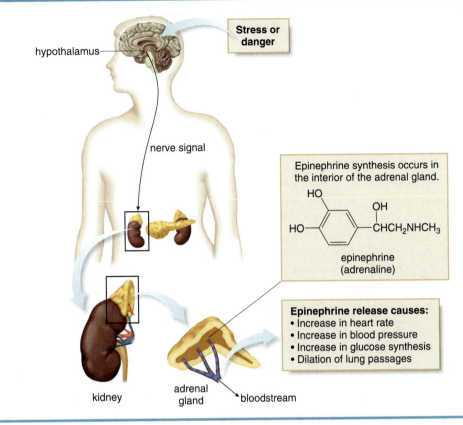

▼ FIGURE 18.6   Epinephrine—The "Fight or Flight" Hormone

respiratory rate. Since they increase glucose concentration in the bloodstream, they decrease hunger and fatigue. While amphetamine and methamphetamine may be prescribed for attention deficit hyperactivity disorder (ADHD) and weight reduction under the trade names Adderall and Desoxyn, respectively, they are highly addictive and widely abused drugs that must be used with extreme care. Both amphetamine and methamphetamine increase the level of dopamine in the brain, causing a pleasurable "high."

The atoms of 2-phenylethylamine are shown in red.

amphetamine

methamphetamine

Pseudoephedrine, the decongestant in Sudafed, is another derivative of 2-phenylethylamine. Since pseudoephedrine can be readily converted to methamphetamine, it has been replaced in many products by a related amine, **phenylephrine.** For example, Pfizer markets its pseudoephedrine-free decongestant as Sudafed PE. Products that contain pseudoephedrine are still over-the-counter remedies, but they are now stocked behind the pharmacy counter so their sale can be more closely monitored.

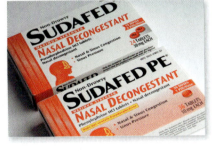

pseudoephedrine
(decongestant in Sudafed)

methamphetamine

phenylephrine
(decongestant in Sudafed PE)

PROBLEM 18.27   Pseudoephedrine can be readily converted to methamphetamine by replacing one OH group by a hydrogen atom. Examine the structure of phenylephrine, and suggest a reason why it cannot be readily converted to methamphetamine.

PROBLEM 18.28   Label the atoms of 2-phenylethylamine in each compound. Mescaline is the active hallucinogen in peyote, a cactus native to the southwestern United States and Mexico. LSD, commonly called "acid," was first synthesized in 1938 by Swiss chemist Albert Hoffman. Its hallucinogenic properties were first discovered when Hoffman himself ingested small amounts of the drug.

a.

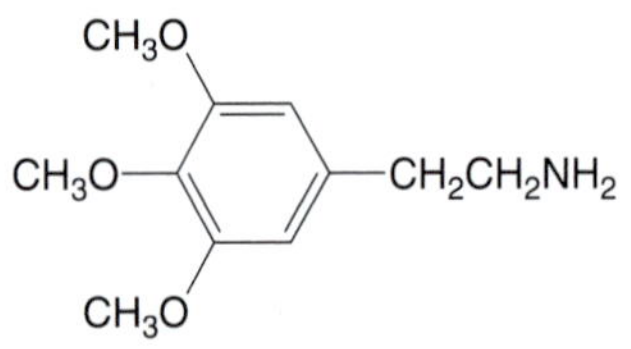

mescaline

b.

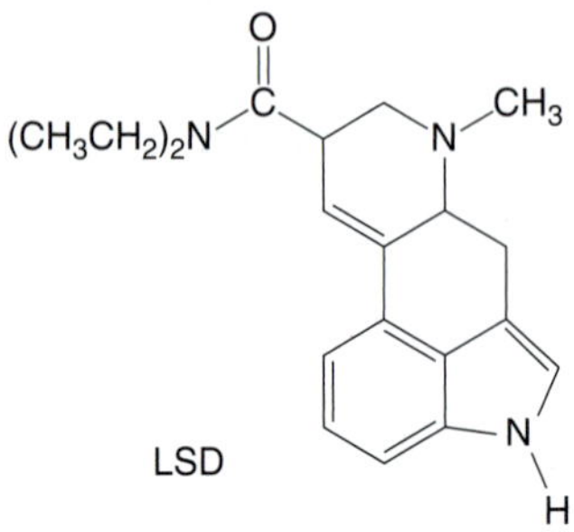

LSD

18.9B   DRUGS TO TREAT ASTHMA

The search for drugs that were structurally related to epinephrine but exhibited only some components of its wide range of biological activities led to the discovery of some useful medications. Both **albuterol** and **salmeterol** are derivatives of 2-phenylethylamine that dilate lung passages; that is, they are **bronchodilators.** They do not, however, stimulate the heart. This makes both compounds useful for the treatment of asthma. Albuterol is a short-acting drug used to relieve the wheezing associated with asthma. Salmeterol is much longer acting, and thus it is often used before bedtime to keep an individual symptom free overnight.

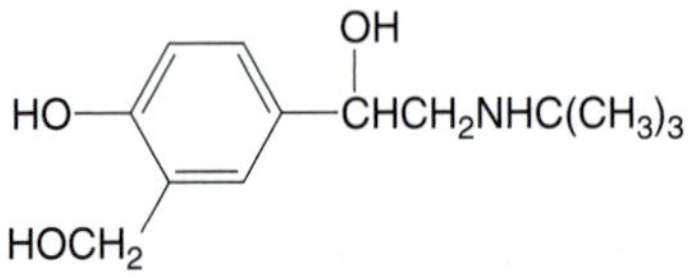

albuterol
(Trade names: Ventolin, Proventil)

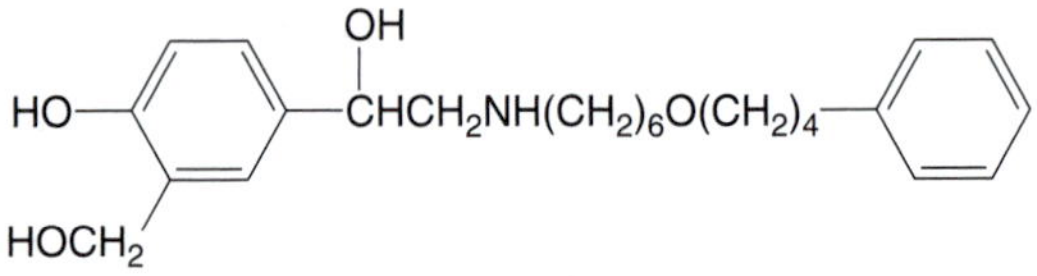

salmeterol
(Trade name: Serevent)

PROBLEM 18.29   Classify the amines and the alcohols (not the phenols) in albuterol and salmeterol as 1°, 2°, or 3°.

18.10   FOCUS ON HEALTH & MEDICINE
HISTAMINE AND ANTIHISTAMINES

Histamine is a biologically active amine formed in many tissues from the amino acid histidine.

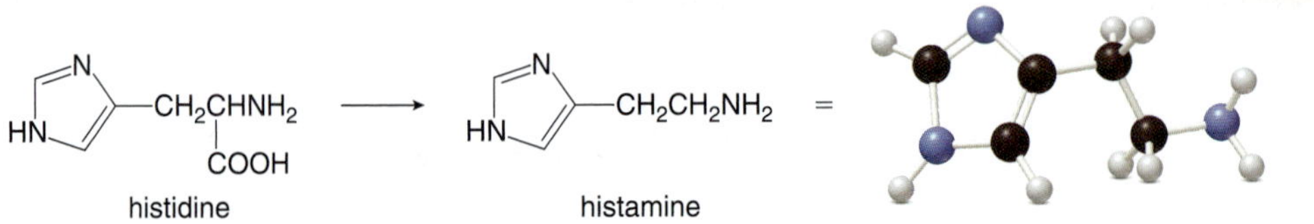

histidine   histamine

Histamine is a neurotransmitter that binds to two different receptors, the H1 and H2 receptors, and produces a wide range of physiological effects on the body. Histamine is a vasodilator (it dilates capillaries), so it is released at the site of an injury or infection to increase blood flow. It

HEALTH NOTE

Scombroid fish poisoning, associated with facial flushing, hives, and general itching, is caused by the ingestion of inadequately refrigerated fish, typically mahimahi and tuna. Bacteria convert the amino acid histidine to histamine, and when consumed in large amounts, this clinical syndrome results.

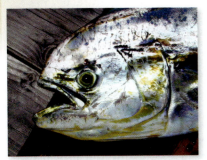

is also responsible for the symptoms of allergies, including a runny nose and watery eyes. In the stomach, histamine stimulates the secretion of acid.

Antihistamines bind to the H1 histamine receptor, but they evoke a different response. An antihistamine like chlorpheniramine or diphenhydramine, for example, inhibits vasodilation, so it is used to treat the symptoms of the common cold and environmental allergies. These antihistamines also cause drowsiness by crossing the blood–brain barrier and binding to H1 receptors in the central nervous system. Newer antihistamines like fexofenodine (trade name: Allegra) also bind to H1 receptors, but they do not cause drowsiness because they cannot cross the blood–brain barrier and bind to the H1 receptors in the central nervous system.

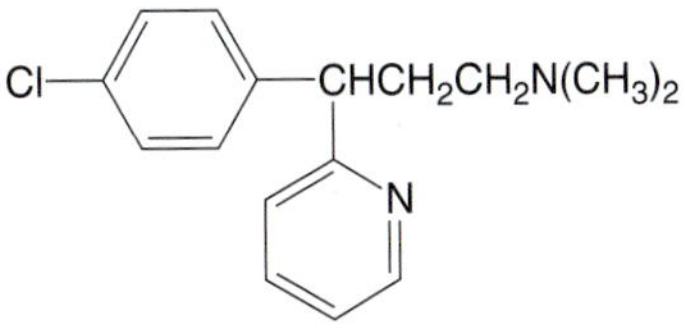

chlorpheniramine
(antihistamine)

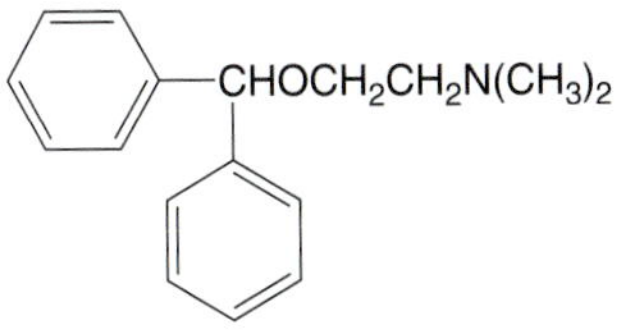

diphenhydramine
(antihistamine)

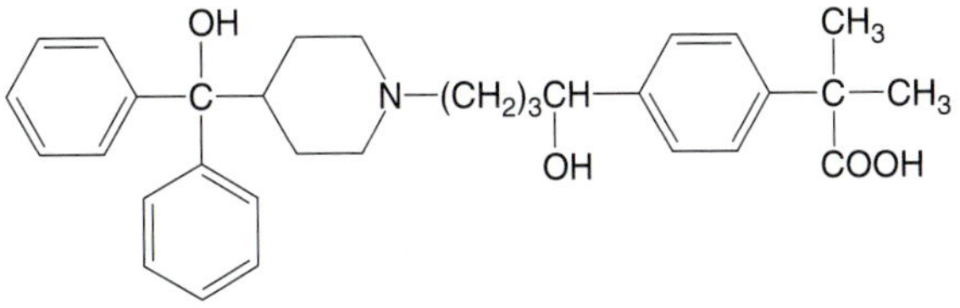

fexophenodine
(nonsedating antihistamine)

Drugs that bind to the H2 receptor produce a different effect. For example, cimetidine (trade name: Tagamet) binds to the H2 histamine receptor and, as a result, reduces acid secretion in the stomach. Thus, cimetidine is an effective **anti-ulcer** medication.

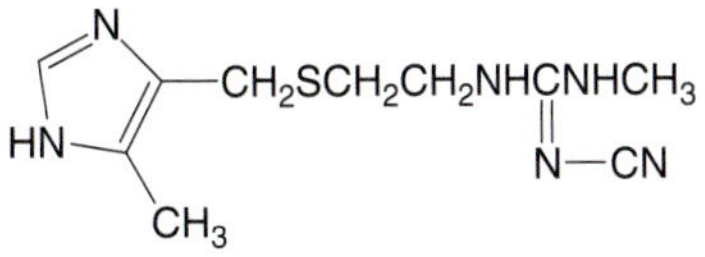

cimetidine
(Trade name: Tagamet)
anti-ulcer drug

PROBLEM 18.30 Draw a complete structure for cimetidine including all lone pairs on heteroatoms.

CHAPTER HIGHLIGHTS

KEY TERMS

Alkaloid (18.4)
Amine (18.1)
Ammonium salt (18.6)
Antihistamine (18.10)
Anti-ulcer drug (18.10)
Axon (18.8)
Dendrite (18.8)
Heterocycle (18.1)
Hormone (18.9)
Neuron (18.8)
Neurotransmitter (18.8)
Postsynaptic neuron (18.8)
Presynaptic neuron (18.8)
Primary (1°) amine (18.1)
Quaternary ammonium ion (18.1)
Secondary (2°) amine (18.1)
SSRI (18.8)
Synapse (18.8)
Tertiary (3°) amine (18.1)
Vesicle (18.8)

KEY REACTIONS

Amines react as bases with inorganic and organic acids (18.6).

$$R-\overset{\overset{\displaystyle H}{|}}{\underset{\underset{\displaystyle H}{|}}{\ddot{N}}}-H \; + \; H-Cl \; \rightleftharpoons \; R-\overset{\overset{\displaystyle H}{|}}{\underset{\underset{\displaystyle H}{|}}{\overset{\pm}{N}}}-H \; + \; Cl^{-}$$

KEY CONCEPTS

❶ What are the characteristics of amines? (18.1, 18.3)

- Amines are organic nitrogen compounds, formed by replacing one or more hydrogen atoms of NH_3 by alkyl groups.
- Primary (1°) amines have one C—N bond; 2° amines have two C—N bonds; 3° amines have three C—N bonds.
- An amine has a lone pair of electrons on the N atom, and is trigonal pyramidal in shape.
- Amines contain polar C—N and N—H bonds. Primary (1°) and 2° amines can hydrogen bond. Intermolecular hydrogen bonds between N and H are weaker than those between O and H.
- Tertiary (3°) amines have lower boiling points than 1° and 2° amines since they cannot hydrogen bond.

❷ How are amines named? (18.2)

- Primary (1°) amines are identified by the suffix *-amine*.
- Secondary (2°) and 3° amines with identical alkyl groups are named by adding the prefix *di-* or *tri-* to the name of the 1° amine.
- Secondary (2°) and 3° amines with different alkyl groups are named as *N*-substituted 1° amines.

❸ What are alkaloids? Give examples of common alkaloids. (18.4, 18.5)

- Alkaloids are naturally occurring amines isolated from plant sources.
- Examples of alkaloids include caffeine (from coffee and tea), nicotine (from tobacco), morphine and codeine (from the opium poppy), quinine (from the cinchona tree), and atropine (from the deadly nightshade plant).

❹ What products are formed when an amine is treated with acid? (18.6)

- Amines act as proton acceptors in water and acid. For example, the reaction of RNH_2 with HCl forms the water-soluble ammonium salt $RNH_3^+ Cl^-$.

❺ What are the characteristics of ammonium salts and how are they named? (18.6, 18.7)

- An ammonium salt consists of a positively charged ammonium ion and an anion.
- An ammonium salt is named by changing the suffix *-amine* of the parent amine to the suffix *-ammonium* followed by the name of the anion.

- Ammonium salts are water-soluble solids.
- Water-insoluble amine drugs are sold as their ammonium salts to increase their solubility in the aqueous environment of the blood.

❻ What are neurotransmitters and how do they differ from hormones? (18.8, 18.9)

- A neurotransmitter is a chemical messenger that transmits a nerve impulse from a neuron to another cell.
- A hormone is a compound produced by an endocrine gland that travels through the bloodstream to a target tissue or organ.

❼ What roles do dopamine and serotonin play in the body? (18.8)

- Dopamine affects movement, emotions, and pleasure. Too little dopamine causes Parkinson's disease. Too much dopamine causes schizophrenia. Dopamine plays a role in addiction.
- Serotonin is important in mood, sleep, perception, and temperature regulation. A deficiency of serotonin causes depression. SSRIs are antidepressants that effectively increase the concentration of serotonin.

❽ Give examples of important derivatives of 2-phenylethylamine. (18.9)

- Derivatives of 2-phenylethylamine contain a benzene ring bonded to a two-carbon chain that is bonded to a nitrogen atom.
- Examples of 2-phenylethylamine derivatives include epinephrine, norepinephrine, amphetamine, and methamphetamine. Albuterol and salmeterol are also derivatives of 2-phenylethylamine that are used to treat asthma.

❾ What is histamine, and how do antihistamines and anti-ulcer drugs work? (18.10)

- Histamine is an amine with a wide range of physiological effects. Histamine dilates capillaries, is responsible for the runny nose and watery eyes of allergies, and stimulates the secretion of stomach acid.
- Antihistamines bind to the H1 histamine receptor and inhibit vasodilation, so they are used to treat the symptoms of colds and allergies.
- Anti-ulcer drugs bind to the H2 histamine receptor and reduce the production of stomach acid.

PROBLEMS

Selected in-chapter and end-of-chapter problems have brief answers provided in Appendix B.

Structure and Bonding

18.31 Classify each amine as 1°, 2°, or 3°.

a. $CH_3CHCH_2CH_2NHCH_3$
   $\overset{|}{CH_3}$

b. [cyclohexyl ring]—N—CH$_3$

c. $CH_3\overset{\overset{\displaystyle CH_3}{|}}{\underset{\underset{\displaystyle CH_3}{|}}{C}}-NH_2$

18.32 Classify each amine as 1°, 2°, or 3°.

a. $CH_3\overset{\overset{\displaystyle CH_3}{|}}{\underset{\underset{\displaystyle CH_3}{|}}{C}}-CH_2NH_2$

b. [cyclohexyl ring]—NH

c. [benzene ring]—N(CH$_3$)$_2$

18.33 Draw the structure of a compound that fits each description:
a. a 1° amine with molecular formula $C_5H_{13}N$
b. a 2° amine with molecular formula $C_6H_{15}N$
c. a 3° amine with molecular formula $C_6H_{13}N$
d. a quaternary ammonium ion with molecular formula $C_8H_{20}N^+$

18.34 Draw the structure of a compound of molecular formula $C_4H_{11}NO$ that fits each description:
a. a compound that contains a 1° amine and a 1° alcohol
b. a compound that contains a 2° amine and a 2° alcohol
c. a compound that contains a 1° amine and a 3° alcohol
d. a compound that contains a 3° amine and a 1° alcohol

18.35 What is the difference between a 3° amine and a 3° alcohol? Give an example of each compound that contains six carbons.

18.36 What is the difference between a 3° amine and a 3° amide? Give an example of each compound that contains six carbons.

Nomenclature

18.37 Give an acceptable name for each amine.

a. $CH_3CH_2-\overset{\overset{\displaystyle H}{|}}{N}-CH_2CH_3$

b. [cyclohexyl ring]—NH$_2$

c. [benzene ring]—NCH$_2$CH$_3$
   $\overset{|}{CH_3}$

d. $CH_3CH_2\overset{\overset{\displaystyle NH_2}{|}}{C}HCH_2CH_2CH_2CH_2CH_3$

18.38 Give an acceptable name for each amine.

e. $CH_3CH_2CH_2\overset{\overset{\displaystyle NHCH_3}{|}}{C}HCH_2CH_3$

f. $CH_3\overset{}{C}HCH_2CH_2CH_2CH_2CH_3$
   $\overset{|}{NHCH_2CH_3}$

18.38 Give an acceptable name for each amine.
a. $CH_3(CH_2)_6NH_2$
b. $CH_3CH_2CH_2CH_2CH_2N(CH_3)_2$
c. [cyclopentyl ring]—NHCH$_2$CH$_3$
d. [pyrrolidine ring]N—CH$_2$CH$_3$
e. [benzene ring with Cl and NH$_2$]
f. [benzene ring]$\overset{\overset{\displaystyle CH_3CHCH_3}{|}}{N}$—CH$_2CH_3$
   $\overset{|}{CH_3}$

18.39 Draw the structure corresponding to each name.
a. 1-decanamine
b. tricyclohexylamine
c. p-bromoaniline
d. 3-aminobutanoic acid
e. N,N-dipropyl-2-octanamine
f. N-ethylhexylamine

18.40 Draw the structure corresponding to each name.
a. cyclobutylamine
b. 4-nonanamine
c. N-pentylaniline
d. 3-amino-2-butanone
e. 3-methyl-1-butanamine
f. N-ethyl-N-propylcyclopentylamine

18.41 What is the difference between N,N-dimethylaniline and 2,4-diethylaniline? Draw structures to illustrate the difference.

18.42 What is the difference between N-butylaniline and o-butylaniline? Draw structures to illustrate the difference.

18.43 Draw the structure of each ammonium salt.
a. dipropylammonium chloride
b. butylammonium bromide
c. ethyldimethylammonium hydroxide

18.44 Draw the structure of each ammonium salt.
a. cyclohexylammonium fluoride
b. triethylammonium iodide
c. ethylmethylammonium chloride

18.45 Draw the structures for the four amines of molecular formula C_3H_9N. Give the systematic name for each amine.

18.46   (a) Draw the structures for the eight constitutional isomers of molecular formula $C_4H_{11}N$. (b) Give the systematic name for each amine. (c) Identify the chirality center present in one of the amines.

Physical Properties and Intermolecular Forces

18.47   Explain why pyridine is more water soluble than benzene. How would you expect the boiling point of pyridine and benzene to compare?

18.48   Explain why $CH_3CH_2CH_2NH_2$ is water soluble but $CH_3CH_2CH_2CH_3$ is not water soluble.

18.49   Which compound in each pair has the higher boiling point?

a. $CH_3CH_2CH_2NH_2$ or $CH_3(CH_2)_7NH_2$

b. $CH_3(CH_2)_6OH$ or $CH_3(CH_2)_6NH_2$

c. ⬡—$CH_2N(CH_3)_2$ or   ⬡—$CH_2CH_2CH_2NH_2$

18.50   Which compound in each pair has the higher boiling point?

a. $(CH_3CH_2)_2C{=}O$ or $(CH_3CH_2)_2CHNH_2$

b. ⬡   or   ⬡NH

c. $CH_3\overset{\underset{|}{Cl}}{C}HCH_2\overset{\underset{|}{NH_2}}{C}HCH_3$ or $CH_3CH_2CH_2\overset{\underset{|}{\overset{+}{N}H_3\ Cl^-}}{C}HCH_3$

18.51   Draw the hydrogen bonding interactions that occur between two molecules of diethylamine [$(CH_3CH_2)_2NH$].

18.52   Draw the hydrogen bonding interactions that occur between diethylamine [$(CH_3CH_2)_2NH$] and water.

18.53   Explain why a 1° amine and a 3° amine having the same number of carbons are soluble in water to a similar extent, but the 1° amine has a higher boiling point.

18.54   Which compound has the higher water solubility: $CH_3(CH_2)_5NH_2$ or $CH_3(CH_2)_5NH_3^+ Cl^-$? Explain your choice.

Acid–Base Reactions of Amines

18.55   Draw the acid–base reaction that occurs when each amine dissolves in water: (a) $CH_3CH_2NH_2$; (b) $(CH_3CH_2)_2NH$; (c) $(CH_3CH_2)_3N$.

18.56   What ammonium salt is formed when each amine is treated with HCl?

a. ⬡—NH_2

b. ⬡—CH_2NHCH_3

c. ⬡—$N(CH_2CH_3)_2$

18.57   Draw the products of each acid–base reaction.

a. $CH_3CH_2CH_2N(CH_3)_2$ +   HCl $\longrightarrow$

b. $CH_3CH_2\overset{\underset{|}{NH_2}}{C}HCH_2CH_3$ + H_2SO_4 $\longrightarrow$

c. ⬡(pyridine)   +   HBr $\longrightarrow$

d. $CH_3CH_2{-}\overset{\underset{|}{\overset{H}{|}}}{\overset{+}{N}}{-}CH_2CH_3$ +   NaOH $\longrightarrow$
    with CH_2CH_3 below N

18.58   Draw the products of each acid–base reaction.

a. ⬡(quinoline)   +   HCl $\longrightarrow$

b. $CH_3CH_2\overset{\underset{|}{NHCH_3}}{C}HCH_2CH_3$ + H_2SO_4 $\longrightarrow$

c. CH_3NH_2 + CH_3COOH $\longrightarrow$

d. $CH_3(CH_2)_5\overset{+}{N}H_3$ +   NaOH $\longrightarrow$

18.59   What ammonium salt is formed when each alkaloid is treated with HCl?

a.

coniine

b.

morphine

18.60   What ammonium salt is formed when each biologically active amine is treated with H_2SO_4?

a. ⬡—$CH_2\overset{\underset{|}{CH_3}}{C}HNH_2$

amphetamine

b.

dopamine

Neurotransmitters, Alkaloids, and Other Biologically Active Amines

18.61 What type of nitrogen heterocycle occurs in both coniine and morphine (Problem 18.59)?

18.62 Only one of the N atoms in nicotine has a trigonal pyramidal shape. Identify which N is trigonal pyramidal and explain why the other N atom has a different molecular shape.

18.63 What physiological effects does caffeine produce on the body?

18.64 Why should caffeine intake be limited in pregnant and nursing mothers?

18.65 Why are aqueous solutions of an alkaloid slightly basic?

18.66 For each alkaloid, list its natural source and give one of its uses in medicine: (a) morphine; (b) quinine; (c) atropine.

18.67 What is the difference between a presynaptic neuron and a postsynaptic neuron?

18.68 Besides the cell body, what are the important parts of a neuron?

18.69 What are the physiological effects of dopamine? What conditions result when dopamine levels are too high or too low?

18.70 Explain why patients with Parkinson's disease cannot be treated by giving them dopamine, even though the disease results from low dopamine levels.

18.71 What are the physiological effects of serotonin? What condition results when there is too little serotonin?

18.72 Give two examples of SSRIs and explain how they combat depression.

18.73 In the body, what amino acid is used to synthesize each compound: (a) dopamine; (b) norepinephrine; (c) serotonin; (d) histamine?

18.74 What is the difference between a neurotransmitter and a hormone?

18.75 Which of the following amines are derivatives of 2-phenylethylamine? Label the atoms of 2-phenylethylamine in those amines that contain this unit.

a.
b.
c.

18.76 Which of the following amines are derivatives of 2-phenylethylamine? Label the atoms of 2-phenylethylamine in those amines that contain this unit.

a.
b.
c.

18.77 Locate the atoms of 2-phenylethylamine in each drug.

a. phentermine
(one component of fen–phen, a banned diet drug)

b. fentanyl
(a narcotic pain reliever)

18.78 Locate the atoms of 2-phenylethylamine in each drug.

a. albuterol
(asthma drug)

b. 3-methylfentanyl
(street drug, 3,000 times more potent than morphine)

18.79 Give an example of an antihistamine. Explain how an antihistamine relieves the runny nose and watery eyes of a cold or allergies.

18.80 Give an example of an anti-ulcer drug, and explain how it is used to treat a stomach ulcer.

General Questions

18.81 Benzphetamine (trade name: Didrex) is a habit-forming diet pill with many of the same properties as amphetamine.

benzphetamine

a. Label the amine as 1°, 2°, or 3°.

b. Label the chirality center in benzphetamine.

c. Draw the two enantiomers of benzphetamine.

d. Draw a constitutional isomer that contains a 1° amine.

e. Draw a constitutional isomer that contains a 3° amine.

f. Draw the structure of benzphetamine hydrochloride.

g. What products are formed when benzphetamine is treated with acetic acid, CH_3COOH?

18.82   Phentermine is one component of the banned diet drug fen–phen.

phentermine

a. Label the amine as 1°, 2°, or 3°.

b. Indicate the molecular shape around each atom.

c. Draw a constitutional isomer that contains a 1° amine.

d. Draw a constitutional isomer that contains a 2° amine.

e. Draw the structure of phentermine hydrobromide.

f. What products are formed when phentermine is treated with benzoic acid, C_6H_5COOH?

18.83   Ritalin is the trade name for methylphenidate, a drug used to treat attention deficit hyperactivity disorder (ADHD).

methylphenidate
(Trade name: Ritalin)

a. Identify the functional groups.

b. Label the amine as 1°, 2°, or 3°.

c. Locate the chirality centers.

d. Locate the atoms of 2-phenylethylamine.

e. Draw the structure of methylphenidate hydrochloride.

18.84   Seldane is the trade name for terfenadine, an antihistamine once used in the United States but withdrawn from the market because of cardiac side effects observed in some patients.

terfenadine
(Trade name: Seldane)

a. Identify the functional groups.

b. Label the amine as 1°, 2°, or 3°.

c. Label any alcohol as 1°, 2°, or 3°.

d. Label any 4° carbon.

e. Locate any chirality centers.

f. Draw the structure of terfenadine hydrochloride.

Applications

18.85   The antihistamine in the over-the-counter product Chlortrimeton is chlorpheniramine maleate. Draw the structure of the amine and carboxylic acid that are used to form this ammonium salt.

chlorpheniramine maleate

18.86   Dextromethorphan hydrobromide is a cough suppressant used in many over-the-counter cough and cold medications. Given the structure of the parent amine dextromethorphan, what is the structure of its hydrobromide salt?

dextromethorphan

18.87   What is the difference between a vasodilator and a bronchodilator? Give an example of a drug that fits each category.

18.88   Why do some antihistamines cause drowsiness while others do not?

18.89   In Section 17.7 we learned that aspirin exists in different forms (neutral or ionic) in the stomach and the intestines. Similar behavior occurs with amines that are ingested, since they undergo acid–base reactions. In what form does the drug albuterol (Section 18.9B) exist in the stomach? In the intestines?

18.90   Explain why eating a poppy seed bagel can produce a positive test in a drug screening.

CHALLENGE QUESTIONS

18.91   Compare the structures of morphine and heroin. Suggest a method to convert morphine to heroin.

18.92   Heroin and the morphine from which it is made are both odorless solids. Drug-sniffing dogs can sometimes locate heroin because of a biting odor that is present in a small enough concentration that humans cannot detect, but the keen nose of a canine can. Suggest what compound might cause the odor that allows for heroin detection. (Hint: What products would be formed by the hydrolysis of heroin in moist air?)

19

CHAPTER GOALS

In this chapter you will learn how to:

1. Describe the general characteristics of lipids

2. Classify fatty acids and describe the relationship between melting point and the number of double bonds

3. Draw the structure of a wax and identify the carboxylic acid and alcohol components

4. Draw the structure of triacylglycerols and describe the difference between a fat and an oil

5. Draw the hydrolysis products of triacylglycerols

6. Identify the two major classes of phospholipids

7. Describe the structure of a cell membrane, as well as different mechanisms of transport across the membrane

8. Recognize the main structural features of steroids like cholesterol, and describe the relationship between blood cholesterol level and cardiovascular disease

9. Define what a hormone is and list several examples of steroid hormones

10. Identify fat-soluble vitamins

11. Discuss the general structural features and biological activity of prostaglandins and leukotrienes

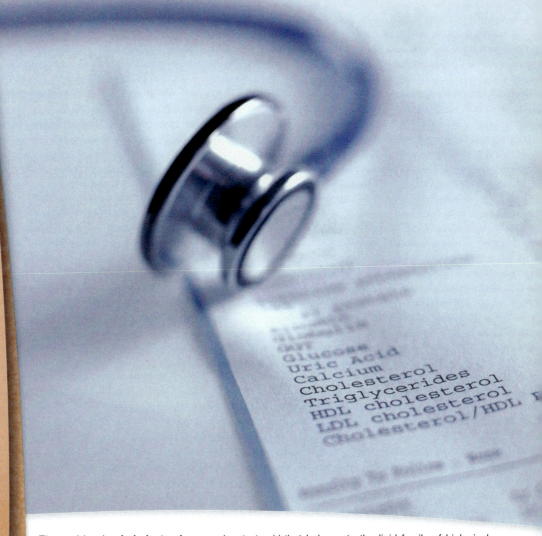

Elevated levels of **cholesterol,** a prominent steroid that belongs to the lipid family of biological molecules, are associated with coronary artery disease.

LIPIDS

CHAPTER 19 is the first of four chapters that deal with the chemistry of *biomolecules,* **organic molecules found in biological systems.** Chapter 19 focuses on lipids, biomolecules that contain many carbon–carbon and carbon–hydrogen bonds, making them soluble in organic solvents and insoluble in water. Chapter 20 discusses carbohydrates, the largest group of biomolecules in nature, while Chapter 21 focuses on proteins and the amino acids that compose them. Finally, the properties of DNA, the polymer responsible for the storage of genetic information in the chromosomes of cells, is presented in Chapter 22. These compounds are all organic molecules, so many of the principles and chemical reactions that you have already learned will be examined once again. But, as you will see, each class of compound has its own unique features that we will discuss as well.

19.1 INTRODUCTION TO LIPIDS

- *Lipids* are biomolecules that are soluble in organic solvents and insoluble in water.

Lipids are unique among organic molecules because their identity is defined on the basis of a *physical property* and not by the presence of a particular functional group. Because of this, lipids come in a wide variety of structures and they have many different functions. Common lipids include triacylglycerols in vegetable oils, cholesterol in egg yolk, and vitamin E in leafy greens.

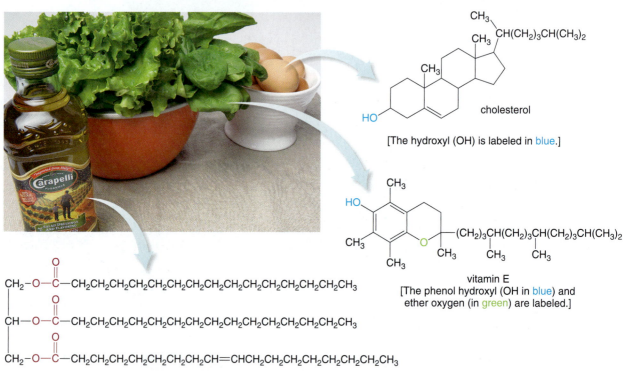

[The hydroxyl (OH) is labeled in blue.]

cholesterol

vitamin E
[The phenol hydroxyl (OH in blue) and ether oxygen (in green) are labeled.]

triacylglycerol
[The three ester groups are labeled in red.]

The word *lipid* comes from the Greek word *lipos* for *fat*.

Lipids contain a large number of nonpolar carbon–carbon and carbon–hydrogen bonds. In addition, most lipids have a few polar bonds that may be found in a variety of functional groups. For example, a triacylglycerol contains three esters, cholesterol possesses a hydroxyl group, and vitamin E has both a phenol (an OH group on a benzene ring) and an ether. The size of these functional groups is small compared to the overall size of the molecules. As a result, lipids are nonpolar or weakly polar molecules that are very soluble in organic solvents like hexane (C_6H_{14}) and carbon tetrachloride (CCl_4), and insoluble in a polar medium like water.

SAMPLE PROBLEM 19.1   Which compounds are likely to be lipids? Sorbitol is a sweetener used in sugar-free mints and gum. β-Carotene is the orange pigment in carrots, and the precursor of vitamin A.

a.

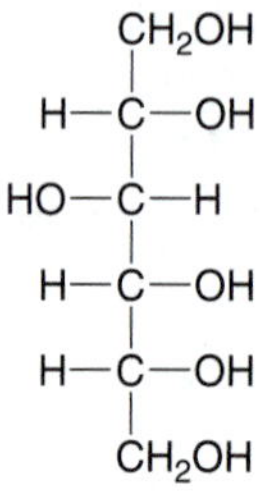

sorbitol

b.

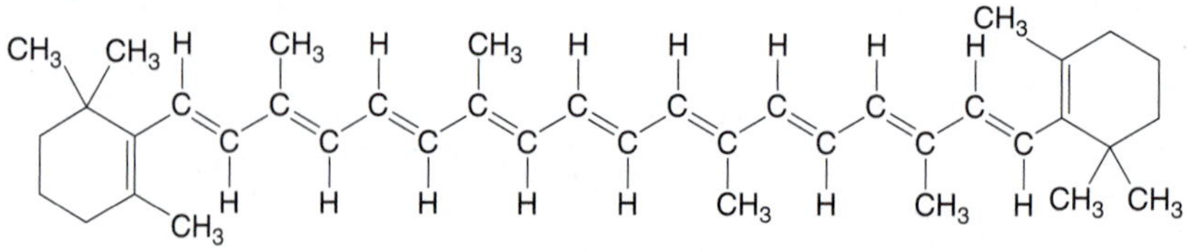

β-carotene

ANALYSIS   Lipids contain many nonpolar C—C and C—H bonds and few polar bonds.

SOLUTION   a. Sorbitol is not likely to be a lipid since it contains six polar OH groups bonded to only six carbon atoms.

b. β-Carotene is likely to be a lipid since it contains only nonpolar C—C and C—H bonds.

PROBLEM 19.1

Which compounds are likely to be lipids?

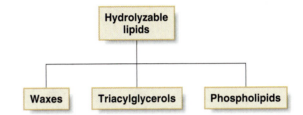

a. 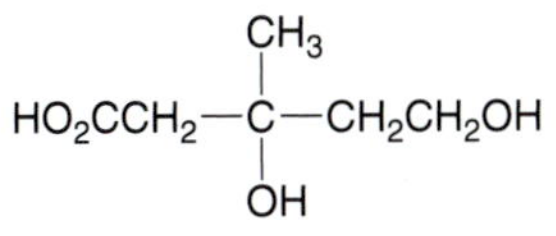 mevalonic acid   b. 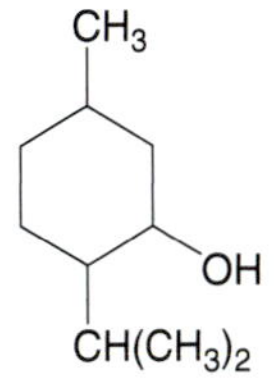 menthol   c. 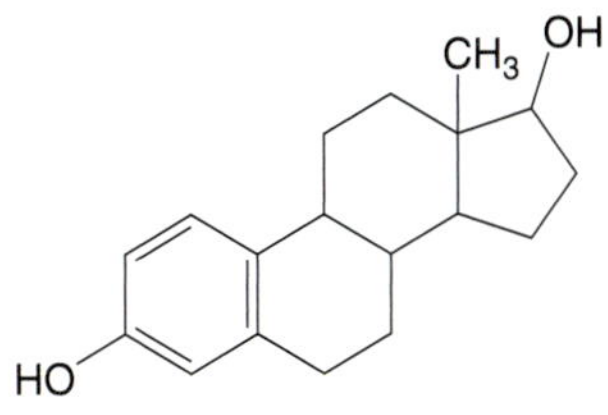 estradiol

TABLE 19.1 Summary of Lipid Chemistry Prior to Chapter 19

Topic	Section
Vitamin A	11.7A
Fatty acids	13.3B
Oral contraceptives	13.5
Margarine and butter	13.7
Aspirin and prostaglandins	17.4B
Soap	17.6C
Olestra, a synthetic fat	17.9C

Because lipids share many properties with hydrocarbons, several features of lipid structure and properties have been discussed in previous chapters, as summarized in Table 19.1.

Lipids can be categorized as hydrolyzable or nonhydrolyzable.

1. *Hydrolyzable lipids* can be converted into smaller molecules by hydrolysis with water. We will examine three subgroups: waxes, triacylglycerols, and phospholipids.

Hydrolyzable lipids

Waxes   **Triacylglycerols**   **Phospholipids**

Most hydrolyzable lipids contain an ester. In the presence of acid, base, or an enzyme, the C—O bond of the ester is cleaved, and a carboxylic acid and an alcohol are formed.

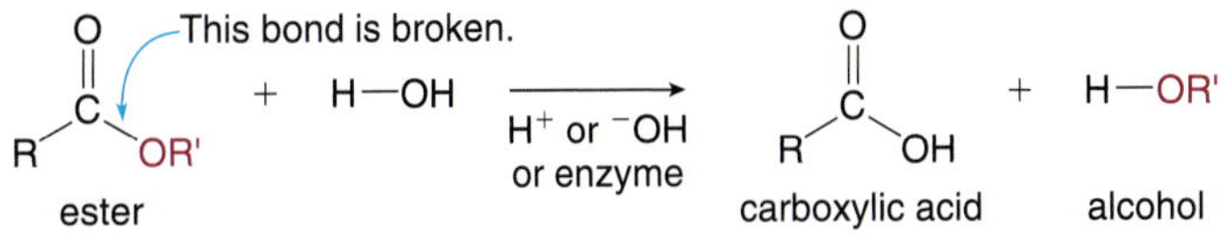

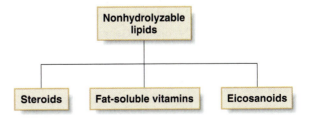

2. *Nonhydrolyzable lipids* cannot be cleaved into smaller units by aqueous hydrolysis. Nonhydrolyzable lipids tend to be more varied in structure. We will examine three different types: steroids, fat-soluble vitamins, and eicosanoids.

Nonhydrolyzable lipids

Steroids   **Fat-soluble vitamins**   **Eicosanoids**

Lipids have many important roles in biological systems. Since lipids release over twice the amount of energy per gram than carbohydrates or proteins (9 kcal/g for lipids compared to 4 kcal/g for carbohydrates and proteins, Section 6.1), lipids are an excellent source of energy. Moreover, lipids are key components of the cell membrane, and they serve as chemical messengers in the body. A lipid coating protects a leaf's surface and keeps it from drying out. Ducks are insulated by a protective layer of lipids on their feathers.

PROBLEM 19.2 In which solvents or solutions will a lipid be soluble: (a) CH_2Cl_2; (b) 5% aqueous NaCl solution; (c) $CH_3CH_2CH_2CH_2CH_3$?

19.2 FATTY ACIDS

Hydrolyzable lipids are derived from **fatty acids,** carboxylic acids that were first discussed in Section 13.3B.

- *Fatty acids* are carboxylic acids (RCOOH) with long carbon chains of 12–20 carbon atoms.

Because they have many nonpolar C—C and C—H bonds and few polar bonds, fatty acids are lipids, making them soluble in organic solvents and insoluble in water. Palmitic acid is a common 16-carbon fatty acid whose structure is given in condensed, skeletal, and three-dimensional representations.

$$CH_3CH_2CH_2CH_2CH_2CH_2CH_2CH_2CH_2CH_2CH_2CH_2CH_2CH_2CH_2COOH$$
palmitic acid
$$C_{16}H_{32}O_2$$

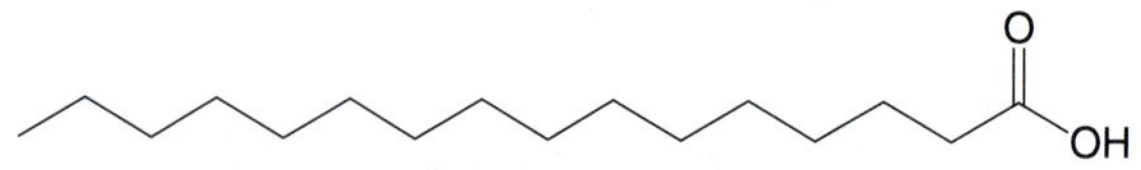

skeletal structure

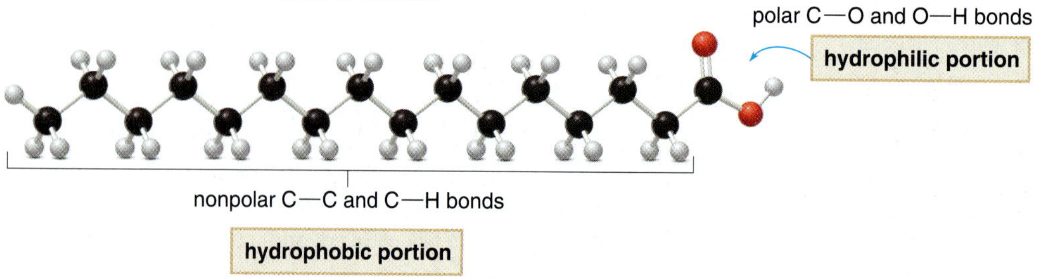

polar C—O and O—H bonds
hydrophilic portion

nonpolar C—C and C—H bonds
hydrophobic portion

The nonpolar part of the molecule (comprised of C—C and C—H bonds) is not attracted to water, so it is said to be *hydrophobic* (water fearing). The polar part of the molecule is attracted to water, so it is said to be *hydrophilic* (water loving). In a lipid, the hydrophobic portion is always much larger than the hydrophilic portion.

Naturally occurring fatty acids have an even number of carbon atoms. There are two types of fatty acids.

- *Saturated* fatty acids have no double bonds in their long hydrocarbon chains.
- *Unsaturated* fatty acids have one or more double bonds in their long hydrocarbon chains. Generally, double bonds in naturally occurring fatty acids are cis.

Recall from Section 13.3 that a **cis** alkene has two alkyl groups on the *same* side of the double bond, while a **trans** alkene has two alkyl groups on *opposite* sides of the double bond.

Three-dimensional models for stearic acid, a saturated fatty acid, and oleic acid, an unsaturated fatty acid with one cis double bond, are shown. Table 19.2 lists the structures of the most common saturated and unsaturated fatty acids.

TABLE 19.2 Common Fatty Acids

Number of C's	Number of C=C's	Structure	Name	Mp (°C)
Saturated Fatty Acids				
12	0	$CH_3(CH_2)_{10}COOH$	Lauric acid	44
14	0	$CH_3(CH_2)_{12}COOH$	Myristic acid	58
16	0	$CH_3(CH_2)_{14}COOH$	Palmitic acid	63
18	0	$CH_3(CH_2)_{16}COOH$	Stearic acid	71
20	0	$CH_3(CH_2)_{18}COOH$	Arachidic acid	77
Unsaturated Fatty Acids				
16	1	$CH_3(CH_2)_5CH{=}CH(CH_2)_7COOH$	Palmitoleic acid	1
18	1	$CH_3(CH_2)_7CH{=}CH(CH_2)_7COOH$	Oleic acid	16
18	2	$CH_3(CH_2)_4CH{=}CHCH_2CH{=}CH(CH_2)_7COOH$	Linoleic acid	−5
18	3	$CH_3CH_2CH{=}CHCH_2CH{=}CHCH_2CH{=}CH(CH_2)_7COOH$	Linolenic acid	−11
20	4	$CH_3(CH_2)_4(CH{=}CHCH_2)_4(CH_2)_2COOH$	Arachidonic acid	−49

cis double bond

stearic acid

a saturated fatty acid
no double bonds in the long chain

oleic acid

an unsaturated fatty acid
one cis double bond in the long chain

The most common saturated fatty acids are palmitic and stearic acid. The most common unsaturated fatty acid is oleic acid. Linoleic and linolenic acids are called **essential fatty acids** because humans cannot synthesize them and must acquire them in our diets.

Oils formed from omega-3 fatty acids may provide health benefits to individuals with cardiovascular disease, as discussed in Section 19.4.

Unsaturated fatty acids are sometimes classified as **omega-n acids,** where n is the carbon at which the first double bond occurs in the carbon chain, beginning at the end of the chain that contains the CH_3 group. Thus, linoleic acid is an omega-6 acid and linolenic acid is an omega-3 acid.

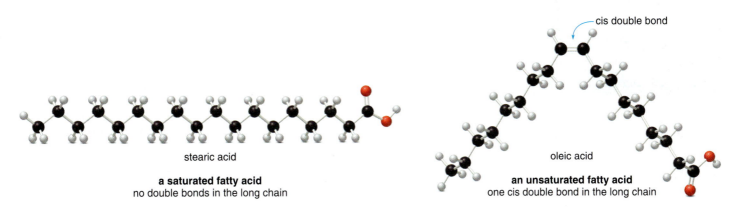

first C=C at C6 ⟶ an **omega-6 acid**

$CH_3CH_2CH_2CH_2CH_2CH{=}CHCH_2CH{=}CHCH_2CH_2CH_2CH_2CH_2CH_2COOH$

linoleic acid

1 2 3 4 5 6

first C=C at C3 ⟶ an **omega-3 acid**

$CH_3CH_2CH{=}CHCH_2CH{=}CHCH_2CH{=}CHCH_2CH_2CH_2CH_2CH_2CH_2COOH$

linolenic acid

1 2 3

As we learned in Section 13.3B, the presence of cis double bonds affects the melting point of these fatty acids greatly.

• As the number of double bonds in the fatty acid *increases,* the melting point *decreases.*

The cis double bonds introduce kinks in the long hydrocarbon chain, which makes it difficult for the molecules to pack closely together in a solid. **The larger the number of cis double bonds, the more kinks in the hydrocarbon chain, and the lower the melting point.**

SAMPLE PROBLEM 19.2

(a) Draw a skeletal structure of gadoleic acid, $CH_3(CH_2)_9CH=CH(CH_2)_7COOH$, a 20-carbon fatty acid obtained from fish oils. (b) Label the hydrophobic and hydrophilic portions. (c) Predict how its melting point compares to the melting points of arachidic acid and arachidonic acid (Table 19.2).

ANALYSIS

• Skeletal structures have a carbon at the intersection of two lines and at the end of every line. The double bond must have the cis arrangement in an unsaturated fatty acid.
• The nonpolar C—C and C—H bonds comprise the hydrophobic portion of a molecule and the polar bonds comprise the hydrophilic portion.
• For the same number of carbons, increasing the number of double bonds decreases the melting point of a fatty acid.

SOLUTION

a. and b. The skeletal structure for gadoleic acid is drawn below. The hydrophilic portion is the COOH group and the hydrophobic portion is the rest of the molecule.

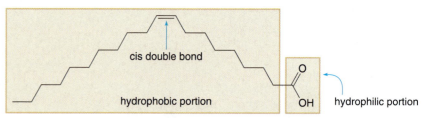

Gadoleic acid (Sample Problem 19.2) and DHA (Problem 19.4) are two fatty acids derived from tuna fish oil.

c. Gadoleic acid has one cis double bond, giving it a lower melting point than arachidic acid (no double bonds), but a higher melting point than arachidonic acid (four double bonds).

PROBLEM 19.3

(a) Draw a skeletal structure for each fatty acid. (b) Label the hydrophobic and hydrophilic portions of each molecule. (c) Without referring to Table 19.2, which fatty acid has the higher melting point, **A** or **B?** Explain your choice.

$CH_3(CH_2)_{16}COOH$ $CH_3(CH_2)_4CH=CHCH_2CH=CH(CH_2)_7COOH$

A   **B**

PROBLEM 19.4

DHA (4,7,10,13,16,19-docosahexaenoic acid) is a common fatty acid present in tuna fish oil. (a) Draw a skeletal structure for DHA. (b) Give the omega-*n* designation for DHA.

$CH_3CH_2CH=CHCH_2CH=CHCH_2CH=CHCH_2CH=CHCH_2CH=CHCH_2CH=CHCH_2CH_2COOH$

DHA

PROBLEM 19.5

Give the omega-*n* designation for (a) oleic acid; (b) arachidonic acid (Table 19.2).

19.3 WAXES

Waxes are the simplest hydrolyzable lipids.

When commercial whaling was commonplace, spermaceti wax obtained from sperm whales was used extensively in cosmetics and candles.

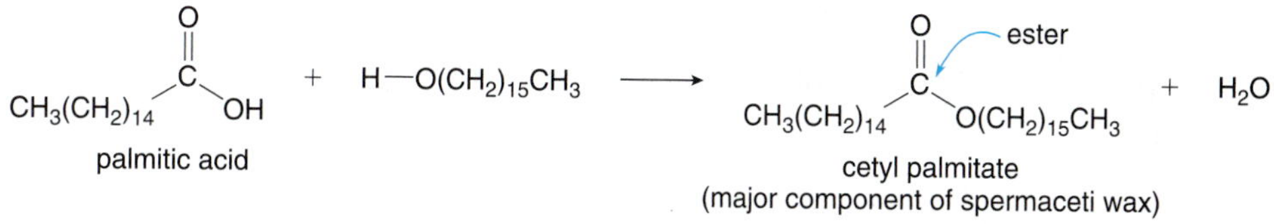

Water beads up on the surface of a leaf because of its waxy coating.

- **Waxes** are esters (RCOOR') formed from a fatty acid (RCOOH) and a high molecular weight alcohol (R'OH).

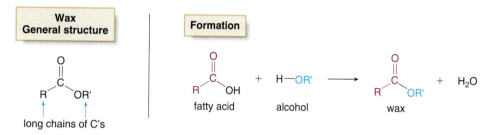

For example, spermaceti wax, isolated from the heads of sperm whales, is largely cetyl palmitate, an ester with the structure $CH_3(CH_2)_{14}COO(CH_2)_{15}CH_3$. Cetyl palmitate is formed from a 16-carbon fatty acid [$CH_3(CH_2)_{14}COOH$] and a 16-carbon alcohol [$CH_3(CH_2)_{15}OH$].

$$CH_3(CH_2)_{14} \overset{O}{\underset{}{C}} OH \;+\; H-O(CH_2)_{15}CH_3 \longrightarrow CH_3(CH_2)_{14} \overset{O}{\underset{}{C}} O(CH_2)_{15}CH_3 \;+\; H_2O$$

palmitic acid

cetyl palmitate
(major component of spermaceti wax)

Because of their long nonpolar carbon chains, **waxes are very hydrophobic.** They form a protective coating on the feathers of birds to make them water repellent, and on leaves to prevent water evaporation. Lanolin is a wax that coats the wool fibers of sheep. Beeswax, a complex mixture of over 200 different compounds, contains the wax myricyl palmitate as its major component. The three-dimensional structure of this compound shows how small the polar ester group is compared to the long hydrocarbon chains.

$$CH_3(CH_2)_{14} \overset{O}{\underset{}{C}} O(CH_2)_{29}CH_3$$

myricyl palmitate

hydrophobic region   hydrophobic region

SAMPLE PROBLEM 19.3

Draw the structure of cetyl myristate, a minor component of spermaceti wax, formed from a 14-carbon fatty acid and a 16-carbon straight chain alcohol.

ANALYSIS To draw the wax, arrange the carboxyl group of the fatty acid (RCOOH) next to the OH group of the alcohol (R'OH) with which it reacts. Then, replace the OH group of the fatty acid with the OR' group of the alcohol, forming an ester RCOOR' with a new C—O bond at the carbonyl carbon.

SOLUTION Draw the structures of the fatty acid and the alcohol, and replace the OH group of the 14-carbon acid with the $O(CH_2)_{15}CH_3$ group of the alcohol.

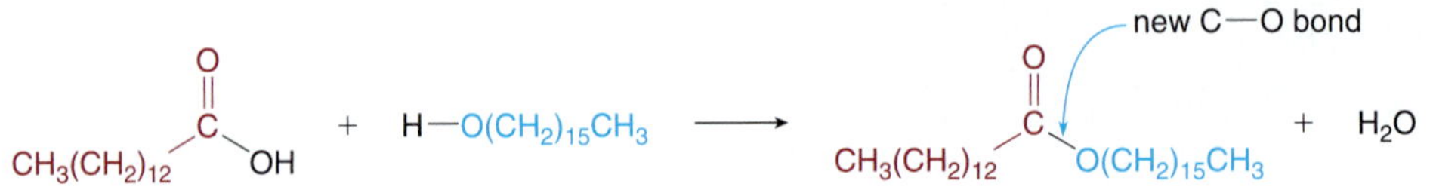

14-carbon fatty acid 16-carbon alcohol cetyl myristate

PROBLEM 19.6

Draw the structure of a wax formed from stearic acid $[CH_3(CH_2)_{16}COOH]$ and each alcohol.

a. $CH_3(CH_2)_9OH$ b. $CH_3(CH_2)_{11}OH$ c. $CH_3(CH_2)_{29}OH$

PROBLEM 19.7

Carnauba wax, a wax that coats the leaves of the Brazilian palm tree, is used for hard, high-gloss finishes for floors, boats, and automobiles. Draw the structure of one component of carnauba wax, formed from a 32-carbon carboxylic acid and a straight chain 34-carbon alcohol.

PROBLEM 19.8

Explain why beeswax is insoluble in water, slightly soluble in ethanol (CH_3CH_2OH), and very soluble in chloroform ($CHCl_3$).

Like other esters, **waxes (RCOOR') are hydrolyzed with water in the presence of acid or base to re-form the carboxylic acid (RCOOH) and alcohol (R'OH)** from which they are prepared (Section 17.9). Thus, hydrolysis of cetyl palmitate in the presence of H_2SO_4 forms a fatty acid and a long chain alcohol by cleaving the carbon–oxygen single bond of the ester.

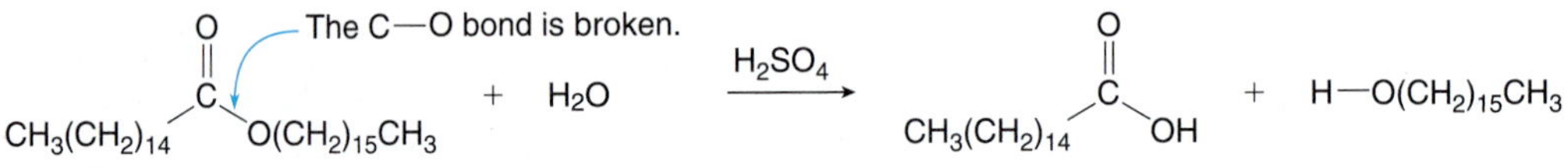

cetyl palmitate fatty acid alcohol

PROBLEM 19.9

What hydrolysis products are formed when cetyl myristate (Sample Problem 19.3) is treated with aqueous acid?

19.4 TRIACYLGLYCEROLS—FATS AND OILS

Animal fats and vegetable oils, the most abundant lipids, are composed of **triacylglycerols.**

- ***Triacylglycerols,*** **or triglycerides, are triesters formed from glycerol and three molecules of fatty acids.**

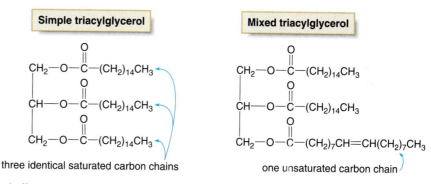

glycerol   fatty acids   triacylglycerol   block diagram

R groups have 11–19 C's.

[Three ester groups are labeled in red.]

19.4A GENERAL FEATURES

Simple triacylglycerols are composed of three identical fatty acid side chains, whereas **mixed triacylglycerols** have two or three different fatty acids. The fatty acids may be saturated or unsaturated. **Monounsaturated triacylglycerols** have a single carbon–carbon double bond, while **polyunsaturated triacylglycerols** have more than one carbon–carbon double bond.

Simple triacylglycerol	Mixed triacylglycerol
$CH_2-O-\overset{\overset{O}{\|\|}}{C}-(CH_2)_{14}CH_3$	$CH_2-O-\overset{\overset{O}{\|\|}}{C}-(CH_2)_{14}CH_3$
$CH-O-\overset{\overset{O}{\|\|}}{C}-(CH_2)_{14}CH_3$	$CH-O-\overset{\overset{O}{\|\|}}{C}-(CH_2)_{14}CH_3$
$CH_2-O-\overset{\overset{O}{\|\|}}{C}-(CH_2)_{14}CH_3$	$CH_2-O-\overset{\overset{O}{\|\|}}{C}-(CH_2)_7CH=CH(CH_2)_7CH_3$
three identical saturated carbon chains	one unsaturated carbon chain

Fats and **oils** are triacylglycerols with different physical properties.

- Fats have higher melting points—they are *solids* at room temperature.
- Oils have lower melting points—they are *liquids* at room temperature.

The identity of the three fatty acids in the triacylglycerol determines whether it is a fat or an oil. *Increasing* **the number of double bonds in the fatty acid side chains** *decreases* **the melting point of the triacylglycerol.**

- Fats are derived from fatty acids having few double bonds.
- Oils are derived from fatty acids having a larger number of double bonds.

Table 19.3 lists the fatty acid composition of some common fats and oils. Solid fats have a relatively high percentage of saturated fatty acids and are generally animal in origin. Thus, lard (hog fat), butter, and whale blubber contain a high percentage of saturated fats. With no double bonds, the three side chains of the saturated lipid lie parallel with each other, making it possible for the triacylglycerol molecules to pack relatively efficiently in a crystalline lattice, thus leading to a high melting point.

no double bonds in the long carbon chains

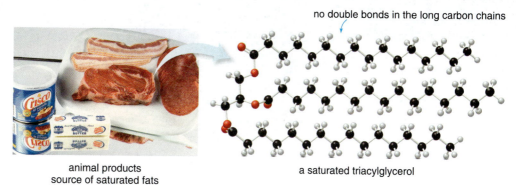

animal products
source of saturated fats

a saturated triacylglycerol

TABLE 19.3 Fatty Acid Composition of Some Fats and Oils

Substance	% Saturated	% Monounsaturated	% Polyunsaturated	% Trans[a]
Cooking Fat				
Lard	39	44	11	1
Butter	60	26	5	5
Shortening	22	29	29	18
Oils				
Canola	7	58	29	0
Safflower	9	12	24	0
Sunflower	10	20	66	0
Corn	13	24	60	0
Olive	13	72	8	0
Soybean	16	44	37	0
Peanut	17	49	32	0
Palm	50	37	10	0
Coconut	87	6	2	0
Margarine/Spreads				
70% Soybean oil, stick	18	2	29	23
60% Vegetable oil spread	18	22	54	5

[a]Trans double bonds are formed when triacylglycerols are partially hydrogenated, as described in Section 13.7.
[Source: Harvard School of Public Health website at www.hsph.harvard.edu/nutritionsource/fats.html, but similar data is available from many other sources.]

Liquid oils have a higher percentage of unsaturated fatty acids and are generally vegetable in origin. Thus, oils derived from corn, soybeans, and olives contain more unsaturated lipids. In the unsaturated lipid, a cis double bond places a kink in the side chain, making it more difficult to pack efficiently in the solid state, thus leading to a lower melting point.

HEALTH NOTE

Oil in the coconuts from plantations like this one is high in saturated fats, which are believed to contribute to a greater risk of heart disease.

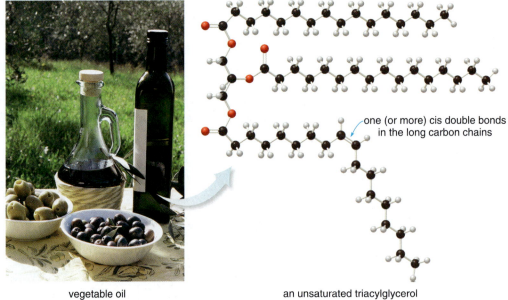

one (or more) cis double bonds in the long carbon chains

vegetable oil
source of unsaturated oils

an unsaturated triacylglycerol

Fish oils are high in unsaturated triacylglycerols derived from omega-3 fatty acids.

Unlike other vegetable oils, oils from palm and coconut trees are very high in saturated fats. Considerable evidence currently suggests that diets high in saturated fats lead to a greater risk of heart disease (Sections 19.4B and 19.8). For this reason, the demand for coconut and palm oil has decreased considerably in recent years, and many coconut plantations previously farmed in the South Pacific are no longer in commercial operation.

Oils derived from fish such as salmon, herring, mackerel, and sardines are very rich in polyunsaturated triacylglycerols. These triacylglycerols pack so poorly that they have very low melting points, and they remain liquids even in very cold water. Fish oils derived from omega-3 fatty acids are thought to be especially beneficial for individuals at risk for developing coronary artery disease.

SAMPLE PROBLEM 19.4

Draw the structure of a triacylglycerol formed from glycerol, one molecule of stearic acid, and two molecules of oleic acid. Bond the stearic acid to the 2° OH group (OH on the middle carbon atom) of glycerol.

ANALYSIS

To draw the triacylglycerol, arrange each OH group of glycerol next to the carboxyl group of a fatty acid. Then join each O atom of glycerol to a carbonyl carbon of a fatty acid, to form three new C—O bonds.

SOLUTION

Form three new ester bonds (RCOOR') from OH groups of glycerol and the three fatty acids (RCOOH).

$$CH_2\text{—}OH$$
$$CH\text{—}OH$$
$$CH_2\text{—}OH$$

glycerol
2° OH

$$HO\text{—}\overset{\overset{O}{\|}}{C}\text{—}(CH_2)_7CH{=}CH(CH_2)_7CH_3$$
$$HO\text{—}\overset{\overset{O}{\|}}{C}\text{—}(CH_2)_{16}CH_3$$
$$HO\text{—}\overset{\overset{O}{\|}}{C}\text{—}(CH_2)_7CH{=}CH(CH_2)_7CH_3$$

oleic acid (R group in blue)
stearic acid (R group in red)

$$CH_2\text{—}O\text{—}\overset{\overset{O}{\|}}{C}\text{—}(CH_2)_7CH{=}CH(CH_2)_7CH_3$$
$$CH\text{—}O\text{—}\overset{\overset{O}{\|}}{C}\text{—}(CH_2)_{16}CH_3$$
$$CH_2\text{—}O\text{—}\overset{\overset{O}{\|}}{C}\text{—}(CH_2)_7CH{=}CH(CH_2)_7CH_3$$

triacylglycerol

$$+ \quad 3\,H_2O$$

PROBLEM 19.10

Draw the structure of a triacylglycerol that contains (a) three molecules of stearic acid; (b) three molecules of oleic acid.

PROBLEM 19.11

Draw the structure of two different triacylglycerols that contain two molecules of stearic acid and one molecule of palmitic acid.

19.4B FOCUS ON HEALTH & MEDICINE
FATS AND OILS IN THE DIET

Fats and oils in our diet come from a variety of sources—meat, dairy products, seeds and nuts, salad dressing, fried foods, and any baked good or packaged food made with oil. Some fat is required in the diet. Fats are the building blocks of cell membranes, and stored body fat insulates an organism and serves as an energy source that can be used at a later time.

Currently, the United States Food and Drug Administration recommends that no more than 20–35% of an individual's calorie intake come from lipids. Moreover, a high intake of *saturated* triacylglycerols is linked to an increased incidence of heart disease. Saturated fats stimulate cholesterol synthesis in the liver and transport to the tissues, resulting in an increase in cholesterol concentration in the blood. An elevated cholesterol level in the blood can lead to cholesterol deposits or plaques on arteries, causing a narrowing of blood vessels, heart attack, and stroke (Section 19.8).

In contrast, *unsaturated* triacylglycerols lower the risk of heart disease by decreasing the amount of cholesterol in the blood. Unsaturated triacylglycerols from omega-3 fatty acids appear to reduce the risk of heart attack, especially in individuals who already have heart disease.

Dietary recommendations on fat intake must also take into account trans triacylglycerols, so-called *trans fats*. As we learned in Section 13.7, trans fats are formed when liquid oils are partially hydrogenated to form semi-solid triacylglycerols. The three-dimensional structure of a trans triacylglycerol shows its similarity to saturated triacylglycerols. Like saturated fats, trans fats also *increase* the amount of cholesterol in the bloodstream, thus increasing an individual's risk of developing coronary artery disease.

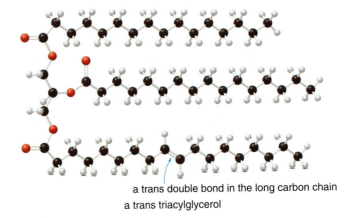

a trans double bond in the long carbon chain
a trans triacylglycerol

Thus, while lipids are a necessary part of anyone's diet, the amount of saturated fat and trans fat should be limited. Calories from lipids should instead be obtained from unsaturated oils. Figure 19.1 illustrates some foods high in saturated, unsaturated, and trans triacylglycerols.

PROBLEM 19.12	Draw the structure of a triacylglycerol that fits each description:

  a. a saturated triacylglycerol formed from three 12-carbon fatty acids

  b. an unsaturated triacylglycerol that contains three cis double bonds

  c. a trans triacylglycerol that contains a trans double bond in each hydrocarbon chain

19.5 HYDROLYSIS OF TRIACYLGLYCEROLS

Like other esters, **triacylglycerols are hydrolyzed with water in the presence of acid, base, or enzymes (in biological systems) to form glycerol and three molecules of fatty acids.** Thus, hydrolysis of tristearin with aqueous sulfuric acid forms glycerol and three molecules of stearic acid.

The three bonds drawn in red are broken in hydrolysis.

$$CH_2-O-C(=O)-(CH_2)_{16}CH_3$$
$$CH-O-C(=O)-(CH_2)_{16}CH_3 \; + \; 3\,H_2O \xrightarrow{H_2SO_4} $$
$$CH_2-O-C(=O)-(CH_2)_{16}CH_3$$

tristearin

$$CH_2-OH$$
$$CH-OH \; + \; 3\,HO-C(=O)-(CH_2)_{16}CH_3$$
$$CH_2-OH$$

glycerol   stearic acid

Hydrolysis cleaves the three single bonds between the carbonyl carbons and the oxygen atoms of the esters (Section 17.9). Since tristearin contains three identical R groups on the carbonyl carbons, three molecules of a single fatty acid, stearic acid, are formed. Triacylglycerols that contain different R groups bonded to the carbonyl carbons form mixtures of fatty acids, as shown in Sample Problem 19.5.

▼ FIGURE 19.1 Saturated, Unsaturated, and Trans Triacylglycerols in the Diet

Food source

Triacylglycerol

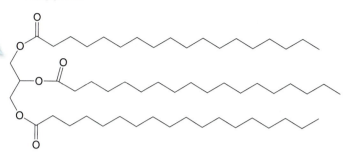

Foods rich in saturated triacylglycerols:
fatty red meat, cheese, butter, fried foods,
ice cream

- no double bonds in the carbon chains
- solid at room temperature
- increases blood cholesterol level

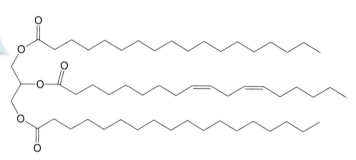

Foods rich in unsaturated triacylglycerols:
plant oils, nuts, soybeans, fish (salmon,
herring, mackerel)

- one (or more) double bonds in the carbon chains
- liquid at room temperature
- decreases blood cholesterol level

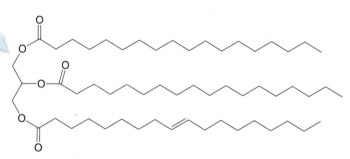

Foods rich in trans triacylglycerols:
margarine, processed foods, fried foods,
some baked goods

- one (or more) trans double bonds in the carbon chains
- semi-solid at room temperature
- increases blood cholesterol level

SAMPLE PROBLEM 19.5   Draw the products formed when the given triacylglycerol is hydrolyzed with water in the presence of sulfuric acid.

$$CH_2-O-\overset{\displaystyle O}{\overset{\|}{C}}-(CH_2)_{16}CH_3$$

$$CH-O-\overset{\displaystyle O}{\overset{\|}{C}}-(CH_2)_{14}CH_3$$

$$CH_2-O-\overset{\displaystyle O}{\overset{\|}{C}}-(CH_2)_7CH=CH(CH_2)_5CH_3$$

ANALYSIS   To draw the products of ester hydrolysis, cleave the three C—O single bonds at the carbonyl carbons to form glycerol and three fatty acids (RCOOH).

SOLUTION   Hydrolysis forms glycerol and stearic, palmitic, and palmitoleic acids.

Bonds broken
during hydrolysis

$$CH_2-O-\overset{\overset{\displaystyle O}{\|}}{C}-(CH_2)_{16}CH_3$$

$$CH-O-\overset{\overset{\displaystyle O}{\|}}{C}-(CH_2)_{14}CH_3$$

$$CH_2-O-\overset{\overset{\displaystyle O}{\|}}{C}-(CH_2)_7CH=CH(CH_2)_5CH_3$$

$$\xrightarrow[H_2SO_4]{3\ H-OH}$$

$$CH_2-OH$$
$$CH-OH$$
$$CH_2-OH$$

glycerol

+

$$HO-\overset{\overset{\displaystyle O}{\|}}{C}-(CH_2)_{16}CH_3$$
stearic acid

$$HO-\overset{\overset{\displaystyle O}{\|}}{C}-(CH_2)_{14}CH_3$$
palmitic acid

$$HO-\overset{\overset{\displaystyle O}{\|}}{C}-(CH_2)_7CH=CH(CH_2)_5CH_3$$
palmitoleic acid

PROBLEM 19.13   Draw the products formed from hydrolysis of each triacylglycerol.

a. $$CH_2-O-\overset{\overset{\displaystyle O}{\|}}{C}-(CH_2)_{12}CH_3$$

$$CH-O-\overset{\overset{\displaystyle O}{\|}}{C}-(CH_2)_{12}CH_3$$

$$CH_2-O-\overset{\overset{\displaystyle O}{\|}}{C}-(CH_2)_{12}CH_3$$

b. $$CH_2-O-\overset{\overset{\displaystyle O}{\|}}{C}-(CH_2)_{12}CH_3$$

$$CH-O-\overset{\overset{\displaystyle O}{\|}}{C}-(CH_2)_7CH=CH(CH_2)_5CH_3$$

$$CH_2-O-\overset{\overset{\displaystyle O}{\|}}{C}-(CH_2)_7CH=CH(CH_2)_7CH_3$$

19.5A   FOCUS ON THE HUMAN BODY
METABOLISM OF TRIACYLGLYCEROLS

Humans store energy in the form of triacylglycerols, kept in a layer of fat cells, called **adipose cells,** below the surface of the skin, in bone marrow, in the breast area (of women), around the kidneys, and in the pelvis (Figure 19.2). Adipose tissue contains large groups of adipose cells, some of which may be so laden with lipid molecules that the remainder of the cell—the cytoplasm and the nucleus—occupies only a small volume on one side. In adulthood, the number of adipose cells is constant. When weight is lost or gained the amount of stored lipid in each cell changes, but the number of adipose cells does not change.

▼ **FIGURE 19.2**   The Storage and Metabolism of Triacylglycerols

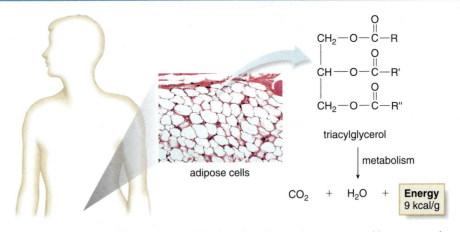

Triacylglycerols are stored in adipose cells below the skin and concentrated in some regions of the body. The average fat content of men and women is ~20% and ~25%, respectively. This stored fat provides two to three months of the body's energy needs.

Adipose tissue serves to insulate the organism, as well as provide energy for its metabolic needs for long periods of time. The first step in the metabolism of a triacylglycerol is hydrolysis of the ester bonds to form glycerol and three fatty acids. **This reaction is simply ester hydrolysis.** In cells, this reaction is carried out with enzymes called **lipases.**

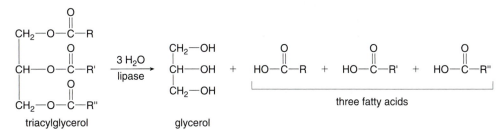

triacylglycerol   glycerol   three fatty acids

Complete metabolism of a triacylglycerol yields CO_2 and H_2O, and a great deal of energy (Section 24.7). This overall reaction is reminiscent of the combustion of alkanes in fossil fuels, a process that also yields CO_2 and H_2O and provides energy to heat homes and power automobiles (Section 12.8). Fundamentally, both processes convert C—C and C—H bonds to C—O and O—H bonds, a highly exothermic reaction. Carbohydrates provide an energy boost, but only for the short term, such as during strenuous exercise. Our long-term energy needs are provided by triacylglycerols, because they store 9 kcal/g, whereas carbohydrates and proteins store only 4 kcal/g.

Because triacylglycerols release heat on combustion, they can in principle be used as fuels for vehicles. In fact, coconut oil was used as a fuel during both World Wars I and II, when gasoline and diesel supplies ran short. Since coconut oil is more viscous than petroleum products and freezes at 24 °C, engines must be modified to use it and it can't be used in cold climates. Nonetheless, a limited number of trucks and boats can now use vegetable oils, sometimes blended with diesel—*biodiesel*—as a fuel source.

ENVIRONMENTAL NOTE

When the price of crude oil is high, the use of **biofuels** such as biodiesel becomes economically attractive. Although biofuels are prepared from renewable resources such as vegetable oils and animal fats, they still burn to form CO_2, a gas that contributes to global warming.

PROBLEM 19.14

Review Section 5.2 on balancing chemical equations. Then, write a balanced equation for the complete combustion (or metabolism) of tristearin to CO_2 and H_2O.

$$\text{tristearin} + O_2 \xrightarrow{\text{enzymes}} CO_2 + H_2O$$

tristearin

19.5B SOAP SYNTHESIS

As we learned in Section 17.6, soaps are metal salts of carboxylic acids that contain many carbon atoms in a long hydrocarbon chain; that is, **soaps are metal salts of fatty acids.** For example, sodium stearate is the sodium salt of stearic acid, an 18-carbon saturated fatty acid.

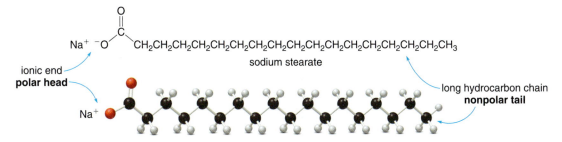

sodium stearate

ionic end
polar head

long hydrocarbon chain
nonpolar tail

Soap is prepared by the basic hydrolysis (saponification) of a triacylglycerol. Heating an animal fat or vegetable oil with aqueous base hydrolyzes the three esters to form glycerol and sodium salts of three fatty acids.

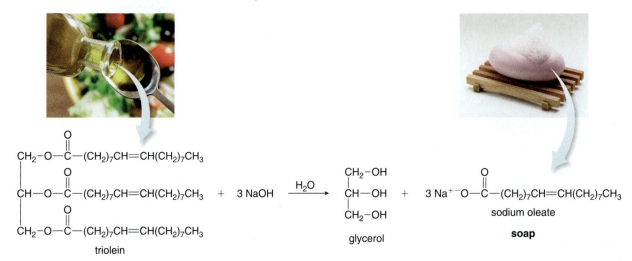

triolein

glycerol

sodium oleate

soap

Saponification comes from the Latin *sapo* meaning *soap*.

These carboxylate salts are **soaps,** which clean away dirt as shown in Figure 17.3. The nonpolar tail dissolves grease and oil and the polar head makes it soluble in water. Most triacylglycerols have two or three different R groups in their hydrocarbon chains, so soaps are usually mixtures of two or three different carboxylate salts.

Soaps are typically made from lard (from hogs), tallow (from cattle or sheep), coconut oil, or palm oil. All soaps work in the same way, but have somewhat different properties depending on the lipid source. The length of the carbon chain in the fatty acids and the number of sites of unsaturation affect the properties of the soap to a limited extent.

PROBLEM 19.15   What is the composition of the soap formed by basic hydrolysis of each triacylglycerol?

a. 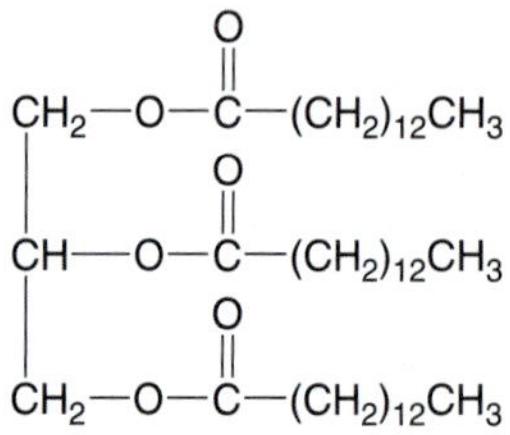
$$CH_2-O-\overset{\overset{\displaystyle O}{\|}}{C}-(CH_2)_{12}CH_3$$
$$CH-O-\overset{\overset{\displaystyle O}{\|}}{C}-(CH_2)_{12}CH_3$$
$$CH_2-O-\overset{\overset{\displaystyle O}{\|}}{C}-(CH_2)_{12}CH_3$$

b. 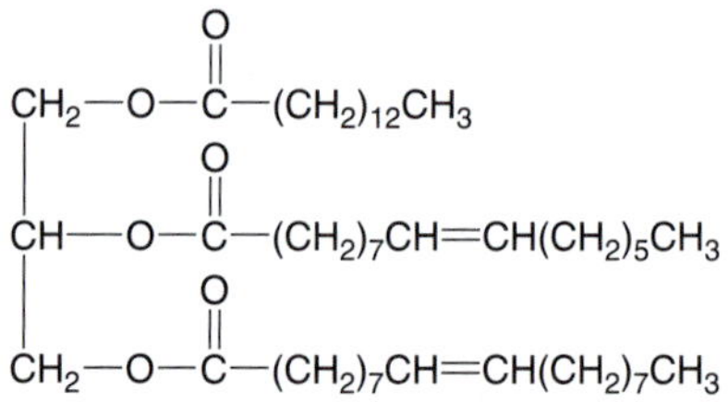
$$CH_2-O-\overset{\overset{\displaystyle O}{\|}}{C}-(CH_2)_{12}CH_3$$
$$CH-O-\overset{\overset{\displaystyle O}{\|}}{C}-(CH_2)_7CH=CH(CH_2)_5CH_3$$
$$CH_2-O-\overset{\overset{\displaystyle O}{\|}}{C}-(CH_2)_7CH=CH(CH_2)_7CH_3$$

19.6 PHOSPHOLIPIDS

Phospholipids are lipids that contain a phosphorus atom. Two common types of phospholipids are **phosphoacylglycerols** and **sphingomyelins.** Both classes are found almost exclusively in the cell membranes of plants and animals as discussed in Section 19.7. Block diagrams that compare the general structural features of triacylglycerols, phosphoacylglycerols, and sphingomyelins are drawn.

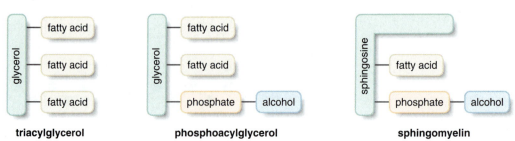

triacylglycerol   **phosphoacylglycerol**   **sphingomyelin**

Phospholipids can be considered organic derivatives of phosphoric acid (H_3PO_4), formed by replacing two of the H atoms by R groups. This type of functional group is called a **phosphodiester.** In cells, the remaining –OH group on phosphorus loses its proton, giving the phosphodiester a net negative charge.

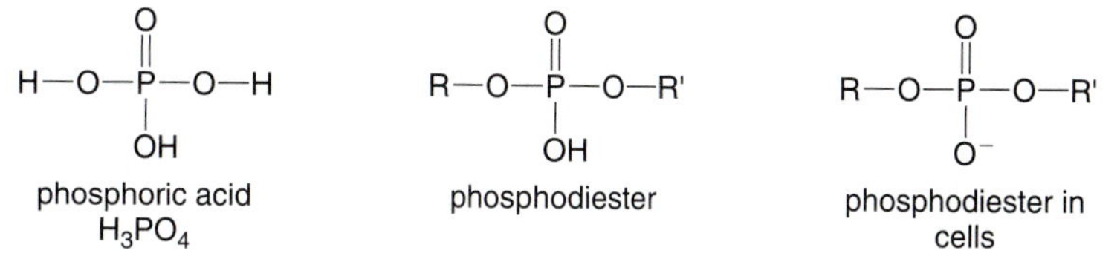

phosphoric acid
H_3PO_4

phosphodiester

phosphodiester in cells

19.6A PHOSPHOACYLGLYCEROLS

Phosphoacylglycerols (or **phosphoglycerides**) are the second most abundant type of lipid. They form the principal lipid component of most cell membranes. Their structure resembles the triacylglycerols of the preceding section with one important difference. Only *two* of the hydroxyl groups of glycerol are esterified with fatty acids. The third OH group is part of a phosphodiester, which is also bonded to an alkyl group (R") derived from a low molecular weight alcohol.

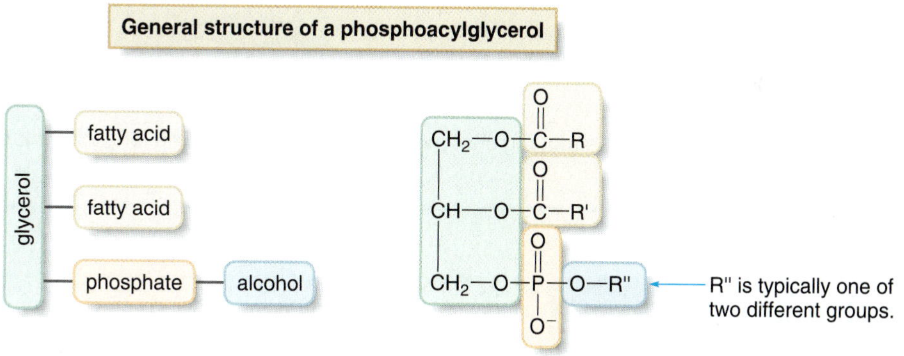

General structure of a phosphoacylglycerol

R" is typically one of two different groups.

There are two prominent types of phosphoacylglycerols. They differ in the identity of the R" group in the phosphodiester.

- When R" = $CH_2CH_2NH_3^+$, the compound is called a *cephalin* or phosphatidylethanolamine.
- When R" = $CH_2CH_2N(CH_3)_3^+$, the compound is called a *lecithin* or phosphatidylcholine.

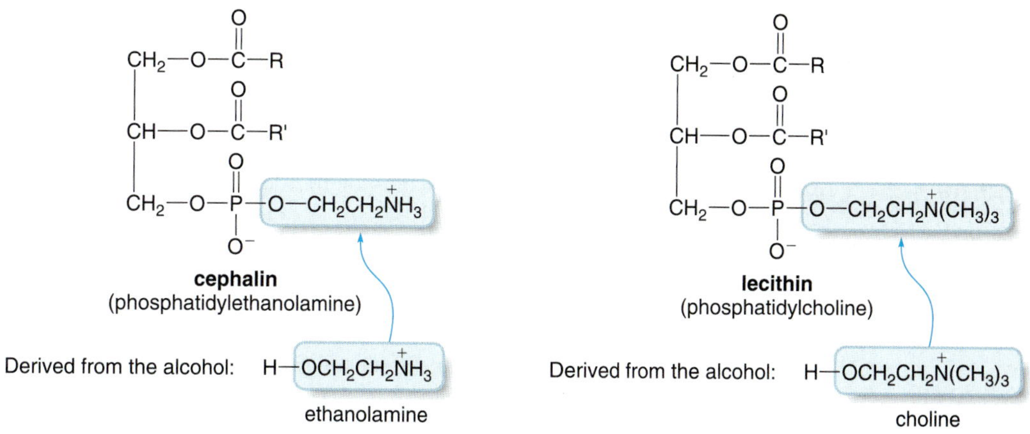

cephalin
(phosphatidylethanolamine)

Derived from the alcohol: $H-OCH_2CH_2\overset{+}{N}H_3$

ethanolamine

lecithin
(phosphatidylcholine)

Derived from the alcohol: $H-OCH_2CH_2\overset{+}{N}(CH_3)_3$

choline

The phosphorus side chain of a phosphoacylglycerol makes it different from a triacylglycerol. **The two fatty acid side chains form two nonpolar "tails" that lie parallel to each other, while the phosphodiester end of the molecule is a charged or polar "head."** A three-dimensional structure of a phosphoacylglycerol is shown in Figure 19.3.

▼ **FIGURE 19.3**   **Three-Dimensional Structure of a Phosphoacylglycerol**

▼ **FIGURE 19.3**   **Three-Dimensional Structure of a Phosphoacylglycerol**

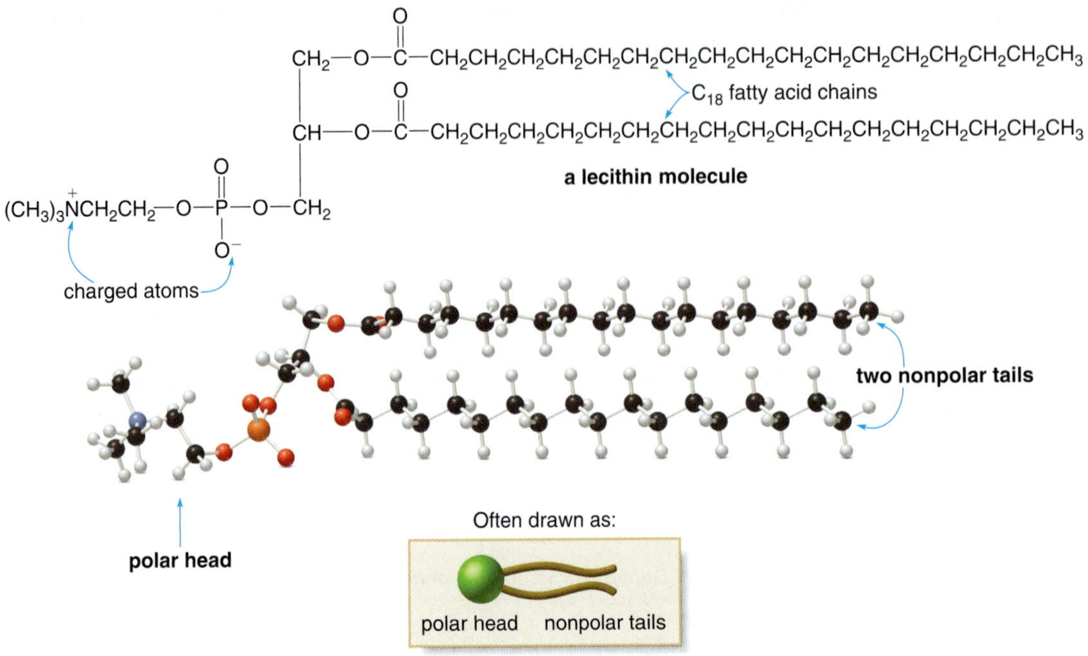

A phosphoacylglycerol has two distinct regions: two nonpolar tails due to the long-chain fatty acids, and a very polar head from the charged phosphodiester.

SAMPLE PROBLEM 19.6

Draw the structure of a cephalin formed from two molecules of stearic acid.

ANALYSIS

Substitute the 18-carbon saturated fatty acid stearic acid for the R and R' groups in the general structure of a cephalin molecule. In a cephalin, $-CH_2CH_2NH_3^+$ forms part of the phosphodiester.

SOLUTION

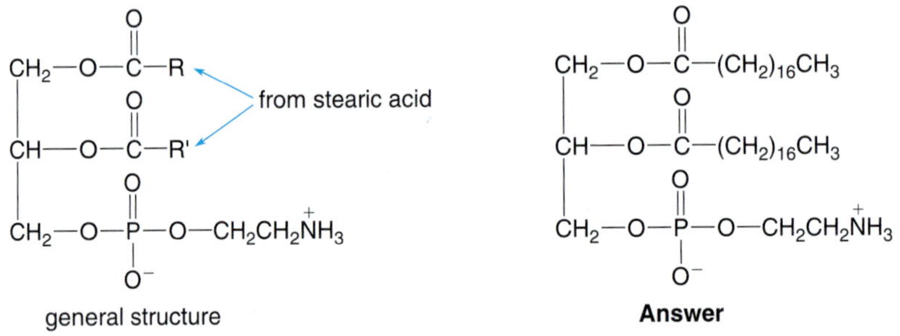

general structure   **Answer**

PROBLEM 19.16

Draw the structure of two different cephalins containing oleic acid and palmitic acid as fatty acid side chains.

19.6B   SPHINGOMYELINS

Sphingomyelins, the second major class of phospholipids, differ in two important ways from the triacylglycerols and the phosphoacylglycerols.

- Sphingomyelins do *not* contain a glycerol backbone. Instead, sphingomyelins are derived from sphingosine.
- Sphingomyelins do *not* contain an ester. Instead, their single fatty acid is bonded to the carbon backbone by an amide bond.

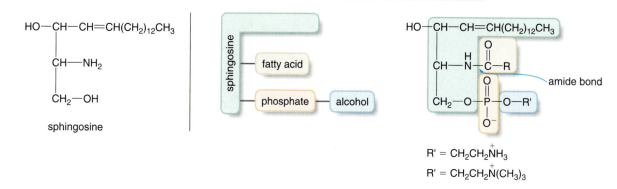

General structure of a sphingomyelin

HO—CH—CH=CH(CH$_2$)$_{12}$CH$_3$

CH—NH$_2$

CH$_2$—OH

sphingosine

sphingosine — fatty acid

phosphate — alcohol

HO—CH—CH=CH(CH$_2$)$_{12}$CH$_3$

CH—N—C—R

CH$_2$—O—P—O—R'

amide bond

R' = CH$_2$CH$_2$$\overset{+}{N}H_3$

R' = CH$_2$CH$_2$$\overset{+}{N}$(CH$_3$)$_3$

Like the phosphoacylglycerols in Section 19.6A, sphingomyelins contain a phosphodiester derived from ethanolamine or choline. As a result, sphingomyelins also contain a polar (ionic) head and two nonpolar tails. An example of a sphingomyelin is drawn.

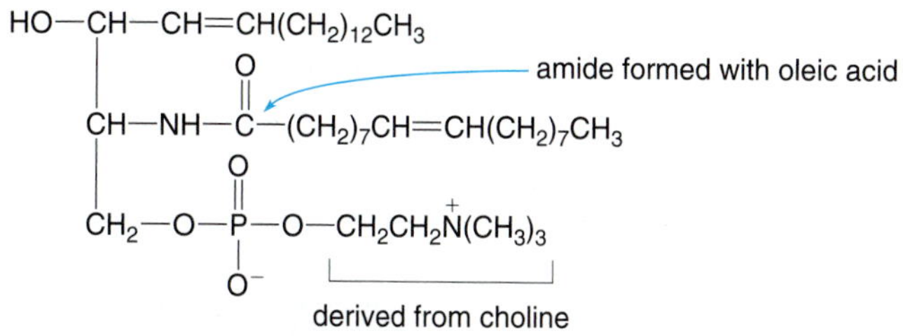

HO—CH—CH=CH(CH$_2$)$_{12}$CH$_3$

CH—NH—C—(CH$_2$)$_7$CH=CH(CH$_2$)$_7$CH$_3$ — amide formed with oleic acid

CH$_2$—O—P—O—CH$_2$CH$_2$$\overset{+}{N}$(CH$_3$)$_3$

O$^-$

derived from choline

a sphingomyelin

The coating that surrounds and insulates nerve cells, the **myelin sheath,** is particularly rich in sphingomyelins, and is vital for proper nerve function (Figure 19.4). Deterioration of the myelin sheath, as seen in multiple sclerosis, leads to disabling neurological disorders.

Figure 19.5 compares the structural features of the most common hydrolyzable lipids: a triacylglycerol, a phosphoacylglycerol, and a sphingomyelin.

PROBLEM 19.17

Identify the components of each lipid and classify it as a triacylglycerol, a phosphoacylglycerol, or a sphingomyelin. Classify any phosphoacylglycerol as a cephalin or lecithin.

a. CH$_2$—O—C—(CH$_2$)$_{12}$CH$_3$

CH—O—C—(CH$_2$)$_{12}$CH$_3$

CH$_2$—O—P—O—CH$_2$CH$_2$$\overset{+}{N}H_3$

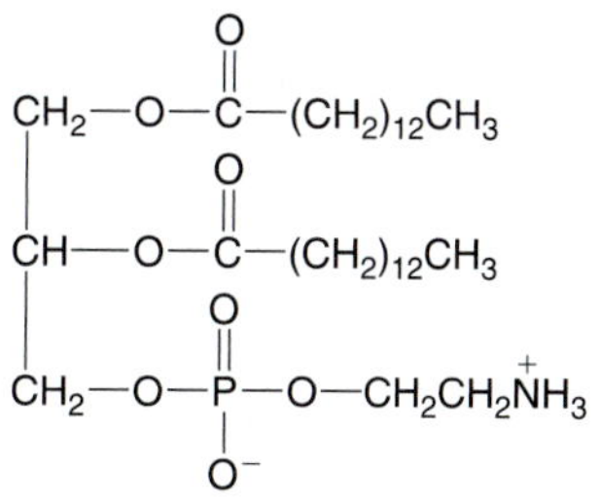

O$^-$

b. CH$_2$—O—C—(CH$_2$)$_{12}$CH$_3$

CH—O—C—(CH$_2$)$_7$CH=CH(CH$_2$)$_5$CH$_3$

CH$_2$—O—C—(CH$_2$)$_{12}$CH$_3$

c. HO—CH—CH=CH(CH$_2$)$_{12}$CH$_3$

CH—NH—C—(CH$_2$)$_{12}$CH$_3$

CH$_2$—O—P—O—CH$_2$CH$_2$$\overset{+}{N}$(CH$_3$)$_3$

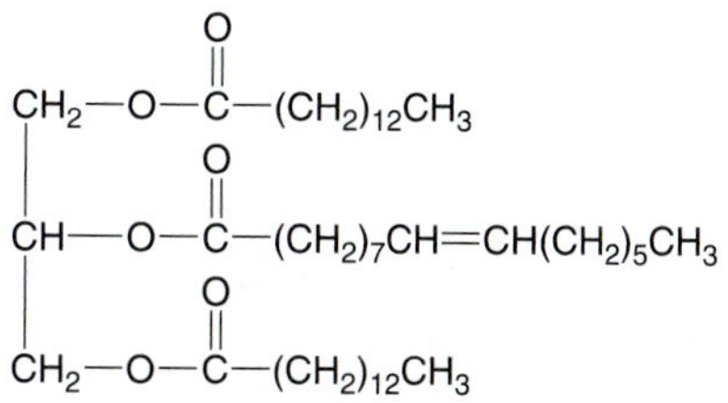

O$^-$

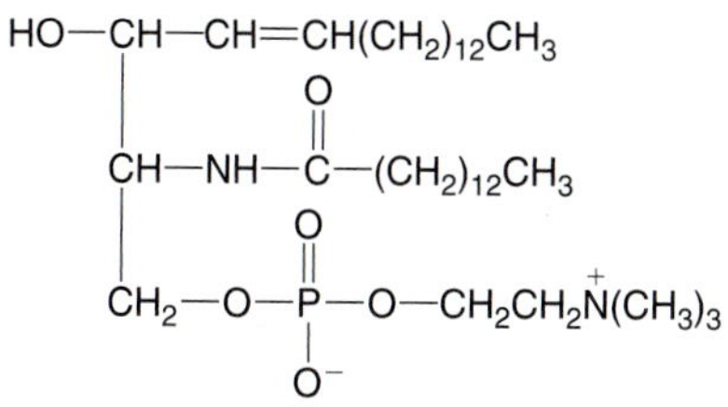

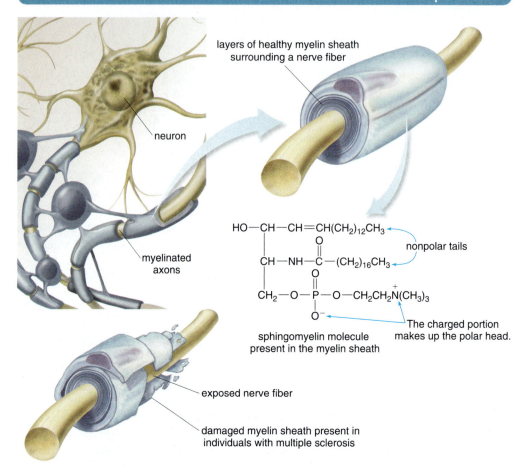

▼ FIGURE 19.4   Focus on Health & Medicine: Nerve Cells and Multiple Sclerosis

layers of healthy myelin sheath surrounding a nerve fiber

neuron

myelinated axons

$$HO-CH-CH=CH(CH_2)_{12}CH_3$$

nonpolar tails

$$CH-NH-\overset{\overset{\displaystyle O}{\|}}{C}-(CH_2)_{16}CH_3$$

$$CH_2-O-\overset{\overset{\displaystyle O}{}}{\underset{\underset{\displaystyle O^-}{}}{P}}-O-CH_2CH_2\overset{+}{N}(CH_3)_3$$

The charged portion makes up the polar head.

sphingomyelin molecule present in the myelin sheath

exposed nerve fiber

damaged myelin sheath present in individuals with multiple sclerosis

Multiple sclerosis (MS) is a degenerative disease characterized by scarring of the myelin sheath, the insulating layer that surrounds a nerve fiber. Without the protective myelin sheath, normal nerve transmission is disrupted and a variety of effects—numbness, blindness, speech disorders, and tremors—can result.

▼ FIGURE 19.5   A Comparison of a Triacylglycerol, a Phosphoacylglycerol, and a Sphingomyelin

$$CH_2-O-\overset{\overset{\displaystyle O}{\|}}{C}-R$$
$$CH-O-\overset{\overset{\displaystyle O}{\|}}{C}-R'$$
$$CH_2-O-\overset{\overset{\displaystyle O}{\|}}{C}-R''$$

$$CH_2-O-\overset{\overset{\displaystyle O}{\|}}{C}-R$$
$$CH-O-\overset{\overset{\displaystyle O}{\|}}{C}-R'$$
$$CH_2-O-\overset{\overset{\displaystyle O}{}}{\underset{\underset{\displaystyle O^-}{}}{P}}-O-CH_2CH_2\overset{+}{N}R''_3$$

R" = H, CH₃

$$HO-CH-CH=CH(CH_2)_{12}CH_3$$
$$CH-NH-\overset{\overset{\displaystyle O}{\|}}{C}-R$$
$$CH_2-O-\overset{\overset{\displaystyle O}{}}{\underset{\underset{\displaystyle O^-}{}}{P}}-O-CH_2CH_2\overset{+}{N}R''_3$$

R" = H, CH₃

- A **triacylglycerol** has three nonpolar side chains.
- Glycerol is the backbone.
- There are three esters formed from three fatty acids.

- A **phosphoacylglycerol** has two nonpolar side chain tails and one ionic head.
- Glycerol is the backbone.
- There are two esters formed with the two fatty acids.
- A phosphodiester is located on a terminal carbon of glycerol.

- A **sphingomyelin** has two nonpolar side chain tails and one ionic head.
- Sphingosine is the backbone.
- A sphingomyelin contains an amide, not an ester.
- A phosphodiester is located on a terminal carbon of sphingosine.

19.7 CELL MEMBRANES

The cell membrane is a beautifully complex example of how chemistry comes into play in a biological system.

19.7A STRUCTURE OF THE CELL MEMBRANE

The basic unit of living organisms is the **cell.** The cytoplasm is the aqueous medium inside the cell, separated from water outside the cell by the **cell membrane.** The cell membrane serves two apparently contradictory functions. It acts as a barrier to the passage of ions and molecules into and out of the cell, but it is also selectively permeable, allowing nutrients in and waste out.

Phospholipids, especially phosphoacylglycerols, are the major component of the cell membrane. **Phospholipids contain a hydrophilic polar head and two nonpolar tails composed of C—C and C—H bonds.** When phospholipids are mixed with water, they assemble in an arrangement called a **lipid bilayer,** with the ionic heads oriented on the outside and the nonpolar tails on the inside. The polar heads electrostatically interact with the polar solvent H_2O, while the nonpolar tails are held in close proximity by numerous London dispersion forces.

To review London dispersion forces, re-read Section 7.7A.

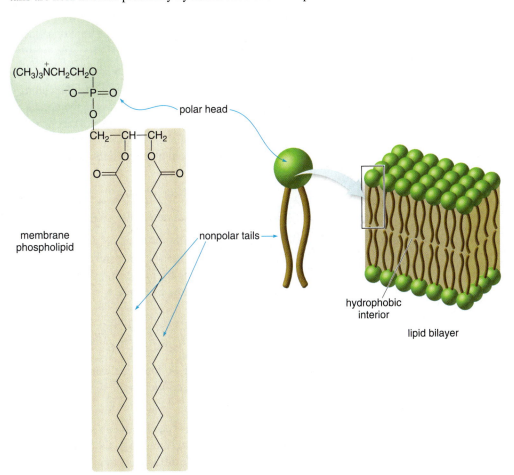

Cell membranes are composed of these lipid bilayers (Figure 19.6). The charged heads of the phospholipids are oriented towards the aqueous interior and exterior of the cell. The nonpolar tails form the hydrophobic interior of the membrane, thus serving as an insoluble barrier that protects the cell from the outside. The identity of the fatty acids in the phospholipid determines the rigidity of this bilayer. When the fatty acids are saturated, they pack well in the interior of the lipid bilayer, and the membrane is less fluid. When there are many unsaturated fatty acids, the nonpolar tails cannot pack as well and the bilayer is more fluid.

▼ FIGURE 19.6 Composition of the Cell Membrane

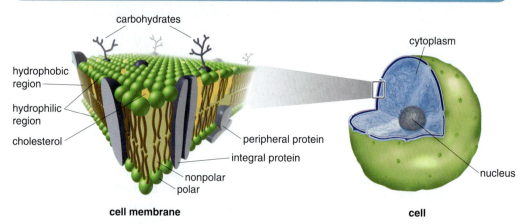

FIGURE 19.6 Composition of the Cell Membrane

cell membrane   cell

Cell membranes are composed of a lipid bilayer having the hydrophilic polar heads of phospholipids arranged on the exterior of the bilayer, where they can interact with the polar aqueous environment inside and outside the cell. The hydrophobic tails of the phospholipid are arranged in the interior of the bilayer, forming a "greasy" layer that is only selectively permeable to the passage of species from one side to the other.

While the phospholipid bilayer forms the main fabric of the cell membrane, proteins and cholesterol (Section 19.8) are embedded in the membrane as well. **Peripheral proteins** are embedded within the membrane and extend outward on one side only. **Integral proteins** extend through the entire bilayer. Some lipids and proteins on the membrane surface are bonded to carbohydrates, forming **glycolipids** and **glycoproteins.** The polar carbohydrate chains extend into the surrounding aqueous environment.

PROBLEM 19.18   Why are phospholipids rather than triacylglycerols present in cell membranes?

PROBLEM 19.19   A cell membrane from one source (**A**) contains phospholipids formed from linoleic and oleic acid. A cell membrane from a second source (**B**) contains phospholipids formed from stearic and palmitic acids. How will these two cell membranes differ?

19.7B TRANSPORT ACROSS A CELL MEMBRANE

How does a molecule or ion in the water on one side of a cell membrane pass through the nonpolar interior of the cell membrane to the other side? A variety of transport mechanisms occur (Figure 19.7).

Small molecules like O_2 and CO_2 can simply diffuse through the cell membrane, traveling from the side of higher concentration to the side of lower concentration. With larger polar molecules and some ions, simple diffusion is too slow or not possible, so a process of **facilitated transport** occurs. Ions such as Cl^- or HCO_3^- and glucose molecules travel through the channels created by integral proteins.

Some ions, notably Na^+, K^+, and Ca^{2+}, must move across the cell membrane against the concentration gradient—that is, from a region of lower concentration to a region of higher concentration. To move an ion across the membrane in this fashion requires energy input, and the process is called **active transport.** Active transport occurs whenever a nerve impulse causes a muscle to contract. In this process, energy is supplied to move K^+ ions from outside to inside a cell, against a concentration gradient.

▼ **FIGURE 19.7** **How Substances Cross a Cell Membrane**

outside

integral membrane protein

energy input needed

inside

cell membrane

simple diffusion

facilitated transport

active transport

PROBLEM 19.20

Why don't ions readily diffuse through the interior of the cell membrane?

19.8 FOCUS ON HEALTH & MEDICINE
CHOLESTEROL, THE MOST PROMINENT STEROID

The steroids are a group of lipids whose carbon skeletons contain three six-membered rings and one five-membered ring. This tetracyclic carbon skeleton is drawn below.

steroid skeleton

numbering the steroid skeleton

Many steroids also contain two methyl groups that are bonded to the rings. The steroid rings are lettered **A, B, C,** and **D,** and the 17 ring carbons are numbered as shown. The two methyl groups are numbered C18 and C19. Steroids differ in the identity and location of the substituents attached to the skeleton.

Cholesterol, the most prominent member of the steroid family, is synthesized in the liver and found in almost all body tissues. It is a vital component for healthy cell membranes, and it serves as the starting material for the synthesis of all other steroids and vitamin D.

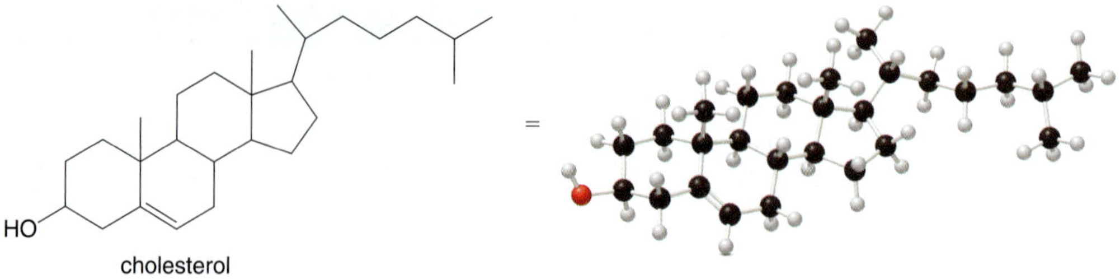

cholesterol

Cholesterol is obtained in the diet from a variety of sources, including meat, cheese, butter, and eggs. Table 19.4 lists the cholesterol content in some foods. While the American Heart Association currently recommends that the daily intake of cholesterol should be less than 300 mg, the average American diet includes 400–500 mg of cholesterol each day.

PROBLEM 19.21   Why is cholesterol classified as a lipid?

PROBLEM 19.22   (a) Label the rings of the steroid nucleus in cholesterol. (b) Give the number of the carbon to which the OH group is bonded. (c) Between which two carbons is the double bond located? (d) Label the polar bonds in cholesterol and explain why it is insoluble in water.

TABLE 19.4   Cholesterol Content in Some Foods

Food	Serving Size	Cholesterol (mg)
Boiled egg	1	225
Cream cheese	1 oz	27
Cheddar cheese	1 oz	19
Butter	3.5 oz	250
Beefsteak	3.5 oz	70
Chicken	3.5 oz	60
Ice cream	3.5 oz	45
Sponge cake	3.5 oz	260

While health experts agree that the amount of cholesterol in the diet should be limited, it is also now clear that elevated *blood* cholesterol (serum cholesterol) can lead to coronary artery disease. It is estimated that only 25% of the cholesterol in the blood comes from dietary sources, with the remainder synthesized in the liver. High blood cholesterol levels are associated with an increased risk of developing coronary artery disease, heart attack, and stroke. To understand the relationship between cholesterol and heart disease, we must learn about how cholesterol is transported through the bloodstream.

Like other lipids, cholesterol is insoluble in the aqueous medium of the blood, since it has only one polar OH group and many nonpolar C—C and C—H bonds. In order for it to be transported from the liver where it is synthesized, to the tissues, cholesterol combines with phospholipids and proteins to form small water-soluble spherical particles called **lipoproteins.**

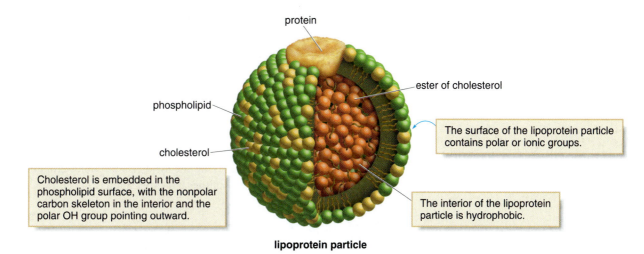

protein

ester of cholesterol

phospholipid

The surface of the lipoprotein particle contains polar or ionic groups.

cholesterol

Cholesterol is embedded in the phospholipid surface, with the nonpolar carbon skeleton in the interior and the polar OH group pointing outward.

The interior of the lipoprotein particle is hydrophobic.

lipoprotein particle

In a lipoprotein, the polar heads of phospholipids and the polar portions of protein molecules are arranged on the surface. The nonpolar molecules are buried in the interior of the particle. In this way, the nonpolar material is "dissolved" in an aqueous environment.

Lipoproteins are classified on the basis of their density, with two types being especially important in determining serum cholesterol levels.

- **Low-density lipoproteins (LDLs) transport cholesterol from the liver to the tissues.**
- **High-density lipoproteins (HDLs) transport cholesterol from the tissues back to the liver.**

LDL particles transport cholesterol to tissues where it is incorporated in cell membranes. When LDLs supply more cholesterol than is needed, LDLs deposit cholesterol on the wall of arteries, forming plaque (Figure 19.8). Atherosclerosis is a disease that results from the buildup of these fatty deposits, restricting the flow of blood, increasing blood pressure, and increasing the likelihood of a heart attack or stroke. As a result, LDL cholesterol is often called "bad" cholesterol.

▼ **FIGURE 19.8   Plaque Formation in an Artery**

a. Open artery

b. Blocked artery

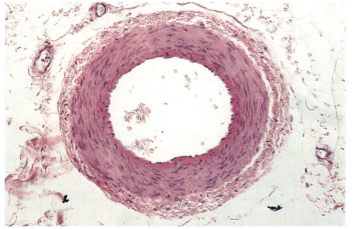

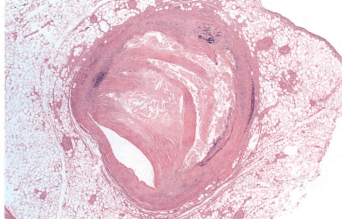

a. Cross-section of a clear artery with no buildup of plaque

b. Artery almost completely blocked by the buildup of plaque

HDL particles transport excess cholesterol from the tissues back to the liver, where it is converted to other substances or eliminated. Thus, HDLs reduce the level of serum cholesterol, so HDL cholesterol is often called "good" cholesterol.

Thus, a physical examination by a physician includes blood work that measures three quantities: total serum cholesterol, HDL cholesterol, and LDL cholesterol. Current recommendations for these values and the role of HDLs and LDLs in determining serum cholesterol levels are shown in Figure 19.9.

Several drugs called **statins** are now available to reduce the level of cholesterol in the bloodstream. These compounds act by blocking the synthesis of cholesterol at its very early stages. Two examples include atorvastatin (Lipitor) and simvastatin (Zocor).

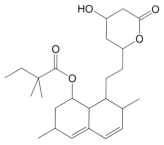

Generic name: atorvastatin
Trade name: Lipitor

Generic name: simvastatin
Trade bame: Zocor

▼ **FIGURE 19.9   HDLs, LDLs, and Cholesterol Level**

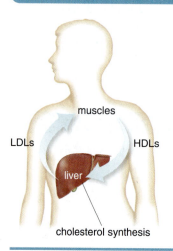

muscles

LDLs

HDLs

liver

cholesterol synthesis

Current recommendations by the National Cholesterol Education Program:

- Total serum cholesterol: < 200 mg/dL

- HDL cholesterol: > 40 mg/dL

- LDL cholesterol: < 100 mg/dL

Cholesterol is synthesized in the liver. High HDL levels are considered desirable since HDL transports cholesterol back to the liver. Low LDLs are considered desirable to avoid the buildup of plaque in the arteries.

PROBLEM 19.23   Would you expect triacylglycerols to be contained in the interior of a lipoprotein particle or on the surface with the phospholipids? Explain your choice.

PROBLEM 19.24   Identify the functional groups in (a) atorvastatin; (b) simvastatin.

19.9 STEROID HORMONES

Many biologically active steroids are hormones secreted by the endocrine glands. **A *hormone* is a molecule that is synthesized in one part of an organism, which then elicits a response at a different site.** Two important classes of steroid hormones are the **sex hormones** and the **adrenal cortical steroids.**

There are two types of female sex hormones, **estrogens** and **progestins.**

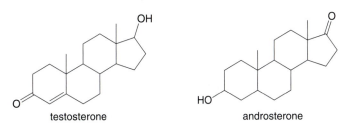

estradiol · estrone · progesterone

Estradiol, progesterone, and oral contraceptives were discussed in Section 13.5.

- Estradiol and estrone are estrogens synthesized in the ovaries. They control the development of secondary sex characteristics in females and regulate the menstrual cycle.
- Progesterone is a progestin often called the "pregnancy hormone." It is responsible for the preparation of the uterus for implantation of a fertilized egg.

The male sex hormones are called **androgens.**

testosterone · androsterone

- Testosterone and androsterone are androgens synthesized in the testes. They control the development of secondary sex characteristics in males—growth of facial hair, increase in muscle mass, and deepening of the voice.

Synthetic androgen analogues, called **anabolic steroids,** promote muscle growth. They were first developed to help individuals whose muscles had atrophied from lack of use following surgery. They have since come to be used by athletes and body builders, although their use is not permitted in competitive sports. Many physical and psychological problems result from their prolonged use.

Anabolic steroids, such as stanozolol, nandrolone, and tetrahydrogestrinone have the same effect on the body as testosterone, but they are more stable, so they are not metabolized as quickly. Tetrahydrogestrinone (also called THG or The Clear), the performance-enhancing drug used by track star Marion Jones during the 2000 Sydney Olympics, was considered a "designer steroid" because it was initially undetected in urine tests for doping. After its chemical structure and properties were determined, it was added to the list of banned anabolic steroids in 2004.

stanozolol · nandrolone · tetrahydrogestrinone

A second group of steroid hormones includes the **adrenal cortical steroids.** Three examples of these hormones are aldosterone, cortisone, and cortisol. All of these compounds are synthesized in the outer layer of the adrenal gland. Aldosterone regulates blood pressure and volume by controlling the concentration of Na^+ and K^+ in body fluids. Cortisone and cortisol serve as anti-inflammatory agents and they regulate carbohydrate metabolism.

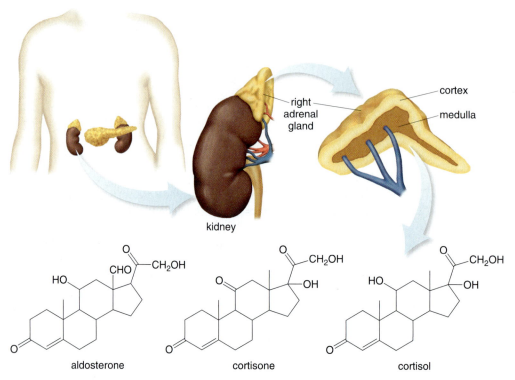

aldosterone   cortisone   cortisol

Cortisone and related compounds are used to suppress organ rejection after transplant surgery and to treat many allergic and autoimmune disorders. Prolonged use of these steroids can have undesired side effects, including bone loss and high blood pressure. Prednisone, a widely used synthetic alternative, has similar anti-inflammatory properties but can be taken orally.

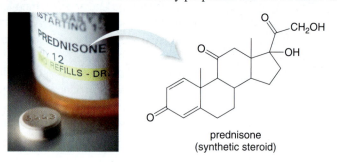

prednisone
(synthetic steroid)

PROBLEM 19.25	Compare the structures of estrone and progesterone. (a) Identify the differences in the A ring of these hormones. (b) How do these hormones differ in functionality at C17?
PROBLEM 19.26	Point out three structural differences between the female sex hormone estrone and the male sex hormone androsterone.
PROBLEM 19.27	Identify the functional groups in aldosterone. Classify each alcohol as 1°, 2°, or 3°.
PROBLEM 19.28	The male sex hormone testosterone and the anabolic steroid nandrolone have very similar structures and, as you might expect, similar biological activity. Point out the single structural difference in these two compounds.

19.10 FOCUS ON HEALTH & MEDICINE
FAT-SOLUBLE VITAMINS

Vitamins are organic compounds required in small quantities for normal metabolism (Section 11.7). Since our cells cannot synthesize these compounds, they must be obtained in the diet. Vitamins can be categorized as fat soluble or water soluble. **The fat-soluble vitamins are lipids.**

The four fat-soluble vitamins—**A, D, E,** and **K**—are found in fruits and vegetables, fish, liver, and dairy products. Although fat-soluble vitamins must be obtained from the diet, they do not have to be ingested every day. Excess vitamins are stored in adipose cells, and then used when needed. Table 19.5 summarizes the dietary sources and recommended daily intake of the fat-soluble vitamins.

TABLE 19.5	Fat-Soluble Vitamins	
Vitamin	**Food Source**	**Recommended Daily Intake**
A	Liver, kidney, oily fish, dairy products, eggs, fortified breakfast cereals	900 μg (men) 700 μg (women)
D	Fortified milk and breakfast cereals	5 μg
E	Sunflower and safflower oils, nuts, beans, whole grains, leafy greens	15 mg
K	Cauliflower, soybeans, broccoli, leafy greens, green tea	120 μg (men) 90 μg (women)

Source: Data from Harvard School of Public Health.

Vitamin A (Section 11.7) is obtained from liver, oily fish, and dairy products, and is synthesized from β-carotene, the orange pigment in carrots. In the body, vitamin A is converted to 11-*cis*-retinal, the light-sensitive compound responsible for vision in all vertebrates (Section 16.7). It is also needed for healthy mucous membranes. A deficiency of vitamin A causes night blindness, as well as dry eyes and skin.

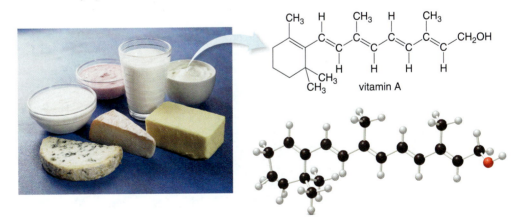

vitamin A

Vitamin D, strictly speaking, is not a vitamin because it can be synthesized in the body from cholesterol. Nevertheless, it is classified as such, and many foods (particularly milk) are fortified with vitamin D so that we get enough of this vital nutrient. Vitamin D helps regulate both calcium and phosphorus metabolism. A deficiency of vitamin D causes rickets, a bone disease characterized by knock-knees, spinal curvature, and other skeletal deformities.

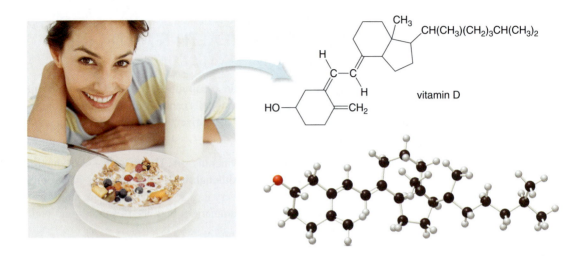

vitamin D

Vitamin E is an antioxidant, and in this way it protects unsaturated side chains in fatty acids from unwanted oxidation (Section 13.12). A deficiency of vitamin E causes numerous neurological problems, although it is rare for vitamin E deficiency to occur.

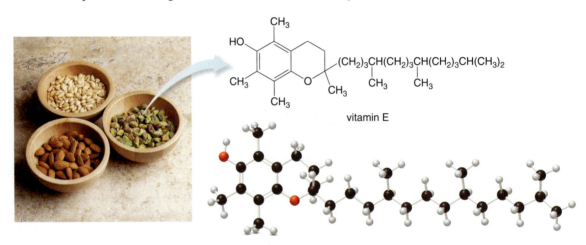

vitamin E

Vitamin K regulates the synthesis of prothrombin and other proteins needed for blood to clot. A severe deficiency of vitamin K leads to excessive and sometimes fatal bleeding because of inadequate blood clotting.

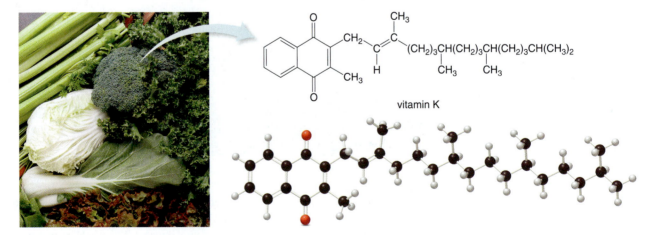

vitamin K

PROBLEM 19.29	Why is it much easier to overdose on a fat-soluble vitamin than a water-soluble vitamin?
PROBLEM 19.30	Vitamin D is synthesized in the body from a steroid. Which of the steroid rings—A, B, C, and D—are intact in vitamin D and which ring has been cleaved?

19.11 FOCUS ON HEALTH & MEDICINE PROSTAGLANDINS AND LEUKOTRIENES

Prostaglandins (Section 17.4) and **leukotrienes** are two types of **eicosanoids,** a group of biologically active compounds containing 20 carbon atoms derived from the fatty acid, arachidonic acid.

$$CH_3(CH_2)_4(CH=CHCH_2)_4(CH_2)_2COOH$$

arachidonic acid
(four cis double bonds)

a fatty acid

PGF$_{2\alpha}$
a prostaglandin

LTC$_4$
a leukotriene

The word *eicosanoid* is derived from the Greek word *eikosi,* meaning 20.

All eicosanoids are very potent compounds present in low concentration in cells. Eicosanoids are not stored in cells; rather, they are synthesized from arachidonic acid in response to an external stimulus. They are **local mediators,** meaning that they perform their function in the environment in which they are synthesized. This distinguishes them from hormones, which are first synthesized and then transported to their site of action.

19.11A PROSTAGLANDINS

Prostaglandins are a group of carboxylic acids that contain a five-membered ring and have a wide range of biological activities. PGF$_{2\alpha}$ and PGE$_1$ are two examples.

PGF$_{2\alpha}$

PGE$_1$

As we learned in Section 17.4, prostaglandins are responsible for inflammation. Aspirin and ibuprofen relieve pain and decrease inflammation because they block the synthesis of prostaglandins by inhibiting the cyclooxygenase enzyme needed for their synthesis.

Prostaglandins also decrease gastric secretions, inhibit blood platelet aggregation, stimulate uterine contractions, and relax the smooth muscles of the uterus. Because of these many biological functions, prostaglandins and their analogues have found several clinical uses. For example, dinoprostone, the generic name for PGE$_2$, is administered to relax the smooth muscles of the uterus when labor is induced, and to terminate early pregnancies. Misoprostol, a synthetic prostaglandin, prevents gastric ulcers in patients who are at high risk of developing them.

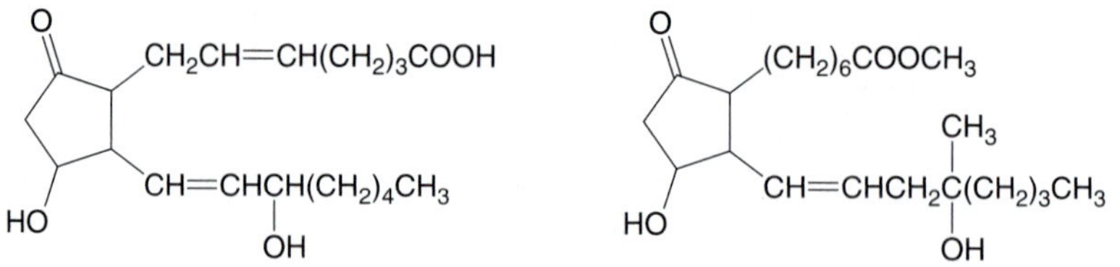

PGE₂
(dinoprostone)

misoprostol

More recently, it has been discovered that two *different* cyclooxygenase enzymes, called COX-1 and COX-2, are responsible for prostaglandin synthesis. COX-1 is involved with the usual production of prostaglandins, but COX-2 is responsible for the synthesis of additional prostaglandins in inflammatory diseases like arthritis. **NSAIDs—nonsteroidal anti-inflammatory drugs**—like aspirin and ibuprofen, inactivate both the COX-1 and COX-2 enzymes. This activity also results in an increase in gastric secretions, making an individual more susceptible to ulcer formation.

A group of anti-inflammatory drugs that block only the COX-2 enzyme was developed in the 1990s. These drugs—rofecoxib, valdecoxib, and celecoxib—do not cause an increase in gastric secretions, and thus were touted as an especially effective treatment for patients with arthritis, who need daily doses of these medications. Unfortunately, both rofecoxib and valdecoxib have been removed from the market, since their use has been associated with an increased risk of heart attack and stroke. Celecoxib, on the other hand, has fewer side effects, so it is still available and widely used.

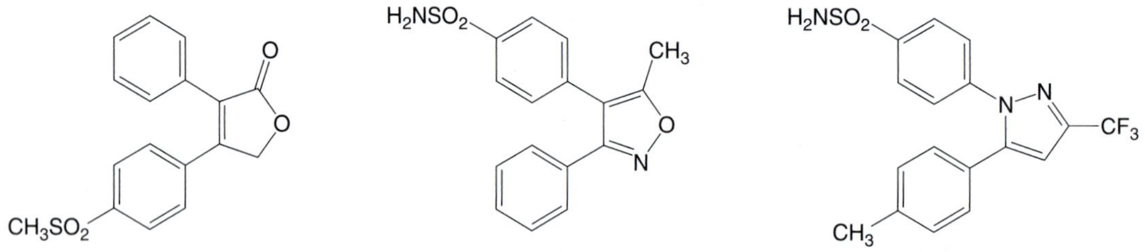

Generic name: rofecoxib
Trade name: Vioxx

Generic name: valdecoxib
Trade name: Bextra

Generic name: celecoxib
Trade name: Celebrex

PROBLEM 19.31

(a) What functional groups are present in PGE₂? (b) Draw a skeletal structure for PGE₂. The double bond closest to the COOH group is cis and the remaining double bond is trans.

19.11B ASTHMA AND LEUKOTRIENES

Asthma is an obstructive lung disease that affects millions of Americans. Because it involves episodic constriction of small airways, bronchodilators such as albuterol (Section 15.3) are used to treat symptoms by widening airways. Since asthma is also characterized by chronic inflammation, inhaled steroids that reduce inflammation are also commonly used.

The **leukotrienes** are molecules that contribute to the asthmatic response by constricting smooth muscles, especially in the lungs. Leukotrienes, such as leukotriene C₄ (LTC₄), are synthesized in several steps from arachidonic acid. The first step involves an oxidation reaction catalyzed by an enzyme called a lipoxygenase.

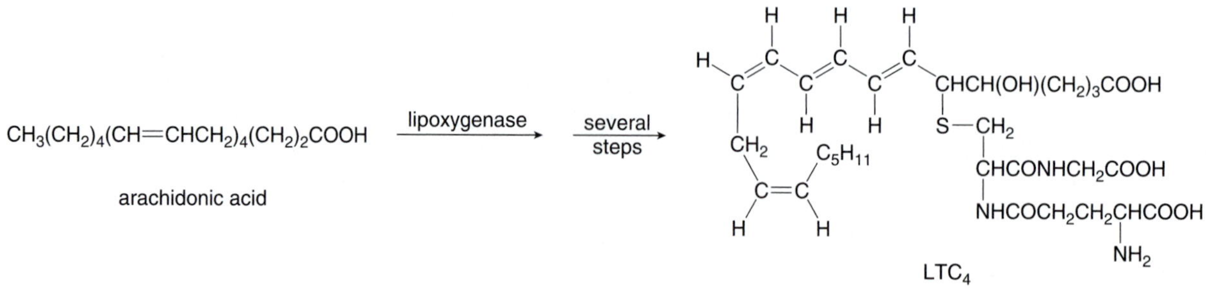

$CH_3(CH_2)_4(CH=CHCH_2)_4(CH_2)_2COOH$

arachidonic acid

lipoxygenase → several steps →

LTC₄

New asthma drugs act by blocking the synthesis of leukotriene C_4 from arachidonic acid. For example, zileuton (trade name: Zyflo) inhibits the lipoxygenase enzyme needed in the first step. By blocking the synthesis of leukotriene C_4, a compound responsible for the disease, zileuton treats the cause of asthma, not just its symptoms.

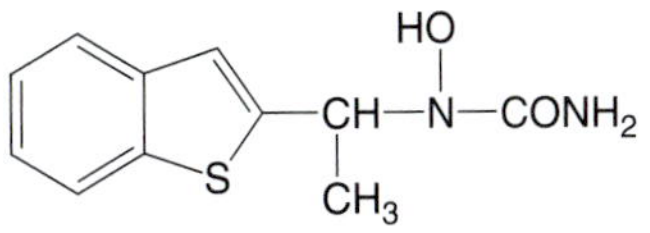

Generic name: zileuton
Trade name: Zyflo

PROBLEM 19.32

(a) A sulfur atom between two alkyl groups (RSR) is called a sulfide. Identify all other functional groups in LTC_4. (b) Label the double bonds in LTC_4 as cis or trans.

CHAPTER HIGHLIGHTS

KEY TERMS

Active transport (19.7)
Adrenal cortical steroid (19.9)
Anabolic steroid (19.9)
Androgen (19.9)
Cell membrane (19.7)
Cephalin (19.6)
Eicosanoid (19.11)
Estrogen (19.9)
Facilitated transport (19.7)
Fat (19.4)
Fat-soluble vitamin (19.10)
Fatty acid (19.2)
High-density lipoprotein (19.8)
Hormone (19.9)

Hydrolyzable lipid (19.1)
Hydrophilic (19.2)
Hydrophobic (19.2)
Lecithin (19.6)
Leukotriene (19.11)
Lipid (19.1)
Lipid bilayer (19.7)
Lipoprotein (19.8)
Low-density lipoprotein (19.8)
Nonhydrolyzable lipid (19.1)
Oil (19.4)
Omega-*n* acid (19.2)
Phosphatidylcholine (19.6)
Phosphatidylethanolamine (19.6)

Phosphoacylglycerol (19.6)
Phosphodiester (19.6)
Phospholipid (19.6)
Progestin (19.9)
Prostaglandin (19.11)
Saponification (19.5)
Saturated fatty acid (19.2)
Soap (19.5)
Sphingomyelin (19.6)
Steroid (19.8)
Triacylglycerol (19.4)
Unsaturated fatty acid (19.2)
Wax (19.3)

KEY REACTIONS

[1] Hydrolysis of waxes (19.3)

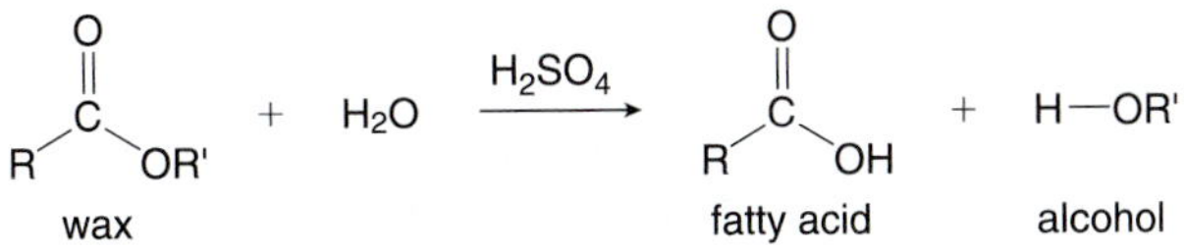

[2] Hydrolysis of triacylglycerols in the presence of acid or enzymes (19.5)

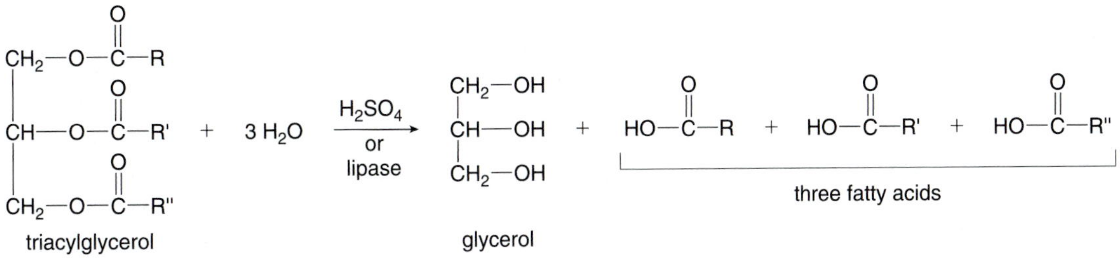

[3] Hydrolysis of triacylglycerols in the presence of base—Saponification (19.5)

KEY CONCEPTS

❶ What are the general characteristics of lipids? (19.1)

- Lipids are biomolecules that contain many nonpolar C—C and C—H bonds, making them soluble in organic solvents and insoluble in water.
- Hydrolyzable lipids, including waxes, triacylglycerols, and phospholipids, can be converted to smaller molecules on reaction with water.
- Nonhydrolyzable lipids, including steroids, fat-soluble vitamins, and eicosanoids, cannot be cleaved into smaller units by hydrolysis.

❷ How are fatty acids classified and what is the relationship between their melting points and the number of double bonds they contain? (19.2)

- Fatty acids are saturated if they contain no carbon–carbon double bonds and unsaturated if they contain one or more double bonds. Unsaturated fatty acids generally contain cis double bonds.
- As the number of double bonds in the fatty acid increases, its melting point decreases.

❸ What are waxes? (19.3)

- A wax is an ester (RCOOR') formed from a fatty acid (RCOOH) and a high molecular weight alcohol (R'OH). Since waxes contain many nonpolar C—C and C—H bonds, they are hydrophobic.
- Waxes (RCOOR') are hydrolyzed to fatty acids (RCOOH) and alcohols (R'OH).

❹ What are triacylglycerols, and how do the triacylglycerols in a fat and oil differ? (19.4)

- Triacylglycerols, or triglycerides, are triesters formed from glycerol and three molecules of fatty acids. A monounsaturated triacylglycerol contains one carbon–carbon double bond, whereas polyunsaturated triacylglycerols have more than one carbon–carbon double bond.
- Fats are triacylglycerols derived from fatty acids having few double bonds, making them solids at room temperature. Fats are generally obtained from animal sources.
- Oils are triacylglycerols derived from fatty acids having a larger number of double bonds, making them liquid at room temperature. Oils are generally obtained from plant sources.

❺ What hydrolysis products are formed from a triacylglycerol? (19.5)

- Triacylglycerols are hydrolyzed in acid or with enzymes (in biological systems) to form glycerol and three molecules of fatty acids. Base hydrolysis of a triacylglycerol forms glycerol and sodium salts of fatty acids—soaps.

❻ What are the major types of phospholipids? (19.6)

- All phospholipids contain a phosphorus atom, and have a polar (ionic) head and two nonpolar tails. Phosphoacylglycerols are derived from glycerol, two molecules of fatty acids, phosphate, and an alcohol (either ethanolamine or choline). Sphingomyelins are derived from sphingosine, a fatty acid (that forms an amide), a phosphate, and an alcohol (either ethanolamine or choline).

❼ Describe the structure of the cell membrane. How do molecules and ions cross the cell membrane? (19.7)

- The main component of the cell membrane is phospholipids, arranged in a lipid bilayer with the ionic heads oriented towards the outside of the bilayer, and the nonpolar tails on the interior.
- Small molecules like O_2 and CO_2 diffuse through the membrane from the side of higher concentration to the side of lower concentration. Larger polar molecules and some ions travel through channels created by integral membrane proteins (facilitated diffusion). Some cations (Na^+, K^+, and Ca^{2+}) must travel against the concentration gradient, a process called active transport, which requires energy input.

❽ What are the main structural features of steroids? What is the relationship between the steroid cholesterol and cardiovascular disease? (19.8)

- Steroids are tetracyclic lipids that contain three six-membered rings and one five-membered ring. Because cholesterol is insoluble in the aqueous medium of the blood, it is transported through the bloodstream in water-soluble particles called lipoproteins.
- Low-density lipoprotein particles (LDLs) transport cholesterol from the liver to the tissues. If the blood cholesterol level is high, it forms plaque on the walls of arteries, increasing the risk of heart attack and stroke. High-density lipoprotein particles (HDLs) transport cholesterol from the tissues to the liver, where it is metabolized or eliminated.

9 **What is a hormone? Give examples of steroid hormones. (19.9)**

- A hormone is a molecule that is synthesized in one part of an organism, and elicits a response at a different site. Steroid hormones include estrogens and progestins (female sex hormones), androgens (male sex hormones), and adrenal cortical steroids such as cortisone, which are synthesized in the adrenal gland.

10 **Which vitamins are fat soluble? (19.10)**

- Fat-soluble vitamins are lipids required in small quantities for normal cell function, and which cannot be synthesized in the body. Vitamins A, D, E, and K are fat soluble.

11 **What are the general characteristics of prostaglandins and leukotrienes? (19.11)**

- Prostaglandins are a group of carboxylic acids that contain a five-membered ring and are derived from arachidonic acid. Prostaglandins cause inflammation, decrease gastric secretions, inhibit blood platelet aggregation, stimulate uterine contractions, and relax the smooth muscle of the uterus.
- Leukotrienes, acyclic molecules derived from arachidonic acid, contribute to the asthmatic response by constricting smooth muscles in the lungs.

PROBLEMS

Selected in-chapter and end-of-chapter problems have brief answers provided in Appendix B.

General Characteristics of Lipids

19.33 Label each compound as a hydrolyzable or nonhydrolyzable lipid.
   a. prostaglandin
   b. triacylglycerol
   c. leukotriene
   d. vitamin A
   e. phosphoacylglycerol
   f. lecithin
   g. cholesterol

19.34 Label each compound as a hydrolyzable or nonhydrolyzable lipid.
   a. eicosanoid
   b. oleic acid
   c. phospholipid
   d. cephalin
   e. wax
   f. estrogen
   g. PGE_1

19.35 In which solvents might a wax be soluble: (a) H_2O; (b) CH_2Cl_2; (c) $CH_3CH_2OCH_2CH_3$?

19.36 In which solvents or solutions might a steroid be soluble: (a) blood plasma; (b) CCl_4; (c) 5% NaCl solution?

Fatty Acids, Waxes, and Triacylglycerols

19.37 Rank the fatty acids in each group in order of increasing melting point.
   a. $CH_3(CH_2)_{14}COOH$, $CH_3(CH_2)_3CH=CH(CH_2)_7COOH$, $CH_3(CH_2)_{12}COOH$
   b. $CH_3(CH_2)_{16}COOH$, $CH_3(CH_2)_7CH=CH(CH_2)_7COOH$, $CH_3(CH_2)_5CH=CH(CH_2)_7COOH$

19.38 (a) What is the difference between a saturated, monounsaturated, and polyunsaturated fatty acid? (b) Give an example of each having the same number of carbons. (c) Rank these compounds in order of increasing melting point.

19.39 How does each of the following affect the melting point of a fatty acid: (a) increasing the number of carbon atoms; (b) increasing the number of double bonds?

19.40 How would you expect the melting points of the following fatty acids to compare? Explain your choice.

19.41 Is a fatty acid a hydrolyzable lipid? Explain your choice.

19.42 Why are soaps water soluble, but the fatty acids from which they are derived, water insoluble?

19.43 Draw the structure of a wax formed from palmitic acid $[CH_3(CH_2)_{14}COOH]$ and each alcohol.
   a. $CH_3(CH_2)_{21}OH$
   b. $CH_3(CH_2)_{11}OH$
   c. $CH_3(CH_2)_9OH$

19.44 Draw the structure of a wax formed from a 30-carbon straight chain alcohol and each carboxylic acid.
   a. lauric acid
   b. myristic acid
   c. $CH_3(CH_2)_{22}COOH$

19.45 What hydrolysis products are formed when each wax is treated with aqueous sulfuric acid?
   a. $CH_3(CH_2)_{16}COO(CH_2)_{17}CH_3$
   b. $CH_3(CH_2)_{12}COO(CH_2)_{25}CH_3$
   c. $CH_3(CH_2)_{14}COO(CH_2)_{27}CH_3$
   d. $CH_3(CH_2)_{22}COO(CH_2)_{13}CH_3$

19.46 What hydrolysis products are formed when each wax is treated with aqueous sulfuric acid?
   a. $CH_3(CH_2)_{18}COO(CH_2)_{29}CH_3$
   b. $CH_3(CH_2)_{24}COO(CH_2)_{23}CH_3$
   c. $CH_3(CH_2)_{14}COO(CH_2)_{17}CH_3$
   d. $CH_3(CH_2)_{12}COO(CH_2)_{15}CH_3$

19.47 Draw a triacylglycerol that fits each description:
a. a triacylglycerol formed from lauric, myristic, and linoleic acids
b. an unsaturated triacylglycerol that contains two cis double bonds in one fatty acid side chain
c. a saturated triacylglycerol formed from three 14-carbon fatty acids
d. a monounsaturated triacylglycerol

19.48 Draw a triacylglycerol that fits each description:
a. a triacylglycerol formed from two molecules of lauric acid and one molecule of palmitic acid
b. a polyunsaturated triacylglycerol formed from three molecules of linoleic acid
c. a trans triacylglycerol that contains two trans double bonds
d. an unsaturated triacylglycerol formed from linolenic acid

19.49 Consider the following four types of compounds: [1] fatty acids; [2] soaps; [3] waxes; [4] triacylglycerols. For each type of compound: (a) give the general structure; (b) draw the structure of a specific example; (c) label the compound as water soluble or water insoluble; (d) label the compound as soluble or insoluble in the organic solvent hexane [$CH_3(CH_2)_4CH_3$].

19.50 How do fats and oils compare with respect to each of the following features?
a. identity and number of functional groups present
b. number of carbon–carbon double bonds present
c. melting point
d. natural source

19.51 Answer the following questions about the given triacylglycerol.

$CH_2-O-C(=O)-(CH_2)_{18}CH_3$
$CH-O-C(=O)-(CH_2)_{16}CH_3$
$CH_2-O-C(=O)-(CH_2)_{10}CH_3$

a. What fatty acids are used to form this triacylglycerol?
b. Would you expect this triacylglycerol to be a solid or a liquid at room temperature?
c. What regions are hydrophobic?
d. What regions are hydrophilic?
e. What hydrolysis products are formed when the triacylglycerol is treated with aqueous sulfuric acid?

19.52 Answer the following questions about the given triacylglycerol.

$CH_2-O-C(=O)-(CH_2)_{14}CH_3$
$CH-O-C(=O)-(CH_2)_7(CH=CHCH_2)_2(CH_2)_3CH_3$
$CH_2-O-C(=O)-(CH_2)_7CH=CH(CH_2)_5CH_3$

a. What fatty acids are used to form this triacylglycerol?
b. Would you expect this triacylglycerol to be a solid or a liquid at room temperature?
c. What regions are hydrophobic?
d. What regions are hydrophilic?
e. What hydrolysis products are formed when the triacylglycerol is treated with aqueous sulfuric acid?

19.53 Draw the products formed when each triacylglycerol is hydrolyzed under each of the following conditions: [1] water and H_2SO_4; [2] water and NaOH.

a. $CH_2-O-C(=O)-(CH_2)_{14}CH_3$
$CH-O-C(=O)-(CH_2)_{14}CH_3$
$CH_2-O-C(=O)-(CH_2)_{16}CH_3$

b. $CH_2-O-C(=O)-(CH_2)_{14}CH_3$
$CH-O-C(=O)-(CH_2)_7CH=CH(CH_2)_7CH_3$
$CH_2-O-C(=O)-(CH_2)_7CH=CH(CH_2)_5CH_3$

19.54 Draw the products formed when each triacylglycerol is hydrolyzed under each of the following conditions: [1] water and H_2SO_4; [2] water and NaOH.

a. $CH_2-O-C(=O)-(CH_2)_{10}CH_3$
$CH-O-C(=O)-(CH_2)_{14}CH_3$
$CH_2-O-C(=O)-(CH_2)_{16}CH_3$

b. $CH_2-O-C(=O)-(CH_2)_7CH=CH(CH_2)_7CH_3$
$CH-O-C(=O)-(CH_2)_{16}CH_3$
$CH_2-O-C(=O)-(CH_2)_7CH=CH(CH_2)_5CH_3$

Phospholipids and Cell Membranes

19.55 What distinguishes a cephalin from a lecithin? Draw a general structure of each type of compound.

19.56 In what ways are phosphoacylglycerols and sphingomyelins similar? In what ways are they different?

19.57 Which of the following are phospholipids: (a) triacylglycerols; (b) leukotrienes; (c) sphingomyelins; (d) fatty acids?

19.58 Which of the following are phospholipids: (a) prostaglandins; (b) cephalins; (c) lecithins; (d) steroids?

19.59 Draw a phospholipid that fits each description.
   a. a cephalin formed from two molecules of palmitoleic acid
   b. a phosphatidylcholine formed from two molecules of lauric acid
   c. a sphingomyelin formed from stearic acid and ethanolamine

19.60 Draw a phospholipid that fits each description.
   a. a lecithin formed from two molecules of oleic acid
   b. a phosphatidylethanolamine formed from two molecules of myristic acid
   c. a sphingomyelin formed from palmitic acid and choline

19.61 Why don't triacylglycerols form lipid bilayers?

19.62 How would a cell membrane having phospholipids that contain a high percentage of oleic acid differ from a cell membrane having phospholipids that contain a high percentage of stearic acid?

19.63 In transporting molecules or ions across a cell membrane, what is the difference between diffusion and facilitated transport? Give an example of a molecule or ion that crosses the membrane by each method.

19.64 In transporting molecules or ions across a cell membrane, what is the difference between facilitated transport and active transport? Give an example of a molecule or ion that crosses the membrane by each method.

Steroids

19.65 Draw the structure of the anabolic steroid 4-androstene-3,17-dione, also called "andro," from the following description. Andro contains the tetracyclic steroid skeleton with carbonyl groups at C3 and C17, a double bond between C4 and C5, and methyl groups bonded to C10 and C13.

19.66 Draw the structure of the anabolic steroid methenolone from the following description. Methenolone contains the tetracyclic steroid skeleton with a carbonyl group at C3, a hydroxyl at C17, a double bond between C1 and C2, and methyl groups bonded to C1, C10, and C13.

19.67 Why must cholesterol be transported through the bloodstream in lipoprotein particles?

19.68 Why are LDLs soluble in the blood?

19.69 Describe the role of HDLs and LDLs in cholesterol transport in the blood. What is the relationship of HDL and LDL levels to cardiovascular disease?

19.70 What are anabolic steroids? Give an example. What adverse effects arise from using anabolic steroids?

19.71 (a) Draw the structure of an estrogen and an androgen. (b) What structural features are similar in the two steroids? (c) What structural features are different? (d) Describe the biological activity of each steroid.

19.72 (a) Draw the structure of an androgen and a progestin. (b) What structural features are similar in the two steroids? (c) What structural features are different? (d) Describe the biological activity of each steroid.

Prostaglandins and Leukotrienes

19.73 What are the similarities and differences between prostaglandins and leukotrienes?

19.74 Why aren't prostaglandins classified as hormones?

19.75 What two structural features characterize all prostaglandins?

19.76 List three biological functions of prostaglandins in the body.

19.77 Explain why aspirin and celecoxib differ in how they act as anti-inflammatory agents.

19.78 How does zileuton treat the cause of asthma, not just relieve its symptoms?

Vitamins

19.79 What are vitamins and why must they be present in the diet?

19.80 Why is vitamin D technically not a vitamin?

19.81 Answer each question with regards to vitamins A and D.
   a. How many tetrahedral carbons does the vitamin contain?
   b. How many trigonal planar carbons does the vitamin contain?
   c. Identify the functional groups.
   d. Label all polar bonds.
   e. What function does the vitamin serve in the body?
   f. What problems result when there is a deficiency of the vitamin?
   g. Give a dietary source.

19.82 Answer each question in Problem 19.81 for vitamins E and K.

General Questions

19.83 Give an example of each type of lipid.
   a. a monounsaturated fatty acid
   b. a wax that contains a total of 30 carbons
   c. a saturated triacylglycerol
   d. a sphingomyelin derived from ethanolamine

19.84   Give an example of each type of lipid.
    a. a polyunsaturated fatty acid
    b. a wax derived from a 12-carbon fatty acid
    c. a polyunsaturated triacylglycerol
    d. a cephalin

19.85   Block diagrams representing the general structures of three types of lipids are drawn. Which terms describe each diagram: (a) phospholipid; (b) triacylglycerol; (c) hydrolyzable lipid; (d) sphingomyelin; (e) phosphoacylglycerol? More than one term may apply to a diagram.

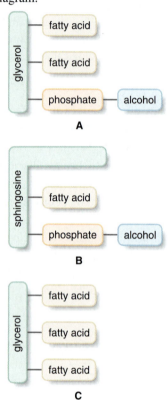

19.86   Which "cartoon" represents a soap and which represents a phosphoacylglycerol? What structural features are present in the polar head and nonpolar tails of each compound?

19.87   Why are phosphoacylglycerols more water soluble than triacylglycerols?

19.88   How are soaps and phosphoacylglycerols similar in structure? How do they differ?

19.89   Some fish oils contain triacylglycerols formed from the polyunsaturated fatty acid, 7,10,13,16,19-docosapentaenoic acid.

$$CH_3CH_2CH{=}CHCH_2CH{=}CHCH_2CH{=}CHCH_2CH{=}CHCH_2CH{=}CH(CH_2)_5COOH$$

7,10,13,16,19-docosapentaenoic acid

    a. Draw a skeletal structure showing the cis arrangement at each double bond.
    b. Label the hydrophobic and hydrophilic portions of the fatty acid.
    c. How does the melting point of this fatty acid compare to its all trans isomer?
    d. Would you expect this fatty acid to be a solid or a liquid at room temperature?
    e. What type of omega-*n* acid is this fatty acid?

19.90   Some marine plankton contain triacylglycerols formed from the polyunsaturated fatty acid, 3,6,9,12,15-octadecapentaenoic acid.

$$CH_3CH_2CH{=}CHCH_2CH{=}CHCH_2CH{=}CHCH_2CH{=}CHCH_2CH{=}CHCH_2COOH$$

3,6,9,12,15-octadecapentaenoic acid

    a. Draw a skeletal structure showing the cis arrangement at each double bond.
    b. Label the hydrophobic and hydrophilic portions of the fatty acid.
    c. How does the melting point of this fatty acid compare to the melting point of oleic acid?
    d. Would you expect this fatty acid to be a solid or a liquid at room temperature?
    e. What type of omega-*n* acid is this fatty acid?

Applications

19.91   The main fatty acid component of the triacylglycerols in coconut oil is lauric acid, $CH_3(CH_2)_{10}COOH$. Explain why coconut oil is a liquid at room temperature despite the fact that it contains a large fraction of this saturated fatty acid.

19.92   Unlike many fats and oils, the cocoa butter used to make chocolate is remarkably uniform in composition. All triacylglycerols contain oleic acid bonded to the 2° OH group of glycerol, and either palmitic acid or stearic acid esterified to the 1° OH groups. Draw the structures of two possible triacylglycerols that compose cocoa butter.

19.93   What is the difference between vegetable oil from plants and motor oil, obtained from crude oil (Section 12.6)?

19.94   Why do laboratory blood results now report the levels of cholesterol, HDLs, and LDLs? What are the current recommendations for these values?

19.95   Can an individual survive on a completely fat-free diet?

19.96   Can an individual survive on a cholesterol-free diet?

19.97   Why should saturated fats in the diet be avoided?

19.98   Why is it recommended that polyunsaturated oils be substituted for saturated fats in the diet?

19.99   Why do animals that live in cold climates have triacylglycerols that contain a higher percentage of unsaturated fatty acid side chains?

19.100 If the serum cholesterol level in an adult is 167 mg/dL, how much cholesterol is contained in 5.0 L of blood? (Recall from Chapter 1 that 1 L = 10 dL.)

CHALLENGE QUESTIONS

19.101 Sometimes it is possible to convert one lipid molecule to another by a simple organic reaction that you learned in previous chapters. What type of reaction converts each starting material to the given product?
a. cortisone $\rightarrow$ cortisol (Section 19.9)
b. estradiol $\rightarrow$ estrone (Section 19.9)
c. $PGE_2 \rightarrow PGE_1$ (Section 19.11)

19.102 How many triacylglycerols can be prepared from three different fatty acids? Draw all possible structures, excluding stereoisomers.

20

- - - - - - - - - - - - - - -

CHAPTER GOALS

In this chapter you will learn how to:

1. Identify the three major types of carbohydrates

2. Recognize the major structural features of monosaccharides

3. Draw the cyclic forms of monosaccharides and classify them as α or β anomers

4. Draw reduction and oxidation products of monosaccharides

5. Recognize the major structural features of disaccharides

6. Describe the characteristics of cellulose, starch, and glycogen

7. Give examples of some carbohydrate derivatives that contain amino groups, amides, or carboxylate anions

8. Describe the role that carbohydrates play in determining blood type

Milk contains **lactose,** a carbohydrate formed from two simple sugars, glucose and galactose.

CARBOHYDRATES

IN Chapter 20, we turn our attention to **carbohydrates,** the largest group of organic molecules in nature, comprising approximately 50% of the earth's biomass. Carbohydrates can be simple or complex, having as few as three or as many as thousands of carbon atoms. The glucose metabolized for energy in cells, the sucrose of table sugar, and the cellulose of plant stems and tree trunks are all examples of carbohydrates. Carbohydrates on cell surfaces determine blood type, and carbohydrates form the backbone of DNA, the carrier of all genetic information in the cell. Unlike the lipids of Chapter 19, which are composed of mainly carbon–carbon and carbon–hydrogen bonds and few functional groups, carbohydrates have many polar functional groups, whose structure and properties can be understood by applying the basic principles of organic chemistry.

20.1 INTRODUCTION

Carbohydrates, commonly referred to as sugars and starches, are polyhydroxy aldehydes and ketones, or compounds that can be hydrolyzed to them. Historically, the word *carbohydrate* was given to this group of compounds because the molecular formula of simple carbohydrates could be written as $C_n(H_2O)_n$, making them *hydrates of carbon.*

Carbohydrates are classified into three groups:

- Monosaccharides (Sections 20.2–20.4)
- Disaccharides (Section 20.5)
- Polysaccharides (Sections 20.6–20.7)

Monosaccharides or simple sugars are the simplest carbohydrates. Glucose and fructose, the two major constituents of honey, are monosaccharides. Glucose contains an aldehyde at one end of a six-carbon chain, and fructose contains a ketone. Every other carbon atom has a hydroxyl group bonded to it. Monosaccharides cannot be converted to simpler compounds by hydrolysis.

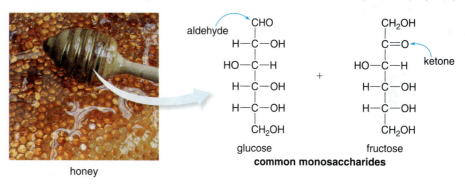

honey

common monosaccharides

Disaccharides are composed of two monosaccharides joined together. Lactose, the principal carbohydrate in milk, is a disaccharide. Disaccharides contain at least one acetal carbon—a carbon atom singly bonded to two OR (alkoxy) groups. Although disaccharides contain no carbonyl groups, they are hydrolyzed to simple monosaccharides that contain an aldehyde or ketone, as we will learn in Section 20.5.

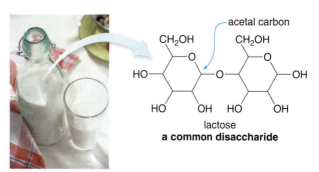

lactose
a common disaccharide

Polysaccharides have three or more monosaccharides joined together. Starch, the main carbohydrate found in the seeds and roots of plants, is a polysaccharide composed of hundreds of glucose molecules joined together. Like disaccharides, polysaccharides contain acetals but no carbonyl groups, and they are hydrolyzed to simple monosaccharides that contain carbonyl groups. Pasta, bread, rice, and potatoes are foods that contain a great deal of starch.

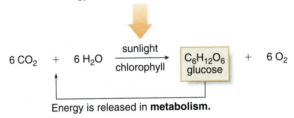

starch
a polysaccharide
[Acetal carbons are labeled with *.]

Carbohydrates are storehouses of chemical energy. Carbohydrates are synthesized in green plants and algae through **photosynthesis,** a process that uses the energy from the sun to convert carbon dioxide and water into glucose and oxygen. Almost all of the oxygen in the atmosphere results from photosynthesis. Plants store glucose in the form of polysaccharides like starch and cellulose (Section 20.6).

Energy is stored in **photosynthesis.**

$$6\ CO_2\ +\ 6\ H_2O\ \xrightarrow[\text{chlorophyll}]{\text{sunlight}}\ \boxed{\begin{array}{c} C_6H_{12}O_6 \\ \text{glucose} \end{array}}\ +\ 6\ O_2$$

Energy is released in **metabolism.**

The energy stored in glucose bonds is released when glucose is metabolized. The oxidation of glucose is a multistep process that forms carbon dioxide, water, and a great deal of energy. Although the metabolism of lipids provides more energy per gram than the metabolism of carbohydrates, glucose is the preferred source when a burst of energy is needed during exercise. Glucose is water soluble, so it can be quickly and easily transported through the bloodstream to tissues.

Chlorophyll in green leaves converts CO_2 and H_2O to glucose and O_2 during photosynthesis.

PROBLEM 20.1	Draw a Lewis structure for glucose that clearly shows the aldehyde carbonyl group and all lone pairs on the oxygen atoms.
PROBLEM 20.2	Label the hemiacetal carbon—the carbon bonded to one OH and one OR group—in lactose (Section 16.8). Label all other hydroxyls as 1°, 2°, or 3°.

20.2 MONOSACCHARIDES

Monosaccharides, the simplest carbohydrates, generally have three to six carbon atoms in a chain, with a **carbonyl group** at either the terminal carbon, numbered C1, or the carbon adjacent to it, numbered C2. In most carbohydrates, each of the remaining carbon atoms has a **hydroxyl group.** Monosaccharides are drawn vertically, with the carbonyl group at (or near) the top.

carbonyl at C1 ----→ aldehyde ----→ **aldose**

carbonyl at C2 ----→ ketone ----→ **ketose**

3–6 C's

OH on all (or most) other C's

monosaccharide

- Monosaccharides with a carbonyl group at C1 are aldehydes called *aldoses.*
- Monosaccharides with a carbonyl group at C2 are ketones called *ketoses.*

Dihydroxyacetone (DHA) is the active ingredient in many artificial tanning agents.

Glyceraldehyde is the simplest aldose and dihydroxyacetone is the simplest ketose. Glyceraldehyde and dihydroxyacetone both have molecular formula $C_3H_6O_3$, so they are **constitutional isomers;** that is, they have the same molecular formula but a different arrangement of atoms. Glucose is the most prevalent aldose.

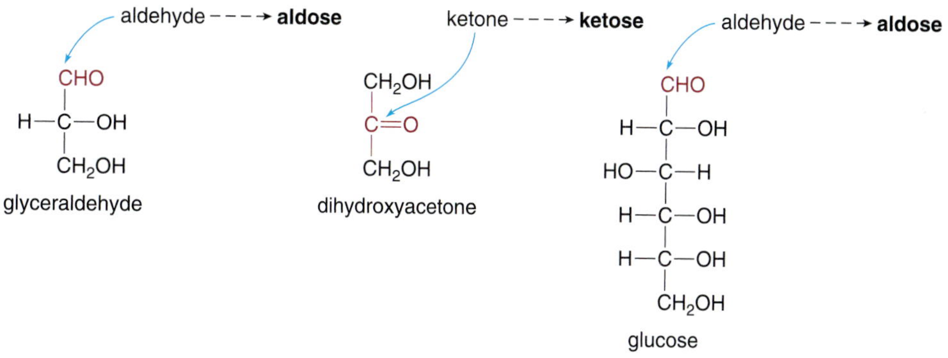

aldehyde ----→ **aldose**   ketone ----→ **ketose**   aldehyde ----→ **aldose**

glyceraldehyde   dihydroxyacetone

glucose

All carbohydrates have common names. The simplest aldehyde, glyceraldehyde, and the simplest ketone, dihydroxyacetone, are the only monosaccharides whose names do not end in the suffix *-ose.*

A monosaccharide is characterized by the number of carbons in its chain.

- A triose has three carbons.
- A tetrose has four carbons.
- A pentose has five carbons.
- A hexose has six carbons.

These terms are then combined with the words *aldose* and *ketose* to indicate both the number of carbon atoms in the monosaccharide and whether it contains an aldehyde or ketone. Thus, glyceraldehyde is an aldotriose (three carbons and an aldehyde), dihydroxyacetone is a ketotriose (three carbons and a ketone), and glucose is an aldohexose (six carbons and an aldehyde).

SAMPLE PROBLEM 20.1

Classify each monosaccharide by the type of carbonyl group and the number of carbons in the chain.

a.   ribose

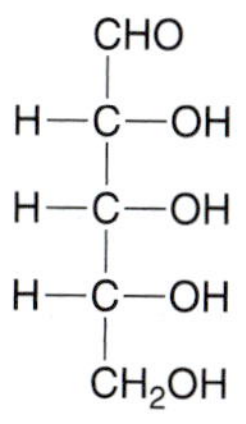

b.   fructose

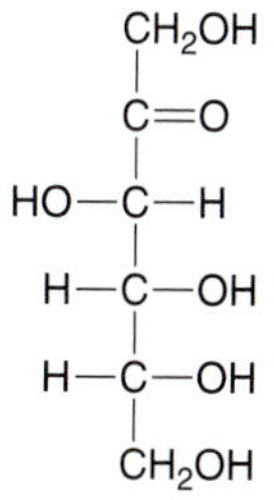

ANALYSIS   Identify the type of carbonyl group to label the monosaccharide as an aldose or ketose. An aldose has the C=O at C1 so that a hydrogen atom is bonded to the carbonyl carbon. A ketose has two carbons bonded to the carbonyl carbon. Count the number of carbons in the chain to determine the suffix—namely, *-triose, -tetrose,* and so forth.

SOLUTION

a.

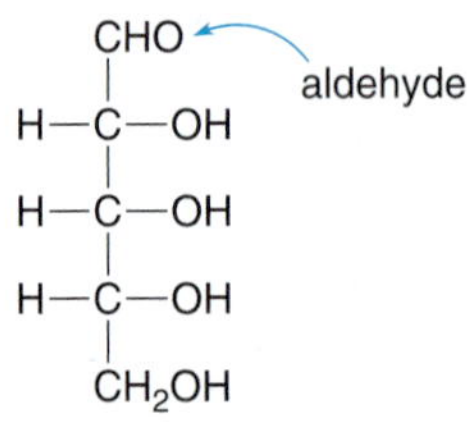

CHO
H—C—OH
H—C—OH
H—C—OH
CH₂OH

aldehyde

ribose
5 C's in the chain
Answer: aldopentose

b.

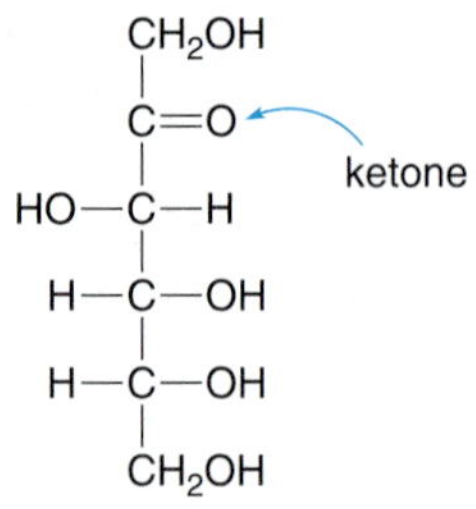

CH₂OH
C=O
HO—C—H
H—C—OH
H—C—OH
CH₂OH

ketone

fructose
6 C's in the chain
Answer: ketohexose

PROBLEM 20.3

Classify each monosaccharide by the type of carbonyl group and the number of carbons in the chain.

a.

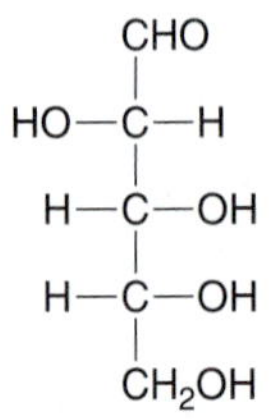

CHO
HO—C—H
H—C—OH
H—C—OH
CH₂OH

arabinose

b.

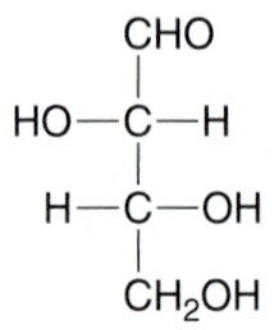

CHO
HO—C—H
H—C—OH
CH₂OH

threose

c.

CH₂OH
C=O
H—C—OH
CH₂OH

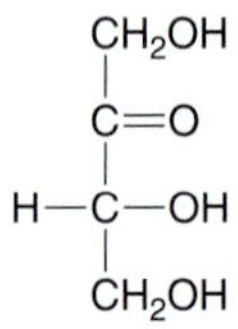

erythrulose

PROBLEM 20.4

Draw the structure of (a) an aldotetrose; (b) a ketopentose; (c) an aldohexose.

Monosaccharides are all sweet tasting, but their relative sweetness varies a great deal. Monosaccharides are polar compounds with high melting points. The presence of so many polar functional groups capable of hydrogen bonding makes them very water soluble.

PROBLEM 20.5

Rank the following compounds in order of increasing water solubility: glucose, hexane [CH₃(CH₂)₄CH₃], and 1-decanol [(CH₃(CH₂)₉OH]. Explain your choice.

20.2A FISCHER PROJECTION FORMULAS

A striking feature of carbohydrate structure is the presence of chirality centers. **All carbohydrates except for dihydroxyacetone contain one or more chirality centers.** The simplest aldose, glyceraldehyde, has one chirality center—one carbon atom bonded to four different groups. Thus, there are two possible enantiomers—mirror images that are not superimposable.

Chirality centers and enantiomers were first encountered in Section 15.2.

Two enantiomers of glyceraldehyde

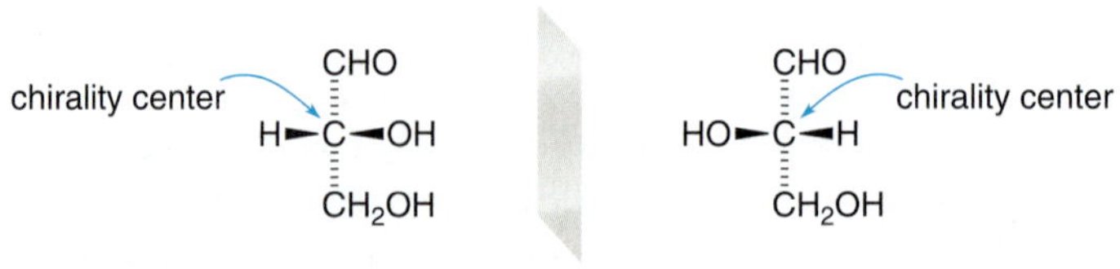

chirality center

CHO
H—C—OH
CH₂OH

naturally occurring isomer
D-glyceraldehyde

chirality center

CHO
HO—C—H
CH₂OH

unnatural isomer
L-glyceraldehyde

Only one enantiomer of glyceraldehyde occurs in nature. When the carbon chain is drawn vertically with the aldehyde at the top, the naturally occurring enantiomer has the OH group drawn on the right side of the carbon chain. **To distinguish the two enantiomers, the prefixes D and L precede the name.** Thus, the naturally occurring enantiomer is labeled D-glyceraldehyde, while the unnatural isomer is L-glyceraldehyde.

Fischer projection formulas are commonly used to depict the chirality centers in monosaccharides. Recall from Section 15.6 that a Fischer projection formula uses a cross to represent a tetrahedral carbon in which the horizontal bonds come forward (on wedges) and vertical bonds go behind (on dashes). In a Fischer projection formula:

- A carbon atom is located at the intersection of the two lines of the cross.
- The horizontal bonds come forward, on wedges.
- The vertical bonds go back, on dashed lines.

Using a Fischer projection formula, D-glyceraldehyde becomes:

D-glyceraldehyde   Fischer projection formula

PROBLEM 20.6

Draw L-glyceraldehyde using a Fischer projection formula.

20.2B MONOSACCHARIDES WITH MORE THAN ONE CHIRALITY CENTER

Fischer projection formulas are also used for compounds like aldohexoses that contain several chirality centers. Glucose, for example, contains four chirality centers labeled in the structure below. To convert the molecule to a Fischer projection, the molecule is drawn with a vertical carbon skeleton with the aldehyde at the top, and the horizontal bonds are assumed to come forward (on wedges). In the Fischer projection, each chirality center is replaced by a cross.

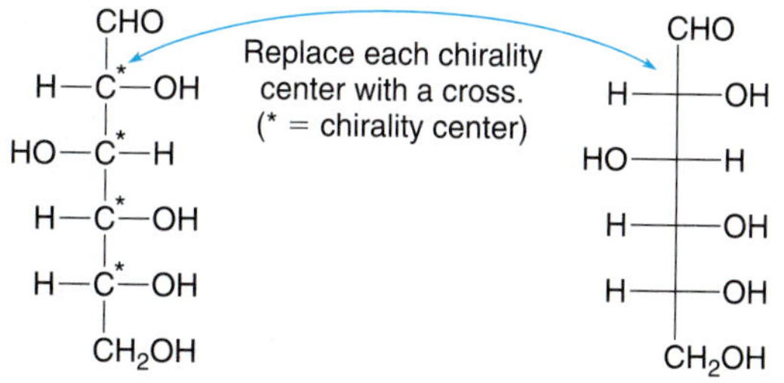

Replace each chirality center with a cross.
(* = chirality center)

glucose   Fischer projection formula

The letters **D** and **L** are used to label all monosaccharides, even those with many chirality centers. **The configuration of the chirality center _farthest_ from the carbonyl group determines whether a monosaccharide is D or L.**

- A D monosaccharide has the OH group on the chirality center farthest from the carbonyl on the right (like D-glyceraldehyde).
- An L monosaccharide has the OH group on the chirality center farthest from the carbonyl on the left (like L-glyceraldehyde).

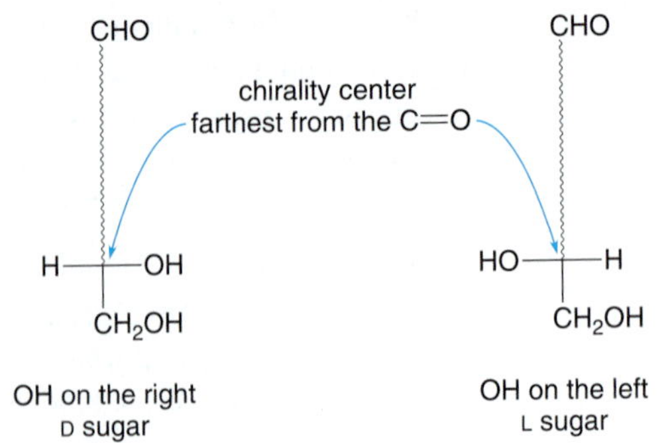

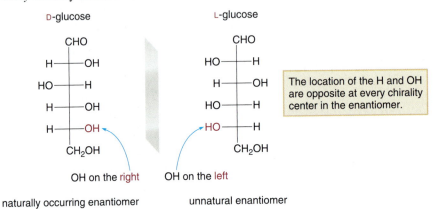

OH on the right
D sugar

OH on the left
L sugar

Glucose and all other naturally occurring sugars are D sugars. L-Glucose, a compound that does not occur in nature, is the enantiomer of D-glucose. L-Glucose has the opposite configuration at *every* chirality center.

D-glucose

CHO
H——OH
HO——H
H——OH
H——OH
CH₂OH

OH on the right

naturally occurring enantiomer

L-glucose

CHO
HO——H
H——OH
HO——H
HO——H
CH₂OH

OH on the left

unnatural enantiomer

The location of the H and OH are opposite at every chirality center in the enantiomer.

SAMPLE PROBLEM 20.2

Consider the aldopentose ribose. (a) Label all chirality centers. (b) Classify ribose as a D or L monosaccharide. (c) Draw the enantiomer.

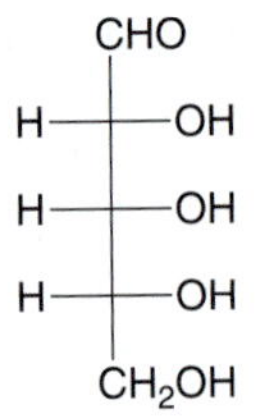

CHO
H——OH
H——OH
H——OH
CH₂OH

ribose

ANALYSIS

- A chirality center has four different groups around a carbon atom.
- The labels D and L are determined by the position of the OH group on the chirality center farthest from the carbonyl group: a D sugar has the OH group on the right and an L sugar has the OH group on the left.
- To draw an enantiomer, draw the mirror image so that each group is a reflection of the group in the original compound.

SOLUTION

a. The three carbons that contain both H and OH groups in ribose are chirality centers.

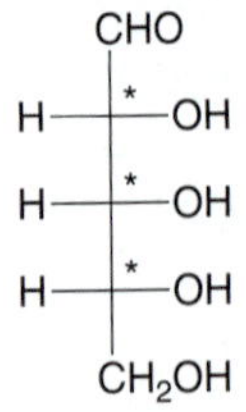

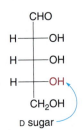

* = chirality center

b. Ribose is a D sugar since the OH group on the chirality center farthest from the carbonyl is on the right.

CHO
H——OH
H——OH
H——OH
CH₂OH

D sugar

c. The enantiomer of D-ribose, L-ribose, has all three OH groups on the left side of the carbon chain.

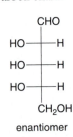

enantiomer

PROBLEM 20.7

For each monosaccharide: [1] label all chirality centers; [2] classify the monosaccharide as D or L; [3] draw the enantiomer.

a.
```
      CHO
   H──┼──OH
   H──┼──OH
      CH₂OH
```

b.
```
      CHO
  HO──┼──H
   H──┼──OH
  HO──┼──H
      CH₂OH
```

c.
```
      CHO
  HO──┼──H
  HO──┼──H
   H──┼──OH
      CH₂OH
```

PROBLEM 20.8

For each type of monosaccharide, [1] give an example of a D sugar; [2] label each chirality center: (a) an aldopentose; (b) a ketohexose; (c) a ketotetrose.

HEALTH NOTE

A 5% intravenous glucose (dextrose) solution provides a patient with calories and hydration.

20.2C COMMON MONOSACCHARIDES

The most common monosaccharides in nature are the aldohexoses D-glucose and D-galactose, and the ketohexose D-fructose.

```
      CHO                     CHO                    CH₂OH
   H─C─OH                  H─C─OH                    C═O
  HO─C─H                  HO─C─H                  HO─C─H
   H─C─OH                 HO─C─H                   H─C─OH
   H─C─OH                  H─C─OH                   H─C─OH
      CH₂OH                   CH₂OH                    CH₂OH

   D-glucose               D-galactose              D-fructose
```

Glucose, also called dextrose, is the sugar referred to when blood sugar is measured. It is the most abundant monosaccharide. Glucose is the building block for the polysaccharides starch and cellulose. Glucose, the carbohydrate that is transported in the bloodstream, provides energy for cells when it is metabolized. Normal blood glucose levels are in the range of 70–110 mg/dL. Excess glucose is converted to the polysaccharide glycogen (Section 20.6) or fat.

Insulin, a protein produced in the pancreas, regulates blood glucose levels. When glucose concentration increases after eating, insulin stimulates the uptake of glucose in tissues and its conversion to glycogen. Patients with diabetes produce insufficient insulin to adequately regulate blood glucose levels, and the concentration of glucose rises. With close attention to diet and daily insulin injections or other medications, a normal level of glucose can be maintained in most diabetic patients. Individuals with poorly controlled diabetes can develop many other significant complications, including cardiovascular disease, chronic renal failure, and blindness.

HEALTH NOTE

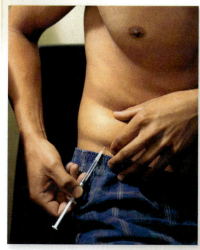

Insulin injections taken by diabetic patients help to maintain a proper blood glucose level.

Galactose is one of the two monosaccharides that form the disaccharide lactose (Section 20.5). Galactose is a stereoisomer of glucose, since the position of a hydrogen atom and hydroxyl group at a single chirality center are different in the two monosaccharides.

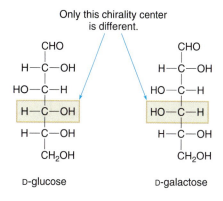

Only this chirality center is different.

```
      CHO                     CHO
   H─C─OH                  H─C─OH
  HO─C─H                  HO─C─H
   H─C─OH                 HO─C─H
   H─C─OH                  H─C─OH
      CH₂OH                   CH₂OH

   D-glucose               D-galactose
```

Individuals with galactosemia, a rare inherited disease, lack an enzyme needed to metabolize galactose. Galactose accumulates, causing a variety of physical problems, including cataracts, cirrhosis, and mental retardation. Galactosemia can be detected in newborn screening, and affected infants must be given soy-based formula to avoid all milk products with lactose. When diagnosed early, individuals who eliminate all lactose- and galactose-containing foods can lead a normal life.

Fructose is one of two monosaccharides that form the disaccharide sucrose (Section 20.5). Fructose is a ketohexose found in honey and is almost twice as sweet as normal table sugar with about the same number of calories per gram.

PROBLEM 20.9 D-Glucose and D-galactose are stereoisomers because they have the same molecular formula but a different three-dimensional geometry. How are D-glucose and D-fructose related? How are D-galactose and D-fructose related?

PROBLEM 20.10 (a) Draw a Fischer projection formula for D-galactose. (b) Draw a Fischer projection formula for the enantiomer of D-galactose.

20.3 THE CYCLIC FORMS OF MONOSACCHARIDES

HEALTH NOTE

An individual with galactosemia must avoid cow's milk and all products derived from cow's milk (Section 20.2C).

In Section 16.8 we learned that an aldehyde or ketone reacts with one equivalent of an alcohol to form a **hemiacetal,** a compound that contains a hydroxyl group (OH) and an alkoxy group (OR) on the same carbon.

OH and OR' on the same C

aldehyde   alcohol   hemiacetal

Generally, hemiacetals from acyclic reactants are unstable. When a compound contains *both* a hydroxyl group and an aldehyde or ketone, however, an intramolecular cyclization reaction forms a stable cyclic hemiacetal.

$HOCH_2CH_2CH_2CH_2$ re-draw   hemiacetal

5-hydroxypentanal   The C=O and OH groups react.

The carbon atom that is part of the hemiacetal is a new chirality center, called the *anomeric carbon*. Thus, two different products called **anomers** are formed.

new chirality center the anomeric carbon

anomeric carbon   anomeric carbon

α anomer   +   β anomer

- The α anomer has the OH group drawn down, below the ring (shown in red).
- The β anomer has the OH group drawn up, above the ring (shown in blue).

PROBLEM 20.11

Locate the hemiacetal carbon in each compound and label the compound as an α or β anomer.

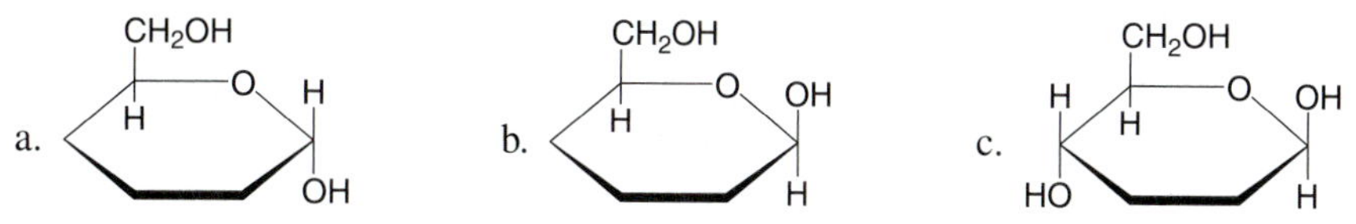

a. b. c.

20.3A THE CYCLIC FORMS OF D-GLUCOSE

Although the monosaccharides in Section 20.2 were drawn as acyclic carbonyl compounds, the hydroxyl and carbonyl groups of monosaccharides undergo intramolecular cyclization to form **hemiacetals.** Let's illustrate the process with D-glucose, and then learn a general method for drawing the cyclic forms of any aldohexose.

Which of the five OH groups in glucose is at the right distance from the carbonyl group to form a six-membered ring? The **O atom on the chirality center farthest from the carbonyl** (C5) is six atoms from the carbonyl carbon, placing it at the proper distance for cyclization to form a six-membered ring with a hemiacetal.

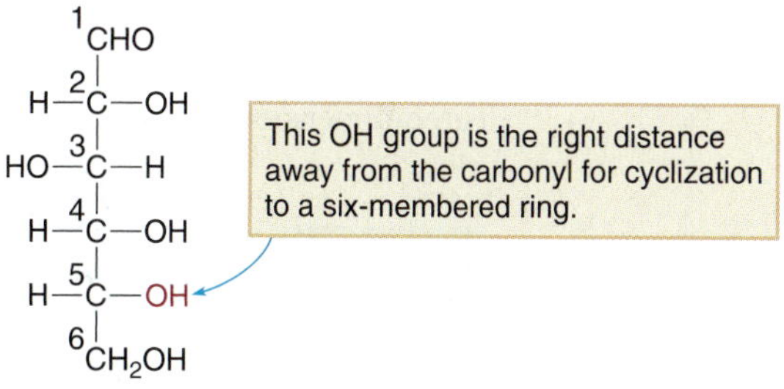

This OH group is the right distance away from the carbonyl for cyclization to a six-membered ring.

D-glucose

To convert this acyclic form (labeled **A**) into a cyclic hemiacetal, first rotate the carbon skeleton clockwise 90° to form **B.** Note that groups that were drawn on the right side of the carbon skeleton in **A** end up *below* the carbon chain in **B.** Then twist the chain to put the OH group on C5 close to the aldehyde carbonyl, forming **C.** In this process, the CH₂OH group at the end of the chain ends up *above* the carbon skeleton. Realize that **A, B,** and **C** are all different representations of D-glucose. The molecule has been rotated and twisted but the carbon skeleton and relative arrangement of the substituents have not changed.

D-glucose

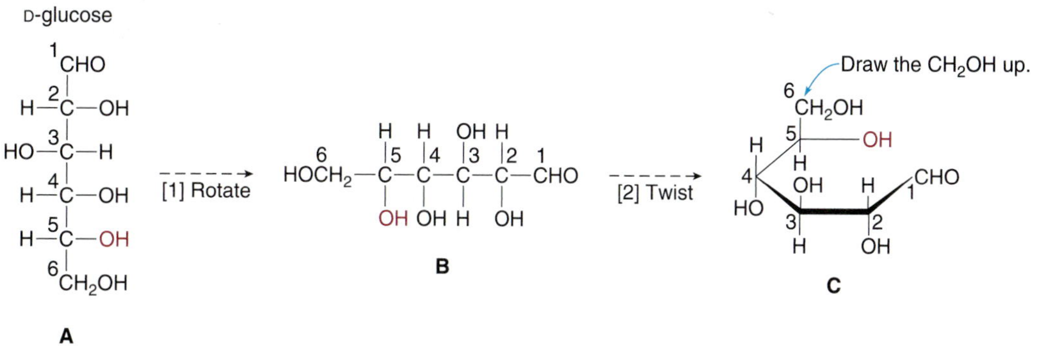

[1] Rotate — [2] Twist — Draw the CH₂OH up.

A B C

We are now set to draw the cyclic hemiacetals formed by reaction of the OH group on C5 with the aldehyde carbonyl. Since cyclization creates a new chirality center, there are *two cyclic forms* **of D-glucose, an α anomer and a β anomer.**

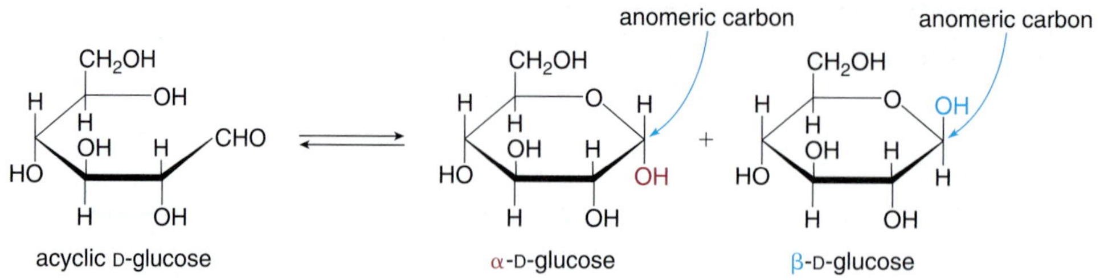

- The α anomer, called α-D-glucose, has the OH group on the anomeric carbon drawn down (shown in red).
- The β anomer, called β-D-glucose, has the OH group on the anomeric carbon drawn up (shown in blue).

These flat, six-membered rings used to represent the cyclic hemiacetals of glucose and other sugars are called **Haworth projections.**

Thus, D-glucose really exists in three different forms—an acyclic aldehyde and two cyclic hemiacetals—all of which are in equilibrium. Each cyclic hemiacetal can be isolated and crystallized separately, but when any one compound is placed in solution, an equilibrium mixture of all three forms results. This process is called **mutarotation.** At equilibrium, the mixture has 37% of the α anomer, 63% of the β anomer, and only a trace amount of the acyclic aldehyde. Three-dimensional models for the three forms of D-glucose are shown in Figure 20.1.

▼ **FIGURE 20.1** **The Three Forms of D-Glucose**

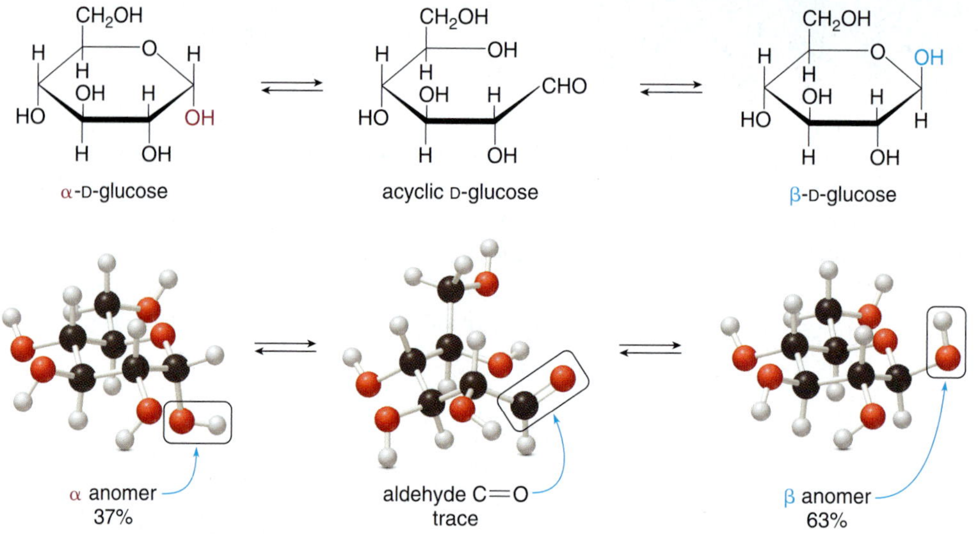

Although Haworth projections represent the six-membered ring of glucose as a flat hexagon, in reality, the ring is puckered as shown in the three-dimensional models (c.f. the ball-and-stick model of cyclohexane in Section 12.5).

20.3B HAWORTH PROJECTIONS

All aldohexoses exist primarily as cyclic hemiacetals typically drawn in Haworth projections. To convert an acyclic monosaccharide to a Haworth projection, follow the stepwise *How To* procedure.

HOW TO   Draw a Haworth Projection from an Acyclic Aldohexose

EXAMPLE   Draw both anomers of D-mannose in a Haworth projection.

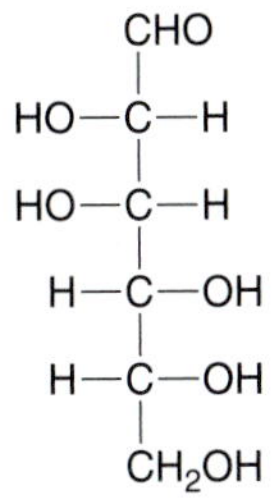

D-mannose

Step [1]   Place the O atom in the upper right corner of a hexagon, and add the CH$_2$OH group on the first carbon to the left of the O atom.

- For naturally occurring D sugars, the CH$_2$OH group is drawn *up*, above the plane of the six-membered ring.

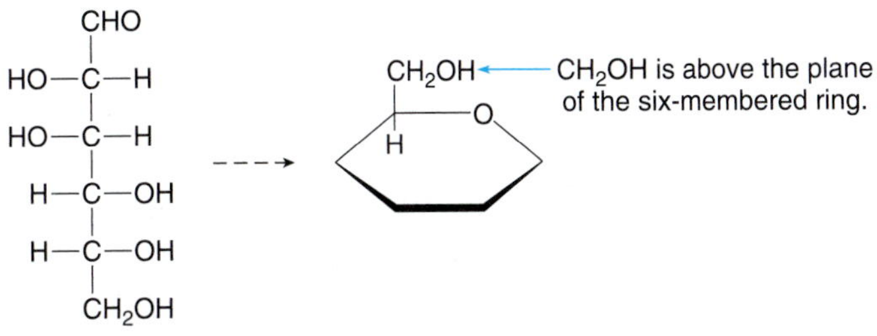

Step [2]   Place the anomeric carbon on the first carbon clockwise from the O atom.

- For an α anomer, the OH is drawn *down* in a D sugar.
- For a β anomer, the OH is drawn *up* in a D sugar.

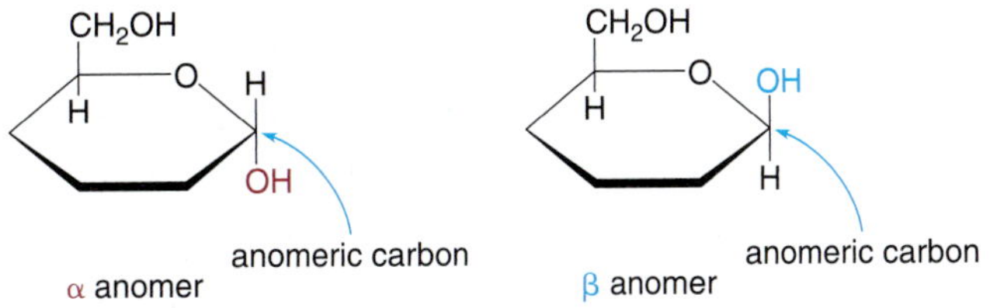

Step [3]   Add the substituents (OH and H) to the three remaining carbons (C2–C4), clockwise around the ring.

- The substituents on the *right* side of the carbon skeleton (shown in red) are drawn *down* in the Haworth projection.
- The substituents on the *left* side of the carbon skeleton (shown in blue) are drawn *up* in the Haworth projection.

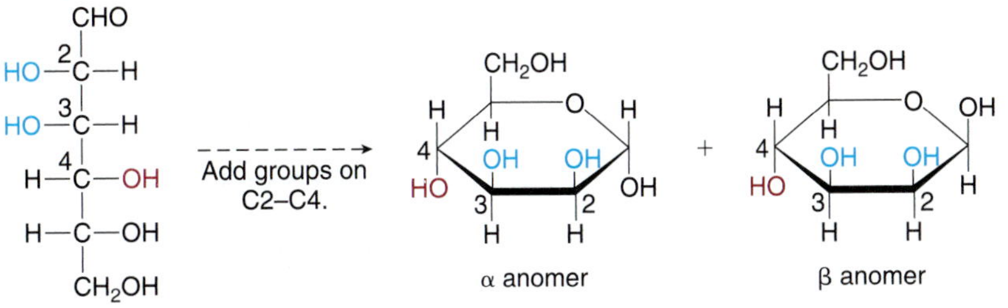

SAMPLE PROBLEM 20.3

Draw the α anomer of D-galactose.

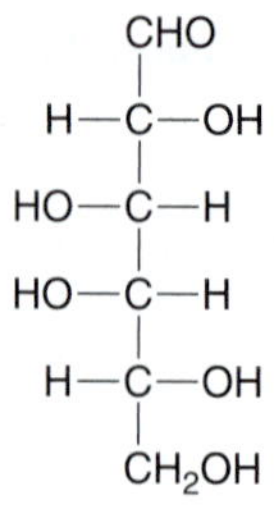

D-galactose

ANALYSIS AND SOLUTION

[1] **Draw a hexagon with an O atom in the upper right corner. Add the CH₂OH above the ring on the first carbon to the left of the O atom.**

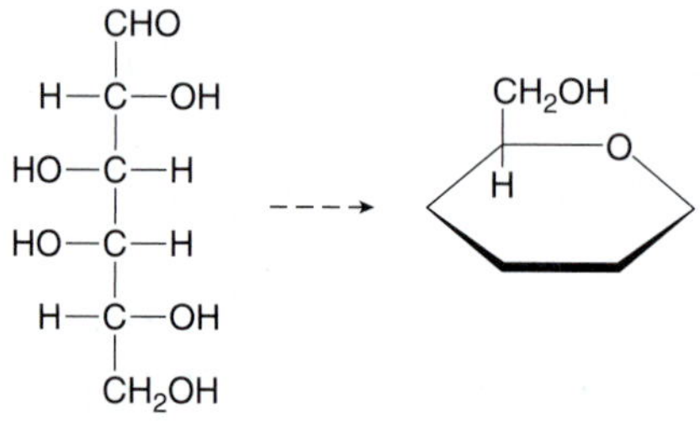

D-galactose

[2] **Draw the anomeric carbon on the first carbon clockwise from the O atom.**

- The α anomer has the OH group drawn down.

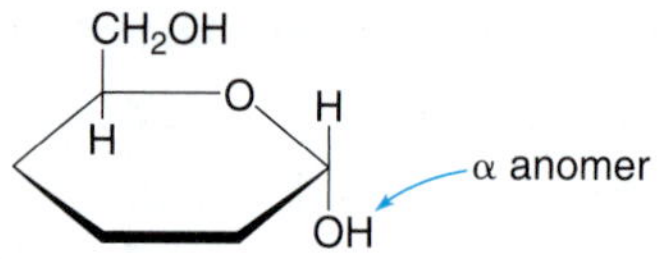

α anomer

[3] **Add the OH groups and H atoms to the three remaining carbons (C2–C4).**

- Groups on the *right* side in the acyclic form are drawn *down,* below the six-membered ring, and groups on the *left* side in the acyclic form are drawn *up,* above the six-membered ring.

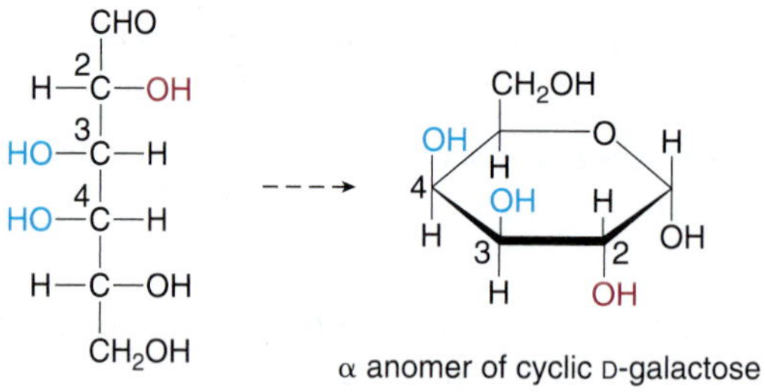

α anomer of cyclic D-galactose

PROBLEM 20.12

Convert each aldohexose to the indicated anomer using a Haworth projection.

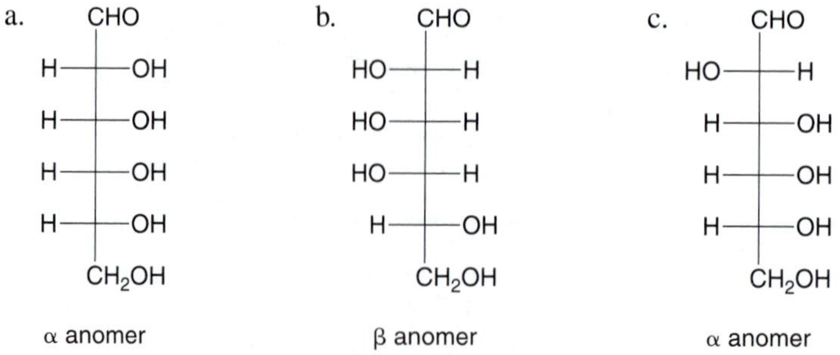

a. α anomer b. β anomer c. α anomer

20.3C THE CYCLIC FORMS OF FRUCTOSE, A KETOHEXOSE

Certain monosaccharides—notably aldopentoses and ketohexoses—form five-membered rings, *not* six-membered rings, in solution. The same principles apply to drawing these structures as for drawing six-membered rings, except the ring size is one atom smaller.

- Cyclization forms two anomers at the hemiacetal carbon. For a D sugar, the OH group is drawn down in the α anomer and up in the β anomer.

For example, D-fructose forms a five-membered ring when it cyclizes to a hemiacetal because the carbonyl group is a ketone at C2, instead of an aldehyde at C1. Note that the cyclic structures of fructose contain two CH_2OH groups bonded to the five-membered ring, one of which is located at the anomeric carbon.

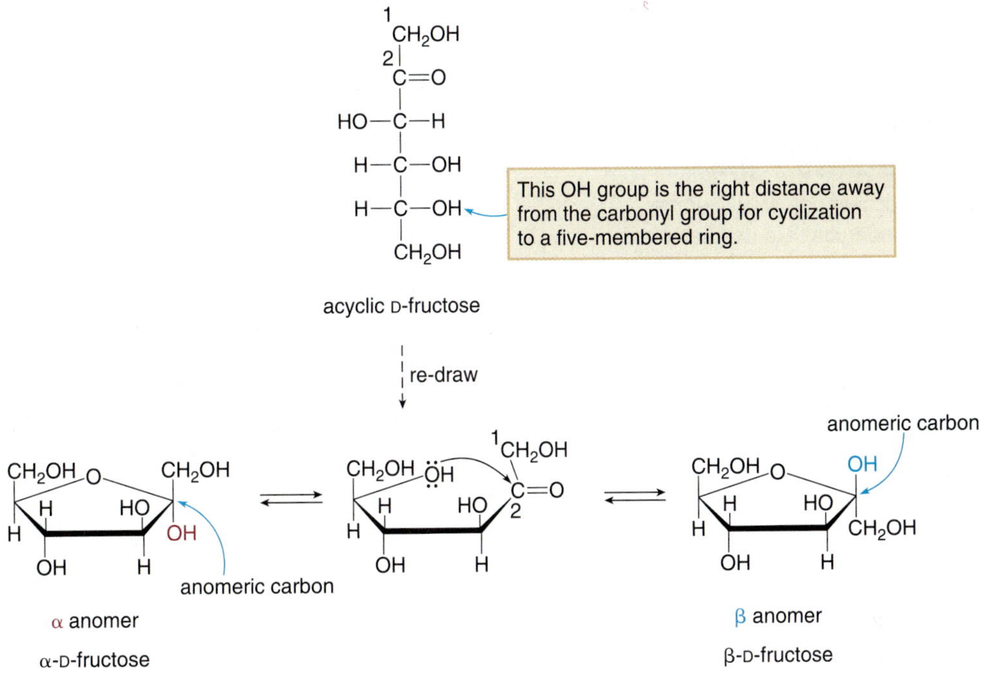

acyclic D-fructose

This OH group is the right distance away from the carbonyl group for cyclization to a five-membered ring.

re-draw

anomeric carbon

α anomer

α-D-fructose

anomeric carbon

β anomer

β-D-fructose

The two cyclic forms of D-fructose contain a five-membered ring.

PROBLEM 20.13

Locate the anomeric carbon in each compound and label the compound as an α or β anomer.

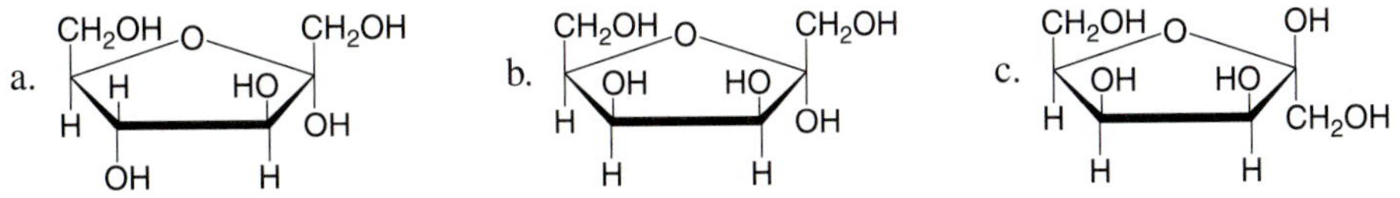

20.4 REDUCTION AND OXIDATION OF MONOSACCHARIDES

The aldehyde carbonyl group of the acyclic form of a monosaccharide undergoes two common reactions—**reduction to an alcohol** and **oxidation to a carboxylic acid.** Even though the acyclic form may be present in only trace amounts, the equilibrium can be tipped in its favor by Le Châtelier's principle (Section 6.6). When the carbonyl group of the acyclic form reacts with a reagent, thus depleting its equilibrium concentration, the equilibrium will shift to compensate for the loss. More of the acyclic form will be produced, which can react further.

20.4A   REDUCTION OF THE ALDEHYDE CARBONYL GROUP

Like other aldehydes, the **carbonyl group of an aldose is reduced to a 1° alcohol** using hydrogen (H_2) in the presence of palladium (Pd) metal (Section 16.6). This alcohol is called an **alditol,** sometimes referred to as a "sugar alcohol." For example, reduction of D-glucose with H_2 and Pd yields glucitol, commonly called sorbitol.

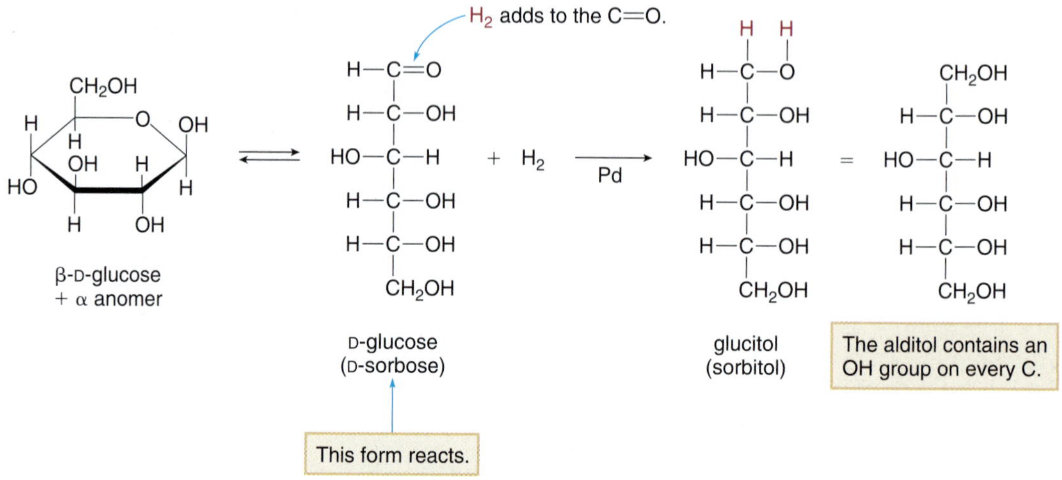

H_2 adds to the C=O.

β-D-glucose
+ α anomer

D-glucose
(D-sorbose)

This form reacts.

glucitol
(sorbitol)

The alditol contains an OH group on every C.

Sorbitol is 60% as sweet as table sugar (sucrose) and contains two-thirds the calories per gram. Sorbitol is used as a sweetening agent in a variety of sugar-free candies and gum.

SAMPLE PROBLEM 20.4

Draw the structure of the compound formed when D-xylose is treated with H_2 in the presence of a Pd catalyst. The product, xylitol, is a sweetener used in sugarless chewing gum and other products.

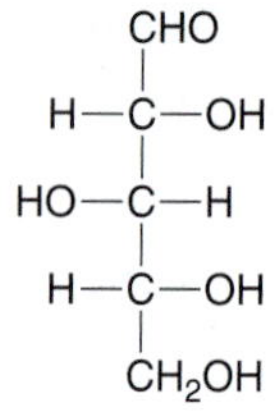

CHO
H—C—OH
HO—C—H
H—C—OH
CH₂OH

D-xylose

ANALYSIS
- Locate the C=O and mentally break one bond in the double bond.
- Mentally break the H—H bond of the reagent and add one H atom to each atom of the C=O, forming new C—H and O—H bonds.

SOLUTION

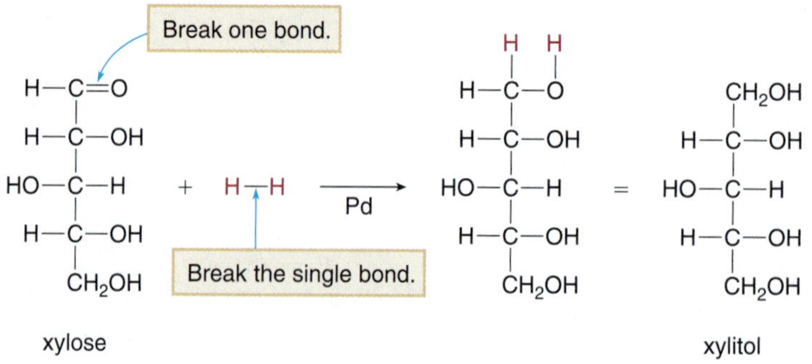

Break one bond.

Break the single bond.

xylose

xylitol

What compound is formed when each aldose is treated with H_2 in the presence of a Pd catalyst?

a.
```
      CHO
       |
  H — C — OH
       |
  H — C — OH
       |
  H — C — OH
       |
     CH₂OH
```

b.
```
      CHO
       |
  H — C — OH
       |
 HO — C — H
       |
 HO — C — H
       |
  H — C — OH
       |
     CH₂OH
```

c.
```
      CHO
       |
  H — C — OH
       |
  H — C — OH
       |
     CH₂OH
```

20.4B OXIDATION OF THE ALDEHYDE CARBONYL GROUP

The aldehyde carbonyl of an aldose is easily oxidized with a variety of reagents to form a carboxyl group, yielding an **aldonic acid.** For example, D-glucose is oxidized with a Cu^{2+} reagent called **Benedict's reagent** to form gluconic acid. In the process, a characteristic color change occurs as the blue Cu^{2+} is reduced to Cu^+, forming brick-red Cu_2O.

```
   H — C = O                          HO — C = O
        |                                  |
   H — C — OH                          H — C — OH
        |                                  |
  HO — C — H     + 2 Cu²⁺   ⎯⎯OH⎯→  HO — C — H     +   Cu₂O
        |          (blue)               |            (brick-red)
   H — C — OH                          H — C — OH
        |                                  |
   H — C — OH                          H — C — OH
        |                                  |
      CH₂OH                             CH₂OH
     D-glucose                        gluconic acid
```

This reaction is an oxidation since the product carboxylic acid has one more C—O bond than the starting aldehyde. Carbohydrates that are oxidized with Benedict's reagent are called **reducing sugars,** because the Cu^{2+} in Benedict's reagent is reduced to Cu^+ during the reaction. Those that do not react with Benedict's reagent are called **nonreducing sugars. All aldoses are reducing sugars.**

Perhaps surprisingly, **ketoses are also reducing sugars.** Although ketoses do not contain an easily oxidized aldehyde, they do contain a ketone carbonyl adjacent to a CH_2OH group that undergoes rearrangement under basic reaction conditions to form an aldose, which is then oxidized with Benedict's reagent. For example, fructose rearranges to an aldose, which is then oxidized to an aldonic acid. Thus, fructose is a reducing sugar.

```
C=O at C2                           C=O at C1
a ketone      CH₂OH                 an aldehyde
                |                        CHO                        COOH
              C=O                         |                          |
                |                    H — C — OH                  H — C — OH
          HO — C — H  ⇌ rearrangement HO — C — H   2 Cu²⁺   HO — C — H    +  Cu₂O
                |                         |          ⎯⎯OH⎯→       |
           H — C — OH                H — C — OH                  H — C — OH
                |                         |                          |
           H — C — OH                H — C — OH                  H — C — OH
                |                         |                          |
              CH₂OH                     CH₂OH                      CH₂OH
             fructose                   aldose                  aldonic acid
```

SAMPLE PROBLEM 20.5

Draw the product formed when each monosaccharide is oxidized with Benedict's reagent.

a. 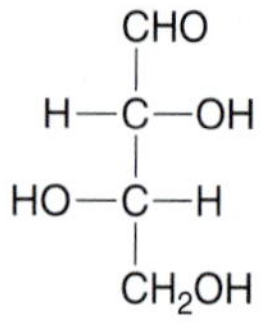
```
        CHO
         |
   H —— C —— OH
         |
  HO —— C —— H
         |
       CH₂OH
```
L-threose

b. 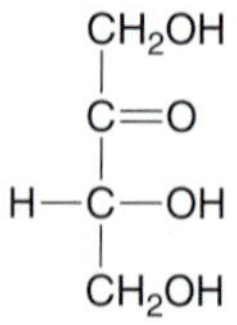
```
       CH₂OH
         |
        C=O
         |
   H —— C —— OH
         |
       CH₂OH
```
D-erythrulose

ANALYSIS   To draw the oxidation product of an aldose, convert the CHO group to COOH. To oxidize a ketose, first rearrange the ketose to an aldose by moving the carbonyl group to C1 to give an aldehyde. Then convert the CHO group to COOH.

SOLUTION   L-Threose in part (a) is an aldose, so its oxidation product can be drawn directly. D-Erythrulose in part (b) is a ketose, so it undergoes rearrangement prior to oxidation.

a. 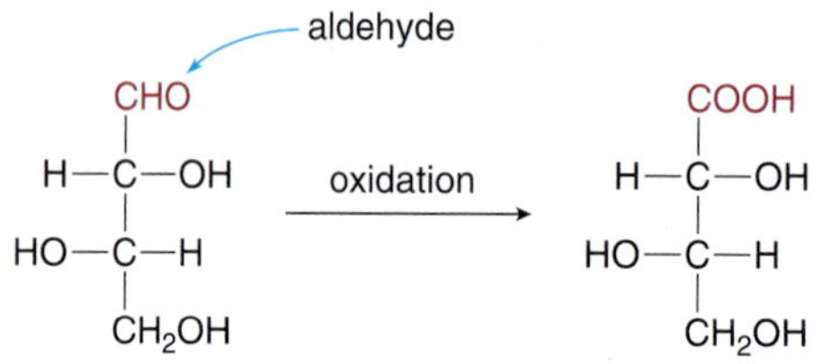
```
                aldehyde
        CHO                              COOH
         |                                |
   H —— C —— OH      oxidation      H —— C —— OH
         |          ───────►             |
  HO —— C —— H                    HO —— C —— H
         |                                |
       CH₂OH                            CH₂OH
```
L-threose

b. 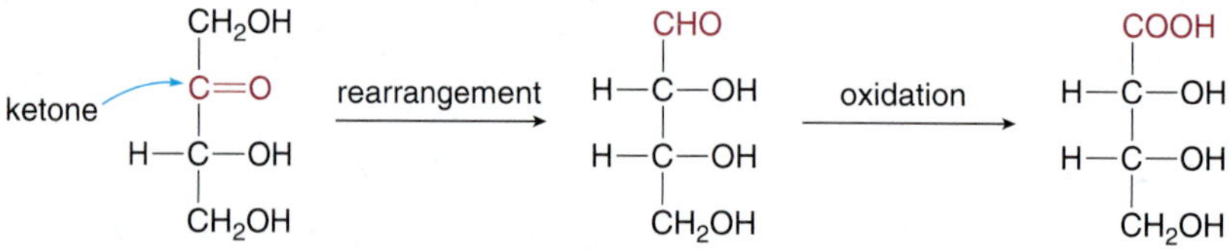
```
        CH₂OH                       CHO                          COOH
         |                           |                            |
   ketone  C=O    rearrangement  H —— C —— OH   oxidation    H —— C —— OH
         |         ───────►           |         ───────►         |
   H —— C —— OH                 H —— C —— OH                H —— C —— OH
         |                           |                            |
       CH₂OH                       CH₂OH                        CH₂OH
```
D-erythrulose

PROBLEM 20.15

What aldonic acid is formed by oxidation of each monosaccharide?

a. 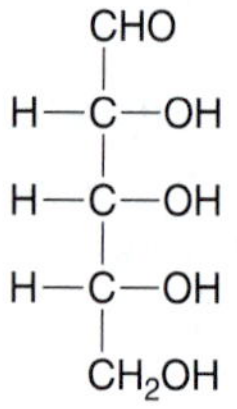
```
        CHO
         |
   H —— C —— OH
         |
   H —— C —— OH
         |
   H —— C —— OH
         |
       CH₂OH
```

b.
```
        CHO
         |
   H —— C —— OH
         |
   H —— C —— OH
         |
       CH₂OH
```

c.
```
       CH₂OH
         |
        C=O
         |
  HO —— C —— H
         |
  HO —— C —— H
         |
   H —— C —— OH
         |
       CH₂OH
```

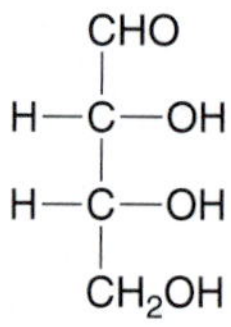

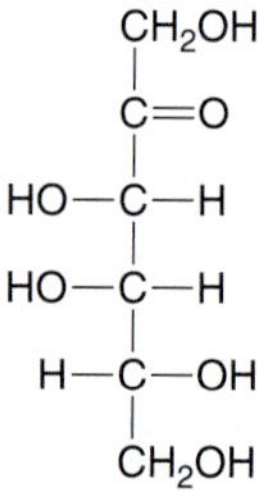

PROBLEM 20.16

What product is formed when D-arabinose is treated with each reagent: (a) H₂, Pd; (b) Benedict's reagent?

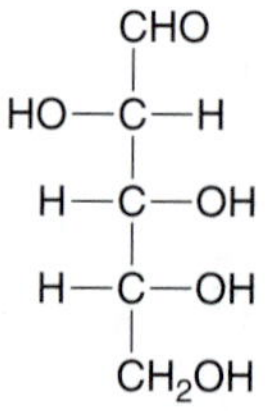

```
        CHO
         |
  HO —— C —— H
         |
   H —— C —— OH
         |
   H —— C —— OH
         |
       CH₂OH
```
D-arabinose

HEALTH NOTE

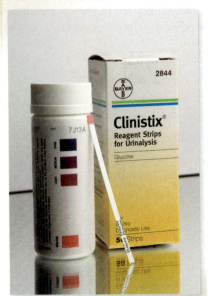

Test strips that contain glucose oxidase are used to measure glucose concentration in urine.

20.4C FOCUS ON HEALTH & MEDICINE
MONITORING GLUCOSE LEVELS

In order to make sure that their blood glucose levels are in the proper range, individuals with diabetes frequently measure the concentration of glucose in their blood. A common method for carrying out this procedure today involves the oxidation of glucose to gluconic acid using the enzyme glucose oxidase.

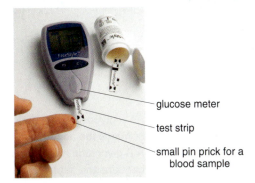

$$
\begin{array}{ccc}
\text{CHO} & & \text{COOH} \\
\text{H—C—OH} & & \text{H—C—OH} \\
\text{HO—C—H} & +\ O_2 \xrightarrow{\text{glucose oxidase}} & \text{HO—C—H} \quad +\ H_2O_2 \\
\text{H—C—OH} & & \text{H—C—OH} \\
\text{H—C—OH} & & \text{H—C—OH} \\
\text{CH}_2\text{OH} & & \text{CH}_2\text{OH} \\
\text{D-glucose} & & \text{gluconic acid}
\end{array}
$$

In the presence of glucose oxidase, oxygen (O_2) in the air oxidizes the aldehyde of glucose to a carboxyl group. The O_2, in turn, is reduced to hydrogen peroxide, H_2O_2. In the first generation of meters for glucose monitoring, the H_2O_2 produced in this reaction was allowed to react with another organic compound to produce a colored product. The intensity of the colored product was then correlated to the amount of glucose in the blood. Test strips used for measuring glucose concentration in the urine are still based on this technology.

Modern glucose meters are electronic devices that measure the amount of oxidizing agent that reacts with a known amount of blood (Figure 20.2). This value is correlated with blood glucose concentration and the result is displayed digitally. A high blood glucose level may mean that an individual needs more insulin, while a low level may mean that it is time to ingest some calories.

▼ **FIGURE 20.2 Monitoring Blood Glucose Levels**

- glucose meter
- test strip
- small pin prick for a blood sample

A small drop of blood is placed on a disposable test strip that is inserted in an electronic blood glucose meter, and the glucose concentration is read on a digital display.

20.5 DISACCHARIDES

Disaccharides are carbohydrates composed of two monosaccharides. Disaccharides are acetals, compounds that contain two alkoxy groups (OR groups) bonded to the same carbon. Recall from Section 16.8 that reaction of a hemiacetal with an alcohol forms an **acetal.**

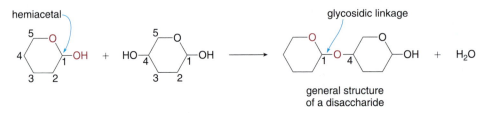

In a similar fashion, a disaccharide results when a hemiacetal of one monosaccharide reacts with a hydroxyl group of a second monosaccharide to form an acetal. The new C—O bond that joins the two rings together is called a **glycosidic linkage.**

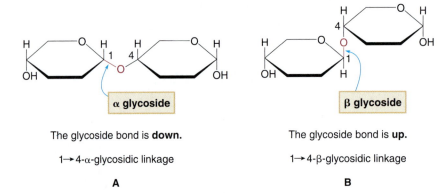

[The acetal O's are labeled in red.]

The two monosaccharide rings may be five-membered or six-membered. All disaccharides contain at least one acetal that joins the rings together. Each ring is numbered beginning at the anomeric carbon, the carbon in each ring bonded to two oxygen atoms.

The glycosidic linkage that joins the two monosaccharides in a disaccharide can be oriented in two different ways, shown with Haworth projections in structures **A** and **B**.

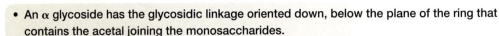

- An α glycoside has the glycosidic linkage oriented down, below the plane of the ring that contains the acetal joining the monosaccharides.
- A β glycoside has the glycosidic linkage oriented up, above the plane of the ring that contains the acetal joining the monosaccharides.

Numbers are used to designate which ring atoms are joined in the disaccharide. Disaccharide **A** has a **1→4-α-glycosidic linkage** since the glycoside bond is oriented down and joins C1 of one ring to C4 of the other. Disaccharide **B** has a **1→4-β-glycosidic linkage** since the glycoside bond is oriented up and joins C1 of one ring to C4 of the other.

Sample Problem 20.6 illustrates these structural features in the disaccharide **maltose.** Maltose, which is formed by the hydrolysis of starch, is found in grains such as barley. Maltose is formed from two molecules of glucose.

Maltose gets its name from malt, the liquid obtained from barley used in the brewing of beer.

maltose

SAMPLE PROBLEM 20.6   (a) Locate the glycosidic linkage in maltose. (b) Number the carbon atoms in both rings. (c) Classify the glycosidic linkage as α or β, and use numbers to designate its location.

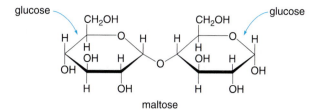

maltose

ANALYSIS AND SOLUTION   **a. and b.** The glycosidic linkage is the acetal C—O bond that joins the two monosaccharides, labeled in red. Each ring is numbered beginning at the anomeric carbon, the carbon bonded to two oxygen atoms.

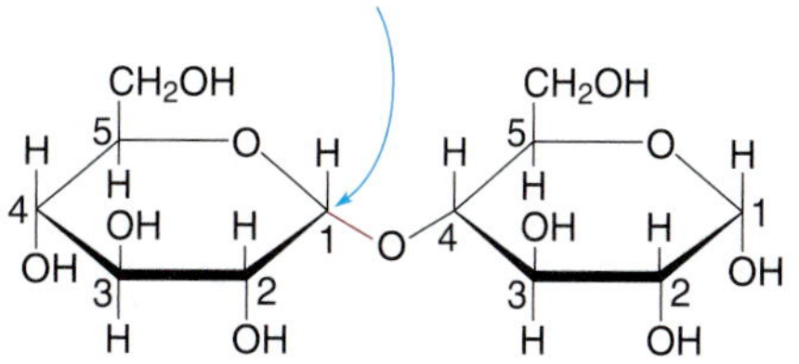

acetal carbon

c. Maltose has an α glycosidic linkage since the C—O bond is drawn down. The glycosidic linkage joins C1 of one ring to C4 of the other ring, so maltose contains a 1→4-α-glycosidic linkage.

PROBLEM 20.17   (a) Locate the glycosidic linkage in cellobiose. (b) Number the carbon atoms in both rings. (c) Classify the glycosidic linkage as α or β, and use numbers to designate its location.

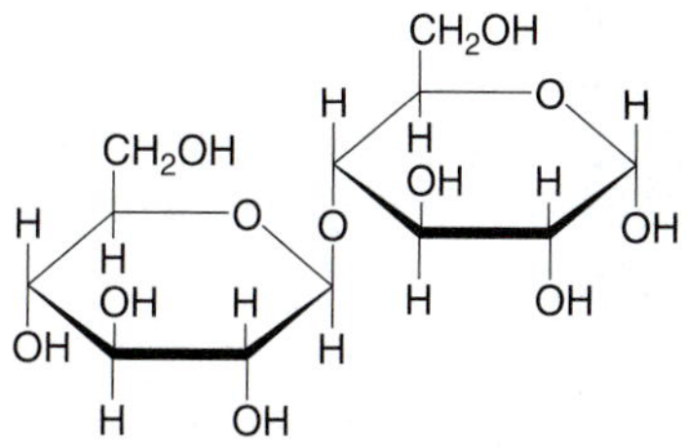

cellobiose

The hydrolysis of a disaccharide cleaves the C—O glycosidic linkage and forms two monosaccharides. For example, hydrolysis of maltose yields two molecules of glucose.

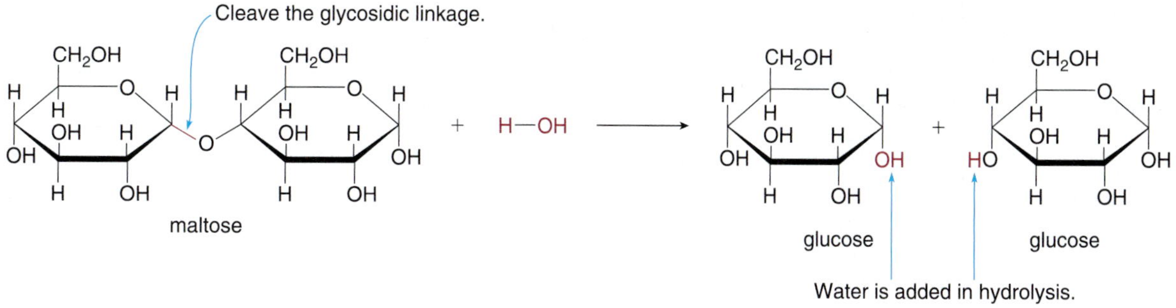

Cleave the glycosidic linkage.

maltose   +   H—OH   ⟶   glucose   +   glucose

Water is added in hydrolysis.

SAMPLE PROBLEM 20.7   Draw the products formed when the following disaccharide is hydrolyzed with water.

ANALYSIS   Locate the glycosidic linkage, the acetal C—O bond that joins the two monosaccharides. Cleave the C—O bond by adding the elements of H_2O across the bond.

SOLUTION   Two monosaccharides are formed by cleaving the glycosidic linkage.

PROBLEM 20.18   What monosaccharides are formed when cellobiose (Problem 20.17) is hydrolyzed with water?

Individuals who are lactose intolerant can drink lactose-free milk. Tablets that contain the lactase enzyme can also be taken when ice cream or other milk products are ingested.

20.5A   FOCUS ON HEALTH & MEDICINE
LACTOSE INTOLERANCE

Lactose is the principal disaccharide found in milk from both humans and cows. Unlike many mono- and disaccharides, lactose is not appreciably sweet. Lactose consists of one galactose ring and one glucose ring, joined by a 1→4-β-glycoside bond from the anomeric carbon of galactose to C4 of glucose.

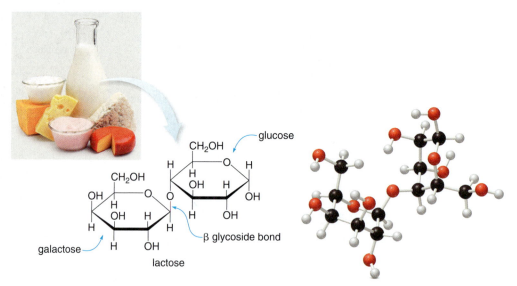

lactose

Lactose is digested in the body by first cleaving the 1→4-β-glycoside bond using the enzyme *lactase*. **Individuals who are lactose intolerant no longer produce this enzyme,** and so lactose cannot be properly digested, causing abdominal cramps and diarrhea. Lactose intolerance is especially prevalent in Asian and African populations whose diets have not traditionally included milk beyond infancy.

PROBLEM 20.19

Lactose contains both an acetal and a hemiacetal. Label both functional groups. What products are formed when lactose is hydrolyzed with water?

20.5B FOCUS ON HEALTH & MEDICINE
SUCROSE AND ARTIFICIAL SWEETENERS

Sucrose, the disaccharide found in sugarcane and the compound generally referred to as "sugar," is the most common disaccharide in nature. It contains one glucose ring and one fructose ring. Unlike maltose and lactose, which contain only six-membered rings, sucrose contains one six-membered and one five-membered ring.

sugarcane

two varieties of refined sugar

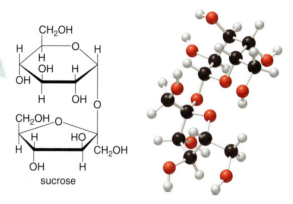

sucrose

TABLE 20.1 Relative Sweetness of Some Carbohydrates and Artificial Sweeteners

Compound	Relative Sweetness
Sorbitol	0.60
Glucose	0.75
Sucrose	1.00
Fructose	1.75
Aspartame	150
Saccharin	350
Sucralose	600

Sucrose's pleasant sweetness has made it a widely used ingredient in baked goods, cereals, bread, and many other products. It is estimated that the average American ingests 100 lb of sucrose annually. Like other carbohydrates, however, sucrose contains many calories. To reduce caloric intake while maintaining sweetness, a variety of artificial sweeteners have been developed. These include aspartame, saccharin, and sucralose (Figure 20.3). These compounds are much sweeter than sucrose, so only a small amount of each compound is needed to achieve the same level of perceived sweetness.

A relative sweetness scale ranks the sweetness of carbohydrates and synthetic sweeteners, as shown in Table 20.1.

Aspartame (trade name: Nutrasweet) is used in the artificial sweetener Equal and many diet beverages. While harmless to most individuals, aspartame is hydrolyzed in the body to the amino acid phenylalanine (and other products). Infants afflicted with phenylketonuria cannot metabolize this amino acid, so it accumulates, causing mental retardation. When this condition is identified early, a diet limiting the consumption of phenylalanine (and compounds like aspartame that are converted to it) can make a normal life possible.

PROBLEM 20.20

Unlike lactose, which contains one acetal and one hemiacetal, sucrose contains two acetals. Locate the acetals in sucrose. Classify each alcohol as 1°, 2°, or 3°.

PROBLEM 20.21

Sucralose is prepared from sucrose. Locate the acetals in sucralose. Label each alkyl halide as 1°, 2°, or 3°.

PROBLEM 20.22

Identify the functional groups in aspartame.

▼ **FIGURE 20.3** Artificial Sweeteners

sucralose
(Trade name: Splenda)

aspartame
(Trade name: Equal)

saccharin
(Trade name: Sweet'n Low)

The sweetness of these artificial sweeteners was discovered accidentally. The sweetness of sucralose was discovered in 1976 when a chemist misunderstood his superior, and so he *tasted* rather than *tested* his compound. Aspartame was discovered in 1965 when a chemist licked his dirty fingers in the lab and tasted its sweetness. Saccharin, the oldest known artificial sweetener, was discovered in 1879 by a chemist who failed to wash his hands after working in the lab. Saccharin was not used extensively until sugar shortages occurred during World War I. Although there were concerns in the 1970s that saccharin causes cancer, there is no proven link between cancer occurrence and saccharin intake at normal levels.

20.6 POLYSACCHARIDES

Polysaccharides contain three or more monosaccharides joined together. Three prevalent polysaccharides in nature are **cellulose, starch,** and **glycogen,** each of which consists of repeating glucose units joined by glycosidic bonds.

β glycosidic linkage

cellulose—repeating structure

α glycosidic linkage

starch and glycogen—repeating structure

- Cellulose contains glucose rings joined in 1→4-β-glycosidic linkages.
- Starch and glycogen contain glucose rings joined in 1→4-α-glycosidic linkages.

20.6A   CELLULOSE

Cellulose is found in the cell walls of nearly all plants, where it gives support and rigidity to wood, plant stems, and grass (Figure 20.4). Wood, cotton, and flax are composed largely of cellulose.

Cellulose is an unbranched polymer composed of repeating glucose units joined in a 1→4-β-glycosidic linkage. The β glycosidic linkages create long linear chains of cellulose molecules that stack in sheets, making an extensive three-dimensional array.

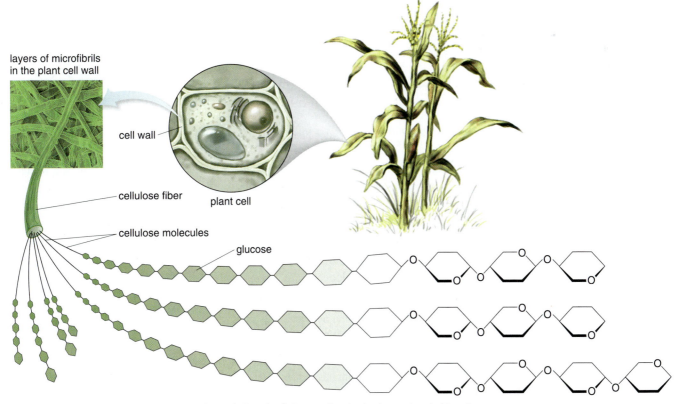

cellulose

β glycosidic linkage
(shown in blue)

In some cells, cellulose is hydrolyzed by an enzyme called a **β-glycosidase,** which cleaves all of the β glycoside bonds, forming glucose. **Humans do not possess this enzyme, and therefore**

▼ FIGURE 20.4   Cellulose

layers of microfibrils
in the plant cell wall

cell wall

cellulose fiber   plant cell

cellulose molecules

glucose

long chains of cellulose molecules hydrogen bonded together

cannot **digest cellulose.** Ruminant animals, on the other hand, such as cattle, deer, and camels, have bacteria containing this enzyme in their digestive systems, so they can derive nutritional benefit from eating grass and leaves.

Much of the insoluble fiber in our diet is cellulose, which passes through the digestive system without being metabolized. Foods rich in cellulose include whole wheat bread, brown rice, and bran cereals. Fiber is an important component of the diet even though it gives us no nutrition; fiber adds bulk to solid waste, so that it is eliminated more readily.

20.6B STARCH

Starch is the main carbohydrate found in the seeds and roots of plants. Corn, rice, wheat, and potatoes are common foods that contain a great deal of starch. **Starch is a polymer composed of repeating glucose units joined in α glycosidic linkages.** The two common forms of starch are **amylose** and **amylopectin.**

α glycosidic linkage
(shown in red)

amylose
(the linear form of starch)

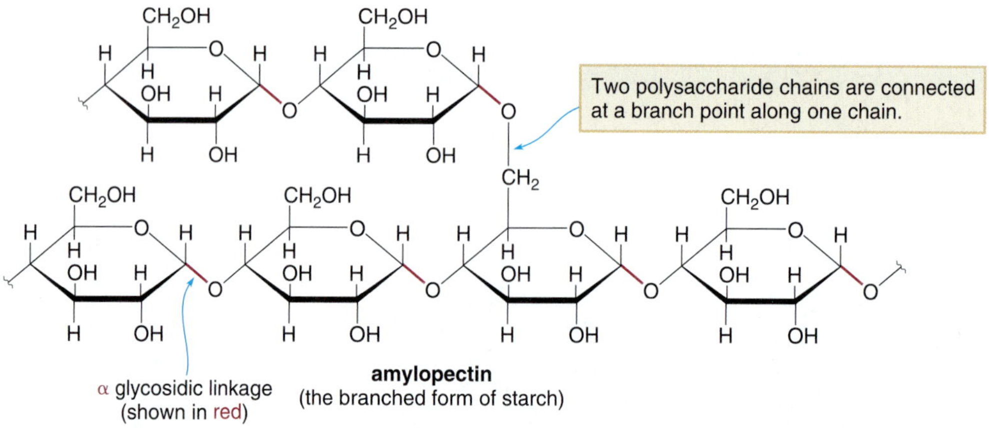

Two polysaccharide chains are connected at a branch point along one chain.

α glycosidic linkage
(shown in red)

amylopectin
(the branched form of starch)

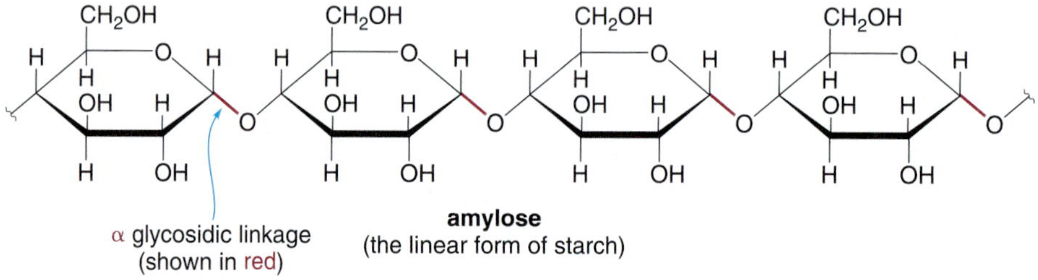

Amylose, which comprises about 20% of starch molecules, has an unbranched skeleton of glucose molecules with **1→4-α-glycoside bonds.** Because of this linkage, an amylose chain adopts a helical arrangement, giving it a very different three-dimensional shape from the linear chains of cellulose (Figure 20.5).

Amylopectin, which comprises about 80% of starch molecules, consists of a backbone of glucose units joined in **α glycosidic bonds,** but it also contains considerable branching along the chain. The linear linkages of amylopectin are formed by **1→4-α-glycoside bonds,** similar to amylose.

Both forms of starch are water soluble. Since the OH groups in these starch molecules are not buried in a three-dimensional network, they are available for hydrogen bonding with water molecules, leading to greater water solubility than cellulose.

Both amylose and amylopectin are hydrolyzed to glucose with cleavage of the glycosidic bonds. The human digestive system has the necessary amylase enzymes needed to catalyze this process. Bread and pasta made from wheat flour, rice, and corn tortillas are all sources of starch that are readily digested.

▼ **FIGURE 20.5** Starch—Amylose and Amylopectin

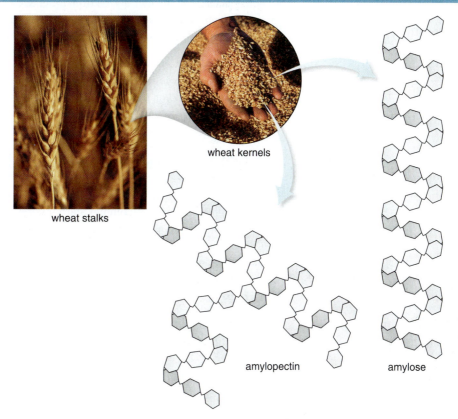

wheat kernels

wheat stalks

amylopectin

amylose

PROBLEM 20.23

Label all acetal carbons in the structure of amylopectin drawn at the beginning of Section 20.6B.

20.6C GLYCOGEN

Glycogen is the major form in which polysaccharides are stored in animals. Glycogen, a polymer of glucose containing **α glycosidic bonds,** has a branched structure similar to amylopectin, but the branching is much more extensive (Figure 20.6).

▼ **FIGURE 20.6** Glycogen

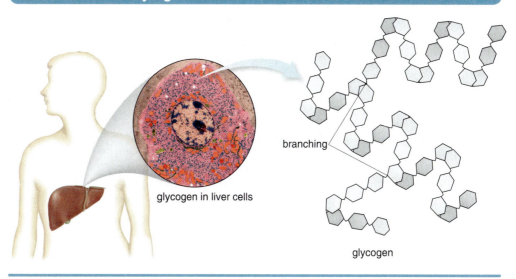

branching

glycogen in liver cells

glycogen

Glycogen is stored principally in the liver and muscle. When glucose is needed for energy in the cell, glucose units are hydrolyzed from the ends of the glycogen polymer, and then further metabolized with the release of energy. Because glycogen has a highly branched structure, there are many glucose units at the ends of the branches that can be cleaved whenever the body needs them.

PROBLEM 20.24

Cellulose is water *in*soluble, despite its many OH groups. Based on its three-dimensional structure, why do you think this is so?

20.7 FOCUS ON THE HUMAN BODY
USEFUL CARBOHYDRATE DERIVATIVES

Many other simple and complex carbohydrates with useful properties exist in the biological world. Several are derived from monosaccharides that contain an amino (NH_2) or amide ($NHCOCH_3$) group in place of an OH group. Examples include D-glucosamine, the most abundant amino sugar in nature, and *N*-acetyl-D-glucosamine (NAG). Other carbohydrates are derived from D-glucuronate, which contains a carboxylate anion, COO^-, in place of the CH_2OH group of the typical monosaccharide skeleton.

HEALTH NOTE

Injections of heparin in an intravenous line keep the line from clogging with blood.

D-glucosamine   *N*-acetyl-D-glucosamine   D-glucuronate

20.7A GLYCOSAMINOGLYCANS

Glycosaminoglycans (GAGs) are a group of unbranched carbohydrates derived from alternating amino sugar and glucuronate units. Glycosaminoglycans form a gel-like matrix that acts as a lubricant, making them key components in connective tissue and joints.

Examples include **hyaluronate,** which is found in the extracellular fluid that lubricates joints and the vitreous humor of the eye; **chondroitin,** a component of cartilage and tendons; and **heparin,** which is stored in the mast cells of the liver and other organs and prevents blood clotting (Figure 20.7). While the monosaccharide rings of hyaluronate and chondroitin are joined by β glycosidic linkages (shown in blue), those of heparin are joined by α glycosidic linkages (shown in red).

PROBLEM 20.25

Classify the glycosidic linkages in chondroitin and heparin (Figure 20.7) as α or β, and use numbers to designate their location.

20.7B CHITIN

Chitin, the second most abundant carbohydrate polymer, is a polysaccharide formed from *N*-acetyl-D-glucosamine units joined together in **1→4-β-glycosidic linkages.** Chitin is identical in structure to cellulose, except that each OH group at C2 is now replaced by $NHCOCH_3$. The exoskeletons of lobsters, crabs, and shrimp are composed of chitin. Like cellulose, chitin chains are held together by an extensive network of hydrogen bonds, forming water-insoluble sheets.

HEALTH NOTE

Both chondroitin and glucosamine are sold as dietary supplements for individuals suffering from osteoarthritis. On-going research is examining the role of these supplements in replacing and rebuilding lost joint cartilage in the hope of halting or reversing the progression of arthritis.

▼ FIGURE 20.7 Glycosaminoglycans

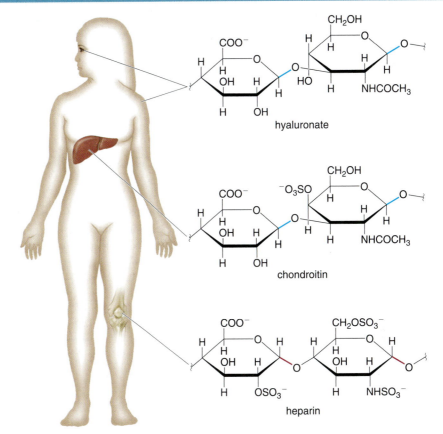

Hyaluronates form viscous solutions that act as lubricants in the fluid around joints. They also give the vitreous humor of the eye its gelatin-like consistency. Chondroitin strengthens cartilage, tendons, and the walls of blood vessels. Heparin is an anticoagulant.

The rigidity of a crab shell is due to chitin, a high molecular weight carbohydrate molecule. Much of the D-glucosamine sold in over-the-counter dietary supplements comes from shellfish.

Chitin-based coatings have found several commercial applications, such as extending the shelf life of fruits. Processing plants now convert the shells of crabs, lobsters, and shrimp to chitin and various derivatives for use in many consumer products. Complete hydrolysis of the glycoside and amide bonds in chitin forms the dietary supplement D-glucosamine.

PROBLEM 20.26

Suppose that chitin contained α glycosidic linkages. Draw a portion of the resulting polysaccharide that contains four *N*-acetyl-D-glucosamine units joined together.

20.8 FOCUS ON THE HUMAN BODY
BLOOD TYPE

Human blood is classified into one of four types using the ABO system discovered in the early 1900s by Karl Landsteiner. There are four blood types—A, B, AB, and O. An individual's blood type is determined by three or four monosaccharides attached to a membrane protein of red blood cells. These monosaccharides include:

D-galactose L-fucose *N*-acetyl-D-glucosamine *N*-acetyl-D-galactosamine

Each blood type is associated with a different carbohydrate structure, as shown in Figure 20.8. Three monosaccharides occur in all blood types—*N*-acetyl-D-glucosamine, D-galactose, and L-fucose. Type A blood contains a fourth monosaccharide, *N*-acetyl-D-galactosamine, and type B blood contains an additional D-galactose unit. Type AB blood has both type A and type B carbohydrates.

The short polysaccharide chains distinguish one type of red blood cell from another, and signal the cells about foreign viruses, bacteria, and other agents. When a foreign substance enters the blood, the body's immune system uses antibodies to attack and destroy the invading substance so that it does the host organism no harm.

Knowing an individual's blood type is necessary before receiving a blood transfusion. Because the blood of an individual may contain antibodies to another blood type, the types of blood that can be given to a patient are often limited. An individual with blood type A produces antibodies to type B blood, and an individual with blood type B produces antibodies to type A blood. Type AB blood contains no antibodies to other blood types, while type O blood contains antibodies to both types A and B. As a result:

- Individuals with type O blood are called *universal donors* because no antibodies to type O are produced by those with types A, B, and AB blood. Type O blood can be given to individuals of any blood type.
- Individuals with type AB blood are called *universal recipients* because their blood contains no antibodies to blood types A, B, or O. Individuals with type AB blood can receive blood of any type.

Table 20.2 lists what blood types can be safely given to an individual. Blood must be carefully screened to make sure that the blood types of the donor and recipient are compatible. Should the wrong blood type be administered, antibodies of the immune system will attack the foreign red blood cells, causing them to clump together, which can block blood vessels and even result in death.

HEALTH NOTE

The blood type of a blood donor and recipient must be compatible, so donated blood is clearly labeled with the donor's blood type.

▼ **FIGURE 20.8** Carbohydrates and Blood Types

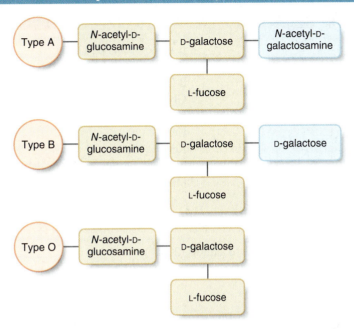

Each blood type is characterized by a different polysaccharide that is covalently bonded to a membrane protein of the red blood cell. There are three different carbohydrate sequences, one each for A, B, and O blood types. Blood type AB contains the sequences for both blood type A and blood type B.

TABLE 20.2   Compatibility Chart of Blood Types

Blood Type	Can Receive Blood Type:	Can Donate to Someone of Blood Type:
A	A, O	A, AB
B	B, O	B, AB
AB	A, B, AB, O	AB
O	O	A, B, AB, O

PROBLEM 20.27 List two structural features that distinguish L-fucose from the other monosaccharides we have seen in naturally occurring molecules.

PROBLEM 20.28 How do *N*-acetyl-D-glucosamine and *N*-acetyl-D-galactosamine differ in structure? Are these two compounds constitutional isomers or stereoisomers? Explain your choice.

CHAPTER HIGHLIGHTS

KEY TERMS

Alditol (20.4)
Aldonic acid (20.4)
Aldose (20.2)
α Anomer (20.3)
β Anomer (20.3)
Anomeric carbon (20.3)
Benedict's reagent (20.4)
Carbohydrate (20.1)
Disaccharide (20.1)

D Monosaccharide (20.2)
α Glycoside (20.5)
β Glycoside (20.5)
Glycosidic linkage (20.5)
Haworth projection (20.3)
Hexose (20.2)
Ketose (20.2)
L Monosaccharide (20.2)
Monosaccharide (20.1)

Mutarotation (20.3)
Nonreducing sugar (20.4)
Pentose (20.2)
Polysaccharide (20.1)
Reducing sugar (20.4)
Tetrose (20.2)
Triose (20.2)

KEY REACTIONS

[1] Reduction of monosaccharides to alditols (20.4A)

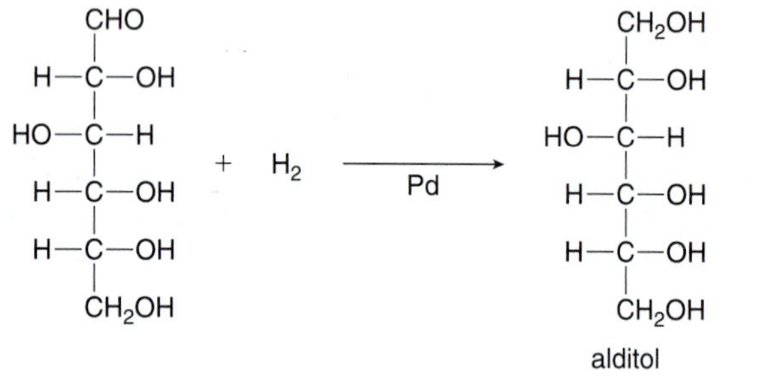

[2] Oxidation of monosaccharides to aldonic acids (20.4B)

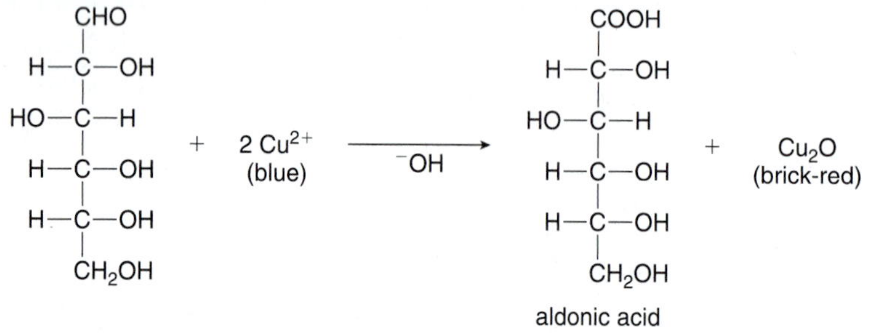

[3] Hydrolysis of disaccharides (20.5)

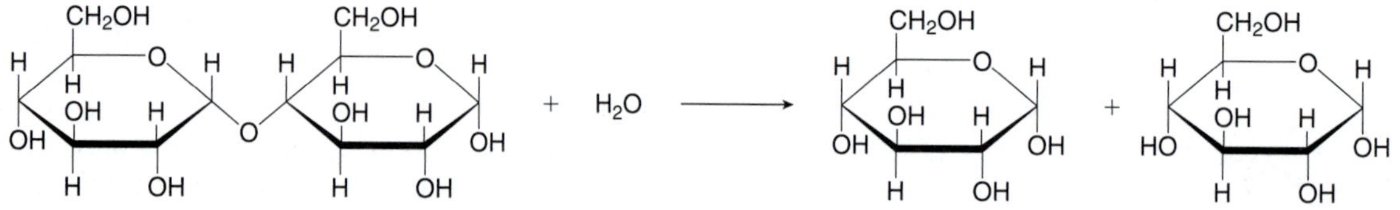

KEY CONCEPTS

① What are the three major types of carbohydrates? (20.1)

- Monosaccharides, which cannot be hydrolyzed to simpler compounds, generally have three to six carbons with a carbonyl group at either the terminal carbon or the carbon adjacent to it. Generally, all other carbons have OH groups bonded to them.
- Disaccharides are composed of two monosaccharides.
- Polysaccharides are composed of three or more monosaccharides.

② What are the major structural features of monosaccharides? (20.2)

- Monosaccharides with a carbonyl group at C1 are called aldoses and those with a carbonyl at C2 are called ketoses. Generally, OH groups are bonded to every other carbon. The terms triose, tetrose, and so forth are used to indicate the number of carbons in the chain.
- The acyclic form of monosaccharides is drawn with Fischer projection formulas. A D sugar has the OH group of the chirality center farthest from the carbonyl on the right side. An L sugar has the OH group of the chirality center farthest from the carbonyl on the left side.

③ How are the cyclic forms of monosaccharides drawn? (20.3)

- In aldohexoses the OH group on C5 reacts with the aldehyde carbonyl to give two cyclic hemiacetals called anomers. The acetal carbon is called the anomeric carbon. The α anomer has the OH group drawn down for a D sugar and the β anomer has the OH group drawn up.

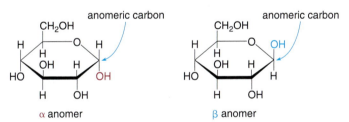

④ What reduction and oxidation products are formed from monosaccharides? (20.4)

- Monosaccharides are reduced to alditols with H_2 and Pd.
- Monosaccharides are oxidized to aldonic acids with Benedict's reagent. Sugars that are oxidized with Benedict's reagent are called reducing sugars.

⑤ What are the major structural features of disaccharides? (20.5)

- Disaccharides contain two monosaccharides joined by an acetal C—O bond called a glycosidic linkage. An α glycoside has the glycosidic linkage oriented down and a β glycoside has the glycosidic linkage oriented up.

- Disaccharides are hydrolyzed to two monosaccharides by the cleavage of the glycosidic C—O bond.
- Lactose (the principal disaccharide in milk) and sucrose (table sugar) are common disaccharides.

⑥ What are the differences in the polysaccharides cellulose, starch, and glycogen? (20.6)

- Cellulose, starch, and glycogen are all polymers of the monosaccharide glucose.
- Cellulose is an unbranched polymer composed of repeating glucose units joined in 1→4-β-glycosidic linkages. Cellulose forms long chains that stack in three-dimensional sheets. The human digestive system does not contain the needed enzyme to metabolize cellulose.
- There are two forms of starch—amylose, which is an unbranched polymer, and amylopectin, which is a branched polysaccharide polymer. Both forms contain 1→4-α-glycosidic linkages, and the polymer winds in a helical arrangement. Starch is digestible since the human digestive system has the needed amylase enzyme to catalyze hydrolysis.
- Glycogen resembles amylopectin but is more extensively branched. Glycogen is the major form in which polysaccharides are stored in animals.

⑦ Give examples of some carbohydrate derivatives that contain amino groups, amides, or carboxylate anions. (20.7)

- Glycosaminoglycans are a group of unbranched carbohydrates derived from amino sugar and glucuronate units. Examples include hyaluronate, which forms a gel-like matrix in joints and the vitreous humor of the eye; chondroitin, which is a component of cartilage and tendons; and heparin, an anticoagulant.
- Chitin, a polymer of N-acetyl-D-glucosamine, forms the hard exoskeletons of crabs, lobsters, and shrimp.

⑧ What role do carbohydrates play in determining blood type? (20.8)

- Human blood type—A, B, AB, or O—is determined by three or four monosaccharides attached to a membrane protein on the surface of red blood cells. There are three different carbohydrate sequences, one for each of the A, B, and O blood types. Blood type AB contains the sequences for both blood type A and blood type B. Since the blood of an individual may contain antibodies to another blood type, blood type must be known before receiving a transfusion.

PROBLEMS

Selected in-chapter and end-of-chapter problems have brief answers provided in Appendix B.

Monosaccharides

20.29   What is the difference between an aldose and a ketose? Give an example of each type of carbohydrate.

20.30   What is the difference between a tetrose and a pentose? Give an example of each type of carbohydrate.

20.31   Draw the structure of each type of carbohydrate.
　　　a. an L-aldopentose　　　c. a five-carbon alditol
　　　b. a D-aldotetrose

20.32   Draw the structure of each type of carbohydrate.
　　　a. a D-aldotriose　　　c. a four-carbon aldonic acid
　　　b. an L-ketohexose

20.33   What is the difference between a Fischer projection and a Haworth projection?

20.34   What is the difference between an α anomer and a β anomer?

20.35   Are α-D-glucose and β-D-glucose enantiomers? Explain your choice.

20.36   Are D-fructose and L-fructose enantiomers? Explain your choice.

20.37   Classify each monosaccharide by the type of carbonyl group and the number of carbons in the chain.

a.
```
       CHO
        |
  HO—C—H
        |
  HO—C—H
        |
      CH₂OH
```

b.
```
       CHO
        |
  HO—C—H
        |
  HO—C—H
        |
  HO—C—H
        |
   H—C—OH
        |
      CH₂OH
```

c.
```
      CH₂OH
        |
       C=O
        |
   H—C—OH
        |
  HO—C—H
        |
      CH₂OH
```

20.38   Classify each monosaccharide by the type of carbonyl group and the number of carbons in the chain.

a.
```
       CHO
        |
  HO—C—H
        |
      CH₂OH
```

b.
```
       CHO
        |
   H—C—OH
        |
  HO—C—H
        |
  HO—C—H
        |
      CH₂OH
```

c.
```
       CHO
        |
   H—C—OH
        |
   H—C—OH
        |
  HO—C—H
        |
  HO—C—H
        |
      CH₂OH
```

20.39   For each compound in Problem 20.37: [1] label all the chirality centers; [2] classify the compound as a D or L monosaccharide; [3] draw the enantiomer; [4] draw a Fischer projection.

20.40   For each compound in Problem 20.38: [1] label all the chirality centers; [2] classify the compound as a D or L monosaccharide; [3] draw the enantiomer; [4] draw a Fischer projection.

20.41   Consider monosaccharides **A**, **B**, and **C**.

A
```
       CHO
        |
   H—C—OH
        |
   H—C—OH
        |
      CH₂OH
```

B
```
       CHO
        |
   H—C—OH
        |
  HO—C—H
        |
      CH₂OH
```

C
```
      CH₂OH
        |
       C=O
        |
   H—C—OH
        |
      CH₂OH
```

a. Which two monosaccharides are stereoisomers?
b. Identify two compounds that are constitutional isomers.
c. Draw the enantiomer of **B**.
d. Draw a Fischer projection for **A**.

20.42   Consider monosaccharides **D**, **E**, and **F**.

D
```
       CHO
        |
   H—C—OH
        |
   H—C—OH
        |
   H—C—OH
        |
      CH₂OH
```

E
```
      CH₂OH
        |
       C=O
        |
  HO—C—H
        |
  HO—C—H
        |
      CH₂OH
```

F
```
       CHO
        |
  HO—C—H
        |
  HO—C—H
        |
   H—C—OH
        |
      CH₂OH
```

a. Which two monosaccharides are stereoisomers?
b. Identify two compounds that are constitutional isomers.
c. Draw the enantiomer of **F**.
d. Draw a Fischer projection for **D**.

20.43   Using Haworth projections, draw the α and β anomers of the following D monosaccharide.

```
       CHO
        |
  HO—C—H
        |
   H—C—OH
        |
  HO—C—H
        |
   H—C—OH
        |
      CH₂OH
```

20.44   Using Haworth projections, draw the α and β anomers of the following D monosaccharide.

```
       CHO
        |
   H—C—OH
        |
   H—C—OH
        |
  HO—C—H
        |
   H—C—OH
        |
      CH₂OH
```

20.45 Consider the following cyclic monosaccharide.

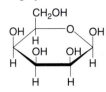

a. Label the hemiacetal carbon.
b. Label the monosaccharide as an α or β anomer.
c. Draw the other anomer.
d. Draw a stereoisomer that differs in the arrangement of substituents at C2.

20.46 Consider the following cyclic monosaccharide.

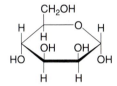

a. Label the hemiacetal carbon.
b. Label the monosaccharide as an α or β anomer.
c. Draw the other anomer.
d. Draw a stereoisomer that differs in the arrangement of substituents at C3.

20.47 For each monosaccharide: [1] label all acetal and hemiacetal carbons; [2] label the anomeric carbon; [3] designate the compound as an α or β anomer.

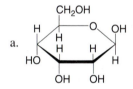 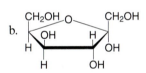

20.48 For each monosaccharide: [1] label all acetal and hemiacetal carbons; [2] label the anomeric carbon; [3] designate the compound as an α or β anomer.

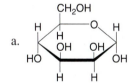 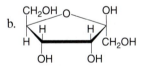

20.49 Why are no D, L labels used for a ketotriose?

20.50 Explain how the labels D and L are assigned in a monosaccharide with three chirality centers.

20.51 Aldopentoses exist as five-membered rings in their cyclic forms. For each aldopentose: [1] label the hemiacetal carbon; [2] designate the monosaccharide as an α or β anomer.

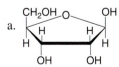 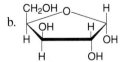

20.52 Aldopentoses exist as five-membered rings in their cyclic forms. For each aldopentose: [1] label the hemiacetal carbon; [2] designate the monosaccharide as an α or β anomer.

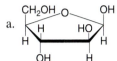 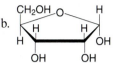

Reactions of Carbohydrates

20.53 Draw the structures of an alditol and an aldonic acid that contain four carbons.

20.54 Draw the structures of an alditol and an aldonic acid that contain six carbons.

20.55 Draw the organic products formed when each monosaccharide is treated with each of the following reagents: [1] H_2, Pd; [2] Cu^{2+}, ^-OH.

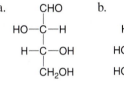

20.56 Draw the organic products formed when each monosaccharide is treated with each of the following reagents: [1] H_2, Pd; [2] Cu^{2+}, ^-OH.

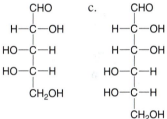

20.57 What product is formed when each compound is treated with Benedict's reagent?

a.
```
      CHO
   H—C—OH
   H—C—OH
  HO—C—H
     CH₂OH
```
b.
```
     CH₂OH
      C=O
   H—C—OH
   H—C—OH
     CH₂OH
```

20.58 What product is formed when each compound is treated with Benedict's reagent?

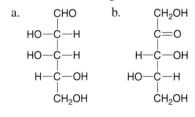

20.59 What does the term *reducing sugar* mean? Give an example.

20.60 What does the term *mutarotation* mean?

20.61 What monosaccharides are formed when each disaccharide is hydrolyzed?

a.

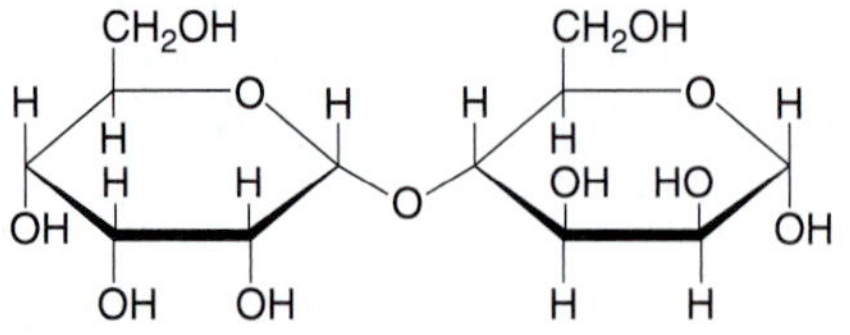

b.

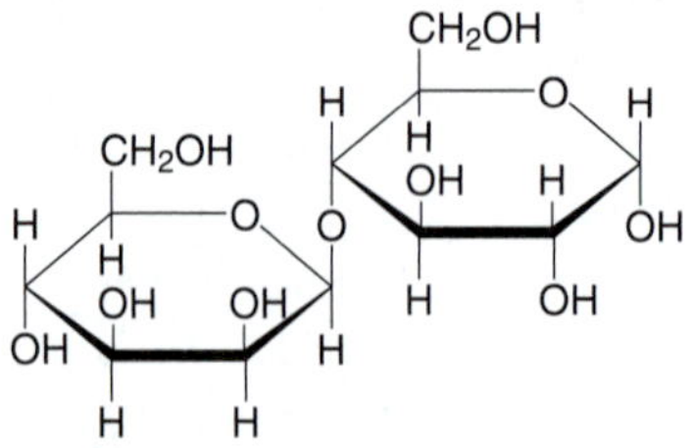

20.62 What monosaccharides are formed when each disaccharide is hydrolyzed?

a.

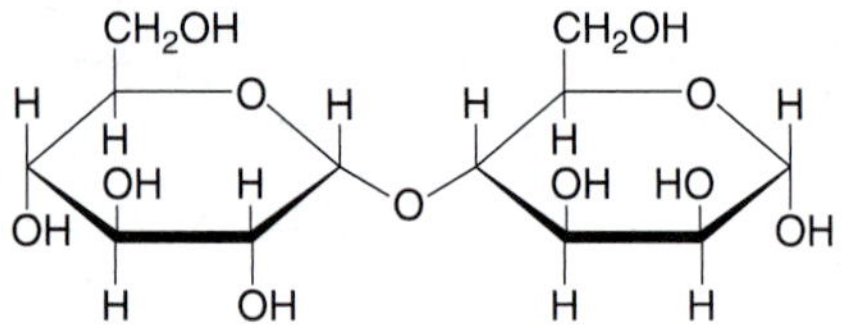

b.

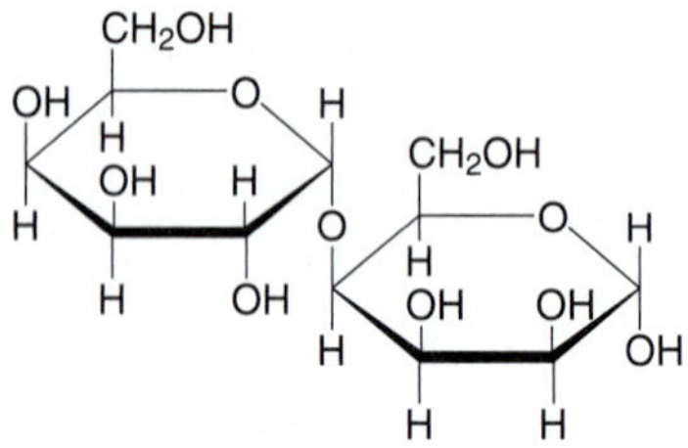

Disaccharides and Polysaccharides

20.63 What is the difference between an α anomer and an α glycoside?

20.64 What is the difference between a β anomer and a β glycoside?

20.65 Draw the structure of a disaccharide that contains two six-membered rings and an α glycosidic linkage.

20.66 Draw the structure of a disaccharide that contains one six-membered ring and one five-membered ring as well as a β glycosidic linkage.

20.67 Describe the similarities and differences in the structures of maltose and lactose.

20.68 Describe the similarities and differences in the structures of lactose and sucrose.

20.69 Consider the disaccharide isomaltose.

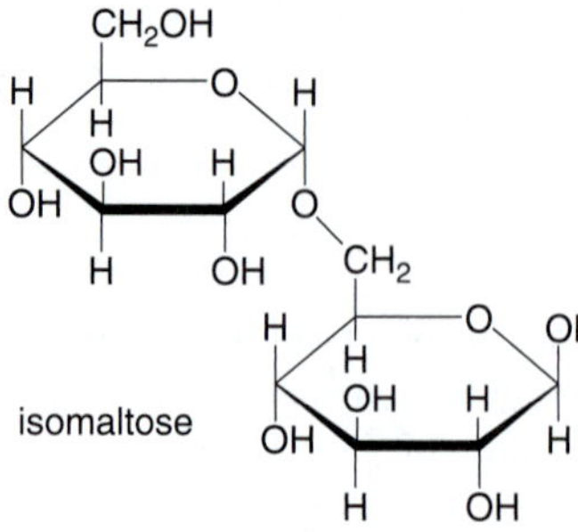

isomaltose

a. Label any acetal and hemiacetal in isomaltose.
b. Number each monosaccharide ring.
c. Classify the glycosidic linkage as α or β, and use numbers to designate its location.
d. Is the hemiacetal drawn as an α or β anomer?
e. What monosaccharides are formed when isomaltose is hydrolyzed?

20.70 Consider the disaccharide sophorose.

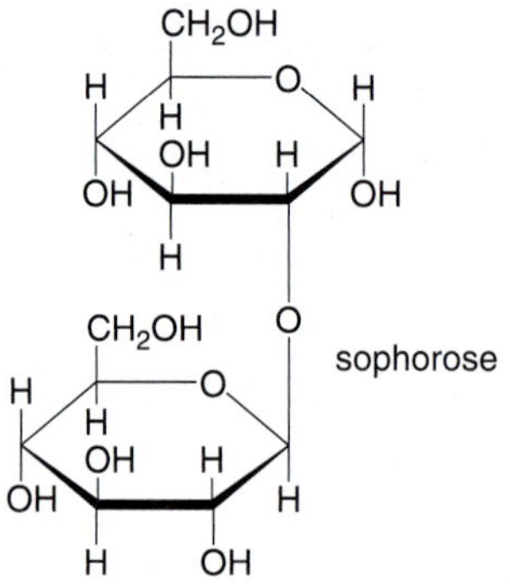

sophorose

a. Label any acetal and hemiacetal in sophorose.
b. Number each monosaccharide ring.
c. Classify the glycosidic linkage as α or β, and use numbers to designate its location.
d. Is the hemiacetal drawn as an α or β anomer?
e. What monosaccharides are formed when sophorose is hydrolyzed?

20.71 Draw the structure of a disaccharide formed from two galactose units joined by a 1→4-β-glycosidic linkage. (The cyclic structure of galactose appears in Sample Problem 20.3.)

20.72 Draw the structure of a disaccharide formed from two mannose units joined by a 1→4-α-glycosidic linkage. (The cyclic structure of mannose appears in Section 20.3B.)

20.73 In what ways are cellulose and amylose similar? How do the structures of cellulose and amylose differ?

20.74 In what ways are amylopectin and glycogen similar? How do the structures of amylopectin and glycogen differ?

20.75 In what ways are cellulose and chitin similar? In what ways are cellulose and chitin different?

20.76 In what ways are amylose and amylopectin similar? How do the structures of amylose and amylopectin differ?

20.77 Draw a short segment of a polysaccharide that contains three galactose units joined together in 1→4-α-glycosidic linkages. (The cyclic structure of galactose appears in Sample Problem 20.3.)

20.78 Draw a short segment of a polysaccharide that contains three mannose units joined together in 1→4-β-glycosidic linkages. (The cyclic structure of mannose appears in Section 20.3B.)

General Questions

20.79 Consider the following monosaccharide.

CHO
HO—C—H
HO—C—H
HO—C—H
H—C—OH
CH₂OH

a. Is the monosaccharide a D or L sugar?
b. Classify the monosaccharide by the type of carbonyl and the number of atoms in the chain.
c. Draw the enantiomer.
d. Label the chirality centers.
e. Draw the α anomer of the cyclic form.
f. What product is formed when the monosaccharide is treated with Benedict's reagent?
g. What product is formed when the monosaccharide is treated with H₂ and Pd?
h. Label the monosaccharide as a reducing or nonreducing sugar.

20.80 Consider the following monosaccharide.

CHO
HO—C—H
HO—C—H
H—C—OH
H—C—OH
CH₂OH

a. Is the monosaccharide a D or L sugar?
b. Classify the monosaccharide by the type of carbonyl and the number of atoms in the chain.
c. Draw the enantiomer.
d. Label the chirality centers.
e. Draw the β anomer of the cyclic form.
f. What product is formed when the monosaccharide is treated with Benedict's reagent?
g. What product is formed when the monosaccharide is treated with H₂ and Pd?
h. Label the monosaccharide as a reducing or nonreducing sugar.

Applications

20.81 Describe the difference between lactose intolerance and galactosemia.

20.82 Explain why cellulose is a necessary component of our diet even though we don't digest it.

20.83 Explain why fructose is called a reduced calorie sweetener while sucralose is called an artificial sweetener.

20.84 How do oxidation reactions help an individual with diabetes monitor blood glucose levels?

20.85 Why can an individual with type A blood receive only blood types A and O, but he or she can donate to individuals with either type A or AB blood?

20.86 Why can an individual with type B blood receive only blood types B and O, but he or she can donate to individuals with either type B or AB blood?

20.87 Would you predict chitin to be digestible by humans? Explain why or why not.

20.88 Explain, with reference to the structures, why starch can be digested by humans but cellulose cannot.

20.89 Name three glycosaminoglycans and their roles in the body.

20.90 Why is chitin water insoluble but the N-acetyl-D-glucosamine from which it is made water soluble?

CHALLENGE QUESTIONS

20.91 In much the same way that aldoses are reduced with H₂ and Pd, 2-ketohexoses are also reduced. One difference occurs, however, in that two products are formed on reduction of 2-ketohexoses. Draw the two products formed when fructose (Section 20.2C) is reduced, and label the products as stereoisomers, constitutional isomers, or anomers. Explain your choice.

20.92 Convert each cyclic form to an acyclic monosaccharide.

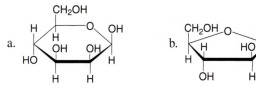

21

CHAPTER GOALS

In this chapter you will learn how to:

1. Identify the general structural features of amino acids

2. Describe the acid–base properties of amino acids

3. Draw simple peptides from individual amino acids, and label the N- and C-terminal amino acids

4. Give examples of simple biologically active peptides

5. Describe the characteristics of the primary, secondary, tertiary, and quaternary structure of proteins

6. Describe the features of fibrous proteins like α-keratin and collagen

7. Describe the features of globular proteins like hemoglobin and myoglobin

8. Draw the products of protein hydrolysis

9. Describe protein denaturation

10. Describe the main features of enzymes

11. Describe the use of enzymes to diagnose and treat disease

Hair, whether it is black, brunette, blond, or red, is composed of the same protein—**keratin.**

AMINO ACIDS, PROTEINS, AND ENZYMES

OF the four major groups of biomolecules—lipids, carbohydrates, proteins, and nucleic acids—proteins have the widest array of functions. **Keratin** and **collagen,** for example, form long insoluble fibers, giving strength and support to tissues. Hair, horns, hooves, and fingernails are all made up of keratin. **Collagen** is found in bone, connective tissue, tendons, and cartilage. **Membrane proteins** transport small organic molecules and ions across cell membranes. **Insulin,** the hormone that regulates blood glucose levels, and **hemoglobin,** which transports oxygen from the lungs to tissues, are proteins. **Enzymes** are proteins that catalyze and regulate all aspects of cellular function. In Chapter 21 we discuss proteins and their primary components, the amino acids.

21.1 INTRODUCTION

Proteins **are biomolecules that contain many amide bonds, formed by joining amino acids together.**

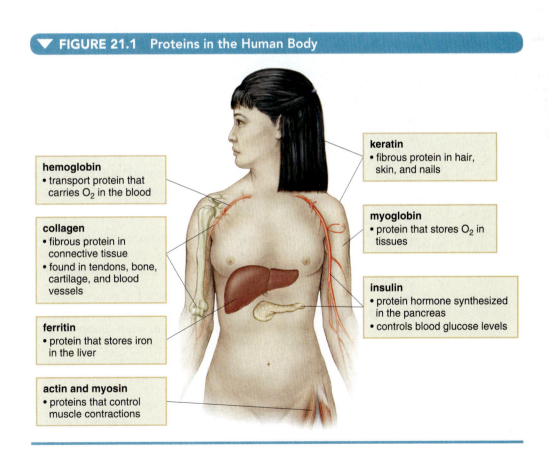

amino acid

protein
[Amide bonds are shown in red.]

The word *protein* comes from the Greek *proteios* meaning "of first importance." Proteins occur widely in the human body, accounting for approximately 50% of its dry weight (Figure 21.1). Fibrous proteins, like keratin in hair, skin, and nails and collagen in connective tissue, give support and structure to tissues and cells. Protein hormones and enzymes regulate the body's metabolism. Transport proteins carry substances through the blood, and storage proteins store elements and ions in organs. Contractile proteins control muscle movements, and immunoglobulins are proteins that defend the body against foreign substances.

Unlike lipids and carbohydrates, which the body stores for use when needed, protein is not stored so it must be consumed on a daily basis. The current recommended daily intake for adults is 0.8 grams of protein per kilogram of body weight. Since children need protein for both growth and maintenance, the recommended daily intake is higher, as shown in Table 21.1. The average protein intake in the United States is often twice what is necessary. Most protein in an American diet comes from eating meat and milk products.

▼ **FIGURE 21.1   Proteins in the Human Body**

keratin
- fibrous protein in hair, skin, and nails

hemoglobin
- transport protein that carries O_2 in the blood

myoglobin
- protein that stores O_2 in tissues

collagen
- fibrous protein in connective tissue
- found in tendons, bone, cartilage, and blood vessels

insulin
- protein hormone synthesized in the pancreas
- controls blood glucose levels

ferritin
- protein that stores iron in the liver

actin and myosin
- proteins that control muscle contractions

TABLE 21.1 Recommended Daily Protein Intake

Group	Daily Protein Intake (g protein/kg body weight)
Children (1–3 years)	1.1
Children (4–13 years)	0.95
Children (14–18 years)	0.85
Adult	0.8

Source: Data from U.S. Food and Drug Administration.

21.2 AMINO ACIDS

To understand protein properties and structure, we must first learn about the amino acids that compose them.

21.2A GENERAL FEATURES OF AMINO ACIDS

Amino acids contain two functional groups—an amino group (NH_2) and a carboxyl group ($COOH$). In most naturally occurring amino acids, the amino group is bonded to the α carbon, the carbon adjacent to the carbonyl group (Section 17.2), making them **α-amino acids.**

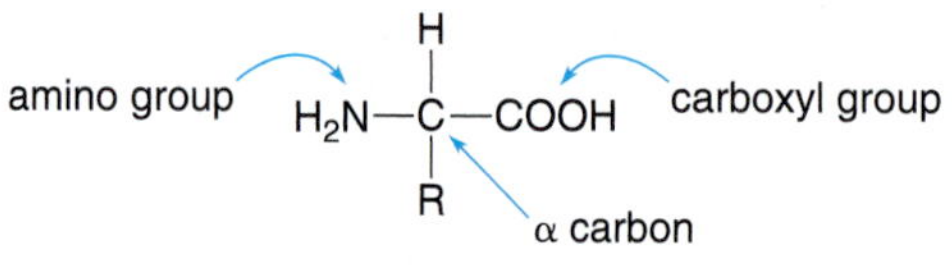

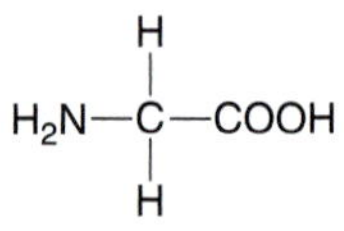

The 20 amino acids that occur naturally in proteins differ in the identity of the R group bonded to the α carbon. The R group is called the **side chain** of the amino acid. The simplest amino acid, called **glycine,** has R = H. Other side chains may be simple alkyl groups, or have additional functional groups such as OH, SH, COOH, or NH_2. Table 21.2 lists the structures of the 20 common amino acids that occur in proteins.

- Amino acids with an additional COOH group in the side chain are called acidic amino acids.
- Those with an additional basic N atom in the side chain are called basic amino acids.
- All others are neutral amino acids.

All amino acids have common names, which are abbreviated by a three-letter or one-letter designation. For example, glycine is often written as the three-letter abbreviation **Gly,** or the one-letter abbreviation **G.** These abbreviations are also given in Table 21.2.

Amino acids never exist in nature as neutral molecules with all uncharged atoms. Since amino acids contain a base (NH_2 group) and an acid (COOH), proton transfer from the acid to the base forms a salt called a **zwitterion,** which contains both a positive and a negative charge. These salts have high melting points and are water soluble.

The acid–base chemistry of amino acids is discussed in greater detail in Section 21.3. The structures in Table 21.2 show the charged form of the amino acids at the physiological pH of the blood.

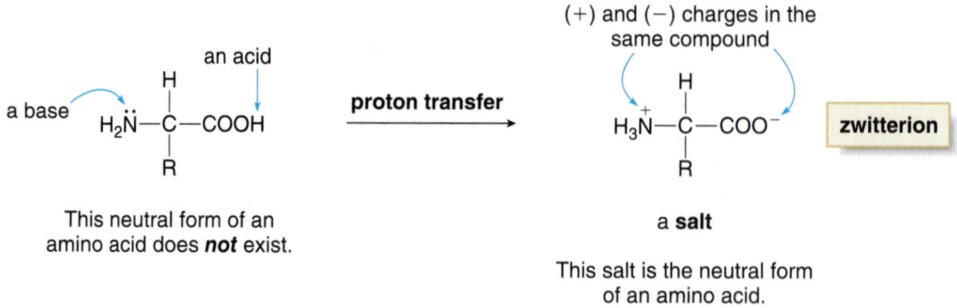

Humans can synthesize only 10 of the 20 amino acids needed for proteins. The remaining 10, called **essential amino acids,** must be obtained from the diet and consumed on a regular, almost daily basis. Diets that include animal products readily supply all of the needed amino acids. Since no one plant source has sufficient amounts of all of the essential amino acids, vegetarian diets must be carefully balanced. Grains—wheat, rice, and corn—are low in lysine, and legumes—beans, peas, and peanuts—are low in methionine, but a combination of these foods provides all the needed amino acids.

PROBLEM 21.1	In addition to the amino and carboxyl groups, what other functional groups are present in each amino acid: (a) asparagine; (b) serine; (c) cysteine?
PROBLEM 21.2	How do the OH groups in Ser, Thr, and Tyr differ?
PROBLEM 21.3	Draw a Lewis structure for arginine that clearly shows all of the lone pairs.

HEALTH NOTE

A diet of rice and tofu provides all essential amino acids. A peanut butter sandwich on wheat bread does the same.

21.2B   STEREOCHEMISTRY OF AMINO ACIDS

Except for the simplest amino acid, glycine, **all other amino acids have a chirality center—a carbon bonded to four different groups—on the α carbon.** Thus, an amino acid like alanine (R = CH$_3$) has two possible enantiomers, drawn below in both three-dimensional representations with wedges and dashed bonds, and Fischer projections.

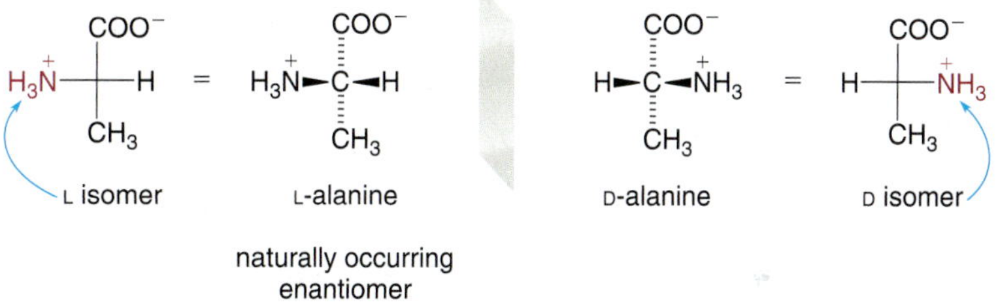

$$
\begin{array}{ccc}
\text{L isomer} & \text{L-alanine} & \text{D-alanine} & \text{D isomer}
\end{array}
$$

naturally occurring
enantiomer

Like monosaccharides, the prefixes **D** and **L** are used to designate the arrangement of groups on the chirality center of amino acids. When drawn with a vertical carbon chain having the –COO⁻ group at the top and the R group at the bottom,

- **L** Amino acids have the –NH$_3^+$ group on the *left* side in the Fischer projection. Common naturally occurring amino acids are **L** isomers.
- **D** Amino acids have the –NH$_3^+$ group on the *right* side in the Fischer projection. **D** Amino acids occur infrequently in nature.

SAMPLE PROBLEM 21.1	Draw the Fischer projection for each amino acid: (a) L-leucine; (b) D-cysteine.
ANALYSIS	To draw an amino acid in a Fischer projection, place the –COO⁻ group at the top and the R group at the bottom. The L isomer has the –NH$_3^+$ on the left side and the D isomer has the –NH$_3^+$ on the right side.
SOLUTION	

a. For leucine, R = CH$_2$CH(CH$_3$)$_2$

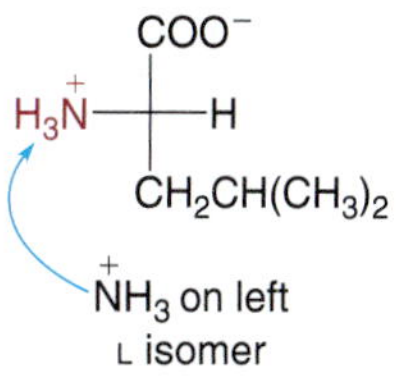

$$
\begin{array}{c}
\text{COO}^- \\
\text{H}_3\text{N} \!\!-\!\!\!-\!\!\!- \text{H} \\
\text{CH}_2\text{CH(CH}_3)_2
\end{array}
$$

⁺NH$_3$ on left
L isomer

b. For cysteine, R = CH$_2$SH

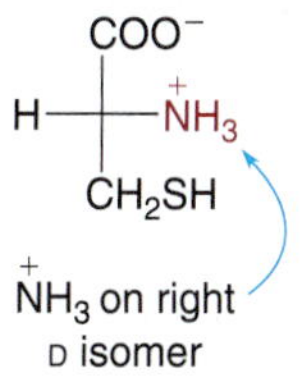

$$
\begin{array}{c}
\text{COO}^- \\
\text{H} \!\!-\!\!\!-\!\!\!- \text{NH}_3 \\
\text{CH}_2\text{SH}
\end{array}
$$

⁺NH$_3$ on right
D isomer

TABLE 21.2 The 20 Common Naturally Occurring Amino Acids

Neutral Amino Acids					
Name	Structure	Abbreviations	Name	Structure	Abbreviations
Alanine	$H_3\overset{+}{N}-\underset{\underset{CH_3}{\vert}}{\overset{\overset{H}{\vert}}{C}}-COO^-$	Ala A	Phenylalanine*	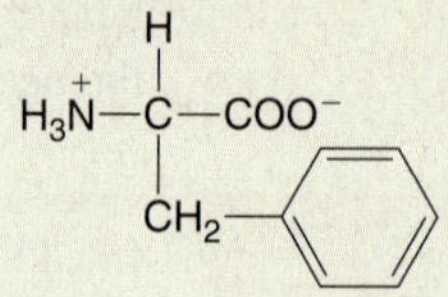 $H_3\overset{+}{N}-\underset{\underset{CH_2-\text{(phenyl)}}{\vert}}{\overset{\overset{H}{\vert}}{C}}-COO^-$	Phe F
Asparagine	$H_3\overset{+}{N}-\underset{\underset{CH_2CONH_2}{\vert}}{\overset{\overset{H}{\vert}}{C}}-COO^-$	Asn N	Proline	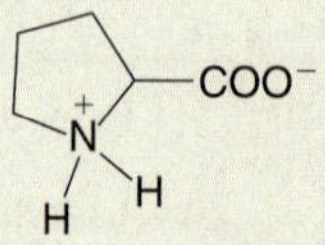 (pyrrolidine ring)—COO⁻	Pro P
Cysteine	$H_3\overset{+}{N}-\underset{\underset{CH_2SH}{\vert}}{\overset{\overset{H}{\vert}}{C}}-COO^-$	Cys C	Serine	$H_3\overset{+}{N}-\underset{\underset{CH_2OH}{\vert}}{\overset{\overset{H}{\vert}}{C}}-COO^-$	Ser S
Glutamine	$H_3\overset{+}{N}-\underset{\underset{CH_2CH_2CONH_2}{\vert}}{\overset{\overset{H}{\vert}}{C}}-COO^-$	Gln Q	Threonine*	$H_3\overset{+}{N}-\underset{\underset{CH(OH)CH_3}{\vert}}{\overset{\overset{H}{\vert}}{C}}-COO^-$	Thr T
Glycine	$H_3\overset{+}{N}-\underset{\underset{H}{\vert}}{\overset{\overset{H}{\vert}}{C}}-COO^-$	Gly G	Tryptophan*	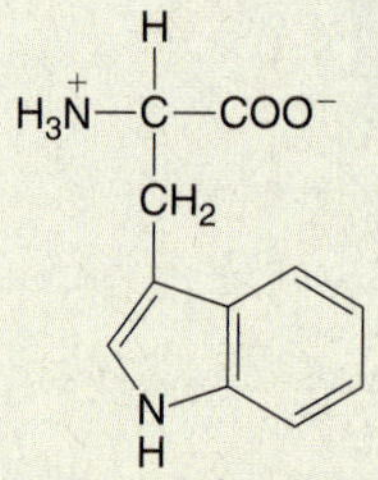 $H_3\overset{+}{N}-\underset{\underset{CH_2-\text{(indole)}}{\vert}}{\overset{\overset{H}{\vert}}{C}}-COO^-$	Trp W
Isoleucine*	$H_3\overset{+}{N}-\underset{\underset{CH(CH_3)CH_2CH_3}{\vert}}{\overset{\overset{H}{\vert}}{C}}-COO^-$	Ile I	Tyrosine	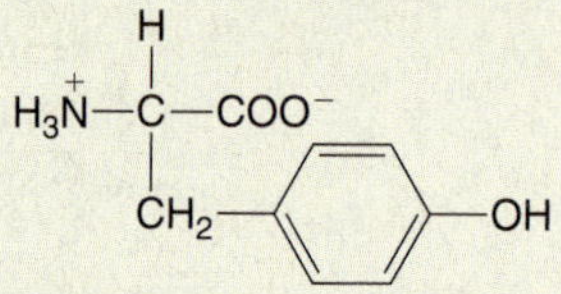 $H_3\overset{+}{N}-\underset{\underset{CH_2-\text{(phenol)}-OH}{\vert}}{\overset{\overset{H}{\vert}}{C}}-COO^-$	Tyr Y
Leucine*	$H_3\overset{+}{N}-\underset{\underset{CH_2CH(CH_3)_2}{\vert}}{\overset{\overset{H}{\vert}}{C}}-COO^-$	Leu L	Valine*	$H_3\overset{+}{N}-\underset{\underset{CH(CH_3)_2}{\vert}}{\overset{\overset{H}{\vert}}{C}}-COO^-$	Val V
Methionine*	$H_3\overset{+}{N}-\underset{\underset{CH_2CH_2SCH_3}{\vert}}{\overset{\overset{H}{\vert}}{C}}-COO^-$	Met M			

Essential amino acids are labeled with an asterisk (*).

(continued on next page)

TABLE 21.2 (continued)

Acidic Amino Acids			Basic Amino Acids		
Name	Structure	Abbreviations	Name	Structure	Abbreviations
Aspartic acid	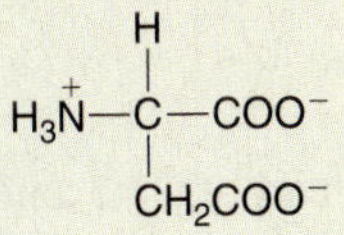 H$_3$N$^+$—C(H)—COO$^-$ with CH$_2$COO$^-$	Asp D	Arginine*	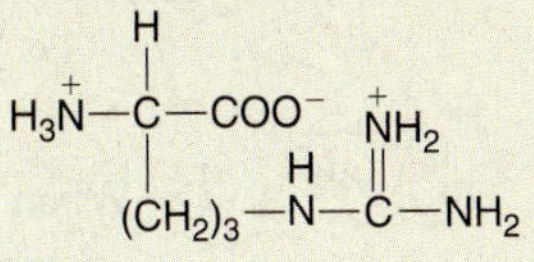 H$_3$N$^+$—C(H)—COO$^-$ with (CH$_2$)$_3$—N(H)—C(NH$_2$)=NH$_2^+$	Arg R
Glutamic acid	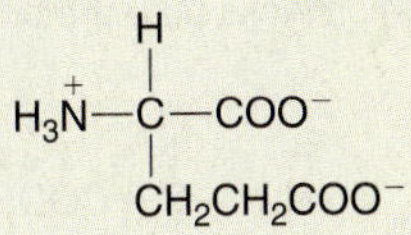 H$_3$N$^+$—C(H)—COO$^-$ with CH$_2$CH$_2$COO$^-$	Glu E	Histidine*	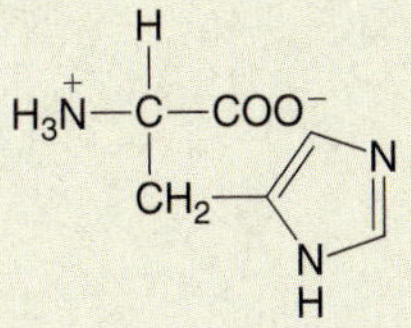 H$_3$N$^+$—C(H)—COO$^-$ with CH$_2$—(imidazole)	His H
			Lysine*	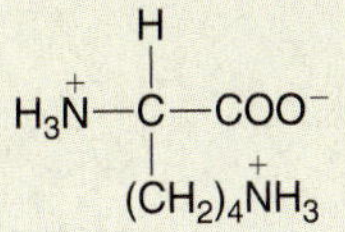 H$_3$N$^+$—C(H)—COO$^-$ with (CH$_2$)$_4$NH$_3^+$	Lys K

Essential amino acids are labeled with an asterisk (*).

PROBLEM 21.4

Draw both enantiomers of each amino acid in Fischer projections and label them as D or L: (a) phenylalanine; (b) methionine.

PROBLEM 21.5

Which of the following amino acids is naturally occurring? By referring to the structures in Table 21.2, name each amino acid and include its D or L designation in the name.

a. 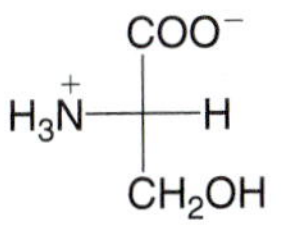
COO$^-$
H$_3$N$^+$——H
CH$_2$OH

b. 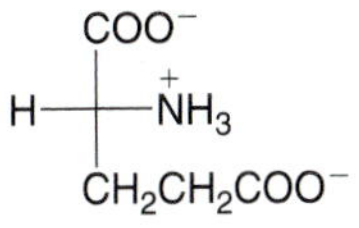
COO$^-$
H——NH$_3^+$
CH$_2$CH$_2$COO$^-$

c. 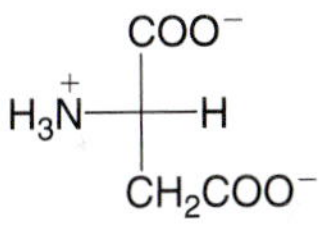
COO$^-$
H$_3$N$^+$——H
CH$_2$COO$^-$

21.3 ACID–BASE BEHAVIOR OF AMINO ACIDS

As mentioned in Section 21.2, an amino acid contains both a basic amino group (NH$_2$) and an acidic carboxyl group (COOH). As a result, proton transfer from the acid to the base forms a **zwitterion,** a salt that contains both a positive and a negative charge. The zwitterion is *neutral;* that is, the net charge on the salt is zero.

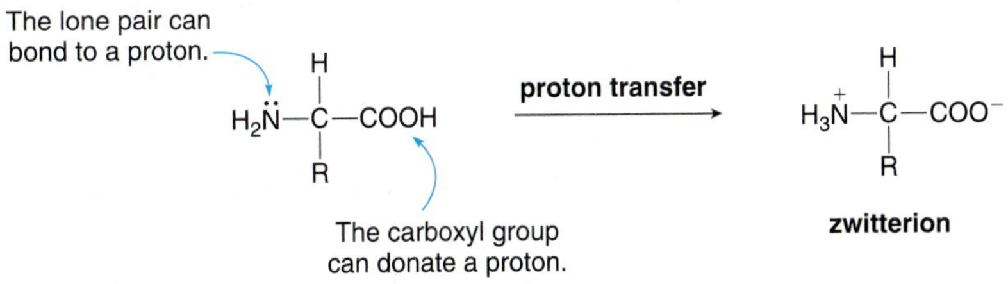

The lone pair can bond to a proton.

H$_2$N—C(H)(R)—COOH →(proton transfer)→ H$_3$N$^+$—C(H)(R)—COO$^-$

The carboxyl group can donate a proton.

zwitterion

In actuality, **an amino acid can exist in different forms, depending on the pH of the aqueous solution in which it is dissolved.** When the pH of a solution is around 6, alanine (R = CH$_3$) and other neutral amino acids exist in their zwitterionic form (**A**), having no net charge. In this form, the carboxyl group bears a net negative charge—it is a **carboxylate anion**—and the amino group bears a net positive charge (an **ammonium cation**).

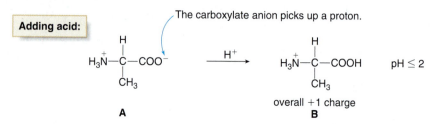

alanine
no net charge
A
pH ≈ 6

When strong acid is added to lower the pH to 2 or less, the carboxylate anion gains a proton and the **amino acid has a net positive charge** (form **B**).

Adding acid:

The carboxylate anion picks up a proton.

$$H_3\overset{+}{N}-\underset{\underset{CH_3}{|}}{\overset{\overset{H}{|}}{C}}-COO^- \xrightarrow{\;H^+\;} H_3\overset{+}{N}-\underset{\underset{CH_3}{|}}{\overset{\overset{H}{|}}{C}}-COOH \qquad pH \le 2$$

A   overall +1 charge
                              B

When strong base is added to **A** to raise the pH to 10 or higher, the ammonium cation loses a proton and the **amino acid has a net negative charge** (form **C**).

Adding base:

The ammonium cation loses a proton.

$$H_3\overset{+}{N}-\underset{\underset{CH_3}{|}}{\overset{\overset{H}{|}}{C}}-COO^- \xrightarrow{\;{}^-OH\;} H_2N-\underset{\underset{CH_3}{|}}{\overset{\overset{H}{|}}{C}}-COO^- \;+\; H_2O \quad pH \ge 10$$

A   overall −1 charge
                              C

Thus, **alanine exists in one of three different forms depending on the pH of the solution in which it is dissolved.** At the physiological pH of 7.4, neutral amino acids are primarily in their zwitterionic forms.

- **The pH at which the amino acid exists primarily in its neutral form is called its *isoelectric point,* abbreviated as p*I*.**

The isoelectric points of neutral amino acids are generally around 6. Acidic amino acids (Table 21.2), which have an additional carboxyl group that can lose a proton, have lower p*I* values (around 3). The three basic amino acids, which have an additional basic nitrogen atom that can accept a proton, have higher p*I* values (7.6–10.8).

SAMPLE PROBLEM 21.2

Draw the structure of the amino acid glycine at each pH: (a) 6; (b) 2; (c) 11.

ANALYSIS

A neutral amino acid exists in its zwitterionic form (no net charge) at its isoelectric point, which is pH ≈ 6. The zwitterionic forms of neutral amino acids appear in Table 21.2. At low pH (≤ 2), the carboxylate anion is protonated and the amino acid has a net positive (+1) charge. At high pH (≥ 10), the ammonium cation loses a proton and the amino acid has a net negative (−1) charge.

SOLUTION

a. At pH = 6, the neutral, zwitterionic form of glycine predominates.

$$H_3\overset{+}{N}-\underset{\underset{H}{|}}{\overset{\overset{H}{|}}{C}}-COO^-$$

neutral
pH = 6

b. At pH = 2, glycine has a net +1 charge.

$$H_3\overset{+}{N}-\underset{\underset{H}{|}}{\overset{\overset{H}{|}}{C}}-COOH$$

+1 charge
pH = 2

c. At pH = 11, glycine has a net −1 charge.

$$H_2N-\underset{\underset{H}{|}}{\overset{\overset{H}{|}}{C}}-COO^-$$

−1 charge
pH = 11

PROBLEM 21.6   Draw the structure of the amino acid valine at each pH: (a) 6; (b) 2; (c) 11. Which form predominates at valine's isoelectric point?

PROBLEM 21.7   Draw the positively charged, neutral, and negatively charged forms for the amino acid phenylalanine. Which species predominates at pH 11? Which species predominates at pH 1?

21.4 PEPTIDES

When amino acids are joined together by amide bonds, they form larger molecules called **peptides** and **proteins.**

- A *dipeptide* has two amino acids joined together by *one* amide bond.
- A *tripeptide* has three amino acids joined together by *two* amide bonds.

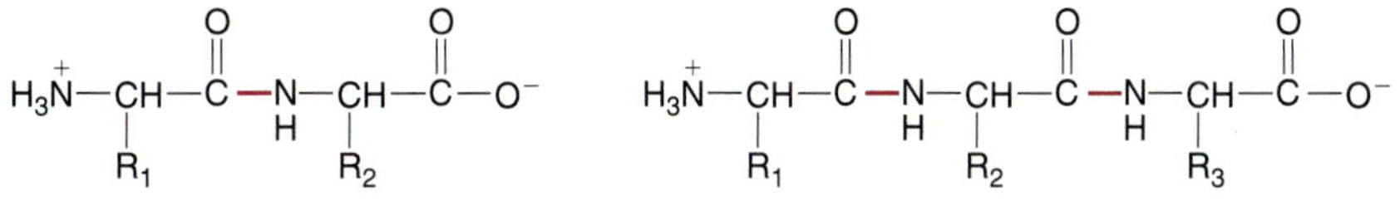

dipeptide   tripeptide

[Amide bonds are shown in red.]

Polypeptides and **proteins** both have many amino acids joined together in long linear chains, but the term **protein** is usually reserved for polymers of more than 40 amino acids.

- The amide bonds in peptides and proteins are called *peptide bonds.*
- The individual amino acids are called *amino acid residues.*

To form a dipeptide, **the –NH$_3^+$ group of one amino acid forms an amide bond with the carboxylate (–COO$^-$) of another amino acid, and the elements of H$_2$O are removed.** Because each amino acid has both functional groups, two different dipeptides can be formed. This is illustrated with alanine (Ala) and serine (Ser) to form dipeptides **A** and **B.**

1. The –COO$^-$ group of alanine can combine with the –NH$_3^+$ group of serine.

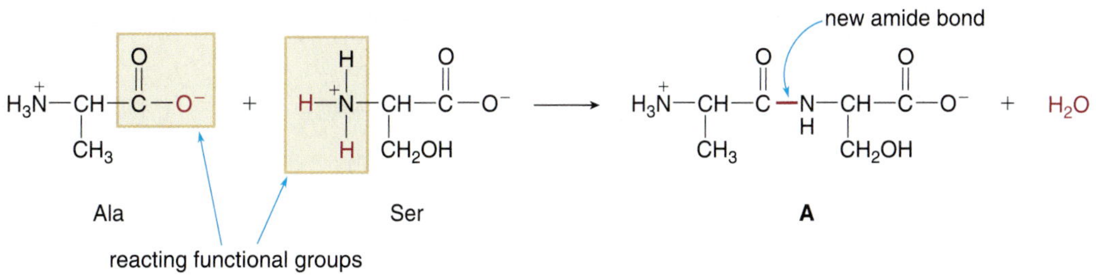

Ala   Ser   A

reacting functional groups

2. The –COO$^-$ group of serine can combine with the –NH$_3^+$ group of alanine.

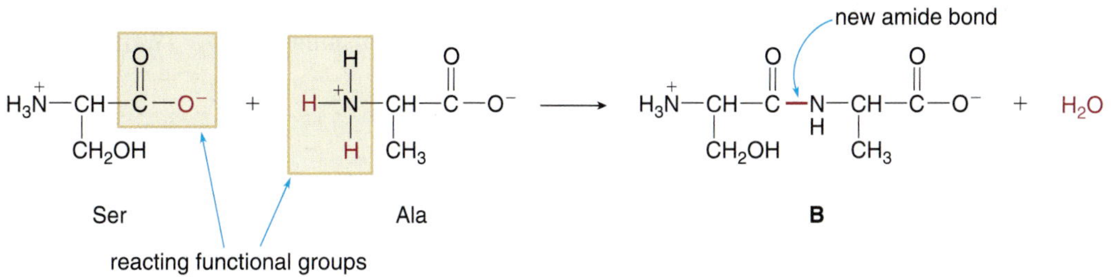

Ser   Ala   B

reacting functional groups

Dipeptides **A** and **B** are **constitutional isomers** of each other. Both have an ammonium cation ($-NH_3^+$) at one end of their chains and a carboxylate anion ($-COO^-$) at the other.

- The amino acid with the free $-NH_3^+$ group on the α carbon is called the N-terminal amino acid.
- The amino acid with the free $-COO^-$ group on the α carbon is called the C-terminal amino acid.

By convention, **the N-terminal amino acid is always written at the *left* end of the chain and the C-terminal amino acid at the *right*.**

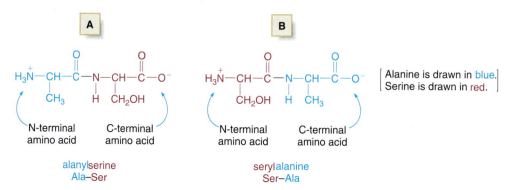

Alanine is drawn in blue.
Serine is drawn in red.

alanylserine
Ala–Ser

serylalanine
Ser–Ala

Peptides are named as derivatives of the C-terminal amino acid. To name a peptide:

- Name the C-terminal amino acid using the names in Table 21.2.
- Name all other amino acids from left to right as substituents of the C-terminal amino acid. Change the *-ine* or *-ic acid* ending of the amino acid name to the suffix *-yl*.

Thus, peptide **A**, which has serine as its C-terminal amino acid, is named as *alanylserine.* Peptide **B**, which has alanine as its C-terminal amino acid, is named as *serylalanine.*

The peptide can be abbreviated by writing the one- or three-letter symbols for the amino acids in the chain from the N-terminal to the C-terminal end. Thus, Ala–Ser has alanine at the N-terminal end and serine at the C-terminal end, whereas Ser–Ala has serine at the N-terminal end and alanine at the C-terminal end.

PROBLEM 21.8

Identify the N-terminal and C-terminal amino acid in each peptide.

a. 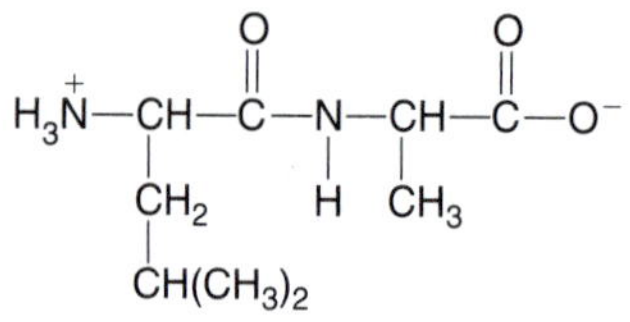
b. Arg–His–Asn–Tyr

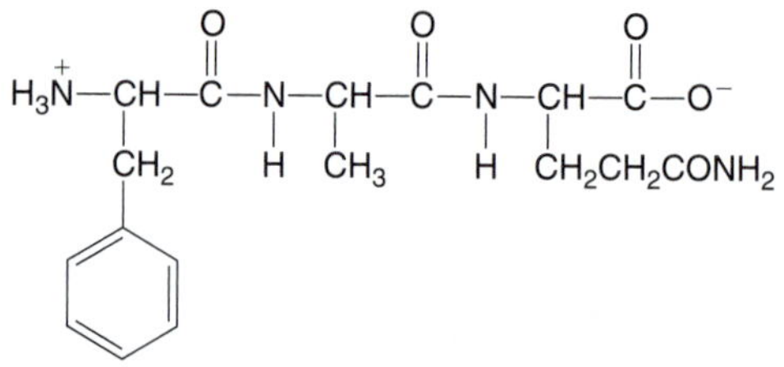

c.
d. Val–Thr–Pro–Phe

PROBLEM 21.9

(a) Identify the N-terminal amino acid in the tetrapeptide alanylglycylleucylmethionine. (b) What is the C-terminal amino acid? (c) Write the peptide using three-letter symbols for the amino acids.

Drawing a dipeptide from component amino acids is illustrated in the stepwise *How To* procedure. Sample Problem 21.3 illustrates how to identify and name the component amino acids from a given tripeptide.

HOW TO   Draw a Dipeptide from Two Amino Acids

EXAMPLE   **Draw the structure of the dipeptide Val–Gly, and label the N-terminal and C-terminal amino acids.**

Step [1]   **Draw the structures of the individual amino acids from left to right.**
- Use the three-letter symbols to identify the amino acids in the peptide. In this example, Val is valine and Gly is glycine.
- Draw the structures of the amino acids by placing the $-COO^-$ of one amino acid next to the $-NH_3^+$ group of the adjacent amino acid.
- Always draw the $-NH_3^+$ group on the left and the $-COO^-$ group on the right.

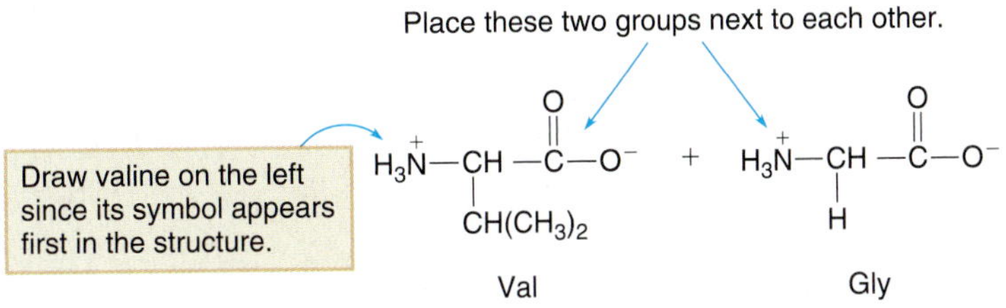

Step [2]   **Join the adjacent $-COO^-$ and $-NH_3^+$ groups together.**
- Form a new amide bond by joining the carbonyl carbon of one amino acid to the N atom of the second amino acid.

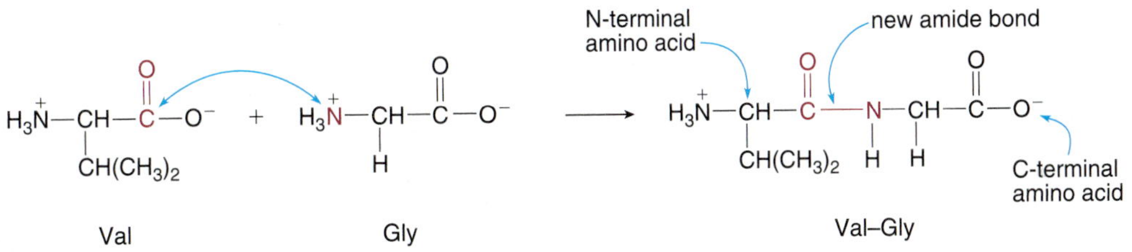

- Valine is the N-terminal amino acid since it contains a free $-NH_3^+$ group on the α carbon.
- Glycine is the C-terminal amino acid since it contains a free $-COO^-$ group on the α carbon.

PROBLEM 21.10   Draw the structure of each dipeptide: (a) Gly–Phe; (b) Gln–Ile; (c) Leu–Cys.

PROBLEM 21.11   (a) Draw the structures of the two possible dipeptides that can be formed by combining leucine and asparagine. (b) In each dipeptide label the N- and C-terminal amino acids. (c) Name each peptide using three-letter abbreviations.

SAMPLE PROBLEM 21.3   Identify the individual amino acids used to form the following tripeptide. What is the name of the tripeptide?

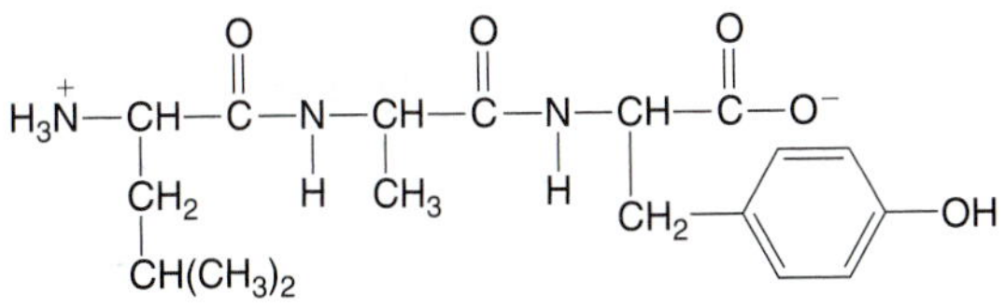

ANALYSIS
- Locate the amide bonds in the peptide. To draw the structures of the individual amino acids, break the amide bonds by adding water. Add O⁻ to the carbonyl carbon to form a carboxylate anion –COO⁻ and 2 H's to the N atom to form –NH₃⁺.
- Identify the amino acids by comparing with Table 21.2.
- To name the peptide: [1] Name the C-terminal amino acid. [2] Name the other amino acids as substituents by changing the *-ine* (or *-ic acid*) ending to the suffix *-yl*. Place the names of the substituent amino acids in order from left to right.

SOLUTION
To determine the amino acids that form the peptide, work backwards. Break the amide bonds in red that join the amino acids together. This forms leucine, alanine, and tyrosine.

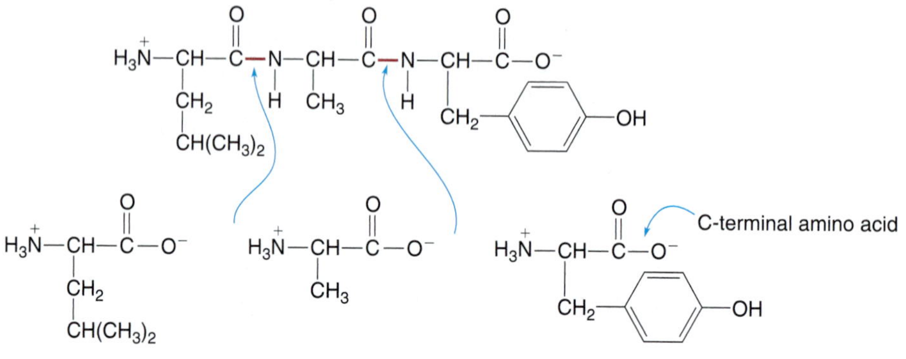

leucine

alanine

tyrosine

The tripeptide is named as a derivative of the C-terminal amino acid, tyrosine, with leucine and alanine as substituents; thus, the tripeptide is named: **leucylalanyltyrosine.**

PROBLEM 21.12

Identify the individual amino acids in each dipeptide, and then name the dipeptide using three-letter abbreviations.

a.

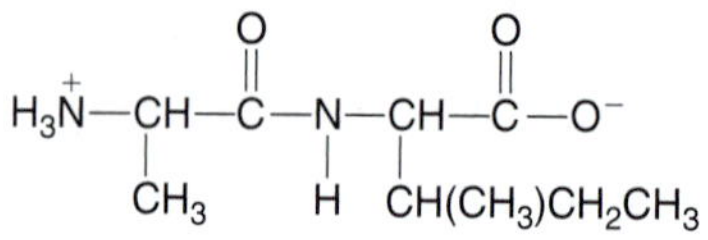

b.

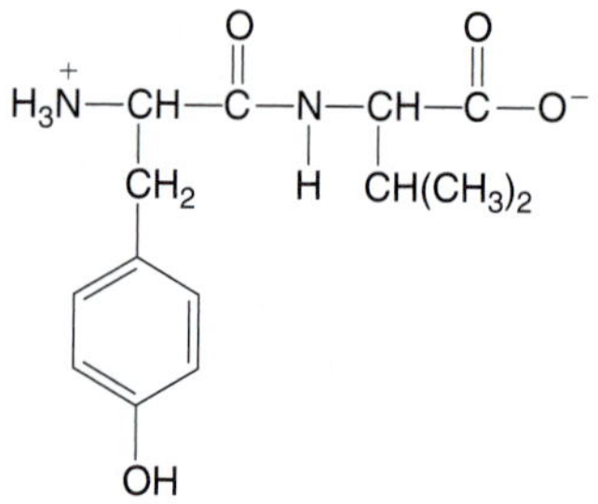

21.5 FOCUS ON THE HUMAN BODY
BIOLOGICALLY ACTIVE PEPTIDES

Many relatively simple peptides have important biological functions.

21.5A NEUROPEPTIDES—ENKEPHALINS AND PAIN RELIEF

Enkephalins, pentapeptides synthesized in the brain, act as pain killers and sedatives by binding to pain receptors. Two enkephalins that differ in the identity of only one amino acid are known. Met-enkephalin contains a C-terminal methionine residue, while leu-enkephalin contains a C-terminal leucine.

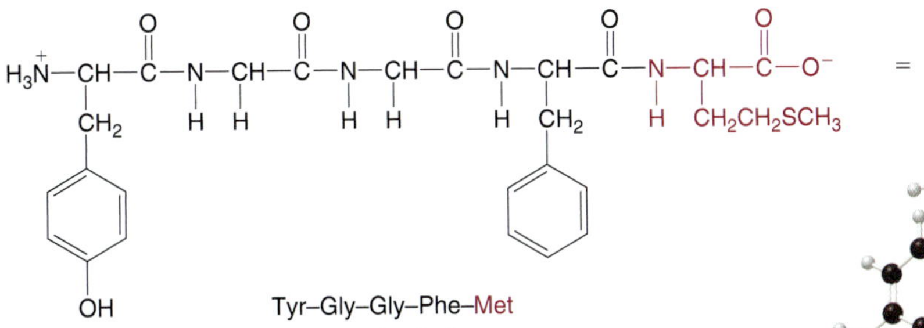

Tyr–Gly–Gly–Phe–Met
met-enkephalin

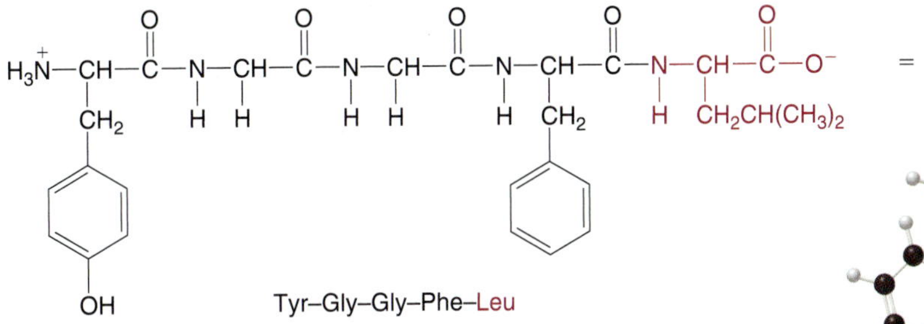

Tyr–Gly–Gly–Phe–Leu
leu-enkephalin

The addictive narcotic analgesics morphine and heroin (Section 18.5) bind to the same receptors as the enkephalins, and thus produce a similar physiological response. Enkephalins, however, produce only short-term pain-relieving effects because their peptide bonds are readily hydrolyzed by enzymes in the brain. Morphine and heroin, on the other hand, are not readily hydrolyzed, so they are biologically active for a much longer time period.

Enkephalins are related to a group of larger polypeptides called **endorphins** that contain 16–31 amino acids. Endorphins also block pain and are thought to produce the feeling of well-being experienced by an athlete after excessive or strenuous exercise.

PROBLEM 21.13	(a) Label the four amide bonds in met-enkephalin. (b) What N-terminal amino acid is present in both enkephalins? (c) How many chirality centers are present in met-enkephalin?

21.5B PEPTIDE HORMONES—OXYTOCIN AND VASOPRESSIN

Oxytocin and **vasopressin** are cyclic nonapeptide hormones secreted by the pituitary gland. Their sequences are identical except for two amino acids, yet this is enough to give them very different biological activities.

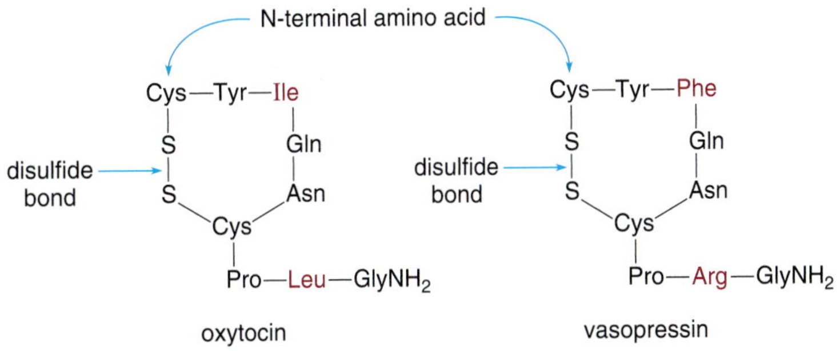

oxytocin   vasopressin

Oxytocin stimulates the contraction of uterine muscles, and it initiates the flow of milk in nursing mothers (Figure 21.2). Oxytocin, sold under the trade names Pitocin and Syntocinon, is used to induce labor.

Vasopressin, also called antidiuretic hormone (ADH), targets the kidneys and helps to keep the electrolytes in body fluids in the normal range. Vasopressin is secreted when the body is dehydrated and causes the kidneys to retain fluid, thus decreasing the volume of the urine (Figure 21.3).

The N-terminal amino acid in both hormones is a cysteine residue, and the C-terminal residue is glycine. Instead of a free carboxylate ($-COO^-$), both peptides have an amide ($-CONH_2$) at the C-terminal end, so this is indicated with the additional NH_2 group drawn at the end of the chain.

The structure of both peptides includes a **disulfide bond,** a form of covalent bonding in which the $-SH$ groups from two cysteine residues are oxidized to form a sulfur–sulfur bond (Section 14.10). In oxytocin and vasopressin, the disulfide bonds make the peptides cyclic.

$$2 \ R\!-\!S\!-\!H \ \xrightarrow{[O]} \ R\!-\!S\!-\!S\!-\!R$$

thiol   disulfide bond

PROBLEM 21.14   Draw the structure of the disulfide formed when cysteine is oxidized.

▼ **FIGURE 21.2   Physiological Effects of Oxytocin**

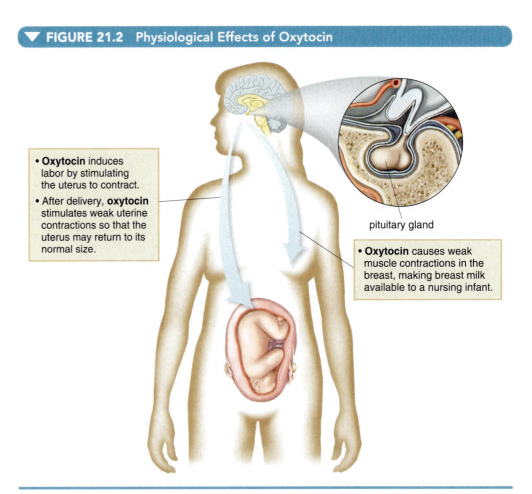

- **Oxytocin** induces labor by stimulating the uterus to contract.
- After delivery, **oxytocin** stimulates weak uterine contractions so that the uterus may return to its normal size.

pituitary gland

- **Oxytocin** causes weak muscle contractions in the breast, making breast milk available to a nursing infant.

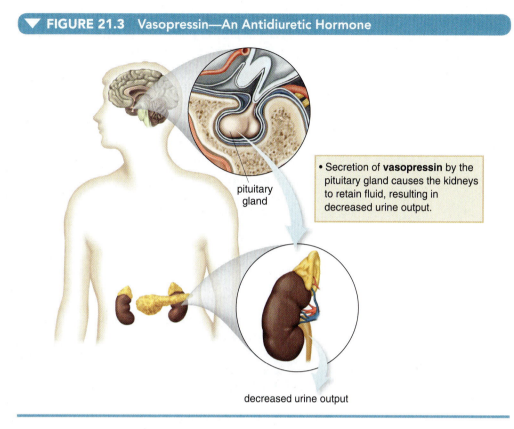

▼ **FIGURE 21.3** Vasopressin—An Antidiuretic Hormone

pituitary gland

• Secretion of **vasopressin** by the pituitary gland causes the kidneys to retain fluid, resulting in decreased urine output.

decreased urine output

21.6 PROTEINS

To understand proteins, the large polymers of amino acids that are responsible for so much of the structure and function of all living cells, we must learn about four levels of structure, called the **primary, secondary, tertiary, and quaternary structure** of proteins.

21.6A PRIMARY STRUCTURE

The *primary structure* of a protein is the particular sequence of amino acids that is joined together by peptide bonds. The most important element of this primary structure is the amide bond that joins the amino acids.

The carbonyl carbon of the amide has **trigonal planar** geometry. All six atoms involved in the peptide bond lie in the same plane. All bond angles are 120° and the C=O and N—H bonds are oriented 180° from each other. As a result, the backbone of the protein adopts a zigzag arrangement as shown in the three-dimensional structure of a portion of a protein molecule.

peptide bond

[Amide bonds are shown in red.]

These six atoms lie in a plane.

The primary structure of a protein—the exact sequence of amino acids—determines all properties and function of a protein. As we will see in Section 21.7, substitution of a single amino acid by a different amino acid can result in very different properties.

none

PROBLEM 21.15

Can two *different* proteins be composed of the same number and type of amino acids?

21.6B SECONDARY STRUCTURE

The three-dimensional arrangement of localized regions of a protein is called its secondary structure. These regions arise due to hydrogen bonding between the N—H proton of one amide and the C=O oxygen of another. Two arrangements that are particularly stable are called the **α-helix** and the **β-pleated sheet.**

α-Helix

The **α-helix** forms when a peptide chain twists into a right-handed or clockwise spiral, as shown in Figure 21.4. Several important features characterize an α-helix.

- Each turn of the helix has 3.6 amino acids.
- The N—H and C=O bonds point along the axis of the helix in opposite directions.
- The C=O group of one amino acid is hydrogen bonded to an N—H group four amino acid residues farther along the chain.
- The R groups of the amino acids extend outward from the core of the helix.

Both the myosin in muscle and the α-keratin in hair are proteins composed almost entirely of α-helices.

▼ **FIGURE 21.4 Two Different Illustrations of the α-Helix**

a. The right-handed α-helix

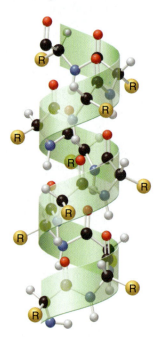

b. The backbone of the α-helix

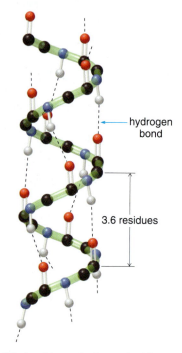

hydrogen bond

3.6 residues

All atoms of the α-helix are drawn in this representation. All C=O bonds are pointing up and all N—H bonds are pointing down.

Only the peptide backbone is drawn in this representation. The hydrogen bonds between the C=O and N—H of amino acids four residues away from each other are shown.

β-Pleated Sheet

The **β-pleated sheet** forms when two or more peptide chains, called **strands,** line up side-by-side, as shown in Figure 21.5. All β-pleated sheets have the following characteristics:

- The C=O and N—H bonds lie in the plane of the sheet.
- Hydrogen bonding often occurs between the N—H and C=O groups of nearby amino acid residues.
- The R groups are oriented above and below the plane of the sheet, and alternate from one side to the other along a given strand.

▼ **FIGURE 21.5**   Three-Dimensional Structure of the β-Pleated Sheet

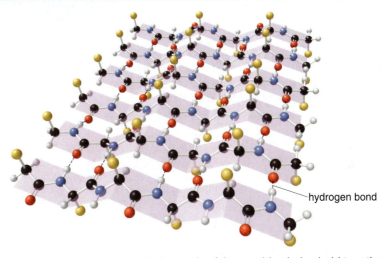

hydrogen bond

The β-pleated sheet consists of extended strands of the peptide chains held together by hydrogen bonding. The C=O and N—H bonds lie in the plane of the sheet, while the R groups (shown as yellow balls) alternate above and below the plane.

The β-pleated sheet arrangement is favored by amino acids with small R groups, like alanine and glycine. With larger R groups, steric interactions prevent the chains from getting close together, so the sheet cannot be stabilized by hydrogen bonding.

Most proteins have regions of α-helix and β-pleated sheet, in addition to other regions that cannot be characterized by either of these arrangements. Shorthand symbols are often used to indicate these regions of secondary structure, as well as disulfide bonds that are sometimes present (Figure 21.6). In particular, a flat helical ribbon is used for the α-helix, while a flat wide arrow is used for the β-pleated sheet. These representations are often used in **ribbon diagrams** to illustrate protein structure.

Proteins are drawn in a variety of ways to show different aspects of their structure. Figure 21.7 illustrates three different representations of the protein lysozyme, an enzyme found in both plants

▼ **FIGURE 21.6**   Shorthand Symbols Representing Protein Structure

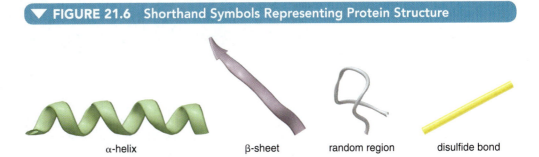

α-helix   β-sheet   random region   disulfide bond

▼ **FIGURE 21.7** Lysozyme

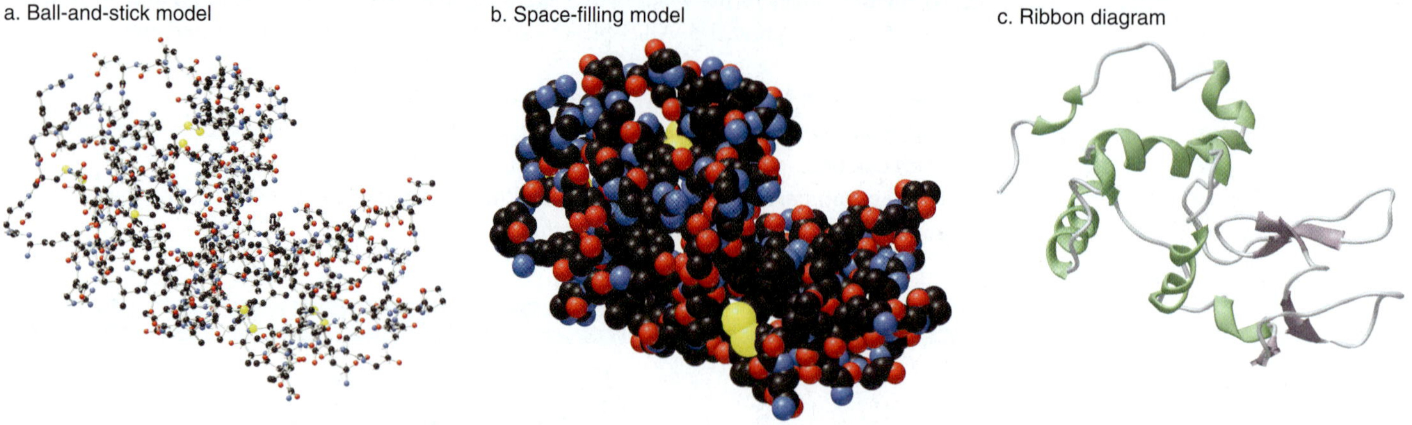

a. Ball-and-stick model

b. Space-filling model

c. Ribbon diagram

(a) The ball-and-stick model of lysozyme shows the protein backbone with color-coded C, N, O, and S atoms. Individual amino acids are most clearly located using this representation. (b) The space-filling model uses color-coded balls for each atom in the backbone of the enzyme and illustrates how the atoms fill the space they occupy. (c) The ribbon diagram shows regions of α-helix and β-sheet that are not clearly in evidence in the other two representations.

and animals. Lysozyme catalyzes the hydrolysis of bonds in bacterial cell walls, weakening them, often causing the bacteria to burst.

Spider dragline silk is a strong yet elastic protein because it has regions of β-pleated sheet and regions of α-helix (Figure 21.8). α-Helical regions impart elasticity to the silk because the peptide chain is twisted (not fully extended), so it can stretch. β-Pleated sheet regions are almost fully extended, so they can't be stretched further, but their highly ordered three-dimensional structure imparts strength to the silk. Thus, spider silk suits the spider by having both types of secondary structure with beneficial properties.

▼ **FIGURE 21.8** Spider Dragline Silk

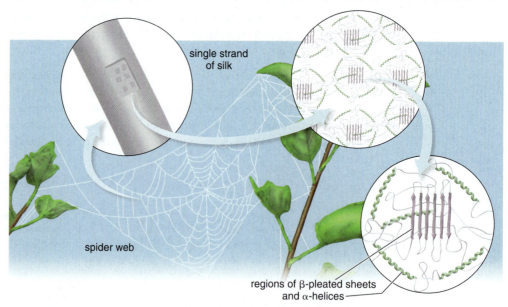

single strand
of silk

spider web

regions of β-pleated sheets
and α-helices

Spider dragline silk has regions of α-helix and β-pleated sheet that make it both elastic and strong. The green coils represent the α-helical regions, while the purple arrows represent the β-pleated sheet regions. The thin gray lines represents other areas of the protein that are neither α-helix nor β-pleated sheet.

21.6C TERTIARY AND QUATERNARY STRUCTURE

The three-dimensional shape adopted by the entire peptide chain is called its tertiary structure. A peptide generally folds into a shape that maximizes its stability. In the aqueous environment of the cell, proteins often fold in such a way as to place a large number of polar and charged groups on their outer surface, to maximize the dipole–dipole and hydrogen bonding interactions with water. This generally places most of the nonpolar side chains in the interior of the protein, where **London dispersion forces** between these hydrophobic groups help stabilize the molecule, too.

In addition, polar functional groups hydrogen bond with each other (not just water), and amino acids with charged side chains like –COO⁻ and –NH₃⁺ can stabilize tertiary structure by **electrostatic interactions.**

Finally, **disulfide bonds are the only covalent bonds that stabilize tertiary structure.** As mentioned in Section 21.5, these strong bonds form by the oxidation of two cysteine residues on either the same polypeptide chain or another polypeptide chain of the same protein.

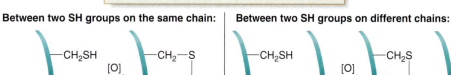

Insulin, for example, consists of two separate polypeptide chains (labeled the **A** and **B** chains) that are covalently linked by two intermolecular disulfide bonds, as shown in Figure 21.9. The **A** chain, which also has an intramolecular disulfide bond, has 21 amino acid residues, whereas the **B** chain has 30.

Figure 21.10 schematically illustrates the many different kinds of intramolecular forces that stabilize the secondary and tertiary structures of polypeptide chains. Nearby amino acid residues that have only nonpolar carbon–carbon and carbon–hydrogen bonds are stabilized by London dispersion forces. Amino acids that contain hydroxyl (OH) and amino groups (NH₂) in their side chains can intermolecularly hydrogen bond to each other.

The shape adopted when two or more folded polypeptide chains come together into one protein complex is called the **quaternary structure** of the protein. Each individual polypeptide chain is called a **subunit** of the overall protein. **Hemoglobin,** for example, consists of two α and two β subunits held together by intermolecular forces in a compact three-dimensional shape. The unique function of hemoglobin is possible only when all four subunits are together.

The four levels of protein structure are summarized in Figure 21.11.

PROBLEM 21.16

Draw the structures of each pair of amino acids and indicate what types of intermolecular forces are present between the side chains. Use Figure 21.10 as a guide.

a. Ser and Tyr   b. Val and Leu   c. 2 Phe residues

PROBLEM 21.17

The fibroin proteins found in silk fibers consist of large regions of β-pleated sheets stacked one on top of another. The polypeptide sequence in these regions has the amino acid glycine at every other residue. Explain how this allows the β-pleated sheets to stack on top of each other.

▼ **FIGURE 21.9**   Insulin

Insulin is a small protein consisting of two polypeptide chains (designated as the **A** and **B** chains), held together by two disulfide bonds. An additional disulfide bond joins two cysteine residues within the **A** chain.

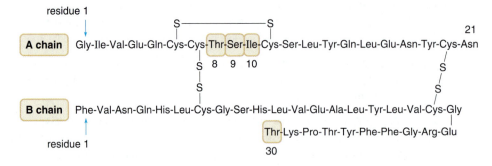

Synthesized by groups of cells in the pancreas called the islets of Langerhans, insulin is the protein that regulates blood glucose levels.

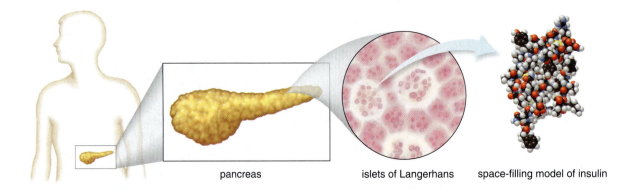

pancreas   islets of Langerhans   space-filling model of insulin

A relative or complete lack of insulin results in diabetes. Many of the abnormalities associated with this disease can be controlled by the injection of insulin. Until the recent availability of human insulin through genetic engineering techniques (Section 22.10), all insulin used by diabetics was obtained from pigs and cattle. The amino acid sequences of these insulin proteins are slightly different from that of human insulin. Pig insulin differs in one amino acid only, while bovine insulin has three different amino acids. This is shown in the accompanying table.

	Chain A			Chain B
Position of Residue →	8	9	10	30
Human insulin	Thr	Ser	Ile	Thr
Pig insulin	Thr	Ser	Ile	Ala
Bovine insulin	Ala	Ser	Val	Ala

▼ **FIGURE 21.10** The Stabilizing Interactions in Secondary and Tertiary Protein Structure

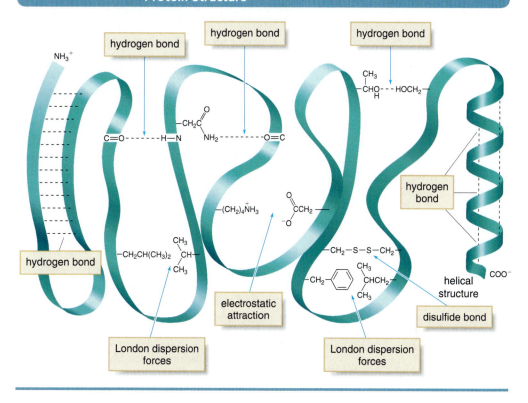

hydrogen bond

hydrogen bond

hydrogen bond

hydrogen bond

hydrogen bond

electrostatic attraction

London dispersion forces

London dispersion forces

disulfide bond

helical structure

▼ **FIGURE 21.11** The Primary, Secondary, Tertiary, and Quaternary Structure of Proteins

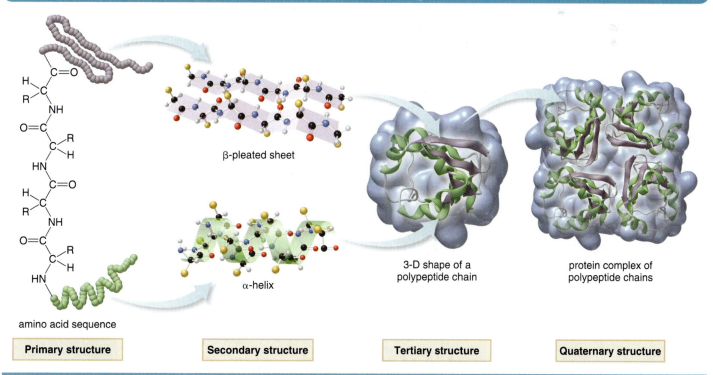

β-pleated sheet

α-helix

amino acid sequence

3-D shape of a polypeptide chain

protein complex of polypeptide chains

Primary structure   **Secondary structure**   **Tertiary structure**   **Quaternary structure**

21.7 FOCUS ON THE HUMAN BODY
COMMON PROTEINS

Proteins are generally classified according to their three-dimensional shapes.

> • *Fibrous proteins* are composed of long linear polypeptide chains that are bundled together to form rods or sheets. These proteins are insoluble in water and serve structural roles, giving strength and protection to tissues and cells.
> • *Globular proteins* are coiled into compact shapes with hydrophilic outer surfaces that make them water soluble. Enzymes and transport proteins are globular to make them soluble in blood and other aqueous environments.

21.7A α-KERATINS

α-Keratins are the proteins found in hair, hooves, nails, skin, and wool. They are composed almost exclusively of long sections of α-helix units, having large numbers of alanine and leucine residues. Since these nonpolar amino acids extend outward from the α-helix, these proteins are very insoluble in water. Two α-keratin helices coil around each other, forming a structure called a **supercoil** or **superhelix.** These, in turn, form larger and larger bundles of fibers, ultimately forming a strand of hair, as shown schematically in Figure 21.12.

α-Keratins also have a number of cysteine residues, and because of this, disulfide bonds are formed between adjacent helices. The number of disulfide bridges determines the strength of the material. Claws, horns, and fingernails have extensive networks of disulfide bonds, making them extremely hard.

21.7B COLLAGEN

Collagen, the most abundant protein in vertebrates, is found in connective tissues such as bone, cartilage, tendons, teeth, and blood vessels. Glycine and proline account for a large fraction of its amino acid residues. Collagen forms an elongated left-handed helix, and then three of these helices wind around each other to form a right-handed **superhelix** or **triple helix.** The side chain of glycine is only a hydrogen atom, so the high glycine content allows the collagen superhelices to lie compactly next to each other, thus stabilizing the superhelices via hydrogen bonding. Two views of the collagen superhelix are shown in Figure 21.13.

▼ **FIGURE 21.12** Anatomy of a Hair

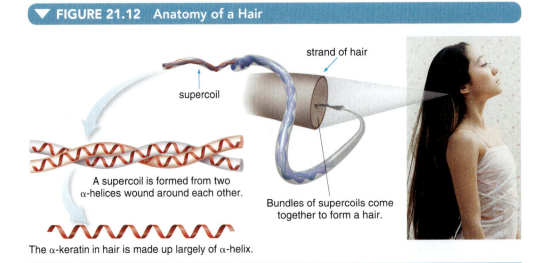

strand of hair

supercoil

A supercoil is formed from two α-helices wound around each other.

Bundles of supercoils come together to form a hair.

The α-keratin in hair is made up largely of α-helix.

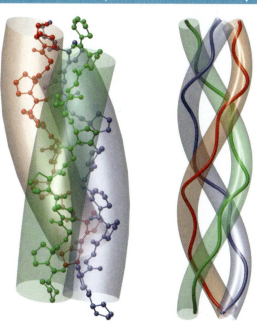

In collagen, three polypeptide chains having an unusual left-handed helix wind around each other in a right-handed triple helix. The high content of small glycine residues allows the chains to lie close to each other, permitting hydrogen bonding between the chains.

Vitamin C is required for a reaction that modifies the original amino acids incorporated into the collagen chain so that strong hydrogen bonds form between the helices. When there is a deficiency of vitamin C in the diet, the collagen fibers do not form properly and scurvy results. Weakened blood vessels and poorly formed cartilage lead to spongy and bloody gums and dark purple skin lesions.

21.7C   HEMOGLOBIN AND MYOGLOBIN

Hemoglobin and **myoglobin,** two globular proteins, are called **conjugated proteins** because they are composed of a protein unit and a nonprotein molecule. In hemoglobin and myoglobin, the nonprotein unit is called **heme,** a complex organic compound containing the Fe^{2+} ion complexed with a large nitrogen-containing ring system. The Fe^{2+} ion of hemoglobin and myoglobin binds oxygen. Hemoglobin, which is present in red blood cells, transports oxygen to wherever it is needed in the body, whereas myoglobin stores oxygen in tissues.

Whales have a particularly high myoglobin concentration in their muscles. It serves as an oxygen reservoir for the whale while it is submerged for long periods of time.

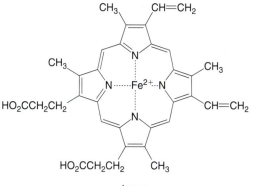

heme

▼ FIGURE 21.14   Protein Ribbon Diagrams for Myoglobin and Hemoglobin

a. Myoglobin

b. Hemoglobin

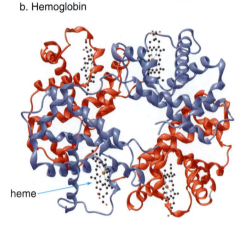

heme

heme

Myoglobin consists of a single polypeptide chain with a heme unit shown in a ball-and-stick model.

Hemoglobin consists of two α and two β chains shown in red and blue, respectively, and four heme units shown in ball-and-stick models.

Myoglobin has 153 amino acid residues in a single polypeptide chain (Figure 21.14a). It has eight separate α-helical sections and a heme group held in a cavity inside the polypeptide. Myoglobin gives cardiac muscle its characteristic red color.

Hemoglobin consists of four polypeptide chains (two α subunits and two β subunits), each of which carries a heme unit (Figure 21.14b). Hemoglobin has more nonpolar amino acid residues than myoglobin. When each subunit is folded, some of these remain on the surface. The London dispersion forces between these hydrophobic groups are what stabilize the quaternary structure of the four subunits.

Carbon monoxide is poisonous because it binds to the Fe^{2+} of hemoglobin 200 times more strongly than does oxygen. Hemoglobin complexed with CO cannot carry O_2 from the lungs to the tissues. Without O_2 available to the tissues for metabolism, cells cannot function, and they die.

The properties of all proteins depend on their three-dimensional shape, and their shape depends on their primary structure—that is, their amino acid sequence. This is particularly well exemplified by comparing normal hemoglobin with **sickle cell hemoglobin,** a mutant variation in which a single amino acid of both β subunits is changed from glutamic acid to valine. The replacement of one acidic amino acid (Glu) with one nonpolar amino acid (Val) changes the shape of hemoglobin, which has profound effects on its function. Red blood cells with sickle cell hemoglobin become elongated and crescent shaped, and they are unusually fragile. As a result, they rupture and occlude capillaries, causing pain and inflammation, leading to severe anemia and organ damage. The end result is often a painful and premature death.

This disease, called **sickle cell anemia,** is found almost exclusively among people originating from central and western Africa, where malaria is an enormous health problem. Sickle cell hemoglobin results from a genetic mutation in the DNA sequence that is responsible for the synthesis of hemoglobin. Individuals who inherit this mutation from both parents develop sickle cell anemia, whereas those who inherit it from only one parent are said to have the sickle cell trait.

They do not develop sickle cell anemia and they are more resistant to malaria than individuals without the mutation. The relative benefit of this mutation apparently accounts for this detrimental gene being passed on from generation to generation.

PROBLEM 21.18	Why is hemoglobin more water soluble than α-keratin?
PROBLEM 21.19	Why is it possible to discuss the quaternary structure of hemoglobin but not the quaternary structure of myoglobin?

21.8 PROTEIN HYDROLYSIS AND DENATURATION

The properties of a protein are greatly altered and often entirely destroyed when any level of protein structure is disturbed.

21.8A PROTEIN HYDROLYSIS

Like other amide bonds, the peptide bonds in proteins are hydrolyzed by treatment with aqueous acid, base, or certain enzymes.

- **The hydrolysis of the amide bonds in a protein forms the individual amino acids that comprise the primary structure.**

For example, hydrolysis of the amide bonds in the tripeptide Ile–Gly–Phe forms the amino acids isoleucine, glycine, and phenylalanine. When each amide bond is broken, the elements of H_2O are added, forming a carboxylate anion ($-COO^-$) in one amino acid and an ammonium cation ($-NH_3^+$) in the other.

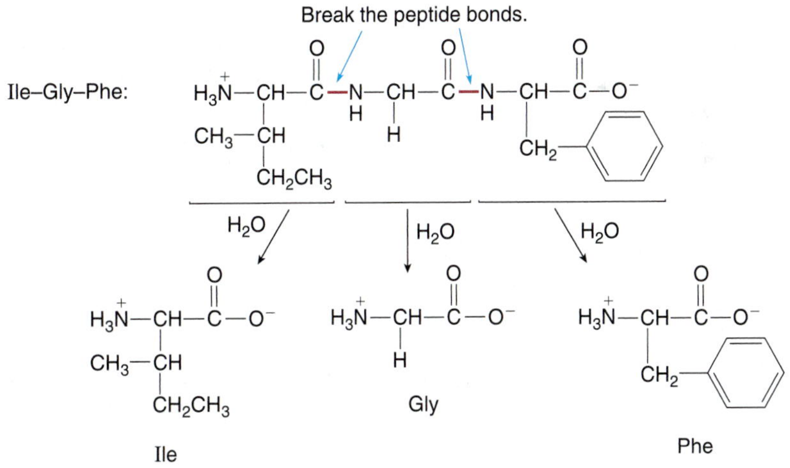

The first step in the digestion of dietary protein is hydrolysis of the amide bonds of the protein backbone. The enzyme pepsin in the acidic gastric juices of the stomach cleaves some of the amide bonds to form smaller peptides, which pass into the small intestines and are further broken down into individual amino acids by the enzymes trypsin and chymotrypsin.

Proteins in the diet serve a variety of nutritional needs. Like carbohydrates and lipids, proteins can be metabolized for energy. Moreover, the individual amino acids formed by hydrolysis are used as starting materials to make new proteins that the body needs. Likewise the N atoms in the amino acids are incorporated into other biomolecules that contain nitrogen.

SAMPLE PROBLEM 21.4

Draw the structures of the amino acids formed by hydrolysis of the neuropeptide leu-enkephalin (Section 21.5).

ANALYSIS

Locate each amide bond in the protein or peptide backbone. To draw the hydrolysis products, break each amide bond by adding the elements of H_2O to form a carboxylate anion ($-COO^-$) in one amino acid and an ammonium cation ($-NH_3^+$) in the other.

SOLUTION

[1] Locate the amide bonds in the peptide backbone.
[2] Break each bond by adding H_2O.

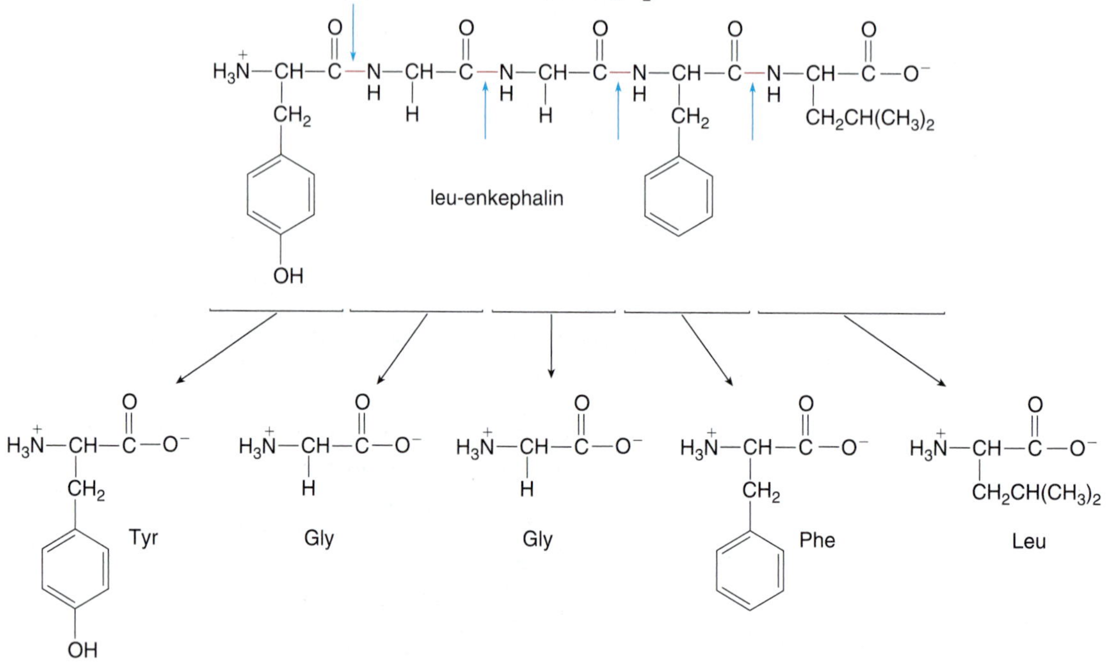

leu-enkephalin

Tyr   Gly   Gly   Phe   Leu

PROBLEM 21.20

Draw the structure of the products formed by hydrolysis of each tripeptide: (a) Ala–Leu–Gly; (b) Ser–Thr–Phe; (c) Leu–Tyr–Asn.

PROBLEM 21.21

What hydrolysis products are formed from the neuropeptide met-enkephalin (Section 21.5)?

21.8B   PROTEIN DENATURATION

When the secondary, tertiary, or quaternary structure of a protein is disturbed, the properties of a protein are also altered and the biological activity is often lost.

- *Denaturation* is the process of altering the shape of a protein without breaking the amide bonds that form the primary structure.

High temperature, acid, base, and even agitation can disrupt the noncovalent interactions that hold a protein in a specific shape. Heat breaks up weak London forces between nonpolar amino acids. Heat, acid, and base disrupt hydrogen bonding interactions between polar amino acids, which account for much of the secondary and tertiary structure. As a result, denaturation causes a globular protein to uncoil into an undefined randomly looped structure.

Cooking or whipping egg whites denatures the globular proteins they contain, forming insoluble protein.

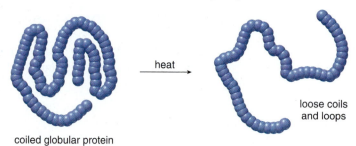

heat

coiled globular protein

loose coils and loops

Denaturation often makes globular proteins less water soluble. Globular proteins are typically folded with hydrophobic regions in the interior to maximize the interaction of polar residues on the outside surface with water. This makes them water soluble. When the protein is denatured, more hydrophobic regions are exposed and the protein often loses water solubility.

We witness many examples of protein denaturation in the kitchen. As milk ages it becomes sour from enzymes that produce lactic acid. The acid also denatures milk proteins, which precipitate as an insoluble curd. Ovalbumin, the major protein in egg white, is denatured when an egg is boiled or fried, forming a solid. Even vigorously whipping egg whites denatures its protein, forming the stiff meringue used to top a lemon meringue pie.

PROBLEM 21.22

Heating collagen, a water-insoluble fibrous protein, forms the jelly-like substance called gelatin. Explain how this process may occur.

21.9 ENZYMES

We conclude the discussion of proteins with *enzymes,* **proteins that serve as biological catalysts for reactions in all living organisms.** Like all catalysts (Section 6.4), enzymes increase the rate of reactions, but they themselves are not permanently changed in the process. Moreover, the position of equilibrium is not altered by an enzyme, nor are the relative energies of the starting material and product.

Enzymes are crucial to the biological reactions that occur in the body, which would otherwise often proceed too slowly to be of any use. In humans, enzymes must catalyze reactions under very specific physiological conditions, usually a pH around 7.4 and a temperature of 37 °C.

21.9A CHARACTERISTICS OF ENZYMES

Enzymes are generally water-soluble, globular proteins that exhibit two characteristic features.

- Enzymes greatly enhance reaction rates. An enzyme-catalyzed reaction can be 10^6 to 10^{12} times faster than a similar uncatalyzed reaction.
- Enzymes are very specific.

The specificity of an enzyme varies. Some enzymes, such as catalase, catalyze a single reaction. Catalase catalyzes the conversion of hydrogen peroxide (H_2O_2) to O_2 and H_2O (Section 5.1).

$$2\ H_2O_2(l) \xrightarrow{\text{catalase}} 2\ H_2O(l)\ +\ O_2(g)$$

Other enzymes, such as carboxypeptidase A, catalyze a particular type of reaction with a variety of substrates. Carboxypeptidase A, a digestive enzyme that breaks down proteins, catalyzes the hydrolysis of a specific type of peptide bond—the amide bond closest to the C-terminal end of the protein.

HEALTH NOTE

When a bloody wound is cleaned with hydrogen peroxide, the enzyme catalase in the blood converts the H_2O_2 to H_2O and O_2, forming a white foam of oxygen bubbles. When hydrogen peroxide is applied to the skin alone, no oxygen bubbles are seen because the skin contains no catalase to catalyze the reaction.

Only this amide bond is broken.

As is the case with catal*ase* and carboxypeptid*ase* A, **the names of most enzymes end in the suffix -ase.** Enzymes are classified into groups depending on the type of reaction they catalyze. For example, **an enzyme that catalyzes a hydrolysis reaction is called a *hydrolase.*** Hydrolases can be further subdivided into **lipases,** which catalyze the hydrolysis of the ester bonds in lipids (Section 19.5), or **proteases,** which catalyze the hydrolysis of proteins. Carboxypeptidase A is a protease.

Additional classes of enzymes are listed in Table 24.1.

The conversion of lactate to pyruvate illustrates other important features of enzyme-catalyzed reactions.

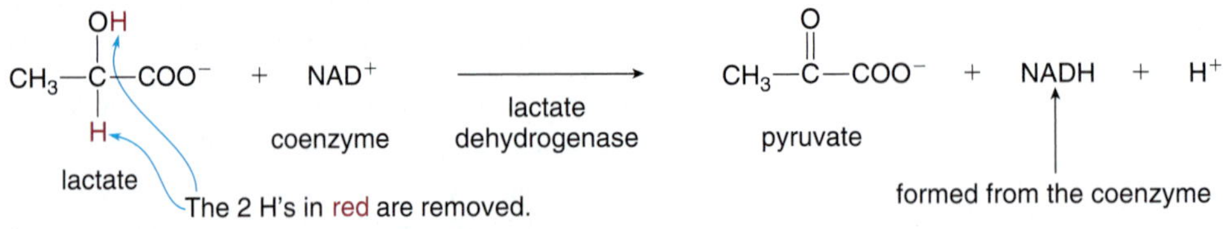

The 2 H's in red are removed.

formed from the coenzyme

The enzyme for this process is *lactate dehydrogenase,* a name derived from the substrate (lactate) and the type of reaction, a **dehydrogenation**—that is, the removal of two hydrogen atoms. To carry out this reaction a **cofactor** is needed.

A **dehydrogenase** catalyzes the removal of two hydrogen atoms from a substrate to form a double bond.

- A *cofactor* is a metal ion or a nonprotein organic molecule needed for an enzyme-catalyzed reaction to occur.

NAD^+ (nicotinamide adenine dinucleotide, Section 16.6) is the cofactor that oxidizes lactate to pyruvate. **An organic compound that serves as an enzyme cofactor is called a *coenzyme.*** While lactate dehydrogenase greatly speeds up this oxidation reaction, NAD^+ is the coenzyme that actually oxidizes the substrate, lactate. NAD^+ is a common biological oxidizing agent used as a coenzyme (Section 23.4).

The need for metal ions as enzyme cofactors explains why trace amounts of certain metals must be present in our diet. For example, certain **oxidases,** enzymes that catalyze oxidation reactions, require either Fe^{2+} or Cu^{2+} as a cofactor needed for transferring electrons.

PROBLEM 21.23

From the name alone, decide which of the following might be enzymes: (a) sucrose; (b) sucrase; (c) lactose; (d) lactase; (e) phosphofructokinase.

PROBLEM 21.24

Lyase is the general name for an enzyme that catalyses either the addition of elements to a double bond or the elimination of elements from a double bond. Fumarase is a lyase that catalyzes the conversion of fumarate to malate.

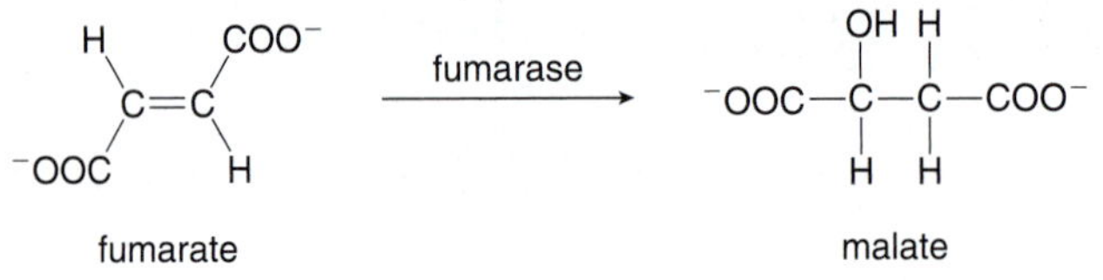

fumarate

malate

a. What elements are added to fumarate to form malate?

b. Would you expect this enzyme to be subclassified as a dehydrase or hydratase? Explain your choice.

21.9B   HOW ENZYMES WORK

An enzyme contains a region called the **active site** that binds the substrate, forming an **enzyme–substrate complex.** The active site is often a small cavity that contains amino acids that are attracted to the substrate with various types of intermolecular forces. Sometimes polar amino acids of the enzyme hydrogen bond to the substrate or nonpolar amino acids have stabilizing hydrophobic interactions.

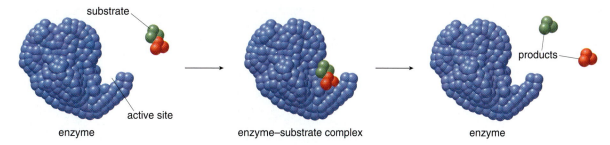

Two models have been proposed to explain the specificity of a substrate for an enzyme's active site: the **lock-and-key model** and the **induced-fit model.**

In the lock-and-key model, the shape of the active site is rigid. The three-dimensional geometry of the substrate must exactly match the shape of the active site for catalysis to occur. The lock-and-key model explains the high specificity observed in many enzyme reactions.

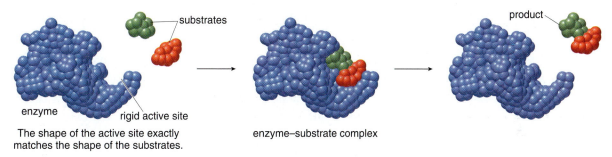

In the induced-fit model, the shape of the active site is more flexible. It is thought that when the substrate and the enzyme interact, the shape of the active site can adjust to fit the shape of the substrate.

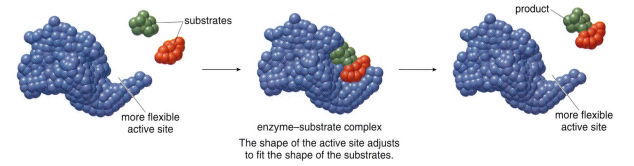

This model is often used to explain why some enzymes catalyze reactions with a wider variety of substrates. The shape of the active site and the substrate must still be reasonably similar, but once bound, the shape of the active site is different from the active site in the unbound enzyme. A well-characterized example of the induced-fit model is seen in the binding of glucose to the enzyme hexokinase, shown in Figure 21.15.

21.9C   ENZYME INHIBITORS

Some substances bind to enzymes and in the process, greatly alter or destroy the enzyme's activity.

- An *inhibitor* is a molecule that causes an enzyme to lose activity.

▼ **FIGURE 21.15   Hexokinase—An Example of the Induced-Fit Model**

a. Shape of the active site before binding

b. Shape of the active site after binding

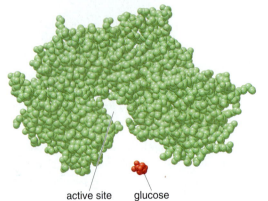

active site   glucose

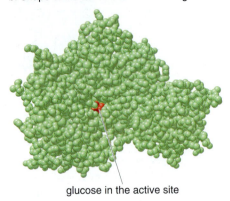

glucose in the active site

The open cavity of the active site of the enzyme closes around the substrate for a tighter fit.

(a) The free enzyme has a small open cavity that fits a glucose molecule. (b) Once bound, the cavity encloses the glucose molecule more tightly in the enzyme–substrate complex.

An inhibitor can bind to an enzyme reversibly or irreversibly.

- A *reversible* inhibitor binds to an enzyme but then enzyme activity is restored when the inhibitor is released.
- An *irreversible* inhibitor covalently binds to an enzyme, permanently destroying its activity.

Penicillin is an antibiotic that kills bacteria because it irreversibly binds to glycopeptide transpeptidase, an enzyme required for the synthesis of a bacterial cell wall (Section 17.11). Penicillin binds to a hydroxyl group (OH) of the enzyme, thus inactivating it, halting cell wall construction, and killing the bacterium. Penicillin has no effect on human cells because they are surrounded by a flexible cell membrane and not a cell wall.

penicillin G

Penicillin is covalently bound to the enzyme, inactivating it.

enzyme

inactive enzyme

Reversible inhibition can be **competitive** or **noncompetitive.**

A *noncompetitive* inhibitor binds to the enzyme but does not bind at the active site. The inhibitor causes the enzyme to change shape so that the active site can no longer bind the substrate. When the noncompetitive inhibitor is no longer bound to the enzyme, normal enzyme activity resumes.

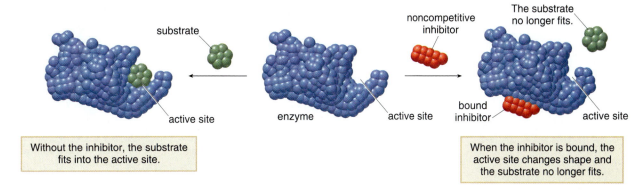

substrate

active site

Without the inhibitor, the substrate fits into the active site.

enzyme active site

noncompetitive inhibitor

The substrate no longer fits.

bound inhibitor active site

When the inhibitor is bound, the active site changes shape and the substrate no longer fits.

A *competitive* inhibitor has a shape and structure similar to the substrate, so it competes with the substrate for binding to the active site. The antibiotic **sulfanilamide** (Section 13.13B) is a competitive inhibitor of the enzyme needed to synthesize the vitamin folic acid from *p*-aminobenzoic acid in bacteria. When sulfanilamide binds to the active site, *p*-aminobenzoic acid cannot be converted to folic acid, so a bacterium cannot grow and reproduce. Sulfanilamide does not affect human cells because humans do not synthesize folic acid, and must obtain it in the diet.

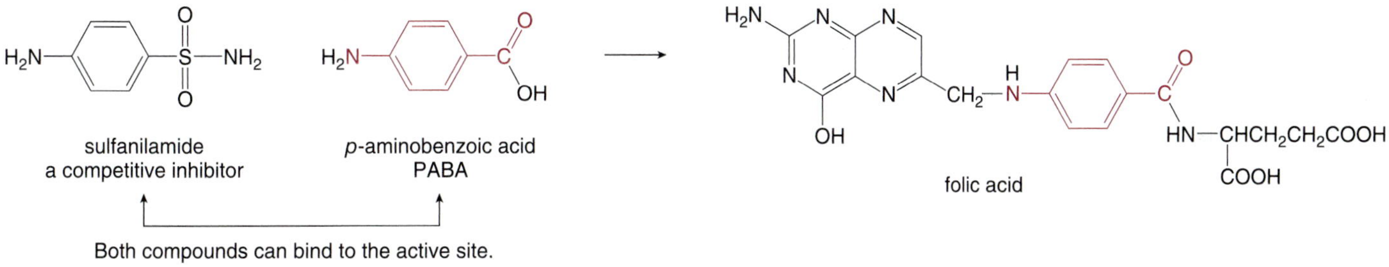

sulfanilamide
a competitive inhibitor

p-aminobenzoic acid
PABA

folic acid

Both compounds can bind to the active site.

PROBLEM 21.25

Use the given representations for an enzyme, substrate, and inhibitor to illustrate the process of competitive inhibition.

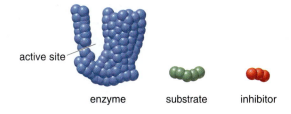

active site

enzyme substrate inhibitor

PROBLEM 21.26

The nerve gas sarin acts as a poison by covalently bonding to a hydroxyl group in the active site of the enzyme acetylcholinesterase. This binding results in a higher-than-normal amount of acetylcholine at a nerve synapse, resulting in muscle spasms. From this description, would you expect sarin to be a competitive, noncompetitive, or irreversible inhibitor?

21.9D ZYMOGENS

Sometimes an enzyme is synthesized in an inactive form, and then it is converted to its active form when it is needed. The inactive precursor of an enzyme is called a **zymogen** or **proenzyme.** Often the zymogen contains additional amino acids in its peptide chain, which are then cleaved when activation is desired.

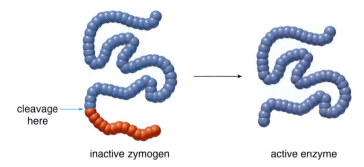

cleavage here

inactive zymogen   active enzyme

For example, the digestive enzymes trypsin and chymotrypsin are synthesized in the pancreas as the zymogens trypsinogen and chymotrypsinogen, respectively. As zymogens, these proteins are inactive and thus they do not digest the proteins that make up the pancreas. When they pass into the small intestines, however, some of the amino acids are removed from the peptide chain to form the active enzymes, which then catalyze the hydrolysis of peptide bonds during protein digestion.

PROBLEM 21.27 Explain why the proteins involved in blood clotting, fibrin and thrombin, are first synthesized as the zymogens fibrinogen and prothrombin, respectively.

21.10 FOCUS ON HEALTH & MEDICINE USING ENZYMES TO DIAGNOSE AND TREAT DISEASES

Measuring enzyme levels in the blood and understanding the key role of enzymes in biological reactions have aided greatly in both diagnosing and treating diseases.

21.10A ENZYME LEVELS AS DIAGNOSTIC TOOLS

Certain enzymes are present in a higher concentration in particular cells. When the cells are damaged by disease or injury, the cells rupture and die, releasing the enzymes into the bloodstream. Measuring the activity of the enzymes in the blood then becomes a powerful tool to diagnose the presence of disease or injury in some organs.

Measuring enzyme levels is now routine for many different conditions. For example, when a patient comes into an emergency room with chest pain, measuring the level of creatine phosphokinase (CPK) and other enzymes indicates whether a heart attack, which results in damage to some portion of the heart, has occurred. Common enzymes used for diagnosis are listed in Table 21.3.

TABLE 21.3 Common Enzymes Used for Diagnosis	
Enzyme	**Condition**
Creatine phosphokinase	Heart attack
Alkaline phosphatase	Liver or bone disease
Acid phosphatase	Prostate cancer
Amylase, lipase	Diseases of the pancreas

21.10B TREATING DISEASE WITH DRUGS THAT INTERACT WITH ENZYMES

Molecules that inhibit an enzyme can be useful drugs. The antibiotics penicillin and sulfanil-amide are two examples discussed in Section 21.9. Drugs used to treat high blood pressure and acquired immune deficiency syndrome (AIDS) are also enzyme inhibitors.

ACE inhibitors are a group of drugs used to treat individuals with high blood pressure. Angio-tensin is an octapeptide that narrows blood vessels, thus increasing blood pressure. Angiotensin is formed from the zymogen angiotensinogen by action of **ACE, the angiotensin-converting enzyme,** which cleaves two amino acids from the inactive decapeptide.

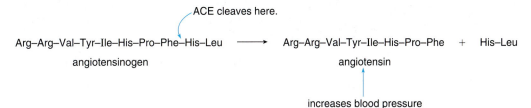

By blocking the conversion of angiotensinogen to angiotensin, blood pressure is decreased. Several effective ACE inhibitors are currently available, including captopril and enalapril.

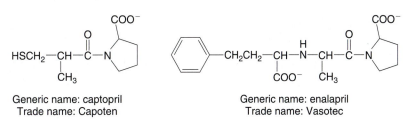

Several enzyme inhibitors are also available to treat human immunodeficiency virus (HIV), the virus that causes AIDS. The most effective treatments are **HIV protease inhibitors.** These drugs inhibit the action of the HIV protease enzyme, an essential enzyme needed by HIV to make copies of itself that go on to infect other cells. Deactivating the HIV protease enzyme decreases the virus population, bringing the disease under control. Several protease inhibitors are currently available, and often an individual takes a "cocktail" of several drugs to keep the disease in check. Amprenavir (trade name: Agenerase) is a protease inhibitor taken twice daily by individuals who are HIV positive. The three-dimensional structure of the HIV-1 protease enzyme is shown in Figure 21.16.

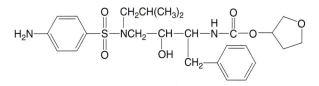

Generic name: amprenavir
Trade name: Agenerase

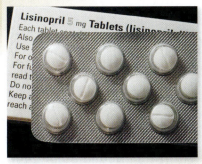
PROBLEM 21.28	How are the structures of the ACE inhibitors captopril and enalapril similar? How are they different?

▼ **FIGURE 21.16** The HIV Protease Enzyme

a. Ball-and-stick model of the HIV protease enzyme

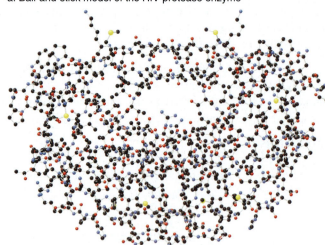

b. Ribbon diagram with the protease inhibitor amprenavir in the active site

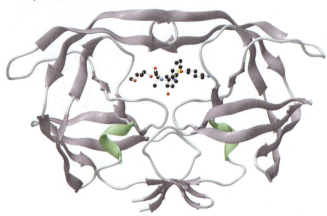

CHAPTER HIGHLIGHTS

KEY TERMS

Active site (21.9)
Amino acid (21.2)
Coenzyme (21.9)
Cofactor (21.9)
Competitive inhibitor (21.9)
Conjugated protein (21.7)
C-Terminal amino acid (21.4)
Denaturation (21.8)
Dipeptide (21.4)
Enzyme (21.9)
Enzyme–substrate complex (21.9)
Fibrous protein (21.7)

Globular protein (21.7)
α-Helix (21.6)
Heme (21.7)
Induced-fit model (21.9)
Inhibitor (21.9)
Irreversible inhibitor (21.9)
Isoelectric point (21.3)
Lock-and-key model (21.9)
Noncompetitive inhibitor (21.9)
N-Terminal amino acid (21.4)
Peptide (21.4)
Peptide bond (21.4)

β-Pleated sheet (21.6)
Primary structure (21.6)
Protein (21.1)
Quaternary structure (21.6)
Reversible inhibitor (21.9)
Secondary structure (21.6)
Tertiary structure (21.6)
Tripeptide (21.4)
Zwitterion (21.2)
Zymogen (21.9)

KEY CONCEPTS

❶ **What are the main structural features of an amino acid? (21.2)**

- Amino acids contain an amino group (NH_2) on the α carbon to the carboxyl group (COOH). Because they contain both an acid and a base, amino acids exist in their neutral form as zwitterions having the general structure $^+H_3NCH(R)COO^-$. Because they are salts, amino acids are water soluble and have high melting points.

- All amino acids except glycine (R = H) have a chirality center on the α carbon. L Amino acids are naturally occurring.

- Amino acids are subclassified as neutral, acidic, or basic by the functional groups present in the R group, as shown in Table 21.2.

❷ **Describe the acid–base properties of amino acids. (21.3)**

- Neutral, uncharged amino acids exist as zwitterions containing an ammonium cation ($-NH_3^+$) and a carboxylate anion ($-COO^-$).

- When strong acid is added, the carboxylate anion gains a proton and the amino acid has a net +1 charge. When strong base is added, the ammonium cation loses a proton and the amino acid has a net −1 charge.

$$H_3\overset{+}{N}-\underset{\underset{CH_3}{|}}{\overset{\overset{H}{|}}{C}}-COOH \quad \xleftarrow{H^+} \quad H_3\overset{+}{N}-\underset{\underset{CH_3}{|}}{\overset{\overset{H}{|}}{C}}-COO^- \quad \xrightarrow{^-OH} \quad H_2N-\underset{\underset{CH_3}{|}}{\overset{\overset{H}{|}}{C}}-COO^-$$

overall +1 charge   overall −1 charge

pH ≤ 2   pH ≈ 6   pH ≥ 10

❸ What are the main structural features of peptides? (21.4)

- Peptides contain amino acids, called amino acid residues, joined together by amide (peptide) bonds. The amino acid that contains the free $-NH_3^+$ group on the α carbon is called the N-terminal amino acid, and the amino acid that contains the free $-COO^-$ group on the α carbon is the C-terminal amino acid.
- Peptides are written from left to right, from the N-terminal to the C-terminal end, using the one- or three-letter abbreviations for the amino acids listed in Table 21.2.

❹ Give examples of simple biologically active peptides. (21.5)

- Enkephalins are pentapeptides that act as sedatives and pain killers by binding to pain receptors.
- Oxytocin is a nonapeptide hormone that stimulates the contraction of uterine muscles and initiates the flow of milk in nursing mothers.
- Vasopressin is a nonapeptide hormone that serves as an antidiuretic; that is, vasopressin causes the kidneys to retain water.

❺ What are the general characteristics of the primary, secondary, tertiary, and quaternary structure of proteins? (21.6)

- The primary structure of a protein is the particular sequence of amino acids joined together by amide bonds.
- The two most common types of secondary structure are the α-helix and the β-pleated sheet. Both structures are stabilized by hydrogen bonds between the N—H and C=O groups.
- The tertiary structure is the three-dimensional shape adopted by the entire peptide chain. London dispersion forces stabilize hydrophobic interactions between nonpolar amino acids. Hydrogen bonding and ionic interactions occur between polar or charged amino acid residues. Disulfide bonds are covalent sulfur–sulfur bonds that occur between cysteine residues in different parts of the peptide chain.
- When a protein contains more than one polypeptide chain, the quaternary structure describes the shape of the protein complex formed by two or more chains.

❻ What are the basic features of fibrous proteins like α-keratin and collagen? (21.7)

- Fibrous proteins are composed of long linear polypeptide chains that serve structural roles and are water insoluble.
- α-Keratin in hair is a fibrous protein composed almost exclusively of α-helix units that wind together to form a superhelix. Disulfide bonds between chains make the resulting bundles of protein chains strong.
- Collagen, found in connective tissue, is composed of a superhelix formed from three elongated left-handed helices.

❼ What are the basic features of globular proteins like hemoglobin and myoglobin? (21.7)

- Globular proteins have compact shapes and are folded to place polar amino acids on the outside to make them water soluble. Hemoglobin and myoglobin are both conjugated proteins composed of a protein unit and a heme molecule. The Fe^{2+} ion of the heme binds oxygen. While myoglobin has a single polypeptide chain, hemoglobin contains four peptide chains that form a single protein molecule.

❽ What products are formed when a protein is hydrolyzed? (21.8)

- Hydrolysis breaks up the primary structure of a protein to form the amino acids that compose it. All of the amide bonds are broken by the addition of water, forming a carboxylate anion ($-COO^-$) in one amino acid and an ammonium cation ($-NH_3^+$) in the other.

❾ What is denaturation? (21.8)

- Denaturation is a process that alters the shape of a protein by disrupting the secondary, tertiary, or quaternary structure. High temperature, acid, base, and agitation can denature a protein. Compact water-soluble proteins uncoil and become less water soluble.

❿ What are the main structural features of enzymes? (21.9)

- Enzymes are biological catalysts that greatly increase the rate of biological reactions and are highly specific for a substrate or a type of substrate. An enzyme binds a substrate at its active site, forming an enzyme–substrate complex by either the lock-and-key model or the induced-fit model.
- Enzyme inhibitors cause an enzyme to lose activity. Irreversible inhibition occurs when an inhibitor covalently binds the enzyme and permanently destroys its activity. Competitive reversible inhibition occurs when the inhibitor is structurally similar to the substrate and competes with it for occupation of the active site. Noncompetitive reversible inhibition occurs when an inhibitor binds to a location other than the active site, altering the shape of the active site.

⓫ How are enzymes used in medicine? (21.10)

- Measuring blood enzyme levels is used to diagnose heart attacks and diseases that cause higher-than-normal concentrations of certain enzymes to enter the blood.
- Drugs that inhibit the action of an enzyme can be used to kill bacteria. ACE inhibitors are used to treat high blood pressure. HIV protease inhibitors are used to treat HIV by binding to an enzyme needed by the virus to replicate itself.

PROBLEMS

Selected in-chapter and end-of-chapter problems have brief answers provided in Appendix B.

Amino Acids

21.29 Naturally occurring amino acids are L-α-amino acids. What do the L and α designations represent?

21.30 Why do neutral amino acids exist as zwitterions with no net charge?

21.31 The amino acid alanine is a solid at room temperature and has a melting point of 315 °C, while pyruvic acid (CH_3COCO_2H) has a similar molecular weight but is a liquid at room temperature with a boiling point of 165 °C. Account for the difference.

21.32 Why is phenylalanine water soluble but 4-phenylbutanoic acid ($C_6H_5CH_2CH_2CH_2COOH$), a compound of similar molecular weight, water insoluble?

21.33 Draw the structure of a naturally occurring amino acid that:
  a. contains a 1° alcohol
  b. contains an amide
  c. is an essential amino acid with an aromatic ring
  d. is a neutral amino acid with a 3° carbon atom in its side chain

21.34 Draw the structure of a naturally occurring amino acid that:
  a. contains a 2° alcohol
  b. contains a thiol
  c. is an acidic amino acid
  d. is a neutral amino acid with a phenol in the side chain

21.35 What two amino acids contain two chirality centers?

21.36 What makes proline different from all of the other amino acids in Table 21.2?

21.37 For each amino acid: [1] draw the L enantiomer in a Fischer projection; [2] classify the amino acid as neutral, acidic, or basic; [3] give the three-letter symbol; [4] give the one-letter symbol.
  a. leucine   c. lysine
  b. tryptophan   d. aspartic acid

21.38 For each amino acid: [1] draw the L enantiomer in a Fischer projection; [2] classify the amino acid as neutral, acidic, or basic; [3] give the three-letter symbol; [4] give the one-letter symbol.
  a. arginine   c. glutamic acid
  b. tyrosine   d. valine

21.39 Draw both enantiomers of each amino acid and label them as D or L: (a) methionine; (b) asparagine.

21.40 Which of the following Fischer projections represent naturally occurring amino acids? Name each amino acid and designate it as a D or L isomer.

a.

b.

c.

21.41 For each amino acid: [1] give the name; [2] give the three-letter abbreviation; [3] give the one-letter abbreviation; [4] classify the amino acid as neutral, acidic, or basic.

a.

b.

c.

21.42 For each amino acid: [1] give the name; [2] give the three-letter abbreviation; [3] give the one-letter abbreviation; [4] classify the amino acid as neutral, acidic, or basic.

a.

b.

c.

Acid–Base Properties of Amino Acids

21.43 Draw the amino acid leucine at each pH: (a) 6; (b) 10; (c) 2. Which form predominates at leucine's isoelectric point?

21.44 Draw the amino acid isoleucine at each pH: (a) 6; (b) 10; (c) 2. Which form predominates at isoleucine's isoelectric point?

21.45 Draw the structure of the neutral, positively charged, and negatively charged forms of the amino acid tyrosine. Which form predominates at pH 1? Which form predominates at pH 11? Which form predominates at the isoelectric point?

21.46 Draw the structure of the neutral, positively charged, and negatively charged forms of the amino acid valine. Which form predominates at pH 1? Which form predominates at pH 11? Which form predominates at the isoelectric point?

Peptides

21.47 (a) Draw the structure of the two possible dipeptides that can be formed by combining valine and phenylalanine. (b) In each dipeptide, label the N- and C-terminal amino acids. (c) Name each peptide using three-letter symbols.

21.48 (a) Draw the structure of the two possible dipeptides that can be formed by combining glycine and asparagine. (b) In each dipeptide, label the N- and C-terminal amino acids. (c) Name each peptide using three-letter symbols.

21.49 For each tripeptide: [1] draw the structure of the tripeptide; [2] label the amide bonds; [3] identify the N-terminal and C-terminal amino acids; [4] name the peptide using three-letter symbols for the amino acids.
a. leucylvalyltryptophan
b. alanylglycylvaline
c. phenylalanylserylthreonine

21.50 For each tripeptide: [1] draw the structure of the tripeptide; [2] label the amide bonds; [3] identify the N-terminal and C-terminal amino acids; [4] name the peptide using three-letter symbols for the amino acids.
a. tyrosylleucylisoleucine
b. histidylglutamyllysine
c. methionylisoleucylcysteine

21.51 For each tripeptide: [1] identify the amino acids that form the peptide; [2] label the N- and C-terminal amino acids; [3] name the tripeptide using three-letter symbols.

21.52 For each tripeptide: [1] identify the amino acids that form the peptide; [2] label the N- and C-terminal amino acids; [3] name the tripeptide using three-letter symbols.

21.53 Draw the structure of the three different tripeptides formed from two moles of serine and one mole of alanine.

21.54 How many different tripeptides can be formed from three different amino acids, methionine, histidine, and arginine? Using three-letter abbreviations, give the names for all of the possible tripeptides.

21.55 Draw the structures of the amino acids formed when the tripeptides in Problem 21.51 are hydrolyzed.

21.56 Draw the structures of the amino acids formed when the tripeptides in Problem 21.52 are hydrolyzed.

21.57 Bradykinin is a nonapeptide that stimulates smooth muscle contraction, dilates blood vessels, causes pain, and is a component of bee venom. Draw the structure of the amino acids formed on hydrolysis of one mole of bradykinin, and tell how many moles of each amino acid are formed.

Arg–Pro–Pro–Gly–Phe–Ser–Pro–Phe–Arg
bradykinin

21.58 Aspartame, the methyl ester of a dipeptide, is the common artificial sweetener NutraSweet (Section 20.5B). Draw the structure of the products formed when the amide and ester bonds are hydrolyzed. Which amino acid contains the carbonyl group of the peptide bond? Which amino acid contains the nitrogen atom of the amide bond?

aspartame

21.59 Chymotrypsin is a digestive enzyme that hydrolyzes only certain peptide bonds: the carbonyl group in the amide bond must come from phenylalanine, tyrosine, tryptophan, or methionine. Draw the structure of the amino acids and peptides formed by hydrolysis of the following peptide with chymotrypsin: Gly–Tyr–Gly–Ala–Phe–Val.

21.60 Trypsin is a digestive enzyme that hydrolyzes peptide bonds only when the carbonyl group in the amide bond comes from lysine or arginine. Draw the structure of the amino acids and peptides formed by hydrolysis of the following peptide with trypsin: Gly–Lys–Arg–Ala–Ala–Arg.

Proteins

21.61 What is the difference between the primary and secondary structure of a protein?

21.62 What is the difference between the tertiary and quaternary structure of a protein?

21.63 What type of intermolecular forces exist between the side chains of each of the following pairs of amino acids?
    a. isoleucine and valine   c. Lys and Glu
    b. threonine and phenylalanine   d. Arg and Asp

21.64 Which of the following pairs of amino acids can have intermolecular hydrogen bonding between the functional groups in their side chains?
    a. two tyrosine residues   c. phenylalanine and tyrosine
    b. serine and threonine   d. alanine and threonine

21.65 Draw the structures of the amino acids asparagine and serine and show how hydrogen bonding is possible between the two side chains.

21.66 Draw the structures of the amino acids tyrosine and threonine and show how hydrogen bonding is possible between the two side chains.

21.67 List four amino acids that would probably be located in the interior of a globular protein.

21.68 List four amino acids that would probably be located on the exterior of a globular protein.

21.69 Label the regions of secondary structure in the following protein ribbon diagram.

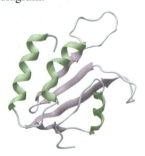

21.70 Label the regions of secondary structure in the following protein ribbon diagram.

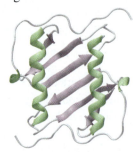

21.71 Compare α-keratin and hemoglobin with regards to each of the following: (a) secondary structure; (b) water solubility; (c) function; (d) location in the body.

21.72 Compare collagen and myoglobin with regards to each of the following: (a) secondary structure; (b) water solubility; (c) function; (d) location in the body.

21.73 When a protein is denatured, how is its primary, secondary, tertiary, and quaternary structure affected?

21.74 Hydrogen bonding stabilizes both the secondary and tertiary structures of a protein. (a) What functional groups hydrogen bond to stabilize secondary structure? (b) What functional groups hydrogen bond to stabilize tertiary structure?

21.75 Describe the function or biological activity of each protein or peptide: (a) insulin; (b) myoglobin; (c) α-keratin; (d) chymotrypsin; (e) oxytocin.

21.76 Describe the function or biological activity of each protein or peptide: (a) collagen; (b) hemoglobin; (c) vasopressin; (d) pepsin; (e) met-enkephalin.

Enzymes

21.77 What is the difference between reversible and irreversible inhibition?

21.78 What is the difference between competitive and noncompetitive inhibition?

21.79 Explain the role of zymogens in (a) controlling blood pressure; (b) the action of digestive enzymes.

21.80 What is the difference between a coenzyme and a cofactor?

21.81 How are enzyme inhibitors used to treat high blood pressure? Give a specific example of a drug used and an enzyme inhibited.

21.82 How are enzyme inhibitors used to treat HIV? Give a specific example of a drug used and an enzyme inhibited.

21.83 Use the given representations for an enzyme, substrate, and inhibitor to illustrate the process of noncompetitive inhibition.

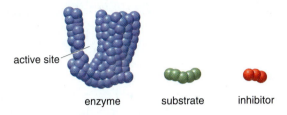

21.84 Use the given representations for an enzyme and substrate to illustrate the difference between the lock-and-key model and the induced-fit model of enzyme specificity.

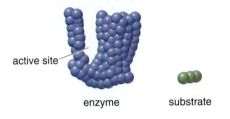

Applications

21.85 What structural feature in α-keratin makes fingernails harder than skin?

21.86 Why does the α-keratin in hair contain many cysteine residues?

21.87 Why must vegetarian diets be carefully balanced?

21.88 Why does cooking meat make it easier to digest?

21.89 Sometimes an incision is cauterized (burned) to close the wound and prevent bleeding. What does cauterization do to protein structure?

21.90 Why is insulin administered by injection instead of taken in tablet form?

21.91 How is sickle cell disease related to hemoglobin structure?

21.92 The silk produced by a silkworm is a protein with a high glycine and alanine content. With reference to the structure, how does this make the silk fiber strong?

21.93 Explain the difference in the mechanism of action of penicillin and sulfanilamide. How is enzyme inhibition involved in both mechanisms? Why don't penicillin and sulfanilamide kill both human and bacterial cells?

21.94 How are blood enzyme levels used to diagnose certain diseases? Give an example of a specific condition and enzyme used for diagnosis.

CHALLENGE QUESTIONS

21.95 Explain why two amino acids—aspartic acid and glutamic acid—have a +1 net charge at low pH, but a –2 net charge at high pH.

21.96 Suggest a reason for the following observation. Proteins like collagen that are high in proline content do not contain an α-helix in their secondary structure.

22

• • • • • • • • • • • • • •

CHAPTER GOALS

In this chapter you will learn how to:

① Draw the structure of nucleosides and nucleotides

② Draw short segments of the nucleic acids DNA and RNA

③ Describe the basic features of the DNA double helix

④ Outline the main steps of replication

⑤ List the three types and functions of RNA molecules

⑥ Explain the process of transcription

⑦ Describe the basic elements of the genetic code

⑧ Explain the process of translation

⑨ Define the terms "mutation" and "genetic disease"

⑩ Describe the basic features of recombinant DNA, the polymerase chain reaction, and DNA fingerprinting

⑪ Describe the main characteristics of viruses

Techniques to analyze DNA samples are commonly utilized in criminal investigations and to screen for genetic diseases.

NUCLEIC ACIDS AND PROTEIN SYNTHESIS

WHETHER you are tall or short, fair-skinned or dark-complexioned, blue-eyed or brown-eyed, your unique characteristics are determined by the nucleic acid polymers that reside in the chromosomes of your cells. The nucleic acid **DNA** stores the genetic information of a particular organism, while the nucleic acid **RNA** translates this genetic information into the synthesis of proteins needed by cells for proper function and development. Even minor alterations in the nucleic acid sequence can have significant effects on an organism, sometimes resulting in devastating diseases like sickle cell anemia and cystic fibrosis. In Chapter 22, we study nucleic acids and learn how the genetic information stored in DNA is translated into protein synthesis.

22.1 NUCLEOSIDES AND NUCLEOTIDES

Nucleic acids are unbranched polymers composed of repeating monomers called *nucleotides*. There are two types of nucleic acids.

- DNA, deoxyribonucleic acid, stores the genetic information of an organism and transmits that information from one generation to another.
- RNA, ribonucleic acid, translates the genetic information contained in DNA into proteins needed for all cellular functions.

The nucleotide monomers that compose DNA and RNA consist of three components—a **monosaccharide, a nitrogen-containing base, and a phosphate group.**

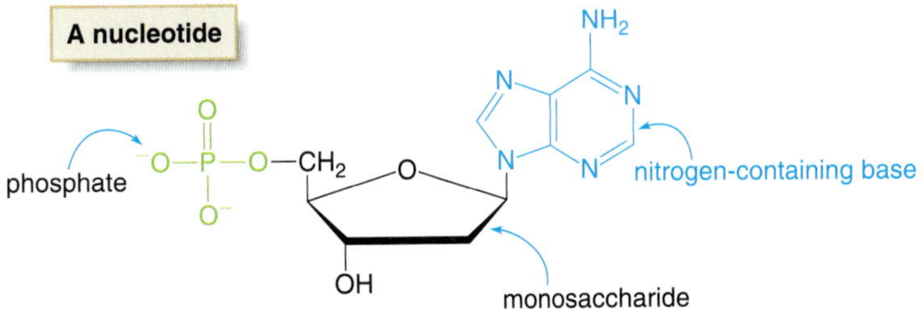

DNA molecules contain several million nucleotides while RNA molecules are much smaller, containing perhaps a few thousand nucleotides. DNA is contained in the chromosomes of the nucleus, each chromosome having a different type of DNA. The number of chromosomes differs from species to species. Humans have 46 chromosomes (23 pairs). An individual chromosome is composed of many genes. **A *gene* is a portion of the DNA molecule responsible for the synthesis of a single protein.**

We begin our study of nucleic acids with a look at the structure and formation of the nucleotide monomers.

22.1A NUCLEOSIDES—JOINING A MONOSACCHARIDE AND A BASE

The nucleotides of both DNA and RNA contain a five-membered ring monosaccharide, often called simply the *sugar* component.

- In RNA, the monosaccharide is the aldopentose D-ribose.
- In DNA the monosaccharide is D-2-deoxyribose, an aldopentose that lacks a hydroxyl group at C2.

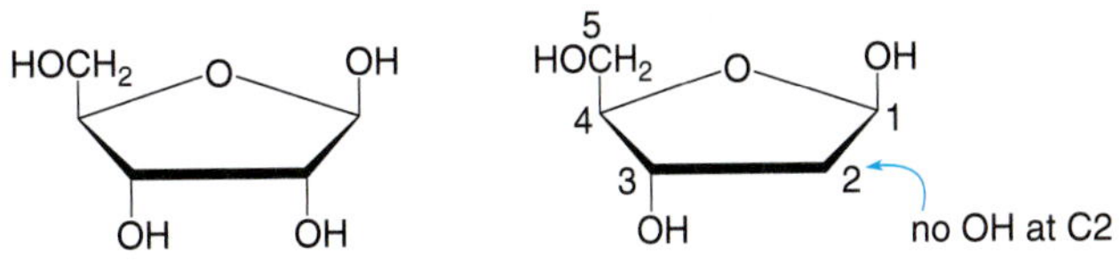

D-ribose
(present in RNA)

D-2-deoxyribose
(present in DNA)

The prefix *deoxy* means *without oxygen.*

Only five common nitrogen-containing bases are present in nucleic acids. Three bases with one ring (**cytosine, uracil,** and **thymine**) are derived from the parent compound pyrimidine. Two bases with two rings (**adenine** and **guanine**) are derived from the parent compound purine. Each base is designated by a one-letter abbreviation as shown.

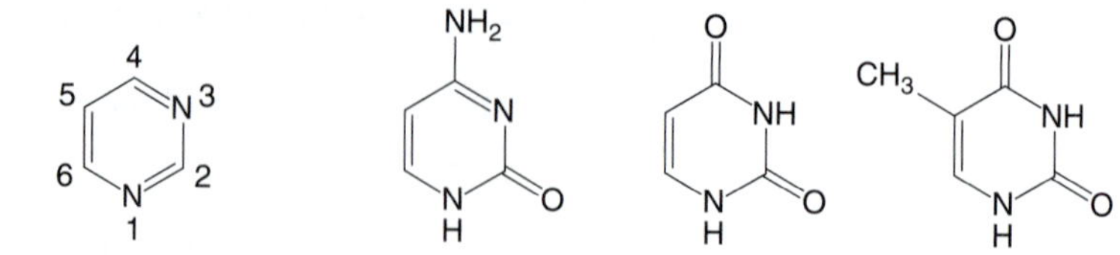

pyrimidine
(parent compound)

cytosine
C

uracil
U

thymine
T

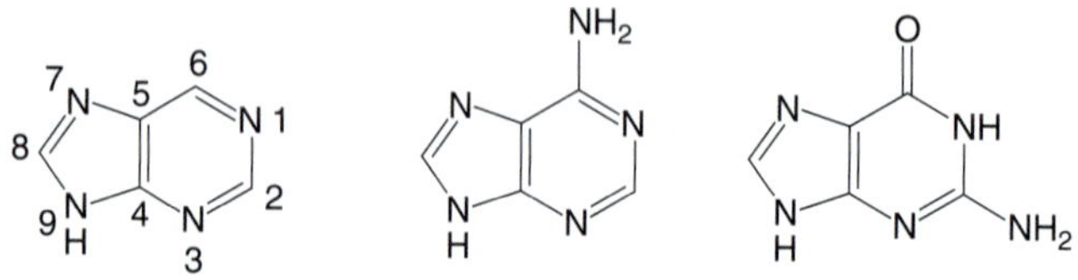

purine
(parent compound)

adenine
A

guanine
G

Uracil (U) occurs only in RNA, while thymine (T) occurs only in DNA. As a result:

- DNA contains the bases A, G, C, and T.
- RNA contains the bases A, G, C, and U.

A *nucleoside* is formed by joining the anomeric carbon of the monosaccharide with a nitrogen atom of the base. A nucleoside is an *N*-glycoside, because it is a nitrogen derivative of the glycosides discussed in Chapter 20. Primes (') are used to number the carbons of the monosaccharide in a nucleoside, to distinguish them from the atoms in the rings of the base.

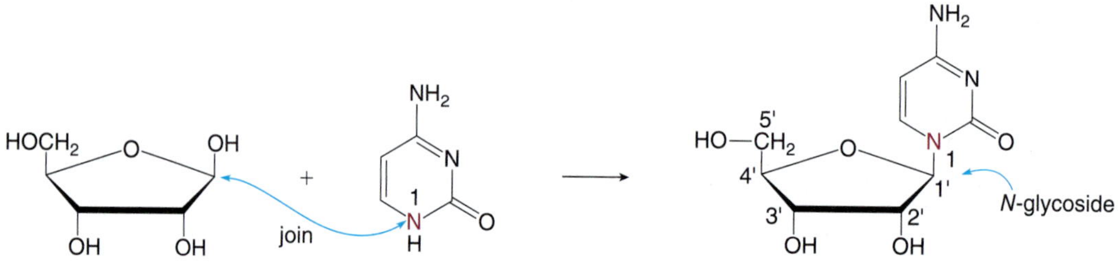

D-ribose

cytosine
C

cytidine
a ribonucleoside

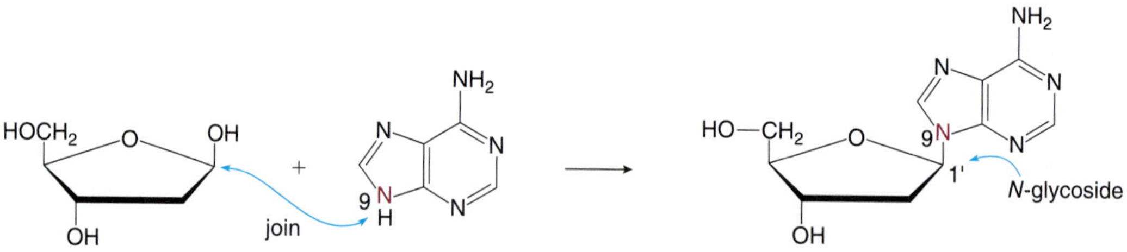

D-2-deoxyribose

adenine
A

deoxyadenosine
a deoxyribonucleoside

With pyrimidine bases, the nitrogen atom at the 1 position bonds with the 1' carbon of the sugar. With purine bases, the nitrogen atom at the 9 position bonds with the 1' carbon of the sugar. For example, joining cytosine with ribose forms the ribonucleoside **cytidine.** Joining adenine with 2-deoxyribose forms the deoxyribonucleoside **deoxyadenosine.**

Nucleosides are named as derivatives of the bases from which they are formed.

- To name a nucleoside derived from a pyrimidine base, use the suffix *-idine* (cytosine → cyt*idine*).
- To name a nucleoside derived from a purine base, use the suffix *-osine* (adenine → aden*osine*).
- For deoxyribonucleosides, add the prefix *deoxy-,* as in *deoxy*adenosine.

SAMPLE PROBLEM 22.1

Identify the base and monosaccharide used to form the following nucleoside, and then name it.

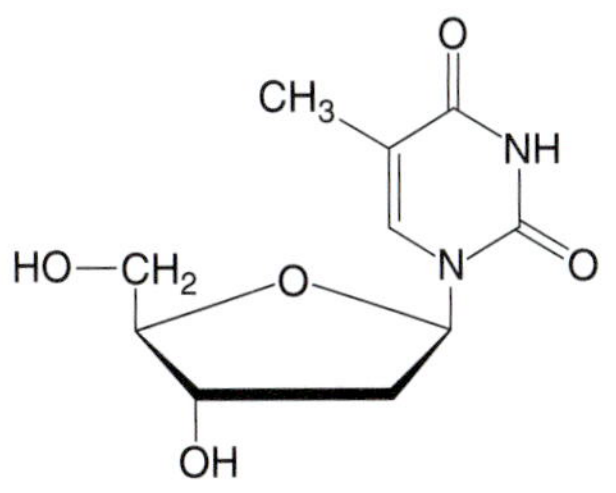

ANALYSIS

- The sugar portion of a nucleoside contains the five-membered ring. If there is an OH group at C2', the sugar is ribose, and if there is no OH group at C2', the sugar is deoxyribose.
- The base is joined to the five-membered ring as an *N*-glycoside. A pyrimidine base has one ring, and is derived from either cytosine, uracil, or thymine. A purine base has two rings, and is derived from either adenine or guanine.
- Nucleosides derived from pyrimidines end in the suffix *-idine.* Nucleosides derived from purines end in the suffix *-osine.*

SOLUTION

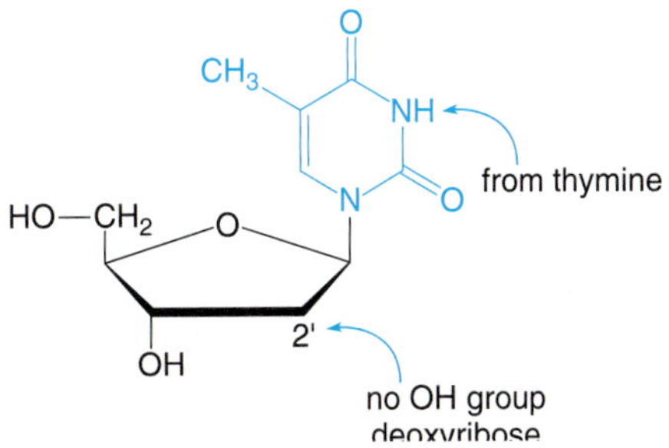

The sugar contains no OH at C2', so it is derived from deoxyribose. The base is thymine. To name the deoxyribonucleoside, change the suffix of the base to *-idine* and add the prefix *deoxy*; thus, thymine → deoxythymidine.

PROBLEM 22.1

Identify the base and monosaccharide used to form the following nucleosides, and then assign names.

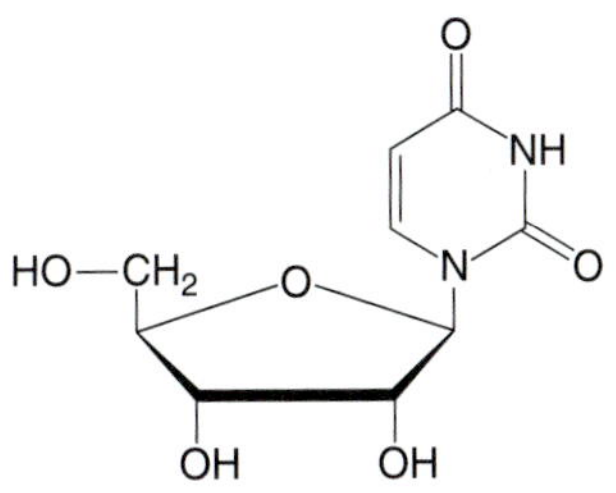

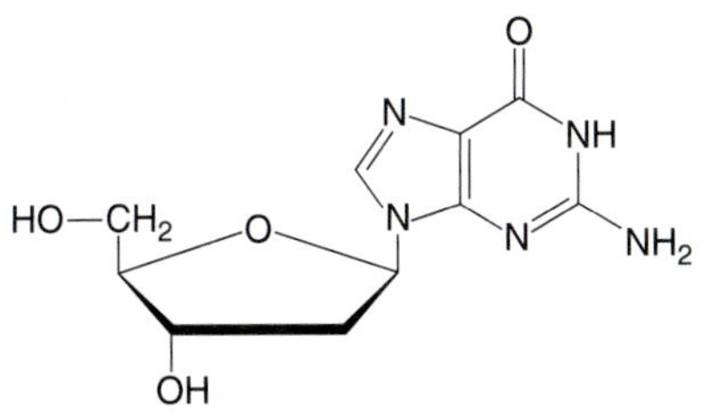

PROBLEM 22.2

Draw the structure of guanosine. Number the carbon atoms in the monosaccharide. Classify the compound as a ribonucleoside or a deoxyribonucleoside.

22.1B   NUCLEOTIDES—JOINING A NUCLEOSIDE WITH A PHOSPHATE

Nucleotides are formed by adding a phosphate group to the 5'-OH of a nucleoside. Nucleotides are named by adding the term *5'-monophosphate* to the name of the nucleoside from which they are derived. **Ribonucleotides** are derived from ribose, while **deoxyribonucleotides** are derived from 2-deoxyribose.

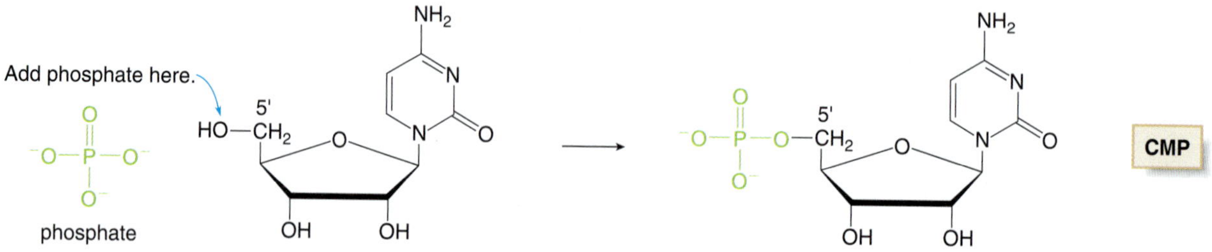

cytidine
a ribonucleoside

cytidine 5'-monophosphate
a ribonucleotide

CMP

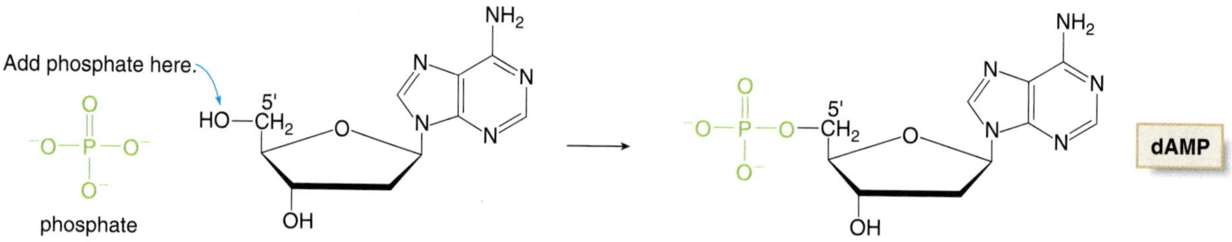

deoxyadenosine
a deoxyribonucleoside

deoxyadenosine 5'-monophosphate
a deoxyribonucleotide

dAMP

Because of the lengthy names of nucleotides, three- or four-letter abbreviations are commonly used instead. Thus, **c**ytidine 5'-**m**ono**p**hosphate is **CMP** and **d**eoxy**a**denosine 5'-**m**ono**p**hosphate is **dAMP.**

Figure 22.1 summarizes the information about nucleic acids and their components learned thus far. Table 22.1 summarizes the names and abbreviations used for the bases, nucleosides, and nucleotides needed in nucleic acid chemistry.

▼ **FIGURE 22.1**   **Summary of the Components of Nucleosides, Nucleotides, and Nucleic Acids**

Type of Compound	Components
Nucleoside	• A monosaccharide + a base • A ribonucleoside contains the monosaccharide ribose. • A deoxyribonucleoside contains the monosaccharide 2-deoxyribose.
Nucleotide	• A nucleoside + phosphate = a monosaccharide + a base + phosphate • A ribonucleotide contains the monosaccharide ribose. • A deoxyribonucleotide contains the monosaccharide 2-deoxyribose.
DNA	• A polymer of deoxyribonucleotides • The monosaccharide is 2-deoxyribose. • The bases are A, G, C, and T.
RNA	• A polymer of ribonucleotides • The monosaccharide is ribose. • The bases are A, G, C, and U.

TABLE 22.1 Names of Bases, Nucleosides, and Nucleotides in Nucleic Acids

Base	Abbreviation	Nucleoside	Nucleotide	Abbreviation
DNA				
Adenine	A	Deoxyadenosine	Deoxyadenosine 5'-monophosphate	dAMP
Guanine	G	Deoxyguanosine	Deoxyguanosine 5'-monophosphate	dGMP
Cytosine	C	Deoxycytidine	Deoxycytidine 5'-monophosphate	dCMP
Thymine	T	Deoxythymidine	Deoxythymidine 5'-monophosphate	dTMP
RNA				
Adenine	A	Adenosine	Adenosine 5'-monophosphate	AMP
Guanine	G	Guanosine	Guanosine 5'-monophosphate	GMP
Cytosine	C	Cytidine	Cytidine 5'-monophosphate	CMP
Uracil	U	Uridine	Uridine 5'-monophosphate	UMP

Di- and triphosphates can also be prepared from nucleosides by adding two and three phosphate groups, respectively, to the 5'-OH. For example, adenosine can be converted to adenosine 5'-diphosphate and adenosine 5'-triphosphate, abbreviated as ADP and ATP, respectively. We will learn about the central role of these phosphates, especially ATP, in energy production in Chapter 23.

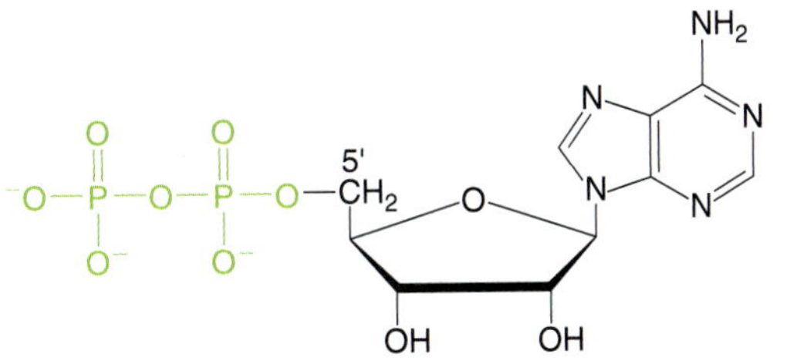

adenosine 5'-diphosphate
ADP

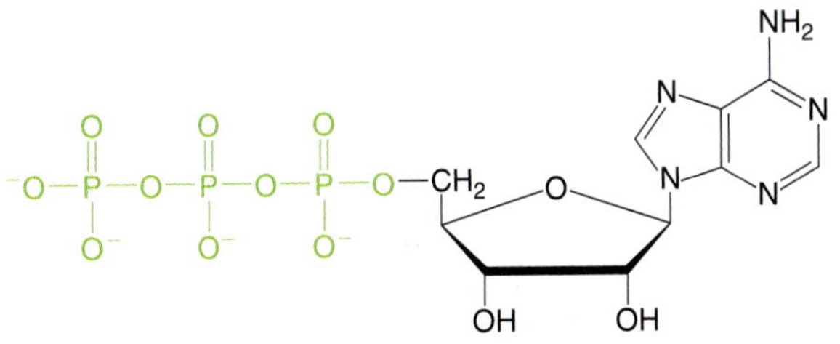

adenosine 5'-triphosphate
ATP

SAMPLE PROBLEM 22.2

Draw the structure of the nucleotide GMP.

ANALYSIS

Translate the abbreviation to a name; GMP is guanosine 5'-monophosphate. First, draw the sugar. Since there is no *deoxy* prefix in the name, GMP is a ribonucleotide and the sugar is ribose. Then draw the base, in this case guanine, bonded to C1' of the sugar ring. Finally, add the phosphate. GMP has one phosphate group bonded to the 5'-OH of the nucleoside.

SOLUTION

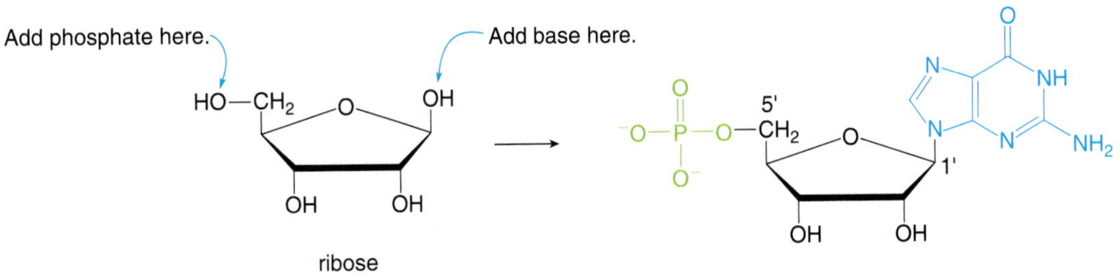

Add phosphate here.

Add base here.

ribose

guanosine 5'-monophosphate
GMP

PROBLEM 22.3	Draw the structure of each nucleotide: (a) UMP; (b) dTMP; (c) AMP.
PROBLEM 22.4	Give the name that corresponds to each abbreviation: (a) GTP; (b) dCDP; (c) dTTP; (d) UDP.
PROBLEM 22.5	Draw the structure of dCTP.

22.2 NUCLEIC ACIDS

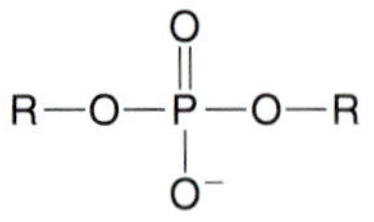

phosphodiester

Nucleic acids—both DNA and RNA—are polymers of nucleotides, formed by joining the 3'-OH group of one nucleotide with the 5'-phosphate of a second nucleotide in a **phosphodiester** linkage (Section 19.6).

For example, joining the 3'-OH group of dCMP (deoxycytidine 5'-monophosphate) and the 5'-phosphate of dAMP (deoxyadenosine 5'-monophosphate) forms a dinucleotide that contains a 5'-phosphate on one end (called the 5' end) and a 3'-OH group on the other end (called the 3' end).

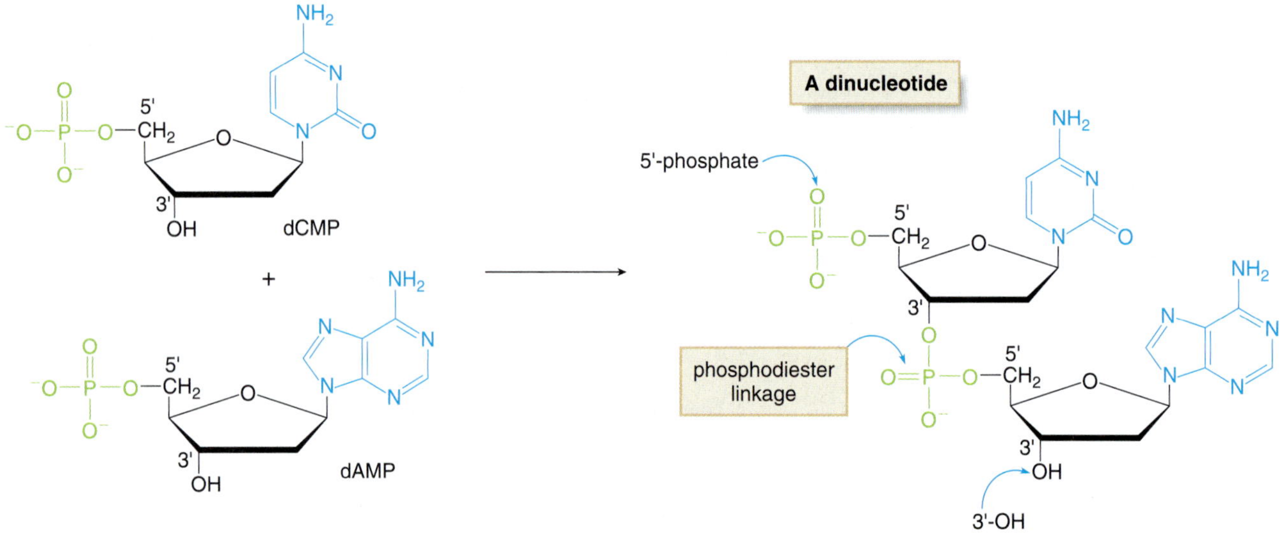

As additional nucleotides are added, the nucleic acid grows, each time forming a new phosphodiester linkage that holds the nucleotides together. Figure 22.2 illustrates the structure of a polynucleotide formed from four different nucleotides. Several features are noteworthy.

- A polynucleotide contains a backbone consisting of alternating sugar and phosphate groups. All polynucleotides contain the same sugar–phosphate backbone.
- The identity and order of the bases distinguish one polynucleotide from another.
- A polynucleotide has one free phosphate group at the 5' end.
- A polynucleotide has a free OH group at the 3' end.

The **primary structure** of a polynucleotide is the sequence of nucleotides that it contains. This sequence, which is determined by the identity of the bases, is unique to a nucleic acid. **In DNA, the sequence of bases carries the genetic information of the organism.**

Polynucleotides are named by the sequence of the bases they contain, beginning at the 5' end and using the one-letter abbreviation for the bases. Thus, the polynucleotide in Figure 22.2 contains the bases cytosine, adenine, thymine, and guanine, in order from the 5' end; thus, it is named CATG.

▼ **FIGURE 22.2** Primary Structure of a Polynucleotide

▼ **FIGURE 22.2** Primary Structure of a Polynucleotide

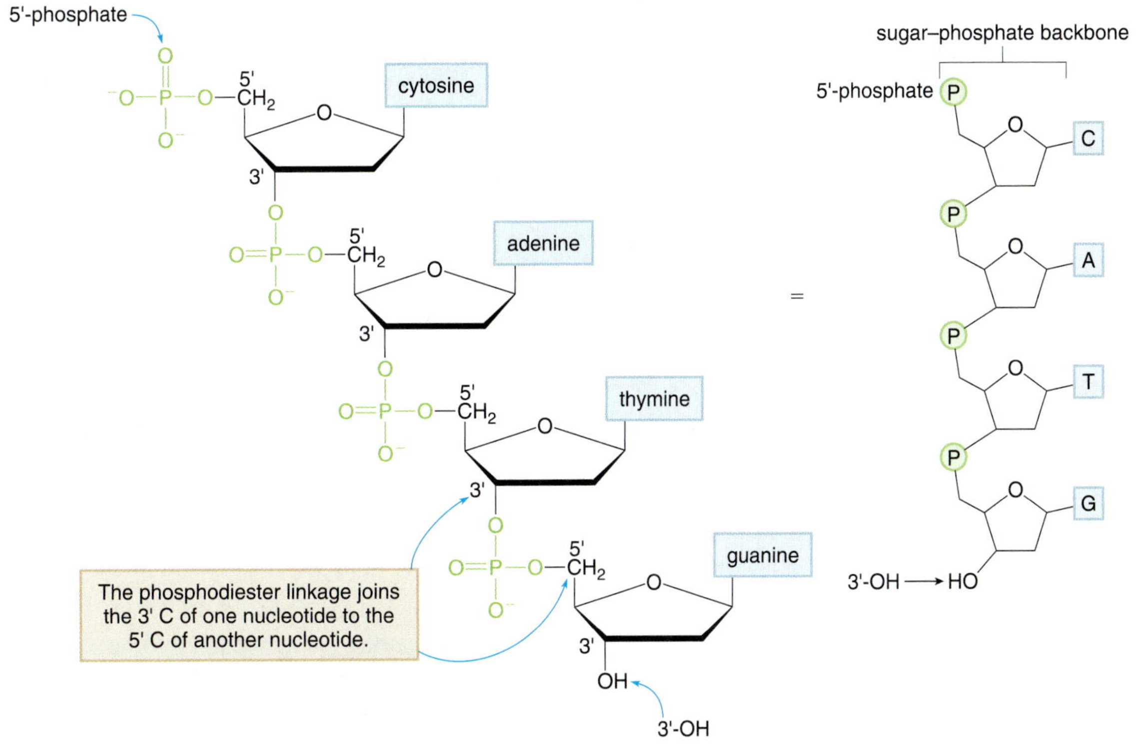

In a polynucleotide, phosphodiester bonds join the 3'-carbon of one nucleotide to the 5'-carbon of another. The name of a polynucleotide is read from the 5' end to the 3' end, using the one-letter abbreviations for the bases it contains. Drawn is the structure of the polynucleotide CATG.

SAMPLE PROBLEM 22.3

(a) Draw the structure of a dinucleotide formed by joining the 3'-OH group of AMP to the 5'-phosphate in GMP. (b) Label the 5' and 3' ends. (c) Name the dinucleotide.

ANALYSIS

Draw the structure of each nucleotide, including the sugar, the phosphate bonded to C5', and the base at C1'. In this case the sugar is ribose since the names of the mononucleotides do not contain the prefix *deoxy*. Bond the 3'-OH group to the 5'-phosphate to form the phosphodiester bond. The name of the dinucleotide begins with the nucleotide that contains the free phosphate at the 5' end.

SOLUTION

a. and b.

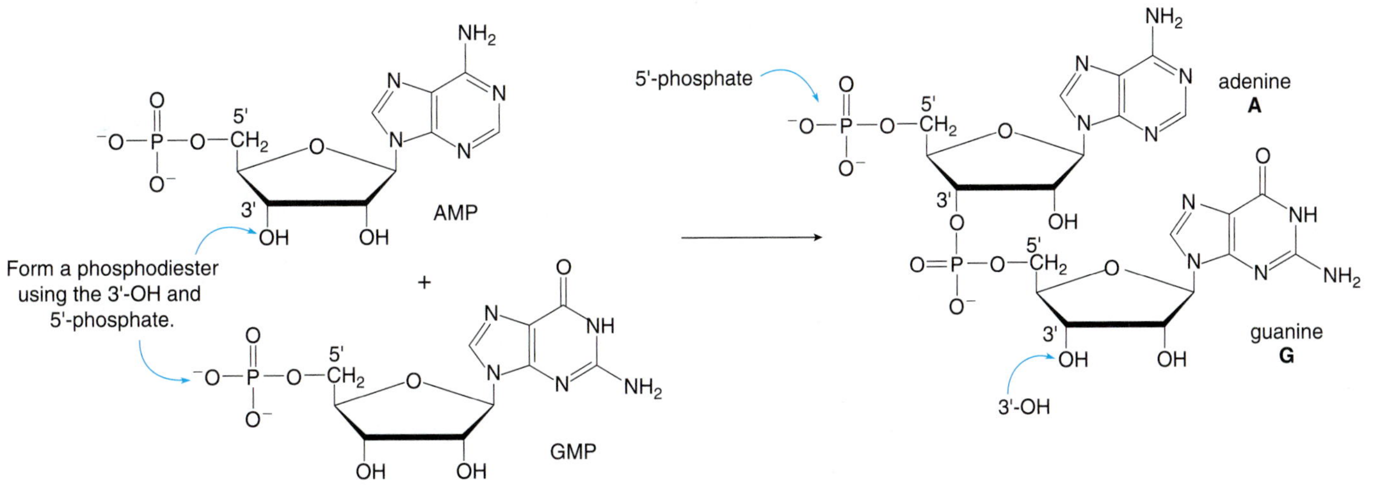

c. Since polynucleotides are named beginning at the 5' end, this dinucleotide is named AG.

PROBLEM 22.6 Draw the structure of a dinucleotide formed by joining the 3'-OH group of dTMP to the 5'-phosphate in dGMP.

PROBLEM 22.7 Draw the structure of each polynucleotide: (a) CU; (b) TAG.

PROBLEM 22.8 Label the 5' end and the 3' end in each polynucleotide: (a) ATTTG; (b) CGCGUU; (c) GGACTT.

22.3 THE DNA DOUBLE HELIX

Our current understanding of the structure of DNA is based on the model proposed initially by James Watson and Francis Crick in 1953 (Figure 22.3).

- **DNA consists of two polynucleotide strands that wind into a right-handed double helix.**

The sugar–phosphate backbone lies on the outside of the helix and the bases lie on the inside, perpendicular to the axis of the helix. The two strands of DNA run in *opposite* directions; that is, one strand runs from the 5' end to the 3' end, while the other runs from the 3' end to the 5' end.

▼ **FIGURE 22.3   The Three-Dimensional Structure of DNA—A Double Helix**

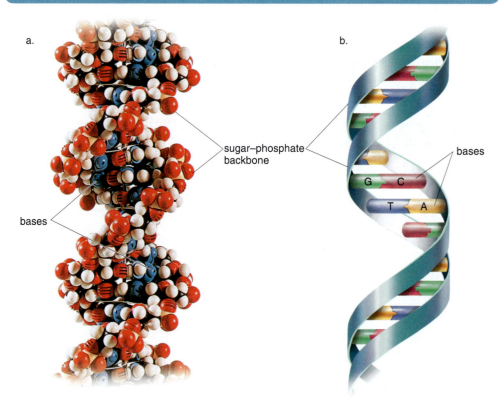

DNA consists of a double helix of polynucleotide chains. In view (a), the three-dimensional molecular model shows the sugar–phosphate backbone with the red (O), black (C), and white (H) atoms visible on the outside of the helix. In view (b), the bases on the interior of the helix are labeled.

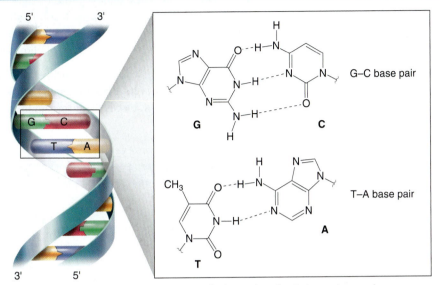

▼ **FIGURE 22.4 Hydrogen Bonding in the DNA Double Helix**

hydrogen bonding between base pairs

Hydrogen bonding of base pairs (A–T and C–G) holds the two strands of DNA together.

The double helix is stabilized by hydrogen bonding between the bases of the two DNA strands as shown in Figure 22.4. A purine base on one strand always hydrogen bonds with a pyrimidine base on the other strand. Two bases hydrogen bond together in a predictable manner, forming **complementary base pairs.**

- Adenine pairs with thymine using two hydrogen bonds, forming an A–T base pair.
- Cytosine pairs with guanine using three hydrogen bonds, forming a C–G base pair.

Because of this consistent pairing of bases, knowing the sequence of one strand of DNA allows us to write the sequence of the other strand, as shown in Sample Problem 22.4.

SAMPLE PROBLEM 22.4

Write the sequence of the complementary strand of the following portion of a DNA molecule: 5'–TAGGCTA–3'.

ANALYSIS

The complementary strand runs in the opposite direction, from the 3' to the 5' end. Use base pairing to determine the corresponding sequence on the complementary strand: A pairs with T and C pairs with G.

SOLUTION

Original strand: 5'–T A G G C T A–3'

Complementary strand: 3'–A T C C G A T–5'

PROBLEM 22.9

Write the complementary strand for each of the following strands of DNA.

a. 5'–AAACGTCC–3'   c. 5'–ATTGCACCCGC–3'
b. 5'–TATACGCC–3'   d. 5'–CACTTGATCGG–3'

The enormously large DNA molecules that compose the **human genome**—the total DNA content of an individual—pack tightly into the nucleus of the cell. The double-stranded DNA helices wind around a core of protein molecules called **histones** to form a chain of **nucleosomes,** as shown in Figure 22.5. The chain of nucleosomes winds into a supercoiled fiber called **chromatin,** which composes each of the 23 pairs of chromosomes in humans.

In Section 22.2 we learned that the **genetic information of an organism is stored in the sequence of bases of its DNA molecules.** How is this information transferred from one generation to another? How, too, is the information stored in DNA molecules used to direct the synthesis of proteins?

▼ **FIGURE 22.5   The Structure of a Chromosome**

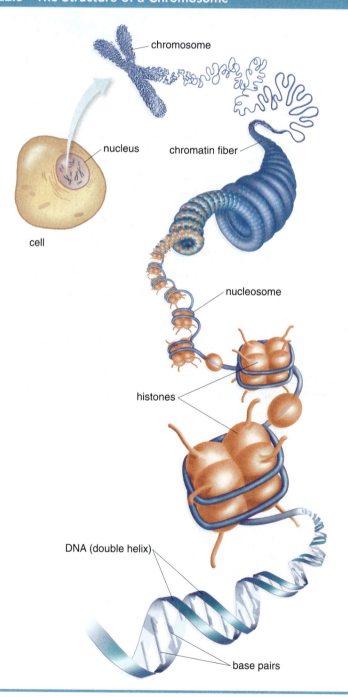

chromosome

nucleus   chromatin fiber

cell

nucleosome

histones

DNA (double helix)

base pairs

To answer these questions we must understand three key processes.

- *Replication* is the process by which DNA makes a copy of itself when a cell divides.
- *Transcription* is the ordered synthesis of RNA from DNA. In this process, the genetic information stored in DNA is passed onto RNA.
- *Translation* is the synthesis of proteins from RNA. In this process, the genetic message contained in RNA determines the specific amino acid sequence of a protein.

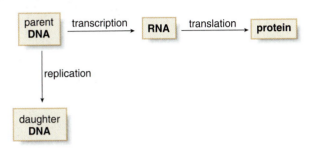

Each chromosome contains many **genes,** those portions of the DNA molecules that result in the synthesis of specific proteins. We say that the genetic message of the DNA molecule is *expressed* in the protein. Only a small fraction (1–2%) of the DNA in a chromosome contains genetic messages or genes that result in protein synthesis.

22.4 REPLICATION

How is the genetic information in the DNA of a parent cell passed onto new daughter cells during replication? The structure of the double helix and the presence of complementary base pairs are central to the replication process.

During replication, the strands of DNA separate and each serves as a template for a new strand. Thus, **the original DNA molecule forms two DNA molecules, each of which contains one strand from the parent DNA and one new strand.** This process is called **semiconservative replication.** The sequence of both strands of the daughter DNA molecules exactly matches the sequence in the parent DNA.

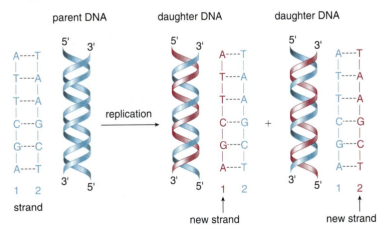

Although the semiconservative nature of replication has been known since the elegant experiments of Matthew Meselson and Franklin Stahl were reported in 1958, the details of replication have only slowly been determined over the last 50 years. The first step in replication is the unwinding of the DNA helix to expose the bases on each strand. Unwinding occurs at many places simultaneously along the helix, creating "bubbles" where replication can occur. The point at which unwinding occurs is called the **replication fork.** Unwinding breaks the hydrogen bonds that hold the two strands of the double helix together.

Once bases have been exposed on the unwound strands of DNA, the enzyme DNA polymerase catalyzes the replication process using the four nucleoside triphosphates (derived from the bases A, T, G, and C) that are available in the nucleus. Three features are key and each is illustrated in Figure 22.6.

▼ **FIGURE 22.6   DNA Replication**

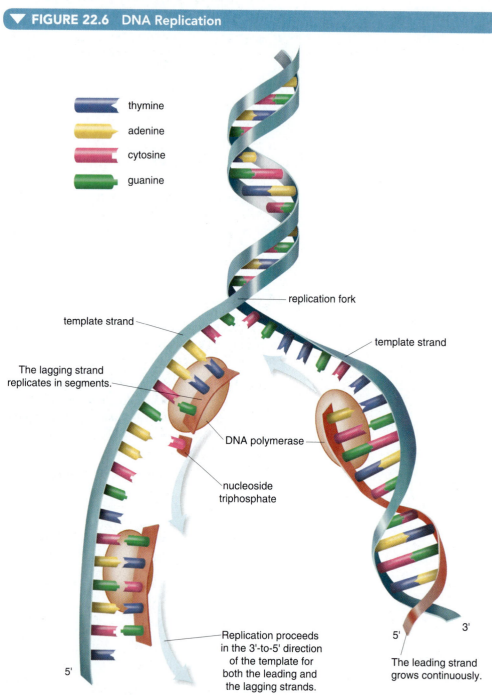

thymine
adenine
cytosine
guanine

replication fork

template strand

The lagging strand replicates in segments.

DNA polymerase

nucleoside triphosphate

template strand

Replication proceeds in the 3'-to-5' direction of the template for both the leading and the lagging strands.

5'

5'   3'

The leading strand grows continuously.

Replication proceeds along both strands of unwound DNA. Replication always occurs in the same direction, from the 3' to the 5' end of the template strand. The leading strand grows continuously, while the lagging strand must be synthesized in fragments that are joined together by a DNA ligase enzyme.

- The identity of the bases on the template strand determines the order of the bases on the new strand: A must pair with T, and G must pair with C.
- A new phosphodiester bond is formed between the 5'-phosphate of the nucleoside triphosphate and the 3'-OH group of the new DNA strand.
- Replication occurs in only one direction on the template strand, from the 3' end to the 5' end.

Since replication proceeds in only one direction—that is, from the 3' end to the 5' end of the template—the two new strands of DNA must be synthesized by somewhat different techniques. One strand, called the **leading strand,** grows continuously. Since its sequence is complementary to the template, its nucleotide sequence grows in the 5' to 3' direction. The other strand, called the **lagging strand,** is synthesized in small fragments, which are then joined together by a DNA ligase enzyme. The end result is two new strands of DNA, one in each of the daughter DNA molecules, both with complementary base pairs joining the two DNA strands together.

SAMPLE PROBLEM 22.5	What is the sequence of a newly synthesized DNA segment if the template strand has the sequence 3'–TGCACC–5'?
ANALYSIS	The newly synthesized strand runs in the opposite direction, from the 5' end to the 3' end in this example. Use base pairing to determine the corresponding sequence on the new strand: A pairs with T and C pairs with G.
SOLUTION	Template strand: 3'–T G C A C C–5' New strand: 5'–A C G T G G–3'

PROBLEM 22.10	What is the sequence of a newly synthesized DNA segment if the template strand has each of the following sequences?

a. 3'–AGAGTCTC–5'   c. 3'–ATCCTGTAC–5'

b. 5'–ATTGCTC–3'   d. 5'–GGCCATACTC–3'

22.5 RNA

While RNA is also composed of nucleotides, there are important differences between DNA and RNA. In RNA,

- The sugar is ribose.
- U (uracil) replaces T (thymine) as one of the bases.
- RNA is single stranded.

RNA molecules are much smaller than DNA molecules. Although RNA contains a single strand, the chain can fold back on itself, forming loops, and intermolecular hydrogen bonding between paired bases on a single strand can form helical regions. When base pairing occurs within an RNA molecule (or between RNA and DNA), **C and G form base pairs,** and **A and U form base pairs.**

There are three different types of RNA molecules.

- Ribosomal RNA (rRNA)
- Messenger RNA (mRNA)
- Transfer RNA (tRNA)

Ribosomal RNA, the most abundant type of RNA, is found in the ribosomes in the cytoplasm of the cell. Each ribosome is composed of one large subunit and one small subunit that contain both RNA and protein. rRNA provides the site where polypeptides are assembled during protein synthesis.

Messenger RNA is the carrier of information from DNA (in the cell nucleus) to the ribosomes (in the cytoplasm). Each gene of a DNA molecule corresponds to a specific mRNA molecule. The sequence of nucleotides in the mRNA molecule determines the amino acid sequence in a particular protein. mRNA is synthesized from DNA on an as-needed basis, and then rapidly degraded after a particular protein is synthesized.

Transfer RNA, the smallest type of RNA, interprets the genetic information in mRNA and brings specific amino acids to the site of protein synthesis in the ribosome. Each amino acid is recognized by one or more tRNA molecules, which contain 70–90 nucleotides. tRNAs have two important sites. The 3' end, called the **acceptor stem,** always contains the nucleotides ACC and has a free OH group that binds a specific amino acid. Each tRNA also contains a sequence of three nucleotides called an **anticodon,** which is complementary to three bases in an mRNA molecule, and identifies what amino acid must be added to a growing polypeptide chain.

tRNA molecules are often drawn in the cloverleaf fashion shown in Figure 22.7a. The acceptor stem and anticodon region are labeled. Folding creates regions of the tRNA in which nearby complementary bases hydrogen bond to each other. A model that more accurately depicts the three-dimensional structure of a tRNA molecule is shown in Figure 22.7b.

Table 22.2 summarizes the characteristics of the three types of RNAs.

▼ **FIGURE 22.7   Transfer RNA**

a. tRNA–Cloverleaf representation

- A — amino acid
- C
- C — 3'
- acceptor stem
- 5'
- hydrogen bonding between complementary base pairs
- anticodon

b. tRNA–Three-dimensional representation

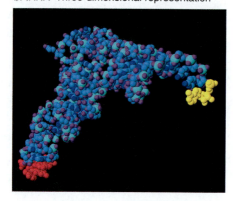

Folding of the tRNA molecule creates regions in which complementary base pairs hydrogen bond to each other. Each tRNA binds a specific amino acid to its 3' end and contains an anticodon that identifies that amino acid for protein synthesis.

In the three-dimensional model of a tRNA, the binding site for the amino acid is shown in yellow and the anticodon is shown in red.

TABLE 22.2   Three Types of RNA Molecules

Type of RNA	Abbreviation	Function
Ribosomal RNA	rRNA	The site of protein synthesis, found in the ribosomes
Messenger RNA	mRNA	Carries the information from DNA to the ribosomes
Transfer RNA	tRNA	Brings specific amino acids to the ribosomes for protein synthesis

22.6   TRANSCRIPTION

The conversion of the information in DNA to the synthesis of proteins begins with *transcription—* **that is, the synthesis of messenger RNA from DNA.**

RNA synthesis begins in the same manner as DNA replication: the double helix of DNA unwinds (Figure 22.8). Since RNA is single stranded, however, only one strand of DNA is needed for RNA synthesis.

- The *template* strand is the strand of DNA used for RNA synthesis.
- The *informational* strand (the non-template strand) is the strand of DNA not used for RNA synthesis.

Each mRNA molecule corresponds to a small segment of a DNA molecule. Transcription begins at a particular sequence of bases on the DNA template using an RNA polymerase enzyme, and proceeds from the 3' end to the 5' end of the template strand. Complementary base pairing determines what RNA nucleotides are added to the growing RNA chain: C pairs with G, T pairs with A, and A pairs with U. Thus, the RNA chain grows from the 5' to 3' direction. Transcription is completed when a particular sequence of bases on the DNA template is reached. The new mRNA molecule is released and the double helix of the DNA molecule re-forms.

▼ **FIGURE 22.8   Transcription**

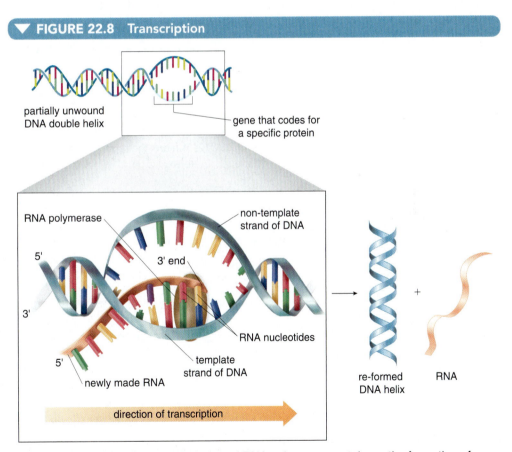

In transcription, the DNA helix unwinds and RNA polymerase catalyzes the formation of mRNA along the DNA template strand. Transcription proceeds from the 3' end to the 5' end of the template, forming an mRNA molecule with base pairs complementary to the DNA template strand.

In bacteria, the new mRNA molecule is ready for protein synthesis immediately after it is prepared. In humans, the mRNA molecule first formed is modified before it is ready for protein synthesis. Portions of the mRNA molecule are removed and pieces of mRNA are spliced together by mechanisms that will not be discussed here.

- Transcription forms a messenger RNA molecule with a sequence that is *complementary* to the DNA template from which it is prepared.
- Since the informational strand of DNA is complementary to the template strand, the mRNA molecule is an *exact copy of the informational strand,* except that the base U replaces T on the RNA strand.

How the base sequence of an RNA molecule compares with the informational and template strands of a DNA molecule is shown in Sample Problem 22.6.

SAMPLE PROBLEM 22.6

If a portion of the template strand of a DNA molecule has the sequence 3'–CTAGGATAC–5', what is the sequence of the mRNA molecule produced from this template? What is the sequence of the informational strand of this segment of the DNA molecule?

ANALYSIS

mRNA has a base sequence that is complementary to the template from which it is prepared. mRNA has a base sequence that is identical to the informational strand of DNA, except that it contains the base U instead of T.

SOLUTION

complementary

Template strand of DNA: 3'–C T A G G A T A C–5'

mRNA sequence: 5'–G A U C C U A U G–3'   complementary

Informational strand of DNA: 5'–G A T C C T A T G–3'

PROBLEM 22.11

For each DNA segment: [1] What is the sequence of the mRNA molecule synthesized from each DNA template? [2] What is the sequence of the informational strand of the DNA molecule?

a. 3'–TGCCTAACG–5'   c. 3'–TTAACGCGA–5'

b. 3'–GACTCC–5'   d. 3'–CAGTGACCGTAC–5'

PROBLEM 22.12

What is the sequence of the DNA template strand from which each of the following mRNA strands were synthesized?

a. 5'–UGGGGCAUU–3'   c. 5'–CCGACGAUG–3'

b. 5'–GUACCU–3'   d. 5'–GUAGUCACG–3'

22.7 THE GENETIC CODE

Once the genetic information of DNA has been transcribed in a messenger RNA molecule, RNA can direct the synthesis of an individual protein. How can RNA, which is composed of only four different nucleotides, direct the synthesis of polypeptides that are formed from 20 different amino acids? The answer lies in the **genetic code.**

- A sequence of three nucleotides (a triplet) codes for a specific amino acid. Each triplet is called a *codon.*

For example, the codon UAC in an mRNA molecule codes for the amino acid serine, and the codon UGC codes for the amino acid cysteine. The same genetic code occurs in almost all organisms, from bacteria to whales to humans.

UAC	→	serine	UGC	→	cysteine
mRNA		amino acid	mRNA		amino acid

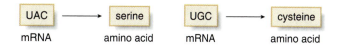

Given four different nucleotides (A, C, G, and U), there are 64 different ways to combine them into groups of three, so there are 64 different codons. Sixty-one codons code for specific amino acids, so many amino acids correspond to more than one codon, as shown in Table 22.3. For example, the codons GGU, GGC, GGA, and GGG all code for the amino acid glycine. Three codons—UAA, UAG, and UGA—do not correspond to any amino acids; they are called **stop codons** because they signal the termination of protein synthesis.

TABLE 22.3 The Genetic Code—Triplets in Messenger RNA

First Base (5' end)	Second Base								Third Base (3' end)
	U		C		A		G		
U	UUU	Phe	UCU	Ser	UAU	Tyr	UGU	Cys	U
	UUC	Phe	UCC	Ser	UAC	Tyr	UGC	Cys	C
	UUA	Leu	UCA	Ser	UAA	Stop	UGA	Stop	A
	UUG	Leu	UCG	Ser	UAG	Stop	UGG	Trp	G
C	CUU	Leu	CCU	Pro	CAU	His	CGU	Arg	U
	CUC	Leu	CCC	Pro	CAC	His	CGC	Arg	C
	CUA	Leu	CCA	Pro	CAA	Gln	CGA	Arg	A
	CUG	Leu	CCG	Pro	CAG	Gln	CGG	Arg	G
A	AUU	Ile	ACU	Thr	AAU	Asn	AGU	Ser	U
	AUC	Ile	ACC	Thr	AAC	Asn	AGC	Ser	C
	AUA	Ile	ACA	Thr	AAA	Lys	AGA	Arg	A
	AUG	Met	ACG	Thr	AAG	Lys	AGG	Arg	G
G	GUU	Val	GCU	Ala	GAU	Asp	GGU	Gly	U
	GUC	Val	GCC	Ala	GAC	Asp	GGC	Gly	C
	GUA	Val	GCA	Ala	GAA	Glu	GGA	Gly	A
	GUG	Val	GCG	Ala	GAG	Glu	GGG	Gly	G

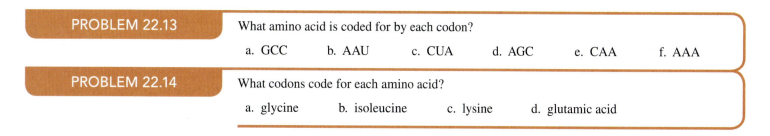

PROBLEM 22.13 What amino acid is coded for by each codon?

a. GCC b. AAU c. CUA d. AGC e. CAA f. AAA

PROBLEM 22.14 What codons code for each amino acid?

a. glycine b. isoleucine c. lysine d. glutamic acid

Codons are written from the 5' to 3' end of mRNA. The 5' end of the mRNA molecule codes for the N-terminal amino acid in a protein, and the 3' end of the mRNA molecule codes for the C-terminal amino acid. Sample Problem 22.7 illustrates the conversion of a sequence of bases in mRNA to a sequence of amino acids in a peptide.

SAMPLE PROBLEM 22.7

Derive the amino acid sequence that is coded for by the following mRNA sequence.

5' CAU AAA ACG GUG UUA AUA 3'

ANALYSIS

Use Table 22.3 to identify the codons that correspond to each amino acid. Codons are written from the 5' to 3' end of an mRNA molecule and correspond to a peptide written from the N-terminal to C-terminal end.

SOLUTION

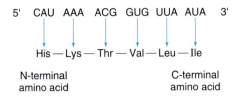

PROBLEM 22.15

Derive the amino acid sequence that is coded for by each mRNA sequence.

a. 5' CAA GAG GUA UCC UAC AGA 3'

b. 5' GUC AUC UGG AGG GGC AUU 3'

c. 5' CUA UGC AGU AGG ACA CCC 3'

PROBLEM 22.16

Write a possible mRNA sequence that codes for each of the following peptides.

a. Met–Arg–His–Phe   c. Gln–Asn–Gly–Ile–Val

b. Gly–Ala–Glu–Gln   d. Thr–His–Asp–Cys–Trp

PROBLEM 22.17

Considering the given sequence of nucleotides in an mRNA molecule, (a) what is the sequence of the DNA template strand from which the RNA was synthesized? (b) What peptide is synthesized by this mRNA sequence?

5' GAG CCC GUA UAC GCC ACG 3'

22.8 TRANSLATION AND PROTEIN SYNTHESIS

The translation of the information in messenger RNA to protein synthesis occurs in the ribosomes. Each type of RNA plays a role in protein synthesis.

- mRNA contains the sequence of codons that determines the order of amino acids in the protein.
- Individual tRNAs bring specific amino acids to add to the peptide chain.
- rRNA contains binding sites that provide the platform on which protein synthesis occurs.

Each individual tRNA contains an **anticodon** of three nucleotides that is complementary to the codon in mRNA and identifies individual amino acids (Section 22.5). For example, a codon of UCA in mRNA corresponds to an anticodon of AGU in a tRNA molecule, which identifies serine as the amino acid. Other examples are shown in Table 22.4.

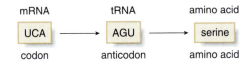

TABLE 22.4 Relating Codons, Anticodons, and Amino Acids

mRNA Codon		tRNA Anticodon		Amino Acid
ACA	$\longrightarrow$	UGU	$\longrightarrow$	threonine
GCG	$\longrightarrow$	CGC	$\longrightarrow$	alanine
AGA	$\longrightarrow$	UCU	$\longrightarrow$	arginine
UCC	$\longrightarrow$	AGG	$\longrightarrow$	serine

PROBLEM 22.18

For each codon: [1] Write the anticodon. [2] What amino acid does each codon represent?

a. CGG   b. GGG   c. UCC   d. AUA   e. CCU   f. GCC

There are three stages in translation: initiation, elongation, and termination. Figure 22.9 depicts the main features of translation.

▼ **FIGURE 22.9 Translation—The Synthesis of Proteins from RNA**

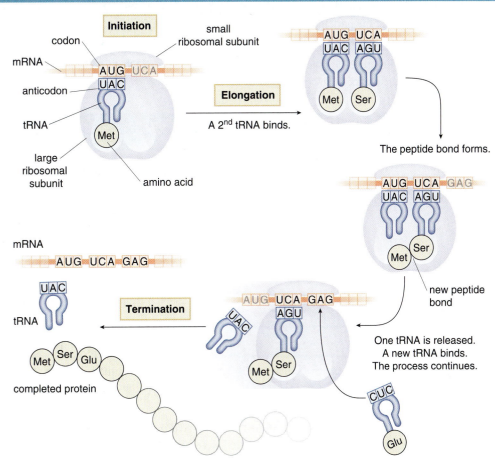

- **Initiation** consists of the binding of the ribosomal subunits to mRNA and the arrival of the first tRNA carrying its amino acid. Hydrogen bonding occurs between the complementary bases of the codon of mRNA and the anticodon of the tRNA.

- The protein is synthesized during **elongation**. One by one a tRNA with its designated amino acid binds to a site on the ribosome adjacent to the first tRNA. A peptide bond forms and a tRNA is released. The ribosome shifts to the next codon and the process continues.

- **Termination** occurs when a stop codon is reached. The synthesis is complete and the protein is released from the complex.

[1] INITIATION

Translation begins when an mRNA molecule binds to the smaller subunit of the ribosome and a tRNA molecule carries the first amino acid of the peptide chain to the binding site. Translation always begins at the codon AUG, which codes for the amino acid methionine. The arriving tRNA contains an anticodon with the complementary base sequence UAC. The large and small subunits of the ribosome combine to form a tight complex where protein synthesis takes place.

[2] ELONGATION

The next tRNA molecule containing an anticodon for the second codon binds to mRNA, delivering its amino acid, and a peptide bond forms between the two amino acids. The first tRNA molecule, which has delivered its amino acid and is no longer needed, dissociates from the complex. The ribosome shifts to the next codon along the mRNA strand and the process continues when a new tRNA molecule binds to the mRNA. Protein synthesis always occurs on two adjacent sites of the ribosome.

[3] TERMINATION

Translation continues until a stop codon is reached. There is no tRNA that contains an anticodon complementary to any of the three stop codons (UAA, UAG, and UGA), so protein synthesis ends and the protein is released from the ribosome. Often the first amino acid in the chain, methionine, is not needed in the final protein and so it is removed after protein synthesis is complete.

Figure 22.10 shows a representative segment of DNA, and the mRNA, tRNA, and amino acid sequences that correspond to it.

▼ **FIGURE 22.10   Comparing the Sequence of DNA, mRNA, tRNA, and a Polypeptide**

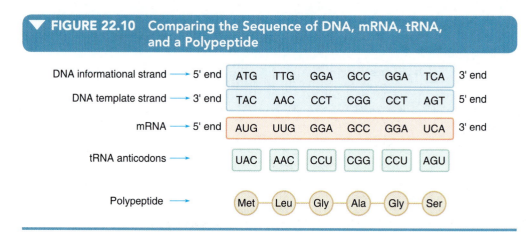

SAMPLE PROBLEM 22.8

What sequence of amino acids would be formed from the following mRNA sequence: 5' CAA AAG ACG UAC CGA 3'? List the anticodons contained in each of the needed tRNA molecules.

ANALYSIS   Use Table 22.3 to determine the amino acid that is coded for by each codon. The anticodons contain complementary bases to the codons: A pairs with U, and C pairs with G.

SOLUTION

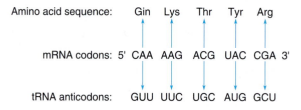

PROBLEM 22.19

What sequence of amino acids would be formed from each mRNA sequence? List the anticodons contained in each of the needed tRNA molecules.

    a. 5' CCA CCG GCA AAC GAA GCA 3'   b. 5' GCA CCA CUA AGA GAC 3'

SAMPLE PROBLEM 22.9

What polypeptide would be synthesized from the following template strand of DNA: 3' CGG TGT CTT TTA 5'?

ANALYSIS

To determine what polypeptide is synthesized from a DNA template, two steps are needed. First use the DNA sequence to determine the transcribed mRNA sequence: C pairs with G, T pairs with A, and A (on DNA) pairs with U (on mRNA). Then use the codons in Table 22.3 to determine what amino acids are coded for by a given codon in mRNA.

SOLUTION

DNA template strand: ⟶ 3' CGG TGT CTT TTA 5'

mRNA: ⟶ 5' GCC ACA GAA AAU 3'

Polypeptide: ⟶ Ala — Thr — Glu — Asn

PROBLEM 22.20

What polypeptide would be synthesized from each of the following template strands of DNA?

    a. 3' TCT CAT CGT AAT GAT TCG 5'   b. 3' GCT CCT AAA TAA CAC TTA 5'

22.9 MUTATIONS AND GENETIC DISEASES

Although replication provides a highly reliable mechanism for making an exact copy of DNA, occasionally an error occurs, thus producing a DNA molecule with a slightly different nucleotide sequence.

- A *mutation* is a change in the nucleotide sequence in a molecule of DNA.

If the mutation occurs in a nonreproductive cell, the mutation is passed on to daughter cells within the organism, but is not transmitted to the next generation. If the mutation occurs in an egg or sperm cell, it is passed on to the next generation of an organism. Some mutations are random events, while others are caused by **mutagens,** chemical substances that alter the structure of DNA. Exposure to high-energy radiation such as X-rays or ultraviolet light can also produce mutations.

Mutations can be classified according to the change that results in a DNA molecule.

- A *point* mutation is the *substitution* of one nucleotide for another.

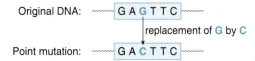

Original DNA: ⁓⁓ G A G T T C ⁓⁓

replacement of G by C

Point mutation: ⁓⁓ G A C T T C ⁓⁓

- A *deletion* mutation occurs when one or more nucleotides is *lost* from a DNA molecule.

Original DNA: ⁓⁓ G A G T T C ⁓⁓

loss of G

Deletion mutation: ⁓⁓ G A T T C ⁓⁓

- An *insertion* mutation occurs when one or more nucleotides is *added* to a DNA molecule.

Original DNA: ⁓⁓ G A G T T C ⁓⁓

addition of C

Insertion mutation: ⁓⁓ G A G C T T C ⁓⁓

A mutation can have a negligible, minimal, or catastrophic effect on an organism. To understand the effect of a mutation, we must determine the mRNA sequence that is transcribed from the DNA sequence as well as the resulting amino acid for which it codes. A point mutation in the three-base sequence CTT in a gene that codes for a particular protein illustrates some possible outcomes of a mutation.

The sequence CTT in DNA is transcribed to the codon GAA in mRNA, and using Table 22.3, this triplet codes for the amino acid glutamic acid. If a point mutation replaces CTT by CTC in DNA, CTC is transcribed to the codon GAG in mRNA. Since GAG codes for the *same* amino acid—glutamic acid—this mutation does not affect the protein synthesized by this segment of DNA. Such a mutation is said to be **silent.**

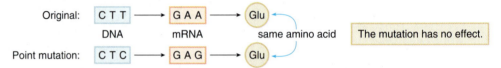

Alternatively, suppose a point mutation replaces CTT by CAT in DNA. CAT is transcribed to the codon GUA in mRNA, and GUA codes for the amino acid valine. Now, the mutation produces a protein with one *different* amino acid—namely, valine instead of glutamic acid. In some proteins this alteration of the primary sequence may have little effect on the protein's secondary and tertiary structure. In other proteins, such as hemoglobin, the substitution of valine for glutamic acid produces a protein with vastly different properties, resulting in the fatal disease sickle cell anemia (Section 21.7).

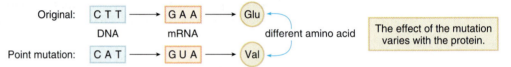

Finally, suppose a point mutation replaces CTT by ATT in DNA. ATT is transcribed to the codon UAA in mRNA, and UAA is a stop codon. This terminates protein synthesis and no more amino acids are added to the protein chain. In this case, a needed protein is not synthesized and depending on the protein's role, the organism may die.

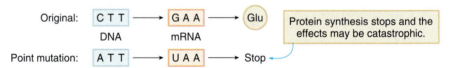

SAMPLE PROBLEM 22.10

(a) What dipeptide is produced from the following segment of DNA: AGAGAT? (b) What happens to the dipeptide when a point mutation occurs and the DNA segment contains the sequence ATAGAT instead?

ANALYSIS   Transcribe the DNA sequence to an mRNA sequence with complementary base pairs. Then use Table 22.3 to determine what amino acids are coded for by each codon.

SOLUTION   a. Since UCU codes for serine and CUA codes for leucine, the dipeptide Ser–Leu results.

b. Since UAU codes for tyrosine, the point mutation results in the synthesis of the dipeptide Tyr–Leu.

PROBLEM 22.21	Consider the following sequence of DNA: AACTGA. (a) What dipeptide is formed from this DNA after transcription and translation? (b) How is the amino acid sequence affected when point mutations produce each of the following DNA sequences: [1] AACGGA; [2] ATCTGA; [3] AATTGA?
PROBLEM 22.22	Suppose a deletion mutation occurs and the DNA segment TATGCACTT is converted to TAACTT. How do the peptides synthesized from each of these DNA segments compare?

When a mutation causes a protein deficiency or results in a defective protein and the condition is inherited from one generation to another, a **genetic disease** results. For example, cystic fibrosis, the most common genetic disease in Caucasians, is caused by a mutation resulting in the synthesis of a defective protein—cystic fibrosis transmembrane conductance regulator (CFTR)—needed for proper passage of ions across cell membranes. Individuals with cystic fibrosis have decreased secretions of pancreatic enzymes, resulting in a failure to thrive (poor growth), and they produce thick sticky mucous in the lungs that leads to devastating lung infections and a shortened life span. Other genetic diseases and their molecular bases are listed in Table 22.5.

TABLE 22.5	Genetic Diseases
Disease	**Characteristics**
Tay–Sachs disease	Mental retardation; caused by a defective hexosaminidase A enzyme
Sickle cell anemia	Anemia; occlusion and inflammation of blood capillaries, caused by defective hemoglobin
Phenylketonuria	Mental retardation; caused by a deficiency of the enzyme phenylalanine hydroxylase needed to convert the amino acid phenylalanine to tyrosine
Galactosemia	Mental retardation; caused by a deficiency of an enzyme needed for galactose metabolism
Huntington's disease	Progressive physical disability; caused by a defect in the gene that codes for the Htt protein, resulting in degeneration in the neurons in certain areas of the brain.

22.10 RECOMBINANT DNA

Laboratory techniques devised over the last 30 years allow scientists to manipulate DNA molecules and produce new forms of DNA that do not occur in nature.

- *Recombinant DNA* is synthetic DNA that contains segments from more than one source.

The term **genetic engineering** is often used to describe the process of manipulating DNA in the laboratory that allows a gene (a segment of DNA) in one organism to be spliced (inserted) into the DNA of another organism, thus forming recombinant DNA.

22.10A GENERAL PRINCIPLES

Three key elements are needed to form recombinant DNA:

- A **DNA molecule** into which a new DNA segment will be inserted
- An **enzyme** that cleaves DNA at specific locations
- A **gene from a second organism** that will be inserted in the original DNA molecule

Forming recombinant DNA often begins with circular, double-stranded DNA molecules called **plasmids** in the bacteria *Escherichia coli* (*E. coli*) (Figure 22.11). Plasmid DNA is isolated from the bacteria and then cleaved (cut) at a specific site with an enzyme called a **restriction endonuclease.** Many different restriction endonucleases are known. Each enzyme recognizes a specific sequence of bases in a DNA molecule and cuts the double-stranded DNA at a particular position.

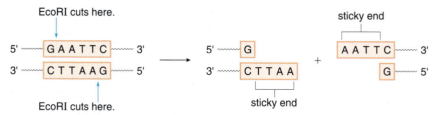

For example, the enzyme EcoRI recognizes the sequence GAATTC, and cuts the DNA molecule between G and A on *both* strands. Since the plasmid DNA is circular to begin with, this forms linear double-stranded DNA with two ends of known sequence. These ends are called **sticky ends** because one strand contains unpaired bases that are ready to hydrogen bond with another DNA segment having the complementary base sequence.

▼ **FIGURE 22.11   Preparing Recombinant DNA**

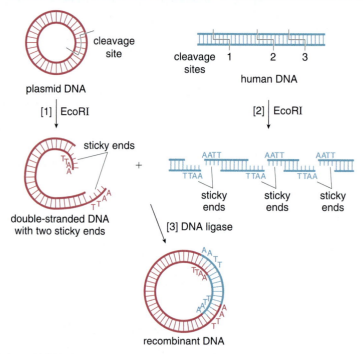

[1] Bacterial plasmid DNA is cut by the restriction endonuclease EcoRI to form double-stranded linear DNA with two sticky ends.

[2] A second sample of human DNA is cut with the same restriction endonuclease to form DNA segments having sticky ends that are complementary to those of the plasmid DNA.

[3] Combining the two pieces of DNA in the presence of a DNA ligase enzyme forms DNA containing the new segment. The recombinant DNA is larger than the original plasmid because of the inserted segment of DNA.

If the same restriction endonuclease is now used to cut DNA from another source, say human DNA, a second segment of DNA is now available having the needed complementary base sequence. When the two segments of DNA are combined together in the presence of a DNA ligase enzyme, the DNA segment is joined to the open plasmid DNA and a recombinant DNA molecule is formed.

The newly engineered recombinant DNA molecule can then be re-inserted into a bacterial cell. Transcription converts the recombinant DNA into RNA and RNA directs the synthesis of the protein that corresponds to the inserted DNA sequence. Since bacteria rapidly grow and divide, large amounts of protein can be synthesized by this method in a short period of time.

Insulin, a hormone used by millions of individuals with diabetes, was the first genetically engineered protein available in significant quantities by these methods. Prior to its availability in 1983, diabetics used insulin isolated from cows or pigs. This insulin has a similar but not identical amino acid sequence to human insulin, as we learned in Section 21.6. These small differences can lead to allergic reactions in some individuals.

PROBLEM 22.23	The restriction endonuclease HindIII recognizes the DNA sequence AAGCTT and cuts the DNA between A and A. Given the following segment of double-stranded DNA, draw the sticky ends that result from cleavage with this enzyme.

$$5' \sim\sim\sim C\ C\ A\ A\ G\ C\ T\ T\ G\ G\ A\ T\ T \sim\sim\sim 3'$$

$$3' \sim\sim\sim G\ G\ T\ T\ C\ G\ A\ A\ C\ C\ T\ A\ A \sim\sim\sim 5'$$

22.10B POLYMERASE CHAIN REACTION

In order to study a specific gene, such as the gene that causes the fatal inherited disorder cystic fibrosis, millions of copies of pure gene are needed. In fact, virtually an unlimited number of copies of any gene can be synthesized in just a few hours using a technique called the **polymerase chain reaction (PCR)**. PCR *clones* a segment of DNA; that is, PCR produces exact copies of a fragment of DNA.

- In this way, PCR *amplifies* a specific portion of a DNA molecule, producing millions of copies of a single molecule.

Four elements are needed to amplify DNA by PCR:

- The **segment of DNA** that must be copied
- **Two primers**—short polynucleotides that are complementary to the two ends of the segment to be amplified
- A **DNA polymerase enzyme** that will catalyze the synthesis of a complementary strand of DNA from a template strand
- **Nucleoside triphosphates** that serve as the source of the nucleotides A, T, C, and G needed in the synthesis of the new strands of DNA

Each cycle of the polymerase chain reaction involves three steps, illustrated in the stepwise *How To* procedure.

HOW TO   Use the Polymerase Chain Reaction to Amplify a Sample of DNA

Step [1] **Heat the DNA segment to unwind the double helix to form single strands.**

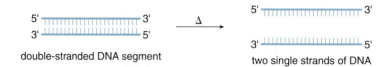

double-stranded DNA segment   two single strands of DNA

Step [2] **Add primers that are complementary to the DNA sequence at either end of the DNA segment to be amplified.**

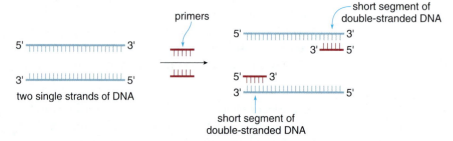

primers

short segment of double-stranded DNA

two single strands of DNA

short segment of double-stranded DNA

- The primers must hydrogen bond to a single strand of DNA and thus they must have the complementary sequence to the template strand.
- The primers form a short segment of double-stranded DNA on each strand, to which the DNA polymerase can add new nucleotides to the 3' end.

Step [3] **Use a DNA polymerase and added nucleotides to lengthen the DNA segment.**

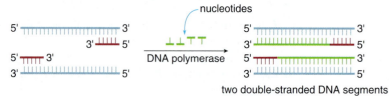

nucleotides

DNA polymerase

two double-stranded DNA segments

- DNA polymerase catalyzes the synthesis of a new strand of DNA complementary to the existing strand using nucleoside triphosphates (dATP, dCTP, dGTP, and dTTP) available in the reaction mixture.
- After one three-step cycle, *one* molecule of double-stranded DNA has formed *two* molecules of double-stranded DNA. Each double-stranded DNA molecule contains one original strand and one newly synthesized strand.
- **After each cycle, the amount of DNA *doubles*.** After 20 cycles, about 1,000,000 copies have been made.

Each step of a PCR cycle is carried out at a different temperature. PCR is now a completely automated process using a **thermal cycler,** an apparatus that controls the heating and cooling needed for each step. A heat-tolerant DNA polymerase called **Taq polymerase** is also typically used, so that new enzyme need not be added as each new cycle begins.

22.10C   FOCUS ON THE HUMAN BODY
DNA FINGERPRINTING

Because the DNA of each individual is unique, DNA fingerprinting is now routinely used as a method of identification.

Almost any type of cell—skin, saliva, semen, blood, and so forth—can be used to obtain a DNA fingerprint. After the DNA in a sample is amplified using PCR techniques, the DNA is cut into fragments with various restriction enzymes and the fragments are separated by size using a technique called gel electrophoresis. DNA fragments can be visualized on X-ray film after they react with a radioactive probe. The result is an image consisting of a set of horizontal bands, each band corresponding to a segment of DNA, sorted from low to high molecular weight.

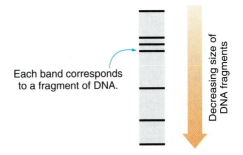

Each band corresponds to a fragment of DNA.

Decreasing size of DNA fragments

To compare the DNA of different individuals, samples are placed next to each other on the same gel and the position of the horizontal bands compared. DNA fingerprinting is now routinely used in criminal cases to establish the guilt or innocence of a suspect (Figure 22.12). Only identical twins have identical DNA, but related individuals have several similar DNA fragments. Thus,

▼ FIGURE 22.12   DNA Fingerprinting in Forensic Analysis

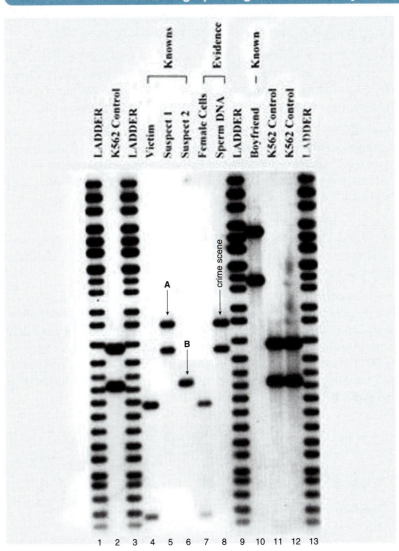

Each vertical lane (numbered 1–13) corresponds to a DNA sample.

- Lanes 1, 3, 9, and 13 are called DNA ladders. They correspond to DNA fragments of known size and are used to show the approximate size of the DNA fragments of the fingerprints.

- Lane 4 is the DNA obtained from a female assault victim and lane 7 is the female DNA from the crime scene. The two lanes match, indicating that they are from the same individual.

- Two individuals (**A** and **B**), whose DNA appears in lanes 5 and 6, were considered suspects.

- The male DNA obtained at the crime scene is shown in lane 8. The horizontal bands correspond to those of suspect **A,** incriminating him and eliminating individual **B** as a suspect.

DNA fingerprinting can be used to establish paternity by comparing the DNA of a child with that of each parent. DNA is also used to identify a body when no other means of identification is possible. DNA analysis was instrumental in identifying human remains found in the rubble of the World Trade Center after the towers collapsed on September 11, 2001.

22.11 FOCUS ON HEALTH & MEDICINE VIRUSES

A *virus* is an infectious agent consisting of a DNA or RNA molecule that is contained within a protein coating. Since a virus contains no enzymes or free nucleotides of its own, it is incapable of replicating. When it invades a host organism, however, it takes over the biochemical machinery of the host.

A virus that contains DNA uses the materials in the host organism to replicate its DNA, transcribe DNA to RNA, and synthesize a protein coating, thus forming new virus particles. These new virus particles leave the host cell and infect new cells and the process continues. Many prevalent diseases, including the common cold, influenza, and herpes are viral in origin.

A vaccine is an inactive form of a virus that causes an individual's immune system to produce antibodies to the virus to ward off infection. Many childhood diseases that were once very common, including mumps, measles, and chickenpox, are now prevented by vaccination. Polio has been almost completely eradicated, even in remote areas worldwide, by vaccination.

A virus that contains a core of RNA is called a **retrovirus.** Once a retrovirus invades a host organism, it must first make DNA by a process called **reverse transcription** (Figure 22.13). Once viral DNA has been synthesized, the DNA can transcribe RNA, which can direct protein synthesis. New retrovirus particles are thus synthesized and released to infect other cells.

AIDS (acquired immune deficiency syndrome) is caused by HIV (human immunodeficiency virus), a retrovirus that attacks lymphocytes central to the body's immune response against invading organisms. As a result, an individual infected with HIV becomes susceptible to life-threatening bacterial infections. HIV is spread by direct contact with the blood or other body fluids of an infected individual.

Major progress in battling the AIDS epidemic has occurred in recent years. HIV is currently best treated with a "cocktail" of drugs designed to destroy the virus at different stages of its reproductive cycle. One group of drugs, the protease inhibitors such as amprenavir (Section 21.10), act as enzyme inhibitors that prevent viral RNA from synthesizing needed proteins.

Other drugs are designed to interfere with reverse transcription, an essential biochemical process unique to the virus. Two drugs in this category are **AZT** (azidodeoxythymidine) and **ddI** (dideoxyinosine). These drugs are synthetic analogues of nucleosides. The structure of each drug closely resembles the nucleotides that must be incorporated in viral DNA, and thus they are inserted into a growing DNA strand after a phosphate is introduced at the 5'-OH. Each drug lacks a hydroxyl group at the 3' position, however, so no additional nucleotide can be added to the DNA chain, thus halting DNA synthesis.

HEALTH NOTE

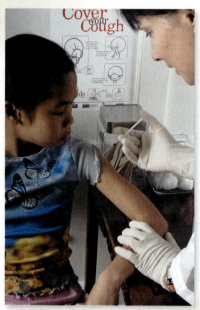

Childhood vaccinations have significantly decreased the incidence of once common diseases such as chickenpox, measles, and mumps.

no OH group → 3'
N₃

azidodeoxythymidine
AZT

no OH group → 3'

dideoxyinosine
ddI

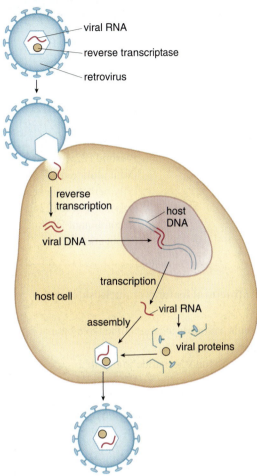

▼ FIGURE 22.13   How a Retrovirus Infects an Organism

A retrovirus (containing viral RNA and a reverse transcriptase) binds and infects a host cell. Reverse transcription forms viral DNA from RNA. One strand of viral DNA becomes a template to transcribe viral RNA, which is then translated into viral proteins. A new virus is assembled and the virus leaves the host cell to infect other cells.

Thus, HIV protease inhibitors and nucleoside analogues both disrupt the life cycle of HIV but they do so by very different mechanisms.

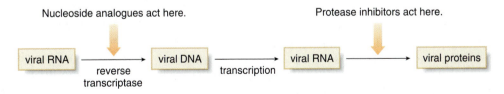

Nucleoside analogues act here.   Protease inhibitors act here.

| viral RNA | reverse transcriptase | viral DNA | transcription | viral RNA | | viral proteins |

PROBLEM 22.24   Explain why antibiotics such as sulfanilamide and penicillin (Sections 21.9) are effective in treating bacterial infections but are completely ineffective in treating viral infections.

CHAPTER HIGHLIGHTS

KEY TERMS

Anticodon (22.8)
Codon (22.7)
Complementary base pairs (22.3)
Deletion mutation (22.9)
Deoxyribonucleic acid (DNA, 22.1, 22.3)
Deoxyribonucleoside (22.1)
Deoxyribonucleotide (22.1)
DNA fingerprinting (22.10)
Gene (22.1)
Genetic code (22.7)
Genetic disease (22.9)
Genetic engineering (22.10)
Insertion mutation (22.9)

Lagging strand (22.4)
Leading strand (22.4)
Messenger RNA (mRNA, 22.5)
Mutation (22.9)
Nucleic acid (22.1, 22.2)
Nucleoside (22.1)
Nucleotide (22.1)
Plasmid (22.10)
Point mutation (22.9)
Polymerase chain reaction (PCR, 22.10)
Polynucleotide (22.2)
Recombinant DNA (22.10)
Replication (22.4)

Restriction endonuclease (22.10)
Retrovirus (22.11)
Reverse transcription (22.11)
Ribonucleic acid (RNA, 22.1, 22.5)
Ribonucleoside (22.1)
Ribonucleotide (22.1)
Ribosomal RNA (rRNA, 22.5)
Transcription (22.6)
Transfer RNA (tRNA, 22.5)
Translation (22.8)
Virus (22.11)

KEY CONCEPTS

❶ What are the main structural features of nucleosides and nucleotides? (22.1)

- A nucleoside contains a monosaccharide joined to a base at the anomeric carbon.
- A nucleotide contains a monosaccharide joined to a base, and a phosphate bonded to the 5'-OH group of the monosaccharide.
- The monosaccharide is either ribose or 2-deoxyribose, and the bases are abbreviated as A, G, C, T, and U.

❷ How do the nucleic acids DNA and RNA differ in structure? (22.2)

- DNA is a polymer of deoxyribonucleotides, where the sugar is 2-deoxyribose and the bases are A, G, C, and T. DNA is double stranded.
- RNA is a polymer of ribonucleotides, where the sugar is ribose and the bases are A, G, C, and U. RNA is single stranded.

❸ Describe the basic features of the DNA double helix. (22.3)

- DNA consists of two polynucleotide strands that wind into a right-handed double helix. The sugar–phosphate backbone lies on the outside of the helix and the bases lie on the inside. The two strands run in opposite directions; that is, one strand runs from the 5' end to the 3' end and the other runs from the 3' end to the 5' end. The double helix is stabilized by hydrogen bonding between complementary base pairs; A pairs with T and C pairs with G.

❹ Outline the main steps in the replication of DNA. (22.4)

- Replication of DNA is semiconservative; an original DNA molecule forms two DNA molecules, each of which has one strand from the parent DNA and one new strand.

- In replication, DNA unwinds and the enzyme DNA polymerase catalyzes replication on both strands. The identity of the bases on the template strand determines the order of the bases on the new strand, with A pairing with T and C pairing with G. Replication occurs from the 3' end to the 5' end of each strand. One strand, the leading strand, grows continuously, while the other strand, the lagging strand, is synthesized in fragments and then joined together with a DNA ligase enzyme.

❺ List the three types of RNA molecules and describe their functions. (22.5)

- Ribosomal RNA (rRNA), which consists of one large subunit and one small subunit, provides the site where proteins are assembled.
- Messenger RNA (mRNA) contains the sequence of nucleotides that determines the amino acid sequence in a protein. mRNA is transcribed from DNA such that each DNA gene corresponds to a different mRNA molecule.
- Transfer RNA (tRNA) contains an anticodon that identifies the amino acid that it carries on its acceptor stem and delivers that amino acid to a growing polypeptide.

❻ What is transcription? (22.6)

- Transcription is the synthesis of mRNA from DNA. The DNA helix unwinds and RNA polymerase catalyzes RNA synthesis from the 3' to 5' end of the template strand, forming mRNA with complementary bases.

❼ What are the main features of the genetic code? (22.7)

- mRNAs contain sequences of three bases called codons that code for individual amino acids. There are 61 codons that correspond to the 20 amino acids, as well as three stop codons that signal the end of protein synthesis.

8 How are proteins synthesized by the process of translation? (22.8)

- Translation begins with initiation, the binding of the ribosomal subunits to mRNA and the arrival of the first tRNA with an amino acid. During elongation, tRNAs bring individual amino acids to the ribosome one after another, and new peptide bonds are formed. Termination occurs when a stop codon is reached.

9 What is a mutation and how are mutations related to genetic diseases? (22.9)

- Mutations are changes in the nucleotide sequence in a DNA molecule. A point mutation results in the substitution of one nucleotide for another. Deletion and insertion mutations result in the loss or addition of nucleotides, respectively. A mutation that causes an inherited condition may result in a genetic disease.

10 What are the principal features of three techniques using DNA in the laboratory—recombinant DNA, the polymerase chain reaction (PCR), and DNA fingerprinting? (22.10)

- Recombinant DNA is synthetic DNA formed when a segment of DNA from one source is inserted into the DNA of another source. When inserted into a bacterium, recombinant DNA can be used to prepare large quantities of useful proteins.
- The polymerase chain reaction (PCR) is used to amplify a portion of a DNA molecule to produce millions of copies of a single gene.
- DNA fingerprinting is used to identify an individual by cutting DNA with restriction endonucleases to give a unique set of fragments.

11 What are the main characteristics of viruses? (22.11)

- A virus is an infectious agent that contains either DNA or RNA within a protein coat. When the virus invades a host cell, it uses the biochemical machinery of the host to replicate. A retrovirus contains RNA and a reverse transcriptase that allow the RNA to synthesize viral DNA, which then transcribes RNA that directs protein synthesis.

PROBLEMS

Selected in-chapter and end-of-chapter problems have brief answers provided in Appendix B.

Nucleosides, Nucleotides, and Nucleic Acids

22.25 What is the difference between a ribonucleoside and a ribonucleotide? Give an example of each.

22.26 What is the difference between a ribonucleotide and a deoxyribonucleotide? Give an example of each.

22.27 What is the difference between a gene and a chromosome?

22.28 What is the difference between uracil and uridine?

22.29 List three structural differences between DNA and RNA.

22.30 List three structural similarities in DNA and RNA.

22.31 Identify the base and monosaccharide in each compound and then name it.

22.32 Identify the base and monosaccharide in each compound and then name it.

22.33 Draw the structure of each of the following:
   a. a purine base
   b. a nucleoside that contains 2-deoxyribose and a pyrimidine base
   c. a nucleotide that contains ribose and a purine base
   d. a triphosphate of guanosine

22.34 Draw the structure of each of the following:
   a. a pyrimidine base
   b. a nucleoside that contains 2-deoxyribose and a purine base
   c. a nucleotide that contains ribose and a pyrimidine base
   d. a diphosphate of uridine

22.35 Draw the structure of each nucleoside or nucleotide.
   a. adenosine   c. GDP
   b. deoxyguanosine   d. dTDP

22.36 Draw the structure of each nucleoside or nucleotide.
   a. uridine   c. dGMP
   b. deoxycytidine   d. UTP

22.37 Draw the structures of the two possible dinucleotides formed from uridine 5'-monophosphate and adenosine 5'-monophosphate.

22.38 Draw the structures of the two possible dinucleotides formed from deoxythymidine 5'-monophosphate and deoxycytidine 5'-monophosphate.

22.39 Draw the structure of each dinucleotide and identify the 5' and 3' ends.
   a. the deoxyribonucleotide formed by joining the 3'-OH group of dTMP with the 5'-phosphate of dAMP
   b. the ribonucleotide formed by joining the 5'-phosphate of GMP with the 3'-OH of CMP

22.40 Draw the structure of each dinucleotide and identify the 5' and 3' ends.
   a. the deoxyribonucleotide formed by joining two deoxyguanosine 5'-phosphates together
   b. the ribonucleotide formed by joining the 5'-phosphate of UMP with the 3'-OH of AMP

22.41 Draw the deoxyribonucleotide TGA. Label the 5' and 3' ends.

22.42 Draw the ribonucleotide CGU. Label the 5' and 3' ends.

22.43 Describe in detail the DNA double helix with reference to each of the following features: (a) the sugar–phosphate backbone; (b) the functional groups at the end of each strand; (c) the hydrogen bonding between strands.

22.44 Describe in detail the DNA double helix with reference to each of the following features: (a) the location of the bases; (b) the complementary base pairing; (c) the phosphodiester linkages.

22.45 Write the sequence of the complementary strand of each segment of a DNA molecule.
   a. 5'–AAATAAC–3'   c. 5'–CGATATCCCG–3'
   b. 5'–ACTGGACT–3'   d. 5'–TTCCCGGGATA–3'

22.46 Write the sequence of the complementary strand of each segment of a DNA molecule.
   a. 5'–TTGCGA–3'   c. 5'–ACTTCAGGT–3'
   b. 5'–CGCGTAAT–3'   d. 5'–CCGGTTAATACGGC–3'

22.47 If 27% of the nucleotides in a sample of DNA contain the base adenine (A), what are the percentages of bases T, G, and C?

22.48 If 19% of the nucleotides in a sample of DNA contain the base cytosine (C), what are the percentages of bases G, A, and T?

22.49 Complementary base pairing occurs between two strands of DNA, between two different sites on an RNA molecule, and between one strand of DNA and one strand of RNA. Which base pairs are the same in all three settings?

22.50 Complementary base pairing occurs between two strands of DNA, and between one strand of DNA and one strand of RNA. Give an example of a base pair that is unique to each setting.

Replication, Transcription, Translation, and Protein Synthesis

22.51 What is the sequence of a newly synthesized DNA segment if the template strand has the sequence 3'–ATGGCCTATGCGAT–5'?

22.52 What is the sequence of a newly synthesized DNA segment if the template strand has the sequence 3'–CGCGATTAGATATTGCCGC–5'?

22.53 During replication, why is there a leading strand and a lagging strand?

22.54 Why is replication said to be semiconservative?

22.55 Explain the roles of messenger RNA and transfer RNA in converting the genetic information coded in DNA into protein synthesis.

22.56 Explain the different roles of DNA polymerase and RNA polymerase in converting the information in DNA to other nucleic acids.

22.57 In what ways are replication and transcription similar? In what ways are they different?

22.58 What are the two main structural features of transfer RNA molecules?

22.59 What mRNA is transcribed from each DNA sequence in Problem 22.45?

22.60 What mRNA is transcribed from each DNA sequence in Problem 22.46?

22.61 For each DNA segment: [1] What is the sequence of the mRNA molecule synthesized from each DNA template? [2] What is the sequence of the informational strand of the DNA molecule?
   a. 3'–ATGGCTTA–5'   c. 3'–GGTATACCG–5'
   b. 3'–CGGCGCTTA–5'   d. 3'–TAGGCCGTA–5'

22.62 For each DNA segment: [1] What is the sequence of the mRNA molecule synthesized from each DNA template? [2] What is the sequence of the informational strand of the DNA molecule?
   a. 3'–GGCCTATA–5'   c. 3'–CTACTG–5'
   b. 3'–GCCGAT–5'   d. 3'–ATTAGAGC–5'

22.63 What is the difference between a codon and an anticodon?

22.64 What is the difference between transcription and translation?

22.65 For each codon, give its anticodon and the amino acid for which it codes.
   a. CUG   b. UUU   c. AAG   d. GCA

22.66 For each codon, give its anticodon and the amino acid for which it codes.
   a. GUU   b. AUA   c. CCC   d. GCG

22.67 Derive the amino acid sequence that is coded for by each mRNA sequence.
    a. 5' CCA ACC UGG GUA GAA 3'
    b. 5' AUG UUU UUA UGG UGG 3'
    c. 5' GUC GAC GAA CCG CAA 3'

22.68 Derive the amino acid sequence that is coded for by each mRNA sequence.
    a. 5' AAA CCC UUU UGU 3'
    b. 5' CCU UUG GAA GUA CUU 3'
    c. 5' GGG UGU AUG CAC CGA UUG 3'

22.69 Write a possible mRNA sequence that codes for each peptide.
    a. Ile–Met–Lys–Ser–Tyr
    b. Pro–Gln–Glu–Asp–Phe
    c. Thr–Ser–Asn–Arg

22.70 Write a possible mRNA sequence that codes for each peptide.
    a. Phe–Phe–Leu–Lys
    b. Val–Gly–Gln–Asp–Asn
    c. Arg–His–Ile–Ser

22.71 Considering each nucleotide sequence in an mRNA molecule: [1] write the sequence of the DNA template strand from which the mRNA was synthesized; [2] give the peptide synthesized by the mRNA.
    a. 5' UAU UCA AUA AAA AAC 3'
    b. 5' GAU GUA AAC AAG CCG 3'

22.72 Considering each nucleotide sequence in an mRNA molecule: [1] write the sequence of the DNA template strand from which the mRNA was synthesized; [2] give the peptide synthesized by the mRNA.
    a. 5' UUG CUC AAC CAA 3'
    b. 5' AUU GUA CCA CAA CCC 3'

Mutations

22.73 What is the difference between a point mutation and a silent mutation?

22.74 Explain why some mutations cause no effect on a cell whereas others have a major effect.

22.75 Consider the following mRNA sequence:
5'–CUU CAG CAC–3'.
    a. What amino acid sequence is coded for by this mRNA?
    b. What is the amino acid sequence if a mutation converts CAC to AAC?
    c. What is the amino acid sequence if a mutation converts CUU to CUC?
    d. What occurs when a mutation converts CAG to UAG?
    e. What occurs if CU is deleted from the beginning of the chain?

22.76 Consider the following mRNA sequence:
5'–ACC UUA CGA–3'.
    a. What amino acid sequence is coded for by this mRNA?
    b. What is the amino acid sequence if a mutation converts ACC to ACU?
    c. What is the amino acid sequence if a mutation converts UUA to UCA?
    d. What is the amino acid sequence if a mutation converts CGA to AGA?
    e. What occurs if C is added to the beginning of the chain?

22.77 Consider the following sequence of DNA:
3'–TTA CGG–5'.
    a. What dipeptide is formed from this DNA after transcription and translation?
    b. If a mutation converts CGG to CGT in DNA, what dipeptide is formed?
    c. If a mutation converts CGG to CCG in DNA, what dipeptide is formed?
    d. If a mutation converts CGG to AGG in DNA, what dipeptide is formed?

22.78 Consider the following sequence of DNA:
3'–ATA GGG–5'.
    a. What dipeptide is formed from this DNA after transcription and translation?
    b. If a mutation converts ATA to ATG in DNA, what dipeptide is formed?
    c. If a mutation converts ATA to AGA in DNA, what dipeptide is formed?
    d. What occurs when a mutation converts ATA to ATT in DNA?

Recombinant DNA and Related Techniques

22.79 What is a restriction endonuclease?

22.80 What is meant by the term "sticky end"?

22.81 The restriction enzyme BamHI recognizes the sequence GGATCC and cuts the DNA between the two G's. Given the following segment of double-stranded DNA, draw the sticky ends that result from cleavage with this enzyme.

    5' ⌇⌇⌇⌇ C C G G T T G G A T C C T T ⌇⌇⌇⌇ 3'

    3' ⌇⌇⌇⌇ G G C C A A C C T A G G A A ⌇⌇⌇⌇ 5'

22.82 The restriction enzyme PstI recognizes the sequence GACGTC and cuts the DNA between the G and A. Given the following segment of double-stranded DNA, draw the sticky ends that result from cleavage with this enzyme.

    5' ⌇⌇⌇⌇ G G A T G A C G T C A A T T ⌇⌇⌇⌇ 3'

    3' ⌇⌇⌇⌇ C C T A C T G C A G T T A A ⌇⌇⌇⌇ 5'

22.83 Describe the main steps in preparing recombinant DNA from plasmid DNA.

22.84 Describe the main steps in using the polymerase chain reaction to amplify a small amount of DNA.

22.85 The given gel contains the DNA fingerprint of a mother (lane 1), father (lane 2), and four children (lanes 3–6). One child is adopted, two children are the offspring of both parents, and one child is the offspring of one parent only.

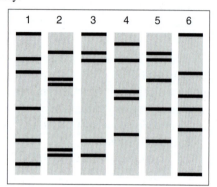

a. Which lanes, if any, represent a biological child of both parents? Explain your choice.
b. Which lanes, if any, represent an adopted child? Explain your choice.

22.86 With reference to the gel in Problem 22.85: (a) which lanes (if any) represent a biological child of the mother only; (b) which lanes (if any) represent a biological child of the father only; (c) which lanes (if any) represent twins? Explain each choice.

Viruses

22.87 What is a retrovirus?

22.88 What is reverse transcription?

22.89 Lamivudine, also called 3TC, is a nucleoside analogue used in the treatment of HIV and the virus that causes hepatitis B. (a) What nucleoside does lamivudine resemble? (b) How does lamivudine inhibit reverse transcription?

lamivudine
3TC

22.90 How does a vaccine protect an individual against certain viral infections?

General Questions

22.91 Consider a template strand of DNA with the following sequence: 3'–CAA TGT ATT TTT GCT–5'. (a) What is the informational strand of DNA that corresponds to this template? (b) What mRNA is prepared from this template? (c) What polypeptide is prepared from the mRNA?

22.92 Consider a template strand of DNA with the following sequence: 3'–ATT AAA GCC TTC TGT–5'. (a) What is the informational strand of DNA that corresponds to this template? (b) What mRNA is prepared from this template? (c) What polypeptide is prepared from the mRNA?

22.93 Fill in the needed bases, codon, anticodon, or amino acid needed to complete the following table that relates the sequences of DNA, mRNA, tRNA, and the resulting polypeptide.

DNA informational strand:	5' end						3' end
DNA template strand:	3' end	TTG	ATA	GGT	TGC	TTC	TAC	5' end
mRNA codons:	5' end						3' end
tRNA anticodons:							
Polypeptide:							

22.94 Fill in the needed bases, codon, anticodon, or amino acid needed to complete the following table that relates the sequences of DNA, mRNA, tRNA, and the resulting polypeptide.

DNA informational strand:	5' end						3' end
DNA template strand:	3' end	TCC	GAC	TTG	TGC	CAT	CAC	5' end
mRNA codons:	5' end						3' end
tRNA anticodons:							
Polypeptide:							

22.95 Fill in the needed bases, codon, anticodon, or amino acid needed to complete the following table that relates the sequences of DNA, mRNA, tRNA, and the resulting polypeptide.

DNA informational strand:	5' end	AAC					3' end
DNA template strand:	3' end		CAT				5' end
mRNA codons:	5' end			UCA		AUG	3' end
tRNA anticodons:					GUG		
Polypeptide:				Thr			

22.96 Fill in the needed bases, codon, anticodon, or amino acid needed to complete the following table that relates the sequences of DNA, mRNA, tRNA, and the resulting polypeptide.

DNA informational strand:	5' end		ACA					3' end
DNA template strand:	3' end	CAA					GTC	5' end
mRNA codons:	5' end			UAC				3' end
tRNA anticodons:					ACA			
Polypeptide:						Lys		

22.97 If there are 325 amino acids in a polypeptide, how many bases are present in a single strand of the gene that codes for it, assuming that every base is transcribed and then translated to polypeptide?

22.98 If a single strand of a gene contains 678 bases, how many amino acids result in the polypeptide prepared from it, assuming every base of the gene is transcribed and then translated?

22.99 Met-enkephalin (Tyr–Gly–Gly–Phe–Met) is a pain killer and sedative (Section 21.5). What is a possible nucleotide sequence in the template strand of the gene that codes for met-enkephalin, assuming that every base of the gene is transcribed and then translated?

22.100 Leu-enkephalin (Tyr–Gly–Gly–Phe–Leu) is a pain killer and sedative (Section 21.5). What is a possible nucleotide sequence in the template strand of the gene that codes for leu-enkephalin, assuming that every base of the gene is transcribed and then translated?

CHALLENGE QUESTIONS

22.101 Give a possible nucleotide sequence in the template strand of the gene that codes for the following tripeptide.

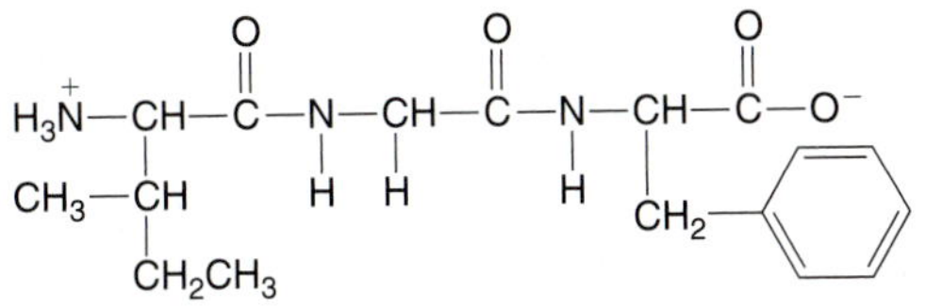

22.102 Give a possible nucleotide sequence in the template strand of the gene that codes for the following tripeptide.

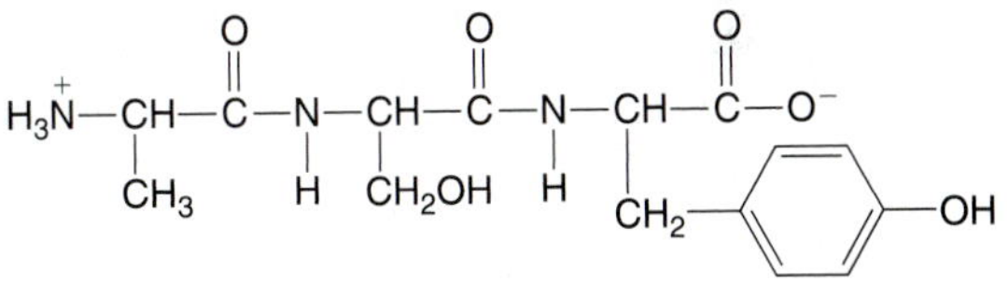

23

CHAPTER GOALS

In this chapter you will learn how to:

1. Define metabolism and explain where energy production occurs in cells

2. Describe the main characteristics of the four stages of catabolism

3. Explain the role of ATP in energy production and how coupled reactions drive energetically unfavorable reactions

4. Describe the roles of the main coenzymes used in metabolism

5. List the main features of the citric acid cycle

6. Describe the main components of the electron transport chain and oxidative phosphorylation

7. Explain why compounds such as cyanide that disrupt the electron transport chain are poisons

The metabolism of carbohydrates, proteins, and lipids provides the raw materials and energy for the growth and maintenance of the human body.

DIGESTION AND THE CONVERSION OF FOOD INTO ENERGY

DESPITE the wide diversity among living organisms, virtually all organisms contain the same types of biomolecules—lipids, carbohydrates, proteins, and nucleic acids— and use the same biochemical reactions. Chapter 23 is the first of two chapters that deal with the complex biochemical pathways that convert one molecule into another in cells. In Chapter 23 we concentrate on the fundamental reactions that produce energy, while in Chapter 24 we examine specific pathways for the metabolism of lipids, carbohydrates, and proteins. Although the pathways often involve many steps, individual reactions can be understood by applying the same chemical principles you have already learned.

23.1 INTRODUCTION

Each moment thousands of reactions occur in a living cell: large molecules are broken down into smaller components, small molecules are converted into larger molecules, and energy changes occur.

- *Metabolism* is the sum of all of the chemical reactions that take place in an organism.

There are two types of metabolic processes called **catabolism** and **anabolism.**

- *Catabolism* is the breakdown of large molecules into smaller ones. Energy is generally released during catabolism.
- *Anabolism* is the synthesis of large molecules from smaller ones. Energy is generally absorbed during anabolism.

The oxidation of glucose ($C_6H_{12}O_6$) to carbon dioxide (CO_2) and water (H_2O) is an example of catabolism, while the synthesis of a protein from component amino acids is an example of anabolism. Often the conversion of a starting material to a final product occurs by an organized series of consecutive reactions called a **metabolic pathway.** A metabolic pathway may be **linear** or **cyclic.**

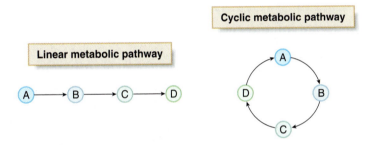

- A *linear* pathway is a series of reactions that generates a final product different from any of the reactants.
- A *cyclic* pathway is a series of reactions that regenerates the first reactant.

In this chapter we concentrate on energy changes that occur during metabolism. The human body needs energy to move, think, talk, laugh, and breathe. It also requires energy to maintain a constant temperature and to synthesize new cells and tissues.

While the reactions in living organisms follow the same principles as reactions in the laboratory, the energy needs of the body come with some special requirements. The body usually can't use the calories of a big meal all at once, so the energy must be stored and then gradually released. Moreover, energy must be stored in molecules that are readily accessible for use anywhere in the body at any time the energy is needed.

Where does energy production occur in cells? A typical animal cell, such as the one shown in Figure 23.1, is surrounded by a cell membrane (Section 19.7) and has a nucleus that contains DNA in chromosomes (Section 22.3). The **cytoplasm,** the region of the cell between the cell membrane and the nucleus, contains various specialized structures called **organelles,** each of which has a specific function. **Mitochondria** are small sausage-shaped organelles in which energy production takes place. Mitochondria contain an outer membrane and an inner membrane with many folds. The area between these two membranes is called the **intermembrane space.** Energy production occurs within the **matrix,** the area surrounded by the inner membrane of the mitochondrion. The number of mitochondria in a cell varies depending on its energy needs. Cells in the heart, brain, and muscles of the human body typically contain many mitochondria.

Mitochondri*on* is the singular form of the plural mitochondria.

▼ **FIGURE 23.1   The Mitochondrion in a Typical Animal Cell**

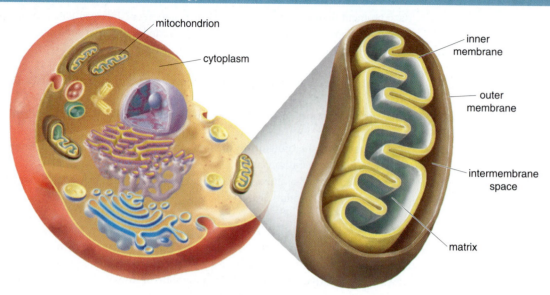

A mitochondrion is a small organelle located in the cytoplasm of the cell. Cellular energy production occurs in the mitochondria.

23.2   AN OVERVIEW OF METABOLISM

Just as gasoline is the fuel that powers automobiles, so, too, food is the fuel that is metabolized by the body to provide energy. Catabolism breaks down the carbohydrates, proteins, and lipids in food into smaller molecules, releasing energy to supply the body's needs. This process involves numerous catabolic pathways that can be organized into four stages, as shown in Figure 23.2.

23.2A   STAGE [1]—DIGESTION

The catabolism of food begins with **digestion,** which is catalyzed by enzymes in the saliva, stomach, and small intestines (Figure 23.3). Digestion converts large molecules into smaller components. The hydrolysis of carbohydrates to monosaccharides begins with amylase enzymes in the saliva, and continues in the small intestines. Protein digestion begins in the stomach, where acid denatures the protein, and the protease pepsin begins to cleave the protein backbone into smaller polypeptides and amino acids. Digestion continues in the small intestines, where trypsin and chymotrypsin further cleave the protein backbone to form amino acids. Triacylglycerols, the most common lipids, are first emulsified by bile secreted by the liver, and then hydrolyzed to glycerol and fatty acids by lipases in the small intestines.

Once these small molecules are formed during digestion, they are each absorbed through the intestinal wall into the bloodstream and transported to other cells in the body. Some substances such as cellulose are not metabolized as they travel through the digestive tract. When these materials can't be absorbed, they pass into the large intestines and are excreted.

23.2B   STAGES [2]–[4] OF CATABOLISM

Once small molecules are formed, catabolism continues to break down each type of molecule to smaller units releasing energy in the process.

FIGURE 23.2 The Four Stages in Biochemical Catabolism and Energy Production

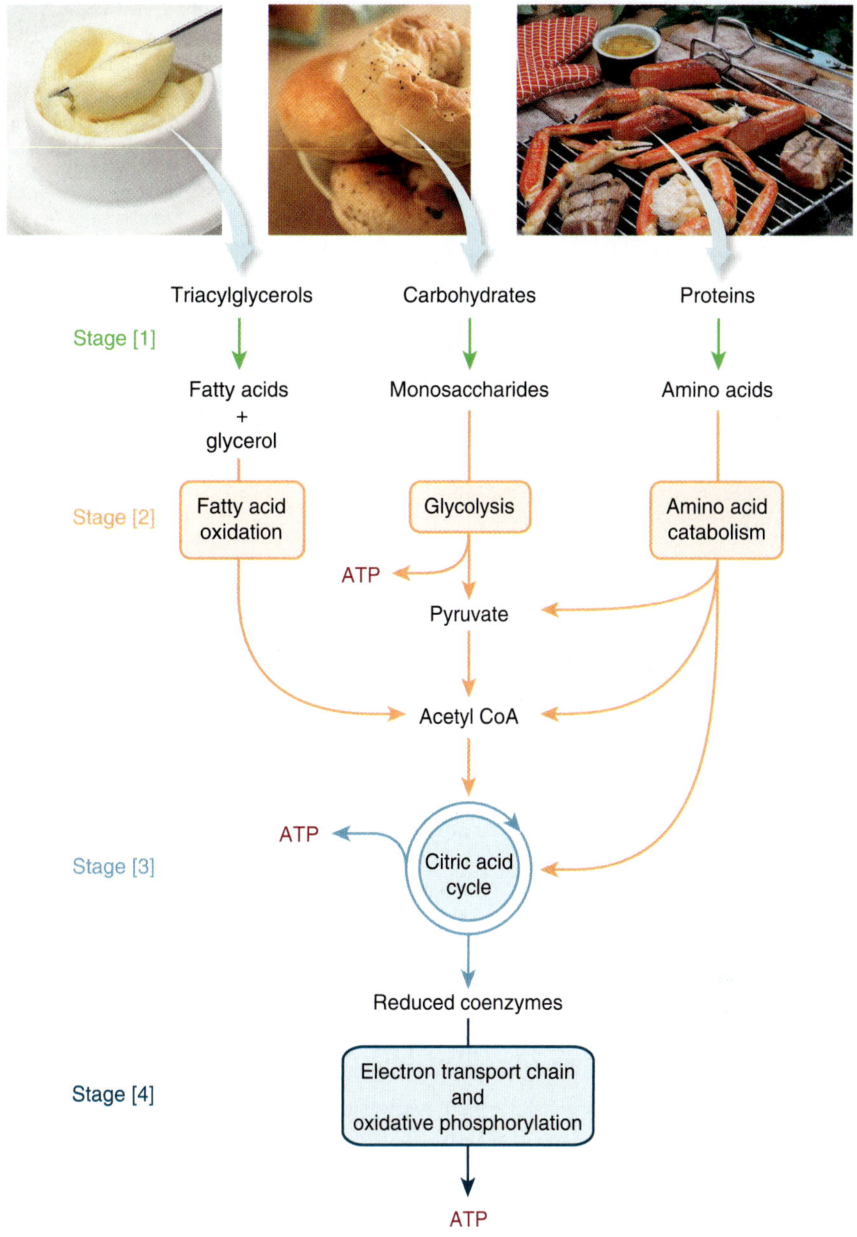

Stage [2]—Formation of Acetyl CoA

Monosaccharides, amino acids, and fatty acids are degraded into **acetyl groups (CH₃CO–),** two-carbon units that are bonded to coenzyme A (a coenzyme), forming **acetyl CoA.** Details of the structure of acetyl CoA are given in Section 23.4.

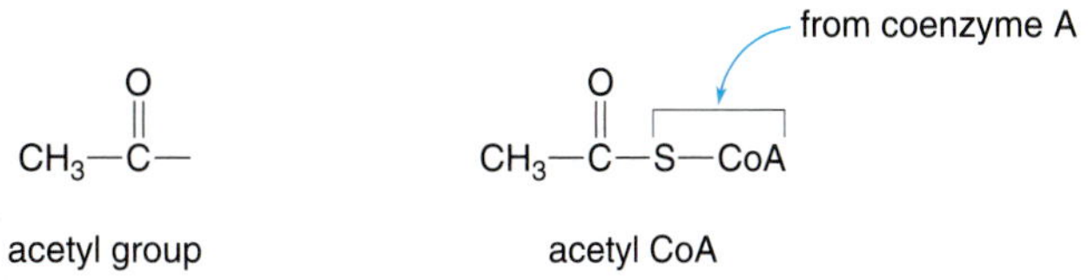

▼ **FIGURE 23.3**   **Digestion of Carbohydrates, Proteins, and Triacylglycerols**

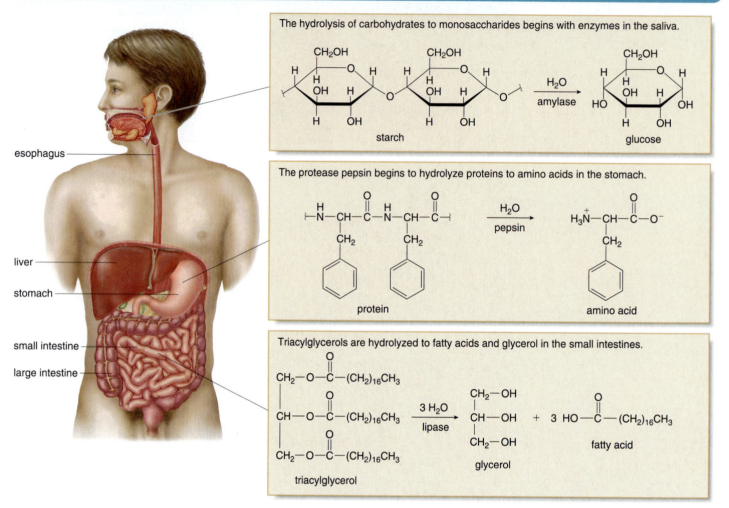

The first stage of catabolism, **digestion,** is the hydrolysis of large molecules to small molecules: polysaccharides such as starch are hydrolyzed to monosaccharides (Section 20.6), proteins are hydrolyzed to their component amino acids (Section 21.8), and triacylglycerols are hydrolyzed to glycerol and fatty acids (Section 19.5). Each of these molecules enters its own metabolic pathway to be further broken down into smaller components, releasing energy.

The product of the catabolic pathways is the *same* for all three types of molecules. As a result, a common catabolic pathway, the **citric acid cycle,** continues the processing of all types of biomolecules to generate energy.

Stage [3]—The Citric Acid Cycle

The citric acid cycle is based in the mitochondria. In this biochemical cycle, the acetyl groups of acetyl CoA are oxidized to carbon dioxide. Some energy produced by this process is stored in the bonds of a nucleoside triphosphate (Section 22.1) and reduced coenzymes, whose structures are shown in Section 23.4.

Stage [4]—The Electron Transport Chain and Oxidative Phosphorylation

Within the mitochondria, the electron transport chain and oxidative phosphorylation produce **ATP**—adenosine 5'-triphosphate—the primary energy-carrying molecule in metabolic pathways. Oxygen combines with hydrogen ions and electrons from the reduced coenzymes to form water. The result of catabolism is that biomolecules are converted to CO_2 and H_2O and energy is produced and stored in ATP molecules.

Since all catabolic pathways converge in the citric acid cycle, we begin our discussion of catabolism with this central common pathway. In order to understand the details, however, we must first learn about the structure and properties of some of the molecules involved. In particular, Section 23.3 is devoted to a discussion of ATP and how it is used to supply energy in reactions. In Section 23.4 we examine the structure and reactions of the key coenzymes nicotinamide adenine dinucleotide (NAD^+), flavin adenine dinucleotide (FAD), and coenzyme A.

PROBLEM 23.1	What advantage might there be to funneling all catabolic pathways into a single common pathway, the citric acid cycle?

23.3 ATP AND ENERGY PRODUCTION

As we learned in Section 22.1, **ATP,** adenosine 5'-triphosphate, is a nucleoside triphosphate formed by adding three phosphates to the 5'-OH group of adenosine, a nucleoside composed of the sugar ribose and the base adenine. Similarly, **ADP,** adenosine 5'-diphosphate, is a nucleoside diphosphate formed by adding two phosphates to the 5'-OH group of adenosine.

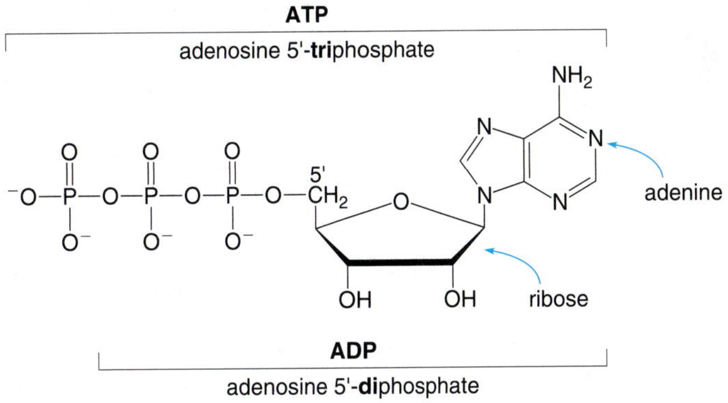

23.3A   GENERAL FEATURES OF ATP HYDROLYSIS AND FORMATION

In metabolic pathways, the interconversion of ATP and ADP is the most important process for the storage and release of energy.

- Hydrolysis of ATP cleaves one phosphate group, forming ADP and hydrogen phosphate, HPO_4^{2-}. This reaction *releases* 7.3 kcal/mol of energy.

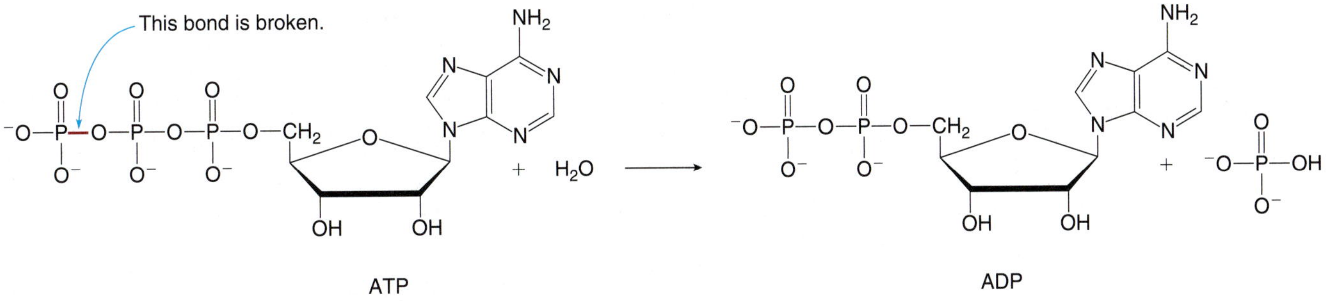

Any process, such as walking, running, swallowing, or breathing, is fueled by the release of energy from the hydrolysis of ATP to ADP. ATP is the most prominent member of a group of "high-energy" molecules, reactive molecules that release energy by cleaving a bond during hydrolysis. Because ATP contains four negatively charged oxygen atoms in close proximity, the electronic repulsion of the like charges drives the hydrolysis to ADP, which has fewer negatively charged oxygens and therefore less electronic repulsion.

- The reverse reaction, phosphorylation, adds a phosphate group to ADP, forming ATP. Phosphorylation *requires* 7.3 kcal/mol of energy.

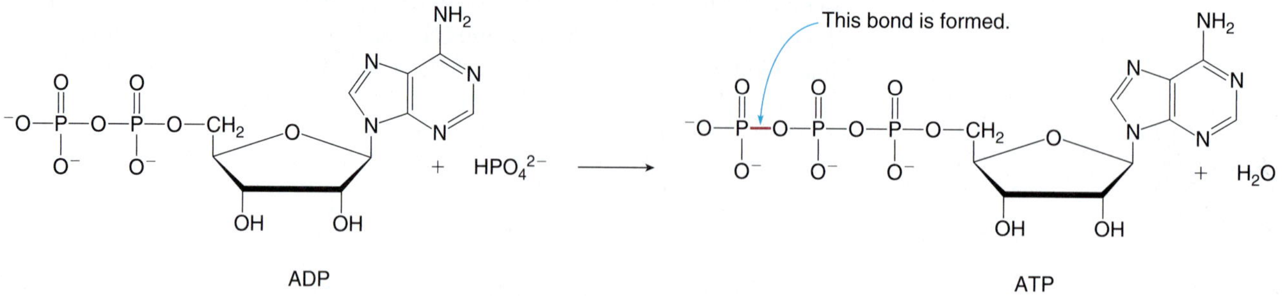

ADP $\quad + \quad HPO_4^{2-} \quad \longrightarrow \quad$ ATP $\quad + \quad H_2O$

By way of these reactions, ATP is a transporter of energy.

- Energy is absorbed and stored in ATP when it is synthesized from ADP.
- Energy is released when ATP is hydrolyzed to ADP.

Figure 23.4 summarizes the reactions and energy changes that occur when ATP is synthesized and hydrolyzed. As we learned in Section 6.2, when energy is *released* in a reaction, the energy change is reported as a *negative* (–) value, so the energy change for ATP hydrolysis is –7.3 kcal/mol. When energy is *absorbed* in a reaction, the energy change is reported as a *positive* (+) value, so the energy change for the phosphorylation of ADP is +7.3 kcal/mol. The synthesis of ATP with HPO_4^{2-} is the reverse of ATP hydrolysis, and the energy changes for such reactions are *equal* in value but *opposite* in sign.

ATP is constantly synthesized and hydrolyzed. It is estimated that each ATP molecule exists for about a minute before it is hydrolyzed and its energy released. Even though the body contains only about one gram of ATP at a given time, the energy needs of the body are such that an average individual synthesizes 40 kg of ATP daily!

▼ FIGURE 23.4 ATP Hydrolysis and Synthesis

							Energy change
ATP Hydrolysis	ATP	+	H_2O	$\longrightarrow$	ADP	+ HPO_4^{2-}	–7.3 kcal/mol
ATP Synthesis	ADP	+	HPO_4^{2-}	$\longrightarrow$	ATP	+ H_2O	+7.3 kcal/mol

Energy is released.   Energy is absorbed.

PROBLEM 23.2

GTP, guanosine 5'-triphosphate, is another high-energy molecule that releases 7.3 kcal of energy when it is hydrolyzed to GDP. Write the equation for the hydrolysis of GTP to GDP.

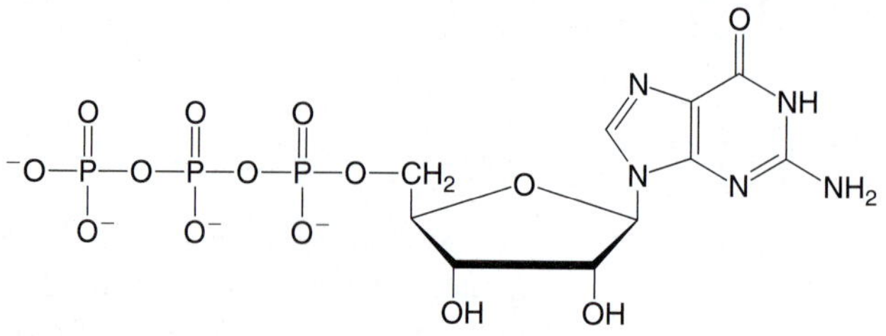

GTP

23.3B COUPLED REACTIONS IN METABOLIC PATHWAYS

As we learned in Section 6.5, a reaction that releases energy is a favorable process because it forms products that are lower in energy than the reactants. On the other hand, a reaction that absorbs energy is an unfavorable process because it forms products that are higher in energy than the reactants.

The hydrolysis of ATP to ADP is a favorable reaction: energy is released and some of the electronic repulsion due to many closely situated negative charges is relieved. The energy released in ATP hydrolysis can be used to drive a reaction that has an unfavorable energy change by **coupling** the two reactions together.

- Coupled reactions are pairs of reactions that occur together. The energy released by one reaction provides the energy to drive the other reaction.

Let's consider the catabolism of glucose to CO_2 and H_2O, a process that requires a large number of steps in various biochemical pathways. Although the overall conversion releases a great deal of energy, the energy change associated with each individual step is much smaller. In some reactions energy is released, while in others energy is absorbed. The hydrolysis of ATP provides the energy to drive the reactions that require energy.

In the first step in the metabolism of glucose, for example, glucose reacts with hydrogen phosphate (HPO_4^{2-}) to form glucose 6-phosphate, a process that requires 3.3 kcal/mol of energy. This process is energetically unfavorable because energy is absorbed (+3.3 kcal/mol). However, by coupling this reaction with ATP hydrolysis, an energetically favorable reaction (−7.3 kcal/mol of energy released), the coupled reaction becomes an energetically favorable process (−4.0 kcal/mol of energy released).

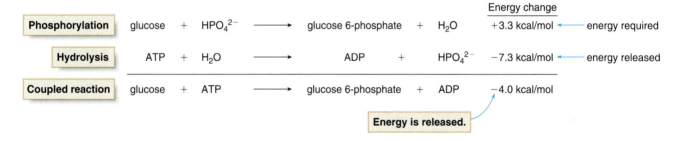

- The coupled reaction is the net reaction, written by summing the substances in both equations and eliminating those compounds that appear on both sides of the reaction arrows.
- The overall energy change is found by summing the energies for the individual steps.

Thus, in this example:

- The hydrolysis of ATP provides the energy for the phosphorylation of glucose.

Although the coupled reactions are written as two separate equations for emphasis in the preceding example, in reality, a single reaction takes place. ATP transfers a phosphate to glucose, forming glucose 6-phosphate and ADP, while both molecules are held in close proximity at the active site of hexokinase, as shown in Figure 23.5.

- Coupling an energetically unfavorable reaction with an energetically favorable reaction that releases *more* energy than the amount required is a common phenomenon in biochemical reactions.

▼ **FIGURE 23.5   A Coupled Reaction—Phosphorylation of Glucose by ATP**

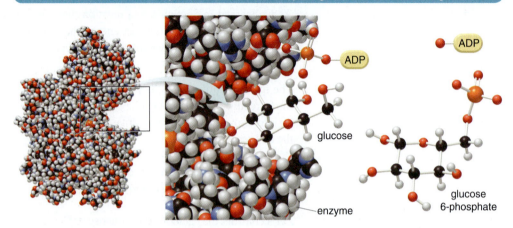

The enzyme hexokinase binds glucose at its active site and ATP is also bound in close proximity. ATP transfers a phosphate directly to the glucose molecule to form ADP and glucose 6-phosphate. This coupled reaction is energetically favorable.

Coupled reactions explain how ATP can be synthesized by the energetically unfavorable process of phosphorylation. Phosphorylation must be coupled with a reaction that releases *more* than 7.3 kcal/mol of energy. Such a reaction is the hydrolysis of phosphoenolpyruvate to form pyruvate.

Phosphorylation	$ADP + HPO_4^{2-} \longrightarrow ATP + H_2O$ $+7.3$ kcal/mol ← energy required
Hydrolysis	$\underset{\text{phosphoenolpyruvate}}{CH_2{=}\overset{\overset{\displaystyle OPO_3^{2-}}{\mid}}{C}{-}COO^-} + H_2O \longrightarrow \underset{\text{pyruvate}}{CH_3{-}\overset{\overset{\displaystyle O}{\|}}{C}{-}COO^-} + HPO_4^{2-}$ -14.8 kcal/mol ← energy released
Coupled reaction	$CH_2{=}\overset{\overset{\displaystyle OPO_3^{2-}}{\mid}}{C}{-}COO^- + ADP \longrightarrow CH_3{-}\overset{\overset{\displaystyle O}{\|}}{C}{-}COO^- + ATP$ -7.5 kcal/mol

Energy change

Energy is released.

Since the hydrolysis of phosphoenolpyruvate releases more energy than the phosphorylation of ADP (–14.8 kcal versus +7.3 kcal), the coupled reaction becomes energetically favorable (–7.5 kcal/mol of energy released).

SAMPLE PROBLEM 23.1

The phosphorylation of fructose to fructose 6-phosphate requires 3.8 kcal/mol of energy. This unfavorable reaction can be driven by the hydrolysis of ATP to ADP. (a) Write the equation for the coupled reaction. (b) How much energy is released in the coupled reaction?

Energy change

$$fructose + HPO_4^{2-} \longrightarrow fructose\ 6\text{-phosphate} + H_2O \qquad +3.8 \text{ kcal/mol}$$

ANALYSIS   Write the equation for the hydrolysis of ATP to ADP (Section 23.3A). Add together the substances in this equation and the given equation to give the net equation—that is, the equation for the coupled reaction. To determine the overall energy change, add together the energy changes for each step.

SOLUTION

a. The coupled equation shows the overall reaction that combines the phosphorylation of fructose and the hydrolysis of ATP.

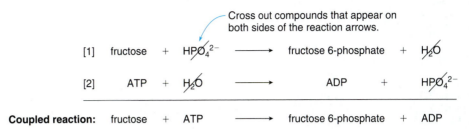

Cross out compounds that appear on both sides of the reaction arrows.

[1] fructose + HPO_4^{2-} ⟶ fructose 6-phosphate + H_2O

[2] ATP + H_2O ⟶ ADP + HPO_4^{2-}

Coupled reaction: fructose + ATP ⟶ fructose 6-phosphate + ADP

b. The overall energy change is the sum of the energy changes for each step:
+3.8 kcal/mol + (−7.3) kcal/mol = −3.5 kcal/mol.

PROBLEM 23.3

The phosphorylation of glucose to glucose 1-phosphate requires 5.0 kcal/mol of energy. This unfavorable reaction can be driven by the hydrolysis of ATP to ADP. (a) Write the equation for the coupled reaction. (b) How much energy is released in the coupled reaction?

glucose + HPO_4^{2-} ⟶ glucose 1-phosphate + H_2O

Coupled reactions that use ATP or coenzymes are often drawn using a combination of horizontal and curved arrows. The principal organic reactants and products are drawn from left to right with a reaction arrow as usual, but additional compounds like ATP and ADP are drawn on a curved arrow. This technique is meant to emphasize the organic substrates of the reaction, while making it clear that other materials are needed for the reaction to occur.

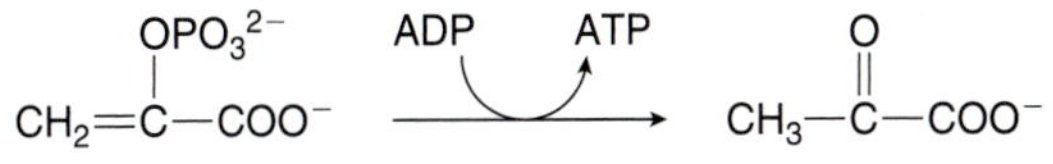

$$CH_2\!=\!\underset{\underset{OPO_3^{2-}}{|}}{C}\!-\!COO^- \quad \overset{ADP \quad ATP}{\longrightarrow} \quad CH_3\!-\!\underset{\overset{\parallel}{O}}{C}\!-\!COO^-$$

phosphoenolpyruvate   pyruvate

PROBLEM 23.4

Write each of the following reactions using curved arrow symbolism.

a. reaction of fructose with ATP to form fructose 6-phosphate and ADP (Sample Problem 23.1)

b. reaction of glucose with ATP to form glucose 6-phosphate and ADP

23.3C FOCUS ON THE HUMAN BODY
CREATINE AND ATHLETIC PERFORMANCE

Many athletes take creatine supplements to boost their performance. Creatine is synthesized in the body from the amino acid arginine, and is present in the diet in meat and fish. Creatine is stored in muscles as creatine phosphate, a high-energy molecule that releases energy by cleavage of its P—N bond on hydrolysis.

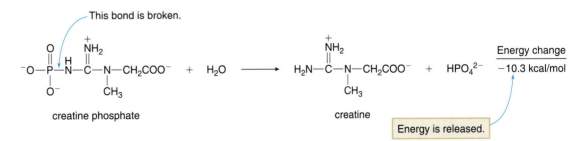

This bond is broken.

creatine phosphate + H_2O ⟶ creatine + HPO_4^{2-}

Energy change
−10.3 kcal/mol

Energy is released.

The immediate energy needs of a sprinter are supplied by the hydrolysis of ATP and creatine phosphate.

Since more energy is released from the hydrolysis of creatine phosphate than is needed for the phosphorylation of ADP, the coupling of these two reactions results in the formation of ATP from ADP.

$$\text{creatine phosphate} \xrightarrow{\quad \text{ADP} \quad \text{ATP} \quad} \text{creatine}$$

Creatine phosphate is stored in muscles. When existing ATP supplies are depleted during strenuous exercise, creatine phosphate reacts with ADP to form a new supply of ATP as a source of more energy. Once existing supplies of both ATP and creatine phosphate have been depleted—which occurs in as little as 10 seconds—other catabolic processes must supply more ATP to continue an activity. Energy-generating pathways that need oxygen input—**aerobic reactions**—are discussed in Sections 23.5–23.6. Energy-generating pathways that do not need oxygen input—**anaerobic reactions**—are discussed in Sections 24.3–24.4.

Do creatine supplements increase athletic performance? Some evidence suggests that in athletic events that require short bursts of intense energy, increased creatine intake increases the amount of creatine phosphate in the muscle. This, in turn, allows an athlete to maintain a high rate of activity for a longer period of time. It is questionable whether creatine alone adds body mass. It is thought, however, that since an athlete has a greater energy reserve, an individual can train harder and longer, resulting in greater muscle mass.

PROBLEM 23.5

Use the values for the hydrolysis of creatine phosphate to creatine (–10.3 kcal/mol) and the phosphorylation of ADP to ATP (+7.3 kcal/mol) to calculate the energy change in the following reaction. Is energy released or required in this reaction?

$$\text{creatine phosphate} + \text{ADP} \longrightarrow \text{creatine} + \text{ATP}$$

23.4 COENZYMES IN METABOLISM

Many reactions in metabolic pathways involve coenzymes. As we learned in Section 21.9, a *coenzyme* **is an organic compound needed for an enzyme-catalyzed reaction to occur.** Some coenzymes serve as important oxidizing and reducing agents (Sections 23.4A and 23.4B), while coenzyme A activates acetyl groups (CH_3CO-), resulting in the transfer of a two-carbon unit to other substrates (Section 23.4C).

23.4A COENZYMES NAD$^+$ AND NADH

Many coenzymes are involved in oxidation and reduction reactions. As we learned in Section 5.8, oxidation and reduction can be defined in terms of changes in electrons, hydrogen atoms, or oxygen atoms.

- Oxidation results in . . . a loss of electrons, *or*
  a loss of hydrogen, *or*
  a gain of oxygen.
- Reduction results in . . . a gain of electrons, *or*
  a gain of hydrogen, *or*
  a loss of oxygen.

A coenzyme may serve as an oxidizing agent or a reducing agent in a biochemical pathway.

- An oxidizing agent causes an oxidation reaction to occur, so the oxidizing agent is reduced.
- A reducing agent causes a reduction reaction to occur, so the reducing agent is oxidized.

In examining what happens to a coenzyme during oxidation and reduction, it is convenient to think in terms of hydrogen atoms being composed of protons (H^+) and electrons (e^-).

- When a coenzyme gains hydrogen atoms—that is, H^+ and e^-—the coenzyme is reduced; thus, the coenzyme is an *oxidizing* agent.
- When a coenzyme loses hydrogen atoms—that is, H^+ and e^-—the coenzyme is oxidized; thus, the coenzyme is a *reducing* agent.

The coenzyme **nicotinamide adenine dinucleotide, NAD^+,** is a common biological oxidizing agent. Although its structure is complex, it is the six-membered ring containing the positively charged nitrogen atom (shown in red) that participates in oxidation reactions. When NAD^+ reacts with two hydrogen atoms, this ring gains one proton and two electrons and one proton is left over. Thus, the ring is reduced and a new C—H bond is formed in the product, written as **NADH,** and referred to as the *reduced form* of nicotinamide adenine dinucleotide.

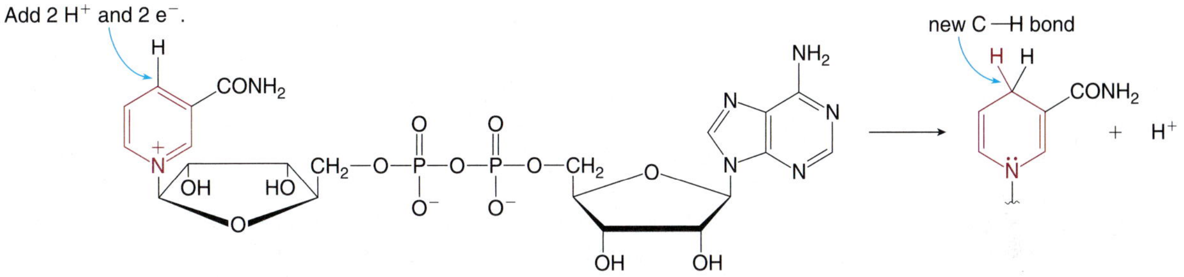

NAD^+
nicotinamide adenine dinucleotide

NADH
(reduced form of NAD^+)

Curved arrow symbolism is often used to depict reactions with coenzymes.

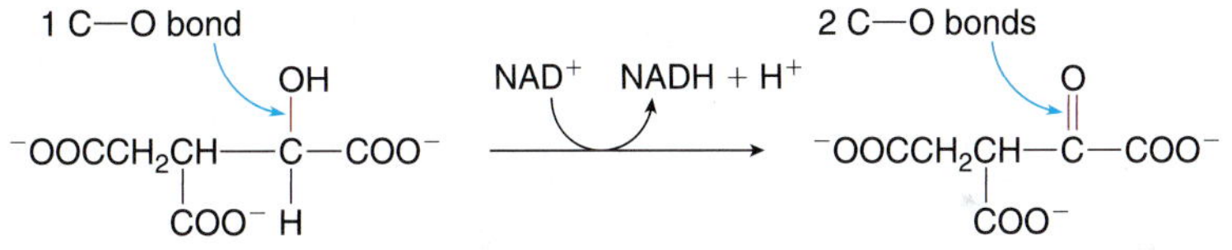

isocitrate

oxalosuccinate

The conversion of isocitrate to oxalosuccinate is an **oxidation** since the number of C—O bonds in the substrate *increases*. **NAD^+ serves as the oxidizing agent, and in the process, is reduced to NADH.** The reduced form of the coenzyme, NADH, is a biological reducing agent, as we learned in Section 16.6. When NADH reacts, it forms NAD^+ as a product. Thus, **NAD^+ and NADH are interconverted by oxidation and reduction reactions.**

SAMPLE PROBLEM 23.2

Label each reaction as an oxidation or reduction, and give the reagent, NAD^+ or NADH, that would be used to carry out the reaction.

a.

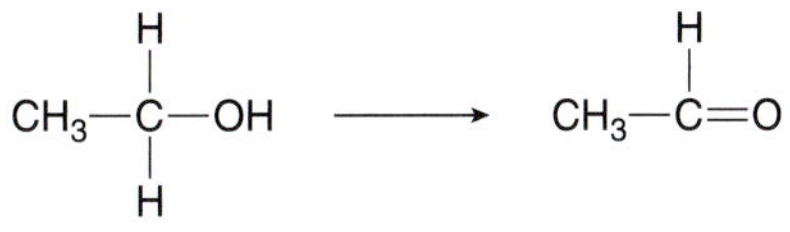

ethanol   acetaldehyde

b.

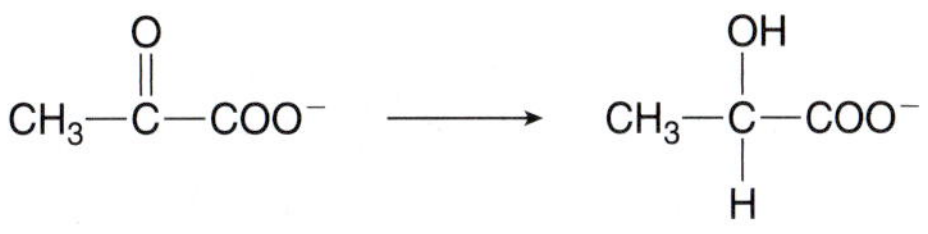

pyruvate   lactate

ANALYSIS   Count the number of C—O bonds in the starting material and product. Oxidation increases the number of C—O bonds and reduction decreases the number of C—O bonds. NAD⁺ is the coenzyme needed for an oxidation, and NADH is the coenzyme needed for a reduction.

SOLUTION   a. The conversion of ethanol to acetaldehyde is an oxidation, since the product has two C—O bonds and the reactant has only one. To carry out the oxidation, the oxidizing agent NAD^+ could be used.

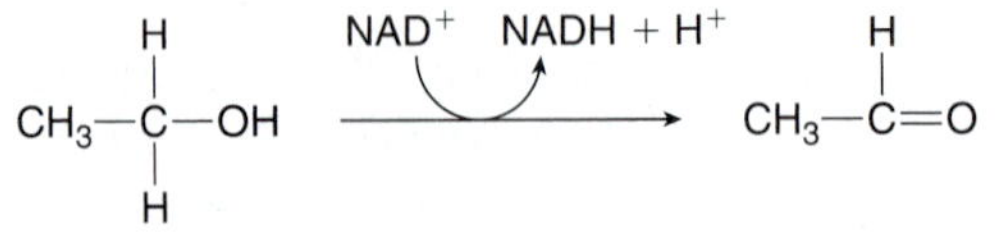

ethanol   acetaldehyde

b. The conversion of pyruvate to lactate is a reduction, since the product has one fewer C—O bond than the reactant. To carry out the reduction, the reducing agent NADH could be used.

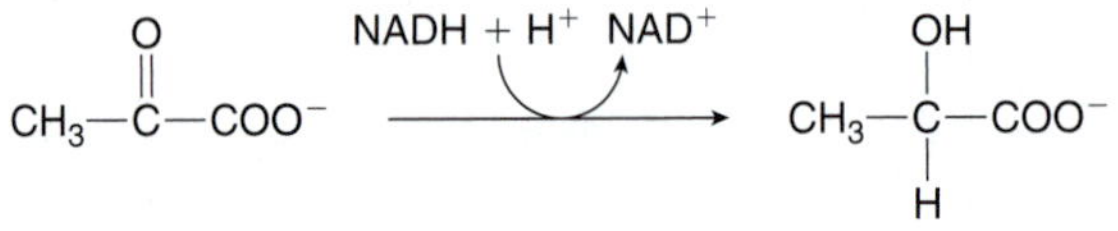

pyruvate   lactate

PROBLEM 23.6   Label each reaction as an oxidation or reduction, and give the reagent, NAD^+ or NADH, that would be used to carry out the reaction.

a. $H_2C=O \longrightarrow CH_3OH$

b.

23.4B   COENZYMES FAD AND FADH₂

Flavin adenine dinucleotide, FAD, is another common biological oxidizing agent. Although its structure is complex, just four atoms of the tricyclic ring system (2 N's and 2 C's shown in red) participate in redox reactions. When it acts as an oxidizing agent, FAD is reduced by adding two hydrogen atoms, forming **FADH₂,** the *reduced form* of flavin adenine dinucleotide.

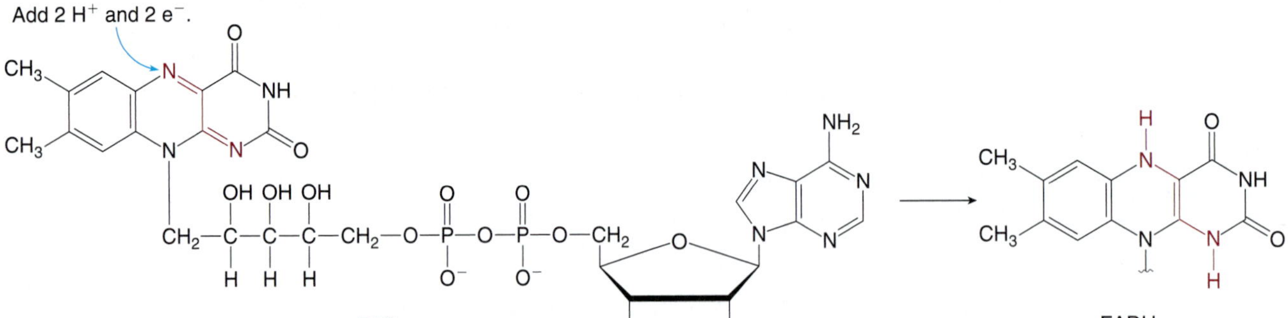

FAD
flavin adenine dinucleotide

FADH₂
(reduced form of FAD)

Table 23.1 summarizes the common coenzymes used in oxidation and reduction reactions.

FAD is synthesized in cells from vitamin B₂, **riboflavin.** Riboflavin is a yellow, water-soluble vitamin obtained in the diet from leafy green vegetables, soybeans, almonds, and liver. When large quantities of riboflavin are ingested, excess amounts are excreted in the urine, giving it a bright yellow appearance.

HEALTH NOTE

Leafy green vegetables, soybeans, and almonds are good sources of riboflavin, vitamin B_2. Since this vitamin is light sensitive, riboflavin-fortified milk contained in glass or clear plastic bottles should be stored in the dark.

TABLE 23.1 Coenzymes Used for Oxidation and Reduction

Coenzyme Name	Abbreviation	Role
Nicotinamide adenine dinucleotide	NAD^+	Oxidizing agent
Nicotinamide adenine dinucleotide (reduced form)	NADH	Reducing agent
Flavin adenine dinucleotide	FAD	Oxidizing agent
Flavin adenine dinucleotide (reduced form)	$FADH_2$	Reducing agent

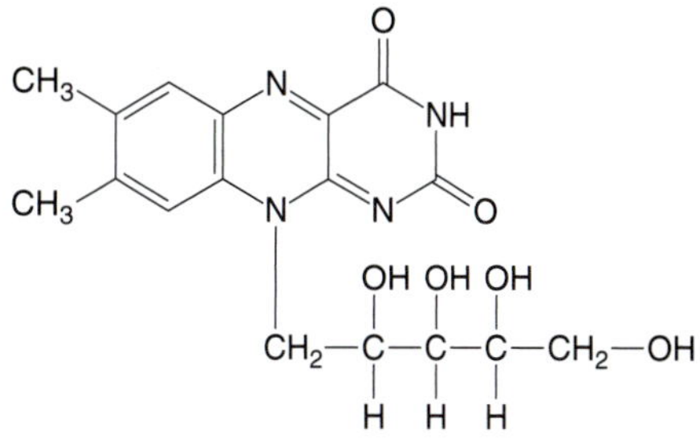

riboflavin
vitamin B_2

PROBLEM 23.7 What makes riboflavin a water-soluble vitamin?

23.4C COENZYME A

Coenzyme A differs from other coenzymes in this section because it is not an oxidizing or a reducing agent. In addition to many other functional groups, coenzyme A contains a **sulfhydryl group (SH group),** making it a **thiol (RSH).** To emphasize this functional group, we sometimes abbreviate the structure as **HS–CoA.**

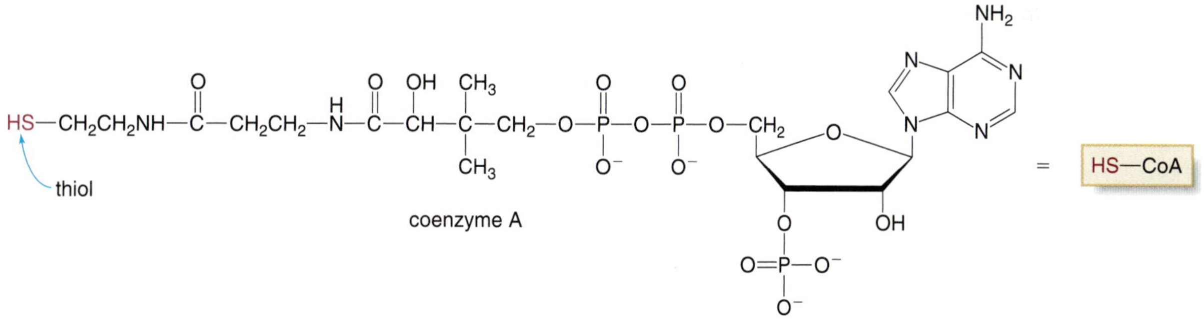

coenzyme A

= HS—CoA

The sulfhydryl group of coenzyme A reacts with acetyl groups (CH_3CO–) or other acyl groups (RCO—) to form thioesters, RCOSR'. When an acetyl group is bonded to coenzyme A, the product is called **acetyl coenzyme A,** or simply **acetyl CoA.**

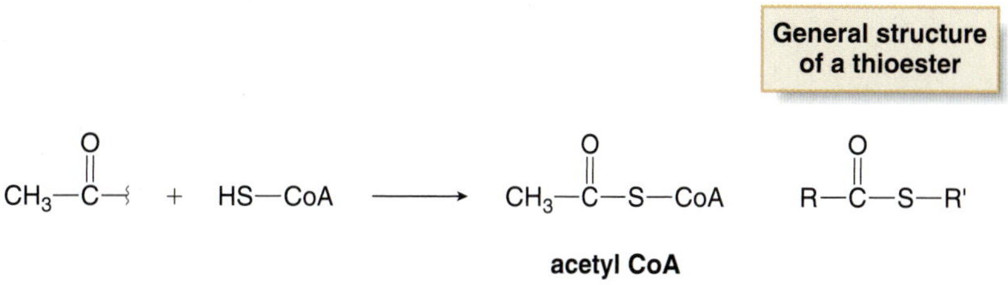

General structure of a thioester

$CH_3-\overset{O}{\overset{\|}{C}}-$ + HS—CoA ⟶ $CH_3-\overset{O}{\overset{\|}{C}}-S-CoA$ $R-\overset{O}{\overset{\|}{C}}-S-R'$

acetyl CoA

Thioesters such as acetyl CoA are another group of high-energy compounds that release energy on reaction with water. In addition, acetyl CoA reacts with other substrates in metabolic pathways to deliver its two-carbon acetyl group, as in the citric acid cycle in Section 23.5.

This bond is broken.

$$CH_3-\overset{O}{\underset{}{C}}-S-CoA + H_2O \longrightarrow CH_3-\overset{O}{\underset{}{C}}-O^- + HS-CoA$$

	Energy change
	−7.5 kcal/mol

acetyl CoA   coenzyme A

Coenzyme A is synthesized in cells from **pantothenic acid, vitamin B$_5$,** which is obtained in the diet from a variety of sources, especially whole grains and eggs.

$$HO-\overset{O}{\underset{}{C}}-CH_2CH_2-\overset{H}{\underset{}{N}}-\overset{O}{\underset{}{C}}-\overset{OH}{\underset{}{CH}}-\overset{CH_3}{\underset{CH_3}{C}}-CH_2-OH$$

pantothenic acid
vitamin B$_5$

PROBLEM 23.8

Predict the water solubility of vitamin B$_5$.

PROBLEM 23.9

What products are formed when the thioester $CH_3CH_2CH_2COSCoA$ is hydrolyzed with water?

23.5   THE CITRIC ACID CYCLE

The **citric acid cycle,** a series of enzyme-catalyzed reactions that occur in mitochondria, comprises the third stage of the catabolism of biomolecules—carbohydrates, lipids, and amino acids—to carbon dioxide, water, and energy.

- The citric acid cycle is a cyclic metabolic pathway that begins with the addition of acetyl CoA to a four-carbon substrate and ends when the same four-carbon compound is formed as a product eight steps later.
- The citric acid cycle produces high-energy compounds for ATP synthesis in stage [4] of catabolism.

23.5A   OVERVIEW OF THE CITRIC ACID CYCLE

The citric acid cycle is also called the tricarboxylic acid cycle (TCA cycle) or the **Krebs cycle,** named for Hans Krebs, a German chemist and Nobel Laureate who worked out the details of these reactions in 1937. A general scheme of the citric acid cycle, shown in Figure 23.6, illustrates the key features. All intermediates in the citric acid cycle are carboxylate anions derived from di- and tricarboxylic acids.

- The citric acid cycle begins when two carbons of acetyl CoA ($CH_3COSCoA$) react with a four-carbon organic substrate to form a six-carbon product (step [1]).
- Two carbon atoms are sequentially removed to form two molecules of CO_2 (steps [3] and [4]).
- Four molecules of reduced coenzymes (NADH and FADH$_2$) are formed in steps [3], [4], [6], and [8]. These molecules serve as carriers of electrons to the electron transport chain in stage [4] of catabolism, which ultimately results in the synthesis of a great deal of ATP.
- One mole of GTP is synthesized in step [5]. GTP is a high-energy nucleoside triphosphate similar to ATP.

Triacylglycerols   Carbohydrates   Proteins
Fatty acids + glycerol   Monosaccharides   Amino acids
Fatty acid oxidation   Glycolysis   Amino acid catabolism
Pyruvate
Acetyl CoA
Citric acid cycle
Reduced coenzymes
Electron transport chain and oxidative phosphorylation

Glycolysis, fatty acid oxidation, and amino acid catabolism—three processes that generate acetyl CoA—are discussed in detail in Chapter 24.

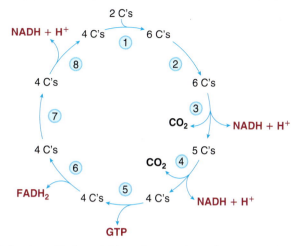

▼ **FIGURE 23.6** **General Features of the Citric Acid Cycle**

The citric acid cycle begins with the addition of two carbons from acetyl CoA in step [1] to a four-carbon organic substrate, drawn at the top of the cyclic pathway. Each turn of the citric acid cycle forms two molecules of CO_2, four molecules of reduced coenzymes (3 NADH and 1 $FADH_2$), and one high-energy GTP molecule.

23.5B SPECIFIC STEPS OF THE CITRIC ACID CYCLE

The eight reactions of the citric acid cycle, which can be conceptually divided into two parts, are shown in Figure 23.7. The first part of the cycle includes the addition of acetyl CoA to oxaloacetate to form the six-carbon product citrate, which undergoes two separate **decarboxylations**—reactions that give off CO_2. In part [2], functional groups are added and oxidized to re-form oxaloacetate, the substrate needed to begin the cycle again.

Part [1] of the Citric Acid Cycle

As shown in Figure 23.7, acetyl CoA enters the cycle by reaction with oxaloacetate at step [1] of the pathway. This reaction, catalyzed by citrate synthase, adds two carbons to oxaloacetate, forming citrate. In step [2], the 3° alcohol of citrate is isomerized to the 2° alcohol isocitrate using aconitase. These first two steps add carbon atoms and rearrange functional groups.

Loss of two carbon atoms begins in step [3], by decarboxylation of isocitrate using isocitrate dehydrogenase. The oxidizing agent NAD^+ also converts the 2° alcohol to a ketone to form α-ketoglutarate, which now contains one fewer carbon atom. This reaction forms NADH and H^+, which will carry electrons and protons gained in this reaction to the electron transport chain. In step [4], decarboxylation releases a second molecule of CO_2. Also, oxidation with NAD^+ in the presence of coenzyme A forms the thioester succinyl CoA using the enzyme α-ketoglutarate dehydrogenase. **By the end of step [4], two carbons are lost as CO_2 and two molecules of NADH are formed.**

Although it might be convenient to think of the two molecules of CO_2 as originating in the acetyl CoA added in step [1], the carbon atoms in CO_2 were *not* added during this cycle of the pathway.

Part [2] of the Citric Acid Cycle

Part [2] consists of four reactions that manipulate the functional groups of succinyl CoA to re-form oxaloacetate. In step [5], the thioester bond of succinyl CoA is hydrolyzed to form succinate, releasing energy to convert GDP to GTP. GTP, guanosine 5'-triphosphate, is similar to ATP:

▼ FIGURE 23.7   Steps in the Citric Acid Cycle

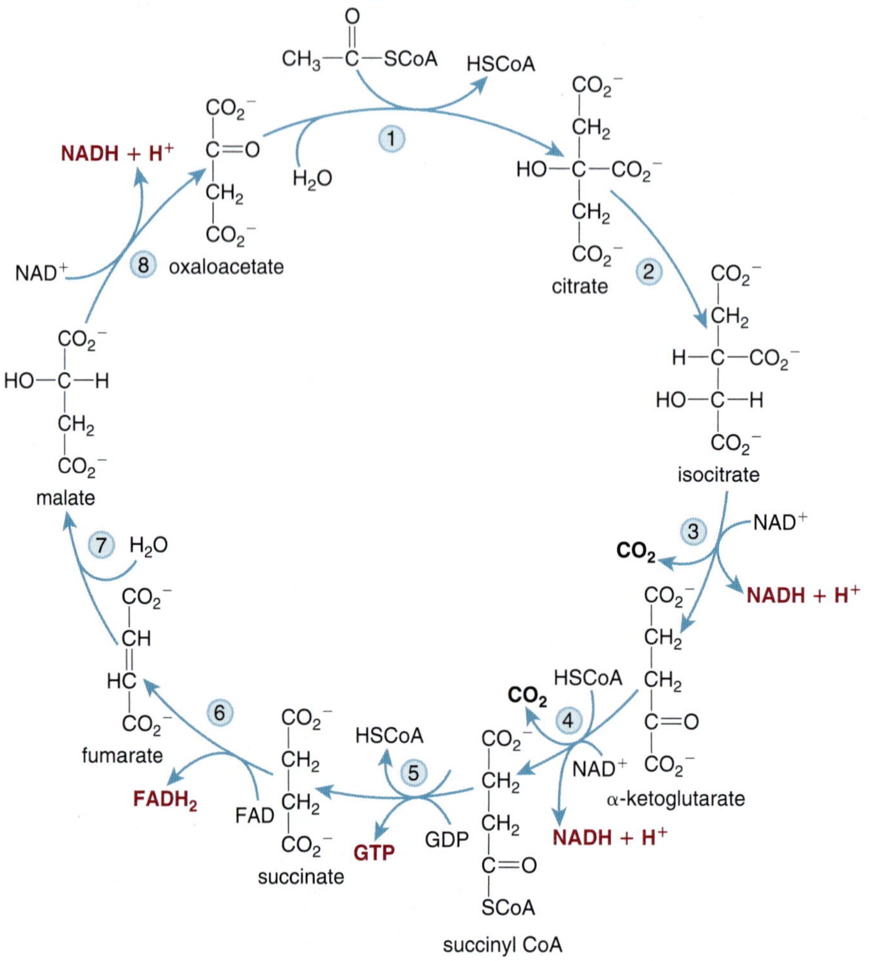

Each step of the citric acid cycle is enzyme catalyzed. The net result of the eight-step cycle is the conversion of the two carbons added to oxaloacetate to two molecules of CO_2. Reduced coenzymes (NADH and $FADH_2$) are also formed, which carry electrons to the electron transport chain to synthesize ATP. One molecule of high-energy GTP is synthesized in step [5].

GTP is a high-energy molecule that releases energy when a P—O bond is hydrolyzed. This is the only step of the citric acid cycle that directly generates a triphosphate.

In step [6], succinate is converted to fumarate with FAD and succinate dehydrogenase. This reaction forms the reduced coenzyme $FADH_2$, which will carry electrons and protons to the electron transport chain. Addition of water in step [7] forms malate and oxidation of the 2° alcohol in malate with NAD^+ forms oxaloacetate in step [8]. Another molecule of NADH is also formed in step [8]. By the end of step [8], two more molecules of reduced coenzymes ($FADH_2$ and NADH) are formed. **Since the product of step [8] is the starting material of step [1], the cycle can continue as long as additional acetyl CoA is available for step [1].**

Overall, the citric acid cycle results in formation of

- two molecules of CO_2
- four molecules of reduced coenzymes (3 NADH and 1 $FADH_2$)
- one molecule of GTP

The net equation for the citric acid cycle can be written as shown. The ultimate fate of each product is also indicated.

$$CH_3-\overset{\overset{\text{O}}{\|}}{C}-SCoA \quad + \quad 2\,H_2O \quad + \quad 3\,NAD^+ \quad + \quad FAD \quad + \quad GDP \quad + \quad HPO_4^{2-}$$

overall reaction

$$2\,CO_2 \quad + \quad HSCoA \quad + \quad 3\,NADH \quad + \quad 3\,H^+ \quad + \quad FADH_2 \quad + \quad GTP$$

exhaled gas   The coenzyme re-enters the cycle.   The reduced coenzymes enter the electron transport chain.   energy source

- **The main function of the citric acid cycle is to produce reduced coenzymes that enter the electron transport chain and ultimately produce ATP.**

The rate of the citric acid cycle depends on the body's need for energy. When energy demands are high and the amount of available ATP is low, the cycle is activated. When energy demands are low and NADH concentration is high, the cycle is inhibited.

Although the citric acid cycle is complex, many individual reactions can be understood by applying the basic principles of organic chemistry learned in previous chapters.

SAMPLE PROBLEM 23.3

(a) Write out the reaction that converts succinate to fumarate with FAD using curved arrow symbolism. (b) Classify the reaction as an oxidation, reduction, or decarboxylation.

ANALYSIS

Use Figure 23.7 to draw the structures for succinate and fumarate. Draw the organic reactant and product on the horizontal arrow and the oxidizing reagent FAD, which is converted to FADH$_2$, on the curved arrow. Oxidation reactions result in a loss of electrons, a loss of hydrogen, or a gain of oxygen. Reduction reactions result in a gain of electrons, a gain of hydrogen, or a loss of oxygen. A decarboxylation results in the loss of CO_2.

SOLUTION

a. Equation:

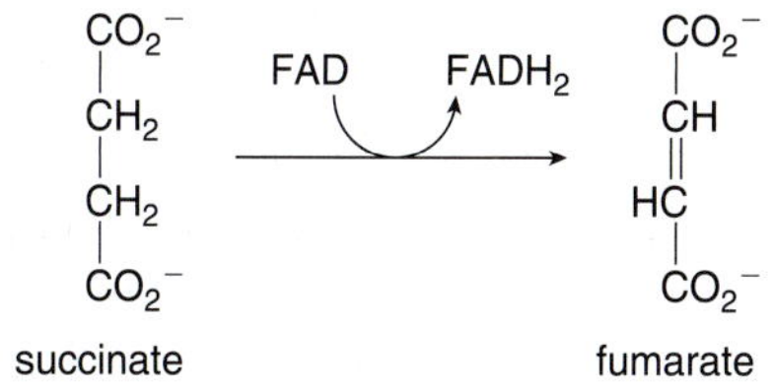

succinate   fumarate

b. Since succinate contains four C—H bonds and fumarate contains only two C—H bonds, hydrogen atoms have been lost, making this reaction an oxidation. In the process FAD is reduced to FADH$_2$.

PROBLEM 23.10

(a) Write out the reaction that converts malate to oxaloacetate with NAD$^+$ using curved arrow symbolism. (b) Classify the reaction as an oxidation, reduction, or decarboxylation.

PROBLEM 23.11

From what you learned about classifying enzymes in Section 21.9, explain why the enzyme used to convert succinate to fumarate is called succinate dehydrogenase.

23.6 THE ELECTRON TRANSPORT CHAIN AND OXIDATIVE PHOSPHORYLATION

Triacylglycerols   Carbohydrates   Proteins

Fatty acids   Monosaccharides   Amino acids
+
glycerol

Fatty acid oxidation   Glycolysis   Amino acid catabolism

Pyruvate

Acetyl CoA

Citric acid cycle

Reduced coenzymes

Electron transport chain and oxidative phosphorylation

Most of the energy generated during the breakdown of biomolecules is formed during stage [4] of catabolism. Because oxygen is required, this process is called **aerobic respiration.** There are two facets to this stage:

- the electron transport chain, or the respiratory chain
- oxidative phosphorylation

The reduced coenzymes formed in the citric acid cycle enter the **electron transport chain** and the electrons they carry are transferred from one molecule to another by a series of oxidation–reduction reactions. Each reaction releases energy until electrons and protons react with oxygen to form water. Electron transfer also causes H^+ ions to be pumped across the inner mitochondrial cell membrane, creating an energy reservoir that is used to synthesize ATP by the **phosphorylation** of ADP.

In contrast to the combustion of gasoline, which releases energy all at once in a single reaction, the energy generated during metabolism is released in small portions as the result of many reactions.

23.6A THE ELECTRON TRANSPORT CHAIN

The electron transport chain is a multistep process that relies on four enzyme systems, called **complexes I, II, III, and IV,** as well as mobile electron carriers. Each complex is composed of enzymes, additional protein molecules, and metal ions that can gain and lose electrons in oxidation and reduction reactions. The complexes are situated in the inner membrane of the mitochondria, arranged so that electrons can be passed to progressively stronger oxidizing agents (Figure 23.8).

▼ **FIGURE 23.8   The Electron Transport Chain and ATP Synthesis in a Mitochondrion**

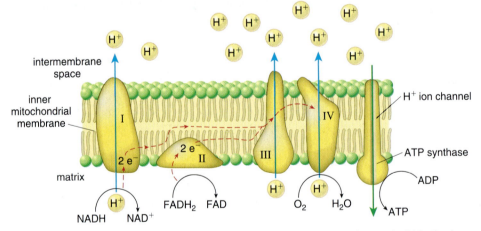

The four enzyme complexes (I–IV) of the electron transport chain are located within the inner membrane of a mitochondrion, between the matrix and the intermembrane space. Electrons enter the chain when NADH and $FADH_2$ are oxidized and then transported through a series of complexes along the pathway shown in red. The electrons ultimately combine with O_2 to form H_2O. Protons (H^+) are pumped across the inner membrane into the intermembrane space at three locations shown by blue arrows. The energy released when protons return to the matrix by traveling through a channel (in green) in the ATP synthase enzyme is used to convert ADP to ATP.

The electron transport chain begins with the reduced coenzymes—3 NADH and 1 $FADH_2$—formed during the citric acid cycle. These reduced coenzymes are electron rich and as such, they are capable of donating electrons to other species. Thus, **NADH and $FADH_2$ are *reducing agents* and when they donate electrons, they are oxidized.** When NADH donates two electrons, it is oxidized to NAD^+, which can re-enter the citric acid cycle. Likewise, when $FADH_2$ donates two electrons, it is oxidized to FAD, which can be used as an oxidant in step [6] of the citric acid cycle once again.

Once in the electron transport chain, the electrons are passed down from complex to complex in a series of redox reactions, and small packets of energy are released along the way. At the end of the chain, **the electrons and protons (obtained from the reduced coenzymes or the matrix of the mitochondrion) react with inhaled oxygen to form water** and this facet of the process is complete.

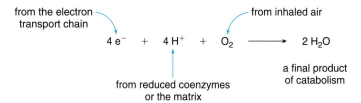

$$4\,e^- \; + \; 4\,H^+ \; + \; O_2 \longrightarrow 2\,H_2O$$

from the electron transport chain

from inhaled air

from reduced coenzymes or the matrix

a final product of catabolism

Because oxygen is needed for the final stage of electron transport, this process is **aerobic.**

PROBLEM 23.12

If NADH and $FADH_2$ were not oxidized in the electron transport chain, what would happen to the citric acid cycle?

PROBLEM 23.13

At several points in the electron transport chain, iron cations gain or lose electrons by reactions that interconvert Fe^{2+} and Fe^{3+} cations. (a) When Fe^{3+} is converted to Fe^{2+}, is the reaction an oxidation or a reduction? (b) Is Fe^{3+} an oxidizing agent or a reducing agent?

23.6B ATP SYNTHESIS BY OXIDATIVE PHOSPHORYLATION

Although the electron transport chain illustrates how electrons carried by reduced coenzymes ultimately react with oxygen to form water, we have still not learned how ATP, the high-energy triphosphate needed for energy transport, is synthesized. The answer lies in what happens to the H^+ ions in the mitochondrion.

H^+ ions generated by reactions in the electron transport chain, as well as H^+ ions present in the matrix of the mitochondria, are pumped across the inner mitochondrial membrane into the intermembrane space at three different sites (Figure 23.8). This process requires energy, since it moves protons against the concentration gradient. The energy comes from redox reactions in the electron transport chain. Since the concentration of H^+ ions is then higher on one side of the membrane, this creates a potential energy gradient, much like the potential energy of water that is stored behind a dam.

To return to the matrix, the H^+ ions travel through a channel in the ATP synthase enzyme. ATP synthase is the enzyme that catalyzes the phosphorylation of ADP to form ATP. The energy released as the protons return to the matrix converts ADP to ATP. This process is called **oxidative phosphorylation,** since the energy that results from the oxidation of the reduced coenzymes is used to transfer a phosphate group.

Energy released from H^+ movement fuels phosphorylation.

$$ADP \; + \; HPO_4^{2-} \longrightarrow ATP \; + \; H_2O$$

PROBLEM 23.14	In which region of the mitochondrion—the matrix or the intermembrane space—would the pH be lower? Explain your choice.

23.6C   ATP YIELD FROM OXIDATIVE PHOSPHORYLATION

How much ATP is generated during stage [4] of catabolism?

- Each NADH enters the electron transport chain at complex I in the inner mitochondrial membrane and the resulting cascade of reactions produces enough energy to synthesize 2.5 ATPs.
- $FADH_2$ enters the electron transport chain at complex II, producing energy for the synthesis of 1.5 ATPs.

How much ATP is generated for each acetyl CoA fragment that enters the entire common catabolic pathway—that is, stages [3] and [4] of catabolism?

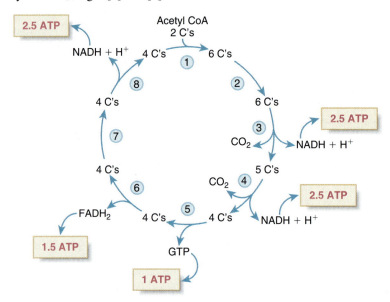

For each turn of the citric acid cycle, three NADH molecules and one $FADH_2$ molecule are formed. In addition, one GTP molecule is produced directly during the citric acid cycle (step [5]); one GTP molecule is equivalent in energy to one ATP molecule. These facts allow us to calculate the total number of ATP molecules formed for each acetyl CoA.

$$\begin{array}{rcl} 3\ NADH \times 2.5\ ATP/NADH & = & 7.5\ ATP \\ 1\ FADH_2 \times 1.5\ ATP/FADH_2 & = & 1.5\ ATP \\ 1\ GTP & = & \underline{1\ \ ATP} \\ & & 10\ \ ATP \end{array}$$

- Complete catabolism of each acetyl CoA molecule that enters the citric acid cycle forms 10 ATP molecules.

Each ATP molecule can now provide the energy to drive energetically unfavorable reactions.

23.7 FOCUS ON HEALTH & MEDICINE
HYDROGEN CYANIDE

The reactions detailed in the electron transport chain and oxidative phosphorylation are so crucial for the health of an organism that disruption of even a single step can have disastrous consequences. When any one step is inhibited, energy production ceases, and an organism cannot survive.

Hydrogen cyanide (HCN, Section 4.6) is a lethal gas used as a chemical weapon during World War I and in the gas chambers of Nazi Germany during World War II. HCN is a poison because it produces cyanide ($^-$CN), which irreversibly binds to the Fe^{3+} ion of cytochrome oxidase. Cytochrome oxidase is the key enzyme in complex IV of the electron transport chain. When bonded to cyanide, Fe^{3+} cannot be reduced to Fe^{2+} and the transfer of electrons to oxygen to form water cannot occur. ATP synthesis is halted and cell death ensues.

A few naturally occurring compounds that contain CN groups form HCN when they are hydrolyzed in the presence of certain enzymes. As an example, **amygdalin** is present in the seeds and pits of apricots, peaches, and wild cherries. Commonly known as laetrile, amygdalin was once touted as an anticancer drug, and is still available in some countries for this purpose. Amygdalin's effectiveness in targeting cancer cells is unproven, and the HCN produced by hydrolysis kills all types of cells in the body.

ENVIRONMENTAL NOTE

The golden bamboo lemur of Madagascar subsists primarily on giant bamboo, despite the high levels of toxic cyanide it contains.

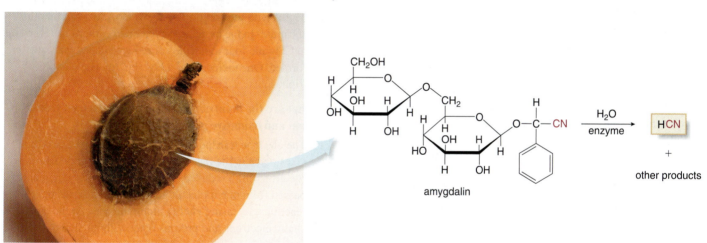

amygdalin

The side effects that result from taking amygdalin resemble those of chronic exposure to cyanide and include headache, vomiting, coma, and in some cases, death.

Interestingly, the diet of the golden bamboo lemur, a critically endangered species that inhabits Madagascar, contains shoots of giant bamboo as its main component. Giant bamboo contains such high levels of cyanide that most mammals would die after ingesting it. Evidently, the golden bamboo lemur has developed a mechanism for detoxifying cyanide, although that mechanism is currently unknown.

CHAPTER HIGHLIGHTS

KEY TERMS

Acetyl CoA (23.4)
Adenosine 5'-diphosphate (ADP, 23.3)
Adenosine 5'-triphosphate (ATP, 23.3)
Anabolism (23.1)
Catabolism (23.1)
Citric acid cycle (23.5)

Coenzyme A (23.4)
Coupled reactions (23.3)
Electron transport chain (23.6)
Flavin adenine dinucleotide (FAD, 23.4)
Metabolism (23.1)
Mitochondrion (23.1)

Nicotinamide adenine dinucleotide
   (NAD$^+$, 23.4)
Oxidative phosphorylation (23.6)
Phosphorylation (23.3)

KEY CONCEPTS

❶ What is metabolism and where is energy produced in cells? (23.1)

- Metabolism is the sum of all of the chemical reactions that take place in an organism. Catabolic reactions break down large molecules and release energy, while anabolic reactions synthesize larger molecules and require energy.
- Energy is produced in the mitochondria, sausage-shaped organelles that contain an outer and inner cell membrane. Energy is produced in the matrix, the area surrounded by the inner membrane.

❷ What are the four stages of metabolism? (23.2)

- Metabolism begins with digestion in stage [1], in which large molecules—polysaccharides, proteins, and triacylglycerols—are hydrolyzed to smaller molecules— monosaccharides, amino acids, fatty acids, and glycerol.
- In stage [2], biomolecules are degraded into two-carbon acetyl units.
- The citric acid cycle comprises stage [3]. The citric acid cycle converts two carbon atoms to two molecules of CO_2, and forms reduced coenzymes, NADH and $FADH_2$, which carry electrons and energy to the electron transport chain.
- In stage [4], the electron transport chain and oxidative phosphorylation produce ATP, and oxygen is converted to water.

❸ What is ATP and how do coupled reactions with ATP drive energetically unfavorable reactions? (23.3)

- ATP is the primary energy-carrying molecule in metabolic pathways. The hydrolysis of ATP cleaves one phosphate group and releases 7.3 kcal/mol of energy.
- The hydrolysis of ATP provides the energy to drive a reaction that requires energy. A pair of reactions of this sort is said to be coupled.

❹ List the main coenzymes in metabolism and describe their roles. (23.4)

- Nicotinamide adenine dinucleotide (NAD$^+$) is a biological oxidizing agent that accepts electrons and protons, thus generating its reduced form NADH. NADH is a reducing agent that donates electrons and protons, re-forming NAD$^+$.
- Flavin adenine dinucleotide (FAD) is a biological oxidizing agent that accepts electrons and protons, thus yielding its

reduced form, $FADH_2$. $FADH_2$ is a reducing agent that donates electrons and protons, re-forming FAD.

- Coenzyme A reacts with acetyl groups (CH_3CO-) to form high-energy thioesters that deliver two-carbon acetyl groups to other substrates.

❺ What are the main features of the citric acid cycle? (23.5)

- The citric acid cycle is an eight-step cyclic pathway that begins with the addition of acetyl CoA to a four-carbon substrate. In the citric acid cycle, two carbons are converted to CO_2 and four molecules of reduced coenzymes are formed. One molecule of a high-energy nucleoside triphosphate is also formed.

❻ What are the main components of the electron transport chain and oxidative phosphorylation? (23.6)

- The electron transport chain is a multistep process that takes place in the inner membrane of mitochondria. Electrons from reduced coenzymes enter the chain and are passed from one molecule to another in a series of redox reactions, releasing energy along the way. At the end of the chain, electrons and protons react with inhaled oxygen to form water.
- H^+ ions are pumped across the inner membrane of the mitochondrion, forming a high concentration of H^+ ions in the intermembrane space, thus creating a potential energy gradient. When the H^+ ions travel through the channel in the ATP synthase enzyme, this energy is used to convert ADP to ATP—a process called oxidative phosphorylation.

❼ Why do compounds such as cyanide act as poisons when they disrupt the electron transport chain? (23.7)

- Since all catabolic pathways converge at the electron transport chain, these steps are needed to produce energy for normal cellular processes. Compounds that disrupt a single step can halt ATP synthesis so that an organism cannot survive.
- Cyanide from HCN irreversibly binds to the Fe^{3+} ion of the enzyme cytochrome oxidase and, as a result, Fe^{3+} cannot be reduced to Fe^{2+} and water cannot be formed from oxygen. Since the electron transport chain is disrupted, energy is not generated for oxidative phosphorylation, ATP is not synthesized, and cell death often follows.

PROBLEMS

Selected in-chapter and end-of-chapter problems have brief answers provided in Appendix B.

Metabolism

23.15 What is the difference between catabolism and anabolism?

23.16 What is the difference between metabolism and digestion?

23.17 Describe the main features of a mitochondrion. Where does energy production occur in a mitochondrion?

23.18 Explain why mitochondria are called the powerhouses of the cell.

23.19 Place the following steps in the catabolism of carbohydrates in order: the electron transport chain, the conversion of glucose to acetyl CoA, the hydrolysis of starch, oxidative phosphorylation, and the citric acid cycle.

23.20 Describe the main features of the four stages of catabolism.

23.21 In what stage of catabolism does each of the following processes occur?
a. cleavage of a protein with chymotrypsin
b. oxidation of a fatty acid to acetyl CoA
c. oxidation of malate to oxaloacetate with NAD^+
d. conversion of ADP to ATP with ATP synthase
e. hydrolysis of starch to glucose with amylase

23.22 In what stage of catabolism does each of the following processes occur?
a. conversion of a monosaccharide to acetyl CoA
b. hydrolysis of a triacylglycerol with lipase
c. reaction of oxygen with protons and electrons to form water
d. conversion of succinate to fumarate with FAD
e. degradation of a fatty acid to acetyl CoA

ATP and Coupled Reactions

23.23 What are coupled reactions and why does coupling occur?

23.24 Explain why ATP and GTP are called "high-energy" compounds.

23.25 Acetyl phosphate is another example of a high-energy phosphate. What products would be formed when acetyl phosphate is hydrolyzed with water?

$$CH_3-\overset{\overset{\displaystyle O}{\|}}{C}-O-\overset{\overset{\displaystyle O}{\|}}{\underset{\underset{\displaystyle O^-}{|}}{P}}-O^-$$

acetyl phosphate

23.26 ADP undergoes hydrolysis in a similar manner to ATP; that is, a P—O bond is cleaved. Draw the structure of the product of ADP hydrolysis.

23.27 The reaction of succinate with coenzyme A to form succinyl CoA requires energy, and the hydrolysis of ATP releases energy as shown in the given equations.

	Energy change
succinate + HSCoA $\longrightarrow$ succinyl CoA + H_2O	+9.4 kcal/mol
ATP + H_2O $\longrightarrow$ ADP + HPO_4^{2-}	−7.3 kcal/mol

a. Combine the equations and write the coupled reaction.
b. Is the coupled reaction energetically favorable? Explain.

23.28 The hydrolysis of succinyl CoA to succinate and coenzyme A releases energy, and the phosphorylation of ADP requires energy as shown in the given equations.

	Energy change
succinyl CoA + H_2O $\longrightarrow$ succinate + HSCoA	−9.4 kcal/mol
ADP + HPO_4^{2-} $\longrightarrow$ ATP + H_2O	+7.3 kcal/mol

a. Combine the equations and write the coupled reaction.
b. Is the coupled reaction energetically favorable? Explain.

23.29 The phosphorylation of glucose with HPO_4^{2-} forms glucose 1-phosphate and water and requires 5.0 kcal/mol of energy.
a. Write the equation for this reaction.
b. This unfavorable reaction can be driven by the hydrolysis of ATP to ADP. Write an equation for the coupled reaction.
c. Calculate the energy change for the coupled reaction.

23.30 Write an equation for the reverse reaction described in Problem 23.29—that is, the hydrolysis of glucose 1-phosphate to form glucose and HPO_4^{2-}. What is the energy change in this reaction?

23.31 The phosphorylation of fructose 6-phosphate to form fructose 1,6-bisphosphate and water requires 3.9 kcal/mol of energy.

fructose 6-phosphate + HPO_4^{2-} $\longrightarrow$

fructose 1,6-bisphosphate + H_2O

a. What is the energy change for the reverse reaction—that is, the hydrolysis of fructose 1,6-bisphosphate to form fructose 6-phosphate?
b. Write a coupled reaction that shows how the energy from ATP hydrolysis (−7.3 kcal/mol) can be used to drive the given reaction.
c. What is the energy change for the coupled reaction?

23.32   a. Use the values in Problem 23.31 to calculate the energy change in the following reaction.

fructose 1,6-bisphosphate + ADP ⟶ fructose 6-phosphate + ATP

   b. Is this reaction energetically favorable or unfavorable?
   c. Write this reaction using curved arrow symbolism.
   d. Can this reaction be used to synthesize ATP from ADP? Explain.

23.33   Refer to the following equations to answer the questions.

	Energy change

1,3-bisphosphoglycerate + H_2O ⟶

3-phosphoglycerate + HPO_4^{2-} −11.8 kcal/mol

ATP + H_2O ⟶ ADP + HPO_4^{2-} −7.3 kcal/mol

   a. Write the equation for the coupled reaction of 1,3-bisphosphoglycerate with ADP to form 3-phosphoglycerate and ATP.
   b. Calculate the energy change for this coupled reaction.
   c. Can this reaction be used to synthesize ATP from ADP? Explain.

23.34   Refer to the equations in Problem 23.33 to answer the following questions.
   a. Write the equation for the coupled reaction of 3-phosphoglycerate with ATP to form 1,3-bisphosphoglycerate and ADP.
   b. Calculate the energy change for this coupled reaction.
   c. Is this reaction energetically favorable? Explain.

23.35   (a) Draw the structure of the high-energy nucleoside triphosphate GTP. (b) Draw the structure of the hydrolysis product formed when one phosphate is removed.

23.36   If the phosphorylation of GDP to form GTP requires 7.3 kcal/mol of energy, how much energy is released when GTP is hydrolyzed to GDP?

Coenzymes

23.37   Classify each substance as an oxidizing agent, a reducing agent, or neither: (a) $FADH_2$; (b) ATP; (c) NAD^+.

23.38   Classify each substance as an oxidizing agent, a reducing agent, or neither: (a) NADH; (b) ADP; (c) FAD.

23.39   When a substrate is oxidized, is NAD^+ oxidized or reduced? Is NAD^+ an oxidizing agent or a reducing agent?

23.40   When a substrate is reduced, is $FADH_2$ oxidized or reduced? Is $FADH_2$ an oxidizing agent or a reducing agent?

23.41   Label each reaction as an oxidation or a reduction and give the coenzyme, NAD^+ or NADH, which might be used to carry out the reaction. Write each reaction using curved arrow symbolism.

a.

b.

23.42   Label each reaction as an oxidation or a reduction and give the coenzyme, NAD^+ or NADH, which might be used to carry out the reaction. Write each reaction using curved arrow symbolism.
   a. CH_3CH_2CHO ⟶ $CH_3CH_2CH_2OH$

   b.

Citric Acid Cycle

23.43   What reactions in the citric acid cycle have each of the following characteristics?
   a. The reaction generates NADH.
   b. CO_2 is removed.
   c. The reaction utilizes FAD.
   d. The reaction forms a new carbon–carbon single bond.

23.44   What reactions in the citric acid cycle have each of the following characteristics?
   a. The reaction generates $FADH_2$.
   b. The organic substrate is oxidized.
   c. The reaction utilizes NAD^+.
   d. The reaction breaks a carbon–carbon bond.

23.45   (a) Which intermediate(s) in the citric acid cycle contain two chirality centers? (b) Which intermediate(s) contain a 2° alcohol?

23.46   (a) Which intermediate(s) in the citric acid cycle contain one chirality center? (b) Which intermediate(s) contain a 3° alcohol?

23.47   The conversion of isocitrate to α-ketoglutarate in step [3] of the citric acid cycle actually occurs by a two-step process: isocitrate is converted first to oxalosuccinate, which then goes on to form α-ketoglutarate.

isocitrate   oxalosuccinate   α-ketoglutarate

   a. Classify reaction [3a] as an oxidation, reduction, or decarboxylation.
   b. Classify reaction [3b] as an oxidation, reduction, or decarboxylation.
   c. For which step is NAD^+ necessary?
   d. Why is the enzyme that catalyzes this reaction called isocitrate dehydrogenase?

23.48 The conversion of citrate to isocitrate in step [2] of the citric acid cycle actually occurs by a two-step process (steps [2a] and [2b]) with aconitate formed as an intermediate.

citrate   aconitate   isocitrate

a. What atoms are added or removed in each step?

b. Steps [2a] and [2b] are not considered oxidation or reduction reactions despite the fact that the number of C—H and C—O bonds changes in these reactions. Explain why this is so.

c. From your knowledge of organic chemistry learned in previous chapters, what type of reaction occurs in step [2a]? In step [2b]?

23.49 In what step does a hydration reaction occur in the citric acid cycle?

23.50 What products of the citric acid cycle are funneled into the electron transport chain?

Electron Transport Chain and Oxidative Phosphorylation

23.51 Why are the reactions that occur in stage [4] of catabolism sometimes called aerobic respiration?

23.52 What is the role of each of the following in the electron transport chain: (a) NADH; (b) O_2; (c) complexes I–IV; (d) H$^+$ ion channel?

23.53 What is the role of each of the following in the electron transport chain: (a) FADH$_2$; (b) ADP; (c) ATP synthase; (d) the inner mitochondrial membrane?

23.54 Explain the importance of the movement of H$^+$ ions across the inner mitochondrial membrane and then their return passage through the H$^+$ ion channel in ATP synthase.

23.55 What are the final products of the electron transport chain?

23.56 What product is formed from each of the following compounds during the electron transport chain: (a) NADH; (b) FADH$_2$; (c) ADP; (d) O_2?

23.57 Why does one NADH that enters the electron transport chain ultimately produce 2.5 ATPs, while one FADH$_2$ produces 1.5 ATPs?

23.58 How does the energy from the proton gradient result in ATP synthesis?

General Questions and Applications

23.59 How does the metabolism of the two carbons of acetyl CoA form 10 molecules of ATP?

23.60 What is the difference between phosphorylation and oxidative phosphorylation?

23.61 What is the structural difference between ATP and ADP?

23.62 What is the structural difference between coenzyme A and acetyl CoA?

23.63 From what you learned about monosaccharides in Chapter 20 and phosphates in Chapter 23: (a) Draw the structure of glucose 1-phosphate. (b) Using structures, write the equation for the hydrolysis of glucose 1-phosphate to glucose and HPO_4^{2-}.

23.64 From what you learned about monosaccharides in Chapter 20 and phosphates in Chapter 23: (a) Draw the structure of glucose 6-phosphate. (b) Using structures, write the equation for the hydrolysis of glucose 6-phosphate to glucose and HPO_4^{2-}.

23.65 How are the citric acid cycle and the electron transport chain interrelated?

23.66 Why is the citric acid cycle considered to be part of the aerobic catabolic pathways, even though oxygen is not directly involved in any step in the cycle?

23.67 In which tissue would a cell likely have more mitochondria, the heart or the bone? Explain your choice.

23.68 What is the fate of an acetyl CoA molecule after several turns of the citric acid cycle?

23.69 What is the role of stored creatine phosphate in muscles?

23.70 Explain how HCN acts as a poison by interfering with the synthesis of ATP.

24

- - - - - - - - - - - - - - - - -

CHAPTER GOALS

In this chapter you will learn how to:

1. Understand the basic features of biochemical reactions
2. Describe the main aspects of glycolysis
3. List the pathways for pyruvate metabolism
4. Calculate the energy yield from glucose metabolism
5. Describe the main features of gluconeogenesis
6. Summarize the process of the β-oxidation of fatty acids
7. Calculate the energy yield from fatty acid oxidation
8. Identify the structures of ketone bodies and describe their role in metabolism
9. Describe the main components of amino acid catabolism

A complex set of biochemical pathways converts ingested carbohydrates, lipids, and proteins to usable materials and energy to meet the body's needs.

CARBOHYDRATE, LIPID, AND PROTEIN METABOLISM

THE metabolism of ingested food begins with the hydrolysis of large biomolecules into small compounds that can be absorbed through the intestinal wall. In Chapter 23 we learned that the last stages of catabolism, which produce energy from acetyl coenzyme A (acetyl CoA) by means of the citric acid cycle, electron transport chain, and oxidative phosphorylation, are the same for all types of biomolecules. The catabolic pathways that form acetyl CoA are different, however, depending on the particular type of biomolecule. In Chapter 24 we examine the specific metabolic pathways for carbohydrates, lipids, and proteins.

24.1   INTRODUCTION

The four stages in metabolism:

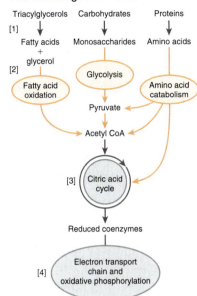

Recall from Section 23.2 that we can conceptually consider catabolism as the sum of four stages. Catabolism begins with digestion in stage [1] in which polysaccharides, triacylglycerols, and proteins are hydrolyzed to smaller compounds that can be absorbed by the bloodstream and delivered to individual cells. Each type of biomolecule is then converted to acetyl CoA by different pathways in stage [2]. Acetyl CoA enters the citric acid cycle and produces reduced coenzymes, whose energy is stored in ATP (stages [3] and [4]). Since we already learned many important facts about stages [1], [3], and [4] of catabolism, we now consider the metabolic pathways that convert monosaccharides, fatty acids and glycerol, and amino acids to acetyl CoA in stage [2].

- **Carbohydrates: Glycolysis** converts glucose, the most common monosaccharide, to pyruvate, which is then metabolized to acetyl CoA (Sections 24.3–24.4).

$$\text{glucose} \xrightarrow{\text{glycolysis}} 2\ CH_3-\underset{O}{\overset{O}{C}}-\underset{O}{\overset{O}{C}}-O^- \longrightarrow 2\ CH_3-\overset{O}{C}-SCoA$$

glucose   pyruvate   acetyl CoA

- **Lipids:** Fatty acids are converted to thioesters, which are oxidized by a stepwise procedure that sequentially cleaves two-carbon units from the carbonyl end to form acetyl CoA (Section 24.7).

$$CH_3(CH_2)_{16}-\overset{O}{C}-OH \longrightarrow CH_3(CH_2)_{16}-\overset{O}{C}-SCoA \longrightarrow 9\ CH_3-\overset{O}{C}-SCoA$$

fatty acid   thioester   acetyl CoA

- **Amino acids:** The amino acids formed from protein hydrolysis are often assembled into new proteins without any other modification. Since excess amino acids are not stored in the body, they are catabolized for energy as discussed in Section 24.9. The amino groups (NH_2) of amino acids are converted to urea [$(NH_2)_2C=O$], which is excreted in urine.

24.2   UNDERSTANDING BIOCHEMICAL REACTIONS

Before examining the specific reactions of glycolysis, let's pause to consider some principles that will help us better understand these processes.

The biochemical reactions we learned in Chapter 23 and those we will examine in Chapter 24 are often challenging for a student to tackle. On the one hand, cells use the basic principles of organic chemistry in oxidation, reduction, acid–base, and other reactions. On the other hand, the use of enzymes and coenzymes allows cells to carry out transformations that are not easily replicated or even possible in the lab, so many reactions take on a new and different appearance. Moreover, many substrates have several functional groups and these reactions often proceed with complete selectivity at just one of them.

While it is often difficult to predict the exact product of some reactions, it should be possible to *understand* and *analyze* a reaction by examining the reactants and products, the reagents (coenzymes and other materials), and the enzymes. The name of an enzyme is often a clue as to the type of reaction. Common types of enzymes encountered in metabolism, such as kinases and isomerases, are listed in Table 24.1.

TABLE 24.1 Common Enzymes in Metabolism	
Type of Enzyme	**Type of Reaction**
Carboxylase	Addition of a carboxylate ($-COO^-$)
Decarboxylase	Removal of carbon dioxide (CO_2)
Dehydrogenase	Removal of two hydrogen atoms
Isomerase	Isomerization of one isomer to another
Kinase	Transfer of a phosphate

- A *kinase* catalyzes the transfer of a phosphate group from one substrate to another.
- An *isomerase* catalyzes the conversion of one isomer to another.

As an example, consider the conversion of fructose 6-phosphate to fructose 1,6-bisphosphate, step [3] in glycolysis (Section 24.3). In this reaction, a phosphate group is transferred from ATP to fructose 6-phosphate.

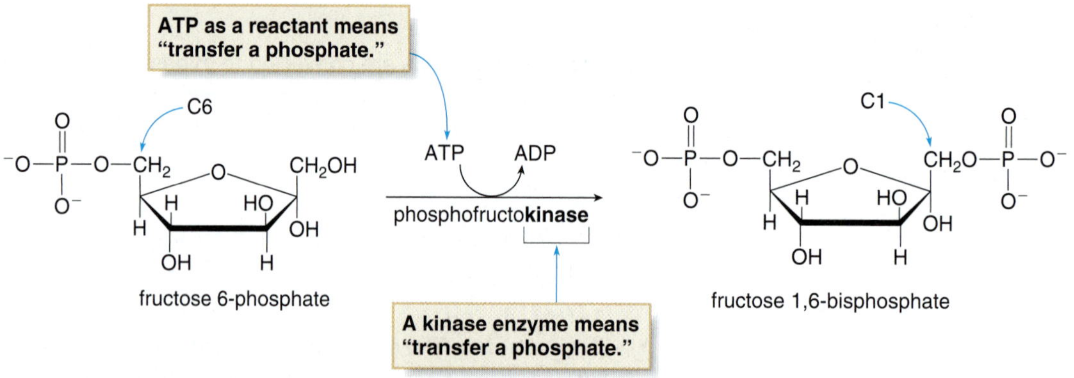

It would probably be hard for you to predict that this reaction converts only the OH group at C1 to a phosphate, given that the substrate has four OH groups. Nonetheless, it is possible to *understand* the reaction by looking at all of its components.

- A phosphate is added to fructose 6-phosphate to form fructose 1,6-bisphosphate, a product with two phosphate groups.
- The new phosphate in the product comes from ATP, a high-energy nucleoside triphosphate that transfers a phosphate to other compounds—in this case, fructose 6-phosphate.
- Loss of a phosphate from ATP forms ADP.
- The reaction is catalyzed by a kinase enzyme, also indicating that a phosphate is transferred.

Sample Problems 24.1 and 24.2 illustrate the process in other examples.

SAMPLE PROBLEM 24.1

Analyze the following reaction by considering the functional groups that change and the name of the enzyme.

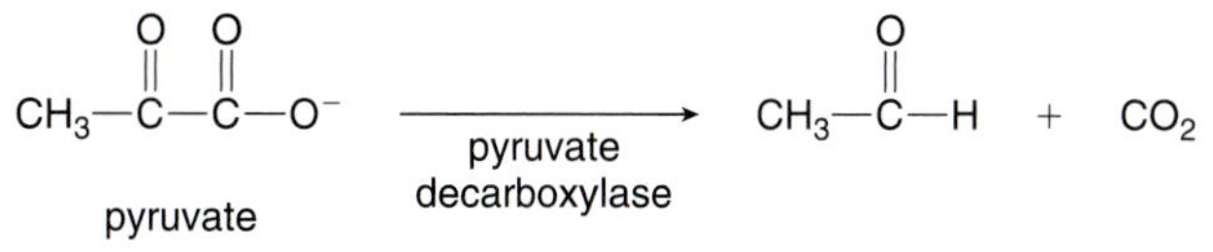

$$CH_3-\underset{\underset{pyruvate}{O}}{\overset{O}{\underset{\|}{C}}}-\underset{\overset{O}{\underset{\|}{C}}}{}-O^- \xrightarrow[\text{decarboxylase}]{\text{pyruvate}} CH_3-\underset{\overset{O}{\underset{\|}{C}}}{}-H + CO_2$$

ANALYSIS
- Consider the functional groups that are added or removed as a clue to the type of reaction.
- Classify the reagent (when one is given) as to the type of reaction it undergoes.
- Use the name of the enzyme as a clue to the reaction type.

SOLUTION
The –COO$^-$ of pyruvate is lost as CO_2, meaning that a **decarboxylation** has occurred. This is supported by the type of enzyme, a *decarboxylase*, which typically catalyzes decarboxylations. A carbon–carbon bond is broken in the process, giving CO_2 as one of the products. No coenzyme is used here; the reaction does not involve oxidation, reduction, or thioester synthesis.

SAMPLE PROBLEM 24.2

Analyze the following reaction by considering the functional groups that change, the coenzyme utilized, and the name of the enzyme.

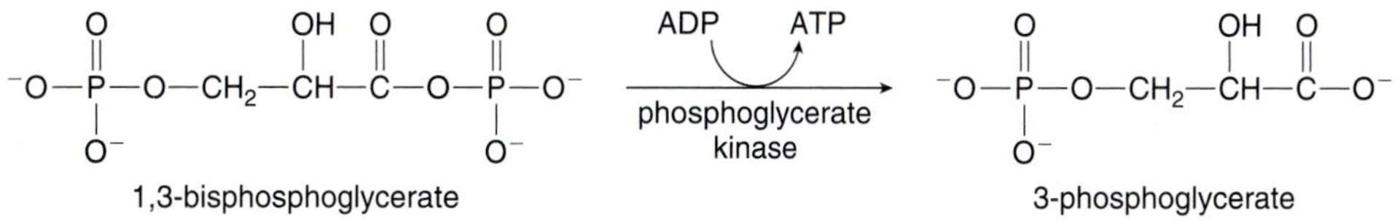

1,3-bisphosphoglycerate $\xrightarrow[\text{kinase}]{\text{phosphoglycerate}}$ 3-phosphoglycerate

(ADP → ATP)

ANALYSIS
- Consider the functional groups that are added or removed as a clue to the type of reaction.
- Classify the reagent—that is, any coenzyme or other reactant—as to the type of reaction it undergoes. For example, NAD$^+$ is an oxidizing agent, while NADH is a reducing agent. ATP transfers phosphate groups to substrates, while ADP accepts phosphate groups from substrates.
- Use the name of the enzyme as a clue to the reaction type.

SOLUTION
A phosphate group is removed from the reactant to give 3-phosphoglycerate. This phosphate is transferred to the nucleoside diphosphate ADP to form the nucleoside triphosphate ATP. Since kinase enzymes catalyze the addition or removal of phosphates, the enzyme used for the reaction is phosphoglycerate *kinase.* No coenzyme is used here; the reaction does not involve oxidation, reduction, or thioester synthesis.

PROBLEM 24.1

Analyze the following reaction by considering the functional groups that change and the name of the enzyme.

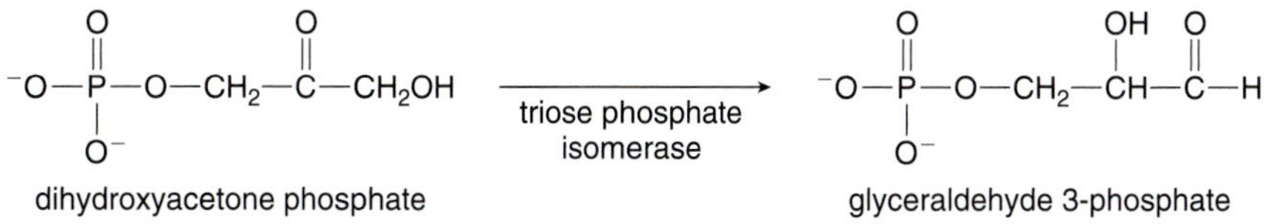

dihydroxyacetone phosphate $\xrightarrow[\text{isomerase}]{\text{triose phosphate}}$ glyceraldehyde 3-phosphate

PROBLEM 24.2

Analyze the following reaction by considering the functional groups that change, the coenzyme utilized, and the name of the enzyme.

glyceraldehyde 3-phosphate $\xrightarrow[\text{dehydrogenase}]{\text{glyceraldehyde 3-phosphate}}$ 3-phosphoglycerate

(NAD$^+$ → NADH + H$^+$)

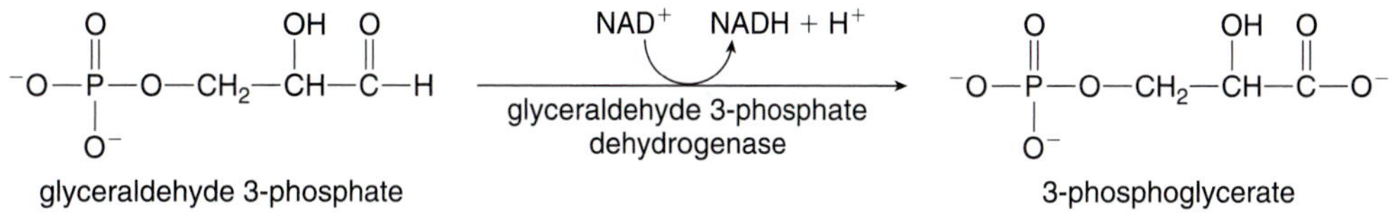

24.3 GLYCOLYSIS

The metabolism of monosaccharides centers around glucose. Whether it is obtained by the hydrolysis of ingested polysaccharides or stored glycogen (Section 20.6), glucose is the principal monosaccharide used for energy in the human body.

The word *glycolysis* is derived from the Greek *glykys* meaning *sweet* and *lysis* meaning *splitting*.

- *Glycolysis* is a linear, 10-step pathway that converts glucose, a six-carbon monosaccharide, to two molecules of pyruvate ($CH_3COCO_2^-$).

Glycolysis is an anaerobic pathway that takes place in the cytoplasm and can be conceptually divided into two parts (Figure 24.1).

- Steps [1]–[5] comprise the *energy-investment phase*. The addition of two phosphate groups requires the energy stored in two ATP molecules. Cleavage of a carbon–carbon bond forms two three-carbon products.
- Steps [6]–[10] comprise the *energy-generating phase*. Each of the three-carbon products is ultimately oxidized, forming NADH, and two high-energy phosphate bonds are broken to form two ATP molecules.

The specific reactions of glycolysis are discussed in Section 24.3A. To better understand these reactions, recall a few facts from Chapter 23.

▼ **FIGURE 24.1 An Overview of Glycolysis**

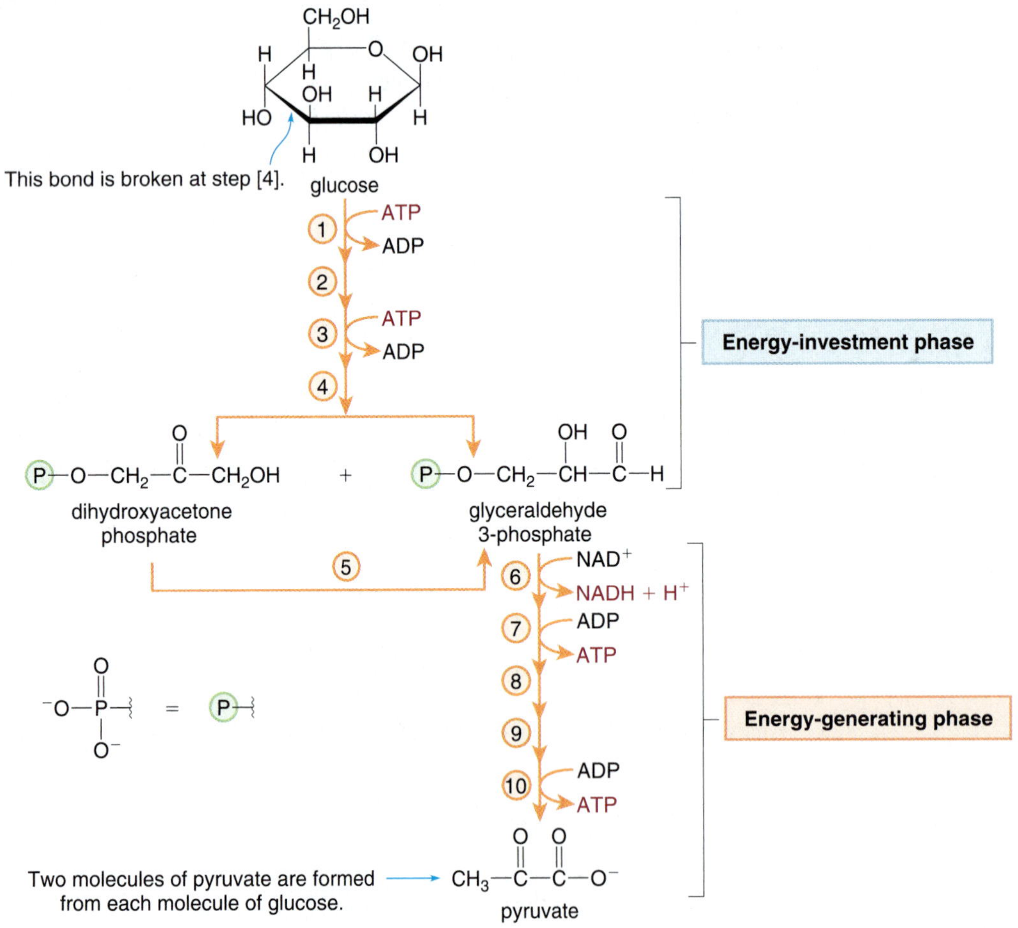

In the energy-investment phase of glycolysis, ATP supplies energy needed for steps [1] and [3]. In the energy-generating phase, each three-carbon product from step [5] forms one NADH and two ATP molecules. Since two glyceraldehyde 3-phosphate molecules are formed from each glucose molecule, a total of two NADH and four ATP molecules are formed in the energy-generating phase.

- Coenzyme NAD$^+$ is a biological *oxidizing* agent that converts C—H bonds to C—O bonds. In the process, NAD$^+$ is reduced to NADH + H$^+$.
- The phosphorylation of ADP *requires* energy and forms ATP, a high-energy nucleoside triphosphate.
- The hydrolysis of ATP *releases* energy and forms ADP.

24.3A THE STEPS IN GLYCOLYSIS

The specific steps and all needed enzymes in glycolysis are shown in Figures 24.2–24.4.

Glycolysis: Steps [1]–[3]

Glycolysis begins with the phosphorylation of glucose to form glucose 6-phosphate (Figure 24.2). This energetically unfavorable reaction is coupled with the hydrolysis of ATP to ADP to make the reaction energetically favorable. Isomerization of glucose 6-phosphate to fructose 6-phosphate takes place with an isomerase enzyme in step [2]. Phosphorylation in step [3], an energy-absorbing reaction, can be driven by the hydrolysis of ATP, yielding fructose 1,6-bisphosphate.

▼ **FIGURE 24.2 Glycolysis: Steps [1]–[3]**

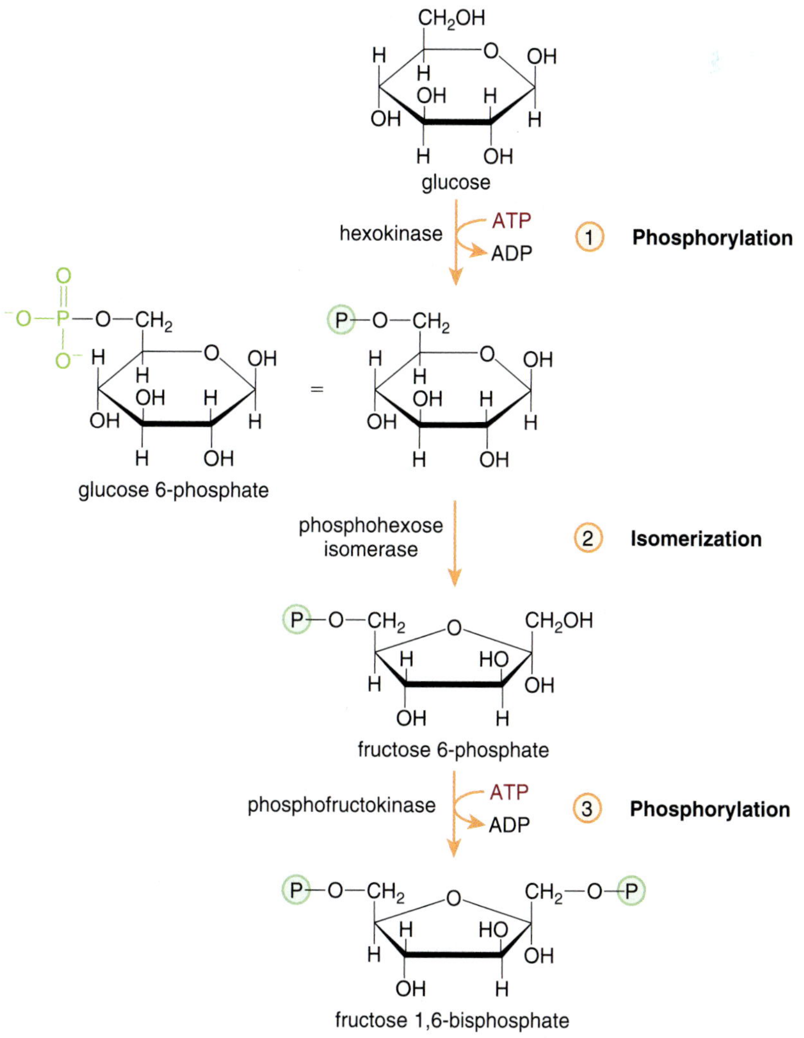

- All —PO$_3^{2-}$ groups in glycolysis are abbreviated as —Ⓟ.

- The energy from two ATP molecules is used for phosphorylation in steps [1] and [3].

- Overall, the first three steps of glycolysis add two phosphate groups and isomerize a six-membered glucose ring to a five-membered fructose ring.
- The energy stored in two ATP molecules is utilized to modify the structure of glucose for the later steps that generate energy.

PROBLEM 24.3

What type of isomers—constitutional isomers or stereoisomers—do glucose 6-phosphate and fructose 6-phosphate represent? Explain your choice.

Glycolysis: Steps [4]–[5]

Cleavage of the six-carbon chain of fructose 1,6-bisphosphate forms two three-carbon products—dihydroxyacetone phosphate and glyceraldehyde 3-phosphate—as shown in Figure 24.3. These two products have the same molecular formula but a different arrangement of atoms; that is, they are *constitutional isomers* of each other. Since only glyceraldehyde 3-phosphate continues on in glycolysis, dihydroxyacetone phosphate is isomerized to glyceraldehyde 3-phosphate in step [5], completing the energy-investment phase of glycolysis.

In summary:

- The first phase of glycolysis converts glucose to *two* molecules of glyceraldehyde 3-phosphate.
- The energy from two ATP molecules is utilized.

PROBLEM 24.4

Identify the type of carbonyl groups present in dihydroxyacetone phosphate and glyceraldehyde 3-phosphate. Classify the OH groups in each compound as 1°, 2°, or 3°.

PROBLEM 24.5

As we learned in Section 23.3, the hydrolysis of ATP to ADP releases 7.3 kcal/mol of energy. If the coupled reaction, fructose 6-phosphate + ATP → fructose 1,6-bisphosphate + ADP, releases 3.4 kcal/mol of energy, how much energy is required for the phosphorylation of fructose 6-phosphate?

$$\text{fructose 6-phosphate} + \text{HPO}_4{}^{2-} \longrightarrow \text{fructose 1,6-bisphosphate} + \text{H}_2\text{O}$$

▼ **FIGURE 24.3   Glycolysis: Steps [4] and [5]**

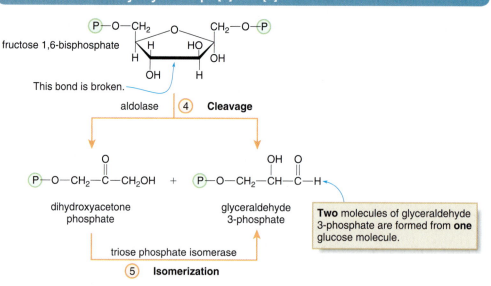

Cleavage of a carbon–carbon bond and isomerization form two molecules of glyceraldehyde 3-phosphate from glucose, completing the energy-investment phase of glycolysis.

Glycolysis: Steps [6]–[10]

Each three-carbon aldehyde (glyceraldehyde 3-phosphate) produced in step [5] of glycolysis is carried through a series of five reactions that ultimately form pyruvate (Figure 24.4).

In step [6], oxidation of the –CHO group of glyceraldehyde 3-phosphate and phosphorylation with HPO_4^{2-} form 1,3-bisphosphoglycerate. In this process, the oxidizing agent NAD^+ is reduced to NADH. Transfer of a phosphate group from 1,3-bisphosphoglycerate to ADP forms 3-phosphoglycerate and generates ATP in step [7]. Isomerization of the phosphate group in step [8]

▼ **FIGURE 24.4** Glycolysis: Steps [6]–[10]

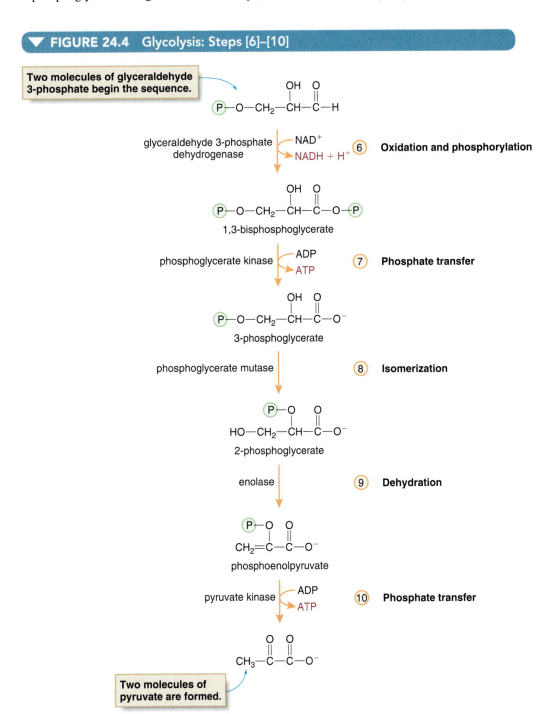

In the energy-generating phase of glycolysis, NADH is formed in step [6] and ATP is formed in steps [7] and [10]. Since one glucose molecule yields two molecules of glyceraldehyde that begin the sequence, 2 $CH_3COCO_2^-$, 2 NADH, and 4 ATP are ultimately formed in steps [6]–[10].

and loss of water in step [9] form phosphoenolpyruvate. Finally, transfer of a phosphate to ADP forms ATP and pyruvate in step [10]. Thus, **one NADH molecule is produced in step [6] and two ATPs are formed in steps [7] and [10] for each glyceraldehyde 3-phosphate.**

- Since each glucose molecule yielded *two* glyceraldehyde 3-phosphate molecules in step [5], overall *two* NADH molecules and *four* ATP molecules are formed in the energy-generating phase of glycolysis.

PROBLEM 24.6 A kinase is an enzyme that catalyzes the transfer of a phosphate group from one substrate to another. Identify all of the reactions of glycolysis that utilize kinases. What species is phosphorylated in each reaction?

PROBLEM 24.7 Identify the three reactions in glycolysis that involve the conversion of one constitutional isomer to another. Name the enzymes involved in each reaction.

PROBLEM 24.8 Why is the enzyme responsible for step [6] of glycolysis called a dehydrogenase?

24.3B THE NET RESULT OF GLYCOLYSIS

Three major products are formed in glycolysis—**ATP, NADH,** and **pyruvate.**

- Two ATP molecules are used in the energy-investment phase (steps [1] and [3]), and four molecules of ATP are formed in the energy-generating phase (steps [7] and [10]). The net result is the *synthesis of two molecules of ATP from glycolysis.*

- Two molecules of NADH are formed from two glyceraldehyde 3-phosphate molecules in step [6]. The NADH formed in glycolysis must be transported from the cytoplasm to the mitochondria for use in the electron transport chain to generate more ATP.

- Two three-carbon molecules of pyruvate ($CH_3COCO_2^-$) are formed from the six carbon atoms of glucose. The fate of pyruvate depends on oxygen availability, as discussed in Section 24.4.

Thus, the ATP production of glycolysis is small compared to that produced from the electron transport chain and oxidative phosphorylation. The overall process of glycolysis can be summarized in the following equation.

$$C_6H_{12}O_6 + 2\ NAD^+ \xrightarrow[2\ ADP]{\quad 2\ ATP \quad} 2\ CH_3-\overset{O}{\overset{\|}{C}}-\overset{O}{\overset{\|}{C}}-O^- + 2\ NADH + 2\ H^+$$

glucose pyruvate

Although glycolysis is an ongoing pathway in cells, the rate of glycolysis depends on the body's need for the products it forms—that is, pyruvate, ATP, and NADH. When ATP levels are high, glycolysis is inhibited at various stages. When ATP levels are depleted, such as during strenuous exercise, glycolysis is activated so that more ATP is synthesized.

PROBLEM 24.9 The level of citrate, one intermediate in the citric acid cycle, also regulates the rate of glycolysis. How might a high level of citrate affect the rate of glycolysis?

24.3C GLYCOLYSIS AND OTHER HEXOSES

Although glucose is the major monosaccharide obtained by the hydrolysis of ingested carbohydrates, three other six-carbon monosaccharides are also obtained in lesser amounts in the diet.

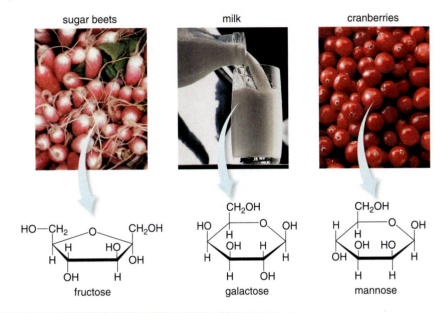

sugar beets milk cranberries

fructose galactose mannose

- **Fructose** is obtained by the hydrolysis of the disaccharide sucrose, found in sugar beets and sugarcane.
- **Galactose** is obtained by the hydrolysis of the disaccharide lactose in milk.
- **Mannose** is obtained from polysaccharides in fruits such as cranberries and currants.

These hexoses enter glycolysis at various points along the metabolic pathway. Fructose enters glycolysis in two different ways. In the muscle and kidneys, it is phosphorylated with ATP to form fructose 6-phosphate, the substrate for step [3]. In the liver, fructose is converted by a multistep route to glyceraldehyde 3-phosphate, which enters the pathway at step [6].

Galactose is phosphorylated with ATP to form galactose 1-phosphate, which is converted to glucose 6-phosphate in several steps. Individuals with **galactosemia** lack one of the necessary enzymes required for this process, resulting in an abnormally high galactose concentration, with disabling and sometimes fatal results (Section 20.2). Galactosemia can be detected in newborn screening, and is treated by eliminating galactose from the diet.

Mannose is converted to mannose 6-phosphate and is then isomerized to fructose 6-phosphate, the substrate in step [3] of glycolysis. In this way, all of the other common hexoses are converted to intermediates in the glycolytic pathway, and as a result, they are metabolized to pyruvate.

PROBLEM 24.10

(a) Are galactose and mannose constitutional isomers or stereoisomers? (b) Draw the structure of galactose 1-phosphate and mannose 6-phosphate. (c) Are these two phosphates constitutional isomers or stereoisomers?

PROBLEM 24.11

Would you expect that the phosphorylation of mannose to form mannose 6-phosphate requires ATP to make it an energetically favorable reaction? Explain your choice.

24.3D FOCUS ON HEALTH & MEDICINE
GLYCOLYSIS AND CANCER CELLS

It has been known for a long time that many cancerous tumors depend largely on glycolysis to supply their energy needs, even though the yield of ATP from glycolysis is small—only two molecules of ATP for each glucose molecule. This occurs in part because glycolysis is an anaerobic pathway and often tumor cells are not located in proximity to oxygen-carrying blood vessels. As a result, glycolysis often proceeds ten times faster in cancer cells compared to healthy cells.

Although many cancer chemotherapies target derailing DNA synthesis as a method of killing fast-growing cancer cells, some recent research has aimed at attacking specific points on the glycolytic pathway. This strategy was pinned on the theory that without the ATP generated from glycolysis, cancer cells would be energy starved and unable to survive. In fact, promising results have been obtained by shutting down glycolysis in breast cancer cells by targeting a specific enzyme, thus opening the way for new therapies with perhaps fewer side effects than traditional chemotherapy.

24.4 THE FATE OF PYRUVATE

While pyruvate is the end product of glycolysis, it is not the final product of glucose metabolism. What happens to pyruvate depends on the existing conditions and the organism. In particular, there are three possible products:

$$CH_3-\underset{\underset{\text{pyruvate}}{}}{\overset{O}{\underset{\|}{C}}}-\overset{O}{\underset{\|}{C}}-O^- \longrightarrow CH_3-\overset{O}{\underset{\underset{\text{acetyl CoA}}{\|}}{C}}-SCoA \quad \text{or} \quad CH_3-\underset{\underset{\underset{\text{lactate}}{H}}{|}}{\overset{OH}{\underset{|}{C}}}-\overset{O}{\underset{\|}{C}}-O^- \quad \text{or} \quad \underset{\text{ethanol}}{CH_3CH_2OH}$$

- **Acetyl CoA, $CH_3COSCoA$, is formed under aerobic conditions.**
- **Lactate, $CH_3CH(OH)CO_2^-$, is formed under anaerobic conditions.**
- **Ethanol, CH_3CH_2OH, is formed in fermentation.**

24.4A CONVERSION TO ACETYL CoA

When oxygen is plentiful, oxidation of pyruvate by NAD^+ in the presence of coenzyme A and a dehydrogenase enzyme forms acetyl CoA and carbon dioxide. For this process to occur, pyruvate must diffuse across the outer membrane of a mitochondrion, and then be transported across the inner mitochondrial membrane into the matrix.

$$CH_3-\overset{O}{\underset{\|}{C}}-\overset{O}{\underset{\underset{\text{pyruvate}}{\|}}{C}}-O^- + HS-CoA \xrightarrow{\overset{NAD^+ \quad NADH + H^+}{\curvearrowright}} CH_3-\overset{O}{\underset{\underset{\text{acetyl CoA}}{\|}}{C}}-SCoA + CO_2$$

Although oxygen is not needed for this specific reaction, oxygen is needed to oxidize NADH back to NAD^+. Without an adequate supply of NAD^+, this pathway cannot occur. The acetyl CoA formed in this process then enters the common metabolic pathways—the citric acid cycle, the electron transport chain, and oxidative phosphorylation—to generate a great deal of ATP.

24.4B FOCUS ON HEALTH & MEDICINE
CONVERSION TO LACTATE

When oxygen levels are low, the metabolism of pyruvate must follow a different course. Oxygen is needed to oxidize the NADH formed in step [6] of glycolysis back to NAD^+. If there is not enough O_2 to re-oxidize NADH, cells must get NAD^+ in a different way. The conversion of pyruvate to lactate provides the solution.

Reduction of pyruvate with NADH forms lactate and NAD^+, which can now re-enter glycolysis and oxidize glyceraldehyde 3-phosphate at step [6].

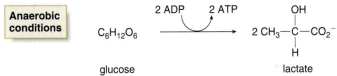

At rest, the O_2 level in cells is sufficient for the aerobic oxidation of pyruvate to acetyl CoA in humans, so lactate production occurs regularly only in red blood cells, which have no mitochondria and therefore no machinery of aerobic catabolism. During periods of strenuous exercise, when ATP needs are high, however, there is an inadequate level of oxygen to re-oxidize NADH in the electron transport chain. At these times, pyruvate is reduced to lactate for the sole purpose of re-oxidizing NADH to NAD^+ to maintain glycolysis. Under anaerobic conditions, therefore, lactate becomes the main product of glucose metabolism and only two ATP molecules are formed per glucose.

HEALTH NOTE

Muscle fatigue is caused by lactate buildup due to the anaerobic metabolism of glucose. Deep breaths after exercise replenish O_2 in oxygen-depleted tissues and allow the re-oxidation of lactate to pyruvate.

Anaerobic metabolism leads to an increase in lactate levels in muscles, which in turn is associated with soreness and cramping. During these periods an "oxygen debt" is created. When vigorous activity ceases, an individual inhales deep breaths of air to repay the oxygen debt caused by heavy exercise. Lactate is then gradually re-oxidized to pyruvate, which can once again be converted to acetyl CoA, and muscle soreness, fatigue, and shortness of breath resolve.

In any tissue deprived of oxygen, pyruvate is converted to lactate rather than acetyl CoA. The pain produced during a heart attack, for example, is caused by an increase in lactate concentration that results when the blood supply to part of the heart muscle is blocked (Figure 24.5). The lack of oxygen delivery to heart tissue results in the anaerobic metabolism of glucose to lactate rather than acetyl CoA.

Measuring lactate levels in the blood is a common diagnostic tool used by physicians to assess how severely ill an individual is. A higher-than-normal lactate concentration generally indicates inadequate oxygen delivery to some tissues. Lactate levels increase transiently during exercise, but can remain elevated because of lung disease, congestive heart failure, or the presence of a serious infection.

▼ **FIGURE 24.5** **Lactate Production During a Heart Attack**

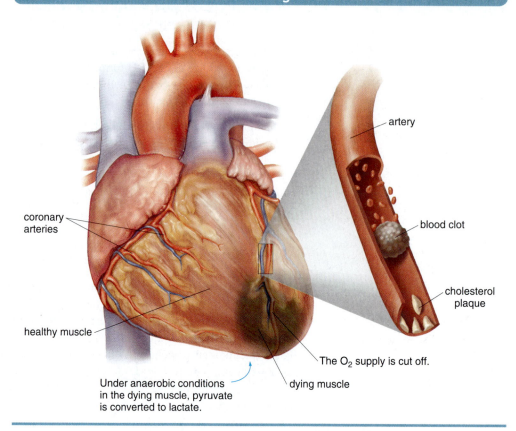

Under anaerobic conditions in the dying muscle, pyruvate is converted to lactate.

24.4C FOCUS ON HEALTH & MEDICINE
CONVERSION TO ETHANOL

In yeast and other microorganisms, pyruvate is converted to ethanol (CH_3CH_2OH) and carbon dioxide (CO_2) by a two-step process: decarboxylation (loss of CO_2) to acetaldehyde followed by reduction to ethanol.

The NAD^+ generated during reduction can enter glycolysis as an oxidizing agent in step [6]. As a result, glucose can be metabolized by yeast under anaerobic conditions: glycolysis forms pyruvate, which is further metabolized to ethanol and carbon dioxide. Two molecules of ATP are generated during glycolysis.

- *Fermentation* is the anaerobic conversion of glucose to ethanol and CO_2.

Fermentation plays a key role in the production of bread, beer, and cheese.

The ethanol in beer, wine, and other alcoholic beverages is obtained by the fermentation of sugar, quite possibly the oldest example of chemical synthesis. The carbohydrate source determines the type of alcoholic beverage formed. The sugars in grapes are fermented to form wine, barley malt is used for beer, and the carbohydrates in corn or rye form whiskey. When the fermented ethanol reaches a concentration of about 12%, the yeast cells that provide the enzymes for the process die and fermentation stops. Any alcoholic beverage with an ethanol concentration of greater than 12% must be distilled to increase the ethanol content. While ethanol consumption is considered socially acceptable in most societies, the irresponsible consumption of alcoholic beverages is a major health and social problem in many parts of the world (Section 14.6).

Fermentation plays a role in forming other food products. Cheese is produced by fermenting curdled milk, while yogurt is prepared by fermenting fresh milk. When yeast is mixed with flour, water, and sugar, the enzymes in yeast carry out fermentation to produce the CO_2 that causes bread to rise. Some of the characteristic and "intoxicating" odor associated with freshly baked bread is due to the ethanol that has evaporated during the baking process.

PROBLEM 24.12	What role does NADH play in the conversion of pyruvate to lactate? What role does NADH play in the conversion of pyruvate to ethanol?
PROBLEM 24.13	(a) In what way(s) is the conversion of pyruvate to acetyl CoA similar to the conversion of pyruvate to ethanol? (b) In what way(s) are the two processes different?
PROBLEM 24.14	(a) In what way(s) is the conversion of pyruvate to lactate similar to the conversion of pyruvate to ethanol? (b) In what way(s) are the two processes different?

24.5 THE ATP YIELD FROM GLUCOSE

How much ATP is generated from the complete catabolism of glucose ($C_6H_{12}O_6$) to carbon dioxide (CO_2)? To answer this question we must take into account the number of ATP molecules formed in the following sequential pathways:

- the glycolysis of glucose to two pyruvate molecules
- the oxidation of two pyruvate molecules to two molecules of acetyl CoA
- the citric acid cycle
- the electron transport chain and oxidative phosphorylation

To calculate how much ATP is generated, we must consider both the ATP formed directly in reactions, as well as the ATP produced from reduced coenzymes after oxidative phosphorylation.

As we learned in Section 23.6, each NADH formed in the mitochondrial matrix provides the energy to yield 2.5 ATPs, while each $FADH_2$ yields 1.5 ATPs. In contrast, the NADH formed during glycolysis is generated in the cytoplasm *outside* the mitochondria. Although it cannot pass through the mitochondrial membrane, its electrons and energy can be transferred to other molecules—NADH or $FADH_2$—ultimately yielding 1.5 or 2.5 ATPs for each NADH. To simplify calculations, we will use the value 2.5 ATP/NADH consistently in calculations.

Finally, any calculation must also take into account that glucose is split into *two* three-carbon molecules at step [4] of glycolysis, so the **ATP yield in each reaction must be *doubled* after this step.** With this information and Figure 24.6 in hand, we can now determine the total yield of ATP from the complete catabolism of glucose.

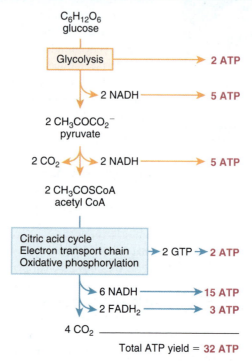

▼ FIGURE 24.6   The ATP Yield from the Aerobic Metabolism of Glucose to CO_2

The complete catabolism of glucose forms six CO_2 molecules and 32 ATP molecules.

- Glycolysis yields a net of two ATP molecules. The two molecules of NADH formed during step [6] of glycolysis yield five additional ATPs.

$$C_6H_{12}O_6 \longrightarrow 2\ CH_3COCO_2^- + 2\ ATP + 2\ NADH$$
glucose   pyruvate

$$2 \times (2.5\ ATP/NADH) = 5\ ATP$$

- Oxidation of two molecules of pyruvate to acetyl CoA in the mitochondria forms two NADH molecules that yield five ATP molecules.

$$2\ CH_3COCO_2^- \longrightarrow 2\ CH_3COSCoA + 2\ CO_2 + 2\ NADH$$
pyruvate   acetyl CoA

$$2 \times (2.5\ ATP/NADH) = 5\ ATP$$

- Beginning with two acetyl CoA molecules, the citric acid cycle (Figure 23.7) forms two GTP molecules, the energy equivalent of two ATPs, in step [5]. The six NADH molecules and two $FADH_2$ molecules also formed yield an additional 18 ATPs from the electron transport chain and oxidative phosphorylation. Thus, 20 ATPs are formed from two acetyl CoA molecules.

$$2 \times (1.5\ ATP/FADH_2) = 3\ ATP$$

$$2\ CH_3COSCoA \longrightarrow 4\ CO_2 + 2\ GTP + 6\ NADH + 2\ FADH_2$$
acetyl CoA

$$2\ ATP$$

$$6 \times (2.5\ ATP/NADH) = 15\ ATP$$

Adding up the ATP formed in each pathway gives a **total of 32 molecules of ATP for the complete catabolism of each glucose molecule.** Most of the ATP generated from glucose metabolism results from the citric acid cycle, electron transport chain, and oxidative phosphorylation.

Glucose is the main source of energy for cells and the only source of energy used by the brain. When energy demands are low, glucose is stored as the polymer glycogen in the liver and muscles. When blood levels of glucose are low, glycogen is hydrolyzed to keep adequate blood glucose levels to satisfy the body's energy needs.

Blood glucose levels are carefully regulated by two hormones. When blood glucose concentration rises after a meal, **insulin** stimulates the passage of glucose into cells for metabolism. When blood glucose levels are low, the hormone **glucagon** stimulates the conversion of stored glycogen to glucose.

PROBLEM 24.15	How much ATP results from each transformation?

a. glucose $\rightarrow$ 2 pyruvate   c. glucose $\rightarrow$ 2 acetyl CoA

b. pyruvate $\rightarrow$ acetyl CoA   d. 2 acetyl CoA $\rightarrow$ 4 CO_2

PROBLEM 24.16	What three reactions form CO_2 when glucose is completely catabolized?

PROBLEM 24.17	What three reactions form a nucleoside triphosphate (GTP or ATP) directly when glucose is completely catabolized?

PROBLEM 24.18	What is the difference in ATP generation between the aerobic oxidation of glucose to CO_2 and the anaerobic conversion of glucose to lactate?

24.6 GLUCONEOGENESIS

Now that we have learned how glucose is converted to CO_2 and energy, we can also consider one other metabolic pathway that involves glucose.

- *Gluconeogenesis* is the synthesis of glucose from noncarbohydrate sources—lactate, amino acids, or glycerol.

As such, **gluconeogenesis is an *anabolic* pathway rather than a catabolic pathway since it results in the synthesis of glucose from smaller molecules.** Gluconeogenesis occurs in the liver when the body has depleted both available glucose and stored glycogen reserves. Such a condition results during sustained vigorous exercise or fasting.

For example, the lactate that builds up in muscles as a result of vigorous exercise can be transported to the liver where it is oxidized to pyruvate, which is then converted to glucose by means of gluconeogenesis. The newly synthesized glucose is then utilized for energy or stored in the muscle as glycogen.

Conceptually, gluconeogenesis is the reverse of glycolysis; that is, **two molecules of pyruvate are converted to glucose by a stepwise pathway that passes through all of the same intermediates encountered in glycolysis.** In fact, seven of the 10 steps of gluconeogenesis use the same enzymes as glycolysis. Three steps in glycolysis (steps [1], [3], and [10]), however, must use different enzymes to be energetically feasible for gluconeogenesis to occur. As a result, **gluconeogenesis provides a mechanism for synthesizing new glucose molecules from other substrates.**

- Glycolysis is a catabolic process that converts glucose to pyruvate.
- Gluconeogenesis is an anabolic process that synthesizes glucose from pyruvate.

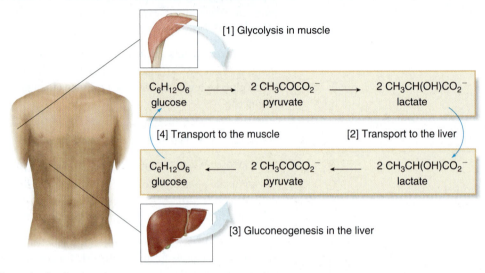

▼ **FIGURE 24.7** The Cori Cycle—Glycolysis and Gluconeogenesis

Steps in the Cori cycle:

[1] The catabolism of glucose in muscle forms pyruvate, which is reduced to lactate when the oxygen supply is limited.

[2] Lactate is transported to the liver.

[3] Oxidation of lactate forms pyruvate, which is then converted to glucose by the 10-step process of gluconeogenesis.

[4] Glucose is transported back to the muscle.

Cycling compounds from the muscle to the liver and back to the muscle is called the **Cori cycle.** The Cori cycle involves the four operations shown in Figure 24.7.

Because the brain relies solely on glucose as an energy source, gluconeogenesis is a mechanism that ensures that the brain has a supply of glucose even when a diet is low in carbohydrates and glycogen reserves are low. Gluconeogenesis is not a commonly used metabolic pathway when carbohydrate intake is high. On the other hand, a low carbohydrate diet makes gluconeogenesis an important pathway that converts noncarbohydrate substrates—fats and amino acids—to needed glucose.

PROBLEM 24.19

From what you learned about glycolysis in Section 24.3, what is the starting material and product of each of the following steps in gluconeogenesis: (a) step [1]; (b) step [3]; (c) step [10]?

24.7 THE CATABOLISM OF TRIACYLGLYCEROLS

The first step in the catabolism of triacylglycerols, the most common lipids, is the hydrolysis of the three ester bonds to form glycerol and fatty acids, which are then metabolized in separate pathways.

24.7A   GLYCEROL CATABOLISM

The glycerol formed from triacylglycerol hydrolysis is converted in two steps to dihydroxy-acetone phosphate. Phosphorylation of glycerol with ATP forms glycerol 3-phosphate, which is then oxidized with NAD^+.

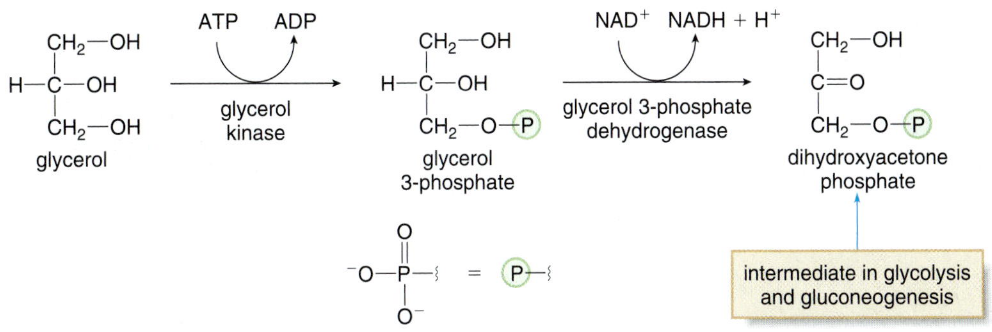

Since dihydroxyacetone phosphate is an intermediate in both glycolysis and gluconeogenesis, two pathways are available, depending on the energy needs of the organism.

PROBLEM 24.20   Analyze the phosphorylation of glycerol in step [1] of glycerol metabolism by looking at the functional groups, the reagent, and the name of the enzyme, as discussed in Section 24.2.

24.7B   FATTY ACID CATABOLISM BY β-OXIDATION

Fatty acids are catabolized by **β-oxidation,** a process in which two-carbon acetyl CoA units are sequentially cleaved from the fatty acid. Key to this process is the oxidation of the β carbon to the carbonyl group of a thioester (RCOSR'), which then undergoes cleavage between the α and β carbons.

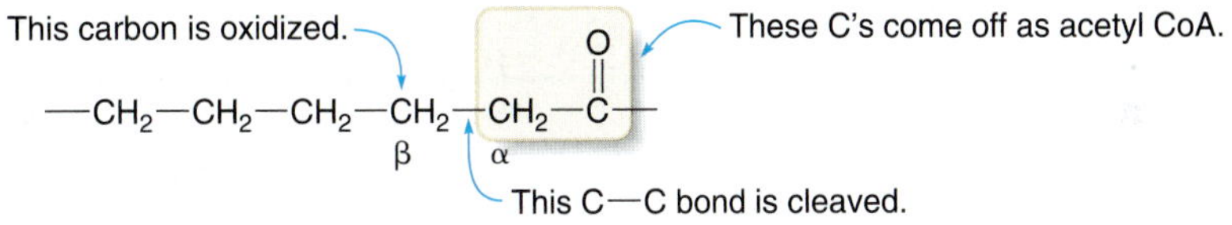

Fatty acid oxidation begins with conversion of the fatty acid to a thioester with coenzyme A. This process requires energy, which comes from the hydrolysis of two P—O bonds in ATP to form AMP, adenosine *mono*phosphate. Much like the beginning of glycolysis requires an energy investment, so, too, the initial step of fatty acid oxidation requires energy input.

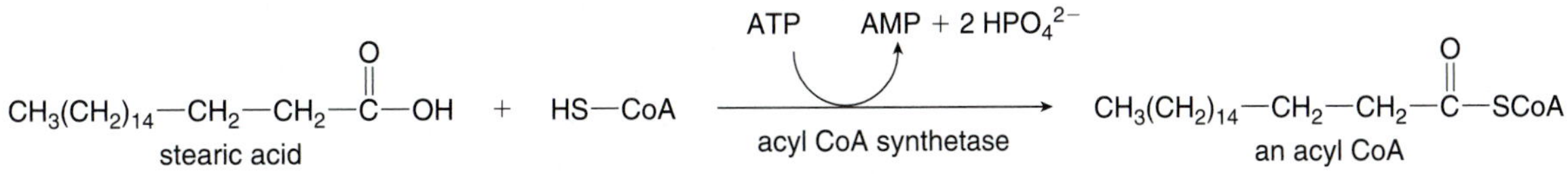

Once the product, an **acyl CoA,** is inside the mitochondrion, the process of β-oxidation is set to begin. β-Oxidation requires four steps to cleave a two-carbon acetyl CoA unit from the acyl CoA, as shown with the 18-carbon fatty acid stearic acid in Figure 24.8.

In step [1], FAD removes two hydrogen atoms to form $FADH_2$ and a double bond between the α and β carbons of the thioester. Water is added to the double bond in step [2] to place an OH group on the β carbon to the carbonyl group, which is then oxidized in step [3] to form a carbonyl group. The NAD^+ oxidizing agent is reduced to NADH in step [3] as well. Finally, cleavage of the bond between the α and β carbons forms acetyl CoA and a 16-carbon acyl CoA in step [4].

• As a result, a new acyl CoA having two carbons fewer than the original acyl CoA is formed.

▼ FIGURE 24.8   β-Oxidation of a Fatty Acid

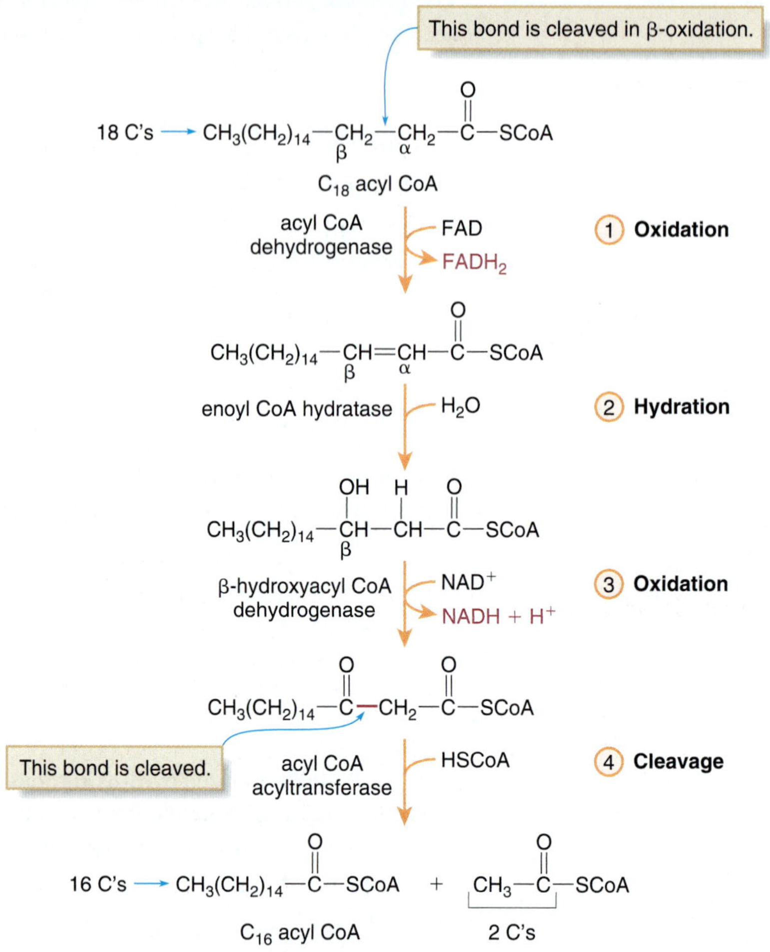

The following equation summarizes the important components of β-oxidation for a general acyl CoA, $RCH_2CH_2COSCoA$. **Each four-step sequence forms one molecule each of acetyl CoA, NADH, and FADH$_2$.**

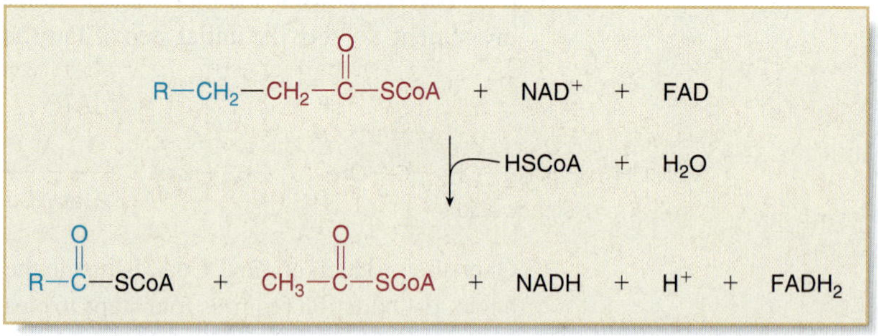

Once the 16-carbon acyl CoA is formed in step [4], it becomes the substrate for a new four-step β-oxidation sequence. This type of metabolic pathway is called a **spiral pathway,** because the same set of reactions is repeated on increasingly smaller substrates. The process continues until a four-carbon acyl CoA is cleaved to form *two* acetyl CoA molecules. As a result:

- An 18-carbon acyl CoA is cleaved to *nine* two-carbon acetyl CoA molecules.
- A total of *eight* cycles of β-oxidation are needed to cleave the eight carbon–carbon bonds.

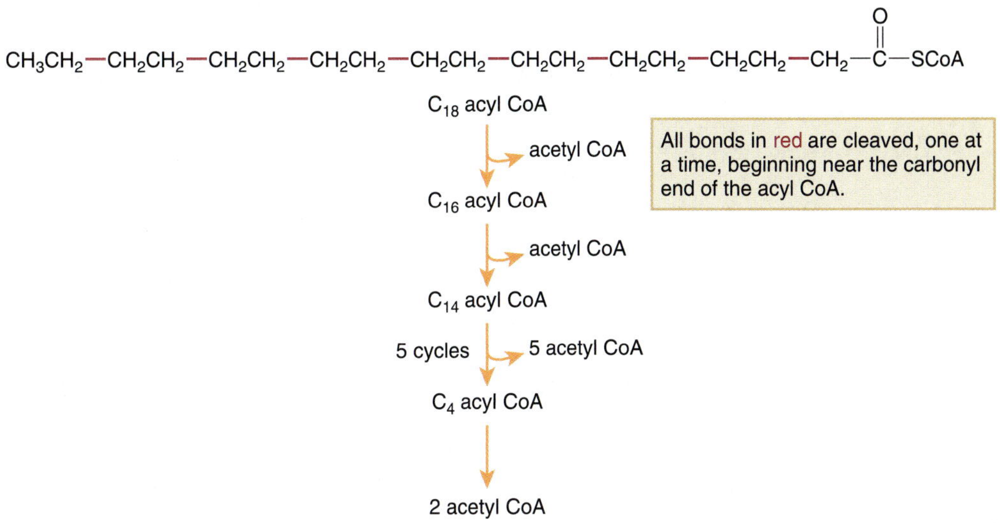

$$CH_3CH_2-CH_2CH_2-CH_2CH_2-CH_2CH_2-CH_2CH_2-CH_2CH_2-CH_2CH_2-CH_2CH_2-CH_2\overset{\overset{\displaystyle O}{\|}}{C}-SCoA$$

C_{18} acyl CoA

→ acetyl CoA

C_{16} acyl CoA

All bonds in red are cleaved, one at a time, beginning near the carbonyl end of the acyl CoA.

→ acetyl CoA

C_{14} acyl CoA

5 cycles → 5 acetyl CoA

C_4 acyl CoA

2 acetyl CoA

Thus, complete β-oxidation of the acyl CoA derived from stearic acid forms:

- **9 CH₃COSCoA molecules** (from the 18-carbon fatty acid)
- **8 NADH** (from eight cycles of β-oxidation)
- **8 FADH₂** (from eight cycles of β-oxidation)

β-Oxidation of unsaturated fatty acids proceeds in a similar fashion, although an additional step is required. **Ultimately every carbon in the original fatty acid ends up as a carbon atom of acetyl CoA.**

SAMPLE PROBLEM 24.3

Consider lauric acid, $CH_3(CH_2)_{10}CO_2H$. (a) How many molecules of acetyl CoA are formed from complete β-oxidation? (b) How many cycles of β-oxidation are needed for complete catabolism?

ANALYSIS

The number of carbons in the fatty acid determines the number of molecules of acetyl CoA formed and the number of times β-oxidation occurs.

- The number of molecules of acetyl CoA equals one-half the number of carbons in the original fatty acid.
- Because the final turn of the cycle forms *two* molecules of acetyl CoA, the number of cycles is one fewer than the number of acetyl CoA molecules formed.

SOLUTION

Since lauric acid has 12 carbons, it forms six molecules of acetyl CoA from five cycles of β-oxidation.

PROBLEM 24.21

For each fatty acid: [1] How many molecules of acetyl CoA are formed from complete catabolism? [2] How many cycles of β-oxidation are needed for complete oxidation?

a. arachidic acid ($C_{20}H_{40}O_2$)   b. palmitoleic acid ($C_{16}H_{30}O_2$)

24.7C   THE ENERGY YIELD FROM FATTY ACID OXIDATION

How much energy—in terms of the number of molecules of ATP formed—results from the complete catabolism of a fatty acid? To determine this quantity, we must take into account the ATP cost for the conversion of the fatty acid to the acyl CoA, as well as the ATP production from the coenzymes (NADH and FADH$_2$) and acetyl CoA formed during β-oxidation. The steps are shown in the accompanying *How To* procedure.

HOW TO   Determine the Number of Molecules of ATP Formed from a Fatty Acid

EXAMPLE   **How much ATP is formed by the complete catabolism of stearic acid, C$_{18}$H$_{36}$O$_2$?**

Step [1]   **Determine the amount of ATP required to synthesize the acyl CoA from the fatty acid.**
- Since the conversion of stearic acid (C$_{17}$H$_{35}$COOH) to an acyl CoA (C$_{17}$H$_{35}$COSCoA) requires the hydrolysis of two P—O bonds, this is equivalent to the energy released when 2 ATPs are converted to 2 ADPs.
- Thus, the **first step in catabolism costs the equivalent of 2 ATPs—that is, –2 ATPs.**

Step [2]   **Add up the ATP generated from the coenzymes produced during β-oxidation.**
- As we learned in Section 24.7B, each cycle of β-oxidation produces one molecule each of NADH and FADH$_2$. To cleave eight carbon–carbon bonds in stearic acid requires eight cycles of β-oxidation, so 8 NADH and 8 FADH$_2$ are produced.

$$8 \text{ NADH } \times 2.5 \text{ ATP/NADH} = 20 \text{ ATP}$$
$$8 \text{ FADH}_2 \times 1.5 \text{ ATP/FADH}_2 = \underline{12 \text{ ATP}}$$
$$\text{From reduced coenzymes: } \quad 32 \text{ ATP}$$

- Thus, 32 ATPs would be produced from oxidative phosphorylation after the reduced coenzymes enter the electron transport chain.

Step [3]   **Determine the amount of ATP that results from each acetyl CoA, and add the results for steps [1]–[3].**
- From Section 24.7B, stearic acid generates nine molecules of acetyl CoA, which then enter the citric acid cycle and go on to produce ATP by the electron transport chain and oxidative phosphorylation. As we learned in Section 23.6, each acetyl CoA results in 10 ATPs.

$$9 \text{ acetyl CoA } \times 10 \text{ ATP/acetyl CoA} = 90 \text{ ATP}$$

- Totaling the values obtained in steps [1]–[3]:

$$(-2) + 32 + 90 = 120 \text{ ATP molecules from stearic acid}$$

Answer

PROBLEM 24.22   Calculate the number of molecules of ATP formed by the complete catabolism of palmitic acid, C$_{16}$H$_{32}$O$_2$.

PROBLEM 24.23   Calculate the number of molecules of ATP formed by the complete catabolism of arachidic acid, C$_{20}$H$_{40}$O$_2$.

How does the energy generated in fatty acid catabolism compare to the energy generated from glucose catabolism? In Section 24.5, we determined that one molecule (or one mole) of glucose generates 32 molecules (or moles) of ATP, while the *How To* calculation demonstrates that one molecule (or one mole) of stearic acid generates 120 molecules (or moles) of ATP.

To compare these values, we need to compare the amount of ATP generated *per gram* of each material. To carry out a calculation of this sort we need to use the molar masses of glucose (180 g/mol) and stearic acid (284 g/mol).

A grizzly bear uses stored body fat as its sole energy source during its many months of hibernation. β-Oxidation of fatty acids releases sufficient energy to maintain a constant body temperature between 32 and 35 °C and operate all necessary cellular processes.

molar mass conversion factor

For glucose: $\dfrac{32\ \text{ATP}}{\text{mol glucose}} \times \dfrac{1\ \text{mol}}{180\ \text{g}} = \dfrac{0.18\ \text{mol ATP}}{\text{g glucose}}$

For stearic acid: $\dfrac{120\ \text{ATP}}{\text{mol stearic acid}} \times \dfrac{1\ \text{mol}}{284\ \text{g}} = \dfrac{0.42\ \text{mol ATP}}{\text{g stearic acid}}$

over twice as much energy per gram

This calculation demonstrates that **a fatty acid produces over *twice as much energy per gram* (in terms of moles of ATP generated) than glucose.** This is why lipids are so much more effective as energy-storing molecules than carbohydrates.

PROBLEM 24.24

Use the number of molecules of ATP formed from the complete catabolism of palmitic acid, $C_{16}H_{32}O_2$, in Problem 24.22 to calculate the molecules (or moles) of ATP formed per gram of palmitic acid (molar mass 256 g/mol).

PROBLEM 24.25

Another way to compare the energy content of molecules is to compare the number of ATPs they generate *per carbon* atom they contain. (a) How many ATPs are formed per carbon when glucose is completely catabolized? (b) How many ATPs are formed per carbon when stearic acid is completely catabolized? (c) Do these data support or refute the fact that lipids are more effective energy-storing molecules than carbohydrates?

24.8 KETONE BODIES

When carbohydrates do not meet energy needs, the body turns to catabolizing stored triacylglycerols, which generate acetyl CoA by β-oxidation of fatty acids. Normally the acetyl CoA is metabolized in the citric acid cycle. When acetyl CoA levels exceed the capacity of the citric acid cycle, however, acetyl CoA is converted to three compounds that are collectively called **ketone bodies**—acetoacetate, β-hydroxybutyrate, and acetone.

$$2\ CH_3-\overset{\overset{\textstyle O}{\|}}{C}-SCoA \xrightarrow{\text{several steps}} CH_3-\overset{\overset{\textstyle O}{\|}}{C}-CH_2-\overset{\overset{\textstyle O}{\|}}{C}-O^- \xrightarrow{\substack{NADH + H^+ \quad NAD^+}} CH_3-\overset{\overset{\textstyle OH}{|}}{\underset{\underset{\textstyle H}{|}}{C}}-CH_2-\overset{\overset{\textstyle O}{\|}}{C}-O^-$$

acetoacetate

β-hydroxybutyrate

$$\downarrow CO_2$$

$$CH_3-\overset{\overset{\textstyle O}{\|}}{C}-CH_3$$

acetone

- *Ketogenesis* is the synthesis of ketone bodies from acetyl CoA.

Acetoacetate is first formed from two molecules of acetyl CoA by a multistep process. Acetoacetate is either reduced to β-hydroxybutyrate with NADH or undergoes decarboxylation to form acetone. Ketone bodies are produced in the liver, and since they are small molecules that can hydrogen bond with water, they are readily soluble in blood and urine. Once they reach tissues, β-hydroxybutyrate and acetoacetate can be re-converted to acetyl CoA and metabolized for energy.

PROBLEM 24.26   Are any structural features common to the three ketone bodies?

PROBLEM 24.27   Why is the term "ketone body" a misleading name for β-hydroxybutyrate?

HEALTH NOTE

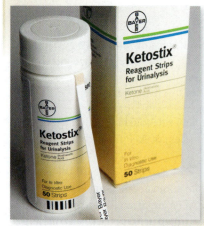

Ketostix are a brand of test strips that detect ketone bodies in urine.

Under some circumstances—notably starvation, vigorous dieting, and uncontrolled diabetes—when glucose is unavailable or cannot pass into a cell for use as fuel, ketone bodies accumulate, a condition called **ketosis.** As a result, ketone bodies are eliminated in urine and the sweet odor of acetone can be detected in exhaled breath. Sometimes the first indication of diabetes in a patient is the detection of excess ketone bodies in a urine test.

An abnormally high concentration of ketone bodies can lead to **ketoacidosis**—that is, a lowering of the blood pH caused by the increased level of β-hydroxybutyrate and acetoacetate. Although the carbonic acid/bicarbonate buffer in the blood prevents drastic pH changes (Section 9.11), even a small drop in pH can alter many critical biochemical processes.

Low carbohydrate diets, popularized by Dr. Robert Atkins in a series of diet books published in the 1990s, restrict carbohydrate intake to induce the utilization of the body's stored fat as its main energy source. This increased level of fatty acid metabolism leads to an increased concentration of ketone bodies in the blood. Increased levels of ketone bodies in the urine—which can be detected using ketone test strips—can be used as an indicator that the body has switched to metabolic machinery that relies on fat rather than carbohydrates as its principal energy source. Some physicians recommend that the level of ketone bodies be monitored to avoid the risk of ketoacidosis.

PROBLEM 24.28   What type of enzymes catalyzes the conversion of acetoacetate to acetone?

PROBLEM 24.29   How is the concentration of acetyl CoA related to the production of ketone bodies?

24.9   AMINO ACID METABOLISM

HEALTH NOTE

Low carbohydrate diets such as the Atkins plan induce the use of stored fat for energy production to assist weight loss.

After proteins are hydrolyzed in the stomach and intestines, the individual amino acids are reassembled to new proteins or converted to intermediates in other metabolic pathways. The catabolism of amino acids provides energy when the supply of carbohydrates and lipids is exhausted.

The catabolism of amino acids can be conceptually divided into two parts: the fate of the amino group and the fate of the carbon atoms. As shown in Figure 24.9, amino acid carbon skeletons are converted to pyruvate, acetyl CoA, or various carbon compounds that are part of the citric acid cycle.

24.9A   DEGRADATION OF AMINO ACIDS—THE FATE OF THE AMINO GROUP

The catabolism of carbohydrates and triacylglycerols deals with the oxidation of carbon atoms only. With amino acids, an **amino group (–NH$_2$)** must be metabolized, as well. The catabolism of amino acids begins with the removal of the amino group from the carbon skeleton and the formation of NH$_4^+$ by a two-step pathway: **transamination followed by oxidative deamination.**

▼ **FIGURE 24.9** An Overview of the Catabolism of Amino Acids

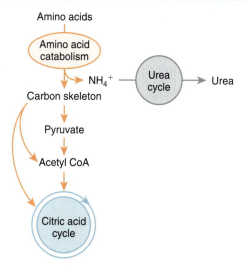

The breakdown of amino acids forms NH_4^+, which enters the urea cycle to form urea, and a carbon skeleton that is metabolized to either pyruvate, acetyl CoA, or an intermediate in the citric acid cycle.

- *Transamination* is the transfer of an amino group from an amino acid to an α-keto acid, usually α-ketoglutarate.

The C–H and C–NH_3^+ groups are replaced by a C=O.

$$R-\overset{\overset{+}{NH_3}}{\underset{H}{C}}-CO_2^- \ + \ R'-\overset{O}{\overset{\parallel}{C}}-CO_2^- \ \xrightarrow{\text{transaminase}} \ R-\overset{O}{\overset{\parallel}{C}}-CO_2^- \ + \ R'-\overset{\overset{+}{NH_3}}{\underset{H}{C}}-CO_2^-$$

amino acid   α-keto acid   α-keto acid   amino acid

The C=O is replaced by C–H and C–NH_3^+ groups.

In transamination, the amino group of the amino acid and the ketone carbonyl oxygen of the α-keto acid are interchanged to form a new amino acid and a new α-keto acid. As an example, transfer of an amino group from alanine to α-ketoglutarate forms pyruvate and glutamate, the completely ionized form of the amino acid glutamic acid.

$$CH_3-\overset{\overset{+}{NH_3}}{\underset{H}{C}}-CO_2^- \ + \ {}^-O_2CCH_2CH_2-\overset{O}{\overset{\parallel}{C}}-CO_2^- \ \xrightarrow{\text{transaminase}} \ CH_3-\overset{O}{\overset{\parallel}{C}}-CO_2^- \ + \ {}^-O_2CCH_2CH_2-\overset{\overset{+}{NH_3}}{\underset{H}{C}}-CO_2^-$$

alanine   α-ketoglutarate   pyruvate   glutamate (ionic form of glutamic acid)

The amino group is removed from the amino acid.

Transamination removes the amino group to form a carbon skeleton that contains only carbon, hydrogen, and oxygen atoms. These products are then degraded along other catabolic pathways as described in Section 24.9B.

SAMPLE PROBLEM 24.4   What products are formed in the following transamination reaction?

$$(CH_3)_2CH-\overset{\overset{+}{NH_3}}{\underset{H}{C}}-CO_2^- \;+\; {}^-O_2CCH_2CH_2-\overset{O}{\overset{\|}{C}}-CO_2^- \;\xrightarrow{\text{transaminase}}$$

$$\text{valine} \qquad\qquad\qquad \text{α-ketoglutarate}$$

ANALYSIS   To draw the products of transamination, convert the C—H and C—NH$_3^+$ groups of the amino acid to a C=O, forming an α-keto acid.

SOLUTION

new C—NH$_3^+$ bond

$$(CH_3)_2CH-\overset{\overset{+}{NH_3}}{\underset{H}{C}}-CO_2^- \;+\; {}^-O_2CCH_2CH_2-\overset{O}{\overset{\|}{C}}-CO_2^- \;\xrightarrow{\text{transaminase}}\; (CH_3)_2CH-\overset{O}{\overset{\|}{C}}-CO_2^- \;+\; {}^-O_2CCH_2CH_2-\overset{\overset{+}{NH_3}}{\underset{H}{C}}-CO_2^-$$

$$\text{valine} \qquad\qquad \text{α-ketoglutarate} \qquad\qquad\qquad\qquad\qquad \text{α-keto acid} \qquad\qquad\qquad\qquad \text{glutamate}$$

The amino group is removed from the amino acid.

PROBLEM 24.30   Draw the products formed when each amino acid undergoes transamination with α-ketoglutarate.

a. $HOCH_2-\overset{\overset{+}{NH_3}}{\underset{H}{C}}-CO_2^-$ b. $CH_3SCH_2CH_2-\overset{\overset{+}{NH_3}}{\underset{H}{C}}-CO_2^-$ c. $HSCH_2-\overset{\overset{+}{NH_3}}{\underset{H}{C}}-CO_2^-$

serine   methionine   cysteine

The glutamate formed by transamination is then degraded by oxidative deamination using NAD$^+$.

- In oxidative deamination, the C—H and C—NH$_3^+$ bonds on the α carbon of glutamate are converted to C=O and an ammonium ion, NH$_4^+$.

These bonds are broken.

$$\underset{\text{glutamate}}{{}^-O_2CCH_2CH_2-\overset{\overset{+}{NH_3}}{\underset{H}{\underset{\alpha}{C}}}-CO_2^-} \;+\; H_2O \;\xrightarrow[\substack{\text{glutamate}\\\text{dehydrogenase}}]{NAD^+ \quad NADH+H^+}\; \underset{\text{α-ketoglutarate}}{{}^-O_2CCH_2CH_2-\overset{O}{\overset{\|}{C}}-CO_2^-} \;+\; NH_4^+$$

enters the urea cycle

This product can be recycled for another transamination.

In oxidative deamination, glutamate is re-converted to α-ketoglutarate, which can undergo transamination with another molecule of an amino acid and the cycle repeats. In this way, α-ketoglutarate removes an amino group from an amino acid and then loses that amino group as NH$_4^+$ in a second step.

The ammonium ion then enters the urea cycle, where it is converted by a multistep pathway to **urea, (NH$_2$)$_2$C=O**, in the liver. Urea is then transported to the kidneys and excreted in urine.

- **The overall result of transamination and oxidative deamination is to remove an amino group from an amino acid and form an ammonium ion, NH$_4^+$.**

$$\underset{\text{amino acid}}{R-\overset{\overset{+}{NH_3}}{\underset{H}{C}}-CO_2^-} \;\xrightarrow[\text{[2] oxidative deamination}]{\text{[1] transamination}}\; \underset{\text{α-keto acid}}{R-\overset{O}{\overset{\|}{C}}-CO_2^-} \;+\; NH_4^+$$

SAMPLE PROBLEM 24.5

What final products are formed when leucine is subjected to transamination followed by oxidative deamination?

$$(CH_3)_2CHCH_2-\overset{\overset{+}{N}H_3}{\underset{H}{C}}-CO_2^-$$

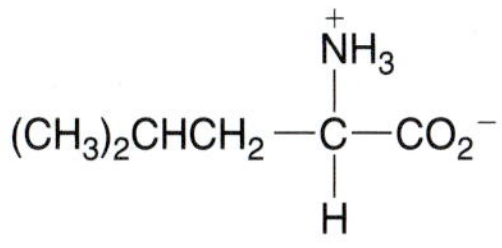

leucine

ANALYSIS To draw the organic product, replace the C—H and C—NH_3^+ on the α carbon of the amino acid by C=O. NH_4^+ is formed from the amino group.

SOLUTION

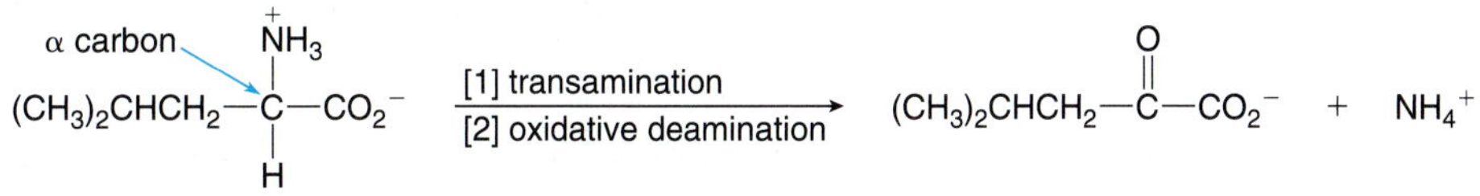

α carbon
$$(CH_3)_2CHCH_2-\overset{\overset{+}{N}H_3}{\underset{H}{C}}-CO_2^- \xrightarrow[\text{[2] oxidative deamination}]{\text{[1] transamination}} (CH_3)_2CHCH_2-\overset{O}{\overset{\|}{C}}-CO_2^- + NH_4^+$$

PROBLEM 24.31

What products are formed when each amino acid is subjected to transamination followed by oxidative deamination: (a) threonine; (b) glycine; (c) isoleucine? Use the structures in Table 21.2.

24.9B DEGRADATION OF AMINO ACIDS—THE FATE OF THE CARBON SKELETON

Once the nitrogen has been removed from an amino acid, the carbon skeletons of individual amino acids are catabolized in a variety of ways. There are three common fates of the carbon skeletons of amino acids, shown in Figure 24.10:

- conversion to pyruvate, $CH_3COCO_2^-$
- conversion to acetyl CoA, $CH_3COSCoA$
- conversion to an intermediate in the citric acid cycle

▼ **FIGURE 24.10 Amino Acid Catabolism**

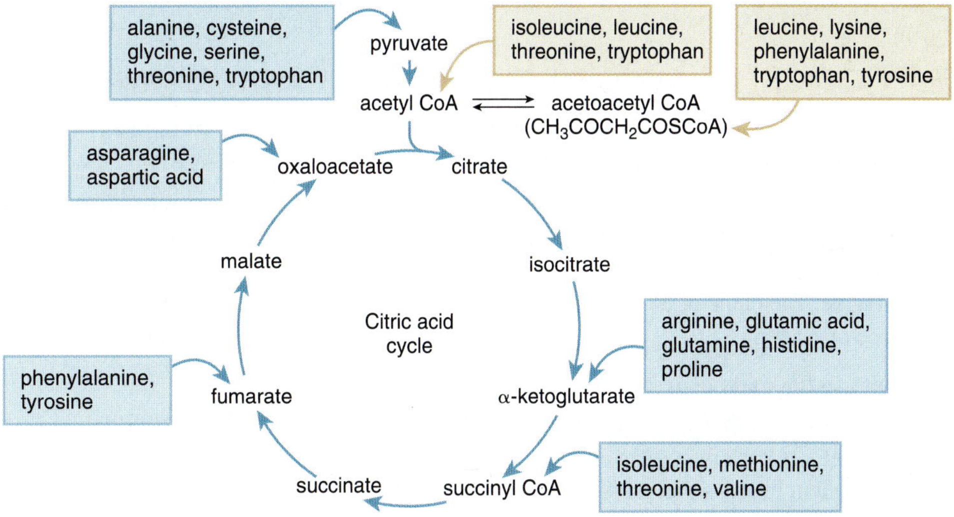

Glucogenic amino acids are highlighted in blue, while ketogenic amino acids are highlighted in tan. Amino acids that appear more than once in the scheme can be degraded by multiple routes.

Some amino acids such as alanine (Section 24.9A) are catabolized to pyruvate. Since pyruvate is an intermediate in both glycolysis and gluconeogenesis, pyruvate can be broken down for energy or used to synthesize glucose. In considering catabolism, amino acids are often divided into two groups.

- *Glucogenic* amino acids are catabolized to pyruvate or an intermediate in the citric acid cycle. Since these catabolic products can be used for gluconeogenesis, glucogenic amino acids can be used to synthesize glucose.

- *Ketogenic* amino acids are converted to acetyl CoA, or the related thioester acetoacetyl CoA, $CH_3COCH_2COSCoA$. These catabolic products cannot be used to synthesize glucose, but they can be converted to ketone bodies and yield energy by this path.

We will not examine the specific pathways that convert the carbon skeletons of individual amino acids into other products. Figure 24.10 illustrates where each amino acid feeds into the metabolic pathways we have already discussed.

PROBLEM 24.32   What metabolic intermediate is produced from the carbon atoms of each amino acid?

a. cysteine   b. aspartic acid   c. valine   d. threonine

CHAPTER HIGHLIGHTS

KEY TERMS

Acyl CoA (24.7)
Cori cycle (24.6)
Decarboxylation (24.2)
Fermentation (24.4)
Glucogenic amino acid (24.9)
Gluconeogenesis (24.6)

Glycolysis (24.3)
Isomerase (24.2)
Ketogenesis (24.8)
Ketogenic amino acid (24.9)
Ketone bodies (24.8)
Ketosis (24.8)

Kinase (24.2)
β-Oxidation (24.7)
Oxidative deamination (24.9)
Transamination (24.9)
Urea cycle (24.9)

KEY CONCEPTS

❶ **What are the main elements that provide clues to the outcome of a biochemical reaction? (24.2)**

- To understand the course of a biochemical reaction, examine the functional groups that are added or removed, the reagents (coenzymes or other materials), and the enzyme. The name of an enzyme is often a clue as to the type of reaction.

❷ **What are the main aspects of glycolysis? (24.3)**

- Glycolysis is a linear, 10-step pathway that converts glucose to two three-carbon pyruvate molecules. In the energy-investment phase, steps [1]–[5], the energy from two ATP molecules is used for phosphorylation and two three-carbon products are formed. In the energy-generating phase, steps [6]–[10], the following species are generated: two pyruvate molecules ($CH_3COCO_2^-$), 2 NADHs, and 4 ATPs.
- The net result of glycolysis, considering both phases, is 2 $CH_3COCO_2^-$, 2 NADHs, and 2 ATPs.

❸ **What are the major pathways for pyruvate metabolism? (24.4)**

- When oxygen is plentiful, pyruvate is converted to acetyl CoA, which can enter the citric acid cycle.
- When the oxygen level is low, the anaerobic metabolism of pyruvate forms lactate and NAD^+.
- In yeast and other microorganisms, pyruvate is converted to ethanol and CO_2 by fermentation.

❹ **How much ATP is formed by the complete catabolism of glucose? (24.5)**

- To calculate the amount of ATP formed in the catabolism of glucose, we must take into account the ATP yield from glycolysis, the oxidation of two molecules of pyruvate to two molecules of acetyl CoA, the citric acid cycle, and oxidative phosphorylation.
- As shown in Figure 24.6, the complete catabolism of glucose forms six CO_2 molecules and 32 molecules of ATP.

⑤ **What are the main features of gluconeogenesis? (24.6)**

- Gluconeogenesis is the synthesis of glucose from noncarbohydrate sources—lactate, amino acids, or glycerol. Gluconeogenesis converts two molecules of pyruvate to glucose. Conceptually, gluconeogenesis is the reverse of glycolysis, but three steps in gluconeogenesis require different enzymes.
- Gluconeogenesis occurs when the body has depleted its supplies of glucose and stored glycogen, and occurs during sustained physical exercise and fasting.

⑥ **Describe the main features of the β-oxidation of fatty acids. (24.7)**

- β-Oxidation is a spiral metabolic pathway that sequentially cleaves two-carbon acetyl CoA units from an acyl CoA derived from a fatty acid. Each cycle of β-oxidation consists of a four-step sequence that forms one molecule each of acetyl CoA, NADH, and FADH$_2$.

⑦ **How much ATP is formed from complete fatty acid oxidation? (24.7)**

- To determine the ATP yield from the complete catabolism of a fatty acid, we must consider the ATP used up in the synthesis of the acyl CoA, the ATP generated from

coenzymes produced during β-oxidation, and the ATP that results from the catabolism of each acetyl CoA.

- As an example, the complete catabolism of stearic acid, C$_{18}$H$_{36}$O$_2$, yields 120 ATPs.

⑧ **What are ketone bodies and how do they play a role in metabolism? (24.8)**

- Ketone bodies—acetoacetate, β-hydroxybutyrate, and acetone—are formed when acetyl CoA levels exceed the capacity of the citric acid cycle. Ketone bodies can be reconverted to acetyl CoA and metabolized for energy. When the level of ketone bodies is high, the pH of the blood can be lowered, causing ketoacidosis.

⑨ **What are the main features of amino acid catabolism? (24.9)**

- The catabolism of amino acids involves two parts. First, the amino group is removed by transamination followed by oxidative deamination. The NH$_4^+$ ion formed enters the urea cycle where it is converted to urea and eliminated in urine. The carbon skeletons of the amino acids are catabolized by a variety of pathways to yield pyruvate, acetyl CoA, or an intermediate in the citric acid cycle.

PROBLEMS

Selected in-chapter and end-of-chapter problems have brief answers provided in Appendix B.

Analyzing Biochemical Reactions

24.33 Analyze each reaction by considering the functional groups that change, the coenzyme or other reactant, and the name of the enzyme.

a.
$$CH_3-\overset{OH}{\underset{}{CH}}-CH_2-\overset{O}{\underset{}{C}}-SCoA \xrightarrow[\substack{\beta\text{-hydroxyacyl CoA} \\ \text{dehydrogenase}}]{NAD^+ \quad NADH + H^+} CH_3-\overset{O}{\underset{}{C}}-CH_2-\overset{O}{\underset{}{C}}-SCoA$$

b.
$$^-O_2CCH_2-\overset{O}{\underset{}{C}}-CO_2^- \xrightarrow[\substack{\text{phosphoenolpyruvate} \\ \text{carboxykinase}}]{GTP \quad GDP} CH_2=\overset{PO_3^{2-}}{\underset{}{C}}-\overset{O}{\underset{}{C}}-O^- + CO_2$$

24.34 Analyze each reaction by considering the functional groups that change, the coenzyme or other reactant, and the name of the enzyme.

a.
$$\underset{\text{3-phosphoglycerate}}{\overset{PO_3^{2-}\quad OH\quad O}{O-CH_2-CH-C-O^-}} \xrightarrow[\substack{\text{phosphoglycerate} \\ \text{kinase}}]{ATP \quad ADP} \underset{\text{1,3-bisphosphoglycerate}}{\overset{PO_3^{2-}\quad OH\quad O}{O-CH_2-CH-C-O-PO_3^{2-}}}$$

b.
$$CH_3-\overset{O}{\underset{}{C}}-CO_2^- + CO_2 \xrightarrow[\text{pyruvate carboxylase}]{ATP \quad ADP + HPO_4^{2-}} \ ^-O_2CCH_2-\overset{O}{\underset{}{C}}-CO_2^-$$

24.35   In the conversion of glucose to stored glycogen, glucose 6-phosphate is converted to glucose 1-phosphate by the enzyme phosphoglucomutase. Suggest a reason for the name of this enzyme.

24.36   When mannose is metabolized to an intermediate in glycolysis, mannose 6-phosphate is converted to fructose 6-phosphate by phosphomannose isomerase. (a) Why is an isomerase enzyme used for this reaction? (b) What type of isomers do mannose 6-phosphate and fructose 6-phosphate represent?

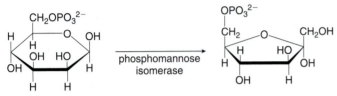

mannose 6-phosphate   fructose 6-phosphate

Glucose Metabolism

24.37   Compare the energy-investment phase and the energy-generating phase of glycolysis with regards to each of the following: (a) the reactant that begins the phase and the final product formed; (b) the amount of ATP used or formed; (c) the number of reduced coenzymes used or formed.

24.38   Compare glycolysis and gluconeogenesis with regards to the reactant that begins the pathway and the product formed at the end of the pathway.

24.39   Considering the individual steps in glycolysis:
a. Which steps form ATP?
b. Which steps use ATP?
c. Which steps form a reduced coenzyme?
d. Which step breaks a C—C bond?

24.40   Explain the role of the coenzymes NAD^+ and NADH in each reaction.
a. pyruvate $\rightarrow$ acetyl CoA
b. pyruvate $\rightarrow$ lactate
c. pyruvate $\rightarrow$ ethanol

24.41   Glucose is completely metabolized to six molecules of CO_2. What specific reactions generate each molecule of CO_2?

24.42   Why is glycolysis described as an anaerobic process?

24.43   Write the overall equation with key coenzymes for each process.
a. glucose $\rightarrow$ pyruvate
b. glucose $\rightarrow$ ethanol
c. pyruvate $\rightarrow$ lactate

24.44   Write the overall equation with key coenzymes for each process.
a. glucose $\rightarrow$ lactate
b. pyruvate $\rightarrow$ ethanol
c. pyruvate $\rightarrow$ acetyl CoA

24.45   Consider the aerobic and anaerobic avenues of pyruvate metabolism in the human body.
a. Where do the carbon atoms of pyruvate end up in each pathway?
b. What coenzymes are used and formed?

24.46   What is the main purpose for the conversion of pyruvate to lactate under anaerobic conditions?

24.47   What is the difference between glycolysis and gluconeogenesis?

24.48   Why is it incorrect to say that gluconeogenesis is simply the reverse of glycolysis?

24.49   What effect is observed on the rate of glycolysis under each of the following conditions?
a. low ATP concentration   c. high carbohydrate diet
b. low ADP concentration   d. low carbohydrate diet

24.50   What effect is observed on the rate of gluconeogenesis under each of the following conditions?
a. low ATP concentration   c. high carbohydrate diet
b. low ADP concentration   d. low carbohydrate diet

24.51   Describe the steps in the Cori cycle. In what organs do key reactions of the cycle occur?

24.52   How does a defect in monosaccharide metabolism result in galactosemia?

24.53   What metabolic products are formed from pyruvate in each case: (a) anaerobic conditions in the body; (b) anaerobic conditions in yeast; (c) aerobic conditions?

24.54   Why must the NADH produced in glycolysis be re-oxidized to NAD^+? How is this accomplished aerobically? How is this accomplished anaerobically?

24.55   Explain in detail how 32 ATP molecules are generated during the complete catabolism of glucose to CO_2.

24.56   What are the starting material and product of each step of gluconeogenesis: (a) step [2]; (b) step [5]; (c) step [8]?

24.57   In fermentation, where do the six carbon atoms of glucose end up?

24.58   In what ways is alcohol fermentation similar to lactate production in muscle?

Triacylglycerol Metabolism

24.59   How much ATP is used or formed when a fatty acid is converted to an acyl CoA? Explain your reasoning.

24.60   How much ATP is ultimately generated from each cycle of β-oxidation of a fatty acid?

24.61   For each carboxylic acid: [1] How many molecules of acetyl CoA are formed from complete β-oxidation? [2] How many cycles of β-oxidation are needed for complete oxidation?
a. hexanoic acid, $CH_3(CH_2)_4CO_2H$
b. myristic acid, $C_{13}H_{27}CO_2H$

24.62 For each carboxylic acid: [1] How many molecules of acetyl CoA are formed from complete β-oxidation? [2] How many cycles of β-oxidation are needed for complete oxidation?
   a. octanoic acid, $CH_3(CH_2)_6CO_2H$
   b. oleic acid, $C_{17}H_{33}CO_2H$

24.63 How much ATP is generated by the complete catabolism of the carboxylic acids in Problem 24.61?

24.64 How much ATP is generated by the complete catabolism of the carboxylic acids in Problem 24.62?

24.65 Draw the structure of the acyl CoA formed from palmitic acid, $C_{15}H_{31}CO_2H$. Use this acyl CoA to write out the four steps of β-oxidation.

24.66 Draw the structure of the acyl CoA formed from hexanoic acid, $C_5H_{11}CO_2H$. Use this acyl CoA to write out the four steps of β-oxidation.

24.67 Why is step [1] of β-oxidation considered an oxidation reaction even though the product has no additional C—O bonds?

24.68 Why is the enzyme for step [2] of β-oxidation called a hydratase? What type of reaction does this enzyme catalyze?

24.69 Consider decanoic acid, $C_9H_{19}CO_2H$.
   a. Label the α and β carbons.
   b. Draw the acyl CoA derived from this fatty acid.
   c. How many acetyl CoA molecules are formed by complete β-oxidation?
   d. How many cycles of β-oxidation are needed for complete oxidation?
   e. How many molecules of ATP are formed from the complete catabolism of this fatty acid?

24.70 Consider docosanoic acid, $C_{21}H_{43}CO_2H$.
   a. Label the α and β carbons.
   b. Draw the acyl CoA derived from this fatty acid.
   c. How many acetyl CoA molecules are formed by complete β-oxidation?
   d. How many cycles of β-oxidation are needed for complete oxidation?
   e. How many molecules of ATP are formed from the complete catabolism of this fatty acid?

24.71 How many moles of ATP per gram of fatty acid are formed from the complete catabolism of decanoic acid (molar mass 172 g/mol) in Problem 24.69?

24.72 How many moles of ATP per gram of fatty acid are formed from the complete catabolism of docosanoic acid (molar mass 341 g/mol) in Problem 24.70?

Ketone Bodies

24.73 What is the difference between ketosis and ketogenesis?

24.74 How is the production of ketone bodies related to ketoacidosis?

24.75 Why are more ketone bodies produced in an individual whose diabetes is poorly managed?

24.76 Why do some individuals use test strips to measure the presence of ketone bodies in their urine?

Amino Acid Metabolism

24.77 What is the difference between ketogenic and glucogenic amino acids?

24.78 What is the difference between transamination and oxidative deamination?

24.79 Draw the structure of the α-keto acid formed by the transamination of each amino acid: (a) glycine; (b) phenylalanine.

24.80 Draw the structure of the α-keto acid formed by the transamination of each amino acid: (a) tyrosine; (b) asparagine.

24.81 Draw the products formed in each transamination reaction.

a. $CH_3-\overset{\overset{+}{N}H_3}{\underset{H}{C}}-CO_2^-$ + $^-O_2CCH_2-\overset{O}{\overset{\|}{C}}-CO_2^-$

b. $(CH_3)_2CHCH_2-\overset{\overset{+}{N}H_3}{\underset{H}{C}}-CO_2^-$ + $^-O_2CCH_2-\overset{O}{\overset{\|}{C}}-CO_2^-$

24.82 Draw the products formed in each transamination reaction.

a. $(CH_3)_2CH-\overset{\overset{+}{N}H_3}{\underset{H}{C}}-CO_2^-$ + $^-O_2CCH_2-\overset{O}{\overset{\|}{C}}-CO_2^-$

b. $CH_3CH_2CH(CH_3)-\overset{\overset{+}{N}H_3}{\underset{H}{C}}-CO_2^-$ + $^-O_2CCH_2CH_2-\overset{O}{\overset{\|}{C}}-CO_2^-$

24.83 What products are formed when each amino acid undergoes oxidative deamination with NAD^+ and a dehydrogenase enzyme: (a) leucine; (b) phenylalanine?

24.84 What products are formed when each amino acid undergoes oxidative deamination with NAD^+ and a dehydrogenase enzyme: (a) tyrosine; (b) tryptophan?

24.85 What metabolic intermediate is formed from the carbon skeleton of each amino acid?
   a. phenylalanine   c. asparagine
   b. glutamic acid   d. glycine

24.86 What metabolic intermediate is formed from the carbon skeleton of each amino acid?
   a. lysine   c. methionine
   b. tryptophan   d. serine

24.87   Can an amino acid be considered both glucogenic and ketogenic? Explain your choice.

24.88   Which amino acids are classified as only glucogenic?

General Questions and Applications

24.89   What is the difference between a spiral and a cyclic metabolic pathway? Give an example of each.

24.90   What is the cause of the pain and cramping in a runner's muscles?

24.91   Explain the reaction that occurs during the baking of bread that causes the bread to rise.

24.92   Why is the concentration of ethanol in wine typically 12% or less?

24.93   How might pyruvate be metabolized in the cornea, tissue that has little blood supply?

24.94   How is pyruvate metabolized in red blood cells, which contain no mitochondria?

24.95   Why is the Atkins low carbohydrate diet called a ketogenic diet?

24.96   What metabolic conditions induce ketogenesis?

24.97   What type of enzyme would catalyze the conversion of hydroxyacetone (CH_3COCH_2OH) to each compound?

a. 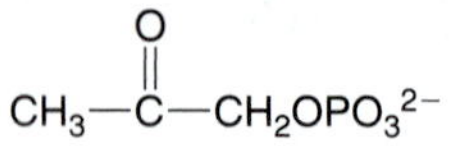 $CH_3-\overset{\overset{\displaystyle O}{\|}}{C}-CH_2OPO_3^{2-}$

b. $CH_3-\overset{\overset{\displaystyle O}{\|}}{C}-CHO$

c. $HO_2CCH_2-\overset{\overset{\displaystyle O}{\|}}{C}-CH_2OH$

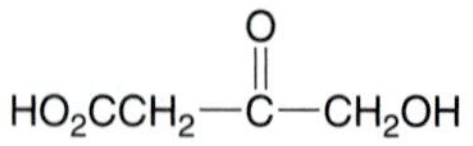

24.98   What coenzyme would convert 6-hydroxy-2-hexanone ($CH_3COCH_2CH_2CH_2OH$) to each compound?

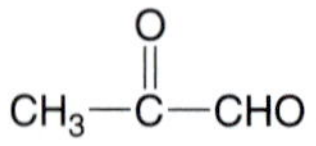

a. 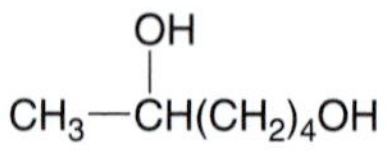 $CH_3-\overset{\overset{\displaystyle OH}{|}}{CH}(CH_2)_4OH$

b. 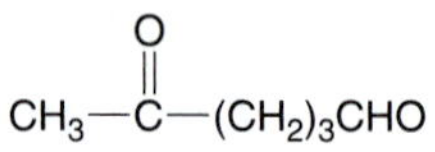 $CH_3-\overset{\overset{\displaystyle O}{\|}}{C}-(CH_2)_3CHO$

c. $CH_3-\overset{\overset{\displaystyle O}{\|}}{C}-CH=CHCH_2CH_2OH$

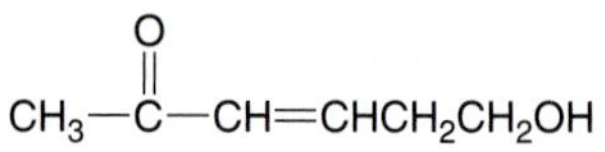

CHALLENGE QUESTIONS

24.99   How much ATP is produced from the complete catabolism of one molecule of glycerol, $HOCH_2CH(OH)CH_2OH$?

24.100 How much ATP is produced from the complete catabolism of one molecule of the given triacylglycerol?

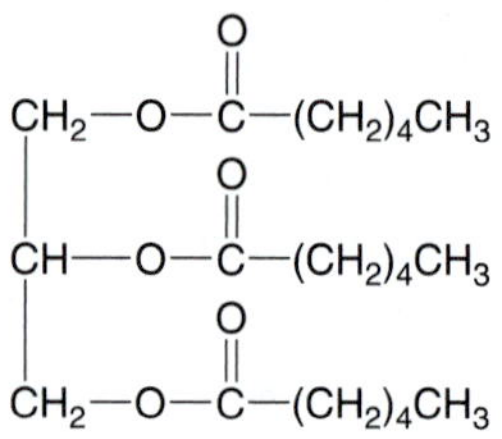

$$CH_2-O-\overset{\overset{\displaystyle O}{\|}}{C}-(CH_2)_4CH_3$$
$$CH-O-\overset{\overset{\displaystyle O}{\|}}{C}-(CH_2)_4CH_3$$
$$CH_2-O-\overset{\overset{\displaystyle O}{\|}}{C}-(CH_2)_4CH_3$$

Appendix A

Useful Mathematical Concepts

Three common mathematical concepts are needed to solve many problems in chemistry:

- Using scientific notation
- Determining the number of significant figures
- Using a scientific calculator

SCIENTIFIC NOTATION

To write numbers that contain many leading zeros (at the beginning) or trailing zeros (at the end), scientists use **scientific notation.**

- In scientific notation, a number is written as $y \times 10^x$, where y (the coefficient) is a number between 1 and 10, and x is an exponent, which can be any positive or negative whole number.

To convert a standard number to scientific notation:

1. Move the decimal point to give a number between 1 and 10.
2. Multiply the result by 10^x, where x is the number of places the decimal point was moved.
   - If the decimal point is moved to the **left,** x is **positive.**
   - If the decimal point is moved to the **right,** x is **negative.**

$$2822. \quad = \quad 2.822 \times 10^3 \qquad \text{the number of places the decimal point was moved to the left}$$

Move the decimal point three places to the left.

$$0.000\ 004\ 5 \quad = \quad 4.5 \times 10^{-6} \qquad \text{the number of places the decimal point was moved to the right}$$

Move the decimal point six places to the right.

To convert a number in scientific notation to a standard number, use the value of x in 10^x to indicate the number of places to move the decimal point in the coefficient.

- Move the decimal point to the **right** when x is **positive.**
- Move the decimal point to the **left** when x is **negative.**

$$2.521 \times 10^2 \qquad\qquad 2.521 \quad ----\rightarrow \quad 252.1$$

Move the decimal point to the right two places.

$$2.68 \times 10^{-2} \qquad\qquad 00268 \quad ----\rightarrow \quad 0.0268$$

Move the decimal point to the left two places.

Table A.1 shows how several numbers are written in scientific notation.

TABLE A.1 Numbers in Standard Form and Scientific Notation

Number	Scientific Notation
26,200	2.62×10^4
0.006 40	6.40×10^{-3}
3,000,000	3×10^6
0.000 000 139	1.39×10^{-7}
2,000.20	2.00020×10^3

Often, numbers written in scientific notation must be multiplied or divided.

- **To multiply two numbers in scientific notation, *multiply* the coefficients together and *add* the exponents in the powers of 10.**

Add exponents. (8 + 3)

$$(3.0 \times 10^8) \times (2.0 \times 10^3) = 6.0 \times 10^{11}$$

Multiply coefficients. (3.0 × 2.0)

- **To divide two numbers in scientific notation, *divide* the coefficients and *subtract* the exponents in the powers of 10.**

Divide coefficients. (6.0/2.0)

$$\frac{6.0 \times 10^6}{2.0 \times 10^{10}} = 3.0 \times 10^{-4}$$

Subtract exponents. (6 − 10)

Table A.2 shows the result of multiplying or dividing several numbers written in scientific notation.

TABLE A.2 Calculations Using Numbers Written in Scientific Notation

Calculation	Answer
$(3.5 \times 10^3) \times (2.2 \times 10^{22}) =$	7.7×10^{25}
$(3.5 \times 10^3)/(2.2 \times 10^{22}) =$	1.6×10^{-19}
$(3.5 \times 10^3) \times (2.2 \times 10^{-10}) =$	7.7×10^{-7}
$(3.5 \times 10^3)/(2.2 \times 10^{-10}) =$	1.6×10^{13}

SIGNIFICANT FIGURES

Whenever we measure a number, there is a degree of uncertainty associated with the result. The last number (furthest to the right) is an estimate. **Significant figures** are all of the digits in a measured number including one estimated digit. How many significant figures are contained in a number?

- All nonzero digits are always significant.
- A zero *counts* as a significant figure when it occurs between two nonzero digits, or at the end of a number with a decimal point.
- A zero does *not* count as a significant figure when it occurs at the beginning of a number, or at the end of a number that does not have a decimal point.

Table A.3 lists the number of significant figures in several quantities.

TABLE A.3	Examples Illustrating Significant Figures		
Quantity	Number of Significant Figures	Quantity	Number of Significant Figures
1,267 g	Four	203 L	Three
24,345 km	Five	6.10 atm	Three
1.200 mg	Four	0.3040 g	Four
0.000 001 mL	One	1,200 m	Two

The number of significant figures must also be taken into account in calculations. To avoid reporting a value with too many digits, we must often **round off the number** to give the correct number of significant figures. Two rules are used in rounding off numbers.

- If the first number that must be dropped is 4 or less, drop it and all remaining numbers.
- If the first number that must be dropped is 5 or greater, *round the number up* by adding one to the last digit that will be retained.

To round 63.854 to two significant figures:

These digits must be retained.

$\overline{63}.854$ first digit to be dropped

These digits must be dropped.

- Since the first digit to be dropped is 8 (a 5 or greater), add 1 to the first digit to its left.
- The number 63.854 rounded to two digits is **64**.

Table A.4 gives other examples of rounding off numbers.

TABLE A.4	Rounding Off Numbers	
Original Number	Rounded to	Rounded Number
15.2538	Two places	15
15.2538	Three places	15.3
15.2538	Four places	15.25
15.2538	Five places	15.254

The first number to be dropped is indicated in red in each original number.

The number of significant figures in the answer of a problem depends on the type of mathematical calculation—multiplication (and division) or addition (and subtraction).

- **In multiplication and division, the answer has the same number of significant figures as the original number with the *fewest* significant figures.**

five significant figures   first digit to be dropped

$$5.5067 \quad \times \quad 2.6 \quad = \quad 14.31742 \text{ rounded to } \mathbf{14}$$

two significant figures

The answer must contain only **two** significant figures.

- **In addition and subtraction, the answer has the same number of decimal places as the original number with the *fewest* decimal places.**

two digits after the decimal point

$$10.17 \quad + \quad 3.5 \quad = \quad 13.67 \text{ rounded to } \mathbf{13.7}$$

one digit after the decimal point

last significant digit   The answer can have only **one** digit after the decimal point.

Table A.5 lists other examples of calculations that take into account the number of significant figures.

TABLE A.5   Calculations Using Significant Figures

Calculation	Answer
$3.2 \times 699 =$	2,236.8 rounded to **2,200**
$4.66892/2.13 =$	2.191981221 rounded to **2.19**
$25.3 + 3.668 + 29.1004 =$	58.0684 rounded to **58.1**
$95.1 - 26.335 =$	68.765 rounded to **68.8**

USING A SCIENTIFIC CALCULATOR

A scientific calculator is capable of carrying out more complicated mathematical functions than simple addition, subtraction, multiplication, and division. For example, these calculators allow the user to convert a standard number to scientific notation, as well as readily determine the logarithm (log) or antilogarithm (antilog) of a value. Carrying out these operations is especially important in determining pH or hydronium ion concentration in Chapter 9.

Described in this section are the steps that can be followed in calculations with some types of calculators. Consult your manual if these steps do not produce the stated result.

CONVERTING A NUMBER TO SCIENTIFIC NOTATION

To convert a number, such as 1,200, from its standard form to scientific notation:

- Enter 1200.
- Press 2nd and then SCI.
- The number will appear as 1.2^{03}, indicating that $1{,}200 = 1.2 \times 10^3$.

ENTERING A NUMBER WRITTEN IN SCIENTIFIC NOTATION

To enter a number written in scientific notation with a positive exponent, such as 1.5×10^8:

- Enter 1.5.
- Press EE.
- Enter 8.
- The number will appear as 1.5^{08}, indicating that it is equal to 1.5×10^8.

To enter a number written in scientific notation with a negative exponent, such as 3.5×10^{-4}:

- Enter 3.5.
- Press EE.
- Enter 4.
- Press CHANGE SIGN ($+ \rightarrow -$).
- The number will appear as 3.5^{-04}, indicating that it is equal to 3.5×10^{-4}.

TAKING THE LOGARITHM OF A NUMBER: CALCULATING pH FROM A KNOWN [H₃O⁺]

Since $pH = -\log [H_3O^+]$, we must learn how to calculate logarithms on a calculator in order to determine pH values. To determine the pH from a known hydronium ion concentration, say $[H_3O^+] = 1.8 \times 10^{-5}$, carry out the following steps:

- Enter 1.8×10^{-5} (Enter 1.8; press EE; enter 5; press CHANGE SIGN). The number 1.8^{-05} will appear.
- Press LOG.
- Press CHANGE SIGN ($+ \rightarrow -$).
- The number 4.744 727 495 will appear. Since the coefficient, 1.8, contains two significant figures, round the logarithm to 4.74, which has two digits to the right of the decimal point. Thus, the pH of the solution is 4.74.

TAKING THE ANTILOGARITHM OF A NUMBER: CALCULATING [H₃O⁺] FROM A KNOWN pH

Since $[H_3O^+] = \text{antilog}(-pH)$, we must learn how to calculate an antilogarithm—that is, the number that has a given logarithm value—using a calculator. To determine the hydronium ion concentration from a given pH, say 3.91, carry out the following steps:

- Enter 3.91.
- Press CHANGE SIGN ($+ \rightarrow -$).
- Press 2nd and then LOG.
- The number 0.000 123 027 will appear. To convert this number to scientific notation, press 2nd and SCI.
- The number $1.230\ 268\ 771^{-04}$ will appear, indicating that $[H_3O^+] = 1.230\ 268\ 771 \times 10^{-4}$. Since the original pH (a logarithm) had two digits to the right of the decimal point, the answer must have two significant figures in the coefficient in scientific notation. As a result, $[H_3O^+] = 1.2 \times 10^{-4}$.

Table A.6 lists pH values that correspond to given $[H_3O^+]$ values. You can practice using a calculator to determine pH or $[H_3O^+]$ by entering a value in one column, following the listed steps, and then checking to see if you obtain the corresponding value in the other column.

TABLE A.6	The pH of a Solution from a Given Hydronium Ion Concentration $[H_3O^+]$		
$[H_3O^+]$	pH	$[H_3O^+]$	pH
1.8×10^{-10}	9.74	4.0×10^{-13}	12.40
3.8×10^{-2}	1.42	6.6×10^{-4}	3.18
5.0×10^{-12}	11.30	2.6×10^{-9}	8.59
4.2×10^{-7}	6.38	7.3×10^{-8}	7.14

Appendix B

Selected Answers to In-Chapter and End-of-Chapter Problems

CHAPTER 1

1.1 water, rock, tree: natural
    tennis ball, CD, plastic bag: synthetic

1.3 b,c,e: chemical   a,d: physical

1.5 a,b,c: mixture   d: pure substance

1.7 a. megaliter   b. millisecond   c. centigram   d. deciliter

1.9 1,000,000,000

1.11 100

1.12 a. 4   b. 5   c. 2   d. 4   e. 3   f. 3   g. 7   h. 2

1.13 a. 4   b. 5   c. 6   d. 4   e. 5   f. 4   g. 4   h. 6

1.15 a. 1.3   b. 0.0025   c. 3,800

1.16 a. 37   b. 0.000 007 93   c. 32   d. 3,100

1.17 a. 28.0 cm   b. 127.5 mL   c. 30 mg   d. 2.08 s

1.18 9.8×10^{-5} g/dL

1.19 a. 9.32×10^{4} c. 6.78×10^{6} e. 4.52×10^{12}
    b. 7.25×10^{-4} d. 3.0×10^{-5} f. 2.8×10^{-11}

1.20 6,020,000,000,000,000,000,000 molecules

1.21 a. 6,500   c. 0.037 80   e. 2,221,000
    b. 0.000 032 6   d. 104,000,000   f. 0.000 000 000 45

1.22 a. $\dfrac{0.621 \text{ mi}}{1 \text{ km}}$ $\dfrac{1 \text{ km}}{0.621 \text{ mi}}$ c. $\dfrac{454 \text{ g}}{1 \text{ lb}}$ $\dfrac{1 \text{ lb}}{454 \text{ g}}$
    b. $\dfrac{1000 \text{ mm}}{1 \text{ m}}$ $\dfrac{1 \text{ m}}{1000 \text{ mm}}$ d. $\dfrac{1000 \text{ µg}}{1 \text{ mg}}$ $\dfrac{1 \text{ mg}}{1000 \text{ µg}}$

1.23 2,560 mi

1.25 a. 250 dL   b. 1,140 g   c. 81 cm   d. 100 mm

1.26 a. 1.91 km   b. 0.7 L   c. 140 cm

1.27 a. 0.018 mi/s   b. 29 m/s

1.28 1.5 tsp

1.29 4 tablets

1.31 83.3 °F or 301.5 K

1.33 7.13 g

1.35 chloroform

1.37 An element is a pure substance that cannot be broken down into simpler substances by a chemical reaction. A compound is a pure substance formed by combining two or more elements.

1.39

Phase	a. Volume	b. Shape	c. Organization	d. Particle Proximity
Solid	Definite	Definite	Very organized	Very close
Liquid	Definite	Assumes shape of container	Less organized	Close
Gas	Not fixed	None	Disorganized	Far apart

1.41 a. physical   b. chemical   c. physical

1.43 This is a physical change since the compound CO_2 is unchanged in this transition. The same "particles" exist at the beginning and end of the process.

1.45 An exact number results from counting objects or is part of a definition, such as having 20 people in a class. An inexact number results from a measurement or observation and contains some uncertainty, such as the distance from the earth to the sun, 9.3×10^{7} miles.

1.47 a. 5 dL   b. 10 mg   c. 5 cm   d. 10 Ms

1.49 a. 4   b. 2   c. 3   d. 2   e. 3   f. 4   g. 4   h. 2

1.51 a. 25,400   c. 0.001 27   e. 195
    b. 1,250,000   d. 0.123   f. 197

1.53 a. 22   b. 21.97   c. 5.4   d. 50.1   e. 100   f. 1,044

1.55 a. 1.234×10^{3} g   c. 5.244×10^{6} L   e. 4.4×10^{4} km
    b. 1.62×10^{-5} m   d. 5.62×10^{-3} g

1.57 a. 340,000,000   c. 300
    b. 0.000 058 22   d. 0.000 000 068 6

1.59 a. 4.44×10^{3} c. 1.3×10^{8}
    b. 5.6×10^{-5} d. 9.8×10^{-4}

1.61 a. 4.00×10^{-4} g   c. 8.0×10^{-5} g
    b. 2×10^{-3} g   d. 3.4×10^{3} mg

1.63 a. 1×10^{-12} b. one trillion, 1×10^{12}

1.65 a. 300,000 mg   d. 10 oz
    b. 2,000,000 µL   e. 0.6 m
    c. 0.050 m   f. 3.2 m

1.67 a. 106 kg   b. 130 cm   c. 1.4 L   d. 99.9 °F

1.69 a. 946 mL   b. 33.9 fl oz

1.71 a. 1,500 g   b. 3.3 lb   c. 53 oz

1.73 a. 127 °F or 326 K   b. 177 °C or 450 K

1.75 a. −10 °C   b. −50 °F

1.77 Density is the mass per unit volume, usually reported in g/mL or g/cc. Specific gravity is the ratio of the density of a substance to the density of water and has no units.

1.79 1.01 g/mL

1.81 152 g rounded to 150 g

1.83 0.974 kg

1.85 a. heptane   b. olive oil   c. water   d. water

1.87 a. 13.6   b. 0.789 g/mL

1.89 All units of measurement in the metric system are related by a factor of ten.

1.91 a. 100 µL, 100 mL, 100 dL   c. 100 mg, 10 g, 0.1 kg
    b. 1,000 µL, 10 mL, 1 dL   d. 1,000 cm, 100 m, 1 km

1.93   a. 0.186 g/dL   b. 1,860 mg/L
1.95   three
1.97   7.8 (or 8)
1.99   a. $0.67   b. $1.00
1.101   784.6 doses, so 784 full doses
1.103   2 tablets (2.38)
1.105   0.150 g
1.107   0.905 mg rounded to 0.9 mg

CHAPTER 2

2.1   a. Ca   b. Rn   c. N   d. Au
2.3   a. neon   c. iodine   e. boron
      b. sulfur   d. silicon   f. mercury
2.5   As, B, Si: metalloids
      Cr, Co, Cu, Fe, Mn, Mo, Ni, Zn: metals
      F, I, Se: nonmetals
2.7   a. 4 hydrogens, 1 carbon
      b. 3 hydrogens, 1 nitrogen
      c. 6 hydrogens, 2 carbons, 1 oxygen
2.8

	Atomic Number	Element	Protons	Electrons
a.	2	Helium	2	2
b.	11	Sodium	11	11
c.	20	Calcium	20	20
d.	47	Silver	47	47
e.	78	Platinum	78	78

2.9

	Protons	Neutrons	Electrons
a.	17	18	17
b.	14	14	14
c.	92	146	92

2.11   a. 95   b. 52
2.12

	Atomic Number	Mass Number	Protons	Neutrons	Electrons
a.	6	13	6	7	6
b.	51	121	51	70	51

2.13

	Protons	Electrons	Atomic Number	Mass Number
$^{24}_{12}$Mg	12	12	12	24
$^{25}_{12}$Mg	12	12	12	25
$^{26}_{12}$Mg	12	12	12	26

2.14   a. 24.31 amu   b. 50.94 amu
2.15

Element	Period	Group
a. Oxygen	2	6A (or 16)
b. Calcium	4	2A (or 2)
c. Phosphorus	3	5A (or 15)
d. Platinum	6	8B (or 10)
e. Iodine	5	7A (or 17)

2.17   a. titanium, Ti, group 4B (or 4), period 4, transition metal
       b. phosphorus, P, group 5A (or 15), period 3, main group element
       c. dysprosium, Dy, no group number, period 6, inner transition element
2.19   a. neon   b. carbon   c. magnesium
2.21   a. magnesium

1s	2s	2p	3s
↑↓	↑↓	↑↓ ↑↓ ↑↓	↑↓

b. aluminum

1s	2s	2p	3s	3p
↑↓	↑↓	↑↓ ↑↓ ↑↓	↑↓	↑ □ □

c. bromine

1s	2s	2p	3s	3p
↑↓	↑↓	↑↓ ↑↓ ↑↓	↑↓	↑↓ ↑↓ ↑↓

4s	3d	4p
↑↓	↑↓ ↑↓ ↑↓ ↑↓ ↑↓	↑↓ ↑↓ ↑

2.23   a. 12 electrons, 2 valence electrons, magnesium
       b. 15 electrons, 5 valence electrons, phosphorus
       c. 40 electrons, 2 valence electrons, zirconium
       d. 26 electrons, 2 valence electrons, iron
2.24   a. 7 valence electrons: $2s^2 2p^5$
       b. 8 valence electrons: $4s^2 4p^6$
       c. 2 valence electrons: $3s^2$
       d. 4 valence electrons: $4s^2 4p^2$
2.25   Se, selenium: $4s^2 4p^4$
       Te, tellurium: $5s^2 5p^4$
       Po, polonium: $6s^2 6p^4$
2.26   a. :Br·   b. Li·   c. ·Al·   d. ·S̈·   e. :N̈e:
2.27   a. neon, carbon, boron
       b. beryllium, magnesium, calcium
       c. sulfur, silicon, magnesium
       d. neon, krypton, xenon
       e. oxygen, sulfur, silicon
       f. fluorine, sulfur, aluminum
2.29   a. gold, astatine, silver
       b. nitrogen, sodium, nickel
       c. sulfur, silicon, tin
       d. calcium, chromium, chlorine
       e. phosphorus, lead, platinum
       f. titanium, tantalum, thallium
2.31   a,c,e: element   b,d: compound
2.33   a. cesium   d. beryllium
       b. ruthenium   e. fluorine
       c. chlorine   f. cerium
2.35   a. sodium: metal, alkali metal, main group element
       b. silver: metal, transition metal
       c. xenon: nonmetal, noble gas, main group element
       d. platinum: metal, transition metal
       e. uranium: metal, inner transition metal
       f. tellurium: metalloid, main group element

2.37

	Element Symbol	Atomic Number	Mass Number	Number of Protons	Number of Neutrons	Number of Electrons
a.	C	6	12	6	6	6
b.	P	15	31	15	16	15
c.	Zn	30	65	30	35	30
d.	Mg	12	24	12	12	12
e.	I	53	127	53	74	53
f.	Be	4	9	4	5	4
g.	Zr	40	91	40	51	40
h.	S	16	32	16	16	16

2.39 0.000 000 000 000 000 000 000 001 672 6 g

2.41
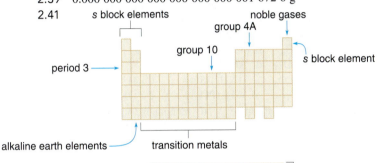

2.43 hydrogen

2.45 silicon, germanium, tin, and lead

2.47 K, Ca, Sc, Ti, V, Cr, Mn, Fe, Co, Ni, Cu, Zn, Ga: metals
Ge, As: metalloids
Se, Br, Kr: nonmetals

2.49 8A

2.51

Mass	Protons	Neutrons	Electrons	Group	Symbol
16	8	8	8	6A	$^{16}_{8}O$
17	8	9	8	6A	$^{17}_{8}O$
18	8	10	8	6A	$^{18}_{8}O$

2.53

	Protons	Neutrons	Electrons
a.	13	14	13
b.	17	18	17
c.	16	18	16

2.55 a. $^{127}_{53}I$ b. $^{79}_{35}Br$ c. $^{107}_{47}Ag$

2.57 107.87 amu

2.59 No, two different elements must have a different number of protons and so, in the neutral atom, they must have a different number of electrons.

2.61 A shell includes all the subshells, such as s, p, d, f.

2.63 Both the $1s$ and $2s$ orbitals are spherical, but the $2s$ orbital is larger.

2.65 a. 1   b. 4   c. 9   d. 16

2.67 The transition metals result from filling d orbitals. There are five d orbitals and each orbital takes two electrons, making 10 columns.

2.69 a. B
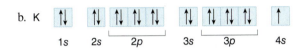

b. K
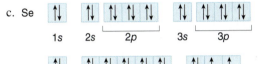

c. Se

d. Ar
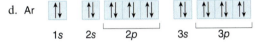

e. Zn
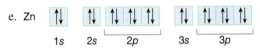

2.71 a. B: $1s^2 2s^2 2p^1$ [He]$2s^2 2p^1$
b. K: $1s^2 2s^2 2p^6 3s^2 3p^6 4s^1$ [Ar]$4s^1$
c. Se: $1s^2 2s^2 2p^6 3s^2 3p^6 4s^2 3d^{10} 4p^4$ [Ar]$4s^2 3d^{10} 4p^4$
d. Ar: $1s^2 2s^2 2p^6 3s^2 3p^6$ [Ne]$3s^2 3p^6$ or [Ar]
e. Zn: $1s^2 2s^2 2p^6 3s^2 3p^6 4s^2 3d^{10}$ [Ar]$4s^2 3d^{10}$

2.73 a. 1   b. 3   c. 1

2.75 a. 38 electrons, 2 valence electrons, strontium
b. 16 electrons, 6 valence electrons, sulfur
c. 11 electrons, 1 valence electron, sodium
d. 17 electrons, 7 valence electrons, chlorine

2.77 An alkali metal has one valence electron and an alkaline earth element has two valence electrons.

2.79

	Electrons	Group Number	Valence Electrons	Period	Valence Shell
a. Carbon	6	4A	4	2	2
b. Calcium	20	2A	2	4	4
c. Krypton	36	8A	8	4	4

2.81   a. carbon: $1s^2 2s^2 2p^2$; $2s^2 2p^2$
      b. calcium: $1s^2 2s^2 2p^6 3s^2 3p^6 4s^2$; $4s^2$
      c. krypton: $1s^2 2s^2 2p^6 3s^2 3p^6 4s^2 3d^{10} 4p^6$; $4s^2 4p^6$

2.83

Element	Valence Electrons
C	4
P	5
Zn	2
Mg	2
I	7
Be	2
Zr	2
S	6

2.85   a. 2   b. 4   c. 7

2.87   Yes, any transition metal element in the fourth period, for example, has electrons in the 4s and 3d orbitals. The valence electrons thus occupy the 4s orbital.

2.89   a. sulfur: 6, $3s^2 3p^4$
      b. chlorine: 7, $3s^2 3p^5$
      c. barium: 2, $6s^2$
      d. titanium: 2, $4s^2$
      e. tin: 4, $5s^2 5p^2$

2.91   a. Be·   b. ·S̈i·   c. ·Ï: d. ·Mg·   e. :Är:

2.93   a. iodine   b. carbon   c. potassium   d. selenium

2.95   a. bromine   b. nitrogen   c. silicon   d. chlorine

2.97   fluorine, oxygen, sulfur, silicon, magnesium

2.99   sodium, magnesium, phosphorus, nitrogen, fluorine

2.101 There is only one 4s electron.

2.103

	a. Type	b. Block	c,d: Radius	e,f: Ionization E	g. Valence Electrons
Sodium	Metal	s			1
Potassium	Metal	s	Largest	Lowest	1
Chlorine	Nonmetal	p	Smallest	Highest	7

2.105 Carbon-11 has the same number of protons and electrons as carbon-12—that is, six. Carbon-11 has only five neutrons, whereas carbon-12 has six neutrons. The symbol for carbon-11 is $^{11}_{6}C$.

CHAPTER 3

3.1   a,d,e,f: covalent   b,c: ionic

3.3   covalent

3.4   a. K^+ b. N^{3-} c. Br^- d. V^{2+}

3.5   a. 28 protons, 26 electrons   c. 30 protons, 28 electrons
     b. 34 protons, 36 electrons   d. 26 protons, 23 electrons

3.6   a. +2   b. −1   c. −2   d. +1

3.7   3

3.9   a. 79 protons, 78 electrons   c. 50 protons, 48 electrons
     b. 79 protons, 76 electrons   d. 50 protons, 46 electrons

3.11 a. 25 protons, 25 electrons
     b. 25 protons, 23 electrons
     c. $1s^2 2s^2 2p^6 3s^2 3p^6 4s^2 3d^5$ The two $4s^2$ valence electrons would be lost to form Mn^{2+}.

3.13 a. NaBr   b. BaO   c. MgI_2 d. Li_2O

3.15 a. sulfide   d. aluminum
     b. copper(I), cuprous   e. tin(IV), stannic
     c. cesium

3.17 hydride

3.18 a. sodium fluoride   e. titanium oxide
     b. magnesium oxide   f. aluminum chloride
     c. strontium bromide   g. calcium iodide
     d. lithium oxide   h. cobalt chloride

3.19 a. chromium(III) chloride, chromic chloride
     b. lead(II) sulfide
     c. tin(IV) fluoride, stannic fluoride
     d. lead(IV) oxide
     e. iron(II) bromide, ferrous bromide
     f. gold(III) chloride

3.21 iron(III) oxide, ferric oxide

3.22 a. $CaBr_2$ c. $FeBr_3$ e. $CrCl_2$
     b. CuI   d. MgS   f. Na_2O

3.23 Ionic compounds have high melting points and high boiling points. They usually dissolve in water. Their solutions conduct electricity and they form crystalline solids.

3.24 a. $MgSO_4$ c. $NiSO_4$ e. Li_2SO_4
     b. Na_2SO_4 d. $Al_2(SO_4)_3$

3.25 a. $NaHCO_3$ c. $(NH_4)_2SO_4$ e. $Ca(HSO_4)_2$
     b. KNO_3 d. $Mg_3(PO_4)_2$ f. $Ba(OH)_2$

3.27 a. sodium carbonate   d. manganese acetate
     b. calcium hydroxide   e. iron(III) hydrogen sulfite, ferric bisulfite
     c. magnesium nitrate   f. magnesium phosphate

3.29 a,b,d: covalent   c: ionic

3.31 a: ionic   b,c,d: covalent

3.33 There is a transfer of electrons from the metal to the nonmetal. A metal gains a noble gas configuration of electrons by giving up electrons. A nonmetal can gain a noble gas configuration by gaining electrons.

3.35 a. Be^{2+} c. S^{2-} e. Cl^-
     b. Ti^{2+} d. Al^{3+} f. Ca^{2+}

3.37 a. Be   b. Cl^- c. Rb^+

3.39 By gaining two electrons they have a filled valence shell.

3.41 a. Na^+ b. Se^{2-} c. Mn^{2+} d. Au^{3+} e. Sn^{4+}

3.43 a,b,c,e: Ne   d: Ar   f: He

3.45 a. lose 1 electron   c. gain 2 electrons
     b. gain 1 electron   d. lose 2 electrons

3.47 b,d: likely to form
     a,e: only seven electrons in outer shell
     c: one electron in outer shell
     f: two electrons in outer shell

3.49

	Number of Valence Electrons	Group Number	Number of Electrons Gained or Lost	Charge	Example
a.	1	1A	Lose 1	1+	Li
b.	2	2A	Lose 2	2+	Mg
c.	6	6A	Gain 2	2−	S
d.	7	7A	Gain 1	1−	Cl

3.51 a. SO_4^{2-} b. NH_4^+ c. HCO_3^- d. ^-CN

3.53 a. 9 protons, 10 electrons
     b. 11 protons, 10 electrons
     c. 47 protons, 50 electrons

3.55 Elements in group 4A would have to gain or lose four electrons to achieve a filled valence shell; that's too many electrons to easily form an ion.

3.57 Transition metals have one or more d electrons. All of these electrons would have to be lost to follow the octet rule, and most transition metals do not lose that many electrons.

3.59 Na donates an electron to F; then each atom has eight electrons in its outer shell.

3.61 a. CaS b. AlBr$_3$ c. LiI d. NiCl$_2$ e. Na$_2$Se

3.63 a. LiNO$_2$ c. NaHSO$_3$ e. Mg(HSO$_3$)$_2$
     b. Ca(CH$_3$COO)$_2$ d. Mn$_3$(PO$_4$)$_2$

3.65

	Br$^-$	$^-$OH	HCO$_3^-$	SO$_3^{2-}$	PO$_4^{3-}$
Na$^+$	NaBr	NaOH	NaHCO$_3$	Na$_2$SO$_3$	Na$_3$PO$_4$
Co^{2+}	CoBr$_2$	Co(OH)$_2$	Co(HCO$_3$)$_2$	CoSO$_3$	Co$_3$(PO$_4$)$_2$
Al^{3+}	AlBr$_3$	Al(OH)$_3$	Al(HCO$_3$)$_3$	Al$_2$(SO$_3$)$_3$	AlPO$_4$

3.67 a. KHSO$_4$ c. Al(HSO$_4$)$_3$
     b. Ba(HSO$_4$)$_2$ d. Zn(HSO$_4$)$_2$

3.69 a. Ba(CN)$_2$ c. BaHPO$_4$
     b. Ba$_3$(PO$_4$)$_2$ d. Ba(H$_2$PO$_4$)$_2$

3.71 a. sodium oxide e. cobalt bromide
     b. barium sulfide f. rubidium bromide
     c. lead(IV) sulfide g. lead(II) bromide
     d. silver chloride

3.73 a. iron(II) chloride, ferrous chloride
     b. iron(III) bromide, ferric bromide
     c. iron(II) sulfide, ferrous sulfide
     d. iron(III) sulfide, ferric sulfide

3.75 Copper cations can be 1+ or 2+, so the Roman numeral designation is required. Ca only exists as 2+.
     CuBr$_2$: copper(II) bromide or cupric bromide
     CaBr$_2$: calcium bromide

3.77 a. Na$_2$S Na$_2$SO$_4$
     b. MgO Mg(OH)$_2$
     c. MgSO$_4$ Mg(HSO$_4$)$_2$

3.79 a. ammonium chloride
     b. lead(II) sulfate
     c. copper(II) nitrate, cupric nitrate
     d. calcium bicarbonate, calcium hydrogen carbonate
     e. iron(II) nitrate, ferrous nitrate

3.81 a. MgCO$_3$ e. Au(NO$_3$)$_3$
     b. NiSO$_4$ f. Li$_3$PO$_4$
     c. Cu(OH)$_2$ g. Al(HCO$_3$)$_3$
     d. K$_2$HPO$_4$ h. Cr(CN)$_2$

3.83 a. Pb(OH)$_4$ lead(IV) hydroxide
     b. Pb(SO$_4$)$_2$ lead(IV) sulfate
     c. Pb(HCO$_3$)$_4$ lead(IV) bicarbonate
     d. Pb(NO$_3$)$_4$ lead(IV) nitrate
     e. Pb$_3$(PO$_4$)$_4$ lead(IV) phosphate
     f. Pb(CH$_3$CO$_2$)$_4$ lead(IV) acetate

3.85 a. True.
     b. False—ionic compounds are solids at room temperature.
     c. False—most ionic compounds are soluble in water.
     d. False—ionic solids exist as crystalline lattices with the ions arranged to maximize the electrostatic interactions of anions and cations.

3.87 Ionic solids have very strong attractive forces so they have high melting points. It takes a great deal of energy to separate the ions from each other.

3.89 NaCl

3.91 a. 30 protons, 30 electrons
     b. 30 protons, 28 electrons
     c. $1s^22s^22p^63s^23p^64s^23d^{10}$ The $4s^2$ electrons are lost to form Zn^{2+}.

3.93

Cation	a. Number of Protons	b. Number of Electrons	c. Noble Gas	d. Role
Na$^+$	11	10	Ne	Major cation in extracellular fluids and blood; maintains blood volume and blood pressure
K$^+$	19	18	Ar	Major intracellular cation
Ca^{2+}	20	18	Ar	Major cation in solid tissues like bone and teeth; required for normal muscle contraction and nerve function
Mg^{2+}	12	10	Ne	Required for normal muscle contraction and nerve function

3.95 AgNO$_3$
3.97 calcium sulfite
3.99 NH$_4$NO$_3$
3.101 a. MgO, KI
      b. calcium hydrogen phosphate
      c. iron(III) phosphate, ferric phosphate
      d. Na$_2$SeO$_3$
      e. The name chromium chloride is ambiguous. Without a designation as chromium(II) or chromium(III), one does not know the ratio of chromium cations to chloride anions.

CHAPTER 4

4.1 H· + ·Cl: ⟶ H:Cl:
    H is surrounded by two electrons, giving it the noble gas configuration of He. Cl is surrounded by eight valence electrons, giving it the noble gas configuration of Ar.

4.3 a. 1 b. 4 c. 1 d. 2 e. 3 f. 2

4.5 Ionic bonding is observed in CaO since Ca is a metal and readily transfers electrons to a nonmetal. Covalent bonding is observed in CO$_2$ since carbon is a nonmetal and does not readily transfer electrons.

4.6 (Lewis structures a–f)

4.7 (Lewis structure)

4.8 a. H—C≡N b. H—C=O (with H) c. H—C=C—Cl

4.9 There are 10 electrons, four from each C = 8 and one from each H = 2.
Each H is surrounded by two electrons and each C by eight.

4.11 :Br—B—Br: (with :Br: above B)

Boron has only six electrons around it, so it does not follow the octet rule.

4.13 a. [H—C(H)—C=Ö: with :Ö: above]⁻ b. [H—C=N—H with :Ö: above]⁻

4.15 :Ö—S̈=Ö: ⟷ Ö=S̈—Ö:

4.16 a. carbon disulfide   c. phosphorus pentachloride
b. sulfur dioxide   d. boron trifluoride

4.17 a. SiO_2 b. PCl_3 c. SO_3 d. N_2O_3

4.18 trigonal planar

4.19 a. bent   c. trigonal pyramidal
b. tetrahedral   d. trigonal planar

4.21 a. polar   c. nonpolar   e. polar
b. ionic   d. polar   f. polar

4.23 a. H—Cl (δ^+ δ^-), polar
b. H—C(H)(H)—C(H)(H)—H nonpolar
c. CH₂F₂ type, polar
d. H—C≡N (δ^+ δ^-) polar
e. Cl₂C=CCl₂ type nonpolar

4.24 a. H—C(H)(H)—C(H)(H)—Ö—H (δ^+ δ^- ... δ^+) polar C–O, O–H. All C's are tetrahedral.
b. H—C(H)(H)—C(=Ö)—H trigonal planar; tetrahedral

4.25 a. LiCl: ionic; the metal Li donates electrons to chlorine.
HCl: covalent; H and Cl share electrons since both are nonmetals and the electronegativity difference is not large enough for electron transfer to occur.
b. KBr: ionic; the metal K donates electrons to bromine.
HBr: covalent; H and Br share electrons since the electronegativity difference is not large enough for electron transfer to occur.

4.27 a. 4 bonds, 0 lone pairs   c. 1 bond, 3 lone pairs
b. 2 bonds, 2 lone pairs   d. 3 bonds, 1 lone pair

4.29 a. H—C=C—C≡N̈ (with H, H on the C=C)
b. H—S̈—C(H)(H)—C(:N—H below, N—H, H)—C—Ö—H (H, :O: above)

4.31 a. H—Ï: c. H—S̈e—H   e. :Cl—C(:Cl: :Cl:)—C(:Cl: :Cl:)—Cl:
b. H—C(H)(:F: below)—F: d. C≡Ö

4.33 a. H—C(H)(H)—N̈(H)—H (H above)
b. H—Ö—N̈=Ö
c. H—C(H)(H)—C≡C—H

4.35 :F:C=C:F: with :F: and :F: (tetrafluoroethylene structure)

4.37 a. [H—N̈:—H with H below]⁻
b. [H—Ö—H with H below]⁺

4.39 a. :Cl—B—Cl: with :Cl: above
b. Ö=S=Ö with :O: below

4.41 A resonance structure is one representation of an electron arrangement with a given placement of atoms. A resonance hybrid is a composite of two or more resonance structures.

4.43 [H—C(H)—C=C—H with :Ö: above and H's]⁻

4.45 a. resonance structures   b. not resonance structures

4.47 [Ö=C—Ö: with :O: below]²⁻ ⟷ [:Ö—C—Ö: with :O: below]²⁻ ⟷ [:Ö—C=Ö: with :O: below]²⁻

4.49 a. phosphorus tribromide   c. nitrogen trichloride
b. sulfur trioxide   d. diphosphorus pentasulfide

4.51 a. SeO_2 b. CCl_4 c. N_2O_5

4.53 water

4.55 a. H—C(H)(H)—Ö—H, tetrahedral (C), bent (O)
b. NF_3, trigonal pyramid
c. H—C(H)(H)—C=C—C(H)(H)—H, tetrahedral, trigonal planar

4.57 a. H—N̈(H)—Ö—H, trigonal pyramid, bent
b. H—C(H)(H)—C(=Ö, :O: above)—H, tetrahedral, trigonal planar
c. [H—C(H)(H)—N(H)(H)—H]⁺, tetrahedral

4.59 BCl_3 is trigonal planar because it has three chlorines bonded to B but no lone pairs. NCl_3 is trigonal pyramidal because it has three Cl's around N as well as a lone pair.

4.61 a. H—C(H)(H)—F:, both 109.5°
b. H—C≡C—C(H)(H)—Ö—H, 180°, 109.5°, 109.5°
c. H—C=C=Ö (H below), 120° 180°, 120°

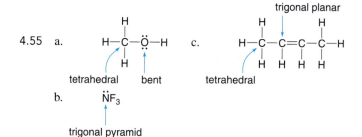

4.63

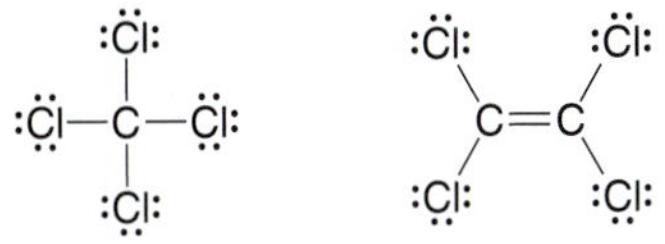

CCl$_4$ is tetrahedral since carbon has four substituents. Each C in C$_2$Cl$_4$ has trigonal planar geometry since each carbon has three substituents.

4.65 a. Se, S, O   b. Na, P, Cl   c. S, Cl, F   d. P, N, O

4.67 A nonpolar bond such as H—H has no net dipole and joins two atoms of similar electronegativity. A polar bond such as H—Br joins atoms with different electronegativities and therefore has a net dipole.

4.69 a,b,c: covalent   d: ionic

4.71 a. nonpolar   b. polar   c. polar   d. polar   e. nonpolar

4.73 a. C—O   b. C—F   c. Si—C

4.75
$\overset{\delta^+}{C}\!-\!\overset{\delta^-}{O}$ $\overset{\delta^+}{C}\!-\!\overset{\delta^-}{N}$ $\overset{\delta^+}{C}\!-\!\overset{\delta^-}{F}$ $\overset{\delta^+}{C}\!-\!\overset{\delta^-}{Cl}$ $\overset{\delta^+}{Si}\!-\!\overset{\delta^-}{C}$ P—H
$\longmapsto$ $\longmapsto$ $\longmapsto$ $\longmapsto$ $\longmapsto$ no dipole

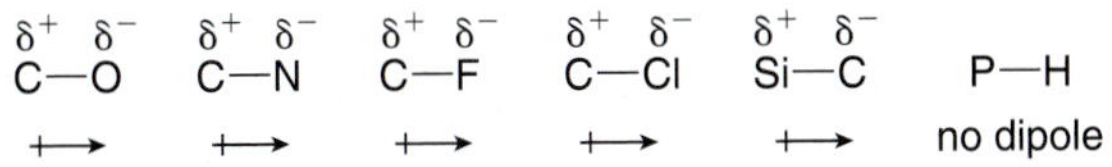

4.77 In CH$_3$NH$_2$ carbon bears a partial positive charge since nitrogen is more electronegative than carbon. In CH$_3$MgBr carbon bears a partial negative charge since carbon is more electronegative than magnesium.

4.79 No, there must be differences in electronegativity to make a compound polar.

4.81 No, a compound with only one polar bond must be polar. The single bond dipole is not cancelled by another bond dipole.

4.83 a.

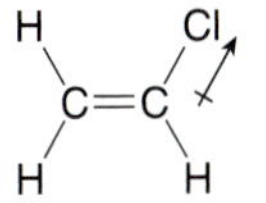

polar bond, polar compound

b.

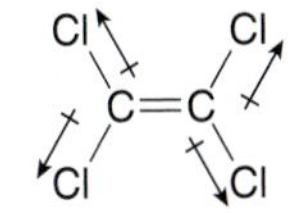

polar bonds, nonpolar compound

c.

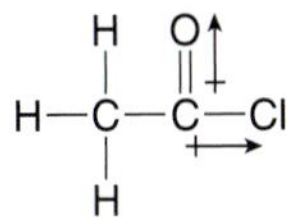

polar bonds, polar compound

4.85 CHCl$_3$ has a net dipole, making it polar. CCl$_4$ has four polar bonds but no net dipole because the four bonds have dipoles that cancel one another.

4.87 a. 20 valence electrons

b,c:

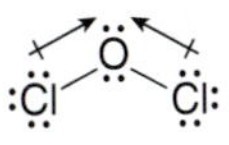

d. bent

e. The compound is polar since the two bond dipoles do not cancel.

4.89 a.-c.

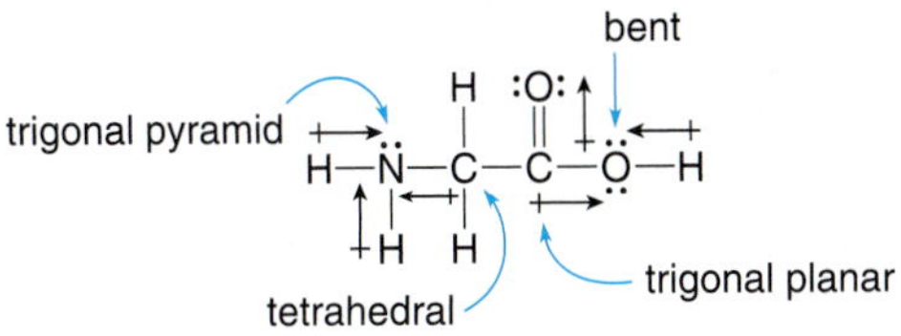

d. Glycine is a polar molecule since the many bond dipoles do not cancel.

4.91 One possibility:

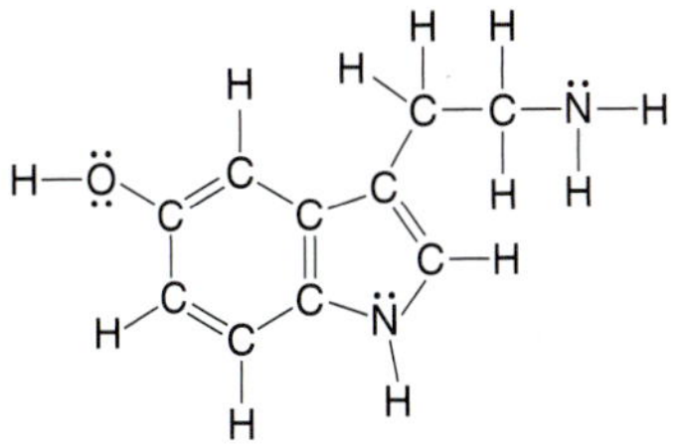

4.93 a. The predicted shape around each carbon is tetrahedral.
b. 60°
c. The 60° bond angle deviates greatly from the theoretical tetrahedral bond angle, making the molecule unstable.

CHAPTER 5

5.1
reactants   products
a. 2 H$_2$O$_2$(aq) $\longrightarrow$ 2 H$_2$O(l) + O$_2$(g)   (4 H, 4 O)
b. 2 C$_8$H$_{18}$ + 25 O$_2$ $\longrightarrow$ 16 CO$_2$ + 18 H$_2$O
(16 C, 50 O, 36 H)
c. 2 Na$_3$PO$_4$(aq) + 3 MgCl$_2$(aq) $\longrightarrow$ Mg$_3$(PO$_4$)$_2$(s)
+ 6 NaCl(aq)   (3 Mg, 2 P, 8 O, 6 Na, 6 Cl)

5.3 a. CH$_4$(g) + 4 Cl$_2$(g) $\overset{\Delta}{\longrightarrow}$ CCl$_4$(l) + 4 HCl(g)
b. C$_3$H$_6$O$_2$(l) + 2 H$_2$(g) $\longrightarrow$ C$_2$H$_6$O(l) + CH$_4$O(l)

5.4 a. 2 H$_2$ + O$_2$ $\longrightarrow$ 2 H$_2$O
b. 2 NO + O$_2$ $\longrightarrow$ 2 NO$_2$
c. 4 Fe + 3 O$_2$ $\longrightarrow$ 2 Fe$_2$O$_3$
d. CH$_4$ + 2 Cl$_2$ $\longrightarrow$ CH$_2$Cl$_2$ + 2 HCl

5.5 2 CO + O$_2$ $\longrightarrow$ 2 CO$_2$

5.6 2 C$_2$H$_6$ + 7 O$_2$ $\longrightarrow$ 4 CO$_2$ + 6 H$_2$O

5.7 N$_2$ + 3 H$_2$ $\longrightarrow$ 2 NH$_3$

5.8 a. 2 Al + 3 H$_2$SO$_4$ $\longrightarrow$ Al$_2$(SO$_4$)$_3$ + 3 H$_2$
b. 3 Na$_2$SO$_3$ + 2 H$_3$PO$_4$ $\longrightarrow$ 3 H$_2$SO$_3$ + 2 Na$_3$PO$_4$

5.9 a,b,c,d: 6.02 × 10^{23}

5.10 a. 1.20 × 10^{24} c. 3.01 × 10^{23}
b. 3.61 × 10^{24} d. 1.51 × 10^{25}

5.11 a. 1.5 × 10^{24} c. 2.4 × 10^{23}
b. 1.5 × 10^{23} d. 3.33 × 10^{25}

5.12 a. 100. mol   b. 0.0500 mol   c. 15 mol

5.13 a. 100.09 amu   b. 166.00 amu

5.15 a. 100.09 g/mol   b. 166.00 g/mol

5.17 a. 29.2 g   b. 332 g   c. 101 g   d. 26.3 g

5.18 a. 1.71 mol   c. 1.39 × 10^{-3} mol
b. 1.59 mol   d. 1.39 mol

5.19 1.80 × 10^{21}

5.21 a. 6.6 mol   b. 1.0 mol   c. 1.2 mol

5.23 a. 51 g   b. 22 g   c. 90. g

5.25 a. 72.4 g   b. 24.2 g   c. 7.24 g

5.27 No. The theoretical yield is the maximum amount of product that can be produced.

5.28 a. 154 g   b. 34.7%

5.29 a. 250 g   b. 62%

5.31 a. 111 g   b. 59.0%

5.33 a. 35%   b. 11%   c. 19%   d. 1.0%

5.34   (oxidized) (reduced)
    a. $Zn(s) + 2 H^+(aq) \longrightarrow Zn^{2+}(aq) + H_2(g)$
       $Zn \longrightarrow Zn^{2+} + 2 e^-$
       $2 H^+ + 2 e^- \longrightarrow H_2$
       (reduced)   (oxidized)
    b. $Fe^{3+}(aq) + Al(s) \longrightarrow Al^{3+}(aq) + Fe(s)$
       $Al \longrightarrow Al^{3+} + 3 e^-$
       $Fe^{3+} + 3 e^- \longrightarrow Fe$
      (oxidized) (reduced)
    c. $2 I^- + Br_2 \longrightarrow I_2 + 2 Br^-$
       $2 I^- \longrightarrow I_2 + 2 e^-$
       $Br_2 + 2 e^- \longrightarrow 2 Br^-$
    d. $2 AgBr \longrightarrow 2 Ag + Br_2$
       $2 Br^-$ (oxidized) $2 Ag^+$ (reduced)
       $2 Br^- \longrightarrow Br_2 + 2 e^-$
       $2 Ag^+ + 2 e^- \longrightarrow 2 Ag$

5.35   a. Zn reducing agent, H^+ oxidizing agent
    b. Fe^{3+} oxidizing agent, Al reducing agent
    c. I^- reducing agent, Br_2 oxidizing agent
    d. Br^- reducing agent, Ag^+ oxidizing agent

5.37 H_2 is oxidized since it gains an O atom and $C_2H_4O_2$ is reduced since it gains hydrogen.

5.38   Zn is the reducing agent and Hg^{2+} is the oxidizing agent.

5.39   The coefficient indicates the number of molecules or moles undergoing reaction, whereas the subscript indicates the number of atoms of each element in a chemical formula.

5.41   A chemical reaction involves the breaking and forming of chemical bonds. A chemical equation is an expression that indicates the reactants on the left side of an arrow and the products formed on the right side.

5.43   a. 2 H, 2 Cl, 1 Ca on both sides; therefore balanced
    b. 1 Ti, 4 Cl, 4 H, 2 O on left side, and 1 Ti, 1 Cl, 1 H, 2 O on right side; therefore NOT balanced
    c. 1 Al, 1 P, 7 O, 6 H on both sides; therefore balanced

5.45   a. $Ni(s) + 2 HCl(aq) \longrightarrow NiCl_2(aq) + H_2(g)$
    b. $CH_4(g) + 4 Cl_2(g) \longrightarrow CCl_4(g) + 4 HCl(g)$
    c. $2 KClO_3 \longrightarrow 2 KCl + 3 O_2$
    d. $Al_2O_3 + 6 HCl \longrightarrow 2 AlCl_3 + 3 H_2O$
    e. $4 Al(OH)_3 + 6 H_2SO_4 \longrightarrow 2 Al_2(SO_4)_3 + 12 H_2O$

5.47   a. $2 C_6H_6 + 15 O_2 \longrightarrow 12 CO_2 + 6 H_2O$
    b. $C_7H_8 + 9 O_2 \longrightarrow 7 CO_2 + 4 H_2O$
    c. $2 C_8H_{18} + 25 O_2 \longrightarrow 16 CO_2 + 18 H_2O$

5.49 $2 S(s) + 3 O_2(g) + 2 H_2O(l) \longrightarrow H_2SO_4(l)$

5.51

5.53   The formula weight is the sum of atomic weights in a compound reported in amu. The molecular weight is the formula weight for a covalent compound.

5.55   a. 69.00 amu, 69.00 g/mol
    b. 28.06 amu, 28.06 g/mol
    c. 342.17 amu, 342.17 g/mol

5.57   a. 176.1 amu, 176.1 g/mol
    b. 167.2 amu, 167.2 g/mol
    c. 321.9 amu, 321.9 g/mol

5.59   a. $C_9H_{11}NO_4$ b. 197.2 amu   c. 197.2 g/mol
5.61   a. 1 mol Sn
    b. 6.02×10^{23} N atoms
    c. 1 mol N_2 molecules
    d. 1 mol CO_2
5.63   a. 182 g   b. 710. g   c. 130. g   d. 390. g
5.65   a. 1.46×10^{-3} mol   c. 0.0730 mol
    b. 0.0146 mol   d. 7.30×10^{-5} mol
5.67   10.0 g of aspirin
5.69   a. 1.20×10^{24} c. 1.60×10^{25} e. 3.01×10^{29}
    b. 1.51×10^{23} d. 1.34×10^{26}
5.71   a. 324 g   b. 52.3 g   c. 1.1×10^3 g   d. 9.82 g
5.73 1.6×10^{17} molecules
5.75   a. 12.5 mol   b. 12 mol   c. 0.50 mol   d. 0.40 mol
5.77   a. 220 g   b. 44 g   c. 4.5 g   d. 240 g
5.79   a. 56.0 g   b. 0.28 g
5.81   The theoretical yield is the predicted amount of product if all reactant is converted to the desired product. The actual yield is the amount actually obtained after reaction occurs.
5.83   75%
5.85   a. 23.9 g b. 62.8%
5.87   A substance that is oxidized loses electrons. A substance that gains electrons acts as an oxidizing agent.
5.89   a. Fe (oxidized) Cu^{2+} (reduced)
      $Fe \longrightarrow Fe^{2+} + 2 e^-$
      $Cu^{2+} + 2 e^- \longrightarrow Cu$
    b. Cl_2 (reduced) $2 I^-$ (oxidized)
      $2 I^- \longrightarrow I_2 + 2 e^-$
      $Cl_2 + 2 e^- \longrightarrow 2 Cl^-$
    c. 2 Na (oxidized) Cl_2 (reduced)
      $2 Na \longrightarrow 2 Na^+ + 2 e^-$
      $Cl_2 + 2 e^- \longrightarrow 2 Cl^-$
5.91   Zn Ag^+
    oxidized   reduced
    reducing agent   oxidizing agent
5.93   Acetylene is reduced because it gains hydrogen atoms.
5.95 $2 Mg + O_2 \longrightarrow 2 MgO$
    $2 Mg \longrightarrow 2 Mg^{2+} + 4 e^-$ $O_2 + 4 e^- \longrightarrow 2 O^{2-}$
5.97   a. 342.3 g/mol
    b. $C_{12}H_{22}O_{11}(s) + H_2O(l) \longrightarrow 4 C_2H_6O(l) + 4 CO_2(g)$
    c. 8 mol   e. 101 g   g. 9.21 g
    d. 10 mol   f. 18.4 g   h. 13.6%
5.99   a. 0.485 mol   b. 2.92×10^{23}
5.101   a. 1.43×10^{20} molecules   b. 2.38×10^{-4} mol
5.103 6.3×10^{22} ions
5.105   a. 354.5 g/mol   b. 18 g   c. 17.8 g   d. 84.3%
5.107   a. 2.1×10^{-3} g   b. 3.9×10^{18} molecules

CHAPTER 6

6.1   a. 10. cal   b. 55,600 cal   c. 1,360 kJ   d. 107,000 J
6.3   126 Cal, rounded to 100 Cal
6.5   a. +88 kcal/mol; endothermic
    b. −136 kcal/mol; exothermic
    c. +119 kcal/mol; endothermic
6.6   The indicated bond has the higher bond dissociation energy, making it the stronger bond.

    a. H—C—Br: b. H—OH

6.7 a. absorbed   c. reactants
     b. reactants   d. endothermic

6.8 106 kcal

6.9 a. 96 kcal   b. 8.0 kcal   c. 1.8 kcal

6.10

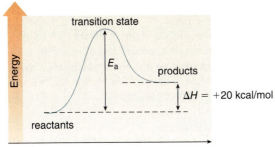

6.11

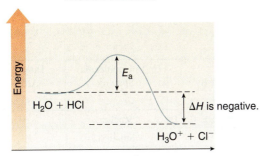

6.13 Yes, H_2SO_4 is a catalyst because it speeds up the rate of the reaction. It does not affect the relative energies of reactants and products.

6.15 a. $2\,SO_2(g) + O_2(g) \longrightarrow 2\,SO_3(g)$ forward reaction
      $2\,SO_3(g) \longrightarrow 2\,SO_2(g) + O_2(g)$ reverse reaction
      b. $N_2(g) + O_2(g) \longrightarrow 2\,NO(g)$ forward reaction
      $2\,NO(g) \longrightarrow N_2(g) + O_2(g)$ reverse reaction
      c. $C_2H_4O_2 + CH_4O \longrightarrow C_3H_6O_2 + H_2O$
      forward reaction
      $C_3H_6O_2 + H_2O \longrightarrow C_2H_4O_2 + CH_4O$
      reverse reaction

6.16 a. $\dfrac{[PCl_5]}{[PCl_3][Cl_2]}$ c. $\dfrac{[HBr]^2}{[Br_2][H_2]}$

      b. $\dfrac{[SO_3]^2}{[SO_2]^2[O_2]}$ d. $\dfrac{[HCl]^3[CHCl_3]}{[CH_4][Cl_2]^3}$

6.17 a. reactants favored   c. reactants and products present
      b. products favored   d. reactants and products present

6.19 a. $\dfrac{[H_2S]^2}{[H_2]^2[S_2]}$

      b. products favored
      c. ΔH negative
      d. products lower in energy
      e. One cannot predict the rate of reaction without knowing the energy of activation.

6.20 a. 4.75   b. 218

6.21 a,d: right   b,c: left

6.22 a. right   b. left

6.23 a. left   b. right

6.24 a. right   b. left

6.25 Potential energy is stored energy, while kinetic energy is the energy of motion. A stationary object on a hill has potential energy, but as it moves down the hill this potential energy is converted to kinetic energy.

6.27 A calorie (cal) is the amount of energy needed to raise the temperature of 1 g of water 1 °C. A joule is also a unit of energy. Joules (J) and calories are related in the following way: 1 cal = 4.184 J.

6.29 a. 563,000 cal/h   c. 2.36×10^6 J/h
      b. 563 kcal/h   d. 2,360 kJ/h

6.31 a. 0.05 kcal   b. 0.23 kJ   c. 230 cal   d. 1.01×10^6 cal

6.33 236 Calories, rounded to 200 Calories

6.35 30 g

6.37 salmon, 113 Calories vs. chicken, 107 Calories

6.39 When energy is absorbed, the reaction is endothermic and ΔH is positive (+).
      When energy is released, the reaction is exothermic and ΔH is negative (−).

6.41 a. Cl_2 b. Cl_2 c. HF

6.43 a,d: exothermic   b,c: endothermic

6.45 a. 240 kcal   b. 280 kcal   c. 2.0×10^2 kcal

6.47 a. stronger bonds in products   c. 339 kcal
      b. 2,710 kcal   d. 37.6 kcal

6.49 The transition state is located at the top of the energy hill that separates reactants from products.
      The difference in energy between the reactants and the transition state is the energy of activation, symbolized by E_a.

6.51 a. X   b. Z   c. Y   d. X,Y   e. X,Z   f. Y   g. Z

6.53 a.

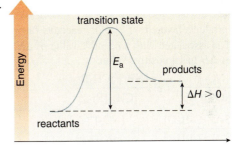

      b.

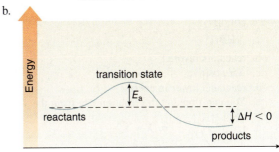

      c.

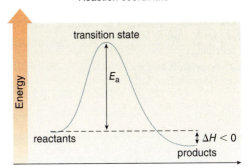

6.55 The reaction is exothermic.

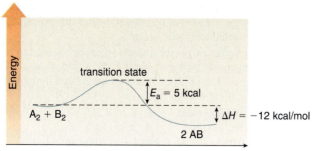

Reaction coordinate

6.57 Reacting molecules must have the proper orientation for new bonds to form.

6.59 Increasing temperature increases the number of collisions. Since the average kinetic energy of the colliding molecules is larger, more collisions are effective at causing reaction.

6.61 a. The reaction with $E_a = 1$ kcal will proceed faster because the energy of activation is lower.
b. K doesn't affect the reaction rate, so we cannot predict which reaction is faster.
c. One cannot predict which reaction will proceed faster from the ΔH.

6.63 b and c

6.65 A catalyst increases the reaction rate (a) and lowers the E_a (c). It has no effect on ΔH (b) or K (d), or the relative energies of the reactants and products (e).

6.67 The forward reaction proceeds from left to right as drawn and the reverse reaction proceeds from right to left as drawn.

6.69 a,b: products favored   c,d: reactants favored

6.71 $K > 1$ is associated with a negative value of ΔH. A $K < 1$ means ΔH has a positive value.

6.73 a. $\dfrac{[NO_2]^2}{[NO]^2[O_2]}$ b. $\dfrac{[HBr]^2[CH_2Br_2]}{[CH_4][Br_2]^2}$

6.75 a. $\dfrac{[H_2O][C_2H_6O]}{[C_2H_4O_2][H_2]^2}$ b. $\dfrac{[NO_2]^4[O_2]}{[N_2O_5]^2}$

6.77 a. $2\,A \rightleftarrows A_2$ b. $A_2 + 3\,B_2 \rightleftarrows 2\,AB_3$

6.79 a. $\dfrac{[Br_2][H_2]}{[HBr]^2}$

b. reactants favored
c. ΔH positive
d. reactants lower in energy
e. You can't predict the reaction rate from the value of K.

6.81 a. $\dfrac{[CO_2][H_2]}{[CO][H_2O]}$ b. $K = 4.2$

6.83 a. increases NO and O_2, decreases N_2
b. increases N_2, NO, and O_2

6.85 a,f: favor shift to right
b,c,d: favor shift to left
e. no change

6.87 a,c,f: favor shift to right
b,d,e: favor shift to left

6.89 a. $\dfrac{[C_2H_6]}{[H_2][C_2H_4]}$

b. reactants higher in energy
c. $K > 1$
d. 20. kcal
e. increase in rate
f. 1,2,4: favor shift to right; 3: favors shift to left; 5: no change, but reduces rate of reaction

6.91 Lactase is an enzyme that converts lactose, a naturally occurring sugar in dairy products, into the two simple sugars, glucose and galactose.

6.93 400 Calories

6.95 1.9 h, rounded to 2 h

6.97 propane 12.0 kcal/g   butane 11.8 kcal/g

6.99 30,200 kcal

CHAPTER 7

7.1

	a. Density	b. Intermolecular Spacing	c. Intermolecular Attraction
Gas	Lowest	Greatest	Lowest
Liquid	Higher	Smaller	Higher
Solid	Highest	Smallest	Highest

7.3 a. 0.83 atm   b. 12 psi

7.5 a. 2,300 mm Hg   b. 14 psi   c. 0.558 atm

7.7 14/9

7.8 a. 1.6 L   b. 3.2 L   c. 0.80 L   d. 16 L

7.9 a. 0.25 atm   b. 1.5 atm   c. 0.075 atm   d. 0.0075 atm

7.10 0.44 L

7.11 a. 250 L   b. 24 mL

7.13 392 K or 119 °C

7.15 As a sealed container is heated, the gases inside expand and increase the container's internal pressure, eventually popping the lid off.

7.16 290 mm Hg

7.17 0.18 mol

7.19 a. 1.0×10^2 L   b. 7.8 L   c. 12.6 L

7.21 36 mol

7.23 53 L

7.24 O_2: 2.5 atm CO_2: 1.5 atm

7.25 methane: 640 mm Hg   ethane: 75 mm Hg
propane: 38 mm Hg

7.27 all: a–d

7.29

	London Dispersion	Dipole–Dipole	Hydrogen Bonding
a. Cl_2	+		
b. HCN	+	+	
c. HF	+	+	+
d. CH_3Cl	+	+	
e. H_2	+		

7.31 a. C_2H_6 b. CH_3OH c. HBr   d. CH_3Br

7.33 Water has stronger intermolecular forces since it can hydrogen bond. This explains why it is a liquid at room temperature, whereas CO_2 is a gas.

7.34 a. CH_4 b. CH_4 c. C_2H_6

7.35 When you get out of a pool, the water on your body evaporates and this cools your skin. When you re-enter the water, the water feels warmer because the skin is cooler.

7.37 Benzene cannot hydrogen bond, whereas water can, so benzene has much weaker intermolecular forces than water and is thus less viscous. Ethylene glycol, on the other hand, has two OH groups capable of hydrogen bonding, so it has stronger intermolecular forces than water and is thus more viscous.

7.39 a. ionic   b. metallic   c. molecular   d. molecular

7.40 a. 3,990 cal   b. 2,790 cal   c. 2.79 kcal   d. 1,430 cal

7.41 a. 23,000 cal   b. 23,000 cal   c. 9.7 kcal   d. 34 kcal

7.43 Both are measurements of pressure. They are related by the conversion factor: 760 mm Hg = 1 atm.

7.45 1.071 atm

7.47 a. 41 psi   b. 0.68 atm   c. 15,200 torr   d. 13,300 Pa

7.49 a. 0.75 L   b. 1.1 L   c. 99.7 mm Hg

7.51 35 mL

7.53 a. 4.0 L   b. 1.1 L   c. 820 K

7.55 1.4 L

7.57 a. 4.34 atm   b. 350 mm Hg   c. 1,300 K

7.59 1.1 atm

7.61 a. 1.4 atm   b. 110 L   c. 740 K

7.63 1,200 L

7.65 STP is "standard temperature and pressure," or 0 °C at 760 mm Hg. The standard molar volume is the volume that one mole of a gas occupies at STP, or 22.4 liters.

7.67 a. 0.22 mol   b. 0.500 mol   c. 0.002 23 mol

7.69 a. 94 L   b. 1.8 L   c. 1.7 L

7.71 0.112 L or 112 mL

7.73 0.017 mol

7.75 0.19 mol or 1.1×10^{23} molecules

7.77 O_2 has more moles and also weighs more.

7.79 97 mm Hg for O_2 360 mm Hg for N_2

7.81 1,780 mm Hg

7.83 London dispersion forces are very weak interactions due to the momentary changes in electron density in a molecule. The larger the molecule, the larger the attractive force between two molecules, and the stronger the intermolecular forces. Dipole–dipole interactions are the stronger attractive forces between the permanent dipoles of two polar molecules.

7.85 Water is capable of hydrogen bonding and these strong intermolecular attractive forces give it a higher boiling point than H_2S.

7.87 a. London forces, dipole–dipole   b. London forces only

7.89 d.

7.91 No, $H_2C{=}O$ has no H on the O atom.

7.93 a. Ethylene has London forces only, whereas methanol has London forces, dipole–dipole forces, and hydrogen bonding.
   b. Methanol has a higher boiling point.
   c. Ethylene has a higher vapor pressure at any given temperature.

7.95 Vapor pressure is the pressure exerted by gas molecules in equilibrium with the liquid phase. A mixture of gases behaves like a pure gas and each component of a gas mixture exerts a pressure called its partial pressure.

7.97 butane, acetaldehyde, Freon-113

7.99 Glycerol is more viscous than water since it has three OH groups and has many opportunities for hydrogen bonding. Acetone cannot hydrogen bond so its intermolecular forces are weaker and thus it has low viscosity.

7.101 An ionic solid is composed of oppositely charged ions. A metallic solid can be considered as a lattice of metal cations surrounded by a cloud of electrons that move freely. Because of their loosely held, delocalized electrons, metals conduct electricity.

7.103 a. ionic   c. metallic   e. amorphous
   b. molecular   d. network

7.105 Evaporation is an endothermic process by which a liquid enters the gas phase. Condensation is an exothermic process that occurs when a gas enters the liquid phase.

7.107 Sublimation is an endothermic process by which a solid transforms directly to the gas phase. Deposition is an exothermic process that occurs when a gas transforms directly to the solid phase.

7.109 a,d: absorbed   b,c: released

7.111 Vaporizing 50.0 g of water takes more energy; 27,000 cal vs. 20,000 cal.

7.113 Charles's law relates volume and temperature. For a fixed amount of gas at constant pressure, the volume of a gas is proportional to its Kelvin temperature.
   Gay–Lussac's law relates pressure and temperature. For a fixed amount of gas at constant volume, the pressure of a gas is proportional to its Kelvin temperature.

7.115 Boyle's law relates volume and pressure. For a fixed amount of gas at constant temperature, the pressure and volume of a gas are inversely related.
   Gay–Lussac's law relates pressure and temperature. For a fixed amount of gas at constant volume, the pressure of a gas is proportional to its Kelvin temperature.

7.117 a. Volume increases as outside atmospheric pressure decreases.
   b. Volume decreases at the lower temperature.
   c. Volume decreases as external pressure increases.
   d. Volume increases as temperature increases.

7.119 The systolic pressure is the maximum pressure generated with each heartbeat. The diastolic pressure is the lowest pressure recorded between heartbeats.

7.121 290 mL
   The gases inside the bag had a volume of 250 mL at 760 mm Hg and take up a greater volume at the reduced pressure of 650 mm Hg.

7.123 Water is one of the few substances that expands as it enters the solid phase. This causes the bottle to crack as the water occupies a larger volume as it freezes.

7.125 a. As a person breathes faster, he eliminates more CO_2 from the lungs; therefore, the measured value of CO_2 is lower than the normal value of 40 mm Hg.
   b. Many people with advanced lung disease have lost lung tissue over time and therefore cannot exchange adequate amounts of oxygen through the lungs and into the

blood, leading to a lower-than-normal partial pressure of oxygen. In addition, they often breathe more slowly and with lower volumes than normal, so they cannot eliminate enough CO_2 and therefore the partial pressure of CO_2 climbs.

7.127 The molar mass is 4 g/mol and the gas is helium.

CHAPTER 8

8.1 a. heterogeneous mixture   b. colloid   c,d,e: solution

8.3 a,c,d,e: water soluble

8.5 Only (a) and (b) would form solutions.

8.6 a,c,e,f: water soluble   b,d: water insoluble

8.7 a,d: no   b,c: yes

8.9 A soft drink becomes "flat" when CO_2 escapes from the solution. CO_2 comes out of solution faster at room temperature than at the cooler temperature of the refrigerator because the solubility of gases in liquids decreases as the temperature increases.

8.11 3.5%

8.13 8.4% (v/v)

8.14 2.4 mL

8.15 2.0×10^2 mL

8.17 a. 0.030 ppm   b. 0.5 ppm   c. 2.0 ppm   d. 0.23 ppm

8.18 a. 2.0 M   c. 10. M   e. 1.2 M
     b. 8.0 M   d. 0.10 M   f. 1.4 M

8.19 10.0 g NaOH in 150 mL gives a concentration of 1.7 M vs. only 1.5 M for 15.0 g in 250 mL.

8.20 a. 1.0×10^2 mL   c. 2.0 mL
     b. 13 mL   d. 2.0×10^3 mL

8.21 a. 4.0 moles   c. 0.050 moles
     b. 0.63 moles   d. 0.063 moles

8.22 a. 7.3 g   b. 150 g   c. 40. g   d. 3.7 g

8.23 a. 5.8 mL   b. 23 mL   c. 15 mL   d. 0.58 mL

8.24 0.35 M

8.25 a. 220 mL   b. 5.0×10^2 mL   c. 7.5 mL   d. 15 mL

8.27 a. 101.0 °C   b. 102.0 °C   c. 103.1 °C   d. 100.4 °C

8.28 a. −3.7 °C   b. −7.4 °C   c. −11.2 °C   d. −1.3 °C

8.29 −7.4 °C

8.30 a. 5.0% sugar solution   b. 4.0 M NaCl   c. 0.75 M NaCl

8.31 a. Water flows across the membrane.
     b. More water initially flows from the 1.0 M side to the more concentrated 1.5 M side. When equilibrium is reached there is equal flow of water in both directions.
     c. The height of the 1.5 M side will be higher and the height of the 1.0 M NaCl will be lower.

8.33 A solution is a homogeneous mixture that contains small particles. A colloid is a homogeneous mixture with larger particles, often having an opaque appearance.

8.35 a,b,d,e: solution   c. heterogeneous mixture   f. colloid

8.37 The substance present in the lesser amount in the solution is the solute, while the substance present in the larger amount in the solution is the solvent.

8.39 a. unsaturated   b,c: saturated   d. supersaturated

8.41 a,c,d

8.43 Water-soluble compounds are ionic or are small polar molecules that can hydrogen bond with the water solvent, but nonpolar compounds, such as oil, are soluble in nonpolar solvents.

8.45 Iodine would not be soluble in water but is soluble in CCl_4 since I_2 is nonpolar and CCl_4 is a nonpolar solvent.

8.47 Cholesterol is not water soluble because it is a large nonpolar molecule with a single OH group.

8.49 a. increased   b. decreased   c,d: no change

8.51 A decrease in temperature (a) increases the solubility of a gas and (b) decreases the solubility of a solid.

8.53 The ion–dipole interactions between ions and water provide the energy needed to break apart the ions from the crystal lattice. The water molecules form a loose shell of solvent around each ion.

8.55 a,b,d,e,g: water soluble

8.57 Weight/volume percent concentration is the number of grams of solute dissolved in 100 mL of solution. Molarity is the number of moles of solute per liter of solution.

8.59 5% (w/v) $\dfrac{5 \text{ g}}{100 \text{ mL}}$ $\dfrac{100 \text{ mL}}{5 \text{ g}}$

     6.0 M $\dfrac{6.0 \text{ mol}}{1.0 \text{ L}}$ $\dfrac{1.0 \text{ L}}{6.0 \text{ mol}}$

     10 ppm $\dfrac{10 \text{ g}}{10^6 \text{ g}}$ $\dfrac{10^6 \text{ g}}{10 \text{ g}}$

8.61 a. 1.3% (w/v)   b. 17% (w/v)   c. 8.00% (w/v)

8.63 17% (v/v)

8.65 a. 2.3 M   b. 0.51 M   c. 0.66 M   d. 0.036 M

8.67 a. Add 12 g of acetic acid to the flask and then water to bring the volume to 250 mL.
     b. Add 55 mL of ethyl acetate to the flask and then water to bring the volume to 250 mL.
     c. Add 37 g of NaCl to the flask and then water to bring the volume to 250 mL.

8.69 a. 0.038 mol   b. 0.090 mol   c. 3.8 mol

8.71 a. 3.2 g   b. 5.7 g   c. 140 g

8.73 83 mL

8.75 a. 95 g   b. 1.6 mol   c. 0.85 M

8.77 a. 0.08 ppm   b. 0.7 ppm

8.79 Concentration is the amount of solute per unit volume in a solution. Dilution is the addition of solvent to decrease the concentration of solute.

8.81 a. 15.0% (w/v)   c. 5.0% (w/v)
     b. 6.00% (w/v)   d. 14% (w/v)

8.83 1.8 M

8.85 a. 10. mL   b. 450 mL   c. 1.5 mL   d. 2.5 mL

8.87 A volatile solute readily escapes into the vapor phase. A nonvolatile solute does not readily escape into the vapor phase, and thus it has a negligible vapor pressure at a given temperature.

8.89 a. 101.5 °C   b. 101.2 °C   c. 103.1 °C

8.91 −4.5 °C

8.93 a. NaCl   b. glucose   c. NaCl   d. glucose

8.95 a. 0.10 M glucose   c. 0.10 M Na_2SO_4
     b. 0.20 M NaCl   d. 0.10 M glucose

8.97 Osmosis is the selective diffusion of water and small molecules across a semipermeable membrane. Osmotic pressure is the pressure that prevents the flow of additional solvent into a solution on one side of a semipermeable membrane.

8.99 Two solutions with the same osmotic pressure are said to be isotonic. A hypotonic solution has a lower concentration of solute relative to another solution.

8.101 a. A > B   c. no change   e. no change
b. B > A   d. A > B

8.103 At warmer temperatures, CO_2 is less soluble in water and more is in the gas phase and escapes as the can is opened and pressure is reduced.

8.105 0.09% (w/v)   0.005 M

8.107 a. 280 mL would have to be given.
b. The hypertonic mannitol draws water out of swollen brain cells and thus reduces the pressure on the brain.

8.109 Water moves out of the cells of the cucumber to the hypertonic salt solution, so the cucumber shrinks and loses its crispness.

8.111 NaCl, KCl, and glucose are found in the bloodstream. If they weren't in the dialyzer fluid, they would move out of the bloodstream into the dialyzer, and their concentrations in the bloodstream would fall.

8.113 6.3% (w/v)

8.115 7.3 g

8.117 4,000 mg

8.119 a. yes   b. 0.0005 mol

CHAPTER 9

9.1 a,c

9.2 a,b,d

9.3
      acid   base   base   acid
a. $HCl(g)$ + $NH_3(g)$ c. $^-OH(aq)$ + $HSO_4^-(aq)$
      acid   base
b. $CH_3COOH(l)$ + $H_2O(l)$

9.4 a. H_3O^+ b. HI   c. H_2CO_3

9.5 a. HS^- b. ^-CN c. SO_4^{2-}

9.6
                        conjugate conjugate
    base   acid   base   acid
a. $H_2O(l)$ + $HI(g)$ ⇌ $I^-(aq)$ + $H_3O^+(aq)$

    acid   base   conjugate conjugate
b. $CH_3COOH(l)$ + $NH_3(g)$ base   acid
    ⇌ $CH_3COO^-(aq)$ + $NH_4^+(aq)$

                    conjugate conjugate
    base   acid   acid   base
c. $Br^-(aq)$ + $HNO_3(aq)$ ⇌ $HBr(aq)$ + $NO_3^-(aq)$

9.7 a. NH_4^+ b. NH_2^-

9.9 a. H_2SO_4 is the stronger acid; H_3PO_4 has the stronger conjugate base.
b. HCl is the stronger acid; HF has the stronger conjugate base.
c. H_2CO_3 is the stronger acid; NH_4^+ has the stronger conjugate base.
d. HF is the stronger acid; HCN has the stronger conjugate base.

9.11 a. HNO_2 HNO_3 b. HNO_3 is the stronger acid.

9.12 a,c: products   b. reactants

9.13 a: reactants   b: products

9.14 a. HPO_4^{2-}, $H_2PO_4^-$, H_3PO_4 b. HCN, CH_3COOH, HF

9.15 a. H_3PO_4 b. $H_2PO_4^-$ CH_3COO^- (stronger base)

9.16 Reactants favored.

9.17 a,b,d,e: H_2CO_3 c. HCN

9.18 a. 10^{-11} M acidic   c. 3.6×10^{-5} M basic
b. 10^{-3} M basic   d. 1.8×10^{-11} M acidic

9.19 a. 10^{-8} M basic   c. 1.9×10^{-4} M acidic
b. 10^{-5} M acidic   d. 1.4×10^{-11} M basic

9.20

	$[H_3O^+]$	$[^-OH]$
a.	10^{-11} M	10^{-3} M
b.	10^{-3} M	10^{-11} M
c.	1.5 M	6.7×10^{-15} M
d.	3.3×10^{-14} M	3.0×10^{-1} M

9.21 a. 6   b. 12   c. 5   d. 11

9.22 a. 1×10^{-13} M   b. 1×10^{-7} M   c. 1×10^{-3} M

9.23 a. basic   b. neutral   c. acidic

9.24 a. 6×10^{-11} M   b. 2×10^{-8} M   c. 5×10^{-5} M

9.25 a. 5.74   b. 11.036   c. 4.06   d. 10.118

9.27 a. $HNO_3(aq)$ + $NaOH(aq)$ ⟶ $H_2O(l)$ + $NaNO_3(aq)$
b. $H_2SO_4(aq)$ + 2 $KOH(aq)$ ⟶ 2 $H_2O(l)$ + $K_2SO_4(aq)$

9.29 $H_2SO_4(aq)$ + $CaCO_3(s)$
    ⟶ $CaSO_4(aq)$ + $\underline{H_2O(l) + CO_2(g)}$
                           from H_2CO_3

9.31 a,c,e: neutral   b,f: basic   d. acidic

9.33 0.41 M

9.35 a. Not a buffer since it contains a strong acid, HBr.
b. A buffer since HF is a weak acid and F^- is its conjugate base.
c. Not a buffer since it contains a weak acid only.

9.37 a,b,c: pH 7.21
The pH remains the same if equal amounts of the weak acid and conjugate base are present.

9.39 By the Arrhenius definition, an acid contains a hydrogen atom and dissolves in water to form a hydrogen ion, H^+, and a base contains hydroxide and dissolves in water to form ^-OH. A base must contain ^-OH. By the Brønsted–Lowry definition, acids and bases are classified according to whether they can donate or accept a proton—a positively charged hydrogen ion, H^+. A Brønsted–Lowry acid is a proton donor and a Brønsted–Lowry base is a proton acceptor. NH_3 is a base by the Brønsted–Lowry definition but not by the Arrhenius definition.

9.41 a,d,f

9.43 a,d,e,f

9.45 NH_3 can accept a proton because the N atom has a lone pair, but CH_4 does not have a lone pair, so it cannot.

9.47 a. H_2S b. HCO_3^- c. HNO_2 d. $\left[\begin{array}{c} H \quad H \\ | \quad\ | \\ H-C-N-H \\ | \quad\ | \\ H \quad H \end{array} \right]^+$

9.49 a. NO_2^- b. NH_3 c. HO_2^-

9.51 a. $HI(g)$ acid $I^-(aq)$ conjugate base
    $NH_3(g)$ base $NH_4^+(aq)$ conjugate acid
b. $HCOOH(l)$ acid $HCOO^-(aq)$ conjugate base
    $H_2O(l)$ base $H_3O^+(aq)$ conjugate acid
c. $HSO_4^-(aq)$ base $H_2SO_4(aq)$ conjugate acid
    $H_2O(l)$ acid $^-OH(aq)$ conjugate base

9.53 a. H_2CO_3 b. CO_3^{2-}

9.55 $HNO_3(aq)$ + $H_2O(l)$ ⟶ $H_3O^+(aq)$ + $NO_3^-(aq)$

9.57 A strong acid fully dissociates in water, whereas a weak acid only partially dissociates.

9.59 **A** represents HCl because it shows a fully dissociated acid.
B represents HF because it is only partially dissociated.

9.61 a. CH_3COOH b. H_3PO_4 c. H_2SO_4

9.63 a. H_2O b. HCO_3^- c. HSO_4^-

9.65 A strong acid readily donates a proton and forms a weak conjugate base that has little ability to accept a proton.

9.67 a. **A**   b. **B**   c. **B**

9.69 The equilibrium constant (K) shows the ratio of the concentrations of the products to the concentrations of the reactants. Since water serves as both the base and the solvent in acid–base reactions, its concentration is essentially constant, and the equation can be rearranged by multiplying both sides by $[H_2O]$. This forms a new constant called the acid dissociation constant, K_a. Thus, K and K_a are related by the concentration of H_2O, and $K_a = [H_2O]K$.

9.71 a. HSO_4^- stronger acid
    SO_4^{2-} conjugate base
    $H_2PO_4^-$
    HPO_4^{2-} conjugate base, stronger base

  b. CH_3COOH stronger acid
    CH_3COO^- conjugate base
    CH_3CH_2COOH
    $CH_3CH_2COO^-$ conjugate base, stronger base

9.73   acid   conjugate acid

  a. $H_3PO_4(aq) + {}^-CN(aq) \rightleftharpoons H_2PO_4^-(aq) + HCN(aq)$
    products favored

    acid   conjugate acid

  b. $Br^-(aq) + HSO_4^-(aq) \rightleftharpoons SO_4^{2-}(aq) + HBr(g)$
    reactants favored

    acid

  c. $CH_3COO^-(aq) + H_2CO_3(aq)$
    reactants favored   conjugate acid
    $\rightleftharpoons CH_3COOH(aq) + HCO_3^-(aq)$

9.75 a. 10^{-6} M basic   c. 3.3×10^{-11} M acidic
  b. 10^{-4} M basic   d. 4.0×10^{-4} M basic

9.77 a. 10^{-12} M basic   c. 1.6×10^{-8} M basic
  b. 2.5×10^{-7} M acidic   d. 1.2×10^{-2} M acidic

9.79 a. 12   b. 6.60   c. 7.80   d. 1.92

9.81 a. 1×10^{-12} M   c. 1.6×10^{-2} M
  b. 1×10^{-1} M   d. 1.3×10^{-9} M

9.83 $[H_3O^+] = 1.3 \times 10^{-6}$ M $[^-OH] = 7.7 \times 10^{-9}$ M

9.85 $[H_3O^+] = 7.9 \times 10^{-5}$ M $[^-OH] = 1.3 \times 10^{-10}$ M

9.87 a. 2.60   b. 12.17

9.89 HCl is fully dissociated to H_3O^+ and Cl^- and CH_3COOH is not fully dissociated.

9.91 a. $HBr(aq) + KOH(aq) \longrightarrow KBr(aq) + H_2O(l)$
  b. $2 HNO_3(aq) + Ca(OH)_2(aq)$
    $\longrightarrow 2 H_2O(l) + Ca(NO_3)_2(aq)$
  c. $HCl(aq) + NaHCO_3(aq)$
    $\longrightarrow NaCl(aq) + H_2O(l) + CO_2(g)$
  d. $H_2SO_4(aq) + Mg(OH)_2(aq)$
    $\longrightarrow 2 H_2O(l) + MgSO_4(aq)$

9.93 $2 HNO_3(aq) + CaCO_3(s)$
    $\longrightarrow Ca(NO_3)_2(aq) + CO_2(g) + H_2O(l)$

9.95 a,e: neutral   b,d,f: basic   c. acidic

9.97 0.14 M

9.99 0.12 M

9.101 25 mL

9.103 If the concentrations of the weak acid and its conjugate base are equal, then a proton donor and a proton acceptor are available and the H_3O^+ concentration will change very little when acid or base is added to a buffer.

9.105 Yes. HCN is a weak acid and ^-CN is its conjugate base, so in equal amounts they form a buffer.

9.107 a. HNO_2 is a proton donor that will react with added base and NO_2^- is a proton acceptor that will react with added acid.
  b. The concentration of HNO_2 increases and that of NO_2^- decreases when a small amount of acid is added to the buffer.
  c. The concentration of HNO_2 decreases and that of NO_2^- increases when a small amount of base is added to the buffer.

9.109 a. 12.66   b. 10.25

9.111 a. 4.74   b. 5.05   c. 4.44

9.113 The strength of an acid relates to how readily it donates a proton, whereas the concentration of an acid refers to how much of the acid is contained in a given volume of solution.

9.115 CO_2 combines with rainwater to form H_2CO_3, which is acidic, lowering the pH.

9.117 $[H_3O^+] = 3.2 \times 10^{-8}$ M $[^-OH] = 3.1 \times 10^{-7}$ M

9.119 By breathing into a bag, the individual breathes in air with a higher CO_2 concentration. Thus, the CO_2 concentration in the lungs and the blood increases, thereby lowering the pH.

9.121 A rise in CO_2 concentration leads to an increase in $H^+(aq)$ concentration by the following equilibria:
    $CO_2(g) + H_2O(l) \rightleftharpoons H_2CO_3(aq) \rightleftharpoons HCO_3^-(aq) + H^+(aq)$

9.123 $^-OCl(aq) + H_2O(l) \longrightarrow HOCl(aq) + {}^-OH(aq)$
    ^-OH is formed by this reaction, so the water is basic.

CHAPTER 10

10.1

	Atomic Number	Mass Number	Number of Protons	Number of Neutrons	Isotope Symbol
Cobalt-59	27	59	27	32	$^{59}_{27}Co$
Cobalt-60	27	60	27	33	$^{60}_{27}Co$

10.3 An α particle has no electrons around the nucleus.

10.5 a. β particle   b. α particle   c. positron

10.6 $^{222}_{86}Rn \longrightarrow {}^4_2He + {}^{218}_{84}Po$

10.7 $^{226}_{88}Ra \longrightarrow {}^4_2He + {}^{222}_{86}Rn$

10.9 $^{131}_{53}I \longrightarrow {}^0_{-1}e + {}^{131}_{54}Xe$

10.11 a. $^{74}_{33}As \longrightarrow {}^0_{+1}e + {}^{74}_{32}Ge$
  b. $^{15}_8O \longrightarrow {}^0_{+1}e + {}^{15}_7N$

10.13 a. $^{11}_5B \longrightarrow {}^{11}_5B + \gamma$
  b. $^{40}_{19}K \longrightarrow {}^{40}_{20}Ca + {}^0_{-1}e + \gamma$

10.14 a. 0.250 g   b. 0.0625 g   c. 0.00391 g   d. 9.54×10^{-7} g

10.15 a. 80.0 mg   b. 20.0 mg   c. 10.0 mg   d. 0.6 mg

10.17 4.4 mL

10.19 1/8

10.21 a. $^{235}_{92}U + {}^1_0n \longrightarrow {}^{90}_{38}Sr + {}^{143}_{54}Xe + 3 {}^1_0n$
  b. $^{235}_{92}U + {}^1_0n \longrightarrow {}^{133}_{51}Sb + {}^{100}_{41}Nb + 3 {}^1_0n$

10.23

	a. Atomic Number	b. Number of Protons	c. Number of Neutrons	d. Mass Number	Isotope Symbol
Fluorine-18	9	9	9	18	$^{18}_{9}F$
Fluorine-19	9	9	10	19	$^{19}_{9}F$

10.25

	Atomic Number	Mass Number	Number of Protons	Number of Neutrons	Isotope Symbol
a. Chromium-51	24	51	24	27	$^{51}_{24}Cr$
b. Palladium-103	46	103	46	57	$^{103}_{46}Pd$
c. Potassium-42	19	42	19	23	$^{42}_{19}K$
d. Xenon-133	54	133	54	79	$^{133}_{54}Xe$

10.27

	Change in Mass	Change in Charge
a. α particle	−4	−2
b. β particle	0	+1
c. γ ray	0	0
d. Positron	0	−1

10.29

	Mass	Charge
a. α	4	+2
b. n	1	0
c. γ	0	0
d. β	0	−1

10.31

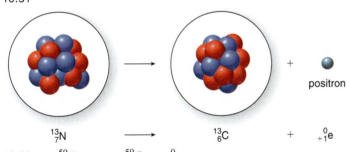

$^{13}_{7}N \longrightarrow ^{13}_{6}C + ^{0}_{+1}e$

10.33 a. $^{59}_{26}Fe \longrightarrow ^{59}_{27}Co + ^{0}_{-1}e$

b. $^{190}_{78}Pt \longrightarrow ^{186}_{76}Os + ^{4}_{2}He$

c. $^{178}_{80}Hg \longrightarrow ^{178}_{79}Au + ^{0}_{+1}e$

10.35 a. $^{90}_{39}Y \longrightarrow ^{90}_{40}Zr + ^{0}_{-1}e$

b. $^{135}_{60}Nd \longrightarrow ^{135}_{59}Pr + ^{0}_{+1}e$

c. $^{210}_{83}Bi \longrightarrow ^{206}_{81}Tl + ^{4}_{2}He$

10.37 $^{214}_{83}Bi \longrightarrow ^{214}_{84}Po + ^{0}_{-1}e$

$^{214}_{83}Bi \longrightarrow ^{210}_{81}Tl + ^{4}_{2}He$

10.39 a. $^{232}_{90}Th \longrightarrow ^{4}_{2}He + ^{228}_{88}Ra$

b. $^{25}_{11}Na \longrightarrow ^{25}_{12}Mg + ^{0}_{-1}e$

c. $^{118}_{54}Xe \longrightarrow ^{118}_{53}I + ^{0}_{+1}e$

d. $^{243}_{96}Cm \longrightarrow ^{4}_{2}He + ^{239}_{94}Pu$

10.41 4.0 days

10.43

	Iodine-131	Xenon-131
a.	32 mg	32 mg
b.	16 mg	48 mg
c.	8.0 mg	56 mg
d.	4.0 mg	60. mg

10.45 No, 25% remains.

10.47 In artifacts over 50,000 years old, the percentage of carbon-14 is too small to accurately measure.

10.49 a. 10 mCi   b. 5 mCi   c. 1 mCi

10.51 1.9×10^8 disintegrations/second

10.53 2.3 mL

10.55 2.6 mL

10.57 Curie refers to the number of disintegrations per second and the number of rads indicates the radiation absorbed by one gram of a substance.

10.59 20 Sv = 2,000 rem
This represents a fatal dose because 600 rem is uniformly fatal.

10.61 Nuclear fission refers to the splitting of nuclei and fusion refers to the joining of small nuclei to form larger ones.

10.63 a. fusion
b. fission
c. both
d. fusion

10.65 a. $^{235}_{92}U + ^{1}_{0}n \longrightarrow ^{137}_{50}Sn + ^{97}_{42}Mo + 2\,^{1}_{0}n$

b. $^{235}_{92}U + ^{1}_{0}n \longrightarrow ^{93}_{36}Kr + ^{140}_{56}Ba + 3\,^{1}_{0}n$

10.67 tritium; $^{3}_{1}H$

10.69 The containment of radiation leaks and disposal of radioactive waste are two problems that are associated with nuclear power production.

10.71 $^{209}_{83}Bi + ^{58}_{26}Fe \longrightarrow ^{266}_{109}Mt + ^{1}_{0}n$
meitnerium-266

10.73 a. $^{74}_{33}As \longrightarrow ^{0}_{+1}e + ^{74}_{32}Ge$
b. 4 mg
c. 1.5 mL

10.75 a. $^{192}_{77}Ir \longrightarrow ^{192}_{78}Pt + ^{0}_{-1}e + \gamma$
b. 30. mg
c. 2 Ci

10.77 a. used for the treatment and diagnosis of thyroid diseases
b. used for the treatment of breast cancer
c. used for the diagnosis of heart disease

10.79 a. Iodine-125, with its longer half-life, is used for the treatment of prostate cancers using implanted radioactive seeds.
b. Iodine-131, with its shorter half-life, is used for the diagnostic and therapeutic treatment of thyroid diseases and tumors. A patient is administered radioactive iodine-131, which is then incorporated into the thyroid hormone, thyroxine. Since its half-life is short, the radioactive iodine isotope decays so that little remains after a month or so.

10.81 Radiology technicians must be shielded to avoid exposure to excessive and dangerous doses of radiation.

10.83 High doses of stable iodine will prevent the absorption and uptake of the radioactive iodine-131.

10.85 370 kcal

CHAPTER 11

11.1 organic (a), (e)
      inorganic (b), (c), (d), (f)

11.2 a. 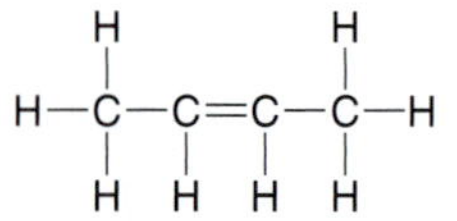 [structure] d. 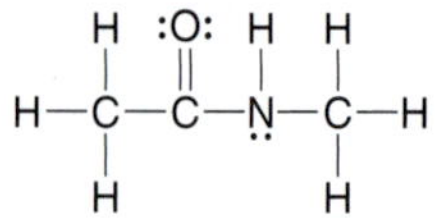 [structure]

      b. 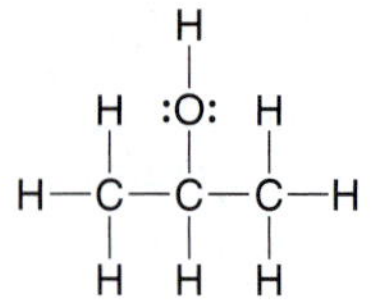 [structure] e. 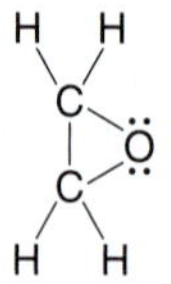 [structure]

      c. 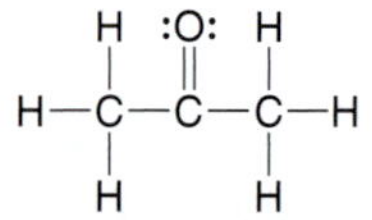 [structure]

11.3 a. (1), (2) both trigonal planar
      b. (1) linear; (2) tetrahedral; (3) bent
      c. (1) trigonal planar; (2) bent
      d. (1) trigonal planar; (2) linear

11.5 a. 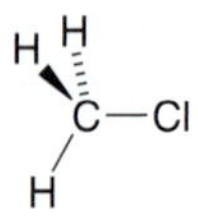 [structure] b. 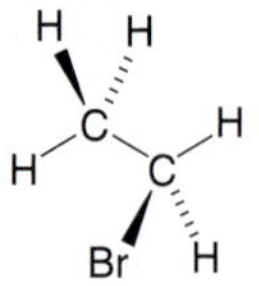 [structure]

11.6 (1) bent; (2) linear; (3),(4) trigonal planar; (5) bent;
      (6) tetrahedral

11.7 a. $CH_3(CH_2)_3CH_3$
      b. $Br(CH_2)_2Br$
      c. $CH_3CH_2O(CH_2)_2OH$
      d. $CH_3CH_2C(CH_3)_3$

11.9 a. 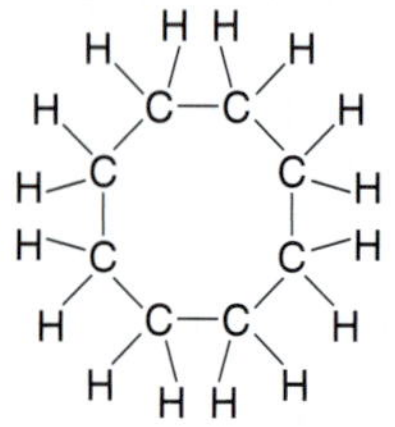 [structure] c. 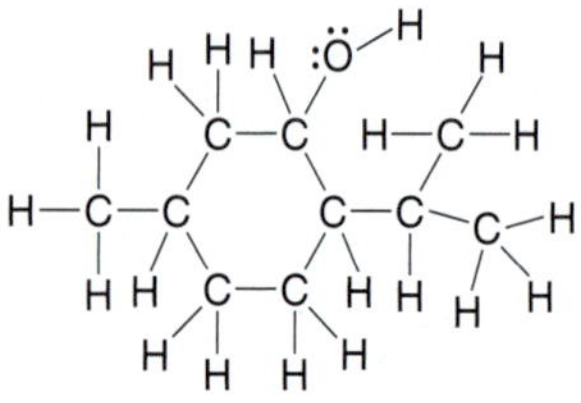 [structure]

      b. 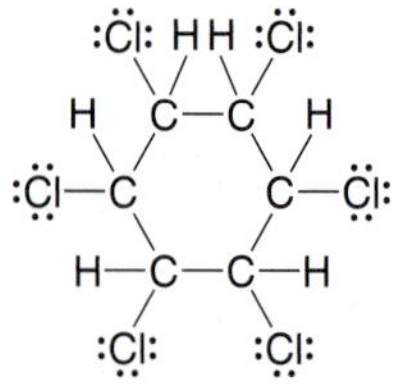 [structure]

11.10 C1: no H; C2: 1 H; C3: 2 H; C4: 1 H; C5: 1 H

11.11 a. hydroxyl
      b. alkene; aromatic ring
      c. two amine groups

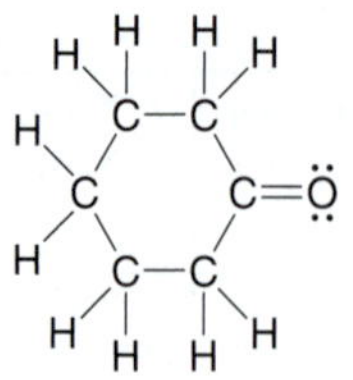

11.12 a. 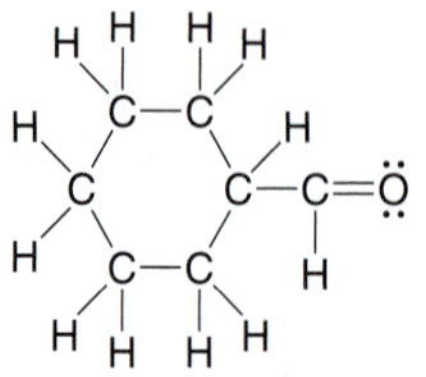 [structure] aldehyde

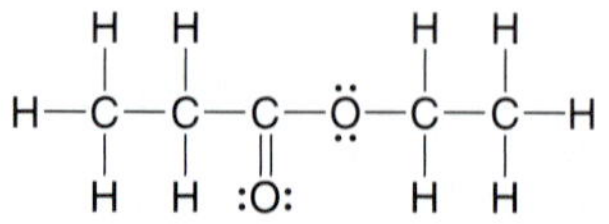

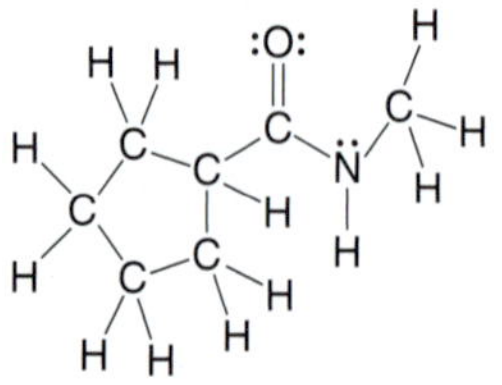

      b. 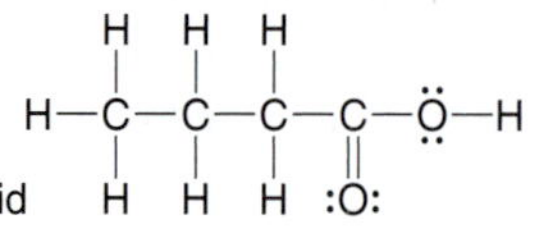 [structure] carboxylic acid

11.13 benzaldehyde: aldehyde
      citric acid: three carboxylic acid groups
      isoamyl acetate: ester
      ethyl butanoate: ester

11.14 hydroxyl group, alkene, aromatic ring

11.15 a. amide, aromatic ring, ether, hydroxyl, amine
      b. ether, aromatic ring, hydroxyl, alkyne

11.17 a. nonpolar; all nonpolar bonds
      b. polar; one polar C=O bond
      c. nonpolar; the four bond dipoles cancel
      d. polar; the polar C—N and N—H bond dipoles don't
         cancel

11.19 a. insoluble
      b. soluble
      c. insoluble

11.21 Niacin is soluble in water.

11.23 inorganic: (a), (b)
      organic: (c)

11.25 a. 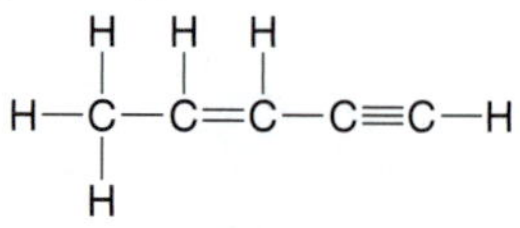 [structure] c. 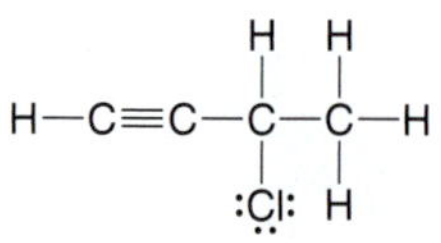 [structure]

      b. 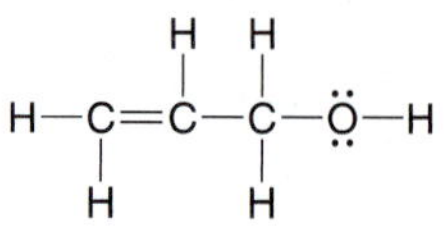 [structure] d. 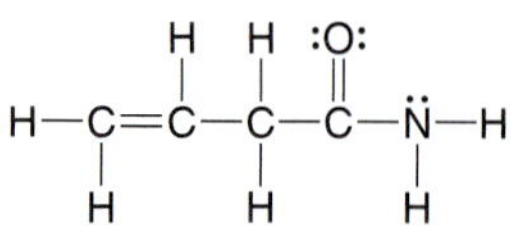 [structure]

11.27 a. (1) trigonal planar; (2) bent
      b. (1) trigonal planar; (2) linear
      c. (1) tetrahedral; (2) bent; (3) tetrahedral

11.29 a. (1) and (2) 180°
      b. (1) and (2) 120°
      c. (1) and (2) 109.5°

11.31 (1) bent; (2) trigonal planar; (3) tetrahedral; (4) trigonal
      pyramidal

11.33 Each C is surrounded by three groups, so each carbon is
      trigonal planar giving 120° bond angles.

11.35 a. $CH_3(CH_2)_2C(CH_3)_3$ c. $(CH_3)_2CHCH_2O(CH_2)_2CH_3$
      b. $BrCH_2CH_2CHCH_2CH_3$ d. $CH_3(CH_2)_2CHO$
                          |
                          Cl

11.37 a. 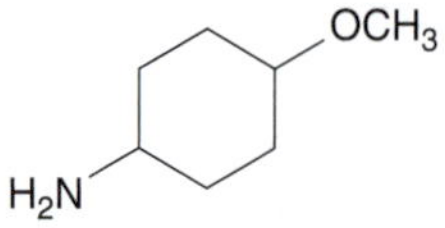 [structure] b. 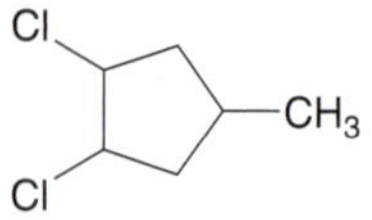 [structure]

c. ketone [structure]

d. ester [structure]

e. amide [structure]

11.39 a. [structure: a 9-carbon chain with substituents, all H's shown]

b. [structure with two CH_3 branches and an $\ddot{O}-H$ group]

c. [structure with $:O:$ carbonyl and $\ddot{O}$ ester linkage on a carbon chain]

d. [ring structure with $:\ddot{C}l:$ and $\ddot{C}l:$ substituents, ester group, and $H-\ddot{O}:$ group]

11.41 a. $(CH_3)_3CHCH_2CH_3$ — Five bonds to C

b. $CH_3(CH_2)_4CH_3OH$ — Five bonds to C

c. $(CH_3)_2C=CH_3$ — Five bonds to C

d. [cyclohexene ring with two CH_3 groups] — Five bonds to C

e. [cyclopentane ring with three CH_3 groups] — Five bonds to C

11.43 a. [aromatic ring with $H-\ddot{O}-$ substituent, a $-C-C-C-\ddot{O}-H$ chain with $:O:$ carbonyl and $:N-H$ amine]

b. [ring structure with multiple $:O:$, $\ddot{O}:$, $H-\ddot{O}:$, and $-C-$ groups]

11.45 H$-\ddot{O}-$C [aromatic ring with $:\ddot{O}:$ H group, $-C-C-\ddot{N}-C-C-H$ chain, $H-\ddot{O}-C-H$ and $H-C-H$ substituents]

11.47 a. two alkenes
       b. ketone
       c. ester

11.49 a. two hydroxyl groups
       b. ester
       c. two alkenes, two alkynes, one hydroxyl
       d. hydroxyl, ether, aldehyde, aromatic ring

11.51 In an alcohol, O is bonded to C and H. In an ether, O is bonded to 2 C's.
       Examples: CH_3CH_2OH (alcohol) and CH_3OCH_3 (ether)

11.53 In a carboxylic acid, the C=O is bonded to a hydroxyl group. In an ester, the C=O is bonded to an OR group.
       Examples: CH_3CH_2COOH (carboxylic acid) and $CH_3CO_2CH_3$ (ester)

11.55 a. $HC\equiv C-CH_3$

       b. $CH_3CH_2CH_2OH$

       c. CH_3CH_2CHO

       d. $CH_3CH_2\overset{O}{\overset{\|}{C}}CH_3$

       e. $CH_3O\overset{O}{\overset{\|}{C}}CH_3$

11.57 alkane $CH_3CH_2CH_2CH_2CH_3$ C_5H_{12}

       alkene $CH_3-\overset{H}{\underset{H}{C}}-\overset{H}{C}=\overset{H}{C}-CH_3$ C_5H_{10}

       alkyne $CH_3-\overset{H}{\underset{H}{C}}-C\equiv C-CH_3$ C_5H_8

11.59 carboxylic acid, amine, amide, ester, aromatic ring

11.61 NaCl dissolves in water but not in dichloromethane and cholesterol dissolves in dichloromethane but not in water.

11.63 Net dipoles and molecular geometry will determine whether a molecule is polar or nonpolar. When a molecule has polar bonds, we must determine whether the bond dipoles cancel or reinforce.

11.65 a. $CH_3\overset{\delta^+}{C}H_2-\overset{\delta^-}{O}-\overset{\delta^+}{C}H_2CH_3$

       b. $\underset{\delta^-}{F}-\overset{\overset{\delta^-}{Cl}}{\underset{\underset{\delta^-}{Cl}}{\overset{\delta^+}{C}}}\overset{\delta^-}{Cl}$

       c. $\overset{\delta^+}{H}-\overset{\delta^-}{O}-CH_2\overset{\overset{\delta^-\ \delta^+}{OH}}{\underset{\delta^+\ \delta^+\ \delta^+}{C}HCH_2}-\overset{\delta^-}{O}-\overset{\delta^+}{H}$

11.67 a. nonpolar
      b. polar
      c. polar

11.69 Sucrose has multiple OH groups and this makes it water soluble. 1-Dodecanol has a long hydrocarbon chain and a single OH group so it is not water soluble.

11.71 a. Spermaceti wax is an ester.
      b. The very long hydrocarbon chains would make it insoluble in water and soluble in organic solvents.

11.73 Fat-soluble vitamins will persist in tissues and accumulate, whereas any excess of water-soluble vitamins will be excreted in the urine.

11.75 Vitamin E is a fat-soluble vitamin, making it soluble in organic solvents and insoluble in water.

11.77 a. Yes, there are just enough H's to give each C four bonds.
      b. No, there are too many H's.
      c. Yes, if you put the three C's in a ring, each C gets two H's.

11.79 a. $CH_3CH_2CH_2NHCH_3$

      b.

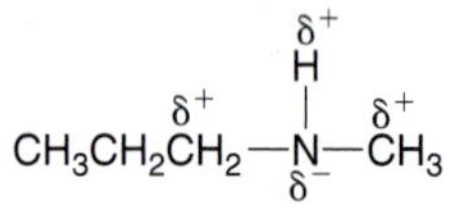

      c. trigonal pyramid around N
      d. This is a polar molecule. The bond dipoles don't cancel.

11.81

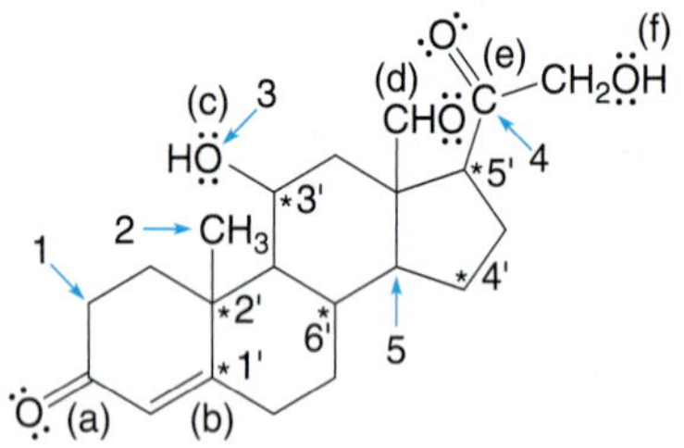

      a. functional groups:
        (a) ketone
        (b) alkene
        (c) hydroxyl
        (d) aldehyde
        (e) ketone
        (f) hydroxyl
      b. Each O needs two lone pairs.
      c. 21 C's
      d. number of H's at carbon:
        1'- 0 H's
        2'- 0 H's
        3'- 1 H
        4'- 2 H's
        5'- 1 H
        6'- 1 H
      e. shape at each carbon:
        1- tetrahedral
        2- tetrahedral
        3- bent
        4- trigonal planar
        5- tetrahedral

f.

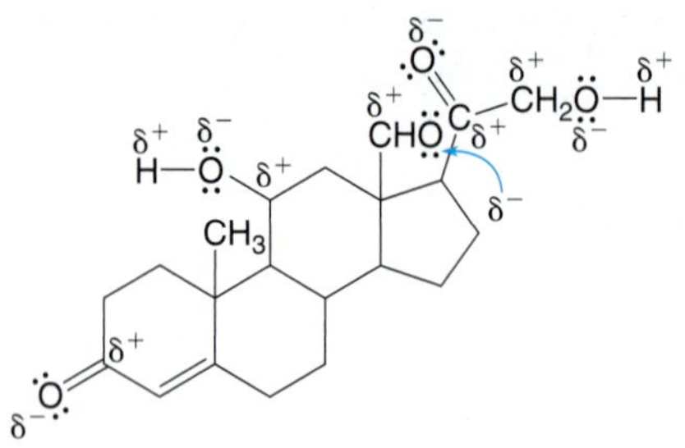

11.83 The waxy coating will prevent loss of water from leaves and will keep leaves crisp.

11.85 THC is not water soluble. THC is fat soluble and will therefore persist in tissues for an extended period of time. Ethanol is water soluble and will be quickly excreted in the urine.

CHAPTER 12

12.1 a. 8   b. 8   c. 18   d. 16

12.3 a. not isomers   b. isomers   c. isomers

12.5

12.7

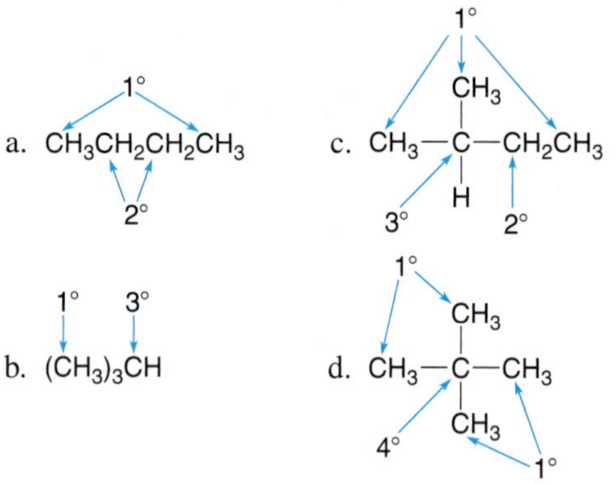

12.9 a.    b.

12.11 a. 3-methylpentane
      b. 4-methyl-5-propylnonane
      c. 2,4-dimethylhexane

12.13

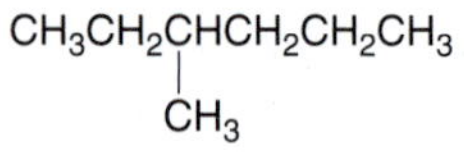

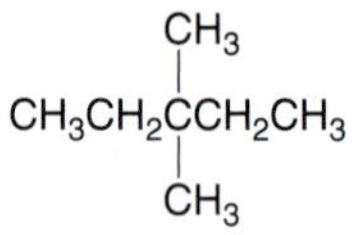

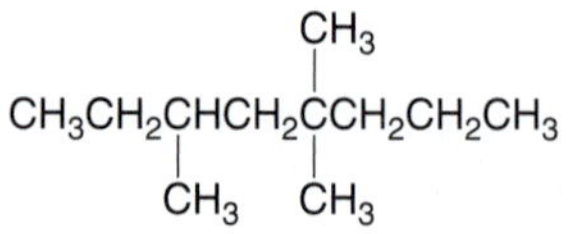

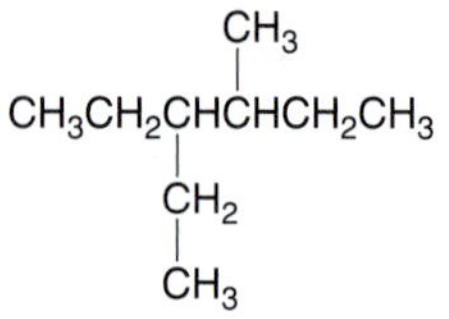

12.15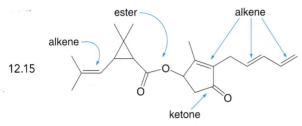

12.17 a. methylcyclobutane
      b. 1,1-dimethylcyclohexane
      c. 1-ethyl-3-propylcyclopentane
      d. 1-ethyl-4-methylcyclohexane

12.19 gasoline < kerosene < diesel

12.21 a. $CH_3CH_2CH_3 + 5\ O_2 \longrightarrow 3\ CO_2 + 4\ H_2O$
      b. $2\ CH_3CH_2CH_2CH_3 + 13\ O_2 \longrightarrow 8\ CO_2 + 10\ H_2O$

12.23 64

12.25 acyclic alkane

12.27 26 C has 54 H's
      27 C has 56 H's
      28 C has 58 H's
      29 C has 60 H's
      30 C has 62 H's

12.29 To form a cycloalkane, you need to form a C—C bond between 2 C's in a chain. To do this, 1 H from each C must be removed. This means a cycloalkane always has two fewer H's than an acyclic alkane of the same number of C's.

12.31 a. $CH_3(CH_2)_3CH_3$
      c. $(CH_3)_3CC(CH_3)_3$
      b. $CH_3CH_2CHCHCH_3$
      d.

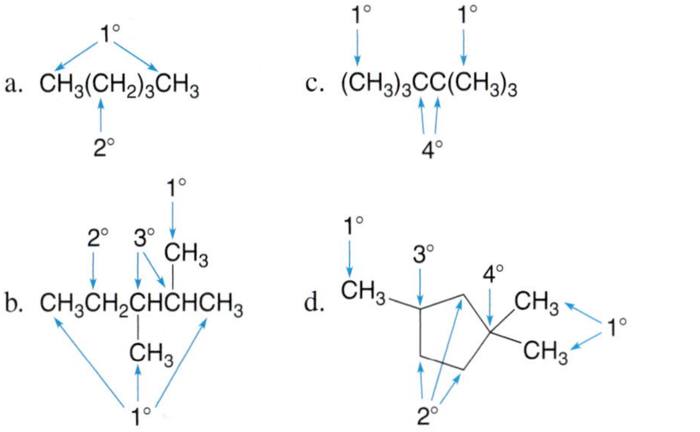

12.33 a. $CH_3CH_2CH_2C{-}CH_3$ (with CH_3 groups)
      b. $CH_3CH_2CH_2CH_2CH_2CH_2CH_3$

12.35 a,c,d: constitutional isomers
      b. identical

12.37 a,d: constitutional isomers
      b,c,e: identical

12.39 a. 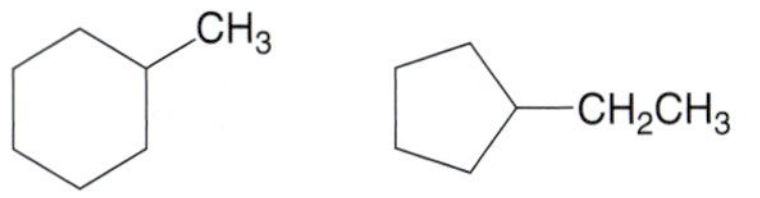
      b. $CH_3CH_2CH_2CH_2CH_2OH$ $CH_3OCH_2CH_2CH_2CH_3$
      c. $CH_3CH_2CH_2Cl$ CH_3CHCH_3 (Cl)

12.41 $CH_3CHCH_2CH_2CH_2CH_2CH_3$ (with CH_3)
      $CH_3CH_2CHCH_2CH_2CH_2CH_3$ (with CH_3)
      $CH_3CH_2CH_2CHCH_2CH_2CH_3$ (with CH_3)

12.43 ▷—OH
      $CH_3\overset{O}{\overset{\|}{C}}CH_3$
      (epoxide)—CH_3

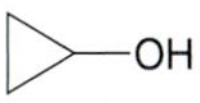

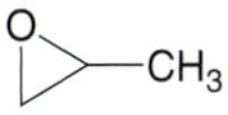

12.45 hexane; 2-methylpentane; 3-methylpentane; 2,2-dimethylbutane; 2,3-dimethylbutane

12.47 a. 3-methylheptane
      b. 3,5-dimethyloctane
      c. 3,3-diethylhexane
      d. 4-ethyl-2,6-dimethyloctane
      e. 6-ethyl-2-methyloctane
      f. 2,2,8,8-tetramethyldecane

12.49 a. cyclooctane
      b. methylcyclopropane
      c. 1,1,2,2-tetramethylcyclobutane
      d. 1,3-diethylcyclohexane

12.51 a. $CH_3CH_2CHCH_2CH_2CH_3$
               CH_2CH_3
      b. $CH_3CH_2CCH_2CH_2CH_2CH_3$ (with CH_3 and CH_2CH_3)
      c. $CH_3CHCHCHCH_2CH_2CH_2CH_3$ (with CH_3, CH_3, CH_3, CH_3)
      d. 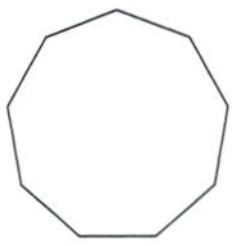
      e. 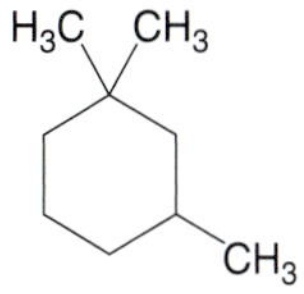 (H_3C, CH_3, CH_3)
      f. 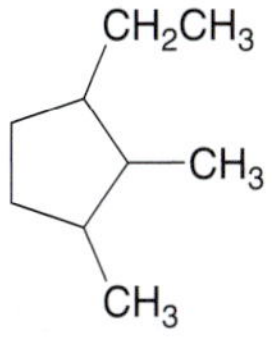 (CH_2CH_3, CH_3, CH_3)

12.53 a. 2-methylbutane: Number to give CH_3 the lower number, 2 not 3.

b. methylcyclopentane: no number assigned if only one substituent

c. 2-methylpentane: five-carbon chain

d. 2,5-dimethylheptane: longest chain not chosen

e. 1,3-dimethylcyclohexane: Number to give the second substituent the lower number.

f. 1-ethyl-2-propylcyclopentane: lower number assigned alphabetically

12.55 a.

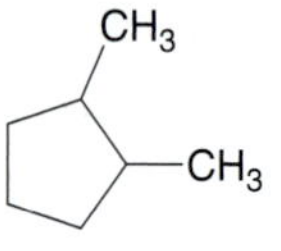

1,2-dimethylcyclopentane

b.

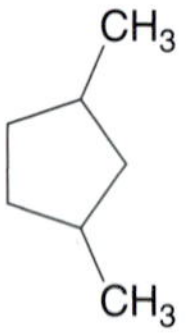

1,3-dimethylcyclopentane

c.

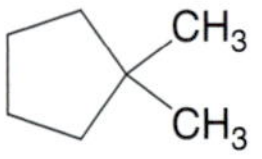

1,1-dimethylcyclopentane

12.57 a.

b.

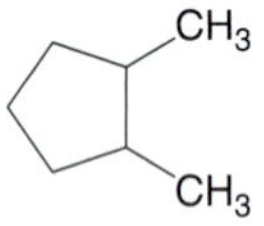

c.

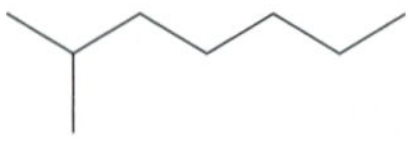

12.59 a.

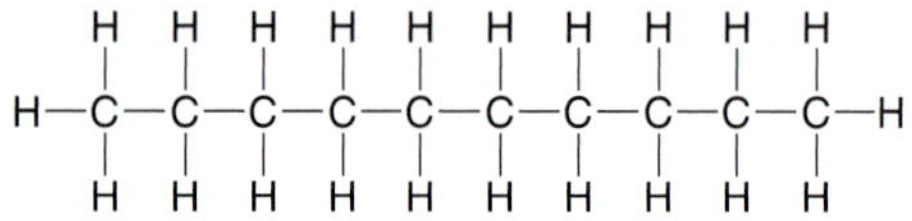

b.

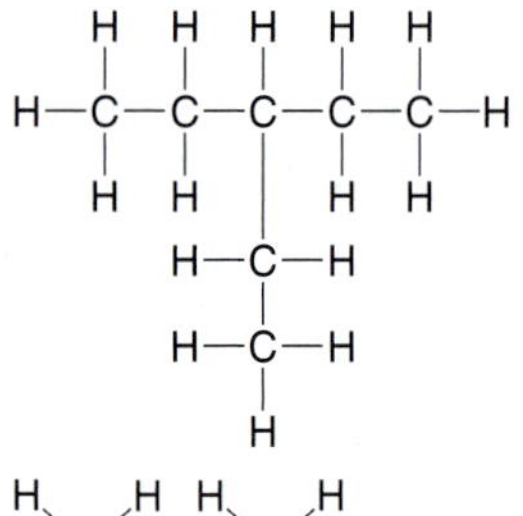

c.

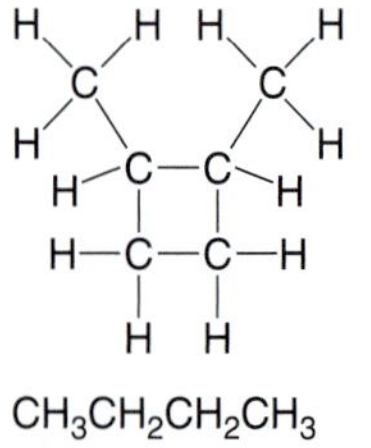

12.61 a. $CH_3CH_2CH_2CH_3$

b.

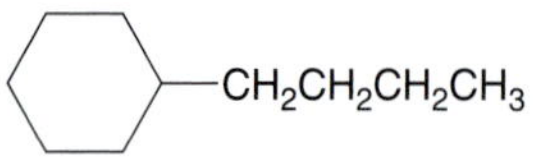

12.63 a. $(CH_3)_4C$, $(CH_3)_2CHCH_2CH_3$, $CH_3CH_2CH_2CH_2CH_3$

b. $(CH_3)_2CHCH(CH_3)_2$, $CH_3CH_2CH_2CH(CH_3)_2$, $CH_3(CH_2)_4CH_3$

12.65 Hexane is a nonpolar hydrocarbon and is soluble in organic solvents but not in water.

12.67 CO_2 and H_2O

12.69 a. $2\ CH_3CH_3 + 7\ O_2 \longrightarrow 4\ CO_2 + 6\ H_2O$

b. $(CH_3)_2CHCH_2CH_3 + 8\ O_2 \longrightarrow 5\ CO_2 + 6\ H_2O$

12.71 a. $2\ CH_3CH_2CH_3 + 7\ O_2 \longrightarrow 6\ CO + 8\ H_2O$

b. $2\ CH_3CH_2CH_2CH_3 + 9\ O_2 \longrightarrow 8\ CO + 10\ H_2O$

12.73 $C_6H_{12}O_6 + 6\ O_2 \longrightarrow 6\ CO_2 + 6\ H_2O$

12.75 Higher molecular weight alkanes in warmer weather means less evaporation. Lower molecular weight alkanes in colder weather means the gasoline won't freeze.

12.77 The mineral oil can prevent the body's absorption of important fat-soluble vitamins, which are expelled with the mineral oil.

12.79 c. The nonpolar asphalt will be most soluble in the paint thinner because "like dissolves like."

12.81 a. 3-ethyl-4,4-dimethylheptane

b.

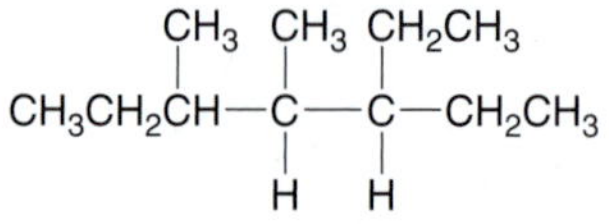

c. not water soluble

d. soluble in organic solvents

e. $C_{11}H_{24} + 17\ O_2 \longrightarrow 11\ CO_2 + 12\ H_2O$

f.

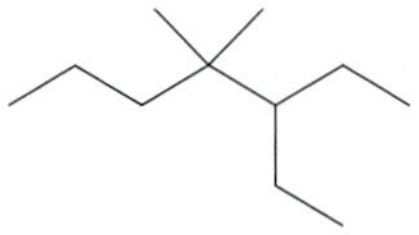

12.83 $C_{10}H_{18}$

12.85 Cyclopentane has a more rigid structure. The rings can get closer together since they are not floppy, resulting in an increased force of attraction.

CHAPTER 13

13.1 a.

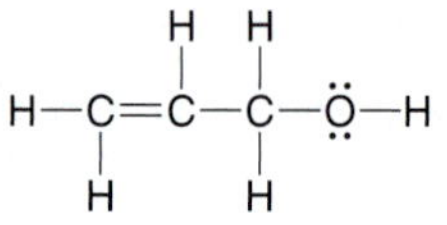

b.

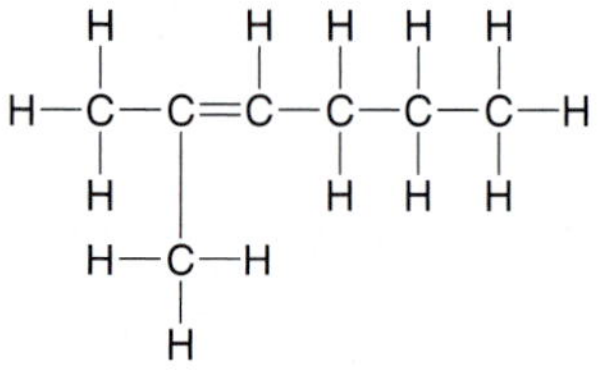

c.

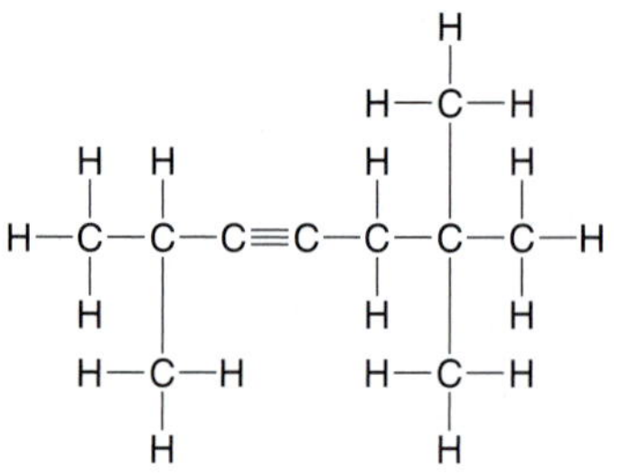

13.3 a. alkene

b. saturated hydrocarbon

c. alkyne

d. alkene

13.5 a. 3-methyl-1-pentene

b. 3-ethyl-3-heptene

c. 2,4-heptadiene

d. 3-ethylcyclopentene

13.7 a. $CH_2=CHCH_2\overset{\overset{\displaystyle CH_3}{|}}{C}HCH_2CH_3$

c. $CH_3\overset{\overset{\displaystyle CH_3}{|}}{C}HC\equiv CCHCH_3$ with CH_3 below

b. $CH_3C=CHCH_2\overset{\overset{\displaystyle CH_2CH_3}{|}}{C}HCH_2CH_3$ with CH_3 below

d. cyclobutene with $CH_2CH_2CH_3$ substituent

13.9 a.

$$\begin{array}{c} H \quad\quad H \\ C=C \\ CH_3 \quad CH_2CH_2CH_2CH_2CH_3 \end{array}$$

b.

$$\begin{array}{c} H \quad\quad CH_2CH_2CH_3 \\ C=C \\ CH_3CH_2 \quad\quad H \end{array}$$

c.

$$\begin{array}{c} \quad\quad CH_3 \\ H \quad CHCH_3 \\ C=C \\ CH_3 \quad H \end{array}$$

13.11

cis

$$\begin{array}{c} H \quad H \\ C=C \\ H \quad C=C \quad CH_2CH_2CH_3 \\ HOCH_2(CH_2)_7CH_2 \quad H \end{array}$$

trans

13.13

$$\begin{array}{c} H \quad H\ H \quad H\ H \quad H\ H \quad H \\ C=C \quad C=C \quad C=C \quad C=C \\ CH_3CH_2CH_2CH_2CH_2 \quad CH_2 \quad CH_2 \quad CH_2 \quad CH_2CH_2CH_2\overset{O}{\overset{||}{C}}OH \end{array}$$

13.15 three aromatic rings, alkene, ether, amine

13.17 a. $CH_3CH_2CH_2CH_2CH_2CH_3$

c. cyclohexane with CH_3 substituent

b. $CH_3\overset{\overset{\displaystyle |}{}}{C}HCH_2CH_2CH(CH_3)_2$ with CH_3 below

13.19 a. $CH_3\overset{\overset{\displaystyle |}{}}{\underset{\underset{\displaystyle Br}{|}}{C}}\overset{\overset{\displaystyle |}{}}{\underset{\underset{\displaystyle H}{|}}{C}}CH_3$

c. $CH_3\overset{\overset{\displaystyle CH_3}{|}}{\underset{\underset{\displaystyle Cl}{|}}{C}}\overset{\overset{\displaystyle H}{|}}{\underset{\underset{\displaystyle H}{|}}{C}}CH_3$

b. cyclohexane with Br, CH_3, H, H

d. cyclopentane with H, Cl, H, H

13.21 a. $CH_3CH_2CH_2\overset{\overset{\displaystyle H}{|}}{\underset{\underset{\displaystyle H}{|}}{C}}\overset{\overset{\displaystyle H}{|}}{\underset{\underset{\displaystyle H}{|}}{C}}H$

d. $CH_3CH_2CH_2\overset{\overset{\displaystyle H}{|}}{\underset{\underset{\displaystyle Br}{|}}{C}}\overset{\overset{\displaystyle H}{|}}{\underset{\underset{\displaystyle H}{|}}{C}}H$

b. $CH_3CH_2CH_2\overset{\overset{\displaystyle H}{|}}{\underset{\underset{\displaystyle Cl}{|}}{C}}\overset{\overset{\displaystyle H}{|}}{\underset{\underset{\displaystyle Cl}{|}}{C}}H$

e. $CH_3CH_2CH_2\overset{\overset{\displaystyle H}{|}}{\underset{\underset{\displaystyle Cl}{|}}{C}}\overset{\overset{\displaystyle H}{|}}{\underset{\underset{\displaystyle H}{|}}{C}}H$

c. $CH_3CH_2CH_2\overset{\overset{\displaystyle Br}{|}}{\underset{\underset{\displaystyle H}{|}}{C}}\overset{\overset{\displaystyle Br}{|}}{\underset{\underset{\displaystyle H}{|}}{C}}H$

f. $CH_3CH_2CH_2\overset{\overset{\displaystyle HO}{|}}{\underset{\underset{\displaystyle H}{|}}{C}}\overset{\overset{\displaystyle H}{|}}{\underset{\underset{\displaystyle H}{|}}{C}}H$

13.23

$$\begin{array}{c} H \quad CH_2 \quad H \\ C=C \quad\quad C=C \\ CH_3CH_2CH_2CH_2CH_2 \quad H\ H \quad CH_2CH_2CH_2CH_2CH_2CH_2CH_2COOH \end{array}$$

The trans stereoisomer will have a higher melting point because the molecules stack together better.

13.24 a.

$$\begin{array}{c} H \quad CH_3\ H \quad CH_3\ H \quad CH_3 \\ -C-C-C-C-C-C- \\ H \quad CH_2\ H \quad CH_2\ H \quad CH_2 \\ CH_3 \quad\quad CH_3 \quad\quad CH_3 \end{array}$$

b.

$$\begin{array}{c} H \quad CH_3\ H \quad CH_3\ H \quad CH_3 \\ -C-C-C-C-C-C- \\ H \quad CN\ H \quad CN\ H \quad CN \end{array}$$

c.

$$\begin{array}{c} H \quad H\ H \quad H\ H \quad H \\ -C-C-C-C-C-C- \\ H \quad\quad H \quad\quad H \end{array}$$
(with chlorobenzene rings on alternate carbons, Cl at meta positions)

13.25 $CH_2=\overset{\overset{\displaystyle H}{|}}{C}\underset{\underset{\displaystyle \overset{\displaystyle O}{|}}{}}{}$ ... $O-\overset{O}{\overset{||}{C}}OCH_3$

13.26 a. propylbenzene
b. *p*-ethyliodobenzene
c. *m*-butylphenol
d. 2-bromo-5-chlorotoluene

13.27 a. benzene ring with $CH_2CH_2CH_2CH_2CH_3$

c. benzene ring with NH_2 and Br

b. benzene ring with two Cl (ortho)

d. benzene ring with CH_2CH_3, CH_2CH_3, and Cl

13.29 CO_2 and H_2O

13.31 a. benzene ring with three Cl

c. benzene ring with two Cl and SO_3H

b. benzene ring with Cl, NO_2, Cl

13.33 a. $C_{10}H_{22}$
b. $C_{10}H_{20}$
c. $C_{10}H_{18}$

13.35 $HC{\equiv}CCH_2CH_2CH_3$

       $CH_3C{\equiv}CCH_2CH_3$

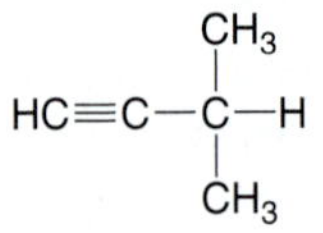

$$HC{\equiv}C-\underset{\underset{\displaystyle CH_3}{|}}{\overset{\overset{\displaystyle CH_3}{|}}{C}}-H$$

13.37   a.

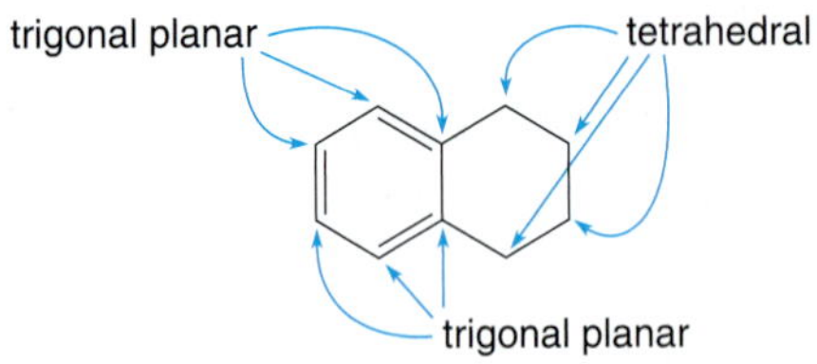

trigonal planar / tetrahedral / trigonal planar

  b.

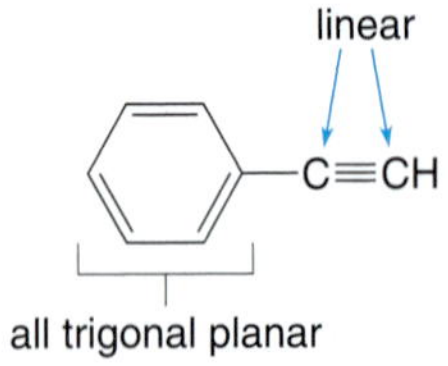

linear / all trigonal planar

  c. tetrahedral

trigonal planar

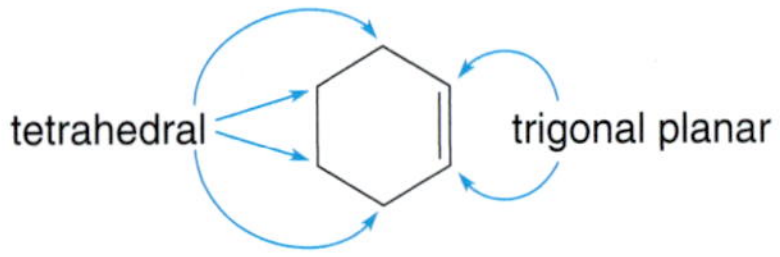

13.39   a. 5,5-dimethyl-1-hexene
       b. 3-ethyl-5,7-dimethyl-3-octene
       c. 2-ethyl-1-heptene
       d. 5,5-dimethyl-2-hexyne
       e. 6-ethyl-5,6-dimethyl-2-octyne
       f. 3,3-dimethyl-1,5-hexadiene

13.41 $HC{\equiv}CCH_2CH_2CH_3$ 1-pentyne

       $CH_3C{\equiv}CCH_2CH_3$ 2-pentyne

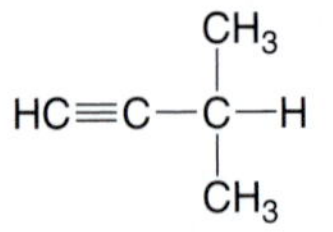

$$HC{\equiv}C-\underset{\underset{\displaystyle CH_3}{|}}{\overset{\overset{\displaystyle CH_3}{|}}{C}}-H \quad \text{3-methyl-1-butyne}$$

13.43   a. 4-methylcyclohexene
       b. 3,3-diethylcyclobutene
       c. 3,5-dimethylcyclohexene

13.45   a. $CH_2{=}CHCH_2CH_2CH_2CH_2CH_3$

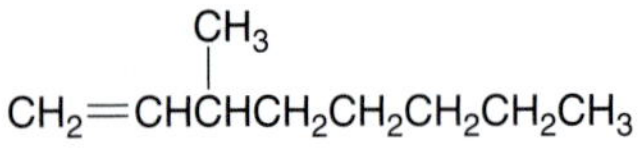

       b.

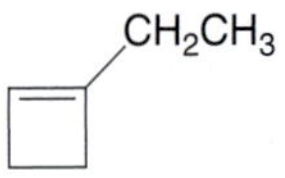

       c. $CH_3CH_2C{\equiv}CCH_3$ (with CH_3 and H)

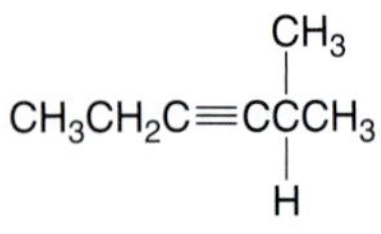

       d. $CH_3CH_2\underset{\underset{\displaystyle H}{|}}{\overset{\overset{\displaystyle CH_2CH_3}{|}}{C}}HC{=}\underset{\underset{\displaystyle CH_2CH_3}{|}}{\overset{\overset{\displaystyle CH_3}{|}}{C}}-CHCH_3$

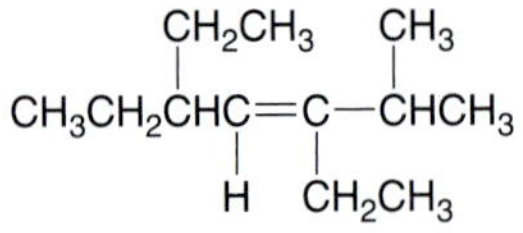

       e. $CH_2{=}CH-CH{=}CHCH_2CH_2CH_3$

       f.

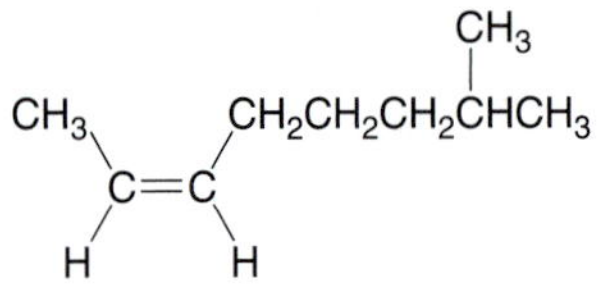

13.47   a. Assign lower number to alkene: 2-methyl-2-hexene.
       b. five-carbon chain: 2-pentene
       c. Number to put the C=C between C1 and C2, and then give the first substituent the lower number: 1,6-dimethylcyclohexene.
       d. seven-carbon chain: 3-methyl-1-heptyne

13.49

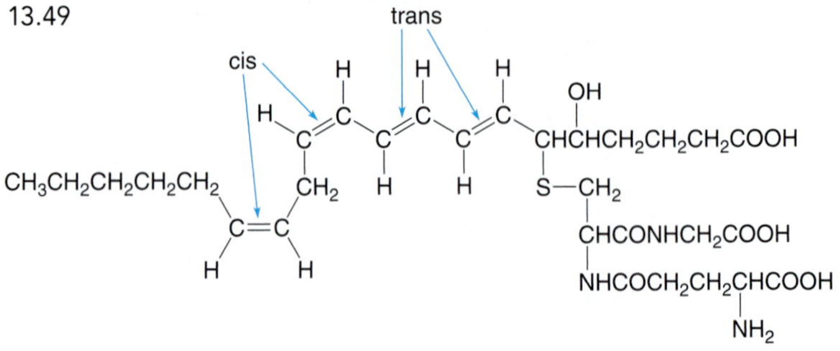

13.51

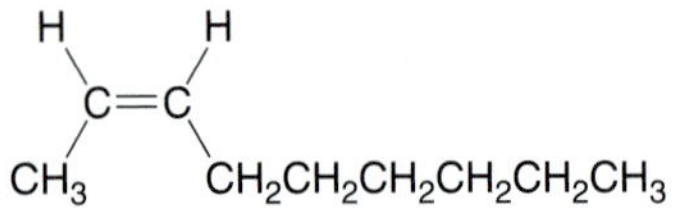

cis-2-nonene

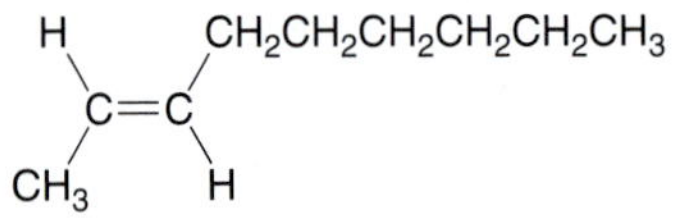

trans-2-nonene

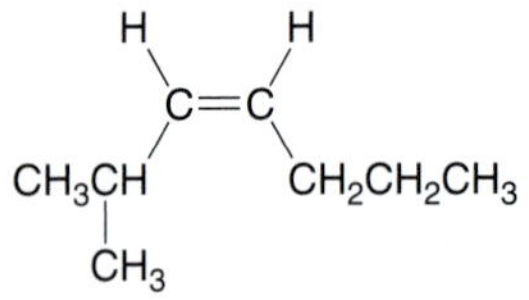

cis-2-methyl-3-heptene

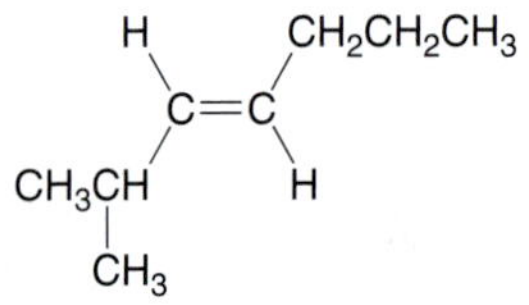

trans-2-methyl-3-heptene

13.53   Constitutional isomers have the same molecular formula but have the atoms bonded to different atoms. Stereoisomers have atoms bonded to the same atoms but in a different three-dimensional arrangement.

13.55   a. constitutional isomers
       b. identical
       c. identical
       d. stereoisomers

13.57   When two like groups are bonded to the same end of a double bond, switching the positions of the two groups results in the same compound.

13.59   a. $CH_3CH_2CH_2CH_2CH_2CH_3$
       b. $(CH_3)_2CHCH_2CH_2CH_2CH_3$
       c.

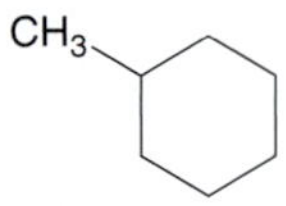

       d.

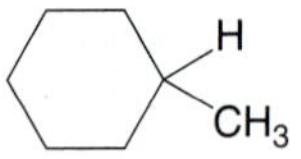

13.61 a. 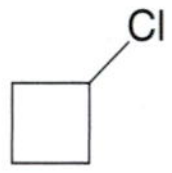 [structure: cyclobutane with Cl]

c. 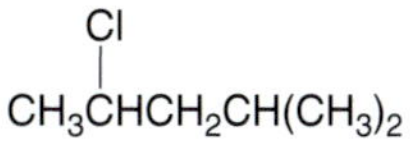 $CH_3CHClCH_2CH(CH_3)_2$

b. 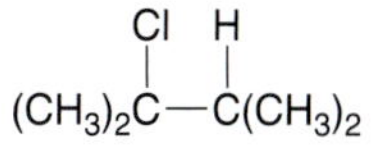 $(CH_3)_2C(Cl)-C(H)(CH_3)_2$

d. 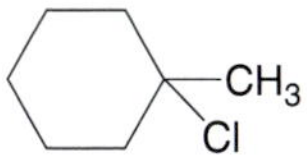 [structure: cyclohexane with CH_3 and Cl]

13.63 a. 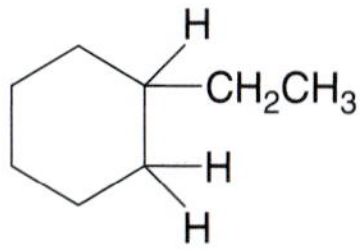 [cyclohexane with CH_2CH_3, H, H]

d. 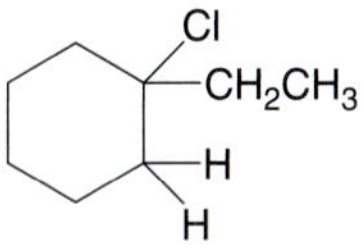 [cyclohexane with Cl, CH_2CH_3, H, H]

b. 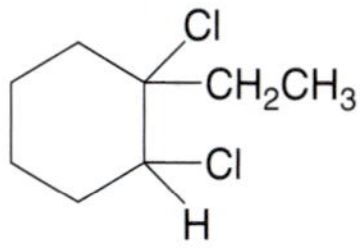 [cyclohexane with Cl, CH_2CH_3, Cl, H]

e. 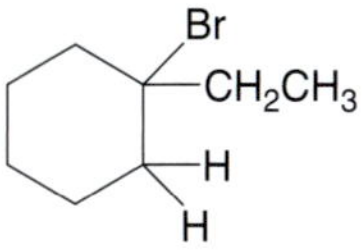 [cyclohexane with Br, CH_2CH_3, H, H]

c. 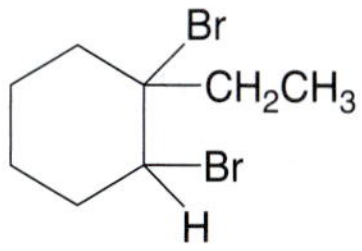 [cyclohexane with Br, CH_2CH_3, Br, H]

f. 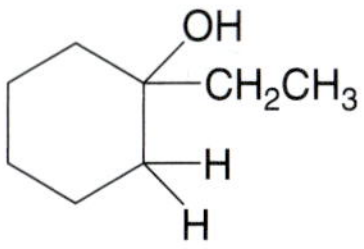 [cyclohexane with OH, CH_2CH_3, H, H]

13.65

$CH_3-C(H)(H)-C(OH)(H)-CH_2CH_3$

$CH_3-C(OH)(H)-C(H)(H)-CH_2CH_3$

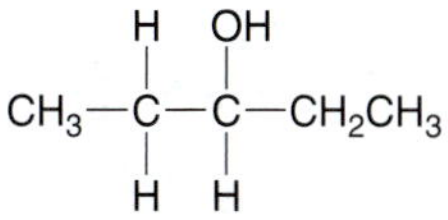

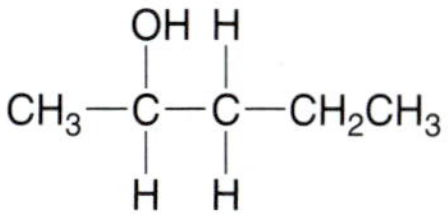

OH adds to either side of the C=C since both carbons have an equal number of H's.

13.67 $CH_2{=}CH_2$ reagent: HBr

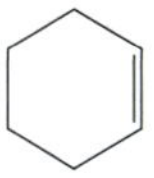

[cyclohexene]   reagent: HCl

[cyclopentene with CH_3]   reagent: Cl_2

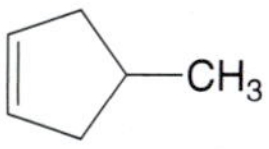

$CH_2{=}CHCH_2CH(CH_3)_2$ reagent: Br_2

13.69 a. HCl   d. HBr
b. H_2, Pd   e. Br_2
c. H_2O, H_2SO_4 f. Cl_2

13.71 A monomer is the basic unit that when combined with similar units connected end-to-end creates a polymer.

13.73 $-CH_2C(COOH)(H)-CH_2C(COOH)(H)-CH_2C(COOH)(H)-$

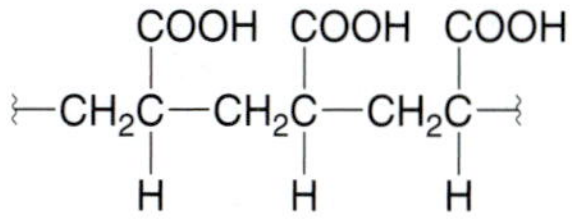

13.75 a. 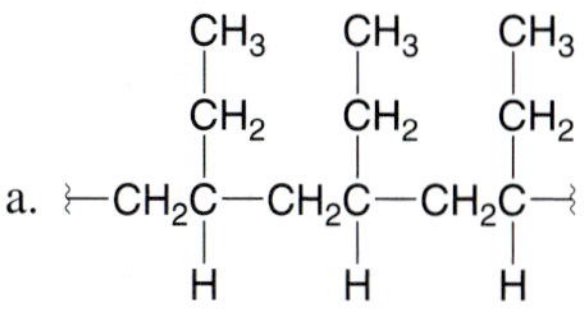 $-CH_2C(CH_2CH_3)(H)-CH_2C(CH_2CH_3)(H)-CH_2C(CH_2CH_3)(H)-$

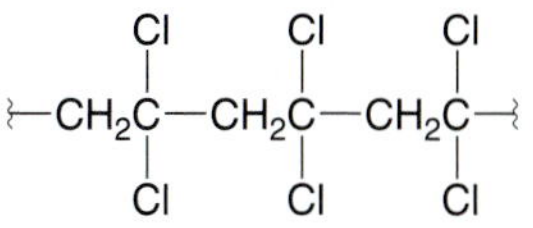

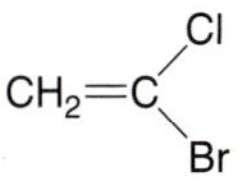

b. 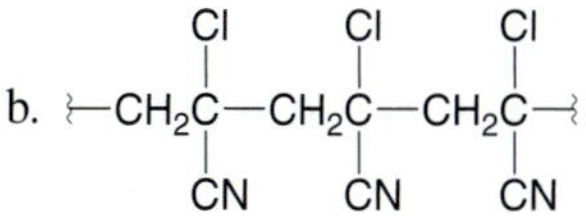 $-CH_2C(CN)(Cl)-CH_2C(CN)(Cl)-CH_2C(CN)(Cl)-$

c. $-CH_2C(Cl)(Cl)-CH_2C(Cl)(Cl)-CH_2C(Cl)(Cl)-$

13.77 $CH_2{=}C(Cl)(Br)$

13.79 [chlorobenzene resonance structure] ↔ [chlorobenzene resonance structure]

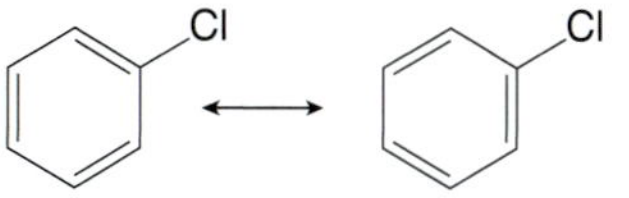

13.81 Two isomers differ in the placement of both atoms and electrons. Two resonance structures differ only in the placement of electrons. The atom placement is the same.

13.83 a. *m*-chloronitrobenzene
b. *p*-nitroaniline
c. *o*-butylethylbenzene
d. 2,5-dichlorophenol

13.85 [structure: benzene with NH_2 and Cl ortho]   o-chloroaniline

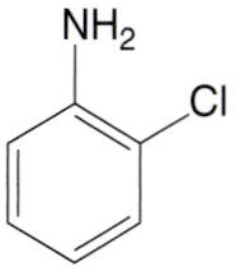

[structure: benzene with NH_2 and Cl meta]   m-chloroaniline

[structure: benzene with NH_2 and Cl para]   p-chloroaniline

13.87 a. 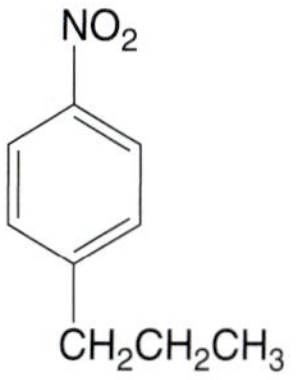 [benzene with NO_2 and $CH_2CH_2CH_3$ para]

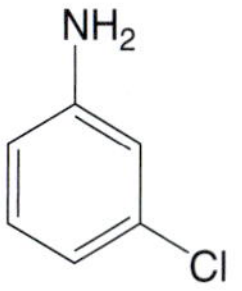

b. 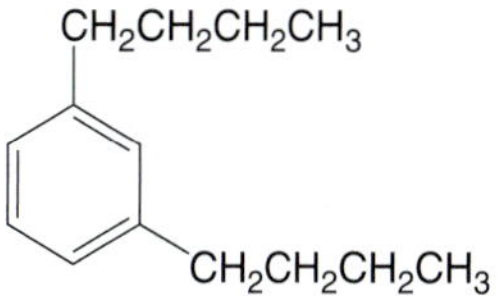 [benzene with $CH_2CH_2CH_2CH_3$ substituents]

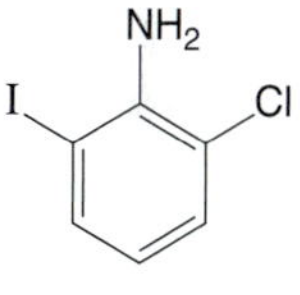

c. 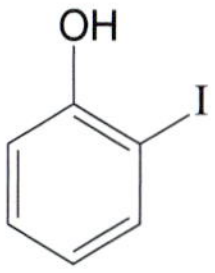 [phenol with OH and I ortho]

d. [benzene with CH_3, Br, Cl]

e. [benzene with I, NH_2, Cl]

13.89 a. Assign the lowest numbers: 2,3-dichlorophenol.
b. You can only use ortho, meta, para, for a disubstituted benzene: 3,5-dibromoaniline.

13.91 [structure: benzene with Cl, CH_3, CH_3]

[structure: benzene with NO_2, CH_3, CH_3]

[structure: benzene with SO_3H, CH_3, CH_3]

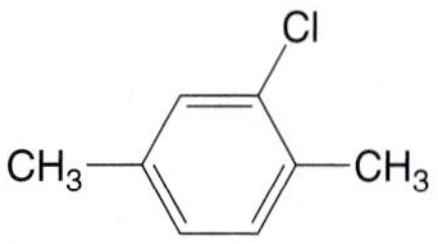

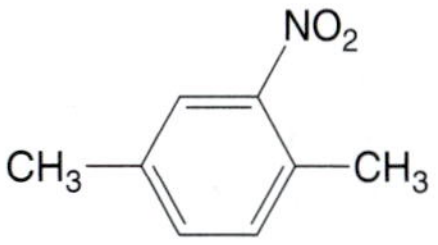

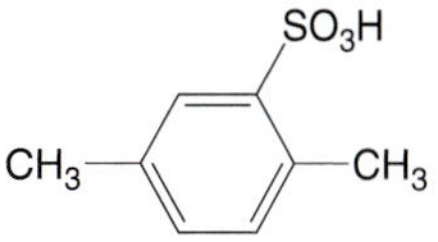

13.93

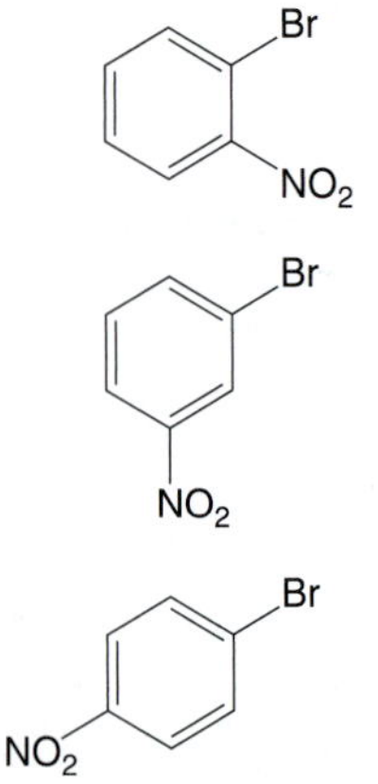

(structure: benzene ring with Br and NO_2 in ortho positions)

(structure: benzene ring with Br and NO_2 in meta positions)

(structure: benzene ring with Br and NO_2 in para positions)

13.95 Vitamin E is an antioxidant.

13.97 Methoxychlor is more water soluble. The OCH_3 groups can hydrogen bond to water. This increase in water solubility makes methoxychlor more biodegradable.

13.99 a.

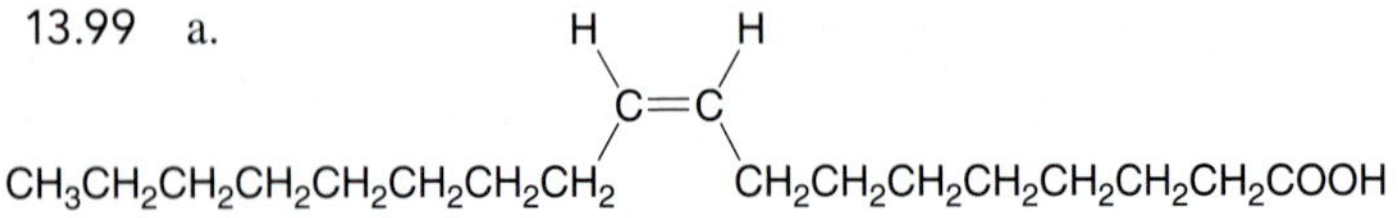

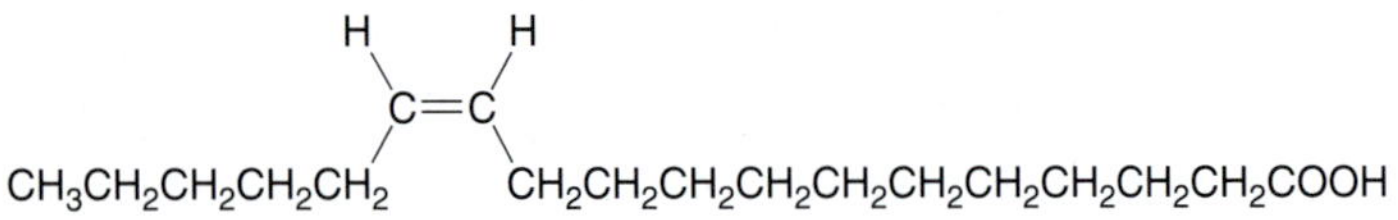

(structure: cis alkene)
$$CH_3CH_2CH_2CH_2CH_2CH_2CH_2CH_2 \quad C{=}C \quad CH_2CH_2CH_2CH_2CH_2CH_2CH_2COOH$$

(structure: cis alkene)
$$CH_3CH_2CH_2CH_2CH_2 \quad C{=}C \quad CH_2CH_2CH_2CH_2CH_2CH_2CH_2CH_2CH_2CH_2COOH$$

b.
$$CH_3CH_2CH_2CH_2CH_2CH_2CH_2CH_2CH_2CH_2CH_2CH_2CH_2CH_2CH_2CH_2CH_2COOH$$

c.
one possibility: (trans alkene)
$$CH_3CH_2CH_2CH_2CH_2CH_2CH_2CH_2 \quad C{=}C \quad CH_2CH_2CH_2CH_2CH_2CH_2CH_2COOH$$

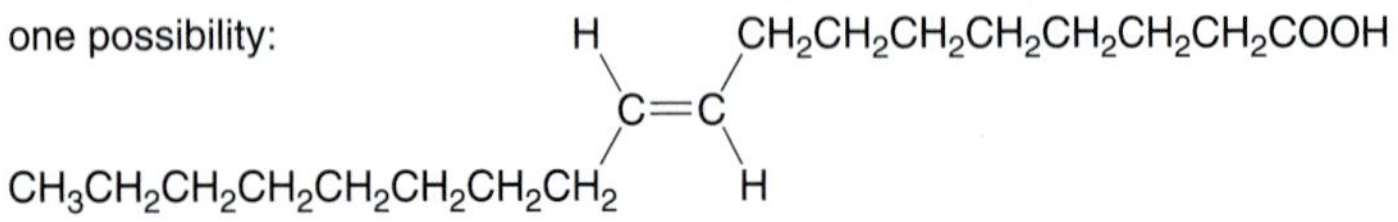

13.101 (c)

13.103 Benzene is converted to a more water-soluble compound that can then be excreted in the urine.

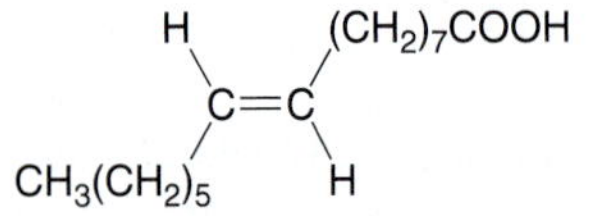

13.105 a.
(cis alkene)
$$CH_3(CH_2)_5 \quad C{=}C \quad (CH_2)_7COOH$$

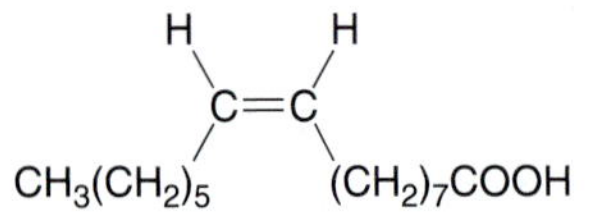

b.
(trans alkene)
$$CH_3(CH_2)_5 \quad C{=}C \quad (CH_2)_7COOH$$

c.
(cis alkene)
$$CH_3(CH_2)_4 \quad C{=}C \quad (CH_2)_8COOH \quad \text{is one possibility}$$

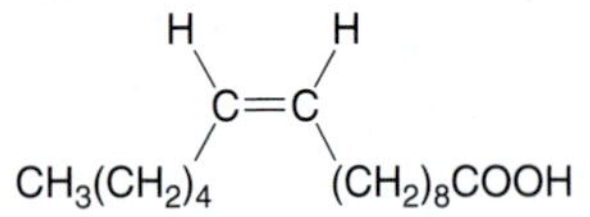

13.107 In addition reactions, two new bonds are formed at both ends of a double bond when a reagent is added. In a substitution reaction, one atom replaces another—one single bond is broken and one single bond is formed.

13.109 All the carbons in benzene are trigonal planar with 120° bond angles, resulting in a flat ring. In cyclohexane, all the carbons are tetrahedral so the ring is puckered.

13.111 They are constitutional isomers because the double bond is located in a different place on the carbon chain.

CHAPTER 14

14.1 a.
(structure with labels: halide → Cl; CH_3, CH_3; Br ← halide; ether → O; ring system)

(structure with labels: CH_3; C—SH ← thiol; CH_3; CH_3-cyclohexene)

(structure of salmeterol with labels: hydroxyl → OH; hydroxyl → HO; ether; $CHCH_2NH(CH_2)_6O(CH_2)_4$—phenyl; $HOCH_2$; hydroxyl)

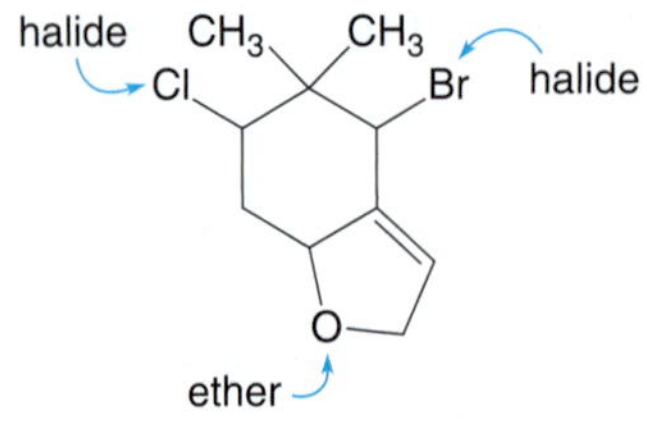

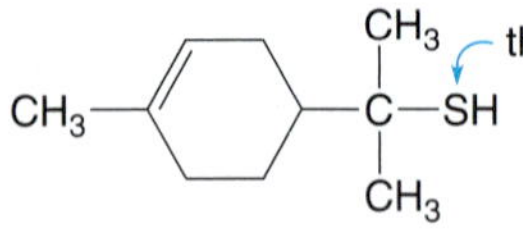

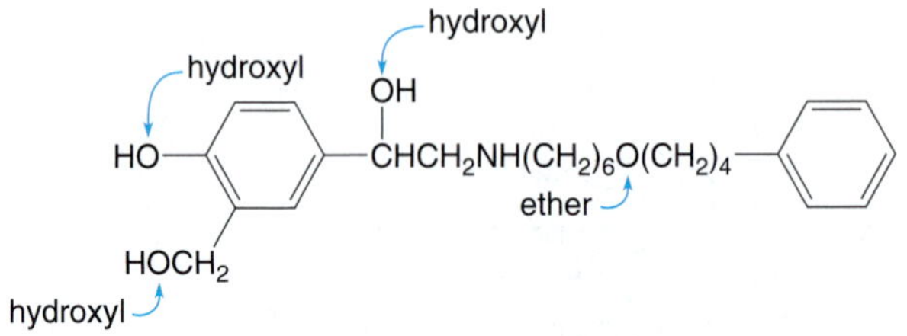

b. The OH on the benzene ring of salmeterol is part of a phenol.

14.3 a. 2°   b. 2°

(steroid structure with labels: O; C—CH_2OH ← 1°; 2°; HO; CH_3; OH; 3°; CH_3; O)

c.

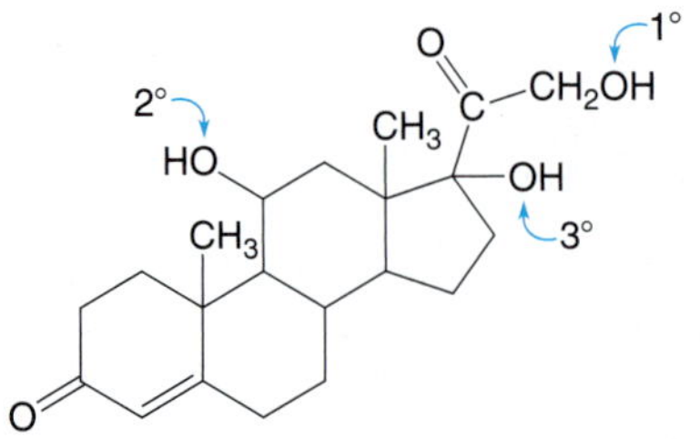

14.5 a. insoluble   b. soluble   c. insoluble   d. insoluble

14.6 a. 2-heptanol   c. 2-methylcyclohexanol
     b. 4-ethyl-3-hexanol   d. 5-ethyl-6-methyl-3-nonanol

14.7 a. $CH_3CCH_2CH_2CHCH_2CH_2CH_3$ (with CH_3 and OH substituents; CH_3)

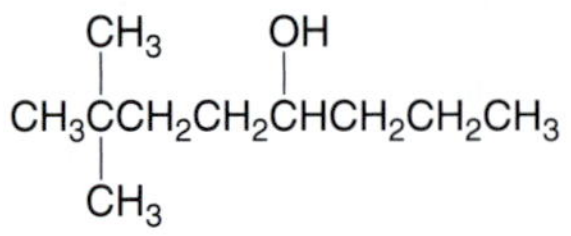

c. (cyclohexane ring with OH, CH_2CH_3, CH_3)

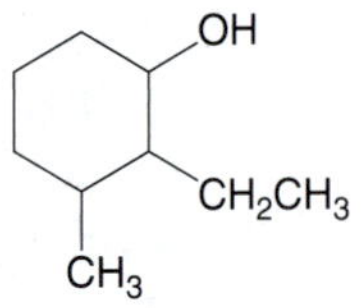

b. $CH_3CH_2CHCHCHCH_2CH_3$ (with CH_3, OH, and $CH_2CH_2CH_3$)

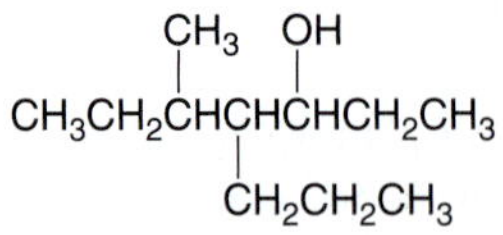

d. (cyclohexane ring with OH and OH)

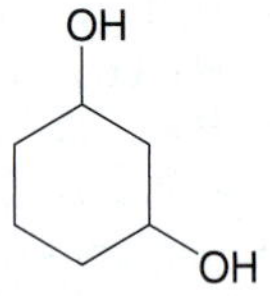

14.9 a. $CH_3{-}C{=}CH_2$ (with H)   b. (benzene ring)$-CH{=}CH_2$ c. (cyclopentene)

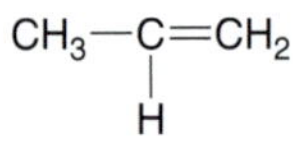

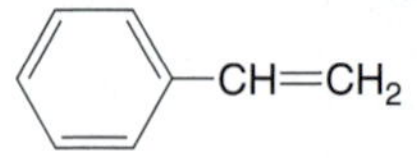

14.11 a. $CH_3CCH_2CH_2CH_3$ (with =O)   c. (cyclopentanone, ring with =O)

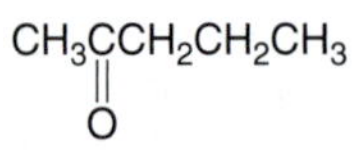

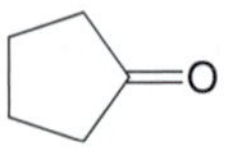

b. $(CH_3)_3CCH_2C{-}OH$ (with =O)   d. no reaction

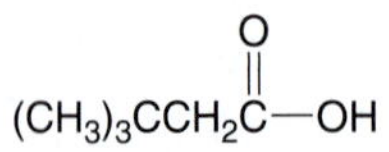

14.13 aldehyde, 3 alkenes, 2 hydroxyl groups

14.15 a. $CH_3(CH_2)_5OCH_3$ c. $CH_3(CH_2)_6OH$

b. (tetrahydropyran ring with O)   d. $CH_3(CH_2)_5OCH_3$

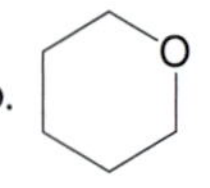

14.17 a. 1-methoxybutane (or butyl methyl ether)
b. methoxycyclohexane
c. dipropyl ether

14.19 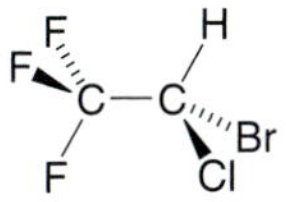

14.21 a. 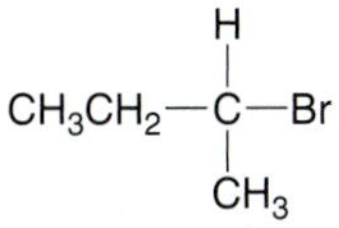 b. 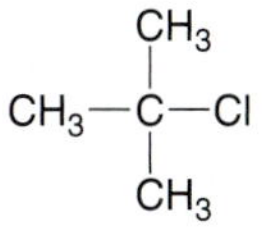 c. 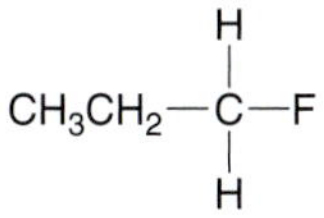

14.23 a. 2-bromohexane
b. 3-chloro-2-methylpentane
c. 1-bromo-2-methylcyclohexane

14.25 a. CFC b. HCFC c. HFC

14.26 a. 3-pentanethiol
b. cyclohexanethiol
c. 5-ethyl-1-heptanethiol

14.27 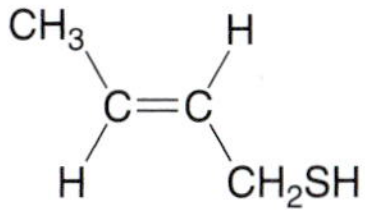

14.29 a. 1° b. 3° c. 2° d. 2°

14.31 a. 2° b. 3° c. 2° d. 1°

14.33 a. 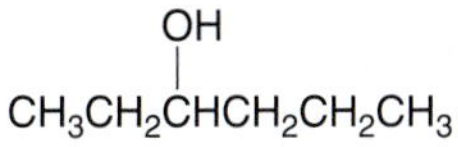 c. 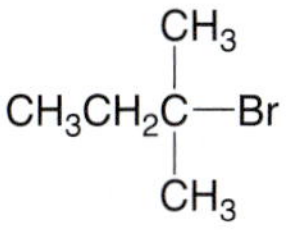

b. $CH_3—O—CH_2CH_2CH_2CH_2CH_3$

14.35

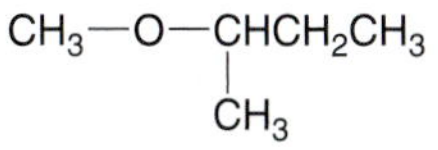 $CH_3—O—CH_2CH_2CH_2CH_3$ 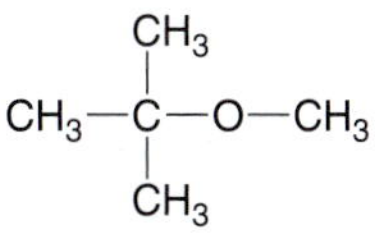

$CH_3CH_2CH_2OCH_2CH_3$ 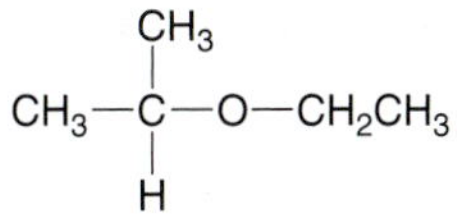 $(CH_3)_2CHCH_2OCH_3$

14.37 a. 2-pentanol
b. 3,3-dimethyl-2-hexanol
c. 4-ethyl-6-methyl-1-heptanol
d. 4-butylcyclohexanol
e. 2,2-dimethylcyclobutanol
f. 2,5-diethylcyclopentanol

14.39 a. 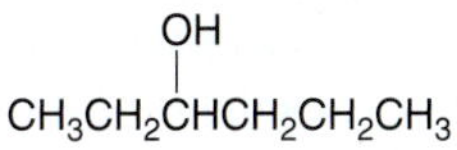
b. $CH_3CH_2CH_2OH$
c. 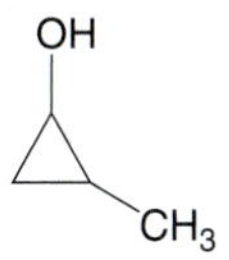
d. 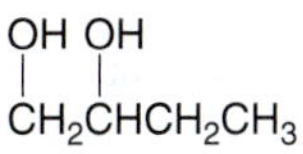
e. 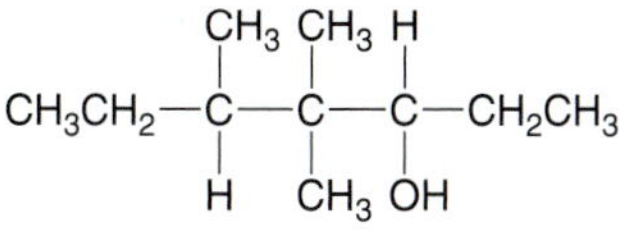
f. 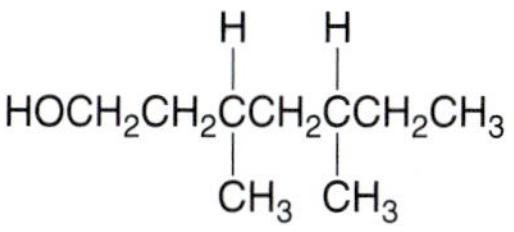

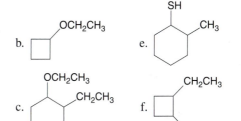

14.41 a. 1-ethoxybutane (or butyl ethyl ether)
b. 2-ethoxypentane
c. ethoxycyclopentane

14.43 2-methoxy-2-methylpropane

14.45 $CH_3CH_2CH_2CH_2CH_2CH_2CH_2OH$ 1-heptanol

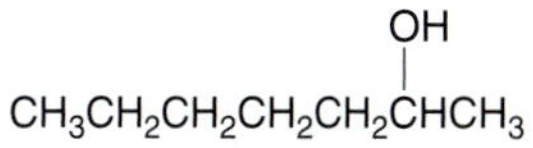 2-heptanol

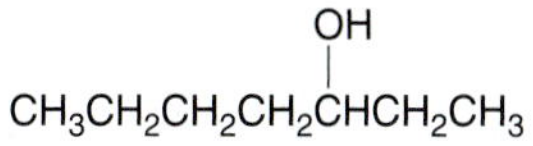 3-heptanol

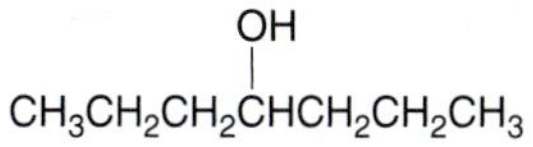 4-heptanol

14.47 a. 4-bromoheptane
b. 1-chloro-3-methylcyclobutane
c. 1-pentanethiol

14.49 a. 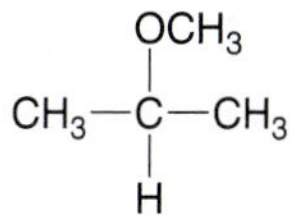 d. $CH_3CH_2CH_2CH_2Cl$

b. 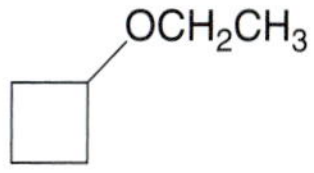 e. 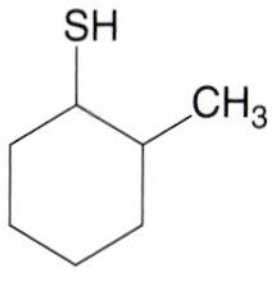

c. 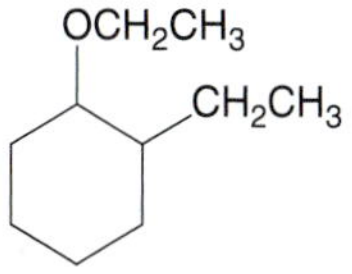 f. 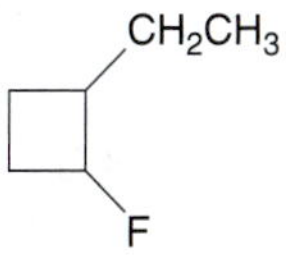

14.51 a. $CH_3CH_2CH_2I$ b. $HOCH_2CH_2OH$ c. $CH_3CH_2CH_2OH$

14.53 Ethanol is a polar molecule capable of hydrogen bonding with itself and water. The greater intermolecular forces make its boiling point higher. Both ethanol and dimethyl ether have only two carbons and can hydrogen bond to H_2O, so both are water soluble.

14.55 1-Butanol is capable of hydrogen bonding with H_2O and is therefore water soluble. 1-Butene is not polar and cannot hydrogen bond with H_2O so it is not water soluble.

14.57 a. 

b. 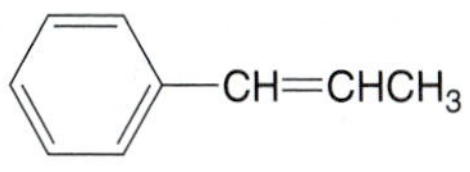

c. $CH_3CH=CHCH_2CH_2CH_3$ (major product) + $CH_2=CH(CH_2)_3CH_3$

d. 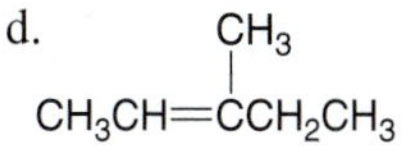 (major product) + 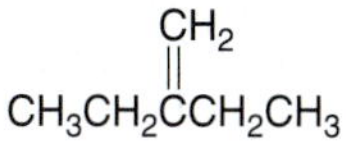

14.59 a. $CH_3CH{=}CHCH_2CH_2CH_3$ and $CH_3CH_2CH{=}CHCH_2CH_3$

b. The carbons of both double bonds are bonded to an equal number of H's; therefore, equal amounts of both isomers are formed.

14.61 a. b.

14.63 $CH_3CH_2CH_2OH$ and CH_3CHCH_3
$\quad\quad\quad\quad\quad\quad\quad\quad\quad\quad\quad\quad\;$ OH

14.65 a. $CH_3(CH_2)_6\overset{O}{\overset{\|}{C}}OH$ c. $CH_3CH_2\overset{O}{\overset{\|}{C}}HCOH$
$\quad\quad\quad\quad\quad\quad\quad\quad\quad\quad\quad\quad\quad\quad\quad\quad\quad\quad\quad\quad\;$ CH_3

b. $CH_3CH_2\overset{O}{\overset{\|}{C}}CH_2CH_3$ d. No reaction

14.67

14.69 a. b. $CH_3{-}\overset{CH_3}{\underset{CH_3}{\overset{|}{C}}}{-}\overset{OH}{\underset{CH_3}{\overset{|}{C}}}H$ c.

14.71 a. $CH_3CH_2CH_2CH{=}CHCH_2CH_3$

b. $CH_3CH_2CH_2\overset{O}{\overset{\|}{C}}CH_2CH_2CH_3$

14.73 **A:** H_2O and H_2SO_4
B: $K_2Cr_2O_7$

14.75 a.

b. $CH_3(CH_2)_4S{-}S(CH_2)_4CH_3$

14.77 a. CH_3SH and $HSCH_2CH_2CH_3$

b. $CH_2{=}CHCH_2SH$ and $HSCH_2CH_2CH_3$

14.79 $C_4H_{10}O + 6\,O_2 \longrightarrow 4\,CO_2 + 5\,H_2O$

14.81 a. CFCs are chlorofluorocarbons with a general formula of CF_xCl_{4-x}, HCFCs have fluorine, chlorine, and hydrogen bonded to carbons, and HFCs have only hydrogen and fluorine bonded to carbon.

b. CFCs interact and destroy the ozone layer, whereas HCFCs and HFCs decompose before ascending to the ozone layer.

14.83 PEG is capable of hydrogen bonding with water and this makes it water soluble. PVC cannot hydrogen bond to water so it is water insoluble.

14.85

$CH_3CH_2OH \;+\; \underset{\text{(orange)}}{K_2Cr_2O_7} \longrightarrow CH_3\overset{O}{\overset{\|}{C}}OH \;+\; Cr^{3+}$ (green color)

The greater the blood alcohol level, the greater the color change from red-orange to green on the Breathalyzer.

14.87 $CH_3\overset{O}{\overset{\|}{C}}{-}\overset{O}{\overset{\|}{C}}OH$
pyruvic acid

14.89

$CH_3CH_2OH \longrightarrow$ $\longrightarrow$
$\quad\quad\quad\quad\quad\quad\quad\quad\quad\quad$ acetaldehyde $\quad\quad\quad\quad\quad\quad$ acetic acid
$\quad\quad\quad\quad\quad\quad\quad\quad\quad\quad\quad\quad\quad$ Antabuse
$\quad\quad\quad\quad\quad\quad\quad\quad\quad\quad$ (blocks conversion)

Antabuse blocks the conversion of acetaldehyde to acetic acid and the accumulation of acetaldehyde makes people ill.

14.91 $(CH_3)_3COH$

14.93 Ether molecules have no H's on O's capable of hydrogen bonding to other ether molecules, but the hydrogens in water can hydrogen bond to the oxygen of the ether.

14.95 a. 3-ethylcyclopentanol
b. 2°

c. and

d.

e.

f.

14.97 a. $CH_3CH_2\overset{CH_3}{\underset{H}{\overset{|}{C}}}CH_2OH$ **A** b. $CH_3CH{=}\overset{CH_3}{\overset{|}{C}}{-}CH_3$

$\quad\quad\quad$ $CH_3CH_2\overset{CH_3}{\underset{OH}{\overset{|}{C}}}CH_3$ **B**

CHAPTER 15

15.1 a. constitutional isomers c. identical
b. stereoisomers d. constitutional isomers

15.3 A molecule that is chiral is not superimposable on its mirror image. A chirality center is a carbon with four different groups bonded to it.

15.4 a. $CH_3\overset{\downarrow}{C}HCH_2CH_3$ c. $CH_2{=}CHCHCH_3$
$\quad\quad\quad\;$ Cl $\quad\quad\quad\quad\quad\quad\quad\quad\quad\quad$ OH

b. none d.

15.5 a. Br—⟨C₆H₄⟩—C(H)(pyridyl)—CH₂CH₂N(CH₃)₂

b. ⟨C₆H₅⟩—C(H)—O—⟨C₆H₄⟩—CF₃, with CH₂CH₂NHCH₃

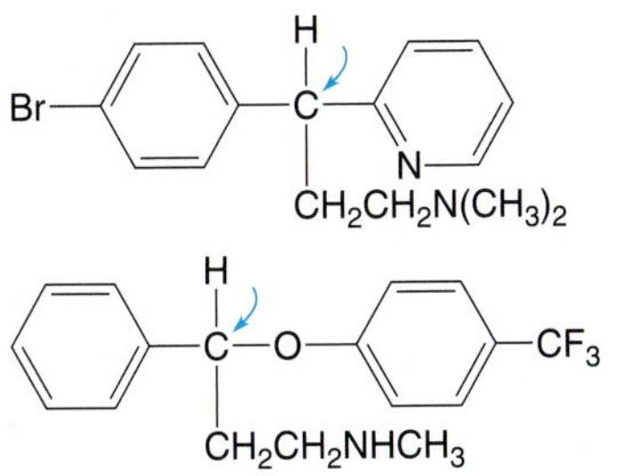

15.7

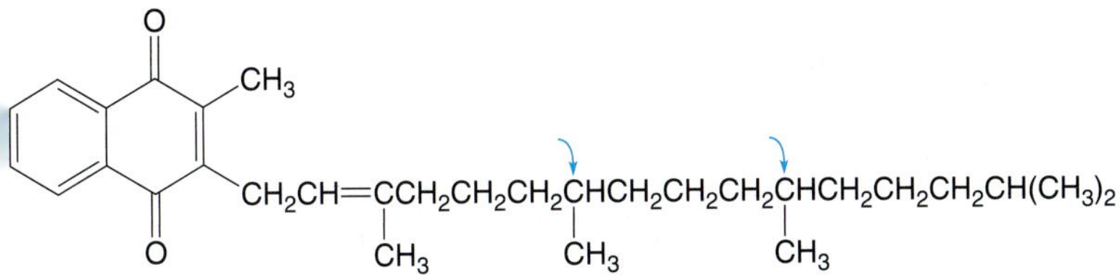

CH₃

CH₂CH=CCH₂CH₂CH₂CHCH₂CH₂CH₂CHCH₂CH₂CH₂CH(CH₃)₂

CH₃   CH₃   CH₃

15.8 a.
CH₃—C(Cl)(H)—CH₂CH₃   CH₃—C(H)(Cl)—CH₃CH₂

b.
OH—C(H)(CH₂CH₃)—CH₂OH   HO—C(H)(CH₃CH₂)—HOCH₂

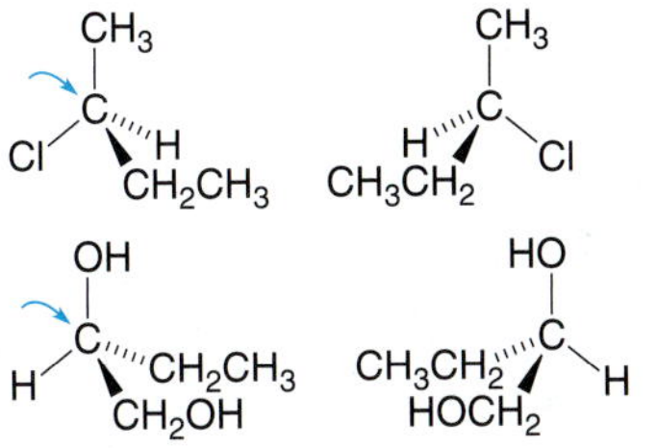

15.9

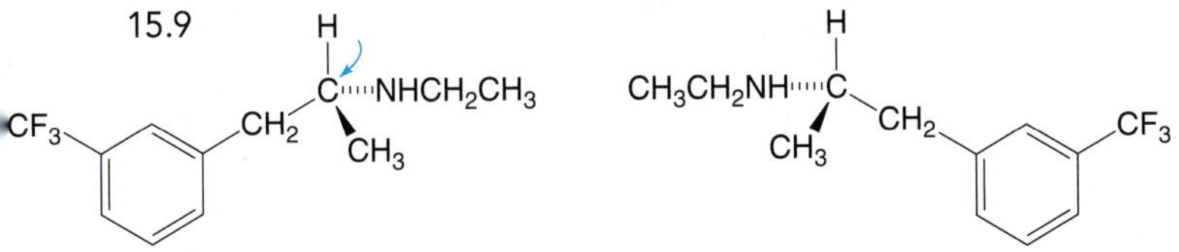

CF₃—⟨C₆H₄⟩—CH₂—C(H)(CH₃)—NHCH₂CH₃   CH₃CH₂NH—C(H)(CH₃)—CH₂—⟨C₆H₄⟩—CF₃

15.11 CH₃NH—(tetralin ring)—⟨C₆H₃⟩(Cl)(Cl)

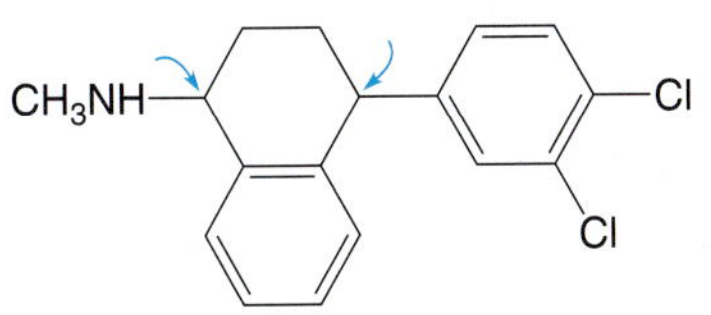

15.13 a.
CH₃
H—Br
Cl

b.
COOH
H—OH
CH₂Cl

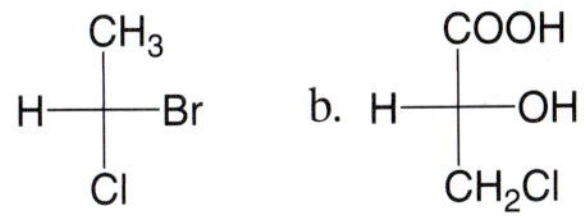

15.15 a.
H
CH₃—CH₂CH₂CH₃   CH₃CH₂CH₂—CH₃
OH   OH

b.
H
ClCH₂—CH₂CH₃   CH₃CH₂—CH₂Cl
Cl   Cl

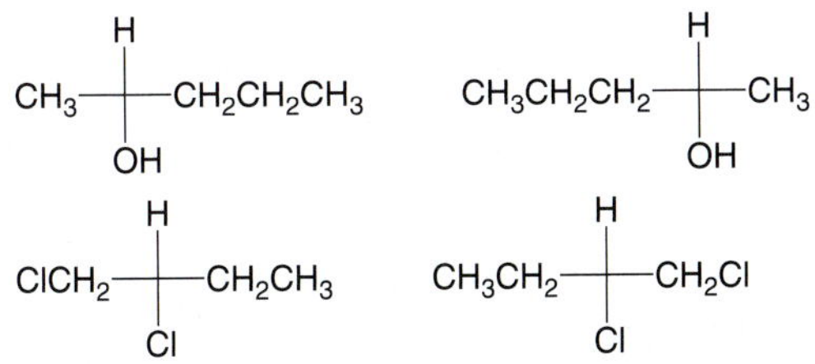

15.16 a. **Z**   b. **Y**   c. **Z** and **W**   d. **Y** and **X**

15.17

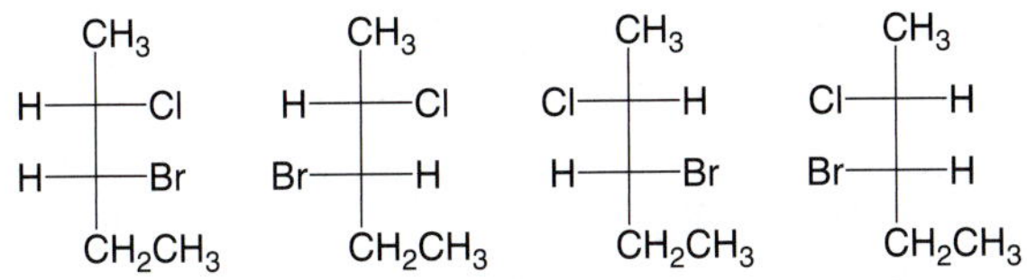

CH₃	CH₃	CH₃	CH₃
H—Cl	H—Cl	Cl—H	Cl—H
H—Br	Br—H	H—Br	Br—H
CH₂CH₃	CH₂CH₃	CH₂CH₃	CH₂CH₃
W	**X**	**Y**	**Z**

15.19 a,d: achiral   b,c: chiral

15.21 Butane does not have an enantiomer since the molecule and its mirror image *are* superimposable and therefore not enantiomers.

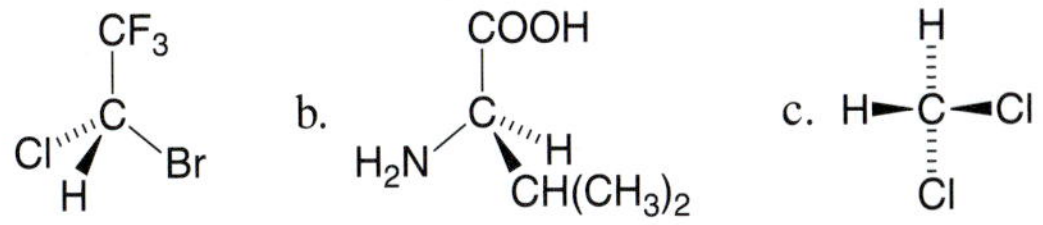

15.23 a.
CF₃
Cl—C—Br
H
chiral

b.
COOH
H₂N—C—H
CH(CH₃)₂
chiral

c.
H
H—C—Cl
Cl
achiral

15.25 a. none   c. none

b. 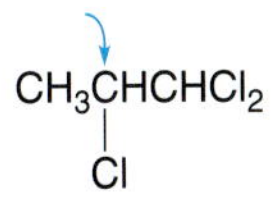 CH₃CHCHCl₂
        Cl

d. 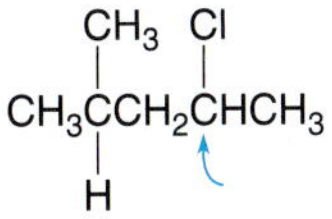 CH₃   Cl
   CH₃CCH₂CHCH₃
        H

15.27 a. none   c. 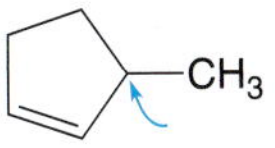 ⟨cyclopentene⟩—CH₃

b. 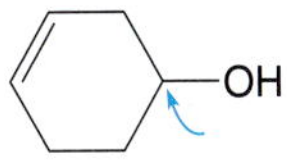 ⟨cyclohexene⟩—OH   d. none

15.29 a.
CH₃
CH₃CH₂—C—CH₂CH₂CH₃
        H

b.
OH
CH₃CH₂—C—CH₂CH₂CH₃
        H

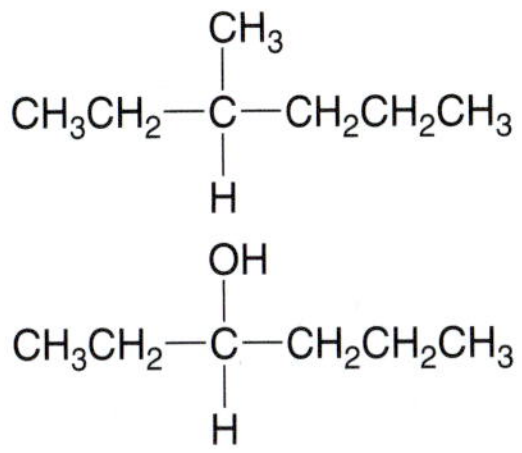

15.31 A chirality center must have a carbon bonded to four groups and a carbonyl carbon has only three groups.

15.33 a. 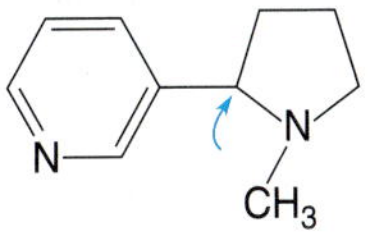 ⟨pyridyl⟩—⟨pyrrolidine⟩
                    CH₃

b. 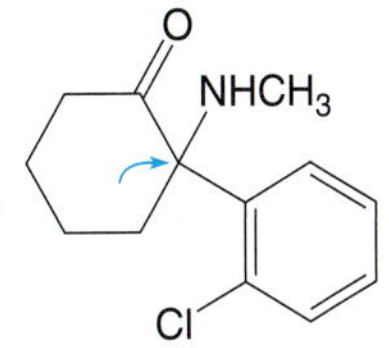 ⟨cyclohexanone⟩—NHCH₃, ⟨C₆H₄⟩—Cl

c. 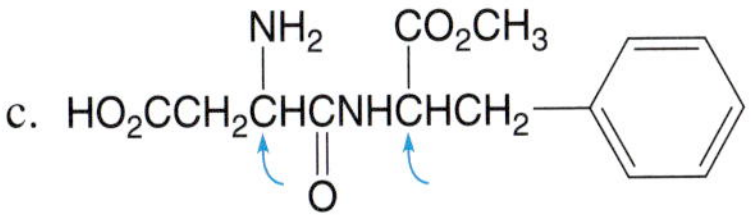
NH₂   CO₂CH₃
HO₂CCH₂CHCNHCHCH₂—⟨C₆H₅⟩
          O

15.35 ⟨C₆H₅⟩—CH₂CHNCH₃
                    CH₃
                    H

CH₃   CH₃
⟨C₆H₅⟩—CH₂—C(H)—NHCH₃   H—C—CH₂—⟨C₆H₅⟩
                              CH₃NH

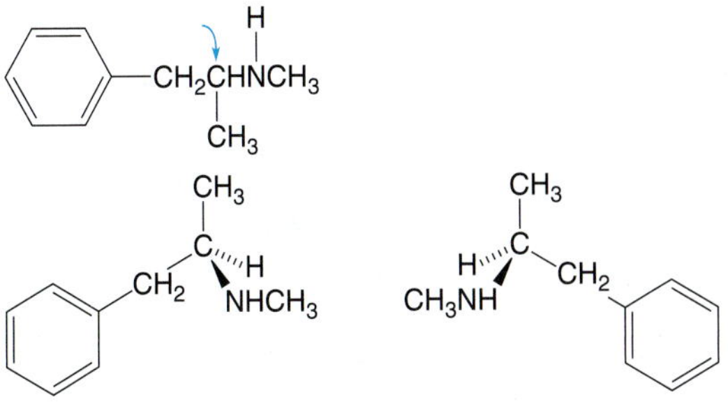

15.37 a. identical   b,c: enantiomers

15.39 a. constitutional isomers   c. identical
b. enantiomers   d. diastereomers

15.41 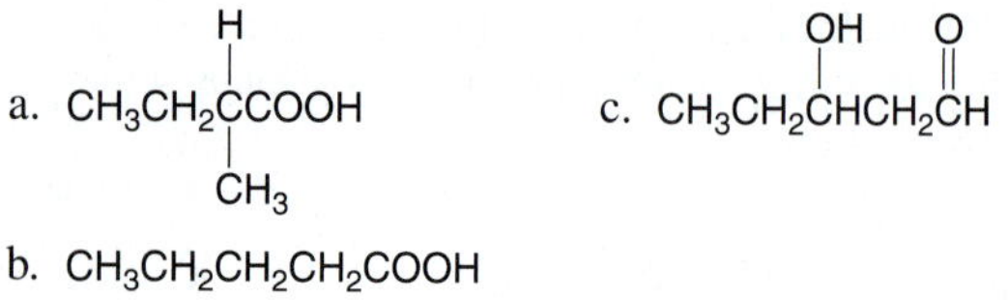

15.43 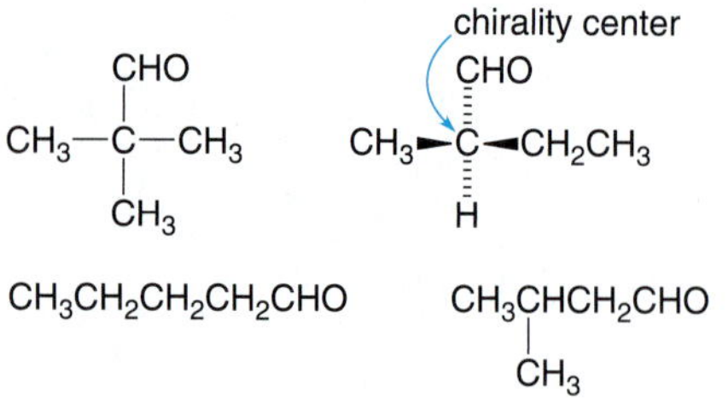

15.45 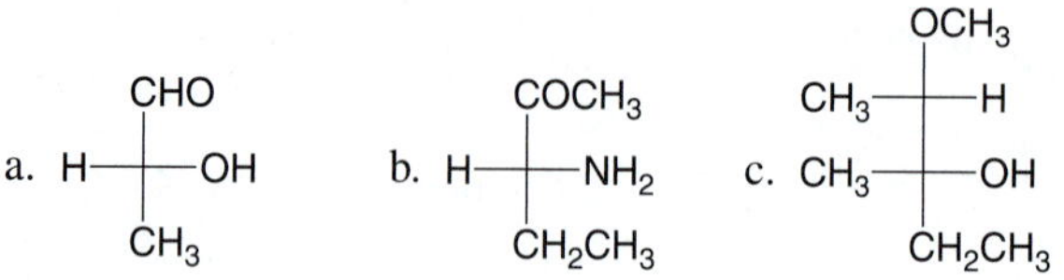

15.47 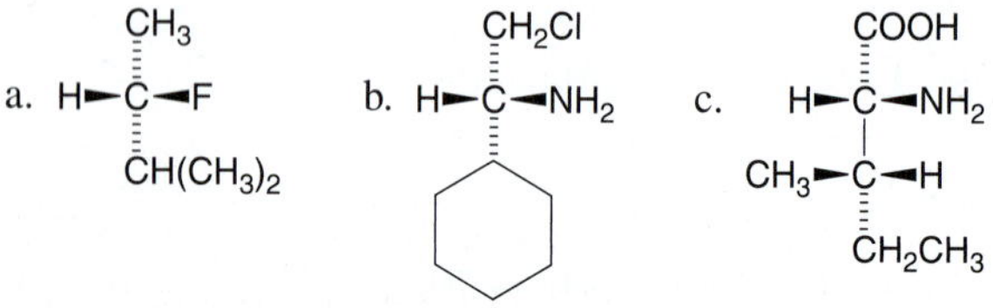

15.49 a,d: identical b,c: enantiomers

15.51 Enantiomers are stereoisomers that are nonsuperimposable mirror images of one another, whereas diastereomers are stereoisomers that are not mirror images of one another.

15.53 a. 1,2,5,6: diastereomers 3,4: enantiomers

b. 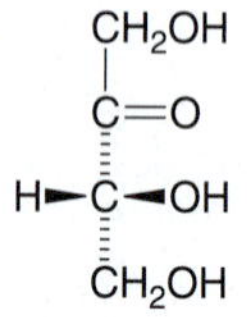

15.55 They are diastereomers because they are not mirror images of one another.

15.57 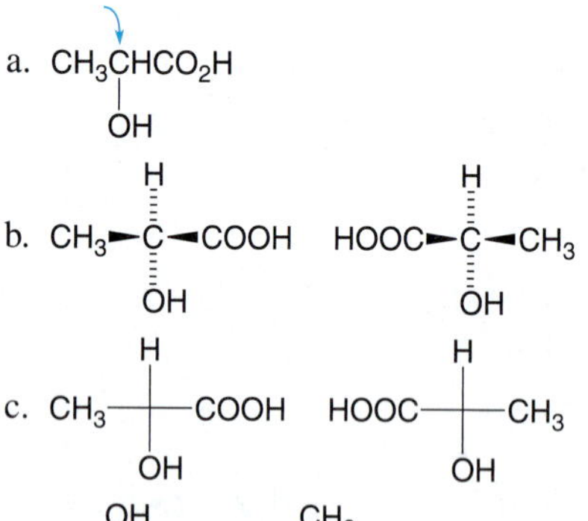

15.59 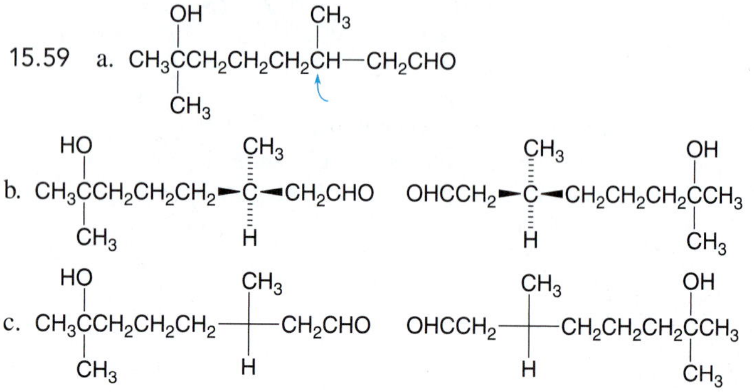

15.61 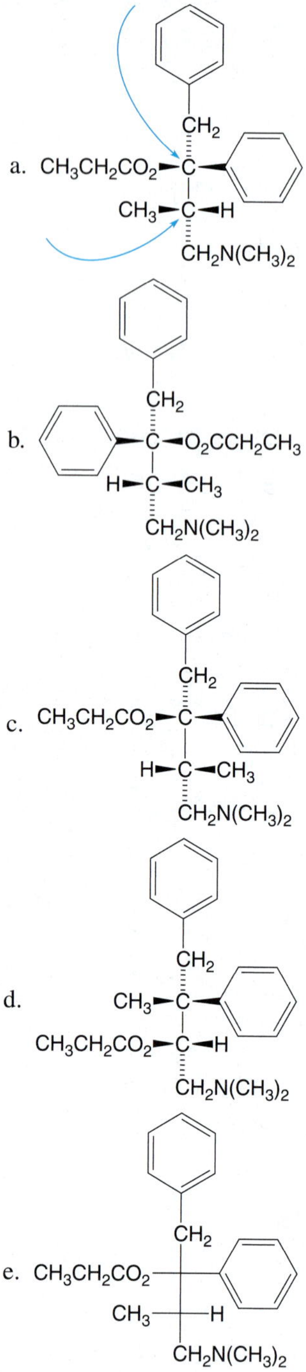

15.63 * = chirality center

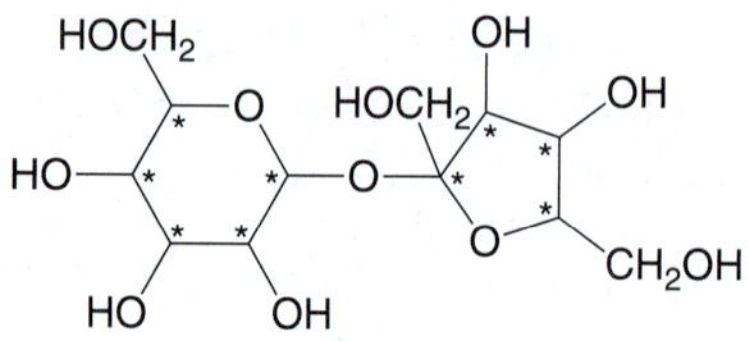

CHAPTER 16

16.1 a. $CH_3CH_2\overset{\displaystyle O}{\underset{}{C}}H$ aldehyde

b. $CH_3CH_2\overset{\displaystyle O}{\underset{}{C}}CH_3$ ketone

c. $(CH_3)_3C\overset{\displaystyle O}{\underset{}{C}}CH_3$ ketone

d. $(CH_3CH_2)_2CH\overset{\displaystyle O}{\underset{}{C}}H$ aldehyde

16.3

16.4 a. 5-methylhexanal   c. 2,5,6-trimethyloctanal
b. 3,3,4,4-tetramethylpentanal

16.5 a. $CH_3\overset{\displaystyle Cl}{\underset{}{C}}HCHO$

b. $CH_3CH_2\overset{\displaystyle CH_3CH_2}{\underset{\displaystyle CH_2CH_3}{C}}HCHCHCH_2CHO$

c. $CH_3CH_2CH_2\overset{}{C}HCH_2CH_2\overset{\displaystyle CH_2CH_3}{\underset{\displaystyle CH_2CH_3}{C}}HCH_2CHO$

d. (benzene ring with CHO and CH_2CH_3)

16.6 a. 4-methyl-3-heptanone   c. 2,2-dimethyl-3-heptanone
b. 2-methylcyclopentanone

16.7 $CH_3\overset{\displaystyle O}{\underset{}{C}}CH_2CH_2CH_2CH_2CH_2CH_3$

16.9 a. (cyclohexanone)   c. (cyclohexanol)
b. $(CH_3CH_2)_2CO$ d. $CH_3(CH_2)_5CHO$

16.11 Hexane is soluble in acetone because both compounds are organic and "like dissolves like." Water is soluble in acetone because acetone has a short hydrocarbon chain and is capable of hydrogen bonding with water.

16.13 a. $CH_3CH_2\overset{\displaystyle O}{\underset{}{C}}OH$ c. $CH_3C=CHCH_2CH_2\overset{\displaystyle CH_3}{\underset{}{C}}HCH_2\overset{\displaystyle O}{\underset{}{C}}-OH$
b. No reaction

16.14 a. $CH_3(CH_2)_6\overset{\displaystyle O}{\underset{}{C}}-OH$ c. (cyclopentane with $\overset{\displaystyle O}{\underset{}{C}}-OH$)

b and d. No reaction

16.15 a. $CH_3CH_2CH_2\overset{\displaystyle OH}{\underset{}{C}}H_2$ c. $CH_3\overset{\displaystyle OH}{\underset{}{C}}HCH_2CH_3$

b. (cyclopentane with OH and CH_3)   d. (benzene ring with CH_2OH)

16.17 a. stereoisomers   b. not isomers   c. stereoisomers

16.18 a. $CH_3\overset{\displaystyle OCH_3}{\underset{\displaystyle H}{C}}OH$ $CH_3\overset{\displaystyle OCH_3}{\underset{\displaystyle H}{C}}OCH_3$
hemiacetal   acetal

b. $(CH_3CH_2)_2\overset{\displaystyle OCH_3}{\underset{}{C}}OH$ $(CH_3CH_2)_2\overset{\displaystyle OCH_3}{\underset{}{C}}OCH_3$
hemiacetal   acetal

c. (benzene ring with $\overset{\displaystyle OCH_2CH_3}{\underset{}{C}}HOH$)   (benzene ring with $\overset{\displaystyle OCH_2CH_3}{\underset{}{C}}HOCH_2CH_3$)
hemiacetal   acetal

16.19 a. ether   b. acetal   c. hemiacetal   d. acetal

16.20 a. (ring structure) HOCH_2, HO, OH, HO, NH_2 — hemiacetal

b. (ring structure) HOCH_2, HO, OH, OH, CH_2OH (benzene ring) — acetal

16.21 a. (tetrahydropyran ring with OCH_2CH_3)   b. (tetrahydrofuran ring with O–cyclohexane)

16.22 a. $CH_3\overset{\displaystyle O}{\underset{}{C}}CH_2CH_2CH_3$ +   2 CH_3OH

b. (cyclohexanone)   +   2 CH_3CH_2OH

c. (cyclohexane with $\overset{\displaystyle O}{\underset{\displaystyle H}{C}}$)   +   2 CH_3OH

16.23 a. $CH_3CH_2CH_2\overset{\displaystyle CH_2CH_3}{\underset{}{C}}HCH_2CHO$ c. (cyclopentanone)

b. $CH_3CH_2\overset{\displaystyle O}{\underset{\displaystyle CH_3}{C}}CHCH_3$ d. (cyclopentane with $\overset{\displaystyle O}{\underset{\displaystyle H}{C}}$)

16.25   a. Both are trigonal planar.
        b. C=O is polar and C=C is not.
        c. Both functional groups undergo addition reactions.

16.27   No, it cannot. C_5H_{12} has too many H's. Since an aldehyde
        has a double bond, the number of C's and H's resembles an
        alkene, not an alkane. A compound with 5 C's would have
        to be $C_5H_{10}O$.

16.29   a. 3-methylhexanal
        b. 3,5-dimethylheptanal
        c. 3-propylhexanal
        d. 6,6-diethyl-2,2-dimethyloctanal
        e. 5-ethyl-2,6-dimethyldecanal
        f. p-chlorobenzaldehyde

16.31   a. $CH_3CH_2\overset{\underset{|}{Cl}}{\underset{\underset{Cl}{|}}{C}}CH_2CHO$
        c.
        b. $CH_3CH_2\underset{\underset{CH_3}{|}}{CH}\overset{\overset{CH_3}{|}}{CH}CH_2CHO$
        d. $CH_3CH_2CH_2\overset{\overset{OH}{|}}{CH}CH_2CH_2CHO$

16.33   a. 4-methyl-2-pentanone
        b. 2,6-dimethylcyclohexanone
        c. o-butylacetophenone
        d. 2,4-dimethyl-3-hexanone
        e. 3,5-diethyl-2-heptanone
        f. 3-chlorocyclopentanone

16.35   a. $CH_3CH_2CH_2\overset{\overset{CH_3}{|}}{\underset{\underset{CH_3}{|}}{C}}\overset{\overset{O}{\parallel}}{C}CH_3$
        c.
        b.
        d.

16.37

        $CH_3CH_2\overset{\overset{CH_3}{|}}{\underset{\underset{CH_3}{|}}{C}}CHO$ $CH_3\overset{\overset{CH_3}{|}}{\underset{\underset{CH_3}{|}}{C}}CH_2CHO$

        2,2-dimethylbutanal   3,3-dimethylbutanal

        $CH_3\overset{\overset{CH_3}{|}}{CH}\underset{\underset{CH_3}{|}}{CH}CHO$ $CH_3CH_2\overset{\overset{CH_2CH_3}{|}}{CH}CHO$

        2,3-dimethylbutanal   2-ethylbutanal

16.39   a. $CH_3CH_2CH_2CH_2\overset{\overset{O}{\parallel}}{C}H$
        pentanal

        c. $CH_3CH_2CH_2\overset{\overset{O}{\parallel}}{\underset{\underset{CH_3}{|}}{C}H}CCH_3$
        3-methyl-2-hexanone

        b. $CH_3CH_2CH_2\overset{\overset{O}{\parallel}}{C}CH_2CH_3$
        3-hexanone

        d. $CH_3CH_2CH_2CH_2CH_2\overset{\overset{O}{\parallel}}{\underset{\underset{CH_3}{|}}{C}H}CH$
        2-methyloctanal

16.41

16.43   a. $(CH_3)_3CCH_2CHO$
        c.
        b.

16.45

16.47   a. insoluble   b. soluble   c. insoluble

16.49   2,3-Butanedione has two carbonyl groups capable of
        hydrogen bonding, whereas acetone has one carbonyl group.
        It would be soluble in diethyl ether.

16.51   a. $CH_3(CH_2)_4COOH$ c. No reaction
        b.
        d. $CH_3(CH_2)_4COOH$

16.53   a. $CH_3(CH_2)_4COOH$ c. No reaction
        b.
        d. No reaction

16.55   a.
        b.

16.57   a. $CH_3CH_2\overset{\overset{CH_3}{|}}{CH}CH_2CHO$
        c. $CH_3CH_2\overset{\overset{CHO}{|}}{CH}CH_2CH_3$
        b.

16.59   a. CH_3CH_2—
        —CH_2OH
        c.
        b.

16.61 a. (methanal / formaldehyde: H–C(=O)–H)   b. $CH_3CH_2CH_2CH_2$–C(=O)–H   c. (3-methylcyclohexanone: cyclohexanone ring with CH_3)

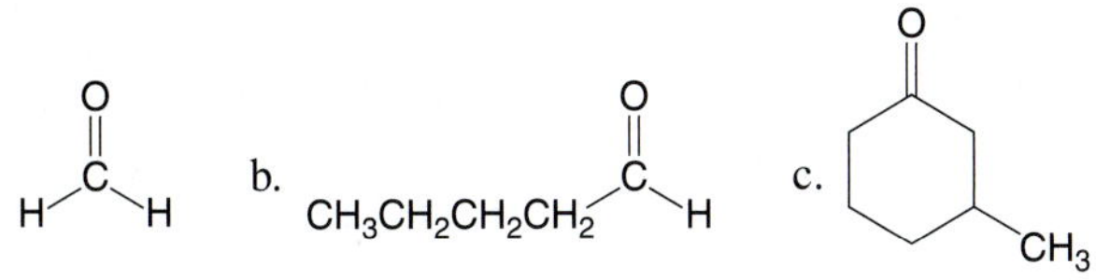

16.63 1-Methylcyclohexanol is a 3° alcohol and cannot be produced from the reduction of a carbonyl compound.

16.65 a. CH_3CH_2–O–CH(H)–O–CH_2CH_3

b. CH_3–C(H)(OH)–O–$CH_2CH_2CH_3$

c. CH_3CH_2–O–CH(H)–CH(H)–O–CH_3

d. CH_3–CH(H)–O–CH(H)–CH(OH)–CH_3

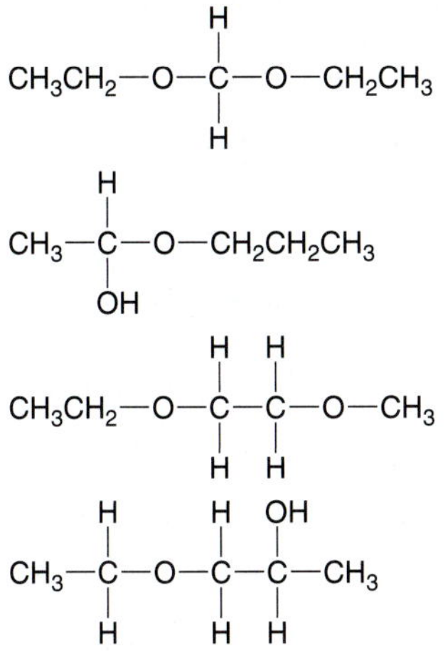

16.67 a. CH_3–C(OCH_3)(H)–OCH_3   (acetal)

b. CH_3–C(OCH_2CH_3)(H)–OH   (hemiacetal)

c. $HOCH_2CH(OCH_3)CH_2CH_3$ (OCH_3 = ether, alcohol at OH)

d. (tetrahydropyran ring, O = ether)

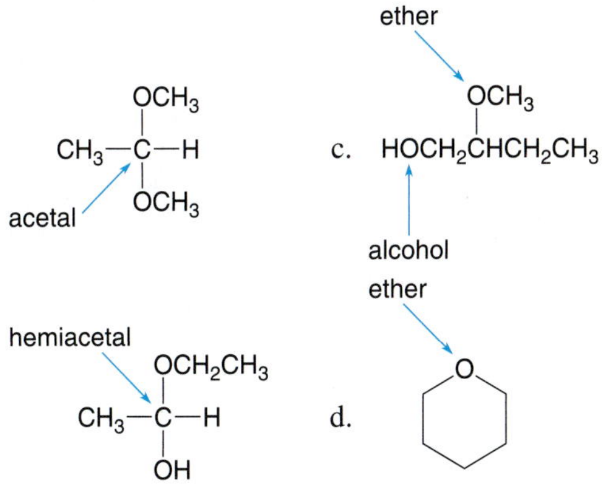

16.69 a. (cyclopentane ring with two CH_3 and OCH_3, OCH_3)

b. $CH_2(OCH_3)_2$

c. CH_3–C(OCH_3)(CH_3O)–$CH_2CH_2CH_3$

d. (benzene)–$CH_2CH(OCH_3)OCH_3$

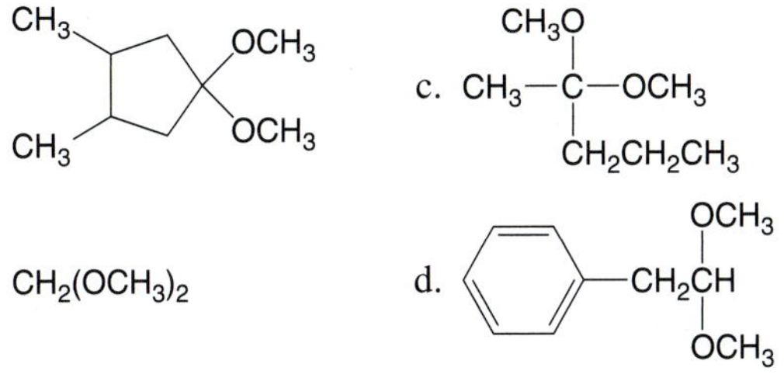

16.71 a. CH_3–C(OH)(CH_3)–O–cyclopentyl

b. CH_3–C(CH_3)–O–cyclopentyl (with =O and cyclopentyl ether)

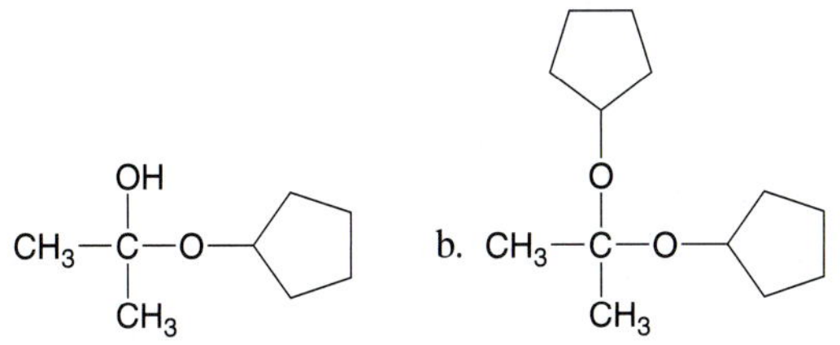

16.73 a. (tetrahydrofuran ring)–OH (hemiacetal carbon)

b. $HOCH_2CH_2CH_2$–C(=O)H

c. (tetrahydrofuran ring)–OCH_3

16.75 (CH_3-substituted tetrahydropyran with OH)

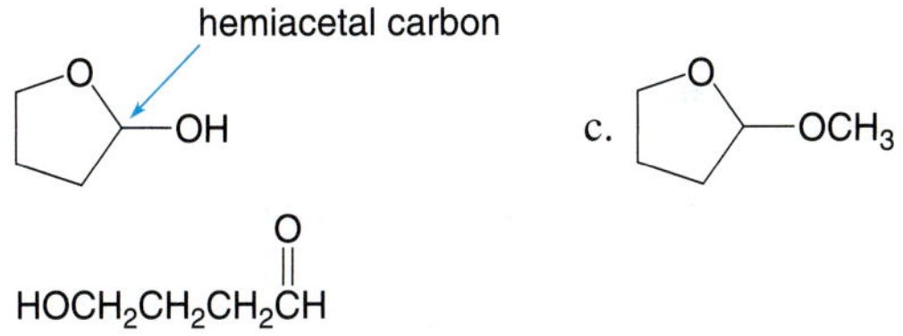

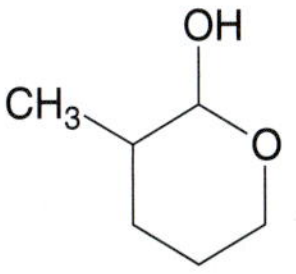

16.77 a. (cyclohexanone) $+$ 2 $HOCH_2CH_2CH_3$

b. H–C(=O)–$CH_2CH_2CH_3$ $+$ 2 $HOCH_3$

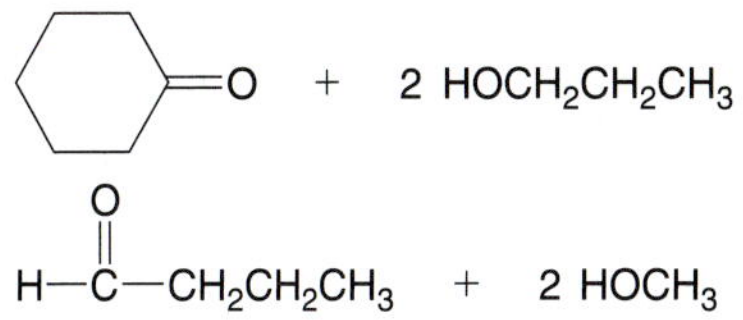

16.79 a. (benzene)–CH_2OH

b. (benzene)–C(=O)OH

c. (benzene)–C(=O)OH

d. (benzene)–C(OCH_3)(H)–OCH_3

e. (benzene)–C(OCH_2CH_3)(H)–OCH_2CH_3

f. (benzene)–C(=O)–H

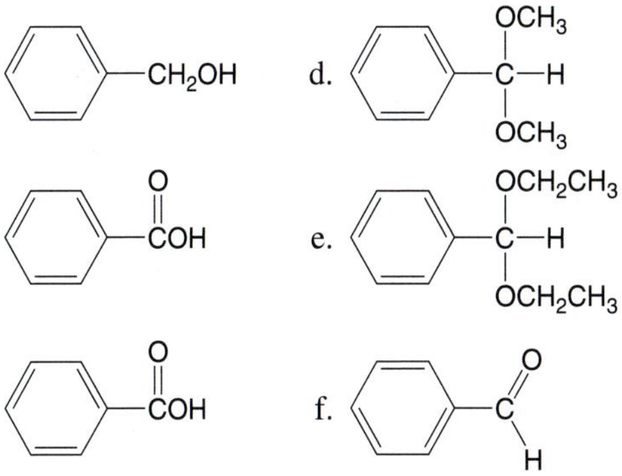

16.81 a. CH_3–C(H)–(CH_2)_4CH_3 (with OH)

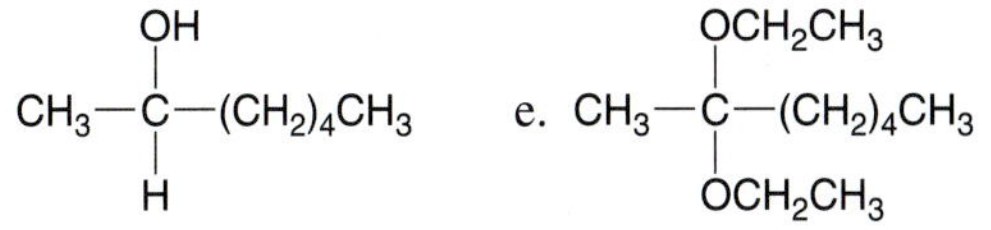

b,c. No reaction

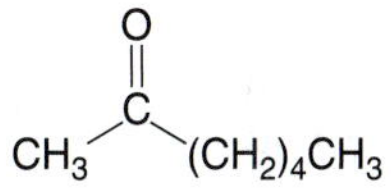

d. CH_3–C(OCH_3)(OCH_3)–(CH_2)_4CH_3

e. CH_3–C(OCH_2CH_3)(OCH_2CH_3)–(CH_2)_4CH_3

f. CH_3–C(=O)–(CH_2)_4CH_3

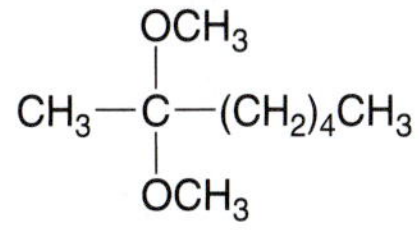

16.83
$CH_2=CHCH_2CH_2CHO$ $CH_3CH=CHCH_2CHO$ $CH_3CH_2CH=CHCHO$

16.85 a. (steroid ring system with CH_3, CH_3, OH, HO)

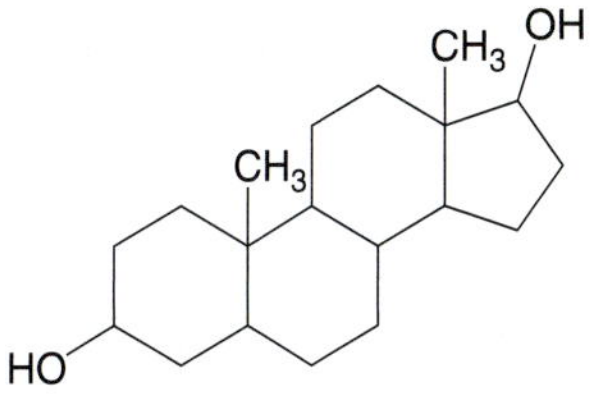

b. (steroid ring system with CH_3, CH_3, =O, O)

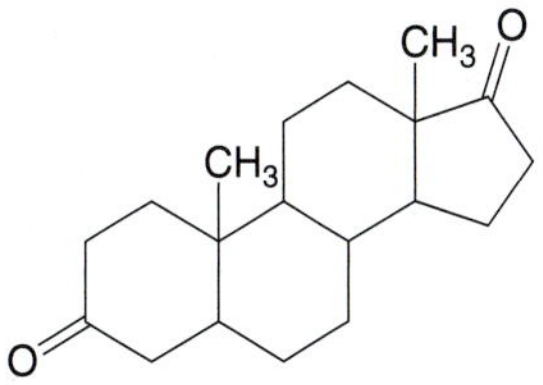

c. (steroid ring system with CH_3, CH_3, OCH_3, OCH_3, HO)

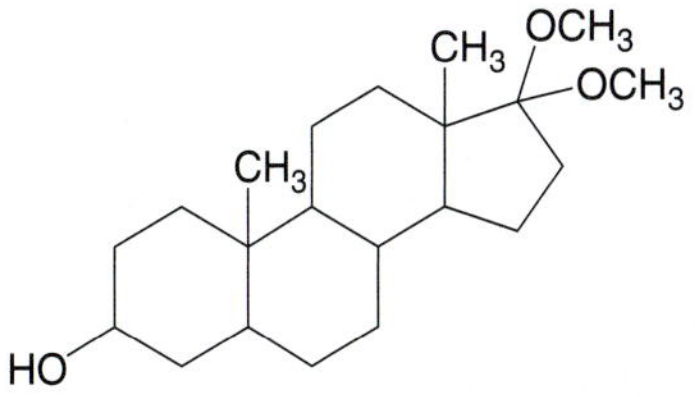

d. (steroid ring system with CH_3, CH_3, OCH_2CH_3, OCH_2CH_3, HO)

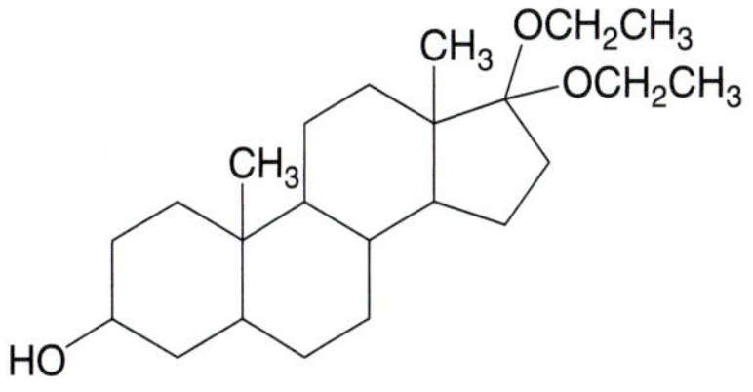

16.87 (benzene)–$CH=CH$–C(=O)–OH

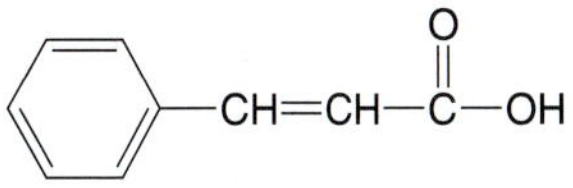

16.89

hemiacetal

$HOCH_2$ — (furanose ring) — OH

alcohol → HO

16.91 The cis double bond in 11-*cis*-retinal produces crowding, making the molecule unstable. Light energy converts this to the more stable trans isomer, and with this conversion an electrical impulse is generated in the optic nerve.

16.93

alcohol

carboxylic acid

CH_3

CH_3

HO_2CCHCH—CH

CH_3O CH_3 acetal   ether

CH_2CH_3 CH_3

CH_3

ether

$HOCH_2$ OH CH_3

hemiacetal

alcohol

CHAPTER 17

17.1

a.
$$CH_3CH_2-\overset{\overset{\displaystyle O}{\|}}{C}-OCH_2CH_3$$
ester

c.
$$(CH_3)_3C-\overset{\overset{\displaystyle O}{\|}}{C}-OH$$
carboxylic acid

b.
$$CH_3-\overset{\overset{\displaystyle O}{\|}}{C}-\underset{H}{N}-CH_3$$
amide

d.
$$(CH_3)_2CH-\overset{\overset{\displaystyle O}{\|}}{C}-N(CH_3)_2$$
amide

17.3
a. 3,3-dimethylhexanoic acid
b. 4-chloropentanoic acid
c. 2,4-diethylhexanoic acid

17.5
a. methyl hexanoate
b. ethyl benzoate
c. propyl pentanoate

17.7
a. pentanamide
b. *N*-methylbenzamide
c. *N,N*-dipropylformamide

17.9 a. CH_3COOH b. $CH_3CH_2CH_2CONH_2$ c. (benzene ring)—COOH

17.11 CH_3CONH_2 is a 1° amide and has two H's available for hydrogen bonding; therefore, it has stronger intermolecular forces and a higher boiling point.

17.13 a. (large lactone ring structure)

17.15

17.17 a.
$$(cyclohexyl)-\overset{\overset{\displaystyle O}{\|}}{C}-O^-Na^+ \quad + \quad H_2O$$

b.
$$CH_3CH_2CH_2-\overset{\overset{\displaystyle O}{\|}}{C}-O^-Na^+ \quad + \quad Na^+HCO_3^-$$

17.19 a. sodium butanoate   b. lithium benzoate

17.21
$$CH_3CH_2CH_2CH_2CH_2CH_2CH_2CH_2CH_2CH_2CH_2CH_2CH_2CH_2CH_2-\overset{\overset{\displaystyle O}{\|}}{C}-O^-K^+$$

17.23 a.
$$(CH_3)_2CHCH_2-\text{(benzene ring)}-\overset{\overset{\displaystyle CH_3}{}}{C}HCOOH \quad + \quad NaOH$$

$$\longrightarrow (CH_3)_2CHCH_2-\text{(benzene ring)}-\overset{\overset{\displaystyle CH_3}{}}{C}HCOO^-Na^+ \quad + \quad H_2O$$

b. The neutral form is present in the stomach.
c. The ionized form is present in the small intestines.

17.24 a.
$$(CH_3)_2CH-\overset{\overset{\displaystyle O}{\|}}{C}-OCH_2CH_3$$

c.
$$CH_3(CH_2)_6-\overset{\overset{\displaystyle O}{\|}}{C}-OCH_2CH_3$$

b.
$$H-\overset{\overset{\displaystyle O}{\|}}{C}-OCH_2CH_3$$

d.
$$(cyclohexyl)-\overset{\overset{\displaystyle O}{\|}}{C}-OCH_2CH_3$$

17.25
$$\text{(benzene ring)}-\overset{\overset{\displaystyle O}{\|}}{C}-OH \quad + \quad CH_3OH$$
with OH on ring

17.27 a.
$$CH_3CH_2CH_2CH_2-\overset{\overset{\displaystyle O}{\|}}{C}-NH_2$$

b.
$$CH_3CH_2CH_2CH_2-\overset{\overset{\displaystyle O}{\|}}{C}-\underset{CH_3}{N}-CH_3$$

c.
$$CH_3CH_2CH_2CH_2-\overset{\overset{\displaystyle O}{\|}}{C}-\underset{H}{N}-(cyclohexyl)$$

d.
$$CH_3CH_2CH_2CH_2-\overset{\overset{\displaystyle O}{\|}}{C}-\underset{H}{N}-CH_3$$

17.29 a.
$$CH_3(CH_2)_8-\overset{\overset{\displaystyle O}{\|}}{C}-OH \quad + \quad CH_3OH$$

b.
$$CH_3\underset{CH_3}{C}HCH_2-\overset{\overset{\displaystyle O}{\|}}{C}-OH \quad + \quad CH_3CH_2OH$$

c.
$$(cyclohexyl)-\overset{\overset{\displaystyle O}{\|}}{C}OH \quad + \quad CH_3CH_2CH_2OH$$

17.31 salicylic acid structure (COOH, OH) $+$ CH_3COH (with C=O)

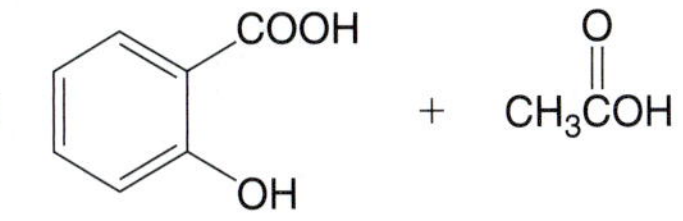

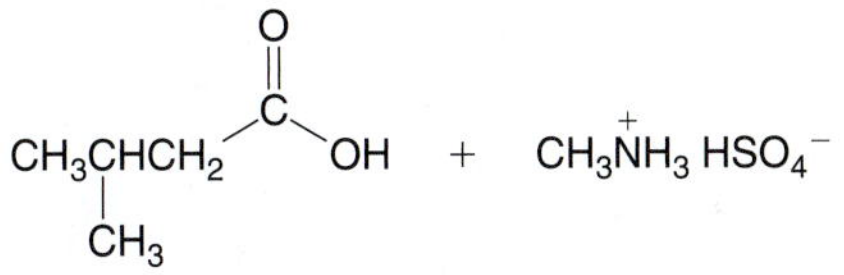

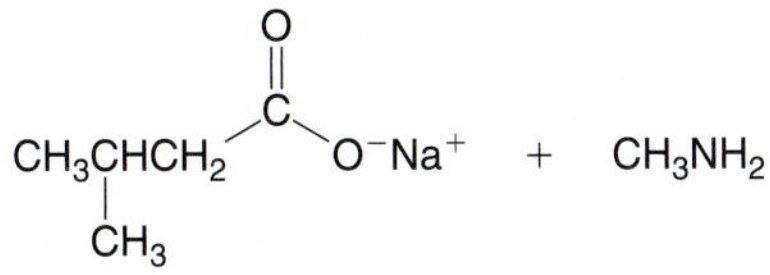

17.32
a. 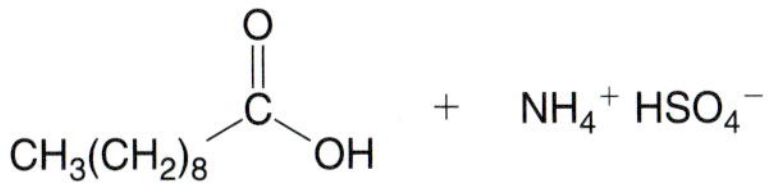 $CH_3(CH_2)_8-C(=O)-OH$ $+$ $NH_4^+ HSO_4^-$

b. $CH_3CHCH_2-C(=O)-OH$ (with CH_3 branch) $+$ $CH_3\overset{+}{N}H_3\ HSO_4^-$

c. cyclohexyl$-C(=O)OH$ $+$ $(CH_3CH_2CH_2)_2\overset{+}{N}H_2\ HSO_4^-$

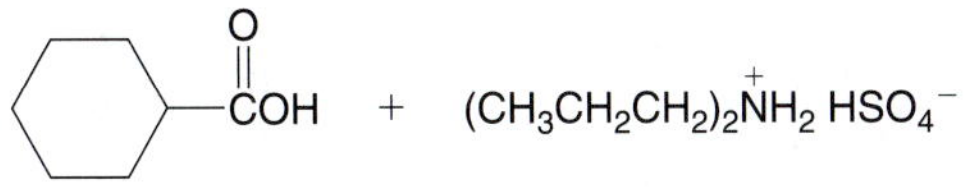

17.33
a. 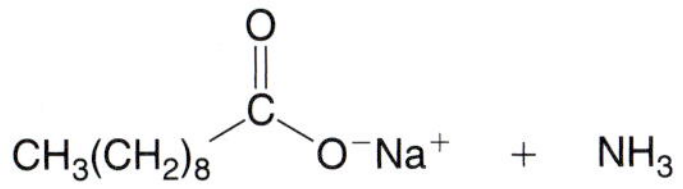 $CH_3(CH_2)_8-C(=O)-O^-Na^+$ $+$ NH_3

b. $CH_3CHCH_2-C(=O)-O^-Na^+$ (with CH_3 branch) $+$ CH_3NH_2

c. cyclohexyl$-C(=O)O^-Na^+$ $+$ $(CH_3CH_2CH_2)_2NH$

17.35 $H_2N-(CH_2)_6-NH_2$ $+$ $HO-C(=O)-(CH_2)_8-C(=O)-OH$

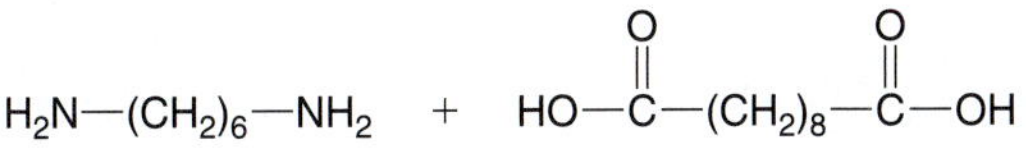

17.37 Polyesters can be converted back to their monomers by acid hydrolysis. The strong C—C bonds in polyethylene are not easily broken.

17.39
a. 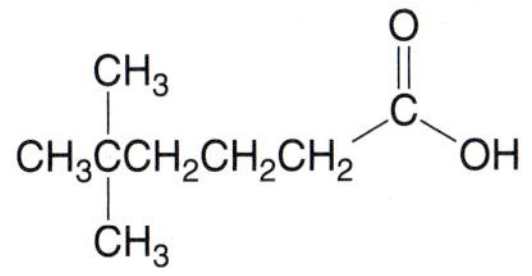 $CH_3CCH_2CH_2CH_2-C(=O)-OH$ (with two CH_3 groups)
c. cyclobutyl$-C(=O)-OCH_3$

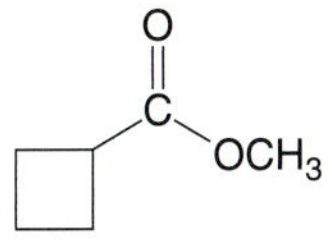

b. 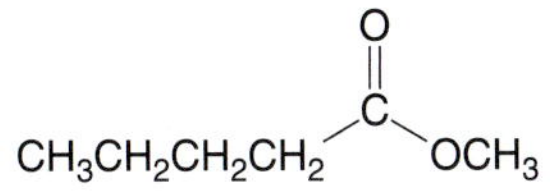 $CH_3CH_2CH_2CH_2-C(=O)-OCH_3$
d. 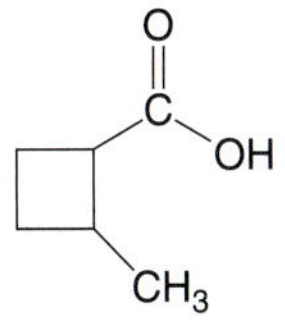 cyclobutyl$-C(=O)-OH$ (with CH_3)

17.41
a. 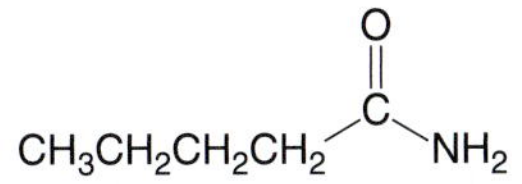 $CH_3CH_2CH_2CH_2-C(=O)-NH_2$
c. $CH_3CH_2-C(=O)-N-CH_3$ (with CH_3)

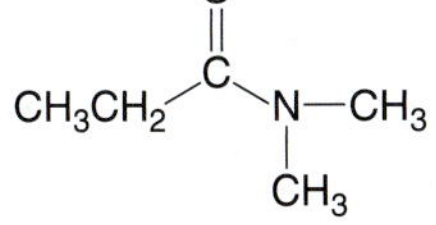

b. 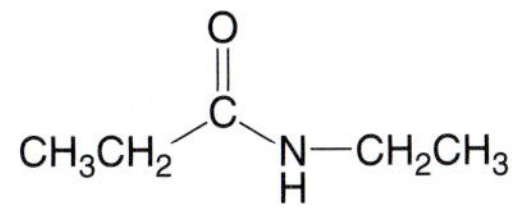 $CH_3CH_2-C(=O)-N-CH_2CH_3$ (with H)

17.43 $CH_3CH_2-C(=O)-O-CH_3$ $CH_3-C(=O)-O-CH_2CH_3$

$H-C(=O)-OCH_2CH_2CH_3$ $H-C(=O)-OCH(CH_3)_2$

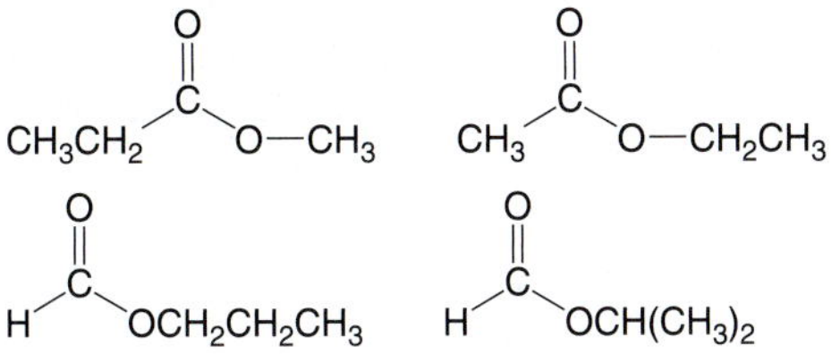

17.45 No, the N atom of the amide must be bonded to a C atom in the ring, so it can't be 1°.

17.47 A lactone is a cyclic ester and a lactam is a cyclic amide.

17.49
a. 4-methylpentanoic acid   c. 2-ethylheptanoic acid
b. 4,5-diethyloctanoic acid   d. o-chlorobenzoic acid

17.51 a. propyl butanoate   b. ethyl formate   c. butyl benzoate

17.53
a. hexanamide
b. N-butylformamide
c. N-ethyl-N-methylbenzamide

17.55
a. formamide   c. 2-hydroxybutanoic acid
b. lithium acetate   d. propyl propanoate

17.57
a. 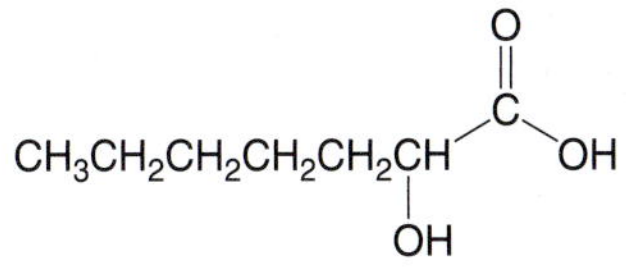 $CH_3CH_2CH_2CH_2CH_2CH-C(=O)-OH$ (with OH)

b. $CH_3CH_2CH_2CH_2CH_2CHCH_2CH_2-C(=O)-OH$ (with Cl)

c. benzene ring with $-C(=O)OH$ and two Br

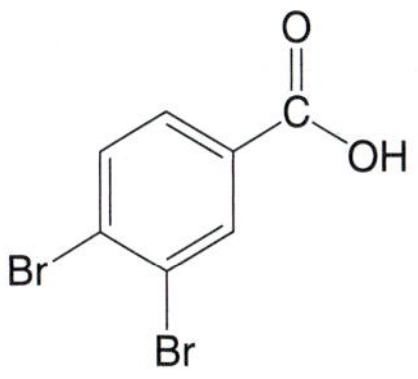

d. 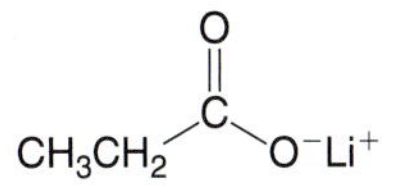 $CH_3CH_2-C(=O)-O^-Li^+$

e. 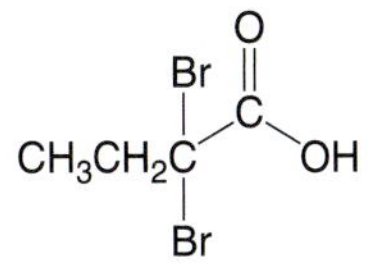 $CH_3CH_2C-C(=O)-OH$ (with two Br)

f. 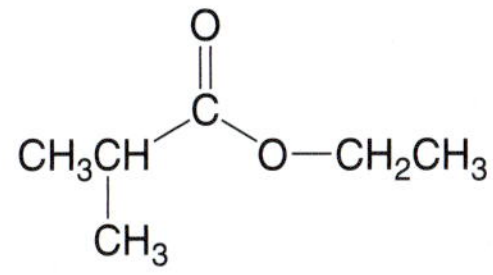 $CH_3CH-C(=O)-O-CH_2CH_3$ (with CH_3)

g. 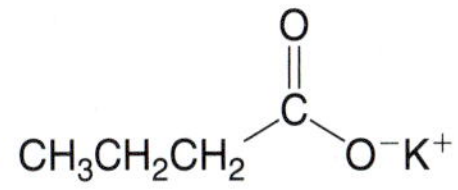 $CH_3CH_2CH_2-C(=O)-O^-K^+$

17.59
a. 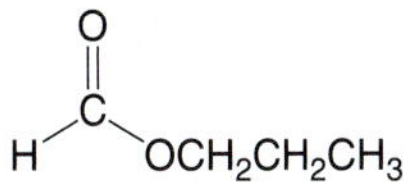 $H-C(=O)-OCH_2CH_2CH_3$

b. $CH_3CH_2CH_2-C(=O)-OCH_2CH_2CH_2CH_3$

c. benzene$-C(=O)-OCH_2CH_2CH_2CH_2CH_2CH_2CH_3$

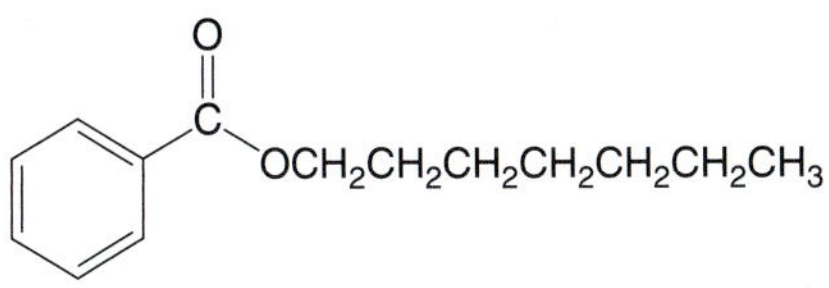

d. 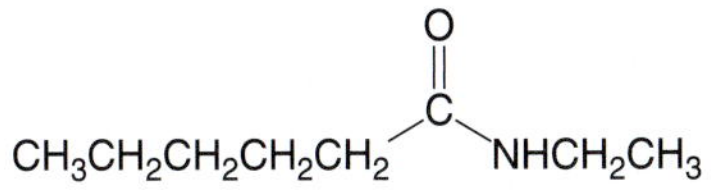 $CH_3CH_2CH_2CH_2-C(=O)-NHCH_2CH_3$

e. 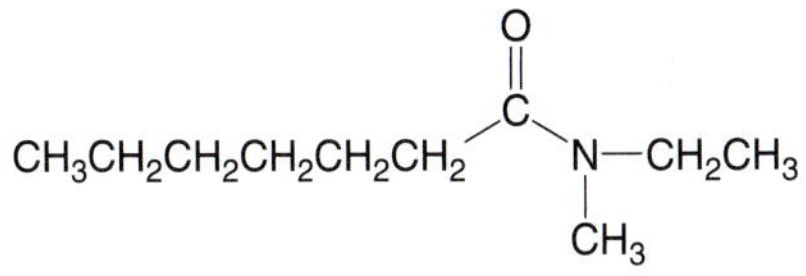 $CH_3CH_2CH_2CH_2CH_2CH_2-C(=O)-N-CH_2CH_3$ (with CH_3)

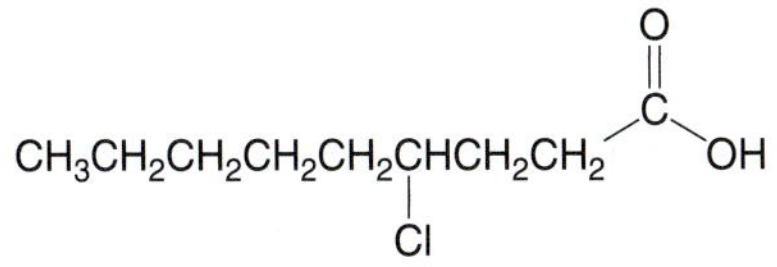

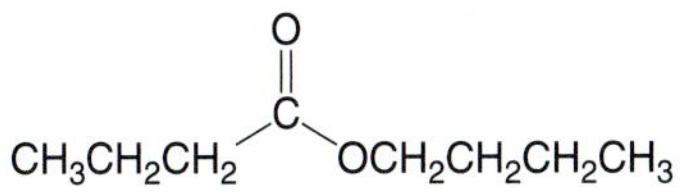

17.61 $CH_3CH_2CH_2\!-\!C(=O)\!-\!NHCH_3$ $CH_3CH_2CH(CH_3)\!-\!C(=O)\!-\!NH_2$

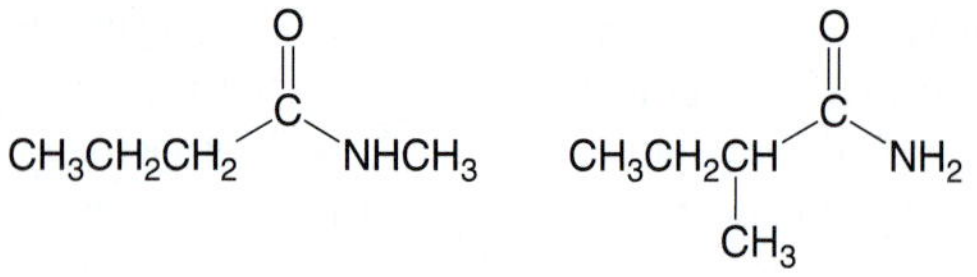

 N-methylbutanamide   2-methylbutanamide

17.63 $CH_3CH_2CH_2CH_2\!-\!C(=O)\!-\!OH$ $CH_3CH(CH_3)CH_2\!-\!C(=O)\!-\!OH$

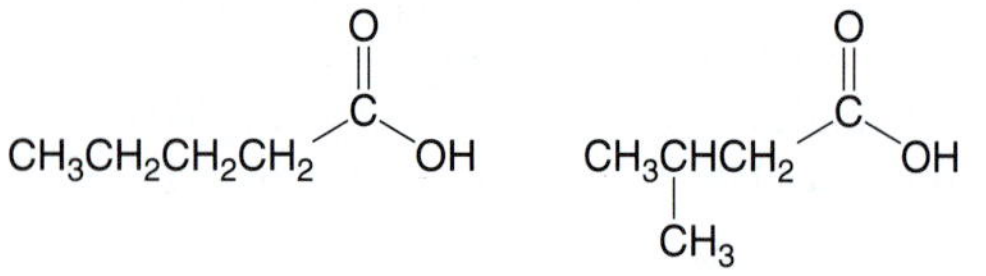

 pentanoic acid   3-methylbutanoic acid

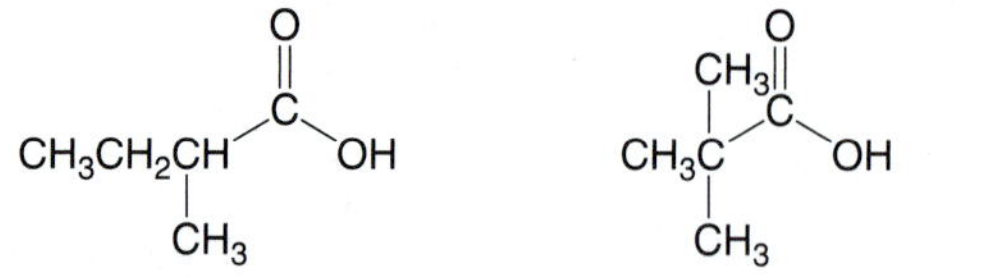

 $CH_3CH_2CH(CH_3)\!-\!C(=O)\!-\!OH$ $CH_3C(CH_3)_2\!-\!C(=O)\!-\!OH$

 2-methylbutanoic acid   2,2-dimethylpropanoic acid

17.65 (a) HCO_2CH_3 can hydrogen bond to water.
     (b) CH_3CH_2COOH can hydrogen bond to itself and to water.

17.67 $CH_3CH_2CH(CH_3)_2 < CH_3CH_2COCH_3 < CH_3CH_2CO_2H$

17.69 $CH_3CH_2CONH_2$ can intermolecularly hydrogen bond, so it has stronger intermolecular forces.

17.71 a. $CH_3(CH_2)_3\!-\!C(=O)\!-\!O^-K^+$ + H_2O

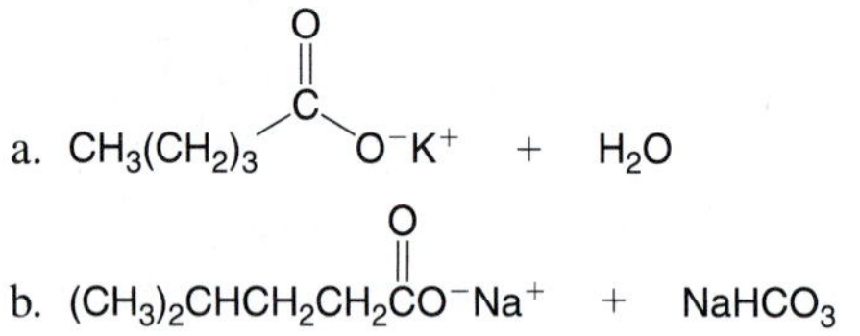

 b. $(CH_3)_2CHCH_2CH_2CO^-Na^+$ + $NaHCO_3$

17.73 a. $CH_3CH_2CH_2\!-\!C(=O)\!-\!OCH_3$

 b. $CH_3CH_2CH_2\!-\!C(=O)\!-\!OCH_2CH_2CH_3$

 c. $CH_3CH_2CH_2\!-\!C(=O)\!-\!O\!-\!$cyclohexyl

 d. $CH_3CH_2CH_2\!-\!C(=O)\!-\!O\!-\!CH_2CH_2\!-\!$phenyl

17.75 $HO\!-\!$C$_6$H$_4$$\!-\!CO_2H$ and CH_3OH

17.77 a. $CH_3CH_2CH_2\!-\!C(=O)\!-\!NH_2$

 b. $CH_3CH_2CH_2\!-\!C(=O)\!-\!NH\!-\!CH_2CH_3$

 c. $CH_3CH_2CH_2\!-\!C(=O)\!-\!N(CH_2CH_3)\!-\!CH_2CH_3$

 d. $CH_3CH_2CH_2\!-\!C(=O)\!-\!N(CH_2CH_3)\!-\!CH_3$

17.79 a. cyclohexyl$\!-\!C(=O)\!-\!OH$ + CH_3OH

 b. $HO\!-\!C(=O)\!-\!CH_3$ + cyclohexyl$\!-\!OH$

17.81 $CH_3CH_2O\!-\!$C$_6$H$_4$$\!-\!NH_2$ + $HO\!-\!C(=O)\!-\!CH_3$

17.83 a. $CH_3CH_2CH_2\!-\!C(=O)\!-\!OH$ + $HOCH(CH_3)_2$

 b. decalinyl$\!-\!OH$ + $HO\!-\!C(=O)\!-\!CH_3$

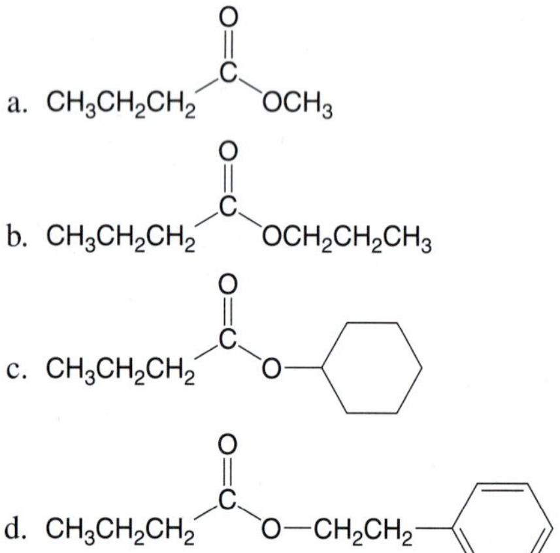

 c. cyclohexyl$\!-\!CH_2CH_2\!-\!OH$ + $HO\!-\!C(=O)\!-\!H$

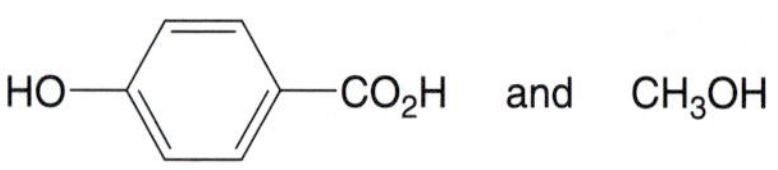

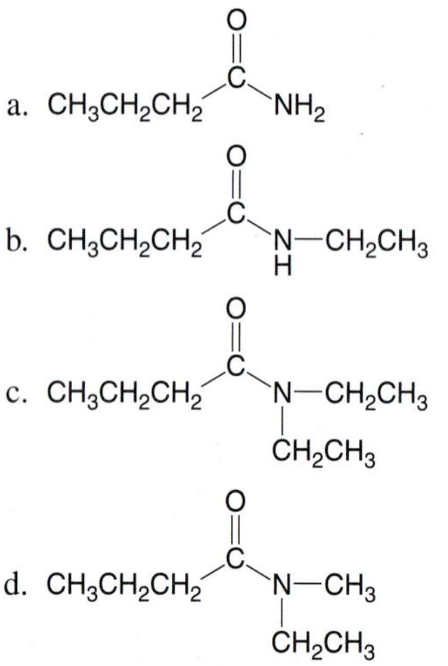

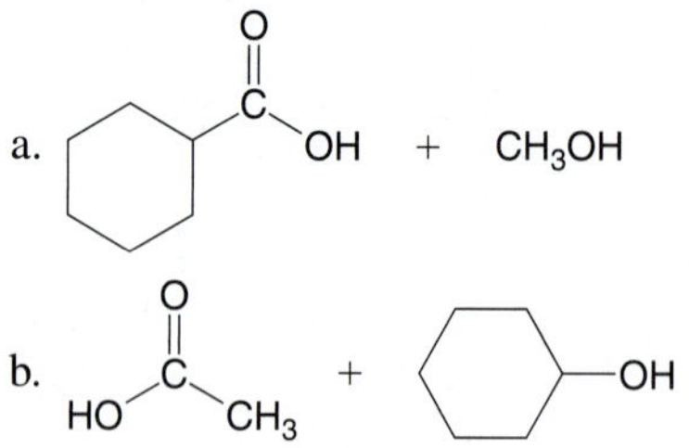

 d. cyclopentyl$\!-\!OH$ + $HO\!-\!C(=O)\!-\!CH_2CH_2CH_3$

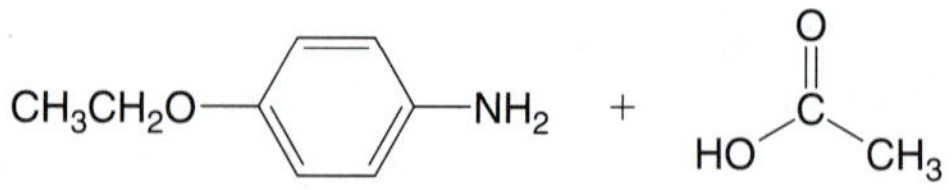

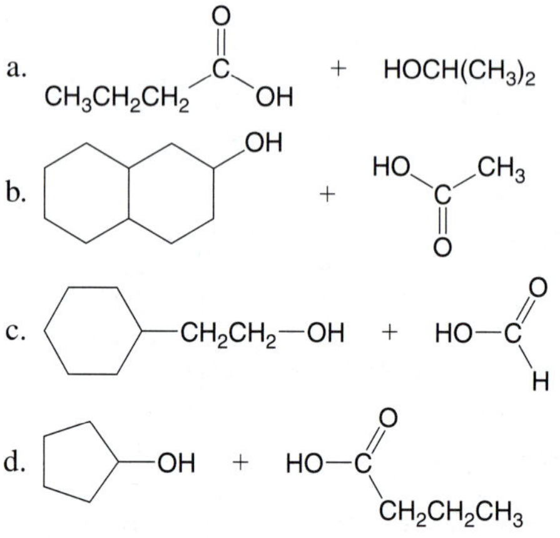

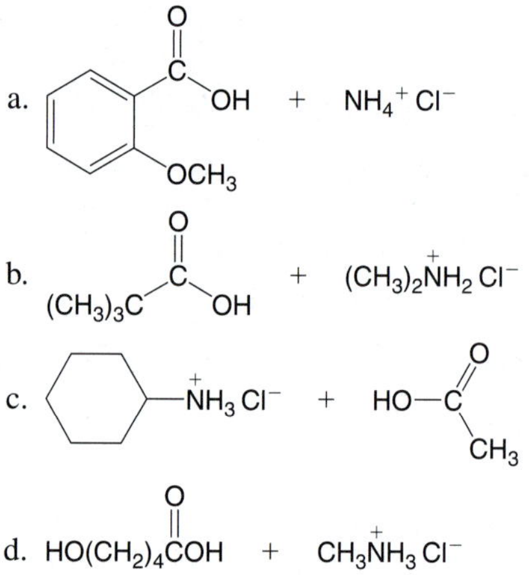

17.85 a. (o-)$C_6H_4(OCH_3)\!-\!C(=O)\!-\!OH$ + $NH_4^+\,Cl^-$

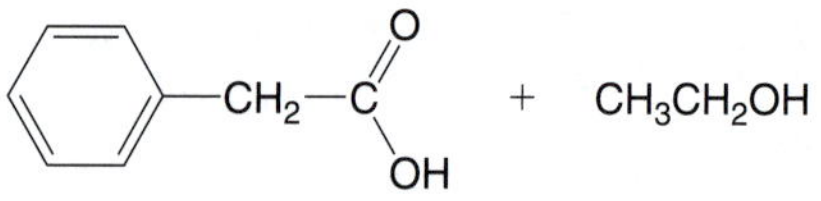

 b. $(CH_3)_3C\!-\!C(=O)\!-\!OH$ + $(CH_3)_2NH_2^+\,Cl^-$

 c. cyclohexyl$\!-\!NH_3^+\,Cl^-$ + $HO\!-\!C(=O)\!-\!CH_3$

 d. $HO(CH_2)_4\!\overset{O}{C}OH$ + $CH_3NH_3^+\,Cl^-$

17.87 phenyl$\!-\!CH_2\!-\!C(=O)\!-\!OH$ + CH_3CH_2OH

17.89 $HO\!-\!C(=O)\!-\!$C$_6$H$_4$$\!-\!C(=O)\!-\!OH$ $NH_2\!-\!(CH_2)_4\!-\!NH_2$

17.91

 $\{\!-\!O\!-\!C(=O)\!-\!$C$_6$H$_4$$\!-\!C(=O)\!-\!OCH_2CH_2CH_2O\!-\!C(=O)\!-\!C_6H_4$$\!-\!C(=O)\!-\!O\!-\!CH_2CH_2CH_2\!\}$

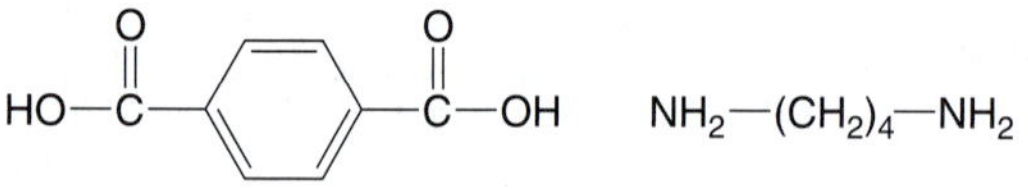

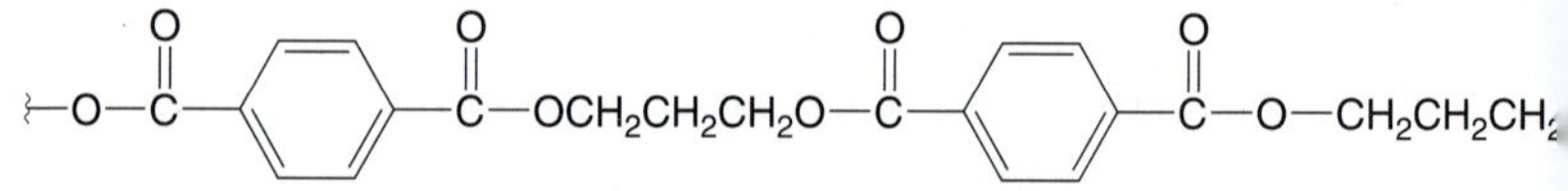

17.93 Saponification is the hydrolysis of an ester with strong base and forms a metal salt, $RCOO^-\,M^+$. Esterification forms a new ester, $RCOOR'$.

17.95 a. $CH_3-C(=O)-OCH_3$

b. $(CH_3)_2CHOH$ + $HO-C(=O)-CH_3$

c. $Na^+ \ ^-O-C(=O)-CH(CH_3)_2$ + $(CH_3)_2CHCH_2OH$

d. $CH_3(CH_2)_4-C(=O)-O^-Na^+$ + H_2O

17.97 a. 5-methylhexanoic acid

b. $CH_3CH_2CHCH_2CH_2-C(=O)-OH$
$\quad\quad\quad |$
$\quad\quad CH_3$

c. $CH_3CH_2CH_2CH_2CH_2-C(=O)-OCH_3$

d. $CH_3CHCH_2CH_2CH_2-C(=O)-O^-Na^+$ + H_2O
$\quad\quad |$
$\quad CH_3$

e. **A** is insoluble in H_2O, but soluble in organic solvent.

f. $CH_3CHCH_2CH_2CH_2-C(=O)-OCH_2CH_3$
$\quad\quad |$
$\quad CH_3$

g. $CH_3CHCH_2CH_2CH_2-C(=O)-NHCH_2CH_3$
$\quad\quad |$
$\quad CH_3$

17.99 **A** = $K_2Cr_2O_7$ **B** = [benzyl alcohol structure with CH$_2$OH] + H_2SO_4

17.101 b. This is a sodium salt of a long-chain carboxylic acid.

17.103 a. [naphthalene structure with CHCOOH, CH$_3$, and CH$_3$O groups] + NaOH

$\longrightarrow$ [naphthalene structure with CHCOO$^-$Na$^+$, CH$_3$, and CH$_3$O groups] + H_2O

b. In the stomach naproxen exists as the neutral carboxylic acid.

c. In the intestines naproxen exists as the ionized carboxylate anion.

17.105 Aspirin inhibits the production of prostaglandins.

17.107 The nonpolar end of a soap molecule binds to dirt and hydrocarbons while the polar end then hydrogen bonds with water, thereby rendering the dirt soluble in water.

17.109 [HO-C(=O)-CH$_2$-CH(NH$_2$)-C(=O)-OH structure] + [benzyl-CH$_2$-CH(NH$_2$)-C(=O)-OH structure] + CH_3OH

17.111 a. [six-membered lactone ring structure]   b. [five-membered lactone ring structure]

CHAPTER 18

18.1 a. $\overset{1°}{H_2N}(CH_2)_3\overset{2°}{N}H(CH_2)_4\overset{2°}{N}H(CH_2)_3\overset{1°}{N}H_2$

b. $CH_3CH_2O-C(=O)-$ [piperidine ring with C_6H_5 and $\overset{3°}{N}-CH_3$]

18.3 a. 2° amine

b. (1) trigonal planar (2) tetrahedral (3) trigonal pyramid

18.4 a. 2-butanamine

b. N-methyl-1-propanamine

c. N,N-dimethylcyclohexanamine

d. dibutylamine

18.5 a. [benzene ring with NHCH$_3$]

b. [benzene ring with NH$_2$ and CH$_2$CH$_3$]

c. [benzene ring with CH$_3$CH$_2$, NH$_2$, and CH$_2$CH$_3$]

d. [benzene ring with CH$_3$CH$_2$ and N(CH$_2$CH$_3$)]

18.7 a. $(CH_3)_2CHCH_2NH_2$ c. [cyclohexane ring with NH$_2$]

b. $(CH_3)_2CHCH_2OH$

18.9 Caffeine is soluble in the organic solvent CH_2Cl_2 because it is organic and "like dissolves like."

18.11 chirality center

[three nicotine structures]

18.13

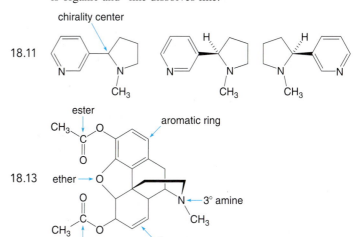

18.15

(structure labeled: 3° amine, alcohol, aromatic ring, ester)

18.16 a. $CH_3CH_2\overset{+}{N}H_3$ + Cl^-

b. $(CH_3CH_2)_2\overset{+}{N}H_2$ + Cl^-

c. $(CH_3CH_2)_3\overset{+}{N}H$ + Cl^-

18.17 a. $CH_3CH_2CH_2CH_2-\overset{+}{N}H_3$ + Cl^-

b. $(CH_3)_2\overset{+}{N}H_2$ + $C_6H_5COO^-$

c.
(piperidinium cation) + ^-OH

18.19 a. methylammonium chloride
b. dipropylammonium bromide
c. ethyldimethylammonium acetate

18.21 a. $(CH_3CH_2)_3N$ b. $CH_3CH_2NH_2$ c. (piperidine NH)

18.23 A quaternary ammonium salt has four R groups bonded to N so there is no proton available that can be removed.

18.25 COOH must be removed and OH must be added.

18.27 This would require addition of a methyl group and removal of two OH groups.

18.29 albuterol: 1° and 2° alcohols, 2° amine
salmeterol: 1° and 2° alcohols, 2° amine

18.31 a. 2°   b. 3°   c. 1°

18.33 a. $CH_3CH_2CH_2CH_2CH_2NH_2$ c. (N—CH₃ piperidine)

b. $CH_3CH_2CH_2\underset{H}{N}CH_2CH_3$

d. $CH_3CH_2\overset{+}{\underset{CH_2CH_3}{N}}CH_2CH_3$

18.35 A 3° amine has three carbon–nitrogen bonds. A 3° alcohol has a hydroxyl on a carbon bonded to three carbon atoms.

(triethylamine and 3-methyl-3-pentanol structures)

18.37 a. diethylamine
b. cyclohexanamine
c. N-ethyl-N-methylaniline
d. 3-octanamine
e. N-methyl-3-hexanamine
f. N-ethyl-2-heptanamine

18.39 a. $CH_3CH_2CH_2CH_2CH_2CH_2CH_2CH_2CH_2CH_2NH_2$

b.
(tricyclohexylamine structure)

18.41

c. (4-bromoaniline)

d. $CH_3\underset{NH_2}{CH}CH_2COOH$

e. $CH_3CH_2CH_2CH_2CH_2CH_2\underset{N(CH_2CH_3)_2}{CH}CH_3$

f. $CH_3CH_2NHCH_2CH_2CH_2CH_2CH_2CH_3$

N,N-dimethylaniline   2,4-diethylaniline

18.43 a. $(CH_3CH_2CH_2)_2\overset{+}{N}H_2$ Cl^-

b. $CH_3CH_2CH_2CH_2\overset{+}{N}H_3$ Br^-

c. $CH_3CH_2\overset{+}{\underset{H}{N}}(CH_3)_2$ ^-OH

18.45

trimethylamine $CH_3CH_2CH_2NH_2$ $CH_3CH_2NHCH_3$ $CH_3\underset{CH_3}{CH}NH_2$
           1-propanamine   N-methylethanamine   2-propanamine

18.47 Pyridine is capable of hydrogen bonding with water, so it is more water soluble. Pyridine has a higher boiling point than benzene.

18.49 a. $CH_3(CH_2)_7NH_2$ c. (cyclohexyl-CH₂CH₂CH₂NH₂)

b. $CH_3(CH_2)_6OH$

18.51
(hydrogen bonding of triethylamine)

18.53 Primary amines can hydrogen bond to each other, whereas 3° amines cannot. Therefore, 1° amines will have a higher boiling point than 3° amines of similar size. Any amine can hydrogen bond to water so both 1° and 3° amines have similar solubility properties.

18.55 a. $CH_3CH_2NH_2 + H_2O \longrightarrow CH_3CH_2\overset{+}{N}H_3 + \ ^-OH$

b. $(CH_3CH_2)_2NH + H_2O \longrightarrow (CH_3CH_2)_2\overset{+}{N}H_2 + \ ^-OH$

c. $(CH_3CH_2)_3N + H_2O \longrightarrow (CH_3CH_2)_3\overset{+}{N}H + \ ^-OH$

18.57 a. $CH_3CH_2CH_2\overset{+}{N}H(CH_3)_2$ + Cl^-

b. $CH_3CH_2\underset{\overset{|}{\overset{+}{N}H_3}}{CH}CH_2CH_3$ + HSO_4^-

c. (pyridinium) + Br^-

d. $CH_3CH_2-\underset{CH_2CH_3}{N}-CH_2CH_3$ + H_2O + Na^+

18.59 a. [structure: piperidinium with $CH_2CH_2CH_3$, Cl^-]   b. [morphine-type structure with HO groups, O, N^+, H, CH_3, Cl^-]

18.61 piperidine

18.63 Caffeine is a mild stimulant, imparting a feeling of alertness after consumption. It also increases heart rate, dilates airways, and stimulates the secretion of stomach acid. These effects are observed because caffeine increases glucose production, making an individual feel energetic.

18.65 An alkaloid solution is slightly basic since its amine pulls off a proton from water, forming ^-OH.

18.67 A presynaptic neuron releases a neurotransmitter, which is taken up by receptors in the postsynaptic neuron.

18.69 Dopamine affects brain processes that control movement, emotions, and pleasure. Normal dopamine levels give an individual a pleasurable, satisfied feeling. Increased levels result in an intense "high." Drugs such as heroin, cocaine, and alcohol increase dopamine levels. When there is too little dopamine in the brain, an individual loses control of fine motor skills and Parkinson's disease results.

18.71 Serotonin plays an important role in mood, sleep, perception, and temperature regulation. We get sleepy after eating a turkey dinner on Thanksgiving because the unusually high level of tryptophan in turkey is converted to serotonin. A deficiency of serotonin causes depression.

18.73 Dopamine and norepinephrine are derived from tyrosine, serotonin from tryptophan, and histamine from histidine.

18.75 a and c

a. [structure: benzene ring—C(CH_3)(CH_3)—$CH_2N(CH_3)_2$]
c. [structure: benzene ring—CH_2—piperidine ring with HN]

18.77 a. [structure: benzene ring—$CH_2C(CH_3)(CH_3)NH_2$]

b. [structure: benzene ring—CH_2CH_2—N piperidine ring—N—benzene ring, with $COCH_2CH_3$]

18.79 Chlorpheniramine is an example of an antihistamine. Antihistamines bind to the H1 histamine receptor, but they evoke a different response. An antihistamine like chlorpheniramine or diphenhydramine, for example, inhibits vasodilation, so it is used to treat the symptoms of the common cold and environmental allergies.

18.81 a,b. [structure: benzene ring—CH_2CHNCH_2—benzene ring, with CH_3 groups; labeled: 3° amine, chirality center]

18.83 a,b,c,d. [structure with labels: aromatic ring, chirality centers, 2° amine, ester; CH—COOCH_3—piperidine ring with HN; caption: 2-phenylethylamine in bold]

c. [two structures with phenyl groups, CH_3, NCH$_2$ groups]

d. [structure: dimethylbenzene ring—CH_2CHNH_2 with CH_3; CH_2—benzene ring]

e. [structure: benzene ring—$CH_2CH_2N(CH_3)CH_3$; CH_2—benzene ring]

f. [structure: benzene ring—$CH_2CHNHCH_2$—benzene ring, with CH_3 groups, Cl^-]

g. [structure: benzene ring—CH_2CHNCH_2—benzene ring, with CH_3 groups] + CH_3COOH

→ [structure: benzene ring—$CH_2CHNHCH_2$—benzene ring, with CH_3 groups] + CH_3COO^-

e. [structure: benzene ring—CH—piperidine ring with N^+H, Cl^-, $COOCH_3$]

18.85 [structure: Cl—benzene ring—$CHCH_2CH_2N(CH_3)_2$ with pyridine ring (N)]   [structure: HOOC—C=C—COOH with H, H]

18.87 A vasodilator dilates blood vessels and a bronchodilator dilates airways in the lungs. Histamine is a vasodilator and albuterol is a bronchodilator.

18.89 Albuterol will exist in the ionic form in the stomach and as the neutral form in the intestines.

18.91 Heroin has two esters, which can be made from the two OH groups in morphine. Add acetic acid (CH_3COOH) and H_2SO_4 to make the two esters in heroin.

CHAPTER 19

19.1 b and c

19.3

A =

hydrophilic portion (points to $-C(=O)OH$)

hydrophobic portion

$$CH_3(CH_2)_{14}-C(=O)-OH$$

B =

hydrophilic portion (points to $-C(=O)OH$)

hydrophobic portion

(unsaturated fatty acid with two cis double bonds, $-C(=O)-OH$)

c. **A** will have the higher melting point because the molecules can pack together better.

19.5 a. omega-9   b. omega-6

19.6 a. $CH_3(CH_2)_{16}CO(CH_2)_9CH_3$ c. $CH_3(CH_2)_{16}CO(CH_2)_{29}CH_3$

b. $CH_3(CH_2)_{16}CO(CH_2)_{11}CH_3$

19.7 $CH_3(CH_2)_{30}CO(CH_2)_{33}CH_3$

19.9 $CH_3(CH_2)_{12}-C(=O)-OH$ + $HO(CH_2)_{15}CH_3$

19.10 a.

$$\begin{array}{l} CH_2-O-C(=O)-(CH_2)_{16}CH_3 \\ CH-O-C(=O)-(CH_2)_{16}CH_3 \\ CH_2-O-C(=O)-(CH_2)_{16}CH_3 \end{array}$$

b.

$$\begin{array}{l} CH_2-O-C(=O)-(CH_2)_7C(H)=C(H)(CH_2)_7CH_3 \\ CH-O-C(=O)-(CH_2)_7C(H)=C(H)(CH_2)_7CH_3 \\ CH_2-O-C(=O)-(CH_2)_7C(H)=C(H)(CH_2)_7CH_3 \end{array}$$

19.11 a.

$$\begin{array}{l} CH_2-O-C(=O)-(CH_2)_{16}CH_3 \\ CH-O-C(=O)-(CH_2)_{16}CH_3 \\ CH_2-O-C(=O)-(CH_2)_{14}CH_3 \end{array}$$

b.

$$\begin{array}{l} CH_2-O-C(=O)-(CH_2)_{16}CH_3 \\ CH-O-C(=O)-(CH_2)_{14}CH_3 \\ CH_2-O-C(=O)-(CH_2)_{16}CH_3 \end{array}$$

19.13 a.

$$\begin{array}{l} CH_2-OH \\ CH-OH \\ CH_2-OH \end{array}$$ + $3\ HO-C(=O)-(CH_2)_{12}CH_3$

b.

$$\begin{array}{l} CH_2-OH \\ CH-OH \\ CH_2-OH \end{array}$$ +

$$HO-C(=O)-(CH_2)_{12}CH_3$$
$$HO-C(=O)-(CH_2)_7CH=CH(CH_2)_5CH_3$$
$$HO-C(=O)-(CH_2)_7CH=CH(CH_2)_7CH_3$$

19.15 a. $3\ Na^+\ {}^-O-C(=O)-(CH_2)_{12}CH_3$

b. $Na^+\ {}^-O-C(=O)-(CH_2)_{12}CH_3$

$Na^+\ {}^-O-C(=O)-(CH_2)_7CH=CH(CH_2)_5CH_3$

$Na^+\ {}^-O-C(=O)-(CH_2)_7CH=CH(CH_2)_7CH_3$

19.16

$$\begin{array}{l} CH_2-O-C(=O)-(CH_2)_{14}CH_3 \\ CH-O-C(=O)-(CH_2)_7C(H)=C(H)(CH_2)_7CH_3 \\ CH_2-O-P(=O)(O^-)-O-CH_2CH_2\overset{+}{N}H_3 \end{array}$$

$$\begin{array}{l} CH_2-O-C(=O)-(CH_2)_7C(H)=C(H)(CH_2)_7CH_3 \\ CH-O-C(=O)-(CH_2)_{14}CH_3 \\ CH_2-O-P(=O)(O^-)-O-CH_2CH_2\overset{+}{N}H_3 \end{array}$$

19.17 a. phosphoacylglycerol, cephalin

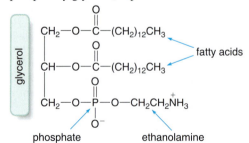

$$\begin{array}{l} CH_2-O-C(=O)-(CH_2)_{12}CH_3 \\ CH-O-C(=O)-(CH_2)_{12}CH_3 \\ CH_2-O-P(=O)(O^-)-O-CH_2CH_2\overset{+}{N}H_3 \end{array}$$

fatty acids

glycerol   phosphate   ethanolamine

b. triacylglycerol

$$\begin{array}{l} CH_2-O-C(=O)-(CH_2)_{12}CH_3 \\ CH-O-C(=O)-(CH_2)_7CH=CH(CH_2)_5CH_3 \\ CH_2-O-C(=O)-(CH_2)_{12}CH_3 \end{array}$$

fatty acids

glycerol

c. sphingomyelin

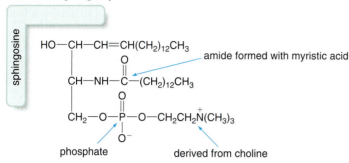

19.19 Membrane **A** will be more fluid or pliable as it contains unsaturated fatty acids.

19.21 Cholesterol is a lipid since it contains many C—C and C—H bonds and it is not water soluble.

19.23 Triacylglycerols would be found in the interior, hydrophobic portion of lipoproteins.

19.25 a. Estrone has a phenol (a benzene ring with a hydroxyl group) and progesterone has a ketone and C=C in ring A. Progesterone also has a methyl group bonded to C10.
b. Estrone has a ketone at C17 and progesterone has a C—C bond, which is attached to a ketone.

19.27

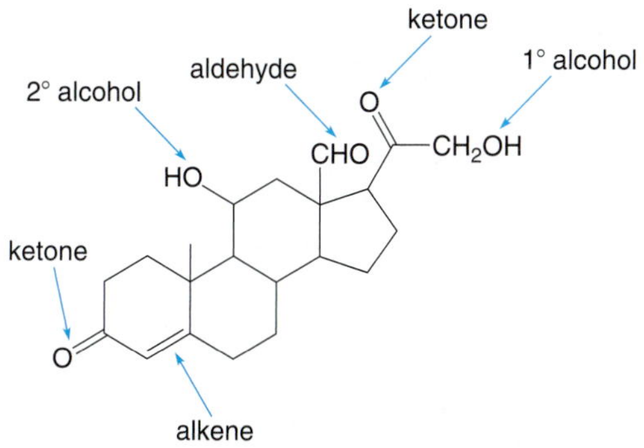

19.29 Water-soluble vitamins are excreted in the urine, whereas fat-soluble vitamins are stored in the body.

19.31 a.

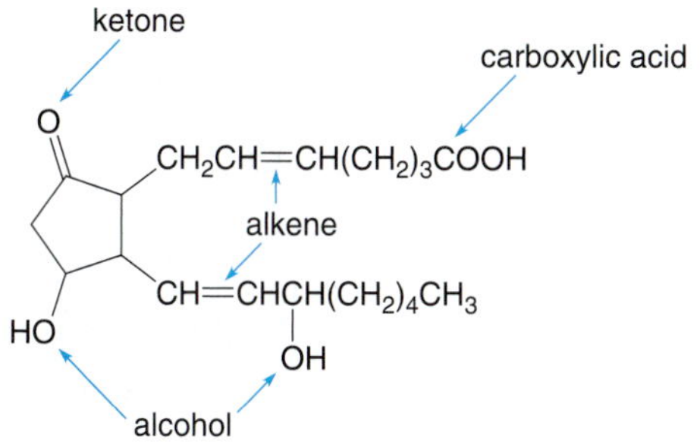

b.

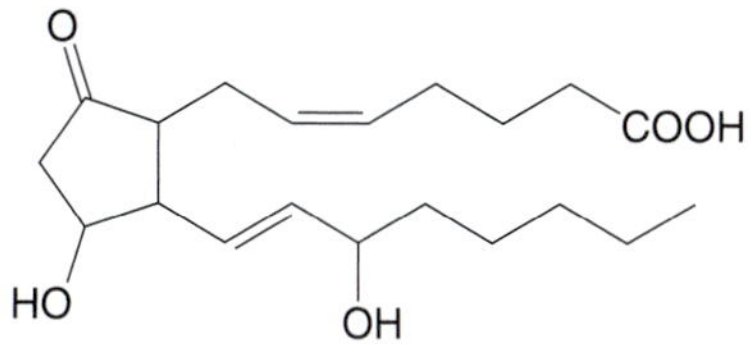

19.33 b,e,f: hydrolyzable   a,c,d,g: nonhydrolyzable

19.35 b and c

19.37 a. $CH_3(CH_2)_3CH=CH(CH_2)_7COOH$, $CH_3(CH_2)_{12}COOH$, $CH_3(CH_2)_{14}COOH$
b. $CH_3(CH_2)_5CH=CH(CH_2)_7COOH$, $CH_3(CH_2)_7CH=CH(CH_2)_7COOH$, $CH_3(CH_2)_{16}COOH$

19.39 a. Increasing the number of carbon atoms increases the melting point.
b. Increasing the number of double bonds decreases the melting point.

19.41 Fatty acids are not hydrolyzable because they contain a very long hydrocarbon chain attached to a carboxylic acid group.

19.43 a. $CH_3(CH_2)_{14}\overset{O}{\overset{\|}{C}}O(CH_2)_{21}CH_3$ c. $CH_3(CH_2)_{14}\overset{O}{\overset{\|}{C}}O(CH_2)_9CH_3$
b. $CH_3(CH_2)_{14}\overset{O}{\overset{\|}{C}}O(CH_2)_{11}CH_3$

19.45 a. $CH_3(CH_2)_{16}COOH + HO(CH_2)_{17}CH_3$
b. $CH_3(CH_2)_{12}COOH + HO(CH_2)_{25}CH_3$
c. $CH_3(CH_2)_{14}COOH + HO(CH_2)_{27}CH_3$
d. $CH_3(CH_2)_{22}COOH + HO(CH_2)_{13}CH_3$

19.47 a.

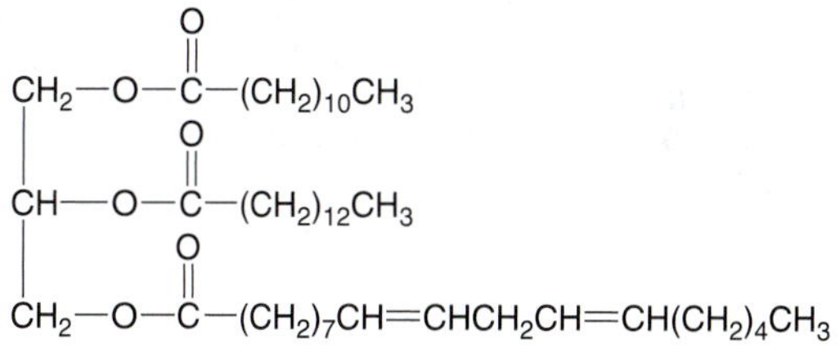

b.

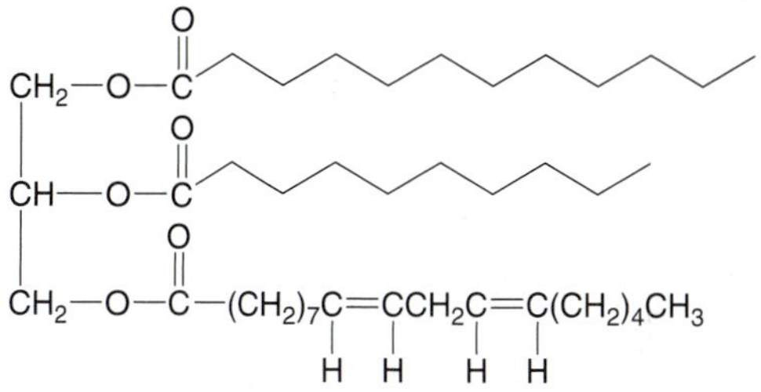

c.

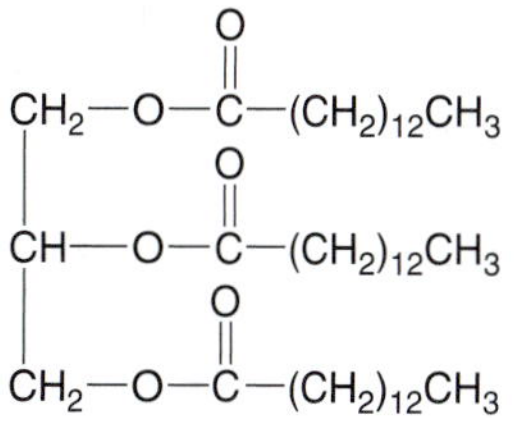

d.

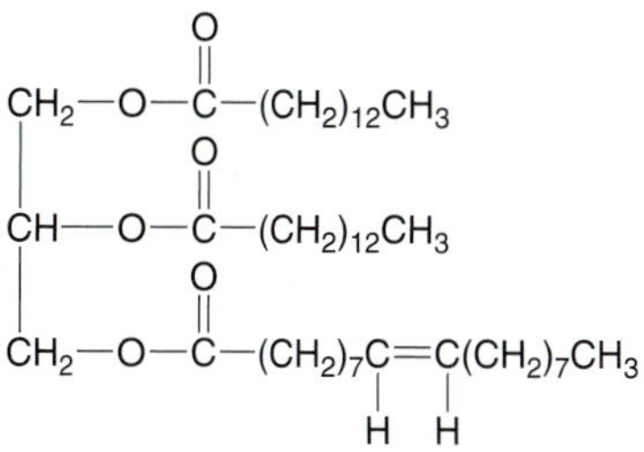

19.49

Compound	a. General structure	b. Example	c. Water soluble (Y/N)	d. Hexane soluble (Y/N)
[1] Fatty acid	$RCOOH$	(chain)COOH	N	Y
[2] Soap	$RCOO^- Na^+$	(chain)$COO^- Na^+$	Y	N
[3] Wax	$RCOOR'$	(chain)COO(chain)	N	Y
[4] Triacylglycerol	$CH_2-O-\overset{O}{\overset{\|}{C}}-R$ $CH-O-\overset{O}{\overset{\|}{C}}-R'$ $CH_2-O-\overset{O}{\overset{\|}{C}}-R''$	$CH_2-O-\overset{O}{\overset{\|}{C}}-(CH_2)_{12}CH_3$ $CH-O-\overset{O}{\overset{\|}{C}}-(CH_2)_{12}CH_3$ $CH_2-O-\overset{O}{\overset{\|}{C}}-(CH_2)_{12}CH_3$	N	Y

19.51 a. arachidic acid, stearic acid, and lauric acid
  b. solid
  c. The long hydrocarbon chains are hydrophobic.
  d. The ester linkages are hydrophilic.
  e. CH_2-OH $HOOC(CH_2)_{10}CH_3$
      $CH-OH$ $+$ $HOOC(CH_2)_{16}CH_3$
      CH_2-OH $HOOC(CH_2)_{18}CH_3$

19.53 a.   [1] water and H_2SO_4

CH_2-OH
$CH-OH$ $+$ $2\ HO\overset{O}{\overset{\|}{C}}-(CH_2)_{14}CH_3$
CH_2-OH $HO\overset{O}{\overset{\|}{C}}-(CH_2)_{16}CH_3$

[2] water and NaOH

CH_2-OH
$CH-OH$ $+$ $2\ Na^{+-}O\overset{O}{\overset{\|}{C}}-(CH_2)_{14}CH_3$
CH_2-OH $Na^{+-}O\overset{O}{\overset{\|}{C}}-(CH_2)_{16}CH_3$

  b.   [1] water and H_2SO_4

CH_2-OH $HO\overset{O}{\overset{\|}{C}}-(CH_2)_7CH=CH(CH_2)_7CH_3$
$CH-OH$ $+$ $HO\overset{O}{\overset{\|}{C}}-(CH_2)_{14}CH_3$
CH_2-OH $HO\overset{O}{\overset{\|}{C}}-(CH_2)_7CH=CH(CH_2)_5CH_3$

[2] water and NaOH

CH_2-OH $Na^{+-}O\overset{O}{\overset{\|}{C}}-(CH_2)_7CH=CH(CH_2)_7CH_3$
$CH-OH$ $+$ $Na^{+-}O\overset{O}{\overset{\|}{C}}-(CH_2)_{14}CH_3$
CH_2-OH $Na^{+-}O\overset{O}{\overset{\|}{C}}-(CH_2)_7CH=CH(CH_2)_5CH_3$

19.55

$CH_2-O-\overset{O}{\overset{\|}{C}}-R$
$CH-O-\overset{O}{\overset{\|}{C}}-R'$ a primary amine salt
$CH_2-O-\overset{O}{\overset{\|}{P}}-O-CH_2CH_2\overset{+}{N}H_3$
     $|$
     O^-

cephalin
(phosphatidylethanolamine)

derived from the alcohol: $H-OCH_2CH_2\overset{+}{N}H_3$

ethanolamine

$CH_2-O-\overset{O}{\overset{\|}{C}}-R$
$CH-O-\overset{O}{\overset{\|}{C}}-R'$ a quaternary amine salt
$CH_2-O-\overset{O}{\overset{\|}{P}}-O-CH_2CH_2\overset{+}{N}(CH_3)_3$
     $|$
     O^-

lecithin
(phosphatidylcholine)

derived from the alcohol: $H-OCH_2CH_2\overset{+}{N}(CH_3)_3$

choline

19.57 c

$CH_2-O-\overset{O}{\overset{\|}{C}}-(CH_2)_7CH=CH(CH_2)_5CH_3$

19.59 a. $CH-O-\overset{O}{\overset{\|}{C}}-(CH_2)_7CH=CH(CH_2)_5CH_3$
$CH_2-O-\overset{O}{\overset{\|}{P}}-O-CH_2CH_2\overset{+}{N}H_3$
     $|$
     O^-

19.61 Triacylglycerols do not have a strongly hydrophilic region contained in a polar head.

19.63 Diffusion is the movement of small molecules through a membrane along a concentration gradient. Facilitated transport is the transport of molecules through channels in a cell membrane. O_2 and CO_2 move by diffusion whereas glucose and Cl^- move by facilitated transport.

19.65

19.67 Cholesterol is insoluble in the aqueous medium of the bloodstream. By being bound to a lipoprotein particle, it can be transported in the aqueous solution.

19.69 Low-density lipoproteins (LDLs) transport cholesterol from the liver to the tissues where it is incorporated in cell membranes. High-density lipoproteins (HDLs) transport cholesterol from the tissues back to the liver. When LDLs supply more cholesterol than is needed, LDLs deposit cholesterol on the wall of arteries, forming plaque. Atherosclerosis is a disease that results from the buildup of these fatty deposits, restricting the flow of blood, increasing blood pressure, and increasing the likelihood of a heart attack or stroke. As a result, LDL cholesterol is often called "bad" cholesterol.

19.71 a.

estrone   testosterone

b. The estrogen (left) and androgen (right) both contain the four rings of the steroid skeleton. Both contain a methyl group bonded to C13.

c. The estrogen has an aromatic A ring and a hydroxyl group on this ring. The androgen has a carbonyl on the A ring but does not contain an aromatic ring. The androgen also contains a C=C in the A ring and an additional CH_3 group at C10. The D rings are also different. The estrogen contains a carbonyl at C17 and the androgen has an OH group.

d. Estrogens, synthesized in the ovaries, control the menstrual cycle and secondary sexual characteristics of females. Androgens, synthesized in the testes, control the development of male secondary sexual characteristics.

19.73 Prostaglandins and leukotrienes are two types of eicosanoids, a group of biologically active compounds containing 20 carbon atoms derived from the fatty acid arachidonic acid. Prostaglandins are a group of carboxylic acids that contain a five-membered ring. Leukotrienes do not contain a ring. They both mediate biological activity at the site where they are formed. Prostaglandins mediate inflammation and uterine contractions. Leukotrienes stimulate smooth muscle contraction in the lungs and lead to the narrowing of airways as associated with asthma.

19.75 Prostaglandins contain a five-membered ring and a carboxyl group (COOH).

19.77 Aspirin inhibits the activity of both COX-1 and COX-2 enzymes, whereas celecoxib inhibits the activity of only COX-2.

19.79 Vitamins are organic compounds required in small quantities for normal metabolism and must be obtained in the diet since our cells cannot synthesize these compounds.

19.81

	Vitamin A	Vitamin D
a.	10	21
b.	10	6
e.	Required for normal vision	Regulates calcium and phosphorus metabolism
f.	Night blindness	Rickets and skeletal deformities
g.	Liver, kidney, oily fish, dairy	Milk and breakfast cereals

c. and d.

19.83 a.

[structure: long hydrocarbon chain with cis double bond ending in COOH]

b. $CH_3(CH_2)_{12}COO(CH_2)_{15}CH_3$

c.

$$CH_2-O-\overset{\overset{O}{\|}}{C}-(CH_2)_{12}CH_3$$
$$CH-O-\overset{\overset{O}{\|}}{C}-(CH_2)_{12}CH_3$$
$$CH_2-O-\overset{\overset{O}{\|}}{C}-(CH_2)_{12}CH_3$$

d.

$$HO-CH-CH=CH(CH_2)_{12}CH_3$$
$$CH-NH-\overset{\overset{O}{\|}}{C}-(CH_2)_{12}CH_3$$
$$CH_2-O-\overset{\overset{O}{\|}}{\underset{\underset{O^-}{|}}{P}}-O-CH_2CH_2\overset{+}{NH_3}$$

19.85 A: a,c,e   **B:** a,c,d   **C:** b,c

19.87 They contain an ionic head, making them more polar than triacylglycerols.

19.89 a, b:

hydrophobic   hydrophilic

[structure: long polyunsaturated hydrocarbon chain ending in COOH labeled hydrophobic (chain) and hydrophilic (OH end)]

c. The melting point would be lower than the melting point of the trans isomer.

d. liquid

e. omega-3 fatty acid

19.91 The hydrocarbon chains have only 12 carbons in them, making them short enough so that the triacylglycerol remains a liquid at room temperature.

19.93 Vegetable oils are composed of triacylglycerols while motor oil, derived from petroleum, is mostly alkanes and other long-chain hydrocarbons.

19.95 No, certain fatty acids and fat-soluble vitamins are required in the diet.

19.97 Saturated fats are more likely to lead to atherosclerosis and heart disease.

19.99 The unsaturated fats have lower melting points, thus remaining liquid at low temperatures. These unsaturated fats allow the cells to remain more fluid with less rigid cell membranes than saturated triacylglycerols would allow for.

19.101 a. reduction   **b.** oxidation   **c.** reduction

CHAPTER 20

20.1

$$H-C=\overset{..}{\overset{..}{O}}$$
$$H-\overset{|}{C}-\overset{..}{\overset{..}{O}}-H$$
$$H-\overset{..}{\overset{..}{O}}-\overset{|}{C}-H$$
$$H-\overset{|}{C}-\overset{..}{\overset{..}{O}}-H$$
$$H-\overset{|}{C}-\overset{..}{\overset{..}{O}}-H$$
$$H-\overset{|}{\underset{\underset{H}{|}}{C}}-\overset{..}{\overset{..}{O}}-H$$

20.3 a. aldopentose   **b.** aldotetrose   **c.** ketotetrose

20.5 Hexane < 1-decanol < glucose. Hexane is a nonpolar hydrocarbon, insoluble in water. 1-Decanol is a long-chain alcohol and minimally soluble in water, due to the polar OH group. Glucose, with multiple hydroxyl groups, is water soluble.

20.7 a.

CHO	CHO
H —*— OH	HO —— H
H —*— OH	HO —— H
CH₂OH	CH₂OH
D	enantiomer

b.

CHO	CHO
HO —*— H	H —— OH
H —*— OH	HO —— H
HO —*— H	H —— OH
CH₂OH	CH₂OH
L	enantiomer

c.

CHO	CHO
HO —*— H	H —— OH
HO —*— H	H —— OH
H —*— OH	HO —— H
CH₂OH	CH₂OH
D	enantiomer

* chirality center

20.9 D-Glucose and D-fructose are constitutional isomers. D-Galactose and D-fructose are also constitutional isomers.

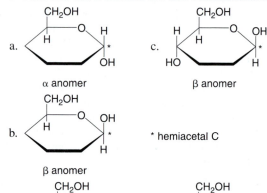

20.11 a. α anomer   **b.** β anomer   **c.** β anomer

* hemiacetal C

20.12 a., b., c.

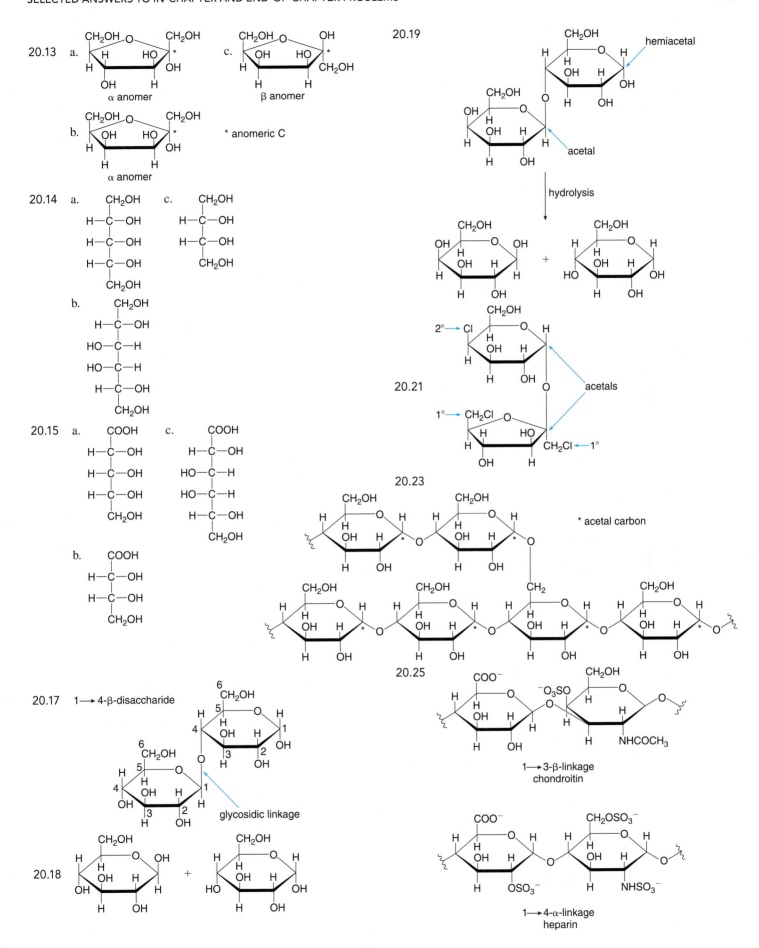

20.27 L-Fucose does not have a CH_2OH at C5 and it is an L enantiomer, whereas other naturally occurring sugars are D enantiomers.

20.29 Aldoses are monosaccharides with a carbonyl group at C1, forming an aldehyde, and ketoses are monosaccharides with a carbonyl group at C2, forming a ketone.

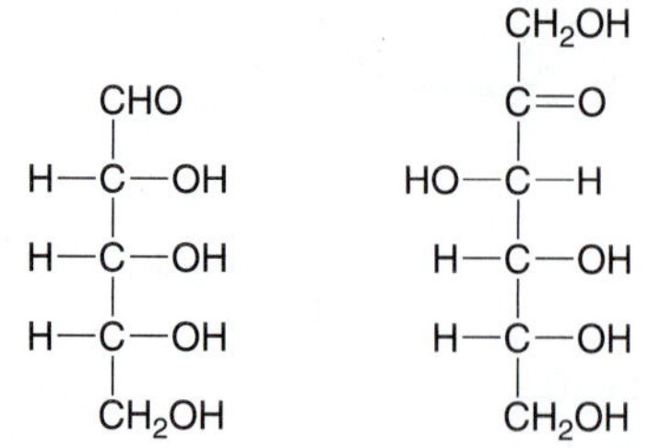

ribose, an aldose   fructose, a ketose

20.31 a. b. c.

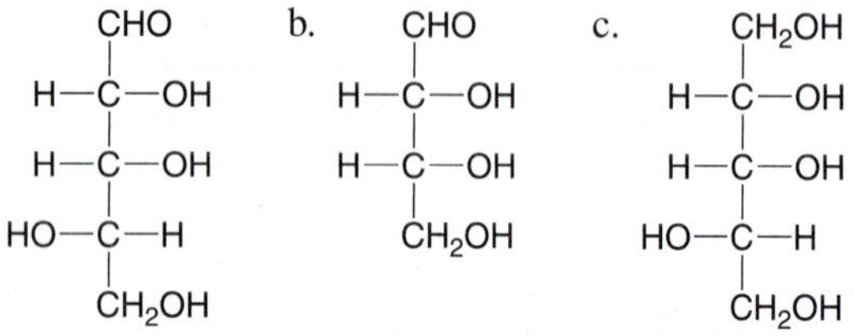

20.33 Haworth projections are representations of the cyclic form of monosaccharides using pentagons for five-membered rings and hexagons for six-membered rings. Fischer projections are linear representations of the acyclic forms of sugars. Each chirality center is located at the center of a cross and the four groups bonded to the chirality center are drawn on the lines of the cross.

20.35 α-D-Glucose and β-D-glucose are not enantiomers because they differ in the orientation of only one OH at C1.

20.37 a. aldotetrose   b. aldohexose   c. ketopentose

20.39 a.

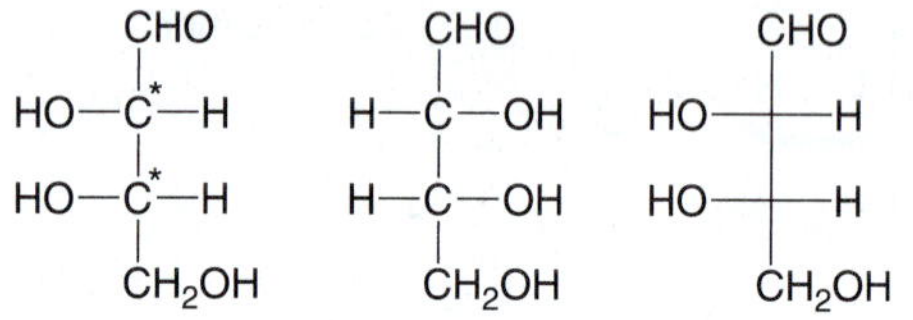

L-tetrose   enantiomer   Fischer projection

b.

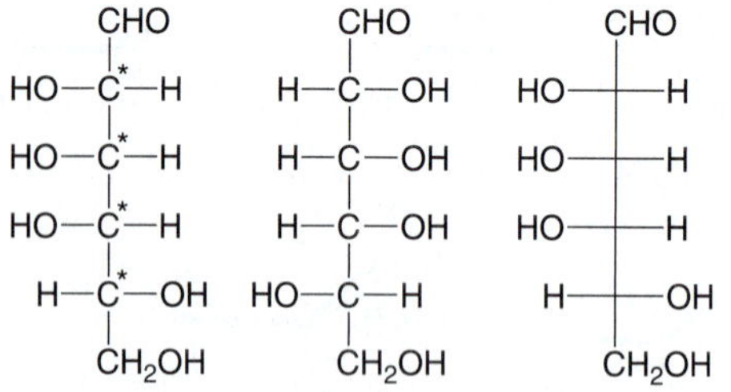

D-hexose   enantiomer   Fischer projection

c.

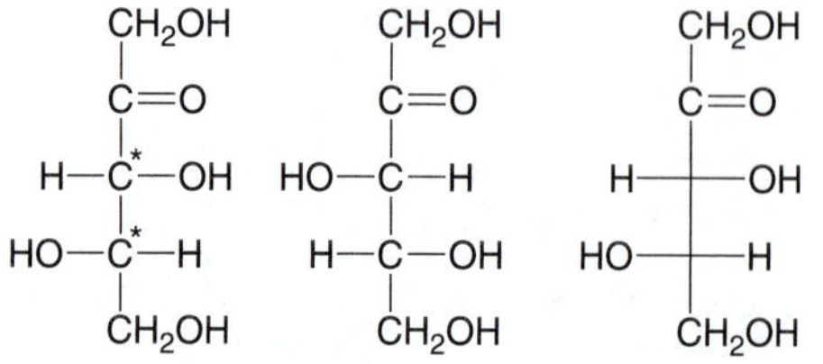

L-pentose   enantiomer   Fischer projection

* chirality center

20.41 a. **A** and **B**
b. **B** and **C** (or **A** and **C**)
c.

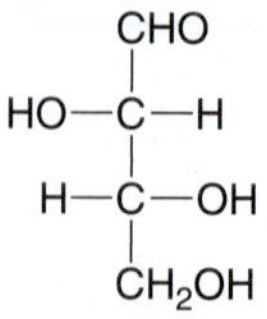

d.

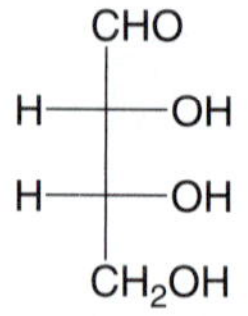

20.43

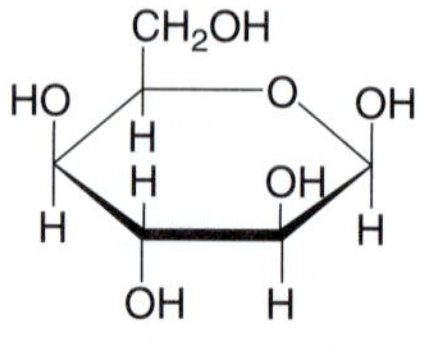

β anomer 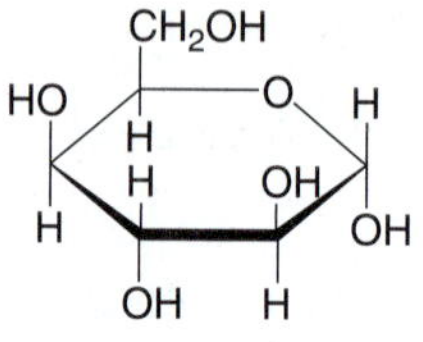   α anomer

20.45

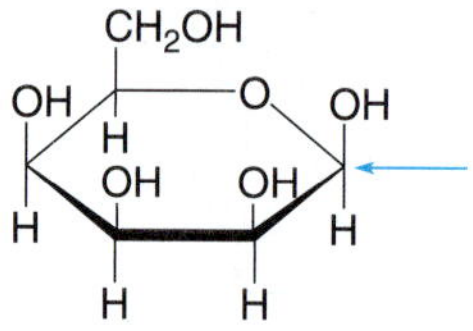

a. hemiacetal carbon
b. β anomer

c.

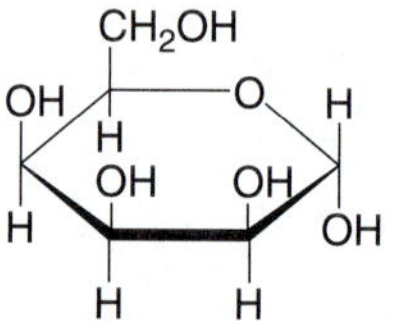

d.

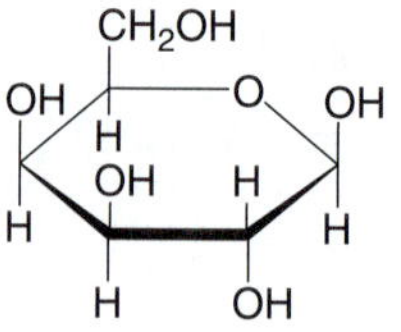

20.47 a.

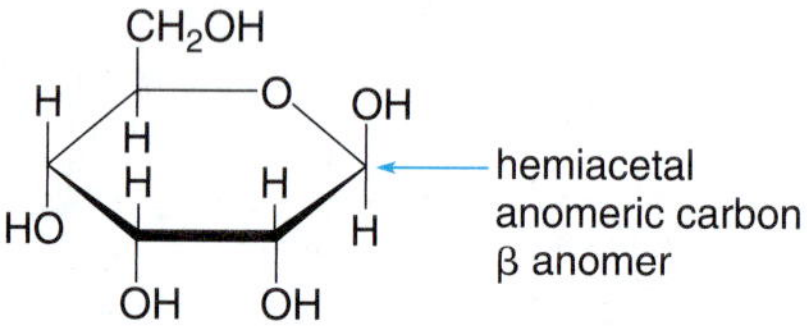

hemiacetal anomeric carbon β anomer

b.

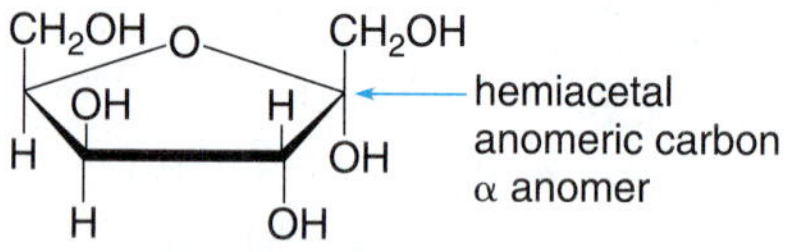

hemiacetal anomeric carbon α anomer

20.49 A ketotriose has no stereoisomers.

20.51 a.

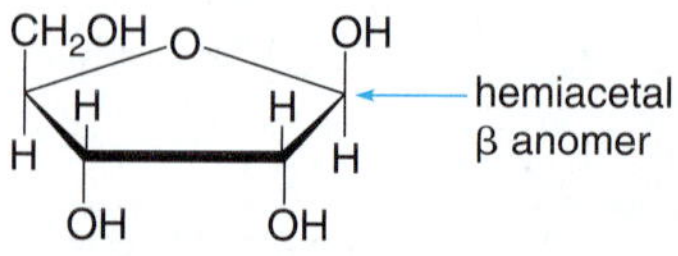

hemiacetal β anomer

b.

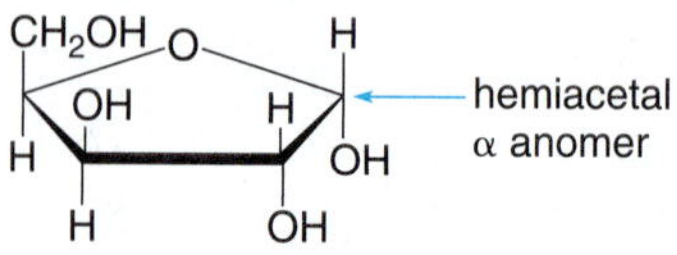

hemiacetal α anomer

20.53

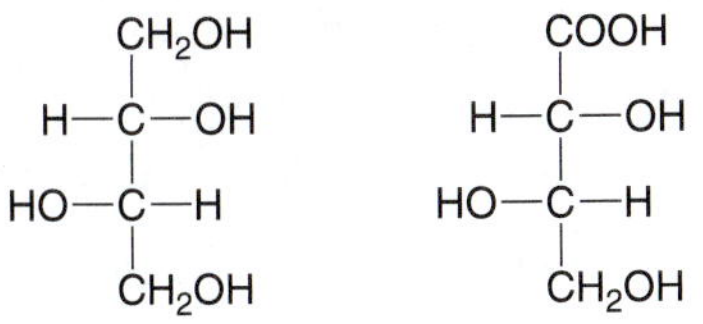

alditol   aldonic acid

20.55 a. [1] [2]

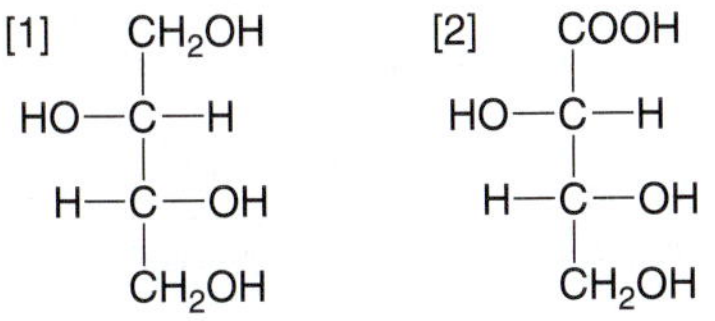

b. [1] [2]

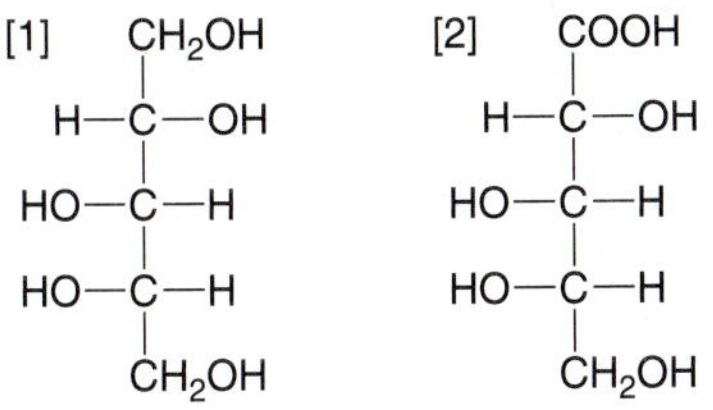

c. [1] [2]

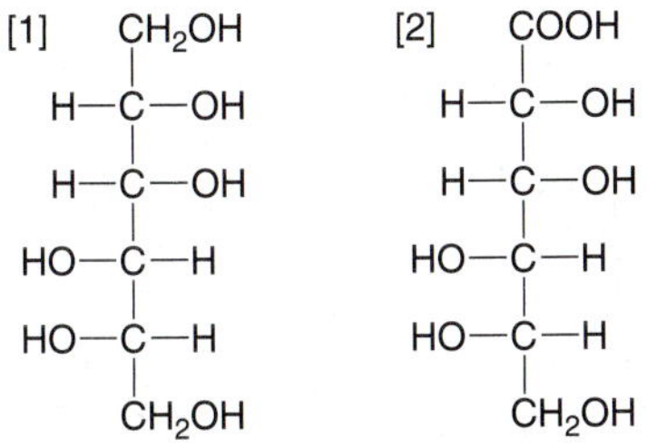

20.57 a.   b.

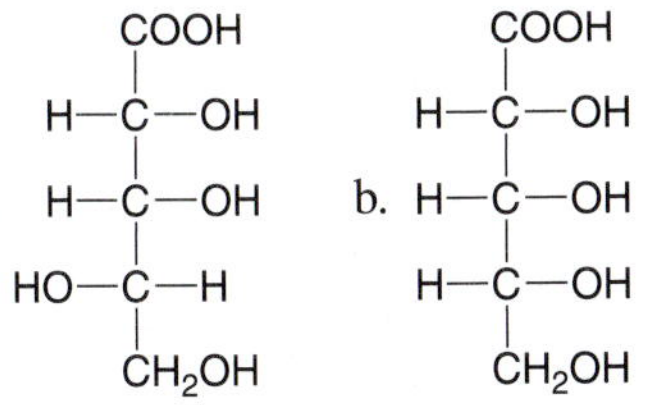

20.59 Reducing sugars are carbohydrates that are oxidized with Benedict's reagent (Cu^{2+}). Glucose, an aldohexose, is a reducing sugar.

20.61 a.

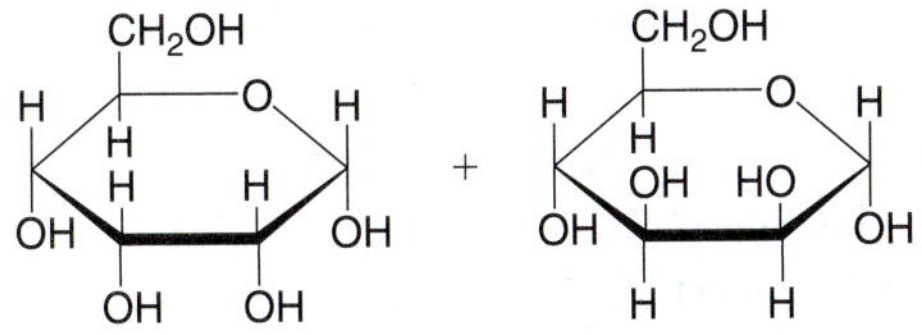

b.

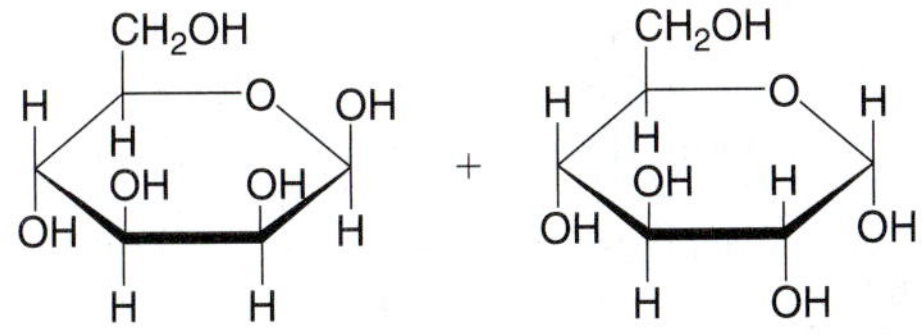

20.63 An α anomer has a hydroxyl at C1, the hemiacetal C, in the down position. An α glycoside has the glycosidic linkage in the down position.

20.65

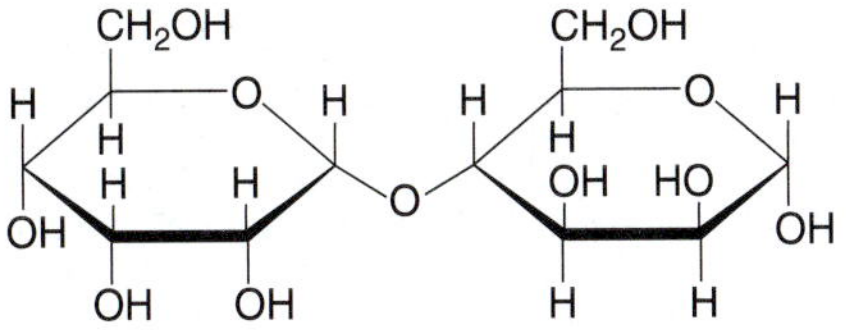

20.67 Maltose and lactose are both disaccharides composed of two hexoses. Maltose has an α glycosidic linkage and lactose has a β glycosidic linkage. Maltose contains two glucose molecules and lactose has one glucose molecule linked with galactose.

20.69 a,b.

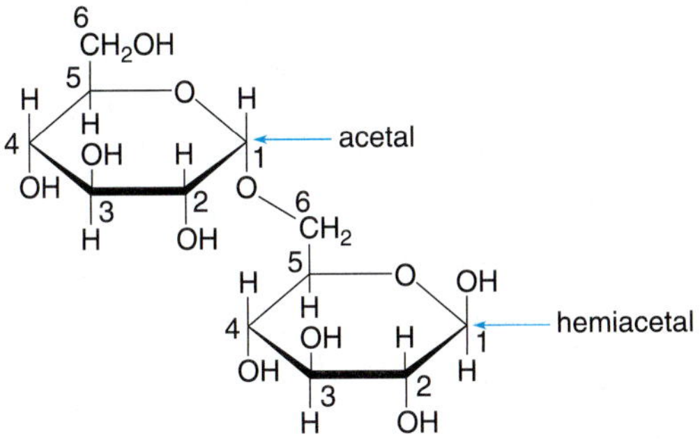

acetal

hemiacetal

c. 1→6-α-glycosidic linkage

d. β anomer

e.

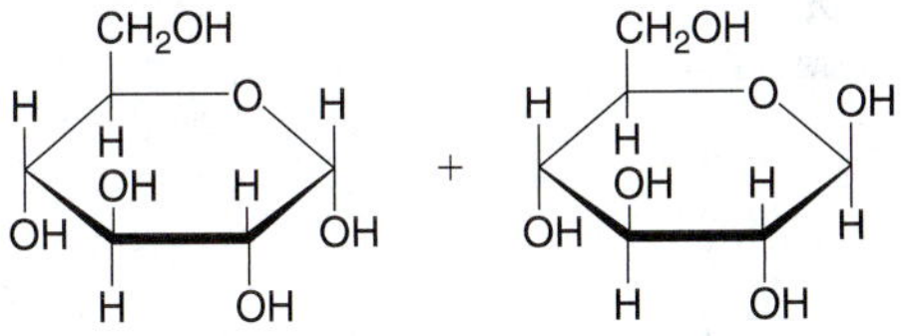

α anomer   +   β anomer

20.71

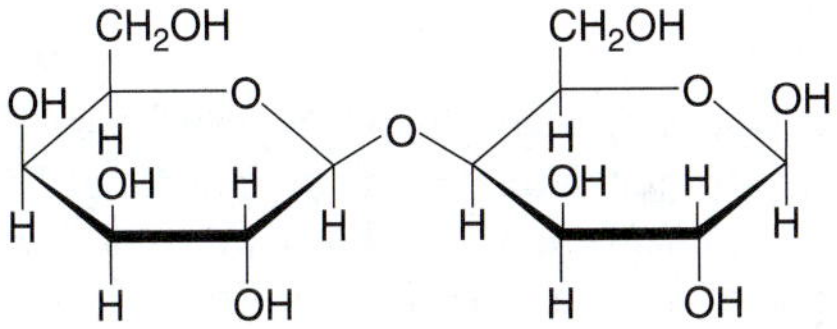

20.73 Cellulose and amylose are both composed of repeating glucose units. In cellulose the glucose units are joined by a 1→4-β-glycosidic linkage, but in amylose they are joined by an α-(1→4) linkage. This leads to very different three-dimensional shapes, with cellulose forming sheets and amylose forming helices.

20.75 Cellulose and chitin both consist of polysaccharides joined with 1→4-β-glycosidic linkages. Cellulose is composed solely of repeating glucose units, whereas chitin is formed from repeating N-acetyl-D-glucosamine units.

20.77

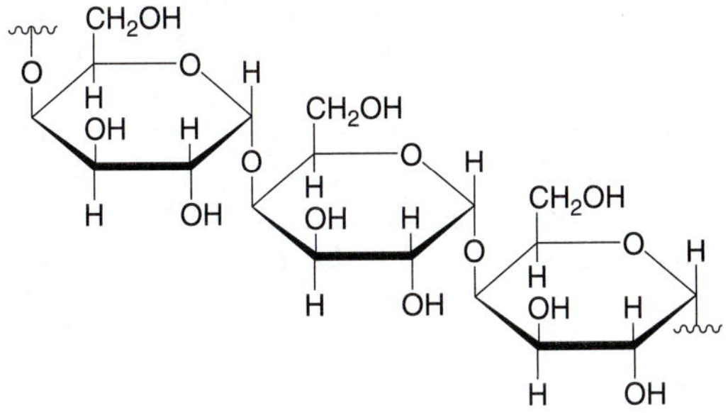

20.79 a. D sugar

b. aldohexose

c.

$$
\begin{array}{c}
\text{CHO} \\
\text{H}-\text{C}-\text{OH} \\
\text{H}-\text{C}-\text{OH} \\
\text{H}-\text{C}-\text{OH} \\
\text{HO}-\text{C}-\text{H} \\
\text{CH}_2\text{OH}
\end{array}
$$

enantiomer

d.

$$
\begin{array}{c}
\text{CHO} \\
\text{HO}-\overset{*}{\text{C}}-\text{H} \\
\text{HO}-\overset{*}{\text{C}}-\text{H} \\
\text{HO}-\overset{*}{\text{C}}-\text{H} \\
\text{H}-\overset{*}{\text{C}}-\text{OH} \\
\text{CH}_2\text{OH}
\end{array}
$$

* chirality center

e.

$$
\begin{array}{c}
\text{CH}_2\text{OH} \\
\text{(cyclic hexose structure)}
\end{array}
$$

f.

$$
\begin{array}{c}
\text{COOH} \\
\text{HO}-\text{C}-\text{H} \\
\text{HO}-\text{C}-\text{H} \\
\text{HO}-\text{C}-\text{H} \\
\text{H}-\text{C}-\text{OH} \\
\text{CH}_2\text{OH}
\end{array}
$$

g.

$$
\begin{array}{c}
\text{CH}_2\text{OH} \\
\text{HO}-\text{C}-\text{H} \\
\text{HO}-\text{C}-\text{H} \\
\text{HO}-\text{C}-\text{H} \\
\text{H}-\text{C}-\text{OH} \\
\text{CH}_2\text{OH}
\end{array}
$$

h. a reducing sugar

20.81 Lactose intolerance results from a lack of the enzyme lactase. It results in abdominal cramping and diarrhea. Galactosemia results from the inability to metabolize galactose. Therefore, it accumulates in the liver, causing cirrhosis, and in the brain, leading to mental retardation.

20.83 Fructose is a naturally occurring sugar with more perceived sweetness per gram than sucrose. Sucralose is a synthetic sweetener; that is, it is not naturally occurring.

20.85 An individual with type A blood can receive only blood types A and O, because he or she will produce antibodies and an immune response to B or AB blood. He or she can donate to individuals with either type A or AB blood as the type A polysaccharides are common to both and no immune response will be generated.

20.87 The long sheets of polysaccharides in chitin, similar to cellulose, have β glycosidic linkages and will not be digestible.

20.89 Hyaluronate is found in the extracellular fluid that lubricates joints and the vitreous humor of the eye. Chondroitin is a component of cartilage and tendons. Heparin is stored in the mast cells of the liver and other organs and prevents blood clotting.

20.91

$$
\begin{array}{c}
\text{CH}_2\text{OH} \\
\text{C}=\text{O} \\
\text{HO}-\text{C}-\text{H} \\
\text{H}-\text{C}-\text{OH} \\
\text{H}-\text{C}-\text{OH} \\
\text{CH}_2\text{OH}
\end{array}
\quad
\xrightarrow{\text{reduction}}
\quad
\begin{array}{c}
\text{CH}_2\text{OH} \\
\text{HO}-\text{C}-\text{H} \\
\text{HO}-\text{C}-\text{H} \\
\text{H}-\text{C}-\text{OH} \\
\text{H}-\text{C}-\text{OH} \\
\text{CH}_2\text{OH}
\end{array}
\quad + \quad
\begin{array}{c}
\text{CH}_2\text{OH} \\
\text{H}-\text{C}-\text{OH} \\
\text{HO}-\text{C}-\text{H} \\
\text{H}-\text{C}-\text{OH} \\
\text{H}-\text{C}-\text{OH} \\
\text{CH}_2\text{OH}
\end{array}
$$

fructose   stereoisomers
They differ in configuration at a single carbon.

CHAPTER 21

21.1 a. amide   b. alcohol   c. thiol

21.3

(Lewis structure of amino acid with side chain)

21.4 a.

$$
\begin{array}{c}
\text{COO}^- \\
\text{H}_3\overset{+}{\text{N}}-\rule{0.7cm}{0.4pt}-\text{H} \\
\text{CH}_2 \\
\text{(phenyl ring)}
\end{array}
\qquad
\begin{array}{c}
\text{COO}^- \\
\text{H}-\rule{0.7cm}{0.4pt}-\overset{+}{\text{N}}\text{H}_3 \\
\text{CH}_2 \\
\text{(phenyl ring)}
\end{array}
$$

L   D

b.

$$
\begin{array}{c}
\text{COO}^- \\
\text{H}_3\overset{+}{\text{N}}-\rule{0.7cm}{0.4pt}-\text{H} \\
\text{CH}_2\text{CH}_2\text{SCH}_3
\end{array}
\qquad
\begin{array}{c}
\text{COO}^- \\
\text{H}-\rule{0.7cm}{0.4pt}-\overset{+}{\text{N}}\text{H}_3 \\
\text{CH}_2\text{CH}_2\text{SCH}_3
\end{array}
$$

L   D

21.5 a. L-serine, naturally occurring
b. D-glutamic acid
c. L-aspartic acid, naturally occurring

21.6 a.

$$
\begin{array}{c}
\text{H} \\
\text{H}_3\overset{+}{\text{N}}-\text{C}-\text{COO}^- \\
\text{CH(CH}_3)_2
\end{array}
$$

predominant at pI

c.

$$
\begin{array}{c}
\text{H} \\
\text{H}_2\text{N}-\text{C}-\text{COO}^- \\
\text{CH(CH}_3)_2
\end{array}
$$

b.

$$
\begin{array}{c}
\text{H} \\
\text{H}_3\overset{+}{\text{N}}-\text{C}-\text{COOH} \\
\text{CH(CH}_3)_2
\end{array}
$$

21.7

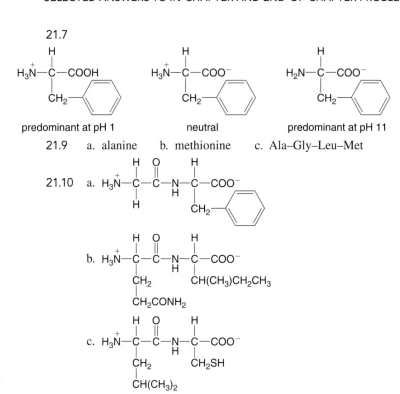

$$H_3\overset{+}{N}-\underset{\underset{CH_2}{|}}{\overset{\overset{H}{|}}{C}}-COOH$$

predominant at pH 1

$$H_3\overset{+}{N}-\underset{\underset{CH_2}{|}}{\overset{\overset{H}{|}}{C}}-COO^-$$

neutral

$$H_2N-\underset{\underset{CH_2}{|}}{\overset{\overset{H}{|}}{C}}-COO^-$$

predominant at pH 11

21.9   a. alanine   b. methionine   c. Ala–Gly–Leu–Met

21.10   a., b., c. (structures)

21.11   a. (structures)

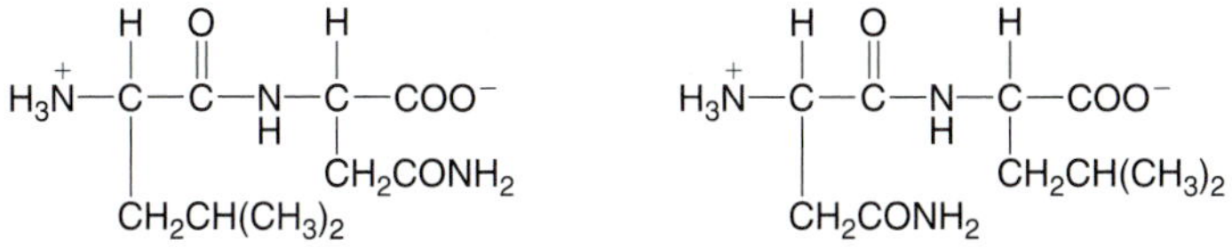

b. N-terminal: leucine   N-terminal: asparagine
C-terminal: asparagine   C-terminal: leucine
c. Leu–Asn   Asn–Leu

21.12   a. alanine and isoleucine: Ala–Ile
b. tyrosine and valine: Tyr–Val

21.13   a. (structure)

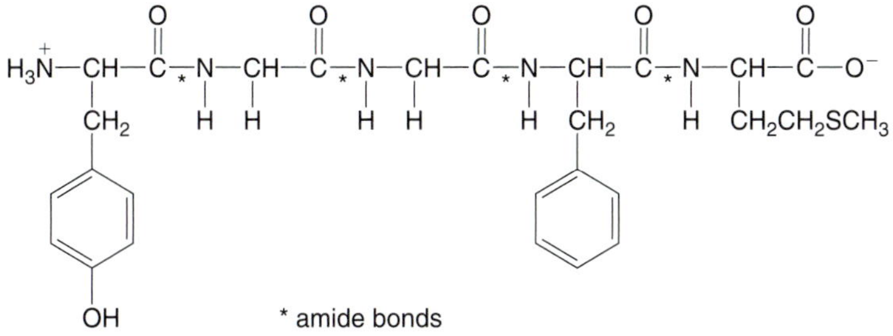

* amide bonds

b. tyrosine
c. 3

21.15   Yes, the amino acids may be ordered differently.

21.17   Glycine has no large side chain and this allows for the β-sheets to stack well together.

21.19   Hemoglobin has four chains that combine in a quaternary structure. Myoglobin has a single protein chain, so it has no quaternary structure.

21.20   a. (structures)   b. (structures)   c. (structures)

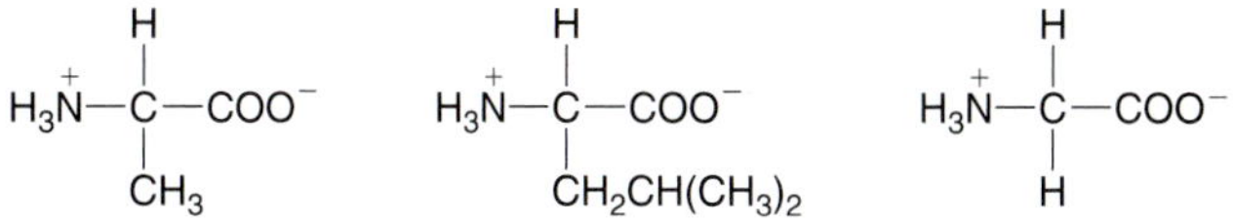

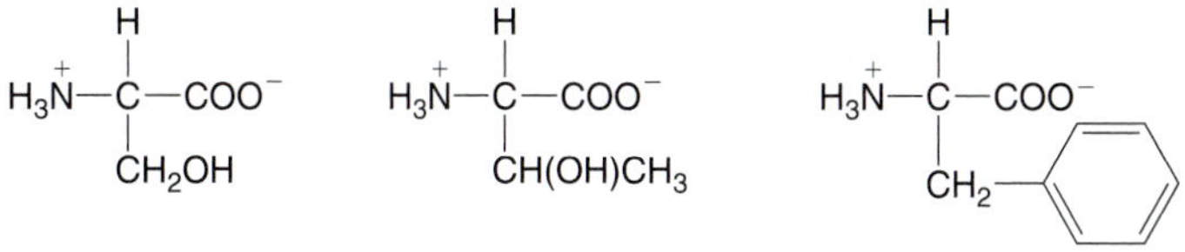

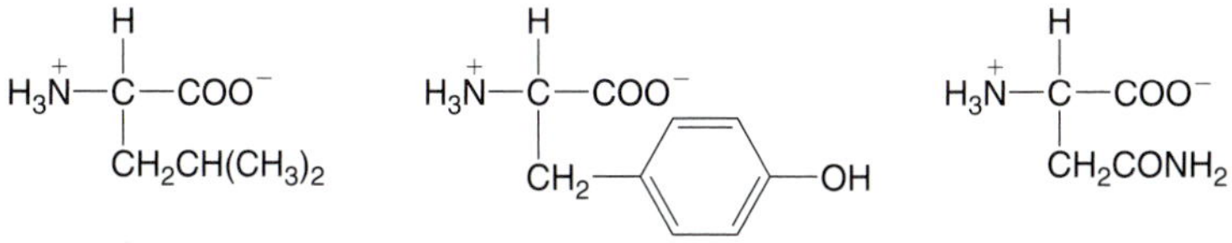

21.21   tyrosine, glycine (2 equivalents), phenylalanine, methionine

21.23   b,d,e

21.25

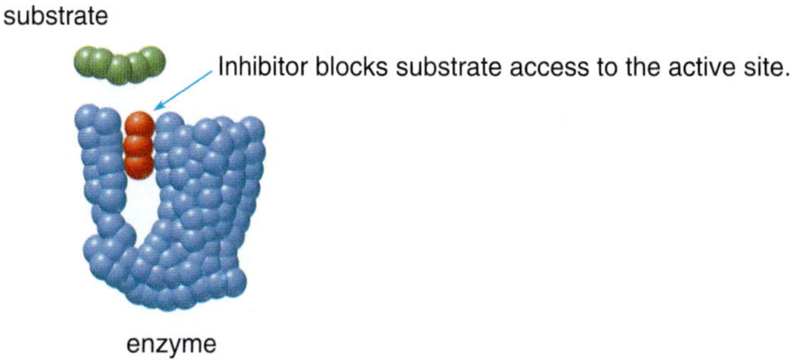

substrate

Inhibitor blocks substrate access to the active site.

enzyme

21.27   Fibrin and thrombin circulate as inactive zymogens so that the blood does not clot unnecessarily. They are activated as required at a bleeding point to form a clot.

21.29   The L designation refers to the configuration at the chirality center. With a vertical carbon chain in the Fischer projection, the L isomer has the $-NH_3^+$ drawn on the left side. The α amino acid designation indicates that the amino group is bonded to the carbon adjacent to the carbonyl group.

21.31   Alanine is an ionic salt with strong electrostatic forces, leading to its high melting point, and making it a solid at room temperature. Pyruvic acid has much weaker intermolecular forces so it is a liquid at room temperature.

21.33   a., b., c., d. (structures)

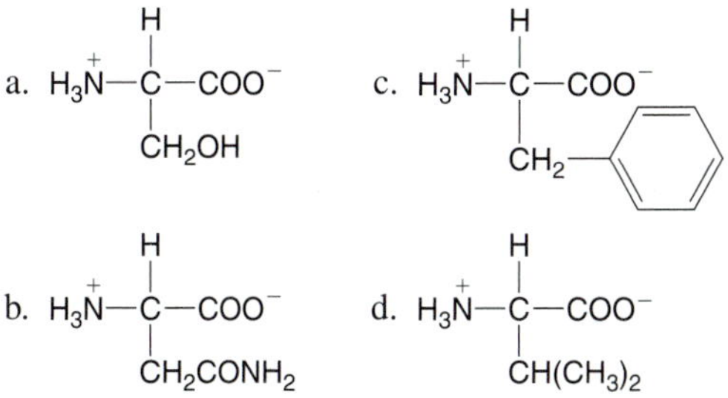

21.35   isoleucine and threonine

21.37

	[1]	[2]	[3]	[4]
a.	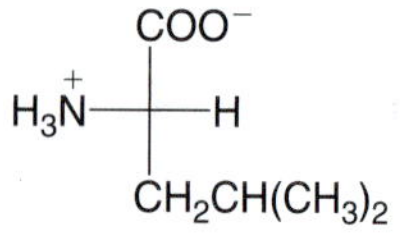 $H_3\overset{+}{N}$—C—H with COO⁻ above and $CH_2CH(CH_3)_2$ below	neutral	Leu	L
b.	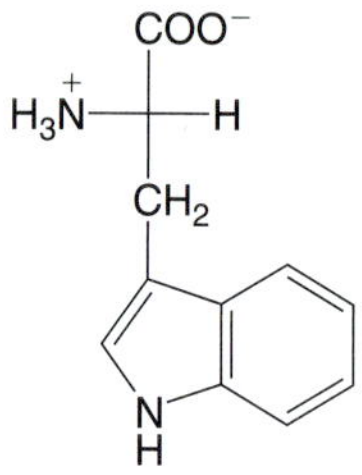 $H_3\overset{+}{N}$—C—H with COO⁻ above and CH_2–(indole) below	neutral	Trp	W
c.	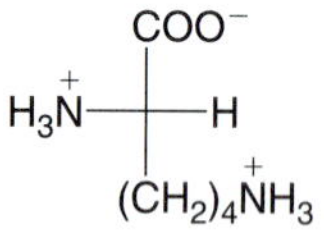 $H_3\overset{+}{N}$—C—H with COO⁻ above and $(CH_2)_4\overset{+}{N}H_3$ below	basic	Lys	K
d.	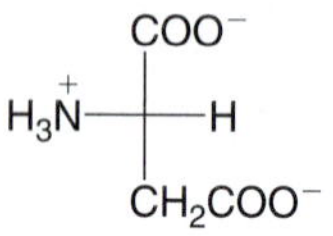 $H_3\overset{+}{N}$—C—H with COO⁻ above and CH_2COO^- below	acidic	Asp	D

21.39

a. 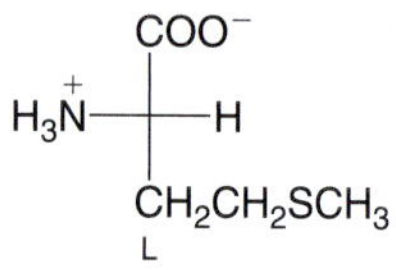 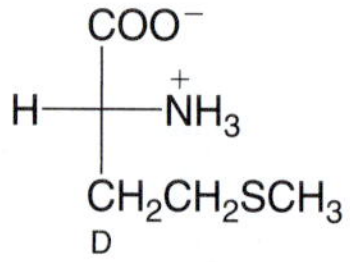 [L] $H_3\overset{+}{N}$—C—H, COO⁻ above, $CH_2CH_2SCH_3$ below (L)   [D] H—C—$\overset{+}{N}H_3$, COO⁻ above, $CH_2CH_2SCH_3$ below (D)

b. 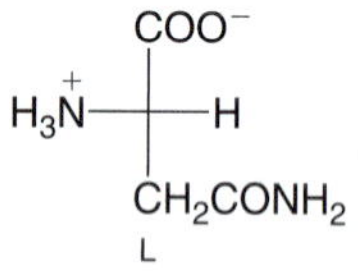 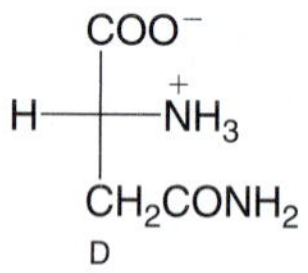 [L] $H_3\overset{+}{N}$—C—H, COO⁻ above, CH_2CONH_2 below (L)   [D] H—C—$\overset{+}{N}H_3$, COO⁻ above, CH_2CONH_2 below (D)

21.41

	[1]	[2]	[3]	[4]
a.	glutamine	Gln	Q	neutral
b.	threonine	Thr	T	neutral
c.	tyrosine	Tyr	Y	neutral

21.43

a. 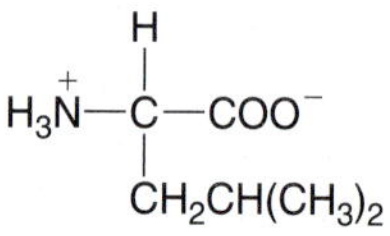 $H_3\overset{+}{N}$—C—COO⁻ (H above, $CH_2CH(CH_3)_2$ below)
predominant form at pI

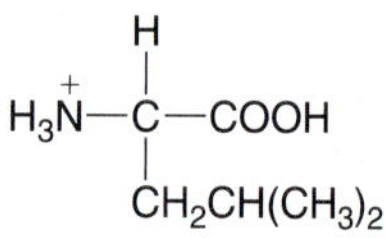

b. 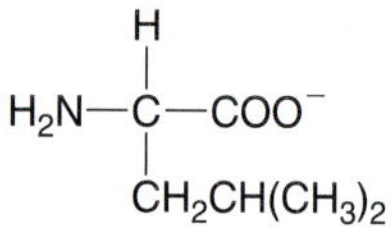 H_2N—C—COO⁻ (H above, $CH_2CH(CH_3)_2$ below)

c. $H_3\overset{+}{N}$—C—COOH (H above, $CH_2CH(CH_3)_2$ below)

21.45

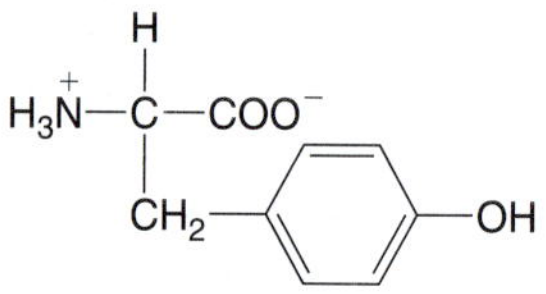

$H_3\overset{+}{N}$—C—COO⁻ (H above, CH_2–C₆H₄–OH below)
neutral
predominant form at pI

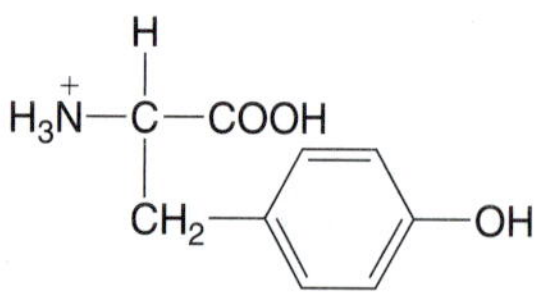

$H_3\overset{+}{N}$—C—COOH (H above, CH_2–C₆H₄–OH below)
positive charge
pH = 1

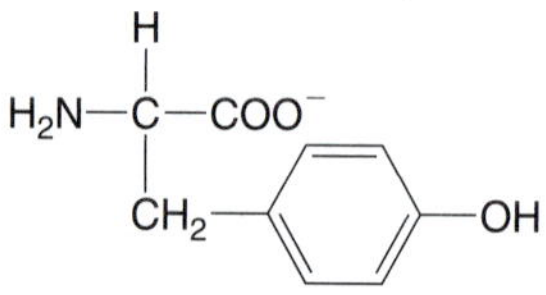

H_2N—C—COO⁻ (H above, CH_2–C₆H₄–OH below)
negative charge
pH = 11

21.47 a,b,c.

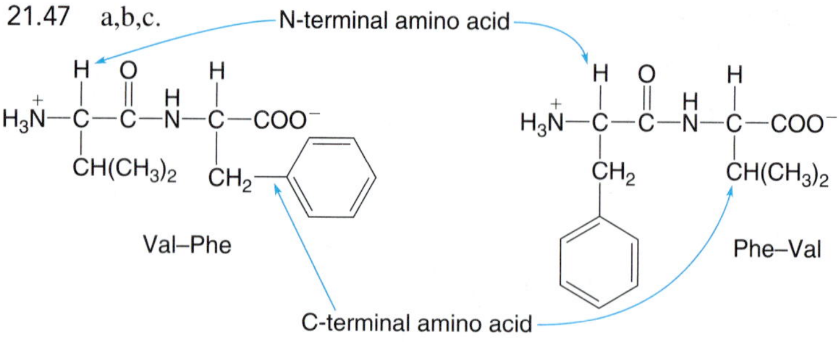

N-terminal amino acid / C-terminal amino acid

Val–Phe

$H_3\overset{+}{N}$—C—C(=O)—N—C—COO⁻ (CH(CH₃)₂ and CH₂–C₆H₅ side chains)

Phe–Val

$H_3\overset{+}{N}$—C—C(=O)—N—C—COO⁻ (CH₂–C₆H₅ and CH(CH₃)₂ side chains)

21.49

a. [1] $H_3\overset{+}{N}$—C—C*—N—C—C*—N—C—COO⁻ with side chains $CH_2CH(CH_3)_2$, $CH(CH_3)_2$, and CH_2–(indole)

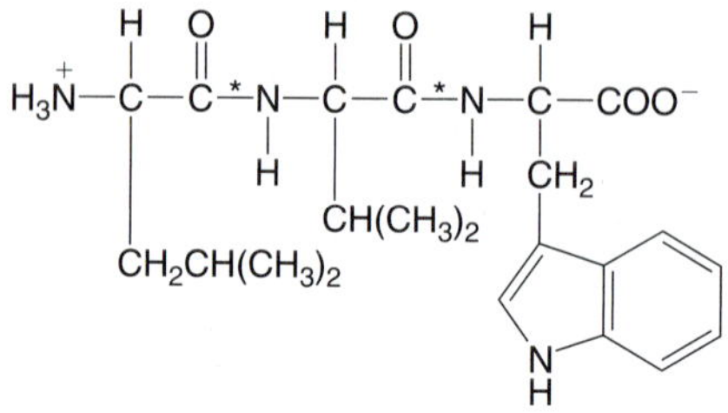

[2] * amide bonds labeled

[3] N-terminal: leucine
C-terminal: tryptophan

[4] Leu–Val–Trp

b. [1] $H_3\overset{+}{N}$—C—C*—N—C—C*—N—C—COO⁻ with side chains CH_3, H, and $CH(CH_3)_2$

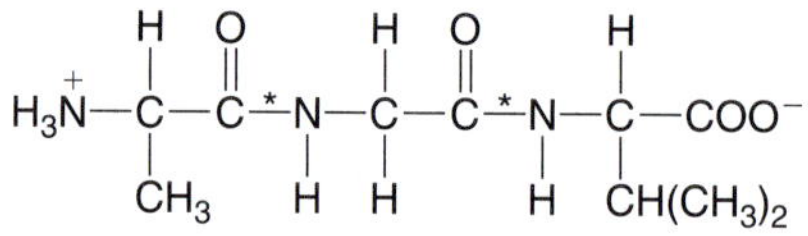

[2] * amide bonds labeled

[3] N-terminal: alanine
C-terminal: valine

[4] Ala–Gly–Val

c. [1] $H_3\overset{+}{N}$—C—C*—N—C—C*—N—C—COO⁻ with side chains CH_2–C₆H₅, CH_2OH, and $CH(OH)CH_3$

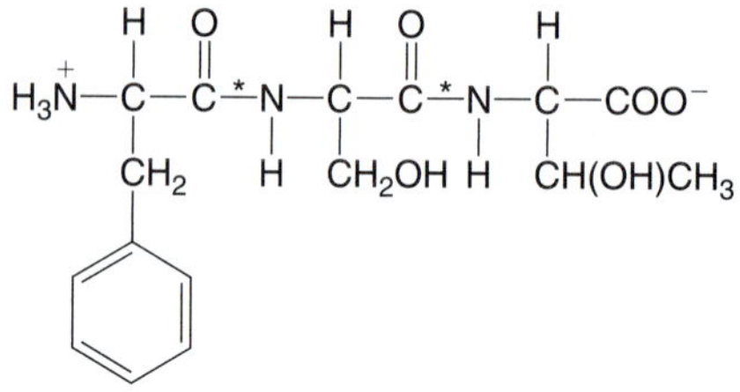

[2] * amide bonds labeled

[3] N-terminal: phenylalanine
C-terminal: threonine

[4] Phe–Ser–Thr

21.51 a. [1], [2] valine: N-terminal, glycine, phenylalanine: C-terminal [3] Val–Gly–Phe

b. [1], [2] leucine: N-terminal, tyrosine, methionine: C-terminal [3] Leu–Tyr–Met

21.53

$H_3N^+-C(H)(CH_2OH)-C(=O)-N(H)-C(H)(CH_3)-C(=O)-N(H)-C(H)(CH_2OH)-COO^-$ Ser–Ala–Ser

$H_3N^+-C(H)(CH_3)-C(=O)-N(H)-C(H)(CH_2OH)-C(=O)-N(H)-C(H)(CH_2OH)-COO^-$ Ala–Ser–Ser

$H_3N^+-C(H)(CH_2OH)-C(=O)-N(H)-C(H)(CH_2OH)-C(=O)-N(H)-C(H)(CH_3)-COO^-$ Ser–Ser–Ala

21.55 a.

$H_3N^+-C(H)(CH(CH_3)_2)-COO^-$ $H_3N^+-C(H)(H)-COO^-$ $H_3N^+-C(H)(CH_2C_6H_5)-COO^-$

b.

$H_3N^+-C(H)(CH_2CH(CH_3)_2)-COO^-$ $H_3N^+-C(H)(CH_2-C_6H_4-OH)-COO^-$ $H_3N^+-C(H)(CH_2CH_2SCH_3)-COO^-$

21.57

Proline structure (pyrrolidine ring with $-COO^-$, NH_2^+)   3 moles Pro

$H_3N^+-C(H)((CH_2)_3-NH-C(NH_2)=NH_2^+)-COO^-$ $H_3N^+-C(H)(CH_2C_6H_5)-COO^-$ 2 moles each, Arg and Phe

$H_3N^+-C(H)(H)-COO^-$ $H_3N^+-C(H)(CH_2OH)-COO^-$ 1 mole each, Gly and Ser

21.59

$H_3N^+-C(H)(H)-C(=O)-N(H)-C(H)(CH_2-C_6H_4-OH)-COO^-$

Gly–Tyr

$H_3N^+-C(H)(CH(CH_3)_2)-COO^-$

Val

$H_3N^+-C(H)(H)-C(=O)-N(H)-C(H)(CH_3)-C(=O)-N(H)-C(H)(CH_2-C_6H_5)-COO^-$

Gly–Ala–Phe

21.61 The primary structure of a protein is the order of its amino acids. The secondary structure refers to the three-dimensional arrangements of regions within the protein.

21.63 a. London dispersion forces   c. electrostatic
b. London dispersion forces   d. electrostatic

21.65

$H_3N^+-C(H)(CH_2-C(=O)-N(H)-H \cdots H-O-CH_2-C(H)(NH_3^+)-COO^-)-COO^-$ hydrogen bonding

21.67 Any four of the following amino acids: valine, alanine, phenylalanine, leucine, glycine, and isoleucine.

21.69

21.71

	a. Secondary Structure	b. H$_2$O Solubility	c. Function	d. Location
Hemoglobin	Globular with much α-helix	Soluble	Carries oxygen to tissues	Blood
Keratin	α-helix	Insoluble	Firm tissues	Nail, hair

21.73 When heated, a protein's primary structure is unaffected. The 2°, 3°, and 4° structures may be altered.

21.75 a. Insulin is a hormone that controls glucose levels.
b. Myoglobin stores oxygen in muscle.
c. α-Keratin forms hard tissues such as hair and nails.
d. Chymotrypsin is a protease that hydrolyzes peptide bonds.
e. Oxytocin is a hormone that stimulates uterine contractions and induces the release of breast milk.

21.77 Reversible enzyme inhibition occurs when an enzyme's activity is restored when an inhibitor is released. Irreversible inhibition renders an enzyme incapable of further activity.

21.79 a. The zymogen angiotensinogen, when activated to angiotensin, acts to increase blood pressure.
b. Digestive enzymes such as trypsin and chymotrypsin are formed as zymogens and are converted to the active form in the intestines.

21.81 Captopril inhibits the angiotensin-converting enzyme, blocking the conversion of angiotensinogen to angiotensin. This reduces the concentration of angiotensin, which in turn lowers blood pressure.

21.83

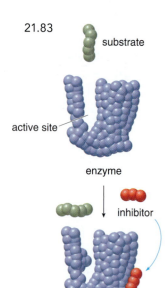

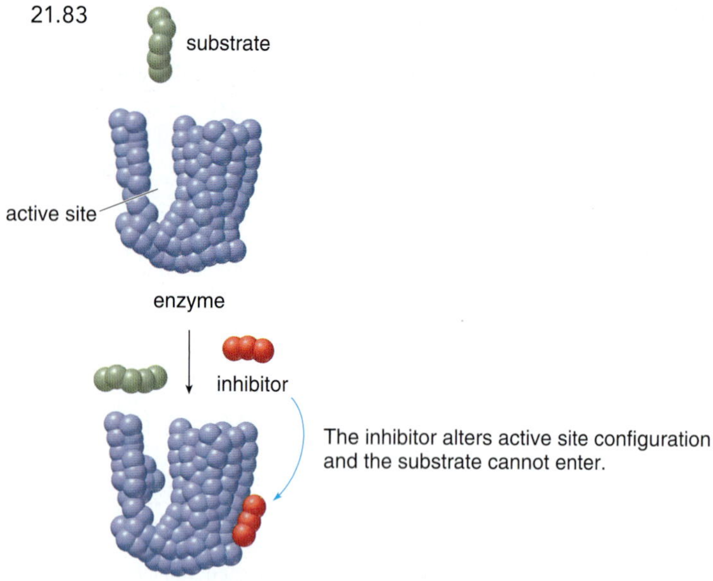

active site

enzyme

inhibitor

The inhibitor alters active site configuration and the substrate cannot enter.

21.85 The α-keratin in nails has more cysteine residues to form disulfide bonds. The larger the number of disulfide bonds, the harder the substance.

21.87 Humans cannot synthesize the amino acids methionine and lysine. Diets that include animal products readily supply all the needed amino acids, but no one plant source has sufficient amounts of all the essential amino acids. Grains—wheat, rice, and corn—are low in lysine, and legumes—beans, peas, and peanuts—are low in methionine, but a combination of these foods provides all the needed amino acids.

21.89 Cauterization denatures the proteins in a wound.

21.91 In sickle hemoglobin there is a substitution of a single amino acid, valine for glutamic acid.

21.93 Penicillin inhibits the formation of the bacterial cell wall by irreversibly binding to an enzyme needed for its construction. It does not affect humans since human cells have a cell membrane, but not a rigid cell wall. Sulfanilamide inhibits the production of folic acid and therefore reproduction in bacteria, but humans do not synthesize folic acid (they must ingest it instead), so it does not affect humans.

21.95 Both aspartic acid and glutamic acid have two carboxylic acid groups. At low pH they have a +1 charge with both acid groups protonated, but at a high pH both acid groups are ionized, leading to a net charge of –2.

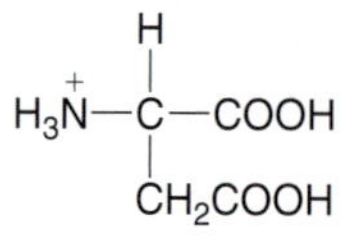

Form of Asp at low pH

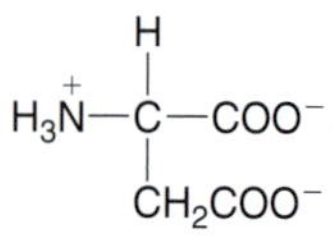

Form of Asp at high pH

CHAPTER 22

22.1 a. base = uracil, sugar = D-ribose, nuclesoside = uridine
b. base = guanine, sugar = D-2-deoxyribose, nucleoside = deoxyguanosine

22.3 a.

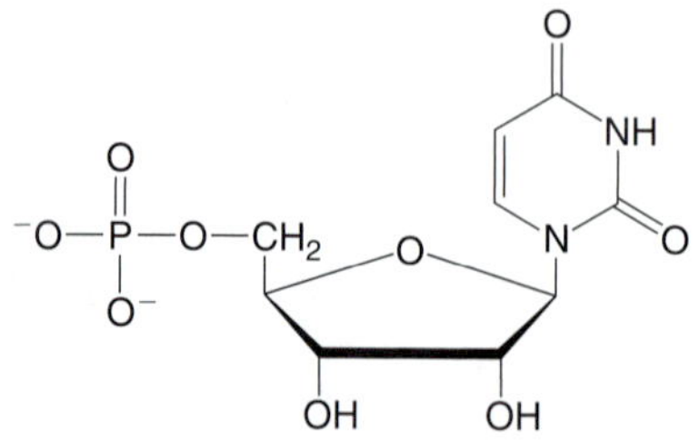

UMP

b.

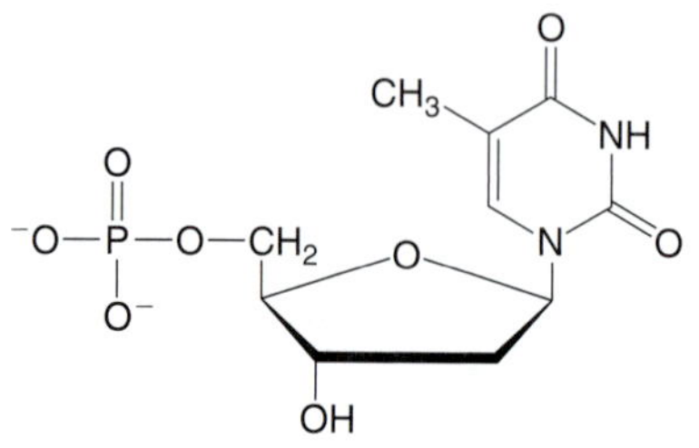

dTMP

c.

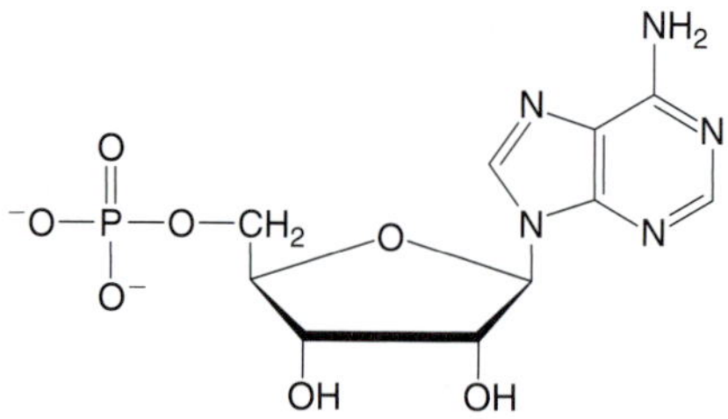

AMP

22.5

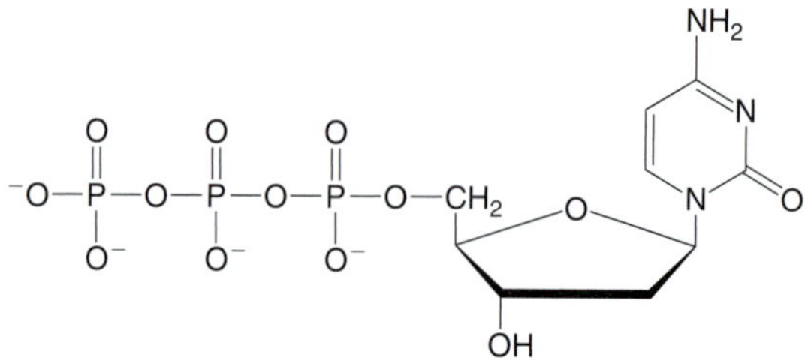

22.7 a.

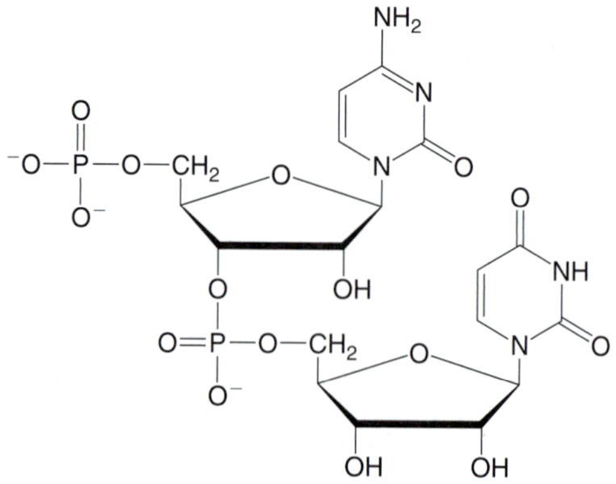

b.

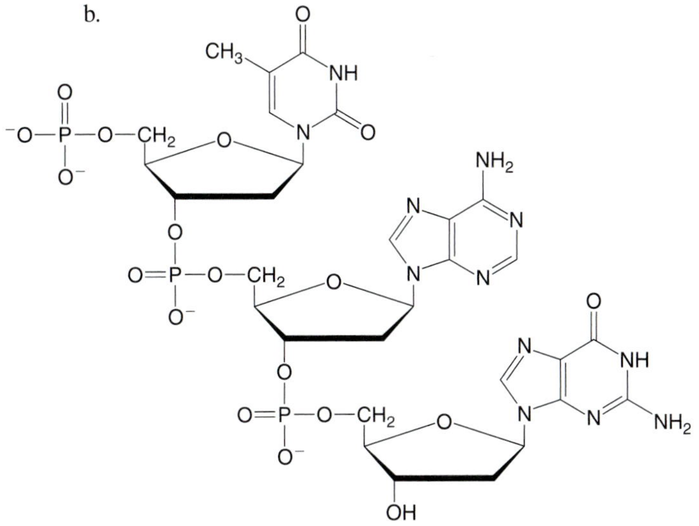

22.9 a. 3'–TTTGCAGG–5'   c. 3'–TAACGTGGGCG–5'
      b. 3'–ATATGCGG–5'   d. 3'–GTGAACTAGCC–5'

22.10 a. 5'–TCTCAGAG–3'   c. 5'–TAGGACATG–3'
      b. 3'–TAACGAG–5'   d. 3'–CCGGTATGAG–5'

22.11 a. [1] 5'–ACGGAUUGC–3'   [2] 5'–ACGGATTGC–3'
      b. [1] 5'–CUGAGG–3'   [2] 5'–CTGAGG–3'
      c. [1] 5'–AAUUGCGCU–3'   [2] 5'–AATTGCGCT–3'
      d. [1] 5'–GUCACUGGCAUG–3'
         [2] 5'–GTCACTGGCATG–3'

22.13 a. Ala   c. Leu   e. Gln
      b. Asn   d. Ser   f. Lys

22.15 a. Gln–Glu–Val–Ser–Tyr–Arg
      b. Val–Ile–Trp–Arg–Gly–Ile
      c. Leu–Cys–Ser–Arg–Thr–Pro

22.17 a. 3'–CTC GGG CAT ATG CGG TGC–5'
      b. Glu–Pro–Val–Tyr–Ala–Thr

22.19 a. Pro–Pro–Ala–Asn–Glu–Ala
         GGU GGC CGU UUG CUU CGU
      b. Ala–Pro–Leu–Arg–Asp
         CGU GGU GAU UCU CUG

22.20 a. Arg–Val–Ala–Leu–Leu–Ser
      b. Arg–Gly–Phe–Ile–Val–Asn

22.21 a. Leu–Thr
      b. [1] Leu–Pro
         [2] Stop codon, so no dipeptide is formed.
         [3] Leu–Thr

22.23

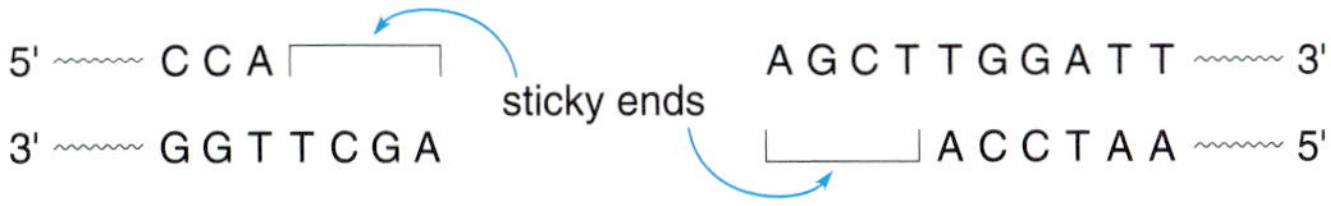

22.25 A ribonucleoside contains ribose (a monosaccharide) and a nitrogen-containing base (A, G, C, or U), and a ribonucleotide has these components plus a phosphate group attached to the 5' position of the ribose.

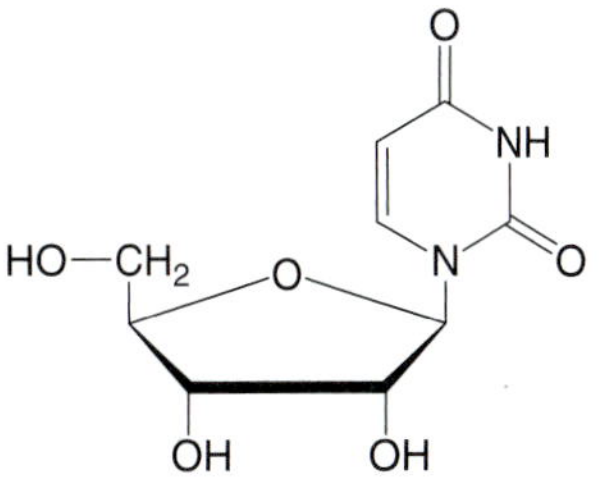

uridine
a ribonucleoside

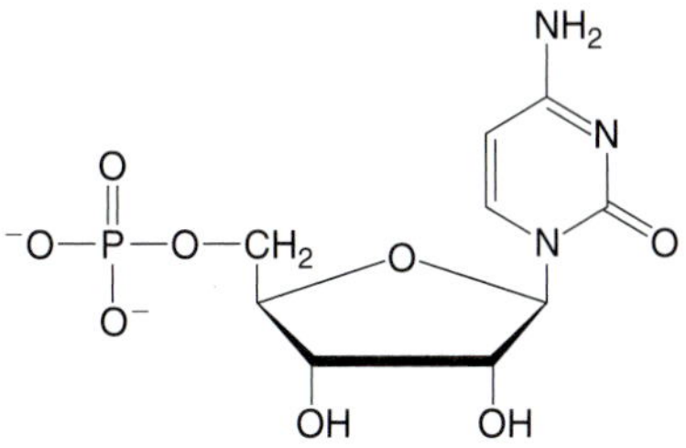

cytidine 5'-monophosphate
a ribonucleotide

22.27 A gene is a portion of the DNA molecule responsible for the synthesis of a single protein. Many genes form each chromosome.

22.29 In RNA, the monosaccharide is the aldopentose D-ribose. In DNA, the monosaccharide is 2-deoxy-D-ribose, an aldopentose that lacks a hydroxyl group at C2.

Uracil (U) occurs only in RNA, while thymine (T) occurs only in DNA. As a result:
    DNA contains the bases A, G, C, and T.
    RNA contains the bases A, G, C, and U.

DNA forms a double helix with complementary base pairs. RNA is a single chain composed of nucleotides.

22.31 a. base: cytosine
         monosaccharide: 2'-deoxyribose
         deoxycytidine
      b. base: guanine
         monosaccharide: ribose
         guanosine 5'-monophosphate

22.33 a.

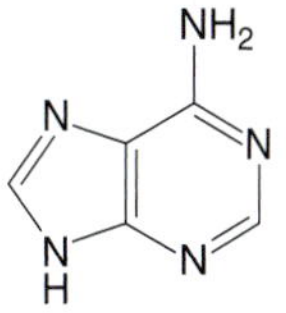

b.

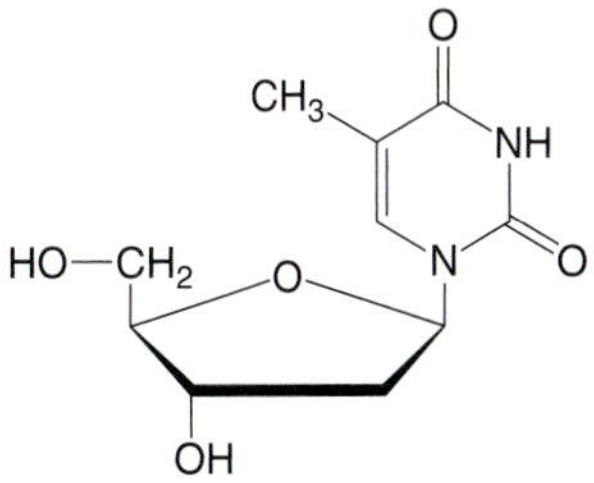

c.

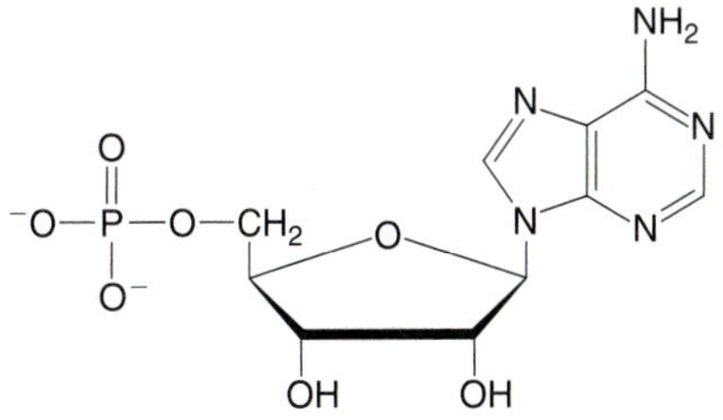

d.

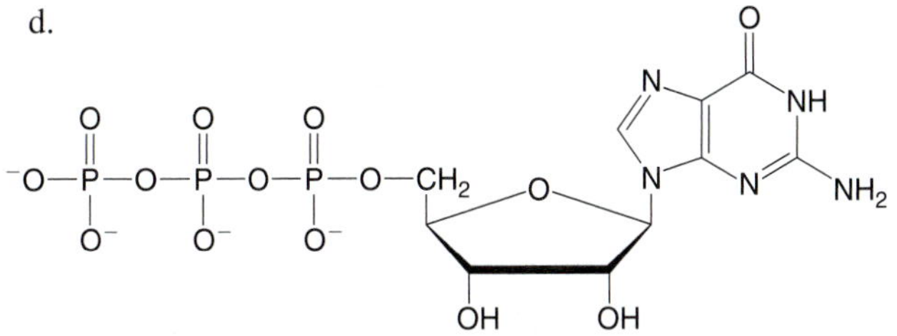

22.35 a.

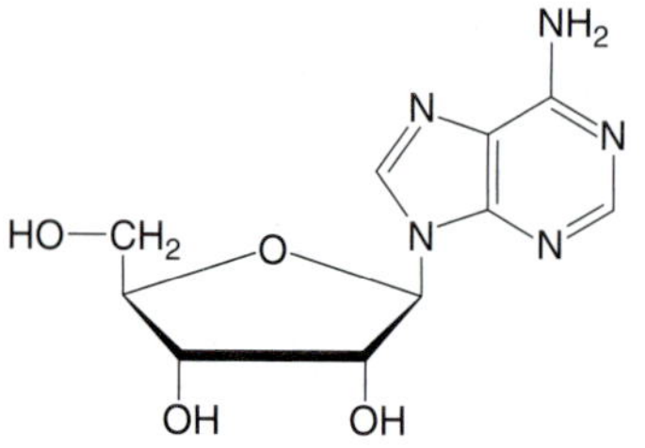

b.

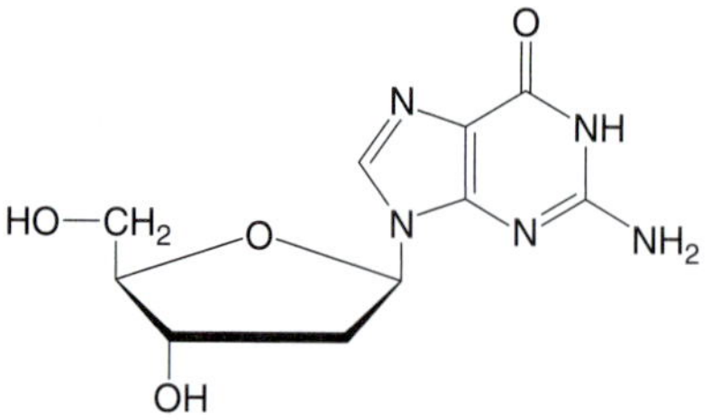

c.

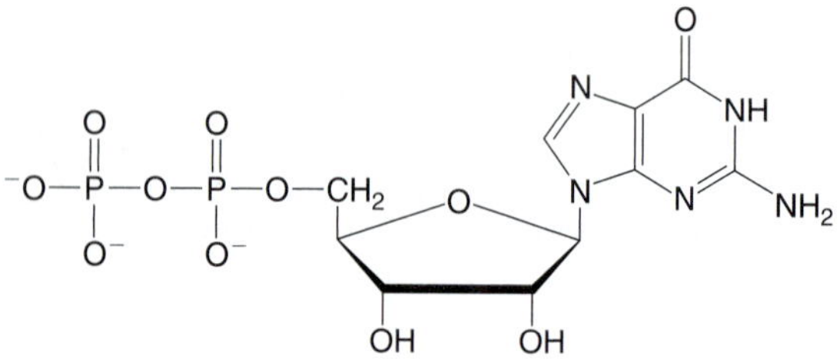

d.

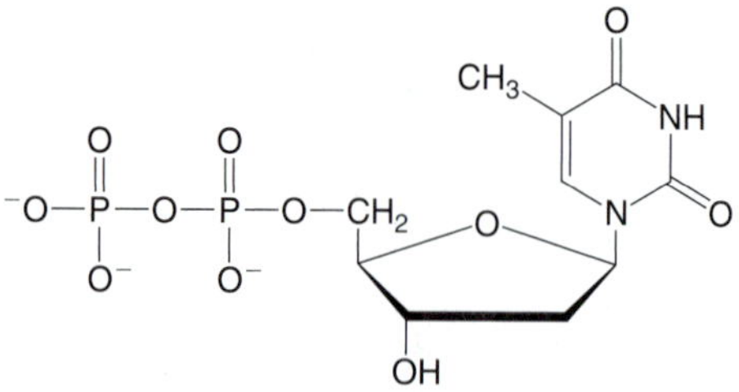

22.37

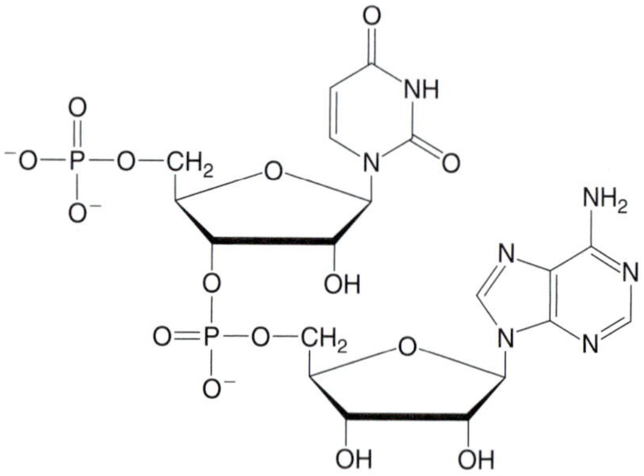

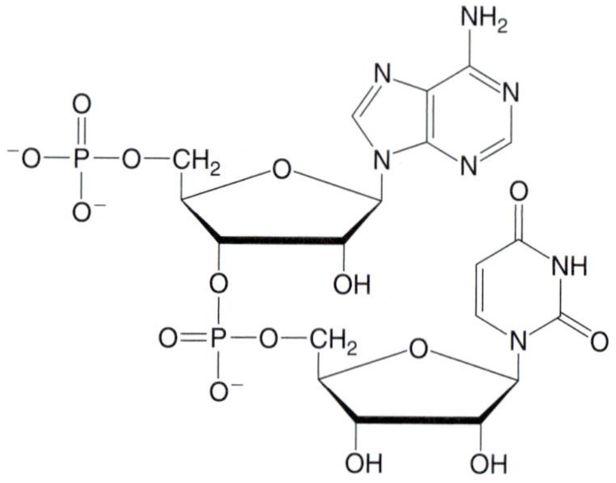

22.39 a.

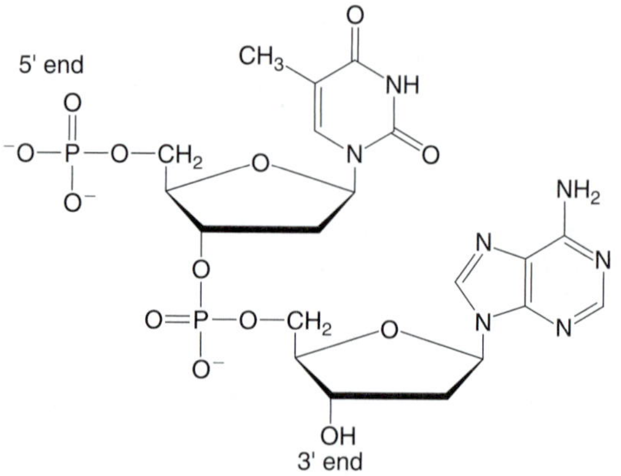

b.

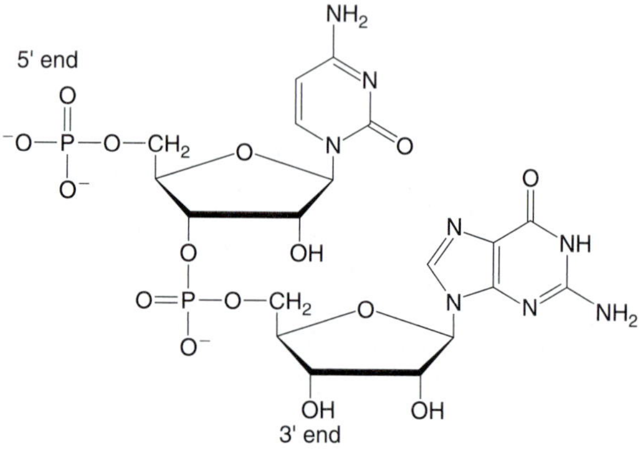

22.41

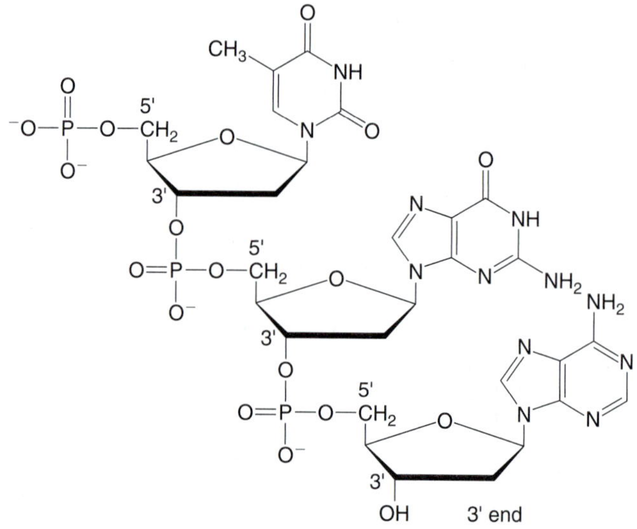

22.43 a. The DNA double helix has 2-deoxyribose as the only sugar. The sugar–phosphate groups are on the outside of the helix.

b. The 5' end has a phosphate and the 3' end has an OH group.

c. Hydrogen bonding occurs in the interior of the helix between base pairs: A pairs with T and G pairs with C.

22.45 a. 3'–TTTATTG–5'   c. 3'–GCTATAGGGC–5'

b. 3'–TGACCTGA–5'   d. 3'–AAGGGCCCTAT–5'

22.47 27% T, 23% G, 23% C

22.49 G and C

22.51 5'–TACCGGATACGCTA–3'

22.53 Replication occurs in only one direction, from the 3' end to the 5' end of the template strand, but the strands of DNA run in opposite directions. The leading strand can be formed continuously from the 3' end as it is unwound. The lagging strand runs in the opposite direction, so it must be formed in segments that are then joined by DNA ligase.

22.55 Messenger RNA carries the specific sequence of the DNA code from the cell nucleus to the ribosomes in the cytoplasm to make a protein. Each transfer RNA brings a specific amino acid to the growing protein chain on the ribosome according to the sequence specified by the mRNA.

22.57 Replication and transcription both result in a strand complementary to the DNA template, but DNA has the nucleotide thymidine and RNA has uridine in corresponding positions. In replication, both strands of DNA are copied. In transcription, only one strand of DNA serves as a template to synthesize mRNA.

22.59 a. 3'–UUUAUUG–5' c. 3'–GCUAUAGGGC–5'
      b. 3'–UGACCUGA–5' d. 3'–AAGGGCCCUAU–5'

22.61 a. [1] 5'-UACCGAAU–3' [2] 5'–TACCGAAT–3'
      b. [1] 5'–GCCGCGAAU–3' [2] 5'–GCCGCGAAT–3'
      c. [1] 5'–CCAUAUGGC–3' [2] 5'–CCATATGGC–3'
      d. [1] 5'–AUCCGGCAU–3' [2] 5'–ATCCGGCAT–3'

22.63 A codon is a group of three mRNA nucleotides that is specific for an amino acid. The anticodon is the complementary sequence of nucleotides on a tRNA.

22.65 a. GAC:Leu b. AAA:Phe c. UUC:Lys d. CGU:Ala

22.67 a. Pro–Thr–Trp–Val–Glu
      b. Met–Phe–Leu–Trp–Trp
      c. Val–Asp–Glu–Pro–Gln

22.69 a. AUU AUG AAA AGU UAU
      b. CCU CAA GAA GAU UUU
      c. ACU AGU AAU CGU

22.71 a. [1] 3' ATA AGT TAT TTT TTG 5'
         [2] Tyr–Ser–Ile–Lys–Asn
      b. [1] 3' CTA CAT TTG TTC GGC 5'
         [2] Asp–Val–Asn–Lys–Pro

22.73 A point mutation results in the substitution of one nucleotide for another in a DNA molecule. A silent mutation is a point mutation in DNA that results in no change in an amino acid sequence.

22.75 a. Leu–Gln–His
      b. Leu–Gln–Asn
      c. Leu–Gln–His, no change
      d. This is a stop codon so the chain is terminated.
      e. A dipeptide, Ser–Ala, is synthesized.

22.77 a. Asn–Ala b. Asn–Ala c. Asn–Gly d. Asn–Ser

22.79 A restriction endonuclease is an enzyme that cleaves plasmid DNA at a specific site.

22.81

5' ~~~ C C G G T T G ⌐‾‾‾‾¬   G A T C C T T ~~~ 3'
                        sticky ends
3' ~~~ G G C C A A C C T A G   ⌊_____⌋ G A A ~~~ 5'

22.83 Circular plasmid DNA is isolated from bacteria and cleaved at a specific site with a restriction endonuclease. This forms "sticky ends" of DNA that are then combined with another DNA source, perhaps human, with complementary "sticky ends" to the plasmid DNA. These DNA strands come together and then a DNA ligase is used to re-form a circular DNA that is then reproduced by the bacteria.

22.85 a. Lanes 3 and 5 represent DNA of children that share both parents because they both have DNA fragments common to both parents.
      b. Lane 4 represents DNA from an adopted child because the DNA fragments have little relationship to the parental DNA fragments.

22.87 A retrovirus is a virus that has RNA rather than DNA in its core. Once it invades a cell it uses reverse transcriptase to synthesize DNA for replication.

22.89 a. cytidine
      b. Lamivudine is a nucleoside analogue that gets incorporated into a DNA chain, but since it does not contain a 3' hydroxyl group, synthesis is terminated.

22.91 a. 5' GTT ACA TAA AAA CGA 3'
      b. 5' GUU ACA UAA AAA CGA 3'
      c. Val–Thr

22.93

DNA informational strand:	5' end	AAC	TAT	CCA	ACG	AAG	ATG	3' end
DNA template strand:	3' end	TTG	ATA	GGT	TGC	TTC	TAC	5' end
mRNA codons:	5' end	AAC	UAU	CCA	ACG	AAG	AUG	3' end
tRNA anticodons:		UUG	AUA	GGU	UGC	UUC	UAC	
Polypeptide:		Asn	Tyr	Pro	Thr	Lys	Met	

22.95

DNA informational strand:	5' end	AAC	GTA	TCA	ACT	CAC	ATG	3' end
DNA template strand:	3' end	TTG	CAT	AGT	TGA	GTG	TAC	5' end
mRNA codons:	5' end	AAC	GUA	UCA	ACU	CAC	AUG	3' end
tRNA anticodons:		UUG	CAU	AGU	UGA	GUG	UAC	
Polypeptide:		Asn	Val	Ser	Thr	His	Met	

22.97 975
22.99 ATA CCA CCA AAA TAC
22.101 TAA CCT AAA

CHAPTER 23

23.1 This makes it possible to use the same reactions and the same enzyme systems to metabolize all types of biomolecules.

23.3 a. glucose + ATP ⟶ glucose 1-phosphate + ADP
     b. –2.3 kcal/mol

23.5 –3.0 kcal/mol of energy released

23.6 a. reduction using NADH b. oxidation using NAD^+

23.7 Riboflavin has numerous polar groups (OH and NH) that can hydrogen bond to water, making it water soluble.

23.9
$$CH_3CH_2CH_2\overset{\overset{\displaystyle O}{\|}}{C}O^- \ + \ HSCoA$$

23.10 a.
NAD^+ $NADH + H^+$

HO–C–H ⟶ C=O
   | |
   CH₂   CH₂
   | |
   CO₂⁻   CO₂⁻
   CO₂⁻   CO₂⁻

b. oxidation

23.11   Succinate dehydrogenase is so named because the elements
        of hydrogen (H_2) are removed from succinate.

23.13   a. reduction   b. oxidizing agent

23.15   Catabolism is the breakdown of large molecules into smaller
        ones and anabolism is the synthesis of large molecules from
        smaller ones.

23.17   Mitochondria contain an outer membrane and an inner
        membrane with many folds. The area between these two
        membranes is called the intermembrane space. Energy
        production occurs within the matrix, the area surrounded by
        the inner membrane.

23.19   hydrolysis of starch, the conversion of glucose to acetyl
        CoA, the citric acid cycle, electron transport chain,
        oxidative phosphorylation

23.21   a. stage [1]   c. stage [3]   e. stage [1]
        b. stage [2]   d. stage [4]

23.23   Coupled reactions are pairs of reactions that occur together.
        The energy released by one reaction provides the energy to
        drive the other reaction.

23.25 $CH_3\overset{\overset{\displaystyle O}{\|}}{C}O^-$ + HPO_4^{2-}

23.27   a. ATP + succinate + HSCoA   Energy change
        $\longrightarrow$ succinyl CoA + ADP + HPO_4^{2-} +2.1 kcal/mol
        b. The reaction is not energetically favored because an input
        of energy is required.

23.29   a. glucose + HPO_4^{2-} $\longrightarrow$ glucose 1-phosphate + H_2O
        b. glucose + ATP $\longrightarrow$ glucose 1-phosphate + ADP
        c. −2.3 kcal/mol

23.31   a. −3.9 kcal/mol
        b. fructose 6-phosphate + HPO_4^{2-}
        $\longrightarrow$ fructose 1,6-bisphosphate + H_2O
        ATP + H_2O $\longrightarrow$ ADP + HPO_4^{2-}
        ───────────────────────────────
        fructose 6-phosphate + ATP
        $\longrightarrow$ fructose 1,6-bisphosphate + ADP
        c. +3.9 + (−7.3) = −3.4 kcal/mol

23.33   a. 1,3-bisphosphoglycerate + ADP
        $\longrightarrow$ 3-phosphoglycerate + ATP
        b. −4.5 kcal/mol
        c. Yes, this is an energetically favorable coupled reaction
        that could be used to synthesize ATP from ADP.

23.35   a.

GTP

b.

GDP

23.37   a. reducing agent   b. neither   c. oxidizing agent

23.39   When a substrate is oxidized, NAD^+ is reduced. NAD^+ is an
        oxidizing agent.

23.41   a.

        b.

23.43   a. steps [3], [4], and [8]   c. step [6]
        b. steps [3] and [4]   d. step [1]

23.45   a. isocitrate   b. malate, isocitrate

23.47   a. oxidation   b. decarboxylation   c. step [3a]
        d. The elements of H_2 (hydrogen) are removed in
        converting isocitrate to oxalosuccinate.

23.49   step [7]

23.51   They require oxygen.

23.53   a. $FADH_2$ donates electrons to the electron transport chain.
        b. ADP is a substrate for the formation of ATP.
        c. ATP synthase catalyzes the formation of ATP from ADP.
        d. The inner mitochondrial membrane contains the four
        complexes for the electron transport chain. ATP synthase
        is also embedded in the membrane and contains the H^+
        ion channel that allows H^+ to return to the matrix.

23.55 NAD^+, FAD, and H_2O

23.57 $FADH_2$ enters the electron transport chain at complex II,
        whereas NADH enters at complex I.

23.59   Each two-carbon fragment of acetyl CoA that enters the
        citric acid cycle produces three molecules of NADH and
        one molecule of $FADH_2$. These then enter the electron
        transport chain and serve as reducing agents. For each
        NADH that enters the electron transport chain, there are
        2.5 ATPs produced, and for each $FADH_2$, there are 1.5 ATPs
        produced. In addition, there is one GTP (an ATP equivalent)
        produced in the citric acid cycle.
        In total: 3 NADH × 2.5 ATP/NADH = 7.5 ATP
                    1 $FADH_2$ × 1.5 ATP/$FADH_2$ = 1.5 ATP
                    1 GTP (ATP equivalent) = 1   ATP
                                            ─────────
                                            10 ATP

23.61   ATP has three phosphate groups and ADP has two.

23.63 a.

glucose 1-phosphate

b.

$+$ H_2O

glucose 1-phosphate

$+$ HPO_4^{2-}

23.65 The citric acid cycle generates the reducing agents NADH and $FADH_2$ that then enter the electron transport chain.

23.67 There would be more mitochondria in the heart because it has far more metabolic needs than bone.

23.69 Creatine phosphate, stored in muscles, is a high-energy phosphate that generates energy upon hydrolysis.

CHAPTER 24

24.1 The reactant and product are constitutional isomers that differ in the location of a carbonyl and hydroxyl group. Thus, an isomerase enzyme converts one isomer to another.

24.3 They are constitutional isomers with a different arrangement of atoms but the same molecular formula.

24.5 3.9 kcal/mol

24.7 step [2]: phosphohexose isomerase
step [5]: triose phosphate isomerase
step [8]: phosphoglycerate mutase

24.9 The rate of glycolysis would decrease.

24.11 Yes, this is similar to the reaction to form glucose 6-phosphate, which requires ATP.

24.13 a. Both create a two-carbon compound from the three-carbon pyruvate with loss of CO_2.
b. Production of acetyl CoA generates NADH and production of ethanol consumes NADH.

24.15 a. 7   b. 2.5 for each pyruvate   c. 12   d. 20

24.17 Steps [7] and [10] of the glycolysis pathway and step [5] of the citric acid cycle form ATP or GTP directly.

24.19

	Starting Material	Product
a.	Pyruvate	Phosphoenolpyruvate
b.	2-Phosphoglycerate	3-Phosphoglycerate
c.	Glucose 6-phosphate	Glucose

24.21 a. [1] 10; [2] 9 cycles   b. [1] 8; [2] 7 cycles

24.22 106

24.23 134

24.25 a. 5.3 ATP/carbon in glucose
b. 6.7 ATP/carbon in stearic acid
c. supports

24.27 β-Hydroxybutyrate contains an alcohol and a carboxylate anion, but no ketone.

24.29 An increased concentration of acetyl CoA increases the production of ketone bodies.

24.30 a.

b.

c.

24.31 a.

b.

c.

24.33 a. Oxidation occurs with loss of hydrogen by a dehydrogenase enzyme. The coenzyme NAD^+ serves as an oxidant.
b. Transfer of phosphate from GTP to the reactant occurs, forming GDP. The loss of carbon dioxide takes place with the carboxykinase enzyme.

24.35 This reaction converts one isomer to another so there is a change in position of the phosphate; thus, the enzyme is called a "mutase."

24.37 Energy-investment phase:
a. glucose, glyceraldehyde 3-phosphate
b. 2 ATP utilized
c. no coenzymes used or formed
Energy-generating phase:
a. glyceraldehyde 3-phosphate, pyruvate
b. 4 ATP/glucose produced
c. 2 NAD^+ used and 2 NADH produced per glucose

24.39 a. 7,10   b. 1,3   c. 6   d. 4

24.41 The conversion of pyruvate to acetyl CoA forms one CO_2 for each pyruvate.
The conversion of isocitrate to α-ketoglutarate and α-ketoglutarate to succinyl CoA in the citric acid cycle forms one CO_2 for each acetyl CoA.

24.43 a.

$$C_6H_{12}O_6 + 2\,NAD^+ \xrightarrow[\text{2 ADP}\quad\text{2 ATP}]{} 2\,CH_3COCO_2^- + 2\,NADH + 2\,H^+$$

glucose   pyruvate

b.

$$C_6H_{12}O_6 \xrightarrow[\text{2 ADP}\quad\text{2 ATP}]{} 2\,CH_3CH_2OH + 2\,CO_2$$

glucose   ethanol

c.

pyruvate $\xrightarrow[\text{NADH + H}^+ \quad \text{NAD}^+]{}$ lactate

24.45 a. aerobic: CO_2
anaerobic: lactate
b. aerobic: NAD^+ and FAD are used and NADH and $FADH_2$ are formed.
anaerobic: NADH is used to convert pyruvate to lactate and NAD^+ is formed.

24.47 Glycolysis takes glucose and breaks it into smaller units, producing pyruvate. Gluconeogenesis is the synthesis of glucose from lactate, amino acids, and glycerol.

24.49 a. increased   c. increased
b. decreased   d. decreased

24.51 In the Cori cycle, glucose in muscle is catabolized to pyruvate and then converted to lactate, which is transported to the liver. In the liver, lactate is converted to glucose by gluconeogenesis and then transported back to the muscle.

24.53 a. lactate   b. ethanol   c. CO_2

24.55 During glycolysis: 2 ATP and 2 NADH generated.
Each NADH in turn leads to 2.5 ATP,
so glycolysis leads to: total 7 ATP
Conversion of 2 pyruvate to 2 acetyl CoA
yields 2 NADH, which leads to: 5 ATP
In the citric acid cycle, 20 additional ATP
are formed for 2 $CH_3COSCoA$:
    2 GTP (ATP), 6 NADH $\longrightarrow$ 15 ATP,
    2 $FADH_2$ $\longrightarrow$ 3 ATP: total 20 ATP
                                        = 32 ATP

24.57 two carbons in two molecules of CO_2 and four in two molecules of CH_3CH_2OH

24.59 Two phosphate bonds of ATP are hydrolyzed forming AMP; therefore, 2 ATP equivalents are used.

24.61 a. [1] 3   [2] 2   b. [1] 7   [2] 6

24.63 a. 36   b. 92

24.65

$$C_{15}H_{31}CO_2H \longrightarrow CH_3(CH_2)_{12}-CH_2-CH_2-\overset{\overset{\displaystyle O}{\|}}{C}-SCoA$$

C_{16} acyl CoA

acyl CoA dehydrogenase $\xrightarrow[\text{FADH}_2]{\text{FAD}}$ ① **Oxidation**

$$CH_3(CH_2)_{12}-CH=CH-\overset{\overset{\displaystyle O}{\|}}{C}-SCoA$$

enoyl CoA hydratase $\xrightarrow{\text{H}_2\text{O}}$ ② **Hydration**

$$CH_3(CH_2)_{12}-\overset{\overset{\displaystyle OH}{|}}{CH}-\overset{\overset{\displaystyle H}{|}}{CH}-\overset{\overset{\displaystyle O}{\|}}{C}-SCoA$$

$\xrightarrow[\text{NADH + H}^+]{\text{NAD}^+}$ ③ **Oxidation**

$$CH_3(CH_2)_{12}-\overset{\overset{\displaystyle O}{\|}}{C}-CH_2-\overset{\overset{\displaystyle O}{\|}}{C}-SCoA$$

acyl CoA acyltransferase $\xrightarrow{\text{HSCoA}}$ ④ **Cleavage**

$$CH_3(CH_2)_{12}-\overset{\overset{\displaystyle O}{\|}}{C}-SCoA + CH_3-\overset{\overset{\displaystyle O}{\|}}{C}-SCoA$$

C_{14} acyl CoA   2 C's

24.67 Two H's are removed. A decrease in the number of C—H bonds = oxidation.

24.69 a. $CH_3(CH_2)_6-\underset{\beta}{CH_2}-\underset{\alpha}{CH_2}-\overset{\overset{\displaystyle O}{\|}}{C}-OH$

b. $CH_3(CH_2)_6-\underset{\beta}{CH_2}-\underset{\alpha}{CH_2}-\overset{\overset{\displaystyle O}{\|}}{C}-SCoA$

c. 5   d. 4   e. 64

24.71 0.37 mol ATP/g decanoic acid

24.73 Ketosis is the condition under which ketone bodies accumulate. Ketogenesis is the synthesis of ketone bodies from acetyl CoA.

24.75 In poorly controlled diabetes, glucose cannot be metabolized. Fatty acids are used for metabolism and ketone bodies are formed to a greater extent.

24.77 Ketogenic amino acids are converted to acetyl CoA or a related thioester and can be converted to ketone bodies. Glucogenic amino acids are catabolized to pyruvate or another intermediate in the citric acid cycle.

24.79 a. 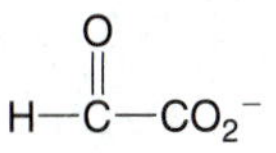 $H-\overset{\overset{\displaystyle O}{\|}}{C}-CO_2^-$ b. 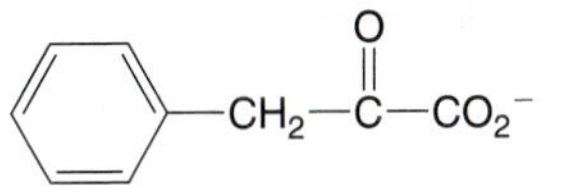 ⌬$-CH_2-\overset{\overset{\displaystyle O}{\|}}{C}-CO_2^-$

24.81 a. 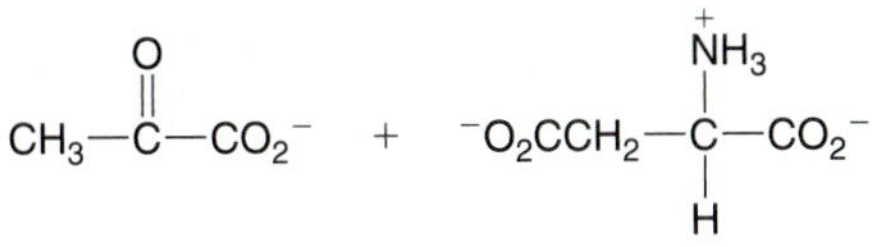 $CH_3-\overset{\overset{\displaystyle O}{\|}}{C}-CO_2^-$ + $^-O_2CCH_2-\overset{\overset{\displaystyle \overset{+}{N}H_3}{|}}{\underset{\underset{\displaystyle H}{|}}{C}}-CO_2^-$

b. 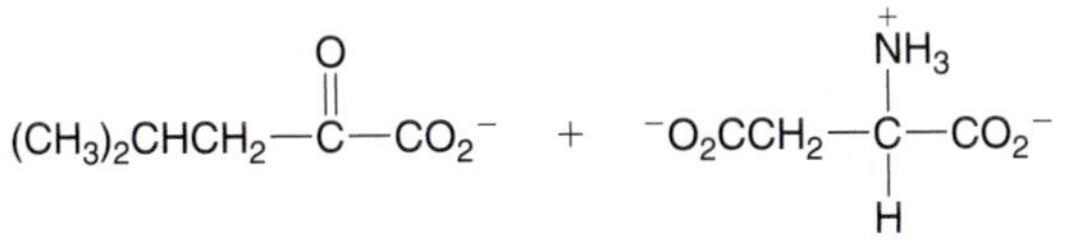 $(CH_3)_2CHCH_2-\overset{\overset{\displaystyle O}{\|}}{C}-CO_2^-$ + $^-O_2CCH_2-\overset{\overset{\displaystyle \overset{+}{N}H_3}{|}}{\underset{\underset{\displaystyle H}{|}}{C}}-CO_2^-$

24.83 a. 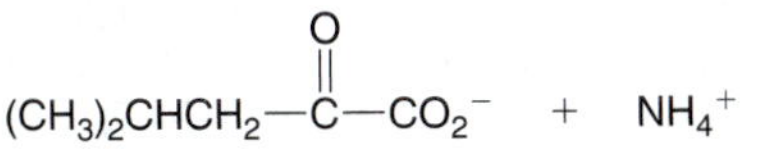 $(CH_3)_2CHCH_2-\overset{\overset{\displaystyle O}{\|}}{C}-CO_2^-$ + NH_4^+

b. 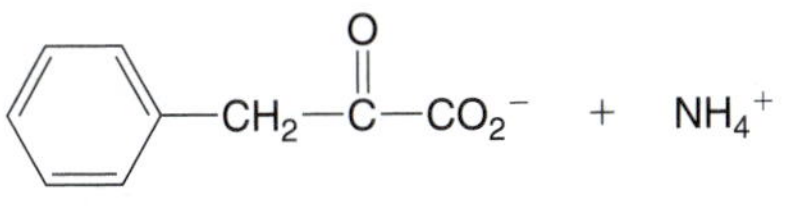 ⌬$-CH_2-\overset{\overset{\displaystyle O}{\|}}{C}-CO_2^-$ + NH_4^+

24.85 a. acetoacetyl CoA, fumarate
b. α-ketoglutarate
c. oxaloacetate
d. pyruvate

24.87 Yes, phenylalanine, for example, can be degraded by multiple routes and can be both ketogenic and glucogenic.

24.89 A spiral pathway involves the same set of reactions repeated on increasingly smaller substrates; example: β-oxidation of fatty acids. A cyclic pathway is a series of reactions that regenerates the first reactant; example: citric acid cycle.

24.91 Glucose is metabolized anaerobically to form CO_2 and ethanol. The CO_2 causes bread to rise.

24.93 Pyruvate is metabolized anaerobically to lactate.

24.95 The diet calls for ingestion of protein and fat, so in the absence of carbohydrates to be metabolized, ketone bodies are formed.

24.97 a. kinase   b. dehydrogenase   c. carboxylase

24.99 18.5 ATP

Glossary

A

Acetal (16.8) A compound that contains two OR groups (alkoxy groups) bonded to the same carbon; general formula $R_2C(OR)_2$.

Acetyl CoA (23.4) A compound formed when an acetyl group ($CH_3CO–$) is bonded to coenzyme A (HS–CoA); $CH_3COSCoA$.

Achiral (15.2) Being superimposable on a mirror image.

Acid (9.1) In the Arrhenius definition, a substance that contains a hydrogen atom and dissolves in water to form a hydrogen ion, H^+.

Acid dissociation constant (9.4) A constant, K_a, related to acid strength. The stronger the acid (HA), the larger the value of K_a; $K_a = [H_3O^+][A:^-]/[HA]$.

Acidic solution (9.5) A solution in which $[H_3O^+] > [^-OH]$; thus, $[H_3O^+] > 10^{-7}$ M.

Actinides (2.4) A group of elements in the periodic table beginning with thorium ($Z = 90$), and immediately following the element actinium (Ac).

Active site (6.4, 21.9) The region in an enzyme that binds a substrate, which then undergoes a very specific reaction with an enhanced rate.

Active transport (19.7) The process of moving an ion across a cell membrane that requires energy input.

Actual yield (5.7) The amount of product formed in a reaction, determined by weighing the product on a balance.

Acyclic alkane (12.1) An alkane with molecular formula C_nH_{2n+2}, which contains a chain of carbon atoms but no rings. An acyclic alkane is also called a saturated hydrocarbon.

Acyl CoA (24.7) The thioester formed from a fatty acid and coenzyme A that undergoes β-oxidation in mitochondria; general structure RCOSCoA.

Acyl group (17.1) A substituent having the formula RCO–.

Addition reaction (13.6) A reaction in which elements are added to a compound.

Adenosine 5'-diphosphate (ADP, 23.3) A nucleoside diphosphate formed by adding two phosphates to the 5'-OH group of adenosine.

Adenosine 5'-triphosphate (ATP, 23.3) A nucleoside triphosphate formed by adding three phosphates to the 5'-OH group of adenosine. ATP is the most prominent member of a group of "high-energy" molecules that release energy by cleavage of a P–O bond during hydrolysis.

Adrenal cortical steroid (19.9) A steroid hormone synthesized in the outer layer of the adrenal gland.

Alcohol (11.5) A compound containing a hydroxyl group (OH) bonded to a tetrahedral carbon atom; general formula ROH.

Aldehyde (11.5) A compound that has a hydrogen atom bonded directly to a carbonyl carbon; general formula RCHO.

Alditol (20.4) A compound produced when the carbonyl group of an aldose is reduced to a 1° alcohol.

Aldonic acid (20.4) A compound produced when the aldehyde carbonyl of an aldose is oxidized to a carboxyl group.

Aldose (20.2) A monosaccharide with an aldehyde carbonyl group at C1.

Alkali metal (2.4) An element located in group 1A (group 1) of the periodic table. Alkali metals include lithium (Li), sodium (Na), potassium (K), rubidium (Rb), cesium (Cs), and francium (Fr).

Alkaline earth element (2.4) An element located in group 2A (group 2) of the periodic table. Alkaline earth elements include beryllium (Be), magnesium (Mg), calcium (Ca), strontium (Sr), barium (Ba), and radium (Ra).

Alkaloid (18.4) A naturally occurring amine derived from a plant source.

Alkane (11.5) A compound having only C–C and C–H single bonds.

Alkene (11.5) A compound having a carbon–carbon double bond.

Alkoxy group (14.7) An OR group. An alkoxy group is named by changing the -yl ending of an alkyl group to -oxy.

Alkyl group (12.4) A group formed by removing one hydrogen from an alkane.

Alkyl halide (11.5) A compound with the general structure R–X that contains a halogen atom (X = F, Cl, Br, or I) bonded to a tetrahedral carbon.

Alkyne (11.5) A compound with a carbon–carbon triple bond.

Alpha (α) carbon (17.2) The carbon adjacent to a carbonyl group.

Alpha (α) particle (10.1) A high-energy particle that is emitted from a radioactive nucleus and contains two protons and two neutrons.

Amide (11.5) A compound that contains a nitrogen atom bonded directly to a carbonyl carbon; general structure $RCONR'_2$, where R' = H or alkyl.

Amine (11.5) An organic compound that contains a nitrogen atom bonded to one, two, or three alkyl groups; general structure RNH_2, R_2NH, or R_3N.

Amino acid (21.2) A compound that contains two functional groups—an amino group (NH_2) and a carboxyl group (COOH) bonded to the same carbon.

Amino group (11.5) An $–NH_2$ group.

Ammonium ion (3.6) An NH_4^+ ion.

Ammonium salt (18.6) An ionic compound that contains a positively charged ammonium ion and an anion. $(CH_3CH_2CH_2NH_3)^+Cl^-$ is an ammonium salt.

Amorphous solid (7.9) A solid with no regular arrangement of its particles.

Amphoteric compound (9.2) A compound that contains both a hydrogen atom and a lone pair of electrons so that it can be either an acid or a base.

Anabolic steroid (19.9) A synthetic androgen analogue that promotes muscle growth.

Anabolism (23.1) The synthesis of large molecules from smaller ones in a metabolic pathway.

Androgen (19.9) A hormone that controls the development of secondary sex characteristics in males.

Anion (3.2) A negatively charged ion with more electrons than protons.

α Anomer (20.3) The cyclic form of a monosaccharide that contains a hemiacetal with the OH group drawn down, below the ring.

β Anomer (20.3) The cyclic form of a monosaccharide that contains a hemiacetal with the OH group drawn up, above the ring.

Anomeric carbon (20.3) The carbon atom that is part of the hemiacetal in a cyclic monosaccharide.

Anticodon (22.8) Three nucleotides in a tRNA molecule that are complementary to the codon in mRNA and identify an individual amino acid.

Antihistamine (18.10) A compound that binds to the H1 histamine receptor and prevents the action of histamine.

Antioxidant (13.4) A compound that prevents an unwanted oxidation reaction from occurring.

Anti-ulcer drug (18.10) A drug that binds to the H2 histamine receptor, thus reducing acid secretion in the stomach.

Aqueous solution (8.1) A solution with water as the solvent.

Aromatic compound (11.5) A compound that contains a benzene ring, a six-membered ring with three double bonds.

Aryl halide (13.13) An organic halide with the halogen bonded directly to a benzene ring.

Atmosphere (7.2) A unit used to measure pressure; 1 atm = 760 mm Hg.

Atom (2.1) The basic building block of matter composed of a nucleus and an electron cloud.

Atomic mass unit (2.2) A unit abbreviated as amu, which equals one-twelfth the mass of a carbon atom that has six protons and six neutrons; 1 amu = 1.661×10^{-24} g.

Atomic number (2.2) The number of protons in the nucleus of an atom; symbolized as Z.

Atomic weight (2.3) The weighted average of the mass of all naturally occurring isotopes of a particular element, reported in atomic mass units.

Avogadro's law (6.4) A gas law that states that the volume of a gas is proportional to the number of moles present when the pressure and temperature are held constant.

Avogadro's number (5.3) A quantity that contains 6.02×10^{23} items—usually atoms, molecules, or ions.

Axon (18.8) The long stem that protrudes from the cell body of a neuron.

B

Balanced chemical equation (5.2) An equation written so that an equal number of atoms of each element is present on both sides.

Barometer (7.2) A device for measuring atmospheric pressure.

Base (9.1) In the Arrhenius definition, a substance that contains hydroxide and dissolves in water to form $^-$OH.

Basic solution (9.5) A solution in which $[^-OH] > [H_3O^+]$; thus, $[^-OH] > 10^{-7}$ M.

Becquerel (10.4) An SI unit used to measure radioactivity, abbreviated as Bq; 1 Bq = 1 disintegration/s.

Benedict's reagent (20.4) A Cu^{2+} reagent that oxidizes the aldehyde carbonyl of an aldose to a carboxyl group, yielding an aldonic acid.

Beta (β) carbon (17.2) The carbon bonded to the α (alpha) carbon and located two carbons from a carbonyl group.

Beta (β) particle (10.1) A high-energy electron emitted from a radioactive nucleus.

Boiling point (7.7) The temperature at which a liquid is converted to the gas phase.

Boiling point elevation (8.7) The increase in the boiling point of a liquid solution due to the presence of a dissolved nonvolatile solute.

Bond dissociation energy (6.2) The energy needed to break a covalent bond by equally dividing the electrons between the two atoms in the bond.

Bonding (3.1) The joining of two atoms in a stable arrangement.

Boyle's law (7.3) A gas law that relates pressure and volume. Boyle's law states that for a fixed amount of gas at constant temperature, the pressure and volume of the gas are inversely related.

Branched-chain alkane (12.2) An alkane that contains one or more carbon branches bonded to a carbon chain.

Brønsted–Lowry acid (9.1) A proton donor.

Brønsted–Lowry base (9.1) A proton acceptor.

Buffer (9.10) A solution whose pH changes very little when acid or base is added. Most buffers are solutions composed of approximately equal amounts of a weak acid and the salt of its conjugate base.

Building-block element (2.1) One of the four nonmetals—oxygen, carbon, hydrogen, and nitrogen—that comprise 96% of the mass of the human body.

C

Calorie (6.1) A unit of energy that equals the amount of energy needed to raise the temperature of 1 g of water by 1 °C; abbreviated as cal, where 1 cal = 4.184 J.

Carbohydrate (20.1) A polyhydroxy aldehyde or ketone, or a compound that can be hydrolyzed to a polyhydroxy aldehyde or ketone.

Carbonate (3.6) A polyatomic anion with the structure CO_3^{2-}.

Carbonyl group (11.5) A carbon–oxygen double bond (C=O).

Carboxyl group (17.1) A COOH group.

Carboxylate anion (17.6) The conjugate base of a carboxylic acid; general structure $RCOO^-$.

Carboxylic acid (11.5) A compound that contains an OH group bonded directly to the carbonyl carbon; general structure RCOOH or RCO_2H.

Catabolism (23.1) The breakdown of large molecules into smaller ones during metabolism.

Catalyst (6.4) A substance that increases the rate of a reaction but is recovered unchanged at the end of the reaction.

Cation (2.8) A positively charged particle with fewer electrons than protons.

Cell membrane (19.7) The semipermeable membrane that surrounds the cell, composed of a lipid bilayer.

Celsius scale (1.9) One of three temperature scales in which water freezes at 0 °C and boils at 100 °C.

Cephalin (19.6) A phosphoacylglycerol in which the identity of the R group esterified to the phosphodiester is $-CH_2CH_2NH_3^+$. A cephalin is also called a phosphatidylethanolamine.

Chain reaction (10.6) The process by which each neutron produced during fission can go on to bombard three other nuclei to produce more nuclei and more neutrons.

Charles's law (7.3) A gas law that states that for a fixed amount of gas at constant pressure, the volume of the gas is proportional to its Kelvin temperature.

Chemical equation (5.1) An expression that uses chemical formulas and other symbols to illustrate what reactants constitute the starting materials in a reaction and what products are formed.

Chemical formula (2.1) A representation that uses element symbols to show the identity of elements in a compound, and subscripts to show the number of atoms of each element contained in the compound.

Chemical properties (1.2) Those properties that determine how a substance can be converted to another substance by a chemical reaction.

Chemistry (1.1) The study of matter—its composition, properties, and transformations.

Chiral (15.2) Not superimposable on a mirror image.

Chirality center (15.2) A carbon atom bonded to four different groups.

Chlorination (13.13) A reaction with chlorine in which a Cl atom substitutes for a hydrogen atom.

Chlorofluorocarbon (14.9) A halogen-containing compound having the general molecular formula CF_xCl_{4-x}.

Cis isomer (13.3) An alkene with two R groups on the same side of the double bond.

Citric acid cycle (23.5) A cyclic metabolic pathway that begins with the addition of acetyl CoA to a four-carbon substrate and ends when the same four-carbon compound is formed as a product eight steps later.

Codon (22.7) A sequence of three nucleotides (triplet) in mRNA that codes for a specific amino acid.

Coenzyme (14.6, 21.9) An organic molecule needed for an enzyme-catalyzed reaction to occur.

Coenzyme A (23.4) A coenzyme that contains a sulfhydryl group (SH group) making it a thiol (RSH), and abbreviated as HS–CoA.

Cofactor (21.9) A metal ion or a nonprotein organic molecule needed for an enzyme-catalyzed reaction to occur.

Colligative properties (8.7) The properties of a solution that depend on the concentration of the solute but not its identity.

Colloid (8.1) A homogeneous mixture with large particles, often having an opaque appearance.

Combined gas law (7.3) A gas law that relates pressure, volume, and temperature. For a constant number of moles, the product of pressure and volume divided by temperature is a constant.

Combustion (12.8) An oxidation reaction in which carbon-containing compounds react with oxygen to form carbon dioxide (CO_2) and water.

Competitive inhibitor (21.9) An enzyme inhibitor that has a shape and structure similar to the substrate, and competes with the substrate for binding to the active site.

Complementary base pairs (22.3) The predictable pairing of bases between two strands of DNA. Adenine pairs with thymine using two hydrogen bonds, forming an A–T base pair, and cytosine pairs with guanine using three hydrogen bonds, forming a C–G base pair.

Compound (1.3) A pure substance formed by chemically combining two or more elements.

Concentration (8.4) The amount of solute dissolved in a given amount of solution.

Condensation (7.8, 7.10) The conversion of a gas to a liquid.

Condensation polymer (17.10) A polymer formed when monomers containing two functional groups come together with loss of a small molecule such as water.

Condensed structure (11.4) A representation used for a compound having a chain of atoms bonded together. The atoms are drawn in, but the two-electron bond lines and lone pairs on heteroatoms are generally omitted.

Conjugate acid (9.2) The product formed by the gain of a proton by a base.

Conjugate acid–base pair (9.2) Two species that differ by the presence of a proton.

Conjugate base (9.2) The product formed by loss of a proton from an acid.

Conjugated protein (21.7) A compound composed of a protein unit and a nonprotein molecule.

Constitutional isomers (12.2) Isomers that differ in the way the atoms are connected to each other.

Conversion factor (1.7) A term that converts a quantity in one unit to a quantity in another unit.

Cori cycle (24.6) The cycling of lactate from the muscle to the liver, where it is re-oxidized to pyruvate and converted to glucose. Glucose is then transported back to the muscle, where it undergoes glycolysis.

Coupled reactions (23.3) Pairs of reactions that occur together. The energy released by an energetically favorable reaction drives an energetically unfavorable reaction.

Covalent bond (3.1) A chemical bond that results from the sharing of electrons between two atoms.

Critical mass (10.6) The amount of a radioactive element required to sustain a chain reaction.

Cross formula (15.6) A Fischer projection formula that replaces a chirality center with a cross. The horizontal lines represent wedged bonds and the vertical lines represent dashed bonds.

Crystalline solid (7.9) A solid with a regular arrangement of particles—atoms, molecules, or ions—with a repeating structure.

C-Terminal amino acid (21.4) In a peptide, the amino acid with the free –COO^- group on the α carbon.

Cubic centimeter (1.4) A unit of volume equal to one milliliter; one cubic centimeter = 1 cm^3 = 1 cc.

Curie (10.4) A unit used to measure radioactivity and equal to 3.7×10^{10} disintegrations/s. A curie corresponds to the decay rate of one gram of the element radium.

Cyclic hemiacetal (16.8) A compound formed by an intramolecular reaction of a hydroxyl group and an aldehyde or ketone.

Cycloalkane (12.1) A compound with the general formula C_nH_{2n} that contains carbons joined in one or more rings.

D

Dalton's law (7.6) A law that states that the total pressure (P_{total}) of a gas mixture is equal to the sum of the partial pressures of its component gases.

***d* Block** (2.7) A group of elements consisting of the 10 columns of transition metals in the periodic table. The *d* subshell is filled last in these elements.

Decarboxylation (24.2) The loss of carbon dioxide (CO_2) from a compound.

Dehydration (14.5) The loss of water (H_2O) from a starting material.

Deletion mutation (22.9) The loss of one or more nucleotides from a DNA molecule.

Denaturation (21.8) The process of altering the shape of a protein without breaking the amide bonds that form the primary structure.

Dendrites (18.8) Numerous small filaments at the end of a neuron.

Density (1.10) A physical property that relates the mass of a substance to its volume; density = g/(mL or cc).

Deoxyribonucleic acid (DNA, 22.1, 22.3) A polymer of deoxyribonucleotides that stores the genetic information of an organism and transmits that information from one generation to another.

Deoxyribonucleoside (22.1) A compound that contains the monosaccharide 2-deoxyribose and a purine or pyrimidine base.

Deoxyribonucleotide (22.1) A compound that contains the monosaccharide 2-deoxyribose bonded to a purine or pyrimidine base, as well as a phosphate at the 5'-OH group.

Deposition (7.10) The conversion of a gas directly to a solid.

Deuterium (2.3) A hydrogen atom having one proton and one neutron, giving it a mass number of two.

Dialysis (8.8) A process that involves the selective passage of substances across a semipermeable membrane, called a dialyzing membrane.

Diastereomers (15.7) Stereoisomers that are not mirror images of each other.

Diatomic molecule (4.1) A molecule that contains two atoms. Hydrogen (H_2) is a diatomic molecule.

Dilution (8.6) The addition of solvent to a solution to decrease the concentration of solute.

Diol (14.3) A compound with two hydroxyl groups, also called a glycol.

Dipeptide (21.4) A peptide formed from two amino acids joined together by one amide bond.

Dipole (4.7) The separation of charge in a bond or molecule.

Dipole–dipole interactions (7.7) The attractive intermolecular forces between the permanent dipoles of two polar molecules.

Diprotic acid (9.1) An acid that contains two acidic protons.

Disaccharide (20.1) A carbohydrate composed of two monosaccharides joined together.

Dissociation (9.3) The process that occurs when an acid or base dissolves in water to form ions.

Disulfide (14.10) A compound that contains a sulfur–sulfur bond.

D Monosaccharide (20.2) A carbohydrate that has the OH group on the chirality center farthest from the carbonyl on the right.

DNA fingerprinting (22.10) A technique in which DNA is amplified using PCR and cut into fragments that are separated by size using gel electrophoresis. This forms a set of horizontal bands, each band corresponding to a segment of DNA, sorted from low to high molecular weight.

Double bond (4.2) A multiple bond that contains four electrons—that is, two two-electron bonds.

Double-headed arrow (13.9) An arrow drawn between resonance structures.

E

Eicosanoids (19.11) A group of biologically active compounds containing 20 carbon atoms derived from arachidonic acid.

Electrolyte (8.1) A substance that conducts an electric current in water.

Electron (2.2) A negatively charged subatomic particle.

Electron cloud (2.2) The space surrounding the nucleus of an atom, which contains electrons and comprises most of the volume of an atom.

Electron transport chain (23.6) A series of reactions that transfers electrons from reduced coenzymes to progressively stronger oxidizing agents, ultimately converting oxygen to water.

Electron-dot symbol (2.7) A symbol that shows the number of valence electrons around an atom.

Electronegativity (4.7) A measure of an atom's attraction for electrons in a bond.

Electronic configuration (2.6) The arrangement of electrons in an atom's orbitals.

Element (1.3) A pure substance that cannot be broken down into simpler substances by a chemical reaction.

Elimination (14.5) A reaction in which elements of the starting material are "lost" and a new multiple bond is formed.

Enantiomers (15.2) Mirror images that are not superimposable.

Endothermic reaction (6.2) A chemical reaction where ΔH is positive (+) and energy is absorbed.

Energy (6.1) The capacity to do work.

Energy diagram (6.3) A schematic representation of the energy changes in a reaction, which plots energy on the vertical axis and the progress of the reaction—the reaction coordinate—on the horizontal axis.

Energy of activation (6.3) The difference in energy between the reactants and the transition state; symbolized by E_a.

English system of measurement (1.4) A system of measurement used primarily in the United States in which units are not systematically related to each other and require memorization.

Enthalpy (6.2) The energy absorbed or released in any reaction—also called the heat of reaction and symbolized by ΔH.

Enzyme (6.4, 21.9) A biological catalyst composed of one or more chains of amino acids in a very specific three-dimensional shape.

Enzyme–substrate complex (21.9) A structure composed of a substrate bonded to the active site of an enzyme.

Epoxide (14.7) A cyclic ether containing an oxygen atom in a three-membered ring.

Equilibrium (6.5) A reaction that consists of forward and reverse reactions that have equal reaction rates, so the concentration of each species does not change.

Equilibrium constant (6.5) A characteristic value for a reaction at a given temperature and equal to the ratio of the concentrations of the products multiplied together to the concentrations of the reactants multiplied together. Each term is raised to a power equal to its coefficient in the balanced chemical equation.

Ester (11.5) A compound that contains an alkoxy group (OR) bonded directly to the carbonyl carbon; general structure RCOOR.

Estrogen (19.9) A hormone that controls the development of secondary sex characteristics in females and regulates the menstrual cycle.

Ether (14.1) A compound that has two alkyl groups bonded to an oxygen atom; general structure ROR.

Evaporation (7.8) The conversion of liquid molecules to gas molecules.

Exact number (1.5) A number that results from counting objects or is part of a definition.

Exothermic reaction (6.2) A reaction in which energy is released and ΔH is negative (–).

F

Facilitated transport (19.7) The process by which some ions and molecules travel through the channels in a cell membrane created by integral proteins.

Factor–label method (1.7) A method of using conversion factors to convert a quantity in one unit to a quantity in another unit.

FAD (Flavin adenine dinucleotide, 23.4) A biological oxidizing agent synthesized in cells from vitamin B_2, riboflavin. FAD is reduced by adding two hydrogen atoms, forming $FADH_2$.

Fahrenheit scale (1.9) One of three temperature scales in which water freezes at 32 °F and boils at 212 °F.

Fat (13.3, 19.4) A triacylglycerol with few double bonds, making it a solid at room temperature.

Fat-soluble vitamin (11.7, 19.10) A vitamin that dissolves in an organic solvent but is insoluble in water. Vitamins A, D, E, and K are fat soluble.

Fatty acid (13.3, 19.2) A carboxylic acid (RCOOH) with a long carbon chain, usually containing 12–20 carbon atoms.

f Block (2.7) A group of elements consisting of the two rows of inner transition metals. The f subshell is filled last in these elements.

Fermentation (24.4) The anaerobic conversion of glucose to ethanol and CO_2.

Fibrous protein (21.7) A water-insoluble protein composed of long linear polypeptide chains that are bundled together to form rods or sheets.

Fischer esterification (17.8) Treatment of a carboxylic acid (RCOOH) with an alcohol (R'OH) and an acid catalyst to form an ester (RCOOR').

Fischer projection formula (15.6) A method of drawing chiral compounds with the chirality center at the intersection of a cross. The horizontal bonds are assumed to be wedges and the vertical bonds are assumed to be dashed lines.

Formula weight (5.4) The sum of the atomic weights of all the atoms in a compound, reported in atomic mass units (amu).

Forward reaction (6.5) In equilibrium, a reaction that proceeds from left to right as drawn.

Freezing (7.10) The conversion of a liquid to a solid.

Freezing point depression (8.7) The decrease in the melting point of a liquid solution due to the presence of a nonvolatile solute.

Functional group (11.5) An atom or a group of atoms with characteristic chemical and physical properties.

G

Gamma (γ) ray (10.1) High-energy radiation released from a radioactive nucleus.

Gas (1.2) A state of matter that has no definite shape or volume. The particles of a gas move randomly and are separated by a distance much larger than their size.

Gas laws (7.3) A series of laws that relate the pressure, volume, and temperature of a gas.

Gay–Lussac's law (7.3) A gas law that states for a fixed amount of gas at constant volume, the pressure of the gas is proportional to its Kelvin temperature.

Geiger counter (10.4) A small portable device used for measuring radioactivity.

Gene (22.1) A portion of a DNA molecule responsible for the synthesis of a single protein.

Genetic code (22.7) The sequence of nucleotides in mRNA (coded in triplets) that specifies the amino acid sequence of a protein. Each triplet is called a codon.

Genetic disease (22.9) A disease resulting from a mutation that causes a condition to be inherited from one generation to another.

Genetic engineering (22.10) The process of manipulating DNA in the laboratory that allows a gene in one organism to be spliced into the DNA of another organism.

Globular protein (21.7) A protein that is coiled into a compact shape with a hydrophilic outer surface to make it water soluble.

Glucogenic amino acid (24.9) An amino acid that can be used to synthesize glucose.

Gluconeogenesis (24.6) The synthesis of glucose from noncarbohydrate sources—lactate, amino acids, or glycerol.

Glycol (14.3) A compound with two hydroxyl groups, also called a diol.

Glycolysis (24.3) A linear, 10-step pathway that converts glucose, a six-carbon monosaccharide, to two three-carbon pyruvate molecules.

α Glycoside (20.5) A monosaccharide with the OR group on the anomeric carbon oriented below the plane of the ring.

β Glycoside (20.5) A monosaccharide with the OR group on the anomeric carbon oriented above the plane of the ring.

Glycosidic linkage (20.5) The acetal C–O bond that joins two monosaccharides together.

Gram (1.4) The basic unit of mass in the metric system; abbreviated as g.

Gray (10.4) A unit that measures absorbed radiation; abbreviated as Gy.

Greenhouse gas (12.8) A gas that absorbs thermal energy that normally radiates from the earth's surface, and redirects it back to the surface.

Ground state (2.6) The lowest energy arrangement of electrons.

Group (2.4) A column in the periodic table.

Group number (2.4) A number that identifies a particular column in the periodic table.

H

Half reaction (5.8) An equation written for an individual oxidation or reduction that shows how many electrons are gained or lost.

Half-life (10.3) The time it takes for one-half of a sample to decay.

Halogenation (13.6) The reaction of a compound with a halogen, X_2, such as the addition of X_2 to an alkene.

Halogen (2.4) An element located in group 7A (group 17) of the periodic table. Halogens include fluorine (F), chlorine (Cl), bromine (Br), iodine (I), and astatine (At).

Hardening (13.7) The reaction of an unsaturated liquid vegetable oil with hydrogen to form a fat with a higher melting point.

Haworth projection (20.3) A planar, six-membered ring used to represent the cyclic hemiacetal of glucose and other sugars.

Heat of fusion (7.10) The amount of energy needed to melt one gram of a substance.

Heat of reaction (6.2) The energy absorbed or released in any reaction and symbolized by ΔH—also called the enthalpy change.

Heat of vaporization (7.10) The amount of energy needed to vaporize one gram of a substance.

α-Helix (21.6) A secondary structure of a protein formed when a peptide chain twists into a right-handed or clockwise spiral.

Heme (21.7) A complex organic compound containing an Fe^{2+} ion complexed with a large nitrogen-containing ring system.

Hemiacetal (16.8) A compound that contains an OH group (hydroxyl) and an OR group (alkoxy) bonded to the same carbon.

Henry's law (8.3) A law that states that the solubility of a gas in a liquid is proportional to the partial pressure of the gas above the liquid.

Heteroatom (11.2) Any atom in an organic compound that is not carbon or hydrogen.

Heterocycle (14.7) A ring that contains a heteroatom.

Heterogeneous mixture (8.1) A mixture that does not have a uniform composition throughout a sample.

Hexose (20.2) A monosaccharide with six carbons.

High-density lipoprotein (19.8) A spherical particle that transports cholesterol from the tissues to the liver.

Homogeneous mixture (8.1) A mixture that has a uniform composition throughout a sample.

Hormone (18.9, 19.9) A compound synthesized in one part of an organism, which then travels through the bloodstream to elicit a response at a target tissue or organ.

Hybrid (13.9) A composite of two or more resonance forms.

Hydration (13.6) The addition of water to a molecule.

Hydrocarbon (11.5) A compound that contains only the elements of carbon and hydrogen.

Hydrochlorofluorocarbon (14.9) A compound such as CF_3CHCl_2 that contains the elements of H, Cl, and F bonded to carbon.

Hydrofluorocarbon (14.9) A compound such as FCH_2CF_3 that contains the elements of H and F bonded to carbon.

Hydrogen bonding (7.7) An attractive intermolecular force that occurs when a hydrogen atom bonded to O, N, or F is electrostatically attracted to an O, N, or F atom in another molecule.

Hydrogenation (13.6) The addition of hydrogen (H_2) to an alkene.

Hydrohalogenation (13.6) The addition of HX (X = Cl or Br) to an alkene.

Hydrolysis (16.8) A cleavage reaction that uses water.

Hydrolyzable lipid (19.1) A lipid that can be converted to smaller molecules by hydrolysis with water.

Hydronium ion (3.6) The H_3O^+ ion.

Hydrophilic (19.2) The polar part of a molecule that is attracted to water.

Hydrophobic (19.2) The nonpolar part of a molecule (C–C and C–H bonds) that is not attracted to water.

Hydroxide (3.6) The $^-$OH ion.

α-Hydroxy acid (17.4) A compound that contains a hydroxyl group on the α carbon to a carboxyl group.

Hydroxyl group (11.5) An OH group.

Hypertonic solution (8.8) A solution that has a higher osmotic pressure than body fluids.

Hypotonic solution (8.8) A solution that has a lower osmotic pressure than body fluids.

I

Ideal gas law (7.5) A gas law that relates the pressure (*P*), volume (*V*), temperature (*T*), and number of moles (*n*) of a gas in a single equation; $PV = nRT$, where *R* is a constant.

Incomplete combustion (12.8) An oxidation reaction that forms carbon monoxide (CO) instead of carbon dioxide (CO_2) because insufficient oxygen is available.

Induced-fit model (21.9) The binding of a substrate to an enzyme such that the shape of the active site adjusts to fit the shape of the substrate.

Inexact number (1.5) A number that results from a measurement or observation and contains some uncertainty.

Inhibitor (21.9) A molecule that causes an enzyme to lose activity.

Inner transition metal elements (2.4) A group of elements consisting of the lanthanides and actinides.

Insertion mutation (22.9) The addition of one or more nucleotides to a DNA molecule.

Intermolecular forces (7.7) The attractive forces that exist between molecules.

Ion (3.1) A charged species in which the number of protons and electrons in an atom is not equal.

Ion–dipole interaction (8.2) The attraction of an ion to a dipole in another molecule.

Ionic bond (3.1) A bond that results from the transfer of electrons from one element to another.

Ionic solid (7.9) A solid composed of oppositely charged ions in a regular arrangement.

Ionization energy (2.8) The energy needed to remove an electron from a neutral atom.

Ion–product constant (9.5) The product of the concentrations of H_3O^+ and $^-$OH in water or an aqueous solution—symbolized by K_w and equal to 1×10^{-14}.

Irreversible inhibitor (21.9) An inhibitor that covalently binds to an enzyme, permanently destroying its activity.

Isoelectric point (21.3) The pH at which an amino acid exists primarily in its neutral form; abbreviated as p*I*.

Isomerase (24.2) An enzyme that catalyzes the conversion of one isomer to another.

Isomers (12.2) Two different compounds with the same molecular formula.

Isotonic solution (8.8) Two solutions with the same osmotic pressure.

Isotopes (2.3) Atoms of the same element having a different number of neutrons.

IUPAC nomenclature (12.3) A systematic method of naming compounds developed by the International Union of Pure and Applied Chemistry.

J

Joule (6.1) A unit of measurement for energy; abbreviated as J, where 1 cal = 4.184 J.

K

Kelvin scale (1.9) A temperature scale commonly used by scientists. The Kelvin scale is divided into kelvins (K); K = °C + 273.

Ketogenesis (24.8) The synthesis of ketone bodies from acetyl CoA.

Ketogenic amino acid (24.9) An amino acid that cannot be used to synthesize glucose, but can be converted to ketone bodies.

Ketone (11.5) A compound that has two alkyl groups bonded to the carbonyl group; general structure RCOR.

Ketone bodies (24.8) Three compounds—acetoacetate, β-hydroxybutyrate, and acetone—formed when acetyl CoA levels exceed the capacity of the citric acid cycle.

Ketose (20.2) A monosaccharide with a carbonyl group at C2.

Ketosis (24.8) The accumulation of ketone bodies during starvation and uncontrolled diabetes.

Kinase (24.2) An enzyme that catalyzes the transfer of a phosphate group from one substrate to another.

Kinetic energy (6.1) The energy of motion.

Kinetic-molecular theory (7.2) A theory that describes the fundamental characteristics of gas particles.

L

Lactam (17.1) A cyclic amide.

Lactone (17.1) A cyclic ester.

Lagging strand (22.4) The strand of DNA synthesized in small fragments during replication, which are then joined together by a DNA ligase enzyme.

Lanthanides (2.4) A group of 14 elements in the periodic table beginning with the element cerium (*Z* = 58) and immediately following the element lanthanum (La).

Law of conservation of energy (6.1) A law that states that the total energy in a system does not change. Energy cannot be created or destroyed.

Law of conservation of mass (5.1) A law that states that atoms cannot be created or destroyed in a chemical reaction.

LD_{50} (10.4) The lethal dose of radiation (or a poison) that kills 50% of a population.

Le Châtelier's principle (6.6) A principle that states that if a chemical system at equilibrium is disturbed or stressed, the system will react in the direction that counteracts the disturbance or relieves the stress.

Leading strand (22.4) The strand of DNA that grows continuously during replication.

Lecithin (19.6) A phosphoacylglycerol in which the identity of the R group esterified to the phosphodiester is $-CH_2CH_2N(CH_3)_3^+$. A lecithin is also called a phosphatidylcholine.

Leukotriene (19.11) A molecule synthesized in several steps from arachidonic acid, which contributes to the asthmatic response by constricting smooth muscles, especially in the lungs.

Lewis structure (4.1) An electron-dot structure for a molecule that shows the location of all valence electrons in the molecule, both the shared electrons in bonds and the nonbonded electron pairs.

"Like dissolves like" (11.6) The principle in solubility that compounds dissolve in solvents having similar types of intermolecular forces.

Lipid (17.9, 19.1) A biomolecule that is soluble in organic solvents and insoluble in water.

Lipid bilayer (19.7) The basic structure of the cell membrane formed from two layers of phospholipids having their ionic heads oriented on the outside and their nonpolar tails on the inside.

Lipoprotein (19.8) A small water-soluble spherical particle composed of proteins and lipids.

Liquid (1.2) A state of matter that has a definite volume, but takes on the shape of the container it occupies. The particles of a liquid are close together but they can randomly move past each other.

Liter (1.4) The basic unit of volume in the metric system; abbreviated as L.

L Monosaccharide (20.2) A carbohydrate that has the OH group on the chirality center farthest from the carbonyl on the left.

Lock-and-key model (21.9) The binding of a substrate to a rigid active site, such that the three-dimensional geometry of the substrate exactly matches the shape of the active site.

London dispersion forces (7.7) Very weak intermolecular interactions due to the momentary changes in electron density in a molecule.

Lone pair (4.1) An unshared electron pair.

Low-density lipoprotein (19.8) A spherical particle containing proteins and lipids, which transports cholesterol from the liver to the tissues.

M

Main group element (2.4) An element in groups 1A–8A of the periodic table.

Major mineral (macronutrient, 2.1) One of the seven elements present in the body in small amounts (0.1–2% by mass) and needed in the daily diet.

Markovnikov's rule (13.6) The rule that states that in the addition of HX to an unsymmetrical alkene, the H atom bonds to the less substituted carbon atom.

Mass (1.4) A measure of the amount of matter in an object.

Mass number (2.2) The total number of protons and neutrons in a nucleus; symbolized as A.

Matter (1.1) Anything that has mass and takes up volume.

Melting (7.10) The conversion of a solid to a liquid.

Melting point (7.7) The temperature at which a solid is converted to the liquid phase.

Messenger RNA (mRNA, 22.5) The carrier of information from DNA (in the cell nucleus) to the ribosomes (in the cell cytoplasm). Each gene of a DNA molecule corresponds to a specific mRNA molecule.

Meta isomer (13.10) A 1,3-disubstituted benzene.

Metabolism (23.1) The sum of all of the chemical reactions that take place in an organism.

Metal (2.1) A shiny element that is a good conductor of heat and electricity.

Metallic solid (7.9) A lattice of metal cations surrounded by a cloud of electrons that move freely.

Metalloid (2.1) An element with properties intermediate between a metal and a nonmetal. Metalloids include boron (B), silicon (Si), germanium (Ge), arsenic (As), antimony (Sb), tellurium (Te), and astatine (At).

Meter (1.4) A unit used to measure length; abbreviated as m.

Metric system (1.4) A measurement system in which each type of measurement has a base unit and all other units are related to the base unit by a prefix that indicates if the unit is larger or smaller than the base unit.

Micelle (17.6) A spherical droplet formed when soap is dissolved in water. The ionic heads of the soap molecules are oriented on the surface and the nonpolar tails are packed in the interior.

Millimeters mercury (7.2) A unit used to measure pressure; abbreviated as mm Hg and also called "torr."

Mitochondrion (23.1) A small sausage-shaped organelle within a cell in which energy production takes place.

Mixture (1.3) Matter composed of more than one component.

Molar mass (5.4) The mass of one mole of any substance, reported in grams per mole.

Molarity (8.5) The number of moles of solute per liter of solution; abbreviated as M.

Molecular formula (4.2) A formula that shows the number and identity of all of the atoms in a compound, but it does not indicate what atoms are bonded to each other.

Molecular solid (7.9) A solid composed of individual molecules arranged regularly.

Molecular weight (5.4) The formula weight of a covalent compound.

Molecule (4.1) A discrete group of atoms that are held together by covalent bonds.

Monomers (13.8) Small molecules that covalently bond together to form polymers.

Monoprotic acid (9.1) An acid that contains one acidic proton.

Monosaccharide (20.1) A carbohydrate that cannot be hydrolyzed to simpler compounds.

Multiple bond (4.2) A chemical bond that contains four or six electrons—that is, a double or a triple bond.

Mutarotation (20.3) The process by which a single anomer of a monosaccharide equilibrates to a mixture of anomers.

Mutation (22.9) A change in the nucleotide sequence in a molecule of DNA.

N

NAD⁺ (nicotinamide adenine dinucleotide, 16.6, 23.4) A biological oxidizing agent and coenzyme synthesized from the vitamin niacin. NAD^+ and NADH are interconverted by oxidation and reduction reactions.

NADH (16.6, 23.4) A biological reducing agent and coenzyme formed when NAD^+ is reduced.

Net ionic equation (9.7) An equation that contains only the species involved in a reaction.

Network solid (7.9) A solid composed of a vast number of atoms covalently bonded together, forming sheets or three-dimensional arrays.

Neuron (18.8) A nerve cell.

Neurotransmitter (18.8) A chemical messenger that transmits nerve impulses from one nerve cell to another cell.

Neutral solution (9.5) Any solution that has an equal concentration of H_3O^+ and ^-OH ions and a pH = 7.

Neutralization reaction (9.7) An acid–base reaction that produces a salt and water as products.

Neutron (2.2) A neutral subatomic particle in the nucleus.

Nitration (13.13) Substitution of a nitro group (NO_2) for a hydrogen.

Nitro group (13.13) An NO_2 group.

Noble gases (2.4) Elements located in group 8A (group 18) of the periodic table. The noble gases are helium (He), neon (Ne), argon (Ar), krypton (Kr), xenon (Xe), and radon (Rn).

Nomenclature (3.4) The system of assigning an unambiguous name to a compound.

Nonbonded electron pair (4.1) An unshared electron pair or lone pair.

Noncompetitive inhibitor (21.9) An inhibitor that binds to an enzyme but does not bind at the active site.

Nonelectrolyte (8.1) A substance that does not conduct an electric current when dissolved in water.

Nonhydrolyzable lipid (19.1) A lipid that cannot be cleaved into smaller units by aqueous hydrolysis.

Nonmetal (2.1) An element that does not have a shiny appearance and poorly conducts heat and electricity.

Nonpolar bond (4.7) A bond in which electrons are equally shared.

Nonpolar molecule (11.6) A molecule that has no net dipole.

Nonreducing sugar (20.4) A sugar that does not react with Benedict's reagent.

Nonvolatile (8.7) Not readily vaporized.

Normal boiling point (7.8) The temperature at which the vapor pressure above a liquid equals 760 mm Hg.

N-Terminal amino acid (21.4) In a peptide, the amino acid with the free $-NH_3^+$ group on the α carbon.

Nuclear fission (10.6) The splitting apart of a nucleus into lighter nuclei and neutrons.

Nuclear fusion (10.6) The joining together of two nuclei to form a larger nucleus.

Nuclear reaction (10.1) A reaction that involves the subatomic particles of the nucleus.

Nucleic acid (22.1, 22.2) An unbranched polymer composed of nucleotides. DNA and RNA are nucleic acids.

Nucleoside (22.1) A compound formed by joining the anomeric carbon of a monosaccharide with a nitrogen atom of a purine or pyrimidine base.

Nucleotide (22.1) A compound formed by adding a phosphate group to the 5'-OH of a nucleoside.

Nucleus (2.2) The dense core of the atom that contains protons and neutrons.

Nylon (17.10) A condensation polymer that contains many amide bonds.

Octet rule (3.2) The rule in bonding that states that main group elements are especially stable when they possess eight electrons (an octet) in the outer shell.

Oil (13.3, 19.4) A triacylglycerol that is liquid at room temperature.

Omega-*n* acid (19.2) An unsaturated fatty acid where *n* is the carbon at which the first double bond occurs in the carbon chain. The numbering begins at the end of the chain with the CH_3 group.

Orbital (2.5) A region of space where the probability of finding an electron is high.

Organic chemistry (11.1) The study of compounds that contain the element carbon.

Ortho isomer (13.10) A 1,2-disubstituted benzene.

Osmosis (8.8) The selective diffusion of water (and small molecules) across a semipermeable membrane from a less concentrated solution to a more concentrated solution.

Osmotic pressure (8.8) The pressure that prevents the flow of additional solvent into a solution on one side of a semipermeable membrane.

Oxidation (5.8, 12.8) The loss of electrons from an atom. Oxidation may result in a gain of oxygen atoms or a loss of hydrogen atoms.

β-Oxidation (24.7) A process in which two-carbon acetyl CoA units are sequentially cleaved from a fatty acid.

Oxidative deamination (24.9) The process by which the C–H and $C-NH_3^+$ bonds on the α carbon of an amino acid are converted to $C=O$ and an ammonium ion (NH_4^+).

Oxidative phosphorylation (23.6) The process by which the energy released from the oxidation of reduced coenzymes is used to convert ADP to ATP using the enzyme ATP synthase.

Oxidizing agent (5.8) A compound that gains electrons (i.e., is reduced), causing another compound to be oxidized.

Para isomer (13.10) A 1,4-disubstituted benzene.

Parent name (12.4) The root that indicates the number of carbons in the longest continuous carbon chain in a molecule.

Partial hydrogenation (13.7) The hydrogenation of some, but not all, of the double bonds in a molecule.

Partial pressure (7.6) The pressure exerted by one component of a mixture of gases.

Parts per million (8.4) A concentration term (abbreviated ppm)—the number of "parts" in 1,000,000 parts of solution.

***p* Block** (2.7) The elements in groups 3A–8A (except helium) in the periodic table. The *p* subshell is filled last in these elements.

Penicillin (17.11) An antibiotic that contains a β-lactam and interferes with the synthesis of the bacterial cell wall.

Pentose (20.2) A monosaccharide with five carbons.

Peptide (21.4) A compound that contains many amino acids joined together by amide bonds.

Peptide bond (21.4) An amide bond in peptides and proteins.

Percent yield (5.7) The amount of product actually formed in a particular reaction divided by the theoretical yield, multiplied by 100%.

Period (2.4) A row in the periodic table.

Periodic table (2.1) A schematic arrangement of all known elements that groups elements with similar properties.

Petroleum (12.6) A complex mixture of compounds, most of which are hydrocarbons containing 1–40 carbon atoms.

pH scale (9.6) The scale used to report the H_3O^+ concentration; $pH = -\log [H_3O^+]$.

Pheromone (12.1) A chemical substance used for communication in a specific animal species, most commonly an insect population.

Phosphate (3.6) A PO_4^{3-} anion.

Phosphatidylcholine (19.6) A phosphoacylglycerol in which the identity of the R group esterified to the phosphodiester is $-CH_2CH_2N(CH_3)_3^+$; also called a lecithin.

Phosphatidylethanolamine (19.6) A phosphoacylglycerol in which the identity of the R group esterified to the phosphodiester is $-CH_2CH_2NH_3^+$; also called a cephalin.

Phosphoacylglycerol (19.6) A lipid with a glycerol backbone that contains two of the hydroxyls esterified with fatty acids and the third hydroxyl as part of a phosphodiester.

Phosphodiester (19.6) A derivative of phosphoric acid (H_3PO_4) that is formed by replacing two of the H atoms by R groups.

Phospholipid (19.6) A lipid that contains a phosphorus atom.

Phosphorylation (23.3) A reaction that adds a phosphate group to a molecule.

Physical properties (1.2) Those properties of a substance that can be observed or measured without changing the composition of the material.

Plasmid (22.10) A circular, double-stranded DNA molecule isolated from bacteria.

β-Pleated sheet (21.6) A secondary structure formed when two or more peptide chains, called strands, line up side-by-side.

Point mutation (22.9) The substitution of one nucleotide for another.

Polar bond (4.7) A bond in which electrons are unequally shared and pulled towards the more electronegative element.

Polar molecule (11.6) A molecule that contains a net dipole.

Polyamide (17.10) A class of condensation polymer that contains many amide bonds.

Polyatomic ion (3.6) A cation or anion that contains more than one atom.

Polycyclic aromatic hydrocarbon (13.11) A compound containing two or more benzene rings that share a carbon–carbon bond.

Polyester (17.10) A class of condensation polymer that contains many ester bonds.

Polymer (13.8) A large molecule made up of repeating units of smaller molecules—called monomers—covalently bonded together.

Polymerase chain reaction (22.10) A technique that produces exact copies of a fragment of DNA.

Polymerization (13.8) The joining together of monomers to make polymers.

Polynucleotide (22.2) A polymer of nucleotides that contains a sugar–phosphate backbone.

Polysaccharide (20.1) Three or more monosaccharides joined together.

p **Orbital** (2.5) A dumbbell-shaped orbital higher in energy than an *s* orbital in the same shell.

Positron (10.1) A radioactive particle that has a negligible mass and a +1 charge.

Postsynaptic neuron (18.8) A nerve cell that contains the receptors that bind a neurotransmitter.

Potential energy (6.1) Energy that is stored.

Pressure (7.2) The force (*F*) exerted per unit area (*A*); symbolized by *P*.

Presynaptic neuron (18.8) A nerve cell that releases a neurotransmitter.

Primary (1°) alcohol (14.2) An alcohol having the general structure RCH_2OH.

Primary (1°) alkyl halide (14.9) An alkyl halide having the general structure RCH_2X.

Primary (1°) amide (17.1) A compound having the general structure $RCONH_2$.

Primary (1°) amine (18.1) A compound having the general structure RNH_2.

Primary (1°) carbon (12.2) A carbon atom bonded to one other carbon.

Primary structure (21.6) The particular sequence of amino acids that are joined together by peptide bonds in a protein.

Product (5.1) A substance formed in a chemical reaction.

Progestin (19.9) A hormone responsible for the preparation of the uterus for implantation of a fertilized egg.

Prostaglandins (17.4, 19.11) A group of carboxylic acids that contain a five-membered ring, are synthesized from arachidonic acid, and have a wide range of biological activities.

Proteins (21.1) Biomolecules that contain many amide bonds, formed by joining amino acids together.

Proton (2.2) A positively (+) charged subatomic particle that resides in the nucleus of the atom.

Proton transfer reaction (9.2) A Brønsted–Lowry acid–base reaction in which a proton is transferred from an acid to a base.

Pure substance (1.3) A substance that contains a single component, and has a constant composition regardless of the sample size.

 Q

Quaternary ammonium ion (18.1) A cation with the general structure R_4N^+.

Quaternary (4°) carbon (12.2) A carbon atom that is bonded to four other carbons.

Quaternary structure (21.6) The shape adopted when two or more folded polypeptide chains come together into one protein complex.

R

Racemic mixture (15.5) An equal mixture of two enantiomers.

Rad (10.4) The radiation absorbed dose; the amount of radiation absorbed by one gram of a substance.

Radioactive decay (10.2) The process by which an unstable radioactive nucleus emits radiation, forming a nucleus of new composition.

Radioactive isotope (10.1) An isotope that is unstable and spontaneously emits energy to form a more stable nucleus.

Radioactivity (10.1) The energy emitted by a radioactive isotope.

Radiocarbon dating (10.3) A method to date artifacts that is based on the ratio of the radioactive carbon-14 isotope to the stable carbon-12 isotope.

Reactant (5.1) The starting material in a reaction.

Reaction rate (6.3) A measure of how fast a chemical reaction occurs.

Recombinant DNA (22.10) Synthetic DNA that contains segments from more than one source.

Redox reaction (5.8) A reaction that involves the transfer of electrons from one element to another.

Reducing agent (5.8) A compound that loses electrons (i.e., is oxidized), causing another compound to be reduced.

Reducing sugar (20.4) A carbohydrate that is oxidized with Benedict's reagent.

Reduction (5.8, 12.8) The gain of electrons by an atom. Reduction may result in the loss of oxygen atoms or the gain of hydrogen atoms.

Refining (12.6) The distillation of crude petroleum to form usable fractions that differ in boiling point.

Rem (10.4) The radiation equivalent for man; the amount of radiation absorbed by a substance that also factors in its energy and potential to damage tissue.

Replication (22.4) The process by which DNA makes a copy of itself when a cell divides.

Resonance structures (4.4) Two Lewis structures having the same arrangement of atoms but a different arrangement of electrons.

Restriction endonuclease (22.10) An enzyme that cuts DNA at a particular sequence of bases.

Retrovirus (22.11) A virus that contains a core of RNA.

Reverse reaction (6.5) In equilibrium, a reaction that proceeds from right to left as drawn.

Reverse transcription (22.11) A process by which a retrovirus produces DNA from RNA.

Reversible inhibitor (21.9) An inhibitor that binds to an enzyme, but enzyme activity is restored when the inhibitor is released.

Reversible reaction (6.5) A reaction that can occur in either direction, from reactants to products or from products to reactants.

Ribonucleic acid (RNA, 22.1, 22.5) A polymer of ribonucleotides that translates genetic information to protein synthesis.

Ribonucleoside (22.1) A compound that contains the monosaccharide ribose and a purine or pyrimidine base.

Ribonucleotide (22.1) A compound that contains the monosaccharide ribose bonded to either a purine or pyrimidine base as well as a phosphate at the 5'-OH group.

Ribosomal RNA (rRNA, 22.5) The most abundant type of RNA. rRNA is found in the ribosomes of the cell and provides the site where polypeptides are assembled during protein synthesis.

S

Saponification (17.9, 19.5) The basic hydrolysis of an ester.

Saturated fatty acids (19.2) Fatty acids that have no double bonds in their long hydrocarbon chains.

Saturated hydrocarbon (12.1) An alkane with molecular formula C_nH_{2n+2} that contains a chain of carbon atoms but no rings.

Saturated solution (8.2) A solution that has the maximum number of grams of solute that can be dissolved.

s Block (2.7) Elements located in groups 1A and 2A of the periodic table as well as the element helium. The s subshell is filled last in these elements.

Scientific notation (1.6) A system in which numbers are written as $y \times 10^x$, where y is a number between 1 and 10 and x can be either positive or negative.

Secondary (2°) alcohol (14.2) An alcohol having the general structure R_2CHOH.

Secondary (2°) alkyl halide (14.9) An alkyl halide having the general structure R_2CHX.

Secondary (2°) amide (17.1) A compound that has the general structure RCONHR'.

Secondary (2°) amine (18.1) A compound that has the general structure R_2NH.

Secondary (2°) carbon (12.2) A carbon atom that is bonded to two other carbons.

Secondary structure (21.6) The three-dimensional arrangement of localized regions of a protein. The α-helix and β-pleated sheet are two kinds of secondary structure.

Semipermeable membrane (8.8) A membrane that allows only certain molecules to pass through.

Shell (2.5) A region where an electron that surrounds a nucleus is confined. A shell is also called a principal energy level.

SI units (1.4) The International System of Units formally adopted as the uniform system of units for the sciences.

Sievert (10.4) A unit that measures absorbed radiation; abbreviated as Sv.

Significant figures (1.5) All of the digits in a measured number, including one estimated digit.

Skeletal structure (11.4) A shorthand method used to draw organic compounds in which carbon atoms are assumed to be at the junction of any two lines or at the end of a line, and all H's on C's are omitted.

Soap (17.6, 19.5) A salt of a long-chain carboxylic acid.

Solid (1.2) A state of matter that has a definite shape and volume. The particles of a solid lie close together, and are arranged in a regular, three-dimensional array.

Solubility (8.2) The amount of solute that dissolves in a given amount of solvent.

Solute (8.1) The substance present in the lesser amount in a solution.

Solution (8.1) A homogeneous mixture that contains small particles. Liquid solutions are transparent.

Solvation (8.2) The process of surrounding particles of a solute with solvent molecules.

Solvent (8.1) The substance present in the larger amount in a solution.

s Orbital (2.5) A spherical orbital that is lower in energy than other orbitals in the same shell.

Specific gravity (1.10) A unitless quantity that compares the density of a substance with the density of water at the same temperature.

Spectator ion (9.7) An ion that appears on both sides of an equation but undergoes no change in a reaction.

Sphingomyelin (19.6) A phospholipid derived from sphingosine, which contains a single fatty acid bonded with an amide bond to the carbon backbone.

SSRI (18.8) A class of antidepressants that acts by inhibiting the reuptake of serotonin by the presynaptic neuron. SSRI is an abbreviation for "selective serotonin reuptake inhibitor."

Standard molar volume (7.4) The volume of one mole of any gas at STP—22.4 L.

States of matter (1.2) The forms in which most matter exists—that is, gas, liquid, and solid.

Stereochemistry (15.1) The three-dimensional structure of molecules.

Stereoisomers (13.3) Isomers that differ only in their three-dimensional arrangement of atoms.

Steroid (19.8) A lipid whose carbon skeleton contains three six-membered rings and one five-membered ring.

STP (7.4) Standard conditions of temperature and pressure—1 atm (760 mm Hg) for pressure and 273 K (0 °C) for temperature.

Straight-chain alkane (12.2) An alkane that has all of its carbons in one continuous chain.

Sublimation (7.10) A phase change in which the solid phase enters the gas phase without passing through the liquid state.

Subshell (2.5) A sublevel within a shell designated by the letters s, p, d, or f, and containing one type of orbital.

Substitution reaction (13.13) A reaction in which an atom is replaced by another atom or a group of atoms.

Sulfate (3.6) An SO_4^{2-} ion.

Sulfhydryl group (14.1) An SH group.

Sulfonation (13.13) The substitution of SO_3H for a hydrogen.

Supersaturated solution (8.3) A solution that contains more than the predicted maximum amount of solute at a given temperature.

Surface tension (7.8) A measure of the resistance of a liquid to spread out.

Synapse (18.8) The gap between neurons across which neurotransmitters act.

T

Temperature (1.9) A measure of how hot or cold an object is.

Tertiary (3°) alcohol (14.2) An alcohol that has the general structure R_3COH.

Tertiary (3°) alkyl halide (14.9) An alkyl halide that has the general structure R_3CX.

Tertiary (3°) amide (17.1) A compound that has the general structure RCONR'$_2$.

Tertiary (3°) amine (18.1) A compound that has the general structure R$_3$N.

Tertiary (3°) carbon (12.2) A carbon atom that is bonded to three other carbons.

Tertiary structure (21.6) The three-dimensional shape adopted by an entire peptide chain.

Tetrose (20.2) A monosaccharide with four carbons.

Theoretical yield (5.7) The amount of product expected from a given amount of reactant based on the coefficients in the balanced chemical equation.

Thiol (14.1) A compound that contains a sulfhydryl group (SH group) bonded to a tetrahedral carbon atom; general structure RSH.

Titration (9.9) A technique for determining an unknown molarity of an acid by adding a base of known molarity to a known volume of acid.

Tollens reagent (16.5) An oxidizing agent that contains silver(I) oxide (Ag$_2$O) in aqueous ammonium hydroxide (NH$_4$OH).

Trace element (micronutrient, 2.1) An element required in the daily diet in small quantities—usually less than 15 mg.

Trans isomer (13.3) An alkene with two R groups on opposite sides of a double bond.

Transamination (24.9) The transfer of an amino group from an amino acid to an α-keto acid.

Transcription (22.6) The process that synthesizes RNA from DNA.

Transfer RNA (tRNA, 22.5) The smallest type of RNA, which brings a specific amino acid to the site of protein synthesis on a ribosome.

Transition metal element (2.4) An element contained in one of the 10 columns in the periodic table numbered 1B–8B.

Transition state (6.3) The unstable energy maximum located at the top of the energy hill in an energy diagram.

Translation (22.8) The synthesis of proteins from RNA.

Triacylglycerol (17.9, 19.4) A triester formed from glycerol and three molecules of fatty acids.

Triol (14.3) A compound with three hydroxyl groups.

Triose (20.2) A monosaccharide with three carbons.

Tripeptide (21.4) A peptide that contains three amino acids joined together by two amide bonds.

Triple bond (4.2) A multiple bond that contains six electrons—that is, three two-electron bonds.

Triprotic acid (9.1) An acid that contains three acidic protons.

Tritium (2.3) A hydrogen atom that has one proton and two neutrons, giving it a mass number of three.

U

Universal gas constant (7.5) The constant, symbolized by *R*, that equals the product of the pressure and volume of a gas, divided by the product of the number of moles and Kelvin temperature; $R = PV/nT$.

Unpaired electron (2.6) A single electron.

Unsaturated fatty acid (19.2) A fatty acid that has one or more double bonds in its long hydrocarbon chain.

Unsaturated hydrocarbon (13.1) A compound that contains fewer than the maximum number of hydrogen atoms per carbon.

Unsaturated solution (8.2) A solution that has less than the maximum number of grams of solute.

Urea cycle (24.9) The process by which an ammonium ion is converted to urea, (NH$_2$)$_2$C=O.

V

Valence electron (2.7) An electron in the outermost shell that takes part in bonding and chemical reactions.

Valence shell electron pair repulsion theory (4.6) A theory that predicts molecular geometry based on the fact that electron pairs repel each other; thus, the most stable arrangement keeps these groups as far away from each other as possible.

Vapor (7.8) Gas molecules formed from the evaporation of a liquid.

Vapor pressure (7.8) The pressure above a liquid exerted by gas molecules in equilibrium with the liquid phase.

Vaporization (7.10) The conversion of a liquid to a gas.

Vesicle (18.8) A small packet where neurotransmitters are stored, located in the filaments of the axon of a neuron near the synapse.

Virus (22.11) An infectious agent consisting of a DNA or RNA molecule that is contained within a protein coating.

Viscosity (7.8) A measure of a fluid's resistance to flow freely.

Vitamin (11.7) An organic compound that must be obtained in the diet and is needed in small amounts for normal cell function.

Volatile (8.7) Readily vaporized.

Volume/volume percent concentration (8.4) The number of milliliters of solute dissolved in 100 mL of solution.

W

Water-soluble vitamin (11.7) A vitamin with many polar bonds that dissolves in water.

Wax (19.3) An ester (RCOOR') formed from a fatty acid (RCOOH) and a high molecular weight alcohol (R'OH).

Weight (1.4) The force that matter feels due to gravity.

Weight/volume percent concentration (8.4) The number of grams of solute dissolved in 100 mL of solution.

X

X-ray (10.7) A high-energy form of radiation.

Z

Zaitsev rule (14.5) A rule that states that the major product in an elimination reaction is the alkene that has more alkyl groups bonded to the double bond.

Zwitterion (21.2) A neutral compound that contains both a positive and a negative charge.

Zymogen (21.9) The inactive precursor of an enzyme that is then converted to its active form when needed.

Credits

Front Matter
Author photo, p. vi: Daniel C. Smith.

Chapter 1
Opener: Paul Kuroda/SuperStock; 1.1a–b: The McGraw-Hill Companies, Inc./Jill Braaten, photographer; 1.2a: © Bob Krist/Corbis; 1.2b: © Corbis RF; 1.2c: © Gerry Ellis/Minden Pictures; 1.2d: © PhotoDisc/Getty; 1.3a: © Douglas Peebles Photography/Alamy; 1.3b: Daniel C. Smith; 1.3c: © Dr. Parvinder Sethi RF; p. 6(left): Daniel C. Smith; p. 6(middle, right): © The McGraw-Hill Companies, Inc./Jill Braaten, photographer; 1.4a: Daniel C. Smith; 1.4b: © Richard T. Nowitz/Photo Researchers, Inc.; 1.4c: © The McGraw-Hill Companies, Inc./Jill Braaten, photographer; 1.5a: Daniel C. Smith; 1.5b: © Keith Eng, 2008; 1.5c: © The McGraw-Hill Companies, Inc./Jill Braaten, photographer; 1.5d: Daniel C. Smith; p. 8: © Doug Wilson/Corbis; p. 10(top): © Chuck Savage/Corbis; p. 10(bottom), p. 11: © PhotoDisc/Getty RF; p. 12: Daniel C. Smith; p. 16: © Stockbyte/Getty RF; p. 20: © The McGraw-Hill Companies, Inc./Elite Images; p. 21: © PhotoDisc/Getty RF; pp. 22, 24 & 26: © The McGraw-Hill Companies, Inc./Jill Braaten, photographer.

Chapter 2
Opener: © PhotoLink/Photodisc/Getty Images RF; p. 35(silver): © Charles D. Winters/Photo Researchers, Inc.; p. 35(sodium): © Andrew Lambert Photography/Photo Researchers, Inc.; p. 35(mercury): © Dirk Wiersma/Photo Researchers, Inc.; p. 35(silicon): © Andrew Lambert Photography/Photo Researchers, Inc.; p. 35(arsenic): © Charles D. Winters/Photo Researchers, Inc.; p. 35(boron): © The McGraw-Hill Companies, Inc./Stephen Frisch, photographer; p. 35(carbon): © Charles D. Winters/Photo Researchers, Inc.; p. 35(sulfur): Daniel C. Smith; p. 35(bromine): © Charles D. Winters/ Photo Researchers, Inc.; 2.4: © Ingram Publishing/SuperStock RF; 2.5b–c: © Dr. A. Leger/Phototake; p. 46(left column all & right column top & bottom): © The McGraw-Hill Companies, Inc./Jill Braaten, photographer and Anthony Arena, chemistry consultant; p. 46(right column middle): © The McGraw-Hill Companies, Inc./Stephen Frisch, photographer; p. 47(top): © The McGraw-Hill Companies, Inc./photographer Joe Franek; p. 47(middle, bottom): © The McGraw-Hill Companies, Inc./Jill Braaten, photographer and Anthony Arena, chemistry consultant; p. 47(radon detector): © Keith Eng, 2008; 2.7a: © Charles O'Rear/Corbis; 2.7b: © Lester V. Bergman/Corbis; 2.7c: © The McGraw-Hill Companies, Inc./Jill Braaten, photographer and Anthony Arena, chemistry consultant; p. 55: © Dr. Parvinder Sethi RF.

Chapter 3
Opener: © Stockbyte/PunchStock RF; p. 67(sodium metal & chlorine gas): © The McGraw-Hill Companies, Inc./Stephen Frisch, photographer; p. 67 (calcite): © Dane S. Johnson/Visuals Unlimited; p. 75: © The McGraw-Hill Companies, Inc./Jill Braaten, photographer; p. 78(tea pot): © James L. Amos/Photo Researchers, Inc.; p. 78(athlete): © Nicholas Pitt/Alamy; p. 78(table salt & toothpaste): © The McGraw-Hill Companies, Inc./Jill Braaten, photographer; p. 82: © BananaStock/PunchStock RF; 3.4: © The McGraw-Hill Companies, Inc./Jill Braaten, photographer; p. 85(top): © The McGraw-Hill Companies, Inc./Elite Images; p. 85(bottom): © CNRI/ Photo Researchers, Inc.; p. 87(Tums, Maalox, Iron): © The McGraw-Hill Companies, Inc./Jill Braaten, photographer; p. 87(normal bone): © Steve Gschmeissner/Photo Researchers, Inc.; (bone): Steve Gschmeissner/Photo Researchers, Inc.

Chapter 4
Opener: © The McGraw-Hill Companies, Inc./Elite Images; p. 95(top): © Photodisc/Alamy; p. 95(middle): © Grant Heilman Photography/ Alamy; p. 95(bottom): Photo by Tim McCabe, USDA Natural Resources Conservation Service; p. 99: U.S. Dept. of Commerce photoset Hawaii Volcanism: Lava Forms; p. 107: Ernest Beat Basel; p. 108: © Image Source Black/Alamy; p. 114(both): © The McGraw-Hill Companies, Inc./Jill Braaten, photographer.

Chapter 5
Opener: © Image Source/Getty Images; 5.1(left): © The McGraw-Hill Companies, Inc./Jill Braaten, photographer and Anthony Arena, chemistry consultant; 5.1(right): Daniel C. Smith; p. 125: © BananaStock/Punchstock RF; p. 127: © Jim Erickson/Corbis; 5.2(airbag): © Jack Sullivan/ Alamy; pp. 130 & 133: © The McGraw-Hill Companies, Inc./Jill Braaten, photographer and Anthony Arena, chemistry consultant; p. 136: © Hisham F. Ibrahim/Getty Images RF; p. 137: © Keith Eng, 2008; p. 138: © Comstock Images/Punchstock RF; p. 139: © Royalty Free/Corbis; 5.3: © The McGraw-Hill Companies, Inc./Jill Braaten, photographer and Anthony Arena, chemistry consultant; p. 143(top): © Bob Daemmrich/Stock Boston; p. 143(bottom): © The McGraw-Hill Companies, Inc./Jill Braaten, photographer and Anthony Arena, chemistry consultant; p. 145: © Ryan McVay/Getty Images RF; 5.4(albuterol & fluoxetine): © The McGraw-Hill Companies, Inc./Jill Braaten, photographer; (atorvastatin): © Beaconstox/ Alamy; p. 151(left): © Tony Cordoza/Alamy; p. 151(right): © Clair Dunn/ Alamy; 5.7: © Keith Eng, 2008.

Chapter 6
Opener: © Jim Erickson/Corbis; p. 161: © The McGraw-Hill Companies, Inc./Jill Braaten, photographer; p. 165(top): © Digital Vision/PunchStock RF; p. 165(bottom): © Comstock/PunchStock RF; p. 170: © Jack Star/ PhotoLink/Getty Images; p. 172: © The McGraw-Hill Companies, Inc./ Jill Braaten, photographer; p. 173: © Ingram Publishing/SuperStock RF; p. 179: Royalty-Free/Corbis.

Chapter 7
Opener: Daniel C. Smith; Figure 7.1: © Doug Menuez/Getty Images RF; p. 198: © Open Door/Alamy RF; p. 201: © Bill Aron/Photo Edit; p. 208: © National Geographic/Getty Images; p. 209: © Spencer Grant/Photo Edit; p. 210(top): © Martin Harvey/Corbis; p. 210(bottom): © Andrew Syred/ Photo Researchers, Inc.; p. 215(top): © PPP/Popper photo/Robertstock; p. 215(bottom): © The McGraw-Hill Companies, Inc./Jill Braaten, photographer; 7.8a: © The McGraw-Hill Companies/Jill Braaten, photographer and Anthony Arena, chemistry consultant; 7.8b: © blickwinkel/Alamy; 7.9a–d: © The McGraw-Hill Companies, Inc./Jill Braaten, photographer; p. 219(bikers): © Royalty Free/Corbis; p. 219(ice & liquid water): © The McGraw-Hill Companies, Inc./Jill Braaten, photographer; p. 220(adding ice): © The McGraw-Hill Companies, Inc./Suzie Ross, photographer; p. 219(liquid water & vapor): © The McGraw-Hill Companies, Inc./Jill Braaten, photographer; p. 222(freeze-dried food): © The McGraw-Hill Companies, Inc./Jill Braaten, photographer and Anthony Arena, chemistry consultant; p. 222(CO_2 (both)): © The McGraw-Hill Companies, Inc./Jill Braaten, photographer and Anthony Arena, chemistry consultant.

Chapter 8

Opener: © Keith Eng, 2008; 8.1a: Brian Evans/Photo Researchers, Inc.; 8.1b: © Digital Vision/Alamy; 8.1c: © Keith Eng, 2008; p. 230(both) & 8.3(both): © The McGraw-Hill Companies, Inc./Jill Braaten, photographer and Anthony Arena, chemistry consultant; p. 232: © Royalty-Free/Corbis; p. 233(top): © Tom Pantages; p. 233(bottom): © Keith Eng, 2008; 8.5(both): © The McGraw-Hill Companies, Inc./Jill Braaten, photographer; p. 237: © The McGraw-Hill Companies, Inc./Suzie Ross, photographer; 8.6(both): © The McGraw-Hill Companies, Inc./Jill Braaten, photographer and Anthony Arena, chemistry consultant; p. 238: © BananaStock/PunchStock RF; p. 239: © David Hoffman Photo Library/Alamy; p. 240: © Comstock/PunchStock RF; 8.7(both): © The McGraw-Hill Companies, Inc./Jill Braaten, photographer; p. 248(left): Traverse City Record-Eagle, John L. Russell/AP/Wide World Photos; (right): © OSF/Doug Allan/Animals Animals; 8.8(all): Dennis Kunkel Microscopy, Inc.

Chapter 9

Opener & p. 260(both): © The McGraw-Hill Companies/Jill Braaten, photographer and Anthony Arena, chemistry consultant; 9.1a: © Colin Dutton/Grand Tour/Corbis; 9.1b: © Photolink/Getty Images RF; 9.1c: © BananaStock/PunchStock RF; 9.2(all): © The McGraw-Hill Companies, Inc./Jill Braaten, photographer; p. 264: © Charles D. Winters/Photo Researchers, Inc.; 9.4(both), 9.5(both): © McGraw-Hill Companies/Jill Braaten, photographer and Anthony Arena, chemistry consultant; p. 271: © Tim Tadder/Corbis; p. 272: © The McGraw-Hill Companies, Inc./Jill Braaten, photographer; p. 275: © Comstock Images/PicturesQuest RF; p. 276: © Royalty-Free/Corbis; 9.6a: © The McGraw-Hill Companies, Inc./Stephen Frisch, photographer; 9.6b: © sciencephotos/Alamy; 9.6c: © Leslie Garland Picture Library/Alamy; 9.7(lemon): © Humberto Olarte Cupas/Alamy RF; 9.7(strawberries): © Photolink/Getty Images RF; 9.7(tomatoes): © Burke/Triolo/Getty Images RF; 9.7(milk): © Mitch Hrdlicka/Getty Images RF; 9.7(Phillips): © The McGraw-Hill Companies, Inc./Jill Braaten, photographer; 9.7(ammonia): © The McGraw-Hill Companies/Terry Wild Studio; 9.7(bleach): © The McGraw-Hill Companies, Inc./Jill Braaten, photographer; p. 279: © Bruce Heinemann/Getty Images RF; pp. 282, 283 & 9.9(all): © The McGraw-Hill Companies, Inc./Jill Braaten, photographer and Anthony Arena, chemistry consultant; p. 287: © Dr. Parvinder Sethi RF.

Chapter 10

Opener: © Owen Franken/Corbis; p. 301: © AFP/Getty Images; p. 302(top): © Cordelia Molloy/Photo Researchers, Inc.; p. 302(bottom): © S. Wanke/PhotoLink/Getty Images RF; 10.2a: Owen Franken/Corbis; 10.2b–c: © Custom Medical Stock Photo; p. 309: © Science VU/Visuals Unlimited; p. 310(top): © Hank Morgan/Photo Researchers, Inc.; p. 310(bottom): Landauer Corporation; 10.4b: © Living Art Enterprises, LLC/Photo Researchers, Inc.; 10.5a–b & 10.7a–c: Daniel C. Smith; 10.8a: © Martin Bond/Photo Researchers, Inc.; 10.9a: © Royalty-Free/Corbis; 10.9b: © Scott Camazine/Photo Researchers, Inc.; 10.9c: James Rosato, RF (R) (MR) (CT) MRI Applications Specialist Brigham and Women's Hospital, Boston, MA.

Chapter 11

Opener: © Corbis RF; 11.1a: © Comstock/PunchStock RF; 11.1b: © Corbis Premium RF/Alamy; 11.1c: © FRI/Corbis Sygma; 11.1d: © Layne Kennedy/Corbis; p. 324(methane): © Ingram Publishing/SuperStock RF; p. 324(ethanol): © Goodshoot/PunchStock RF; p. 324(capsaicin): © The McGraw-Hill Companies, Inc./Jill Braaten, photographer; p. 324(caffeine): © Comstock/PunchStock RF; p. 327(top): © Vol. 129 PhotoDisc/Getty Images; p. 327(bottom): © PhotoDisc/Getty; p. 328: © Corbis/RF; p. 331: John Somerville; p. 336(top): Czerwinksi/Ontario Ministry of Natu-

ral Resources; p. 336(bottom): Courtesy of Gebauer Company, Cleveland, Ohio; 11.2(cherries): © Brand X Pictures/PunchStock RF; 11.2(bananas): © Ingram Publishing/SuperStock RF; 11.2(oranges): © Brand X Pictures/PunchStock RF; 11.2(pineapple): © Corbis RF; Table 11.6(butane lighter): © PIXTAL/PunchStock RF; Table 11.6(salt shaker): © Royalty-Free/Corbis; p. 345: © The McGraw-Hill Companies, Inc./Jill Braaten, photographer; p. 346(top): © BananaStock/PhotoStock RF; p. 346(bottom): © The McGraw-Hill Companies Inc./John Flournoy, photographer.

Chapter 12

Opener: © Edmond Van Hoorick/Getty Images; p. 356(left): God of Insects; p. 356(right): © Douglas Peebles/Corbis; p. 362: © PhotoDisc Website; p. 363: © Rachel Epstein/Photo Edit; p. 368(left): From the Richard A. Howard Image Collection of the Smithsonian Institution; p. 368(right): © The McGraw-Hill Companies, Inc./Suzie Ross, photographer; 12.1a: © PhotoDisc Website; p. 371(top): © The McGraw-Hill Companies, Inc./Jill Braaten, photographer; p. 371(bottom): © Vince Bevan/Alamy; p. 372: PhotoLink/Getty Images RF.

Chapter 13

Opener & p. 380: © The McGraw-Hill Companies, Inc./Jill Braaten, photographer; p. 387: © Degginger E. R./Animals Animals/Earth Scenes; p. 390(top): © The McGraw-Hill Companies, Inc./Jill Braaten, photographer; p. 390(bottom): Sergio Piumatti; p. 393: © 1998 Richard Megna, Fundamental Photographs, NYC; p. 395: Ethanol Promotion and Information Council (EPIC); p. 396: © Royalty-Free/Corbis; 13.5(both): © The McGraw-Hill Companies, Inc./Jill Braaten, photographer; p. 397(both): © The McGraw-Hill Companies, Inc./Elite Images; 13.6(pvc): © The McGraw-Hill Companies, Inc./John Thoeming, photographer; 13.6(sports balls): © Dynamicgraphics/Jupiterimages; 3.6(blankets): © Fernando Bengoe/Corbis; 3.6(styrofoam): © The McGraw-Hill Companies, Inc./John Thoeming, photographer; p. 399(polyethylene): © The McGraw-Hill Companies, Inc./Jill Braaten, photographer; Table 13.2(blood bags): Image Source/Getty Images RF; Table 13.2(syringes): © Royalty Free/Corbis; Table 13.2(dental floss): © Cristina Fumi/Alamy; 13.7(top): © The McGraw-Hill Companies, Inc./Jill Braaten, photographer; 13.7(middle): © Mikael Karlsson/Arresting Images RF; 13.7(bottom): © Picture Partners/Alamy; p. 406(sunscreen): © The McGraw-Hill Companies, Inc./John Thoeming, photographer; p. 406(tobacco): Photo by Bob Nichols, USDA Natural Resources Conservation Service; p. 406(cigarette): PhotoAlto/PictureQuest RF; p. 407(vanilla bean): © CD16/Author's Image RF; p. 407(cloves): © Peter Arnold, Inc./Alamy; p. 407(turmeric): Mr. Napat Kitipanangkul, Science School, Walailak University with financial support by Biodiversity; Research and training program (BRT); p. 408(top): © Royalty-Free/Corbis; p. 408(bottom): © The McGraw-Hill Companies, Inc./Elite Images.

Chapter 14

Opener: © John A. Rizzo/Getty Images RF; p. 419(left): © Creatas/PunchStock RF; p. 419(middle): © Mira/Alamy; p. 419(right): © Greg Vaughn/Alamy; 14.1(left): © Dynamic Graphics Group/PunchStock RF; 14.1(middle): © Goodshoot/PunchStock RF; 14.1(right): © BananaStock/PunchStock RF; p. 425(top): Ethanol Promotion and Information Council (EPIC); p. 425(wheat): © Vol. 1 PhotoDisc/Getty Images; p. 425(cotton): © Corbis/R-F Website; 14.4: © The McGraw-Hill Companies, Inc./Elite Images; p. 431(party): © BananaStock/PunchStock RF; p. 431(crashed car): Gwinnet County Police Department/Centers for Disease Control and Prevention; p. 433(top): © Ed Reschke; p. 433(bottom): Peter J. S. Franks, Scripps Institution of Oceanography; p. 436: The Boston Medical Library in the Francis A Countway Library of Medicine; p. 439: © The McGraw-Hill Companies, Inc./Jill Braaten, photographer; p. 440: NASA; 14.7(both): © The McGraw-Hill Companies, Inc./Suzie Ross, photographer.

Chapter 15

Opener: © The McGraw-Hill Companies, Inc./Elite Images; p. 451: Daniel C. Smith; 15.1a: © The McGraw-Hill Companies, Inc./Jill Braaten, photographer; 15.1b: © Biosphoto/Thouvenin Claude/Peter Arnold Inc.; p. 455(left): © PhotoDisc/Getty Images; p. 455(right): © Henriette Kress; p. 456: © Comstock/PunchStock RF; p. 458: John Fisher; p. 460: © The McGraw-Hill Companies, Inc./Jill Braaten, photographer; p. 461: © Nigel Cattlin/Alamy; p. 466(left): © The McGraw-Hill Companies, Inc./Elite Images; p. 466(right): © David Sieren/Visuals Unlimited.

Chapter 16

Opener: Daniel C. Smith; p. 475: © Iconotec/Alamer and Cali, photographers RF; p. 480(all): © The McGraw-Hill Companies, Inc./Jill Braaten, photographer; 16.1: © The McGraw-Hill Companies, Inc./Jill Braaten, photographer; p. 482(top): Ben Mills; p. 482(bottom): © The McGraw-Hill Companies, Inc./photographer Joe Franek; p. 486(top): © M. K. Ranjitsinh/Photo Researchers, Inc.; p. 486(bottom): © The McGraw-Hill Companies, Inc./Jill Braaten, photographer; p. 487(top): © PhotoDisc/Getty Images RF; (bottom): p. 493(top): © The McGraw-Hill Companies, Inc. Elite Images; p. 493(bottom): Richo Cech, Horizon Herbs, LLC.

Chapter 17

Opener: Photo by Forest and Kim Starr; p. 505: © The McGraw-Hill Companies, Inc./Jill Braaten, photographer; p. 512(top): Photo by Forest and Kim Starr; p. 512(bottom): © The McGraw-Hill Companies, Inc./Jill Braaten, photographer; p. 514(mangoes): © CD16/Author's Image RF; p. 514(apricots): © Jupiterimages/ImageSource RF; p. 514(wintergreens): © UpperCut Images/Alamy; p. 514(melatonin, orajel, Tylenol): © The McGraw-Hill Companies, Inc./Jill Braaten, photographer; p. 519(willow): © National Geographic/Getty Images; p. 519(meadowsweet): © Biopix. dkhttp://www.biopix.dk; p. 523(cockroach): Coby Schal/NC State University; p. 523(Off): Photo by Scott Bauer/USDA; p. 527: © The McGraw-Hill Companies, Inc./Jill Braaten, photographer; p. 528(top): Image courtesy of The Advertising Archives; p. 528(bottom): © Royalty-Free/Corbis; p. 529(top): © The McGraw-Hill Companies, Inc./John Thoeming, photographer; p. 529(bottom): © The McGraw-Hill Companies, Inc./Jill Braaten, photographer; 17.5a: © SIU/Visuals Unlimited, Inc.; 17.5b: © Antonia Reeve/Photo Researchers, Inc.; p. 531: © The McGraw-Hill Companies, Inc./Jill Braaten, photographer; p. 532: Library of Congress.

Chapter 18

Opener: © Taxi/Getty; p. 541(left): © Maximilian Weinzierl/Alamy; (right): © D. Hurst/Alamy; p. 547: © BananaStock/PunchStock RF; 18.2(plant): © Mediacolor's/Alamy; 18.2(coffee): Nick Koudis/Getty Images RF; 18.2(cherries): © Inga Spence/Visuals Unlimited; p. 549: © Stockdisc/PunchStock RF; 18.3(plant): © Flora Torrance/Life File RF; 18.3(cigarettes): © Nancy R. Cohen/Getty Images RF; 18.3(dried leaves): © Wayne Hutchinson/Alamy; p. 550(poppy seeds): © Stockbyte Silver/Getty Images; p. 550(opium poppy): © Scott Camazine/Photo Researchers, Inc.; p. 551 (cinchona tree): Photo by Forest and Kim Starr; p. 551(nightshade plant): Werner Arnold; p. 555: © The McGraw-Hill Companies, Inc./Jill Braaten, photographer; p. 559(Bufo frog): © Daniel C. Smith; p. 559(Psilocybe mushrooms): © IT Stock/age fotostock RF; p. 561: © The McGraw-Hill Companies, Inc./Jill Braaten, photographer; p. 563: Daniel C. Smith.

Chapter 19

Opener: © Andrew Brookes/Corbis; p. 570: © The McGraw-Hill Companies, Inc./Jill Braaten, photographer; p. 574: © Brand X Pictures/PunchStock RF; p. 575(whale): © Denis Scott/Corbis; p. 575(leaf): Daniel C. Smith; p. 575(lip balm): © The McGraw-Hill Companies, Inc./Jill Braaten, photographer; p. 575(bees): © Photodisc/Getty RF; p. 577: © The McGraw-Hill Companies, Inc./Jill Braaten, photographer; p. 578(left): © Taxi/Getty; p. 578(right): © Jupiterimages/ImageSource RF; p. 579: Daniel C. Smith; p. 580, 19.1 (all): © The McGraw-Hill Companies, Inc./Jill Braaten, photographer; 19.2: © The McGraw-Hill Companies, Inc./Dennis Strete, photographer; p. 583: © AP/Wide World Photos; p. 584(left): © Jupiterimages/ImageSource RF; p. 584(right): David Buffington/Getty Images RF; p. 591: © Getty Images RF; 19.8(both): © Ed Reschke; p. 595: © Comstock/JupiterImages RF; p. 596: © A. T. Willett/Alamy; p. 597: © BananaStock RF; p. 598(top): © Stockbyte/PunchStock RF; p. 598(middle): © IT Stock/PunchStock RF; p. 598(bottom): © AGStockUSA, Inc./Alamy.

Chapter 20

Opener: © Japack Company/Corbis; p. 609(top): © Michelle Garrett/Corbis; p. 609(bottom): ImageSource/Corbis RF; p. 610(top): © Jim Erickson/Corbis; p. 610(bottom): Daniel C. Smith; p. 611: © The McGraw-Hill Companies, Inc./Elite Images; p. 615(top): Alamy Images; p. 615(bottom): © Keith Brofsky/Getty Images RF; p. 616: © Comstock/PunchStock RF; p. 617: © Michael Newman/Photo Edit; pp. 622, 625 & 20.2: © The McGraw-Hill Companies, Inc./Jill Braaten, photographer; p. 626: © Vol. 3/PhotoDisc/Getty Images; p. 628(left): © The McGraw-Hill Companies, Inc./Jill Braaten, photographer; p. 628(right): © D. Hurst/Alamy RF; p. 629(left): © David Muench/Corbis; p. 629(right): © The McGraw-Hill Companies, Inc./Elite Images; 20.3: © The McGraw-Hill Companies, Inc./Jill Braaten, photographer; 20.4: Biophoto Associates/Photo Researchers, Inc.; 20.5: © Vol. 1 PhotoDisc/Getty Images; 20.5(inset): © Borland/PhotoLink/Getty Images RF; 20.6: © Dennis Kunkel/Visuals Unlimited; p. 634: © Geldi/Alamy; p. 635(top): © The McGraw-Hill Companies, Inc./Jill Braaten, photographer; p. 635(bottom): © Lawson Wood/Corbis; p. 636: © Royalty-Free/Corbis.

Chapter 21

Opener: © Purestock/PunchStock RF; p. 647: John A. Raineri; 21.12: © Image Source/PunchStock RF; p. 665: Daniel C. Smith; p. 666: © Eye of Science/Photo Researchers, Inc.; p. 668: © The McGraw-Hill Companies, Inc./Jill Braaten, photographer; p. 669: Daniel C. Smith; p. 675: © Photo Researchers, Inc.; 21.16b: Roger Sayle.

Chapter 22

Opener: © Lawrence Lawry/Getty Images RF; 22.7b: © Kenneth Eward/Photo Researchers, Inc.; 22.12: Courtesy Genelex Corp., www.HealthandDNA.com; p. 710: © The McGraw-Hill Companies, Inc./Jill Braaten, photographer.

Chapter 23

Opener: © BananaStock/PunchStock RF; 23.2(butter): © Royalty-Free/Corbis; 23.2(bagels): © Nancy R. Cohen/Getty Images RF; 23.2(meat): © PhotoLink/Getty Images/RF; p. 728: © David Stoecklein/Corbis; p. 731: © The McGraw-Hill Companies, Inc./Jill Braaten, photographer; p. 739(top): © Connie Bransilver/Photo Researchers, Inc.; p. 739(bottom): © The McGraw-Hill Companies, Inc./Jill Braaten, photographer.

Chapter 24

Opener: © Dennis Gray/Cole Group/Getty Images RF; p. 753(beets): © Corbis RF; p. 753(milk): © Isabelle Rozenbaum & Frederic Cirou/PhotoAlto/PunchStock RF; p. 753(cranberries): © Burke/Triolo/Getty Images RF; p. 755: © Warren Morgan/Corbis; p. 757: © John A. Rizzo/Getty Images RF; p. 765: Daniel C. Smith; p. 766(both): © The McGraw-Hill Companies, Inc./Jill Braaten, photographer.

Index

Page numbers followed by *f*, indicate figures; *mn*, marginal notes; and *t*, tables. Page numbers preceded by "A-" indicate pages in the *Appendix*.

Common Functional Groups in Organic Chemistry

Type of Compound	General Structure	Example	Functional Group
Alcohol	$R-\overset{..}{\underset{..}{O}}H$	$CH_3-\overset{..}{\underset{..}{O}}H$	$-OH$ hydroxyl group
Aldehyde	$\underset{R}{\overset{:O:}{\underset{}{\parallel}}}\overset{}{C}-H$	$\underset{CH_3}{\overset{:O:}{\parallel}}C-H$	$C=O$ carbonyl group
Alkane	$R-H$	CH_3CH_3	$--$
Alkene	$C=C$	$\overset{H\quad H}{\underset{H\quad H}{C=C}}$	double bond
Alkyl halide	$R-\overset{..}{\underset{..}{X}}:$ (X = F, Cl, Br, I)	$CH_3-\overset{..}{\underset{..}{Br}}:$	$-X$
Alkyne	$-C\equiv C-$	$H-C\equiv C-H$	triple bond
Amide	$\underset{R}{\overset{:O:}{\parallel}}C-\overset{H\ (or\ R)}{\underset{H\ (or\ R)}{N}}$	$\underset{CH_3}{\overset{:O:}{\parallel}}C-\overset{..}{N}H_2$	$-CONH_2$, $-CONHR$, $-CONR_2$
Amine	$R-\overset{..}{N}H_2$ or $R_2\overset{..}{N}H$ or $R_3\overset{..}{N}$	$CH_3-\overset{..}{N}H_2$	$-NH_2$ amino group
Aromatic compound	⬡	⬡	benzene ring
Carboxylic acid	$\underset{R}{\overset{:O:}{\parallel}}C-\overset{..}{\underset{..}{O}}H$	$\underset{CH_3}{\overset{:O:}{\parallel}}C-\overset{..}{\underset{..}{O}}H$	$-COOH$ carboxyl group
Ester	$\underset{R}{\overset{:O:}{\parallel}}C-\overset{..}{\underset{..}{O}}R$	$\underset{CH_3}{\overset{:O:}{\parallel}}C-\overset{..}{\underset{..}{O}}CH_3$	$-COOR$
Ether	$R-\overset{..}{\underset{..}{O}}-R$	$CH_3-\overset{..}{\underset{..}{O}}-CH_3$	$-OR$
Ketone	$\underset{R}{\overset{:O:}{\parallel}}C-R$	$\underset{CH_3}{\overset{:O:}{\parallel}}C-CH_3$	$C=O$ carbonyl group